Study and Solutions Guide

PRECALCULUS WITH LIMITS

Larson / Hostetler

Dianna L. Zook

Indiana University – Purdue University
Fort Wayne, Indiana

Houghton Mifflin Company   Boston   New York

Publisher: Richard Stratton
Sponsoring Editor: Cathy Cantin
Development Manager: Maureen Ross
Development Editor: Lisa Collette
Supervising Editor: Karen Carter
Senior Project Editor: Patty Bergin
Art and Design Manager: Gary Crespo
Executive Marketing Manager: Brenda Bravener-Greville
Director of Manufacturing: Priscilla Manchester

Printed in the United States of America

ISBN 13: 978-0-618-66092-6
ISBN 10: 0-618-66092-5

23456789-CS-10 09 08 07

TO THE STUDENT

This *Study and Solutions Guide* for *Precalculus with Limits*, is a supplement to the textbook by Ron Larson and Robert Hostetler.

As a mathematics Instructor, I often have students come to me with questions about assigned homework. When I ask to see their work, the reply often is "I didn't know where to start." The purpose of the study guide is to provide brief summaries of the topics covered in the textbook and enough detailed solutions to problems so that you will be able to work the remaining exercises.

I have made every effort to see that the solutions are correct. However, I would appreciate hearing about any errors or other suggestions for improvement.

I would like to thank the staff at Larson Texts, Inc. for their help in the production of this guide.

Dianna L. Zook
Indiana University – Purdue University
Fort Wayne, IN 46805
(zook@ipfw.edu)

Study Strategies

- Attend all classes and come prepared. Have your homework completed. Bring the text, paper, pen or pencil, and a calculator (scientific or graphing) to each class.

- Read the section in the text that is to be covered before class. Make notes about any questions that you have and if they are not answered during the lecture, ask them at the appropriate time.

- Participate in class. As mentioned above, ask questions,

- Take notes on all definitions, concepts, rules, formulas, and examples. After class, read your notes and fill in any gaps, or make notations of questions that you have.

- DO THE HOMEWORK!!! You learn mathematics by doing it yourself. Allow at least two hours outside of each class for homework. Do not fall behind.

- Seek help when needed. Visit your instructor during office hours and come prepared with specific questions; check with your school's tutoring service; find a study partner in class; check additional books in the library for more examples—just do something before the problem becomes insurmountable.

- Do not cram for exams. Each chapter in the text contains a review chapter and a chapter test, and this Study Guide contains a practice test at the end of each chapter. (The answers are at the end of Part I.) Work these problems a few days before the exam and review any areas of weakness.

CONTENTS

Part I   **Solutions to Odd-Numbered Exercises** **1**

Chapter 1   Functions and Their Graphs. **1**

Chapter 2   Polynomial and Rational Functions **75**

Chapter 3   Exponential and Logarithmic Functions. **147**

Chapter 4   Trigonometry. **185**

Chapter 5   Analytic Trigonometry. **240**

Chapter 6   Additional Topics in Trigonometry **290**

Chapter 7   Systems of Equations and Inequalities. **331**

Chapter 8   Matrices and Determinants **391**

Chapter 9   Sequences, Series, and Probability. **443**

Chapter 10   Topics in Analytic Geometry **489**

Chapter 11   Analytic Geometry in Three Dimensions. **556**

Chapter 12   Limits and an Introduction to Calculus **579**

Appendix A   Review of Fundamental Concepts of Algebra. **614**

                      Solutions to Practice Tests. **656**

Part II   **Solutions to Chapter and Cumulative Tests** **685**

PART I

CHAPTER 1
Functions and Their Graphs

Section 1.1   Rectangular Coordinates 2

Section 1.2   Graphs of Equations 7

Section 1.3   Linear Equations in Two Variables 12

Section 1.4   Functions 19

Section 1.5   Analyzing Graphs of Functions 25

Section 1.6   A Library of Parent Functions 31

Section 1.7   Transformations of Functions 35

Section 1.8   Combinations of Functions: Composite Functions 43

Section 1.9   Inverse Functions 48

Section 1.10 Mathematical Modeling and Variation 56

Review Exercises 60

Problem Solving 70

Practice Test 73

CHAPTER 1
Functions and Their Graphs

Section 1.1   Rectangular Coordinates

- You should be able to use the point-plotting method of graphing.

- You should be able to find x- and y-intercepts.

  (a) To find the x-intercepts, let $y = 0$ and solve for x.

  (b) To find the y-intercepts, let $x = 0$ and solve for y.

- You should be able to test for symmetry.

  (a) To test for x-axis symmetry, replace y with $-y$.

  (b) To test for y-axis symmetry, replace x with $-x$.

  (c) To test for origin symmetry, replace x with $-x$ and y with $-y$.

- You should know the standard equation of a circle with center (h, k) and radius r:

  $$(x - h)^2 + (y - k)^2 = r^2$$

Vocabulary Check

1. (a) **v** horizontal real number line

(b) **vi** vertical real number line

(c) **i** point of intersection of vertical axis and horizontal axis

(d) **iv** four regions of the coordinate plane

(e) **iii** directed distance from the y-axis

(f) **ii** directed distance from the x-axis

2. Cartesian   **3.** Distance Formula   **4.** Midpoint Formula

---

1. $A: (2, 6)$, $B: (-6, -2)$, $C: (4, -4)$, $D: (-3, 2)$

3.

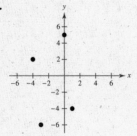

5.

7. $(-3, 4)$ **9.** $(-5, -5)$ **11.** $x > 0$ and $y < 0$ in Quadrant IV.

13. $x = -4$ and $y > 0$ in Quadrant II.   **15.** $y < -5$ in Quadrants III and IV.

17. $(x, -y)$ is in the second Quadrant means that (x, y) is in Quadrant III.

19. (x, y), $xy > 0$ means x and y have the same signs. This occurs in Quadrants I and III.

21.

Year, x	Number of stores, y
1996	3054
1997	3406
1998	3599
1999	3985
2000	4189
2001	4414
2002	4688
2003	4906

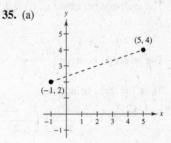

Year (6 ↔ 1996)

23. $d = |5 - (-3)| = 8$

25. $d = |2 - (-3)| = 5$

27. (a) The distance between $(0, 2)$ and $(4, 2)$ is 4.

The distance between $(4, 2)$ and $(4, 5)$ is 3.

The distance between $(0, 2)$ and $(4, 5)$ is

$$\sqrt{(4 - 0)^2 + (5 - 2)^2} = \sqrt{16 + 9} = \sqrt{25} = 5.$$

(b) $4^2 + 3^2 = 16 + 9 = 25 = 5^2$

29. (a) The distance between $(-1, 1)$ and $(9, 1)$ is 10.

The distance between $(9, 1)$ and $(9, 4)$ is 3.

The distance between $(-1, 1)$ and $(9, 4)$ is

$$\sqrt{(9 - (-1))^2 + (4 - 1)^2} = \sqrt{100 + 9} = \sqrt{109}.$$

(b) $10^2 + 3^2 = 109 = \left(\sqrt{109}\right)^2$

31. (a)

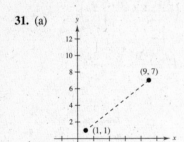

(b) $d = \sqrt{(9 - 1)^2 + (7 - 1)^2}$

$\quad = \sqrt{64 + 36} = 10$

(c) $\left(\dfrac{9 + 1}{2}, \dfrac{7 + 1}{2}\right) = (5, 4)$

33. (a)

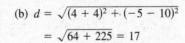

(b) $d = \sqrt{(4 + 4)^2 + (-5 - 10)^2}$

$\quad = \sqrt{64 + 225} = 17$

(c) $\left(\dfrac{4 - 4}{2}, \dfrac{-5 + 10}{2}\right) = \left(0, \dfrac{5}{2}\right)$

35. (a)

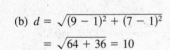

(b) $d = \sqrt{(5 + 1)^2 + (4 - 2)^2}$

$\quad = \sqrt{36 + 4} = 2\sqrt{10}$

(c) $\left(\dfrac{-1 + 5}{2}, \dfrac{2 + 4}{2}\right) = (2, 3)$

37. (a)

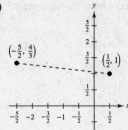

(b) $d = \sqrt{\left(\dfrac{1}{2} + \dfrac{5}{2}\right)^2 + \left(1 - \dfrac{4}{3}\right)^2}$

$= \sqrt{9 + \dfrac{1}{9}} = \dfrac{\sqrt{82}}{3}$

(c) $\left(\dfrac{-(5/2) + (1/2)}{2}, \dfrac{(4/3) + 1}{2}\right) = \left(-1, \dfrac{7}{6}\right)$

39. (a)

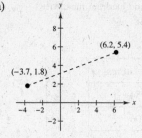

(b) $d = \sqrt{(6.2 + 3.7)^2 + (5.4 - 1.8)^2}$

$= \sqrt{98.01 + 12.96} = \sqrt{110.97}$

(c) $\left(\dfrac{6.2 - 3.7}{2}, \dfrac{5.4 + 1.8}{2}\right) = (1.25, 3.6)$

41. $d_1 = \sqrt{(4 - 2)^2 + (0 - 1)^2} = \sqrt{5}$

$d_2 = \sqrt{(4 + 1)^2 + (0 + 5)^2} = \sqrt{50}$

$d_3 = \sqrt{(2 + 1)^2 + (1 + 5)^2} = \sqrt{45}$

$\left(\sqrt{5}\right)^2 + \left(\sqrt{45}\right)^2 = \left(\sqrt{50}\right)^2$

43. Since $x_m = \dfrac{x_1 + x_2}{2}$ and $y_m = \dfrac{y_1 + y_2}{2}$ we have:

$$2x_m = x_1 + x_2 \qquad\qquad 2y_m = y_1 + y_2$$

$$2x_m - x_1 = x_2 \qquad\qquad 2y_m - y_1 = y_2$$

Thus, $(x_2, y_2) = (2x_m - x_1, 2y_m - y_1)$.

45. The midpoint of the given line segment is $\left(\dfrac{x_1 + x_2}{2}, \dfrac{y_1 + y_2}{2}\right)$.

The midpoint between (x_1, y_1) and $\left(\dfrac{x_1 + x_2}{2}, \dfrac{y_1 + y_2}{2}\right)$ is $\left(\dfrac{x_1 + \dfrac{x_1 + x_2}{2}}{2}, \dfrac{y_1 + \dfrac{y_1 + y_2}{2}}{2}\right) = \left(\dfrac{3x_1 + x_2}{4}, \dfrac{3y_1 + y_2}{4}\right)$.

The midpoint between $\left(\dfrac{x_1 + x_2}{2}, \dfrac{y_1 + y_2}{2}\right)$ and (x_2, y_2) is $\left(\dfrac{\dfrac{x_1 + x_2}{2} + x_2}{2}, \dfrac{\dfrac{y_1 + y_2}{2} + y_2}{2}\right) = \left(\dfrac{x_1 + 3x_2}{4}, \dfrac{y_1 + 3y_2}{4}\right)$.

Thus, the three points are

$$\left(\dfrac{3x_1 + x_2}{4}, \dfrac{3y_1 + y_2}{4}\right), \left(\dfrac{x_1 + x_2}{2}, \dfrac{y_1 + y_2}{2}\right), \text{ and } \left(\dfrac{x_1 + 3x_2}{4}, \dfrac{y_1 + 3y_2}{4}\right).$$

47. $d = \sqrt{(42 - 18)^2 + (50 - 12)^2}$

$= \sqrt{24^2 + 38^2}$

$= \sqrt{2020}$

$= 2\sqrt{505}$

≈ 45 yards

49. $\left(\dfrac{2001 + 2003}{2}, \dfrac{3433 + 4174}{2}\right) = (2002, 3803.5)$

In 2002, the sales for Big Lots was approximately $3803.5 million.

51. $(-2 + 2, -4 + 5) = (0, 1)$

$(2 + 2, -3 + 5) = (4, 2)$

$(-1 + 2, -1 + 5) = (1, 4)$

53. $(-7 + 4, -2 + 8) = (-3, 6)$

$(-2 + 4, 2 + 8) = (2, 10)$

$(-2 + 4, -4 + 8) = (2, 4)$

$(-7 + 4, -4 + 8) = (-3, 4)$

55. The highest price of butter is approximately $3.31 per pound. This occurred in 2001.

57. $\left[\dfrac{2400 - 700}{700}\right](100) \approx 242.9\%$ increase

59. (a) The number of artists elected each year seems to be nearly steady except for the first few years. Between 6 and 8 artists will be elected in 2008.

(b) Elections for inclusion in the Rock and Roll Hall of Fame began in 1986.

61. (1996, 18,546), (2004, 21,900)

By Exercise 45 we have the following:

$\left(\dfrac{3(1996) + 2004}{4}, \dfrac{3(18,546) + 21,900}{4}\right) = (1998, 19,384.5)$

$\left(\dfrac{1996 + 2004}{2}, \dfrac{18,546 + 21,900}{2}\right) = (2000, 20,223)$

$\left(\dfrac{1996 + 3(2004)}{4}, \dfrac{18,546 + 3(21,900)}{4}\right) = (2002, 21,061.5)$

Year	Sales for Coca-Cola Company
1998	\$19,384.5 million
2000	\$20,223 million
2002	\$21,061.5 million

63. $V = \dfrac{4}{3}\pi r^3$

$5.96 = \dfrac{4}{3}\pi r^3$

$17.88 = 4\pi r^3$

$\dfrac{17.88}{4\pi} = r^3$

$r = \sqrt[3]{\dfrac{4.47}{\pi}} \approx 1.12$ inches

65. $3S = 129$

$S = 43$ centimeters

$h^2 + \left(\dfrac{S}{2}\right)^2 = S^2$

$h^2 = \dfrac{3S^2}{4}$

$h = \dfrac{\sqrt{3}S}{2}$

$A = \dfrac{1}{2}bh = \dfrac{1}{2}S\left(\dfrac{\sqrt{3}S}{2}\right) = \dfrac{\sqrt{3}S^2}{4}$

When $S = 43$ centimeters,

$A = \dfrac{\sqrt{3}(43)^2}{4} \approx 800.64$ square centimeters.

67. (a)
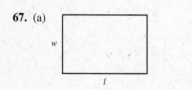

(b) $l = 1.5w$

$P = 2l + 2w$

$= 2(1.5w) + 2w$

$= 5w$

(c) $25 = 5w$

$5 = w$

Width: $w = 5$ meters

Length: $l = 1.5w = 7.5$ meters

Dimensions: 7.5 meters $\times$ 5 meters

69. (a)

Year, x	Pieces of mail, y (in billions)
1996	183
1997	191
1998	197
1999	202
2000	208
2001	207
2002	203
2003	202

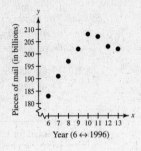

(b) The greatest decrease occurred in 2002.

(c) Answers will vary. Technology now enables us to transport information in ways other than by mail. The internet is one example.

71.

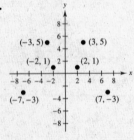

(a) The point is reflected through the y-axis.

(b) The point is reflected through the x-axis.

(c) The point is reflected through the origin.

73. False, you would have to use the Midpoint Formula 15 times.

75. No. It depends on the magnitude of the quantities measured.

77. Since (x_0, y_0) lies in Quadrant II, $(x_0, -y_0)$ must lie in Quadrant III. Matches (b).

79. Since (x_0, y_0) lies in Quadrant II, $\left(x_0, \frac{1}{2}y_0\right)$ must lie in Quadrant II. Matches (d).

81.
$$2x + 1 = 7x - 4$$
$$-5x = -5$$
$$x = 1$$

83.
$$x^2 - 4x - 7 = 0$$
$$x^2 - 4x = 7$$
$$x^2 - 4x + 4 = 7 + 4$$
$$(x - 2)^2 = 11$$
$$x - 2 = \pm\sqrt{11}$$
$$x = 2 \pm \sqrt{11}$$

85.
$$3x + 1 < 2(2 - x)$$
$$3x + 1 < 4 - 2x$$
$$5x < 3$$
$$x < \frac{3}{5}$$

87.
$$|x - 18| < 4$$
$$-4 < x - 18 < 4$$
$$14 < x < 22$$

Section 1.2   Graphs of Equations

You should know the following important facts about lines.

■ The graph of $y = mx + b$ is a straight line. It is called a linear equation in two variables.

   (a) The slope (steepness) is m.

   (b) The y-intercept is $(0, b)$.

■ The slope of the line through (x_1, y_1) and (x_2, y_2) is

$$m = \frac{y_2 - y_1}{x_2 - x_1} = \frac{\text{change in } y}{\text{change in } x} = \frac{\text{rise}}{\text{run}}.$$

■ (a) If $m > 0$, the line rises from left to right.

   (b) If $m = 0$, the line is horizontal.

   (c) If $m < 0$, the line falls from left to right.

   (d) If m is undefined, the line is vertical.

■ Equations of Lines

   (a) Slope-Intercept Form: $y = mx + b$

   (b) Point-Slope Form: $y - y_1 = m(x - x_1)$

   (c) Two-Point Form: $y - y_1 = \dfrac{y_2 - y_1}{x_2 - x_1}(x - x_1)$

   (d) General Form: $Ax + By + C = 0$

   (e) Vertical Line: $x = a$

   (f) Horizontal Line: $y = b$

■ Given two distinct nonvertical lines

   $L_1: y = m_1 x + b_1$ and $L_2: y = m_2 x + b_2$

   (a) L_1 is parallel to L_2 if and only if $m_1 = m_2$ and $b_1 \neq b_2$.

   (b) L_1 is perpendicular to L_2 if and only if $m_1 = -1/m_2$.

Vocabulary Check

1. solution or solution point   **2.** graph   **3.** intercepts

4. y-axis   **5.** circle; (h, k); r **6.** numerical

1. $y = \sqrt{x + 4}$

   (a) $(0, 2)$: $2 \stackrel{?}{=} \sqrt{0 + 4}$

                $2 = 2$

        Yes, the point *is* on the graph.

   (b) $(5, 3)$: $3 \stackrel{?}{=} \sqrt{5 + 4}$

                $3 = \sqrt{9}$

        Yes, the point *is* on the graph.

3. $y = 4 - |x - 2|$

   (a) $(1, 5)$: $5 \stackrel{?}{=} 4 - |1 - 2|$

                $5 \neq 4 - 1$

        No, the point *is not* on the graph.

   (b) $(6, 0)$: $0 \stackrel{?}{=} 4 - |6 - 2|$

                $0 = 4 - 4$

        Yes, the point *is* on the graph.

5. $y = -2x + 5$

x	-1	0	1	2	$\frac{5}{2}$
y	7	5	3	1	0
(x, y)	$(-1, 7)$	$(0, 5)$	$(1, 3)$	$(2, 1)$	$\left(\frac{5}{2}, 0\right)$

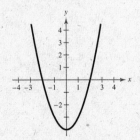

7. $y = x^2 - 3x$

x	-1	0	1	2	3
y	4	0	-2	-2	0
(x, y)	$(-1, 4)$	$(0, 0)$	$(1, -2)$	$(2, -2)$	$(3, 0)$

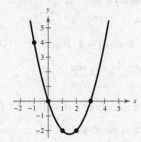

9. $y = 16 - 4x^2$

x-intercepts: $0 = 16 - 4x^2$

$\qquad 4x^2 = 16$

$\qquad x^2 = 4$

$\qquad x = \pm 2$

$\qquad (-2, 0), (2, 0)$

y-intercept: $y = 16 - 4(0)^2 = 16$

$\qquad (0, 16)$

11. $y = 5x - 6$

x-intercept: $0 = 5x - 6$

$\qquad 6 = 5x$

$\qquad \frac{6}{5} = x$

$\qquad \left(\frac{6}{5}, 0\right)$

y-intercept: $y = 5(0) - 6 = -6$

$\qquad (0, -6)$

13. $y = \sqrt{x + 4}$

x-intercept: $0 = \sqrt{x + 4}$

$\qquad 0 = x + 4$

$\qquad -4 = x$

$\qquad (-4, 0)$

y-intercept: $y = \sqrt{0 + 4} = 2$

$\qquad (0, 2)$

15. $y = |3x - 7|$

x-intercept: $0 = |3x - 7|$

$\qquad 0 = 3x - 7$

$\qquad \frac{7}{3} = 0$

$\qquad \left(\frac{7}{3}, 0\right)$

y-intercept: $y = |3(0) - 7| = 7$

$\qquad (0, 7)$

17. $y = 2x^3 - 4x^2$

x-intercepts: $0 = 2x^3 - 4x^2$

$\qquad 0 = 2x^2(x - 2)$

$\qquad x = 0 \quad \text{or} \quad x = 2$

$\qquad (0, 0), (2, 0)$

y-intercept: $y = 2(0)^3 - 4(0)^2$

$\qquad y = 0$

$\qquad (0, 0)$

19. $y^2 = 6 - x$

x-intercept: $0 = 6 - x$

$\qquad x = 6$

$\qquad (6, 0)$

y-intercepts: $y^2 = 6 - 0$

$\qquad y = \pm\sqrt{6}$

$\qquad \left(0, \sqrt{6}\right), \left(0, -\sqrt{6}\right)$

21. y-axis symmetry

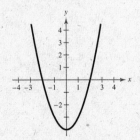

23. Origin symmetry

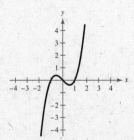

25. $x^2 - y = 0$

$(-x)^2 - y = 0 \implies x^2 - y = 0 \implies$ *y*-axis symmetry

$x^2 - (-y) = 0 \implies x^2 + y = 0 \implies$ No *x*-axis symmetry

$(-x)^2 - (-y) = 0 \implies x^2 + y = 0 \implies$ No origin symmetry

27. $y = x^3$

$y = (-x)^3 \implies y = -x^3 \implies$ No *y*-axis symmetry

$-y = x^3 \implies y = -x^3 \implies$ No *x*-axis symmetry

$-y = (-x)^3 \implies -y = -x^3 \implies y = x^3 \implies$ Origin symmetry

29. $y = \dfrac{x}{x^2 + 1}$

$y = \dfrac{-x}{(-x)^2 + 1} \implies y = \dfrac{-x}{x^2 + 1} \implies$ No *y*-axis symmetry

$-y = \dfrac{x}{x^2 + 1} \implies y = \dfrac{-x}{x^2 + 1} \implies$ No *x*-axis symmetry

$-y = \dfrac{-x}{(-x)^2 + 1} \implies -y = \dfrac{-x}{x^2 + 1} \implies y = \dfrac{x}{x^2 + 1} \implies$ Origin symmetry

31. $xy^2 + 10 = 0$

$(-x)y^2 + 10 = 0 \implies -xy^2 + 10 = 0 \implies$ No *y*-axis symmetry

$x(-y)^2 + 10 = 0 \implies xy^2 + 10 = 0 \implies$ *x*-axis symmetry

$(-x)(-y)^2 + 10 = 0 \implies -xy^2 + 10 = 0 \implies$ No origin symmetry

33. $y = -3x + 1$

x-intercept: $\left(\tfrac{1}{3}, 0\right)$

y-intercept: $(0, 1)$

No axis or origin symmetry

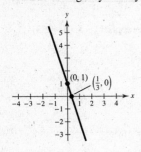

35. $y = x^2 - 2x$

Intercepts: $(0, 0), (2, 0)$

No axis or origin symmetry

x	-1	0	1	2	3
y	3	0	-1	0	3

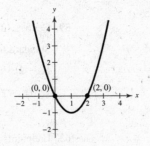

37. $y = x^3 + 3$

Intercepts: $(0, 3), \left(\sqrt[3]{-3}, 0\right)$

No axis or origin symmetry

x	-2	-1	0	1	2
y	-5	2	3	4	11

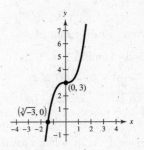

39. $y = \sqrt{x - 3}$

Domain: $[3, \infty)$

Intercept: $(3, 0)$

No axis or origin symmetry

x	3	4	7	12
y	0	1	2	3

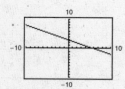

41. $y = |x - 6|$

Intercepts: $(0, 6), (6, 0)$

No axis or origin symmetry

x	-2	0	2	4	6	8	10
y	8	6	4	2	0	2	4

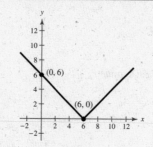

43. $x = y^2 - 1$

Intercepts: $(0, -1), (0, 1), (-1, 0)$

x-axis symmetry

x	-1	0	3
y	0	± 1	± 2

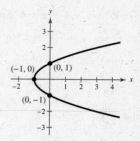

45. $y = 3 - \dfrac{1}{2}x$

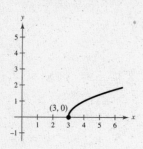

Intercepts: $(6, 0), (0, 3)$

47. $y = x^2 - 4x + 3$

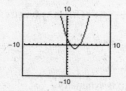

Intercepts: $(3, 0), (1, 0), (0, 3)$

49. $y = \dfrac{2x}{x - 1}$

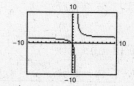

Intercept: $(0, 0)$

51. $y = \sqrt[3]{x}$

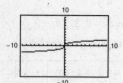

Intercept: $(0, 0)$

53. $y = x\sqrt{x + 6}$

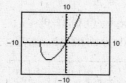

Intercepts: $(0, 0), (-6, 0)$

55. $y = |x + 3|$

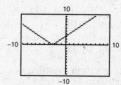

Intercepts: $(-3, 0), (0, 3)$

57. Center: $(0, 0)$; radius: 4

Standard form:

$(x - 0)^2 + (y - 0)^2 = 4^2$

$x^2 + y^2 = 16$

59. Center: $(2, -1)$; radius: 4

Standard form:

$(x - 2)^2 + (y - (-1))^2 = 4^2$

$(x - 2)^2 + (y + 1)^2 = 16$

61. Center: $(-1, 2)$; solution point: $(0, 0)$

$(x - (-1))^2 + (y - 2)^2 = r^2$

$(0 + 1)^2 + (0 - 2)^2 = r^2 \Longrightarrow 5 = r^2$

Standard form: $(x + 1)^2 + (y - 2)^2 = 5$

63. Endpoints of a diameter: $(0, 0), (6, 8)$

Center: $\left(\dfrac{0 + 6}{2}, \dfrac{0 + 8}{2} \right) = (3, 4)$

$(x - 3)^2 + (y - 4)^2 = r^2$

$(0 - 3)^2 + (0 - 4)^2 = r^2 \Longrightarrow 25 = r^2$

Standard form: $(x - 3)^2 + (y - 4)^2 = 25$

65. $x^2 + y^2 = 25$

Center: $(0, 0)$, radius: 5

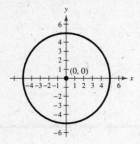

67. $(x - 1)^2 + (y + 3)^2 = 9$

Center: $(1, -3)$, radius: 3

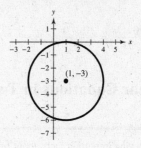

69. $\left(x - \frac{1}{2}\right)^2 + \left(y - \frac{1}{2}\right)^2 = \frac{9}{4}$

Center: $\left(\frac{1}{2}, \frac{1}{2}\right)$, radius: $\frac{3}{2}$

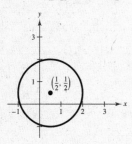

71. $y = 225{,}000 - 20{,}000t, \ 0 \le t \le 8$

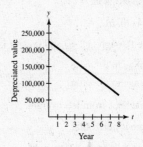

73. (a)

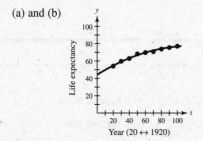

(c) 8000

0   180

(b) $2x + 2y = \frac{1040}{3}$

$2y = \frac{1040}{3} - 2x$

$y = \frac{520}{3} - x$

$A = xy = x\left(\frac{520}{3} - x\right)$

(d) When $x = y = 86\frac{2}{3}$ yards, the area is a maximum of $7511\frac{1}{9}$ square yards.

(e) A regulation NFL playing field is 120 yards long and $53\frac{1}{3}$ yards wide. The actual area is 6400 square yards.

75. $y = -0.0025t^2 + 0.574t + 44.25, \ 20 \le t \le 100$

(a) and (b)

(c) For the year 1948, let $t = 48$: $y \approx 66.0$ years.

(d) For the year 2005, let $t = 105$: $y \approx 77.0$ years.

For the year 2010, let $t = 110$: $y \approx 77.1$ years.

(e) No. The graph reaches a maximum of $y \approx 77.2$ years when $t \approx 114.8$, or during the year 2014. After this time, the model has life expectancy decreasing, which is not realistic.

77. False. A graph is symmetric with respect to the x-axis if, whenever (x, y) is on the graph, $(x, -y)$ is also on the graph.

79. The viewing window is incorrect. Change the viewing window. Examples will vary. For example, $y = x^2 + 20$ will not appear in the standard window setting.

81. $9x^5 + 4x^3 - 7$

     Terms: $9x^5, 4x^3, -7$

83. $\sqrt{18x} - \sqrt{2x} = 3\sqrt{2x} - \sqrt{2x} = 2\sqrt{2x}$

85. $\dfrac{70}{\sqrt{7x}} = \dfrac{70}{\sqrt{7x}} \cdot \dfrac{\sqrt{7x}}{\sqrt{7x}} = \dfrac{70\sqrt{7x}}{7x} = \dfrac{10\sqrt{7x}}{x}$

87. $\sqrt[6]{t^2} = t^{2/6} = |t|^{1/3} = \sqrt[3]{|t|}$

Section 1.3   Linear Equations in Two Variables

You should know the following important facts about lines.

■   The graph of $y = mx + b$ is a straight line. It is called a linear equation in two variables.

     (a) The slope (steepness) is m.

     (b) The y-intercept is $(0, b)$.

■   The slope of the line through (x_1, y_1) and (x_2, y_2) is

$$m = \frac{y_2 - y_1}{x_2 - x_1} = \frac{\text{change in } y}{\text{change in } x} = \frac{\text{rise}}{\text{run}}.$$

     (a) If $m > 0$, the line rises from left to right.

     (b) If $m = 0$, the line is horizontal.

     (c) If $m < 0$, the line falls from left to right.

     (d) If m is undefined, the line is vertical.

■   Equations of Lines

     (a) Slope-Intercept Form: $y = mx + b$

     (b) Point-Slope Form: $y - y_1 = m(x - x_1)$

     (c) Two-Point Form: $y - y_1 = \dfrac{y_2 - y_1}{x_2 - x_1}(x - x_1)$

     (d) General Form: $Ax + By + C = 0$

     (e) Vertical Line: $x = a$

     (f) Horizontal Line: $y = b$

■   Given two distinct nonvertical lines

     $L_1: y = m_1x + b_1$ and $L_2: y = m_2x + b_2$

     (a) L_1 is parallel to L_2 if and only if $m_1 = m_2$ and $b_1 \neq b_2$.

     (b) L_1 is perpendicular to L_2 if and only if $m_1 = -1/m_2$.

Vocabulary Check

1. linear

2. slope

3. parallel

4. perpendicular

5. rate or rate of change

6. linear extrapolation

7. (a) $Ax + By + C = 0$ (iii) general form

     (b) $x = a$ (i)   vertical line

     (c) $y = b$ (v)   horizontal line

     (d) $y = mx + b$ (ii)   slope-intercept form

     (e) $y - y_1 = m(x - x_1)$ (iv)   point-slope form

1. (a) $m = \frac{2}{3}$. Since the slope is positive, the line rises.
Matches L_2.

(b) m is undefined. The line is vertical. Matches L_3.

(c) $m = -2$. The line falls. Matches L_1.

3.

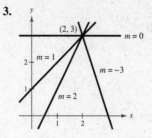

5. Two points on the line: $(0, 0)$ and $(4, 6)$

$$\text{Slope} = \frac{\text{rise}}{\text{run}} = \frac{6}{4} = \frac{3}{2}$$

7. Two points on the line: $(0, 8)$ and $(2, 0)$

$$\text{Slope} = \frac{\text{rise}}{\text{run}} = \frac{-8}{2} = -4$$

9. $y = 5x + 3$

Slope: $m = 5$

y-intercept: $(0, 3)$

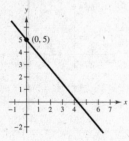

11. $y = -\frac{1}{2}x + 4$

Slope: $m = -\frac{1}{2}$

y-intercept: $(0, 4)$

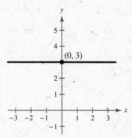

13. $5x - 2 = 0$

$x = \frac{2}{5}$, vertical line

Slope: undefined

No y-intercept

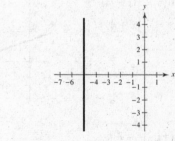

15. $7x + 6y = 30$

$y = -\frac{7}{6}x + 5$

Slope: $m = -\frac{7}{6}$

y-intercept: $(0, 5)$

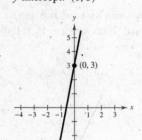

17. $y - 3 = 0$

$y = 3$, horizontal line

Slope: $m = 0$

y-intercept: $(0, 3)$

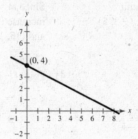

19. $x + 5 = 0$

$x = -5$

Slope: undefined (vertical line)

No y-intercept

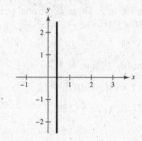

21. $m = \frac{6 - (-2)}{1 - (-3)} = \frac{8}{4} = 2$

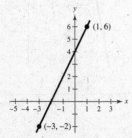

23. $m = \frac{4 - (-1)}{-6 - (-6)} = \frac{5}{0}$

m is undefined.

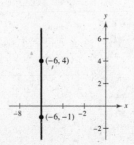

25. $m = \frac{-\frac{1}{3} - \left(-\frac{4}{3}\right)}{-\frac{3}{2} - \frac{11}{2}} = -\frac{1}{7}$

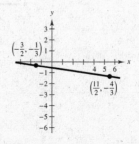

27. $m = \dfrac{1.6 - 3.1}{-5.2 - 4.8} = \dfrac{-1.5}{-10} = 0.15$

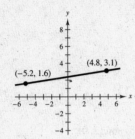

29. Point: $(2, 1)$, Slope: $m = 0$

Since $m = 0$, y does not change. Three points are $(0, 1)$, $(3, 1)$, and $(-1, 1)$.

31. Point: $(5, -6)$, Slope: $m = 1$

Since $m = 1$, y increases by 1 for every one unit increase in x. Three points are $(6, -5)$, $(7, -4)$, and $(8, -3)$.

33. Point: $(-8, 1)$, Slope is undefined.

Since m is undefined, x does not change. Three points are $(-8, 0)$, $(-8, 2)$, and $(-8, 3)$.

35. Point: $(-5, 4)$, Slope: $m = 2$

Since $m = 2 = \frac{2}{1}$, y increases by 2 for every one unit increase in x. Three additional points are $(-4, 6)$, $(-3, 8)$, and $(-2, 10)$.

37. Point: $(7, -2)$, Slope: $m = \frac{1}{2}$

Since $m = \frac{1}{2}$, y increases by 1 unit for every two unit increase in x. Three additional points are $(9, -1)$, $(11, 0)$, and $(13, 1)$.

39. Point $(0, -2)$; $m = 3$

$y + 2 = 3(x - 0)$

$y = 3x - 2$

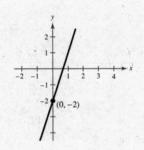

41. Point $(-3, 6)$; $m = -2$

$y - 6 = -2(x + 3)$

$y = -2x$

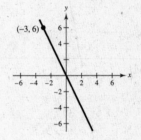

43. Point $(4, 0)$; $m = -\frac{1}{3}$

$y - 0 = -\frac{1}{3}(x - 4)$

$y = -\frac{1}{3}x + \frac{4}{3}$

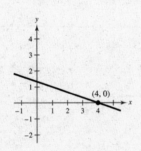

45. Point $(6, -1)$; m is undefined.

The line is vertical.

$x = 6$

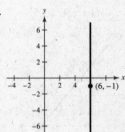

47. Point $\left(4, \frac{5}{2}\right)$; $m = 0$

The line is horizontal.

$y = \frac{5}{2}$

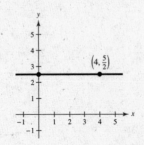

49. Point $(-5.1, 1.8)$; $m = 5$

$y - 1.8 = 5(x - (-5.1))$

$y = 5x + 27.3$

51. $(5, -1)$ and $(-5, 5)$

$$y + 1 = \frac{5 + 1}{-5 - 5}(x - 5)$$

$$y = -\frac{3}{5}(x - 5) - 1$$

$$y = -\frac{3}{5}x + 2$$

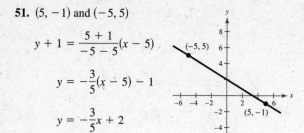

53. $(-8, 1)$ and $(-8, 7)$

Since both points have $x = -8$, the slope is undefined, and the line is vertical.

$$x = -8$$

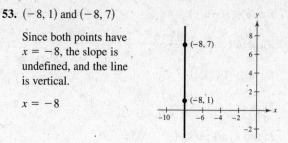

55. $\left(2, \frac{1}{2}\right)$ and $\left(\frac{1}{2}, \frac{5}{4}\right)$

$$y - \frac{1}{2} = \frac{\frac{5}{4} - \frac{1}{2}}{\frac{1}{2} - 2}(x - 2)$$

$$y = -\frac{1}{2}(x - 2) + \frac{1}{2}$$

$$y = -\frac{1}{2}x + \frac{3}{2}$$

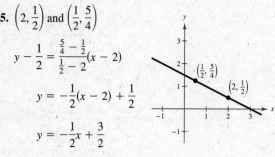

57. $\left(-\frac{1}{10}, -\frac{3}{5}\right)$ and $\left(\frac{9}{10}, -\frac{9}{5}\right)$

$$y - \left(-\frac{3}{5}\right) = \frac{-\frac{9}{5} - \left(-\frac{3}{5}\right)}{\frac{9}{10} - \left(-\frac{1}{10}\right)}\left(x - \left(-\frac{1}{10}\right)\right)$$

$$y = -\frac{6}{5}\left(x + \frac{1}{10}\right) - \frac{3}{5}$$

$$y = -\frac{6}{5}x - \frac{18}{25}$$

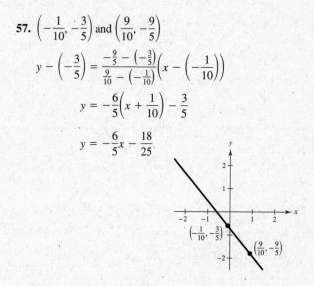

59. $(1, 0.6)$ and $(-2, -0.6)$

$$y - 0.6 = \frac{-0.6 - 0.6}{-2 - 1}(x - 1)$$

$$y = 0.4(x - 1) + 0.6$$

$$y = 0.4x + 0.2$$

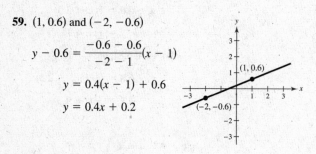

61. $(2, -1)$ and $\left(\frac{1}{3}, -1\right)$

$$y + 1 = \frac{-1 - (-1)}{\frac{1}{3} - 2}(x - 2)$$

$$y + 1 = 0$$

$$y = -1$$

The line is horizontal.

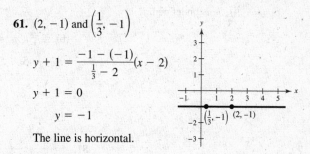

63. $\left(\frac{7}{3}, -8\right)$ and $\left(\frac{7}{3}, 1\right)$

$$m = \frac{1 - (-8)}{\frac{7}{3} - \frac{7}{3}} = \frac{9}{0} \text{ and is undefined.}$$

$$x = \frac{7}{3}$$

The line is vertical.

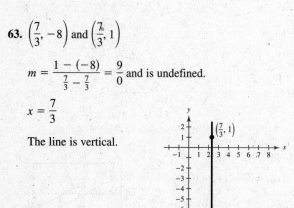

65. L_1: $(0, -1), (5, 9)$

Slope of L_1: $m = \frac{9 + 1}{5 - 0} = 2$

L_2: $(0, 3), (4, 1)$

Slope of L_2: $m = \frac{1 - 3}{4 - 0} = -\frac{1}{2}$

L_1 and L_2 are perpendicular.

67. L_1: $(3, 6)$, $(-6, 0)$

Slope of L_1: $m = \dfrac{0 - 6}{-6 - 3} = \dfrac{2}{3}$

L_2: $(0, -1)$, $\left(5, \frac{7}{3}\right)$

Slope of L_2: $m = \dfrac{\frac{7}{3} + 1}{5 - 0} = \dfrac{2}{3}$

L_1 and L_2 are parallel.

69. $4x - 2y = 3$

$y = 2x - \frac{3}{2}$

Slope: $m = 2$

(a) $(2, 1)$, $m = 2$

$y - 1 = 2(x - 2)$

$y = 2x - 3$

(b) $(2, 1)$, $m = -\dfrac{1}{2}$

$y - 1 = -\dfrac{1}{2}(x - 2)$

$y = -\dfrac{1}{2}x + 2$

71. $3x + 4y = 7$

$y = -\dfrac{3}{4}x + \dfrac{7}{4}$

Slope: $m = -\dfrac{3}{4}$

(a) $\left(-\dfrac{2}{3}, \dfrac{7}{8}\right)$, $m = -\dfrac{3}{4}$

$y - \dfrac{7}{8} = -\dfrac{3}{4}\left(x - \left(-\dfrac{2}{3}\right)\right)$

$y = -\dfrac{3}{4}x + \dfrac{3}{8}$

(b) $\left(-\dfrac{2}{3}, \dfrac{7}{8}\right)$, $m = \dfrac{4}{3}$

$y - \dfrac{7}{8} = \dfrac{4}{3}\left(x - \left(-\dfrac{2}{3}\right)\right)$

$y = \dfrac{4}{3}x + \dfrac{127}{72}$

73. $y = -3$

$m = 0$

(a) $(-1, 0)$ and $m = 0$

$y = 0$

(b) $(-1, 0)$, m is undefined.

$x = -1$

75. $x = 4$

m is undefined.

(a) $(2, 5)$, m is undefined. The line is vertical, passing through $(2, 5)$.

$x = 2$

(b) $(2, 5)$, $m = 0$

$y = 5$

77. $x - y = 4$

$y = x - 4$

Slope: $m = 1$

(a) $(2.5, 6.8)$, $m = 1$

$y - 6.8 = 1(x - 2.5)$

$y = x + 4.3$

(b) $(2.5, 6.8)$, $m = -1$

$y - 6.8 = (-1)(x - 2.5)$

$y = -x + 9.3$

79. $\dfrac{x}{2} + \dfrac{y}{3} = 1$

$3x + 2y - 6 = 0$

81. $\dfrac{x}{-1/6} + \dfrac{y}{-2/3} = 1$

$6x + \dfrac{3}{2}y = -1$

$12x + 3y + 2 = 0$

83. $\dfrac{x}{c} + \dfrac{y}{c} = 1$, $c \neq 0$

$x + y = c$

$1 + 2 = c$

$3 = c$

$x + y = 3$

$x + y - 3 = 0$

85. (a) $y = 2x$

(b) $y = -2x$

(c) $y = \frac{1}{2}x$

(b) and (c) are perpendicular.

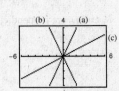

87. (a) $y = -\frac{1}{2}x$

(b) $y = -\frac{1}{2}x + 3$

(c) $y = 2x - 4$

(a) and (b) are parallel. (c) is perpendicular to (a) and (b).

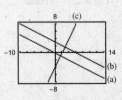

89. Set the distance between $(4, -1)$ and (x, y) equal to the distance between $(-2, 3)$ and (x, y).

$$\sqrt{(x-4)^2 + [y-(-1)]^2} = \sqrt{[x-(-2)]^2 + (y-3)^2}$$

$$(x-4)^2 + (y+1)^2 = (x+2)^2 + (y-3)^2$$

$$x^2 - 8x + 16 + y^2 + 2y + 1 = x^2 + 4x + 4 + y^2 - 6y + 9$$

$$-8x + 2y + 17 = 4x - 6y + 13$$

$$0 = 12x - 8y - 4$$

$$0 = 4(3x - 2y - 1)$$

$$0 = 3x - 2y - 1$$

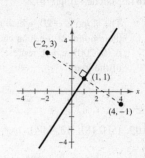

This line is the perpendicular bisector of the line segment connecting $(4, -1)$ and $(-2, 3)$.

91. Set the distance between $\left(3, \frac{5}{2}\right)$ and (x, y) equal to the distance between $(-7, 1)$ and (x, y).

$$\sqrt{(x-3)^2 + \left(y-\frac{5}{2}\right)^2} = \sqrt{[x-(-7)]^2 + (y-1)^2}$$

$$(x-3)^2 + \left(y-\frac{5}{2}\right)^2 = (x+7)^2 + (y-1)^2$$

$$x^2 - 6x + 9 + y^2 - 5y + \frac{25}{4} = x^2 + 14x + 49 + y^2 - 2y + 1$$

$$-6x - 5y + \frac{61}{4} = 14x - 2y + 50$$

$$-24x - 20y + 61 = 56x - 8y + 200$$

$$80x + 12y + 139 = 0$$

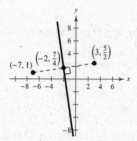

This line is the perpendicular bisector of the line segment connecting $\left(3, \frac{5}{2}\right)$ and $(-7, 1)$.

93. (a) $m = 135$. The sales are increasing 135 units per year.

(b) $m = 0$. There is no change in sales during the year.

(c) $m = -40$. The sales are decreasing 40 units per year.

95. (a) $(0, 55{,}722), (2, 61{,}768)$: $m = \dfrac{61{,}768 - 55{,}722}{2 - 0} = 3023$ $(6, 69{,}277), (8, 74{,}380)$: $m = \dfrac{74{,}380 - 69{,}277}{8 - 6} = 2551.5$

$(2, 61{,}768), (4, 64{,}993)$: $m = \dfrac{64{,}993 - 61{,}768}{4 - 2} = 1612.5$ $(8, 74{,}380), (10, 79{,}839)$: $m = \dfrac{79{,}839 - 74{,}380}{10 - 8} = 2729.5$

$(4, 64{,}993), (6, 69{,}277)$: $m = \dfrac{69{,}277 - 64{,}993}{6 - 4} = 2142$ $(10, 79{,}839), (12, 83{,}944)$: $m = \dfrac{83{,}944 - 79{,}839}{12 - 10} = 2052.5$

The average salary increased the most from 1990 to 1992 and the least from 1992 to 1994.

(b) $(0, 55{,}722), (12, 83{,}944)$: $m = \dfrac{83{,}944 - 55{,}722}{12 - 0} \approx \$2351.83$

(c) The average salary for senior high school principals increased by \$2351.83 per year over the 12 years between 1990 and 2002.

97. $y = \frac{6}{100}x$

$y = \frac{6}{100}(200) = 12$ feet

99. $(5, 2540), m = -125$

$$V - 2540 = -125(t - 5)$$

$$V - 2540 = -125t + 625$$

$$V = -125t + 3165, \ 5 \le t \le 10$$

101. Matches graph (b).

The slope is -20, which represents the decrease in the amount of the loan each week. The y-intercept is $(0, 200)$, which represents the original amount of the loan.

103. Matches graph (a).

The slope is 0.32, which represents the increase in travel cost for each mile driven. The y-intercept is $(0, 30)$, which represents the fixed cost of $30 per day for meals. This amount does not depend on the number of miles driven.

105. $(5, 0.18)$, $(13, 4.04)$: $m = \dfrac{4.04 - 0.18}{13 - 5} = 0.4825$

$$y - 0.18 = 0.4825(t - 5)$$

$$y = 0.4825t - 2.2325$$

For 2008, use $t = 18$: $y(18) \approx \$6.45$

For 2010, use $t = 20$: $y(20) \approx \$7.42$

107. Using the points $(0, 875)$ and $(5, 0)$, where the first coordinate represents the year t and the second coordinate represents the value V, we have

$$m = \dfrac{0 - 875}{5 - 0} = -175$$

$$V = -175t + 875, \ 0 \le t \le 5.$$

109. (a) $(0, 40{,}571)$, $(4, 41{,}289)$:

$$m = \dfrac{41{,}289 - 40{,}571}{4 - 0} = 179.5$$

$$y = 179.5t + 40{,}571$$

(b) For 2008, use $t = 8$: $y(8) = 42{,}007$ students.

For 2010, use $t = 10$: $y(10) = 42{,}366$ students.

(c) The slope is $m = 179.5$, which represents the increase in the number of students each year.

111. Sale price = List price $-$ 15% of the list price

$$S = L - 0.15L$$

$$S = 0.85L$$

113. (a) $C = 36{,}500 + 5.25t + 11.50t$

$$= 16.75t + 36{,}500$$

(b) $R = 27t$

(c) $P = R - C$

$$= 27t - (16.75t + 36{,}500)$$

$$= 10.25t - 36{,}500$$

(d) $\qquad 0 = 10.25t - 36{,}500$

$$36{,}500 = 10.25t$$

$$t \approx 3561 \text{ hours}$$

115. (a)

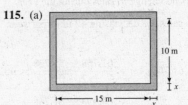

(b) $y = 2(15 + 2x) + 2(10 + 2x) = 8x + 50$

(c)

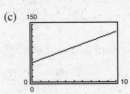

(d) Since $m = 8$, each 1-meter increase in x will increase y by 8 meters.

117. $C = 0.38x + 120$

119. (a) and (b)

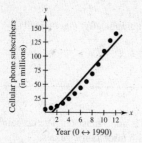

Year (0 ↔ 1990)

(c) Answers will vary. Find two points on your line and then find the equation of the line through your points. Sample answer: $y \approx 11.72x - 14.08$

(d) Answers will vary. Sample answer: The y-intercept should represent the number of initial subscribers. In this case, since b is negative, it cannot be interpreted as such. The slope of 11.72 represents the increase in the number of subscribers per year (in millions).

(e) The model is a fairly good fit to the data.

(f) Answers will vary. Sample answer:

$y(18) \approx 11.72(18) - 14.08$

$= 196.88$ million subscribers in 2008

121. False. The slope with the greatest magnitude corresponds to the steepest line.

123. Using the Distance Formula, we have $AB = 6$, $BC = \sqrt{40}$, and $AC = 2$. Since $6^2 + 2^2 = \left(\sqrt{40}\right)^2$, the triangle is a right triangle.

125. No. The slope cannot be determined without knowing the scale on the y-axis. The slopes will be the same if the scale on the y-axis of (a) is $2\frac{1}{2}$ and the scale on the y-axis of (b) is 1. Then the slope of both is $\frac{5}{4}$.

127. The V-intercept measures the initial cost and the slope measures annual depreciation.

129. $y = 8 - 3x$ is a linear equation with slope $m = -3$ and y-intercept $(0, 8)$. Matches graph (d).

131. $y = \frac{1}{2}x^2 + 2x + 1$ is a quadratic equation. Its graph is a parabola with vertex $(-2, -1)$ and y-intercept $(0, 1)$. Matches graph (a).

133. $-7(3 - x) = 14(x - 1)$

$-21 + 7x = 14x - 14$

$-7x = 7$

$x = -1$

135. $2x^2 - 21x + 49 = 0$

$(2x - 7)(x - 7) = 0$

$2x - 7 = 0$ or $x - 7 = 0$

$x = \frac{7}{2}$ or $x = 7$

137. $\sqrt{x - 9} + 15 = 0$

$\sqrt{x - 9} = -15$

No real solution

The square root of $x - 9$ cannot be negative.

139. Answers will vary.

Section 1.4   Functions

- Given a set or an equation, you should be able to determine if it represents a function.

- Know that functions can be represented in four ways: verbally, numerically, graphically, and algebraically.

- Given a function, you should be able to do the following.

    (a) Find the domain and range.

    (b) Evaluate it at specific values.

- You should be able to use function notation.

Vocabulary Check

1. domain; range; function

2. verbally; numerically; graphically; algebraically

3. independent; dependent

4. piecewise-defined

5. implied domain

6. difference quotient

---

1. Yes, the relationship is a function. Each domain value is matched with only one range value.

3. No, the relationship is not a function. The domain values are each matched with three range values.

5. Yes, it does represent a function. Each input value is matched with only one output value.

7. No, it does not represent a function. The input values of 10 and 7 are each matched with two output values.

9. (a) Each element of A is matched with exactly one element of B, so it does represent a function.

   (b) The element 1 in A is matched with two elements, -2 and 1 of B, so it does not represent a function.

   (c) Each element of A is matched with exactly one element of B, so it does represent a function.

   (d) The element 2 in A is not matched with an element of B, so the relation does not represent a function.

11. Each is a function. For each year there corresponds one and only one circulation.

13. $x^2 + y^2 = 4 \implies y = \pm\sqrt{4 - x^2}$

   No, y *is not* a function of x.

15. $x^2 + y = 4 \implies y = 4 - x^2$

   Yes, y *is* a function of x.

17. $2x + 3y = 4 \implies y = \frac{1}{3}(4 - 2x)$

   Yes, y *is* a function of x.

19. $y^2 = x^2 - 1 \implies y = \pm\sqrt{x^2 - 1}$

   Thus, y *is not* a function of x.

21. $y = |4 - x|$

   Yes, y *is* a function of x.

23. $x = 14$

   Thus, this *is not* a function of x.

25. $f(x) = 2x - 3$

   (a) $f(1) = 2(1) - 3 = -1$

   (b) $f(-3) = 2(-3) - 3 = -9$

   (c) $f(x - 1) = 2(x - 1) - 3 = 2x - 5$

27. $V(r) = \frac{4}{3}\pi r^3$

   (a) $V(3) = \frac{4}{3}\pi(3)^3 = \frac{4}{3}\pi(27) = 36\pi$

   (b) $V\left(\frac{3}{2}\right) = \frac{4}{3}\pi\left(\frac{3}{2}\right)^3 = \frac{4}{3}\pi\left(\frac{27}{8}\right) = \frac{9}{2}\pi$

   (c) $V(2r) = \frac{4}{3}\pi(2r)^3 = \frac{4}{3}\pi(8r^3) = \frac{32}{3}\pi r^3$

29. $f(y) = 3 - \sqrt{y}$

   (a) $f(4) = 3 - \sqrt{4} = 1$

   (b) $f(0.25) = 3 - \sqrt{0.25} = 2.5$

   (c) $f(4x^2) = 3 - \sqrt{4x^2} = 3 - 2|x|$

31. $q(x) = \dfrac{1}{x^2 - 9}$

   (a) $q(0) = \dfrac{1}{0^2 - 9} = -\dfrac{1}{9}$

   (b) $q(3) = \dfrac{1}{3^2 - 9}$ is undefined.

   (c) $q(y + 3) = \dfrac{1}{(y + 3)^2 - 9} = \dfrac{1}{y^2 + 6y}$

33. $f(x) = \dfrac{|x|}{x}$

    (a) $f(2) = \dfrac{|2|}{2} = 1$

    (b) $f(-2) = \dfrac{|-2|}{-2} = -1$

    (c) $f(x-1) = \dfrac{|x-1|}{x-1} = \begin{cases} -1, & \text{if } x < 1 \\ 1, & \text{if } x > 1 \end{cases}$

35. $f(x) = \begin{cases} 2x+1, & x < 0 \\ 2x+2, & x \geq 0 \end{cases}$

    (a) $f(-1) = 2(-1) + 1 = -1$

    (b) $f(0) = 2(0) + 2 = 2$

    (c) $f(2) = 2(2) + 2 = 6$

37. $f(x) = \begin{cases} 3x-1, & x < -1 \\ 4, & -1 \leq x \leq 1 \\ x^2, & x > 1 \end{cases}$

    (a) $f(-2) = 3(-2) - 1 = -7$

    (b) $f\left(-\tfrac{1}{2}\right) = 4$

    (c) $f(3) = 3^2 = 9$

39. $f(x) = x^2 - 3$

$f(-2) = (-2)^2 - 3 = 1$

$f(-1) = (-1)^2 - 3 = -2$

$f(0) = (0)^2 - 3 = -3$

$f(1) = (1)^2 - 3 = -2$

$f(2) = (2)^2 - 3 = 1$

x	-2	-1	0	1	2
$f(x)$	1	-2	-3	-2	1

41. $h(t) = \tfrac{1}{2}|t + 3|$

$h(-5) = \tfrac{1}{2}|-5 + 3| = 1$

$h(-4) = \tfrac{1}{2}|-4 + 3| = \tfrac{1}{2}$

$h(-3) = \tfrac{1}{2}|-3 + 3| = 0$

$h(-2) = \tfrac{1}{2}|-2 + 3| = \tfrac{1}{2}$

$h(-1) = \tfrac{1}{2}|-1 + 3| = 1$

t	-5	-4	-3	-2	-1
$h(t)$	1	$\tfrac{1}{2}$	0	$\tfrac{1}{2}$	1

43. $f(x) = \begin{cases} -\tfrac{1}{2}x + 4, & x \leq 0 \\ (x-2)^2, & x > 0 \end{cases}$

x	-2	-1	0	1	2
$f(x)$	5	$\tfrac{9}{2}$	4	1	0

$f(-2) = -\tfrac{1}{2}(-2) + 4 = 5$

$f(-1) = -\tfrac{1}{2}(-1) + 4 = 4\tfrac{1}{2} = \tfrac{9}{2}$

$f(0) = -\tfrac{1}{2}(0) + 4 = 4$

$f(1) = (1 - 2)^2 = 1$

$f(2) = (2 - 2)^2 = 0$

45. $15 - 3x = 0$

$3x = 15$

$x = 5$

47. $\dfrac{3x - 4}{5} = 0$

$3x - 4 = 0$

$x = \dfrac{4}{3}$

49. $x^2 - 9 = 0$

$x^2 = 9$

$x = \pm 3$

51. $x^3 - x = 0$

$x(x^2 - 1) = 0$

$x(x + 1)(x - 1) = 0$

$x = 0,\ x = -1,\text{ or } x = 1$

53. $f(x) = g(x)$

$x^2 + 2x + 1 = 3x + 3$

$x^2 - x - 2 = 0$

$(x + 1)(x - 2) = 0$

$x = -1\ \text{ or }\ x = 2$

55. $f(x) = g(x)$

$\sqrt{3x} + 1 = x + 1$

$\sqrt{3x} = x$

$3x = x^2$

$0 = x^2 - 3x$

$0 = x(x - 3)$

$x = 0\ \text{ or }\ x = 3$

57. $f(x) = 5x^2 + 2x - 1$

Since $f(x)$ is a polynomial, the domain is all real numbers x.

59. $h(t) = \dfrac{4}{t}$

Domain: All real numbers t except $t = 0$

61. $g(y) = \sqrt{y - 10}$

Domain: $y - 10 \geq 0$

$y \geq 10$

63. $f(x) = \sqrt[4]{1 - x^2}$

Domain: $1 - x^2 \geq 0$

By solving this inequality, we conclude that $-1 \leq x \leq 1$ or $[-1, 1]$.

65. $g(x) = \dfrac{1}{x} - \dfrac{3}{x + 2}$

Domain: All real numbers x except $x = 0$, $x = -2$

67. $f(s) = \dfrac{\sqrt{s - 1}}{s - 4}$

Domain: $s - 1 \geq 0 \Rightarrow s \geq 1$ and $s \neq 4$

The domain consists of all real numbers s, such that $s \geq 1$ and $s \neq 4$.

69. $f(x) = \dfrac{x - 4}{\sqrt{x}}$

The domain is all real numbers such that $x > 0$ or $(0, \infty)$.

71. $f(x) = x^2$

$\{(-2, 4), (-1, 1), (0, 0), (1, 1), (2, 4)\}$

73. $f(x) = |x| + 2$

$\{(-2, 4), (-1, 3), (0, 2), (1, 3), (2, 4)\}$

75. By plotting the points, we have a parabola, so $g(x) = cx^2$. Since $(-4, -32)$ is on the graph, we have $-32 = c(-4)^2 \Rightarrow c = -2$. Thus, $g(x) = -2x^2$.

77. Since the function is undefined at 0, we have $r(x) = c/x$. Since $(-4, -8)$ is on the graph, we have $-8 = c/-4 \Rightarrow c = 32$. Thus, $r(x) = 32/x$.

79.
$$f(x) = x^2 - x + 1$$
$$f(2 + h) = (2 + h)^2 - (2 + h) + 1$$
$$= 4 + 4h + h^2 - 2 - h + 1$$
$$= h^2 + 3h + 3$$
$$f(2) = (2)^2 - 2 + 1 = 3$$
$$f(2 + h) - f(2) = h^2 + 3h$$
$$\frac{f(2 + h) - f(2)}{h} = \frac{h^2 + 3h}{h} = h + 3, \; h \neq 0$$

81.
$$f(x) = x^3 + 3x$$
$$f(x + h) = (x + h)^3 + 3(x + h)$$
$$= x^3 + 3x^2h + 3xh^2 + h^3 + 3x + 3h$$
$$\frac{f(x + h) - f(x)}{h} = \frac{(x^3 + 3x^2h + 3xh^2 + h^3 + 3x + 3h) - (x^3 + 3x)}{h}$$
$$= \frac{h(3x^2 + 3xh + h^2 + 3)}{h}$$
$$= 3x^2 + 3xh + h^2 + 3, \; h \neq 0$$

83. $g(x) = \dfrac{1}{x^2}$
$$\frac{g(x) - g(3)}{x - 3} = \frac{\dfrac{1}{x^2} - \dfrac{1}{9}}{x - 3}$$
$$= \frac{9 - x^2}{9x^2(x - 3)}$$
$$= \frac{-(x + 3)(x - 3)}{9x^2(x - 3)}$$
$$= -\frac{x + 3}{9x^2}, \; x \neq 3$$

85. $f(x) = \sqrt{5x}$

$$\frac{f(x) - f(5)}{x - 5} = \frac{\sqrt{5x} - 5}{x - 5}$$

87. $A = s^2$ and $P = 4s \implies \dfrac{P}{4} = s$

$$A = \left(\frac{P}{4}\right)^2 = \frac{P^2}{16}$$

89. (a)

Height, x	Volume, V
1	484
2	800
3	972
4	1024
5	980
6	864

The volume is maximum when $x = 4$ and $V = 1024$ cubic centimeters.

(b)

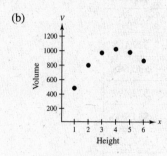

V is a function of x.

(c) $V = x(24 - 2x)^2$

Domain: $0 < x < 12$

91. $A = \dfrac{1}{2}bh = \dfrac{1}{2}xy$

Since $(0, y)$, $(2, 1)$, and $(x, 0)$ all lie on the same line, the slopes between any pair are equal.

$$\frac{1 - y}{2 - 0} = \frac{0 - 1}{x - 2}$$

$$\frac{1 - y}{2} = \frac{-1}{x - 2}$$

$$y = \frac{2}{x - 2} + 1$$

$$y = \frac{x}{x - 2}$$

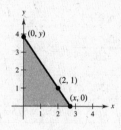

Therefore,

$$A = \frac{1}{2}x\left(\frac{x}{x - 2}\right) = \frac{x^2}{2(x - 2)}.$$

The domain of A includes x-values such that $x^2/[2(x - 2)] > 0$. By solving this inequality, we find that the domain is $x > 2$.

93. $y = -\frac{1}{10}x^2 + 3x + 6$

$y(30) = -\frac{1}{10}(30)^2 + 3(30) + 6 = 6$ feet

If the child holds a glove at a height of 5 feet, then the ball *will* be over the child's head since it will be at a height of 6 feet.

95. $p(t) = \begin{cases} 0.182t^2 + 0.57t + 27.3, & 0 \le t \le 7 \\ 2.50t + 21.3, & 8 \le t \le 12 \end{cases}$

Year	Function Value	Price
1990	$p(0) = 27.3$	$27,300
1991	$p(1) = 28.052$	$28,052
1992	$p(2) = 29.168$	$29,168
1993	$p(3) = 30.648$	$30,648
1994	$p(4) = 32.492$	$32,492
1995	$p(5) = 34.7$	$34,700
1996	$p(6) = 37.272$	$37,272
1997	$p(7) = 40.208$	$40,208
1998	$p(8) = 41.3$	$41,300
1999	$p(9) = 43.8$	$43,800
2000	$p(10) = 46.3$	$46,300
2001	$p(11) = 48.8$	$48,800
2002	$p(12) = 51.3$	$51,300

97. (a) Cost = variable costs + fixed costs

$$C = 12.30x + 98,000$$

(b) Revenue = price per unit × number of units

$$R = 17.98x$$

(c) Profit = Revenue − Cost

$$P = 17.98x - (12.30x + 98,000)$$

$$P = 5.68x - 98,000$$

99. (a) $R = n(\text{rate}) = n[8.00 - 0.05(n - 80)], \ n \ge 80$

$$R = 12.00n - 0.05n^2 = 12n - \frac{n^2}{20} = \frac{240n - n^2}{20}, \ n \ge 80$$

(b)

n	90	100	110	120	130	140	150
$R(n)$	$675	$700	$715	$720	$715	$700	$675

The revenue is maximum when 120 people take the trip.

101. (a)

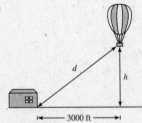

(b) $(3000)^2 + h^2 = d^2$

$$h = \sqrt{d^2 - (3000)^2}$$

Domain: $d \ge 3000$ (since both $d \ge 0$ and $d^2 - (3000)^2 \ge 0$)

103. False. The range is $[-1, \infty)$.

105. The domain is the set of inputs of the function, and the range is the set of outputs.

107. (a) Yes. The amount that you pay in sales tax will increase as the price of the item purchased increases.

(b) No. The length of time that you study the night before an exam does not necessarily determine your score on the exam.

109. $\dfrac{t}{3} + \dfrac{t}{5} = 1$

$$15\left(\frac{t}{3} + \frac{t}{5}\right) = 15(1)$$

$$5t + 3t = 15$$

$$8t = 15$$

$$t = \frac{15}{8}$$

111.
$$\frac{3}{x(x+1)} - \frac{4}{x} = \frac{1}{x+1}$$

$$x(x+1)\left[\frac{3}{x(x+1)} - \frac{4}{x}\right] = x(x+1)\left(\frac{1}{x+1}\right)$$

$$3 - 4(x+1) = x$$

$$3 - 4x - 4 = x$$

$$-1 = 5x$$

$$-\frac{1}{5} = x$$

113. $(-2, -5)$ and $(4, -1)$

$$m = \frac{-1 - (-5)}{4 - (-2)} = \frac{4}{6} = \frac{2}{3}$$

$$y - (-5) = \frac{2}{3}(x - (-2))$$

$$y + 5 = \frac{2}{3}x + \frac{4}{3}$$

$$3y + 15 = 2x + 4$$

$$2x - 3y - 11 = 0$$

115. $(-6, 5)$ and $(3, -5)$

$$m = \frac{-5 - 5}{3 - (-6)} = -\frac{10}{9}$$

$$y - 5 = -\frac{10}{9}(x - (-6))$$

$$9y - 45 = -10x - 60$$

$$10x + 9y + 15 = 0$$

Section 1.5   Analyzing Graphs of Functions

- ■ You should be able to determine the domain and range of a function from its graph.
- ■ You should be able to use the vertical line test for functions.
- ■ You should be able to find the zeros of a function.
- ■ You should be able to determine when a function is constant, increasing, or decreasing.
- ■ You should be able to approximate relative minimums and relative maximums from the graph of a function.
- ■ You should know that f is
  - (a) odd if $f(-x) = -f(x)$.   (b) even if $f(-x) = f(x)$.

Vocabulary Check

1. ordered pairs

2. vertical line test

3. zeros

4. decreasing

5. maximum

6. average rate of change; secant

7. odd

8. even

---

1. Domain: $(-\infty, -1] \cup [1, \infty)$

Range: $[0, \infty)$

3. Domain: $[-4, 4]$

Range: $[0, 4]$

5. (a) $f(-2) = 0$ (b) $f(-1) = -1$

(c) $f\left(\frac{1}{2}\right) = 0$ (d) $f(1) = -3$

7. (a) $f(-2) = -3$ (b) $f(1) = 0$

(c) $f(0) = 1$ (d) $f(2) = -3$

9. $y = \frac{1}{2}x^2$

A vertical line intersects the graph just once, so y *is* a function of x.

11. $x - y^2 = 1 \implies y = \pm\sqrt{x - 1}$

y *is not* a function of x. Some vertical lines cross the graph twice.

13. $x^2 = 2xy - 1$

A vertical line intersects the graph just once, so y *is* a function of x.

15. $2x^2 - 7x - 30 = 0$

$(2x + 5)(x - 6) = 0$

$2x + 5 = 0 \quad$ or $\quad x - 6 = 0$

$x = -\dfrac{5}{2} \quad$ or $\quad x = 6$

17. $\dfrac{x}{9x^2 - 4} = 0$

$x = 0$

19. $\frac{1}{2}x^3 - x = 0$

$x^3 - 2x = 2(0)$

$x(x^2 - 2) = 0$

$x = 0 \quad$ or $\quad x^2 - 2 = 0$

$x^2 = 2$

$x = \pm\sqrt{2}$

21. $\qquad 4x^3 - 24x^2 - x + 6 = 0$

$4x^2(x - 6) - 1(x - 6) = 0$

$(x - 6)(4x^2 - 1) = 0$

$(x - 6)(2x + 1)(2x - 1) = 0$

$x - 6 = 0, \quad 2x + 1 = 0, \quad 2x - 1 = 0$

$x = 6, \qquad x = -\frac{1}{2}, \qquad x = \frac{1}{2}$

23. $\sqrt{2x} - 1 = 0$

$\sqrt{2x} = 1$

$2x = 1$

$x = \dfrac{1}{2}$

25. (a)

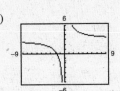

Zero: $x = -\dfrac{5}{3}$

(b) $3 + \dfrac{5}{x} = 0$

$3x + 5 = 0$

$x = -\dfrac{5}{3}$

27. (a)

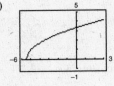

Zero: $x = -\dfrac{11}{2}$

(b) $\sqrt{2x + 11} = 0$

$2x + 11 = 0$

$x = -\frac{11}{2}$

29. (a)

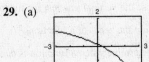

Zero: $x = \dfrac{1}{3}$

(b) $\dfrac{3x - 1}{x - 6} = 0$

$3x - 1 = 0$

$x = \dfrac{1}{3}$

31. $f(x) = \frac{3}{2}x$

f is increasing on $(-\infty, \infty)$.

33. $f(x) = x^3 - 3x^2 + 2$

f is increasing on $(-\infty, 0)$ and $(2, \infty)$.

f is decreasing on $(0, 2)$.

35. $f(x) = \begin{cases} x + 3, & x \le 0 \\ 3, & 0 < x \le 2 \\ 2x + 1, & x > 2 \end{cases}$

f is increasing on $(-\infty, 0)$ and $(2, \infty)$.

f is constant on $(0, 2)$.

37. $f(x) = |x + 1| + |x - 1|$

f is increasing on $(1, \infty)$.

f is constant on $(-1, 1)$.

f is decreasing on $(-\infty, -1)$.

39. $f(x) = 3$

(a)

Constant on $(-\infty, \infty)$

(b)

x	-2	-1	0	1	2
$f(x)$	3	3	3	3	3

41. $g(s) = \dfrac{s^2}{4}$

(a)

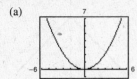

Decreasing on $(-\infty, 0)$; Increasing on $(0, \infty)$

(b)

s	-4	-2	0	2	4
$g(s)$	4	1	0	1	4

43. $f(t) = -t^4$

(a)

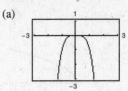

Increasing on $(-\infty, 0)$; Decreasing on $(0, \infty)$

(b)

t	-2	-1	0	1	2
$f(t)$	-16	-1	0	-1	-16

45. $f(x) = \sqrt{1 - x}$

(a)

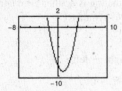   Decreasing on $(-\infty, 1)$

(b)

x	-3	-2	-1	0	1
$f(x)$	2	$\sqrt{3}$	$\sqrt{2}$	1	0

47. $f(x) = x^{3/2}$

(a)

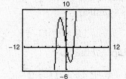   Increasing on $(0, \infty)$

(b)

x	0	1	2	3	4
$f(x)$	0	1	2.8	5.2	8

49. $f(x) = (x - 4)(x + 2)$

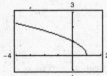

Relative minimum: $(1, -9)$

51. $f(x) = -x^2 + 3x - 2$

Relative maximum: $(1.5, 0.25)$

53. $f(x) = x(x - 2)(x + 3)$

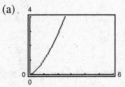

Relative minimum: $(1.12, -4.06)$

Relative maximum: $(-1.79, 8.21)$

55. $f(x) = 4 - x$

$f(x) \geq 0$ on $(-\infty, 4]$.

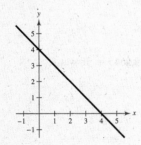

57. $f(x) = x^2 + x$

$f(x) \geq 0$ on $(-\infty, -1]$ and $[0, \infty)$.

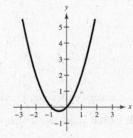

59. $f(x) = \sqrt{x - 1}$

$f(x) \geq 0$ on $[1, \infty)$.

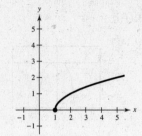

61. $f(x) = -(1 + |x|)$

$f(x)$ is never greater than 0.
$(f(x) < 0$ for all $x.)$

63. $f(x) = -2x + 15$

$\dfrac{f(3) - f(0)}{3 - 0} = \dfrac{9 - 15}{3} = -2$

The average rate of change from $x_1 = 0$ to $x_2 = 3$ is -2.

65. $f(x) = x^2 + 12x - 4$

$\dfrac{f(5) - f(1)}{5 - 1} = \dfrac{81 - 9}{4} = 18$

The average rate of change from $x_1 = 1$ to $x_2 = 5$ is 18.

67. $f(x) = x^3 - 3x^2 - x$

$\dfrac{f(3) - f(1)}{3 - 1} = \dfrac{-3 - (-3)}{2} = 0$

The average rate of change from $x_1 = 1$ to $x_2 = 3$ is 0.

69. $f(x) = -\sqrt{x - 2} + 5$

$\dfrac{f(11) - f(3)}{11 - 3} = \dfrac{2 - 4}{8} = -\dfrac{1}{4}$

The average rate of change from $x_1 = 3$ to $x_2 = 11$ is $-\frac{1}{4}$.

71. $f(x) = x^6 - 2x^2 + 3$

$f(-x) = (-x)^6 - 2(-x)^2 + 3$

$\quad = x^6 - 2x^2 + 3$

$\quad = f(x)$

The function is even.
y-axis symmetry

73. $g(x) = x^3 - 5x$

$g(-x) = (-x)^3 - 5(-x)$

$\quad = -x^3 + 5x$

$\quad = -g(x)$

The function is odd.
Origin symmetry

75. $f(t) = t^2 + 2t - 3$

$f(-t) = (-t)^2 + 2(-t) - 3$

$\quad = t^2 - 2t - 3$

$\quad \neq f(t), \neq -f(t)$

The function is neither even nor
odd. No symmetry

77. $h = \text{top} - \text{bottom}$

$\quad = (-x^2 + 4x - 1) - 2$

$\quad = -x^2 + 4x - 3$

79. $h = \text{top} - \text{bottom}$

$\quad = (4x - x^2) - 2x$

$\quad = 2x - x^2$

81. $L = \text{right} - \text{left}$

$\quad = \frac{1}{2}y^2 - 0 = \frac{1}{2}y^2$

83. $L = \text{right} - \text{left}$

$\quad = 4 - y^2$

85. $L = -0.294x^2 + 97.744x - 664.875,\ 20 \leq x \leq 90$

(a)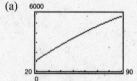

(b) $L = 2000$ when $x \approx 29.9645 \approx 30$ watts.

87. (a) For the average salaries of college professors, a scale of \$10,000 would be appropriate.

(b) For the population of the United States, use a scale of 10,000,000.

(c) For the percent of the civilian workforce that is unemployed, use a scale of 1%.

89. $r = 15.639t^3 - 104.75t^2 + 303.5t - 301, \ 2 \le t \le 7$

(a)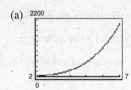

(b) $\dfrac{r(7) - r(2)}{7 - 2} = \dfrac{2054.927 - 12.112}{5} = 408.563$

The average rate of change from 2002 to 2007 is $408.563 billion per year. The estimated revenue is increasing each year at a rapid pace.

91. $s_0 = 6, \ v_0 = 64$

(a) $s = -16t^2 + 64t + 6$

(b)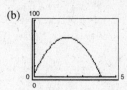

(c) $\dfrac{s(3) - s(0)}{3 - 0} = \dfrac{54 - 6}{3} = 16$

(d) The average rate of change of the height of the object with respect to time over the interval $t_1 = 0$ to $t_2 = 3$ is 16 feet per second.

(e) $s(0) = 6, m = 16$

Secant line: $y - 6 = 16(t - 0)$

$$y = 16t + 6$$

(f)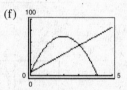

93. $v_0 = 120, \ s_0 = 0$

(a) $s = -16t^2 + 120t$

(b)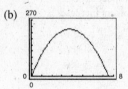

(c) $\dfrac{s(5) - s(3)}{5 - 3} = \dfrac{200 - 216}{2} = -8$

(d) The average decrease in the height of the object over the interval $t_1 = 3$ to $t_2 = 5$ is 8 feet per second.

(e) $s(5) = 200, m = -8$

Secant line: $y - 200 = -8(t - 5)$

$$y = -8t + 240$$

(f)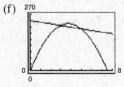

95. $v_0 = 0, \ s_0 = 120$

(a) $s = -16t^2 + 120$

(b)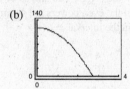

(c) $\dfrac{s(2) - s(0)}{2 - 0} = \dfrac{56 - 120}{2} = -32$

(d) On the interval $t_1 = 0$ to $t_2 = 2$, the height of the object is decreasing at a rate of 32 feet per second.

(e) $s(0) = 120, m = -32$

Secant line: $y - 120 = -32(t - 0)$

$$y = -32t + 120$$

(f)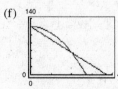

97. False. The function $f(x) = \sqrt{x^2 + 1}$ has a domain of all real numbers.

99. (a) Even. The graph is a reflection in the x-axis.

(b) Even. The graph is a reflection in the y-axis.

(c) Even. The graph is a vertical translation of f.

(d) Neither. The graph is a horizontal translation of f.

101. $\left(-\frac{3}{2}, 4\right)$

(a) If f is even, another point is $\left(\frac{3}{2}, 4\right)$.

(b) If f is odd, another point is $\left(\frac{3}{2}, -4\right)$.

103. $(4, 9)$

(a) If f is even, another point is $(-4, 9)$.

(b) If f is odd, another point is $(-4, -9)$.

105. (a) $y = x$

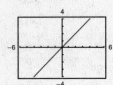

(b) $y = x^2$

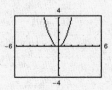

(c) $y = x^3$

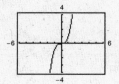

(d) $y = x^4$

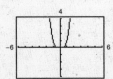

(e) $y = x^5$

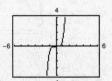

(f) $y = x^6$

All the graphs pass through the origin. The graphs of the odd powers of x are symmetric with respect to the origin and the graphs of the even powers are symmetric with respect to the y-axis. As the powers increase, the graphs become flatter in the interval $-1 < x < 1$.

107. $x^2 - 10x = 0$

$x(x - 10) = 0$

$x = 0$ or $x = 10$

109. $x^3 - x = 0$

$x(x^2 - 1) = 0$

$x = 0$ or $x^2 - 1 = 0$

$x^2 = 1$

$x = \pm 1$

111. $f(x) = 5x - 8$

(a) $f(9) = 5(9) - 8 = 37$

(b) $f(-4) = 5(-4) - 8 = -28$

(c) $f(x - 7) = 5(x - 7) - 8 = 5x - 35 - 8 = 5x - 43$

113. $f(x) = \sqrt{x - 12} - 9$

(a) $f(12) = \sqrt{12 - 12} - 9 = 0 - 9 = -9$

(b) $f(40) = \sqrt{40 - 12} - 9 = \sqrt{28} - 9 = 2\sqrt{7} - 9$

(c) $f\left(-\sqrt{36}\right)$ does not exist. The given value is not in the domain of the function.

115. $f(x) = x^2 - 2x + 9$

$f(3 + h) = (3 + h)^2 - 2(3 + h) + 9$

$= 9 + 6h + h^2 - 6 - 2h + 9$

$= h^2 + 4h + 12$

$f(3) = 3^2 - 2(3) + 9 = 12$

$\dfrac{f(3 + h) - f(3)}{h} = \dfrac{(h^2 + 4h + 12) - (12)}{h} = \dfrac{h^2 + 4h}{h} = \dfrac{h(h + 4)}{h} = h + 4, \ h \neq 0$

Section 1.6   A Library of Parent Functions

■ You should be able to identify and graph the following types of functions:

(a) Linear functions like $f(x) = ax + b$

(b) Squaring functions like $f(x) = x^2$

(c) Cubic functions like $f(x) = x^3$

(d) Square root functions like $f(x) = \sqrt{x}$

(e) Reciprocal functions like $f(x) = \dfrac{1}{x}$

(f) Constant functions like $f(x) = c$

(g) Absolute value functions like $f(x) = |x|$

(h) Step and piecewise-defined functions like $f(x) = [\![x]\!]$

■ You should be able to determine the following about these parent functions:

(a) Domain and range

(b) x-intercept(s) and y-intercept

(c) Symmetries

(d) Where it is increasing, decreasing, or constant

(e) If it is odd, even or neither

(f) Relative maximums and relative minimums

Vocabulary Check

1. $f(x) = [\![x]\!]$

(g) greatest integer function

2. $f(x) = x$

(i) identity function

3. $f(x) = \dfrac{1}{x}$

(h) reciprocal function

4. $f(x) = x^2$

(a) squaring function

5. $f(x) = \sqrt{x}$

(b) square root function

6. $f(x) = c$

(e) constant function

7. $f(x) = |x|$

(f) absolute value function

8. $f(x) = x^3$

(c) cubic function

9. $f(x) = ax + b$

(d) linear function

---

1. (a) $f(1) = 4,\ f(0) = 6$

$(1, 4)$ and $(0, 6)$

$m = \dfrac{6 - 4}{0 - 1} = -2$

$y - 6 = -2(x - 0)$

$y = -2x + 6$

$f(x) = -2x + 6$

(b)

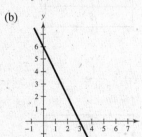

3. (a) $f(5) = -4,\ f(-2) = 17$

$(5, -4)$ and $(-2, 17)$

$m = \dfrac{17 - (-4)}{-2 - 5} = \dfrac{21}{-7} = -3$

$y - (-4) = -3(x - 5)$

$y + 4 = -3x + 15$

$y = -3x + 11$

$f(x) = -3x + 11$

(b)

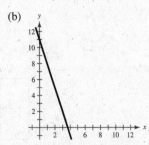

5. (a) $f(-5) = -1$, $f(5) = -1$

$(-5, -1)$ and $(5, -1)$

$$m = \frac{-1 - (-1)}{5 - (-5)} = \frac{0}{10} = 0$$

$$y - (-1) = 0(x - (-5))$$

$$y + 1 = 0$$

$$y = -1$$

$$f(x) = -1$$

(b)

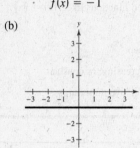

7. (a) $f\left(\frac{1}{2}\right) = -6, f(4) = -3$

$\left(\frac{1}{2}, -6\right)$ and $(4, -3)$

$$m = \frac{-3 - (-6)}{4 - (1/2)} = \frac{3}{7/2} = \frac{6}{7}$$

$$y - (-3) = \frac{6}{7}(x - 4)$$

$$y + 3 = \frac{6}{7}x - \frac{24}{7}$$

$$y = \frac{6}{7}x - \frac{45}{7}$$

$$f(x) = \frac{6}{7}x - \frac{45}{7}$$

(b)

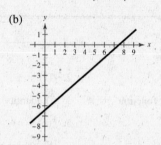

9. $f(x) = -x - \frac{3}{4}$

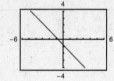

11. $f(x) = -\frac{1}{6}x - \frac{5}{2}$

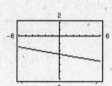

13. $f(x) = x^2 - 2x$

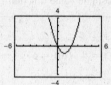

15. $h(x) = -x^2 + 4x + 12$

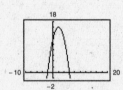

17. $f(x) = x^3 - 1$

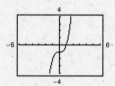

19. $f(x) = (x - 1)^3 + 2$

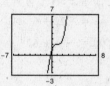

21. $f(x) = 4\sqrt{x}$

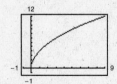

23. $g(x) = 2 - \sqrt{x + 4}$

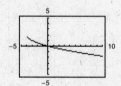

25. $f(x) = -\frac{1}{x}$

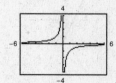

27. $h(x) = \dfrac{1}{x + 2}$

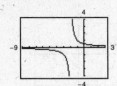

29. $f(x) = [\![x]\!]$

(a) $f(2.1) = 2$

(b) $f(2.9) = 2$

(c) $f(-3.1) = -4$

(d) $f\left(\dfrac{7}{2}\right) = 3$

31. $h(x) = [\![x + 3]\!]$

(a) $h(-2) = [\![1]\!] = 1$

(b) $h\left(\dfrac{1}{2}\right) = [\![3.5]\!] = 3$

(c) $h(4.2) = [\![7.2]\!] = 7$

(d) $h(-21.6) = [\![-18.6]\!] = -19$

33. $h(x) = [\![3x - 1]\!]$

(a) $h(2.5) = [\![6.5]\!] = 6$

(b) $h(-3.2) = [\![-10.6]\!] = -11$

(c) $h\left(\dfrac{7}{3}\right) = [\![6]\!] = 6$

(d) $h\left(-\dfrac{21}{3}\right) = [\![-22]\!] = -22$

35. $g(x) = 3[\![x - 2]\!] + 5$

(a) $g(-2.7) = 3[\![-4.7]\!] + 5 = 3(-5) + 5 = -10$

(b) $g(-1) = 3[\![-3]\!] + 5 = 3(-3) + 5 = -4$

(c) $g(0.8) = 3[\![-1.2]\!] + 5 = 3(-2) + 5 = -1$

(d) $g(14.5) = 3[\![12.5]\!] + 5 = 3(12) + 5 = 41$

37. $g(x) = -[\![x]\!]$

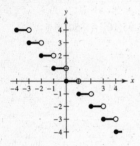

39. $g(x) = [\![x]\!] - 2$

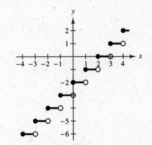

41. $g(x) = [\![x + 1]\!]$

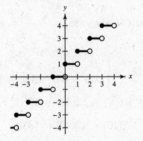

43. $f(x) = \begin{cases} 2x + 3, & x < 0 \\ 3 - x, & x \geq 0 \end{cases}$

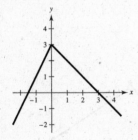

45. $f(x) = \begin{cases} \sqrt{4 + x}, & x < 0 \\ \sqrt{4 - x}, & x \geq 0 \end{cases}$

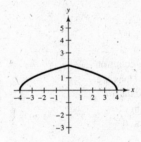

47. $f(x) = \begin{cases} x^2 + 5, & x \leq 1 \\ -x^2 + 4x + 3, & x > 1 \end{cases}$

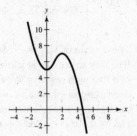

49. $h(x) = \begin{cases} 4 - x^2, & x < -2 \\ 3 + x, & -2 \leq x < 0 \\ x^2 + 1, & x \geq 0 \end{cases}$

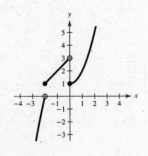

51. $s(x) = 2\left(\dfrac{1}{4}x - \left[\!\left[\dfrac{1}{4}x\right]\!\right]\right)$

(a)

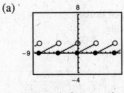

(b) Domain: $(-\infty, \infty)$

Range: $[0, 2)$

(c) Sawtooth pattern

53. (a) Parent function: $f(x) = |x|$

(b) $g(x) = |x + 2| - 1$

(c)

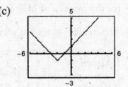

55. (a) Parent function: $f(x) = x^3$

(b) $g(x) = (x - 1)^3 - 2$

(c)

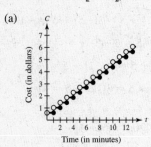

57. (a) Parent function: $f(x) = c$

(b) $g(x) = 2$

(c)

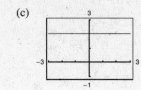

59. (a) Parent function: $f(x) = x$

(b) $g(x) = x - 2$

(c)

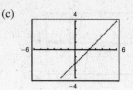

61. $C = 0.60 - 0.42[\![1 - t]\!], \ t > 0$

(a)

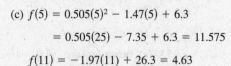

Time (in minutes)

(b) $C(12.5) = \$5.64$

63. $C = 10.75 + 3.95[\![x]\!], \ x > 0$

(a)

Cost of overnight delivery (in dollars)

Weight (in pounds)

(b) $C(10.33) = 10.75 + 3.95(10) = \$50.25$

65. $W(h) = \begin{cases} 12h, & 0 < h \le 40 \\ 18(h - 40) + 480, & h > 40 \end{cases}$

(a) $W(30) = 12(30) = \$360$

$W(40) = 12(40) = \$480$

$W(45) = 18(5) + 480 = \$570$

$W(50) = 18(10) + 480 = \$660$

(b) $W(h) = \begin{cases} 12h, & 0 < h \le 45 \\ 18(h - 45) + 540, & h > 45 \end{cases}$

67. (a) The domain of $f(x) = -1.97x + 26.3$ is $6 < x \le 12$.
One way to see this is to notice that this is the equation
of a line with negative slope, so the function values are
decreasing as x increases, which matches the data for
the corresponding part of the table. The domain of
$f(x) = 0.505x^2 - 1.47x + 6.3$ is then $1 \le x \le 6$.

(c) $f(5) = 0.505(5)^2 - 1.47(5) + 6.3$

$= 0.505(25) - 7.35 + 6.3 = 11.575$

$f(11) = -1.97(11) + 26.3 = 4.63$

These values represent the income in thousands of dollars
for the months of May and November, respectively.

(d) The model values are very close to the actual values.

(b)

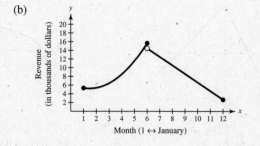

Month (1 ↔ January)

Month, x	1	2	3	4	5	6	7	8	9	10	11	12
Revenue, y	5.2	5.6	6.6	8.3	11.5	15.8	12.8	10.1	8.6	6.9	4.5	2.7
Model, $f(x)$	5.3	5.4	6.4	8.5	11.6	15.7	12.5	10.5	8.6	6.6	4.6	2.7

69. False. A piecewise-defined function is a function that is defined by two or more equations over a specified domain. That domain may or may not include x- and y-intercepts.

71. For the line through $(0, 6)$ and $(3, 2)$: $m = \dfrac{6-2}{0-3} = -\dfrac{4}{3}$

$$y - 6 = -\frac{4}{3}(x - 0) \implies y = -\frac{4}{3}x + 6$$

For the line through $(3, 2)$ and $(8, 0)$: $m = \dfrac{2-0}{3-8} = -\dfrac{2}{5}$

$$y - 0 = -\frac{2}{5}(x - 8) \implies y = -\frac{2}{5}x + \frac{16}{5}$$

$$f(x) = \begin{cases} -\frac{4}{3}x + 6, & 0 \le x \le 3 \\ -\frac{2}{5}x + \frac{16}{5}, & 3 < x \le 8 \end{cases}$$

Note that the respective domains can also be $0 \le x < 3$ and $3 \le x \le 8$.

73. $3x + 4 \le 12 - 5x$

$8x + 4 \le 12$

$8x \le 8$

$x \le 1$

75. L_1: $(-2, -2)$ and $(2, 10)$

$$m_1 = \frac{10 - (-2)}{2 - (-2)} = \frac{12}{4} = 3$$

L_2: $(-1, 3)$ and $(3, 9)$

$$m_2 = \frac{9 - 3}{3 - (-1)} = \frac{6}{4} = \frac{3}{2}$$

The lines are neither parallel nor perpendicular.

Section 1.7 Transformations of Functions

■ You should know the basic types of transformations.

Let $y = f(x)$ and let c be a positive real number.

1. $h(x) = f(x) + c$	Vertical shift c units upward
2. $h(x) = f(x) - c$	Vertical shift c units downward
3. $h(x) = f(x - c)$	Horizontal shift c units to the right
4. $h(x) = f(x + c)$	Horizontal shift c units to the left
5. $h(x) = -f(x)$	Reflection in the x-axis
6. $h(x) = f(-x)$	Reflection in the y-axis
7. $h(x) = cf(x), c > 1$	Vertical stretch
8. $h(x) = cf(x), 0 < c < 1$	Vertical shrink
9. $h(x) = f(cx), c > 1$	Horizontal shrink
10. $h(x) = f(cx), 0 < c < 1$	Horizontal stretch

Vocabulary Check

1. rigid   **2.** $-f(x); f(-x)$ **3.** nonrigid

4. horizontal shrink; horizontal stretch   **5.** vertical stretch; vertical shrink   **6.** (a) iv   (b) ii   (c) iii   (d) i

1. (a) $f(x) = |x| + c$ Vertical shifts

   $c = -1 : f(x) = |x| - 1$ 1 unit down

   $c = 1 : f(x) = |x| + 1$ 1 unit up

   $c = 3 : f(x) = |x| + 3$ 3 units up

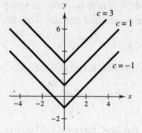

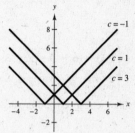

(b) $f(x) = |x - c|$ Horizontal shifts

   $c = -1 : f(x) = |x + 1|$ 1 unit left

   $c = 1 : f(x) = |x - 1|$ 1 unit right

   $c = 3 : f(x) = |x - 3|$ 3 units right

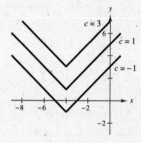

(c) $f(x) = |x + 4| + c$ Horizontal shift four units left and a vertical shift

   $c = -1 : f(x) = |x + 4| - 1$ 1 unit down

   $c = 1 : f(x) = |x + 4| + 1$ 1 unit up

   $c = 3 : f(x) = |x + 4| + 3$ 3 units up

3. (a) $f(x) = [\![x]\!] + c$ Vertical shifts

   $c = -2 : f(x) = [\![x]\!] - 2$ 2 units down

   $c = 0 : f(x) = [\![x]\!]$ Parent function

   $c = 2 : f(x) = [\![x]\!] + 2$ 2 units up

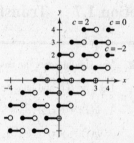

(b) $f(x) = [\![x + c]\!]$ Horizontal shifts

   $c = -2 : f(x) = [\![x - 2]\!]$ 2 units right

   $c = 0 : f(x) = [\![x]\!]$ Parent function

   $c = 2 : f(x) = [\![x + 2]\!]$ 2 units left

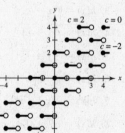

(c) $f(x) = [\![x - 1]\!] + c$ Horizontal shift 1 unit right and a vertical shift

   $c = -2 : f(x) = [\![x - 1]\!] - 2$ 2 units down

   $c = 0 : f(x) = [\![x - 1]\!]$ 2 units up

   $c = 2 : f(x) = [\![x - 1]\!] + 2$ 2 units up

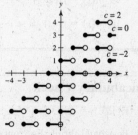

5. (a) $y = f(x) + 2$

Vertical shift 2 units upward

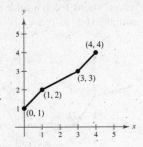

(b) $y = f(x - 2)$

Horizontal shift 2 units to the right

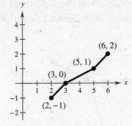

(c) $y = 2f(x)$

Vertical stretch (each y-value is multiplied by 2)

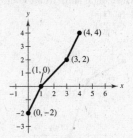

(d) $y = -f(x)$

Reflection in the x-axis

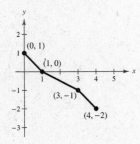

(e) $y = f(x + 3)$

Horizontal shift 3 units to the left

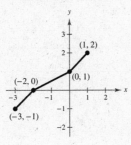

(f) $y = f(-x)$

Reflection in the y-axis

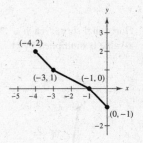

(g) $y = f\left(\frac{1}{2}x\right)$

Horizontal stretch (each x-value is multiplied by 2)

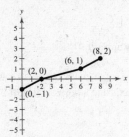
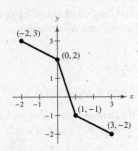

7. (a) $y = f(x) - 1$

Vertical shift 1 unit downward

(b) $y = f(x - 1)$

Horizontal shift 1 unit to the right

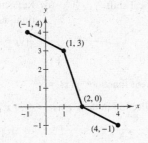

(c) $y = f(-x)$

Reflection about the y-axis

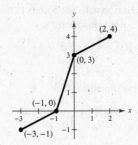

—CONTINUED—

7. —CONTINUED—

(d) $y = f(x + 1)$

Horizontal shift 1 unit to the left

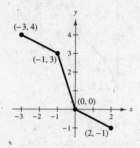

(e) $y = -f(x - 2)$

Reflection about the x-axis and a horizontal shift 2 units to the right

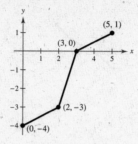

(f) $y = \frac{1}{2}f(x)$

Vertical shrink $\left(\text{each } y\text{-value is multiplied by } \frac{1}{2}\right)$

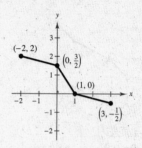

(g) $y = f(2x)$

Horizontal shrink $\left(\text{each } x\text{-value is multiplied by } \frac{1}{2}\right)$

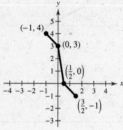
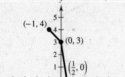

9. Parent function: $f(x) = x^2$

(a) Vertical shift 1 unit downward

$g(x) = x^2 - 1$

(b) Reflection about the x-axis, horizontal shift 1 unit to the left, and a vertical shift 1 unit upward

$g(x) = -(x + 1)^2 + 1$

(c) Reflection about the x-axis, horizontal shift 2 units to the right, and a vertical shift 6 units upward

$g(x) = -(x - 2)^2 + 6$

(d) Horizontal shift 5 units to the right and a vertical shift 3 units downward

$g(x) = (x - 5)^2 - 3$

11. Parent function: $f(x) = |x|$

(a) Vertical shift 5 units upward

$g(x) = |x| + 5$

(b) Reflection in the x-axis and a horizontal shift 3 units to the left

$g(x) = -|x + 3|$

(c) Horizontal shift 2 units to the right and a vertical shift 4 units downward

$g(x) = |x - 2| - 4$

(d) Reflection in the x-axis, horizontal shift 6 units to the right, and a vertical shift 1 unit downward

$g(x) = -|x - 6| - 1$

13. Parent function: $f(x) = x^3$

Horizontal shift 2 units to the right: $y = (x - 2)^3$

15. Parent function: $f(x) = x^2$

Reflection in the x-axis: $y = -x^2$

17. Parent function: $f(x) = \sqrt{x}$

Reflection in the x-axis and a vertical shift 1 unit upward: $y = -\sqrt{x} + 1$

19. $g(x) = 12 - x^2$

(a) Parent function: $f(x) = x^2$

(b) Reflection in the x-axis and a vertical shift 12 units upward

(c)

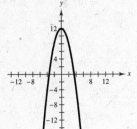

(d) $g(x) = 12 - f(x)$

21. $g(x) = x^3 + 7$

(a) Parent function: $f(x) = x^3$

(b) Vertical shift 7 units upward

(c)

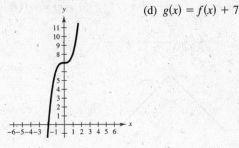

(d) $g(x) = f(x) + 7$

23. $g(x) = \frac{2}{3}x^2 + 4$

(a) Parent function: $f(x) = x^2$

(b) Vertical shrink of two-thirds, and a vertical shift 4 units upward

(c)

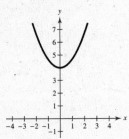

(d) $g(x) = \frac{2}{3}f(x) + 4$

25. $g(x) = 2 - (x + 5)^2$

(a) Parent function: $f(x) = x^2$

(b) Reflection in the x-axis, horizontal shift 5 units to the left, and a vertical shift 2 units upward

(c)

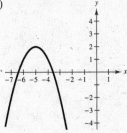

(d) $g(x) = 2 - f(x + 5)$

27. $g(x) = \sqrt{3x}$

(a) Parent function: $f(x) = \sqrt{x}$

(b) Horizontal shrink by $\frac{1}{3}$

(c)

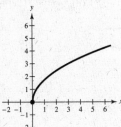

(d) $g(x) = f(3x)$

29. $g(x) = (x - 1)^3 + 2$

(a) Parent function: $f(x) = x^3$

(b) Horizontal shift 1 unit to the right and a vertical shift 2 units upward

(c)

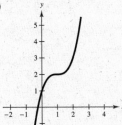

(d) $g(x) = f(x - 1) + 2$

31. $g(x) = -|x| - 2$

   (a) Parent function: $f(x) = |x|$

   (b) Reflection in the x-axis; vertical shift 2 units downward

   (c)   (d) $g(x) = -f(x) - 2$

33. $g(x) = -|x + 4| + 8$

   (a) Parent function: $f(x) = |x|$

   (b) Reflection in the x-axis, horizontal shift 4 units to the left, and a vertical shift 8 units upward

   (c)   (d) $g(x) = -f(x + 4) + 8$

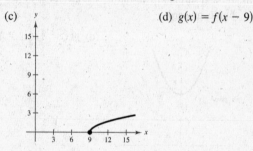

35. $g(x) = 3 - [\![x]\!]$

   (a) Parent function: $f(x) = [\![x]\!]$

   (b) Reflection in the x-axis and a vertical shift 3 units up

   (c)   (d) $g(x) = 3 - f(x)$

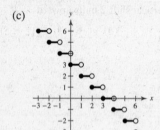

37. $g(x) = \sqrt{x - 9}$

   (a) Parent function: $f(x) = \sqrt{x}$

   (b) Horizontal shift 9 units to the right

   (c)   (d) $g(x) = f(x - 9)$

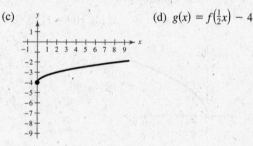

39. $g(x) = \sqrt{7 - x} - 2$ or $g(x) = \sqrt{-(x - 7)} - 2$

   (a) Parent function: $f(x) = \sqrt{x}$

   (b) Reflection in the y-axis, horizontal shift 7 units to the right, and a vertical shift 2 units downward

   (c)   (d) $g(x) = f(7 - x) - 2$

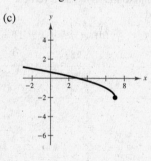

41. $g(x) = \sqrt{\frac{1}{2}x} - 4$

   (a) Parent function: $f(x) = \sqrt{x}$

   (b) Horizontal stretch (each x-value is multiplied by 2) and a vertical shift 4 units down

   (c)   (d) $g(x) = f\left(\frac{1}{2}x\right) - 4$

43. $f(x) = x^2$ moved 2 units to the right and 8 units down.

   $g(x) = (x - 2)^2 - 8$

45. $f(x) = x^3$ moved 13 units to the right.

   $g(x) = (x - 13)^3$

47. $f(x) = |x|$ moved 10 units up and reflected about the x-axis.

   $g(x) = -(|x| + 10) = -|x| - 10$

49. $f(x) = \sqrt{x}$ moved 6 units to the left and reflected in both the x- and y-axes.

   $g(x) = -\sqrt{-x + 6}$

51. $f(x) = x^2$

    (a) Reflection in the x-axis and a vertical stretch (each y-value is multiplied by 3)

       $g(x) = -3x^2$

    (b) Vertical shift 3 units upward and a vertical stretch (each y-value is multiplied by 4)

       $g(x) = 4x^2 + 3$

53. $f(x) = |x|$

    (a) Reflection in the x-axis and a vertical shrink (each y-value is multiplied by $\frac{1}{2}$)

       $g(x) = -\frac{1}{2}|x|$

    (b) Vertical stretch (each y-value is multiplied by 3) and a vertical shift 3 units downward

       $g(x) = 3|x| - 3$

55. Parent function: $f(x) = x^3$

Vertical stretch (each y-value is multiplied by 2)

$g(x) = 2x^3$

57. Parent function: $f(x) = x^2$

Reflection in the x-axis; vertical shrink (each y-value is multiplied by $\frac{1}{2}$)

$g(x) = -\frac{1}{2}x^2$

59. Parent function: $f(x) = \sqrt{x}$

Reflection in the y-axis; vertical shrink (each y-value is multiplied by $\frac{1}{2}$)

$g(x) = \frac{1}{2}\sqrt{-x}$

61. Parent function: $f(x) = x^3$

Reflection in the x-axis, horizontal shift 2 units to the right and a vertical shift 2 units upward

$g(x) = -(x - 2)^3 + 2$

63. Parent function: $f(x) = \sqrt{x}$

Reflection in the x-axis and a vertical shift 3 units downward

$g(x) = -\sqrt{x} - 3$

65. (a) $g(x) = f(x) + 2$

Vertical shift 2 units upward

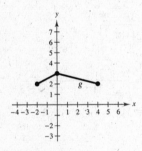

(b) $g(x) = f(x) - 1$

Vertical shift 1 unit downward

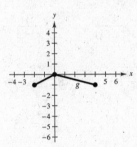

(c) $g(x) = f(-x)$

Reflection in the y-axis

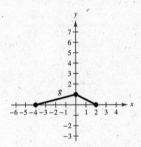

(d) $g(x) = -2f(x)$

Reflection in the x-axis and a vertical stretch (each y-value is multiplied by 2)

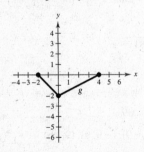

(e) $g(x) = f(4x)$

Horizontal shrink (each x-value is multiplied by $\frac{1}{4}$)

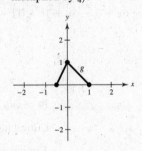

(f) $g(x) = f\left(\frac{1}{2}x\right)$

Horizontal stretch (each x-value is multiplied by 2)

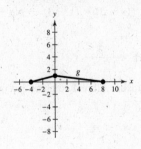

67. $F = f(t) = 20.6 + 0.035t^2$, $0 \le t \le 22$

(a) A vertical shrink by 0.035 and a vertical shift of 20.6 units upward

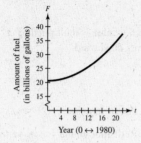

Year (0 ↔ 1980)

(b) $\dfrac{f(22) - f(0)}{22 - 0} = \dfrac{37.54 - 20.6}{22} = 0.77$

The average increase in fuel used by trucks was 0.77 billion gallons per year between 1980 and 2002.

(c) $g(t) = 20.6 + 0.035(t + 10)^2 = f(t + 10)$

This represents a horizontal shift 10 units to the left.

(d) $g(20) = 52.1$ billion gallons

Yes. There are many factors involved here. The number of trucks on the road continues to increase but are more fuel efficient. The availability and the cost of overseas and domestic fuel also plays a role in usage.

69. True, since $|x| = |-x|$, the graphs of $f(x) = |x| + 6$ and $f(x) = |-x| + 6$ are identical.

71. (a) The profits were only $\frac{3}{4}$ as large as expected:

$g(t) = \frac{3}{4}f(t)$

(b) The profits were \$10,000 greater than predicted:

$g(t) = f(t) + 10,000$

(c) There was a two-year delay: $g(t) = f(t - 2)$

73. $y = f(x + 2) - 1$

Horizontal shift 2 units to the left and a vertical shift 1 unit downward

$(0, 1) \to (0 - 2, 1 - 1) = (-2, 0)$

$(1, 2) \to (1 - 2, 2 - 1) = (-1, 1)$

$(2, 3) \to (2 - 2, 3 - 1) = (0, 2)$

75. $\dfrac{4}{x} + \dfrac{4}{1 - x} = \dfrac{4(1 - x) + 4x}{x(1 - x)} = \dfrac{4 - 4x + 4x}{x(1 - x)} = \dfrac{4}{x(1 - x)}$

77. $\dfrac{3}{x - 1} - \dfrac{2}{x(x - 1)} = \dfrac{3x - 2}{x(x - 1)}$

79. $(x - 4)\left(\dfrac{1}{\sqrt{x^2 - 4}}\right) = \dfrac{x - 4}{\sqrt{x^2 - 4}} = \dfrac{(x - 4)\sqrt{x^2 - 4}}{x^2 - 4}$

81. $(x^2 - 9) \div \left(\dfrac{x + 3}{5}\right) = \dfrac{(x + 3)(x - 3)}{1} \cdot \dfrac{5}{x + 3}$

$= 5(x - 3), \ x \ne -3$

83. $f(x) = x^2 - 6x + 11$

(a) $f(-3) = (-3)^2 - 6(-3) + 11 = 38$

(b) $f\left(-\frac{1}{2}\right) = \left(-\frac{1}{2}\right)^2 - 6\left(-\frac{1}{2}\right) + 11 = \frac{1}{4} + 3 + 11 = \frac{57}{4}$

(c) $f(x - 3) = (x - 3)^2 - 6(x - 3) + 11 = x^2 - 6x + 9 - 6x + 18 + 11 = x^2 - 12x + 38$

85. $f(x) = \dfrac{2}{11 - x}$

Domain: All real numbers except $x = 11$

87. $f(x) = \sqrt{81 - x^2}$

$81 - x^2 \ge 0$

$(9 + x)(9 - x) \ge 0$

Critical numbers: $x = \pm 9$

Test intervals: $(-\infty, -9), (-9, 9), (9, \infty)$

Test: Is $81 - x^2 \ge 0$?

Solution: $[-9, 9]$

Domain of $f(x)$: $-9 \le x \le 9$

Section 1.8   Combinations of Functions: Composite Functions

■ Given two functions, *f* and *g*, you should be able to form the following functions (if defined):

1. Sum: $(f + g)(x) = f(x) + g(x)$

2. Difference: $(f - g)(x) = f(x) - g(x)$

3. Product: $(fg)(x) = f(x)g(x)$

4. Quotient: $(f/g)(x) = f(x)/g(x), g(x) \neq 0$

5. Composition of *f* with *g*: $(f \circ g)(x) = f(g(x))$

6. Composition of *g* with *f*: $(g \circ f)(x) = g(f(x))$

Vocabulary Check

1. addition, subtraction, multiplication, division

2. composition

3. $g(x)$

4. inner; outer

1.

x	0	1	2	3
f	2	3	1	2
g	-1	0	$\frac{1}{2}$	0
$f + g$	1	3	$\frac{3}{2}$	2

3.

x	-2	0	1	2	4
f	2	0	1	2	4
g	4	2	1	0	2
$f + g$	6	2	2	2	6

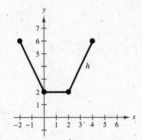

5. $f(x) = x + 2, g(x) = x - 2$

(a) $(f + g)(x) = f(x) + g(x) = (x + 2) + (x - 2) = 2x$

(b) $(f - g)(x) = f(x) - g(x) = (x + 2) - (x - 2) = 4$

(c) $(fg)(x) = f(x) \cdot g(x) = (x + 2)(x - 2) = x^2 - 4$

(d) $\left(\dfrac{f}{g}\right)(x) = \dfrac{f(x)}{g(x)} = \dfrac{x + 2}{x - 2}$

Domain: all real numbers *x* except $x = 2$

7. $f(x) = x^2, g(x) = 4x - 5$

(a) $(f + g)(x) = f(x) + g(x)$

$$= x^2 + (4x - 5) = x^2 + 4x - 5$$

(b) $(f - g)(x) = f(x) - g(x)$

$$= x^2 - (4x - 5) = x^2 - 4x + 5$$

(c) $(fg)(x) = f(x) \cdot g(x) = x^2(4x - 5) = 4x^3 - 5x^2$

(d) $\left(\dfrac{f}{g}\right)(x) = \dfrac{f(x)}{g(x)} = \dfrac{x^2}{4x - 5}$

Domain: all real numbers *x* except $x = \dfrac{5}{4}$

9. $f(x) = x^2 + 6$, $g(x) = \sqrt{1 - x}$

(a) $(f + g)(x) = f(x) + g(x) = (x^2 + 6) + \sqrt{1 - x}$

(b) $(f - g)(x) = f(x) - g(x) = (x^2 + 6) - \sqrt{1 - x}$

(c) $(fg)(x) = f(x) \cdot g(x) = (x^2 + 6)\sqrt{1 - x}$

(d) $\left(\dfrac{f}{g}\right)(x) = \dfrac{f(x)}{g(x)} = \dfrac{x^2 + 6}{\sqrt{1 - x}} = \dfrac{(x^2 + 6)\sqrt{1 - x}}{1 - x}$,

Domain: $x < 1$

11. $f(x) = \dfrac{1}{x}$, $g(x) = \dfrac{1}{x^2}$

(a) $(f + g)(x) = f(x) + g(x) = \dfrac{1}{x} + \dfrac{1}{x^2} = \dfrac{x + 1}{x^2}$

(b) $(f - g)(x) = f(x) - g(x) = \dfrac{1}{x} - \dfrac{1}{x^2} = \dfrac{x - 1}{x^2}$

(c) $(fg)(x) = f(x) \cdot g(x) = \dfrac{1}{x}\left(\dfrac{1}{x^2}\right) = \dfrac{1}{x^3}$

(d) $\left(\dfrac{f}{g}\right)(x) = \dfrac{f(x)}{g(x)} = \dfrac{1/x}{1/x^2} = \dfrac{x^2}{x} = x$

Domain: all real numbers x except $x = 0$

For Exercises 13–24, $f(x) = x^2 + 1$ **and** $g(x) = x - 4$.

13. $(f + g)(2) = f(2) + g(2) = (2^2 + 1) + (2 - 4) = 3$

15. $(f - g)(0) = f(0) - g(0) = (0^2 + 1) - (0 - 4) = 5$

17. $(f - g)(3t) = f(3t) - g(3t) = [(3t)^2 + 1] - (3t - 4)$

$\qquad\qquad = 9t^2 - 3t + 5$

19. $(fg)(6) = f(6)g(6) = (6^2 + 1)(6 - 4) = 74$

21. $\left(\dfrac{f}{g}\right)(5) = \dfrac{f(5)}{g(5)} = \dfrac{5^2 + 1}{5 - 4} = 26$

23. $\left(\dfrac{f}{g}\right)(-1) - g(3) = \dfrac{f(-1)}{g(-1)} - g(3)$

$\qquad\qquad = \dfrac{(-1)^2 + 1}{-1 - 4} - (3 - 4)$

$\qquad\qquad = -\dfrac{2}{5} + 1 = \dfrac{3}{5}$

25. $f(x) = \frac{1}{2}x$, $g(x) = x - 1$, $(f + g)(x) = \frac{3}{2}x - 1$

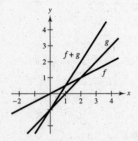

27. $f(x) = x^2$, $g(x) = -2x$, $(f + g)(x) = x^2 - 2x$

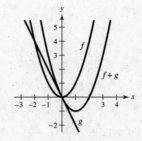

29. $f(x) = 3x$, $g(x) = -\dfrac{x^3}{10}$, $(f + g)(x) = 3x - \dfrac{x^3}{10}$

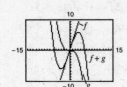

For $0 \le x \le 2$, $f(x)$ contributes most to the magnitude.

For $x > 6$, $g(x)$ contributes most to the magnitude.

31. $f(x) = x^2$, $g(x) = x - 1$

(a) $(f \circ g)(x) = f(g(x)) = f(x - 1) = (x - 1)^2$

(b) $(g \circ f)(x) = g(f(x)) = g(x^2) = x^2 - 1$

(c) $(f \circ f)(x) = f(f(x)) = f(x^2) = (x^2)^2 = x^4$

33. $f(x) = \sqrt[3]{x - 1}$, $g(x) = x^3 + 1$

    (a) $(f \circ g)(x) = f(g(x))$

$$= f(x^3 + 1)$$

$$= \sqrt[3]{(x^3 + 1) - 1}$$

$$= \sqrt[3]{x^3} = x$$

    (b) $(g \circ f)(x) = g(f(x))$

$$= g\left(\sqrt[3]{x - 1}\right)$$

$$= \left(\sqrt[3]{x - 1}\right)^3 + 1$$

$$= (x - 1) + 1 = x$$

    (c) $(f \circ f)(x) = f(f(x))$

$$= f\left(\sqrt[3]{x - 1}\right)$$

$$= \sqrt[3]{\sqrt[3]{x - 1} - 1}$$

35. $f(x) = \sqrt{x + 4}$ Domain: $x \geq -4$

    $g(x) = x^2$ Domain: all real numbers x

    (a) $(f \circ g)(x) = f(g(x)) = f(x^2) = \sqrt{x^2 + 4}$

    Domain: all real numbers x

    (b) $(g \circ f)(x) = g(f(x))$

$$= g\left(\sqrt{x + 4}\right) = \left(\sqrt{x + 4}\right)^2 = x + 4$$

    Domain: $x \geq -4$

37. $f(x) = x^2 + 1$ Domain: all real numbers x

    $g(x) = \sqrt{x}$ Domain: $x \geq 0$

    (a) $(f \circ g)(x) = f(g(x)) = f(\sqrt{x}) = (\sqrt{x})^2 + 1 = x + 1$

    Domain: $x \geq 0$

    (b) $(g \circ f)(x) = g(f(x)) = g(x^2 + 1) = \sqrt{x^2 + 1}$

    Domain: all real numbers x

39. $f(x) = |x|$ Domain: all real numbers x

    $g(x) = x + 6$ Domain: all real numbers x

    (a) $(f \circ g)(x) = f(g(x)) = f(x + 6) = |x + 6|$

    Domain: all real numbers x

    (b) $(g \circ f)(x) = g(f(x)) = g(|x|) = |x| + 6$

    Domain: all real numbers x

41. $f(x) = \dfrac{1}{x}$ Domain: all real numbers x except $x = 0$

    $g(x) = x + 3$ Domain: all real numbers x

    (a) $(f \circ g)(x) = f(g(x)) = f(x + 3) = \dfrac{1}{x + 3}$

    Domain: all real numbers x except $x = -3$

    (b) $(g \circ f)(x) = g(f(x)) = g\left(\dfrac{1}{x}\right) = \dfrac{1}{x} + 3$

    Domain: all real numbers x except $x = 0$

43. (a) $(f + g)(3) = f(3) + g(3) = 2 + 1 = 3$

    (b) $\left(\dfrac{f}{g}\right)(2) = \dfrac{f(2)}{g(2)} = \dfrac{0}{2} = 0$

45. (a) $(f \circ g)(2) = f(g(2)) = f(2) = 0$

    (b) $(g \circ f)(2) = g(f(2)) = g(0) = 4$

47. $h(x) = (2x^2 + 1)^2$

One possibility: Let $f(x) = x^2$ and $g(x) = 2x + 1$, then $(f \circ g)(x) = h(x)$.

49. $h(x) = \sqrt[3]{x^2 - 4}$

One possibility: Let $f(x) = \sqrt[3]{x}$ and $g(x) = x^2 - 4$, then $(f \circ g)(x) = h(x)$.

51. $h(x) = \dfrac{1}{x + 2}$

One possibility: Let $f(x) = 1/x$ and $g(x) = x + 2$, then $(f \circ g)(x) = h(x)$.

53. $h(x) = \dfrac{-x^2 + 3}{4 - x^2}$

One possibility: Let $f(x) = \dfrac{x + 3}{4 + x}$ and $g(x) = -x^2$, then

$(f \circ g)(x) = h(x)$.

55. $T(x) = R(x) + B(x) = \frac{3}{4}x + \frac{1}{15}x^2$

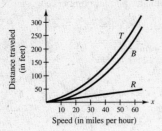

57. (a) $c(t) = \dfrac{p(t) + b(t) - d(t)}{p(t)} \times 100$

(b) $c(5)$ represents the percent change in the population in the year 2005.

59. $A(t) = 3.36t^2 - 59.8t + 735,\ N(t) = 1.95t^2 - 42.2t + 603$

(a) $(A + N)(t) = A(t) + N(t) = 5.31t^2 - 102.0t + 1338$

This represents the combined Army and Navy personnel (in thousands) from 1990 to 2002, where $t = 0$ corresponds to 1990.

$(A + N)(4) = 1014.96$ thousand

$(A + N)(8) = 861.84$ thousand

$(A + N)(12) = 878.64$ thousand

(b) $(A - N)(t) = A(t) - N(t) = 1.41t^2 - 17.6t + 132$

This represents the number of Army personnel (in thousands) more than the number of Navy personnel from 1990 to 2002, where $t = 0$ corresponds to 1990.

$(A - N)(4) = 84.16$ thousand

$(A - N)(8) = 81.44$ thousand

$(A - N)(12) = 123.84$ thousand

61.

Year	y_1	y_2	y_3
1995	146.2	329.1	44.8
1996	152.0	344.1	48.1
1997	162.2	359.9	52.1
1998	175.2	382.0	55.6
1999	184.4	412.1	57.8
2000	194.7	449.0	57.4
2001	205.5	496.1	57.8

(a) $y_1 \approx 10.20t + 92.7$

$y_2 \approx 3.357t^2 - 26.46t + 379.5$

$y_3 \approx -0.465t^2 + 9.71t + 7.4$

(b) $y_1 + y_2 + y_3 \approx 2.892t^2 - 6.55t + 479.6$

This sum represents the total spent on health services and supplies for the years 1995 through 2001. It includes out-of-pocket payments, insurance premiums, and other types of payments.

(c)

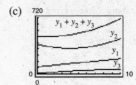

(d) For 2008 use $t = 18$:

$(y_1 + y_2 + y_3)(18) \approx \$1298.708$ billion

For 2010 use $t = 20$:

$(y_1 + y_2 + y_3)(20) \approx \$1505.4$ billion

63. (a) $r(x) = \dfrac{x}{2}$

(b) $A(r) = \pi r^2$

(c) $(A \circ r)(x) = A(r(x)) = A\left(\dfrac{x}{2}\right) = \pi\left(\dfrac{x}{2}\right)^2$

$(A \circ r)(x)$ represents the area of the circular base of the tank on the square foundation with side length x.

65. (a) $N(T(t)) = N(3t + 2)$

$= 10(3t + 2)^2 - 20(3t + 2) + 600$

$= 10(9t^2 + 12t + 4) - 60t - 40 + 600$

$= 90t^2 + 60t + 600$

$= 30(3t^2 + 2t + 20),\ 0 \le t \le 6$

This represents the number of bacteria in the food as a function of time.

(b) $30(3t^2 + 2t + 20) = 1500$

$3t^2 + 2t + 20 = 50$

$3t^2 + 2t - 30 = 0$

By the Quadratic Formula, $t \approx -3.513$ or 2.846. Choosing the positive value for t, we have $t \approx 2.846$ hours.

67. (a) $f(g(x)) = f(0.03x) = 0.03x - 500{,}000$

(b) $g(f(x)) = g(x - 500{,}000) = 0.03(x - 500{,}000)$

$g(f(x))$ represents your bonus of 3% of an amount over $500,000.

69. False. $(f \circ g)(x) = 6x + 1$ and $(g \circ f)(x) = 6x + 6.$

71. Let $f(x)$ and $g(x)$ be two odd functions and define $h(x) = f(x)g(x)$. Then

$$h(-x) = f(-x)g(-x)$$
$$= [-f(x)][-g(x)] \quad \text{since } f \text{ and } g \text{ are odd}$$
$$= f(x)g(x)$$
$$= h(x).$$

Thus, $h(x)$ is even.

Let $f(x)$ and $g(x)$ be two even functions and define $h(x) = f(x)g(x)$. Then

$$h(-x) = f(-x)g(-x)$$
$$= f(x)g(x) \quad \text{since } f \text{ and } g \text{ are even}$$
$$= h(x).$$

Thus, $h(x)$ is even.

73. $f(x) = 3x - 4$

$$\frac{f(x + h) - f(x)}{h} = \frac{[3(x + h) - 4] - (3x - 4)}{h}$$
$$= \frac{3x + 3h - 4 - 3x + 4}{h}$$
$$= \frac{3h}{h}$$
$$= 3, \; h \neq 0$$

75. $f(x) = \dfrac{4}{x}$

$$\frac{f(x + h) - f(x)}{h} = \frac{\dfrac{4}{x + h} - \dfrac{4}{x}}{h} = \frac{\dfrac{4x - 4(x + h)}{x(x + h)}}{\dfrac{h}{1}}$$

$$= \frac{4x - 4x - 4h}{x(x + h)} \cdot \frac{1}{h} = \frac{-4h}{x(x + h)} \cdot \frac{1}{h} = \frac{-4}{x(x + h)}, \; h \neq 0$$

77. Point: $(2, -4)$

Slope: $m = 3$

$$y - (-4) = 3(x - 2)$$
$$y + 4 = 3x - 6$$
$$3x - y - 10 = 0$$

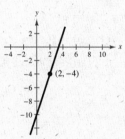

79. Point: $(8, -1)$

Slope: $m = -\dfrac{3}{2}$

$$y - (-1) = -\tfrac{3}{2}(x - 8)$$
$$y + 1 = -\tfrac{3}{2}x + 12$$
$$2y + 2 = -3x + 24$$
$$3x + 2y - 22 = 0$$

Section 1.9   Inverse Functions

■ Two functions f and g are inverses of each other if $f(g(x)) = x$ for every x in the domain of g and $g(f(x)) = x$ for every x in the domain of f.

■ A function f has an inverse function if and only if no **horizontal** line crosses the graph of f at more than one point.

■ The graph of f^{-1} is a reflection of the graph of f about the line $y = x$.

■ Be able to find the inverse of a function, if it exists.

    1. Use the Horizontal Line Test to see if f^{-1} exists.

    2. Replace $f(x)$ with y.

    3. Interchange x and y and solve for y.

    4. Replace y with $f^{-1}(x)$.

Vocabulary Check

  1. inverse; f-inverse   **2.** range; domain   **3.** $y = x$

  4. one-to-one   **5.** Horizontal

1. $f(x) = 6x$

$$f^{-1}(x) = \frac{x}{6} = \frac{1}{6}x$$

$$f(f^{-1}(x)) = f\left(\frac{x}{6}\right) = 6\left(\frac{x}{6}\right) = x$$

$$f^{-1}(f(x)) = f^{-1}(6x) = \frac{6x}{6} = x$$

3. $f(x) = x + 9$

$$f^{-1}(x) = x - 9$$

$$f(f^{-1}(x)) = f(x - 9) = (x - 9) + 9 = x$$

$$f^{-1}(f(x)) = f^{-1}(x + 9) = (x + 9) - 9 = x$$

5. $f(x) = 3x + 1$

$$f^{-1}(x) = \frac{x - 1}{3}$$

$$f(f^{-1}(x)) = f\left(\frac{x - 1}{3}\right) = 3\left(\frac{x - 1}{3}\right) + 1 = x$$

$$f^{-1}(f(x)) = f^{-1}(3x + 1) = \frac{(3x + 1) - 1}{3} = x$$

7. $f(x) = \sqrt[3]{x}$

$$f^{-1}(x) = x^3$$

$$f(f^{-1}(x)) = f(x^3) = \sqrt[3]{x^3} = x$$

$$f^{-1}(f(x)) = f^{-1}(\sqrt[3]{x}) = (\sqrt[3]{x})^3 = x$$

9. The inverse is a line through $(-1, 0)$.
    Matches graph (c).

11. The inverse is half a parabola starting at $(1, 0)$.
    Matches graph (a).

13. $f(x) = 2x$, $g(x) = \dfrac{x}{2}$

    (a) $f(g(x)) = f\left(\dfrac{x}{2}\right) = 2\left(\dfrac{x}{2}\right) = x$

       $g(f(x)) = g(2x) = \dfrac{2x}{2} = x$

    (b)

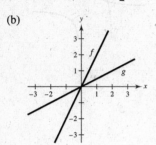

15. $f(x) = 7x + 1$, $g(x) = \dfrac{x - 1}{7}$

    (a) $f(g(x)) = f\left(\dfrac{x-1}{7}\right) = 7\left(\dfrac{x-1}{7}\right) + 1 = x$

       $g(f(x)) = g(7x + 1) = \dfrac{(7x+1) - 1}{7} = x$

    (b)

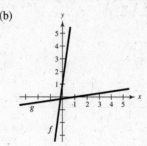

17. $f(x) = \dfrac{x^3}{8}$, $g(x) = \sqrt[3]{8x}$

    (a) $f(g(x)) = f\left(\sqrt[3]{8x}\right) = \dfrac{\left(\sqrt[3]{8x}\right)^3}{8} = \dfrac{8x}{8} = x$

       $g(f(x)) = g\left(\dfrac{x^3}{8}\right) = \sqrt[3]{8\left(\dfrac{x^3}{8}\right)} = \sqrt[3]{x^3} = x$

    (b)

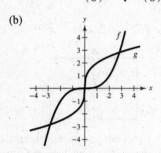

19. $f(x) = \sqrt{x - 4}$, $g(x) = x^2 + 4$, $x \ge 0$

    (a) $f(g(x)) = f(x^2 + 4)$, $x \ge 0$

           $= \sqrt{(x^2 + 4) - 4} = x$

       $g(f(x)) = g\left(\sqrt{x - 4}\right)$

           $= \left(\sqrt{x - 4}\right)^2 + 4 = x$

    (b)

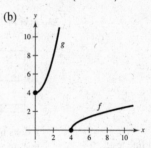

21. $f(x) = 9 - x^2$, $x \ge 0$; $g(x) = \sqrt{9 - x}$, $x \le 9$

    (a) $f(g(x)) = f\left(\sqrt{9 - x}\right)$, $x \le 9$

           $= 9 - \left(\sqrt{9 - x}\right)^2 = x$

       $g(f(x)) = g(9 - x^2)$, $x \ge 0$

           $= \sqrt{9 - (9 - x^2)} = x$

    (b)

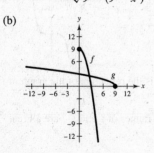

23. $f(x) = \dfrac{x-1}{x+5}$, $g(x) = -\dfrac{5x+1}{x-1}$

(b)

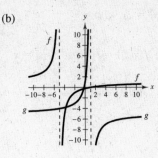

(a) $f(g(x)) = f\left(-\dfrac{5x+1}{x-1}\right)$

$$= \dfrac{\left(-\dfrac{5x+1}{x-1} - 1\right)}{\left(-\dfrac{5x+1}{x-1} + 5\right)} \cdot \dfrac{x-1}{x-1} = \dfrac{-(5x+1)-(x-1)}{-(5x+1)+5(x-1)} = \dfrac{-6x}{-6} = x$$

$g(f(x)) = g\left(\dfrac{x-1}{x+5}\right)$

$$= -\dfrac{\left[5\left(\dfrac{x-1}{x+5}\right) + 1\right]}{\left[\dfrac{x-1}{x+5} - 1\right]} \cdot \dfrac{x+5}{x+5} = -\dfrac{5(x-1)+(x+5)}{(x-1)-(x+5)} = -\dfrac{6x}{-6} = x$$

25. No, $\{(-2, -1), (1, 0), (2, 1), (1, 2), (-2, 3), (-6, 4)\}$ does not represent a function. -2 and 1 are paired with two different values.

27.

x	-2	0	2	4	6	8
$f^{-1}(x)$	-2	-1	0	1	2	3

29. Yes, since no horizontal line crosses the graph of f at more than one point, f has an inverse.

31. No, since some horizontal lines cross the graph of f twice, f *does not* have an inverse.

33. $g(x) = \dfrac{4-x}{6}$

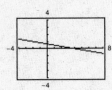

g passes the horizontal line test, so g has an inverse.

35. $h(x) = |x+4| - |x-4|$

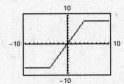

h does not pass the horizontal line test, so h *does not* have an inverse.

37. $f(x) = -2x\sqrt{16 - x^2}$

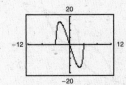

f does not pass the horizontal line test, so f *does not* have an inverse.

39. (a) $f(x) = 2x - 3$

$y = 2x - 3$

$x = 2y - 3$

$y = \dfrac{x+3}{2}$

$f^{-1}(x) = \dfrac{x+3}{2}$

(b)

(c) The graph of f^{-1} is the reflection of the graph of f about the line $y = x$.

(d) The domains and ranges of f and f^{-1} are all real numbers.

41. (a) $f(x) = x^5 - 2$

$y = x^5 - 2$

$x = y^5 - 2$

$y = \sqrt[5]{x+2}$

$f^{-1}(x) = \sqrt[5]{x+2}$

(b)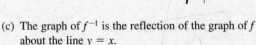

(c) The graph of f^{-1} is the reflection of the graph of f about the line $y = x$.

(d) The domains and ranges of f and f^{-1} are all real numbers.

43. (a) $f(x) = \sqrt{x}$

$y = \sqrt{x}$

$x = \sqrt{y}$

$y = x^2$

$f^{-1}(x) = x^2,\ x \geq 0$

(b)

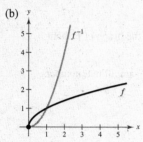

(c) The graph of f^{-1} is the reflection of the graph of f about the line $y = x$.

(d) The domains and ranges of f and f^{-1} are $[0, \infty)$.

45. (a) $f(x) = \sqrt{4 - x^2},\ 0 \leq x \leq 2$

$y = \sqrt{4 - x^2}$

$x = \sqrt{4 - y^2}$

$x^2 = 4 - y^2$

$y^2 = 4 - x^2$

$y = \sqrt{4 - x^2}$

$f^{-1}(x) = \sqrt{4 - x^2},\ 0 \leq x \leq 2$

(b)

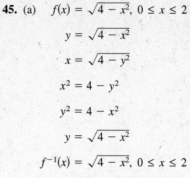

(c) The graph of f^{-1} is the same as the graph of f.

(d) The domains and ranges of f and f^{-1} are $[0, 2]$.

47. (a) $f(x) = \dfrac{4}{x}$

$y = \dfrac{4}{x}$

$x = \dfrac{4}{y}$

$xy = 4$

$y = \dfrac{4}{x}$

$f^{-1}(x) = \dfrac{4}{x}$

(b)

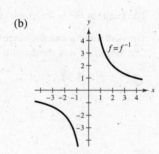

(c) The graph of f^{-1} is the same as the graph of f.

(d) The domains and ranges of f and f^{-1} are all real numbers except for 0.

49. (a) $f(x) = \dfrac{x + 1}{x - 2}$

$y = \dfrac{x + 1}{x - 2}$

$x = \dfrac{y + 1}{y - 2}$

$x(y - 2) = y + 1$

$xy - 2x = y + 1$

$xy - y = 2x + 1$

$y(x - 1) = 2x + 1$

$y = \dfrac{2x + 1}{x - 1}$

$f^{-1}(x) = \dfrac{2x + 1}{x - 1}$

(b)

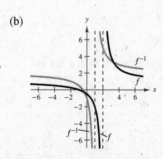

(c) The graph of f^{-1} is the reflection of the graph of f about the line $y = x$.

(d) The domain of f and the range of f^{-1} is all real numbers except 2. The range of f and the domain of f^{-1} is all real numbers except 1.

51. (a) $f(x) = \sqrt[3]{x - 1}$

$$y = \sqrt[3]{x - 1}$$

$$x = \sqrt[3]{y - 1}$$

$$x^3 = y - 1$$

$$y = x^3 + 1$$

$$f^{-1}(x) = x^3 + 1$$

(b)

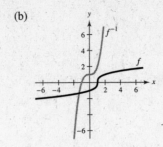

(c) The graph of f^{-1} is the reflection of the graph of f about the line $y = x$.

(d) The domains and ranges of f and f^{-1} are all real numbers.

53. (a) $f(x) = \dfrac{6x + 4}{4x + 5}$

$$y = \dfrac{6x + 4}{4x + 5}$$

$$x = \dfrac{6y + 4}{4y + 5}$$

$$x(4y + 5) = 6y + 4$$

$$4xy + 5x = 6y + 4$$

$$4xy - 6y = -5x + 4$$

$$y(4x - 6) = -5x + 4$$

$$y = \dfrac{-5x + 4}{4x - 6}$$

$$f^{-1}(x) = \dfrac{-5x + 4}{4x - 6} = \dfrac{5x - 4}{6 - 4x}$$

(b)

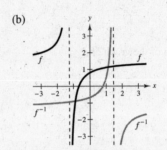

(c) The graph of f^{-1} is the graph of f reflected about the line $y = x$.

(d) The domain of f and the range of f^{-1} is all real numbers except $-\frac{5}{4}$. The range of f and the domain of f^{-1} is all real numbers except $\frac{3}{2}$.

55. $f(x) = x^4$

$$y = x^4$$

$$x = y^4$$

$$y = \pm\sqrt[4]{x}$$

This does not represent y as a function of x. f does not have an inverse.

57. $g(x) = \dfrac{x}{8}$

$$y = \dfrac{x}{8}$$

$$x = \dfrac{y}{8}$$

$$y = 8x$$

This is a function of x, so g has an inverse.

$$g^{-1}(x) = 8x$$

59. $p(x) = -4$

$$y = -4$$

Since $y = -4$ for all x, the graph is a horizontal line and fails the Horizontal Line Test. p does not have an inverse.

61. $f(x) = (x + 3)^2,\ x \geq -3 \implies y \geq 0$

$$y = (x + 3)^2,\ x \geq -3,\ y \geq 0$$

$$x = (y + 3)^2,\ y \geq -3,\ x \geq 0$$

$$\sqrt{x} = y + 3,\ y \geq -3,\ x \geq 0$$

$$y = \sqrt{x} - 3,\ x \geq 0,\ y \geq -3$$

This is a function of x, so f has an inverse.

$$f^{-1}(x) = \sqrt{x} - 3,\ x \geq 0$$

63. $f(x) = \begin{cases} x + 3, & x < 0 \\ 6 - x, & x \geq 0 \end{cases}$

The graph fails the Horizontal Line Test, so $f(x)$ does not have an inverse.

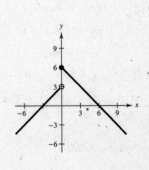

65. $h(x) = -\dfrac{4}{x^2}$

The graph fails the Horizontal Line Test so h does not have an inverse.

67. $f(x) = \sqrt{2x+3} \Rightarrow x \geq -\dfrac{3}{2},\ y \geq 0$

$y = \sqrt{2x+3},\ x \geq -\dfrac{3}{2},\ y \geq 0$

$x = \sqrt{2y+3},\ y \geq -\dfrac{3}{2},\ x \geq 0$

$x^2 = 2y+3,\ x \geq 0,\ y \geq -\dfrac{3}{2}$

$y = \dfrac{x^2-3}{2},\ x \geq 0,\ y \geq -\dfrac{3}{2}$

This is a function of x, so f has an inverse.

$f^{-1}(x) = \dfrac{x^2-3}{2},\ x \geq 0$

In Exercises 69–73, $f(x) = \frac{1}{8}x - 3$, $f^{-1}(x) = 8(x+3)$, $g(x) = x^3$, $g^{-1}(x) = \sqrt[3]{x}$.

69. $(f^{-1} \circ g^{-1})(1) = f^{-1}(g^{-1}(1))$
$= f^{-1}\left(\sqrt[3]{1}\right)$
$= 8\left(\sqrt[3]{1} + 3\right) = 32$

71. $(f^{-1} \circ f^{-1})(6) = f^{-1}(f^{-1}(6))$
$= f^{-1}(8[6+3])$
$= 8[8(6+3)+3] = 600$

73. $(f \circ g)(x) = f(g(x)) = f(x^3) = \frac{1}{8}x^3 - 3$
$y = \frac{1}{8}x^3 - 3$
$x = \frac{1}{8}y^3 - 3$
$x + 3 = \frac{1}{8}y^3$
$8(x+3) = y^3$
$\sqrt[3]{8(x+3)} = y$
$(f \circ g)^{-1}(x) = 2\sqrt[3]{x+3}$

In Exercises 75–77, $f(x) = x + 4$, $f^{-1}(x) = x - 4$, $g(x) = 2x - 5$, $g^{-1}(x) = \dfrac{x+5}{2}$.

75. $(g^{-1} \circ f^{-1})(x) = g^{-1}(f^{-1}(x))$
$= g^{-1}(x-4)$
$= \dfrac{(x-4)+5}{2}$
$= \dfrac{x+1}{2}$

77. $(f \circ g)(x) = f(g(x))$
$= f(2x-5)$
$= (2x-5)+4$
$= 2x-1$
$(f \circ g)^{-1}(x) = \dfrac{x+1}{2}$

Note: Comparing Exercises 75 and 77, we see that $(f \circ g)^{-1}(x) = (g^{-1} \circ f^{-1})(x)$.

79. (a) $f^{-1}(108,209) = 11$

(b) f^{-1} represents the year for a given number of households in the United States.

(c) $y \approx 1578.68t + 90,183.63$

(d)
$$y = 1578.68t + 90,183.63$$
$$t = 1578.68y + 90,183.63$$
$$\frac{t - 90,183.63}{1578.68} = y$$
$$f^{-1}(t) = \frac{t - 90,183.63}{1578.68}$$

(e) $f^{-1}(117,022) \approx 17$

(f) $f^{-1}(108,209) \approx 11.418$

This is close to the value of 11 in the table.

81. (a) Yes. Since the values of f increase each year, no two f-values are paired with the same t-value so f does have an inverse.

(b) f^{-1} would represent the year that a given number of miles was traveled by motor vehicles.

(c) Since $f(8) = 2632, f^{-1}(2632) = 8$.

(d) No. Since the new value is the same as the value given for 2000, f would not pass the Horizontal Line Test and would not have an inverse.

83. (a)
$$y = 0.03x^2 + 245.50, 0 < x < 100$$
$$\Rightarrow 245.50 < y < 545.50$$
$$x = 0.03y^2 + 245.50$$
$$x - 245.50 = 0.03y^2$$
$$\frac{x - 245.50}{0.03} = y^2$$
$$\sqrt{\frac{x - 245.50}{0.03}} = y, \ 245.50 < x < 545.50$$
$$f^{-1}(x) = \sqrt{\frac{x - 245.50}{0.03}}$$
$$x = \text{temperature in degrees Fahrenheit}$$
$$y = \text{percent load for a diesel engine}$$

(b)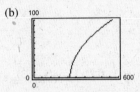

(c)
$$0.03x^2 + 245.50 \le 500$$
$$0.03x^2 \le 254.50$$
$$x^2 \le 8483.33$$
$$x \le 92.10$$

Thus, $0 < x \le 92.10$.

85. False. $f(x) = x^2$ is even and does not have an inverse.

87. Let $(f \circ g)(x) = y$. Then $x = (f \circ g)^{-1}(y)$. Also,
$$(f \circ g)(x) = y \Rightarrow f(g(x)) = y$$
$$g(x) = f^{-1}(y)$$
$$x = g^{-1}(f^{-1}(y))$$
$$x = (g^{-1} \circ f^{-1})(y).$$

Since f and g are both one-to-one functions,
$$(f \circ g)^{-1} = g^{-1} \circ f^{-1}.$$

89.

x	1	3	4	6
f	1	2	6	7

x	1	2	6	7
$f^{-1}(x)$	1	3	4	6

91.

x	-2	-1	3	4
f	6	0	-2	-3

x		-3	-2	0	6
$f^{-1}(x)$		4	3	-1	-2

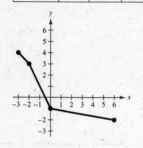

93. If $f(x) = k(2 - x - x^3)$ has an inverse and $f^{-1}(3) = -2$, then $f(-2) = 3$. Thus,

$$f(-2) = k(2 - (-2) - (-2)^3) = 3$$
$$k(2 + 2 + 8) = 3$$
$$12k = 3$$
$$k = \frac{3}{12} = \frac{1}{4}$$

So, $k = \frac{1}{4}$.

95. $x^2 = 64$

$x = \pm\sqrt{64} = \pm 8$

97. $4x^2 - 12x + 9 = 0$

$$(2x - 3)^2 = 0$$
$$2x - 3 = 0$$
$$x = \frac{3}{2}$$

99. $x^2 - 6x + 4 = 0$ Complete the square.

$$x^2 - 6x = -4$$
$$x^2 - 6x + 9 = -4 + 9$$
$$(x - 3)^2 = 5$$
$$x - 3 = \pm\sqrt{5}$$
$$x = 3 \pm \sqrt{5}$$

101. $50 + 5x = 3x^2$

$$0 = 3x^2 - 5x - 50$$
$$0 = (3x + 10)(x - 5)$$
$$3x + 10 = 0 \implies x = -\frac{10}{3}$$
$$x - 5 = 0 \implies x = 5$$

103. Let $2n$ = first positive even integer. Then $2n + 2$ = next positive even integer.

$$2n(2n + 2) = 288$$
$$4n^2 + 4n - 288 = 0$$
$$4(n^2 + n - 72) = 0$$
$$4(n + 9)(n - 8) = 0$$

$n + 9 = 0 \implies n = -9$ Not a solution since the integers are positive.

$n - 8 = 0 \implies n = 8$

So, $2n = 16$ and $2n + 2 = 18$.

Section 1.10   Mathematical Modeling and Variation

You should know the following the following terms and formulas.

- Direct variation (varies directly, directly proportional)

    (a) $y = kx$ (b) $y = kx^n$ (as nth power)

- Inverse variation (varies inversely, inversely proportional)

    (a) $y = k/x$ (b) $y = k/(x^n)$ (as nth power)

- Joint variation (varies jointly, jointly proportional)

    (a) $z = kxy$ (b) $z = kx^n y^m$ (as nth power of x and mth power of y)

- k is called the constant of proportionality.

- Least Squares Regression Line $y = ax + b$. Use your calculator or computer to enter the data points and to find the "best-fitting"linear model.

Vocabulary Check

1. variation; regression   **2.** sum of square differences   **3.** correlation coefficient

4. directly proportional   **5.** constant of variation   **6.** directly proportional

7. inverse   **8.** combined   **9.** jointly proportional

1. $y = 1767.0t + 123,916$

Year	Actual Number (in thousands)	Model (in thousands)
1992	128,105	127,450
1993	129,200	129,217
1994	131,056	130,984
1995	132,304	132,751
1996	133,943	134,518
1997	136,297	136,285
1998	137,673	138,052
1999	139,368	139,819
2000	142,583	141,586
2001	143,734	143,353
2002	144,863	145,120

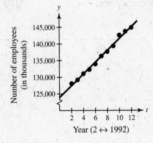

The model is a good fit for the actual data.

3.

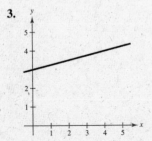

Using the points $(0, 3)$ and $(4, 4)$, we have $y = \frac{1}{4}x + 3$.

5.

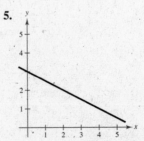

Using the points $(2, 2)$ and $(4, 1)$, we have $y = -\frac{1}{2}x + 3$.

7. (a) and (b)

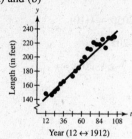

Year (12 ↔ 1912)

$y \approx t + 130$

(c) $y \approx 1.03t + 130.27$

(d) The models are similar.

(e) When $t = 108$, we have:

Model in part (b): 238 feet

Model in part (c): 241.51 feet

(f) Answers will vary.

9. (a) and (c)

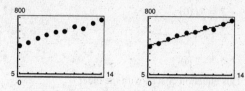

The model is a good fit to the actual data. ($r \approx 0.98$)

(b) $S \approx 38.4t + 224$

(d) For 2005, use $t = 15$: $S \approx \$800.4$ million

For 2007, use $t = 17$: $S \approx \$877.3$ million

(e) Each year the annual gross ticket sales for Broadway shows in New York City increase by approximately $38.4 million.

11. The graph appears to represent $y = 4/x$, so y varies inversely as x.

13. $k = 1$

x	2	4	6	8	10
$y = kx^2$	4	16	36	64	100

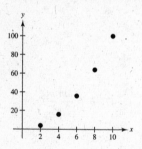

15. $k = \frac{1}{2}$

x	2	4	6	8	10
$y = kx^2$	2	8	18	32	50

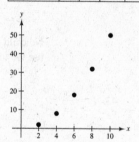

17. $k = 2$

x	2	4	6	8	10
$y = \dfrac{k}{x^2}$	$\frac{1}{2}$	$\frac{1}{8}$	$\frac{1}{18}$	$\frac{1}{32}$	$\frac{1}{50}$

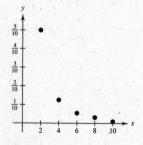

19. $k = 10$

x	2	4	6	8	10
$y = \dfrac{k}{x^2}$	$\frac{5}{2}$	$\frac{5}{8}$	$\frac{5}{18}$	$\frac{5}{32}$	$\frac{1}{10}$

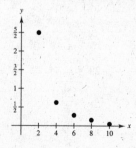

21. The table represents the equation $y = 5/x$.

23.
$$y = kx$$
$$-7 = k(10)$$
$$-\frac{7}{10} = k$$
$$y = -\frac{7}{10}x$$

This equation checks with the other points given in the table.

25.
$$y = kx$$
$$12 = k(5)$$
$$\frac{12}{5} = k$$
$$y = \frac{12}{5}x$$

27.
$$y = kx$$
$$2050 = k(10)$$
$$205 = k$$
$$y = 205x$$

29.
$$I = kP$$
$$87.50 = k(2500)$$
$$0.035 = k$$
$$I = 0.035P$$

31.
$$y = kx$$
$$33 = k(13)$$
$$\frac{33}{13} = k$$
$$y = \frac{33}{13}x$$

When $x = 10$ inches, $y \approx 25.4$ centimeters.

When $x = 20$ inches, $y \approx 50.8$ centimeters.

33.
$$y = kx$$
$$5520 = k(150,000)$$
$$0.0368 = k$$
$$y = 0.0368x$$
$$y = 0.0368(200,000)$$
$$= \$7360$$

The property tax is $7360.

35.
$$d = kF$$
$$0.15 = k(265)$$
$$\frac{3}{5300} = k$$
$$d = \frac{3}{5300}F$$
(a) $d = \frac{3}{5300}(90) \approx 0.05$ meter
(b) $0.1 = \frac{3}{5300}F$
$$\frac{530}{3} = F$$
$$F = 176\frac{2}{3} \text{ newtons}$$

37.
$$d = kF$$
$$1.9 = k(25) \implies k = 0.076$$
$$d = 0.076F$$

When the distance compressed is 3 inches, we have
$$3 = 0.076F$$
$$F \approx 39.47.$$

No child over 39.47 pounds should use the toy.

39. $A = kr^2$

41. $y = \dfrac{k}{x^2}$

43. $F = \dfrac{kg}{r^2}$

45. $P = \dfrac{k}{V}$

47. $F = \dfrac{km_1m_2}{r^2}$

49. $A = \dfrac{1}{2}bh$

The area of a triangle is jointly proportional to its base and height.

51. $V = \dfrac{4}{3}\pi r^3$

The volume of a sphere varies directly as the cube of its radius.

53. $r = \dfrac{d}{t}$

Average speed is directly proportional to the distance and inversely proportional to the time.

55.
$$A = kr^2$$
$$9\pi = k(3)^2$$
$$\pi = k$$
$$A = \pi r^2$$

57.
$$y = \frac{k}{x}$$
$$7 = \frac{k}{4}$$
$$28 = k$$
$$y = \frac{28}{x}$$

59.
$$F = krs^3$$
$$4158 = k(11)(3)^3$$
$$k = 14$$
$$F = 14rs^3$$

61.
$$z = \frac{kx^2}{y}$$
$$6 = \frac{k(6)^2}{4}$$
$$\frac{24}{36} = k$$
$$\frac{2}{3} = k$$
$$z = \frac{2/3x^2}{y} = \frac{2x^2}{3y}$$

63. $d = kv^2$

$$0.02 = k\left(\frac{1}{4}\right)^2$$

$$k = 0.32$$

$$d = 0.32v^2$$

$$0.12 = 0.32v^2$$

$$v^2 = \frac{0.12}{0.32} = \frac{3}{8}$$

$$v = \frac{\sqrt{3}}{2\sqrt{2}} = \frac{\sqrt{6}}{4} \approx 0.61 \text{ mi/hr}$$

65. $r = \dfrac{kl}{A}, \; A = \pi r^2 = \dfrac{\pi d^2}{4}$

$$r = \frac{4kl}{\pi d^2}$$

$$66.17 = \frac{4(1000)k}{\pi\left(\frac{0.0126}{12}\right)^2}$$

$$k \approx 5.73 \times 10^{-8}$$

$$r = \frac{4(5.73 \times 10^{-8})l}{\pi\left(\frac{0.0126}{12}\right)^2}$$

$$33.5 = \frac{4(5.73 \times 10^{-8})l}{\pi\left(\frac{0.0126}{12}\right)^2}$$

$$\frac{33.5 \pi\left(\frac{0.0126}{12}\right)^2}{4(5.73 \times 10^{-8})} = l$$

$$l \approx 506 \text{ feet}$$

67. $W = kmh$

$$2116.8 = k(120)(1.8)$$

$$k = \frac{2116.8}{(120)(1.8)} = 9.8$$

$$W = 9.8mh$$

When $m = 100$ kilograms and $h = 1.5$ meters, we have
$W = 9.8(100)(1.5) = 1470$ joules.

69. $v = \dfrac{k}{A}$

$$v = \frac{k}{0.75A} = \frac{4}{3}\left(\frac{k}{A}\right)$$

The velocity is increased by one-third.

71. (a)

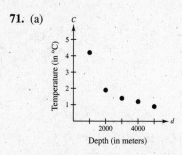

(b) Yes, the data appears to be modeled (approximately) by the inverse proportion model.

$4.2 = \dfrac{k_1}{1000}$ $1.9 = \dfrac{k_2}{2000}$ $1.4 = \dfrac{k_3}{3000}$ $1.2 = \dfrac{k_4}{4000}$ $0.9 = \dfrac{k_5}{5000}$

$4200 = k_1$ $3800 = k_2$ $4200 = k_3$ $4800 = k_4$ $4500 = k_5$

(c) Mean: $k = \dfrac{4200 + 3800 + 4200 + 4800 + 4500}{5} = 4300,$ Model: $C = \dfrac{4300}{d}$

(d)

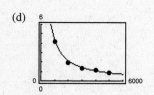

(e) $3 = \dfrac{4300}{d}$

$$d = \frac{4300}{3} = 1433\frac{1}{3} \text{ meters}$$

73. $y = \dfrac{262.76}{x^{2.12}}$

(a)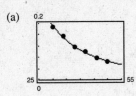

(b) $y = \dfrac{262.76}{(25)^{2.12}}$

≈ 0.2857 microwatts per sq. cm.

75. False. y will increase if k is positive and y will decrease if k is negative.

77. False. The closer the value of $|r|$ is to 1, the better the fit.

79. The accuracy of the model in predicting prize winnings is questionable because the model is based on limited data.

81. $3x + 2 > 17$

$3x > 15$

$x > 5$

83. $|2x - 1| < 9$

$-9 < 2x - 1 < 9$

$-8 < 2x < 10$

$-4 < x < 5$

85. $f(x) = \dfrac{x^2 + 5}{x - 3}$

(a) $f(0) = \dfrac{0^2 + 5}{0 - 3} = -\dfrac{5}{3}$

(b) $f(-3) = \dfrac{(-3)^2 + 5}{-3 - 3} = \dfrac{14}{-6} = -\dfrac{7}{3}$

(c) $f(4) = \dfrac{4^2 + 5}{4 - 3} = 21$

87. Answers will vary.

Review Exercises for Chapter 1

1.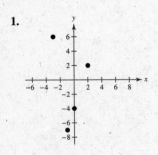

3. $x > 0$ and $y = -2$ in Quadrant IV.

5. (a)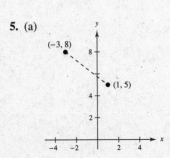

(b) $d = \sqrt{(-3 - 1)^2 + (8 - 5)^2} = \sqrt{16 + 9} = 5$

(c) Midpoint: $\left(\dfrac{-3 + 1}{2}, \dfrac{8 + 5}{2}\right) = \left(-1, \dfrac{13}{2}\right)$

7. (a)

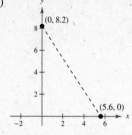

(b) $d = \sqrt{(5.6 - 0)^2 + (0 - 8.2)^2}$

$= \sqrt{31.36 + 67.24} = \sqrt{98.6} \approx 9.9$

(c) Midpoint: $\left(\dfrac{0 + 5.6}{2}, \dfrac{8.2 + 0}{2}\right) = (2.8, 4.1)$

9. $(4 - 2, 8 - 3) = (2, 5)$

$(6 - 2, 8 - 3) = (4, 5)$

$(4 - 2, 3 - 3) = (2, 0)$

$(6 - 2, 3 - 3) = (4, 0)$

11. $(2001, 539.1), (2003, 773.8)$

$\left(\dfrac{2001 + 2003}{2}, \dfrac{539.1 + 773.8}{2}\right) = (2002, 656.45)$

In 2002, the sales were approximately \$656.45 million.

13. $\dfrac{4}{3}\pi r^3 = 47{,}712.94$

$r = \sqrt[3]{\dfrac{47{,}712.94(3)}{4\pi}}$

$r \approx 22.5$ centimeters

15. $y = 3x - 5$

x	-2	-1	0	1	2
y	-11	-8	-5	-2	1

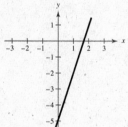

17. $y = x^2 - 3x$

x	-1	0	1	2	3	4
y	4	0	-2	-2	0	4

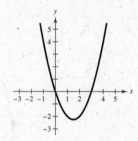

19. $y - 2x - 3 = 0$

$y = 2x + 3$

Line with x-intercept $\left(-\frac{3}{2}, 0\right)$ and y-intercept $(0, 3)$

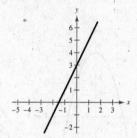

21. $y = \sqrt{5 - x}$

Domain: $(-\infty, 5]$

x	5	4	1	-4
y	0	1	2	3

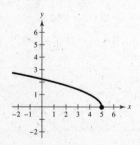

23. $y + 2x^2 = 0$

$y = -2x^2$ is a parabola.

x	0	± 1	± 2
y	0	-2	-8

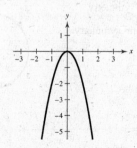

25. $y = 2x + 7$

x-intercept: Let $y = 0$.

$0 = 2x + 7$

$x = -\dfrac{7}{2}$

$\left(-\dfrac{7}{2}, 0\right)$

y-intercept: Let $x = 0$.

$y = 2(0) + 7$

$y = 7$

$(0, 7)$

27. $y = (x - 3)^2 - 4$

    x-intercepts: $0 = (x - 3)^2 - 4 \Rightarrow (x - 3)^2 = 4$

                            $\Rightarrow x - 3 = \pm 2$

                            $\Rightarrow x = 3 \pm 2$

                            $\Rightarrow x = 5 \text{ or } x = 1$

    y-intercept: $y = (0 - 3)^2 - 4 \Rightarrow y = 9 - 4 \Rightarrow y = 5$

    The x-intercepts are $(1, 0)$ and $(5, 0)$. The y-intercept is $(0, 5)$.

29. $y = -4x + 1$

    Intercepts: $\left(\frac{1}{4}, 0\right), (0, 1)$

    $y = -4(-x) + 1 \Rightarrow y = 4x + 1 \Rightarrow$ No y-axis symmetry

    $-y = -4x + 1 \Rightarrow y = 4x - 1 \Rightarrow$ No x-axis symmetry

    $-y = -4(-x) + 1 \Rightarrow y = -4x - 1 \Rightarrow$ No origin symmetry

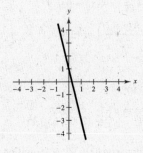

31. $y = 5 - x^2$

    Intercepts: $\left(\pm\sqrt{5}, 0\right), (0, 5)$

    $y = 5 - (-x)^2 \Rightarrow y = 5 - x^2 \Rightarrow y$-axis symmetry

    $-y = 5 - x^2 \Rightarrow y = -5 + x^2 \Rightarrow$ No x-axis symmetry

    $-y = 5 - (-x)^2 \Rightarrow y = -5 + x^2 \Rightarrow$ No origin symmetry

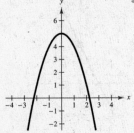

33. $y = x^3 + 3$

    Intercepts: $\left(-\sqrt[3]{3}, 0\right), (0, 3)$

    $y = (-x)^3 + 3 \Rightarrow y = -x^3 + 3 \Rightarrow$ No y-axis symmetry

    $-y = x^3 + 3 \Rightarrow y = -x^3 - 3 \Rightarrow$ No x-axis symmetry

    $-y = (-x)^3 + 3 \Rightarrow y = x^3 - 3 \Rightarrow$ No origin symmetry

35. $y = \sqrt{x + 5}$

    Domain: $[-5, \infty)$

    Intercepts: $(-5, 0), \left(0, \sqrt{5}\right)$

    $y = \sqrt{-x + 5} \Rightarrow$ No y-axis symmetry

    $-y = \sqrt{x + 5} \Rightarrow y = -\sqrt{x + 5} \Rightarrow$ No x-axis symmetry

    $-y = \sqrt{-x + 5} \Rightarrow y = -\sqrt{-x + 5} \Rightarrow$ No origin symmetry

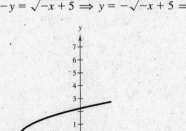

37. $x^2 + y^2 = 9$

    Center: $(0, 0)$

    Radius: 3

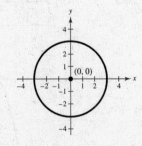

39.
$$(x + 2)^2 + y^2 = 16$$
$$(x - (-2))^2 + (y - 0)^2 = 4^2$$
Center: $(-2, 0)$

Radius: 4

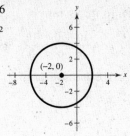

41.
$$\left(x - \tfrac{1}{2}\right)^2 + (y + 1)^2 = 36$$
$$\left(x - \tfrac{1}{2}\right)^2 + (y - (-1))^2 = 6^2$$
Center: $\left(\tfrac{1}{2}, -1\right)$

Radius: 6

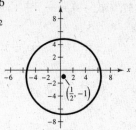

43. Endpoints of a diameter: $(0, 0)$ and $(4, -6)$

Center: $\left(\dfrac{0 + 4}{2}, \dfrac{0 + (-6)}{2}\right) = (2, -3)$

Radius: $r = \sqrt{(2 - 0)^2 + (-3 - 0)^2} = \sqrt{4 + 9} = \sqrt{13}$

Standard form: $(x - 2)^2 + (y - (-3))^2 = \left(\sqrt{13}\right)^2$
$$(x - 2)^2 + (y + 3)^2 = 13$$

45. $F = \tfrac{5}{4}x, \; 0 \le x \le 20$

(a)

x	0	4	8	12	16	20
F	0	5	10	15	20	25

(b)

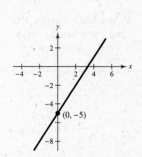

(c) When $x = 10$, $F = \dfrac{50}{4} = 12.5$ pounds.

47. $y = 6$

Horizontal line, $m = 0$

y-intercept: $(0, 6)$

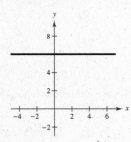

49. $y = 3x + 13$

Slope: $m = 3 = \tfrac{3}{1}$

y-intercept: $(0, 13)$

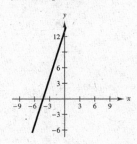

51. $(3, -4), (-7, 1)$

$$m = \frac{1 - (-4)}{-7 - 3} = \frac{5}{-10} = -\frac{1}{2}$$

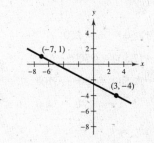

53. $(-4.5, 6), (2.1, 3)$

$$m = \frac{3 - 6}{2.1 - (-4.5)}$$

$$= \frac{-3}{6.6} = -\frac{30}{66} = -\frac{5}{11}$$

55. $(0, -5), \; m = \dfrac{3}{2}$

$$y - (-5) = \frac{3}{2}(x - 0)$$

$$y + 5 = \frac{3}{2}x$$

$$y = \frac{3}{2}x - 5$$

57. $(10, -3), m = -\dfrac{1}{2}$

$$y - (-3) = -\dfrac{1}{2}(x - 10)$$

$$y + 3 = -\dfrac{1}{2}x + 5$$

$$y = -\dfrac{1}{2}x + 2$$

$(10, -3)$

59. $(0, 0), (0, 10)$

$$m = \dfrac{10 - 0}{0 - 0} = \dfrac{10}{0}, \quad \text{undefined}$$

The line is vertical.

$$x = 0$$

61. $(-1, 4), (2, 0)$

$$m = \dfrac{0 - 4}{2 - (-1)} = -\dfrac{4}{3}$$

$$y - 4 = -\dfrac{4}{3}(x - (-1))$$

$$y - 4 = -\dfrac{4}{3}x - \dfrac{4}{3}$$

$$y = -\dfrac{4}{3}x + \dfrac{8}{3}$$

63. Point: $(3, -2)$

$5x - 4y = 8 \implies y = \dfrac{5}{4}x - 2$ and $m = \dfrac{5}{4}$

(a) Parallel slope: $m = \dfrac{5}{4}$

$$y - (-2) = \dfrac{5}{4}(x - 3)$$

$$y + 2 = \dfrac{5}{4}x - \dfrac{15}{4}$$

$$y = \dfrac{5}{4}x - \dfrac{23}{4}$$

(b) Perpendicular slope: $m = -\dfrac{4}{5}$

$$y - (-2) = -\dfrac{4}{5}(x - 3)$$

$$y + 2 = -\dfrac{4}{5}x + \dfrac{12}{5}$$

$$y = -\dfrac{4}{5}x + \dfrac{2}{5}$$

65. $(6, 12{,}500) \quad m = 850$

$$y - 12{,}500 = 850(t - 6)$$

$$y - 12{,}500 = 850t - 5100$$

$$y = 850t + 7400, \quad 6 \le t \le 11$$

67. $16x - y^4 = 0$

$$y^4 = 16x$$

$$y = \pm 2\sqrt[4]{x}$$

No, y is not a function of x. Some x-values correspond to two y-values.

69. $y = \sqrt{1 - x}$

Yes. Each x-value, $x \le 1$, corresponds to only one y-value so y is a function of x.

71. $f(x) = x^2 + 1$

(a) $f(2) = (2)^2 + 1 = 5$

(b) $f(-4) = (-4)^2 + 1 = 17$

(c) $f(t^2) = (t^2)^2 + 1 = t^4 + 1$

(d) $f(t + 1) = (t + 1)^2 + 1$

$$= t^2 + 2t + 2$$

73. $f(x) = \sqrt{25 - x^2}$

Domain: $25 - x^2 \geq 0$

         $(5 + x)(5 - x) \geq 0$

Critical numbers: $x = \pm 5$

Test intervals: $(-\infty, -5), \ (-5, 5), \ (5, \infty)$

Test: Is $25 - x^2 \geq 0$?

Solution set: $-5 \leq x \leq 5$

Thus, the domain is all real numbers x such that $-5 \leq x \leq 5$, or $[-5, 5]$.

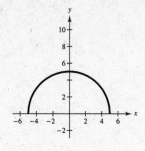

75. $h(x) = \dfrac{x}{x^2 - x - 6}$

     $= \dfrac{x}{(x + 2)(x - 3)}$

Domain: All real numbers x
except $x = -2, 3$

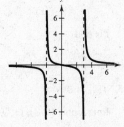

77. $v(t) = -32t + 48$

(a) $v(1) = 16$ feet per second

(b) $0 = -32t + 48$

     $t = \frac{48}{32} = 1.5$ seconds

(c) $v(2) = -16$ feet per second

79. $f(x) = 2x^2 + 3x - 1$

$$\frac{f(x + h) - f(x)}{h} = \frac{[2(x + h)^2 + 3(x + h) - 1] - (2x^2 + 3x - 1)}{h}$$

$$= \frac{2x^2 + 4xh + 2h^2 + 3x + 3h - 1 - 2x^2 - 3x + 1}{h}$$

$$= \frac{h(4x + 2h + 3)}{h}$$

$$= 4x + 2h + 3, \ h \neq 0$$

81. $y = (x - 3)^2$

The graph passes the Vertical Line Test. y *is* a function of x.

83. $x - 4 = y^2$

The graph does not pass the Vertical Line Test. y *is not* a function of x.

85. $f(x) = 3x^2 - 16x + 21$

$3x^2 - 16x + 21 = 0$

$(3x - 7)(x - 3) = 0$

$3x - 7 = 0 \quad$ or $\quad x - 3 = 0$

$x = \dfrac{7}{3} \quad$ or $\qquad x = 3$

87. $f(x) = \dfrac{8x + 3}{11 - x}$

$\dfrac{8x + 3}{11 - x} = 0$

$8x + 3 = 0$

$x = -\dfrac{3}{8}$

89. $f(x) = |x| + |x + 1|$

f is increasing on $(0, \infty)$.

f is decreasing on $(-\infty, -1)$.

f is constant on $(-1, 0)$.

91. $f(x) = -x^2 + 2x + 1$

Relative maximum: $(1, 2)$

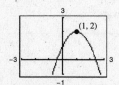

93. $f(x) = x^3 - 6x^4$

Relative maximum: $(0.125, 0.000488) \approx (0.13, 0.00)$

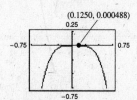

95. $f(x) = -x^2 + 8x - 4$

$$\frac{f(4) - f(0)}{4 - 0} = \frac{12 - (-4)}{4} = 4$$

The average rate of change of f from $x_1 = 0$ to $x_2 = 4$ is 4.

97. $f(x) = 2 - \sqrt{x + 1}$

$$\frac{f(7) - f(3)}{7 - 3} = \frac{(2 - \sqrt{8}) - (2 - 2)}{4}$$

$$= \frac{2 - 2\sqrt{2}}{4} = \frac{1 - \sqrt{2}}{2}$$

The average rate of change of f from $x_1 = 3$ to $x_2 = 7$ is $(1 - \sqrt{2})/2$.

99. $f(x) = x^5 + 4x - 7$

$$f(-x) = (-x)^5 + 4(-x) - 7$$

$$= -x^5 - 4x - 7$$

$$\neq f(x)$$

$$\neq -f(x)$$

Neither even nor odd

101. $f(x) = 2x\sqrt{x^2 + 3}$

$$f(-x) = 2(-x)\sqrt{(-x)^2 + 3}$$

$$= -2x\sqrt{x^2 + 3}$$

$$= -f(x)$$

f is odd.

103. $f(2) = -6, f(-1) = 3$

Points: $(2, -6), (-1, 3)$

$$m = \frac{3 - (-6)}{-1 - 2} = \frac{9}{-3} = -3$$

$$y - (-6) = -3(x - 2)$$

$$y + 6 = -3x + 6$$

$$y = -3x$$

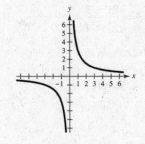

105. $f(x) = 3 - x^2$

Intercepts: $(0, 3), (\pm\sqrt{3}, 0)$

y-axis symmetry

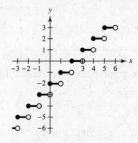

107. $f(x) = -\sqrt{x}$

Domain: $x \geq 0$

Intercepts: $(0, 0)$

x	0	1	4	9
y	0	-1	-2	-3

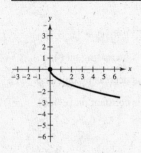

109. $g(x) = \dfrac{3}{x}$

No intercepts

Origin symmetry

x	-3	-1	1	3
y	-1	-3	3	1

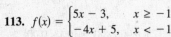

111. $f(x) = [\![x]\!] - 2$

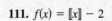

113. $f(x) = \begin{cases} 5x - 3, & x \geq -1 \\ -4x + 5, & x < -1 \end{cases}$

115. Common function: $f(x) = x^3$

Horizontal shift 4 units to the left and a vertical shift 4 units upward

117. (a) $f(x) = x^2$

(b) $h(x) = x^2 - 9$

Vertical shift 9 units downward

(c)

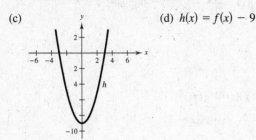

(d) $h(x) = f(x) - 9$

119. (a) $f(x) = \sqrt{x}$

(b) $h(x) = \sqrt{x - 7}$

Horizontal shift 7 units to the right

(c)

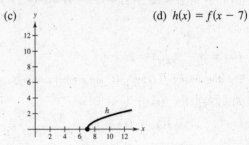

(d) $h(x) = f(x - 7)$

121. (a) $f(x) = x^2$

(b) $h(x) = -(x + 3)^2 + 1$

Reflection in the x-axis, a horizontal shift 3 units to the left, and a vertical shift 1 unit upward

(c)

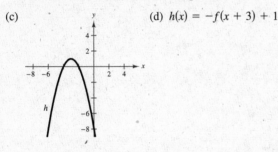

(d) $h(x) = -f(x + 3) + 1$

123. (a) $f(x) = [\![x]\!]$

(b) $h(x) = -[\![x]\!] + 6$

Reflection in the x-axis and a vertical shift 6 units upward

(c)

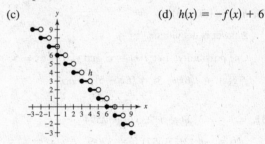

(d) $h(x) = -f(x) + 6$

125. (a) $f(x) = |x|$

(b) $h(x) = -|-x + 4| + 6$

Reflection in both the x- and y-axes; horizontal shift of 4 units to the right; vertical shift of 6 units upward

(c)

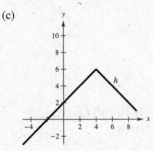

(d) $h(x) = -f(-(x - 4)) + 6 = -f(-x + 4) + 6$

127. (a) $f(x) = [\![x]\!]$

(b) $h(x) = 5[\![x - 9]\!]$

Horizontal shift 9 units to the right and a vertical stretch (each y-value is multiplied by 5)

(c)

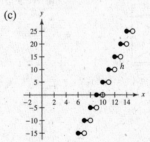

(d) $h(x) = 5f(x - 9)$

129. (a) $f(x) = \sqrt{x}$

(b) $h(x) = -2\sqrt{x - 4}$

Reflection in the x-axis, a vertical stretch (each y-value is multiplied by 2), and a horizontal shift 4 units to the right

(d) $h(x) = -2f(x - 4)$

(c)

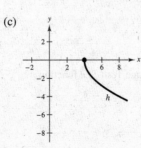

131. $f(x) = x^2 + 3$, $g(x) = 2x - 1$

(a) $(f + g)(x) = (x^2 + 3) + (2x - 1) = x^2 + 2x + 2$

(b) $(f - g)(x) = (x^2 + 3) - (2x - 1) = x^2 - 2x + 4$

(c) $(fg)(x) = (x^2 + 3)(2x - 1) = 2x^3 - x^2 + 6x - 3$

(d) $\left(\dfrac{f}{g}\right)(x) = \dfrac{x^2 + 3}{2x - 1}$,   Domain: $x \neq \dfrac{1}{2}$

133. $f(x) = \frac{1}{3}x - 3$, $g(x) = 3x + 1$

The domains of $f(x)$ and $g(x)$ are all real numbers.

(a) $(f \circ g)(x) = f(g(x))$

$\qquad = f(3x + 1)$

$\qquad = \frac{1}{3}(3x + 1) - 3$

$\qquad = x + \frac{1}{3} - 3$

$\qquad = x - \frac{8}{3}$

Domain: all real numbers

(b) $(g \circ f)(x) = g(f(x))$

$\qquad = g\left(\frac{1}{3}x - 3\right)$

$\qquad = 3\left(\frac{1}{3}x - 3\right) + 1$

$\qquad = x - 9 + 1$

$\qquad = x - 8$

Domain: all real numbers

135. $h(x) = (6x - 5)^3$

Answer is not unique.

One possibility: Let $f(x) = x^3$ and $g(x) = 6x - 5$.

$f(g(x)) = f(6x - 5) = (6x - 5)^3 = h(x)$

137. $v(t) = -31.86t^2 + 233.6t + 2594$

$d(t) = -4.18t^2 + 571.0t - 3706$

(a) $(v + d)(t) = v(t) + d(t)$

$\qquad\quad = -36.04t^2 + 804.6t - 1112$

$(v + d)(t)$ represents the combined factory sales (in millions of dollars) for VCRs and DVD players from 1997 to 2003.

(b)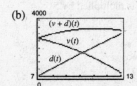

(c) $(v + d)(10) = \$3330$ million

139. $\qquad f(x) = x - 7$

$\qquad f^{-1}(x) = x + 7$

$f(f^{-1}(x)) = f(x + 7) = (x + 7) - 7 = x$

$f^{-1}(f(x)) = f^{-1}(x - 7) = (x - 7) + 7 = x$

141. The graph passes the Horizontal Line Test. The function has an inverse.

143. $f(x) = 4 - \dfrac{1}{3}x$

The graph passes the Horizontal Line Test. The function has an inverse.

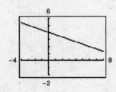

145. $h(t) = \dfrac{2}{t - 3}$

The graph passes the Horizontal Line Test. The function has an inverse.

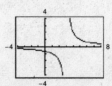

147. (a) $f(x) = \frac{1}{2}x - 3$

$$y = \frac{1}{2}x - 3$$

$$x = \frac{1}{2}y - 3$$

$$x + 3 = \frac{1}{2}y$$

$$2(x + 3) = y$$

$$f^{-1}(x) = 2x + 6$$

(c) The graph of f^{-1} is the reflection of the graph of f about the line $y = x$.

(b)

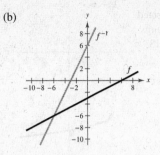

(d) The domains and ranges of f and f^{-1} are the set of all real numbers.

149. (a) $f(x) = \sqrt{x + 1}$

$$y = \sqrt{x + 1}$$

$$x = \sqrt{y + 1}$$

$$x^2 = y + 1$$

$$x^2 - 1 = y$$

$$f^{-1}(x) = x^2 - 1, \; x \geq 0$$

Note: The inverse must have a restricted domain.

(c) The graph of f^{-1} is the reflection of the graph of f about the line $y = x$.

(b)

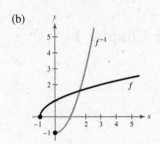

(d) The domain of f and the range of f^{-1} is $[-1, \infty)$. The range of f and the domain of f^{-1} is $[0, \infty)$.

151. $f(x) = 2(x - 4)^2$ is increasing on $[4, \infty)$.

Let $f(x) = 2(x - 4)^2$, $x \geq 4$ and $y \geq 0$.

$$y = 2(x - 4)^2$$

$$x = 2(y - 4)^2, \; x \geq 0, \; y \geq 4$$

$$\frac{x}{2} = (y - 4)^2$$

$$\sqrt{\frac{x}{2}} = y - 4$$

$$\sqrt{\frac{x}{2}} + 4 = y$$

$$f^{-1}(x) = \sqrt{\frac{x}{2}} + 4, \; x \geq 0$$

153. $I = 2.09t + 37.2$

(a)

Median income (in thousands of dollars)

Year (5 ↔ 1995)

(b) The model is a good fit to the actual data.

155. $D = km$

$4 = 2.5k$

$1.6 = k$

In 2 miles:

$D = 1.6(2) = 3.2$ kilometers

In 10 miles:

$D = 1.6(10) = 16$ kilometers

157. $F = ks^2$

If speed is doubled,

$F = k(2s)^2$

$F = 4ks^2$.

Thus, the force will be changed by a factor of 4.

159. $T = \dfrac{k}{r}$

$3 = \dfrac{k}{65}$

$k = 3(65) = 195$

$T = \dfrac{195}{r}$

When $r = 80$ mph,

$T = \dfrac{195}{80} = 2.4375$ hours

≈ 2 hours, 26 minutes.

161. False. The graph is reflected in the x-axis, shifted 9 units to the left, then shifted 13 units down.

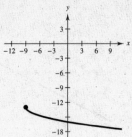

163. True. If $y = kx$, then

$$x = \frac{1}{k}y.$$

165. A function from a Set A to a Set B is a relation that assigns to each element x in the Set A exactly one element y in the Set B.

Problem Solving for Chapter 1

1. (a) $W_1 = 0.07x + 2000$

(b) $W_2 = 0.05x + 2300$

(c)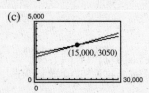

Point of intersection: (15,000, 3050)

Both jobs pay the same, $3050, if you sell $15,000 per month.

(d) If you think you can sell $20,000 per month, keep your current job with the higher commission rate. For sales over $15,000 it pays more than the other job.

3. (a) Let $f(x)$ and $g(x)$ be two even functions. Then define $h(x) = f(x) \pm g(x)$.

$$h(-x) = f(-x) \pm g(-x)$$
$$= f(x) \pm g(x) \text{ since } f \text{ and } g \text{ are even}$$
$$= h(x)$$

So, $h(x)$ is also even.

(b) Let $f(x)$ and $g(x)$ be two odd functions. Then define $h(x) = f(x) \pm g(x)$.

$$h(-x) = f(-x) \pm g(-x)$$
$$= -f(x) \mp g(x) \text{ since } f \text{ and } g \text{ are odd}$$
$$= -h(x)$$

So, $h(x)$ is also odd. (If $f(x) \neq g(x)$)

(c) Let $f(x)$ be odd and $g(x)$ be even. Then define $h(x) = f(x) \pm g(x)$.

$$h(-x) = f(-x) \pm g(-x)$$
$$= -f(x) \pm g(x) \text{ since } f \text{ is odd and } g \text{ is even}$$
$$\neq h(x)$$
$$\neq -h(x)$$

So, $h(x)$ is neither odd nor even.

5. $f(x) = a_{2n}x^{2n} + a_{2n-2}x^{2n-2} + \cdots + a_2x^2 + a_0$

$f(-x) = a_{2n}(-x)^{2n} + a_{2n-2}(-x)^{2n-2} + \cdots + a_2(-x)^2 + a_0$

$\qquad = a_{2n}x^{2n} + a_{2n-2}x^{2n-2} + \cdots + a_2x^2 + a_0$

$\qquad = f(x)$

Therefore, $f(x)$ is even.

7. (a) April 11: 10 hours

April 12: 24 hours

April 13: 24 hours

April 14: $23\frac{2}{3}$ hours

Total: $81\frac{2}{3}$ hours

(b) Speed $= \dfrac{\text{distance}}{\text{time}} = \dfrac{2100}{81\frac{2}{3}} = \dfrac{180}{7} = 25\frac{5}{7}$ mph

(c) $D = -\dfrac{180}{7}t + 3400$

Domain: $0 \le t \le \dfrac{1190}{9}$

Range: $0 \le D \le 3400$

(d)

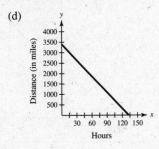

9. (a)–(d) Use $f(x) = 4x$ and $g(x) = x + 6$.

(a) $(fg)(x) = f(x + 6) = 4(x + 6) = 4x + 24$

(b) $(f \circ g)^{-1}(x) = \dfrac{x - 24}{4} = \dfrac{1}{4}x - 6$

(c) $f^{-1}(x) = \dfrac{1}{4}x$

$g^{-1}(x) = x - 6$

(d) $(g^{-1} \circ f^{-1})(x) = g^{-1}\left(\dfrac{1}{4}x\right) = \dfrac{1}{4}x - 6$

(e) $f(x) = x^3 + 1$ and $g(x) = 2x$

$(f \circ g)(x) = f(2x) = (2x)^3 + 1 = 8x^3 + 1$

$(f \circ g)^{-1}(x) = \sqrt[3]{\dfrac{x-1}{8}} = \dfrac{1}{2}\sqrt[3]{x - 1}$

$f^{-1}(x) = \sqrt[3]{x - 1}$

$g^{-1}(x) = \dfrac{1}{2}x$

$(g^{-1} \circ f^{-1})(x) = g^{-1}\left(\sqrt[3]{x - 1}\right) = \dfrac{1}{2}\sqrt[3]{x - 1}$

(f) Answers will vary.

(g) Conjecture: $(f \circ g)^{-1}(x) = (g^{-1} \circ f^{-1})(x)$

11. $H(x) = \begin{cases} 1, & x \ge 0 \\ 0, & x < 0 \end{cases}$

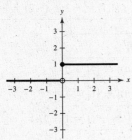

(a) $H(x) - 2$

(b) $H(x - 2)$

(c) $-H(x)$

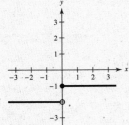

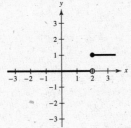

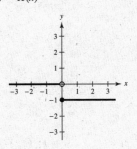

—CONTINUED—

11. —CONTINUED—

(d) $H(-x)$

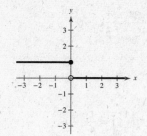

(e) $\frac{1}{2}H(x)$

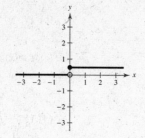

(f) $-H(x-2)+2$

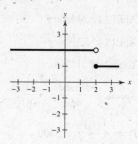

13. $(f \circ (g \circ h))(x) = f((g \circ h)(x))$

$$= f(g(h(x)))$$

$$= (f \circ g \circ h)(x)$$

$((f \circ g) \circ h)(x) = (f \circ g)(h(x))$

$$= f(g(h(x)))$$

$$= (f \circ g \circ h)(x)$$

15.

x	$f(x)$	$f^{-1}(x)$
-4	—	2
-3	4	1
-2	1	0
-1	0	—
0	-2	-1
1	-3	-2
2	-4	—
3	—	—
4	—	-3

(a)

x	$f(f^{-1}(x))$
-4	$f(f^{-1}(-4)) = f(2) = -4$
-2	$f(f^{-1}(-2)) = f(0) = -2$
0	$f(f^{-1}(0)) = f(-1) = 0$
4	$f(f^{-1}(4)) = f(-3) = 4$

(b)

x	$(f + f^{-1})(x)$
-3	$f(-3) + f^{-1}(-3) = 4 + 1 = 5$
-2	$f(-2) + f^{-1}(-2) = 1 + 0 = 1$
0	$f(0) + f^{-1}(0) = -2 + (-1) = -3$
1	$f(1) + f^{-1}(1) = -3 + (-2) = -5$

(c)

x	$(f \cdot f^{-1})(x)$
-3	$f(-3)f^{-1}(-3) = (4)(1) = 4$
-2	$f(-2)f^{-1}(-2) = (1)(0) = 0$
0	$f(0)f^{-1}(0) = (-2)(-1) = 2$
1	$f(1)f^{-1}(1) = (-3)(-2) = 6$

(d)

x	$|f^{-1}(x)|$
-4	$|f^{-1}(-4)| = |2| = 2$
-3	$|f^{-1}(-3)| = |1| = 1$
0	$|f^{-1}(0)| = |-1| = 1$
4	$|f^{-1}(4)| = |-3| = 3$

Chapter 1   Practice Test

1. Given the points $(-3, 4)$ and $(5, -6)$, find (a) the midpoint of the line segment joining the points, and (b) the distance between the points.

2. Graph $y = \sqrt{7 - x}$.

3. Write the standard equation of the circle with center $(-3, 5)$ and radius 6.

4. Find the equation of the line through $(2, 4)$ and $(3, -1)$.

5. Find the equation of the line with slope $m = 4/3$ and y-intercept $b = -3$.

6. Find the equation of the line through $(4, 1)$ perpendicular to the line $2x + 3y = 0$.

7. If it costs a company \$32 to produce 5 units of a product and \$44 to produce 9 units, how much does it cost to produce 20 units? (Assume that the cost function is linear.)

8. Given $f(x) = x^2 - 2x + 1$, find $f(x - 3)$.

9. Given $f(x) = 4x - 11$, find $\dfrac{f(x) - f(3)}{x - 3}$

10. Find the domain and range of $f(x) = \sqrt{36 - x^2}$.

11. Which equations determine y as a function of x?

    (a) $6x - 5y + 4 = 0$

    (b) $x^2 + y^2 = 9$

    (c) $y^3 = x^2 + 6$

12. Sketch the graph of $f(x) = x^2 - 5$.

13. Sketch the graph of $f(x) = |x + 3|$.

14. Sketch the graph of $f(x) = \begin{cases} 2x + 1, & \text{if } x \geq 0, \\ x^2 - x, & \text{if } x < 0. \end{cases}$

15. Use the graph of $f(x) = |x|$ to graph the following:

    (a) $f(x + 2)$

    (b) $-f(x) + 2$

16. Given $f(x) = 3x + 7$ and $g(x) = 2x^2 - 5$, find the following:

(a) $(g - f)(x)$

(b) $(fg)(x)$

17. Given $f(x) = x^2 - 2x + 16$ and $g(x) = 2x + 3$, find $f(g(x))$.

18. Given $f(x) = x^3 + 7$, find $f^{-1}(x)$.

19. Which of the following functions have inverses?

(a) $f(x) = |x - 6|$

(b) $f(x) = ax + b, \ a \neq 0$

(c) $f(x) = x^3 - 19$

20. Given $f(x) = \sqrt{\dfrac{3 - x}{x}}, \ 0 < x \leq 3$, find $f^{-1}(x)$.

Exercises 21–23, true or false?

21. $y = 3x + 7$ and $y = \frac{1}{3}x - 4$ are perpendicular.

22. $(f \circ g)^{-1} = g^{-1} \circ f^{-1}$

23. If a function has an inverse, then it must pass both the Vertical Line Test and the Horizontal Line Test.

24. If z varies directly as the cube of x and inversely as the square root of y, and $z = -1$ when $x = -1$ and $y = 25$, find z in terms of x and y.

25. Use your calculator to find the least square regression line for the data.

x	-2	-1	0	1	2	3
y	1	2.4	3	3.1	4	4.7

C H A P T E R   2
Polynomial and Rational Functions

Section 2.1   Quadratic Functions and Models **76**

Section 2.2   Polynomial Functions of Higher Degree **84**

Section 2.3   Polynomial and Synthetic Division **93**

Section 2.4   Complex Numbers **100**

Section 2.5   Zeros of Polynomial Functions **104**

Section 2.6   Rational Functions **114**

Section 2.7   Nonlinear Inequalities **123**

Review Exercises **132**

Problem Solving **143**

Practice Test **145**

CHAPTER 2
Polynomial and Rational Functions

Section 2.1   Quadratic Functions and Models

> You should know the following facts about parabolas.
>
> ■ $f(x) = ax^2 + bx + c,\ a \neq 0$, is a quadratic function, and its graph is a parabola.
>
> ■ If $a > 0$, the parabola opens upward and the vertex is the point with the minimum y-value.
>   If $a < 0$, the parabola opens downward and the vertex is the point with the maximum y-value.
>
> ■ The vertex is $(-b/2a, f(-b/2a))$.
>
> ■ To find the x-intercepts (if any), solve
> $$ax^2 + bx + c = 0.$$
>
> ■ The standard form of the equation of a parabola is
> $$f(x) = a(x - h)^2 + k$$
>   where $a \neq 0$.
>
>   (a) The vertex is (h, k).
>
>   (b) The axis is the vertical line $x = h$.

Vocabulary Check

1. nonnegative integer; real   **2.** quadratic; parabola   **3.** axis or axis of symmetry

4. positive; minimum   **5.** negative; maximum

---

1. $f(x) = (x - 2)^2$ opens upward and has vertex $(2, 0)$.
Matches graph (g).

3. $f(x) = x^2 - 2$ opens upward and has vertex $(0, -2)$.
Matches graph (b).

5. $f(x) = 4 - (x - 2)^2 = -(x - 2)^2 + 4$ opens downward
and has vertex $(2, 4)$. Matches graph (f).

7. $f(x) = -(x - 3)^2 - 2$ opens downward and has
vertex $(3, -2)$. Matches graph (e).

9. (a) $y = \frac{1}{2}x^2$

(b) $y = -\frac{1}{8}x^2$

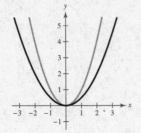

Vertical shrink

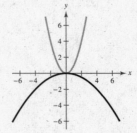

Vertical shrink and reflection in the x-axis

—CONTINUED—

9. **—CONTINUED—**

(c) $y = \frac{3}{2}x^2$

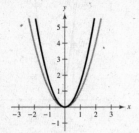

Vertical stretch

(d) $y = -3x^2$

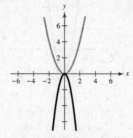

Vertical stretch and reflection in the x-axis

11. (a) $y = (x - 1)^2$

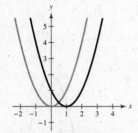

Horizontal translation one unit to the right

(b) $y = (3x)^2 + 1$

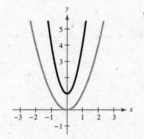

Horizontal shrink and a vertical translation one unit upward

(c) $y = \left(\frac{1}{3}x\right)^2 - 3$

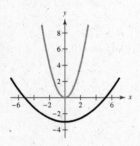

Horizontal stretch and a vertical translation three units downward

(d) $y = (x + 3)^2$

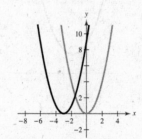

Horizontal translation three units to the left

13. $f(x) = x^2 - 5$

Vertex: $(0, -5)$

Axis of symmetry: $x = 0$ or the y-axis

Find x-intercepts:

$x^2 - 5 = 0$

$x^2 = 5$

$x = \pm\sqrt{5}$

x-intercepts:

$\left(-\sqrt{5}, 0\right), \left(\sqrt{5}, 0\right)$

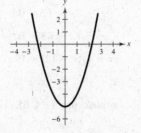

15. $f(x) = \frac{1}{2}x^2 - 4 = \frac{1}{2}(x - 0)^2 - 4$

Vertex: $(0, -4)$

Axis of symmetry: $x = 0$ or the y-axis

Find x-intercepts:

$\frac{1}{2}x^2 - 4 = 0$

$x^2 = 8$

$x = \pm\sqrt{8} = \pm 2\sqrt{2}$

x-intercepts:

$\left(-2\sqrt{2}, 0\right), \left(2\sqrt{2}, 0\right)$

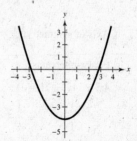

17. $f(x) = (x + 5)^2 - 6$

Vertex: $(-5, -6)$

Axis of symmetry: $x = -5$

Find x-intercepts:

$(x + 5)^2 - 6 = 0$

$(x + 5)^2 = 6$

$x + 5 = \pm\sqrt{6}$

$x = -5 \pm \sqrt{6}$

x-intercepts: $\left(-5 - \sqrt{6}, 0\right), \left(-5 + \sqrt{6}, 0\right)$

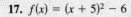

19. $h(x) = x^2 - 8x + 16 = (x - 4)^2$

Vertex: $(4, 0)$

Axis of symmetry: $x = 4$

x-intercept: $(4, 0)$

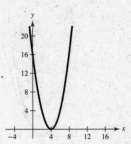

21. $f(x) = x^2 - x + \dfrac{5}{4}$

$= \left(x^2 - x + \dfrac{1}{4}\right) - \dfrac{1}{4} + \dfrac{5}{4}$

$= \left(x - \dfrac{1}{2}\right)^2 + 1$

Vertex: $\left(\dfrac{1}{2}, 1\right)$

Axis of symmetry: $x = \dfrac{1}{2}$

Find x-intercepts:

$x^2 - x + \dfrac{5}{4} = 0$

$x = \dfrac{1 \pm \sqrt{1 - 5}}{2}$

Not a real number

No x-intercepts

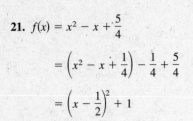

23. $f(x) = -x^2 + 2x + 5$

$= -(x^2 - 2x + 1) - (-1) + 5$

$= -(x - 1)^2 + 6$

Vertex: $(1, 6)$

Axis of symmetry: $x = 1$

Find x-intercepts:

$-x^2 + 2x + 5 = 0$

$x^2 - 2x - 5 = 0$

$x = \dfrac{2 \pm \sqrt{4 + 20}}{2}$

$= 1 \pm \sqrt{6}$

x-intercepts: $\left(1 - \sqrt{6}, 0\right), \left(1 + \sqrt{6}, 0\right)$

25. $h(x) = 4x^2 - 4x + 21$

$= 4\left(x^2 - x + \dfrac{1}{4}\right) - 4\left(\dfrac{1}{4}\right) + 21$

$= 4\left(x - \dfrac{1}{2}\right)^2 + 20$

Vertex: $\left(\dfrac{1}{2}, 20\right)$

Axis of symmetry: $x = \dfrac{1}{2}$

Find x-intercepts:

$4x^2 - 4x + 21 = 0$

$x = \dfrac{4 \pm \sqrt{16 - 336}}{2(4)}$

Not a real number $\implies$ No x-intercepts

27. $f(x) = \dfrac{1}{4}x^2 - 2x - 12$

$= \dfrac{1}{4}(x^2 - 8x + 16) - \dfrac{1}{4}(16) - 12$

$= \dfrac{1}{4}(x - 4)^2 - 16$

Vertex: $(4, -16)$

Axis of symmetry: $x = 4$

Find x-intercepts:

$\dfrac{1}{4}x^2 - 2x - 12 = 0$

$x^2 - 8x - 48 = 0$

$(x + 4)(x - 12) = 0$

$x = -4 \quad \text{or} \quad x = 12$

x-intercepts: $(-4, 0), (12, 0)$

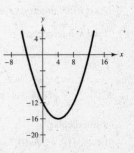

29. $f(x) = -(x^2 + 2x - 3) = -(x + 1)^2 + 4$

Vertex: $(-1, 4)$

Axis of symmetry: $x = -1$

x-intercepts: $(-3, 0), (1, 0)$

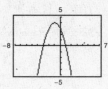

31. $g(x) = x^2 + 8x + 11 = (x + 4)^2 - 5$

Vertex: $(-4, -5)$

Axis of symmetry: $x = -4$

x-intercepts: $\left(-4 \pm \sqrt{5}, 0\right)$

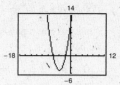

33. $f(x) = 2x^2 - 16x + 31$

$\qquad = 2(x - 4)^2 - 1$

Vertex: $(4, -1)$

Axis of symmetry: $x = 4$

x-intercepts: $\left(4 \pm \frac{1}{2}\sqrt{2}, 0\right)$

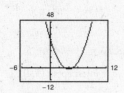

35. $g(x) = \frac{1}{2}(x^2 + 4x - 2) = \frac{1}{2}(x + 2)^2 - 3$

Vertex: $(-2, -3)$

Axis of symmetry: $x = -2$

x-intercepts: $\left(-2 \pm \sqrt{6}, 0\right)$

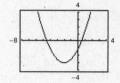

37. $(1, 0)$ is the vertex.

$y = a(x - 1)^2 + 0 = a(x - 1)^2$

Since the graph passes through the point $(0, 1)$, we have:

$1 = a(0 - 1)^2$

$1 = a$

$y = 1(x - 1)^2 = (x - 1)^2$

39. $(-1, 4)$ is the vertex.

$y = a(x + 1)^2 + 4$

Since the graph passes through the point $(1, 0)$, we have:

$0 = a(1 + 1)^2 + 4$

$-4 = 4a$

$-1 = a$

$y = -1(x + 1)^2 + 4 = -(x + 1)^2 + 4$

41. $(-2, 2)$ is the vertex.

$y = a(x + 2)^2 + 2$

Since the graph passes through the point $(-1, 0)$, we have:

$0 = a(-1 + 2)^2 + 2$

$-2 = a$

$y = -2(x + 2)^2 + 2$

43. $(-2, 5)$ is the vertex.

$f(x) = a(x + 2)^2 + 5$

Since the graph passes through the point $(0, 9)$, we have:

$9 = a(0 + 2)^2 + 5$

$4 = 4a$

$1 = a$

$f(x) = 1(x + 2)^2 + 5 = (x + 2)^2 + 5$

45. $(3, 4)$ is the vertex.

$f(x) = a(x - 3)^2 + 4$

Since the graph passes through the point $(1, 2)$, we have:

$2 = a(1 - 3)^2 + 4$

$-2 = 4a$

$-\frac{1}{2} = a$

$f(x) = -\frac{1}{2}(x - 3)^2 + 4$

47. $(5, 12)$ is the vertex.

$f(x) = a(x - 5)^2 + 12$

Since the graph passes through the point $(7, 15)$, we have:

$15 = a(7 - 5)^2 + 12$

$3 = 4a \implies a = \frac{3}{4}$

$f(x) = \frac{3}{4}(x - 5)^2 + 12$

49. $\left(-\frac{1}{4}, \frac{3}{2}\right)$ is the vertex.

$f(x) = a\left(x + \frac{1}{4}\right)^2 + \frac{3}{2}$

Since the graph passes through the point $(-2, 0)$, we have:

$0 = a\left(-2 + \frac{1}{4}\right)^2 + \frac{3}{2}$

$-\frac{3}{2} = \frac{49}{16}a \implies a = -\frac{24}{49}$

$f(x) = -\frac{24}{49}\left(x + \frac{1}{4}\right)^2 + \frac{3}{2}$

51. $\left(-\frac{5}{2}, 0\right)$ is the vertex.

$f(x) = a\left(x + \frac{5}{2}\right)^2$

Since the graph passes through the point $\left(-\frac{7}{2}, -\frac{16}{3}\right)$, we have:

$-\frac{16}{3} = a\left(-\frac{7}{2} + \frac{5}{2}\right)^2$

$-\frac{16}{3} = a$

$f(x) = -\frac{16}{3}\left(x + \frac{5}{2}\right)^2$

53. $y = x^2 - 16$

x-intercepts: $(\pm 4, 0)$

$0 = x^2 - 16$

$x^2 = 16$

$x = \pm 4$

55. $y = x^2 - 4x - 5$

x-intercepts: $(5, 0), (-1, 0)$

$0 = x^2 - 4x - 5$

$0 = (x - 5)(x + 1)$

$x = 5$ or $x = -1$

57. $f(x) = x^2 - 4x$

x-intercepts: $(0, 0), (4, 0)$

$0 = x^2 - 4x$

$0 = x(x - 4)$

$x = 0$ or $x = 4$

The x-intercepts and the solutions of $f(x) = 0$ are the same.

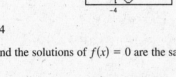

59. $f(x) = x^2 - 9x + 18$

x-intercepts: $(3, 0), (6, 0)$

$0 = x^2 - 9x + 18$

$0 = (x - 3)(x - 6)$

$x = 3$ or $x = 6$

The x-intercepts and the solutions of $f(x) = 0$ are the same.

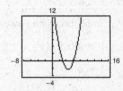

61. $f(x) = 2x^2 - 7x - 30$

x-intercepts: $\left(-\frac{5}{2}, 0\right), (6, 0)$

$0 = 2x^2 - 7x - 30$

$0 = (2x + 5)(x - 6)$

$x = -\frac{5}{2}$ or $x = 6$

The x-intercepts and the solutions of $f(x) = 0$ are the same.

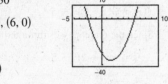

63. $f(x) = -\frac{1}{2}(x^2 - 6x - 7)$

x-intercepts: $(-1, 0), (7, 0)$

$0 = -\frac{1}{2}(x^2 - 6x - 7)$

$0 = x^2 - 6x - 7$

$0 = (x + 1)(x - 7)$

$x = -1$ or $x = 7$

The x-intercepts and the solutions of $f(x) = 0$ are the same.

65. $f(x) = [x - (-1)](x - 3)$ opens upward

$\quad = (x + 1)(x - 3)$

$\quad = x^2 - 2x - 3$

$g(x) = -[x - (-1)](x - 3)$ opens downward

$\quad = -(x + 1)(x - 3)$

$\quad = -(x^2 - 2x - 3)$

$\quad = -x^2 + 2x + 3$

Note: $f(x) = a(x + 1)(x - 3)$ has x-intercepts $(-1, 0)$ and $(3, 0)$ for all real numbers $a \neq 0$.

67. $f(x) = (x - 0)(x - 10)$ opens upward

$\quad = x^2 - 10x$

$g(x) = -(x - 0)(x - 10)$ opens downward

$\quad = -x^2 + 10x$

Note: $f(x) = a(x - 0)(x - 10) = ax(x - 10)$ has x-intercepts $(0, 0)$ and $(10, 0)$ for all real numbers $a \neq 0$.

69. $f(x) = [x - (-3)][x - (-\frac{1}{2})](2)$ opens upward

$\qquad = (x + 3)(x + \frac{1}{2})(2)$

$\qquad = (x + 3)(2x + 1)$

$\qquad = 2x^2 + 7x + 3$

$\quad g(x) = -(2x^2 + 7x + 3)$ opens downward

$\qquad = -2x^2 - 7x - 3$

Note: $f(x) = a(x + 3)(2x + 1)$ has x-intercepts $(-3, 0)$ and $(-\frac{1}{2}, 0)$ for all real numbers $a \neq 0$.

71. Let $x =$ the first number and $y =$ the second number. Then the sum is

$x + y = 110 \implies y = 110 - x.$

The product is $P(x) = xy = x(110 - x) = 110x - x^2.$

$P(x) = -x^2 + 110x$

$\qquad = -(x^2 - 110x + 3025 - 3025)$

$\qquad = -[(x - 55)^2 - 3025]$

$\qquad = -(x - 55)^2 + 3025$

The maximum value of the product occurs at the vertex of $P(x)$ and is 3025. This happens when $x = y = 55.$

73. Let $x =$ the first number and $y =$ the second number. Then the sum is

$x + 2y = 24 \implies y = \dfrac{24 - x}{2}.$

The product is $P(x) = xy = x\left(\dfrac{24 - x}{2}\right).$

$P(x) = \dfrac{1}{2}(-x^2 + 24x)$

$\qquad = -\dfrac{1}{2}(x^2 - 24x + 144 - 144)$

$\qquad = -\dfrac{1}{2}[(x - 12)^2 - 144] = -\dfrac{1}{2}(x - 12)^2 + 72$

The maximum value of the product occurs at the vertex of $P(x)$ and is 72. This happens when $x = 12$ and $y = (24 - 12)/2 = 6.$ Thus, the numbers are 12 and 6.

75. (a)

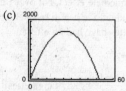

$4x + 3y = 200 \implies y = \dfrac{1}{3}(200 - 4x) = \dfrac{4}{3}(50 - x)$

$A = 2xy = 2x\left[\dfrac{4}{3}(50 - x)\right] = \dfrac{8}{3}x(50 - x) = \dfrac{8x(50 - x)}{3}$

(b)

x	A
5	600
10	$1066\frac{2}{3}$
15	1400
20	1600
25	$1666\frac{2}{3}$
30	1600

This area is maximum when $x = 25$ feet and $y = \frac{100}{3} = 33\frac{1}{3}$ feet.

(c)

This area is maximum when $x = 25$ feet and $y = \frac{100}{3} = 33\frac{1}{3}$ feet.

—**CONTINUED**—

75. —CONTINUED—

(d) $A = \dfrac{8}{3}x(50 - x)$

$\qquad = -\dfrac{8}{3}(x^2 - 50x)$

$\qquad = -\dfrac{8}{3}(x^2 - 50x + 625 - 625)$

$\qquad = -\dfrac{8}{3}[(x - 25)^2 - 625]$

$\qquad = -\dfrac{8}{3}(x - 25)^2 + \dfrac{5000}{3}$

The maximum area occurs at the vertex and is 5000/3 square feet. This happens when $x = 25$ feet and $y = (200 - 4(25))/3 = 100/3$ feet. The dimensions are $2x = 50$ feet by $33\frac{1}{3}$ feet.

(e) They are all identical.

$\qquad x = 25$ feet and $y = 33\frac{1}{3}$ feet

77. $y = -\dfrac{4}{9}x^2 + \dfrac{24}{9}x + 12$

The vertex occurs at $-\dfrac{b}{2a} = \dfrac{-24/9}{2(-4/9)} = 3$. The maximum height is $y(3) = -\dfrac{4}{9}(3)^2 + \dfrac{24}{9}(3) + 12 = 16$ feet.

79. $C = 800 - 10x + 0.25x^2 = 0.25x^2 - 10x + 800$

The vertex occurs at $x = -\dfrac{b}{2a} = -\dfrac{-10}{2(0.25)} = 20$.

The cost is minimum when $x = 20$ fixtures.

81. $P = -0.0002x^2 + 140x - 250,000$

The vertex occurs at $x = -\dfrac{b}{2a} = -\dfrac{140}{2(-0.0002)} = 350,000$.

The profit is maximum when $x = 350,000$ units.

83. $R(p) = -25p^2 + 1200p$

(a) $R(20) = \$14,000$ thousand

$\quad R(25) = \$14,375$ thousand

$\quad R(30) = \$13,500$ thousand

(b) The revenue is a maximum at the vertex.

$\quad -\dfrac{b}{2a} = \dfrac{-1200}{2(-25)} = 24$

$\quad R(24) = 14,400$

The unit price that will yield a maximum revenue of $14,400 thousand is $24.

85. $C = 4299 - 1.8t - 1.36t^2, 0 \le t \le 43$

(a)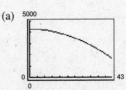

(b) Vertex $\approx (0, 4299)$

The vertex occurs when $y \approx 4299$ which is the maximum average annual consumption. The warnings may not have had an immediate effect, but over time they and other findings about the health risks and the increased cost of cigarettes have had an effect.

(c) $C(40) = 2051$

Annually: $\dfrac{209,128,094(2051)}{48,308,590} \approx 8879$ cigarettes

Daily: $\dfrac{8879}{366} \approx 24$ cigarettes

87. (a)

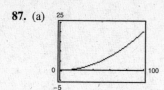

(b) $0.002s^2 + 0.005s - 0.029 = 10$

$$2s^2 + 5s - 29 = 10,000$$

$$2s^2 + 5s - 10,029 = 0$$

$$a = 2, b = 5, c = -10,029$$

$$s = \frac{-5 \pm \sqrt{5^2 - 4(2)(-10,029)}}{2(2)}$$

$$s = \frac{-5 \pm \sqrt{80,257}}{4}$$

$$s \approx -72.1, 69.6$$

The maximum speed if power is not to exceed 10 horsepower is 69.6 miles per hour.

89. True. The equation $-12x^2 - 1 = 0$ has no real solution, so the graph has no x-intercepts.

91. $f(x) = ax^2 + bx + c$

$$= a\left(x^2 + \frac{b}{a}x\right) + c$$

$$= a\left(x^2 + \frac{b}{a}x + \frac{b^2}{4a^2} - \frac{b^2}{4a^2}\right) + c$$

$$= a\left(x + \frac{b}{2a}\right)^2 - \frac{b^2}{4a} + c$$

$$= a\left(x - \left(-\frac{b}{2a}\right)\right)^2 + \frac{4ac - b^2}{4a}$$

$$f\left(-\frac{b}{2a}\right) = a\left(\frac{b^2}{4a^2}\right) + b\left(-\frac{b}{2a}\right) + c$$

$$= \frac{b^2}{4a} - \frac{b^2}{2a} + c$$

$$= \frac{b^2 - 2b^2 + 4ac}{4a} = \frac{4ac - b^2}{4a}$$

So, the vertex occurs at

$$\left(-\frac{b}{2a}, \frac{4ac - b^2}{4a}\right) = \left(-\frac{b}{2a}, f\left(-\frac{b}{2a}\right)\right).$$

93. Yes. A graph of a quadratic equation whose vertex is $(0, 0)$ has only one x-intercept.

95. $(-4, 3)$ and $(2, 1)$

$$m = \frac{1 - 3}{2 - (-4)} = \frac{-2}{6} = -\frac{1}{3}$$

$$y - 1 = -\frac{1}{3}(x - 2)$$

$$y - 1 = -\frac{1}{3}x + \frac{2}{3}$$

$$y = -\frac{1}{3}x + \frac{5}{3}$$

97. $4x + 5y = 10 \implies y = -\frac{4}{5}x + 2$ and $m = -\frac{4}{5}$

The slope of the perpendicular line through $(0, 3)$ is $m = \frac{5}{4}$ and the y-intercept is $b = 3$.

$$y = \frac{5}{4}x + 3$$

For Exercises 99–103, let $f(x) = 14x - 3$, **and** $g(x) = 8x^2$.

99. $(f + g)(-3) = f(-3) + g(-3)$

$$= [14(-3) - 3] + 8(-3)^2 = 27$$

101. $(fg)\left(-\frac{4}{7}\right) = f\left(-\frac{4}{7}\right)g\left(-\frac{4}{7}\right)$

$$= \left[14\left(-\frac{4}{7}\right) - 3\right]\left[8\left(-\frac{4}{7}\right)^2\right]$$

$$= (-11)\left(\frac{128}{49}\right) = -\frac{1408}{49}$$

103. $(f \circ g)(-1) = f(g(-1)) = f(8) = 14(8) - 3 = 109$

105. Answers will vary.

Section 2.2   Polynomial Functions of Higher Degree

You should know the following basic principles about polynomials.

- $f(x) = a_n x^n + a_{n-1}x^{n-1} + \cdots + a_2 x^2 + a_1 x + a_0, a_n \neq 0$, is a polynomial function of degree n.
- If f is of odd degree and
  - (a) $a_n > 0$, then
    1. $f(x) \to \infty$ as $x \to \infty$.
    2. $f(x) \to -\infty$ as $x \to -\infty$.
  - (b) $a_n < 0$, then
    1. $f(x) \to -\infty$ as $x \to \infty$.
    2. $f(x) \to \infty$ as $x \to -\infty$.
- If f is of even degree and
  - (a) $a_n > 0$, then
    1. $f(x) \to \infty$ as $x \to \infty$.
    2. $f(x) \to \infty$ as $x \to -\infty$.
  - (b) $a_n < 0$, then
    1. $f(x) \to -\infty$ as $x \to \infty$.
    2. $f(x) \to -\infty$ as $x \to -\infty$.
- The following are equivalent for a polynomial function.
  - (a) $x = a$ is a zero of a function.
  - (b) $x = a$ is a solution of the polynomial equation $f(x) = 0$.
  - (c) $(x - a)$ is a factor of the polynomial.
  - (d) $(a, 0)$ is an x-intercept of the graph of f.
- A polynomial of degree n has at most n distinct zeros and at most $n - 1$ turning points.
- A factor $(x - a)^k, k > 1$, yields a repeated zero of $x = a$ of multiplicity k.
  - (a) If k is odd, the graph crosses the x-axis at $x = a$.
  - (b) If k is even, the graph just touches the x-axis at $x = a$.
- If f is a polynomial function such that $a < b$ and $f(a) \neq f(b)$, then f takes on every value between $f(a)$ and $f(b)$ in the interval $[a, b]$.
- If you can find a value where a polynomial is positive and another value where it is negative, then there is at least one real zero between the values.

Vocabulary Check

1. continuous

2. Leading Coefficient Test

3. $n; n - 1$

4. solution; $(x - a)$; x-intercept

5. touches; crosses

6. standard

7. Intermediate Value

1. $f(x) = -2x + 3$ is a line with y-intercept $(0, 3)$. Matches graph (c).

3. $f(x) = -2x^2 - 5x$ is a parabola with x-intercepts $(0, 0)$ $\left(-\frac{5}{2}, 0\right)$ and opens downward. Matches graph (h).

5. $f(x) = -\frac{1}{4}x^4 + 3x^2$ has intercepts $(0, 0)$ and $(\pm 2\sqrt{3}, 0)$.
Matches graph (a).

7. $f(x) = x^4 + 2x^3$ has intercepts $(0, 0)$ and $(-2, 0)$.
Matches graph (d).

9. $y = x^3$

(a) $f(x) = (x - 2)^3$

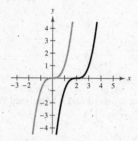

Horizontal shift two units to the right

(b) $f(x) = x^3 - 2$

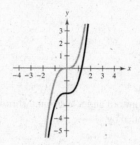

Vertical shift two units downward

(c) $f(x) = -\frac{1}{2}x^3$

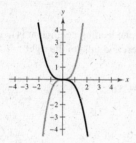

Reflection in the *x*-axis and a vertical shrink

(d) $f(x) = (x - 2)^3 - 2$

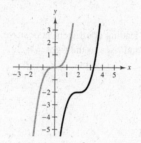

Horizontal shift two units to the right and
a vertical shift two units downward

11. $y = x^4$

(a) $f(x) = (x + 3)^4$

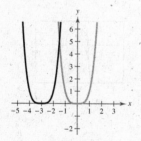

Horizontal shift three units to the left

(b) $f(x) = x^4 - 3$

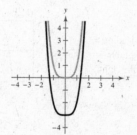

Vertical shift three units downward

(c) $f(x) = 4 - x^4$

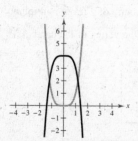

Reflection in the *x*-axis and then a vertical
shift four units upward

(d) $f(x) = \frac{1}{2}(x - 1)^4$

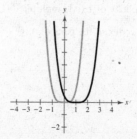

Horizontal shift one unit to the right and a vertical shrink
(each *y*-value is multiplied by $\frac{1}{2}$)

—CONTINUED—

11. —CONTINUED—

(e) $f(x) = (2x)^4 + 1$

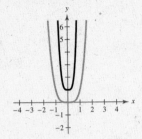

Vertical shift one unit upward and a horizontal shrink
(each y-value is multiplied by $\frac{1}{2}$)

(f) $f(x) = \left(\frac{1}{2}x\right)^4 - 2$

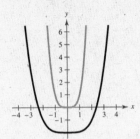

Vertical shift two units downward and a horizontal stretch
(each y-value is multiplied by $\frac{1}{2}$)

13. $f(x) = \frac{1}{3}x^3 + 5x$

Degree: 3

Leading coefficient: $\frac{1}{3}$

The degree is odd and the leading coefficient is positive.
The graph falls to the left and rises to the right.

15. $g(x) = 5 - \frac{7}{2}x - 3x^2$

Degree: 2

Leading coefficient: -3

The degree is even and the leading coefficient is negative.
The graph falls to the left and falls to the right.

17. $f(x) = -2.1x^5 + 4x^3 - 2$

Degree: 5

Leading coefficient: -2.1

The degree is odd and the leading coefficient is negative.
The graph rises to the left and falls to the right.

19. $f(x) = 6 - 2x + 4x^2 - 5x^3$

Degree: 3

Leading coefficient: -5

The degree is odd and the leading coefficient is negative.
The graph rises to the left and falls to the right.

21. $h(t) = -\frac{2}{3}(t^2 - 5t + 3)$

Degree: 2

Leading coefficient: $-\frac{2}{3}$

The degree is even and the leading
coefficient is negative. The graph
falls to the left and falls to the right.

23. $f(x) = 3x^3 - 9x + 1;$

$\quad g(x) = x^3$

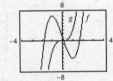

25. $f(x) = -(x^4 - 4x^3 + 16x);$

$\quad g(x) = -x^4$

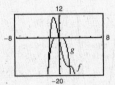

27. $f(x) = x^2 - 25$

(a) $0 = x^2 - 25 = (x + 5)(x - 5)$

Zeros: $x = \pm 5$

(b) Each zero has a multiplicity of 1 (odd multiplicity).

Turning point: 1 (the vertex of the parabola)

(c)

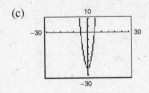

29. $h(t) = t^2 - 6t + 9$

(a) $0 = t^2 - 6t + 9 = (t - 3)^2$

Zero: $t = 3$

(b) $t = 3$ has a multiplicity of 2 (even multiplicity).

Turning point: 1 (the vertex of the parabola)

(c)

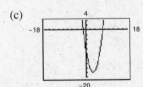

31. $f(x) = \frac{1}{3}x^2 + \frac{1}{3}x - \frac{2}{3}$

    (a) $0 = \frac{1}{3}x^2 + \frac{1}{3}x - \frac{2}{3}$

           $= \frac{1}{3}(x^2 + x - 2)$

           $= \frac{1}{3}(x + 2)(x - 1)$

    Zeros: $x = -2, x = 1$

    (b) Each zero has a multiplicity of 1 (odd multiplicity).

    Turning point: 1 (the vertex of the parabola)

(c)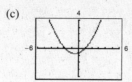

33. $f(x) = 3x^3 - 12x^2 + 3x$

    (a) $0 = 3x^3 - 12x^2 + 3x = 3x(x^2 - 4x + 1)$

    Zeros: $x = 0, x = 2 \pm \sqrt{3}$ (by the Quadratic
    Formula)

    (b) Each zero has a multiplicity of 1 (odd multiplicity).

    Turning points: 2

(c)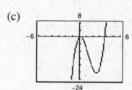

35. $f(t) = t^3 - 4t^2 + 4t$

    (a) $0 = t^3 - 4t^2 + 4t = t(t^2 - 4t + 4) = t(t - 2)^2$

    Zeros: $t = 0, t = 2$

    (b) $t = 0$ has a multiplicity of 1 (odd multiplicity).

    $t = 2$ has a multiplicity of 2 (even multiplicity).

    Turning points: 2

(c)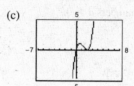

37. $g(t) = t^5 - 6t^3 + 9t$

    (a) $0 = t^5 - 6t^3 + 9t = t(t^4 - 6t^2 + 9) = t(t^2 - 3)^2$

        $= t(t + \sqrt{3})^2(t - \sqrt{3})^2$

    Zeros: $t = 0, t = \pm\sqrt{3}$

    (b) $t = 0$ has a multiplicity of 1 (odd multiplicity).

    $t = \pm\sqrt{3}$ each have a multiplicity of 2
    (even multiplicity).

    Turning points: 4

(c)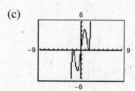

39. $f(x) = 5x^4 + 15x^2 + 10$

    (a) $0 = 5x^4 + 15x^2 + 10$

        $= 5(x^4 + 3x^2 + 2)$

        $= 5(x^2 + 1)(x^2 + 2)$

    No real zeros

    (b) Turning point: 1

    (c)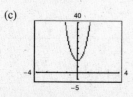

41. $g(x) = x^3 + 3x^2 - 4x - 12$

    (a) $0 = x^3 + 3x^2 - 4x - 12 = x^2(x + 3) - 4(x + 3)$

        $= (x^2 - 4)(x + 3) = (x - 2)(x + 2)(x + 3)$

    Zeros: $x = \pm 2, x = -3$

    (b) Each zero has a multiplicity of 1 (odd multiplicity).

    Turning points: 2

    (c)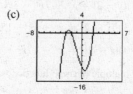

43. $y = 4x^3 - 20x^2 + 25x$

(a)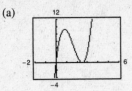

(b) *x*-intercepts: $(0, 0), \left(\frac{5}{2}, 0\right)$

(c) $0 = 4x^3 - 20x^2 + 25x$

$0 = x(2x - 5)^2$

$x = 0 \text{ or } x = \frac{5}{2}$

(d) The solutions are the same as the *x*-coordinates of the *x*-intercepts.

45. $y = x^5 - 5x^3 + 4x$

(a)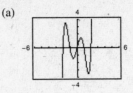

(b) *x*-intercepts: $(0, 0), (\pm 1, 0), (\pm 2, 0)$

(c) $0 = x^5 - 5x^3 + 4x$

$0 = x(x^2 - 1)(x^2 - 4)$

$0 = x(x + 1)(x - 1)(x + 2)(x - 2)$

$x = 0, \pm 1, \pm 2$

(d) The solutions are the same as the *x*-coordinates of the *x*-intercepts.

47. $f(x) = (x - 0)(x - 10)$

$f(x) = x^2 - 10x$

Note: $f(x) = a(x - 0)(x - 10) = ax(x - 10)$ has zeros 0 and 10 for all real numbers $a \neq 0$.

49. $f(x) = (x - 2)(x - (-6))$

$= (x - 2)(x + 6)$

$= x^2 + 4x - 12$

Note: $f(x) = a(x - 2)(x + 6)$ has zeros 2 and -6 for all real numbers $a \neq 0$.

51. $f(x) = (x - 0)(x - (-2))(x - (-3))$

$= x(x + 2)(x + 3)$

$= x^3 + 5x^2 + 6x$

Note: $f(x) = ax(x + 2)(x + 3)$ has zeros $0, -2, -3$ for all real numbers $a \neq 0$.

53. $f(x) = (x - 4)(x + 3)(x - 3)(x - 0)$

$= (x - 4)(x^2 - 9)x$

$= x^4 - 4x^3 - 9x^2 + 36x$

Note: $f(x) = a(x^4 - 4x^3 - 9x^2 + 36x)$ has these zeros for all real numbers $a \neq 0$.

55. $f(x) = \left[x - \left(1 + \sqrt{3}\right)\right]\left[x - \left(1 - \sqrt{3}\right)\right]$

$= \left[(x - 1) - \sqrt{3}\right]\left[(x - 1) + \sqrt{3}\right]$

$= (x - 1)^2 - \left(\sqrt{3}\right)^2$

$= x^2 - 2x + 1 - 3$

$= x^2 - 2x - 2$

Note: $f(x) = a(x^2 - 2x - 2)$ has these zeros for all real numbers $a \neq 0$.

57. $f(x) = (x - (-2))(x - (-2))$

$= (x + 2)^2 = x^2 + 4x + 4$

Note: $f(x) = a(x^2 + 4x + 4), a \neq 0$, has degree 2 and zero $x = -2$.

59. $f(x) = (x - (-3))(x - 0)(x - 1)$

$= x(x + 3)(x - 1) = x^3 + 2x^2 - 3x$

Note: $f(x) = a(x^3 + 2x^2 - 3x), a \neq 0$, has degree 3 and zeros $x = -3, 0, 1$.

61. $f(x) = (x - 0)\left(x - \sqrt{3}\right)\left(x - \left(-\sqrt{3}\right)\right)$

$= x\left(x - \sqrt{3}\right)\left(x + \sqrt{3}\right) = x^3 - 3x$

Note: $f(x) = a(x^3 - 3x), a \neq 0$, has degree 3 and zeros $x = 0, \sqrt{3}, -\sqrt{3}$.

63. $f(x) = (x - (-5))^2(x - 1)(x - 2) = x^4 + 7x^3 - 3x^2 - 55x + 50$

or $f(x) = (x - (-5))(x - 1)^2(x - 2) = x^4 + x^3 - 15x^2 + 23x - 10$

or $f(x) = (x - (-5))(x - 1)(x - 2)^2 = x^4 - 17x^2 + 36x - 20$

Note: Any nonzero scalar multiple of these functions would also have degree 4 and zeros $x = -5, 1, 2$.

65. $f(x) = x^4(x + 4) = x^5 + 4x^4$

    or $f(x) = x^3(x + 4)^2 = x^5 + 8x^4 + 16x^3$

    or $f(x) = x^2(x + 4)^3 = x^5 + 12x^4 + 48x^3 + 64x^2$

    or $f(x) = x(x + 4)^4 = x^5 + 16x^4 + 96x^3 + 256x^2 + 256x$

Note: Any nonzero scalar multiple of these functions would also have degree 5 and zeros $x = 0$ and -4.

67. $f(x) = x^3 - 9x = x(x^2 - 9) = x(x + 3)(x - 3)$

    (a) Falls to the left; rises to the right

    (b) Zeros: $0, -3, 3$

    (c)

x	-3	-2	-1	0	1	2	3
$f(x)$	0	10	8	0	-8	-10	0

    (d)

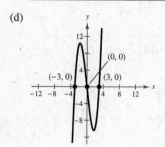

69. $f(t) = \frac{1}{4}(t^2 - 2t + 15) = \frac{1}{4}(t - 1)^2 + \frac{7}{2}$

    (a) Rises to the left; rises to the right

    (b) No real zero (no x-intercepts)

    (c)

t	-1	0	1	2	3
$f(t)$	4.5	3.75	3.5	3.75	4.5

    (d) The graph is a parabola with vertex $\left(1, \frac{7}{2}\right)$.

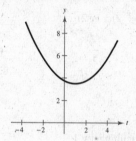

71. $f(x) = x^3 - 3x^2 = x^2(x - 3)$

    (a) Falls to the left; rises to the right

    (b) Zeros: $0, 3$

    (c)

x	-1	0	1	2	3
$f(x)$	-4	0	-2	-4	0

    (d)

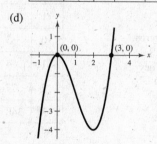

73. $f(x) = 3x^3 - 15x^2 + 18x = 3x(x - 2)(x - 3)$

    (a) Falls to the left; rises to the right

    (b) Zeros: $0, 2, 3$

    (c)

x	0	1	2	2.5	3	3.5
$f(x)$	0	6	0	-1.875	0	7.875

    (d)

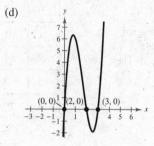

75. $f(x) = -5x^2 - x^3 = -x^2(5 + x)$

    (a) Rises to the left; falls to the right

    (b) Zeros: $0, -5$

    (c)

x	-5	-4	-3	-2	-1	0	1
$f(x)$	0	-16	-18	-12	-4	0	-6

    (d)

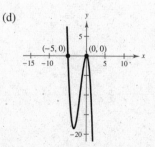

77. $f(x) = x^2(x - 4)$

(a) Falls to the left; rises to the right

(b) Zeros: 0, 4

(c)

x	-1	0	1	2	3	4	5
$f(x)$	-5	0	-3	-8	-9	0	25

(d)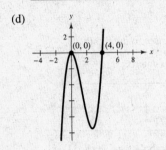

79. $g(t) = -\frac{1}{4}(t - 2)^2(t + 2)^2$

(a) Falls to the left; falls to the right

(b) Zeros: 2, -2

(c)

t	-3	-2	-1	0	1	2	3
$g(t)$	$-\frac{25}{4}$	0	$-\frac{9}{4}$	-4	$-\frac{9}{4}$	0	$-\frac{25}{4}$

(d)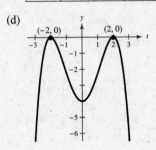

81. $f(x) = x^3 - 4x = x(x + 2)(x - 2)$

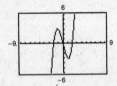

Zeros: 0, -2, 2 all of multiplicity 1

83. $g(x) = \frac{1}{5}(x + 1)^2(x - 3)(2x - 9)$

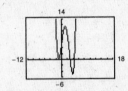

Zeros: -1 of multiplicity 2; 3 of multiplicity 1; $\frac{9}{2}$ of multiplicity 1

85. $f(x) = x^3 - 3x^2 + 3$

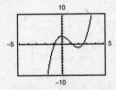

The function has three zeros. They are in the intervals $(-1, 0)$, $(1, 2)$ and $(2, 3)$. They are $x \approx -0.879, 1.347, 2.532$.

x	y_1
-3	-51
-2	-17
-1	-1
0	3
1	1
2	-1
3	3
4	19

87. $g(x) = 3x^4 + 4x^3 - 3$

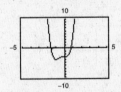

The function has two zeros. They are in the intervals $(-2, -1)$ and $(0, 1)$. They are $x \approx -1.585, 0.779$.

x	y_1
-4	509
-3	132
-2	13
-1	-4
0	-3
1	4
2	77
3	348

89. (a) Volume $= l \cdot w \cdot h$

height $= x$

length $=$ width $= 36 - 2x$

Thus, $V(x) = (36 - 2x)(36 - 2x)(x) = x(36 - 2x)^2$.

(b) Domain: $0 < x < 18$

The length and width must be positive.

(d)
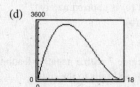

The maximum point on the graph occurs at $x = 6$.
This agrees with the maximum found in part (c).

(c)

Box Height	Box Width	Box Volume, V
1	$36 - 2(1)$	$1[36 - 2(1)]^2 = 1156$
2	$36 - 2(2)$	$2[36 - 2(2)]^2 = 2048$
3	$36 - 2(3)$	$3[36 - 2(3)]^2 = 2700$
4	$36 - 2(4)$	$4[36 - 2(4)]^2 = 3136$
5	$36 - 2(5)$	$5[36 - 2(5)]^2 = 3380$
6	$36 - 2(6)$	$6[36 - 2(6)]^2 = 3456$
7	$36 - 2(7)$	$7[36 - 2(7)]^2 = 3388$

The volume is a maximum of 3456 cubic inches when the height is 6 inches and the length and width are each 24 inches. So the dimensions are $6 \times 24 \times 24$ inches.

91. (a) $A = l \cdot w = (12 - 2x)(x) = -2x^2 + 12x$ square inches

(b) 16 feet $= 192$ inches

$V = l \cdot w \cdot h$

$= (12 - 2x)(x)(192)$

$= -384x^2 + 2304x$ cubic inches

(c) Since x and $12 - 2x$ cannot be negative, we have $0 < x < 6$ inches for the domain.

(d)

x	V
0	0
1	1920
2	3072
3	3456
4	3072
5	1920
6	0

When $x = 3$, the volume is a maximum with $V = 3456$ in.^{3}. The dimensions of the gutter cross-section are 3 inches $\times$ 6 inches $\times$ 3 inches.

(e)
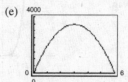

Maximum: $(3, 3456)$

The maximum value is the same.

(f) No. The volume is a product of the constant length and the cross-sectional area. The value of x would remain the same; only the value of V would change if the length was changed.

93. $y_1 = 0.139t^3 - 4.42t^2 + 51.1t - 39$

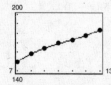

The model is a good fit to the actual data.

95. Midwest: $y_1(18) = \$259.368$ thousand $= \$259,368$

South: $y_2(18) = \$223.472$ thousand $= \$223,472$

Since the models are both cubic functions with positive leading coefficients, both will increase without bound as t increases, thus should only be used for short term projections.

97. $G = -0.003t^3 + 0.137t^2 + 0.458t - 0.839, \ 2 \le t \le 34$

(a)

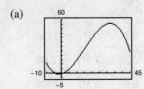

(b) The tree is growing most rapidly at $t \approx 15$.

(c) $y = -0.009t^2 + 0.274t + 0.458$

$$-\frac{b}{2a} = \frac{-0.274}{2(-0.009)} \approx 15.222$$

$y(15.222) \approx 2.543$

Vertex $\approx (15.22, 2.54)$

(d) The x-value of the vertex in part (c) is approximately equal to the value found in part (b).

99. False. A fifth degree polynomial can have at most four turning points.

101. True. A polynomial of degree 7 with a negative leading coefficient rises to the left and falls to the right.

103. $f(x) = x^4;\ f(x)$ is even.

(a) $g(x) = f(x) + 2$

Vertical shift two units upward

$g(-x) = f(-x) + 2$

$\qquad = f(x) + 2$

$\qquad = g(x)$

Even

(b) $g(x) = f(x + 2)$

Horizontal shift two units to the left

Neither odd nor even

(d) $g(x) = -f(x) = -x^4$

Reflection in the x-axis

Even

(f) $g(x) = \frac{1}{2}f(x) = \frac{1}{2}x^4$

Vertical shrink

Even

(h) $g(x) = (f \circ f)(x) = f(f(x)) = f(x^4) = (x^4)^4 = x^{16}$

Even

(c) $g(x) = f(-x) = (-x)^4 = x^4$

Reflection in the y-axis. The graph looks the same.

Even

(e) $g(x) = f\left(\frac{1}{2}x\right) = \frac{1}{16}x^4$

Horizontal stretch

Even

(g) $g(x) = f(x^{3/4}) = (x^{3/4})^4 = x^3,\ x \ge 0$

Neither odd nor even

105. $5x^2 + 7x - 24 = (5x - 8)(x + 3)$

107. $4x^4 - 7x^3 - 15x^2 = x^2(4x^2 - 7x - 15)$

$\qquad\qquad\qquad\quad = x^2(4x + 5)(x - 3)$

109. $\quad 2x^2 - x - 28 = 0$

$\qquad (2x + 7)(x - 4) = 0$

$\qquad\qquad 2x + 7 = 0 \implies x = -\frac{7}{2}$

$\qquad\qquad\ x - 4 = 0 \implies x = 4$

111. $\quad 12x^2 + 11x - 5 = 0$

$\qquad (3x - 1)(4x + 5) = 0$

$\qquad\qquad 3x - 1 = 0 \implies x = \frac{1}{3}$

$\qquad\qquad 4x + 5 = 0 \implies x = -\frac{5}{4}$

113.
$$x^2 - 2x - 21 = 0$$
$$(x^2 - 2x + (-1)^2) - 21 - 1 = 0$$
$$(x - 1)^2 - 22 = 0$$
$$(x - 1)^2 = 22$$
$$x - 1 = \pm\sqrt{22}$$
$$x = 1 \pm \sqrt{22}$$

115.
$$2x^2 + 5x - 20 = 0$$
$$2\left(x^2 + \frac{5}{2}x\right) - 20 = 0$$
$$2\left(x^2 + \frac{5}{2}x + \left(\frac{5}{4}\right)^2\right) - 20 - \frac{25}{8} = 0$$
$$2\left(x + \frac{5}{4}\right)^2 - \frac{185}{8} = 0$$
$$\left(x + \frac{5}{4}\right)^2 = \frac{185}{16}$$
$$x + \frac{5}{4} = \pm\frac{\sqrt{185}}{4}$$
$$x = \frac{-5 \pm \sqrt{185}}{4}$$

117. $f(x) = (x + 4)^2$

Common function: $y = x^2$

Transformation: Horizontal shift four units to the left

119. $f(x) = \sqrt{x + 1} - 5$

Common function: $y = \sqrt{x}$

Transformation: Horizontal shift one unit to the left and a vertical shift five units downward

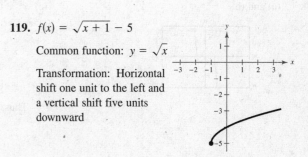

121. $f(x) = 2[\![x]\!] + 9$

Common function: $y = [\![x]\!]$

Transformation: Vertical stretch (each y-value is multiplied by 2), then a vertical shift nine units upward

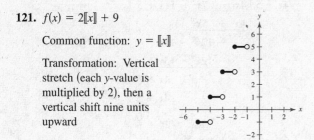

Section 2.3   Polynomial and Synthetic Division

You should know the following basic techniques and principles of polynomial division.

■   The Division Algorithm (Long Division of Polynomials)

■   Synthetic Division

■ $f(k)$ is equal to the remainder of $f(x)$ divided by $(x - k)$ (the Remainder Theorem).

■ $f(k) = 0$ if and only if $(x - k)$ is a factor of $f(x)$.

Vocabulary Check

1. $f(x)$ is the dividend; $d(x)$ is the divisor; $g(x)$ is the quotient; $r(x)$ is the remainder

2. improper; proper   **3.** synthetic division   **4.** factor   **5.** remainder

1. $y_1 = \dfrac{x^2}{x+2}$ and $y_2 = x - 2 + \dfrac{4}{x+2}$

$$
\begin{array}{r}
x - 2 \\
x + 2 \overline{\smash{)}\, x^2 + 0x + 0} \\
\underline{x^2 + 2x} \\
-2x + 0 \\
\underline{-2x - 4} \\
4
\end{array}
$$

Thus, $\dfrac{x^2}{x+2} = x - 2 + \dfrac{4}{x+2}$ and $y_1 = y_2$.

3. $y_1 = \dfrac{x^5 - 3x^3}{x^2 + 1}$ and $y_2 = x^3 - 4x + \dfrac{4x}{x^2 + 1}$

(a) and (b)

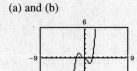

(c)
$$
\begin{array}{r}
x^3 - 4x \\
x^2 + 0x + 1 \overline{\smash{)}\, x^5 + 0x^4 - 3x^3 + 0x^2 + 0x + 0} \\
\underline{x^5 + 0x^4 + x^3} \\
-4x^3 + 0x^2 + 0x \\
\underline{-4x^3 + 0x^2 - 4x} \\
4x + 0
\end{array}
$$

Thus, $\dfrac{x^5 - 3x^3}{x^2 + 1} = x^3 - 4x + \dfrac{4x}{x^2 + 1}$ and $y_1 = y_2$.

5.
$$
\begin{array}{r}
2x + 4 \\
x + 3 \overline{\smash{)}\, 2x^2 + 10x + 12} \\
\underline{2x^2 + 6x} \\
4x + 12 \\
\underline{4x + 12} \\
0
\end{array}
$$

$$\dfrac{2x^2 + 10x + 12}{x + 3} = 2x + 4$$

7.
$$
\begin{array}{r}
x^2 - 3x + 1 \\
4x + 5 \overline{\smash{)}\, 4x^3 - 7x^2 - 11x + 5} \\
\underline{4x^3 + 5x^2} \\
-12x^2 - 11x \\
\underline{-12x^2 - 15x} \\
4x + 5 \\
\underline{4x + 5} \\
0
\end{array}
$$

$$\dfrac{4x^3 - 7x^2 - 11x + 5}{4x + 5} = x^2 - 3x + 1$$

9.
$$
\begin{array}{r}
x^3 + 3x^2 - 1 \\
x + 2 \overline{\smash{)}\, x^4 + 5x^3 + 6x^2 - x - 2} \\
\underline{x^4 + 2x^3} \\
3x^3 + 6x^2 \\
\underline{3x^3 + 6x^2} \\
- x - 2 \\
\underline{- x - 2} \\
0
\end{array}
$$

$$\dfrac{x^4 + 5x^3 + 6x^2 - x - 2}{x + 2} = x^3 + 3x^2 - 1$$

11.
$$
\begin{array}{r}
7 \\
x + 2 \overline{\smash{)}\, 7x + 3} \\
\underline{7x + 14} \\
- 11
\end{array}
$$

$$\dfrac{7x + 3}{x + 2} = 7 - \dfrac{11}{x + 2}$$

13.
$$
\begin{array}{r}
3x + 5 \\
2x^2 + 0x + 1 \overline{)\;6x^3 + 10x^2 + \;\;x + 8} \\
\underline{6x^3 + \;\;0x^2 + 3x} \\
10x^2 - 2x + 8 \\
\underline{10x^2 + 0x + 5} \\
- 2x + 3
\end{array}
$$

$$
\frac{6x^3 + 10x^2 + x + 8}{2x^2 + 1} = 3x + 5 - \frac{2x - 3}{2x^2 + 1}
$$

15.
$$
\begin{array}{r}
x^2 + 2x + \;\;4 \\
x^2 - 2x + 3 \overline{)\;x^4 + 0x^3 + 3x^2 + 0x + \;\;1} \\
\underline{x^4 - 2x^3 + 3x^2} \\
2x^3 + 0x^2 + 0x \\
\underline{2x^3 - 4x^2 + 6x} \\
4x^2 - 6x + \;\;1 \\
\underline{4x^2 - 8x + 12} \\
2x - 11
\end{array}
$$

$$
\Rightarrow \quad \frac{x^4 + 3x^2 + 1}{x^2 - 2x + 3} = x^2 + 2x + 4 + \frac{2x - 11}{x^2 - 2x + 3}
$$

17.
$$
\begin{array}{r}
x + 3 \\
x^3 - 3x^2 + 3x - 1 \overline{)\;x^4 + 0x^3 + 0x^2 + 0x + 0} \\
\underline{x^4 - 3x^3 + 3x^2 - \;\;x} \\
3x^3 - 3x^2 + \;\;x + 0 \\
\underline{3x^3 - 9x^2 + 9x - 3} \\
6x^2 - 8x + 3
\end{array}
$$

$$
\frac{x^4}{(x - 1)^3} = x + 3 + \frac{6x^2 - 8x + 3}{(x - 1)^3}
$$

19.
$$
\begin{array}{r|rrrr}
5 & 3 & -17 & 15 & -25 \\
  &   & 15 & -10 & 25 \\
\hline
  & 3 & -2 & 5 & 0
\end{array}
$$

$$
\frac{3x^3 - 17x^2 + 15x - 25}{x - 5} = 3x^2 - 2x + 5
$$

21.
$$
\begin{array}{r|rrrr}
-2 & 4 & 8 & -9 & -18 \\
   &   & -8 & 0 & 18 \\
\hline
   & 4 & 0 & -9 & 0
\end{array}
$$

$$
\frac{4x^3 + 8x^2 - 9x - 18}{x + 2} = 4x^2 - 9
$$

23.
$$
\begin{array}{r|rrrr}
-10 & -1 & 0 & 75 & -250 \\
    &   & 10 & -100 & 250 \\
\hline
    & -1 & 10 & -25 & 0
\end{array}
$$

$$
\frac{-x^3 + 75x - 250}{x + 10} = -x^2 + 10x - 25
$$

25.
$$
\begin{array}{r|rrrr}
4 & 5 & -6 & 0 & 8 \\
  &   & 20 & 56 & 224 \\
\hline
  & 5 & 14 & 56 & 232
\end{array}
$$

$$
\frac{5x^3 - 6x^2 + 8}{x - 4} = 5x^2 + 14x + 56 + \frac{232}{x - 4}
$$

27.
$$
\begin{array}{r|rrrrr}
6 & 10 & -50 & 0 & 0 & -800 \\
  &   & 60 & 60 & 360 & 2160 \\
\hline
  & 10 & 10 & 60 & 360 & 1360
\end{array}
$$

$$
\frac{10x^4 - 50x^3 - 800}{x - 6} = 10x^3 + 10x^2 + 60x + 360 + \frac{1360}{x - 6}
$$

29.
$$
\begin{array}{r|rrrr}
-8 & 1 & 0 & 0 & 512 \\
   &   & -8 & 64 & -512 \\
\hline
   & 1 & -8 & 64 & 0
\end{array}
$$

$$
\frac{x^3 + 512}{x + 8} = x^2 - 8x + 64
$$

31.
$$
\begin{array}{r|rrrrr}
2 & -3 & 0 & 0 & 0 & 0 \\
  &   & -6 & -12 & -24 & -48 \\
\hline
  & -3 & -6 & -12 & -24 & -48
\end{array}
$$

$$
\frac{-3x^4}{x - 2} = -3x^3 - 6x^2 - 12x - 24 - \frac{48}{x - 2}
$$

33.

$$
\begin{array}{r|rrrrr}
6 & -1 & 0 & 0 & 180 & 0 \\
  &   & -6 & -36 & -216 & -216 \\
\hline
  & -1 & -6 & -36 & -36 & -216
\end{array}
$$

$$\frac{180x - x^4}{x - 6} = -x^3 - 6x^2 - 36x - 36 - \frac{216}{x - 6}$$

35.

$$
\begin{array}{r|rrrr}
-\frac{1}{2} & 4 & 16 & -23 & -15 \\
             &   & -2 & -7 & 15 \\
\hline
             & 4 & 14 & -30 & 0
\end{array}
$$

$$\frac{4x^3 + 16x^2 - 23x - 15}{x + (1/2)} = 4x^2 + 14x - 30$$

37. $f(x) = x^3 - x^2 - 14x + 11, \ k = 4$

$$
\begin{array}{r|rrrr}
4 & 1 & -1 & -14 & 11 \\
  &   & 4 & 12 & -8 \\
\hline
  & 1 & 3 & -2 & 3
\end{array}
$$

$f(x) = (x - 4)(x^2 + 3x - 2) + 3$

$f(4) = 4^3 - 4^2 - 14(4) + 11 = 3$

39. $f(x) = 15x^4 + 10x^3 - 6x^2 + 14, \ k = -\frac{2}{3}$

$$
\begin{array}{r|rrrrr}
-\frac{2}{3} & 15 & 10 & -6 & 0 & 14 \\
             &   & -10 & 0 & 4 & -\frac{8}{3} \\
\hline
             & 15 & 0 & -6 & 4 & \frac{34}{3}
\end{array}
$$

$f(x) = \left(x + \frac{2}{3}\right)(15x^3 - 6x + 4) + \frac{34}{3}$

$f\left(-\frac{2}{3}\right) = 15\left(-\frac{2}{3}\right)^4 + 10\left(-\frac{2}{3}\right)^3 - 6\left(-\frac{2}{3}\right)^2 + 14 = \frac{34}{3}$

41. $f(x) = x^3 + 3x^2 - 2x - 14, \ k = \sqrt{2}$

$$
\begin{array}{r|rrrr}
\sqrt{2} & 1 & 3 & -2 & -14 \\
         &   & \sqrt{2} & 2 + 3\sqrt{2} & 6 \\
\hline
         & 1 & 3 + \sqrt{2} & 3\sqrt{2} & -8
\end{array}
$$

$f(x) = (x - \sqrt{2})\left[x^2 + (3 + \sqrt{2})x + 3\sqrt{2}\right] - 8$

$f(\sqrt{2}) = (\sqrt{2})^3 + 3(\sqrt{2})^2 - 2\sqrt{2} - 14 = -8$

43. $f(x) = -4x^3 + 6x^2 + 12x + 4, \ k = 1 - \sqrt{3}$

$$
\begin{array}{r|rrrr}
1 - \sqrt{3} & -4 & 6 & 12 & 4 \\
             &   & -4 + 4\sqrt{3} & -10 + 2\sqrt{3} & -4 \\
\hline
             & -4 & 2 + 4\sqrt{3} & 2 + 2\sqrt{3} & 0
\end{array}
$$

$f(x) = \left[x - (1 - \sqrt{3})\right]\left[-4x^2 + (2 + 4\sqrt{3})x + (2 + 2\sqrt{3})\right] + 0$

$f(1 - \sqrt{3}) = -4(1 - \sqrt{3})^3 + 6(1 - \sqrt{3})^2 + 12(1 - \sqrt{3}) + 4 = 0$

45. $f(x) = 4x^3 - 13x + 10$

(a)
$$
\begin{array}{r|rrrr}
1 & 4 & 0 & -13 & 10 \\
  &   & 4 & 4   & -9 \\
\hline
  & 4 & 4 & -9 & 1
\end{array}
$$

$f(1) = 1$

(b)
$$
\begin{array}{r|rrrr}
-2 & 4 & 0 & -13 & 10 \\
   &   & -8 & 16 & -6 \\
\hline
   & 4 & -8 & 3 & 4
\end{array}
$$

$f(-2) = 4$

(c)
$$
\begin{array}{r|rrrr}
\frac{1}{2} & 4 & 0 & -13 & 10 \\
            &   & 2 & 1   & -6 \\
\hline
            & 4 & 2 & -12 & 4
\end{array}
$$

$f\left(\frac{1}{2}\right) = 4$

(d)
$$
\begin{array}{r|rrrr}
8 & 4 & 0 & -13 & 10 \\
  &   & 32 & 256 & 1944 \\
\hline
  & 4 & 32 & 243 & 1954
\end{array}
$$

$f(8) = 1954$

47. $h(x) = 3x^3 + 5x^2 - 10x + 1$

(a)
$$
\begin{array}{r|rrrr}
3 & 3 & 5 & -10 & 1 \\
  &   & 9 & 42 & 96 \\
\hline
  & 3 & 14 & 32 & 97
\end{array}
$$

$h(3) = 97$

(b)
$$
\begin{array}{r|rrrr}
\frac{1}{3} & 3 & 5 & -10 & 1 \\
            &   & 1 & 2   & -\frac{8}{3} \\
\hline
            & 3 & 6 & -8 & -\frac{5}{3}
\end{array}
$$

$h\left(\frac{1}{3}\right) = -\frac{5}{3}$

(c)
$$
\begin{array}{r|rrrr}
-2 & 3 & 5 & -10 & 1 \\
   &   & -6 & 2   & 16 \\
\hline
   & 3 & -1 & -8 & 17
\end{array}
$$

$h(-2) = 17$

(d)
$$
\begin{array}{r|rrrr}
-5 & 3 & 5 & -10 & 1 \\
   &   & -15 & 50 & -200 \\
\hline
   & 3 & -10 & 40 & -199
\end{array}
$$

$h(-5) = -199$

49.
$$
\begin{array}{r|rrrr}
2 & 1 & 0 & -7 & 6 \\
  &   & 2 & 4 & -6 \\
\hline
  & 1 & 2 & -3 & 0
\end{array}
$$

$x^3 - 7x + 6 = (x - 2)(x^2 + 2x - 3)$

$\qquad\qquad = (x - 2)(x + 3)(x - 1)$

Zeros: $2, -3, 1$

51.
$$
\begin{array}{r|rrrr}
\frac{1}{2} & 2 & -15 & 27 & -10 \\
  &   & 1 & -7 & 10 \\
\hline
  & 2 & -14 & 20 & 0
\end{array}
$$

$2x^3 - 15x^2 + 27x - 10$

$\qquad = \left(x - \frac{1}{2}\right)(2x^2 - 14x + 20)$

$\qquad = (2x - 1)(x - 2)(x - 5)$

Zeros: $\frac{1}{2}, 2, 5$

53.
$$
\begin{array}{r|rrrr}
\sqrt{3} & 1 & 2 & -3 & -6 \\
  &   & \sqrt{3} & 3 + 2\sqrt{3} & 6 \\
\hline
  & 1 & 2 + \sqrt{3} & 2\sqrt{3} & 0
\end{array}
$$
$$
\begin{array}{r|rrr}
-\sqrt{3} & 1 & 2 + \sqrt{3} & 2\sqrt{3} \\
  &   & -\sqrt{3} & -2\sqrt{3} \\
\hline
  & 1 & 2 & 0
\end{array}
$$

$x^3 + 2x^2 - 3x - 6 = (x - \sqrt{3})(x + \sqrt{3})(x + 2)$

Zeros: $\pm\sqrt{3}, -2$

55.
$$
\begin{array}{r|rrrr}
1 + \sqrt{3} & 1 & -3 & 0 & 2 \\
  &   & 1 + \sqrt{3} & 1 - \sqrt{3} & -2 \\
\hline
  & 1 & -2 + \sqrt{3} & 1 - \sqrt{3} & 0
\end{array}
$$
$$
\begin{array}{r|rrr}
1 - \sqrt{3} & 1 & -2 + \sqrt{3} & 1 - \sqrt{3} \\
  &   & 1 - \sqrt{3} & -1 + \sqrt{3} \\
\hline
  & 1 & -1 & 0
\end{array}
$$

$x^3 - 3x^2 + 2 = \left[x - \left(1 + \sqrt{3}\right)\right]\left[x - \left(1 - \sqrt{3}\right)\right](x - 1)$

$\qquad = (x - 1)(x - 1 - \sqrt{3})(x - 1 + \sqrt{3})$

Zeros: $1, 1 \pm \sqrt{3}$

57. $f(x) = 2x^3 + x^2 - 5x + 2$; Factors: $(x + 2), (x - 1)$

(a)
$$
\begin{array}{r|rrrr}
-2 & 2 & 1 & -5 & 2 \\
  &   & -4 & 6 & -2 \\
\hline
  & 2 & -3 & 1 & 0
\end{array}
$$
$$
\begin{array}{r|rrr}
1 & 2 & -3 & 1 \\
  &   & 2 & -1 \\
\hline
  & 2 & -1 & 0
\end{array}
$$

Both are factors of $f(x)$ since the remainders are zero.

(b) The remaining factor of $f(x)$ is $(2x - 1)$.

(c) $f(x) = (2x - 1)(x + 2)(x - 1)$

(d) Zeros: $\frac{1}{2}, -2, 1$

(e)

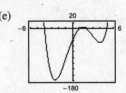

59. $f(x) = x^4 - 4x^3 - 15x^2 + 58x - 40$; Factors: $(x - 5), (x + 4)$

(a)
$$
\begin{array}{r|rrrrr}
5 & 1 & -4 & -15 & 58 & -40 \\
  &   & 5 & 5 & -50 & 40 \\
\hline
  & 1 & 1 & -10 & 8 & 0
\end{array}
$$
$$
\begin{array}{r|rrrr}
-4 & 1 & 1 & -10 & 8 \\
  &   & -4 & 12 & -8 \\
\hline
  & 1 & -3 & 2 & 0
\end{array}
$$

Both are factors of $f(x)$ since the remainders are zero.

(b) $x^2 - 3x + 2 = (x - 1)(x - 2)$

The remaining factors are $(x - 1)$ and $(x - 2)$.

(c) $f(x) = (x - 1)(x - 2)(x - 5)(x + 4)$

(d) Zeros: $1, 2, 5, -4$

(e)

61. $f(x) = 6x^3 + 41x^2 - 9x - 14;$

Factors: $(2x + 1), (3x - 2)$

(a)

$$-\tfrac{1}{2} \begin{array}{|rrrr} 6 & 41 & -9 & -14 \\ & -3 & -19 & 14 \\ \hline 6 & 38 & -28 & 0 \end{array}$$

$$\tfrac{2}{3} \begin{array}{|rrr} 6 & 38 & -28 \\ & 4 & 28 \\ \hline 6 & 42 & 0 \end{array}$$

Both are factors since the remainders are zero.

(b) $6x + 42 = 6(x + 7)$

This shows that $\dfrac{f(x)}{\left(x + \frac{1}{2}\right)\left(x - \frac{2}{3}\right)} = 6(x + 7),$

so $\dfrac{f(x)}{(2x + 1)(3x - 2)} = x + 7.$

The remaining factor is $(x + 7).$

(c) $f(x) = (x + 7)(2x + 1)(3x - 2)$

(d) Zeros: $-7, -\dfrac{1}{2}, \dfrac{2}{3}$

(e)

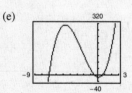

63. $f(x) = 2x^3 - x^2 - 10x + 5;$

Factors: $(2x - 1), \left(x + \sqrt{5}\right)$

(a)

$$\tfrac{1}{2} \begin{array}{|rrrr} 2 & -1 & -10 & 5 \\ & 1 & 0 & -5 \\ \hline 2 & 0 & -10 & 0 \end{array}$$

$$-\sqrt{5} \begin{array}{|rrr} 2 & 0 & -10 \\ & -2\sqrt{5} & 10 \\ \hline 2 & -2\sqrt{5} & 0 \end{array}$$

Both are factors since the remainders are zero.

(b) $2x - 2\sqrt{5} = 2\left(x - \sqrt{5}\right)$

This shows that $\dfrac{f(x)}{\left(x - \frac{1}{2}\right)\left(x + \sqrt{5}\right)} = 2\left(x - \sqrt{5}\right),$

so $\dfrac{f(x)}{(2x - 1)\left(x + \sqrt{5}\right)} = x - \sqrt{5}.$

The remaining factor is $\left(x - \sqrt{5}\right).$

(c) $f(x) = \left(x + \sqrt{5}\right)\left(x - \sqrt{5}\right)(2x - 1)$

(d) Zeros: $-\sqrt{5}, \sqrt{5}, \dfrac{1}{2}$

(e)

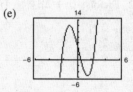

65. $f(x) = x^3 - 2x^2 - 5x + 10$

(a) The zeros of f are 2 and $\approx \pm 2.236.$

(b) An exact zero is $x = 2.$

(c)

$$2 \begin{array}{|rrrr} 1 & -2 & -5 & 10 \\ & 2 & 0 & -10 \\ \hline 1 & 0 & -5 & 0 \end{array}$$

$f(x) = (x - 2)(x^2 - 5)$

$= (x - 2)\left(x - \sqrt{5}\right)\left(x + \sqrt{5}\right)$

67. $h(t) = t^3 - 2t^2 - 7t + 2$

(a) The zeros of h are $t = -2, t \approx 3.732, t \approx 0.268.$

(b) An exact zero is $t = -2.$

(c)

$$-2 \begin{array}{|rrrr} 1 & -2 & -7 & 2 \\ & -2 & 8 & -2 \\ \hline 1 & -4 & 1 & 0 \end{array}$$

$h(t) = (t + 2)(t^2 - 4t + 1)$

By the Quadratic Formula, the zeros of $t^2 - 4t + 1$ are $2 \pm \sqrt{3}$. Thus,

$h(t) = (t + 2)\left[t - \left(2 + \sqrt{3}\right)\right]\left[t - \left(2 - \sqrt{3}\right)\right]$

$= (t + 2)\left(t - 2 - \sqrt{3}\right)\left(t - 2 + \sqrt{3}\right).$

69. $\dfrac{4x^3 - 8x^2 + x + 3}{2x - 3}$

$$\tfrac{3}{2} \begin{array}{|rrrr} 4 & -8 & 1 & 3 \\ & 6 & -3 & -3 \\ \hline 4 & -2 & -2 & 0 \end{array}$$

$\dfrac{4x^3 - 8x^2 + x + 3}{x - \frac{3}{2}} = 4x^2 - 2x - 2 = 2(2x^2 - x - 1)$

Thus, $\dfrac{4x^3 - 8x^2 + x + 3}{2x - 3} = 2x^2 - x - 1, x \neq \dfrac{3}{2}.$

71. $\dfrac{x^4 + 6x^3 + 11x^2 + 6x}{x^2 + 3x + 2} = \dfrac{x^4 + 6x^3 + 11x^2 + 6x}{(x + 1)(x + 2)}$

$$-1 \begin{array}{|rrrrr} 1 & 6 & 11 & 6 & 0 \\ & -1 & -5 & -6 & 0 \\ \hline 1 & 5 & 6 & 0 & 0 \end{array}$$

$$-2 \begin{array}{|rrrr} 1 & 5 & 6 & 0 \\ & -2 & -6 & 0 \\ \hline 1 & 3 & 0 & 0 \end{array}$$

$\dfrac{x^4 + 6x^3 + 11x^2 + 6x}{(x + 1)(x + 2)} = x^2 + 3x, x \neq -2, -1$

73. (a) and (b)

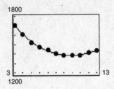

(c) $M \approx -0.242t^3 + 12.43t^2 - 173.4t + 2118$

Year, t	Military Personnel	M
3	1705	1703
4	1611	1608
5	1518	1532
6	1472	1473
7	1439	1430
8	1407	1402
9	1386	1388
10	1384	1385
11	1385	1393
12	1412	1409
13	1434	1433

The model is a good fit to the actual data.

(d)

$$18 \,\big|\ \begin{array}{ccccc} -0.242 & 12.43 & -173.4 & 2118 \\ & -4.356 & 145.332 & -505.224 \\ \hline -0.242 & 8.074 & -28.068 & 1612.776 \end{array}$$

$M(18) \approx 1613$ thousand

No, this model should not be used to predict the number of military personnel in the future. It predicts an increase in military personnel until 2024 and then it decreases and will approach negative infinity quickly.

75. False. If $(7x + 4)$ is a factor of f, then $-\frac{4}{7}$ is a zero of f.

77. True. The degree of the numerator is greater than the degree of the denominator.

79.

$$\begin{array}{r} x^{2n} + 6x^n + 9 \\ x^n + 3 \,\overline{)\, x^{3n} + 9x^{2n} + 27x^n + 27} \\ \underline{x^{3n} + 3x^{2n}} \\ 6x^{2n} + 27x^n \\ \underline{6x^{2n} + 18x^n} \\ 9x^n + 27 \\ \underline{9x^n + 27} \\ 0 \end{array}$$

$$\frac{x^{3n} + 9x^{2n} + 27x^n + 27}{x^n + 3} = x^{2n} + 6x^n + 9$$

81. A divisor divides evenly into a dividend if the remainder is zero.

83.

$$5 \,\big|\ \begin{array}{cccc} 1 & 4 & -3 & c \\ & 5 & 45 & 210 \\ \hline 1 & 9 & 42 & c + 210 \end{array}$$

To divide evenly, $c + 210$ must equal zero. Thus, c must equal -210.

85. $f(x) = (x + 3)^2(x - 3)(x + 1)^3$

The remainder when $k = -3$ is zero since $(x + 3)$ is a factor of $f(x)$.

87.
$$9x^2 - 25 = 0$$
$$(3x - 5)(3x + 5) = 0$$
$$3x - 5 = 0 \implies x = \frac{5}{3}$$
$$3x + 5 = 0 \implies x = -\frac{5}{3}$$

89. $5x^2 - 3x - 14 = 0$
$$(5x + 7)(x - 2) = 0$$
$$5x + 7 = 0 \implies x = -\frac{7}{5}$$
$$x - 2 = 0 \implies x = 2$$

91. $2x^2 + 6x + 3 = 0$
$$x = \frac{-b \pm \sqrt{b^2 - 4ac}}{2a} = \frac{-6 \pm \sqrt{6^2 - 4(2)(3)}}{2(2)} = \frac{-6 \pm \sqrt{12}}{4}$$
$$= \frac{-3 \pm \sqrt{3}}{2}$$

93. $f(x) = (x - 0)(x - 3)(x - 4)$
$$= x(x - 3)(x - 4)$$
$$= x(x^2 - 7x + 12)$$
$$= x^3 - 7x^2 + 12x$$

Note: Any nonzero scalar multiple of $f(x)$ would also have these zeros.

95. $f(x) = [x - (-3)][x - (1 + \sqrt{2})][x - (1 - \sqrt{2})]$
$$= (x + 3)[(x - 1) - \sqrt{2}][(x - 1) + \sqrt{2}]$$
$$= (x + 3)[(x - 1)^2 - (\sqrt{2})^2]$$
$$= (x + 3)(x^2 - 2x - 1)$$
$$= x^3 + x^2 - 7x - 3$$

Note: Any nonzero scalar multiple of $f(x)$ would also have these zeros.

Section 2.4   Complex Numbers

- Standard form: $a + bi$.

  If $b = 0$, then $a + bi$ is a real number.

  If $a = 0$ and $b \neq 0$, then $a + bi$ is a pure imaginary number.

- Equality of Complex Numbers: $a + bi = c + di$ if and only if $a = c$ and $b = d$

- Operations on complex numbers

  (a) Addition: $(a + bi) + (c + di) = (a + c) + (b + d)i$

  (b) Subtraction: $(a + bi) - (c + di) = (a - c) + (b - d)i$

  (c) Multiplication: $(a + bi)(c + di) = (ac - bd) + (ad + bc)i$

  (d) Division: $\dfrac{a + bi}{c + di} = \dfrac{a + bi}{c + di} \cdot \dfrac{c - di}{c - di} = \dfrac{ac + bd}{c^2 + d^2} + \dfrac{bc - ad}{c^2 + d^2}i$

- The complex conjugate of $a + bi$ is $a - bi$:

  $(a + bi)(a - bi) = a^2 + b^2$

- The additive inverse of $a + bi$ is $-a - bi$.

- $\sqrt{-a} = \sqrt{a}\,i$ for $a > 0$.

Vocabulary Check

1. (a) iii   (b) i   (c) ii

2. $\sqrt{-1}$; -1

3. principal square

4. complex conjugates

1. $a + bi = -10 + 6i$

$a = -10$

$b = 6$

3. $(a - 1) + (b + 3)i = 5 + 8i$

$a - 1 = 5 \implies a = 6$

$b + 3 = 8 \implies b = 5$

5. $4 + \sqrt{-9} = 4 + 3i$

7. $2 - \sqrt{-27} = 2 - \sqrt{27}i$

$\qquad = 2 - 3\sqrt{3}i$

9. $\sqrt{-75} = \sqrt{75}i = 5\sqrt{3}i$

11. $8 = 8 + 0i = 8$

13. $-6i + i^2 = -6i - 1$

$\qquad = -1 - 6i$

15. $\sqrt{-0.09} = \sqrt{0.09}i$

$\qquad = 0.3i$

17. $(5 + i) + (6 - 2i) = 11 - i$

19. $(8 - i) - (4 - i) = 8 - i - 4 + i = 4$

21. $\left(-2 + \sqrt{-8}\right) + \left(5 - \sqrt{-50}\right) = -2 + 2\sqrt{2}i + 5 - 5\sqrt{2}i$

$\qquad\qquad = 3 - 3\sqrt{2}i$

23. $13i - (14 - 7i) = 13i - 14 + 7i$

$\qquad\qquad = -14 + 20i$

25. $-\left(\frac{3}{2} + \frac{5}{2}i\right) + \left(\frac{5}{3} + \frac{11}{3}i\right) = -\frac{3}{2} - \frac{5}{2}i + \frac{5}{3} + \frac{11}{3}i$

$\qquad\qquad\qquad = -\frac{9}{6} - \frac{15}{6}i + \frac{10}{6} + \frac{22}{6}i$

$\qquad\qquad\qquad = \frac{1}{6} + \frac{7}{6}i$

27. $(1 + i)(3 - 2i) = 3 - 2i + 3i - 2i^2$

$\qquad\qquad = 3 + i + 2 = 5 + i$

29. $6i(5 - 2i) = 30i - 12i^2 = 30i + 12$

$\qquad\qquad = 12 + 30i$

31. $\left(\sqrt{14} + \sqrt{10}i\right)\left(\sqrt{14} - \sqrt{10}i\right) = 14 - 10i^2$

$\qquad\qquad\qquad = 14 + 10 = 24$

33. $(4 + 5i)^2 = 16 + 40i + 25i^2$

$\qquad\qquad = 16 + 40i - 25$

$\qquad\qquad = -9 + 40i$

35. $(2 + 3i)^2 + (2 - 3i)^2 = 4 + 12i + 9i^2 + 4 - 12i + 9i^2$

$\qquad\qquad\qquad = 4 + 12i - 9 + 4 - 12i - 9$

$\qquad\qquad\qquad = -10$

37. The complex conjugate of $6 + 3i$ is $6 - 3i$.

$(6 + 3i)(6 - 3i) = 36 - (3i)^2 = 36 + 9 = 45$

39. The complex conjugate of $-1 - \sqrt{5}i$ is $-1 + \sqrt{5}i$.

$\left(-1 - \sqrt{5}i\right)\left(-1 + \sqrt{5}i\right) = (-1)^2 - \left(\sqrt{5}i\right)^2$

$\qquad\qquad\qquad = 1 + 5 = 6$

41. The complex conjugate of $\sqrt{-20} = 2\sqrt{5}i$ is $-2\sqrt{5}i$.

$\left(2\sqrt{5}i\right)\left(-2\sqrt{5}i\right) = -20i^2 = 20$

43. The complex conjugate of $\sqrt{8}$ is $\sqrt{8}$.

$\left(\sqrt{8}\right)\left(\sqrt{8}\right) = 8$

45. $\dfrac{5}{i} = \dfrac{5}{i} \cdot \dfrac{-i}{-i} = \dfrac{-5i}{1} = -5i$

47. $\dfrac{2}{4 - 5i} = \dfrac{2}{4 - 5i} \cdot \dfrac{4 + 5i}{4 + 5i}$

$\qquad = \dfrac{2(4 + 5i)}{16 + 25} = \dfrac{8 + 10i}{41} = \dfrac{8}{41} + \dfrac{10}{41}i$

49. $\dfrac{3 + i}{3 - i} = \dfrac{3 + i}{3 - i} \cdot \dfrac{3 + i}{3 + i}$

$\qquad = \dfrac{9 + 6i + i^2}{9 + 1} = \dfrac{8 + 6i}{10} = \dfrac{4}{5} + \dfrac{3}{5}i$

51. $\dfrac{6 - 5i}{i} = \dfrac{6 - 5i}{i} \cdot \dfrac{-i}{-i}$

$\qquad = \dfrac{-6i + 5i^2}{1} = -5 - 6i$

53. $\dfrac{3i}{(4 - 5i)^2} = \dfrac{3i}{16 - 40i + 25i^2} = \dfrac{3i}{-9 - 40i} \cdot \dfrac{-9 + 40i}{-9 + 40i}$

$\qquad = \dfrac{-27i + 120i^2}{81 + 1600} = \dfrac{-120 - 27i}{1681}$

$\qquad = -\dfrac{120}{1681} - \dfrac{27}{1681}i$

55. $\dfrac{2}{1 + i} - \dfrac{3}{1 - i} = \dfrac{2(1 - i) - 3(1 + i)}{(1 + i)(1 - i)}$

$\qquad = \dfrac{2 - 2i - 3 - 3i}{1 + 1}$

$\qquad = \dfrac{-1 - 5i}{2}$

$\qquad = -\dfrac{1}{2} - \dfrac{5}{2}i$

57. $\dfrac{i}{3 - 2i} + \dfrac{2i}{3 + 8i} = \dfrac{i(3 + 8i) + 2i(3 - 2i)}{(3 - 2i)(3 + 8i)}$

$\qquad = \dfrac{3i + 8i^2 + 6i - 4i^2}{9 + 24i - 6i - 16i^2}$

$\qquad = \dfrac{4i^2 + 9i}{9 + 18i + 16}$

$\qquad = \dfrac{-4 + 9i}{25 + 18i} \cdot \dfrac{25 - 18i}{25 - 18i}$

$\qquad = \dfrac{-100 + 72i + 225i - 162i^2}{625 + 324}$

$\qquad = \dfrac{-100 + 297i + 162}{949}$

$\qquad = \dfrac{62 + 297i}{949} = \dfrac{62}{949} + \dfrac{297}{949}i$

59. $\sqrt{-6} \cdot \sqrt{-2} = (\sqrt{6}i)(\sqrt{2}i) = \sqrt{12}i^2 = (2\sqrt{3})(-1)$

$\qquad\qquad\qquad = -2\sqrt{3}$

61. $(\sqrt{-10})^2 = (\sqrt{10}i)^2 = 10i^2 = -10$

63. $(3 + \sqrt{-5})(7 - \sqrt{-10}) = (3 + \sqrt{5}i)(7 - \sqrt{10}i)$

$\qquad\qquad = 21 - 3\sqrt{10}i + 7\sqrt{5}i - \sqrt{50}i^2$

$\qquad\qquad = (21 + \sqrt{50}) + (7\sqrt{5} - 3\sqrt{10})i$

$\qquad\qquad = (21 + 5\sqrt{2}) + (7\sqrt{5} - 3\sqrt{10})i$

65. $x^2 - 2x + 2 = 0; \ a = 1, \ b = -2, \ c = 2$

$\qquad x = \dfrac{-(-2) \pm \sqrt{(-2)^2 - 4(1)(2)}}{2(1)}$

$\qquad = \dfrac{2 \pm \sqrt{-4}}{2}$

$\qquad = \dfrac{2 \pm 2i}{2} = 1 \pm i$

67. $4x^2 + 16x + 17 = 0; \ a = 4, \ b = 16, \ c = 17$

$\qquad x = \dfrac{-16 \pm \sqrt{(16)^2 - 4(4)(17)}}{2(4)}$

$\qquad = \dfrac{-16 \pm \sqrt{-16}}{8}$

$\qquad = \dfrac{-16 \pm 4i}{8} = -2 \pm \dfrac{1}{2}i$

69. $4x^2 + 16x + 15 = 0; \ a = 4, \ b = 16, \ c = 15$

$\qquad x = \dfrac{-16 \pm \sqrt{(16)^2 - 4(4)(15)}}{2(4)}$

$\qquad = \dfrac{-16 \pm \sqrt{16}}{8} = \dfrac{-16 \pm 4}{8}$

$\qquad x = -\dfrac{12}{8} = -\dfrac{3}{2} \quad \text{or} \quad x = -\dfrac{20}{8} = -\dfrac{5}{2}$

71. $\dfrac{3}{2}x^2 - 6x + 9 = 0 \qquad$ Multiply both sides by 2.

$\qquad 3x^2 - 12x + 18 = 0$

$\qquad x = \dfrac{-(-12) \pm \sqrt{(-12)^2 - 4(3)(18)}}{2(3)}$

$\qquad = \dfrac{12 \pm \sqrt{-72}}{6} = \dfrac{12 \pm 6\sqrt{2}i}{6} = 2 \pm \sqrt{2}i$

73. $1.4x^2 - 2x - 10 = 0$ Multiply both sides by 5.

$7x^2 - 10x - 50 = 0$

$x = \dfrac{-(-10) \pm \sqrt{(-10)^2 - 4(7)(-50)}}{2(7)}$

$= \dfrac{10 \pm \sqrt{1500}}{14} = \dfrac{10 \pm 10\sqrt{15}}{14}$

$= \dfrac{5 \pm 5\sqrt{15}}{7} = \dfrac{5}{7} \pm \dfrac{5\sqrt{15}}{7}$

75. $-6i^3 + i^2 = -6i^2i + i^2$

$= -6(-1)i + (-1)$

$= 6i - 1$

$= -1 + 6i$

77. $-5i^5 = -5i^2i^2i$

$= -5(-1)(-1)i$

$= -5i$

79. $\left(\sqrt{-75}\right)^3 = \left(5\sqrt{3}i\right)^3$

$= 5^3\left(\sqrt{3}\right)^3 i^3$

$= 125\left(3\sqrt{3}\right)(-1)i$

$= -375\sqrt{3}i$

81. $\dfrac{1}{i^3} = \dfrac{1}{-i} = \dfrac{1}{-i} \cdot \dfrac{i}{i}$

$= \dfrac{i}{-i^2} = \dfrac{i}{1} = i$

83. (a) $z_1 = 9 + 16i, z_2 = 20 - 10i$

(b) $\dfrac{1}{z} = \dfrac{1}{z_1} + \dfrac{1}{z_2} = \dfrac{1}{9 + 16i} + \dfrac{1}{20 - 10i} = \dfrac{20 - 10i + 9 + 16i}{(9 + 16i)(20 - 10i)} = \dfrac{29 + 6i}{340 + 230i}$

$z = \left(\dfrac{340 + 230i}{29 + 6i}\right)\left(\dfrac{29 - 6i}{29 - 6i}\right) = \dfrac{11{,}240 + 4630i}{877} = \dfrac{11{,}240}{877} + \dfrac{4630}{877}i$

85. (a) $2^4 = 16$

(b) $(-2)^4 = 16$

(c) $(2i)^4 = 2^4i^4 = 16(1) = 16$

(d) $(-2i)^4 = (-2)^4i^4 = 16(1) = 16$

87. False, if $b = 0$ then $a + bi = a - bi = a$.

That is, if the complex number is real, the number equals its conjugate.

89. False

$i^{44} + i^{150} - i^{74} - i^{109} + i^{61} = (i^4)^{11} + (i^4)^{37}(i^2) - (i^4)^{18}(i^2) - (i^4)^{27}(i) + (i^4)^{15}(i)$

$= (1)^{11} + (1)^{37}(-1) - (1)^{18}(-1) - (1)^{27}(i) + (1)^{15}(i)$

$= 1 + (-1) + 1 - i + i = 1$

91. $(a_1 + b_1i)(a_2 + b_2i) = a_1a_2 + a_1b_2i + a_2b_1i + b_1b_2i^2$

$= (a_1a_2 - b_1b_2) + (a_1b_2 + a_2b_1)i$

The complex conjugate of this product is $(a_1a_2 - b_1b_2) - (a_1b_2 + a_2b_1)i$.

The product of the complex conjugates is:

$(a_1 - b_1i)(a_2 - b_2i) = a_1a_2 - a_1b_2i - a_2b_1i + b_1b_2i^2$

$= (a_1a_2 - b_1b_2) - (a_1b_2 + a_2b_1)i$

Thus, the complex conjugate of the product of two complex numbers is the product of their complex conjugates.

93. $(4 + 3x) + (8 - 6x - x^2) = -x^2 - 3x + 12$

95. $\left(3x - \frac{1}{2}\right)(x + 4) = 3x^2 + 12x - \frac{1}{2}x - 2$

$= 3x^2 + \frac{23}{2}x - 2$

97. $-x - 12 = 19$

$\qquad -x = 31$

$\qquad x = -31$

99. $4(5x - 6) - 3(6x + 1) = 0$

$\qquad 20x - 24 - 18x - 3 = 0$

$\qquad 2x - 27 = 0$

$\qquad 2x = 27$

$\qquad x = \frac{27}{2}$

101. $\qquad V = \frac{4}{3}\pi a^2 b$

$\qquad 3V = 4\pi a^2 b$

$\qquad \dfrac{3V}{4\pi b} = a^2$

$\qquad \sqrt{\dfrac{3V}{4\pi b}} = a$

$\qquad a = \dfrac{1}{2}\sqrt{\dfrac{3V}{\pi b}} = \dfrac{\sqrt{3V\pi b}}{2\pi b}$

103. Let $x = $ # liters withdrawn and replaced.

$\qquad 0.50(5 - x) + 1.00x = 0.60(5)$

$\qquad 2.50 - 0.50x + 1.00x = 3.00$

$\qquad 0.50x = 0.50$

$\qquad x = 1$ liter

Section 2.5   Zeros of Polynomial Functions

- You should know that if f is a polynomial of degree $n > 0$, then f has at least one zero in the complex number system.
- You should know the Linear Factorization Theorem.
- You should know the Rational Zero Test.
- You should know shortcuts for the Rational Zero Test. Possible rational zeros $= \dfrac{\text{factors of constant term}}{\text{factors of leading coefficient}}$

  (a) Use a graphing or programmable calculator.

  (b) Sketch a graph.

  (c) After finding a root, use synthetic division to reduce the degree of the polynomial.

- You should know that if $a + bi$ is a complex zero of a polynomial f, with real coefficients, then $a - bi$ is also a complex zero of f.
- You should know the difference between a factor that is irreducible over the rationals (such as $x^2 - 7$) and a factor that is irreducible over the reals (such as $x^2 + 9$).
- You should know Descartes's Rule of Signs. (For a polynomial with real coefficients and a non-zero constant term.)

  (a) The number of positive real zeros of f is either equal to the number of variations of sign of f or is less than that number by an even integer.

  (b) The number of negative real zeros of f is either equal to the number of variations in sign of $f(-x)$ or is less than that number by an even integer.

  (c) When there is only one variation in sign, there is exactly one positive (or negative) real zero.

- You should be able to observe the last row obtained from synthetic division in order to determine upper or lower bounds.

  (a) If the test value is positive and all of the entries in the last row are positive or zero, then the test value is an upper bound.

  (b) If the test value is negative and the entries in the last row alternate from positive to negative, then the test value is a lower bound. (Zero entries count as positive or negative.)

Vocabulary Check

1. Fundamental Theorem of Algebra
2. Linear Factorization Theorem
3. Rational Zero
4. conjugate
5. irreducible; reals
6. Descarte's Rule of Signs
7. lower; upper

---

1. $f(x) = x(x - 6)^2$

The zeros are: $x = 0, x = 6$

3. $g(x) = (x - 2)(x + 4)^3$

The zeros are: $x = 2, x = -4$

5. $f(x) = (x + 6)(x + i)(x - i)$

The three zeros are: $x = -6, x = -i, x = i$

7. $f(x) = x^3 + 3x^2 - x - 3$

Possible rational zeros: $\pm 1, \pm 3$

Zeros shown on graph: $-3, -1, 1$

9. $f(x) = 2x^4 - 17x^3 + 35x^2 + 9x - 45$

Possible rational zeros: $\pm 1, \pm 3, \pm 5, \pm 9, \pm 15, \pm 45,$
$\pm \frac{1}{2}, \pm \frac{3}{2}, \pm \frac{5}{2}, \pm \frac{9}{2}, \pm \frac{15}{2}, \pm \frac{45}{2}$

Zeros shown on graph: $-1, \frac{3}{2}, 3, 5$

11. $f(x) = x^3 - 6x^2 + 11x - 6$

Possible rational zeros: $\pm 1, \pm 2, \pm 3, \pm 6$

$$
\begin{array}{r|rrrr}
1 & 1 & -6 & 11 & -6 \\
  &   & 1 & -5 & 6 \\
\hline
  & 1 & -5 & 6 & 0
\end{array}
$$

$x^3 - 6x^2 + 11x - 6 = (x - 1)(x^2 - 5x + 6)$
$$= (x - 1)(x - 2)(x - 3)$$

Thus, the rational zeros are 1, 2, and 3.

13. $g(x) = x^3 - 4x^2 - x + 4 = x^2(x - 4) - 1(x - 4)$

$\qquad = (x - 4)(x^2 - 1)$

$\qquad = (x - 4)(x - 1)(x + 1)$

Thus, the rational zeros of $g(x)$ are 4 and ± 1.

15. $h(t) = t^3 + 12t^2 + 21t + 10$

Possible rational zeros: $\pm 1, \pm 2, \pm 5, \pm 10$

$$
\begin{array}{r|rrrr}
-1 & 1 & 12 & 21 & 10 \\
   &   & -1 & -11 & -10 \\
\hline
   & 1 & 11 & 10 & 0
\end{array}
$$

$t^3 + 12t^2 + 21t + 10 = (t + 1)(t^2 + 11t + 10)$
$$= (t + 1)(t + 1)(t + 10)$$
$$= (t + 1)^2(t + 10)$$

Thus, the rational zeros are -1 and -10.

17. $C(x) = 2x^3 + 3x^2 - 1$

Possible rational zeros: $\pm 1, \pm \frac{1}{2}$

$$
\begin{array}{r|rrrr}
-1 & 2 & 3 & 0 & -1 \\
   &   & -2 & -1 & 1 \\
\hline
   & 2 & 1 & -1 & 0
\end{array}
$$

$2x^3 + 3x^2 - 1 = (x + 1)(2x^2 + x - 1)$
$$= (x + 1)(x + 1)(2x - 1)$$
$$= (x + 1)^2(2x - 1)$$

Thus, the rational zeros are -1 and $\frac{1}{2}$.

19. $f(x) = 9x^4 - 9x^3 - 58x^2 + 4x + 24$

Possible rational zeros: $\pm 1, \pm 2, \pm 3, \pm 4, \pm 6, \pm 8, \pm 12, \pm 24,$
$\pm \frac{1}{3}, \pm \frac{2}{3}, \pm \frac{4}{3}, \pm \frac{8}{3}, \pm \frac{1}{9}, \pm \frac{2}{9}, \pm \frac{4}{9}, \pm \frac{8}{9}$

$$
\begin{array}{r|rrrrr}
-2 & 9 & -9 & -58 & 4 & 24 \\
   &   & -18 & 54 & 8 & -24 \\
\hline
   & 9 & -27 & -4 & 12 & 0
\end{array}
$$

$$
\begin{array}{r|rrrr}
3 & 9 & -27 & -4 & 12 \\
  &   & 27 & 0 & -12 \\
\hline
  & 9 & 0 & -4 & 0
\end{array}
$$

$9x^4 - 9x^3 - 58x^2 + 4x - 24$

$\qquad = (x + 2)(x - 3)(9x^2 - 4)$

$\qquad = (x + 2)(x - 3)(3x - 2)(3x + 2)$

Thus, the rational zeros are $-2, 3$, and $\pm \frac{2}{3}$.

21. $z^4 - z^3 - 2z - 4 = 0$

Possible rational zeros: $\pm 1, \pm 2, \pm 4$

$$
\begin{array}{r|rrrrr}
-1 & 1 & -1 & 0 & -2 & -4 \\
   &   & -1 & 2 & -2 & 4 \\
\hline
   & 1 & -2 & 2 & -4 & 0
\end{array}
$$

$$
\begin{array}{r|rrrr}
2 & 1 & -2 & 2 & -4 \\
  &   & 2 & 0 & 4 \\
\hline
  & 1 & 0 & 2 & 0
\end{array}
$$

$z^4 - z^3 - 2z - 4 = (z + 1)(z - 2)(z^2 + 2)$

The only real zeros are -1 and 2.

23. $2y^4 + 7y^3 - 26y^2 + 23y - 6 = 0$

Possible rational zeros: $\pm 1, \pm 2, \pm 3, \pm 6, \pm \frac{1}{2}, \pm \frac{3}{2}$

$$
\begin{array}{r|rrrrr}
1 & 2 & 7 & -26 & 23 & -6 \\
  &   & 2 & 9 & -17 & 6 \\
\hline
  & 2 & 9 & -17 & 6 & 0
\end{array}
$$

$$
\begin{array}{r|rrrr}
-6 & 2 & 9 & -17 & 6 \\
   &   & -12 & 18 & -6 \\
\hline
   & 2 & -3 & 1 & 0
\end{array}
$$

$2y^4 + 7y^3 - 26y^2 + 23y - 6 = (y - 1)(y + 6)(2y^2 - 3y + 1)$

$\qquad\qquad = (y - 1)(y + 6)(2y - 1)(y - 1) = (y - 1)^2(y + 6)(2y - 1)$

The only real zeros are $1, -6,$ and $\frac{1}{2}$.

25. $f(x) = x^3 + x^2 - 4x - 4$

(a) Possible rational zeros: $\pm 1, \pm 2, \pm 4$

(b)

(c) The zeros are: $-2, -1, 2$

27. $f(x) = -4x^3 + 15x^2 - 8x - 3$

(a) Possible rational zeros: $\pm 1, \pm 3, \pm \frac{1}{2}, \pm \frac{3}{2}, \pm \frac{1}{4}, \pm \frac{3}{4}$

(b)

(c) The zeros are: $-\frac{1}{4}, 1, 3$

29. $f(x) = -2x^4 + 13x^3 - 21x^2 + 2x + 8$

(a) Possible rational zeros: $\pm 1, \pm 2, \pm 4, \pm 8, \pm \frac{1}{2}$

(b)

(c) The zeros are: $-\frac{1}{2}, 1, 2, 4$

31. $f(x) = 32x^3 - 52x^2 + 17x + 3$

(a) Possible rational zeros: $\pm 1, \pm 3, \pm \frac{1}{2}, \pm \frac{3}{2}, \pm \frac{1}{4}, \pm \frac{3}{4},$
$\qquad \pm \frac{1}{6}, \pm \frac{3}{8}, \pm \frac{1}{16}, \pm \frac{3}{16}, \pm \frac{1}{32}, \pm \frac{3}{32}$

(b)

(c) The zeros are: $-\frac{1}{8}, \frac{3}{4}, 1$

33. $f(x) = x^4 - 3x^2 + 2$

(a) From the calculator we have $x = \pm 1$ and $x \approx \pm 1.414$.

(b) An exact zero is $x = 1$.

$$
\begin{array}{r|rrrrr}
1 & 1 & 0 & -3 & 0 & 2 \\
  &   & 1 & 1 & -2 & -2 \\
\hline
  & 1 & 1 & -2 & -2 & 0
\end{array}
$$

(c)
$$
\begin{array}{r|rrrr}
-1 & 1 & 1 & -2 & -2 \\
   &   & -1 & 0 & 2 \\
\hline
   & 1 & 0 & -2 & 0
\end{array}
$$

$f(x) = (x - 1)(x + 1)(x^2 - 2)$

$\quad = (x - 1)(x + 1)\left(x - \sqrt{2}\right)\left(x + \sqrt{2}\right)$

35. $h(x) = x^5 - 7x^4 + 10x^3 + 14x^2 - 24x$

(a) $h(x) = x(x^4 - 7x^3 + 10x^2 + 14x - 24)$

From the calculator we have $x = 0, 3, 4$ and $x \approx \pm 1.414$.

(b) An exact zero is $x = 3$.

$$
\begin{array}{r|rrrrr}
3 & 1 & -7 & 10 & 14 & -24 \\
  &   & 3 & -12 & -6 & 24 \\
\hline
  & 1 & -4 & -2 & 8 & 0
\end{array}
$$

(c)
$$
\begin{array}{r|rrrr}
4 & 1 & -4 & -2 & 8 \\
  &   & 4 & 0 & -8 \\
\hline
  & 1 & 0 & -2 & 0
\end{array}
$$

$h(x) = x(x - 3)(x - 4)(x^2 - 2)$

$\quad = x(x - 3)(x - 4)\left(x - \sqrt{2}\right)\left(x + \sqrt{2}\right)$

37. $f(x) = (x - 1)(x - 5i)(x + 5i)$

$\quad = (x - 1)(x^2 + 25)$

$\quad = x^3 - x^2 + 25x - 25$

Note: $f(x) = a(x^3 - x^2 + 25x - 25)$, where a is any nonzero real number, has the zeros 1 and $\pm 5i$.

39. $f(x) = (x - 6)[x - (-5 + 2i)][x - (-5 - 2i)]$

$\quad = (x - 6)[(x + 5) - 2i][(x + 5) + 2i]$

$\quad = (x - 6)[(x + 5)^2 - (2i)^2]$

$\quad = (x - 6)(x^2 + 10x + 25 + 4)$

$\quad = (x - 6)(x^2 + 10x + 29)$

$\quad = x^3 + 4x^2 - 31x - 174$

Note: $f(x) = a(x^3 + 4x^2 - 31x - 174)$, where a is any nonzero real number, has the zeros 6, and $-5 \pm 2i$.

41. If $3 + \sqrt{2}i$ is a zero, so is its conjugate, $3 - \sqrt{2}i$.

$f(x) = (3x - 2)(x + 1)\left[x - \left(3 + \sqrt{2}i\right)\right]\left[x - \left(3 - \sqrt{2}i\right)\right]$

$\quad = (3x - 2)(x + 1)\left[(x - 3) - \sqrt{2}i\right]\left[(x - 3) + \sqrt{2}i\right]$

$\quad = (3x^2 + x - 2)\left[(x - 3)^2 - \left(\sqrt{2}i\right)^2\right]$

$\quad = (3x^2 + x - 2)(x^2 - 6x + 9 + 2)$

$\quad = (3x^2 + x - 2)(x^2 - 6x + 11)$

$\quad = 3x^4 - 17x^3 + 25x^2 + 23x - 22$

Note: $f(x) = a(3x^4 - 17x^3 + 25x^2 + 23x - 22)$, where a is any nonzero real number, has the zeros $\frac{2}{3}$, -1, and $3 \pm \sqrt{2}i$.

43. $f(x) = x^4 + 6x^2 - 27$

(a) $f(x) = (x^2 + 9)(x^2 - 3)$

(b) $f(x) = (x^2 + 9)\left(x + \sqrt{3}\right)\left(x - \sqrt{3}\right)$

(c) $f(x) = (x + 3i)(x - 3i)\left(x + \sqrt{3}\right)\left(x - \sqrt{3}\right)$

45. $f(x) = x^4 - 4x^3 + 5x^2 - 2x - 6$

$$
\begin{array}{r}
x^2 - 2x + 3 \\
x^2 - 2x - 2 \overline{\smash{\big)}\ x^4 - 4x^3 + 5x^2 - 2x - 6} \\
\underline{x^4 - 2x^3 - 2x^2} \\
-2x^3 + 7x^2 - 2x \\
\underline{-2x^3 + 4x^2 + 4x} \\
3x^2 - 6x - 6 \\
\underline{3x^2 - 6x - 6} \\
0
\end{array}
$$

$f(x) = (x^2 - 2x - 2)(x^2 - 2x + 3)$

(a) $f(x) = (x^2 - 2x - 2)(x^2 - 2x + 3)$

(b) $f(x) = \left(x - 1 + \sqrt{3}\right)\left(x - 1 - \sqrt{3}\right)(x^2 - 2x + 3)$

(c) $f(x) = \left(x - 1 + \sqrt{3}\right)\left(x - 1 - \sqrt{3}\right)\left(x - 1 + \sqrt{2}\,i\right)\left(x - 1 - \sqrt{2}\,i\right)$

Note: Use the Quadratic Formula for (b) and (c).

47. $f(x) = 2x^3 + 3x^2 + 50x + 75$

Since $5i$ is a zero, so is $-5i$.

$$
\begin{array}{r|rrrr}
5i & 2 & 3 & 50 & 75 \\
   &   & 10i & -50 + 15i & -75 \\
\hline
   & 2 & 3 + 10i & 15i & 0
\end{array}
$$

$$
\begin{array}{r|rrr}
-5i & 2 & 3 + 10i & 15i \\
    &   & -10i & -15i \\
\hline
    & 2 & 3 & 0
\end{array}
$$

The zero of $2x + 3$ is $x = -\frac{3}{2}$. The zeros of $f(x)$ are $x = -\frac{3}{2}$ and $x = \pm 5i$.

Alternate Solution

Since $x = \pm 5i$ are zeros of $f(x)$, $(x + 5i)(x - 5i) = x^2 + 25$ is a factor of $f(x)$. By long division we have:

$$
\begin{array}{r}
2x + 3 \\
x^2 + 0x + 25 \overline{\smash{)}2x^3 + 3x^2 + 50x + 75} \\
\underline{2x^3 + 0x^2 + 50x } \\
3x^2 + 0x + 75 \\
\underline{3x^2 + 0x + 75} \\
0
\end{array}
$$

Thus, $f(x) = (x^2 + 25)(2x + 3)$ and the zeros of f are $x = \pm 5i$ and $x = -\frac{3}{2}$.

49. $f(x) = 2x^4 - x^3 + 7x^2 - 4x - 4$

Since $2i$ is a zero, so is $-2i$.

$$
\begin{array}{r|rrrrr}
2i & 2 & -1 & 7 & -4 & -4 \\
   &   & 4i & -8 - 2i & 4 - 2i & 4 \\
\hline
   & 2 & -1 + 4i & -1 - 2i & -2i & 0
\end{array}
$$

$$
\begin{array}{r|rrrr}
-2i & 2 & -1 + 4i & -1 - 2i & -2i \\
    &   & -4i & 2i & 2i \\
\hline
    & 2 & -1 & -1 & 0
\end{array}
$$

The zeros of $2x^2 - x - 1 = (2x + 1)(x - 1)$ are $x = -\frac{1}{2}$ and $x = 1$. The zeros of $f(x)$ are $x = \pm 2i$, $x = -\frac{1}{2}$, and $x = 1$.

Alternate Solution

Since $x = \pm 2i$ are zeros of $f(x)$, $(x + 2i)(x - 2i) = x^2 + 4$ is a factor of $f(x)$. By long division we have:

$$
\begin{array}{r}
2x^2 - x - 1 \\
x^2 + 0x + 4 \overline{\smash{)}2x^4 - x^3 + 7x^2 - 4x - 4} \\
\underline{2x^4 + 0x^3 + 8x^2 } \\
-x^3 - x^2 - 4x \\
\underline{-x^3 + 0x^2 - 4x } \\
-x^2 + 0x - 4 \\
\underline{-x^2 + 0x - 4} \\
0
\end{array}
$$

Thus, $f(x) = (x^2 + 4)(2x^2 - x - 1)$

$$= (x + 2i)(x - 2i)(2x + 1)(x - 1)$$

and the zeros of $f(x)$ are $x = \pm 2i$, $x = -\frac{1}{2}$, and $x = 1$.

51. $g(x) = 4x^3 + 23x^2 + 34x - 10$

Since $-3 + i$ is a zero, so is $-3 - i$.

$$
\begin{array}{r|rrrr}
-3 + i & 4 & 23 & 34 & -10 \\
       &   & -12 + 4i & -37 - i & 10 \\
\hline
       & 4 & 11 + 4i & -3 - i & 0
\end{array}
$$

$$
\begin{array}{r|rrr}
-3 - i & 4 & 11 + 4i & -3 - i \\
       &   & -12 - 4i & 3 + i \\
\hline
       & 4 & -1 & 0
\end{array}
$$

The zero of $4x - 1$ is $x = \frac{1}{4}$. The zeros of $g(x)$ are $x = -3 \pm i$ and $x = \frac{1}{4}$.

Alternate Solution

Since $-3 \pm i$ are zeros of $g(x)$,

$$[x - (-3 + i)][x - (-3 - i)] = [(x + 3) - i][(x + 3) + i]$$
$$= (x + 3)^2 - i^2$$
$$= x^2 + 6x + 10$$

is a factor of $g(x)$. By long division we have:

$$
\begin{array}{r}
4x - 1 \\
x^2 + 6x + 10 \overline{\smash{)}4x^3 + 23x^2 + 34x - 10} \\
\underline{4x^3 + 24x^2 + 40x } \\
-x^2 - 6x - 10 \\
\underline{-x^2 - 6x - 10} \\
0
\end{array}
$$

Thus, $g(x) = (x^2 + 6x + 10)(4x - 1)$ and the zeros of $g(x)$ are $x = -3 \pm i$ and $x = \frac{1}{4}$.

53. $f(x) = x^4 + 3x^3 - 5x^2 - 21x + 22$

Since $-3 + \sqrt{2}\,i$ is a zero, so is $-3 - \sqrt{2}\,i$, and

$$\left[x - \left(-3 + \sqrt{2}\,i\right)\right]\left[x - \left(-3 - \sqrt{2}\,i\right)\right] = \left[(x + 3) - \sqrt{2}\,i\right]\left[(x + 3) + \sqrt{2}\,i\right]$$

$$= (x + 3)^2 - \left(\sqrt{2}\,i\right)^2$$

$$= x^2 + 6x + 11$$

is a factor of $f(x)$. By long division, we have:

$$
\begin{array}{r}
x^2 - 3x + 2 \\
x^2 + 6x + 11 \overline{\smash{)}\, x^4 + 3x^3 - 5x^2 - 21x + 22} \\
\underline{x^4 + 6x^3 + 11x^2} \\
-3x^3 - 16x^2 - 21x \\
\underline{-3x^3 - 18x^2 - 33x} \\
2x^2 + 12x + 22 \\
\underline{2x^2 + 12x + 22} \\
0
\end{array}
$$

Thus,

$$f(x) = (x^2 + 6x + 11)(x^2 - 3x + 2)$$

$$= (x^2 + 6x + 11)(x - 1)(x - 2)$$

and the zeros of f are $x = -3 \pm \sqrt{2}\,i$, $x = 1$, and $x = 2$.

55. $f(x) = x^2 + 25$

$$= (x + 5i)(x - 5i)$$

The zeros of $f(x)$ are $x = \pm 5i$.

57. $h(x) = x^2 - 4x + 1$

By the Quadratic Formula, the zeros of $h(x)$ are

$$x = \frac{4 \pm \sqrt{16 - 4}}{2} = 2 \pm \sqrt{3}.$$

$$h(x) = \left[x - \left(2 + \sqrt{3}\right)\right]\left[x - \left(2 - \sqrt{3}\right)\right]$$

$$= \left(x - 2 - \sqrt{3}\right)\left(x - 2 + \sqrt{3}\right)$$

59. $f(x) = x^4 - 81$

$$= (x^2 - 9)(x^2 + 9)$$

$$= (x + 3)(x - 3)(x + 3i)(x - 3i)$$

The zeros of $f(x)$ are $x = \pm 3$ and $x = \pm 3i$.

61. $f(z) = z^2 - 2z + 2$

By the Quadratic Formula, the zeros of $f(z)$ are

$$z = \frac{2 \pm \sqrt{4 - 8}}{2} = 1 \pm i.$$

$$f(z) = [z - (1 + i)][z - (1 - i)] = (z - 1 - i)(z - 1 + i)$$

63. $g(x) = x^3 - 6x^2 + 13x - 10$

Possible rational zeros: ± 1, ± 2, ± 5, ± 10

$$
\begin{array}{r|rrrr}
2 & 1 & -6 & 13 & -10 \\
  &   & 2 & -8 & 10 \\
\hline
  & 1 & -4 & 5 & 0
\end{array}
$$

By the Quadratic Formula, the zeros of $x^2 - 4x + 5$ are

$$x = \frac{4 \pm \sqrt{16 - 20}}{2} = 2 \pm i.$$

The zeros of $g(x)$ are $x = 2$ and $x = 2 \pm i$.

$$g(x) = (x - 2)[x - (2 + i)][x - (2 - i)]$$

$$= (x - 2)(x - 2 - i)(x - 2 + i)$$

65. $h(x) = x^3 - x + 6$

Possible rational zeros: ± 1, ± 2, ± 3, ± 6

$$
\begin{array}{r|rrrr}
-2 & 1 & 0 & -1 & 6 \\
   &   & -2 & 4 & -6 \\
\hline
   & 1 & -2 & 3 & 0
\end{array}
$$

By the Quadratic Formula, the zeros of $x^2 - 2x + 3$ are

$$x = \frac{2 \pm \sqrt{4 - 12}}{2} = 1 \pm \sqrt{2}\,i.$$

The zeros of $h(x)$ are $x = -2$ and $x = 1 \pm \sqrt{2}\,i$.

$$h(x) = [x - (-2)]\left[x - \left(1 + \sqrt{2}\,i\right)\right]\left[x - \left(1 - \sqrt{2}\,i\right)\right]$$

$$= (x + 2)\left(x - 1 - \sqrt{2}\,i\right)\left(x - 1 + \sqrt{2}\,i\right)$$

67. $f(x) = 5x^3 - 9x^2 + 28x + 6$

Possible rational zeros:
$\pm 1, \pm 2, \pm 3, \pm 6, \pm \frac{1}{5}, \pm \frac{2}{5}, \pm \frac{3}{5}, \pm \frac{6}{5}$

$$
\begin{array}{r|rrrr}
-\frac{1}{5} & 5 & -9 & 28 & 6 \\
 & & -1 & 2 & -6 \\
\hline
 & 5 & -10 & 30 & 0
\end{array}
$$

By the Quadratic Formula, the zeros of
$5x^2 - 10x + 30 = 5(x^2 - 2x + 6)$ are

$$x = \frac{2 \pm \sqrt{4 - 24}}{2} = 1 \pm \sqrt{5}\,i.$$

The zeros of $f(x)$ are $x = -\frac{1}{5}$ and $x = 1 \pm \sqrt{5}\,i$.

$$f(x) = \left[x - \left(-\tfrac{1}{5}\right)\right](5)\left[x - \left(1 + \sqrt{5}\,i\right)\right]\left[x - \left(1 - \sqrt{5}\,i\right)\right]$$

$$= (5x + 1)\left(x - 1 - \sqrt{5}\,i\right)\left(x - 1 + \sqrt{5}\,i\right)$$

69. $g(x) = x^4 - 4x^3 + 8x^2 - 16x + 16$

Possible rational zeros: $\pm 1, \pm 2, \pm 4, \pm 8, \pm 16$

$$
\begin{array}{r|rrrrr}
2 & 1 & -4 & 8 & -16 & 16 \\
 & & 2 & -4 & 8 & -16 \\
\hline
 & 1 & -2 & 4 & -8 & 0
\end{array}
$$

$$
\begin{array}{r|rrrr}
2 & 1 & -2 & 4 & -8 \\
 & & 2 & 0 & 8 \\
\hline
 & 1 & 0 & 4 & 0
\end{array}
$$

$g(x) = (x - 2)(x - 2)(x^2 + 4) = (x - 2)^2(x + 2i)(x - 2i)$

The zeros of $g(x)$ are 2 and $\pm 2i$.

71. $f(x) = x^4 + 10x^2 + 9$

$$= (x^2 + 1)(x^2 + 9)$$

$$= (x + i)(x - i)(x + 3i)(x - 3i)$$

The zeros of $f(x)$ are $x = \pm i$ and $x = \pm 3i$.

73. $f(x) = x^3 + 24x^2 + 214x + 740$

Possible rational zeros: $\pm 1, \pm 2, \pm 4, \pm 5, \pm 10, \pm 20, \pm 37,$
$\pm 74, \pm 148, \pm 185, \pm 370, \pm 740$

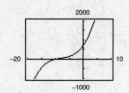

Based on the graph, try $x = -10$.

$$
\begin{array}{r|rrrr}
-10 & 1 & 24 & 214 & 740 \\
 & & -10 & -140 & -740 \\
\hline
 & 1 & 14 & 74 & 0
\end{array}
$$

By the Quadratic Formula, the zeros of $x^2 + 14x + 74$ are

$$x = \frac{-14 \pm \sqrt{196 - 296}}{2} = -7 \pm 5i.$$

The zeros of $f(x)$ are $x = -10$ and $x = -7 \pm 5i$.

75. $f(x) = 16x^3 - 20x^2 - 4x + 15$

Possible rational zeros:
$\pm 1, \pm 3, \pm 5, \pm 15, \pm \frac{1}{2}, \pm \frac{3}{2}, \pm \frac{5}{2}, \pm \frac{15}{2}, \pm \frac{1}{4}, \pm \frac{3}{4},$
$\pm \frac{5}{4}, \pm \frac{15}{4}, \pm \frac{1}{8}, \pm \frac{3}{8}, \pm \frac{5}{8}, \pm \frac{15}{8}, \pm \frac{1}{16}, \pm \frac{3}{16}, \pm \frac{5}{16}, \pm \frac{15}{16}$

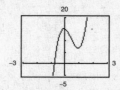

Based on the graph, try $x = -\frac{3}{4}$.

$$
\begin{array}{r|rrrr}
-\frac{3}{4} & 16 & -20 & -4 & 15 \\
 & & -12 & 24 & -15 \\
\hline
 & 16 & -32 & 20 & 0
\end{array}
$$

By the Quadratic Formula, the zeros of

$16x^2 - 32x + 20 = 4(4x^2 - 8x + 5)$ are

$$x = \frac{8 \pm \sqrt{64 - 80}}{8} = 1 \pm \frac{1}{2}i.$$

The zeros of $f(x)$ are $x = -\frac{3}{4}$ and $x = 1 \pm \frac{1}{2}i$.

77. $f(x) = 2x^4 + 5x^3 + 4x^2 + 5x + 2$

Possible rational zeros: $\pm 1, \pm 2, \pm\frac{1}{2}$

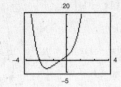

Based on the graph, try $x = -2$ and $x = -\frac{1}{2}$.

$$\begin{array}{r|rrrrr} -2 & 2 & 5 & 4 & 5 & 2 \\ & & -4 & -2 & -4 & -2 \\ \hline & 2 & 1 & 2 & 1 & 0 \end{array}$$

$$\begin{array}{r|rrrr} -\frac{1}{2} & 2 & 1 & 2 & 1 \\ & & -1 & 0 & -1 \\ \hline & 2 & 0 & 2 & 0 \end{array}$$

The zeros of $2x^2 + 2 = 2(x^2 + 1)$ are $x = \pm i$.

The zeros of $f(x)$ are $x = -2$, $x = -\frac{1}{2}$, and $x = \pm i$.

79. $g(x) = 5x^5 + 10x = 5x(x^4 + 2)$

Let $f(x) = x^4 + 2$.

Sign variations: 0, positive zeros: 0

$f(-x) = x^4 + 2$

Sign variations: 0, negative zeros: 0

81. $h(x) = 3x^4 + 2x^2 + 1$

Sign variations: 0, positive zeros: 0

$h(-x) = 3x^4 + 2x^2 + 1$

Sign variations: 0, negative zeros: 0

83. $g(x) = 2x^3 - 3x^2 - 3$

Sign variations: 1, positive zeros: 1

$g(-x) = -2x^3 - 3x^2 - 3$

Sign variations: 0, negative zeros: 0

85. $f(x) = -5x^3 + x^2 - x + 5$

Sign variations: 3, positive zeros: 3 or 1

$f(-x) = 5x^3 + x^2 + x + 5$

Sign variations: 0, negative zeros: 0

87. $f(x) = x^4 - 4x^3 + 15$

(a) $$\begin{array}{r|rrrrr} 4 & 1 & -4 & 0 & 0 & 15 \\ & & 4 & 0 & 0 & 0 \\ \hline & 1 & 0 & 0 & 0 & 15 \end{array}$$

4 is an upper bound.

(b) $$\begin{array}{r|rrrrr} -1 & 1 & -4 & 0 & 0 & 15 \\ & & -1 & 5 & -5 & 5 \\ \hline & 1 & -5 & 5 & -5 & 20 \end{array}$$

-1 is a lower bound.

89. $f(x) = x^4 - 4x^3 + 16x - 16$

(a) $$\begin{array}{r|rrrrr} 5 & 1 & -4 & 0 & 16 & -16 \\ & & 5 & 5 & 25 & 205 \\ \hline & 1 & 1 & 5 & 41 & 189 \end{array}$$

5 is an upper bound.

(b) $$\begin{array}{r|rrrrr} -3 & 1 & -4 & 0 & 16 & -16 \\ & & -3 & 21 & -63 & 141 \\ \hline & 1 & -7 & 21 & -47 & 125 \end{array}$$

-3 is a lower bound.

91. $f(x) = 4x^3 - 3x - 1$

Possible rational zeros: $\pm 1, \pm\frac{1}{2}, \pm\frac{1}{4}$

$$\begin{array}{r|rrrr} 1 & 4 & 0 & -3 & -1 \\ & & 4 & 4 & 1 \\ \hline & 4 & 4 & 1 & 0 \end{array}$$

$4x^3 - 3x - 1 = (x - 1)(4x^2 + 4x + 1)$

$\qquad\qquad\quad = (x - 1)(2x + 1)^2$

Thus, the zeros are 1 and $-\frac{1}{2}$.

93. $f(y) = 4y^3 + 3y^2 + 8y + 6$

Possible rational zeros: $\pm 1, \pm 2, \pm 3, \pm 6, \pm\frac{1}{2}, \pm\frac{3}{2}, \pm\frac{1}{4}, \pm\frac{3}{4}$

$$\begin{array}{r|rrrr} -\frac{3}{4} & 4 & 3 & 8 & 6 \\ & & -3 & 0 & -6 \\ \hline & 4 & 0 & 8 & 0 \end{array}$$

$4y^3 + 3y^2 + 8y + 6 = \left(y + \frac{3}{4}\right)(4y^2 + 8)$

$\qquad\qquad\qquad\quad = \left(y + \frac{3}{4}\right)4(y^2 + 2)$

$\qquad\qquad\qquad\quad = (4y + 3)(y^2 + 2)$

Thus, the only real zero is $-\frac{3}{4}$.

95. $P(x) = x^4 - \frac{25}{4}x^2 + 9$

$\qquad = \frac{1}{4}(4x^4 - 25x^2 + 36)$

$\qquad = \frac{1}{4}(4x^2 - 9)(x^2 - 4)$

$\qquad = \frac{1}{4}(2x + 3)(2x - 3)(x + 2)(x - 2)$

The rational zeros are $\pm\frac{3}{2}$ and ±2.

97. $f(x) = x^3 - \frac{1}{4}x^2 - x + \frac{1}{4}$

$\qquad = \frac{1}{4}(4x^3 - x^2 - 4x + 1)$

$\qquad = \frac{1}{4}[x^2(4x - 1) - 1(4x - 1)]$

$\qquad = \frac{1}{4}(4x - 1)(x^2 - 1)$

$\qquad = \frac{1}{4}(4x - 1)(x + 1)(x - 1)$

The rational zeros are $\frac{1}{4}$ and ±1.

99. $f(x) = x^3 - 1 = (x - 1)(x^2 + x + 1)$

Rational zeros: 1 $(x = 1)$

Irrational zeros: 0

Matches (d).

101. $f(x) = x^3 - x = x(x + 1)(x - 1)$

Rational zeros: 3 $(x = 0, \pm1)$

Irrational zeros: 0

Matches (b).

103. (a)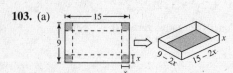

(c)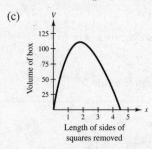

The volume is maximum when $x \approx 1.82$.

The dimensions are: length $\approx 15 - 2(1.82) = 11.36$

$\qquad\qquad\qquad\quad$ width $\approx 9 - 2(1.82) = 5.36$

$\qquad\qquad\qquad\quad$ height $= x \approx 1.82$

$\qquad$ 1.82 cm $\times$ 5.36 cm $\times$ 11.36 cm

(b) $V = l \cdot w \cdot h = (15 - 2x)(9 - 2x)x$

$\qquad = x(9 - 2x)(15 - 2x)$

Since length, width, and height must be positive, we have $0 < x < \frac{9}{2}$ for the domain.

(d) $56 = x(9 - 2x)(15 - 2x)$

$\quad 56 = 135x - 48x^2 + 4x^3$

$\quad\; 0 = 4x^3 - 48x^2 + 135x - 56$

The zeros of this polynomial are $\frac{1}{2}$, $\frac{7}{2}$, and 8. x cannot equal 8 since it is not in the domain of V. [The length cannot equal -1 and the width cannot equal -7. The product of $(8)(-1)(-7) = 56$ so it showed up as an extraneous solution.]

Thus, the volume is 56 cubic centimeters when $x = \frac{1}{2}$ centimeter or $x = \frac{7}{2}$ centimeters.

105. $\qquad\qquad P = -76x^3 + 4830x^2 - 320{,}000, \; 0 \le x \le 60$

$\qquad\qquad 2{,}500{,}000 = -76x^3 + 4830x^2 - 320{,}000$

$76x^3 - 4830x^2 + 2{,}820{,}000 = 0$

The zeros of this equation are $x \approx 46.1$, $x \approx 38.4$, and $x \approx -21.0$. Since $0 \le x \le 60$, we disregard $x \approx -21.0$. The smaller remaining solution is $x \approx 38.4$. The advertising expense is \$384,000.

107. (a) Current bin: $V = 2 \times 3 \times 4 = 24$ cubic feet

$\qquad$ New bin: $V = 5(24) = 120$ cubic feet

$\qquad\qquad V = (2 + x)(3 + x)(4 + x) = 120$

(b) $x^3 + 9x^2 + 26x + 24 = 120$

$\quad x^3 + 9x^2 + 26x - 96 = 0$

The only real zero of this polynomial is $x = 2$. All the dimensions should be increased by 2 feet, so the new bin will have dimensions of 4 feet by 5 feet by 6 feet.

109. $C = 100\left(\dfrac{200}{x^2} + \dfrac{x}{x+30}\right), x \geq 1$

C is minimum when $3x^3 - 40x^2 - 2400x - 36000 = 0$.

The only real zero is $x \approx 40$ or 4000 units.

111.
$$P = R - C = xp - C$$
$$= x(140 - 0.0001x) - (80x + 150{,}000)$$
$$= -0.0001x^2 + 60x - 150{,}000$$
$$9{,}000{,}000 = -0.0001x^2 + 60x - 150{,}000$$

Thus, $0 = 0.0001x^2 - 60x + 9{,}150{,}000$.

$$x = \frac{60 \pm \sqrt{-60}}{0.0002} = 300{,}000 \pm 10{,}000\sqrt{15}\,i$$

Since the solutions are both complex, it is not possible to determine a price p that would yield a profit of 9 million dollars.

113. False. The most nonreal complex zeros it can have is two and the Linear Factorization Theorem guarantees that there are three linear factors, so one zero must be real.

115. $g(x) = -f(x)$. This function would have the same zeros as $f(x)$ so $r_1, r_2,$ and r_3 are also zeros of $g(x)$.

117. $g(x) = f(x - 5)$. The graph of $g(x)$ is a horizontal shift of the graph of $f(x)$ five units to the right so the zeros of $g(x)$ are $5 + r_1, 5 + r_2,$ and $5 + r_3$.

119. $g(x) = 3 + f(x)$. Since $g(x)$ is a vertical shift of the graph of $f(x)$, the zeros of $g(x)$ cannot be determined.

121. $f(x) = x^4 - 4x^2 + k$

$$x^2 = \frac{-(-4) \pm \sqrt{(-4)^2 - 4(1)(k)}}{2(1)} = \frac{4 \pm 2\sqrt{4 - k}}{2} = 2 \pm \sqrt{4 - k}$$
$$x = \pm\sqrt{2 \pm \sqrt{4 - k}}$$

(a) For there to be four distinct real roots, both $4 - k$ and $2 \pm \sqrt{4 - k}$ must be positive. This occurs when $0 < k < 4$. Thus, some possible k-values are $k = 1, k = 2, k = 3, k = \frac{1}{2}, k = \sqrt{2}$, etc.

(b) For there to be two real roots, each of multiplicity 2, $4 - k$ must equal zero. Thus, $k = 4$.

(c) For there to be two real zeros and two complex zeros, $2 + \sqrt{4 - k}$ must be positive and $2 - \sqrt{4 - k}$ must be negative. This occurs when $k < 0$. Thus, some possible k-values are $k = -1, k = -2, k = -\frac{1}{2}$, etc.

(d) For there to be four complex zeros, $2 \pm \sqrt{4 - k}$ must be nonreal. This occurs when $k > 4$. Some possible k-values are $k = 5, k = 6, k = 7.4$, etc.

123. Zeros: $-2, \frac{1}{2}, 3$

$$f(x) = -(x + 2)(2x - 1)(x - 3)$$
$$= -2x^3 + 3x^2 + 11x - 6$$

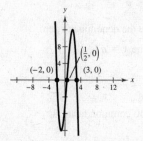

Any nonzero scalar multiple of f would have the same three zeros. Let $g(x) = af(x)$, $a > 0$.

There are infinitely many possible functions for f.

125. Answers will vary. Some of the factoring techniques are:

1. Factor out the greatest common factor.

2. Use special product formulas.

   $a^2 - b^2 = (a + b)(a - b)$

   $a^2 + 2ab + b^2 = (a + b)^2$

   $a^2 - 2ab + b^2 = (a - b)^2$

   $a^3 + b^3 = (a + b)(a^2 - ab + b^2)$

   $a^3 - b^3 = (a - b)(a^2 + ab + b^2)$

3. Factor by grouping, if possible.

4. Factor general trinomials with binomial factors by "guess-and-test" or by the grouping method.

5. Use the Rational Zero Test together with synthetic division to factor a polynomial.

6. Use Descartes's Rule of Signs to determine the number of real zeros. Then find any zeros and use them to factor the polynomial.

7. Find any upper and lower bounds for the real zeros to eliminate some of the possible rational zeros. Then test the remaining candidates by synthetic division and use any zeros to factor the polynomial.

127. (a) $f(x) = \left(x - \sqrt{b}\,i\right)\left(x + \sqrt{b}\,i\right) = x^2 + b$

(b) $f(x) = [x - (a + bi)][x - (a - bi)]$

$= [(x - a) - bi][(x - a) + bi]$

$= (x - a)^2 - (bi)^2$

$= x^2 - 2ax + a^2 + b^2$

129. $(-3 + 6i) - (8 - 3i) = -3 + 6i - 8 + 3i = -11 + 9i$

131. $(6 - 2i)(1 + 7i) = 6 + 42i - 2i - 14i^2 = 20 + 40i$

133. $g(x) = f(x - 2)$

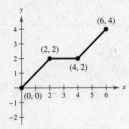

Horizontal shift two units to the right

135. $g(x) = 2f(x)$

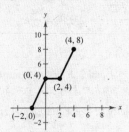

Vertical stretch (each y-value is multiplied by 2)

137. $g(x) = f(2x)$

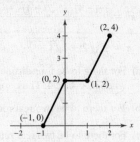

Horizontal shrink $\left(\text{each } x\text{-value is multiplied by } \frac{1}{2}\right)$

Section 2.6   Rational Functions

■ You should know the following basic facts about rational functions.

(a) A function of the form $f(x) = N(x)/D(x)$, $D(x) \neq 0$, where $N(x)$ and $D(x)$ are polynomials, is called a rational function.

(b) The domain of a rational function is the set of all real numbers except those which make the denominator zero.

(c) If $f(x) = N(x)/D(x)$ is in reduced form, and a is a value such that $D(a) = 0$, then the line $x = a$ is a vertical asymptote of the graph of f. $f(x) \to \infty$ or $f(x) \to -\infty$ as $x \to a$.

(d) The line $y = b$ is a horizontal asymptote of the graph of f if $f(x) \to b$ as $x \to \infty$ or $x \to -\infty$.

(e) Let $f(x) = \dfrac{N(x)}{D(x)} = \dfrac{a_n x^n + a_{n-1} x^{n-1} + \cdots + a_1 x + a_0}{b_m x^m + b_{m-1} x^{m-1} + \cdots + b_1 x + b_0}$ where $N(x)$ and $D(x)$ have no common factors.

1. If $n < m$, then the x-axis ($y = 0$) is a horizontal asymptote.

2. If $n = m$, then $y = \dfrac{a_n}{b_m}$ is a horizontal asymptote.

3. If $n > m$, then there are no horizontal asymptotes.

Vocabulary Check

1. rational functions   **2.** vertical asymptote   **3.** horizontal asymptote   **4.** slant asymptote

1. $f(x) = \dfrac{1}{x-1}$

(a)

x	$f(x)$
0.5	-2
0.9	-10
0.99	-100
0.999	-1000

x	$f(x)$
1.5	2
1.1	10
1.01	100
1.001	1000

x	$f(x)$
5	0.25
10	$0.\overline{1}$
100	$0.\overline{01}$
1000	$0.\overline{001}$

(b) The zero of the denominator is $x = 1$, so $x = 1$ is a vertical asymptote. The degree of the numerator is less than the degree of the denominator so the x-axis, or $y = 0$, is a horizontal asymptote.

(c) The domain is all real numbers except $x = 1$.

3. $f(x) = \dfrac{3x^2}{x^2 - 1}$

(a)

x	$f(x)$
0.5	-1
0.9	-12.79
0.99	-147.8
0.999	-1498

x	$f(x)$
1.5	5.4
1.1	17.29
1.01	152.3
1.001	1502

x	$f(x)$
5	3.125
10	$3.\overline{03}$
100	$3.\overline{0003}$
1000	3

(b) The zeros of the denominator are $x = \pm 1$ so both $x = 1$ and $x = -1$ are vertical asymptotes. Since the degree of the numerator equals the degree of the denominator, $y = \frac{3}{1} = 3$ is a horizontal asymptote.

(c) The domain is all real numbers except $x = \pm 1$.

5. $f(x) = \dfrac{1}{x^2}$

Domain: all real numbers except $x = 0$

Vertical asymptote: $x = 0$

Horizontal asymptote: $y = 0$

[Degree of $N(x) <$ degree of $D(x)$]

7. $f(x) = \dfrac{2 + x}{2 - x} = \dfrac{x + 2}{-x + 2}$

Domain: all real numbers except $x = 2$

Vertical asymptote: $x = 2$

Horizontal asymptote: $y = -1$

[Degree of $N(x) =$ degree of $D(x)$]

9. $f(x) = \dfrac{x^3}{x^2 - 1}$

Domain: all real numbers except $x = \pm 1$

Vertical asymptotes: $x = \pm 1$

Horizontal asymptote: None

[Degree of $N(x) >$ degree of $D(x)$]

11. $f(x) = \dfrac{3x^2 + 1}{x^2 + x + 9}$

Domain: All real numbers. The denominator has no real zeros. [Try the Quadratic Formula on the denominator.]

Vertical asymptote: None

Horizontal asymptote: $y = 3$

[Degree of $N(x) =$ degree of $D(x)$]

13. $f(x) = \dfrac{2}{x + 3}$

Vertical asymptote: $y = -3$

Horizontal asymptote: $y = 0$

Matches graph (d).

15. $f(x) = \dfrac{x - 1}{x - 4}$

Vertical asymptote: $x = 4$

Horizontal asymptote: $y = 1$

Matches graph (c).

17. $g(x) = \dfrac{x^2 - 1}{x + 1} = \dfrac{(x - 1)(x + 1)}{x + 1}$

The only zero of $g(x)$ is $x = 1$.

$x = -1$ makes $g(x)$ undefined.

19. $f(x) = 1 - \dfrac{3}{x - 3}$

$$1 - \frac{3}{x - 3} = 0$$

$$1 = \frac{3}{x - 3}$$

$$x - 3 = 3$$

$$x = 6 \text{ is a zero of } f(x).$$

21. $f(x) = \dfrac{x - 4}{x^2 - 16} = \dfrac{1}{x + 4}, \ x \neq 4$

Domain: all real numbers x except $x = \pm 4$

Horizontal asymptote: $y = 0$

[Degree of $N(x)$ < degree of $D(x)$]

Vertical asymptote: $x = -4$ (Since $x - 4$ is a common factor of $N(x)$ and $D(x)$, $x = 4$ is not a vertical asymptote of $f(x)$.)

23. $f(x) = \dfrac{x^2 - 1}{x^2 - 2x - 3} = \dfrac{(x + 1)(x - 1)}{(x + 1)(x - 3)} = \dfrac{x - 1}{x - 3}, \ x \neq -1$

Domain: all real numbers x except $x = -1$ and $x = 3$

Horizontal asymptote: $y = 1$

[Degree of $N(x)$ = degree of $D(x)$]

Vertical asymptote: $x = 3$
(Since $x + 1$ is a common factor of $N(x)$ and $D(x)$, $x = -1$ is not a vertical asymptote of $f(x)$.)

25. $f(x) = \dfrac{x^2 - 3x - 4}{2x^2 + x - 1}$

$$= \frac{(x + 1)(x - 4)}{(2x - 1)(x + 1)} = \frac{x - 4}{2x - 1}, \ x \neq -1$$

Domain: all real numbers x except $x = \dfrac{1}{2}$ and $x = -1$

Horizontal asymptote: $y = \dfrac{1}{2}$

[Degree of $N(x)$ = degree of $D(x)$]

Vertical asymptote: $x = \dfrac{1}{2}$ (Since $x + 1$ is a common factor of $N(x)$ and $D(x)$, $x = -1$ is not a vertical asymptote of $f(x)$.)

27. $f(x) = \dfrac{1}{x + 2}$

(a) Domain: all real numbers x except $x = -2$

(b) y-intercept: $\left(0, \dfrac{1}{2}\right)$

(c) Vertical asymptote: $x = -2$
 Horizontal asymptote: $y = 0$

(d)

x	-4	-3	-1	0	1
y	$-\frac{1}{2}$	-1	1	$\frac{1}{2}$	$\frac{1}{3}$

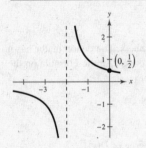

29. $h(x) = \dfrac{-1}{x + 2}$

(a) Domain: all real numbers x except $x = -2$

(b) y-intercept: $\left(0, -\dfrac{1}{2}\right)$

(c) Vertical asymptote: $x = -2$
 Horizontal asymptote: $y = 0$

(d)

x	-4	-3	-1	0
y	$\frac{1}{2}$	1	-1	$-\frac{1}{2}$

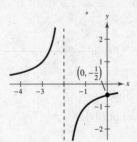

Note: This is the graph of $f(x) = \dfrac{1}{x + 2}$

(Exercise 27) reflected about the x-axis.

31. $C(x) = \dfrac{5 + 2x}{1 + x} = \dfrac{2x + 5}{x + 1}$

(a) Domain: all real numbers x except $x = -1$

(b) x-intercept: $\left(-\dfrac{5}{2}, 0\right)$

y-intercept: $(0, 5)$

(c) Vertical asymptote: $x = -1$
Horizontal asymptote: $y = 2$

(d)

x	-4	-3	-2	0	1	2
$C(x)$	1	$\frac{1}{2}$	-1	5	$\frac{7}{2}$	3

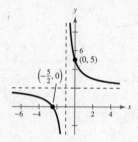

33. $f(x) = \dfrac{x^2}{x^2 + 9}$

(a) Domain: all real numbers x

(b) Intercept: $(0, 0)$

(c) Horizontal asymptote: $y = 1$

(d)

x	± 1	± 2	± 3
y	$\frac{1}{10}$	$\frac{4}{13}$	$\frac{1}{2}$

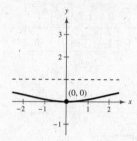

35. $g(s) = \dfrac{s}{s^2 + 1}$

(a) Domain: all real numbers s

(b) Intercept: $(0, 0)$

(c) Horizontal asymptote: $y = 0$

(d)

s	-2	-1	0	1	2
$g(s)$	$-\frac{2}{5}$	$-\frac{1}{2}$	0	$\frac{1}{2}$	$\frac{2}{5}$

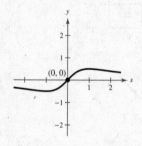

37. $h(x) = \dfrac{x^2 - 5x + 4}{x^2 - 4} = \dfrac{(x - 1)(x - 4)}{(x + 2)(x - 2)}$

(a) Domain: all real numbers x except $x = \pm 2$

(b) x-intercepts: $(1, 0)$, $(4, 0)$
y-intercept: $(0, -1)$

(c) Vertical asymptotes: $x = -2$, $x = 2$
Horizontal asymptote: $y = 1$

(d)

x	-4	-3	-1	0	1	3	4
y	$\frac{10}{3}$	$\frac{28}{5}$	$-\frac{10}{3}$	-1	0	$-\frac{2}{5}$	0

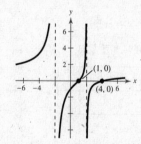

39. $f(x) = \dfrac{2x^2 - 5x - 3}{x^3 - 2x^2 - x + 2} = \dfrac{(2x + 1)(x - 3)}{(x - 2)(x + 1)(x - 1)}$

(a) Domain: all real numbers x except $x = 2$, $x = \pm 1$

(b) x-intercepts: $\left(-\dfrac{1}{2}, 0\right), (3, 0)$

  y-intercept: $\left(0, -\dfrac{3}{2}\right)$

(c) Vertical asymptotes: $x = 2$, $x = -1$ and $x = 1$

  Horizontal asymptotes: $y = 0$

(d)

x	-3	-2	0	1.5	3	4
$f(x)$	$-\dfrac{3}{4}$	$-\dfrac{5}{4}$	$-\dfrac{3}{2}$	$\dfrac{48}{5}$	0	$\dfrac{3}{10}$

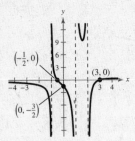

41. $f(x) = \dfrac{x^2 + 3x}{x^2 + x - 6} = \dfrac{x(x + 3)}{(x + 3)(x - 2)} = \dfrac{x}{x - 2}$, $x \ne -3$

(a) Domain: all real numbers x except $x = -3$ and $x = 2$

(b) Intercept: $(0, 0)$

(c) Vertical asymptote: $x = 2$

  Horizontal asymptote: $y = 1$

(d)

x	-1	0	1	3	4
y	$\dfrac{1}{3}$	0	-1	3	2

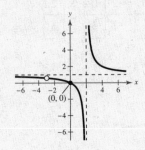

43. $f(x) = \dfrac{2x^2 - 5x + 2}{2x^2 - x - 6}$

  $= \dfrac{(2x - 1)(x - 2)}{(2x + 3)(x - 2)} = \dfrac{2x - 1}{2x + 3}$, $x \ne 2$

(a) Domain: all real numbers x except $x = 2$ and

  $x = -\dfrac{3}{2}$

(b) x-intercept: $\left(\dfrac{1}{2}, 0\right)$

  y-intercept: $\left(0, -\dfrac{1}{3}\right)$

(c) Vertical asymptote: $x = -\dfrac{3}{2}$

  Horizontal asymptote: $y = 1$

(d)

x	-3	-2	-1	0	1
y	$\dfrac{7}{3}$	5	-3	$-\dfrac{1}{3}$	$\dfrac{1}{5}$

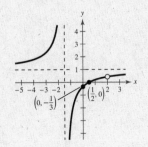

45. $f(t) = \dfrac{t^2 - 1}{t + 1} = \dfrac{(t + 1)(t - 1)}{t + 1} = t - 1$, $t \ne -1$

(a) Domain: all real numbers t except $t = -1$

(b) t-intercept: $(1, 0)$

  y-intercept: $(0, -1)$

(c) Vertical asymptote: none

  Horizontal asymptote: none

(d)

t	-3	-2	0	1
y	-4	-3	-1	0

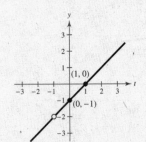

47. $f(x) = \dfrac{x^2 - 1}{x + 1}$, $g(x) = x - 1$

    (a) Domain of f: all real numbers x except $x = -1$

    Domain of g: all real numbers x

    (b) Because $(x + 1)$ is a factor of both the numerator and the denominator of f, $x = -1$ is not a vertical asymptote. f has no vertical asymptotes.

    (c)

x	-3	-2	-1.5	-1	-0.5	0	1
$f(x)$	-4	-3	-2.5	Undef.	-1.5	-1	0
$g(x)$	-4	-3	-2.5	-2	-1.5	-1	0

    (d)

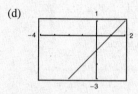

    (e) Because there are only a finite number of pixels, the utility may not attempt to evaluate the function where it does not exist.

51. $h(x) = \dfrac{x^2 - 4}{x} = x - \dfrac{4}{x}$

    (a) Domain: all real numbers x except $x = 0$

    (b) Intercepts: $(2, 0)$, $(-2, 0)$

    (c) Vertical asymptote: $x = 0$
       Slant asymptote: $y = x$

    (d)

x	-3	-1	1	3
y	$-\frac{5}{3}$	3	-3	$\frac{5}{3}$

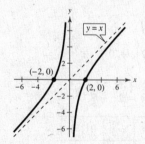

55. $g(x) = \dfrac{x^2 + 1}{x} = x + \dfrac{1}{x}$

    (a) Domain: all real numbers x except $x = 0$

    (b) No intercepts

    (c) Vertical asymptote: $x = 0$
       Slant asymptote: $y = x$

49. $f(x) = \dfrac{x - 2}{x^2 - 2x}$, $g(x) = \dfrac{1}{x}$

    (a) Domain of f: all real numbers x except $x = 0$ and $x = 2$

    Domain of g: all real numbers x except $x = 0$

    (b) Because $(x - 2)$ is a factor of both the numerator and the denominator of f, $x = 2$ is not a vertical asymptote. The only vertical asymptote of f is $x = 0$.

    (c)

x	-0.5	0	0.5	1	1.5	2	3
$f(x)$	-2	Undef.	2	1	$\frac{2}{3}$	Undef.	$\frac{1}{3}$
$g(x)$	-2	Undef.	2	1	$\frac{2}{3}$	$\frac{1}{2}$	$\frac{1}{3}$

    (d)

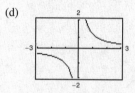

    (e) Because there are only a finite number of pixels, the utility may not attempt to evaluate the function where it does not exist.

53. $f(x) = \dfrac{2x^2 + 1}{x} = 2x + \dfrac{1}{x}$

    (a) Domain: all real numbers x except $x = 0$

    (b) No intercepts

    (c) Vertical asymptote: $x = 0$
       Slant asymptote: $y = 2x$

    (d)

x	-4	-2	2	4	6
$f(x)$	$-\frac{33}{4}$	$-\frac{9}{2}$	$\frac{9}{2}$	$\frac{33}{4}$	$\frac{73}{6}$

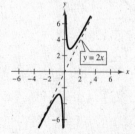

(d)

x	-4	-2	2	4	6
$g(x)$	$-\frac{17}{4}$	$-\frac{5}{2}$	$\frac{5}{2}$	$\frac{17}{4}$	$\frac{37}{6}$

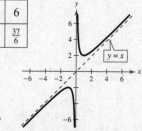

57. $f(t) = -\dfrac{t^2 + 1}{t + 5} = -t + 5 - \dfrac{26}{t + 5}$

   (a) Domain: all real numbers t except $t = -5$

   (b) Intercept: $\left(0, -\dfrac{1}{5}\right)$

   (c) Vertical asymptote: $t = -5$
      Slant asymptote: $y = -t + 5$

   (d)

t	-7	-6	-4	-3	0
y	25	37	-17	-5	$-\dfrac{1}{5}$

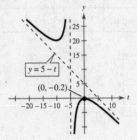

59. $f(x) = \dfrac{x^3}{x^2 - 1} = x + \dfrac{x}{x^2 - 1}$

   (a) Domain: all real numbers x except $x = \pm 1$

   (b) Intercept: $(0, 0)$

   (c) Vertical asymptotes: $x = \pm 1$
      Slant asymptote: $y = x$

   (d)

x	-4	-2	0	2	4
$f(x)$	$-\dfrac{64}{15}$	$-\dfrac{8}{3}$	0	$\dfrac{8}{3}$	$\dfrac{64}{15}$

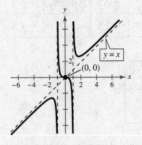

61. $f(x) = \dfrac{x^2 - x + 1}{x - 1} = x + \dfrac{1}{x - 1}$

   (a) Domain: all real numbers x except $x = 1$

   (b) y-intercept: $(0, -1)$

   (c) Vertical asymptote: $x = 1$
      Slant asymptote: $y = x$

   (d)

x	-4	-2	0	2	4
$f(x)$	$-\dfrac{21}{5}$	$-\dfrac{7}{3}$	-1	3	$\dfrac{13}{3}$

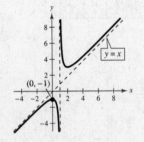

63. $f(x) = \dfrac{2x^3 - x^2 - 2x + 1}{x^2 + 3x + 2}$

$= \dfrac{(2x - 1)(x + 1)(x - 1)}{(x + 1)(x + 2)}$

$= \dfrac{(2x - 1)(x - 1)}{x + 2}, \quad x \neq -1$

$= \dfrac{2x^2 - 3x + 1}{x + 2}$

$= 2x - 7 + \dfrac{15}{x + 2}, \quad x \neq -1$

   (a) Domain: all real numbers x except $x = -1$ and
         $x = -2$

   (b) y-intercept: $\left(0, \dfrac{1}{2}\right)$

      x-intercepts: $\left(\dfrac{1}{2}, 0\right), (1, 0)$

   (c) Vertical asymptote: $x = -2$
      Slant asymptote: $y = 2x - 7$

   (d)

x	-4	-3	$-\dfrac{3}{2}$	0	1
y	$-\dfrac{45}{2}$	-28	20	$\dfrac{1}{2}$	0

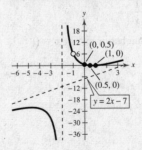

65. $f(x) = \dfrac{x^2 + 5x + 8}{x + 3} = x + 2 + \dfrac{2}{x + 3}$

Domain: all real numbers x except $x = -3$

y-intercept: $\left(0, \dfrac{8}{3}\right)$

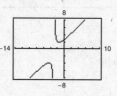

Vertical asymptote: $x = -3$

Slant asymptote: $y = x + 2$

Line: $y = x + 2$

67. $g(x) = \dfrac{1 + 3x^2 - x^3}{x^2} = \dfrac{1}{x^2} + 3 - x = -x + 3 + \dfrac{1}{x^2}$

Domain: all real numbers x except $x = 0$

Vertical asymptote: $x = 0$

Slant asymptote: $y = -x + 3$

Line: $y = -x + 3$

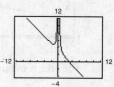

69. $y = \dfrac{x + 1}{x - 3}$

(a) x-intercept: $(-1, 0)$

(b) $\quad 0 = \dfrac{x + 1}{x - 3}$

$\quad 0 = x + 1$

$\quad -1 = x$

71. $y = \dfrac{1}{x} - x$

(a) x-intercepts: $(\pm 1, 0)$ (b) $\quad 0 = \dfrac{1}{x} - x$

$\quad x = \dfrac{1}{x}$

$\quad x^2 = 1$

$\quad x = \pm 1$

73. $C = \dfrac{255p}{100 - p}$, $0 \le p < 100$

(a)

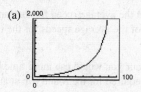

(b) $C(10) = \dfrac{255(10)}{100 - 10} \approx 28.33$ million dollars

$C(40) = \dfrac{255(40)}{100 - 40} = 170$ million dollars

$C(75) = \dfrac{255(75)}{100 - 75} = 765$ million dollars

(c) $C \to \infty$ as $x \to 100$. No, it would not be possible to remove 100% of the pollutants.

75. $N = \dfrac{20(5 + 3t)}{1 + 0.04t}$, $t \ge 0$

(a) $N(5) \approx 333$ deer

$N(10) = 500$ deer

$N(25) = 800$ deer

(b) The herd is limited by the horizontal asymptote:

$N = \dfrac{60}{0.04} = 1500$ deer

77. (a) $A = xy$ and

$(x - 4)(y - 2) = 30$

$y - 2 = \dfrac{30}{x - 4}$

$y = 2 + \dfrac{30}{x - 4} = \dfrac{2x + 22}{x - 4}$

Thus, $A = xy = x\left(\dfrac{2x + 22}{x - 4}\right) = \dfrac{2x(x + 11)}{x - 4}$.

(b) Domain: Since the margins on the left and right are each 2 inches, $x > 4$. In interval notation, the domain is $(4, \infty)$.

—CONTINUED—

77. —CONTINUED—

(c)

x	5	6	7	8	9	10	11	12	13	14	15
y_1 (Area)	160	102	84	76	72	70	69.143	69	69.333	70	70.909

The area is minimum when $x \approx 11.75$ inches and $y \approx 5.87$ inches.

The area is minimum when x is approximately 12.

79. (a) Let $t_1 = $ time from Akron to Columbus and $t_2 = $ time from Columbus back to Akron.

$$xt_1 = 100 \implies t_1 = \frac{100}{x}$$

$$yt_2 = 100 \implies t_2 = \frac{100}{y}$$

$$50(t_1 + t_2) = 200$$

$$t_1 + t_2 = 4$$

$$\frac{100}{x} + \frac{100}{y} = 4$$

$$100y + 100x = 4xy$$

$$25y + 25x = xy$$

$$25x = xy - 25y$$

$$25x = y(x - 25)$$

Thus, $y = \dfrac{25x}{x - 25}$.

(b) Vertical asymptote: $x = 25$

Horizontal asymptote: $y = 25$

(c)

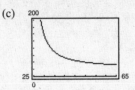

(d)

x	30	35	40	45	50	55	60
y	150	87.5	66.67	56.25	50	45.83	42.86

(e) Yes. You would expect the average speed for the round trip to be the average of the average speeds for the two parts of the trip.

(f) No. At 20 miles per hour you would use more time in one direction than is required for the round trip at an average speed of 50 miles per hour.

81. False. Polynomial functions do not have vertical asymptotes.

83. Vertical asymptote: None $\implies$ The denominator is not zero for any value of x (unless the numerator is also zero there).

Horizontal asymptote: $y = 2 \implies$ The degree of the numerator equals the degree of the denominator.

$f(x) = \dfrac{2x^2}{x^2 + 1}$ is one possible function. There are many correct answers.

85. $x^2 - 15x + 56 = (x - 8)(x - 7)$

87. $x^3 - 5x^2 + 4x - 20 = (x - 5)(x^2 + 4)$
$$= (x - 5)(x + 2i)(x - 2i)$$

89. $10 - 3x \leq 0$

$3x \geq 10$

$x \geq \frac{10}{3}$

91. $|4(x - 2)| < 20$

$-20 < 4x - 8 < 20$

$-12 < 4x < 28$

$-3 < x < 7$

93. Answers will vary.

Section 2.7 Nonlinear Inequalities

■ You should be able to solve inequalities.

(a) Find the critical number.

    1. Values that make the expression zero

    2. Values that make the expression undefined

(b) Test one value in each test interval on the real number line resulting from the critical numbers.

(c) Determine the solution intervals.

Vocabulary Check

1. critical; test intervals   **2.** zeros; undefined values   **3.** $P = R - C$

1. $x^2 - 3 < 0$

(a) $x = 3$

$(3)^2 - 3 \overset{?}{<} 0$

$6 \not< 0$

No, $x = 3$ *is not* a solution.

(b) $x = 0$

$(0)^2 - 3 \overset{?}{<} 0$

$-3 < 0$

Yes, $x = 0$ *is* a solution.

(c) $x = \frac{3}{2}$

$\left(\frac{3}{2}\right)^2 - 3 \overset{?}{<} 0$

$-\frac{3}{4} < 0$

Yes, $x = \frac{3}{2}$ *is* a solution.

(d) $x = -5$

$(-5)^2 - 3 \overset{?}{<} 0$

$22 \not< 0$

No, $x = -5$ *is not* a solution.

3. $\dfrac{x + 2}{x - 4} \geq 3$

(a) $x = 5$

$\dfrac{5 + 2}{5 - 4} \overset{?}{\geq} 3$

$7 \geq 3$

Yes, $x = 5$ *is* a solution.

(b) $x = 4$

$\dfrac{4 + 2}{4 - 4} \overset{?}{\geq} 3$

$\dfrac{6}{0}$ is undefined.

No, $x = 4$ *is not* a solution.

(c) $x = -\dfrac{9}{2}$

$\dfrac{-\frac{9}{2} + 2}{-\frac{9}{2} - 4} \overset{?}{\geq} 3$

$\dfrac{5}{17} \not\geq 3$

No, $x = -\frac{9}{2}$ *is not* a solution.

(d) $x = \dfrac{9}{2}$

$\dfrac{\frac{9}{2} + 2}{\frac{9}{2} - 4} \overset{?}{\geq} 3$

$13 \geq 3$

Yes, $x = \frac{9}{2}$ *is* a solution.

5. $2x^2 - x - 6 = (2x + 3)(x - 2)$

$2x + 3 = 0 \implies x = -\dfrac{3}{2}$

$x - 2 = 0 \implies x = 2$

Critical numbers: $x = -\dfrac{3}{2}, x = 2$

7. $2 + \dfrac{3}{x - 5} = \dfrac{2(x - 5) + 3}{x - 5}$

$= \dfrac{2x - 7}{x - 5}$

$2x - 7 = 0 \implies x = \dfrac{7}{2}$

$x - 5 = 0 \implies x = 5$

Critical numbers: $x = \dfrac{7}{2}, x = 5$

9.
$$x^2 \leq 9$$
$$x^2 - 9 \leq 0$$
$$(x + 3)(x - 3) \leq 0$$

Critical numbers: $x = \pm 3$

Test intervals: $(-\infty, -3), (-3, 3), (3, \infty)$

Test: Is $(x + 3)(x - 3) \leq 0$?

Interval	x-Value	Value of $x^2 - 9$	Conclusion
$(-\infty, -3)$	$x = -4$	$16 - 9 = 7$	Positive
$(-3, 3)$	$x = 0$	$0 - 9 = -9$	Negative
$(3, \infty)$	$x = 4$	$16 - 9 = 7$	Positive

Solution set: $[-3, 3]$

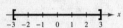

11.
$$(x + 2)^2 < 25$$
$$x^2 + 4x + 4 < 25$$
$$x^2 + 4x - 21 < 0$$
$$(x + 7)(x - 3) < 0$$

Critical numbers: $x = -7, x = 3$

Test intervals: $(-\infty, -7), (-7, 3), (3, \infty)$

Test: Is $(x + 7)(x - 3) < 0$?

Interval	x-Value	Value of $(x + 7)(x - 3)$	Conclusion
$(-\infty, -7)$	$x = -10$	$(-3)(-13) = 39$	Positive
$(-7, 3)$	$x = 0$	$(7)(-3) = -21$	Negative
$(3, \infty)$	$x = 5$	$(12)(2) = 24$	Positive

Solution set: $(-7, 3)$

13.
$$x^2 + 4x + 4 \geq 9$$
$$x^2 + 4x - 5 \geq 0$$
$$(x + 5)(x - 1) \geq 0$$

Critical numbers: $x = -5, x = 1$

Test intervals: $(-\infty, -5), (-5, 1), (1, \infty)$

Test: Is $(x + 5)(x - 1) \geq 0$?

Interval	x-Value	Value of $(x + 5)(x - 1)$	Conclusion
$(-\infty, -5)$	$x = -6$	$(-1)(-7) = 7$	Positive
$(-5, 1)$	$x = 0$	$(5)(-1) = -5$	Negative
$(1, \infty)$	$x = 2$	$(7)(1) = 7$	Positive

Solution set: $(-\infty, -5] \cup [1, \infty)$

15.
$$x^2 + x < 6$$
$$x^2 + x - 6 < 0$$
$$(x + 3)(x - 2) < 0$$

Critical numbers: $x = -3, x = 2$

Test intervals: $(-\infty, -3), (-3, 2), (2, \infty)$

Test: Is $(x + 3)(x - 2) < 0$?

Interval	x-Value	Value of $(x + 3)(x - 2)$	Conclusion
$(-\infty, -3)$	$x = -4$	$(-1)(-6) = 6$	Positive
$(-3, 2)$	$x = 0$	$(3)(-2) = -6$	Negative
$(2, \infty)$	$x = 3$	$(6)(1) = 6$	Positive

Solution set: $(-3, 2)$

17.
$$x^2 + 2x - 3 < 0$$
$$(x + 3)(x - 1) < 0$$

Critical numbers: $x = -3, x = 1$

Test intervals: $(-\infty, -3), (-3, 1), (1, \infty)$

Test: Is $(x + 3)(x - 1) < 0$?

Interval	x-Value	Value of $(x + 3)(x - 1)$	Conclusion
$(-\infty, -3)$	$x = -4$	$(-1)(-5) = 5$	Positive
$(-3, 1)$	$x = 0$	$(3)(-1) = -3$	Negative
$(1, \infty)$	$x = 2$	$(5)(1) = 5$	Positive

Solution set: $(-3, 1)$

19. $x^2 + 8x - 5 \geq 0$

   $x^2 + 8x - 5 = 0$ Complete the square.

   $x^2 + 8x + 16 = 5 + 16$

   $(x + 4)^2 = 21$

   $x + 4 = \pm\sqrt{21}$

   $x = -4 \pm \sqrt{21}$

Critical numbers: $x = -4 \pm \sqrt{21}$

Test intervals: $\left(-\infty, -4 - \sqrt{21}\right), \left(-4 - \sqrt{21}, -4 + \sqrt{21}\right), \left(-4 + \sqrt{21}, \infty\right)$

Test: Is $x^2 + 8x - 5 \geq 0$?

Interval	x-Value	Value of $x^2 + 8x - 5$	Conclusion
$\left(-\infty, -4 - \sqrt{21}\right)$	$x = -10$	$100 - 80 - 5 = 15$	Positive
$\left(-4 - \sqrt{21}, -4 + \sqrt{21}\right)$	$x = 0$	$0 + 0 - 5 = -5$	Negative
$\left(-4 + \sqrt{21}, \infty\right)$	$x = 2$	$4 + 16 - 5 = 15$	Positive

Solution set: $\left(-\infty < -4 - \sqrt{21}\right] \cup \left[-4 + \sqrt{21}, \infty\right)$

21. $x^3 - 3x^2 - x + 3 > 0$

   $x^2(x - 3) - 1(x - 3) > 0$

   $(x^2 - 1)(x - 3) > 0$

   $(x + 1)(x - 1)(x - 3) > 0$

Critical numbers: $x = \pm 1, x = 3$

Test intervals: $(-\infty, -1), (-1, 1), (1, 3), (3, \infty)$

Test: Is $(x + 1)(x - 1)(x - 3) > 0$?

Interval	x-Value	Value of $(x + 1)(x - 1)(x - 3)$	Conclusion
$(-\infty, -1)$	$x = -2$	$(-1)(-3)(-5) = -15$	Negative
$(-1, 1)$	$x = 0$	$(1)(-1)(-3) = 3$	Positive
$(1, 3)$	$x = 2$	$(3)(1)(-1) = -3$	Negative
$(3, \infty)$	$x = 4$	$(5)(3)(1) = 15$	Positive

Solution set: $(-1, 1) \cup (3, \infty)$

23. $x^3 - 2x^2 - 9x - 2 \geq -20$

$\qquad x^3 - 2x^2 - 9x + 18 \geq 0$

$\qquad x^2(x - 2) - 9(x - 2) \geq 0$

$\qquad\qquad (x - 2)(x^2 - 9) \geq 0$

$\qquad (x - 2)(x + 3)(x - 3) \geq 0$

Critical numbers: $x = 2, x = \pm 3$

Test intervals: $(-\infty, -3), (-3, 2), (2, 3), (3, \infty)$

Test: Is $(x - 2)(x + 3)(x - 3) \geq 0$?

Interval	x-Value	Value of $(x - 2)(x + 3)(x - 3)$	Conclusion
$(-\infty, -3)$	$x = -4$	$(-6)(-1)(-7) = -42$	Negative
$(-3, 2)$	$x = 0$	$(-2)(3)(-3) = 18$	Positive
$(2, 3)$	$x = 2.5$	$(0.5)(5.5)(-0.5) = -1.375$	Negative
$(3, \infty)$	$x = 4$	$(2)(7)(1) = 14$	Positive

Solution set: $[-3, 2] \cup [3, \infty)$

25. $4x^2 - 4x + 1 \leq 0$

$\qquad (2x - 1)^2 \leq 0$

Critical number: $x = \frac{1}{2}$

Test intervals: $\left(-\infty, \frac{1}{2}\right), \left(\frac{1}{2}, \infty\right)$

Test: Is $(2x - 1)^2 \leq 0$?

Interval	x-Value	Value of $(2x - 1)^2$	Conclusion
$\left(-\infty, \frac{1}{2}\right)$	$x = 0$	$(-1)^2 = 1$	Positive
$\left(\frac{1}{2}, \infty\right)$	$x = 1$	$(1)^2 = 1$	Positive

Solution set: $x = \frac{1}{2}$

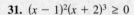

27. $4x^3 - 6x^2 < 0$

$2x^2(2x - 3) < 0$

Critical numbers: $x = 0, x = \frac{3}{2}$

Test intervals: $(-\infty, 0), \left(0, \frac{3}{2}\right), \left(\frac{3}{2}, \infty\right)$

Test: Is $2x^2(2x - 3) < 0$?

By testing an x-value in each test interval in the inequality, we see that the solution set is: $(-\infty, 0) \cup \left(0, \frac{3}{2}\right)$

29. $x^3 - 4x \geq 0$

$x(x + 2)(x - 2) \geq 0$

Critical numbers: $x = 0, x = \pm 2$

Test intervals: $(-\infty, -2), (-2, 0), (0, 2), (2, \infty)$

Test: Is $x(x + 2)(x - 2) \geq 0$?

By testing an x-value in each test interval in the inequality, we see that the solution set is: $[-2, 0] \cup [2, \infty)$

31. $(x - 1)^2(x + 2)^3 \geq 0$

Critical numbers: $x = 1, x = -2$

Test intervals: $(-\infty, -2), (-2, 1), (1, \infty)$

Test: Is $(x - 1)^2(x + 3)^3 \geq 0$?

By testing an x-value in each test interval in the inequality, we see that the solution set is: $[-2, \infty)$

33. $y = -x^2 + 2x + 3$

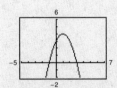

(a) $y \leq 0$ when $x \leq -1$ or $x \geq 3$.

(b) $y \geq 3$ when $0 \leq x \leq 2$.

35. $y = \frac{1}{8}x^3 - \frac{1}{2}x$

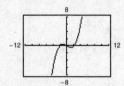

(a) $y \geq 0$ when $-2 \leq x \leq 0, 2 \leq x < \infty$.

(b) $y \leq 6$ when $x \leq 4$.

37. $\frac{1}{x} - x > 0$

$$\frac{1 - x^2}{x} > 0$$

Critical numbers: $x = 0, x = \pm 1$

Test intervals: $(-\infty, -1), (-1, 0), (0, 1), (1, \infty)$

Test: Is $\frac{1 - x^2}{x} > 0$?

By testing an x-value in each test interval in the inequality, we see that the solution set is: $(-\infty, -1) \cup (0, 1)$

39. $\frac{x + 6}{x + 1} - 2 < 0$

$$\frac{x + 6 - 2(x + 1)}{x + 1} < 0$$

$$\frac{4 - x}{x + 1} < 0$$

Critical numbers: $x = -1, x = 4$

Test intervals: $(-\infty, -1), (-1, 4), (4, \infty)$

Test: Is $\frac{4 - x}{x + 1} < 0$?

By testing an x-value in each test interval in the inequality, we see that the solution set is: $(-\infty, -1) \cup (4, \infty)$

41. $\frac{3x - 5}{x - 5} > 4$

$$\frac{3x - 5}{x - 5} - 4 > 0$$

$$\frac{3x - 5 - 4(x - 5)}{x - 5} > 0$$

$$\frac{15 - x}{x - 5} > 0$$

Critical numbers: $x = 5, x = 15$

Test intervals: $(-\infty, 5), (5, 15), (15, \infty)$

Test: Is $\frac{15 - x}{x - 5} > 0$?

By testing an x-value in each test interval in the inequality, we see that the solution set is: $(5, 15)$

43. $\frac{4}{x + 5} > \frac{1}{2x + 3}$

$$\frac{4}{x + 5} - \frac{1}{2x + 3} > 0$$

$$\frac{4(2x + 3) - (x + 5)}{(x + 5)(2x + 3)} > 0$$

$$\frac{7x + 7}{(x + 5)(2x + 3)} > 0$$

Critical numbers: $x = -1, x = -5, x = -\frac{3}{2}$

Test intervals: $(-\infty, -5), \left(-5, -\frac{3}{2}\right),$

$$\left(-\frac{3}{2}, -1\right), (-1, \infty)$$

Test: Is $\frac{7(x + 1)}{(x + 5)(2x + 3)} > 0$?

By testing an x-value in each test interval in the inequality, we see that the solution set is: $\left(-5, -\frac{3}{2}\right) \cup (-1, \infty)$

45.
$$\frac{1}{x-3} \le \frac{9}{4x+3}$$

$$\frac{1}{x-3} - \frac{9}{4x+3} \le 0$$

$$\frac{4x+3 - 9(x-3)}{(x-3)(4x+3)} \le 0$$

$$\frac{30 - 5x}{(x-3)(4x+3)} \le 0$$

Critical numbers: $x = 3, x = -\dfrac{3}{4}, x = 6$

Test intervals: $\left(-\infty, -\dfrac{3}{4}\right), \left(-\dfrac{3}{4}, 3\right), (3, 6), (6, \infty)$

Test: Is $\dfrac{30 - 5x}{(x-3)(4x+3)} \le 0$?

By testing an x-value in each
test interval in the inequality,
we see that the solution set is:
$\left(-\frac{3}{4}, 3\right) \cup [6, \infty)$

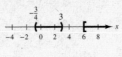

47.
$$\frac{x^2 + 2x}{x^2 - 9} \le 0$$

$$\frac{x(x+2)}{(x+3)(x-3)} \le 0$$

Critical numbers: $x = 0, x = -2, x = \pm 3$

Test intervals:
$(-\infty, -3), (-3, -2), (-2, 0), (0, 3), (3, \infty)$

Test: Is $\dfrac{x(x+2)}{(x+3)(x-3)} \le 0$?

By testing an x-value in each test interval in the inequality,
we see that the solution set is: $(-3, -2] \cup [0, 3)$

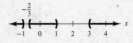

49.
$$\frac{5}{x-1} - \frac{2x}{x+1} < 1$$

$$\frac{5}{x-1} - \frac{2x}{x+1} - 1 < 0$$

$$\frac{5(x+1) - 2x(x-1) - (x-1)(x+1)}{(x-1)(x+1)} < 0$$

$$\frac{5x + 5 - 2x^2 + 2x - x^2 + 1}{(x-1)(x+1)} < 0$$

$$\frac{-3x^2 + 7x + 6}{(x-1)(x+1)} < 0$$

$$\frac{-(3x+2)(x-3)}{(x-1)(x+1)} < 0$$

Critical numbers: $x = -\dfrac{2}{3}, x = 3, x = \pm 1$

Test intervals: $(-\infty, -1), \left(-1, -\dfrac{2}{3}\right), \left(-\dfrac{2}{3}, 1\right), (1, 3), (3, \infty)$

Test: Is $\dfrac{-(3x+2)(x-3)}{(x-1)(x+1)} < 0$?

By testing an x-value in each test interval in the inequality, we see that the solution set is: $(-\infty, -1) \cup \left(-\frac{2}{3}, 1\right) \cup (3, \infty)$

51. $y = \dfrac{3x}{x-2}$

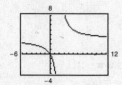

(a) $y \le 0$ when $0 \le x < 2$.

(b) $y \ge 6$ when $2 < x \le 4$.

53. $y = \dfrac{2x^2}{x^2+4}$

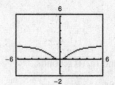

(a) $y \ge 1$ when $x \le -2$ or $x \ge 2$.

This can also be expressed as $|x| \ge 2$.

(b) $y \le 2$ for all real numbers x.

This can also be expressed as $-\infty < x < \infty$.

55. $4 - x^2 \ge 0$

$(2 + x)(2 - x) \ge 0$

Critical numbers: $x = \pm 2$

Test intervals: $(-\infty, -2), (-2, 2), (2, \infty)$

Test: Is $4 - x^2 \ge 0$?

By testing an x-value in each test interval in the inequality, we see that the domain set is: $[-2, 2]$

57. $x^2 - 7x + 12 \ge 0$

$(x - 3)(x - 4) \ge 0$

Critical numbers: $x = 3, x = 4$

Test intervals: $(-\infty, 3), (3, 4), (4, \infty)$

Test: Is $(x - 3)(x - 4) \ge 0$?

By testing an x-value in each test interval in the inequality, we see that the domain set is: $(-\infty, 3] \cup [4, \infty)$

59. $\dfrac{x}{x^2 - 2x - 35} \ge 0$

$\dfrac{x}{(x + 5)(x - 7)} \ge 0$

Critical numbers: $x = 0, x = -5, x = 7$

Test intervals: $(-\infty, -5), (-5, 0), (0, 7), (7, \infty)$

Test: Is $\dfrac{x}{(x + 5)(x - 7)} \ge 0$?

By testing an x-value in each test interval in the inequality, we see that the domain set is: $(-5, 0] \cup (7, \infty)$

61. $0.4x^2 + 5.26 < 10.2$

$0.4x^2 - 4.94 < 0$

$0.4(x^2 - 12.35) < 0$

Critical numbers: $x \approx \pm 3.51$

Test intervals: $(-\infty, -3.51), (-3.51, 3.51), (3.51, \infty)$

By testing an x-value in each test interval in the inequality, we see that the solution set is: $(-3.51, 3.51)$

63. $-0.5x^2 + 12.5x + 1.6 > 0$

The zeros are $x = \dfrac{-12.5 \pm \sqrt{(12.5)^2 - 4(-0.5)(1.6)}}{2(-0.5)}$.

Critical numbers: $x \approx -0.13, x \approx 25.13$

Test intervals: $(-\infty, -0.13), (-0.13, 25.13), (25.13, \infty)$

By testing an x-value in each test interval in the inequality, we see that the solution set is: $(-0.13, 25.13)$

65. $\dfrac{1}{2.3x - 5.2} > 3.4$

$\dfrac{1}{2.3x - 5.2} - 3.4 > 0$

$\dfrac{1 - 3.4(2.3x - 5.2)}{2.3x - 5.2} > 0$

$\dfrac{-7.82x + 18.68}{2.3x - 5.2} > 0$

Critical numbers: $x \approx 2.39, x \approx 2.26$

Test intervals: $(-\infty, 2.26), (2.26, 2.39), (2.39, \infty)$

By testing an x-value in each test interval in the inequality, we see that the solution set is: $(2.26, 2.39)$

67. $s = -16t^2 + v_0t + s_0 = -16t^2 + 160t$

    (a) $-16t^2 + 160t = 0$

        $-16t(t - 10) = 0$

          $t = 0, t = 10$

       It will be back on the ground in 10 seconds.

    (b) $-16t^2 + 160t > 384$

       $-16t^2 + 160t - 384 > 0$

      $-16(t^2 - 10t + 24) > 0$

          $t^2 - 10t + 24 < 0$

         $(t - 4)(t - 6) < 0$

           $4 < t < 6$ seconds

69. $2L + 2W = 100 \Rightarrow W = 50 - L$

               $LW \geq 500$

            $L(50 - L) \geq 500$

        $-L^2 + 50L - 500 \geq 0$

By the Quadratic Formula we have:

Critical numbers: $L = 25 \pm 5\sqrt{5}$

Test: Is $-L^2 + 50L - 500 \geq 0$?

Solution set: $25 - 5\sqrt{5} \leq L \leq 25 + 5\sqrt{5}$

13.8 meters $\leq L \leq$ 36.2 meters

71. $R = x(75 - 0.0005x)$ and $C = 30x + 250,000$

    $P = R - C$

      $= (75x - 0.0005x^2) - (30x + 250,000)$

      $= -0.0005x^2 + 45x - 250,000$

                    $P \geq 750,000$

  $-0.0005x^2 + 45x - 250,000 \geq 750,000$

$-0.0005x^2 + 45x - 1,000,000 \geq 0$

Critical numbers: $x = 40,000, x = 50,000$ (These were obtained by using the Quadratic Formula.)

Test intervals: $(0, 40,000), (40,000, 50,000), (50,000, \infty)$

By testing x-values in each test interval in the inequality, we see that the solution set is $[40,000, 50,000]$ or $40,000 \leq x \leq 50,000$. The price per unit is

$$p = \frac{R}{x} = 75 - 0.0005x.$$

For $x = 40,000$, $p = \$55$. For $x = 50,000$, $p = \$50$. Therefore, for $40,000 \leq x \leq 50,000$, $\$50.00 \leq p \leq \$55.00$.

73. $C = 0.0031t^3 - 0.216t^2 + 5.54t + 19.1, \; 0 \leq t \leq 23$

    (a)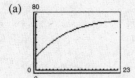

    (b)

t	C
24	70.5
26	71.6
28	72.9
30	74.6
32	76.8
34	79.6

      C will be greater than 75% when $t \approx 31$, which corresponds to 2011.

    (c) $C = 75$ when $t \approx 30.41$.

    (d)

t	C
36	83.2
37	85.4
38	87.8
39	90.5
40	93.5
41	96.8
42	100.4
43	104.4

      C will be between 85% and 100% when t is between 37 and 42. These values correspond to the years 2017 to 2022.

    (e) $85 \leq C \leq 100$ when $36.82 \leq t \leq 41.89$ or $37 \leq t \leq 42$.

    (f) The model is a third-degree polynomial and as $t \to \infty, C \to \infty$.

75.
$$\frac{1}{R} = \frac{1}{R_1} + \frac{1}{2}$$

$$2R_1 = 2R + RR_1$$

$$2R_1 = R(2 + R_1)$$

$$\frac{2R_1}{2 + R_1} = R$$

Since $R \geq 1$, we have

$$\frac{2R_1}{2 + R_1} \geq 1$$

$$\frac{2R_1}{2 + R_1} - 1 \geq 0$$

$$\frac{R_1 - 2}{2 + R_1} \geq 0.$$

Since $R_1 > 0$, the only critical number is $R_1 = 2$.
The inequality is satisfied when $R_1 \geq 2$ ohms.

77. True
$$x^3 - 2x^2 - 11x + 12 = (x + 3)(x - 1)(x - 4)$$

The test intervals are $(-\infty, -3)$, $(-3, 1)$, $(1, 4)$, and $(4, \infty)$.

79. $x^2 + bx + 4 = 0$

To have at least one real solution, $b^2 - 16 \geq 0$. This occurs when $b \leq -4$ or $b \geq 4$. This can be written as $(-\infty, -4] \cup [4, \infty)$.

81. $3x^2 + bx + 10 = 0$

To have at least one real solution, $b^2 - 4(3)(10) \geq 0$.

$$b^2 - 120 \geq 0$$

$$\left(b + \sqrt{120}\right)\left(b - \sqrt{120}\right) \geq 0$$

Critical numbers: $b = \pm\sqrt{120} = \pm 2\sqrt{30}$

Test intervals:
$$\left(-\infty, -2\sqrt{30}\right), \left(-2\sqrt{30}, 2\sqrt{30}\right), \left(2\sqrt{30}, \infty\right)$$

Test: Is $b^2 - 120 \geq 0$?

Solution set: $\left(-\infty, -2\sqrt{30}\right] \cup \left[2\sqrt{30}, \infty\right)$

83. (a) If $a > 0$ and $c \leq 0$, then b can be any real number. If $a > 0$ and $c > 0$, then for $b^2 - 4ac$ to be greater than or equal to zero, b is restricted to $b < -2\sqrt{ac}$ or $b > 2\sqrt{ac}$.

(b) The center of the interval for b in Exercises 79–82 is 0.

85. $4x^2 + 20x + 25 = (2x + 5)^2$

87. $x^2(x + 3) - 4(x + 3) = (x^2 - 4)(x + 3)$
$$= (x + 2)(x - 2)(x + 3)$$

89. Area = (length)(width)
$$= (2x + 1)(x)$$
$$= 2x^2 + x$$

Review Exercises for Chapter 2

1. (a) $y = 2x^2$

Vertical stretch

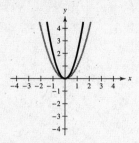

(b) $y = -2x^2$

Vertical stretch and a reflection in the x-axis

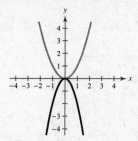

(c) $y = x^2 + 2$

Vertical shift two units upward

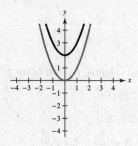

(d) $y = (x + 2)^2$

Horizontal shift two units to the left

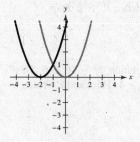

3. $g(x) = x^2 - 2x$

$\quad = x^2 - 2x + 1 - 1$

$\quad = (x - 1)^2 - 1$

Vertex: $(1, -1)$

Axis of symmetry: $x = 1$

$0 = x^2 - 2x = x(x - 2)$

x-intercepts: $(0, 0), (2, 0)$

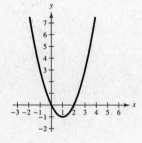

5. $f(x) = x^2 + 8x + 10$

$\quad = x^2 + 8x + 16 - 16 + 10$

$\quad = (x + 4)^2 - 6$

Vertex: $(-4, -6)$

Axis of symmetry: $x = -4$

$\quad 0 = (x + 4)^2 - 6$

$(x + 4)^2 = 6$

$\quad x + 4 = \pm\sqrt{6}$

$\quad\quad x = -4 \pm \sqrt{6}$

x-intercepts: $\left(-4 \pm \sqrt{6}, 0\right)$

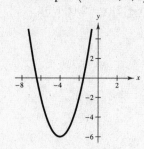

7. $f(t) = -2t^2 + 4t + 1$

$= -2(t^2 - 2t + 1 - 1) + 1$

$= -2[(t-1)^2 - 1] + 1$

$= -2(t-1)^2 + 3$

Vertex: $(1, 3)$

Axis of symmetry: $t = 1$

$0 = -2(t-1)^2 + 3$

$2(t-1)^2 = 3$

$t - 1 = \pm\sqrt{\dfrac{3}{2}}$

$t = 1 \pm \dfrac{\sqrt{6}}{2}$

t-intercepts: $\left(1 \pm \dfrac{\sqrt{6}}{2}, 0\right)$

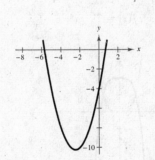

9. $h(x) = 4x^2 + 4x + 13$

$= 4(x^2 + x) + 13$

$= 4\left(x^2 + x + \dfrac{1}{4} - \dfrac{1}{4}\right) + 13$

$= 4\left(x^2 + x + \dfrac{1}{4}\right) - 1 + 13$

$= 4\left(x + \dfrac{1}{2}\right)^2 + 12$

Vertex: $\left(-\dfrac{1}{2}, 12\right)$

Axis of symmetry: $x = -\dfrac{1}{2}$

$0 = 4\left(x + \dfrac{1}{2}\right)^2 + 12$

$\left(x + \dfrac{1}{2}\right)^2 = -3$

No real zeros

x-intercepts: none

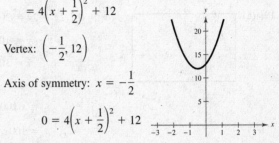

11. $h(x) = x^2 + 5x - 4$

$= x^2 + 5x + \dfrac{25}{4} - \dfrac{25}{4} - 4$

$= \left(x + \dfrac{5}{2}\right)^2 - \dfrac{25}{4} - \dfrac{16}{4}$

$= \left(x + \dfrac{5}{2}\right)^2 - \dfrac{41}{4}$

Vertex: $\left(-\dfrac{5}{2}, -\dfrac{41}{4}\right)$

Axis of symmetry: $x = -\dfrac{5}{2}$

$0 = x^2 + 5x - 4$

By the Quadratic Formula, $x = \dfrac{-5 \pm \sqrt{41}}{2}$.

x-intercepts: $\left(\dfrac{-5 \pm \sqrt{41}}{2}, 0\right)$

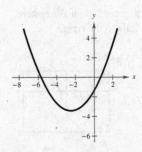

13. $f(x) = \dfrac{1}{3}(x^2 + 5x - 4)$

$= \dfrac{1}{3}\left(x^2 + 5x + \dfrac{25}{4} - \dfrac{25}{4} - 4\right)$

$= \dfrac{1}{3}\left[\left(x + \dfrac{5}{2}\right)^2 - \dfrac{41}{4}\right]$

$= \dfrac{1}{3}\left(x + \dfrac{5}{2}\right)^2 - \dfrac{41}{12}$

Vertex: $\left(-\dfrac{5}{2}, -\dfrac{41}{12}\right)$

Axis of symmetry: $x = -\dfrac{5}{2}$

$0 = x^2 + 5x - 4$

By the Quadratic Formula, $x = \dfrac{-5 \pm \sqrt{41}}{2}$.

x-intercepts: $\left(\dfrac{-5 \pm \sqrt{41}}{2}, 0\right)$

15. Vertex: $(4, 1) \Rightarrow f(x) = a(x - 4)^2 + 1$

Point: $(2, -1) \Rightarrow -1 = a(2 - 4)^2 + 1$

$$-2 = 4a$$

$$-\tfrac{1}{2} = a$$

Thus, $f(x) = -\tfrac{1}{2}(x - 4)^2 + 1$.

17. Vertex: $(1, -4) \Rightarrow f(x) = a(x - 1)^2 - 4$

Point: $(2, -3) \Rightarrow -3 = a(2 - 1)^2 - 4$

$$1 = a$$

Thus, $f(x) = (x - 1)^2 - 4$.

19. (a)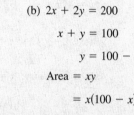

(b) $2x + 2y = 200$

$$x + y = 100$$

$$y = 100 - x$$

$$\text{Area} = xy$$

$$= x(100 - x)$$

$$= 100x - x^2$$

(c) $\text{Area} = 100x - x^2$

$$= -(x^2 - 100x + 2500 - 2500)$$

$$= -[(x - 50)^2 - 2500]$$

$$= -(x - 50)^2 + 2500$$

The maximum area occurs at the vertex when $x = 50$ and $y = 100 - 50 = 50$. The dimensions with the maximum area are $x = 50$ meters and $y = 50$ meters.

21. $C = 70{,}000 - 120x + 0.055x^2$

The minimum cost occurs at the vertex of the parabola.

Vertex: $-\dfrac{b}{2a} = -\dfrac{-120}{2(0.055)} \approx 1091$ units

Approximately 1091 units should be produced each day to yield a minimum cost.

23. $y = x^3, f(x) = -(x - 4)^3$

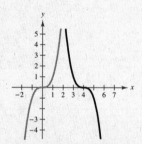

Transformation: Reflection in the x-axis and a horizontal shift four units to the right

25. $y = x^4, f(x) = 2 - x^4$

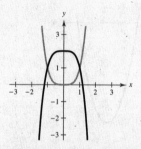

Transformation: Reflection in the x-axis and a vertical shift two units upward

27. $y = x^5, f(x) = (x - 3)^5$

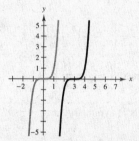

Transformation: Horizontal shift three units to the right

29. $f(x) = -x^2 + 6x + 9$

The degree is even and the leading coefficient is negative. The graph falls to the left and falls to the right.

31. $g(x) = \tfrac{3}{4}(x^4 + 3x^2 + 2)$

The degree is even and the leading coefficient is positive. The graph rises to the left and rises to the right.

33. $f(x) = 2x^2 + 11x - 21$

$$0 = 2x^2 + 11x - 21$$

$$= (2x - 3)(x + 7)$$

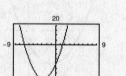

Zeros: $x = \tfrac{3}{2}, -7$, all of multiplicity 1 (odd multiplicity)

Turning points: 1

35. $f(t) = t^3 - 3t$

$$0 = t^3 - 3t$$

$$0 = t(t^2 - 3)$$

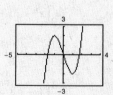

Zeros: $t = 0, \pm\sqrt{3}$ all of multiplicity 1 (odd multiplicity)

Turning points: 2

37. $f(x) = -12x^3 + 20x^2$

$0 = -12x^3 + 20x^2$

$0 = -4x^2(3x - 5)$

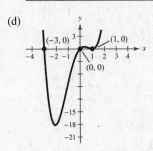

Zeros: $x = 0$ of multiplicity 2
(even multiplicity)

$x = \frac{5}{3}$ of multiplicity 1
(odd multiplicity)

Turning points: 2

39. $f(x) = -x^3 + x^2 - 2$

(a) The degree is odd and the leading coefficient is negative. The graph rises to the left and falls to the right.

(b) Zero: $x = -1$

(c)

x	-3	-2	-1	0	1	2
$f(x)$	34	10	0	-2	-2	-6

(d)

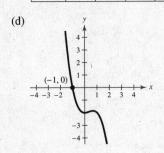

41. $f(x) = x(x^3 + x^2 - 5x + 3)$

(a) The degree is even and the leading coefficient is positive. The graph rises to the left and rises to the right.

(b) Zeros: $x = 0, 1, -3$

(c)

x	-4	-3	-2	-1	0	1	2	3
$f(x)$	100	0	-18	-8	0	0	10	72

(d)

43. (a) $f(x) = 3x^3 - x^2 + 3$

x	-3	-2	-1	0	1	2	3
$f(x)$	-87	-25	-1	3	5	23	75

(b) The zero is in the interval $[-1, 0]$.

Zero: $x \approx -0.900$

45. (a) $f(x) = x^4 - 5x - 1$

x	-3	-2	-1	0	1	2	3
$f(x)$	95	25	5	-1	-5	5	65

(b) There are two zeros, one in the interval $[-1, 0]$ and one in the interval $[1, 2]$

Zeros: $x \approx -0.200, x \approx 1.772$

47.

$$
\begin{array}{r}
8x + 5 \\
3x - 2 \overline{\smash{)}\ 24x^2 - x - 8} \\
\underline{24x^2 - 16x} \\
15x - 8 \\
\underline{15x - 10} \\
2
\end{array}
$$

Thus, $\dfrac{24x^2 - x - 8}{3x - 2} = 8x + 5 + \dfrac{2}{3x - 2}$.

49.

$$
\begin{array}{r}
5x + 2 \\
x^2 - 3x + 1 \overline{\smash{)}\ 5x^3 - 13x^2 - x + 2} \\
\underline{5x^3 - 15x^2 + 5x} \\
2x^2 - 6x + 2 \\
\underline{2x^2 - 6x + 2} \\
0
\end{array}
$$

Thus, $\dfrac{5x^3 - 13x^2 - x + 2}{x^2 - 3x + 1} = 5x + 2$.

51.
$$x^2 + 0x + 2 \overline{)x^4 - 3x^3 + 4x^2 - 6x + 3}$$
$$\begin{array}{r} x^2 - 3x + 2 \\ \hline x^4 + 0x^3 + 2x^2 \\ \hline -3x^3 + 2x^2 - 6x \\ -3x^3 + 0x^2 - 6x \\ \hline 2x^2 + 0x + 3 \\ 2x^2 + 0x + 4 \\ \hline -1 \end{array}$$

Thus, $\dfrac{x^4 - 3x^3 + 4x^2 - 6x + 3}{x^2 + 2} = x^2 - 3x + 2 - \dfrac{1}{x^2 + 2}.$

53.
$$\begin{array}{r|rrrrr} 2 & 6 & -4 & -27 & 18 & 0 \\ & & 12 & 16 & -22 & -8 \\ \hline & 6 & 8 & -11 & -4 & -8 \end{array}$$

Thus,

$\dfrac{6x^4 - 4x^3 - 27x^2 + 18x}{x - 2} = 6x^3 + 8x^2 - 11x - 4 - \dfrac{8}{x - 2}.$

55.
$$\begin{array}{r|rrrr} 4 & 2 & -19 & 38 & 24 \\ & & 8 & -44 & -24 \\ \hline & 2 & -11 & -6 & 0 \end{array}$$

Thus, $\dfrac{2x^3 - 19x^2 + 38x + 24}{x - 4} = 2x^2 - 11x - 6.$

57. $f(x) = 20x^4 + 9x^3 - 14x^2 - 3x$

(a)
$$\begin{array}{r|rrrrr} -1 & 20 & 9 & -14 & -3 & 0 \\ & & -20 & 11 & 3 & 0 \\ \hline & 20 & -11 & -3 & 0 & 0 \end{array}$$

Yes, $x = -1$ is a zero of f.

(c)
$$\begin{array}{r|rrrrr} 0 & 20 & 9 & -14 & -3 & 0 \\ & & 0 & 0 & 0 & 0 \\ \hline & 20 & 9 & -14 & -3 & 0 \end{array}$$

Yes, $x = 0$ is a zero of f.

(b)
$$\begin{array}{r|rrrrr} \frac{3}{4} & 20 & 9 & -14 & -3 & 0 \\ & & 15 & 18 & 3 & 0 \\ \hline & 20 & 24 & 4 & 0 & 0 \end{array}$$

Yes, $x = \frac{3}{4}$ is a zero of f.

(d)
$$\begin{array}{r|rrrrr} 1 & 20 & 9 & -14 & -3 & 0 \\ & & 20 & 29 & 15 & 12 \\ \hline & 20 & 29 & 15 & 12 & 12 \end{array}$$

No, $x = 1$ is not a zero of f.

59. $f(x) = x^4 + 10x^3 - 24x^2 + 20x + 44$

(a)
$$\begin{array}{r|rrrrr} -3 & 1 & 10 & -24 & 20 & 44 \\ & & -3 & -21 & 135 & -465 \\ \hline & 1 & 7 & -45 & 155 & -421 \end{array}$$

Thus, $f(-3) = -421$.

(b)
$$\begin{array}{r|rrrrr} -1 & 1 & 10 & -24 & 20 & 44 \\ & & -1 & -9 & 33 & -53 \\ \hline & 1 & 9 & -33 & 53 & -9 \end{array}$$

$f(-1) = -9$

61. $f(x) = x^3 + 4x^2 - 25x - 28$; Factor: $(x - 4)$

(a)
$$\begin{array}{r|rrrr} 4 & 1 & 4 & -25 & -28 \\ & & 4 & 32 & 28 \\ \hline & 1 & 8 & 7 & 0 \end{array}$$

Yes, $x - 4$ is a factor of $f(x)$.

(b) $x^2 + 8x + 7 = (x + 7)(x + 1)$

The remaining factors of f are $(x + 7)$ and $(x + 1)$.

(c) $f(x) = x^3 + 4x^2 - 25x - 28$

$= (x + 7)(x + 1)(x - 4)$

(d) Zeros: $-7, -1, 4$

(e)
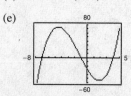

63. $f(x) = x^4 - 4x^3 - 7x^2 + 22x + 24$

Factors: $(x + 2), (x - 3)$

(a)

$$
\begin{array}{r|rrrrr}
-2 & 1 & -4 & -7 & 22 & 24 \\
   &   & -2 & 12 & -10 & -24 \\
\hline
   & 1 & -6 & 5 & 12 & 0
\end{array}
$$

$$
\begin{array}{r|rrrr}
3 & 1 & -6 & 5 & 12 \\
  &   & 3 & -9 & -12 \\
\hline
  & 1 & -3 & -4 & 0
\end{array}
$$

Both are factors since the remainders are zero.

(b) $x^2 - 3x - 4 = (x + 1)(x - 4)$

The remaining factors are $(x + 1)$ and $(x - 4)$.

(c) $f(x) = (x + 1)(x - 4)(x + 2)(x - 3)$

(d) Zeros: $-2, -1, 3, 4$

(e)

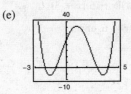

65. $6 + \sqrt{-4} = 6 + 2i$

67. $i^2 + 3i = -1 + 3i$

69. $(7 + 5i) + (-4 + 2i) = (7 - 4) + (5i + 2i) = 3 + 7i$

71. $5i(13 - 8i) = 65i - 40i^2 = 40 + 65i$

73. $(10 - 8i)(2 - 3i) = 20 - 30i - 16i + 24i^2$
$$= -4 - 46i$$

75. $\dfrac{6 + i}{4 - i} = \dfrac{6 + i}{4 - i} \cdot \dfrac{4 + i}{4 + i}$

$= \dfrac{24 + 10i + i^2}{16 + 1}$

$= \dfrac{23 + 10i}{17}$

$= \dfrac{23}{17} + \dfrac{10}{17}i$

77. $\dfrac{4}{2 - 3i} + \dfrac{2}{1 + i} = \dfrac{4}{2 - 3i} \cdot \dfrac{2 + 3i}{2 + 3i} + \dfrac{2}{1 + i} \cdot \dfrac{1 - i}{1 - i}$

$= \dfrac{8 + 12i}{4 + 9} + \dfrac{2 - 2i}{1 + 1}$

$= \dfrac{8}{13} + \dfrac{12}{13}i + 1 - i$

$= \left(\dfrac{8}{13} + 1\right) + \left(\dfrac{12}{13}i - i\right)$

$= \dfrac{21}{13} - \dfrac{1}{13}i$

79. $3x^2 + 1 = 0$

$3x^2 = -1$

$x^2 = -\dfrac{1}{3}$

$x = \pm\sqrt{-\dfrac{1}{3}}$

$= \pm\sqrt{\dfrac{1}{3}}\, i = \pm\dfrac{\sqrt{3}}{3}i$

81. $x^2 - 2x + 10 = 0$

$x^2 - 2x + 1 = -10 + 1$

$(x - 1)^2 = -9$

$x - 1 = \pm\sqrt{-9}$

$x = 1 \pm 3i$

83. $f(x) = 3x(x - 2)^2$

Zeros: $x = 0, x = 2$

85. $f(x) = x^2 - 9x + 8 = (x - 1)(x - 8)$

Zeros: $x = 1, x = 8$

87. $f(x) = (x + 4)(x - 6)(x - 2i)(x + 2i)$

Zeros: $x = -4, x = 6, x = 2i, x = -2i$

89. $f(x) = -4x^3 + 8x^2 - 3x + 15$

Possible rational zeros:

$\pm 1, \pm 3, \pm 5, \pm 15, \pm\frac{1}{2}, \pm\frac{3}{2}, \pm\frac{5}{2}, \pm\frac{15}{2}, \pm\frac{1}{4}, \pm\frac{3}{4}, \pm\frac{5}{4}, \pm\frac{15}{4}$

91. $f(x) = x^3 - 2x^2 - 21x - 18$

Possible rational zeros: $\pm 1, \pm 2, \pm 3, \pm 6, \pm 9, \pm 18$

$$
\begin{array}{r|rrrr}
-1 & 1 & -2 & -21 & -18 \\
   &   & -1 & 3 & 18 \\
\hline
   & 1 & -3 & -18 & 0
\end{array}
$$

$x^3 - 2x^2 - 21x - 18 = (x + 1)(x^2 - 3x - 18)$

$\qquad\qquad\qquad\quad = (x + 1)(x - 6)(x + 3)$

The zeros of $f(x)$ are $x = -1$, $x = 6$, and $x = -3$.

93. $f(x) = x^3 - 10x^2 + 17x - 8$

Possible rational zeros: $\pm 1, \pm 2, \pm 4, \pm 8$

$$
\begin{array}{r|rrrr}
1 & 1 & -10 & 17 & -8 \\
   &   & 1 & -9 & 8 \\
\hline
   & 1 & -9 & 8 & 0
\end{array}
$$

$x^3 - 10x^2 + 17x - 8 = (x - 1)(x^2 - 9x + 8)$

$\qquad\qquad\qquad\quad = (x - 1)(x - 1)(x - 8)$

$\qquad\qquad\qquad\quad = (x - 1)^2(x - 8)$

The zeros of $f(x)$ are $x = 1$ and $x = 8$.

95. $f(x) = x^4 + x^3 - 11x^2 + x - 12$

Possible rational zeros: $\pm 1, \pm 2, \pm 3, \pm 4, \pm 6, \pm 12$

$$
\begin{array}{r|rrrrr}
3 & 1 & 1 & -11 & 1 & -12 \\
   &   & 3 & 12 & 3 & 12 \\
\hline
   & 1 & 4 & 1 & 4 & 0
\end{array}
$$

$$
\begin{array}{r|rrrr}
-4 & 1 & 4 & 1 & 4 \\
   &   & -4 & 0 & -4 \\
\hline
   & 1 & 0 & 1 & 0
\end{array}
$$

$x^4 + x^3 - 11x^2 + x - 12 = (x - 3)(x + 4)(x^2 + 1)$

The real zeros of $f(x)$ are $x = 3$, and $x = -4$.

97. $f(x) = 3\left(x - \frac{2}{3}\right)(x - 4)\left(x - \sqrt{3}i\right)\left(x + \sqrt{3}i\right)$ Since $\sqrt{3}i$ is a zero, so is $-\sqrt{3}i$.

$\qquad = (3x - 2)(x - 4)(x^2 + 3)$ Multiply by 3 to clear the fraction.

$\qquad = (3x^2 - 14x + 8)(x^2 + 3)$

$\qquad = 3x^4 - 14x^3 + 17x^2 - 42x + 24$

Note: $f(x) = a(3x^4 - 14x^3 + 17x^2 - 42x + 24)$, where a is any real nonzero number, has zeros $\frac{2}{3}$, 4, and $\pm\sqrt{3}i$.

99. $f(x) = x^3 - 4x^2 + x - 4$, Zero: i

Since i is a zero, so is $-i$.

$$
\begin{array}{r|rrrr}
i & 1 & -4 & 1 & -4 \\
   &   & i & -1-4i & 4 \\
\hline
   & 1 & -4+i & -4i & 0
\end{array}
$$

$$
\begin{array}{r|rrr}
-i & 1 & -4+i & -4i \\
   &   & -i & 4i \\
\hline
   & 1 & -4 & 0
\end{array}
$$

$f(x) = (x - i)(x + i)(x - 4)$, Zeros: $x = \pm i, 4$

101. $g(x) = 2x^4 - 3x^3 - 13x^2 + 37x - 15$, Zero: $2 + i$

Since $2 + i$ is a zero, so is $2 - i$

$$
\begin{array}{r|rrrrr}
2+i & 2 & -3 & -13 & 37 & -15 \\
   &   & 4+2i & 5i & -31-3i & 15 \\
\hline
   & 2 & 1+2i & -13+5i & 6-3i & 0
\end{array}
$$

$$
\begin{array}{r|rrrr}
2-i & 2 & 1+2i & -13+5i & 6-3i \\
   &   & 4-2i & 10-5i & -6+3i \\
\hline
   & 2 & 5 & -3 & 0
\end{array}
$$

$g(x) = [x - (2 + i)][x - (2 - i)](2x^2 + 5x - 3)$

$\qquad = (x - 2 - i)(x - 2 + i)(2x - 1)(x + 3)$

Zeros: $x = 2 \pm i, \frac{1}{2}, -3$

103. $f(x) = x^3 + 4x^2 - 5x$

$\qquad = x(x^2 + 4x - 5)$

$\qquad = x(x + 5)(x - 1)$

Zeros: $x = 0, -5, 1$

105. $g(x) = x^4 + 4x^3 - 3x^2 + 40x + 208$, Zero: $x = -4$

$$
\begin{array}{r|rrrrr}
-4 & 1 & 4 & -3 & 40 & 208 \\
   &   & -4 & 0 & 12 & -208 \\
\hline
   & 1 & 0 & -3 & 52 & 0
\end{array}
$$

$$
\begin{array}{r|rrrr}
-4 & 1 & 0 & -3 & 52 \\
   &   & -4 & 16 & -52 \\
\hline
   & 1 & -4 & 13 & 0
\end{array}
$$

$g(x) = (x + 4)^2(x^2 - 4x + 13)$

By the Quadratic Formula the zeros of $x^2 - 4x + 13$ are $x = 2 \pm 3i$. The zeros of $g(x)$ are $x = -4$ of multiplicity 2, and $x = 2 \pm 3i$.

$g(x) = (x + 4)^2[x - (2 + 3i)][x - (2 - 3i)]$

$\quad = (x + 4)^2(x - 2 - 3i)(x - 2 + 3i)$

107. $g(x) = 5x^3 + 3x^2 - 6x + 9$

$g(x)$ has two variations in sign, so g has either two or no positive real zeros.

$g(-x) = -5x^3 + 3x^2 + 6x + 9$

$g(-x)$ has one variation in sign, so g has one negative real zero.

109. $f(x) = 4x^3 - 3x^2 + 4x - 3$

(a)
$$
\begin{array}{r|rrrr}
1 & 4 & -3 & 4 & -3 \\
  &   & 4 & 1 & 5 \\
\hline
  & 4 & 1 & 5 & 2
\end{array}
$$

Since the last row has all positive entries, $x = 1$ is an upper bound.

(b)
$$
\begin{array}{r|rrrr}
-\frac{1}{4} & 4 & -3 & 4 & -3 \\
  &   & -1 & 1 & -\frac{5}{4} \\
\hline
  & 4 & -4 & 5 & -\frac{17}{4}
\end{array}
$$

Since the last row entries alternate in sign, $x = -\frac{1}{4}$ is a lower bound.

111. $f(x) = \dfrac{5x}{x + 12}$

Domain: all real numbers x except $x = -12$

113. $f(x) = \dfrac{8}{x^2 - 10x + 24}$

$\quad = \dfrac{8}{(x - 4)(x - 6)}$

Domain: all real numbers x except $x = 4$ and $x = 6$

115. $f(x) = \dfrac{4}{x + 3}$

Vertical asymptote: $x = -3$

Horizontal asymptote: $y = 0$

117. $h(x) = \dfrac{2x - 10}{x^2 - 2x - 15}$

$\quad = \dfrac{2(x - 5)}{(x + 3)(x - 5)}$

$\quad = \dfrac{2}{x + 3}, \quad x \neq 5$

Vertical asymptote: $x = -3$

Horizontal asymptote: $y = 0$

119. $f(x) = \dfrac{-5}{x^2}$

(a) Domain: all real numbers x except $x = 0$

(b) No intercepts

(c) Vertical asymptote: $x = 0$
   Horizontal asymptote: $y = 0$

(d)

x	± 3	± 2	± 1
y	$-\frac{5}{9}$	$-\frac{5}{4}$	-5

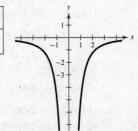

121. $g(x) = \dfrac{2 + x}{1 - x} = -\dfrac{x + 2}{x - 1}$

(a) Domain: all real numbers x except $x = 1$

(b) x-intercept: $(-2, 0)$
y-intercept: $(0, 2)$

(c) Vertical asymptote: $x = 1$
Horizontal asymptote: $y = -1$

(d)

x	-1	0	2	3
y	$\frac{1}{2}$	2	-4	$-\frac{5}{2}$

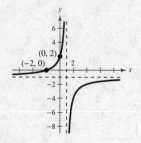

123. $p(x) = \dfrac{x^2}{x^2 + 1}$

(a) Domain: all real numbers x

(b) Intercept: $(0, 0)$

(c) Horizontal asymptote: $y = 1$

(d)

x	± 3	± 2	± 1	0
y	$\frac{9}{10}$	$\frac{4}{5}$	$\frac{1}{2}$	0

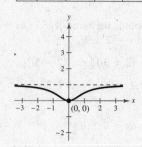

125. $f(x) = \dfrac{x}{x^2 + 1}$

(a) Domain: all real numbers x

(b) Intercept: $(0, 0)$

(c) Horizontal asymptote: $y = 0$

(d)

x	-2	-1	0	1	2
y	$-\frac{2}{5}$	$-\frac{1}{2}$	0	$\frac{1}{2}$	$\frac{2}{5}$

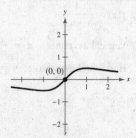

127. $f(x) = \dfrac{-6x^2}{x^2 + 1}$

(a) Domain: all real numbers x

(b) Intercept: $(0, 0)$

(c) Horizontal asymptote: $y = -6$

(d)

x	± 3	± 2	± 1	0
y	$-\frac{27}{5}$	$-\frac{24}{5}$	-3	0

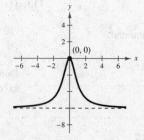

129. $f(x) = \dfrac{6x^2 - 11x + 3}{3x^2 - x}$

$$= \dfrac{(3x - 1)(2x - 3)}{x(3x - 1)} = \dfrac{2x - 3}{x}, \; x \neq \dfrac{1}{3}$$

(a) Domain: all real numbers x except

$x = 0$ and $x = \dfrac{1}{3}$

(b) x-intercept: $\left(\dfrac{3}{2}, 0\right)$
y-intercept: none

(c) Vertical asymptote: $x = 0$
Horizontal asymptote: $y = 2$

(d)

x	-2	-1	1	2	3	4
y	$\frac{7}{2}$	5	-1	$\frac{1}{2}$	1	$\frac{5}{4}$

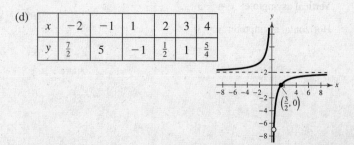

131. $f(x) = \dfrac{2x^3}{x^2 + 1} = 2x - \dfrac{2x}{x^2 + 1}$

(a) Domain: all real numbers x

(b) Intercept: $(0, 0)$

(c) Slant asymptote: $y = 2x$

(d)

x	-2	-1	0	1	2
y	$-\dfrac{16}{5}$	-1	0	1	$\dfrac{16}{5}$

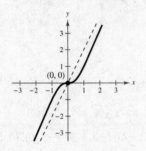

133. $f(x) = \dfrac{3x^3 - 2x^2 - 3x + 2}{3x^2 - x - 4}$

$\qquad = \dfrac{(3x - 2)(x + 1)(x - 1)}{(3x - 4)(x + 1)}$

$\qquad = \dfrac{(3x - 2)(x - 1)}{3x - 4}$

$\qquad = x - \dfrac{1}{3} + \dfrac{2/3}{3x - 4}, \quad x \neq -1$

(a) Domain: all real numbers x except $x = -1$, $x = \dfrac{4}{3}$

(b) x-intercepts: $(1, 0)$ and $\left(\dfrac{2}{3}, 0\right)$

y-intercept: $\left(0, -\dfrac{1}{2}\right)$

(c) Vertical asymptote: $x = \dfrac{4}{3}$

Slant asymptote: $y = x - \dfrac{1}{3}$

(d)

x	-3	-2	0	1	2	3
y	$-\dfrac{44}{13}$	$-\dfrac{12}{5}$	$-\dfrac{1}{2}$	0	2	$\dfrac{14}{5}$

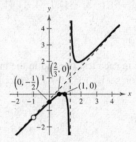

135. $\overline{C} = \dfrac{C}{x} = \dfrac{0.5x + 500}{x}, \quad 0 < x$

Horizontal asymptote: $\overline{C} = \dfrac{0.5}{1} = 0.5$

As x increases, the average cost per unit approaches the horizontal asymptote, $\overline{C} = 0.5 = \$0.50$.

137. (a)

(b) The area of print is $(x - 4)(y - 4)$, which is 30 square inches.

$(x - 4)(y - 4) = 30$

$y - 4 = \dfrac{30}{x - 4}$

$y = \dfrac{30}{x - 4} + 4$

$y = \dfrac{30 + 4(x - 4)}{x - 4}$

$y = \dfrac{4x + 14}{x - 4}$

$y = \dfrac{2(2x + 7)}{x - 4}$

Total area $= xy = x\left[\dfrac{2(2x + 7)}{x - 4}\right] = \dfrac{2x(2x + 7)}{x - 4}$

139.
$$6x^2 + 5x < 4$$
$$6x^2 + 5x - 4 < 0$$
$$(3x + 4)(2x - 1) < 0$$

Critical numbers: $x = -\frac{4}{3}, x = \frac{1}{2}$

Test intervals: $\left(-\infty, -\frac{4}{3}, \right), \left(-\frac{4}{3}, \frac{1}{2}\right), \left(\frac{1}{2}, \infty\right)$

Test: Is $(3x + 4)(2x - 1) < 0$?

By testing an x-value in each test interval in the inequality, we see that the solution set is: $\left(-\frac{4}{3}, \frac{1}{2}\right)$

141.
$$x^3 - 16x \geq 0$$
$$x(x + 4)(x - 4) \geq 0$$

Critical numbers: $x = 0, x = \pm 4$

Test intervals: $(-\infty, -4), (-4, 0), (0, 4), (4, \infty)$

Test: Is $x(x + 4)(x - 4) \geq 0$?

By testing an x-value in each test interval in the inequality, we see that the solution set is: $[-4, 0] \cup [4, \infty)$.

143.
$$\frac{2}{x + 1} \leq \frac{3}{x - 1}$$

$$\frac{2(x - 1) - 3(x + 1)}{(x + 1)(x - 1)} \leq 0$$

$$\frac{2x - 2 - 3x - 3}{(x + 1)(x - 1)} \leq 0$$

$$\frac{-(x + 5)}{(x + 1)(x - 1)} \leq 0$$

Critical numbers: $x = -5, x = \pm 1$

Test intervals: $(-\infty, -5), (-5, -1), (-1, 1), (1, \infty)$

Test: Is $\dfrac{-(x + 5)}{(x + 1)(x - 1)} \leq 0$?

By testing an x-value in each test interval in the inequality, we see that the solution set is: $[-5, -1) \cup (1, \infty)$

145.
$$\frac{x^2 + 7x + 12}{x} \geq 0$$

$$\frac{(x + 4)(x + 3)}{x} \geq 0$$

Critical numbers: $x = -4, x = -3, x = 0$

Test intervals: $(-\infty, -4), (-4, -3), (-3, 0), (0, \infty)$

Test: Is $\dfrac{(x + 4)(x + 3)}{x} \geq 0$?

By testing an x-value in each test interval in the inequality, we see that the solution set is: $[-4, -3] \cup (0, \infty)$

147.
$$5000(1 + r)^2 > 5500$$
$$(1 + r)^2 > 1.1$$
$$1 + r > 1.0488$$
$$r > 0.0488$$
$$r > 4.9\%$$

149. False. A fourth-degree polynomial can have at most four zeros and complex zeros occur in conjugate pairs.

151. The maximum (or minimum) value of a quadratic function is located at its graph's vertex. To find the vertex, either write the equation in standard form or use the formula

$$\left(-\frac{b}{2a}, f\left(-\frac{b}{2a}\right)\right).$$

If the leading coefficient is positive, the vertex is a minimum. If the leading coefficient is negative, the vertex is a maximum.

153. An asymptote of a graph is a line to which the graph becomes arbitrarily close as x increases or decreases without bound.

Problem Solving for Chapter 2

1. $f(x) = ax^3 + bx^2 + cx + d$

$$
\begin{array}{r}
ax^2 + (ak + b)x + (ak^2 + bk + c) \\
x - k \overline{)\, ax^3 + bx^2 \quad\quad + cx \quad\quad\quad\quad + d} \\
\underline{ax^3 - akx^2} \\
(ak + b)x^2 + cx \\
\underline{(ak + b)x^2 - (ak^2 + bk)x} \\
(ak^2 + bk + c)x + d \\
\underline{(ak^2 + bk + c)x - (ak^3 + bk^2 + ck)} \\
(ak^3 + bk^2 + ck + d)
\end{array}
$$

Thus, $f(x) = ax^3 + bx^2 + cx + d = (x - k)[ax^2 + (ak + b)x + (ak^2 + bx + c)] + ak^3 + bk^2 + ck + d$ and $f(k) = ak^3 + bk^2 + ck + d$. Since the remainder $r = ak^3 + bk^2 + ck + d, f(k) = r$.

3. $V = l \cdot w \cdot h = x^2(x + 3)$

$x^2(x + 3) = 20$

$x^3 + 3x^2 - 20 = 0$

Possible rational zeros: $\pm 1, \pm 2, \pm 4, \pm 5, \pm 10, \pm 20$

$$
\begin{array}{r|rrrr}
2 & 1 & 3 & 0 & -20 \\
  &   & 2 & 10 & 20 \\
\hline
  & 1 & 5 & 10 & 0
\end{array}
$$

$(x - 2)(x^2 + 5x + 10) = 0$

$x = 2 \ \text{ or } \ x = \dfrac{-5 \pm \sqrt{15}i}{2}$

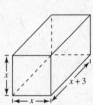

Choosing the real positive value for x we have: $x = 2$ and $x + 3 = 5$. The dimensions of the mold are 2 inches $\times$ 2 inches $\times$ 5 inches.

5. (a) $y = ax^2 + bx + c$

$(0, -4)$: $-4 = a(0)^2 + b(0) + c$

$-4 = c$

$(4, 0)$: $0 = a(4)^2 + b(4) - 4$

$0 = 16a + 4b - 4 = 4(4a + b - 1)$

$0 = 4a + b - 1$ or $b = 1 - 4a$

$(1, 0)$: $0 = a(1)^2 + b(1) - 4$

$4 = a + b$

$4 = a + (1 - 4a)$

$4 = 1 - 3a$

$-3 = -3a$

$a = -1$

$b = 1 - 4(-1) = 5$

$y = -x^2 + 5x - 4$

(b) Enter the data points $(0, -4)$, $(1, 0)$, $(2, 2)$, $(4, 0)$, $(6, -10)$ and use the regression feature to obtain $y = -x^2 + 5x - 4$.

7. $f(x) = (x - k)q(x) + r$

(a) Cubic, passes through $(2, 5)$, rises to the right

One possibility:

$f(x) = (x - 2)x^2 + 5$

$= x^3 - 2x^2 + 5$

(b) Cubic, passes through $(-3, 1)$, falls to the right

One possibility:

$f(x) = -(x + 3)x^2 + 1$

$= -x^3 - 3x^2 + 1$

9. $(a + bi)(a - bi) = a^2 - abi + abi - b^2i^2 = a^2 + b^2$

Since a and b are real numbers, $a^2 + b^2$ is also a real number.

11. $f(x) = \dfrac{ax}{(x - b)^2}$

(a) $b \neq 0 \implies x = b$ is a vertical asymptote.
a causes a vertical stretch if $|a| > 1$ and a vertical shrink if $0 < |a| < 1$. For $|a| > 1$, the graph becomes wider as $|a|$ increases. When a is negative the graph is reflected about the x-axis.

(b) $a \neq 0$. Varying the value of b varies the vertical asymptote of the graph of f. For $b > 0$, the graph is translated to the right. For $b < 0$, the graph is reflected in the x-axis and is translated to the left.

Chapter 2   Practice Test

1. Sketch the graph of $f(x) = x^2 - 6x + 5$ and identify the vertex and the intercepts.

2. Find the number of units x that produce a minimum cost C if
   $C = 0.01x^2 - 90x + 15,000$.

3. Find the quadratic function that has a maximum at $(1, 7)$ and passes through the point $(2, 5)$.

4. Find two quadratic functions that have x-intercepts $(2, 0)$ and $\left(\frac{4}{3}, 0\right)$.

5. Use the leading coefficient test to determine the right and left end behavior of the graph of the polynomial function $f(x) = -3x^5 + 2x^3 - 17$.

6. Find all the real zeros of $f(x) = x^5 - 5x^3 + 4x$.

7. Find a polynomial function with 0, 3, and -2 as zeros.

8. Sketch $f(x) = x^3 - 12x$.

9. Divide $3x^4 - 7x^2 + 2x - 10$ by $x - 3$ using long division.

10. Divide $x^3 - 11$ by $x^2 + 2x - 1$.

11. Use synthetic division to divide $3x^5 + 13x^4 + 12x - 1$ by $x + 5$.

12. Use synthetic division to find $f(-6)$ given $f(x) = 7x^3 + 40x^2 - 12x + 15$.

13. Find the real zeros of $f(x) = x^3 - 19x - 30$.

14. Find the real zeros of $f(x) = x^4 + x^3 - 8x^2 - 9x - 9$.

15. List all possible rational zeros of the function $f(x) = 6x^3 - 5x^2 + 4x - 15$.

16. Find the rational zeros of the polynomial $f(x) = x^3 - \frac{20}{3}x^2 + 9x - \frac{10}{3}$.

17. Write $f(x) = x^4 + x^3 + 5x - 10$ as a product of linear factors.

18. Find a polynomial with real coefficients that has 2, $3 + i$, and $3 - 2i$ as zeros.

19. Use synthetic division to show that $3i$ is a zero of $f(x) = x^3 + 4x^2 + 9x + 36$.

20. Sketch the graph of $f(x) = \dfrac{x - 1}{2x}$ and label all intercepts and asymptotes.

21. Find all the asymptotes of $f(x) = \dfrac{8x^2 - 9}{x^2 + 1}$.

22. Find all the asymptotes of $f(x) = \dfrac{4x^2 - 2x + 7}{x - 1}$.

23. Given $z_1 = 4 - 3i$ and $z_2 = -2 + i$, find the following:

   (a) $z_1 - z_2$

   (b) $z_1 z_2$

   (c) z_1/z_2

24. Solve the inequality: $x^2 - 49 \leq 0$

25. Solve the inequality: $\dfrac{x + 3}{x - 7} \geq 0$

CHAPTER 3
Exponential and Logarithmic Functions

Section 3.1 Exponential Functions and Their Graphs **148**

Section 3.2 Logarithmic Functions and Their Graphs **152**

Section 3.3 Properties of Logarithms **157**

Section 3.4 Exponential and Logarithmic Equations **161**

Section 3.5 Exponential and Logarithmic Models **168**

Review Exercises **175**

Problem Solving **182**

Practice Test **184**

CHAPTER 3
Exponential and Logarithmic Functions

Section 3.1   Exponential Functions and Their Graphs

> ■ You should know that a function of the form $f(x) = a^x$, where $a > 0, a \neq 1$, is called an exponential function with base a.
>
> ■ You should be able to graph exponential functions.
>
> ■ You should know formulas for compound interest.
>
>   (a) For n compoundings per year: $A = P\left(1 + \dfrac{r}{n}\right)^{nt}$.
>
>   (b) For continuous compoundings: $A = Pe^{rt}$.

Vocabulary Check

1. algebraic

2. transcendental

3. natural exponential; natural

4. $A = P\left(1 + \dfrac{r}{n}\right)^{nt}$

5. $A = Pe^{rt}$

---

1. $f(5.6) = (3.4)^{5.6} \approx 946.852$

3. $f(-\pi) = 5^{-\pi} \approx 0.006$

5. $g(x) = 5000(2^x) = 5000(2^{-1.5})$
≈ 1767.767

7. $f(x) = 2^x$

Increasing

Asymptote: $y = 0$

Intercept: $(0, 1)$

Matches graph (d).

9. $f(x) = 2^{-x}$

Decreasing

Asymptote: $y = 0$

Intercept: $(0, 1)$

Matches graph (a).

11. $f(x) = \left(\frac{1}{2}\right)^x$

x	-2	-1	0	1	2
$f(x)$	4	2	1	0.5	0.25

Asymptote: $y = 0$

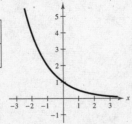

13. $f(x) = 6^{-x}$

x	-2	-1	0	1	2
$f(x)$	36	6	1	0.167	0.028

Asymptote: $y = 0$

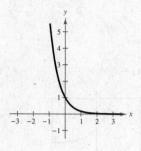

15. $f(x) = 2^{x-1}$

x	-2	-1	0	1	2
$f(x)$	0.125	0.25	0.5	1	2

Asymptote: $y = 0$

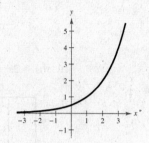

17. $f(x) = 3^x$, $g(x) = 3^{x-4}$

Because $g(x) = f(x - 4)$, the graph of g can be obtained by shifting the graph of f four units to the right.

19. $f(x) = -2^x$, $g(x) = 5 - 2^x$

Because $g(x) = 5 + f(x)$, the graph of g can be obtained by shifting the graph of f five units upward.

21. $f(x) = \left(\frac{7}{2}\right)^x$, $g(x) = -\left(\frac{7}{2}\right)^{-x+6}$

Because $g(x) = -f(-x + 6)$, the graph of g can be obtained by reflecting the graph of f in the x-axis and y-axis and shifting f six units to the right. (**Note:** This is equivalent to shifting f six units to the left and then reflecting the graph in the x-axis and y-axis.)

23. $y = 2^{-x^2}$

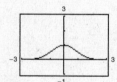

25. $f(x) = 3^{x-2} + 1$

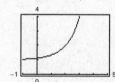

27. $f\left(\frac{3}{4}\right) = e^{-3/4} \approx 0.472$

29. $f(10) = 2e^{-5(10)} \approx 3.857 \times 10^{-22}$

31. $f(6) = 5000e^{0.06(6)} \approx 7166.647$

33. $f(x) = e^x$

x	-2	-1	0	1	2
$f(x)$	0.135	0.368	1	2.718	7.389

Asymptote: $y = 0$

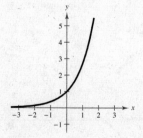

35. $f(x) = 3e^{x+4}$

x	-8	-7	-6	-5	-4
$f(x)$	0.055	0.149	0.406	1.104	3

Asymptote: $y = 0$

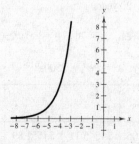

37. $f(x) = 2e^{x-2} + 4$

x	-2	-1	0	1	2
$f(x)$	4.037	4.100	4.271	4.736	6

Asymptote: $y = 4$

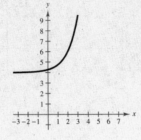

39. $y = 1.08^{-5x}$

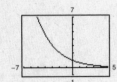

41. $s(t) = 2e^{0.12t}$

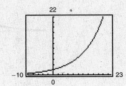

43. $g(x) = 1 + e^{-x}$

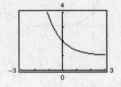

45. $3^{x+1} = 27$

$3^{x+1} = 3^3$

$x + 1 = 3$

$x = 2$

47. $2^{x-2} = \frac{1}{32}$

$2^{x-2} = 2^{-5}$

$x - 2 = -5$

$x = -3$

49. $e^{3x+2} = e^3$

$3x + 2 = 3$

$3x = 1$

$x = \frac{1}{3}$

51. $e^{x^2-3} = e^{2x}$

$x^2 - 3 = 2x$

$x^2 - 2x - 3 = 0$

$(x - 3)(x + 1) = 0$

$x = 3$ or $x = -1$

53. $P = \$2500$, $r = 2.5\%$, $t = 10$ years

Compounded n times per year: $A = P\left(1 + \dfrac{r}{n}\right)^{nt} = 2500\left(1 + \dfrac{0.025}{n}\right)^{10n}$

Compounded continuously: $A = Pe^{rt} = 2500e^{0.025(10)}$

n	1	2	4	12	365	Continuous Compounding
A	\$3200.21	\$3205.09	\$3207.57	\$3209.23	\$3210.04	\$3210.06

55. $P = \$2500$, $r = 3\%$, $t = 20$ years

Compounded n times per year: $A = P\left(1 + \dfrac{r}{n}\right)^{nt} = 2500\left(1 + \dfrac{0.03}{n}\right)^{20n}$

Compounded continuously: $A = Pe^{rt} = 2500e^{0.03(20)}$

n	1	2	4	12	365	Continuous Compounding
A	\$4515.28	\$4535.05	\$4545.11	\$4551.89	\$4555.18	\$4555.30

57. $A = Pe^{rt} = 12,000e^{0.04t}$

t	10	20	30	40	50
A	\$17,901.90	\$26,706.49	\$39,841.40	\$59,436.39	\$88,668.67

59. $A = Pe^{rt} = 12,000e^{0.065t}$

t	10	20	30	40	50
A	\$22,986.49	\$44,031.56	\$84,344.25	\$161,564.86	\$309,484.08

61. $A = 25,000e^{(0.0875)(25)} \approx \$222,822.57$

63. $C(10) = 23.95(1.04)^{10} \approx \$35.45$

65. $V(t) = 100e^{4.6052t}$

  (a) $V(1) \approx 10,000.298$ computers

  (b) $V(1.5) \approx 10,004.472$ computers

  (c) $V(2) \approx 1,000,059.63$ computers

67. $Q = 25\left(\frac{1}{2}\right)^{t/1599}$

  (a) $Q(0) = 25$ grams

  (b) $Q(1000) \approx 16.21$ grams

  (c)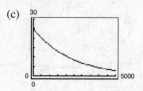

69. $y = \dfrac{100}{1 + 7e^{-0.069x}}$

  (a)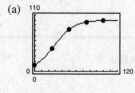

  (b)

x	Sample Data	Model
0	12	12.5
25	44	44.5
50	81	81.82
75	96	96.19
100	99	99.3

  (c) When $x = 36$:

$$y = \frac{100}{1 + 7e^{-0.069(36)}} \approx 63.14\%.$$

  (d) $\frac{2}{3}(100) = \dfrac{100}{1 + 7e^{-0.069x}}$ when

  $x \approx 38$ masses.

71. True. The line $y = -2$ is a horizontal asymptote for the graph of $f(x) = 10^x - 2$.

73. $f(x) = 3^{x-2}$

$= 3^x 3^{-2}$

$= 3^x\left(\dfrac{1}{3^2}\right)$

$= \dfrac{1}{9}(3^x)$

$= h(x)$

Thus, $f(x) \neq g(x)$, but $f(x) = h(x)$.

75. $f(x) = 16(4^{-x})$ and $f(x) = 16(4^{-x})$

$= 4^2(4^{-x})$ $= 16(2^2)^{-x}$

$= 4^{2-x}$ $= 16(2^{-2x})$

$= \left(\dfrac{1}{4}\right)^{-(2-x)}$ $= h(x)$

$= \left(\dfrac{1}{4}\right)^{x-2}$

$= g(x)$

Thus, $f(x) = g(x) = h(x)$.

77. $y = 3^x$ and $y = 4^x$

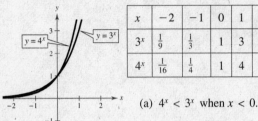

x	-2	-1	0	1	2
3^x	$\frac{1}{9}$	$\frac{1}{3}$	1	3	9
4^x	$\frac{1}{16}$	$\frac{1}{4}$	1	4	16

(a) $4^x < 3^x$ when $x < 0$.

(b) $4^x > 3^x$ when $x > 0$.

79. $f(x) = \left(1 + \dfrac{0.5}{x}\right)^x$ and $g(x) = e^{0.5}$ (Horizontal line)

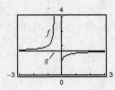

As $x \to \infty$, $f(x) \to g(x)$.

As $x \to -\infty$, $f(x) \to g(x)$.

81. $x^2 + y^2 = 25$

$$y^2 = 25 - x^2$$

$$y = \pm\sqrt{25 - x^2}$$

83. $f(x) = \dfrac{2}{9 + x}$

Vertical asymptote: $x = -9$

Horizontal asymptote: $y = 0$

x	-11	-10	-8	-7
$f(x)$	-1	-2	2	1

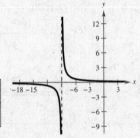

85. Answers will vary.

Section 3.2   Logarithmic Functions and Their Graphs

- You should know that a function of the form $y = \log_a x$, where $a > 0$, $a \neq 1$, and $x > 0$, is called a logarithm of x to base a.

- You should be able to convert from logarithmic form to exponential form and vice versa.

  $y = \log_a x \iff a^y = x$

- You should know the following properties of logarithms.

  (a) $\log_a 1 = 0$ since $a^0 = 1$.

  (b) $\log_a a = 1$ since $a^1 = a$.

  (c) $\log_a a^x = x$ since $a^x = a^x$.

  (d) $a^{\log_a x} = x$ Inverse Property

  (e) If $\log_a x = \log_a y$, then $x = y$.

- You should know the definition of the natural logarithmic function.

  $\log_e x = \ln x,\ x > 0$

- You should know the properties of the natural logarithmic function.

  (a) $\ln 1 = 0$ since $e^0 = 1$.

  (b) $\ln e = 1$ since $e^1 = e$.

  (c) $\ln e^x = x$ since $e^x = e^x$.

  (d) $e^{\ln x} = x$ Inverse Property

  (e) If $\ln x = \ln y$, then $x = y$.

- You should be able to graph logarithmic functions.

Vocabulary Check

1. logarithmic   **2.** 10   **3.** natural; e

4. $a^{\log_a x} = x$ **5.** $x = y$

---

1. $\log_4 64 = 3 \implies 4^3 = 64$

3. $\log_7 \frac{1}{49} = -2 \implies 7^{-2} = \frac{1}{49}$

5. $\log_{32} 4 = \frac{2}{5} \implies 32^{2/5} = 4$

7. $\log_{36} 6 = \frac{1}{2} \implies 36^{1/2} = 6$

9. $5^3 = 125 \implies \log_5 125 = 3$

11. $81^{1/4} = 3 \implies \log_{81} 3 = \frac{1}{4}$

13. $6^{-2} = \frac{1}{36} \implies \log_6 \frac{1}{36} = -2$

15. $7^0 = 1 \implies \log_7 1 = 0$

17. $f(x) = \log_2 x$

$f(16) = \log_2 16 = 4$ since $2^4 = 16$

19. $f(x) = \log_7 x$

$f(1) = \log_7 1 = 0$ since $7^0 = 1$

21. $g(x) = \log_a x$

$g(a^2) = \log_a a^2$

$= 2$ by the Inverse Property

23. $f(x) = \log x$

$f\left(\frac{4}{5}\right) = \log\left(\frac{4}{5}\right) \approx -0.097$

25. $f(x) = \log x$

$f(12.5) \approx 1.097$

27. $\log_3 3^4 = 4$ since $3^4 = 3^4$

29. $\log_\pi \pi = 1$ since $\pi^1 = \pi$

31. $f(x) = \log_4 x$

Domain: $x > 0 \implies$ The domain is $(0, \infty)$.

x-intercept: $(1, 0)$

Vertical asymptote: $x = 0$

$y = \log_4 x \implies 4^y = x$

x	$\frac{1}{4}$	1	4	2
$f(x)$	-1	0	1	$\frac{1}{2}$

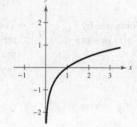

33. $y = -\log_3 x + 2$

Domain: $(0, \infty)$

x-intercept:

$-\log_3 x + 2 = 0$

$2 = \log_3 x$

$3^2 = x$

$9 = x$

The x-intercept is $(9, 0)$.

Vertical asymptote: $x = 0$

$y = -\log_3 x + 2$

$\log_3 x = 2 - y \implies 3^{2-y} = x$

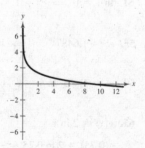

x	27	9	3	1	$\frac{1}{3}$
y	-1	0	1	2	3

35. $f(x) = -\log_6(x + 2)$

Domain: $x + 2 > 0 \implies x > -2$

The domain is $(-2, \infty)$.

x-intercept:

$0 = -\log_6(x + 2)$

$0 = \log_6(x + 2)$

$6^0 = x + 2$

$1 = x + 2$

$-1 = x$

The x-intercept is $(-1, 0)$.

Vertical asymptote: $x + 2 = 0 \implies x = -2$

$y = -\log_6(x + 2)$

$-y = \log_6(x + 2)$

$6^{-y} - 2 = x$

x	4	-1	$-1\frac{5}{6}$	$-1\frac{35}{36}$
$f(x)$	-1	0	1	2

37. $y = \log\left(\dfrac{x}{5}\right)$

x	1	2	3	4	5	6	7
y	-0.70	-0.40	-0.22	-0.10	0	0.08	0.15

Domain: $\dfrac{x}{5} > 0 \implies x > 0$

The domain is $(0, \infty)$.

x-intercept:

$\log\left(\dfrac{x}{5}\right) = 0$

$\quad \dfrac{x}{5} = 10^0$

$\quad \dfrac{x}{5} = 1 \implies x = 5$

The x-intercept is $(5, 0)$.

Vertical asymptote: $\dfrac{x}{5} = 0 \implies x = 0$

The vertical asymptote is the y-axis.

39. $f(x) = \log_3 x + 2$

Asymptote: $x = 0$

Point on graph: $(1, 2)$

Matches graph (c).

The graph of $f(x)$ is obtained by shifting the graph of $g(x)$ upward two units.

41. $f(x) = -\log_3(x + 2)$

Asymptote: $x = -2$

Point on graph: $(-1, 0)$

Matches graph (d).

The graph of $f(x)$ is obtained by reflecting the graph of $g(x)$ about the x-axis and shifting the graph two units to the left.

43. $f(x) = \log_3(1 - x) = \log_3[-(x - 1)]$

Asymptote: $x = 1$

Point on graph: $(0, 0)$

Matches graph (b).

The graph of $f(x)$ is obtained by reflecting the graph of $g(x)$ about the y-axis and shifting the graph one unit to the right.

45. $\ln \frac{1}{2} = -0.693 \ldots \implies e^{-0.693 \ldots} = \frac{1}{2}$

47. $\ln 4 = 1.386 \ldots \implies e^{1.386 \ldots} = 4$

49. $\ln 250 = 5.521 \ldots \implies e^{5.521 \ldots} = 250$

51. $\ln 1 = 0 \implies e^0 = 1$

53. $e^3 = 20.0855 \ldots \implies \ln 20.0855 \ldots = 3$

55. $e^{1/2} = 1.6487 \ldots \implies \ln 1.6487 \ldots = \frac{1}{2}$

57. $e^{-0.5} = 0.6065 \ldots \implies \ln 0.6065 \ldots = -0.5$

59. $e^x = 4 \implies \ln 4 = x$

61. $f(x) = \ln x$

$f(18.42) = \ln 18.42 \approx 2.913$

63. $g(x) = 2 \ln x$

$g(0.75) = 2 \ln 0.75 \approx -0.575$

65. $g(x) = \ln x$

$g(e^3) = \ln e^3 = 3$ by the Inverse Property

67. $g(x) = \ln x$

$g(e^{-2/3}) = \ln e^{-2/3} = -\frac{2}{3}$ by the Inverse Property

69. $f(x) = \ln(x - 1)$

Domain: $x - 1 > 0 \Rightarrow x > 1$

The domain is $(1, \infty)$.

x-intercept:

$0 = \ln(x - 1)$

$e^0 = x - 1$

$2 = x$

The x-intercept is $(2, 0)$.

Vertical asymptote: $x - 1 = 0 \Rightarrow x = 1$

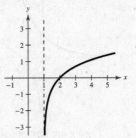

x	1.5	2	3	4
$f(x)$	-0.69	0	0.69	1.10

71. $g(x) = \ln(-x)$

Domain: $-x > 0 \Rightarrow x < 0$

The domain is $(-\infty, 0)$.

x-intercept:

$0 = \ln(-x)$

$e^0 = -x$

$-1 = x$

The x-intercept is $(-1, 0)$.

Vertical asymptote: $-x = 0 \Rightarrow x = 0$

x	-0.5	-1	-2	-3
$g(x)$	-0.69	0	0.69	1.10

73. $y_1 = \log(x + 1)$

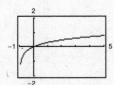

75. $y_1 = \ln(x - 1)$

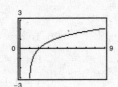

77. $y = \ln x + 2$

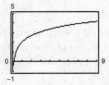

79. $\log_2(x + 1) = \log_2 4$

$x + 1 = 4$

$x = 3$

81. $\log(2x + 1) = \log 15$

$2x + 1 = 15$

$x = 7$

83. $\ln(x + 2) = \ln 6$

$x + 2 = 6$

$x = 4$

85. $\ln(x^2 - 2) = \ln 23$

$x^2 - 2 = 23$

$x^2 = 25$

$x = \pm 5$

87. $t = 12.542 \ln\left(\dfrac{x}{x - 1000}\right), x > 1000$

(a) When $x = \$1100.65$:

$t = 12.542 \ln\left(\dfrac{1100.65}{1100.65 - 1000}\right) \approx 30$ years

When $x = \$1254.68$:

$t = 12.542 \ln\left(\dfrac{1254.68}{1254.68 - 1000}\right) \approx 20$ years

(b) Total amounts: $(1100.65)(12)(30) = \$396{,}234.00$

$(1254.68)(12)(20) = \$301{,}123.20$

(c) Interest charges: $396{,}234 - 150{,}000 = \$246{,}234$

$301{,}123.20 - 150{,}000 = \$151{,}123.20$

(d) The vertical asymptote is $x = 1000$. The closer the payment is to \$1000 per month, the longer the length of the mortgage will be. Also, the monthly payment must be greater than \$1000.

89. $f(t) = 80 - 17 \log(t + 1), 0 \leq t \leq 12$

(a)

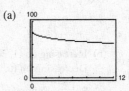

(b) $f(0) = 80 - 17 \log 1 = 80.0$

(c) $f(4) = 80 - 17 \log 5 \approx 68.1$

(d) $f(10) = 80 - 17 \log 11 \approx 62.3$

91. False. Reflecting $g(x)$ about the line $y = x$ will determine the graph of $f(x)$.

93. $f(x) = 3^x$, $g(x) = \log_3 x$

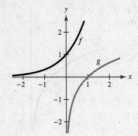

f and g are inverses. Their graphs are reflected about the line $y = x$.

95. $f(x) = e^x$, $g(x) = \ln x$

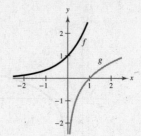

f and g are inverses. Their graphs are reflected about the line $y = x$.

97. (a) $f(x) = \ln x$, $g(x) = \sqrt{x}$

The natural log function grows at a slower rate than the square root function.

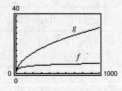

(b) $f(x) = \ln x$, $g(x) = \sqrt[4]{x}$

The natural log function grows at a slower rate than the fourth root function.

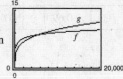

99. (a) False. If y were an exponential function of x, then $y = a^x$, but $a^1 = a$, not 0. Because one point is $(1, 0)$, y is not an exponential function of x.

(c) True. $x = a^y$

For $a = 2$, $x = 2^y$.

$y = 0, 2^0 = 1$

$y = 1, 2^1 = 2$

$y = 3, 2^3 = 8$

(b) True. $y = \log_a x$

For $a = 2$, $y = \log_2 x$.

$x = 1, \log_2 1 = 0$

$x = 2, \log_2 2 = 1$

$x = 8, \log_2 8 = 3$

(d) False. If y were a linear function of x, the slope between $(1, 0)$ and $(2, 1)$ and the slope between $(2, 1)$ and $(8, 3)$ would be the same. However,

$$m_1 = \frac{1 - 0}{2 - 1} = 1 \text{ and } m_2 = \frac{3 - 1}{8 - 2} = \frac{2}{6} = \frac{1}{3}.$$

Therefore, y is not a linear function of x.

101. $f(x) = |\ln x|$

(a)

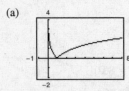

(b) Increasing on $(1, \infty)$
Decreasing on $(0, 1)$

(c) Relative minimum: $(1, 0)$

For Exercises 103–107, use $f(x) = 3x + 2$ and $g(x) = x^3 - 1$.

103. $(f + g)(2) = f(2) + g(2)$

$= [3(2) + 2] + [(2)^3 - 1]$

$= 8 + 7$

$= 15$

105. $(fg)(6) = f(6)g(6)$

$= [3(6) + 2][(6)^3 - 1]$

$= (20)(215)$

$= 4300$

107. $(f \circ g)(7) = f(g(7))$

$= f((7)^3 - 1)$

$= f(342)$

$= 3(342) + 2$

$= 1028$

Section 3.3   Properties of Logarithms

- You should know the following properties of logarithms.

  (a) $\log_a x = \dfrac{\log_b x}{\log_b a}$ $\log_a x = \dfrac{\log_{10} x}{\log_{10} a}$ $\log_a x = \dfrac{\ln x}{\ln a}$

  (b) $\log_a(uv) = \log_a u + \log_a v$ $\ln(uv) = \ln u + \ln v$

  (c) $\log_a\!\left(\dfrac{u}{v}\right) = \log_a u - \log_a v$ $\ln\!\left(\dfrac{u}{v}\right) = \ln u - \ln v$

  (d) $\log_a u^n = n \log_a u$ $\ln u^n = n \ln u$

- You should be able to rewrite logarithmic expressions using these properties.

Vocabulary Check

1. change-of-base

2. $\dfrac{\log x}{\log a} = \dfrac{\ln x}{\ln a}$

3. $\log_a(uv) = \log_a u + \log_a v$
This is the Product Property. Matches (c).

4. $\ln u^n = n \ln u$
This is the Power Property. Matches (a).

5. $\log_a \dfrac{u}{v} = \log_a u - \log_a v$

This is the Quotient Property. Matches (b).

1. (a) $\log_5 x = \dfrac{\log x}{\log 5}$

    (b) $\log_5 x = \dfrac{\ln x}{\ln 5}$

3. (a) $\log_{1/5} x = \dfrac{\log x}{\log(1/5)}$

    (b) $\log_{1/5} x = \dfrac{\ln x}{\ln(1/5)}$

5. (a) $\log_x \dfrac{3}{10} = \dfrac{\log(3/10)}{\log x}$

    (b) $\log_x \dfrac{3}{10} = \dfrac{\ln(3/10)}{\ln x}$

7. (a) $\log_{2.6} x = \dfrac{\log x}{\log 2.6}$

    (b) $\log_{2.6} x = \dfrac{\ln x}{\ln 2.6}$

9. $\log_3 7 = \dfrac{\log 7}{\log 3} = \dfrac{\ln 7}{\ln 3} \approx 1.771$

11. $\log_{1/2} 4 = \dfrac{\log 4}{\log(1/2)}$

         $= \dfrac{\ln 4}{\ln(1/2)} = -2.000$

13. $\log_9(0.4) = \dfrac{\log 0.4}{\log 9} = \dfrac{\ln 0.4}{\ln 9} \approx -0.417$

15. $\log_{15} 1250 = \dfrac{\log 1250}{\log 15} = \dfrac{\ln 1250}{\ln 15} \approx 2.633$

17. $\log_4 8 = \dfrac{\log_2 8}{\log_2 4}$

      $= \dfrac{\log_2 2^3}{\log_2 2^2} = \dfrac{3}{2}$

19. $\log_5 \dfrac{1}{250} = \log_5\!\left(\dfrac{1}{125} \cdot \dfrac{1}{2}\right)$

         $= \log_5 \dfrac{1}{125} + \log_5 \dfrac{1}{2}$

         $= \log_5 5^{-3} + \log_5 2^{-1}$

         $= -3 - \log_5 2$

21. $\ln(5e^6) = \ln 5 + \ln e^6$

         $= \ln 5 + 6$

         $= 6 + \ln 5$

23. $\log_3 9 = 2 \log_3 3 = 2$

25. $\log_2 \sqrt[4]{8} = \frac{1}{4} \log_2 2^3 = \frac{3}{4} \log_2 2 = \frac{3}{4}(1) = \frac{3}{4}$

27. $\log_4 16^{1.2} = 1.2(\log_4 16) = 1.2 \log_4 4^2 = 1.2(2) = 2.4$

29. $\log_3(-9)$ is undefined. -9 is not in the domain of $\log_3 x$.

31. $\ln e^{4.5} = 4.5$

33. $\ln \dfrac{1}{\sqrt{e}} = \ln 1 - \ln \sqrt{e}$

$$= 0 - \frac{1}{2}\ln e$$

$$= 0 - \frac{1}{2}(1)$$

$$= -\frac{1}{2}$$

35. $\ln e^2 + \ln e^5 = 2 + 5 = 7$

37. $\log_5 75 - \log_5 3 = \log_5 \dfrac{75}{3}$

$$= \log_5 25$$

$$= \log_5 5^2$$

$$= 2 \log_5 5$$

$$= 2$$

39. $\log_4 5x = \log_4 5 + \log_4 x$

41. $\log_8 x^4 = 4 \log_8 x$

43. $\log_5 \dfrac{5}{x} = \log_5 5 - \log_5 x$

$$= 1 - \log_5 x$$

45. $\ln \sqrt{z} = \ln z^{1/2} = \dfrac{1}{2}\ln z$

47. $\ln xyz^2 = \ln x + \ln y + \ln z^2$

$$= \ln x + \ln y + 2 \ln z$$

49. $\ln z(z-1)^2 = \ln z + \ln(z-1)^2$

$$= \ln z + 2\ln(z-1),\ z > 1$$

51. $\log_2 \dfrac{\sqrt{a-1}}{9} = \log_2 \sqrt{a-1} - \log_2 9$

$$= \frac{1}{2}\log_2(a-1) - \log_2 3^2$$

$$= \frac{1}{2}\log_2(a-1) - 2\log_2 3,\ a > 1$$

53. $\ln \sqrt[3]{\dfrac{x}{y}} = \dfrac{1}{3}\ln \dfrac{x}{y}$

$$= \frac{1}{3}[\ln x - \ln y]$$

$$= \frac{1}{3}\ln x - \frac{1}{3}\ln y$$

55. $\ln\left(\dfrac{x^4 \sqrt{y}}{z^5}\right) = \ln x^4 \sqrt{y} - \ln z^5$

$$= \ln x^4 + \ln \sqrt{y} - \ln z^5$$

$$= 4 \ln x + \frac{1}{2}\ln y - 5 \ln z$$

57. $\log_5\left(\dfrac{x^2}{y^2 z^3}\right) = \log_5 x^2 - \log_5 y^2 z^3$

$$= \log_5 x^2 - (\log_5 y^2 + \log_5 z^3)$$

$$= 2 \log_5 x - 2 \log_5 y - 3 \log_5 z$$

59. $\ln \sqrt[4]{x^3(x^2+3)} = \dfrac{1}{4}\ln x^3(x^2+3)$

$$= \frac{1}{4}[\ln x^3 + \ln(x^2+3)]$$

$$= \frac{1}{4}[3 \ln x + \ln(x^2+3)]$$

$$= \frac{3}{4}\ln x + \frac{1}{4}\ln(x^2+3)$$

61. $\ln x + \ln 3 = \ln 3x$

63. $\log_4 z - \log_4 y = \log_4 \dfrac{z}{y}$

65. $2 \log_2(x+4) = \log_2(x+4)^2$

67. $\frac{1}{4}\log_3 5x = \log_3(5x)^{1/4} = \log_3 \sqrt[4]{5x}$

69. $\ln x - 3\ln(x+1) = \ln x - \ln(x+1)^3$

$$= \ln \frac{x}{(x+1)^3}$$

71. $\log x - 2\log y + 3\log z = \log x - \log y^2 + \log z^3$

$$= \log \frac{x}{y^2} + \log z^3 = \log \frac{xz^3}{y^2}$$

73. $\ln x - 4[\ln(x+2) + \ln(x-2)] = \ln x - 4\ln(x+2)(x-2)$

$$= \ln x - 4\ln(x^2 - 4)$$

$$= \ln x - \ln(x^2 - 4)^4$$

$$= \ln \frac{x}{(x^2 - 4)^4}$$

75. $\frac{1}{3}[2\ln(x+3) + \ln x - \ln(x^2 - 1)] = \frac{1}{3}[\ln(x+3)^2 + \ln x - \ln(x^2 - 1)]$

$$= \frac{1}{3}[\ln x(x+3)^2 - \ln(x^2 - 1)]$$

$$= \frac{1}{3}\ln \frac{x(x+3)^2}{x^2 - 1}$$

$$= \ln \sqrt[3]{\frac{x(x+3)^2}{x^2 - 1}}$$

77. $\frac{1}{3}[\log_8 y + 2\log_8(y+4)] - \log_8(y-1) = \frac{1}{3}[\log_8 y + \log_8(y+4)^2] - \log_8(y-1)$

$$= \frac{1}{3}\log_8 y(y+4)^2 - \log_8(y-1)$$

$$= \log_8 \sqrt[3]{y(y+4)^2} - \log_8(y-1)$$

$$= \log_8\left(\frac{\sqrt[3]{y(y+4)^2}}{y-1}\right)$$

79. $\log_2 \frac{32}{4} = \log_2 32 - \log_2 4 \ne \frac{\log_2 32}{\log_2 4}$

The second and third expressions are equal by Property 2.

81. $\beta = 10\log\left(\frac{I}{10^{-12}}\right)$

$$= 10[\log I - \log 10^{-12}]$$

$$= 10[\log I + 12]$$

$$= 120 + 10\log I$$

When $I = 10^{-6}$:

$$\beta = 120 + 10\log 10^{-6} = 120 + 10(-6) = 60 \text{ decibels}$$

83. $\beta = 120 + 10\log(2I)$

$$= 120 + 10(\log 2 + \log I)$$

$$= (120 + 10\log I) + 10\log 2$$

With both stereos playing, the music is $10\log 2 \approx 3$ decibels louder.

85. By using the regression feature on a graphing calculator we obtain $y \approx 256.24 - 20.8\ln x$.

87. $f(x) = \ln x$

False, $f(0) \neq 0$ since 0 is not in the domain of $f(x)$.

$f(1) = \ln 1 = 0$

89. False. $f(x) - f(2) = \ln x - \ln 2 = \ln \dfrac{x}{2} \neq \ln(x - 2)$

91. False.

$f(u) = 2f(v) \Rightarrow \ln u = 2 \ln v \Rightarrow \ln u = \ln v^2 \Rightarrow u = v^2$

93. Let $x = \log_b u$ and $y = \log_b v$, then $b^x = u$ and $b^y = v$.

$$\dfrac{u}{v} = \dfrac{b^x}{b^y} = b^{x-y}$$

Then $\log_b(u/v) = \log_b(b^{x-y}) = x - y = \log_b u - \log_b v$.

95. $f(x) = \log_2 x = \dfrac{\log x}{\log 2} = \dfrac{\ln x}{\ln 2}$

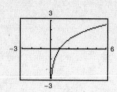

97. $f(x) = \log_{1/2} x$

$= \dfrac{\log x}{\log(1/2)} = \dfrac{\ln x}{\ln(1/2)}$

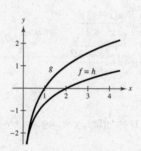

99. $f(x) = \log_{11.8} x$

$= \dfrac{\log x}{\log 11.8} = \dfrac{\ln x}{\ln 11.8}$

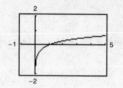

101. $f(x) = \ln \dfrac{x}{2}$, $g(x) = \dfrac{\ln x}{\ln 2}$, $h(x) = \ln x - \ln 2$

$f(x) = h(x)$ by Property 2

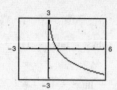

103. $\dfrac{24xy^{-2}}{16x^{-3}y} = \dfrac{24xx^3}{16yy^2} = \dfrac{3x^4}{2y^3}, x \neq 0$

105. $(18x^3y^4)^{-3}(18x^3y^4)^3 = \dfrac{(18x^3y^4)^3}{(18x^3y^4)^3} = 1$ if $x \neq 0, y \neq 0$.

107. $3x^2 + 2x - 1 = 0$

$(3x - 1)(x + 1) = 0$

$3x - 1 = 0 \Rightarrow x = \dfrac{1}{3}$

$x + 1 = 0 \Rightarrow x = -1$

109. $\dfrac{2}{3x + 1} = \dfrac{x}{4}$

$(3x + 1)(x) = (2)(4)$

$3x^2 + x - 8 = 0$

$x = \dfrac{-1 \pm \sqrt{1^2 - 4(3)(-8)}}{2(3)}$

$= \dfrac{-1 \pm \sqrt{97}}{6}$

Section 3.4   Exponential and Logarithmic Equations

■ To solve an exponential equation, isolate the exponential expression, then take the logarithm of both sides. Then solve for the variable.

1. $\log_a a^x = x$ 2. $\ln e^x = x$

■ To solve a logarithmic equation, rewrite it in exponential form. Then solve for the variable.

1. $a^{\log_a x} = x$ 2. $e^{\ln x} = x$

■ If $a > 0$ and $a \neq 1$ we have the following:

1. $\log_a x = \log_a y \iff x = y$

2. $a^x = a^y \iff x = y$

■ Check for extraneous solutions.

Vocabulary Check

1. solve

2. (a) $x = y$ (b) $x = y$
   (c) x (d) x

3. extraneous

---

1. $4^{2x-7} = 64$

  (a) $x = 5$

  $4^{2(5)-7} = 4^3 = 64$

  Yes, $x = 5$ *is* a solution.

  (b) $x = 2$

  $4^{2(2)-7} = 4^{-3} = \frac{1}{64} \neq 64$

  No, $x = 2$ *is not* a solution.

3. $3e^{x+2} = 75$

  (a) $x = -2 + e^{25}$

  $3e^{(-2+e^{25})+2} = 3e^{e^{25}} \neq 75$

  No, $x = -2 + e^{25}$ *is not* a solution.

  (b) $x = -2 + \ln 25$

  $3e^{(-2+\ln 25)+2} = 3e^{\ln 25} = 3(25) = 75$

  Yes, $x = -2 + \ln 25$ *is* a solution.

  (c) $x \approx 1.219$

  $3e^{1.219+2} = 3e^{3.219} \approx 75$

  Yes, $x \approx 1.219$ *is* a solution.

5. $\log_4(3x) = 3 \implies 3x = 4^3 \implies 3x = 64$

  (a) $x \approx 21.333$

  $3(21.333) \approx 64$

  Yes, 21.333 *is* an approximate solution.

  (b) $x = -4$

  $3(-4) = -12 \neq 64$

  No, $x = -4$ *is not* a solution.

  (c) $x = \frac{64}{3}$

  $3\left(\frac{64}{3}\right) = 64$

  Yes, $x = \frac{64}{3}$ *is* a solution.

7. $\ln(2x + 3) = 5.8$

  (a) $x = \frac{1}{2}(-3 + \ln 5.8)$

  $\ln\left[2\left(\frac{1}{2}\right)(-3 + \ln 5.8) + 3\right] = \ln(\ln 5.8) \neq 5.8$

  No, $x = \frac{1}{2}(-3 + \ln 5.8)$ *is not* a solution.

  (b) $x = \frac{1}{2}(-3 + e^{5.8})$

  $\ln\left[2\left(\frac{1}{2}\right)(-3 + e^{5.8}) + 3\right] = \ln(e^{5.8}) = 5.8$

  Yes, $x = \frac{1}{2}(-3 + e^{5.8})$ *is* a solution.

  (c) $x \approx 163.650$

  $\ln[2(163.650) + 3] = \ln 330.3 \approx 5.8$

  Yes, $x \approx 163.650$ *is* an approximate solution.

9. $4^x = 16$

$4^x = 4^2$

$x = 2$

11. $\left(\frac{1}{2}\right)^x = 32$

$2^{-x} = 2^5$

$-x = 5$

$x = -5$

13. $\ln x - \ln 2 = 0$

$\ln x = \ln 2$

$x = 2$

15. $e^x = 2$

$\ln e^x = \ln 2$

$x = \ln 2$

$x \approx 0.693$

17. $\ln x = -1$

$e^{\ln x} = e^{-1}$

$x = e^{-1}$

$x \approx 0.368$

19. $\log_4 x = 3$

$4^{\log_4 x} = 4^3$

$x = 4^3$

$x = 64$

21. $f(x) = g(x)$

$2^x = 8$

$2^x = 2^3$

$x = 3$

Point of intersection:
$(3, 8)$

23. $f(x) = g(x)$

$\log_3 x = 2$

$x = 3^2$

$x = 9$

Point of intersection:
$(9, 2)$

25. $e^x = e^{x^2 - 2}$

$x = x^2 - 2$

$0 = x^2 - x - 2$

$0 = (x + 1)(x - 2)$

$x = -1 \text{ or } x = 2$

27. $e^{x^2 - 3} = e^{x - 2}$

$x^2 - 3 = x - 2$

$x^2 - x - 1 = 0$

By the Quadratic Formula
$x \approx 1.618 \text{ or } x \approx -0.618.$

29. $4(3^x) = 20$

$3^x = 5$

$\log_3 3^x = \log_3 5$

$x = \log_3 5 = \dfrac{\log 5}{\log 3} \text{ or } \dfrac{\ln 5}{\ln 3}$

$x \approx 1.465$

31. $2e^x = 10$

$e^x = 5$

$\ln e^x = \ln 5$

$x = \ln 5 \approx 1.609$

33. $e^x - 9 = 19$

$e^x = 28$

$\ln e^x = \ln 28$

$x = \ln 28 \approx 3.332$

35. $3^{2x} = 80$

$\ln 3^{2x} = \ln 80$

$2x \ln 3 = \ln 80$

$x = \dfrac{\ln 80}{2 \ln 3} \approx 1.994$

37. $5^{-t/2} = 0.20$

$5^{-t/2} = \dfrac{1}{5}$

$5^{-t/2} = 5^{-1}$

$-\dfrac{t}{2} = -1$

$t = 2$

39. $3^{x-1} = 27$

$3^{x-1} = 3^3$

$x - 1 = 3$

$x = 4$

41. $\qquad 2^{3-x} = 565$

$\ln 2^{3-x} = \ln 565$

$(3 - x) \ln 2 = \ln 565$

$3 \ln 2 - x \ln 2 = \ln 565$

$-x \ln 2 = \ln 565 - 3 \ln 2$

$x \ln 2 = 3 \ln 2 - \ln 565$

$x = \dfrac{3 \ln 2 - \ln 565}{\ln 2}$

$= 3 - \dfrac{\ln 565}{\ln 2} \approx -1.642$

43. $8(10^{3x}) = 12$

$10^{3x} = \dfrac{12}{8}$

$\log 10^{3x} = \log\left(\dfrac{3}{2}\right)$

$3x = \log\left(\dfrac{3}{2}\right)$

$x = \dfrac{1}{3} \log\left(\dfrac{3}{2}\right) \approx 0.059$

45. $3(5^{x-1}) = 21$

$5^{x-1} = 7$

$\ln 5^{x-1} = \ln 7$

$(x-1) \ln 5 = \ln 7$

$x - 1 = \dfrac{\ln 7}{\ln 5}$

$x = 1 + \dfrac{\ln 7}{\ln 5} \approx 2.209$

47. $e^{3x} = 12$

$3x = \ln 12$

$x = \dfrac{\ln 12}{3} \approx 0.828$

49. $500e^{-x} = 300$

$e^{-x} = \dfrac{3}{5}$

$-x = \ln \dfrac{3}{5}$

$x = -\ln \dfrac{3}{5}$

$= \ln \dfrac{5}{3} \approx 0.511$

51. $7 - 2e^x = 5$

$-2e^x = -2$

$e^x = 1$

$x = \ln 1 = 0$

53. $6(2^{3x-1}) - 7 = 9$

$6(2^{3x-1}) = 16$

$2^{3x-1} = \dfrac{8}{3}$

$\log_2 2^{3x-1} = \log_2 \left(\dfrac{8}{3} \right)$

$3x - 1 = \log_2 \left(\dfrac{8}{3} \right) = \dfrac{\log(8/3)}{\log 2} \text{ or } \dfrac{\ln(8/3)}{\ln 2}$

$x = \dfrac{1}{3} \left[\dfrac{\log(8/3)}{\log 2} + 1 \right] \approx 0.805$

55. $e^{2x} - 4e^x - 5 = 0$

$(e^x + 1)(e^x - 5) = 0$

$e^x = -1 \quad \text{or} \quad e^x = 5$

(No solution) $\quad\quad x = \ln 5 \approx 1.609$

57. $e^{2x} - 3e^x - 4 = 0$

$(e^x + 1)(e^x - 4) = 0$

$e^x + 1 = 0 \Rightarrow e^x = -1$

Not possible since $e^x > 0$ for all x.

$e^x - 4 = 0 \Rightarrow e^x = 4 \Rightarrow x = \ln 4 \approx 1.386$

59. $\dfrac{500}{100 - e^{x/2}} = 20$

$500 = 20(100 - e^{x/2})$

$25 = 100 - e^{x/2}$

$e^{x/2} = 75$

$\dfrac{x}{2} = \ln 75$

$x = 2 \ln 75 \approx 8.635$

61. $\dfrac{3000}{2 + e^{2x}} = 2$

$3000 = 2(2 + e^{2x})$

$1500 = 2 + e^{2x}$

$1498 = e^{2x}$

$\ln 1498 = 2x$

$x = \dfrac{\ln 1498}{2} \approx 3.656$

63. $\left(1 + \dfrac{0.065}{365} \right)^{365t} = 4$

$\ln \left(1 + \dfrac{0.065}{365} \right)^{365t} = \ln 4$

$365t \ln \left(1 + \dfrac{0.065}{365} \right) = \ln 4$

$t = \dfrac{\ln 4}{365 \ln \left(1 + \frac{0.065}{365} \right)} \approx 21.330$

65. $\left(1 + \dfrac{0.10}{12} \right)^{12t} = 2$

$\ln \left(1 + \dfrac{0.10}{12} \right)^{12t} = \ln 2$

$12t \ln \left(1 + \dfrac{0.10}{12} \right) = \ln 2$

$t = \dfrac{\ln 2}{12 \ln \left(1 + \frac{0.10}{12} \right)} \approx 6.960$

67. $g(x) = 6e^{1-x} - 25$

Algebraically:

$6e^{1-x} = 25$

$e^{1-x} = \dfrac{25}{6}$

$1 - x = \ln\left(\dfrac{25}{6}\right)$

$x = 1 - \ln\left(\dfrac{25}{6}\right)$

$x \approx -0.427$

The zero is $x \approx -0.427$.

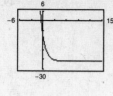

69. $f(x) = 3e^{3x/2} - 962$

Algebraically:

$3e^{3x/2} = 962$

$e^{3x/2} = \dfrac{962}{3}$

$\dfrac{3x}{2} = \ln\left(\dfrac{962}{3}\right)$

$x = \tfrac{2}{3}\ln\left(\dfrac{962}{3}\right)$

$x \approx 3.847$

The zero is $x \approx 3.847$.

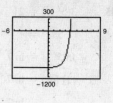

71. $g(t) = e^{0.09t} - 3$

Algebraically:

$e^{0.09t} = 3$

$0.09t = \ln 3$

$t = \dfrac{\ln 3}{0.09}$

$t \approx 12.207$

The zero is $t \approx 12.207$.

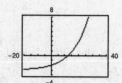

73. $h(t) = e^{0.125t} - 8$

Algebraically:

$e^{0.125t} - 8 = 0$

$e^{0.125t} = 8$

$0.125t = \ln 8$

$t = \dfrac{\ln 8}{0.125}$

$t \approx 16.636$

The zero is $t \approx 16.636$.

75. $\ln x = -3$

$x = e^{-3} \approx 0.050$

77. $\ln 2x = 2.4$

$2x = e^{2.4}$

$x = \dfrac{e^{2.4}}{2} \approx 5.512$

79. $\log x = 6$

$x = 10^6$

$= 1,000,000.000$

81. $3 \ln 5x = 10$

$\ln 5x = \dfrac{10}{3}$

$5x = e^{10/3}$

$x = \dfrac{e^{10/3}}{5} \approx 5.606$

83. $\ln\sqrt{x + 2} = 1$

$\sqrt{x + 2} = e^1$

$x + 2 = e^2$

$x = e^2 - 2 \approx 5.389$

85. $7 + 3 \ln x = 5$

$3 \ln x = -2$

$\ln x = -\tfrac{2}{3}$

$x = e^{-2/3} \approx 0.513$

87. $6 \log_3(0.5x) = 11$

$\log_3(0.5x) = \dfrac{11}{6}$

$3^{\log_3(0.5x)} = 3^{11/6}$

$0.5x = 3^{11/6}$

$x = 2(3^{11/6}) \approx 14.988$

89. $\ln x - \ln(x + 1) = 2$

$$\ln\left(\frac{x}{x+1}\right) = 2$$

$$\frac{x}{x+1} = e^2$$

$$x = e^2(x + 1)$$

$$x = e^2 x + e^2$$

$$x - e^2 x = e^2$$

$$x(1 - e^2) = e^2$$

$$x = \frac{e^2}{1 - e^2} \approx -1.157$$

This negative value is extraneous. The equation has no solution.

93. $\ln(x + 5) = \ln(x - 1) - \ln(x + 1)$

$$\ln(x + 5) = \ln\left(\frac{x-1}{x+1}\right)$$

$$x + 5 = \frac{x-1}{x+1}$$

$$(x + 5)(x + 1) = x - 1$$

$$x^2 + 6x + 5 = x - 1$$

$$x^2 + 5x + 6 = 0$$

$$(x + 2)(x + 3) = 0$$

$$x = -2 \ \text{or} \ x = -3$$

Both of these solutions are extraneous, so the equation has no solution.

97. $\log(x + 4) - \log x = \log(x + 2)$

$$\log\left(\frac{x+4}{x}\right) = \log(x + 2)$$

$$\frac{x+4}{x} = x + 2$$

$$x + 4 = x^2 + 2x$$

$$0 = x^2 + x - 4$$

$$x = \frac{-1 \pm \sqrt{17}}{2} \quad \text{Quadratic Formula}$$

Choosing the positive value of x (the negative value is extraneous), we have

$$x = \frac{-1 + \sqrt{17}}{2} \approx 1.562.$$

91. $\ln x + \ln(x - 2) = 1$

$$\ln[x(x - 2)] = 1$$

$$x(x - 2) = e^1$$

$$x^2 - 2x - e = 0$$

$$x = \frac{2 \pm \sqrt{4 + 4e}}{2}$$

$$= \frac{2 \pm 2\sqrt{1 + e}}{2} = 1 \pm \sqrt{1 + e}$$

The negative value is extraneous. The only solution is $x = 1 + \sqrt{1 + e} \approx 2.928$.

95. $\log_2(2x - 3) = \log_2(x + 4)$

$$2x - 3 = x + 4$$

$$x = 7$$

99. $\log_4 x - \log_4(x - 1) = \frac{1}{2}$

$$\log_4\left(\frac{x}{x-1}\right) = \frac{1}{2}$$

$$4^{\log_4[x/(x-1)]} = 4^{1/2}$$

$$\frac{x}{x-1} = 4^{1/2}$$

$$x = 2(x - 1)$$

$$x = 2x - 2$$

$$-x = -2$$

$$x = 2$$

101. $\log 8x - \log\left(1 + \sqrt{x}\right) = 2$

$$\log \frac{8x}{1 + \sqrt{x}} = 2$$

$$\frac{8x}{1 + \sqrt{x}} = 10^2$$

$$8x = 100\left(1 + \sqrt{x}\right)$$

$$2x = 25\left(1 + \sqrt{x}\right) = 25 + 25\sqrt{x}$$

$$2x - 25 = 25\sqrt{x}$$

$$(2x - 25)^2 = \left(25\sqrt{x}\right)^2$$

$$4x^2 - 100x + 625 = 625x$$

$$4x^2 - 725x + 625 = 0$$

$$x = \frac{725 \pm \sqrt{725^2 - 4(4)(625)}}{2(4)} = \frac{725 \pm \sqrt{515,625}}{8} = \frac{25\left(29 \pm 5\sqrt{33}\right)}{8}$$

$$x \approx 0.866 \text{ (extraneous)} \quad \text{or} \quad x \approx 180.384$$

The only solution is $x = \dfrac{25\left(29 + 5\sqrt{33}\right)}{8} \approx 180.384$.

---

103. $y_1 = 7$

$y_2 = 2^x$

From the graph we have
$x \approx 2.807$ when $y = 7$.
Algebraically:

$$2^x = 7$$

$$\ln 2^x = \ln 7$$

$$x \ln 2 = \ln 7$$

$$x = \frac{\ln 7}{\ln 2} \approx 2.807$$

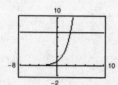

105. $y_1 = 3$

$y_2 = \ln x$

From the graph we have
$x \approx 20.086$ when $y = 3$.
Algebraically:

$$3 - \ln x = 0$$

$$\ln x = 3$$

$$x = e^3 \approx 20.086$$

---

107. (a)

$$A = Pe^{rt}$$

$$5000 = 2500e^{0.085t}$$

$$2 = e^{0.085t}$$

$$\ln 2 = 0.085t$$

$$\frac{\ln 2}{0.085} = t$$

$$t \approx 8.2 \text{ years}$$

(b)

$$A = Pe^{rt}$$

$$7500 = 2500e^{0.085t}$$

$$3 = e^{0.085t}$$

$$\ln 3 = 0.085t$$

$$\frac{\ln 3}{0.085} = t$$

$$t \approx 12.9 \text{ years}$$

109. $p = 500 - 0.5\left(e^{0.004x}\right)$

(a)

$$p = 350$$

$$350 = 500 - 0.5\left(e^{0.004x}\right)$$

$$300 = e^{0.004x}$$

$$0.004x = \ln 300$$

$$x \approx 1426 \text{ units}$$

(b)

$$p = 300$$

$$300 = 500 - 0.5\left(e^{0.004x}\right)$$

$$400 = e^{0.004x}$$

$$0.004x = \ln 400$$

$$x \approx 1498 \text{ units}$$

111. $V = 6.7e^{-48.1/t}$, $t \geq 0$

(a)

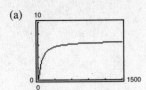

(b) As $t \to \infty$, $V \to 6.7$.

Horizontal asymptote: $V = 6.7$

The yield will approach 6.7 million cubic feet per acre.

(c) $1.3 = 6.7e^{-48.1/t}$

$$\frac{1.3}{6.7} = e^{-48.1/t}$$

$$\ln\left(\frac{13}{67}\right) = \frac{-48.1}{t}$$

$$t = \frac{-48.1}{\ln(13/67)} \approx 29.3 \text{ years}$$

113. $y = 7312 - 630.0 \ln t$, $5 \leq t \leq 12$

$$7312 - 630.0 \ln t = 5800$$

$$-630.0 \ln t = -1512$$

$$\ln t = 2.4$$

$$t = e^{2.4} \approx 11$$

$t \approx 11$ corresponds to the year 2001.

115. (a) From the graph shown in the textbook, we see horizontal asymptotes at $y = 0$ and $y = 100$.
These represent the lower and upper percent bounds; the range falls between 0% and 100%.

(b) Males

$$50 = \frac{100}{1 + e^{-0.6114(x-69.71)}}$$

$$1 + e^{-0.6114(x-69.71)} = 2$$

$$e^{-0.6114(x-69.71)} = 1$$

$$-0.6114(x - 69.71) = \ln 1$$

$$-0.6114(x - 69.71) = 0$$

$$x = 69.71 \text{ inches}$$

Females

$$50 = \frac{100}{1 + e^{-0.66607(x-64.51)}}$$

$$1 + e^{-0.66607(x-64.51)} = 2$$

$$e^{-0.6667(x-64.51)} = 1$$

$$-0.66607(x - 64.51) = \ln 1$$

$$-0.66607(x - 64.51) = 0$$

$$x = 64.51 \text{ inches}$$

117. $y = -3.00 + 11.88 \ln x + \dfrac{36.94}{x}$

(a)

x	0.2	0.4	0.6	0.8	1.0
y	162.6	78.5	52.5	40.5	33.9

(b)

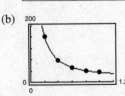

The model seems to fit the data well.

(c) When $y = 30$:

$$30 = -3.00 + 11.88 \ln x + \frac{36.94}{x}$$

Add the graph of $y = 30$ to the graph in part (a) and estimate the point of intersection of the two graphs. We find that $x \approx 1.20$ meters.

(d) No, it is probably not practical to lower the number of gs experienced during impact to less than 23 because the required distance traveled at $y = 23$ is $x \approx 2.27$ meters. It is probably not practical to design a car allowing a passenger to move forward 2.27 meters (or 7.45 feet) during an impact.

119. $\log_a(uv) = \log_a u + \log_a v$

True by Property 1 in Section 3.3.

121. $\log_a(u - v) = \log_a u - \log_a v$

False.

$$1.95 \approx \log(100 - 10)$$

$$\neq \log 100 - \log 10 = 1$$

123. Yes, a logarithmic equation can have more than one extraneous solution. See Exercise 93.

125. Yes.

Time to Double	Time to Quadruple
$2P = Pe^{rt}$	$4P = Pe^{rt}$
$2 = e^{rt}$	$4 = e^{rt}$
$\ln 2 = rt$	$\ln 4 = rt$
$\dfrac{\ln 2}{r} = t$	$\dfrac{2\ln 2}{r} = t$

Thus, the time to quadruple is twice as long as the time to double.

127. $\sqrt{48x^2y^5} = \sqrt{16x^2y^43y} = 4|x|y^2\sqrt{3y}$

129. $\sqrt[3]{25}\sqrt[3]{15} = \sqrt[3]{375} = \sqrt[3]{125 \cdot 3} = 5\sqrt[3]{3}$

131. $f(x) = |x| + 9$

Domain: all real numbers x

y-intercept: $(0, 9)$

y-axis symmetry

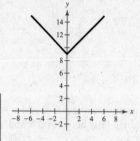

x	0	± 1	± 2	± 3
y	9	10	11	12

133. $g(x) = \begin{cases} 2x, & x < 0 \\ -x^2 + 4, & x \geq 0 \end{cases}$

Domain: all real numbers x

x-intercept: $(2, 0)$

y-intercept: $(0, 4)$

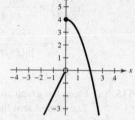

x	-3	-2	-1	-0.5	0	1	2	3
y	-6	-4	-2	-1	4	3	2	-5

135. $\log_6 9 = \dfrac{\log_{10} 9}{\log_{10} 6} = \dfrac{\ln 9}{\ln 6} \approx 1.226$

137. $\log_{3/4} 5 = \dfrac{\log_{10} 5}{\log_{10}(3/4)} = \dfrac{\ln 5}{\ln(3/4)} \approx -5.595$

Section 3.5   Exponential and Logarithmic Models

- You should be able to solve growth and decay problems.
  - (a) Exponential growth if $b > 0$ and $y = ae^{bx}$.
  - (b) Exponential decay if $b > 0$ and $y = ae^{-bx}$.
- You should be able to use the Gaussian model
  $$y = ae^{-(x-b)^2/c}.$$
- You should be able to use the logistic growth model
  $$y = \frac{a}{1 + be^{-rx}}.$$
- You should be able to use the logarithmic models
  $$y = a + b\ln x, \; y = a + b\log x.$$

Vocabulary Check

1. $y = ae^{bx}; \; y = ae^{-bx}$

2. $y = a + b\ln x; \; y = a + b\log x$

3. normally distributed

4. bell; average value

5. sigmoidal

1. $y = 2e^{x/4}$

This is an exponential growth model. Matches graph (c).

3. $y = 6 + \log(x + 2)$

This is a logarithmic function shifted up six units and left two units. Matches graph (b).

5. $y = \ln(x + 1)$

This is a logarithmic model shifted left one unit. Matches graph (d).

7. Since $A = 1000e^{0.035t}$, the time to double is given by $2000 = 1000e^{0.035t}$ and we have

$$2 = e^{0.035t}$$

$$\ln 2 = \ln e^{0.035t}$$

$$\ln 2 = 0.035t$$

$$t = \frac{\ln 2}{0.035} \approx 19.8 \text{ years.}$$

Amount after 10 years: $A = 1000e^{0.35} \approx \$1419.07$

9. Since $A = 750e^{rt}$ and $A = 1500$ when $t = 7.75$, we have the following.

$$1500 = 750e^{7.75r}$$

$$2 = e^{7.75r}$$

$$\ln 2 = \ln e^{7.75r}$$

$$\ln 2 = 7.75r$$

$$r = \frac{\ln 2}{7.75} \approx 0.089438 = 8.9438\%$$

Amount after 10 years: $A = 750e^{0.089438(10)} \approx \$1834.37$

11. Since $A = 500e^{rt}$ and $A = \$1505.00$ when $t = 10$, we have the following.

$$1505.00 = 500e^{10r}$$

$$r = \frac{\ln(1505.00/500)}{10} \approx 0.110 = 11.0\%$$

The time to double is given by

$$1000 = 500e^{0.110t}$$

$$t = \frac{\ln 2}{0.110} \approx 6.3 \text{ years.}$$

13. Since $A = Pe^{0.045t}$ and $A = 10,000.00$ when $t = 10$, we have the following.

$$10,000.00 = Pe^{0.045(10)}$$

$$\frac{10,000.00}{e^{0.045(10)}} = P \approx \$6376.28$$

The time to double is given by $t = \dfrac{\ln 2}{0.045} \approx 15.40 \text{ years.}$

15. $500,000 = P\left(1 + \dfrac{0.075}{12}\right)^{12(20)}$

$$P = \frac{500,000}{\left(1 + \dfrac{0.075}{12}\right)^{12(20)}}$$

$$= \frac{500,000}{1.00625^{240}} \approx \$112,087.09$$

17. $P = 1000, r = 11\%$

(a) $n = 1$

$$(1 + 0.11)^t = 2$$

$$t \ln 1.11 = \ln 2$$

$$t = \frac{\ln 2}{\ln 1.11} \approx 6.642 \text{ years}$$

(c) $n = 365$

$$\left(1 + \frac{0.11}{365}\right)^{365t} = 2$$

$$365t \ln\left(1 + \frac{0.11}{365}\right) = \ln 2$$

$$t = \frac{\ln 2}{365 \ln\left(1 + \frac{0.11}{365}\right)} \approx 6.302 \text{ years}$$

(b) $n = 12$

$$\left(1 + \frac{0.11}{12}\right)^{12t} = 2$$

$$12t \ln\left(1 + \frac{0.11}{12}\right) = \ln 2$$

$$t = \frac{\ln 2}{12 \ln\left(1 + \frac{0.11}{12}\right)} \approx 6.330 \text{ years}$$

(d) Compounded continuously

$$e^{0.11t} = 2$$

$$0.11t = \ln 2$$

$$t = \frac{\ln 2}{0.11} \approx 6.301 \text{ years}$$

19. $3P = Pe^{rt}$

$3 = e^{rt}$

$\ln 3 = rt$

$\dfrac{\ln 3}{r} = t$

r	2%	4%	6%	8%	10%	12%
$t = \dfrac{\ln 3}{r}$ (years)	54.93	27.47	18.31	13.73	10.99	9.16

21. $3P = P(1 + r)^t$

$3 = (1 + r)^t$

$\ln 3 = \ln(1 + r)^t$

$\ln 3 = t\ln(1 + r)$

$\dfrac{\ln 3}{\ln(1 + r)} = t$

r	2%	4%	6%	8%	10%	12%
$t = \dfrac{\ln 3}{\ln(1 + r)}$ (years)	55.48	28.01	18.85	14.27	11.53	9.69

23. Continuous compounding results in faster growth.

$A = 1 + 0.075[\![t]\!]$ and $A = e^{0.07t}$

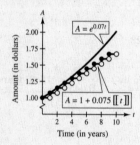

25. $\dfrac{1}{2}C = Ce^{k(1599)}$

$0.5 = e^{k(1599)}$

$\ln 0.5 = \ln e^{k(1599)}$

$\ln 0.5 = k(1599)$

$k = \dfrac{\ln 0.5}{1599}$

Given $C = 10$ grams after 1000 years, we have

$y = 10e^{[(\ln 0.5)/1599](1000)}$

≈ 6.48 grams.

27. $\dfrac{1}{2}C = Ce^{k(5715)}$

$0.5 = e^{k(5715)}$

$\ln 0.5 = \ln e^{k(5715)}$

$\ln 0.5 = k(5715)$

$k = \dfrac{\ln 0.5}{5715}$

Given $y = 2$ grams after 1000 years, we have

$2 = Ce^{[(\ln 0.5)/5715](1000)}$

$C \approx 2.26$ grams.

29. $\dfrac{1}{2}C = Ce^{k(24,100)}$

$0.5 = e^{k(24,100)}$

$\ln 0.5 = \ln e^{k(24,100)}$

$\ln 0.5 = k(24,100)$

$k = \dfrac{\ln 0.5}{24,100}$

Given $y = 2.1$ grams after 1000 years, we have

$2.1 = Ce^{[(\ln 0.5)/24,100](1000)}$

$C \approx 2.16$ grams.

31. $y = ae^{bx}$

$1 = ae^{b(0)} \implies 1 = a$

$10 = e^{b(3)}$

$\ln 10 = 3b$

$\dfrac{\ln 10}{3} = b \implies b \approx 0.7675$

Thus, $y = e^{0.7675x}$.

33. $y = ae^{bx}$

$5 = ae^{b(0)} \implies 5 = a$

$1 = 5e^{b(4)}$

$\dfrac{1}{5} = e^{4b}$

$\ln\left(\dfrac{1}{5}\right) = 4b$

$\dfrac{\ln(1/5)}{4} = b \implies b \approx -0.4024$

Thus, $y = 5e^{-0.4024x}$.

35. $P = 2430e^{-0.0029t}$

  (a) Since the exponent is negative, this is an exponential decay model. The population is decreasing.

  (b) For 2000, let $t = 0$: $P = 2430$ thousand people

     For 2003, let $t = 3$: $P \approx 2408.95$ thousand people

  (c) 2.3 million = 2300 thousand

$$2300 = 2430e^{-0.0029t}$$

$$\frac{2300}{2430} = e^{-0.0029t}$$

$$\ln\left(\frac{2300}{2430}\right) = -0.0029t$$

$$t = \frac{\ln(2300/2430)}{-0.0029} \approx 18.96$$

The population will reach 2.3 million (according to the model) during the later part of the year 2018.

37. $y = 4080e^{kt}$

When $t = 3$, $y = 10{,}000$:

$$10{,}000 = 4080e^{k(3)}$$

$$\frac{10{,}000}{4080} = e^{3k}$$

$$\ln\left(\frac{10{,}000}{4080}\right) = 3k$$

$$k = \frac{\ln(10{,}000/4080)}{3} \approx 0.2988$$

When $t = 24$: $y = 4080e^{0.2988(24)} \approx 5{,}309{,}734$ hits

39. $N = 100e^{kt}$

$$300 = 100e^{5k}$$

$$3 = e^{5k}$$

$$\ln 3 = \ln e^{5k}$$

$$\ln 3 = 5k$$

$$k = \frac{\ln 3}{5} \approx 0.2197$$

$$N = 100e^{0.2197t}$$

$$200 = 100e^{0.2197t}$$

$$t = \frac{\ln 2}{0.2197} \approx 3.15 \text{ hours}$$

41. $R = \dfrac{1}{10^{12}}e^{-t/8223}$

  (a) $R = \dfrac{1}{8^{14}}$

$$\frac{1}{10^{12}}e^{-t/8223} = \frac{1}{8^{14}}$$

$$e^{-t/8223} = \frac{10^{12}}{8^{14}}$$

$$-\frac{t}{8223} = \ln\left(\frac{10^{12}}{8^{14}}\right)$$

$$t = -8223\ln\left(\frac{10^{12}}{8^{14}}\right) \approx 12{,}180 \text{ years old}$$

  (b) $\dfrac{1}{10^{12}}e^{-t/8223} = \dfrac{1}{13^{11}}$

$$e^{-t/8223} = \frac{10^{12}}{13^{11}}$$

$$-\frac{t}{8223} = \ln\left(\frac{10^{12}}{13^{11}}\right)$$

$$t = -8223\ln\left(\frac{10^{12}}{13^{11}}\right) \approx 4797 \text{ years old}$$

43. $(0, 30{,}788), (2, 18{,}000)$

  (a) $m = \dfrac{18{,}000 - 30{,}788}{2 - 0} = -6394$

    $b = 30{,}788$

    Linear model: $V = -6394t + 30{,}788$

—CONTINUED—

43. —CONTINUED—

(b)
$$a = 30{,}788$$

$$18{,}000 = 30{,}788e^{k(2)}$$

$$\frac{4500}{7697} = e^{2k}$$

$$\ln\left(\frac{4500}{7697}\right) = 2k$$

$$k = \frac{1}{2}\ln\left(\frac{4500}{7697}\right) \approx -0.268$$

Exponential model: $V = 30{,}788e^{-0.268t}$

(d)

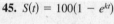

t	1	3
$V = -6394t + 30{,}788$	$24,394	$11,606
$V = 30{,}788e^{-0.268t}$	$23,550	$13,779

(c)
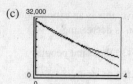

The exponential model depreciates faster in the first two years.

(e) The linear model gives a higher value for the car for the first two years, then the exponential model yields a higher value. If the car is less than two years old, the seller would most likely want to use the linear model and the buyer the exponential model. If it is more than two years old, the opposite is true.

45. $S(t) = 100(1 - e^{kt})$

(a)
$$15 = 100(1 - e^{k(1)})$$

$$-85 = -100e^{k}$$

$$\frac{85}{100} = e^{k}$$

$$0.85 = e^{k}$$

$$\ln 0.85 = \ln e^{k}$$

$$k = \ln 0.85$$

$$k \approx -0.1625$$

$$S(t) = 100(1 - e^{-0.1625t})$$

(b)
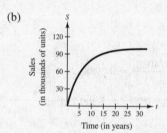

(c) $S(5) = 100(1 - e^{-0.1625(5)}) \approx 55.625 = 55{,}625$ units

47. $y = 0.0266e^{-(x-100)^2/450}, \ 70 \le x \le 116$

(a)
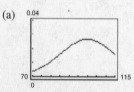

(b) The average IQ score of an adult student is 100.

49. $p(t) = \dfrac{1000}{1 + 9e^{-0.1656t}}$

(a) $p(5) = \dfrac{1000}{1 + 9e^{-0.1656(5)}} \approx 203$ animals

(b)
$$500 = \frac{1000}{1 + 9e^{-0.1656t}}$$

$$1 + 9e^{-0.1656t} = 2$$

$$9e^{-0.1656t} = 1$$

$$e^{-0.1656t} = \frac{1}{9}$$

$$t = -\frac{\ln(1/9)}{0.1656} \approx 13 \text{ months}$$

(c)
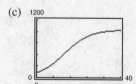

The horizontal asymptotes are $p = 0$ and $p = 1000$. The asymptote with the larger p-value, $p = 1000$, indicates that the population size will approach 1000 as time increases.

51. $R = \log \dfrac{I}{I_0} = \log I$ since $I_0 = 1$.

    (a) $7.9 = \log I \implies I = 10^{7.9} \approx 79{,}432{,}823$

    (b) $8.3 = \log I \implies I = 10^{8.3} \approx 199{,}526{,}231$

    (c) $4.2 = \log I \implies I = 10^{4.2} \approx 15{,}849$

53. $\beta = 10 \log \dfrac{I}{I_0}$ where $I_0 = 10^{-12}$ watt/m^2.

    (a) $\beta = 10 \log \dfrac{10^{-10}}{10^{-12}} = 10 \log 10^2 = 20$ decibels

    (b) $\beta = 10 \log \dfrac{10^{-5}}{10^{-12}} = 10 \log 10^7 = 70$ decibels

    (c) $\beta = 10 \log \dfrac{10^{-8}}{10^{-12}} = 10 \log 10^4 = 40$ decibels

    (d) $\beta = 10 \log \dfrac{1}{10^{-12}} = 10 \log 10^{12} = 120$ decibels

55.
$$\beta = 10 \log \dfrac{I}{I_0}$$
$$\dfrac{\beta}{10} = \log \dfrac{I}{I_0}$$
$$10^{\beta/10} = 10^{\log I/I_0}$$
$$10^{\beta/10} = \dfrac{I}{I_0}$$
$$I = I_0 10^{\beta/10}$$
$$\% \text{ decrease} = \dfrac{I_0 10^{9.3} - I_0 10^{8.0}}{I_0 10^{9.3}} \times 100 \approx 95\%$$

57. $\text{pH} = -\log[\text{H}^+]$

    $-\log(2.3 \times 10^{-5}) \approx 4.64$

59. $5.8 = -\log[\text{H}^+]$

    $-5.8 = \log[\text{H}^+]$

    $10^{-5.8} = 10^{\log[\text{H}^+]}$

    $10^{-5.8} = [\text{H}^+]$

    $[\text{H}^+] \approx 1.58 \times 10^{-6}$ mole per liter

61. $2.9 = -\log[\text{H}^+]$

    $-2.9 = \log[\text{H}^+]$

    $[\text{H}^+] = 10^{-2.9}$ for the apple juice

    $8.0 = -\log[\text{H}^+]$

    $-8.0 = \log[\text{H}^+]$

    $[\text{H}^+] = 10^{-8}$ for the drinking water

    $\dfrac{10^{-2.9}}{10^{-8}} = 10^{5.1}$ times the hydrogen ion concentration of drinking water

63. $t = -10 \ln \dfrac{T - 70}{98.6 - 70}$

At 9:00 A.M. we have:

$t = -10 \ln \dfrac{85.7 - 70}{98.6 - 70} \approx 6$ hours

From this you can conclude that the person died at 3:00 A.M.

65. $u = 120{,}000 \left[\dfrac{0.075t}{1 - \left(\dfrac{1}{1 + 0.075/12}\right)^{12t}} - 1 \right]$

    (a)

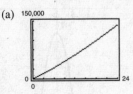

    (b) From the graph, $u = \$120{,}000$ when $t \approx 21$ years. It would take approximately 37.6 years to pay \$240,000 in interest. Yes, it is possible to pay twice as much in interest charges as the size of the mortgage. It is especially likely when the interest rates are higher.

67. False. The domain can be the set of real numbers for a logistic growth function.

71. (a) Logarithmic

(b) Logistic

(c) Exponential (decay)

(d) Linear

(e) None of the above (appears to be a combination of a linear and a quadratic)

(f) Exponential (growth)

69. False. The graph of $f(x)$ is the graph of $g(x)$ shifted upward five units.

73. $(-1, 2), (0, 5)$

(a)

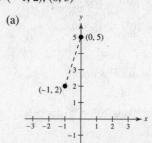

(b) $d = \sqrt{(0 - (-1))^2 + (5 - 2)^2} = \sqrt{1^2 + 3^2} = \sqrt{10}$

(c) Midpoint: $\left(\dfrac{-1 + 0}{2}, \dfrac{2 + 5}{2} \right) = \left(-\dfrac{1}{2}, \dfrac{7}{2} \right)$

(d) $m = \dfrac{5 - 2}{0 - (-1)} = \dfrac{3}{1} = 3$

75. $(3, 3), (14, -2)$

(a)

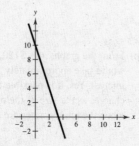

(b) $d = \sqrt{(14 - 3)^2 + (-2 - 3)^2}$

$= \sqrt{11^2 + (-5)^2} = \sqrt{146}$

(c) Midpoint: $\left(\dfrac{3 + 14}{2}, \dfrac{3 + (-2)}{2} \right) = \left(\dfrac{17}{2}, \dfrac{1}{2} \right)$

(d) $m = \dfrac{-2 - 3}{14 - 3} = -\dfrac{5}{11}$

77. $\left(\dfrac{1}{2}, -\dfrac{1}{4} \right), \left(\dfrac{3}{4}, 0 \right)$

(a)

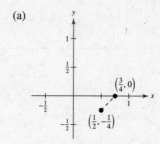

(b) $d = \sqrt{\left(\dfrac{3}{4} - \dfrac{1}{2} \right)^2 + \left(0 - \left(-\dfrac{1}{4} \right) \right)^2}$

$= \sqrt{\left(\dfrac{1}{4} \right)^2 + \left(\dfrac{1}{4} \right)^2} = \sqrt{\dfrac{1}{8}}$

(c) Midpoint: $\left(\dfrac{(1/2) + (3/4)}{2}, \dfrac{(-1/4) + 0}{2} \right) = \left(\dfrac{5}{8}, -\dfrac{1}{8} \right)$

(d) $m = \dfrac{0 - (-1/4)}{(3/4) - (1/2)} = \dfrac{1/4}{1/4} = 1$

79. $y = 10 - 3x$

Line

Slope: $m = -3$

y-intercept: $(0, 10)$

81. $y = -2x^2 - 3$

$y = -2(x - 0)^2 - 3$

Parabola

Vertex: $(0, -3)$

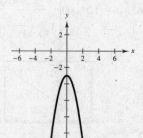

83. $3x^2 - 4y = 0$

$$3x^2 = 4y$$

$$x^2 = \tfrac{4}{3}y$$

Parabola

Vertex: $(0, 0)$

Focus: $\left(0, \tfrac{1}{3}\right)$

Directrix: $y = -\tfrac{1}{3}$

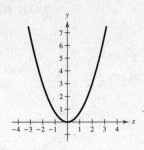

85. $y = \dfrac{4}{1 - 3x}$

Vertical asymptote: $x = \dfrac{1}{3}$

Horizontal asymptote: $y = 0$

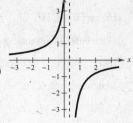

87. $x^2 + (y - 8)^2 = 25$

Circle

Center: $(0, 8)$

Radius: 5

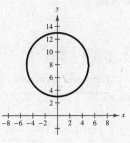

89. $f(x) = 2^{x-1} + 5$

Horizontal asymptote: $y = 5$

x	-5	-3	-1	0	1	3	5
$f(x)$	5.02	5.06	5.3	5.5	6	9	21

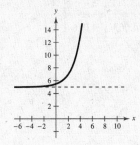

91. $f(x) = 3^x - 4$

Horizontal asymptote: $y = -4$

x	-4	-2	-1	0	1	2
$f(x)$	-3.99	-3.89	-3.67	-3	-1	5

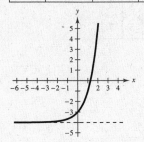

93. Answers will vary.

Review Exercises for Chapter 3

1. $f(x) = 6.1^x$

$f(2.4) = 6.1^{2.4} \approx 76.699$

3. $f(x) = 2^{-0.5x}$

$f(\pi) = 2^{-0.5(\pi)} \approx 0.337$

5. $f(x) = 7(0.2^x)$

$f\left(-\sqrt{11}\right) = 7\left(0.2^{-\sqrt{11}}\right)$

≈ 1456.529

7. $f(x) = 4^x$

Intercept: $(0, 1)$

Horizontal asymptote: x-axis

Increasing on: $(-\infty, \infty)$

Matches graph (c).

9. $f(x) = -4^x$

Intercept: $(0, -1)$

Horizontal asymptote: x-axis

Decreasing on: $(-\infty, \infty)$

Matches graph (a).

11. $f(x) = 5^x$

$g(x) = 5^{x-1}$

Since $g(x) = f(x-1)$, the graph of g can be obtained by shifting the graph of f one unit to the right.

13. $f(x) = \left(\frac{1}{2}\right)^x$

$g(x) = -\left(\frac{1}{2}\right)^{x+2}$

Since $g(x) = -f(x+2)$, the graph of g can be obtained by reflecting the graph of f about the x-axis and shifting $-f$ two units to the left.

15. $f(x) = 4^{-x} + 4$

Horizontal asymptote: $y = 4$

x	-1	0	1	2	3
$f(x)$	8	5	4.25	4.063	4.016

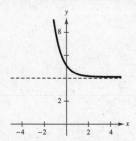

17. $f(x) = -2.65^{x+1}$

Horizontal asymptote: $y = 0$

x	-2	-1	0	1	2
$f(x)$	-0.377	-1	-2.65	-7.023	-18.61

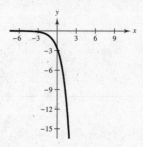

19. $f(x) = 5^{x-2} + 4$

Horizontal asymptote: $y = 4$

x	-1	0	1	2	3
$f(x)$	4.008	4.04	4.2	5	9

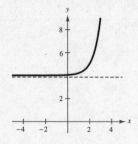

21. $f(x) = \left(\frac{1}{2}\right)^{-x} + 3 = 2^x + 3$

Horizontal asymptote: $y = 3$

x	-2	-1	0	1	2
$f(x)$	3.25	3.5	4	5	7

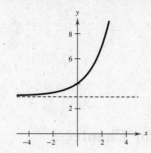

23. $3^{x+2} = \frac{1}{9}$

$3^{x+2} = 3^{-2}$

$x + 2 = -2$

$x = -4$

25. $e^{5x-7} = e^{15}$

$5x - 7 = 15$

$5x = 22$

$x = \frac{22}{5}$

27. $e^8 \approx 2980.958$

29. $e^{-1.7} \approx 0.183$

31. $h(x) = e^{-x/2}$

x	-2	-1	0	1	2
$h(x)$	2.72	1.65	1	0.61	0.37

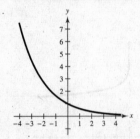

33. $f(x) = e^{x+2}$

x	-3	-2	-1	0	1
$f(x)$	0.37	1	2.72	7.39	20.09

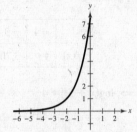

35. $A = 3500\left(1 + \dfrac{0.065}{n}\right)^{10n}$ or $A = 3500e^{(0.065)(10)}$

n	1	2	4	12	365	Continuous Compounding
A	\$6569.98	\$6635.43	\$6669.46	\$6692.64	\$6704.00	\$6704.39

37. $F(t) = 1 - e^{-t/3}$

(a) $F\left(\frac{1}{2}\right) \approx 0.154$

(b) $F(2) \approx 0.487$

(c) $F(5) \approx 0.811$

39. (a) $A = 50{,}000e^{(0.0875)(35)} \approx \$1{,}069{,}047.14$

(b) The doubling time is $\dfrac{\ln 2}{0.0875} \approx 7.9$ years.

41. $4^3 = 64$

$\log_4 64 = 3$

43. $e^{0.8} = 2.2255\ldots$

$\ln 2.2255\ldots = 0.8$

45. $f(x) = \log x$

$f(1000) = \log 1000$

$= \log 10^3 = 3$

47. $g(x) = \log_2 x$

$g\left(\frac{1}{8}\right) = \log_2\left(\frac{1}{8}\right) = \log_2 2^{-3} = -3$

49. $\log_4(x + 7) = \log_4 14$

$x + 7 = 14$

$x = 7$

51. $\ln(x + 9) = \ln 4$

$x + 9 = 4$

$x = -5$

53. $g(x) = \log_7 x \implies x = 7^y$

Domain: $(0, \infty)$

x-intercept: $(1, 0)$

Vertical asymptote: $x = 0$

x	$\frac{1}{7}$	1	7	49
$g(x)$	-1	0	1	2

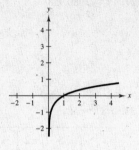

55. $f(x) = \log\left(\dfrac{x}{3}\right) \implies \dfrac{x}{3} = 10^y \implies x = 3(10^y)$

Domain: $(0, \infty)$

x-intercept: $(3, 0)$

Vertical asymptote:
$x = 0$

x	0.03	0.3	3	30
$f(x)$	-2	-1	0	1

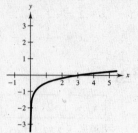

57. $f(x) = 4 - \log(x + 5)$

Domain: $(-5, \infty)$

x-intercept: $(9995, 0)$

Since $4 - \log(x + 5) = 0 \implies \log(x + 5) = 4$

$$x + 5 = 10^4$$

$$x = 10^4 - 5 = 9995.$$

Vertical asymptote: $x = -5$

x	-4	-3	-2	-1	0	1
$f(x)$	4	3.70	3.52	3.40	3.30	3.22

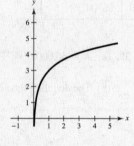

59. $\ln 22.6 \approx 3.118$

61. $\ln e^{-12} = -12$

63. $\ln\left(\sqrt{7} + 5\right) \approx 2.034$

65. $f(x) = \ln x + 3$

Domain: $(0, \infty)$

x-intercept: $\ln x + 3 = 0$

$$\ln x = -3$$

$$x = e^{-3}$$

$$(e^{-3}, 0)$$

Vertical asymptote: $x = 0$

x	1	2	3	$\frac{1}{2}$	$\frac{1}{4}$
$f(x)$	3	3.69	4.10	2.31	1.61

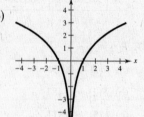

67. $h(x) = \ln(x^2) = 2\ln|x|$

Domain: $(-\infty, 0) \cup (0, \infty)$

x-intercepts: $(\pm 1, 0)$

Vertical asymptote: $x = 0$

x	± 0.5	± 1	± 2	± 3	± 4
y	-1.39	0	1.39	2.20	2.77

69. $h = 116\log(a + 40) - 176$

$h(55) = 116\log(55 + 40) - 176$

≈ 53.4 inches

71. $\log_4 9 = \dfrac{\log 9}{\log 4} \approx 1.585$

$\log_4 9 = \dfrac{\ln 9}{\ln 4} \approx 1.585$

73. $\log_{1/2} 5 = \dfrac{\log 5}{\log(1/2)} \approx -2.322$

$\log_{1/2} 5 = \dfrac{\ln 5}{\ln(1/2)} \approx -2.322$

75. $\log 18 = \log(2 \cdot 3^2)$

$= \log 2 + 2\log 3$

≈ 1.255

77. $\ln 20 = \ln(2^2 \cdot 5)$

$= 2\ln 2 + \ln 5 \approx 2.996$

79. $\log_5 5x^2 = \log_5 5 + \log_5 x^2$

$= 1 + 2\log_5 x$

81. $\log_3 \dfrac{6}{\sqrt[3]{x}} = \log_3 6 - \log_3 \sqrt[3]{x}$

$\qquad = \log_3(3 \cdot 2) - \log_3 x^{1/3}$

$\qquad = \log_3 3 + \log_3 2 - \dfrac{1}{3}\log_3 x$

$\qquad = 1 + \log_3 2 - \dfrac{1}{3}\log_3 x$

83. $\ln x^2 y^2 z = \ln x^2 + \ln y^2 + \ln z$

$\qquad = 2\ln x + 2\ln y + \ln z$

85. $\ln\!\left(\dfrac{x+3}{xy}\right) = \ln(x+3) - \ln xy$

$\qquad = \ln(x+3) - [\ln x + \ln y]$

$\qquad = \ln(x+3) - \ln x - \ln y$

87. $\log_2 5 + \log_2 x = \log_2 5x$

89. $\ln x - \dfrac{1}{4}\ln y = \ln x - \ln \sqrt[4]{y} = \ln\!\left(\dfrac{x}{\sqrt[4]{y}}\right)$

91. $\dfrac{1}{3}\log_8(x+4) + 7\log_8 y = \log_8 \sqrt[3]{x+4} + \log_8 y^7$

$\qquad = \log_8\!\left(y^7 \sqrt[3]{x+4}\right)$

93. $\dfrac{1}{2}\ln(2x-1) - 2\ln(x+1) = \ln\sqrt{2x-1} - \ln(x+1)^2$

$\qquad = \ln\dfrac{\sqrt{2x-1}}{(x+1)^2}$

95. $t = 50\log\dfrac{18,000}{18,000-h}$

(a) Domain: $0 \le h < 18,000$

(b)

Vertical asymptote: $h = 18,000$

(c) As the plane approaches its absolute ceiling, it climbs at a slower rate, so the time required increases.

(d) $50\log\dfrac{18,000}{18,000-4000} \approx 5.46$ minutes

97. $8^x = 512$

$8^x = 8^3$

$x = 3$

99. $e^x = 3$

$x = \ln 3$

101. $\log_4 x = 2$

$x = 4^2 = 16$

103. $\ln x = 4$

$x = e^4$

105. $e^x = 12$

$\ln e^x = \ln 12$

$x = \ln 12 \approx 2.485$

107. $e^{4x} = e^{x^2+3}$

$4x = x^2 + 3$

$0 = x^2 - 4x + 3$

$0 = (x-1)(x-3)$

$x = 1$ or $x = 3$

109. $2^x + 13 = 35$

$\qquad 2^x = 22$

$\qquad x = \log_2 22$

$\qquad\quad = \dfrac{\log 22}{\log 2} \text{ or } \dfrac{\ln 22}{\ln 2}$

$\qquad x \approx 4.459$

111. $-4(5^x) = -68$

$\qquad 5^x = 17$

$\qquad \ln 5^x = \ln 17$

$\qquad x \ln 5 = \ln 17$

$\qquad x = \dfrac{\ln 17}{\ln 5} \approx 1.760$

113. $e^{2x} - 7e^x + 10 = 0$

$\qquad (e^x - 2)(e^x - 5) = 0$

$\qquad e^x = 2 \qquad$ or $\qquad e^x = 5$

$\qquad \ln e^x = \ln 2 \qquad\qquad \ln e^x = \ln 5$

$\qquad x = \ln 2 \approx 0.693 \qquad x = \ln 5 \approx 1.609$

115. $2^{0.6x} - 3x = 0$

Graph $y_1 = 2^{0.6x} - 3x$.

The x-intercepts are at
$x \approx 0.392$ and at $x \approx 7.480$.

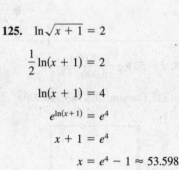

117. $25e^{-0.3x} = 12$

Graph $y_1 = 25e^{-0.3x}$ and
$y_2 = 12$.

The graphs intersect at
$x \approx 2.447$.

119. $\ln 3x = 8.2$

$\qquad e^{\ln 3x} = e^{8.2}$

$\qquad 3x = e^{8.2}$

$\qquad x = \dfrac{e^{8.2}}{3} \approx 1213.650$

121. $2 \ln 4x = 15$

$\qquad \ln 4x = \dfrac{15}{2}$

$\qquad e^{\ln 4x} = e^{7.5}$

$\qquad 4x = e^{7.5}$

$\qquad x = \dfrac{1}{4}e^{7.5} \approx 452.011$

123. $\ln x - \ln 3 = 2$

$\qquad \ln \dfrac{x}{3} = 2$

$\qquad e^{\ln(x/3)} = e^2$

$\qquad \dfrac{x}{3} = e^2$

$\qquad x = 3e^2 \approx 22.167$

125. $\ln \sqrt{x + 1} = 2$

$\qquad \dfrac{1}{2}\ln(x + 1) = 2$

$\qquad \ln(x + 1) = 4$

$\qquad e^{\ln(x+1)} = e^4$

$\qquad x + 1 = e^4$

$\qquad x = e^4 - 1 \approx 53.598$

127. $\qquad \log_8(x - 1) = \log_8(x - 2) - \log_8(x + 2)$

$\qquad\quad \log_8(x - 1) = \log_8\left(\dfrac{x - 2}{x + 2}\right)$

$\qquad\qquad\quad x - 1 = \dfrac{x - 2}{x + 2}$

$\qquad (x - 1)(x + 2) = x - 2$

$\qquad\quad x^2 + x - 2 = x - 2$

$\qquad\qquad\qquad x^2 = 0$

$\qquad\qquad\qquad x = 0$

Since $x = 0$ is not in the domain of $\log_8(x - 1)$ or of $\log_8(x - 2)$, it is an extraneous solution. The equation has no solution.

129. $\log(1 - x) = -1$

$\qquad 1 - x = 10^{-1}$

$\qquad 1 - \dfrac{1}{10} = x$

$\qquad x = 0.900$

131. $2 \ln(x + 3) + 3x = 8$

Graph $y_1 = 2 \ln(x + 3) + 3x$ and $y_2 = 8$.

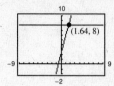

The graphs intersect at approximately $(1.643, 8)$.
The solution of the equation is $x \approx 1.643$.

133. $4 \ln(x + 5) - x = 10$

Graph $y_1 = 4 \ln(x + 5) - x$ and $y_2 = 10$.

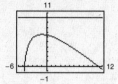

The graphs do not intersect. The equation has no solution.

135. $3(7550) = 7550e^{0.0725t}$

$$3 = e^{0.0725t}$$

$$\ln 3 = \ln e^{0.0725t}$$

$$\ln 3 = 0.0725t$$

$$t = \frac{\ln 3}{0.0725} \approx 15.2 \text{ years}$$

137. $y = 3e^{-2x/3}$

Exponential decay model

Matches graph (e).

139. $y = \ln(x + 3)$

Logarithmic model

Vertical asymptote: $x = -3$

Graph includes $(-2, 0)$

Matches graph (f).

141. $y = 2e^{-(x+4)^2/3}$

Gaussian model

Matches graph (a).

143.
$$y = ae^{bx}$$

$$2 = ae^{b(0)} \implies a = 2$$

$$3 = 2e^{b(4)}$$

$$1.5 = e^{4b}$$

$$\ln 1.5 = 4b \implies b \approx 0.1014$$

Thus, $y \approx 2e^{0.1014x}$.

145.
$$P = 3499e^{0.0135t}$$

$$4.5 \text{ million} = 4500 \text{ thousand}$$

$$4500 = 3499e^{0.0135t}$$

$$\frac{4500}{3499} = e^{0.0135t}$$

$$\ln\left(\frac{4500}{3499}\right) = 0.0135t$$

$$t = \frac{\ln(4500/3499)}{0.0135} \approx 18.6 \text{ years}$$

According to this model, the population of South Carolina will reach 4.5 million during the year 2008.

147. (a) $20,000 = 10,000e^{r(5)}$

$$2 = e^{5r}$$

$$\ln 2 = 5r$$

$$\frac{\ln 2}{5} = r$$

$$r \approx 0.138629$$

$$= 13.8629\%$$

(b) $A = 10,000e^{0.138629}$

$$\approx \$11,486.98$$

149. $y = 0.0499e^{-(x-71)^2/128}$,

$40 \le x \le 100$

(a) Graph $y_1 = 0.0499e^{-(x-71)^2/128}$.

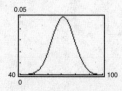

(b) The average test score is 71.

151. $\beta = 10 \log\left(\dfrac{I}{10^{-16}}\right)$

$125 = 10 \log\left(\dfrac{I}{10^{-16}}\right)$

$12.5 = \log\left(\dfrac{I}{10^{-16}}\right)$

$10^{12.5} = \dfrac{I}{10^{-16}}$

$I = 10^{-3.5}$ watt/cm^2

153. True. By the inverse properties, $\log_b b^{2x} = 2x$.

155. Since graphs (b) and (d) represent exponential decay, b and d are negative. Since graphs (a) and (c) represent exponential growth, a and c are positive.

Problem Solving for Chapter 3

1. $y = a^x$

$y_1 = 0.5^x$

$y_2 = 1.2^x$

$y_3 = 2.0^x$

$y_4 = x$

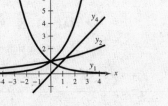

The curves $y = 0.5^x$ and $y = 1.2^x$ cross the line $y = x$. From checking the graphs, it appears that $y = x$ will cross $y = a^x$ for $0 \le a \le 1.44$.

3. The exponential function, $y = e^x$, increases at a faster rate than the polynomial function $y = x^n$.

5. (a) $f(u + v) = a^{u+v}$
$= a^u \cdot a^v$
$= f(u) \cdot f(v)$

(b) $f(2x) = a^{2x}$
$= (a^x)^2$
$= [f(x)]^2$

7. (a)

(b)

(c)

9. $f(x) = e^x - e^{-x}$

$y = e^x - e^{-x}$

$x = e^y - e^{-y}$

$x = \dfrac{e^{2y} - 1}{e^y}$

$xe^y = e^{2y} - 1$

$e^{2y} - xe^y - 1 = 0$

$e^y = \dfrac{x \pm \sqrt{x^2 + 4}}{2}$ Quadratic Formula

Choosing the positive quantity for e^y we have

$y = \ln\left(\dfrac{x + \sqrt{x^2 + 4}}{2}\right)$. Thus, $f^{-1}(x) = \ln\left(\dfrac{x + \sqrt{x^2 + 4}}{2}\right)$.

11. Answer (c). $y = 6(1 - e^{-x^2/2})$

The graph passes through $(0, 0)$ and neither (a) nor (b) pass through the origin. Also, the graph has y-axis symmetry and a horizontal asymptote at $y = 6$.

13. $y_1 = c_1\left(\dfrac{1}{2}\right)^{t/k_1}$ and $y_2 = c_2\left(\dfrac{1}{2}\right)^{t/k_2}$

$$c_1\left(\frac{1}{2}\right)^{t/k_1} = c_2\left(\frac{1}{2}\right)^{t/k_2}$$

$$\frac{c_1}{c_2} = \left(\frac{1}{2}\right)^{(t/k_2 - t/k_1)}$$

$$\ln\left(\frac{c_1}{c_2}\right) = \left(\frac{t}{k_2} - \frac{t}{k_1}\right)\ln\left(\frac{1}{2}\right)$$

$$\ln c_1 - \ln c_2 = t\left(\frac{1}{k_2} - \frac{1}{k_1}\right)\ln\left(\frac{1}{2}\right)$$

$$t = \frac{\ln c_1 - \ln c_2}{[(1/k_2) - (1/k_1)]\ln(1/2)}$$

17. $(\ln x)^2 = \ln x^2$

$(\ln x)^2 - 2\ln x = 0$

$\ln x(\ln x - 2) = 0$

$\ln x = 0$ or $\ln x = 2$

$x = 1$ or $x = e^2$

19. $y_4 = (x-1) - \frac{1}{2}(x-1)^2 + \frac{1}{3}(x-1)^3 - \frac{1}{4}(x-1)^4$

The pattern implies that

$\ln x = (x-1) - \frac{1}{2}(x-1)^2 + \frac{1}{3}(x-1)^3 - \frac{1}{4}(x-1)^4 + \ldots$

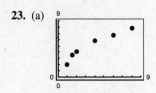

15. (a) $y \approx 252.606(1.0310)^t$

(b) $y \approx 400.88t^2 - 1464.6t + 291{,}782$

(c)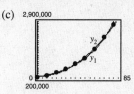

(d) Both models appear to be "good fits" for the data, but neither would be reliable to predict the population of the United States in 2010. The exponential model approaches infinity rapidly.

21. $y = 80.4 - 11\ln x$

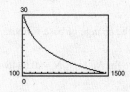

$y(300) = 80.4 - 11\ln 300 \approx 17.7$ ft^3/min

23. (a)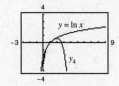

(b) The data could best be modeled by a logarithmic model.

(c) The shape of the curve looks much more logarithmic than linear or exponential.

(d) $y \approx 2.1518 + 2.7044\ln x$

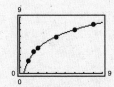

(e) The model is a good fit to the actual data.

25. (a)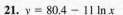

(b) The data could best be modeled by a linear model.

(c) The shape of the curve looks much more linear than exponential or logarithmic.

(d) $y \approx -0.7884x + 8.2566$

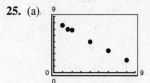

(e) The model is a good fit to the actual data.

Chapter 3   Practice Test

1. Solve for x: $x^{3/5} = 8$.

2. Solve for x: $3^{x-1} = \frac{1}{81}$.

3. Graph $f(x) = 2^{-x}$.

4. Graph $g(x) = e^x + 1$.

5. If \$5000 is invested at 9% interest, find the amount after three years if the interest is compounded

   (a) monthly.   (b) quarterly.   (c) continuously.

6. Write the equation in logarithmic form: $7^{-2} = \frac{1}{49}$.

7. Solve for x: $x - 4 = \log_2 \frac{1}{64}$.

8. Given $\log_b 2 = 0.3562$ and $\log_b 5 = 0.8271$, evaluate $\log_b \sqrt[4]{8/25}$.

9. Write $5 \ln x - \frac{1}{2} \ln y + 6 \ln z$ as a single logarithm.

10. Using your calculator and the change of base formula, evaluate $\log_9 28$.

11. Use your calculator to solve for N: $\log_{10} N = 0.6646$

12. Graph $y = \log_4 x$.

13. Determine the domain of $f(x) = \log_3(x^2 - 9)$.

14. Graph $y = \ln(x - 2)$.

15. True or false: $\dfrac{\ln x}{\ln y} = \ln(x - y)$

16. Solve for x: $5^x = 41$

17. Solve for x: $x - x^2 = \log_5 \frac{1}{25}$

18. Solve for x: $\log_2 x + \log_2(x - 3) = 2$

19. Solve for x: $\dfrac{e^x + e^{-x}}{3} = 4$

20. Six thousand dollars is deposited into a fund at an annual interest rate of 13%. Find the time required for the investment to double if the interest is compounded continuously.

CHAPTER 4
Trigonometry

Section 4.1   Radian and Degree Measure **186**

Section 4.2   Trigonometric Functions: The Unit Circle **191**

Section 4.3   Right Triangle Trigonometry **194**

Section 4.4   Trigonometric Functions of Any Angle **200**

Section 4.5   Graphs of Sine and Cosine Functions **206**

Section 4.6   Graphs of Other Trigonometric Functions **213**

Section 4.7   Inverse Trigonometric Functions **219**

Section 4.8   Applications and Models **225**

Review Exercises **230**

Problem Solving **237**

Practice Test **239**

CHAPTER 4
Trigonometry

Section 4.1   Radian and Degree Measure

You should know the following basic facts about angles, their measurement, and their applications.

■ Types of Angles:
   (a) Acute: Measure between $0°$ and $90°$.
   (b) Right: Measure $90°$.
   (c) Obtuse: Measure between $90°$ and $180°$.
   (d) Straight: Measure $180°$.
■ α and β are complementary if $\alpha + \beta = 90°$. They are supplementary if $\alpha + \beta = 180°$.
■ Two angles in standard position that have the same terminal side are called coterminal angles.
■ To convert degrees to radians, use $1° = \pi/180$ radians.
■ To convert radians to degrees, use 1 radian $= (180/\pi)°$.
■ $1' =$ one minute $= 1/60$ of $1°$.
■ $1'' =$ one second $= 1/60$ of $1' = 1/3600$ of $1°$.
■ The length of a circular arc is $s = r\theta$ where θ is measured in radians.
■ Linear speed $= \dfrac{\text{arc length}}{\text{time}} = \dfrac{s}{t}$
■ Angular speed $= \theta/t = s/rt$

Vocabulary Check

1. Trigonometry

2. angle

3. coterminal

4. radian

5. acute; obtuse

6. complementary; supplementary

7. degree

8. linear

9. angular

10. $A = \frac{1}{2}r^2\theta$

1.
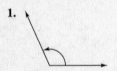

The angle shown is approximately
2 radians.

3.

The angle shown is approximately
-3 radians.

5.
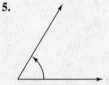

The angle shown is approximately
1 radian.

7. (a) Since $0 < \dfrac{\pi}{5} < \dfrac{\pi}{2}$; $\dfrac{\pi}{5}$ lies in Quadrant I.

   (b) Since $\pi < \dfrac{7\pi}{5} < \dfrac{3\pi}{2}$; $\dfrac{7\pi}{5}$ lies in Quadrant III.

9. (a) Since $-\dfrac{\pi}{2} < -\dfrac{\pi}{12} < 0$; $-\dfrac{\pi}{12}$ lies in Quadrant IV.

   (b) Since $-\pi < -2 < -\dfrac{\pi}{2}$; -2 lies in Quadrant III.

11. (a) Since $\pi < 3.5 < \dfrac{3\pi}{2}$; 3.5 lies in Quadrant III.

   (b) Since $\dfrac{\pi}{2} < 2.25 < \pi$; 2.25 lies in Quadrant II.

13. (a) $\dfrac{5\pi}{4}$

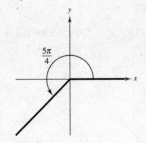

   (b) $-\dfrac{2\pi}{3}$

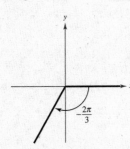

15. (a) $\dfrac{11\pi}{6}$

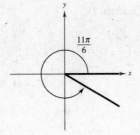

   (b) -3

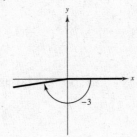

17. (a) Coterminal angles for $\dfrac{\pi}{6}$

$$\dfrac{\pi}{6} + 2\pi = \dfrac{13\pi}{6}$$

$$\dfrac{\pi}{6} - 2\pi = -\dfrac{11\pi}{6}$$

   (b) Coterminal angles for $\dfrac{5\pi}{6}$

$$\dfrac{5\pi}{6} + 2\pi = \dfrac{17\pi}{6}$$

$$\dfrac{5\pi}{6} - 2\pi = -\dfrac{7\pi}{6}$$

19. (a) Coterminal angles for $\dfrac{2\pi}{3}$

$$\dfrac{2\pi}{3} + 2\pi = \dfrac{8\pi}{3}$$

$$\dfrac{2\pi}{3} - 2\pi = -\dfrac{4\pi}{3}$$

   (b) Coterminal angles for $\dfrac{\pi}{12}$

$$\dfrac{\pi}{12} + 2\pi = \dfrac{25\pi}{12}$$

$$\dfrac{\pi}{12} - 2\pi = -\dfrac{23\pi}{12}$$

21. (a) Complement: $\dfrac{\pi}{2} - \dfrac{\pi}{3} = \dfrac{\pi}{6}$

   Supplement: $\pi - \dfrac{\pi}{3} = \dfrac{2\pi}{3}$

   (b) Complement: Not possible, $\dfrac{3\pi}{4}$ is greater than $\dfrac{\pi}{2}$.

   Supplement: $\pi - \dfrac{3\pi}{4} = \dfrac{\pi}{4}$

23. (a) Complement: $\dfrac{\pi}{2} - 1 \approx 0.57$

   Supplement: $\pi - 1 \approx 2.14$

   (b) Complement: Not possible, 2 is greater than $\dfrac{\pi}{2}$.

   Supplement: $\pi - 2 \approx 1.14$

25.

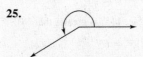

The angle shown is approximately 210°.

27.

The angle shown is approximately −60°.

29.

The angle shown is approximately 165°.

31. (a) Since 90° < 130° < 180°, 130° lies in Quadrant II.

(b) Since 270° < 285° < 360°, 285° lies in Quadrant IV.

33. (a) Since −180° < −132° 50′ < −90°, −132° 50′ lies in Quadrant III.

(b) Since −360° < −336° < −270°, −336° lies in Quadrant I.

35. (a) 30°

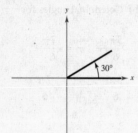

(b) 150°

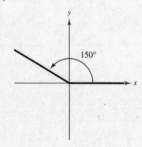

37. (a) 405°

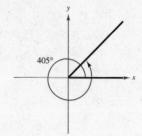

(b) 480°

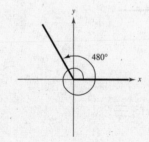

39. (a) Coterminal angles for 45°

$45° + 360° = 405°$

$45° − 360° = −315°$

(b) Coterminal angles for −36°

$−36° + 360° = 324°$

$−36° − 360° = −396°$

41. (a) Coterminal angles for 240°

$240° + 360° = 600°$

$240° − 360° = −120°$

(b) Coterminal angles for −180°

$−180° + 360° = 180°$

$−180° − 360° = −540°$

43. (a) Complement: $90° − 18° = 72°$

Supplement: $180° − 18° = 162°$

(b) Complement: Not possible, 115° is greater than 90°.

Supplement: $1180° − 115° = 65°$

45. (a) Complement: $90° − 79° = 11°$

Supplement: $180° − 79° = 101°$

(b) Complement: Not possible, 150° is greater than 90°.

Supplement: $180° − 150° = 30°$

47. (a) $30° = 30\left(\dfrac{\pi}{180}\right) = \dfrac{\pi}{6}$

(b) $150° = 150\left(\dfrac{\pi}{180}\right) = \dfrac{5\pi}{6}$

49. (a) $−20° = −20\left(\dfrac{\pi}{180}\right) = −\dfrac{\pi}{9}$

(b) $−240° = −240\left(\dfrac{\pi}{180}\right) = −\dfrac{4\pi}{3}$

51. (a) $\dfrac{3\pi}{2} = \dfrac{3\pi}{2}\left(\dfrac{180}{\pi}\right)° = 270°$

(b) $\dfrac{7\pi}{6} = \dfrac{7\pi}{6}\left(\dfrac{180}{\pi}\right)° = 210°$

53. (a) $\dfrac{7\pi}{3} = \dfrac{7\pi}{3}\left(\dfrac{180}{\pi}\right)^{\circ} = 420°$

(b) $-\dfrac{11\pi}{30} = -\dfrac{11\pi}{30}\left(\dfrac{180}{\pi}\right)^{\circ} = -66°$

55. $115° = 115\left(\dfrac{\pi}{180}\right)$

≈ 2.007 radians

57. $-216.35° = -216.35\left(\dfrac{\pi}{180}\right)$

≈ -3.776 radians

59. $532° = 532\left(\dfrac{\pi}{180}\right) \approx 9.285$ radians

61. $-0.83° = -0.83\left(\dfrac{\pi}{180}\right)$

≈ -0.014 radian

63. $\dfrac{\pi}{7} = \dfrac{\pi}{7}\left(\dfrac{180}{\pi}\right)^{\circ} \approx 25.714°$

65. $\dfrac{15\pi}{8} = \dfrac{15\pi}{8}\left(\dfrac{180}{\pi}\right)^{\circ} = 337.500°$

67. $-4.2\pi = -4.2\pi\left(\dfrac{180}{\pi}\right)^{\circ}$

$= -756.000°$

69. $-2 = -2\left(\dfrac{180}{\pi}\right)^{\circ} \approx -114.592°$

71. (a) $54° \, 45' = 54° + \left(\dfrac{45}{60}\right)^{\circ} = 54.75°$

(b) $-128° \, 30' = -128° - \left(\dfrac{30}{60}\right)^{\circ} = -128.5°$

73. (a) $85° \, 18' \, 30'' = \left(85 + \dfrac{18}{60} + \dfrac{30}{3600}\right)^{\circ} \approx 85.308°$

(b) $330° \, 25'' = \left(330 + \dfrac{25}{3600}\right)^{\circ} \approx 330.007°$

75. (a) $240.6° = 240° + 0.6(60)' = 240° \, 36'$

(b) $-145.8° = -[145° + 0.8(60')] = -145° \, 48'$

77. (a) $2.5° = 2° \, 30'$

(b) $-3.58° = -(3° + (0.58)(60'))$

$= -(3° + 34' + 0.8(60''))$

$= -3° \, 34' \, 48''$

79. $s = r\theta$

$6 = 5\theta$

$\theta = \dfrac{6}{5}$ radians

81. $s = r\theta$

$32 = 7\theta$

$\theta = \dfrac{32}{7}$

$= 4\dfrac{4}{7}$ radians

83. $s = r\theta$

$6 = 27\theta$

$\theta = \dfrac{6}{27}$

$= \dfrac{2}{9}$ radians

85. $s = r\theta$

$25 = 14.5\theta$

$\theta = \dfrac{25}{14.5} = \dfrac{50}{29}$ radians

87. $s = r\theta$, θ in radians

$s = 15(180)\left(\dfrac{\pi}{180}\right) = 15\pi$ inches

≈ 47.12 inches

89. $s = r\theta$, θ in radians

$s = 3(1) = 3$ meters

91. $A = \dfrac{1}{2}r^2\theta$

$A = \dfrac{1}{2}(4)^2\left(\dfrac{\pi}{3}\right) = \dfrac{8\pi}{3}$ square inches

≈ 8.38 square inches

93. $A = \dfrac{1}{2}r^2\theta$

$A = \dfrac{1}{2}(2.5)^2(225)\left(\dfrac{\pi}{180}\right) \approx 12.27$ square feet

95. $\theta = 41° \, 15' \, 50'' - 32° \, 47' \, 39''$

$\approx 8.46972° \approx 0.14782$ radian

$s = r\theta \approx 4000(0.14782) \approx 591.3$ miles

97. $\theta = \dfrac{s}{r} = \dfrac{450}{6378} \approx 0.071$ radian $\approx 4.04°$

99. $\theta = \dfrac{s}{r} = \dfrac{2.5}{6} = \dfrac{25}{60} = \dfrac{5}{12}$ radian

101. (a) 65 miles per hour $= \dfrac{65(5280)}{60} = 5720$ feet per minute

The circumference of the tire is $C = 2.5\pi$ feet.

The number of revolutions per minute is
$$r = \frac{5720}{2.5\pi} \approx 728.3 \text{ revolutions per minute}$$

(b) The angular speed is $\dfrac{\theta}{t}$.

$$\theta = \frac{5720}{2.5\pi}(2\pi) = 4576 \text{ radians}$$

$$\text{Angular speed} = \frac{4576 \text{ radians}}{1 \text{ minute}} = 4576 \text{ radians per minute}$$

103. (a) Angular speed $= \dfrac{(5200)(2\pi) \text{ radians}}{1 \text{ minute}}$

$= 10{,}400\pi$ radians per minute

$\approx 32{,}672.56$ radians per minute

(b) Linear speed $= \dfrac{\left(\dfrac{7.25}{2} \text{ in.}\right)\left(\dfrac{1 \text{ ft}}{12 \text{ in.}}\right)(5200)(2\pi)}{1 \text{ minute}}$

$= 3141\dfrac{2}{3}\pi$ feet per minute

≈ 9869.84 feet per minute

105. (a) $(200)(2\pi) \leq$ Angular speed

$\leq (500)(2\pi)$ radians per minute

Interval: $[400\pi,\ 1000\pi]$ radians per minute

(b) $(6)(200)(2\pi) \leq$ Linear speed

$\leq (6)(500)(2\pi)$ centimeters per minute

Interval: $[2400\pi,\ 6000\pi]$ centimeters per minute

107. $A = \dfrac{1}{2}r^2\theta$

$= \dfrac{1}{2}(35)^2(140°)\left(\dfrac{\pi}{180°}\right)$

$\approx 476.39\pi$ square meters

≈ 1496.62 square meters

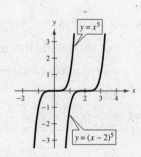

109. False. An angle measure of 4π radians corresponds to two complete revolutions from the initial to the terminal side of an angle.

111. False. The terminal side of $-1260°$ lies on the negative x-axis.

113. Increases, since the linear speed is proportional to the radius.

115. The arc length is increasing. In order for the angle θ to remain constant as the radius r increases, the arc length s must increase in proportion to r, as can be seen from the formula $s = r\theta$.

117. $\dfrac{4}{4\sqrt{2}} = \dfrac{4}{4\sqrt{2}} \cdot \dfrac{\sqrt{2}}{\sqrt{2}} = \dfrac{4\sqrt{2}}{8} = \dfrac{\sqrt{2}}{2}$

119. $\sqrt{2^2 + 6^2} = \sqrt{4 + 36} = \sqrt{40} = \sqrt{4 \cdot 10} = 2\sqrt{10}$

121. $f(x) = (x - 2)^5$

Graph of $y = x^5$ shifted to the right by two units

123. $f(x) = 2 - x^5$

Graph of $y = x^5$ reflected in x-axis and shifted upward by two units

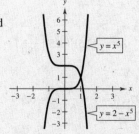

Section 4.2   Trigonometric Functions: The Unit Circle

■ You should know the definition of the trigonometric functions in terms of the unit circle. Let t be a real number and (x, y) the point on the unit circle corresponding to t.

$$\sin t = y \qquad\qquad\qquad \csc t = \frac{1}{y}, \quad y \neq 0$$

$$\cos t = x \qquad\qquad\qquad \sec t = \frac{1}{x}, \quad x \neq 0$$

$$\tan t = \frac{y}{x}, \quad x \neq 0 \qquad\qquad \cot t = \frac{x}{y}, \quad y \neq 0$$

■ The cosine and secant functions are even.

$$\cos(-t) = \cos t \qquad\qquad \sec(-t) = \sec t$$

■ The other four trigonometric functions are odd.

$$\sin(-t) = -\sin t \qquad\qquad \csc(-t) = -\csc t$$

$$\tan(-t) = -\tan t \qquad\qquad \cot(-t) = -\cot t$$

■ Be able to evaluate the trigonometric functions with a calculator.

Vocabulary Check

1. unit circle

2. periodic

3. period

4. odd; even

1. $x = -\dfrac{8}{17}, y = \dfrac{15}{17}$

$$\sin \theta = y = \frac{15}{17} \qquad\qquad \csc \theta = \frac{1}{y} = \frac{17}{15}$$

$$\cos \theta = x = -\frac{8}{17} \qquad\qquad \sec \theta = \frac{1}{x} = -\frac{17}{8}$$

$$\tan \theta = \frac{y}{x} = -\frac{15}{8} \qquad\qquad \cot \theta = \frac{x}{y} = -\frac{8}{15}$$

3. $x = \dfrac{12}{13}, y = -\dfrac{5}{13}$

$$\sin \theta = y = -\frac{5}{13} \qquad\qquad \csc \theta = \frac{1}{y} = -\frac{13}{5}$$

$$\cos \theta = x = \frac{12}{13} \qquad\qquad \sec \theta = \frac{1}{x} = \frac{13}{12}$$

$$\tan \theta = \frac{y}{x} = -\frac{5}{12} \qquad\qquad \cot \theta = \frac{x}{y} = -\frac{12}{5}$$

5. $t = \dfrac{\pi}{4}$ corresponds to $\left(\dfrac{\sqrt{2}}{2}, \dfrac{\sqrt{2}}{2}\right)$.

7. $t = \dfrac{7\pi}{6}$ corresponds to $\left(-\dfrac{\sqrt{3}}{2}, -\dfrac{1}{2}\right)$.

9. $t = \dfrac{4\pi}{3}$ corresponds to $\left(-\dfrac{1}{2}, -\dfrac{\sqrt{3}}{2}\right)$.

11. $t = \dfrac{3\pi}{2}$ corresponds to $(0, -1)$.

13. $t = \dfrac{\pi}{4}$ corresponds to $\left(\dfrac{\sqrt{2}}{2}, \dfrac{\sqrt{2}}{2}\right)$.

$$\sin t = y = \frac{\sqrt{2}}{2}$$

$$\cos t = x = \frac{\sqrt{2}}{2}$$

$$\tan t = \frac{y}{x} = 1$$

15. $t = -\dfrac{\pi}{6}$ corresponds to $\left(\dfrac{\sqrt{3}}{2}, -\dfrac{1}{2}\right)$.

$$\sin t = y = -\frac{1}{2}$$

$$\cos t = x = \frac{\sqrt{3}}{2}$$

$$\tan t = \frac{y}{x} = -\frac{1}{\sqrt{3}} = -\frac{\sqrt{3}}{3}$$

17. $t = -\dfrac{7\pi}{4}$ corresponds to $\left(\dfrac{\sqrt{2}}{2}, \dfrac{\sqrt{2}}{2}\right)$.

$\sin t = y = \dfrac{\sqrt{2}}{2}$

$\cos t = x = \dfrac{\sqrt{2}}{2}$

$\tan t = \dfrac{y}{x} = 1$

19. $t = \dfrac{11\pi}{6}$ corresponds to $\left(\dfrac{\sqrt{3}}{2}, -\dfrac{1}{2}\right)$.

$\sin t = y = -\dfrac{1}{2}$

$\cos t = x = \dfrac{\sqrt{3}}{2}$

$\tan t = \dfrac{y}{x} = -\dfrac{1}{\sqrt{3}} = -\dfrac{\sqrt{3}}{3}$

21. $t = -\dfrac{3\pi}{2}$ corresponds to $(0, 1)$.

$\sin t = y = 1$

$\cos t = x = 0$

$\tan t = \dfrac{y}{x}$ is undefined.

23. $t = \dfrac{3\pi}{4}$ corresponds to $\left(-\dfrac{\sqrt{2}}{2}, \dfrac{\sqrt{2}}{2}\right)$.

$\sin t = y = \dfrac{\sqrt{2}}{2}$ $\qquad$ $\csc t = \dfrac{1}{y} = \sqrt{2}$

$\cos t = x = -\dfrac{\sqrt{2}}{2}$ $\qquad$ $\sec t = \dfrac{1}{x} = -\sqrt{2}$

$\tan t = \dfrac{y}{x} = -1$ $\qquad$ $\cot t = \dfrac{x}{y} = -1$

25. $t = -\dfrac{\pi}{2}$ corresponds to $(0, -1)$.

$\sin t = y = -1$ $\qquad$ $\csc t = \dfrac{1}{y} = -1$

$\cos t = x = 0$ $\qquad$ $\sec t = \dfrac{1}{x}$ is undefined.

$\tan t = \dfrac{y}{x}$ is undefined. $\qquad$ $\cot t = \dfrac{x}{y} = 0$

27. $t = \dfrac{4\pi}{3}$ corresponds to $\left(-\dfrac{1}{2}, -\dfrac{\sqrt{3}}{2}\right)$.

$\sin t = y = -\dfrac{\sqrt{3}}{2}$ $\qquad$ $\csc t = \dfrac{1}{y} = -\dfrac{2\sqrt{3}}{3}$

$\cos t = x = -\dfrac{1}{2}$ $\qquad$ $\sec t = \dfrac{1}{x} = -2$

$\tan t = \dfrac{y}{x} = \sqrt{3}$ $\qquad$ $\cot t = \dfrac{x}{y} = \dfrac{\sqrt{3}}{3}$

29. $\sin 5\pi = \sin \pi = 0$

31. $\cos \dfrac{8\pi}{3} = \cos \dfrac{2\pi}{3} = -\dfrac{1}{2}$

33. $\cos\left(-\dfrac{15\pi}{2}\right) = \cos \dfrac{\pi}{2} = 0$

35. $\sin\left(-\dfrac{9\pi}{4}\right) = \sin\left(\dfrac{7\pi}{4}\right) = -\dfrac{\sqrt{2}}{2}$

37. $\sin t = \dfrac{1}{3}$

(a) $\sin(-t) = -\sin t = -\dfrac{1}{3}$

(b) $\csc(-t) = -\csc t = -3$

39. $\cos(-t) = -\dfrac{1}{5}$

(a) $\cos t = \cos(-t) = -\dfrac{1}{5}$

(b) $\sec(-t) = \dfrac{1}{\cos(-t)} = -5$

41. $\sin t = \dfrac{4}{5}$

(a) $\sin(\pi - t) = \sin t = \dfrac{4}{5}$

(b) $\sin(t + \pi) = -\sin t = -\dfrac{4}{5}$

43. $\sin \dfrac{\pi}{4} \approx 0.7071$

45. $\csc 1.3 = \dfrac{1}{\sin 1.3} \approx 1.0378$

47. $\cos(-1.7) \approx -0.1288$

49. $\csc 0.8 = \dfrac{1}{\sin 0.8} \approx 1.3940$

51. $\sec 22.8 = \dfrac{1}{\cos 22.8} \approx -1.4486$

53. (a) $\sin 5 = y \approx -1$

(b) $\cos 2 = x \approx -0.4$

55. (a) $\sin t = 0.25$

$t \approx 0.25$ or 2.89

(b) $\cos t = -0.25$

$t \approx 1.82$ or 4.46

57. $y(t) = \frac{1}{4}e^{-t}\cos 6t$

(a)

t	0	$\frac{1}{4}$	$\frac{1}{2}$	$\frac{3}{4}$	1
y	0.25	0.0138	-0.1501	-0.0249	0.0883

(b) From the table feature of a graphing utility we see that $y \approx 0$ when $t \approx 5$ seconds.

(c) As t increases, the displacement oscillates but decreases in amplitude.

59. False. $\sin(-t) = -\sin t$ means the function is odd, not that the sine of a negative angle is a negative number.

For example: $\sin\left(-\dfrac{3\pi}{2}\right) = -\sin\left(\dfrac{3\pi}{2}\right) = -(-1) = 1.$

Even though the angle is negative, the sine value is positive.

61. (a) The points have y-axis symmetry.

(b) $\sin t_1 = \sin(\pi - t_1)$ since they have the same y-value.

(c) $\cos(\pi - t_1) = -\cos t_1$ since the x-values have the opposite signs.

63. $f(x) = \frac{1}{2}(3x - 2)$

$y = \frac{1}{2}(3x - 2)$

$x = \frac{1}{2}(3y - 2)$

$2x = 3y - 2$

$\frac{2}{3}x + \frac{2}{3} = y$

$f^{-1}(x) = \frac{2}{3}x + \frac{2}{3} = \frac{2}{3}(x + 1)$

65. $f(x) = \sqrt{x^2 - 4}, x \geq 2$

$y = \sqrt{x^2 - 4}$

$x = \sqrt{y^2 - 4}$

$x^2 = y^2 - 4$

$\pm\sqrt{x^2 + 4} = y$

$f^{-1}(x) = \sqrt{x^2 + 4}, x \geq 0$

67. $f(x) = \dfrac{2x}{x - 3}$

Intercept: $(0, 0)$

Vertical asymptote: $x = 3$

Horizontal asymptote: $y = 2$

x	-1	0	1	2	4	5	6
y	$\frac{1}{2}$	0	-1	-4	8	5	4

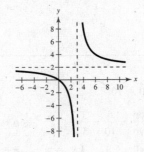

69. $f(x) = \dfrac{x^2 + 3x - 10}{2x^2 - 8} = \dfrac{(x + 5)(x - 2)}{2(x + 2)(x - 2)} = \dfrac{x + 5}{2(x + 2)}, x \neq 2$

Intercepts: $(-5, 0), \left(0, \dfrac{5}{4}\right)$

Vertical asymptote: $x = -2$

Horizontal asymptote: $y = \dfrac{1}{2}$

Hole in the graph at $\left(2, \dfrac{7}{8}\right)$

x	-5	-4	-3	-1	0	1	3
y	0	$-\frac{1}{4}$	-1	2	$\frac{5}{4}$	1	$\frac{4}{5}$

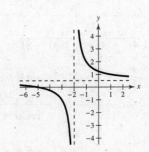

Section 4.3   Right Triangle Trigonometry

■ You should know the right triangle definition of trigonometric functions.

(a) $\sin \theta = \dfrac{\text{opp}}{\text{hyp}}$ (b) $\cos \theta = \dfrac{\text{adj}}{\text{hyp}}$ (c) $\tan \theta = \dfrac{\text{opp}}{\text{adj}}$

(d) $\csc \theta = \dfrac{\text{hyp}}{\text{opp}}$ (e) $\sec \theta = \dfrac{\text{hyp}}{\text{adj}}$ (f) $\cot \theta = \dfrac{\text{adj}}{\text{opp}}$

■ You should know the following identities.

(a) $\sin \theta = \dfrac{1}{\csc \theta}$ (b) $\csc \theta = \dfrac{1}{\sin \theta}$ (c) $\cos \theta = \dfrac{1}{\sec \theta}$

(d) $\sec \theta = \dfrac{1}{\cos \theta}$ (e) $\tan \theta = \dfrac{1}{\cot \theta}$ (f) $\cot \theta = \dfrac{1}{\tan \theta}$

(g) $\tan \theta = \dfrac{\sin \theta}{\cos \theta}$ (h) $\cot \theta = \dfrac{\cos \theta}{\sin \theta}$ (i) $\sin^2 \theta + \cos^2 \theta = 1$

(j) $1 + \tan^2 \theta = \sec^2 \theta$ (k) $1 + \cot^2 \theta = \csc^2 \theta$

■ You should know that two acute angles α and β are complementary if $\alpha + \beta = 90°$, and that cofunctions of complementary angles are equal.

■ You should know the trigonometric function values of 30°, 45°, and 60°, or be able to construct triangles from which you can determine them.

Vocabulary Check

1. (i) $\dfrac{\text{hypotenuse}}{\text{adjacent}} = \sec \theta$ (v)   (ii) $\dfrac{\text{adjacent}}{\text{opposite}} = \cot \theta$ (iv)   (iii) $\dfrac{\text{hypotenuse}}{\text{opposite}} = \csc \theta$ (vi)

   (iv) $\dfrac{\text{adjacent}}{\text{hypotenuse}} = \cos \theta$ (iii)   (v) $\dfrac{\text{opposite}}{\text{hypotenuse}} = \sin \theta$ (i)   (vi) $\dfrac{\text{opposite}}{\text{adjacent}} = \tan \theta$ (ii)

2. opposite; adjacent; hypotenuse

3. elevation; depression

---

1. $\text{hyp} = \sqrt{6^2 + 8^2} = \sqrt{36 + 64} = \sqrt{100} = 10$

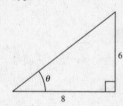

$\sin \theta = \dfrac{\text{opp}}{\text{hyp}} = \dfrac{6}{10} = \dfrac{3}{5}$ $\csc \theta = \dfrac{\text{hyp}}{\text{opp}} = \dfrac{10}{6} = \dfrac{5}{3}$

$\cos \theta = \dfrac{\text{adj}}{\text{hyp}} = \dfrac{8}{10} = \dfrac{4}{5}$ $\sec \theta = \dfrac{\text{hyp}}{\text{adj}} = \dfrac{10}{8} = \dfrac{5}{4}$

$\tan \theta = \dfrac{\text{opp}}{\text{adj}} = \dfrac{6}{8} = \dfrac{3}{4}$ $\cot \theta = \dfrac{\text{adj}}{\text{opp}} = \dfrac{8}{6} = \dfrac{4}{3}$

3. $\text{adj} = \sqrt{41^2 - 9^2} = \sqrt{1681 - 81} = \sqrt{1600} = 40$

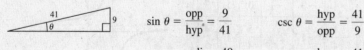

$\sin \theta = \dfrac{\text{opp}}{\text{hyp}} = \dfrac{9}{41}$ $\csc \theta = \dfrac{\text{hyp}}{\text{opp}} = \dfrac{41}{9}$

$\cos \theta = \dfrac{\text{adj}}{\text{hyp}} = \dfrac{40}{41}$ $\sec \theta = \dfrac{\text{hyp}}{\text{adj}} = \dfrac{41}{40}$

$\tan \theta = \dfrac{\text{opp}}{\text{adj}} = \dfrac{9}{40}$ $\cot \theta = \dfrac{\text{adj}}{\text{opp}} = \dfrac{40}{9}$

5. adj $= \sqrt{3^2 - 1^2} = \sqrt{8} = 2\sqrt{2}$

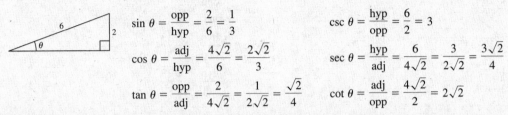

$$\sin\theta = \frac{\text{opp}}{\text{hyp}} = \frac{1}{3}$$

$$\csc\theta = \frac{\text{hyp}}{\text{opp}} = 3$$

$$\cos\theta = \frac{\text{adj}}{\text{hyp}} = \frac{2\sqrt{2}}{3}$$

$$\sec\theta = \frac{\text{hyp}}{\text{adj}} = \frac{3}{2\sqrt{2}} = \frac{3\sqrt{2}}{4}$$

$$\tan\theta = \frac{\text{opp}}{\text{adj}} = \frac{1}{2\sqrt{2}} = \frac{\sqrt{2}}{4}$$

$$\cot\theta = \frac{\text{adj}}{\text{opp}} = 2\sqrt{2}$$

adj $= \sqrt{6^2 - 2^2} = \sqrt{32} = 4\sqrt{2}$

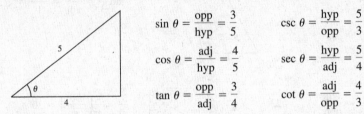

$$\sin\theta = \frac{\text{opp}}{\text{hyp}} = \frac{2}{6} = \frac{1}{3}$$

$$\csc\theta = \frac{\text{hyp}}{\text{opp}} = \frac{6}{2} = 3$$

$$\cos\theta = \frac{\text{adj}}{\text{hyp}} = \frac{4\sqrt{2}}{6} = \frac{2\sqrt{2}}{3}$$

$$\sec\theta = \frac{\text{hyp}}{\text{adj}} = \frac{6}{4\sqrt{2}} = \frac{3}{2\sqrt{2}} = \frac{3\sqrt{2}}{4}$$

$$\tan\theta = \frac{\text{opp}}{\text{adj}} = \frac{2}{4\sqrt{2}} = \frac{1}{2\sqrt{2}} = \frac{\sqrt{2}}{4}$$

$$\cot\theta = \frac{\text{adj}}{\text{opp}} = \frac{4\sqrt{2}}{2} = 2\sqrt{2}$$

The function values are the same since the triangles are similar and the corresponding sides are proportional.

7. opp $= \sqrt{5^2 - 4^2} = 3$

$$\sin\theta = \frac{\text{opp}}{\text{hyp}} = \frac{3}{5}$$

$$\csc\theta = \frac{\text{hyp}}{\text{opp}} = \frac{5}{3}$$

$$\cos\theta = \frac{\text{adj}}{\text{hyp}} = \frac{4}{5}$$

$$\sec\theta = \frac{\text{hyp}}{\text{adj}} = \frac{5}{4}$$

$$\tan\theta = \frac{\text{opp}}{\text{adj}} = \frac{3}{4}$$

$$\cot\theta = \frac{\text{adj}}{\text{opp}} = \frac{4}{3}$$

opp $= \sqrt{1.25^2 - 1^2} = 0.75$

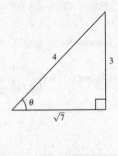

$$\sin\theta = \frac{\text{opp}}{\text{hyp}} = \frac{0.75}{1.25} = \frac{3}{5}$$

$$\csc\theta = \frac{\text{hyp}}{\text{opp}} = \frac{1.25}{0.75} = \frac{5}{3}$$

$$\cos\theta = \frac{\text{adj}}{\text{hyp}} = \frac{1}{1.25} = \frac{4}{5}$$

$$\sec\theta = \frac{\text{hyp}}{\text{adj}} = \frac{1.25}{1} = \frac{5}{4}$$

$$\tan\theta = \frac{\text{opp}}{\text{adj}} = \frac{0.75}{1} = \frac{3}{4}$$

$$\cot\theta = \frac{\text{adj}}{\text{opp}} = \frac{1}{0.75} = \frac{4}{3}$$

The function values are the same since the triangles are similar and the corresponding sides are proportional.

9. Given: $\sin\theta = \dfrac{3}{4} = \dfrac{\text{opp}}{\text{hyp}}$

$3^2 + (\text{adj})^2 = 4^2$

adj $= \sqrt{7}$

$$\cos\theta = \frac{\text{adj}}{\text{hyp}} = \frac{\sqrt{7}}{4}$$

$$\tan\theta = \frac{\text{opp}}{\text{adj}} = \frac{3\sqrt{7}}{7}$$

$$\csc\theta = \frac{\text{hyp}}{\text{opp}} = \frac{4}{3}$$

$$\sec\theta = \frac{\text{hyp}}{\text{adj}} = \frac{4\sqrt{7}}{7}$$

$$\cot\theta = \frac{\text{adj}}{\text{opp}} = \frac{\sqrt{7}}{3}$$

11. Given: $\sec\theta = 2 = \dfrac{2}{1} = \dfrac{\text{hyp}}{\text{adj}}$

$(\text{opp})^2 + 1^2 = 2^2$

opp $= \sqrt{3}$

$$\sin\theta = \frac{\text{opp}}{\text{hyp}} = \frac{\sqrt{3}}{2}$$

$$\cos\theta = \frac{\text{adj}}{\text{hyp}} = \frac{1}{2}$$

$$\tan\theta = \frac{\text{opp}}{\text{adj}} = \sqrt{3}$$

$$\csc\theta = \frac{\text{hyp}}{\text{opp}} = \frac{2\sqrt{3}}{3}$$

$$\cot\theta = \frac{\text{adj}}{\text{opp}} = \frac{\sqrt{3}}{3}$$

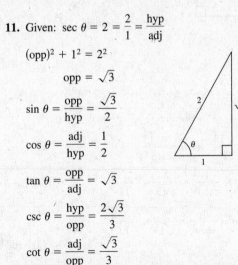

13. Given: $\tan \theta = 3 = \dfrac{3}{1} = \dfrac{\text{opp}}{\text{adj}}$

$\qquad 3^2 + 1^2 = (\text{hyp})^2$

$\qquad \text{hyp} = \sqrt{10}$

$\qquad \sin \theta = \dfrac{\text{opp}}{\text{hyp}} = \dfrac{3\sqrt{10}}{10}$

$\qquad \cos \theta = \dfrac{\text{adj}}{\text{hyp}} = \dfrac{\sqrt{10}}{10}$

$\qquad \csc \theta = \dfrac{\text{hyp}}{\text{opp}} = \dfrac{\sqrt{10}}{3}$

$\qquad \sec \theta = \dfrac{\text{hyp}}{\text{adj}} = \sqrt{10}$

$\qquad \cot \theta = \dfrac{\text{adj}}{\text{opp}} = \dfrac{1}{3}$

15. Given: $\cot \theta = \dfrac{3}{2} = \dfrac{\text{adj}}{\text{opp}}$

$\qquad 2^2 + 3^2 = (\text{hyp})^2$

$\qquad \text{hyp} = \sqrt{13}$

$\qquad \sin \theta = \dfrac{\text{opp}}{\text{hyp}} = \dfrac{2}{\sqrt{13}} = \dfrac{2\sqrt{13}}{13}$

$\qquad \cos \theta = \dfrac{\text{adj}}{\text{hyp}} = \dfrac{3}{\sqrt{13}} = \dfrac{3\sqrt{13}}{13}$

$\qquad \tan \theta = \dfrac{\text{opp}}{\text{adj}} = \dfrac{2}{3}$

$\qquad \csc \theta = \dfrac{\text{hyp}}{\text{opp}} = \dfrac{\sqrt{13}}{2}$

$\qquad \sec \theta = \dfrac{\text{hyp}}{\text{adj}} = \dfrac{\sqrt{13}}{3}$

17.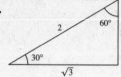

$30° = 30°\left(\dfrac{\pi}{180°}\right) = \dfrac{\pi}{6}$ radian

$\sin 30° = \dfrac{\text{opp}}{\text{hyp}} = \dfrac{1}{2}$

19.

$\dfrac{\pi}{3} = \dfrac{\pi}{3}\left(\dfrac{180°}{\pi}\right) = 60°$

$\tan \dfrac{\pi}{3} = \dfrac{\text{opp}}{\text{adj}} = \dfrac{\sqrt{3}}{1} = \sqrt{3}$

21.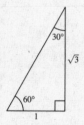

$\cot \theta = \dfrac{\sqrt{3}}{3} = \dfrac{1}{\sqrt{3}} = \dfrac{\text{adj}}{\text{opp}}$

$\theta = 60° = \dfrac{\pi}{3}$ radian

23.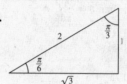

$\dfrac{\pi}{6} = \dfrac{\pi}{6}\left(\dfrac{180°}{\pi}\right) = 30°$

$\cos \dfrac{\pi}{6} = \dfrac{\text{adj}}{\text{hyp}} = \dfrac{\sqrt{3}}{2}$

25.

$\cot \theta = 1 = \dfrac{1}{1} = \dfrac{\text{adj}}{\text{opp}}$

$\theta = 45° = 45°\left(\dfrac{\pi}{180°}\right)$

$\qquad = \dfrac{\pi}{4}$ radian

27. $\sin 60° = \dfrac{\sqrt{3}}{2}$, $\cos 60° = \dfrac{1}{2}$

(a) $\tan 60° = \dfrac{\sin 60°}{\cos 60°} = \sqrt{3}$

(b) $\sin 30° = \cos 60° = \dfrac{1}{2}$

(c) $\cos 30° = \sin 60° = \dfrac{\sqrt{3}}{2}$

(d) $\cot 60° = \dfrac{\cos 60°}{\sin 60°} = \dfrac{1}{\sqrt{3}} = \dfrac{\sqrt{3}}{3}$

29. $\csc \theta = \dfrac{\sqrt{13}}{2}$, $\sec \theta = \dfrac{\sqrt{13}}{3}$

(a) $\sin \theta = \dfrac{1}{\csc \theta} = \dfrac{2}{\sqrt{13}} = \dfrac{2\sqrt{13}}{13}$

(b) $\cos \theta = \dfrac{1}{\sec \theta} = \dfrac{3}{\sqrt{13}} = \dfrac{3\sqrt{13}}{13}$

(c) $\tan \theta = \dfrac{\sin \theta}{\cos \theta} = \dfrac{\dfrac{2\sqrt{13}}{13}}{\dfrac{3\sqrt{13}}{13}} = \dfrac{2}{3}$

(d) $\sec(90° - \theta) = \csc \theta = \dfrac{\sqrt{13}}{2}$

31. $\cos \alpha = \dfrac{1}{3}$

(a) $\sec \alpha = \dfrac{1}{\cos \alpha} = 3$

(c) $\cot \alpha = \dfrac{\cos \alpha}{\sin \alpha} = \dfrac{\frac{1}{3}}{\frac{2\sqrt{2}}{3}} = \dfrac{1}{2\sqrt{2}} = \dfrac{\sqrt{2}}{4}$

(b) $\sin^2 \alpha + \cos^2 \alpha = 1$

$\sin^2 \alpha + \left(\dfrac{1}{3}\right)^2 = 1$

$\sin^2 \alpha = \dfrac{8}{9}$

$\sin \alpha = \dfrac{2\sqrt{2}}{3}$

(d) $\sin(90° - \alpha) = \cos \alpha = \dfrac{1}{3}$

33. $\tan \theta \cot \theta = \tan \theta \left(\dfrac{1}{\tan \theta}\right) = 1$

35. $\tan \alpha \cos \alpha = \left(\dfrac{\sin \alpha}{\cos \alpha}\right) \cos \alpha = \sin \alpha$

37. $(1 + \cos \theta)(1 - \cos \theta) = 1 - \cos^2 \theta$
$= (\sin^2 \theta + \cos^2 \theta) - \cos^2 \theta$
$= \sin^2 \theta$

39. $(\sec \theta + \tan \theta)(\sec \theta - \tan \theta) = \sec^2 \theta - \tan^2 \theta$
$= (1 + \tan^2 \theta) - \tan^2 \theta$
$= 1$

41. $\dfrac{\sin \theta}{\cos \theta} + \dfrac{\cos \theta}{\sin \theta} = \dfrac{\sin^2 \theta + \cos^2 \theta}{\sin \theta \cos \theta}$
$= \dfrac{1}{\sin \theta \cos \theta}$
$= \dfrac{1}{\sin \theta} \cdot \dfrac{1}{\cos \theta}$
$= \csc \theta \sec \theta$

43. (a) $\sin 10° \approx 0.1736$

(b) $\cos 80° \approx 0.1736$

Note: $\cos 80° = \sin(90° - 80°) = \sin 10°$

45. (a) $\sin 16.35° \approx 0.2815$

(b) $\csc 16.35° = \dfrac{1}{\sin 16.35°} \approx 3.5523$

47. (a) $\sec 42° 12' = \sec 42.2° = \dfrac{1}{\cos 42.2°} \approx 1.3499$

(b) $\csc 48° 7' = \dfrac{1}{\sin\left(48 + \frac{7}{60}\right)°} \approx 1.3432$

49. (a) $\cot 11° 15' = \dfrac{1}{\tan 11.25°} \approx 5.0273$

(b) $\tan 11° 15' = \tan 11.25° \approx 0.1989$

51. (a) $\csc 32° 40' 3'' = \dfrac{1}{\sin 32.6675°} \approx 1.8527$

(b) $\tan 44° 28' 16'' \approx \tan 44.4711° \approx 0.9817$

53. (a) $\sin \theta = \dfrac{1}{2} \implies \theta = 30° = \dfrac{\pi}{6}$

(b) $\csc \theta = 2 \implies \theta = 30° = \dfrac{\pi}{6}$

55. (a) $\sec \theta = 2 \implies \theta = 60° = \dfrac{\pi}{3}$

(b) $\cot \theta = 1 \implies \theta = 45° = \dfrac{\pi}{4}$

57. (a) $\csc \theta = \dfrac{2\sqrt{3}}{3} \implies \theta = 60° = \dfrac{\pi}{3}$

(b) $\sin \theta = \dfrac{\sqrt{2}}{2} \implies \theta = 45° = \dfrac{\pi}{4}$

59.

$\tan 30° = \dfrac{30}{x}$
$\dfrac{1}{\sqrt{3}} = \dfrac{30}{x}$
$x = 30\sqrt{3}$

61.

$$\tan 60° = \frac{32}{x}$$

$$\sqrt{3} = \frac{32}{x}$$

$$\sqrt{3}x = 32$$

$$x = \frac{32}{\sqrt{3}} = \frac{32\sqrt{3}}{3}$$

63. $\tan 82° = \dfrac{x}{45}$

$$x = 45 \tan 82°$$

Height of the building:

$$123 + 45 \tan 82° \approx 443.2 \text{ meters}$$

Distance between friends:

$$\cos 82° = \frac{45}{y} \implies y = \frac{45}{\cos 82°}$$

$$\approx 323.34 \text{ meters}$$

65.

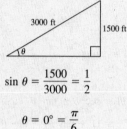

$$\sin \theta = \frac{1500}{3000} = \frac{1}{2}$$

$$\theta = 0° = \frac{\pi}{6}$$

67.

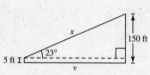

(a) $\sin 23° = \dfrac{145}{x}$

$$x = \frac{145}{\sin 23°} \approx 371.1 \text{ feet}$$

(b) $\tan 23° = \dfrac{145}{y}$

$$y = \frac{145}{\tan 23°} \approx 341.6 \text{ feet}$$

(c) Moving down the line:

$$\frac{145/\sin 23°}{6} \approx 61.85 \text{ feet per second}$$

Dropping vertically:

$$\frac{145}{6} \approx 24.17 \text{ feet per second}$$

69. $\sin 30° = \dfrac{y_1}{56}$

$$y_1 = (\sin 30°)(56) = \left(\frac{1}{2}\right)(56) = 28$$

$$\cos 30° = \frac{x_1}{56}$$

$$x_1 = \cos 30°(56) = \frac{\sqrt{3}}{2}(56) = 28\sqrt{3}$$

$$(x_1, y_1) = \left(28\sqrt{3}, 28\right)$$

$$\sin 60° = \frac{y_2}{56}$$

$$y_2 = \sin 60°(56) = \left(\frac{\sqrt{3}}{2}\right)(56) = 28\sqrt{3}$$

$$\cos 60° = \frac{x_2}{56}$$

$$x_2 = (\cos 60°)(56) = \left(\frac{1}{2}\right)(56) = 28$$

$$(x_2, y_2) = \left(28, 28\sqrt{3}\right)$$

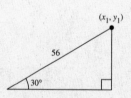

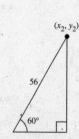

71. (a)

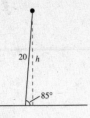

(b) $\sin 85° = \dfrac{h}{20}$

(c) $h = 20 \sin 85° \approx 19.9$ meters

(d) The side of the triangle labeled h will become shorter.

(e)

Angle, θ	Height (in meters)
80°	19.7
70°	18.8
60°	17.3
50°	15.3
40°	12.9
30°	10.0
20°	6.8
10°	3.5

(f) The height of the balloon decreases as θ decreases.

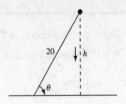

73. True,

$$\csc x = \dfrac{1}{\sin x} \implies \sin 60° \csc 60° = \sin 60°\left(\dfrac{1}{\sin 60°}\right) = 1$$

75. False, $\dfrac{\sqrt{2}}{2} + \dfrac{\sqrt{2}}{2} = \sqrt{2} \neq 1$

77. False, $\dfrac{\sin 60°}{\sin 30°} = \dfrac{\cos 30°}{\sin 30°} = \cot 30° \approx 1.7321$;

$\sin 2° \approx 0.0349$

79. This is true because the corresponding sides of similar triangles are proportional.

81. (a)

θ	0.1	0.2	0.3	0.4	0.5
$\sin \theta$	0.0998	0.1987	0.2955	0.3894	0.4794

(b) In the interval $(0, 0.5]$, $\theta > \sin \theta$.

(c) As θ approaches 0, $\sin \theta$ approaches θ.

83. $\dfrac{x^2 - 6x}{x^2 + 4x - 12} \cdot \dfrac{x^2 + 12x + 36}{x^2 - 36} = \dfrac{x(x - 6)}{(x + 6)(x - 2)} \cdot \dfrac{(x + 6)(x + 6)}{(x + 6)(x - 6)}$

$$= \dfrac{x}{x - 2}, x \neq \pm 6$$

85. $\dfrac{3}{x + 2} - \dfrac{2}{x - 2} + \dfrac{x}{x^2 + 4x + 4} = \dfrac{3(x + 2)(x - 2) - 2(x + 2)^2 + x(x - 2)}{(x - 2)(x + 2)^2}$

$$= \dfrac{3(x^2 - 4) - 2(x^2 + 4x + 4) + x^2 - 2x}{(x - 2)(x + 2)^2}$$

$$= \dfrac{2x^2 - 10x - 20}{(x - 2)(x + 2)^2} = \dfrac{2(x^2 - 5x - 10)}{(x - 2)(x + 2)^2}$$

Section 4.4 Trigonometric Functions of Any Angle

■ Know the Definitions of Trigonometric Functions of Any Angle.

If θ is in standard position, (x, y) a point on the terminal side and $r = \sqrt{x^2 + y^2} \neq 0$, then:

$$\sin \theta = \frac{y}{r} \qquad\qquad \csc \theta = \frac{r}{y}, \ y \neq 0$$

$$\cos \theta = \frac{x}{r} \qquad\qquad \sec \theta = \frac{r}{x}, \ x \neq 0$$

$$\tan \theta = \frac{y}{x}, \ x \neq 0 \qquad \cot \theta = \frac{x}{y}, \ y \neq 0$$

■ You should know the signs of the trigonometric functions in each quadrant.

■ You should know the trigonometric function values of the quadrant angles 0, $\dfrac{\pi}{2}$, π, and $\dfrac{3\pi}{2}$.

■ You should be able to find reference angles.

■ You should be able to evaluate trigonometric functions of any angle. (Use reference angles.)

■ You should know that the period of sine and cosine is 2π.

Vocabulary Check

1. $\sin \theta = \dfrac{y}{r}$ **2.** $\csc \theta$ **3.** $\tan \theta = \dfrac{y}{x}$

4. $\dfrac{r}{x}$ **5.** $\dfrac{x}{r} = \cos \theta$ **6.** $\cot \theta$

7. reference

1. (a) $(x, y) = (4, 3)$

$r = \sqrt{16 + 9} = 5$

$\sin \theta = \dfrac{y}{r} = \dfrac{3}{5}$ $\csc \theta = \dfrac{r}{y} = \dfrac{5}{3}$

$\cos \theta = \dfrac{x}{r} = \dfrac{4}{5}$ $\sec \theta = \dfrac{r}{x} = \dfrac{5}{4}$

$\tan \theta = \dfrac{y}{x} = \dfrac{3}{4}$ $\cot \theta = \dfrac{x}{y} = \dfrac{4}{3}$

(b) $(x, y) = (8, -15)$

$r = \sqrt{64 + 225} = 17$

$\sin \theta = \dfrac{y}{r} = -\dfrac{15}{17}$ $\csc \theta = \dfrac{r}{y} = -\dfrac{17}{15}$

$\cos \theta = \dfrac{x}{r} = \dfrac{8}{17}$ $\sec \theta = \dfrac{r}{x} = \dfrac{17}{8}$

$\tan \theta = \dfrac{y}{x} = -\dfrac{15}{8}$ $\cot \theta = \dfrac{x}{y} = -\dfrac{8}{15}$

3. (a) $(x, y) = \left(-\sqrt{3}, -1\right)$

$r = \sqrt{3 + 1} = 2$

$\sin \theta = \dfrac{y}{r} = -\dfrac{1}{2}$ $\csc \theta = \dfrac{r}{y} = -2$

$\cos \theta = \dfrac{x}{r} = -\dfrac{\sqrt{3}}{2}$ $\sec \theta = \dfrac{r}{x} = -\dfrac{2\sqrt{3}}{3}$

$\tan \theta = \dfrac{y}{x} = \dfrac{\sqrt{3}}{3}$ $\cot \theta = \dfrac{x}{y} = \sqrt{3}$

(b) $(x, y) = (-4, 1)$

$r = \sqrt{16 + 1} = \sqrt{17}$

$\sin \theta = \dfrac{y}{r} = \dfrac{\sqrt{17}}{17}$ $\csc \theta = \dfrac{r}{y} = \sqrt{17}$

$\cos \theta = \dfrac{x}{r} = -\dfrac{4\sqrt{17}}{17}$ $\sec \theta = \dfrac{r}{x} = -\dfrac{\sqrt{17}}{4}$

$\tan \theta = \dfrac{y}{x} = -\dfrac{1}{4}$ $\cot \theta = \dfrac{x}{y} = -4$

5. $(x, y) = (7, 24)$

$r = \sqrt{49 + 576} = 25$

$\sin \theta = \dfrac{y}{r} = \dfrac{24}{25}$

$\cos \theta = \dfrac{x}{r} = \dfrac{7}{25}$

$\tan \theta = \dfrac{y}{x} = \dfrac{24}{7}$

$\csc \theta = \dfrac{r}{y} = \dfrac{25}{24}$

$\sec \theta = \dfrac{r}{x} = \dfrac{25}{7}$

$\cot \theta = \dfrac{x}{y} = \dfrac{7}{24}$

7. $(x, y) = (-4, 10)$

$r = \sqrt{16 + 100} = 2\sqrt{29}$

$\sin \theta = \dfrac{y}{r} = \dfrac{5\sqrt{29}}{29}$

$\cos \theta = \dfrac{x}{r} = -\dfrac{2\sqrt{29}}{29}$

$\tan \theta = \dfrac{y}{x} = -\dfrac{5}{2}$

$\csc \theta = \dfrac{r}{y} = \dfrac{\sqrt{29}}{5}$

$\sec \theta = \dfrac{r}{x} = -\dfrac{\sqrt{29}}{2}$

$\cot \theta = \dfrac{x}{y} = -\dfrac{2}{5}$

9. $(x, y) = (-3.5, 6.8)$

$r = \sqrt{12.25 + 46.24} = \dfrac{\sqrt{5849}}{10}$

$\sin \theta = \dfrac{y}{r} = \dfrac{68\sqrt{5849}}{5849} \approx 0.9$

$\cos \theta = \dfrac{x}{r} = -\dfrac{35\sqrt{5849}}{5849} \approx -0.5$

$\tan \theta = \dfrac{y}{x} = -\dfrac{68}{35} \approx -1.9$

$\csc \theta = \dfrac{r}{y} = \dfrac{\sqrt{5849}}{68} \approx 1.1$

$\sec \theta = \dfrac{r}{x} = -\dfrac{\sqrt{5849}}{35} \approx -2.2$

$\cot \theta = \dfrac{x}{y} = -\dfrac{35}{68} \approx -0.5$

11. $\sin \theta < 0 \implies \theta$ lies in Quadrant III or in Quadrant IV.

$\cos \theta < 0 \implies \theta$ lies in Quadrant II or in Quadrant III.

$\sin \theta < 0$ *and* $\cos \theta < 0 \implies \theta$ lies in Quadrant III.

13. $\sin \theta > 0 \implies \theta$ lies in Quadrant I or in Quadrant II.

$\tan \theta < 0 \implies \theta$ lies in Quadrant II or in Quadrant IV.

$\sin \theta > 0$ *and* $\tan \theta < 0 \implies \theta$ lies in Quadrant II.

15. $\sin \theta = \dfrac{y}{r} = \dfrac{3}{5} \implies x^2 = 25 - 9 = 16$

θ in Quadrant II $\implies x = -4$

$\sin \theta = \dfrac{y}{r} = \dfrac{3}{5}$ $\csc \theta = \dfrac{r}{y} = \dfrac{5}{3}$

$\cos \theta = \dfrac{x}{r} = -\dfrac{4}{5}$ $\sec \theta = \dfrac{r}{x} = -\dfrac{5}{4}$

$\tan \theta = \dfrac{y}{x} = -\dfrac{3}{4}$ $\cot \theta = \dfrac{x}{y} = -\dfrac{4}{3}$

17. $\tan \theta = \dfrac{y}{x} = \dfrac{-15}{8}$

$\sin \theta < 0$ and $\tan \theta < 0 \implies \theta$ is in Quadrant IV $\implies$ $y < 0$ and $x > 0$.

$x = 8, y = -15, r = 17$

$\sin \theta = \dfrac{y}{r} = -\dfrac{15}{17}$ $\csc \theta = \dfrac{r}{y} = -\dfrac{17}{15}$

$\cos \theta = \dfrac{x}{r} = \dfrac{8}{17}$ $\sec \theta = \dfrac{r}{x} = \dfrac{17}{8}$

$\tan \theta = \dfrac{y}{x} = -\dfrac{15}{8}$ $\cot \theta = \dfrac{x}{y} = -\dfrac{8}{15}$

19. $\cot \theta = \dfrac{x}{y} = -\dfrac{3}{1} = \dfrac{3}{-1}$

$\cos \theta > 0 \implies \theta$ is in Quadrant IV $\implies$ x is positive;

$x = 3, y = -1, r = \sqrt{10}$

$\sin \theta = \dfrac{y}{r} = -\dfrac{\sqrt{10}}{10}$ $\csc \theta = \dfrac{r}{y} = -\sqrt{10}$

$\cos \theta = \dfrac{x}{r} = \dfrac{3\sqrt{10}}{10}$ $\sec \theta = \dfrac{r}{x} = \dfrac{\sqrt{10}}{3}$

$\tan \theta = \dfrac{y}{x} = -\dfrac{1}{3}$ $\cot \theta = \dfrac{x}{y} = -3$

21. $\sec \theta = \dfrac{r}{x} = \dfrac{2}{-1} \implies y^2 = 4 - 1 = 3$

$\sin \theta > 0 \implies \theta$ is in Quadrant II $\implies$ $y = \sqrt{3}$

$\sin \theta = \dfrac{y}{r} = \dfrac{\sqrt{3}}{2}$ $\csc \theta = \dfrac{r}{y} = \dfrac{2\sqrt{3}}{3}$

$\cos \theta = \dfrac{x}{r} = -\dfrac{1}{2}$ $\sec \theta = \dfrac{r}{x} = -2$

$\tan \theta = \dfrac{y}{x} = -\sqrt{3}$ $\cot \theta = \dfrac{x}{y} = -\dfrac{\sqrt{3}}{3}$

23. cot θ is undefined, $\dfrac{\pi}{2} \le \theta \le \dfrac{3\pi}{2} \Rightarrow y = 0 \Rightarrow \theta = \pi$

$\sin \pi = 0$ csc π is undefined.

$\cos \pi = -1$ $\sec \pi = -1$

$\tan \pi = 0$ cot π is undefined.

25. To find a point on the terminal side of θ use any point on the line $y = -x$ that lies in Quadrant II. $(-1, 1)$ is one such point.

$x = -1, y = 1, r = \sqrt{2}$

$\sin \theta = \dfrac{1}{\sqrt{2}} = \dfrac{\sqrt{2}}{2}$ $\csc \theta = \sqrt{2}$

$\cos \theta = -\dfrac{1}{\sqrt{2}} = -\dfrac{\sqrt{2}}{2}$ $\sec \theta = -\sqrt{2}$

                          $\cot \theta = -1$

$\tan \theta = -1$

27. To find a point on the terminal side of θ, use any point on the line $y = 2x$ that lies in Quadrant III. $(-1, -2)$ is one such point.

$x = -1, y = -2, r = \sqrt{5}$

$\sin \theta = -\dfrac{2}{\sqrt{5}} = -\dfrac{2\sqrt{5}}{5}$ $\csc \theta = \dfrac{\sqrt{5}}{-2} = -\dfrac{\sqrt{5}}{2}$

$\cos \theta = -\dfrac{1}{\sqrt{5}} = -\dfrac{\sqrt{5}}{5}$ $\sec \theta = \dfrac{\sqrt{5}}{-1} = -\sqrt{5}$

$\tan \theta = \dfrac{-2}{-1} = 2$ $\cot \theta = \dfrac{-1}{-2} = \dfrac{1}{2}$

29. $(x, y) = (-1, 0), r = 1$

$\sin \pi = \dfrac{y}{r} = 0$

31. $(x, y) = (0, -1), r = 1$

$\sec \dfrac{3\pi}{2} = \dfrac{r}{x} = \dfrac{1}{0}$

$\Rightarrow$ undefined

33. $(x, y) = (0, 1), r = 1$

$\sin \dfrac{\pi}{2} = \dfrac{y}{r} = 1$

35. $(x, y) = (-1, 0), r = 1$

$\csc \pi = \dfrac{r}{y} = \dfrac{1}{0}$

$\Rightarrow$ undefined

37. $\theta = 203°$

$\theta' = 203° - 180° = 23°$

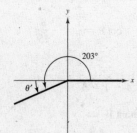

39. $\theta = -245°$

$360° - 245° = 115°$ (coterminal angle)

$\theta' = 180° - 115° = 65°$

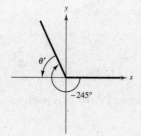

41. $\theta = \dfrac{2\pi}{3}$

$\theta' = \pi - \dfrac{2\pi}{3} = \dfrac{\pi}{3}$

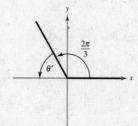

43. $\theta = 3.5$

$\theta' = 3.5 - \pi$

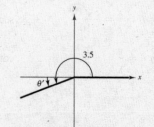

45. $\theta = 225°$, $\theta' = 360° - 225° = 45°$, Quadrant III

$$\sin 225° = -\sin 45° = -\frac{\sqrt{2}}{2}$$

$$\cos 225° = -\cos 45° = -\frac{\sqrt{2}}{2}$$

$$\tan 225° = \tan 45° = 1$$

47. $\theta = 750°$ is coterminal with 30°.

$\theta' = 30°$, Quadrant I

$$\sin 750° = \sin 30° = \frac{1}{2}$$

$$\cos 750° = \cos 30° = \frac{\sqrt{3}}{2}$$

$$\tan 750° = \tan 30° = \frac{\sqrt{3}}{3}$$

49. $\theta = -150°$ is coterminal with 210°.

$\theta' = 210° - 180° = 30°$, Quadrant III

$$\sin(-150°) = -\sin 30° = -\frac{1}{2}$$

$$\cos(-150°) = -\cos 30° = -\frac{\sqrt{3}}{2}$$

$$\tan(-150°) = \tan 30° = \frac{\sqrt{3}}{3}$$

51. $\theta = \frac{4\pi}{3}$, $\theta' = \frac{\pi}{3}$, Quadrant III

$$\sin \frac{4\pi}{3} = -\sin \frac{\pi}{3} = -\frac{\sqrt{3}}{2}$$

$$\cos \frac{4\pi}{3} = -\cos \frac{\pi}{3} = -\frac{1}{2}$$

$$\tan \frac{4\pi}{3} = \tan \frac{\pi}{3} = \sqrt{3}$$

53. $\theta = -\frac{\pi}{6}$, $\theta' = \frac{\pi}{6}$, Quadrant IV

$$\sin\left(-\frac{\pi}{6}\right) = -\sin \frac{\pi}{6} = -\frac{1}{2}$$

$$\cos\left(-\frac{\pi}{6}\right) = \cos \frac{\pi}{6} = \frac{\sqrt{3}}{2}$$

$$\tan\left(-\frac{\pi}{6}\right) = -\tan \frac{\pi}{6} = -\frac{\sqrt{3}}{3}$$

55. $\theta = \frac{11\pi}{4}$ is coterminal with $\frac{3}{4}\pi$.

$\theta' = \pi - \frac{3}{4}\pi = \frac{\pi}{4}$, Quadrant II

$$\sin \frac{11\pi}{4} = \sin \frac{\pi}{4} = \frac{\sqrt{2}}{2}$$

$$\cos \frac{11\pi}{4} = -\cos \frac{\pi}{4} = -\frac{\sqrt{2}}{2}$$

$$\tan \frac{11\pi}{4} = -\tan \frac{\pi}{4} = -1$$

57. $\theta = -\frac{3\pi}{2}$ is coterminal with $\frac{\pi}{2}$, $\theta' = \frac{\pi}{2}$.

$$\sin\left(-\frac{3\pi}{2}\right) = \sin \frac{\pi}{2} = 1$$

$$\cos\left(-\frac{3\pi}{2}\right) = \cos \frac{\pi}{2} = 0$$

$$\tan\left(-\frac{3\pi}{2}\right) = \tan \frac{\pi}{2} \text{ which is undefined.}$$

59.

$$\sin \theta = -\frac{3}{5}$$

$$\sin^2 \theta + \cos^2 \theta = 1$$

$$\cos^2 \theta = 1 - \sin^2 \theta$$

$$\cos^2 \theta = 1 - \left(-\frac{3}{5}\right)^2$$

$$\cos^2 \theta = 1 - \frac{9}{25}$$

$$\cos^2 \theta = \frac{16}{25}$$

$\cos \theta > 0$ in Quadrant IV.

$$\cos \theta = \frac{4}{5}$$

61. $\tan \theta = \dfrac{3}{2}$

$\sec^2 \theta = 1 + \tan^2 \theta$

$\sec^2 \theta = 1 + \left(\dfrac{3}{2}\right)^2$

$\sec^2 \theta = 1 + \dfrac{9}{4}$

$\sec^2 \theta = \dfrac{13}{4}$

$\sec \theta < 0$ in Quadrant III.

$\sec \theta = -\dfrac{\sqrt{13}}{2}$

63. $\cos \theta = \dfrac{5}{8}$

$\cos \theta = \dfrac{1}{\sec \theta} \implies \sec \theta = \dfrac{1}{\cos \theta}$

$\sec \theta = \dfrac{1}{5/8} = \dfrac{8}{5}$

65. $\sin 10° \approx 0.1736$

67. $\cos(-110°) \approx -0.3420$

69. $\tan 304° \approx -1.4826$

71. $\sec 72° = \dfrac{1}{\cos 72°} \approx 3.2361$

73. $\tan 4.5 \approx 4.6373$

75. $\tan \dfrac{\pi}{9} \approx 0.3640$

77. $\sin(-0.65) \approx -0.6052$

79. $\cot\left(-\dfrac{11\pi}{8}\right) = \dfrac{1}{\tan\left(-\dfrac{11\pi}{8}\right)} \approx -0.4142$

81. (a) $\sin \theta = \dfrac{1}{2} \implies$ reference angle is $30°$ or $\dfrac{\pi}{6}$ and θ is in Quadrant I or Quadrant II.

Values in degrees: $30°, 150°$

Values in radians: $\dfrac{\pi}{6}, \dfrac{5\pi}{6}$

(b) $\sin \theta = -\dfrac{1}{2} \implies$ reference angle is $30°$ or $\dfrac{\pi}{6}$ and θ is in Quadrant III or Quadrant IV.

Values in degrees: $210°, 330°$

Values in radians: $\dfrac{7\pi}{6}, \dfrac{11\pi}{6}$

83. (a) $\csc \theta = \dfrac{2\sqrt{3}}{3} \implies$ reference angle is $60°$ or $\dfrac{\pi}{3}$ and θ is in Quadrant I or Quadrant II.

Values in degrees: $60°, 120°$

Values in radians: $\dfrac{\pi}{3}, \dfrac{2\pi}{3}$

(b) $\cot \theta = -1 \implies$ reference angle is $45°$ or $\dfrac{\pi}{4}$ and θ is in Quadrant II or Quadrant IV.

Values in degrees: $135°, 315°$

Values in radians: $\dfrac{3\pi}{4}, \dfrac{7\pi}{4}$

85. (a) $\tan \theta = 1 \implies$ reference angle is 45° or $\dfrac{\pi}{4}$ and θ is in Quadrant I or Quadrant III.

Values in degrees: 45°, 225°

Values in radians: $\dfrac{\pi}{4}, \dfrac{5\pi}{4}$

(b) $\cot \theta = -\sqrt{3} \implies$ reference angle is 30° or $\dfrac{\pi}{6}$ and θ is in Quadrant II or Quadrant IV.

Values in degrees: 150°, 330°

Values in radians: $\dfrac{5\pi}{6}, \dfrac{11\pi}{6}$

87. (a) New York City:

$N \approx 22.099 \sin(0.522t - 2.219) + 55.008$

Fairbanks:

$F \approx 36.641 \sin(0.502t - 1.831) + 25.610$

(b)

Month	New York City	Fairbanks
February	34.6°	$-1.4°$
March	41.6°	13.9°
May	63.4°	48.6°
June	72.5°	59.5°
August	75.5°	55.6°
September	68.6°	41.7°
November	46.8°	6.5°

(c) The periods are about the same for both models, approximately 12 months.

89. $y(t) = 2 \cos 6t$

(a) $y(0) = 2 \cos 0 = 2$ centimeters

(b) $y\left(\dfrac{1}{4}\right) = 2 \cos\left(\dfrac{3}{2}\right) \approx 0.14$ centimeter

(c) $y\left(\dfrac{1}{2}\right) = 2 \cos 3 \approx -1.98$ centimeters

91. $I = 5e^{-2t} \sin t$

$I(0.7) = 5e^{-1.4} \sin 0.7 \approx 0.79$ ampere

93. False. In each of the four quadrants, the sign of the secant function and the cosine function will be the same since they are reciprocals of each other.

95. As θ increases from 0° to 90°, x decreases from 12 cm to 0 cm and y increases from 0 cm to 12 cm.

Therefore, $\sin \theta = \dfrac{y}{12}$ increases from 0 to 1 and $\cos \theta = \dfrac{x}{12}$ decreases from 1 to 0. Thus,

$\tan \theta = \dfrac{y}{x}$ increases without bound, and when $\theta = 90°$ the tangent is undefined.

97. $y = x^2 + 3x - 4 = (x + 4)(x - 1)$

x-intercepts: $(-4, 0), (1, 0)$

y-intercept: $(0, -4)$

No asymptotes

Domain: All real numbers x

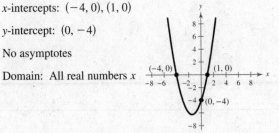

99. $f(x) = x^3 + 8$

x-intercept: $(-2, 0)$

y-intercept: $(0, 8)$

No asymptotes

Domain: All real numbers x

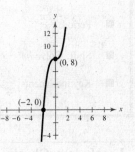

101. $f(x) = \dfrac{x - 7}{x^2 + 4x + 4} = \dfrac{x - 7}{(x + 2)^2}$

x-intercept: $(7, 0)$

y-intercept: $\left(0, -\dfrac{7}{4}\right)$

Vertical asymptote: $x = -2$

Horizontal asymptote: $y = 0$

Domain: All real numbers except $x = -2$

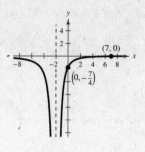

103. $y = 2^{x-1}$

y-intercept: $\left(0, \dfrac{1}{2}\right)$

Horizontal asymptote: $y = 0$

Domain: All real numbers x

x	-1	0	1	2	3
y	$\frac{1}{4}$	$\frac{1}{2}$	1	2	4

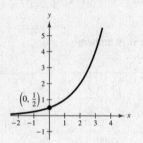

105. $y = \ln x^4$

Domain: All real numbers except $x = 0$

x-intercepts: $(\pm 1, 0)$

Vertical asymptote: $x = 0$

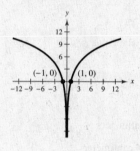

Section 4.5   Graphs of Sine and Cosine Functions

- You should be able to graph $y = a \sin(bx - c)$ and $y = a \cos(bx - c)$. (Assume $b > 0$.)

- Amplitude: $|a|$

- Period: $\dfrac{2\pi}{b}$

- Shift: Solve $bx - c = 0$ and $bx - c = 2\pi$.

- Key increments: $\dfrac{1}{4}$ (period)

Vocabulary Check

1. cycle

2. amplitude

3. $\dfrac{2\pi}{b}$

4. phase shift

5. vertical shift

1. $y = 3 \sin 2x$

Period: $\dfrac{2\pi}{2} = \pi$

Amplitude: $|3| = 3$

3. $y = \dfrac{5}{2} \cos \dfrac{x}{2}$

Period: $\dfrac{2\pi}{1/2} = 4\pi$

Amplitude: $\left|\dfrac{5}{2}\right| = \dfrac{5}{2}$

5. $y = \dfrac{1}{2} \sin \dfrac{\pi x}{3}$

Period: $\dfrac{2\pi}{\pi/3} = 6$

Amplitude: $\left|\dfrac{1}{2}\right| = \dfrac{1}{2}$

7. $y = -2 \sin x$

Period: $\dfrac{2\pi}{1} = 2\pi$

Amplitude: $|-2| = 2$

9. $y = 3 \sin 10x$

Period: $\dfrac{2\pi}{10} = \dfrac{\pi}{5}$

Amplitude: $|3| = 3$

11. $y = \dfrac{1}{2} \cos \dfrac{2x}{3}$

Period: $\dfrac{2\pi}{2/3} = 3\pi$

Amplitude: $\left|\dfrac{1}{2}\right| = \dfrac{1}{2}$

13. $y = \dfrac{1}{4} \sin 2\pi x$

Period: $\dfrac{2\pi}{2\pi} = 1$

Amplitude: $\left|\dfrac{1}{4}\right| = \dfrac{1}{4}$

15. $f(x) = \sin x$

$g(x) = \sin(x - \pi)$

The graph of g is a horizontal shift to the right π units of the graph of f (a phase shift).

17. $f(x) = \cos 2x$

$g(x) = -\cos 2x$

The graph of g is a reflection in the x-axis of the graph of f.

19. $f(x) = \cos x$

$g(x) = \cos 2x$

The period of f is twice that of g.

21. $f(x) = \sin 2x$

$f(x) = 3 + \sin 2x$

The graph of g is a vertical shift three units upward of the graph of f.

23. The graph of g has twice the amplitude as the graph of f. The period is the same.

25. The graph of g is a horizontal shift π units to the right of the graph of f.

27. $f(x) = -2 \sin x$

Period: $\dfrac{2\pi}{b} = \dfrac{2\pi}{1} = 2\pi$

Amplitude: 2

Symmetry: origin

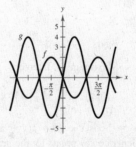

Key points:	Intercept	Minimum	Intercept	Maximum	Intercept
	$(0, 0)$	$\left(\dfrac{\pi}{2}, -2\right)$	$(\pi, 0)$	$\left(\dfrac{3\pi}{2}, 0\right)$	$(2\pi, 0)$

Since $g(x) = 4 \sin x = (-2)f(x)$, generate key points for the graph of $g(x)$ by multiplying the y-coordinate of each key point of $f(x)$ by -2.

29. $f(x) = \cos x$

Period: $\dfrac{2\pi}{b} = \dfrac{2\pi}{1} = 2\pi$

Amplitude: 1

Symmetry: y-axis

Key points: Maximum   Intercept   Minimum   Intercept   Maximum

$(0, 1)$ $\left(\dfrac{\pi}{2}, 0\right)$ $(\pi, -1)$ $\left(\dfrac{3\pi}{2}, 0\right)$ $(2\pi, 1)$

Since $g(x) = 1 + \cos(x) = f(x) + 1$, the graph of $g(x)$ is the graph of $f(x)$, but translated upward by one unit. Generate key points for the graph of $g(x)$ by adding 1 to the y-coordinate of each key point of $f(x)$.

31. $f(x) = -\dfrac{1}{2} \sin \dfrac{x}{2}$

Period: $\dfrac{2\pi}{b} = \dfrac{2\pi}{1/2} = 4\pi$

Amplitude: $\dfrac{1}{2}$

Symmetry: origin

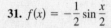

Key points: Intercept   Minimum   Intercept   Maximum   Intercept

$(0, 0)$ $\left(\pi, -\dfrac{1}{2}\right)$ $(2\pi, 0)$ $\left(3\pi, \dfrac{1}{2}\right)$ $(4\pi, 0)$

Since $g(x) = 3 - \dfrac{1}{2} \sin \dfrac{x}{2} = 3 - f(x)$, the graph of $g(x)$ is the graph of $f(x)$, but translated upward by three units. Generate key points for the graph of $g(x)$ by adding 3 to the y-coordinate of each key point of $f(x)$.

33. $f(x) = 2 \cos x$

Period: $\dfrac{2\pi}{b} = \dfrac{2\pi}{1} = 2\pi$

Amplitude: 2

Symmetry: y-axis

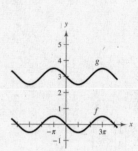

Key points: Maximum   Intercept   Minimum   Intercept   Maximum

$(0, 2)$ $\left(\dfrac{\pi}{2}, 0\right)$ $(\pi, -2)$ $\left(\dfrac{3\pi}{2}, 0\right)$ $(2\pi, 2)$

Since $g(x) = 2\cos(x + \pi) = f(x + \pi)$, the graph of $g(x)$ is the graph of $f(x)$, but with a phase shift (horizontal translation) of $-\pi$. Generate key points for the graph of $g(x)$ by shifting each key point of $f(x)$ π units to the left.

35. $y = 3 \sin x$

Period: 2π

Amplitude: 3

Key points:

$(0, 0), \left(\dfrac{\pi}{2}, 3\right), (\pi, 0),$

$\left(\dfrac{3\pi}{2}, -3\right), (2\pi, 0)$

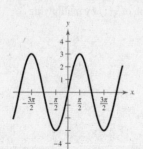

37. $y = \dfrac{1}{3} \cos x$

Period: 2π

Amplitude: $\dfrac{1}{3}$

Key points:

$\left(0, \dfrac{1}{3}\right), \left(\dfrac{\pi}{2}, 0\right), \left(\pi, -\dfrac{1}{3}\right),$

$\left(\dfrac{3\pi}{2}, 0\right), \left(2\pi, \dfrac{1}{3}\right)$

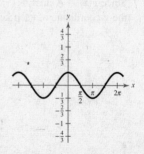

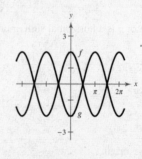

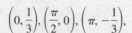

39. $y = \cos \dfrac{x}{2}$

Period: 4π

Amplitude: 1

Key points:

$(0, 1), (\pi, 0), (2\pi, -1),$

$(3\pi, 0), (4\pi, 1)$

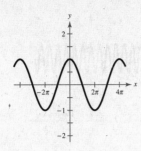

41. $y = \cos 2\pi x$

Period: $\dfrac{2\pi}{2\pi} = 1$

Amplitude: 1

Key points:

$(0, 1), \left(\dfrac{1}{4}, 0\right), \left(\dfrac{1}{2}, -1\right), \left(\dfrac{3}{4}, 0\right)$

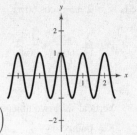

43. $y = -\sin \dfrac{2\pi x}{3}$; $a = -1, b = \dfrac{2\pi}{3}, c = 0$

Period: $\dfrac{2\pi}{2\pi/3} = 3$

Amplitude: 1

Key points: $(0, 0), \left(\dfrac{3}{4}, -1\right), \left(\dfrac{3}{2}, 0\right), \left(\dfrac{9}{4}, 1\right), (3, 0)$

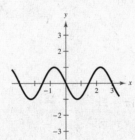

45. $y = \sin\left(x - \dfrac{\pi}{4}\right)$; $a = 1, b = 1, c = \dfrac{\pi}{4}$

Period: 2π

Amplitude: 1

Shift: Set $x - \dfrac{\pi}{4} = 0$ and $x - \dfrac{\pi}{4} = 2\pi$

$\qquad x = \dfrac{\pi}{4} \qquad\qquad x = \dfrac{9\pi}{4}$

Key points: $\left(\dfrac{\pi}{4}, 0\right), \left(\dfrac{3\pi}{4}, 1\right), \left(\dfrac{5\pi}{4}, 0\right), \left(\dfrac{7\pi}{4}, -1\right), \left(\dfrac{9\pi}{4}, 0\right)$

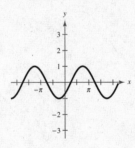

47. $y = 3\cos(x + \pi)$

Period: 2π

Amplitude: 3

Shift: Set $x + \pi = 0$ and $x + \pi = 2\pi$

$\qquad x = -\pi \qquad\qquad x = \pi$

Key points: $(-\pi, 3), \left(-\dfrac{\pi}{2}, 0\right), (0, -3), \left(\dfrac{\pi}{2}, 0\right), (\pi, 3)$

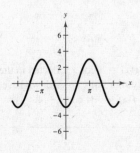

49. $y = 2 - \sin \dfrac{2\pi x}{3}$

Period: 3

Amplitude: 1

Key points: $(0, 2), \left(\dfrac{3}{4}, 1\right), \left(\dfrac{3}{2}, 2\right), \left(\dfrac{9}{4}, 3\right), (3, 2)$

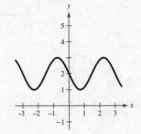

51. $y = 2 + \dfrac{1}{10} \cos 60\pi x$

Period: $\dfrac{2\pi}{60\pi} = \dfrac{1}{30}$

Amplitude: $\dfrac{1}{10}$

Vertical shift two units upward

Key points:

$(0, 2.1), \left(\dfrac{1}{120}, 2\right), \left(\dfrac{1}{60}, 1.9\right), \left(\dfrac{1}{40}, 2\right), \left(\dfrac{1}{30}, 2.1\right)$

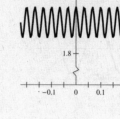

53. $y = 3 \cos(x + \pi) - 3$

Period: 2π

Amplitude: 3

Shift: Set $x + \pi = 0$ and $x + \pi = 2\pi$

$\qquad\qquad x = -\pi \qquad\qquad\quad x = \pi$

Key points: $(-\pi, 0), \left(-\dfrac{\pi}{2}, -3\right), (0, -6), \left(\dfrac{\pi}{2}, -3\right), (\pi, 0)$

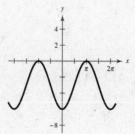

55. $y = \dfrac{2}{3} \cos\left(\dfrac{x}{2} - \dfrac{\pi}{4}\right); \; a = \dfrac{2}{3}, \; b = \dfrac{1}{2}, \; c = \dfrac{\pi}{4}$

Period: 4π

Amplitude: $\dfrac{2}{3}$

Shift: $\dfrac{x}{2} - \dfrac{\pi}{4} = 0$ and $\dfrac{x}{2} - \dfrac{\pi}{4} = 2\pi$

$\qquad\qquad x = \dfrac{\pi}{2} \qquad\qquad x = \dfrac{9\pi}{2}$

Key points: $\left(\dfrac{\pi}{2}, \dfrac{2}{3}\right), \left(\dfrac{3\pi}{2}, 0\right), \left(\dfrac{5\pi}{2}, \dfrac{-2}{3}\right), \left(\dfrac{7\pi}{2}, 0\right), \left(\dfrac{9\pi}{2}, \dfrac{2}{3}\right)$

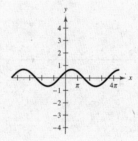

57. $y = -2 \sin(4x + \pi)$

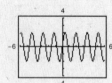

59. $y = \cos\left(2\pi x - \dfrac{\pi}{2}\right) + 1$

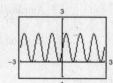

61. $y = -0.1 \sin\left(\dfrac{\pi x}{10} + \pi\right)$

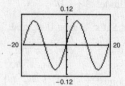

63. $f(x) = a \cos x + d$

Amplitude: $\frac{1}{2}[3 - (-1)] = 2 \implies a = 2$

Vertical shift one unit upward of

$g(x) = 2 \cos x \implies d = 1$

Thus, $f(x) = 2 \cos x + 1$.

65. $f(x) = a \cos x + d$

Amplitude: $\frac{1}{2}[8 - 0] = 4$

Since $f(x)$ is the graph of $g(x) = 4 \cos x$ reflected in the x-axis and shifted vertically four units upward, we have $a = -4$ and $d = 4$. Thus, $f(x) = -4 \cos x + 4$.

67. $y = a\sin(bx - c)$

Amplitude: $|a| = |3|$

Since the graph is reflected in the x-axis, we have $a = -3$.

Period: $\dfrac{2\pi}{b} = \pi \Rightarrow b = 2$

Phase shift: $c = 0$

Thus, $y = -3\sin 2x$.

69. $y = a\sin(bx - c)$

Amplitude: $a = 2$

Period: $2\pi \Rightarrow b = 1$

Phase shift: $bx - c = 0$ when $x = -\dfrac{\pi}{4}$

$$(1)\left(\dfrac{-\pi}{4}\right) - c = 0 \Rightarrow c = -\dfrac{\pi}{4}$$

Thus, $y = 2\sin\left(x + \dfrac{\pi}{4}\right)$.

71. $y_1 = \sin x$

$y_2 = -\dfrac{1}{2}$

In the interval $[-2\pi, 2\pi]$,

$\sin x = -\dfrac{1}{2}$ when $x = -\dfrac{5\pi}{6}, -\dfrac{\pi}{6}, \dfrac{7\pi}{6}, \dfrac{11\pi}{6}$.

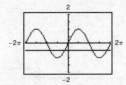

73. $y = 0.85\sin\dfrac{\pi t}{3}$

(a) Time for one cycle $= \dfrac{2\pi}{\pi/3} = 6$ sec

(b) Cycles per min $= \dfrac{60}{6} = 10$ cycles per min

(c) Amplitude: 0.85; Period: 6

Key points: $(0, 0), \left(\dfrac{3}{2}, 0.85\right), (3, 0), \left(\dfrac{9}{2}, -0.85\right), (6, 0)$

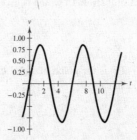

75. $y = 0.001\sin 880\pi t$

(a) Period: $\dfrac{2\pi}{880\pi} = \dfrac{1}{440}$ seconds　　(b) $f = \dfrac{1}{p} = 440$ cycles per second

77. (a) $a = \dfrac{1}{2}[\text{high} - \text{low}] = \dfrac{1}{2}[83.5 - 29.6] = 26.95$

$p = 2[\text{high time} - \text{low time}] = 2[7 - 1] = 12$

$b = \dfrac{2\pi}{p} = \dfrac{2\pi}{12} = \dfrac{\pi}{6}$

$\dfrac{c}{b} = 7 \Rightarrow c = 7\left(\dfrac{\pi}{6}\right) \approx 3.67$

$d = \dfrac{1}{2}[\text{high} + \text{low}] = \dfrac{1}{2}[83.5 + 29.6] = 56.55$

$C(t) = 56.55 + 26.95\cos\left(\dfrac{\pi t}{6} - 3.67\right)$

(b)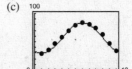

The model is a good fit.

(c)

The model is a good fit.

—CONTINUED—

77. —CONTINUED—

(d) Tallahassee average maximum: 77.90°

Chicago average maximum: 56.55°

The constant term, d, gives the average maximum temperature.

(e) The period for both models is $\dfrac{2\pi}{\pi/6} = 12$ months.

This is as we expected since one full period is one year.

(f) Chicago has the greater variability in temperature throughout the year. The amplitude, a, determines this variability since it is $\frac{1}{2}$[high temp $-$ low temp].

79. $C = 30.3 + 21.6\sin\left(\dfrac{2\pi t}{365} + 10.9\right)$

(a) Period $= \dfrac{2\pi}{2\pi/365} = 365$

Yes, this is what is expected because there are 365 days in a year.

(b) The average daily fuel consumption is given by the amount of the vertical shift (from 0) which is given by the constant 30.3.

(c)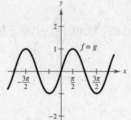

The consumption exceeds 40 gallons per day when $124 < x < 252$.

81. False. The graph of $\sin(x + 2\pi)$ is the graph of $\sin(x)$ translated to the *left* by one period, and the graphs are indeed identical.

83. True. Since

$\cos x = \sin\left(x + \dfrac{\pi}{2}\right),\ y = -\cos x = -\sin\left(x + \dfrac{\pi}{2}\right),$

and so is a reflection in the x-axis of

$y = \sin\left(x + \dfrac{\pi}{2}\right).$

85.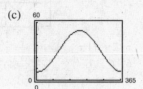

Since the graphs are the same, the conjecture is that

$\sin(x) = \cos\left(x - \dfrac{\pi}{2}\right).$

87. (a)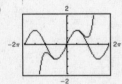

The graphs are nearly the same for $-\dfrac{\pi}{2} < x < \dfrac{\pi}{2}$.

(b)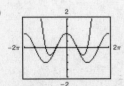

The graphs are nearly the same for $-\dfrac{\pi}{2} < x < \dfrac{\pi}{2}$.

(c) $\sin x \approx x - \dfrac{x^3}{3!} + \dfrac{x^5}{5!} - \dfrac{x^7}{7!}$

$\cos x \approx 1 - \dfrac{x^2}{2!} + \dfrac{x^4}{4!} - \dfrac{x^6}{6!}$

 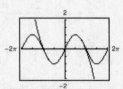

The graphs now agree over a wider range, $-\dfrac{3\pi}{4} < x < \dfrac{3\pi}{4}$.

89. $\log_{10} \sqrt{x-2} = \log_{10}(x-2)^{1/2} = \dfrac{1}{2}\log_{10}(x-2)$

91. $\ln \dfrac{t^3}{t-1} = \ln t^3 - \ln(t-1) = 3\ln t - \ln(t-1)$

93. $\dfrac{1}{2}(\log_{10} x + \log_{10} y) = \dfrac{1}{2}\log_{10}(xy)$

$= \log_{10}\sqrt{xy}$

95. $\ln 3x - 4\ln y = \ln 3x - \ln y^4$

$= \ln\left(\dfrac{3x}{y^4}\right)$

97. Answers will vary.

Section 4.6   Graphs of Other Trigonometric Functions

- ■ You should be able to graph

  $y = a\tan(bx - c)$ $y = a\cot(bx - c)$

  $y = a\sec(bx - c)$ $y = a\csc(bx - c)$

- ■ When graphing $y = a\sec(bx - c)$ or $y = a\csc(bx - c)$ you should first graph $y = a\cos(bx - c)$ or $y = a\sin(bx - c)$ because

  (a) The x-intercepts of sine and cosine are the vertical asymptotes of cosecant and secant.

  (b) The maximums of sine and cosine are the local minimums of cosecant and secant.

  (c) The minimums of sine and cosine are the local maximums of cosecant and secant.

- ■ You should be able to graph using a damping factor.

Vocabulary Check

1. vertical

2. reciprocal

3. damping

4. π

5. $x \neq n\pi$

6. $(-\infty, -1] \cup [1, \infty)$

7. 2π

---

1. $y = \sec 2x$

Period: $\dfrac{2\pi}{2} = \pi$

Matches graph (e).

3. $y = \dfrac{1}{2}\cot \pi x$

Period: $\dfrac{\pi}{\pi} = 1$

Matches graph (a).

5. $y = \dfrac{1}{2}\sec \dfrac{\pi x}{2}$

Period: $\dfrac{2\pi}{b} = \dfrac{2\pi}{\pi/2} = 4$

Asymptotes: $x = -1, x = 1$

Matches graph (f).

7. $y = \dfrac{1}{3}\tan x$

Period: π

Two consecutive asymptotes:

$x = -\dfrac{\pi}{2}$ and $x = \dfrac{\pi}{2}$

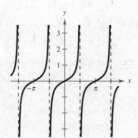

x	$-\dfrac{\pi}{4}$	0	$\dfrac{\pi}{4}$
y	$-\dfrac{1}{3}$	0	$\dfrac{1}{3}$

9. $y = \tan 3x$

Period: $\dfrac{\pi}{3}$

Two consecutive asymptotes:

$3x = -\dfrac{\pi}{2} \Rightarrow x = -\dfrac{\pi}{6}$

$3x - \dfrac{\pi}{2} \Rightarrow x = \dfrac{\pi}{6}$

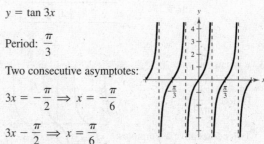

x	$-\dfrac{\pi}{12}$	0	$\dfrac{\pi}{12}$
y	-1	0	1

11. $y = -\dfrac{1}{2}\sec x$

Period: 2π

Two consecutive
asymptotes:

$x = -\dfrac{\pi}{2}, x = \dfrac{\pi}{2}$

x	$-\dfrac{\pi}{3}$	0	$\dfrac{\pi}{3}$
y	-1	$-\dfrac{1}{2}$	-1

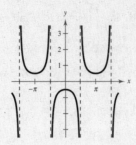

13. $y = \csc \pi x$

Period: $\dfrac{2\pi}{\pi} = 2$

Two consecutive
asymptotes:

$x = 0, x = 1$

x	$\dfrac{1}{6}$	$\dfrac{1}{2}$	$\dfrac{5}{6}$
y	2	1	2

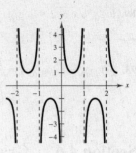

15. $y = \sec \pi x - 1$

Period: $\dfrac{2\pi}{\pi} = 2$

Two consecutive
asymptotes:

$x = -\dfrac{1}{2}, x = \dfrac{1}{2}$

x	$-\dfrac{1}{3}$	0	$\dfrac{1}{3}$
y	1	0	1

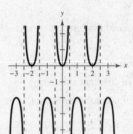

17. $y = \csc \dfrac{x}{2}$

Period: $\dfrac{2\pi}{1/2} = 4\pi$

Two consecutive
asymptotes:

$x = 0, x = 2\pi$

x	$\dfrac{\pi}{3}$	π	$\dfrac{5\pi}{3}$
y	2	1	2

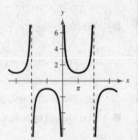

19. $y = \cot \dfrac{x}{2}$

Period: $\dfrac{\pi}{1/2} = 2\pi$

Two consecutive
asymptotes:

$\dfrac{x}{2} = 0 \implies x = 0$

$\dfrac{x}{2} = \pi \implies x = 2\pi$

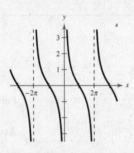

x	$\dfrac{\pi}{2}$	π	$\dfrac{3\pi}{2}$
y	1	0	-1

21. $y = \dfrac{1}{2}\sec 2x$

Period: $\dfrac{2\pi}{2} = \pi$

x	$-\dfrac{\pi}{6}$	0	$\dfrac{\pi}{6}$
y	1	$\dfrac{1}{2}$	1

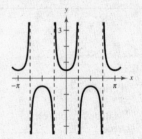

23. $y = \tan \dfrac{\pi x}{4}$

Period: $\dfrac{\pi}{\pi/4} = 4$

Two consecutive asymptotes:

$\dfrac{\pi x}{4} = -\dfrac{\pi}{2} \implies x = -2$

$\dfrac{\pi x}{4} = \dfrac{\pi}{2} \implies x = 2$

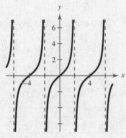

x	-1	0	1
y	-1	0	1

25. $y = \csc(\pi - x)$

Period: 2π

Two consecutive
asymptotes:

$x = 0, x = \pi$

x	$\dfrac{\pi}{6}$	$\dfrac{\pi}{2}$	$\dfrac{5\pi}{6}$
y	2	1	2

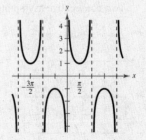

27. $y = 2 \sec(x + \pi)$

Period: 2π

Two consecutive asymptotes:

$x = -\dfrac{\pi}{2}, x = \dfrac{\pi}{2}$

x	$-\dfrac{\pi}{3}$	0	$\dfrac{\pi}{3}$
y	-4	-2	-4

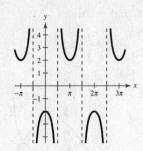

29. $y = \dfrac{1}{4} \csc\left(x + \dfrac{\pi}{4}\right)$

Period: 2π

Two consecutive asymptotes:

$x = -\dfrac{\pi}{4}, x = \dfrac{3\pi}{4}$

x	$-\dfrac{\pi}{12}$	$\dfrac{\pi}{4}$	$\dfrac{7\pi}{12}$
y	$\dfrac{1}{2}$	$\dfrac{1}{4}$	$\dfrac{1}{2}$

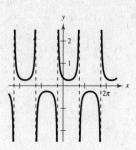

31. $y = \tan \dfrac{x}{3}$

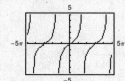

33. $y = -2 \sec 4x = \dfrac{-2}{\cos 4x}$

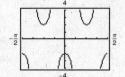

35. $y = \tan\left(x - \dfrac{\pi}{4}\right)$

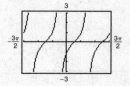

37. $y = -\csc(4x - \pi)$

$y = \dfrac{-1}{\sin(4x - \pi)}$

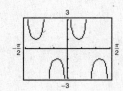

39. $y = 0.1 \tan\left(\dfrac{\pi x}{4} + \dfrac{\pi}{4}\right)$

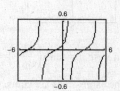

41. $\tan x = 1$

$x = -\dfrac{7\pi}{4}, -\dfrac{3\pi}{4}, \dfrac{\pi}{4}, \dfrac{5\pi}{4}$

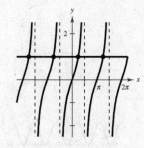

43. $\cot x = -\dfrac{\sqrt{3}}{3}$

$x = -\dfrac{4\pi}{3}, -\dfrac{\pi}{3}, \dfrac{2\pi}{3}, \dfrac{5\pi}{3}$

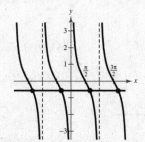

45. $\sec x = -2$

$x = \pm\dfrac{2\pi}{3}, \pm\dfrac{4\pi}{3}$

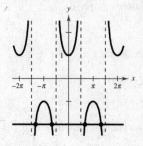

47. $\csc x = \sqrt{2}$

$x = -\dfrac{7\pi}{4}, -\dfrac{5\pi}{4}, \dfrac{\pi}{4}, \dfrac{3\pi}{4}$

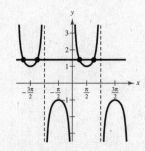

49. $f(x) = \sec x = \dfrac{1}{\cos x}$

$f(-x) = \sec(-x)$

$\qquad = \dfrac{1}{\cos(-x)}$

$\qquad = \dfrac{1}{\cos x}$

$\qquad = f(x)$

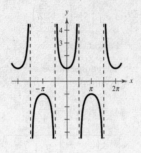

Thus, $f(x) = \sec x$ is an even function and the graph has y-axis symmetry.

51. $f(x) = 2\sin x$

$g(x) = \dfrac{1}{2}\csc x$

(a)

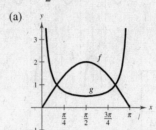

(b) $f > g$ on the interval, $\dfrac{\pi}{6} < x < \dfrac{5\pi}{6}$

(c) As $x \to \pi$, $f(x) = 2\sin x \to 0$ and
$g(x) = \frac{1}{2}\csc x \to \pm\infty$ since $g(x)$ is the reciprocal of $f(x)$.

53. $y_1 = \sin x \csc x$ and $y_2 = 1$

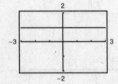

$\sin x \csc x = \sin x\left(\dfrac{1}{\sin x}\right) = 1,\ \sin x \neq 0$

The expressions are equivalent except when $\sin x = 0$ and y_1 is undefined.

55. $y_1 = \dfrac{\cos x}{\sin x}$ and $y_2 = \cot x = \dfrac{1}{\tan x}$

$\cot x = \dfrac{\cos x}{\sin x}$

The expressions are equivalent.

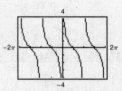

57. $f(x) = |x \cos x|$

As $x \to 0$, $f(x) \to 0$ and $f(x) > 0$.

Matches graph (d).

59. $g(x) = |x| \sin x$

As $x \to 0$, $g(x) \to 0$ and $g(x)$ is odd.

Matches graph (b).

61. $f(x) = \sin x + \cos\left(x + \dfrac{\pi}{2}\right)$

$g(x) = 0$

$f(x) = g(x)$

The graph is the line $y = 0$.

63. $f(x) = \sin^2 x$

$g(x) = \dfrac{1}{2}(1 - \cos 2x)$

$f(x) = g(x)$

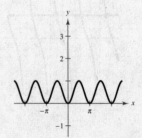

65. $g(x) = e^{-x^2/2}\sin x$

$-e^{-x^2/2} \leq g(x) \leq e^{-x^2/2}$

The damping factor is $y = e^{-x^2/2}$.

As $x \to \infty$, $g(x) \to 0$.

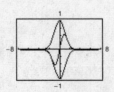

67. $f(x) = 2^{-x/4} \cos \pi x$

$-2^{-x/4} \le f(x) \le 2^{-x/4}$

Damping factor: $y = 2^{-x/4}$

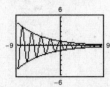

As $x \to \infty, f(x) \to 0$.

69. $y = \dfrac{6}{x} + \cos x, \; x > 0$

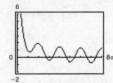

As $x \to 0, \; y \to \infty$.

71. $g(x) = \dfrac{\sin x}{x}$

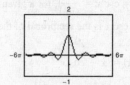

As $x \to 0, \; g(x) \to 1$.

73. $f(x) = \sin \dfrac{1}{x}$

As $x \to 0, f(x)$ oscillates between -1 and 1.

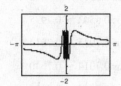

75. $\tan x = \dfrac{7}{d}$

$d = \dfrac{7}{\tan x} = 7 \cot x$

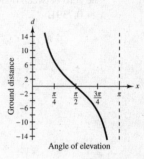

77. $C = 5000 + 2000 \sin \dfrac{\pi t}{12}, \; R = 25{,}000 + 15{,}000 \cos \dfrac{\pi t}{12}$

(a)

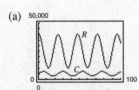

(b) As the predator population increases, the number of prey decreases. When the number of prey is small, the number of predators decreases.

(c) The period for both C and R is:

$$p = \frac{2\pi}{\pi/12} = 24 \text{ months}$$

When the prey population is highest, the predator population is increasing most rapidly.
When the prey population is lowest, the predator population is decreasing most rapidly.
When the predator population is lowest, the prey population is increasing most rapidly.
When the predator population is highest, the prey population is decreasing most rapidly.

In addition, weather, food sources for the prey, hunting, all affect the populations of both the predator and the prey.

79. $H(t) = 54.33 - 20.38 \cos \dfrac{\pi t}{6} - 15.69 \sin \dfrac{\pi t}{6}$

$L(t) = 39.36 - 15.70 \cos \dfrac{\pi t}{6} - 14.16 \sin \dfrac{\pi t}{6}$

(a) Period of $\cos \dfrac{\pi t}{6}: \dfrac{2\pi}{\pi/6} = 12$

Period of $\sin \dfrac{\pi t}{6}: \dfrac{2\pi}{\pi/6} = 12$

Period of $H(t)$: 12 months

Period of $L(t)$: 12 months

(b) From the graph, it appears that the greatest difference between high and low temperatures occurs in summer. The smallest difference occurs in winter.

(c) The highest high and low temperatures appear to occur around the middle of July, roughly one month after the time when the sun is northernmost in the sky.

81. True. Since

$y = \csc x = \dfrac{1}{\sin x}$, for a given value of x, the y-coordinate

of $\csc x$ is the reciprocal of the y-coordinate of $\sin x$.

83. As $x \to \dfrac{\pi}{2}$ from the left, $f(x) = \tan x \to \infty$.

As $x \to \dfrac{\pi}{2}$ from the right, $f(x) = \tan x \to {}^{-}\infty$.

85. $f(x) = x - \cos x$

(a)

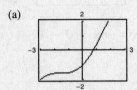

The zero between 0 and 1 occurs at
$x \approx 0.7391$.

(b) $x_n = \cos(x_{n-1})$

$x_0 = 1$

$x_1 = \cos 1 \approx 0.5403$

$x_2 = \cos 0.5403 \approx 0.8576$

$x_3 = \cos 0.8576 \approx 0.6543$

$x_4 = \cos 0.6543 \approx 0.7935$

$x_5 = \cos 0.7935 \approx 0.7014$

$x_6 = \cos 0.7014 \approx 0.7640$

$x_7 = \cos 0.7640 \approx 0.7221$

$x_8 = \cos 0.7221 \approx 0.7504$

$x_9 = \cos 0.7504 \approx 0.7314$

$\vdots$

This sequence appears to be approaching the zero of f: $x \approx 0.7391$.

87. $y_1 = \sec x$

$y_2 = 1 + \dfrac{x^2}{2!} + \dfrac{5x^4}{4!}$

The graph appears to
coincide on the interval
$-1.1 \le x \le 1.1$.

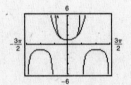

89. $e^{2x} = 54$

$2x = \ln 54$

$x = \dfrac{\ln 54}{2} \approx 1.994$

91. $\dfrac{300}{1 + e^{-x}} = 100$

$\dfrac{300}{100} = 1 + e^{-x}$

$3 = 1 + e^{-x}$

$2 = e^{-x}$

$\ln 2 = -x$

$x = -\ln 2 \approx -0.693$

93. $\ln(3x - 2) = 73$

$3x - 2 = e^{73}$

$3x = 2 + e^{73}$

$x = \dfrac{2 + e^{73}}{3}$

$\approx 1.684 \times 10^{31}$

95. $\ln(x^2 + 1) = 3.2$

$x^2 + 1 = e^{3.2}$

$x^2 = e^{3.2} - 1$

$x = \pm\sqrt{e^{3.2} - 1} \approx \pm 4.851$

97. $\log_8 x + \log_8(x - 1) = \tfrac{1}{3}$

$\log_8[x(x - 1)] = \tfrac{1}{3}$

$x(x - 1) = 8^{1/3}$

$x^2 - x = 2$

$x^2 - x - 2 = 0$

$(x - 2)(x + 1) = 0$

$x = 2, -1$

$x = -1$ is extraneous (not in the domain of $\log_8 x$) so only
$x = 2$ is a solution.

Section 4.7   Inverse Trigonometric Functions

- You should know the definitions, domains, and ranges of $y = \arcsin x$, $y = \arccos x$, and $y = \arctan x$.

Function	Domain	Range
$y = \arcsin x \implies x = \sin y$	$-1 \le x \le 1$	$-\dfrac{\pi}{2} \le y \le \dfrac{\pi}{2}$
$y = \arccos x \implies x = \cos y$	$-1 \le x \le 1$	$0 \le y \le \pi$
$y = \arctan x \implies x = \tan y$	$-\infty < x < \infty$	$-\dfrac{\pi}{2} < x < \dfrac{\pi}{2}$

- You should know the inverse properties of the inverse trigonometric functions.

$$\sin(\arcsin x) = x \quad \text{and} \quad \arcsin(\sin y) = y, \ -\frac{\pi}{2} \le y \le \frac{\pi}{2}$$

$$\cos(\arccos x) = x \quad \text{and} \quad \arccos(\cos y) = y, \ 0 \le y \le \pi$$

$$\tan(\arctan x) = x \quad \text{and} \quad \arctan(\tan y) = y, \ -\frac{\pi}{2} < y < \frac{\pi}{2}$$

- You should be able to use the triangle technique to convert trigonometric functions of inverse trigonometric functions into algebraic expressions.

Vocabulary Check

Function	Alternative Notation	Domain	Range
1. $y = \arcsin x$	$y = \sin^{-1} x$	$-1 \le x \le 1$	$-\dfrac{\pi}{2} \le y \le \dfrac{\pi}{2}$
2. $y = \arccos x$	$y = \cos^{-1} x$	$-1 \le x \le 1$	$0 \le y \le \pi$
3. $y = \arctan x$	$y = \tan^{-1} x$	$-\infty < x < \infty$	$-\dfrac{\pi}{2} < y < \dfrac{\pi}{2}$

1. $y = \arcsin \dfrac{1}{2} \implies \sin y = \dfrac{1}{2}$ for $-\dfrac{\pi}{2} \le y \le \dfrac{\pi}{2} \implies y = \dfrac{\pi}{6}$

3. $y = \arccos \dfrac{1}{2} \implies \cos y = \dfrac{1}{2}$ for $0 \le y \le \pi \implies y = \dfrac{\pi}{3}$

5. $y = \arctan \dfrac{\sqrt{3}}{3} \implies \tan y = \dfrac{\sqrt{3}}{3}$ for

$-\dfrac{\pi}{2} < y < \dfrac{\pi}{2} \implies y = \dfrac{\pi}{6}$

7. $y = \arccos\left(-\dfrac{\sqrt{3}}{2}\right) \implies \cos y = -\dfrac{\sqrt{3}}{2}$ for

$0 \le y \le \pi \implies y = \dfrac{5\pi}{6}$

9. $y = \arctan(-\sqrt{3}) \implies \tan y = -\sqrt{3}$ for

$-\dfrac{\pi}{2} < y < \dfrac{\pi}{2} \implies y = -\dfrac{\pi}{3}$

11. $y = \arccos\left(-\dfrac{1}{2}\right) \implies \cos y = -\dfrac{1}{2}$ for

$0 \le y \le \pi \implies y = \dfrac{2\pi}{3}$

13. $y = \arcsin \dfrac{\sqrt{3}}{2} \implies \sin y = \dfrac{\sqrt{3}}{2}$ for

$-\dfrac{\pi}{2} \le y \le \dfrac{\pi}{2} \implies y = \dfrac{\pi}{3}$

15. $y = \arctan 0 \implies \tan y = 0$ for $-\dfrac{\pi}{2} < y < \dfrac{\pi}{2} \implies y = 0$

17. $f(x) = \sin x$

$g(x) = \arcsin x$

$y = x$

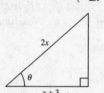

19. $\arccos 0.28 = \cos^{-1} 0.28 \approx 1.29$

21. $\arcsin(-0.75) = \sin^{-1}(-0.75) \approx -0.85$

23. $\arctan(-3) = \tan^{-1}(-3) \approx -1.25$

25. $\arcsin 0.31 = \sin^{-1} 0.31 \approx 0.32$

27. $\arccos(-0.41) = \cos^{-1}(-0.41) \approx 1.99$

29. $\arctan 0.92 = \tan^{-1} 0.92 \approx 0.74$

31. $\arcsin\left(\frac{3}{4}\right) = \sin^{-1}(0.75) \approx 0.85$

33. $\arctan\left(\dfrac{7}{2}\right) = \tan^{-1}(3.5) \approx 1.29$

35. This is the graph of $y = \arctan x$. The coordinates are

$\left(-\sqrt{3}, -\dfrac{\pi}{3}\right), \left(-\dfrac{\sqrt{3}}{3}, -\dfrac{\pi}{6}\right),$ and $\left(1, \dfrac{\pi}{4}\right)$.

37. $\tan \theta = \dfrac{x}{4}$

$\theta = \arctan \dfrac{x}{4}$

39. $\sin \theta = \dfrac{x+2}{5}$

$\theta = \arcsin\left(\dfrac{x+2}{5}\right)$

41. $\cos \theta = \dfrac{x+3}{2x}$

$\theta = \arccos\left(\dfrac{x+3}{2x}\right)$

43. $\sin(\arcsin 0.3) = 0.3$

45. $\cos[\arccos(-0.1)] = -0.1$

47. $\arcsin(\sin 3\pi) = \arcsin(0) = 0$

Note: 3π is not in the range of the arcsine function.

49. Let $y = \arctan \dfrac{3}{4}$,

$\tan y = \dfrac{3}{4}, \; 0 < y < \dfrac{\pi}{2}$,

and $\sin y = \dfrac{3}{5}$.

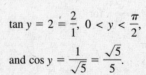

51. Let $y = \arctan 2$,

$\tan y = 2 = \dfrac{2}{1}, \; 0 < y < \dfrac{\pi}{2}$,

and $\cos y = \dfrac{1}{\sqrt{5}} = \dfrac{\sqrt{5}}{5}$.

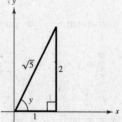

53. Let $y = \arcsin \dfrac{5}{13}$,

$\sin y = \dfrac{5}{13}$, $0 < y < \dfrac{\pi}{2}$, and $\cos y = \dfrac{12}{13}$.

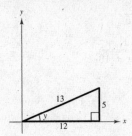

55. Let $y = \arctan\left(-\dfrac{3}{5}\right)$,

$\tan y = -\dfrac{3}{5}$, $-\dfrac{\pi}{2} < y < 0$, and $\sec y = \dfrac{\sqrt{34}}{5}$.

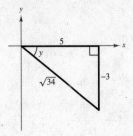

57. Let $y = \arccos\left(-\dfrac{2}{3}\right)$,

$\cos y = -\dfrac{2}{3}$, $\dfrac{\pi}{2} < y < \pi$, and $\sin y = \dfrac{\sqrt{5}}{3}$.

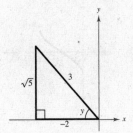

59. Let $y = \arctan x$,

$\tan y = x = \dfrac{x}{1}$,

and $\cot y = \dfrac{1}{x}$.

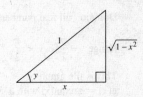

61. Let $y = \arcsin(2x)$,

$\sin y = 2x = \dfrac{2x}{1}$,

and $\cos y = \sqrt{1 - 4x^2}$.

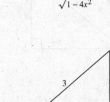

63. Let $y = \arccos x$,

$\cos y = x = \dfrac{x}{1}$,

and $\sin y = \sqrt{1 - x^2}$.

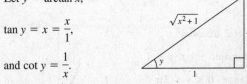

65. Let $y = \arccos\left(\dfrac{x}{3}\right)$,

$\cos y = \dfrac{x}{3}$,

and $\tan y = \dfrac{\sqrt{9 - x^2}}{x}$.

67. Let $y = \arctan \dfrac{x}{\sqrt{2}}$,

$\tan y = \dfrac{x}{\sqrt{2}}$,

and $\csc y = \dfrac{\sqrt{x^2 + 2}}{x}$.

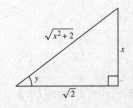

69. $f(x) = \sin(\arctan 2x)$, $g(x) = \dfrac{2x}{\sqrt{1 + 4x^2}}$

They are equal. Let $y = \arctan 2x$,

$\tan y = 2x = \dfrac{2x}{1}$,

and $\sin y = \dfrac{2x}{\sqrt{1 + 4x^2}}$.

$g(x) = \dfrac{2x}{\sqrt{1 + 4x^2}} = f(x)$

The graph has horizontal asymptotes at $y = \pm 1$.

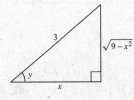

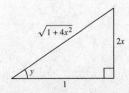

71. Let $y = \arctan \dfrac{9}{x}$.

$\tan y = \dfrac{9}{x}$ and $\sin y = \dfrac{9}{\sqrt{x^2 + 81}}, x > 0; \dfrac{-9}{\sqrt{x^2 + 81}}, x < 0.$

Thus,

$\arcsin y = \dfrac{9}{\sqrt{x^2 + 81}}, x > 0;$

$\arcsin y = \dfrac{-9}{\sqrt{x^2 + 81}}, x < 0.$

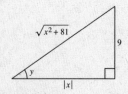

73. Let $y = \arccos \dfrac{3}{\sqrt{x^2 - 2x + 10}}$. Then,

$\cos y = \dfrac{3}{\sqrt{x^2 - 2x + 10}} = \dfrac{3}{\sqrt{(x - 1)^2 + 9}}$

and $\sin y = \dfrac{|x - 1|}{\sqrt{(x - 1)^2 + 9}}$.

Thus, $y = \arcsin \dfrac{|x - 1|}{\sqrt{x^2 - 2x + 10}}$.

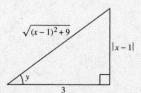

75. $y = 2 \arccos x$

Domain: $-1 \le x \le 1$

Range: $0 \le y \le 2\pi$

This is the graph of $f(x) = \arccos x$ with a factor of 2.

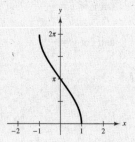

77. $f(x) = \arcsin(x - 1)$

Domain: $0 \le x \le 2$

Range: $-\dfrac{\pi}{2} \le y \le \dfrac{\pi}{2}$

This is the graph of $g(x) = \arcsin(x)$ shifted one unit to the right.

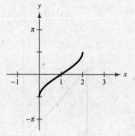

79. $f(x) = \arctan 2x$

Domain: all real numbers

Range: $-\dfrac{\pi}{2} < y < \dfrac{\pi}{2}$

This is the graph of $g(x) = \arctan(x)$ with a horizontal stretch of a factor of 2.

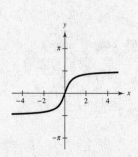

81. $h(v) = \tan(\arccos v) = \dfrac{\sqrt{1 - v^2}}{v}$

Domain: $-1 \le v \le 1, v \ne 0$

Range: all real numbers

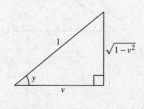 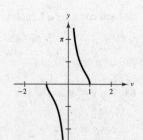

83. $f(x) = 2 \arccos(2x)$

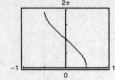

85. $f(x) = \arctan(2x - 3)$

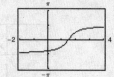

87. $f(x) = \pi - \arcsin\left(\tfrac{2}{3}\right) \approx 2.412$

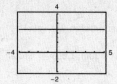

89. $f(t) = 3\cos 2t + 3\sin 2t = \sqrt{3^2 + 3^2}\,\sin\!\left(2t + \arctan\dfrac{3}{3}\right)$

$\qquad = 3\sqrt{2}\,\sin(2t + \arctan 1)$

$\qquad = 3\sqrt{2}\,\sin\!\left(2t + \dfrac{\pi}{4}\right)$

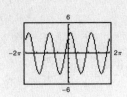

The graph implies the identity is true.

91. (a) $\sin\theta = \dfrac{5}{s}$

$\qquad \theta = \arcsin\dfrac{5}{s}$

(b) $s = 40$: $\theta = \arcsin\dfrac{5}{40} \approx 0.13$

$\quad s = 20$: $\theta = \arcsin\dfrac{5}{20} \approx 0.25$

93. $\beta = \arctan\dfrac{3x}{x^2 + 4}$

(a)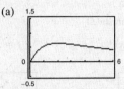

(b) β is maximum when $x = 2$ feet.

(c) The graph has a horizontal asymptote at $\beta = 0$. As x increases, β decreases.

95.

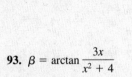

(a) $\tan\theta = \dfrac{20}{41}$

$\qquad \theta = \arctan\!\left(\dfrac{20}{41}\right) \approx 26.0°$

(b) $\tan 26° = \dfrac{h}{50}$

$\qquad h = 50\tan 26° \approx 24.39$ feet

97. (a) $\tan\theta = \dfrac{x}{20}$

$\qquad \theta = \arctan\dfrac{x}{20}$

(b) $x = 5$: $\theta = \arctan\dfrac{5}{20} \approx 14.0°$

$\quad x = 12$: $\theta = \arctan\dfrac{12}{20} \approx 31.0°$

99. False.

$\dfrac{5\pi}{4}$ is not in the range of the arctangent function.

$\arctan 1 = \dfrac{\pi}{4}$

101. $y = \operatorname{arccot} x$ if and only if $\cot y = x$.

Domain: $-\infty < x < \infty$

Range: $0 < x < \pi$

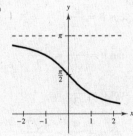

103. $y = \operatorname{arccsc} x$ if and only if $\csc y = x$.

Domain: $(-\infty, -1] \cup [1, \infty)$

Range: $\left[-\dfrac{\pi}{2}, 0\right) \cup \left(0, \dfrac{\pi}{2}\right]$

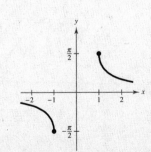

105. Area $= \arctan b - \arctan a$

(a) $a = 0, b = 1$

$\quad$ Area $= \arctan 1 - \arctan 0 = \dfrac{\pi}{4} - 0 = \dfrac{\pi}{4}$

(b) $a = -1, b = 1$

$\quad$ Area $= \arctan 1 - \arctan(-1)$

$\qquad = \dfrac{\pi}{4} - \left(-\dfrac{\pi}{4}\right) = \dfrac{\pi}{2}$

(c) $a = 0, b = 3$

$\quad$ Area $= \arctan 3 - \arctan 0$

$\qquad \approx 1.25 - 0 = 1.25$

(d) $a = -1, b = 3$

$\quad$ Area $= \arctan 3 - \arctan(-1)$

$\qquad \approx 1.25 - \left(-\dfrac{\pi}{4}\right) \approx 2.03$

107. $f(x) = \sin(x), f^{-1}(x) = \arcsin(x)$

(a) $f \cdot f^{-1} = \sin(\arcsin x)$ $\qquad\qquad$ $f^{-1} \cdot f = \arcsin(\sin x)$

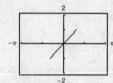
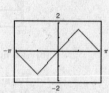

(b) The graphs coincide with the graph of $y = x$ only for certain values of x.

$\quad f \cdot f^{-1} = x$ over its entire domain, $-1 \le x \le 1$.

$\quad f^{-1} \cdot f = x$ over the region $-\dfrac{\pi}{2} \le x \le \dfrac{\pi}{2}$, corresponding to the region where $\sin x$ is one-to-one and thus has an inverse.

109. $(8.2)^{3.4} \approx 1279.284$

111. $(1.1)^{50} \approx 117.391$

113.
$$\sin \theta = \frac{3}{4} = \frac{\text{opp}}{\text{hyp}}$$
$$(\text{adj})^2 + (3)^2 = (4)^2$$
$$(\text{adj})^2 + 9 = 16$$
$$(\text{adj})^2 = 7$$
$$\text{adj} = \sqrt{7}$$
$$\cos \theta = \frac{\sqrt{7}}{4}$$
$$\tan \theta = \frac{3}{\sqrt{7}} = \frac{3\sqrt{7}}{7}$$
$$\cot \theta = \frac{\sqrt{7}}{3}$$
$$\sec \theta = \frac{4}{\sqrt{7}} = \frac{4\sqrt{7}}{7}$$
$$\csc \theta = \frac{4}{3}$$

115.
$$\cos \theta = \frac{5}{6} = \frac{\text{adj}}{\text{hyp}}$$
$$(\text{opp})^2 + (5)^2 = (6)^2$$
$$(\text{opp})^2 + 25 = 36$$
$$(\text{opp})^2 = 11$$
$$\text{opp} = \sqrt{11}$$
$$\sin \theta = \frac{\sqrt{11}}{6}$$
$$\tan \theta = \frac{\sqrt{11}}{5}$$
$$\cot \theta = \frac{5}{\sqrt{11}} = \frac{5\sqrt{11}}{11}$$
$$\sec \theta = \frac{6}{5}$$
$$\csc \theta = \frac{6}{\sqrt{11}} = \frac{6\sqrt{11}}{11}$$

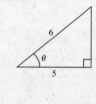

117. Let x = the number of people presently in the group. Each person's share is now $250{,}000/x$.
If two more join the group, each person's share would then be $250{,}000/(x + 2)$.

$$\begin{array}{c} \text{Share per person with} \\ \text{two more people} \end{array} = \begin{array}{c} \text{Original share} \\ \text{per person} \end{array} - 6250$$

$$\frac{250{,}000}{x + 2} = \frac{250{,}000}{x} - 6250$$

$$250{,}000x = 250{,}000(x + 2) - 6250x(x + 2)$$

$$250{,}000x = 250{,}000x + 500{,}000 - 6250x^2 - 12500x$$

$$6250x^2 + 12500x - 500{,}000 = 0$$

$$6250(x^2 + 2x - 80) = 0$$

$$6250(x + 10)(x - 8) = 0$$

$$x = -10 \quad \text{or} \quad x = 8$$

$x = -10$ is not possible.

There were 8 people in the original group.

119. (a) $A = 15{,}000\left(1 + \dfrac{0.035}{4}\right)^{(4)(10)} \approx \$21{,}253.63$

(b) $A = 15{,}000\left(1 + \dfrac{0.035}{12}\right)^{(12)(10)} \approx \$21{,}275.17$

(c) $A = 15{,}000\left(1 + \dfrac{0.035}{365}\right)^{(365)(10)} \approx \$21{,}285.66$

(d) $A = 15{,}000e^{(0.035)(10)} \approx \$21{,}286.01$

Section 4.8   Applications and Models

- You should be able to solve right triangles.
- You should be able to solve right triangle applications.
- You should be able to solve applications of simple harmonic motion.

Vocabulary Check

1. elevation; depression   **2.** bearing   **3.** harmonic motion

1. Given: $A = 20°$, $b = 10$

$\tan A = \dfrac{a}{b} \implies a = b \tan A = 10 \tan 20° \approx 3.64$

$\cos A = \dfrac{b}{c} \implies c = \dfrac{b}{\cos A} = \dfrac{10}{\cos 20°} \approx 10.64$

$B = 90° - 20° = 70°$

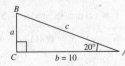

3. Given: $B = 71°$, $b = 24$

$\tan B = \dfrac{b}{a} \implies a = \dfrac{b}{\tan B} = \dfrac{24}{\tan 71°} \approx 8.26$

$\sin B = \dfrac{b}{c} \implies c = \dfrac{b}{\sin B} = \dfrac{24}{\sin 71°} \approx 25.38$

$A = 90° - 71° = 19°$

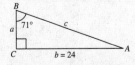

5. Given: $a = 6$, $b = 10$

$c^2 = a^2 + b^2 \implies c = \sqrt{36 + 100}$

$= 2\sqrt{34} \approx 11.66$

$\tan A = \dfrac{a}{b} = \dfrac{6}{10} \implies A = \arctan \dfrac{3}{5} \approx 30.96°$

$B = 90° - 30.96° = 59.04°$

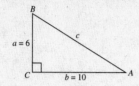

7. Given: $b = 16$, $c = 52$

$a = \sqrt{52^2 - 16^2}$

$= \sqrt{2448} = 12\sqrt{17} \approx 49.48$

$\cos A = \dfrac{16}{52}$

$A = \arccos \dfrac{16}{52} \approx 72.08°$

$B = 90° - 72.08° \approx 17.92°$

9. Given: $A = 12° \, 15'$, $c = 430.5$

$B = 90° - 12° \, 15' = 77° \, 45'$

$\sin 12° \, 15' = \dfrac{a}{430.5}$

$a = 430.5 \sin 12° \, 15' \approx 91.34$

$\cos 12° \, 15' = \dfrac{b}{430.5}$

$b = 430.5 \cos 12° \, 15' \approx 420.70$

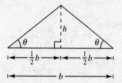

11. $\tan \theta = \dfrac{h}{(1/2)b} \implies h = \dfrac{1}{2} b \tan \theta$

$h = \dfrac{1}{2}(4) \tan 52° \approx 2.56$ inches

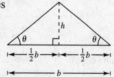

13. $\tan \theta = \dfrac{h}{(1/2)b} \implies h = \dfrac{1}{2} b \tan \theta$

$h = \dfrac{1}{2}(46) \tan 41° \approx 19.99$ inches

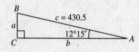

15. $\tan 25° = \dfrac{50}{x}$

$x = \dfrac{50}{\tan 25°}$

≈ 107.2 feet

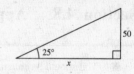

17. $\sin 80° = \dfrac{h}{20}$

$20 \sin 80° = h$

$h \approx 19.7$ feet

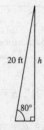

19. (a)

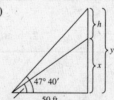

(b) Let the height of the church $= x$ and the height of the church and steeple $= y$. Then,

$\tan 35° = \dfrac{x}{50}$ and $\tan 47° 40' = \dfrac{y}{50}$

$x = 50 \tan 35°$ and $y = 50 \tan 47° 40'$

$h = y - x = 50(\tan 47° 40' - \tan 35°)$.

(c) $h \approx 19.9$ feet

21. $\sin 34° = \dfrac{x}{4000}$

$\quad x = 4000 \sin 34°$

$\quad\quad \approx 2236.8$ feet

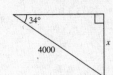

23. (a)

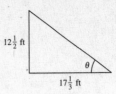

(b) $\tan \theta = \dfrac{12\frac{1}{2}}{17\frac{1}{3}}$

(c) $\theta = \arctan \dfrac{12\frac{1}{2}}{17\frac{1}{3}} \approx 35.8°$

The angle of elevation of the sum is 35.8°.

25. $1200 \text{ feet} + 150 \text{ feet} - 400 \text{ feet} = 950 \text{ feet}$

$5 \text{ miles} = 5 \text{ miles}\left(\dfrac{5280 \text{ feet}}{1 \text{ mile}}\right) = 26{,}400 \text{ feet}$

$\tan \theta = \dfrac{950}{26{,}400}$

$\theta = \arctan\left(\dfrac{950}{26{,}400}\right)$

$\quad \approx 2.06°$

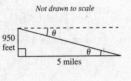

27. $\sin 10.5° = \dfrac{x}{4}$

$x = 4 \sin 10.5° \approx 0.73$ mile

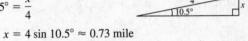

29. The plane has traveled $1.5(600) = 900$ miles.

$\sin 38° = \dfrac{a}{900} \Longrightarrow a \approx 554$ miles north

$\cos 38° = \dfrac{b}{900} \Longrightarrow b \approx 709$ miles east

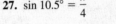

31.

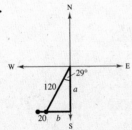

(a) $\cos 29° = \dfrac{a}{120} \Longrightarrow a \approx 104.95$ nautical miles south

$\sin 29° = \dfrac{b}{120} \Longrightarrow b \approx 58.18$ nautical miles west

(b) $\tan \theta = \dfrac{20+b}{a} \approx \dfrac{78.18}{104.95} \Longrightarrow \theta \approx 36.7°$

Bearing: S 36.7° W

Distance: $d \approx \sqrt{104.95^2 + 78.18^2}$

$\quad\quad \approx 130.9$ nautical miles from port

33. $\theta = 32°, \ \phi = 68°$

(a) $\alpha = 90° - 32° = 58°$

Bearing from A to C: N 58° E

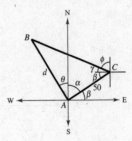

(b) $\beta = \theta = 32°$

$\gamma = 90° - \phi = 22°$

$C = \beta + \gamma = 54°$

$\tan C = \dfrac{d}{50} \Longrightarrow \tan 54°$

$\quad = \dfrac{d}{50} \Longrightarrow d \approx 68.82$ meters

35. $\tan \theta = \dfrac{45}{30} \implies \theta \approx 56.3°$

Bearing: N 56.3° W

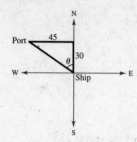

37. $\tan 6.5° = \dfrac{350}{d} \implies d \approx 3071.91$ ft

$\tan 4° = \dfrac{350}{D} \implies D \approx 5005.23$ ft

Distance between ships: $D - d \approx 1933.3$ ft

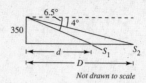

Not drawn to scale

39. $\tan 57° = \dfrac{a}{x} \implies x = a \cot 57°$

$\tan 16° = \dfrac{a}{x + (55/6)}$

$\tan 16° = \dfrac{a}{a \cot 57° + (55/6)}$

$\cot 16° = \dfrac{a \cot 57° + (55/6)}{a}$

$a \cot 16° - a \cot 57° = \dfrac{55}{6} \implies a \approx 3.23$ miles

$\approx 17{,}054$ ft

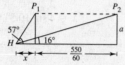

41. $L_1:\ 3x - 2y = 5 \implies y = \dfrac{3}{2}x - \dfrac{5}{2} \implies m_1 = \dfrac{3}{2}$

$L_2:\ x + y = 1 \implies y = -x + 1 \implies m_2 = -1$

$\tan \alpha = \left| \dfrac{-1 - (3/2)}{1 + (-1)(3/2)} \right| = \left| \dfrac{-5/2}{-1/2} \right| = 5$

$\alpha = \arctan 5 \approx 78.7°$

43. The diagonal of the base has a length of $\sqrt{a^2 + a^2} = \sqrt{2}a$. Now, we have

$\tan \theta = \dfrac{a}{\sqrt{2}a} = \dfrac{1}{\sqrt{2}}$

$\theta = \arctan \dfrac{1}{\sqrt{2}}$

$\theta \approx 35.3°.$

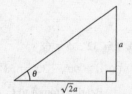

45. $\sin 36° = \dfrac{d}{25} \implies d \approx 14.69$

Length of side: $2d \approx 29.4$ inches

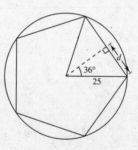

47. $\cos 30° = \dfrac{b}{r}$

$b = r \cos 30°$

$b = \dfrac{\sqrt{3}r}{2}$

$y = 2b = 2\left(\dfrac{\sqrt{3}r}{2} \right)$

$= \sqrt{3}r$

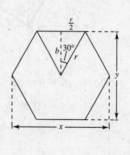

49.

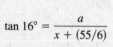

$\tan 35° = \dfrac{b}{10}$

$b = 10 \tan 35° \approx 7$

$\cos 35° = \dfrac{10}{a}$

$a = \dfrac{10}{\cos 35°} \approx 12.2$

51. $d = 0$ when $t = 0$, $a = 4$, period $= 2$

Use $d = a \sin \omega t$ since $d = 0$ when $t = 0$.

$$\frac{2\pi}{\omega} = 2 \implies \omega = \pi$$

Thus, $d = 4 \sin(\pi t)$.

53. $d = 3$ when $t = 0$, $a = 3$, period $= 1.5$

Use $d = a \cos \omega t$ since $d = 3$ when $t = 0$.

$$\frac{2\pi}{\omega} = 1.5 \implies \omega = \frac{4\pi}{3}$$

Thus, $d = 3 \cos\left(\frac{4\pi}{3}t\right) = 3 \cos\left(\frac{4\pi t}{3}\right)$.

55. $d = 4 \cos 8\pi t$

(a) Maximum displacement $=$ amplitude $= 4$

(b) Frequency $= \dfrac{\omega}{2\pi} = \dfrac{8\pi}{2\pi}$

$\qquad = 4$ cycles per unit of time

(c) $d = 4 \cos 40\pi = 4$

(d) $8\pi t = \dfrac{\pi}{2} \implies t = \dfrac{1}{16}$

57. $d = \dfrac{1}{16} \sin 120\pi t$

(a) Maximum displacement $=$ amplitude $= \dfrac{1}{16}$

(b) Frequency $= \dfrac{\omega}{2\pi} = \dfrac{120\pi}{2\pi}$

$\qquad = 60$ cycles per unit of time

(c) $d = \dfrac{1}{16} \sin 600\pi = 0$

(d) $120\pi t = \pi \implies t = \dfrac{1}{120}$

59.

$$d = a \sin \omega t$$

$$\text{Frequency} = \frac{\omega}{2\pi}$$

$$264 = \frac{\omega}{2\pi}$$

$$\omega = 2\pi(264) = 528\pi$$

61. $y = \dfrac{1}{4} \cos 16t$, $t > 0$

(a)

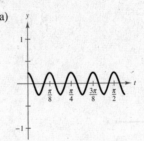

(b) Period: $\dfrac{2\pi}{16} = \dfrac{\pi}{8}$

(c) $\dfrac{1}{4} \cos 16t = 0$ when $16t = \dfrac{\pi}{2} \implies t = \dfrac{\pi}{32}$

63. (a) and (b)

Base 1	Base 2	Altitude	Area
8	$8 + 16 \cos 10°$	$8 \sin 10°$	22.1
8	$8 + 16 \cos 20°$	$8 \sin 20°$	42.5
8	$8 + 16 \cos 30°$	$8 \sin 30°$	59.7
8	$8 + 16 \cos 40°$	$8 \sin 40°$	72.7
8	$8 + 16 \cos 50°$	$8 \sin 50°$	80.5
8	$8 + 16 \cos 60°$	$8 \sin 60°$	83.1
8	$8 + 16 \cos 70°$	$8 \sin 70°$	80.7

The maximum occurs when $\theta = 60°$ and is approximately 83.1 square feet.

(c) $A(\theta) = \left[8 + (8 + 16 \cos \theta) \right] \left[\dfrac{8 \sin \theta}{2} \right]$

$\qquad = (16 + 16 \cos \theta)(4 \sin \theta)$

$\qquad = 64(1 + \cos \theta)(\sin \theta)$

(d)

The maximum of 83.1 square feet occurs when

$$\theta = \frac{\pi}{3} = 60°.$$

65. False. Since the tower is not exactly vertical, a right triangle with sides 191 feet and *d* is not formed.

67. No. N 24° E means 24° east of north.

69. $m = 4$, passes through $(-1, 2)$

$$y - 2 = 4(x - (-1))$$

$$y - 2 = 4x + 4$$

$$y = 4x + 6$$

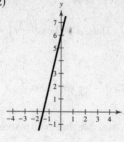

71. Passes through $(-2, 6)$ and $(3, 2)$

$$m = \frac{2 - 6}{3 - (-2)} = -\frac{4}{5}$$

$$y - 6 = -\frac{4}{5}[x - (-2)]$$

$$y - 6 = -\frac{4}{5}x - \frac{8}{5}$$

$$y = -\frac{4}{5}x + \frac{22}{5}$$

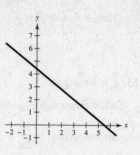

Review Exercises for Chapter 4

1. $\theta \approx 0.5$ radian

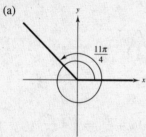

3. $\theta = \dfrac{11\pi}{4}$

(a)

(b) The angle lies in Quadrant II.

(c) Coterminal angles:

$$\frac{11\pi}{4} - 2\pi = \frac{3\pi}{4}$$

$$\frac{3\pi}{4} - 2\pi = -\frac{5\pi}{4}$$

5. $\theta = -\dfrac{4\pi}{3}$

(a)

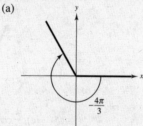

(b) The angle lies in Quadrant II.

(c) Coterminal angles:

$$-\frac{4\pi}{3} + 2\pi = \frac{2\pi}{3}$$

$$-\frac{4\pi}{3} - 2\pi = -\frac{10\pi}{3}$$

7. $\theta = 70°$

(a)

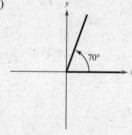

(b) The angle lies in Quadrant I.

(c) Coterminal angles:

$$70° + 360° = 430°$$

$$70° - 360° = -290°$$

9. $\theta = -110°$

(a)

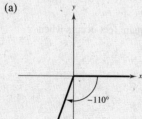

(b) The angle lies in Quadrant III.

(c) Coterminal angles:

$$-110° + 360° = 250°$$

$$-110° - 360° = -470°$$

11. $480° = 480° \cdot \dfrac{\pi \text{ rad}}{180°}$

$$= \frac{8\pi}{3} \text{ radians}$$

$$\approx 8.378 \text{ radians}$$

13. $-33°\ 45' = -33.75° = -33.75° \cdot \dfrac{\pi\ \text{rad}}{180°}$

$= -\dfrac{3\pi}{16}\ \text{radian} \approx -0.589\ \text{radian}$

15. $\dfrac{5\pi\ \text{rad}}{7} = \dfrac{5\pi\ \text{rad}}{7} \cdot \dfrac{180°}{\pi\ \text{rad}} \approx 128.571°$

17. $-3.5\ \text{rad} = -3.5\ \text{rad} \cdot \dfrac{180°}{\pi\ \text{rad}} \approx -200.535°$

19. $138° = \dfrac{138\pi}{180} = \dfrac{23\pi}{30}\ \text{radians}$

$s = r\theta = 20\left(\dfrac{23\pi}{30}\right) \approx 48.17\ \text{inches}$

21. (a) Angular speed $= \dfrac{\left(33\frac{1}{3}\right)(2\pi)\ \text{radians}}{1\ \text{minute}}$

$= 66\tfrac{2}{3}\pi\ \text{radians per minute}$

(b) Linear speed $= \dfrac{6\left(66\frac{2}{3}\pi\right)\ \text{inches}}{1\ \text{minute}}$

$= 400\pi\ \text{inches per minute}$

23. $120° = \dfrac{120\pi}{180} = \dfrac{2\pi}{3}\ \text{radians}$

$A = \dfrac{1}{2}r^2\theta = \dfrac{1}{2}(18)^2\left(\dfrac{2\pi}{3}\right) \approx 339.29\ \text{square inches}$

25. $t = \dfrac{2\pi}{3}$ corresponds to the point $\left(-\dfrac{1}{2}, \dfrac{\sqrt{3}}{2}\right)$.

27. $t = \dfrac{5\pi}{6}$ corresponds to the point $\left(-\dfrac{\sqrt{3}}{2}, \dfrac{1}{2}\right)$.

29. $t = \dfrac{7\pi}{6}$ corresponds to the point $\left(-\dfrac{\sqrt{3}}{2}, -\dfrac{1}{2}\right)$.

$\sin\dfrac{7\pi}{6} = y = -\dfrac{1}{2}$ $\csc\dfrac{7\pi}{6} = \dfrac{1}{y} = -2$

$\cos\dfrac{7\pi}{6} = x = -\dfrac{\sqrt{3}}{2}$ $\sec\dfrac{7\pi}{6} = \dfrac{1}{x} = -\dfrac{2\sqrt{3}}{3}$

$\tan\dfrac{7\pi}{6} = \dfrac{y}{x} = \dfrac{1}{\sqrt{3}} = \dfrac{\sqrt{3}}{3}$ $\cot\dfrac{7\pi}{6} = \dfrac{x}{y} = \sqrt{3}$

31. $t = -\dfrac{2\pi}{3}$ corresponds to the point $\left(-\dfrac{1}{2}, -\dfrac{\sqrt{3}}{2}\right)$.

$\sin\left(-\dfrac{2\pi}{3}\right) = y = -\dfrac{\sqrt{3}}{2}$ $\csc\left(-\dfrac{2\pi}{3}\right) = \dfrac{1}{y} = -\dfrac{2\sqrt{3}}{3}$

$\cos\left(-\dfrac{2\pi}{3}\right) = x = -\dfrac{1}{2}$ $\sec\left(-\dfrac{2\pi}{3}\right) = \dfrac{1}{x} = -2$

$\tan\left(-\dfrac{2\pi}{3}\right) = \dfrac{y}{x} = \sqrt{3}$ $\cot\left(-\dfrac{2\pi}{3}\right) = \dfrac{x}{y} = \dfrac{\sqrt{3}}{3}$

33. $\sin\dfrac{11\pi}{4} = \sin\dfrac{3\pi}{4} = \dfrac{\sqrt{2}}{2}$

35. $\sin\left(-\dfrac{17\pi}{6}\right) = \sin\left(-\dfrac{5\pi}{6}\right) = -\dfrac{1}{2}$

37. $\tan 33 \approx -75.3130$

39. $\sec\left(\dfrac{12\pi}{5}\right) = \dfrac{1}{\cos\left(\dfrac{12\pi}{5}\right)} \approx 3.2361$

41. $\text{opp} = 4,\ \text{adj} = 5,\ \text{hyp} = \sqrt{4^2 + 5^2} = \sqrt{41}$

$\sin\theta = \dfrac{\text{opp}}{\text{hyp}} = \dfrac{4}{\sqrt{41}} = \dfrac{4\sqrt{41}}{41}$ $\csc\theta = \dfrac{\text{hyp}}{\text{opp}} = \dfrac{\sqrt{41}}{4}$

$\cos\theta = \dfrac{\text{adj}}{\text{hyp}} = \dfrac{5}{\sqrt{41}} = \dfrac{5\sqrt{41}}{41}$ $\sec\theta = \dfrac{\text{hyp}}{\text{adj}} = \dfrac{\sqrt{41}}{5}$

$\tan\theta = \dfrac{\text{opp}}{\text{adj}} = \dfrac{4}{5}$ $\cot\theta = \dfrac{\text{adj}}{\text{opp}} = \dfrac{5}{4}$

43. $\text{adj} = 4,\ \text{hyp} = 8,\ \text{opp} = \sqrt{8^2 - 4^2} = \sqrt{48} = 4\sqrt{3}$

$\sin\theta = \dfrac{\text{opp}}{\text{hyp}} = \dfrac{4\sqrt{3}}{8} = \dfrac{\sqrt{3}}{2}$ $\csc\theta = \dfrac{\text{hyp}}{\text{opp}} = \dfrac{8}{4\sqrt{3}} = \dfrac{2\sqrt{3}}{3}$

$\cos\theta = \dfrac{\text{adj}}{\text{hyp}} = \dfrac{4}{8} = \dfrac{1}{2}$ $\sec\theta = \dfrac{\text{hyp}}{\text{adj}} = \dfrac{8}{4} = 2$

$\tan\theta = \dfrac{\text{opp}}{\text{adj}} = \dfrac{4\sqrt{3}}{4} = \sqrt{3}$ $\cot\theta = \dfrac{\text{adj}}{\text{opp}} = \dfrac{4}{4\sqrt{3}} = \dfrac{\sqrt{3}}{3}$

45. $\sin \theta = \dfrac{1}{3}$

(a) $\csc \theta = \dfrac{1}{\sin \theta} = 3$

(b) $\sin^2 \theta + \cos^2 \theta = 1$

$$\left(\dfrac{1}{3}\right)^2 + \cos^2 \theta = 1$$

$$\cos^2 \theta = 1 - \dfrac{1}{9}$$

$$\cos^2 \theta = \dfrac{8}{9}$$

$$\cos \theta = \sqrt{\dfrac{8}{9}}$$

$$\cos \theta = \dfrac{2\sqrt{2}}{3}$$

(c) $\sec \theta = \dfrac{1}{\cos \theta} = \dfrac{3}{2\sqrt{2}} = \dfrac{3\sqrt{2}}{4}$

(d) $\tan \theta = \dfrac{\sin \theta}{\cos \theta} = \dfrac{1/3}{(2\sqrt{2})/3} = \dfrac{1}{2\sqrt{2}} = \dfrac{\sqrt{2}}{4}$

47. $\csc \theta = 4$

(a) $\sin \theta = \dfrac{1}{\csc \theta} = \dfrac{1}{4}$

(b) $\sin^2 \theta + \cos^2 \theta = 1$

$$\left(\dfrac{1}{4}\right)^2 + \cos^2 \theta = 1$$

$$\cos^2 \theta = 1 - \dfrac{1}{16}$$

$$\cos^2 \theta = \dfrac{15}{16}$$

$$\cos \theta = \sqrt{\dfrac{15}{16}}$$

$$\cos \theta = \dfrac{\sqrt{15}}{4}$$

(c) $\sec \theta = \dfrac{1}{\cos \theta} = \dfrac{4}{\sqrt{15}} = \dfrac{4\sqrt{15}}{15}$

(d) $\tan \theta = \dfrac{\sin \theta}{\cos \theta} = \dfrac{1/4}{\sqrt{15}/4} = \dfrac{1}{\sqrt{15}} = \dfrac{\sqrt{15}}{15}$

49. $\tan 33° \approx 0.6494$

51. $\sin 34.2° \approx 0.5621$

53. $\cot 15° \, 14' = \dfrac{1}{\tan\left(15 + \frac{14}{60}\right)}$

$$\approx 3.6722$$

55. $\sin 1° \, 10' = \dfrac{x}{3.5}$

$x = 3.5 \sin 1° \, 10' \approx 0.07$ kilometer or 71.3 meters

3.5 km
1°10'
x
Not drawn to scale

57. $x = 12, \; y = 16, \; r = \sqrt{144 + 256} = \sqrt{400} = 20$

$\sin \theta = \dfrac{y}{r} = \dfrac{4}{5}$ $\csc \theta = \dfrac{r}{y} = \dfrac{5}{4}$

$\cos \theta = \dfrac{x}{r} = \dfrac{3}{5}$ $\sec \theta = \dfrac{r}{x} = \dfrac{5}{3}$

$\tan \theta = \dfrac{y}{x} = \dfrac{4}{3}$ $\cot \theta = \dfrac{x}{y} = \dfrac{3}{4}$

59. $x = \dfrac{2}{3}, \; y = \dfrac{5}{2}$

$$r = \sqrt{\left(\dfrac{2}{3}\right)^2 + \left(\dfrac{5}{2}\right)^2} = \dfrac{\sqrt{241}}{6}$$

$\sin \theta = \dfrac{y}{r} = \dfrac{5/2}{\sqrt{241}/6} = \dfrac{15}{\sqrt{241}} = \dfrac{15\sqrt{241}}{241}$ $\csc \theta = \dfrac{r}{y} = \dfrac{\sqrt{241}/6}{5/2} = \dfrac{2\sqrt{241}}{30} = \dfrac{\sqrt{241}}{15}$

$\cos \theta = \dfrac{x}{r} = \dfrac{2/3}{\sqrt{241}/6} = \dfrac{4}{\sqrt{241}} = \dfrac{4\sqrt{241}}{241}$ $\sec \theta = \dfrac{r}{x} = \dfrac{\sqrt{241}/6}{2/3} = \dfrac{\sqrt{241}}{4}$

$\tan \theta = \dfrac{y}{x} = \dfrac{5/2}{2/3} = \dfrac{15}{4}$ $\cot \theta = \dfrac{x}{y} = \dfrac{2/3}{5/2} = \dfrac{4}{15}$

61. $x = -0.5, y = 4.5$

$$r = \sqrt{(-0.5)^2 + (4.5)^2} = \sqrt{20.5} = \frac{\sqrt{82}}{2}$$

$$\sin\theta = \frac{y}{r} = \frac{4.5}{\sqrt{82}/2} = \frac{9\sqrt{82}}{82} \qquad \csc\theta = \frac{r}{y} = \frac{\sqrt{82}/2}{4.5} = \frac{\sqrt{82}}{9}$$

$$\cos\theta = \frac{x}{r} = \frac{-0.5}{\sqrt{82}/2} = \frac{-\sqrt{82}}{82} \qquad \sec\theta = \frac{r}{x} = \frac{\sqrt{82}/2}{-0.5} = -\sqrt{82}$$

$$\tan\theta = \frac{y}{x} = \frac{4.5}{-0.5} = -9 \qquad \cot\theta = \frac{x}{y} = \frac{-0.5}{4.5} = -\frac{1}{9}$$

63. $(x, 4x), \ x > 0$

$$x' = x, y' = 4x$$

$$r = \sqrt{x^2 + (4x)^2} = \sqrt{17}x$$

$$\sin\theta = \frac{y'}{r} = \frac{4x}{\sqrt{17}x} = \frac{4\sqrt{17}}{17} \qquad \csc\theta = \frac{r}{y'} = \frac{\sqrt{17}x}{4x} = \frac{\sqrt{17}}{4}$$

$$\cos\theta = \frac{x'}{r} = \frac{x}{\sqrt{17}x} = \frac{\sqrt{17}}{17} \qquad \sec\theta = \frac{r}{x'} = \frac{\sqrt{17}x}{x} = \sqrt{17}$$

$$\tan\theta = \frac{y'}{x'} = \frac{4x}{x} = 4 \qquad \cot\theta = \frac{x'}{y'} = \frac{x}{4x} = \frac{1}{4}$$

65. $\sec\theta = \dfrac{6}{5}, \ \tan\theta < 0 \implies \theta$ is in Quadrant IV.

$$r = 6, x = 5, y = -\sqrt{36 - 25} = -\sqrt{11}$$

$$\sin\theta = \frac{y}{r} = -\frac{\sqrt{11}}{6} \qquad \csc\theta = \frac{r}{y} = -\frac{6\sqrt{11}}{11}$$

$$\cos\theta = \frac{x}{r} = \frac{5}{6} \qquad \sec\theta = \frac{6}{5}$$

$$\tan\theta = \frac{y}{x} = -\frac{\sqrt{11}}{5} \qquad \cot\theta = -\frac{5\sqrt{11}}{11}$$

67. $\sin\theta = \dfrac{3}{8}, \ \cos\theta < 0 \implies \theta$ is in Quadrant II.

$$y = 3, r = 8, x = -\sqrt{55}$$

$$\sin\theta = \frac{y}{r} = \frac{3}{8}$$

$$\cos\theta = \frac{x}{r} = -\frac{\sqrt{55}}{8}$$

$$\tan\theta = \frac{y}{x} = -\frac{3}{\sqrt{55}} = -\frac{3\sqrt{55}}{55}$$

$$\csc\theta = \frac{8}{3}$$

$$\sec\theta = -\frac{8}{\sqrt{55}} = -\frac{8\sqrt{55}}{55}$$

$$\cot\theta = -\frac{\sqrt{55}}{3}$$

69. $\cos\theta = \dfrac{x}{r} = \dfrac{-2}{5} \implies y^2 = 21$

$\sin\theta > 0 \implies \theta$ is in Quadrant II $\implies y = \sqrt{21}$

$$\sin\theta = \frac{y}{r} = \frac{\sqrt{21}}{5}$$

$$\tan\theta = \frac{y}{x} = -\frac{\sqrt{21}}{2}$$

$$\csc\theta = \frac{r}{y} = \frac{5}{\sqrt{21}} = \frac{5\sqrt{21}}{21}$$

$$\sec\theta = \frac{r}{x} = \frac{5}{-2} = -\frac{5}{2}$$

$$\cot\theta = \frac{x}{y} = \frac{-2}{\sqrt{21}} = -\frac{2\sqrt{21}}{21}$$

71. $\theta = 264°$

$\theta' = 264° - 180° = 84°$

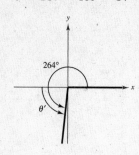

73. $\theta = -\dfrac{6\pi}{5}$

$-\dfrac{6\pi}{5} + 2\pi = \dfrac{4\pi}{5}$

$\theta' = \pi - \dfrac{4\pi}{5} = \dfrac{\pi}{5}$

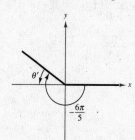

75. $\sin \dfrac{\pi}{3} = \dfrac{\sqrt{3}}{2}$

$\cos \dfrac{\pi}{3} = \dfrac{1}{2}$

$\tan \dfrac{\pi}{3} = \sqrt{3}$

77. $\sin\left(-\dfrac{7\pi}{3}\right) = -\sin\dfrac{\pi}{3} = -\dfrac{\sqrt{3}}{2}$

$\cos\left(-\dfrac{7\pi}{3}\right) = \cos\dfrac{\pi}{3} = \dfrac{1}{2}$

$\tan\left(-\dfrac{7\pi}{3}\right) = -\tan\dfrac{\pi}{3} = -\sqrt{3}$

79. $\sin 495° = \sin 45° = \dfrac{\sqrt{2}}{2}$

$\cos 495° = -\cos 45° = -\dfrac{\sqrt{2}}{2}$

$\tan 495° = -\tan 45° = -1$

81. $\sin(-240°) = \sin 60° = \dfrac{\sqrt{3}}{2}$

$\cos(-240°) = -\cos 60° = -\dfrac{1}{2}$

$\tan(-240°) = -\tan 60° = -\sqrt{3}$

83. $\sin 4 \approx -0.7568$

85. $\sin(-3.2) \approx 0.0584$

87. $\sec\left(\dfrac{12\pi}{5}\right) = \dfrac{1}{\cos\left(\dfrac{12\pi}{5}\right)} \approx 3.2361$

89. $y = \sin x$

Amplitude: 1

Period: 2π

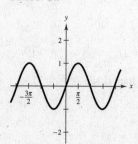

91. $f(x) = 5\sin\dfrac{2x}{5}$

Amplitude: 5

Period: $\dfrac{2\pi}{2/5} = 5\pi$

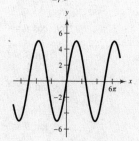

93. $y = 2 + \sin x$

Shift the graph of $y = \sin x$ two units upward.

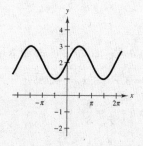

95. $g(t) = \dfrac{5}{2}\sin(t - \pi)$

Amplitude: $\dfrac{5}{2}$

Period: 2π

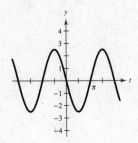

97. $y = a \sin bx$

(a) $a = 2$,

$\dfrac{2\pi}{b} = \dfrac{1}{264} \Rightarrow b = 528\pi$

$y = 2\sin(528\pi x)$

(b) $f = \dfrac{1}{1/264}$

$= 264$ cycles per second.

99. $f(x) = \tan x$

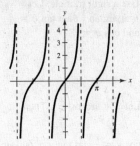

101. $f(x) = \cot x$

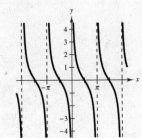

103. $f(x) = \sec x$

Graph $y = \cos x$ first.

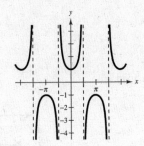

105. $f(x) = \csc x$

Graph $y = \sin x$ first.

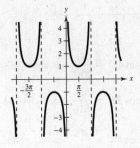

107. $f(x) = x \cos x$

Graph $y = x$ and $y = -x$ first.

As $x \to \infty, f(x) \to \infty$.

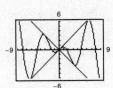

109. $\arcsin\left(-\dfrac{1}{2}\right) = -\arcsin\dfrac{1}{2} = -\dfrac{\pi}{6}$

111. $\arcsin 0.4 \approx 0.41$ radian

113. $\sin^{-1}(-0.44) \approx -0.46$ radian

115. $\arccos\dfrac{\sqrt{3}}{2} = \dfrac{\pi}{6}$

117. $\cos^{-1}(-1) = \pi$

119. $\arccos 0.324 \approx 1.24$ radians

121. $\tan^{-1}(-1.5) \approx -0.98$ radian

123. $f(x) = 2\arcsin x = 2\sin^{-1}(x)$

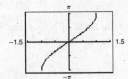

125. $f(x) = \arctan\left(\dfrac{x}{2}\right) = \tan^{-1}\left(\dfrac{x}{2}\right)$

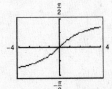

127. $\cos\left(\arctan \frac{3}{4}\right) = \frac{4}{5}$

Use a right triangle. Let
$\theta = \arctan \frac{3}{4}$ then $\tan \theta = \frac{3}{4}$
and $\cos \theta = \frac{4}{5}$.

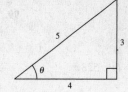

129. $\sec\left(\arctan \frac{12}{5}\right) = \frac{13}{5}$

Use a right triangle. Let $\theta = \arctan \frac{12}{5}$
then $\tan \theta = \frac{12}{5}$ and $\sec \theta = \frac{13}{5}$.

131. Let $y = \arccos\left(\frac{x}{2}\right)$. Then

$\cos y = \dfrac{x}{2}$ and $\tan y = \tan\left(\arccos\left(\dfrac{x}{2}\right)\right) = \dfrac{\sqrt{4-x^2}}{x}$.

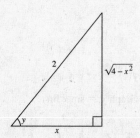

133. $\tan \theta = \dfrac{70}{30}$

$\theta = \arctan\left(\dfrac{70}{30}\right) \approx 66.8°$

135. $\sin 48° = \dfrac{d_1}{650} \implies d_1 \approx 483$

$\cos 25° = \dfrac{d_2}{810} \implies d_2 \approx 734$ $\left.\begin{array}{c} \\ \\ \end{array}\right\}$ $d_1 + d_2 \approx 1217$

$\cos 48° = \dfrac{d_3}{650} \implies d_3 \approx 435$

$\sin 25° = \dfrac{d_4}{810} \implies d_4 \approx 342$ $\left.\begin{array}{c} \\ \\ \end{array}\right\}$ $d_3 - d_4 \approx 93$

$\tan \theta \approx \dfrac{93}{1217} \implies \theta \approx 4.4°$

$\sec 4.4° \approx \dfrac{D}{1217} \implies D \approx 1217 \sec 4.4° \approx 1221$

The distance is 1221 miles and the bearing is 85.6°.

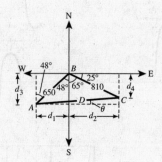

137. False. The sine or cosine
functions are often useful
for modeling simple harmonic
motion.

139. False. For each θ there
corresponds exactly one
value of y.

141. $y = 3 \sin x$

Amplitude: 3

Period: 2π

Matches graph (d).

143. $y = 2 \sin \pi x$

Amplitude: 2

Period: 2

Matches graph (b).

145. $f(\theta) = \sec \theta$ is undefined at the zeros of $g(\theta) = \cos \theta$

since $\sec \theta = \dfrac{1}{\cos \theta}$.

147. The ranges for the other four trigonometric functions are not bounded. For $y = \tan x$ and $y = \cot x$, the range is $(-\infty, \infty)$. For $y = \sec x$ and $y = \csc x$, the range is $(-\infty, -1] \cup [1, \infty)$.

149. $A = \frac{1}{2}r^2\theta, s = r\theta$

(a) $A = \frac{1}{2}r^2(0.8) = 0.4r^2, r > 0$

$s = r(0.8) = 0.8r, r > 0$

As r increases, the area function increases more rapidly.

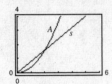

(b) $A = \frac{1}{2}(10)^2\theta = 50\theta, \theta > 0$

$s = 10\theta, \theta > 0$

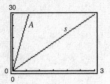

Problem Solving for Chapter 4

1. (a) $8:57 - 6:45 = 2$ hours 12 minutes $= 132$ minutes

$\frac{132}{48} = \frac{11}{4}$ revolutions

$\theta = \left(\frac{11}{4}\right)(2\pi) = \frac{11\pi}{2}$ radians or 990°

(b) $s = r\theta = 47.25(5.5\pi) \approx 816.42$ feet

3. (a) $\sin 39° = \dfrac{3000}{d}$

$d = \dfrac{3000}{\sin 39°} \approx 4767$ feet

(b) $\tan 39° = \dfrac{3000}{x}$

$x = \dfrac{3000}{\tan 39°} \approx 3705$ feet

(c) $\tan 63° = \dfrac{w + 3705}{3000}$

$3000 \tan 63° = w + 3705$

$w = 3000 \tan 63° - 3705 \approx 2183$ feet

5. (a) $h(x) = \cos^2 x$

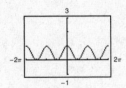

h is even.

(b) $h(x) = \sin^2 x$

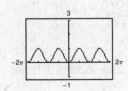

h is even.

7. If we alter the model so that $h = 1$ when $t = 0$, we can use either a sine or a cosine model.

$a = \frac{1}{2}[\text{max} - \text{min}] = \frac{1}{2}[101 - 1] = 50$

$d = \frac{1}{2}[\text{max} + \text{min}] = \frac{1}{2}[101 + 1] = 51$

$b = 8\pi$

For the cosine model we have: $h = 51 - 50\cos(8\pi t)$

For the sine model we have: $h = 51 - 50\sin\left(8\pi t + \dfrac{\pi}{2}\right)$

Notice that we needed the horizontal shift so that the sine value was one when $t = 0$.

Another model would be: $h = 51 + 50\sin\left(8\pi t + \dfrac{3\pi}{2}\right)$

Here we wanted the sine value to be 1 when $t = 0$.

9. Physical (23 days): $P = \sin\dfrac{2\pi t}{23}, t \geq 0$

Emotional (28 days): $E = \sin\dfrac{2\pi t}{28}, t \geq 0$

Intellectual (33 days): $I = \sin\dfrac{2\pi t}{33}, t \geq 0$

(a)

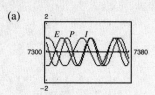

(b) Number of days since birth until September 1, 2006:

$$t = \underbrace{365 \times 20}_{\text{20 years}} + \underbrace{5}_{\text{leap years}} + \underbrace{11}_{\substack{\text{remaining} \\ \text{July days}}} + \underbrace{31}_{\text{August days}} + \underbrace{1}_{\substack{\text{day in} \\ \text{September}}}$$

$t = 7348$

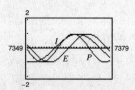

All three drop early in the month, then peak toward the middle of the month, and drop again toward the latter part of the month.

(c) For September 22, 2006, use $t = 7369$.

$P \approx 0.631$

$E \approx 0.901$

$I \approx 0.945$

11. (a) Both graphs have a period of 2 and intersect when $x = 5.35$. They should also intersect when $x = 5.35 - 2 = 3.35$ and $x = 5.35 + 2 = 7.35$.

(b) The graphs intersect when $x = 5.35 - 3(2) = -0.65$.

(c) Since $13.35 = 5.35 + 4(2)$ and $-4.65 = 5.35 - 5(2)$ the graphs will intersect again at these values. Therefore $f(13.35) = g(-4.65)$.

13.

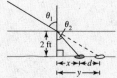

(a) $\dfrac{\sin \theta_1}{\sin \theta_2} = 1.333$

$\sin \theta_2 = \dfrac{\sin \theta_1}{1.333} = \dfrac{\sin 60°}{1.333} \approx 0.6497$

$\theta_2 \approx 40.52°$

(b) $\tan \theta_2 = \dfrac{x}{2} \implies x = 2 \tan 40.52° \approx 1.71$ feet

$\tan \theta_1 = \dfrac{y}{2} \implies y = 2 \tan 60° \approx 3.46$ feet

(c) $d = y - x = 3.46 - 1.71 = 1.75$ feet

(d) As you more closer to the rock, θ_1 decreases, which causes y to decrease, which in turn causes d to decrease.

Chapter 4   Practice Test

1. Express 350° in radian measure.

2. Express $(5\pi)/9$ in degree measure.

3. Convert $135°\,14'\,12''$ to decimal form.

4. Convert $-22.569°$ to D° M′ S″ form.

5. If $\cos\theta = \frac{2}{3}$, use the trigonometric identities to find $\tan\theta$.

6. Find θ given $\sin\theta = 0.9063$.

7. Solve for x in the figure below.

8. Find the reference angle θ' for $\theta = (6\pi)/5$.

9. Evaluate csc 3.92.

10. Find $\sec\theta$ given that θ lies in Quadrant III and $\tan\theta = 6$.

11. Graph $y = 3\sin\dfrac{x}{2}$.

12. Graph $y = -2\cos(x - \pi)$.

13. Graph $y = \tan 2x$.

14. Graph $y = -\csc\left(x + \dfrac{\pi}{4}\right)$.

15. Graph $y = 2x + \sin x$, using a graphing calculator.

16. Graph $y = 3x\cos x$, using a graphing calculator.

17. Evaluate arcsin 1.

18. Evaluate arctan(-3).

19. Evaluate $\sin\left(\arccos\dfrac{4}{\sqrt{35}}\right)$.

20. Write an algebraic expression for $\cos\left(\arcsin\dfrac{x}{4}\right)$.

For Exercises 21–23, solve the right triangle.

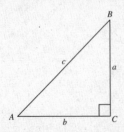

21. $A = 40°$, $c = 12$

22. $B = 6.84°$, $a = 21.3$

23. $a = 5$, $b = 9$

24. A 20-foot ladder leans against the side of a barn. Find the height of the top of the ladder if the angle of elevation of the ladder is 67°.

25. An observer in a lighthouse 250 feet above sea level spots a ship off the shore. If the angle of depression to the ship is 5°, how far out is the ship?

C H A P T E R   5
Analytic Trigonometry

Section 5.1   Using Fundamental Identities 241

Section 5.2   Verifying Trigonometric Identities 248

Section 5.3   Solving Trigonometric Equations 252

Section 5.4   Sum and Difference Formulas 258

Section 5.5   Multiple-Angle and Product-to-Sum Formulas 267

Review Exercises 279

Problem Solving 286

Practice Test 289

CHAPTER 5
Analytic Trigonometry

Section 5.1   Using Fundamental Identities

■ You should know the fundamental trigonometric identities.

(a) Reciprocal Identities

$$\sin u = \frac{1}{\csc u} \qquad\qquad \csc u = \frac{1}{\sin u}$$

$$\cos u = \frac{1}{\sec u} \qquad\qquad \sec u = \frac{1}{\cos u}$$

$$\tan u = \frac{1}{\cot u} = \frac{\sin u}{\cos u} \qquad\qquad \cot u = \frac{1}{\tan u} = \frac{\cos u}{\sin u}$$

(b) Pythagorean Identities

$$\sin^2 u + \cos^2 u = 1$$

$$1 + \tan^2 u = \sec^2 u$$

$$1 + \cot^2 u = \csc^2 u$$

(c) Cofunction Identities

$$\sin\left(\frac{\pi}{2} - u\right) = \cos u \qquad\qquad \cos\left(\frac{\pi}{2} - u\right) = \sin u$$

$$\tan\left(\frac{\pi}{2} - u\right) = \cot u \qquad\qquad \cot\left(\frac{\pi}{2} - u\right) = \tan u$$

$$\sec\left(\frac{\pi}{2} - u\right) = \csc u \qquad\qquad \csc\left(\frac{\pi}{2} - u\right) = \sec u$$

(d) Even/Odd Identities

$$\sin(-x) = -\sin x \qquad\qquad \csc(-x) = -\csc x$$

$$\cos(-x) = \cos x \qquad\qquad \sec(-x) = \sec x$$

$$\tan(-x) = -\tan x \qquad\qquad \cot(-x) = -\cot x$$

■ You should be able to use these fundamental identities to find function values.

■ You should be able to convert trigonometric expressions to equivalent forms by using the fundamental identities.

Vocabulary Check

1. $\tan u$

2. $\cos u$

3. $\cot u$

4. $\csc u$

5. $\cot^2 u$

6. $\sec^2 u$

7. $\cos u$

8. $\csc u$

9. $\cos u$

10. $-\tan u$

1. $\sin x = \dfrac{\sqrt{3}}{2}$, $\cos x = -\dfrac{1}{2} \implies x$ is in Quadrant II.

$\tan x = \dfrac{\sin x}{\cos x} = \dfrac{\sqrt{3}/2}{-1/2} = -\sqrt{3}$

$\cot x = \dfrac{1}{\tan x} = -\dfrac{1}{\sqrt{3}} = -\dfrac{\sqrt{3}}{3}$

$\sec x = \dfrac{1}{\cos x} = \dfrac{1}{-1/2} = -2$

$\csc x = \dfrac{1}{\sin x} = \dfrac{1}{\sqrt{3}/2} = \dfrac{2}{\sqrt{3}} = \dfrac{2\sqrt{3}}{3}$

3. $\sec \theta = \sqrt{2}$, $\sin \theta = -\dfrac{\sqrt{2}}{2} \implies \theta$ is in Quadrant IV.

$\cos \theta = \dfrac{1}{\sec \theta} = \dfrac{1}{\sqrt{2}} = \dfrac{\sqrt{2}}{2}$

$\tan \theta = \dfrac{\sin \theta}{\cos \theta} = \dfrac{-\sqrt{2}/2}{\sqrt{2}/2} = -1$

$\cot \theta = \dfrac{1}{\tan \theta} = -1$

$\csc \theta = \dfrac{1}{\sin \theta} = -\sqrt{2}$

5. $\tan x = \dfrac{5}{12}$, $\sec x = -\dfrac{13}{12} \implies x$ is in

Quadrant III.

$\cos x = \dfrac{1}{\sec x} = -\dfrac{12}{13}$

$\sin x = -\sqrt{1 - \cos^2 x} = -\sqrt{1 - \dfrac{144}{169}} = -\dfrac{5}{13}$

$\cot x = \dfrac{1}{\tan x} = \dfrac{12}{5}$

$\csc x = \dfrac{1}{\sin x} = -\dfrac{13}{5}$

7. $\sec \phi = \dfrac{3}{2}$, $\csc \phi = -\dfrac{3\sqrt{5}}{5} \implies \phi$ is in Quadrant IV.

$\sin \phi = \dfrac{1}{\csc \phi} = \dfrac{1}{-3\sqrt{5}/5} = -\dfrac{\sqrt{5}}{3}$

$\cos \phi = \dfrac{1}{\sec \phi} = \dfrac{1}{3/2} = \dfrac{2}{3}$

$\tan \phi = \dfrac{\sin \phi}{\cos \phi} = \dfrac{-\sqrt{5}/3}{2/3} = -\dfrac{\sqrt{5}}{2}$

$\cot \phi = \dfrac{1}{\tan \phi} = \dfrac{1}{-\sqrt{5}/2} = -\dfrac{2}{\sqrt{5}} = -\dfrac{2\sqrt{5}}{5}$

9. $\sin(-x) = -\dfrac{1}{3} \implies \sin x = \dfrac{1}{3}$, $\tan x = -\dfrac{\sqrt{2}}{4} \implies x$ is

in Quadrant II.

$\cos x = -\sqrt{1 - \sin^2 x} = -\sqrt{1 - \dfrac{1}{9}} = -\dfrac{2\sqrt{2}}{3}$

$\cot x = \dfrac{1}{\tan x} = \dfrac{1}{-\sqrt{2}/4} = -2\sqrt{2}$

$\sec x = \dfrac{1}{\cos x} = \dfrac{1}{-2\sqrt{2}/3} = -\dfrac{3\sqrt{2}}{4}$

$\csc x = \dfrac{1}{\sin x} = \dfrac{1}{1/3} = 3$

11. $\tan \theta = 2$, $\sin \theta < 0 \implies \theta$ is in Quadrant III.

$\sec \theta = -\sqrt{\tan^2 \theta + 1} = -\sqrt{4 + 1} = -\sqrt{5}$

$\cos \theta = \dfrac{1}{\sec \theta} = -\dfrac{1}{\sqrt{5}} = -\dfrac{\sqrt{5}}{5}$

$\sin \theta = -\sqrt{1 - \cos^2 \theta}$

$ = -\sqrt{1 - \dfrac{1}{5}} = -\dfrac{2}{\sqrt{5}} = -\dfrac{2\sqrt{5}}{5}$

$\csc \theta = \dfrac{1}{\sin \theta} = -\dfrac{\sqrt{5}}{2}$

$\cot \theta = \dfrac{1}{\tan \theta} = \dfrac{1}{2}$

13. $\sin \theta = -1$, $\cot \theta = 0 \implies \theta = \dfrac{3\pi}{2}$

$\cos \theta = \sqrt{1 - \sin^2 \theta} = 0$

$\sec \theta$ is undefined.

$\tan \theta$ is undefined.

$\csc \theta = -1$

15. $\sec x \cos x = \sec x \cdot \dfrac{1}{\sec x} = 1$

The expression is matched with (d).

17. $\cot^2 x - \csc^2 x = \cot^2 x - (1 + \cot^2 x) = -1$

The expression is matched with (b).

19. $\dfrac{\sin(-x)}{\cos(-x)} = \dfrac{-\sin x}{\cos x} = -\tan x$

The expression is matched with (e).

21. $\sin x \sec x = \sin x \cdot \dfrac{1}{\cos x} = \tan x$

The expression is matched with (b).

23. $\sec^4 x - \tan^4 x = (\sec^2 x + \tan^2 x)(\sec^2 x - \tan^2 x)$

$$= (\sec^2 x + \tan^2 x)(1) = \sec^2 x + \tan^2 x$$

The expression is matched with (f).

25. $\dfrac{\sec^2 x - 1}{\sin^2 x} = \dfrac{\tan^2 x}{\sin^2 x} = \dfrac{\sin^2 x}{\cos^2 x} \cdot \dfrac{1}{\sin^2 x} = \sec^2 x$

The expression is matched with (e).

27. $\cot \theta \sec \theta = \dfrac{\cos \theta}{\sin \theta} \cdot \dfrac{1}{\cos \theta} = \dfrac{1}{\sin \theta} = \csc \theta$

29. $\sin \phi \, (\csc \phi - \sin \phi) = (\sin \phi)\dfrac{1}{\sin \phi} - \sin^2 \phi$

$$= 1 - \sin^2 \phi = \cos^2 \phi$$

31. $\dfrac{\cot x}{\csc x} = \dfrac{\cos x / \sin x}{1/\sin x}$

$$= \dfrac{\cos x}{\sin x} \cdot \dfrac{\sin x}{1} = \cos x$$

33. $\dfrac{1 - \sin^2 x}{\csc^2 x - 1} = \dfrac{\cos^2 x}{\cot^2 x} = \cos^2 x \tan^2 x = (\cos^2 x)\dfrac{\sin^2 x}{\cos^2 x}$

$$= \sin^2 x$$

35. $\sec \alpha \dfrac{\sin \alpha}{\tan \alpha} = \dfrac{1}{\cos \alpha}(\sin \alpha) \cot \alpha$

$$= \dfrac{1}{\cos \alpha}(\sin \alpha)\left(\dfrac{\cos \alpha}{\sin \alpha}\right) = 1$$

37. $\cos\left(\dfrac{\pi}{2} - x\right) \sec x = (\sin x)(\sec x) = (\sin x)\left(\dfrac{1}{\cos x}\right) = \dfrac{\sin x}{\cos x} = \tan x$

39. $\dfrac{\cos^2 y}{1 - \sin y} = \dfrac{1 - \sin^2 y}{1 - \sin y} = \dfrac{(1 + \sin y)(1 - \sin y)}{1 - \sin y} = 1 + \sin y$

41. $\sin \beta \tan \beta + \cos \beta = (\sin \beta)\dfrac{\sin \beta}{\cos \beta} + \cos \beta$

$$= \dfrac{\sin^2 \beta}{\cos \beta} + \dfrac{\cos^2 \beta}{\cos \beta}$$

$$= \dfrac{\sin^2 \beta + \cos^2 \beta}{\cos \beta}$$

$$= \dfrac{1}{\cos \beta}$$

$$= \sec \beta$$

43. $\cot u \sin u + \tan u \cos u = \dfrac{\cos u}{\sin u}(\sin u) + \dfrac{\sin u}{\cos u}(\cos u)$

$$= \cos u + \sin u$$

45. $\tan^2 x - \tan^2 x \sin^2 x = \tan^2 x(1 - \sin^2 x)$

$$= \tan^2 x \cos^2 x$$

$$= \dfrac{\sin^2 x}{\cos^2 x} \cdot \cos^2 x$$

$$= \sin^2 x$$

47. $\sin^2 x \sec^2 x - \sin^2 x = \sin^2 x(\sec^2 x - 1)$

$$= \sin^2 x \tan^2 x$$

49. $\dfrac{\sec^2 x - 1}{\sec x - 1} = \dfrac{(\sec x + 1)(\sec x - 1)}{\sec x - 1} = \sec x + 1$

51. $\tan^4 x + 2 \tan^2 x + 1 = (\tan^2 x + 1)^2$

$$= (\sec^2 x)^2$$

$$= \sec^4 x$$

53. $\sin^4 x - \cos^4 x = (\sin^2 x + \cos^2 x)(\sin^2 x - \cos^2 x)$

$$= (1)(\sin^2 x - \cos^2 x)$$

$$= \sin^2 x - \cos^2 x$$

55. $\csc^3 x - \csc^2 x - \csc x + 1 = \csc^2 x(\csc x - 1) - 1(\csc x - 1)$

$$= (\csc^2 x - 1)(\csc x - 1)$$

$$= \cot^2 x(\csc x - 1)$$

57. $(\sin x + \cos x)^2 = \sin^2 x + 2 \sin x \cos x + \cos^2 x$

$$= (\sin^2 x + \cos^2 x) + 2 \sin x \cos x$$

$$= 1 + 2 \sin x \cos x$$

59. $(2 \csc x + 2)(2 \csc x - 2) = 4 \csc^2 x - 4$

$$= 4(\csc^2 x - 1)$$

$$= 4 \cot^2 x$$

61. $\dfrac{1}{1 + \cos x} + \dfrac{1}{1 - \cos x} = \dfrac{1 - \cos x + 1 + \cos x}{(1 + \cos x)(1 - \cos x)}$

$$= \dfrac{2}{1 - \cos^2 x}$$

$$= \dfrac{2}{\sin^2 x}$$

$$= 2 \csc^2 x$$

63. $\dfrac{\cos x}{1 + \sin x} + \dfrac{1 + \sin x}{\cos x} = \dfrac{\cos^2 x + (1 + \sin x)^2}{\cos x(1 + \sin x)} = \dfrac{\cos^2 x + 1 + 2 \sin x + \sin^2 x}{\cos x(1 + \sin x)}$

$$= \dfrac{2 + 2 \sin x}{\cos x(1 + \sin x)}$$

$$= \dfrac{2(1 + \sin x)}{\cos x(1 + \sin x)}$$

$$= \dfrac{2}{\cos x}$$

$$= 2 \sec x$$

65. $\dfrac{\sin^2 y}{1 - \cos y} = \dfrac{1 - \cos^2 y}{1 - \cos y}$

$$= \dfrac{(1 + \cos y)(1 - \cos y)}{1 - \cos y}$$

$$= 1 + \cos y$$

67. $\dfrac{3}{\sec x - \tan x} \cdot \dfrac{\sec x + \tan x}{\sec x + \tan x} = \dfrac{3(\sec x + \tan x)}{\sec^2 x - \tan^2 x}$

$$= \dfrac{3(\sec x + \tan x)}{1}$$

$$= 3(\sec x + \tan x)$$

69. $y_1 = \cos\left(\dfrac{\pi}{2} - x\right),\ y_2 = \sin x$

x	0.2	0.4	0.6	0.8	1.0	1.2	1.4
y_1	0.1987	0.3894	0.5646	0.7174	0.8415	0.9320	0.9854
y_2	0.1987	0.3894	0.5646	0.7174	0.8415	0.9320	0.9854

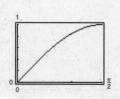

Conclusion: $y_1 = y_2$

71. $y_1 = \dfrac{\cos x}{1 - \sin x}$, $y_2 = \dfrac{1 + \sin x}{\cos x}$

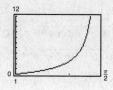

x	0.2	0.4	0.6	0.8	1.0	1.2	1.4
y_1	1.2230	1.5085	1.8958	2.4650	3.4082	5.3319	11.6814
y_2	1.2230	1.5085	1.8958	2.4650	3.4082	5.3319	11.6814

Conclusion: $y_1 = y_2$

73. $y_1 = \cos x \cot x + \sin x = \csc x$

$$\cos x \cot x + \sin x = \cos x\left(\frac{\cos x}{\sin x}\right) + \sin x$$

$$= \frac{\cos^2 x}{\sin x} + \frac{\sin^2 x}{\sin x}$$

$$= \frac{\cos^2 x + \sin^2 x}{\sin x} = \frac{1}{\sin x} = \csc x$$

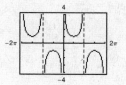

75. $y_1 = \dfrac{1}{\sin x}\left(\dfrac{1}{\cos x} - \cos x\right) = \tan x$

$$\frac{1}{\sin x}\left(\frac{1}{\cos x} - \cos x\right) = \frac{1}{\sin x \cos x} - \frac{\cos x}{\sin x}$$

$$= \frac{1 - \cos^2 x}{\sin x \cos x} = \frac{\sin^2 x}{\sin x \cos x} = \frac{\sin x}{\cos x} = \tan x$$

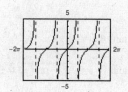

77. Let $x = 3\cos\theta$, then

$$\sqrt{9 - x^2} = \sqrt{9 - (3\cos\theta)^2} = \sqrt{9 - 9\cos^2\theta} = \sqrt{9(1 - \cos^2\theta)}$$
$$= \sqrt{9\sin^2\theta} = 3\sin\theta.$$

79. Let $x = 3\sec\theta$, then

$$\sqrt{x^2 - 9} = \sqrt{(3\sec\theta)^2 - 9}$$
$$= \sqrt{9\sec^2\theta - 9}$$
$$= \sqrt{9(\sec^2\theta - 1)}$$
$$= \sqrt{9\tan^2\theta}$$
$$= 3\tan\theta.$$

81. Let $x = 5\tan\theta$, then

$$\sqrt{x^2 + 25} = \sqrt{(5\tan\theta)^2 + 25}$$
$$= \sqrt{25\tan^2\theta + 25}$$
$$= \sqrt{25(\tan^2\theta + 1)}$$
$$= \sqrt{25\sec^2\theta}$$
$$= 5\sec\theta.$$

83. Let $x = 3\sin\theta$, then $\sqrt{9 - x^2} = 3$ becomes

$$\sqrt{9 - (3\sin\theta)^2} = 3$$
$$\sqrt{9 - 9\sin^2\theta} = 3$$
$$\sqrt{9(1 - \sin^2\theta)} = 3$$
$$\sqrt{9\cos^2\theta} = 3$$
$$3\cos\theta = 3$$
$$\cos\theta = 1$$
$$\sin\theta = \sqrt{1 - \cos^2\theta} = \sqrt{1 - (1)^2} = 0.$$

85. Let $x = 2\cos\theta$, then $\sqrt{16 - 4x^2} = 2\sqrt{2}$ becomes

$$\sqrt{16 - 4(2\cos\theta)^2} = 2\sqrt{2}$$

$$\sqrt{16 - 16\cos^2\theta} = 2\sqrt{2}$$

$$\sqrt{16(1 - \cos^2\theta)} = 2\sqrt{2}$$

$$\sqrt{16\sin^2\theta} = 2\sqrt{2}$$

$$4\sin\theta = 2\sqrt{2}$$

$$\sin\theta = \frac{\sqrt{2}}{2}$$

$$\cos\theta = \sqrt{1 - \sin^2\theta}$$

$$= \sqrt{1 - \frac{1}{2}}$$

$$= \sqrt{\frac{1}{2}}$$

$$= \frac{\sqrt{2}}{2}.$$

87. $\sin\theta = \sqrt{1 - \cos^2\theta}$

Let $y_1 = \sin x$ and $y_2 = \sqrt{1 - \cos^2 x}$, $0 \le x \le 2\pi$.

$y_1 = y_2$ for $0 \le x \le \pi$, so we have

$\sin\theta = \sqrt{1 - \cos^2\theta}$ for $0 \le \theta \le \pi$.

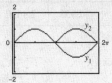

89. $\sec\theta = \sqrt{1 + \tan^2\theta}$

Let $y_1 = \dfrac{1}{\cos x}$ and $y_2 = \sqrt{1 + \tan^2 x}$, $0 \le x \le 2\pi$.

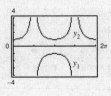

$y_1 = y_2$ for $0 \le x < \dfrac{\pi}{2}$ and $\dfrac{3\pi}{2} < x \le 2\pi$, so we have

$\sec\theta = \sqrt{1 + \tan^2\theta}$ for $0 \le \theta < \dfrac{\pi}{2}$ and $\dfrac{3\pi}{2} < \theta < 2\pi$.

91. $\ln|\cos x| - \ln|\sin x| = \ln\dfrac{|\cos x|}{|\sin x|} = \ln|\cot x|$

93. $\ln|\cot t| + \ln(1 + \tan^2 t) = \ln\left[|\cot t|(1 + \tan^2 t)\right]$

$$= \ln|\cot t \sec^2 t|$$

$$= \ln\left|\frac{\cos t}{\sin t} \cdot \frac{1}{\cos^2 t}\right|$$

$$= \ln\left|\frac{1}{\sin t \cos t}\right| = \ln|\csc t \sec t|$$

95. (a) $\csc^2 132° - \cot^2 132° \approx 1.8107 - 0.8107 = 1$

(b) $\csc^2\dfrac{2\pi}{7} - \cot^2\dfrac{2\pi}{7} \approx 1.6360 - 0.6360 = 1$

97. $\cos\left(\dfrac{\pi}{2} - \theta\right) = \sin\theta$

(a) $\theta = 80°$

$$\cos(90° - 80°) = \sin 80°$$

$$0.9848 = 0.9848$$

(b) $\theta = 0.8$

$$\cos\left(\frac{\pi}{2} - 0.8\right) = \sin 0.8$$

$$0.7174 = 0.7174$$

99. $\mu W\cos\theta = W\sin\theta$

$$\mu = \frac{W\sin\theta}{W\cos\theta} = \tan\theta$$

101. True. For example, $\sin(-x) = -\sin x$ means that the graph of $\sin x$ is symmetric about the origin.

103. As $x \to \dfrac{\pi^-}{2}$, $\sin x \to 1$ and $\csc x \to 1$.

105. As $x \to \dfrac{\pi^-}{2}$, $\tan x \to \infty$ and $\cot x \to 0$.

107. $\cos \theta = \sqrt{1 - \sin^2 \theta}$ *is not* an identity.

$\cos^2 \theta + \sin^2 \theta = 1 \implies \cos \theta = \pm\sqrt{1 - \sin^2 \theta}$

109. $\dfrac{\sin k\theta}{\cos k\theta} = \tan \theta$ *is not* an identity.

$\dfrac{\sin k\theta}{\cos k\theta} = \tan k\theta$

111. $\sin \theta \csc \theta = 1$ *is* an identity.

$\sin \theta \cdot \dfrac{1}{\sin \theta} = 1$, provided $\sin \theta \neq 0$.

113. Let (x, y) be any point on the terminal side of θ.

Then, $r = \sqrt{x^2 + y^2}$ and

$$\sin^2 \theta + \cos^2 \theta = \left(\dfrac{y}{r}\right)^2 + \left(\dfrac{x}{r}\right)^2$$

$$= \dfrac{y^2 + x^2}{r^2}$$

$$= \dfrac{r^2}{r^2}$$

$$= 1.$$

115. $\left(\sqrt{x} + 5\right)\left(\sqrt{x} - 5\right) = \left(\sqrt{x}\right)^2 - (5)^2 = x - 25$

117. $\dfrac{1}{x + 5} + \dfrac{x}{x - 8} = \dfrac{(x - 8) + x(x + 5)}{(x + 5)(x - 8)}$

$$= \dfrac{x^2 + 6x - 8}{(x + 5)(x - 8)}$$

119. $\dfrac{2x}{x^2 - 4} - \dfrac{7}{x + 4} = \dfrac{2x(x + 4) - 7(x^2 - 4)}{(x^2 - 4)(x + 4)}$

$$= \dfrac{2x^2 + 8x - 7x^2 + 28}{(x^2 - 4)(x + 4)}$$

$$= \dfrac{-5x^2 + 8x + 28}{(x^2 - 4)(x + 4)}$$

121. $f(x) = \dfrac{1}{2} \sin(\pi x)$

Amplitude: $\dfrac{1}{2}$

Period: $\dfrac{2\pi}{\pi} = 2$

Key points:

$(0, 0), \left(\dfrac{1}{2}, \dfrac{1}{2}\right), (1, 0), \left(\dfrac{3}{2}, -\dfrac{1}{2}\right), (2, 0)$

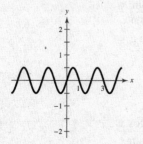

123. $f(x) = \dfrac{1}{2} \sec\left(x + \dfrac{\pi}{4}\right)$

Sketch the graph of

$y = \dfrac{1}{2} \cos\left(x + \dfrac{\pi}{4}\right)$ first.

Amplitude: $\dfrac{1}{2}$

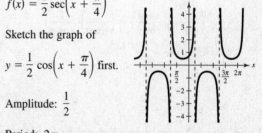

Period: 2π

One cycle: $x + \dfrac{\pi}{4} = 0 \implies x = -\dfrac{\pi}{4}$

$x + \dfrac{\pi}{4} = 2\pi \implies x = \dfrac{7\pi}{4}$

The x-intercepts of $y = \dfrac{1}{2} \cos\left(x + \dfrac{\pi}{4}\right)$ correspond to the

vertical asymptotes of $f(x) = \dfrac{1}{2} \sec\left(x + \dfrac{\pi}{4}\right)$.

$x = \dfrac{\pi}{4}, x = \dfrac{5\pi}{4}, \ldots$

Section 5.2   Verifying Trigonometric Identities

■ You should know the difference between an expression, a conditional equation, and an identity.

■ You should be able to solve trigonometric identities, using the following techniques.

  (a) Work with *one* side at a time. Do not "cross" the equal sign.

  (b) Use algebraic techniques such as combining fractions, factoring expressions, rationalizing denominators, and squaring binomials.

  (c) Use the fundamental identities.

  (d) Convert all the terms into sines and cosines.

Vocabulary Check

1. identity

2. conditional equation

3. $\tan u$

4. $\cot u$

5. $\cos^2 u$

6. $\sin u$

7. $-\csc u$

8. $\sec u$

---

1. $\sin t \csc t = \sin t \left(\dfrac{1}{\sin t} \right) = 1$

3. $(1 + \sin \alpha)(1 - \sin \alpha) = 1 - \sin^2 \alpha = \cos^2 \alpha$

5. $\cos^2 \beta - \sin^2 \beta = (1 - \sin^2 \beta) - \sin^2 \beta$
$$= 1 - 2\sin^2 \beta$$

7. $\sin^2 \alpha - \sin^4 \alpha = \sin^2 \alpha (1 - \sin^2 \alpha)$
$$= (1 - \cos^2 \alpha)(\cos^2 \alpha)$$
$$= \cos^2 \alpha - \cos^4 \alpha$$

9. $\dfrac{\csc^2 \theta}{\cot \theta} = \csc^2 \theta \left(\dfrac{1}{\cot \theta} \right) = \csc^2 \theta \tan \theta$
$$= \left(\dfrac{1}{\sin^2 \theta} \right) \left(\dfrac{\sin \theta}{\cos \theta} \right) = \left(\dfrac{1}{\sin \theta} \right) \left(\dfrac{1}{\cos \theta} \right)$$
$$= \csc \theta \sec \theta$$

11. $\dfrac{\cot^2 t}{\csc t} = \dfrac{\cos^2 t}{\sin^2 t} \cdot \sin t$
$$= \dfrac{\cos^2 t}{\sin t}$$
$$= \dfrac{1 - \sin^2 t}{\sin t} = \dfrac{1}{\sin t} - \dfrac{\sin^2 t}{\sin t}$$
$$= \csc t - \sin t$$

13. $\sin^{1/2} x \cos x - \sin^{5/2} x \cos x = \sin^{1/2} x \cos x (1 - \sin^2 x) = \sin^{1/2} x \cos x \cdot \cos^2 x = \cos^3 x \sqrt{\sin x}$

15. $\dfrac{1}{\sec x \tan x} = \cos x \cot x = \cos x \cdot \dfrac{\cos x}{\sin x}$
$$= \dfrac{\cos^2 x}{\sin x}$$
$$= \dfrac{1 - \sin^2 x}{\sin x}$$
$$= \dfrac{1}{\sin x} - \sin x$$
$$= \csc x - \sin x$$

17. $\csc x - \sin x = \dfrac{1}{\sin x} - \dfrac{\sin^2 x}{\sin x}$
$$= \dfrac{1 - \sin^2 x}{\sin x}$$
$$= \dfrac{\cos^2 x}{\sin x}$$
$$= \dfrac{\cos x}{1} \cdot \dfrac{\cos x}{\sin x}$$
$$= \cos x \cot x$$

19. $\dfrac{1}{\tan x} + \dfrac{1}{\cot x} = \dfrac{\cot x + \tan x}{\tan x \cot x}$

$\qquad = \dfrac{\cot x + \tan x}{1}$

$\qquad = \tan x + \cot x$

21. $\dfrac{\cos \theta \cot \theta}{1 - \sin \theta} - 1 = \dfrac{\cos \theta \cot \theta - (1 - \sin \theta)}{1 - \sin \theta}$

$\qquad = \dfrac{\cos \theta \left(\dfrac{\cos \theta}{\sin \theta} \right) - 1 + \sin \theta}{1 - \sin \theta} \cdot \dfrac{\sin \theta}{\sin \theta}$

$\qquad = \dfrac{\cos^2 \theta - \sin \theta + \sin^2 \theta}{\sin \theta (1 - \sin \theta)}$

$\qquad = \dfrac{1 - \sin \theta}{\sin \theta (1 - \sin \theta)}$

$\qquad = \dfrac{1}{\sin \theta}$

$\qquad = \csc \theta$

23. $\dfrac{1}{\sin x + 1} + \dfrac{1}{\csc x + 1} = \dfrac{\csc x + 1 + \sin x + 1}{(\sin x + 1)(\csc x + 1)}$

$\qquad = \dfrac{\sin x + \csc x + 2}{\sin x \csc x + \sin x + \csc x + 1}$

$\qquad = \dfrac{\sin x + \csc x + 2}{1 + \sin x + \csc x + 1}$

$\qquad = \dfrac{\sin x + \csc x + 2}{\sin x + \csc x + 2}$

$\qquad = 1$

25. $\tan \left(\dfrac{\pi}{2} - \theta \right) \tan \theta = \cot \theta \tan \theta = \left(\dfrac{1}{\tan \theta} \right) \tan \theta = 1$

27. $\dfrac{\csc(-x)}{\sec(-x)} = \dfrac{1/\sin(-x)}{1/\cos(-x)}$

$\qquad = \dfrac{\cos(-x)}{\sin(-x)}$

$\qquad = \dfrac{\cos x}{-\sin x}$

$\qquad = -\cot x$

29. $\dfrac{\tan x \cot x}{\cos x} = \dfrac{1}{\cos x} = \sec x$

31. $\dfrac{\tan x + \cot y}{\tan x \cot y} = \dfrac{\dfrac{1}{\cot x} + \dfrac{1}{\tan y}}{\dfrac{1}{\cot x} \cdot \dfrac{1}{\tan y}} \cdot \dfrac{\cot x \tan y}{\cot x \tan y}$

$\qquad = \tan y + \cot x$

33. $\sqrt{\dfrac{1 + \sin \theta}{1 - \sin \theta}} = \sqrt{\dfrac{1 + \sin \theta}{1 - \sin \theta} \cdot \dfrac{1 + \sin \theta}{1 + \sin \theta}}$

$\qquad = \sqrt{\dfrac{(1 + \sin \theta)^2}{1 - \sin^2 \theta}}$

$\qquad = \sqrt{\dfrac{(1 + \sin \theta)^2}{\cos^2 \theta}}$

$\qquad = \dfrac{1 + \sin \theta}{|\cos \theta|}$

35. $\cos^2 \beta + \cos^2 \left(\dfrac{\pi}{2} - \beta \right) = \cos^2 \beta + \sin^2 \beta = 1$

37. $\sin t \csc \left(\dfrac{\pi}{2} - t \right) = \sin t \sec t = \sin t \left(\dfrac{1}{\cos t} \right)$

$\qquad = \dfrac{\sin t}{\cos t} = \tan t$

39. (a)

(b)

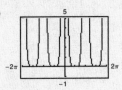

Identity

Let $y_1 = \dfrac{2}{(\cos x)^2} - \dfrac{2(\sin x)^2}{(\cos x)^2} - (\sin x)^2 - (\cos x)^2$ and $y_2 = 1$.

Identity

(c) $2 \sec^2 x - 2 \sec^2 x \sin^2 x - \sin^2 x - \cos^2 x = 2 \sec^2 x(1 - \sin^2 x) - (\sin^2 x + \cos^2 x)$

$$= 2 \sec^2 x(\cos^2 x) - 1$$

$$= 2 \cdot \frac{1}{\cos^2 x} \cdot \cos^2 x - 1$$

$$= 2 - 1$$

$$= 1$$

41. (a)

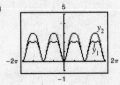

Let $y_1 = 2 + (\cos x)^2 - 3(\cos x)^4$ and

$y_2 = (\sin x)^2(3 + 2(\cos x)^2)$.

Not an identity

(b)

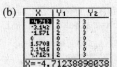

Not an identity

(c) $2 + \cos^2 x - 3 \cos^4 x = (1 - \cos^2 x)(2 + 3 \cos^2 x)$

$$= \sin^2 x(2 + 3 \cos^2 x)$$

$$\neq \sin^2 x(3 + 2 \cos^2 x)$$

43. (a)

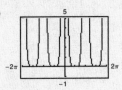

Let $y_1 = \dfrac{1}{(\sin x)^4} - \dfrac{2}{(\sin x)^2} + 1$ and $y_2 = \dfrac{1}{(\tan x)^4}$.

Identity

(b)

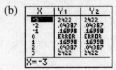

Identity

(c) $\csc^4 x - 2 \csc^2 x + 1 = (\csc^2 x - 1)^2$

$$= (\cot^2 x)^2 = \cot^4 x$$

45. (a)

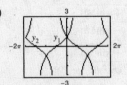

Let $y_1 = \dfrac{\cos x}{(1 - \sin x)}$ and $y_2 = \dfrac{(1 - \sin x)}{\cos x}$.

Not an identity

(b)

Not an identity

(c) $\dfrac{\cos x}{1 - \sin x} = \dfrac{\cos x}{1 - \sin x} \cdot \dfrac{1 + \sin x}{1 + \sin x}$

$$= \frac{\cos x(1 + \sin x)}{1 - \sin^2 x}$$

$$= \frac{\cos x(1 + \sin x)}{\cos^2 x}$$

$$= \frac{1 + \sin x}{\cos x}$$

47. $\tan^3 x \sec^2 x - \tan^3 x = \tan^3 x(\sec^2 x - 1)$
$$= \tan^3 x \tan^2 x$$
$$= \tan^5 x$$

49. $(\sin^2 x - \sin^4 x)\cos x = \sin^2 x(1 - \sin^2 x)\cos x$
$$= \sin^2 x \cos^2 x \cos x$$
$$= \sin^2 x \cos^3 x$$

51. $\sin^2 25° + \sin^2 65° = \sin^2 25° + \cos^2(90° - 65°)$
$$= \sin^2 25° + \cos^2 25°$$
$$= 1$$

53. $\cos^2 20° + \cos^2 52° + \cos^2 38° + \cos^2 70° = \cos^2 20° + \cos^2 52° + \sin^2(90° - 38°) + \sin^2(90° - 70°)$
$$= \cos^2 20° + \cos^2 52° + \sin^2 52° + \sin^2 20°$$
$$= (\cos^2 20° + \sin^2 20°) + (\cos^2 52° + \sin^2 52°)$$
$$= 1 + 1$$
$$= 2$$

55. $\cos x - \csc x \cot x = \cos x - \dfrac{1}{\sin x}\dfrac{\cos x}{\sin x}$
$$= \cos x\left(1 - \dfrac{1}{\sin^2 x}\right)$$
$$= \cos x(1 - \csc^2 x)$$
$$= -\cos x(\csc^2 x - 1)$$
$$= -\cos x \cot^2 x$$

57. False. For the equation to be an identity, it must be true for all values of θ in the domain.

59. Since $\sin^2 \theta = 1 - \cos^2 \theta$, then

$\sin \theta = \pm\sqrt{1 - \cos^2 \theta}$; $\sin \theta \neq \sqrt{1 - \cos^2 \theta}$ if θ lies in

Quadrant III or IV. One such angle is $\theta = \dfrac{7\pi}{4}$.

61. $(2 + 3i) - \sqrt{-26} = 2 + 3i - \sqrt{26}i = 2 + \left(3 - \sqrt{26}\right)i$

63. $\sqrt{-16}\left(1 + \sqrt{-4}\right) = 4i(1 + 2i)$
$$= 4i + 8i^2$$
$$= 4i - 8$$
$$= -8 + 4i$$

65. $x^2 + 6x - 12 = 0$
$$a = 1, b = 6, c = -12$$
$$x = \dfrac{-6 \pm \sqrt{6^2 - 4(1)(-12)}}{2(1)}$$
$$= \dfrac{-6 \pm \sqrt{36 + 48}}{2}$$
$$= \dfrac{-6 \pm \sqrt{84}}{2}$$
$$= \dfrac{-6 \pm 2\sqrt{21}}{2}$$
$$= -3 \pm \sqrt{21}$$

67. $3x^2 - 6x - 12 = 0$
$$3(x^2 - 2x - 4) = 0$$
$$x^2 - 2x - 4 = 0$$
$$a = 1, b = -2, c = -4$$
$$x = \dfrac{-(-2) \pm \sqrt{(-2)^2 - 4(1)(-4)}}{2(1)}$$
$$= \dfrac{2 \pm \sqrt{4 + 16}}{2}$$
$$= \dfrac{2 \pm \sqrt{20}}{2}$$
$$= \dfrac{2 \pm 2\sqrt{5}}{2}$$
$$= 1 \pm \sqrt{5}$$

Section 5.3   Solving Trigonometric Equations

- ■ You should be able to identify and solve trigonometric equations.
- ■ A trigonometric equation is a conditional equation. It is true for a specific set of values.
- ■ To solve trigonometric equations, use algebraic techniques such as collecting like terms, extracting square roots, factoring, squaring, converting to quadratic type, using formulas, and using inverse functions. Study the examples in this section.

Vocabulary Check

1. general   2. quadratic   3. extraneous

---

1. $2 \cos x - 1 = 0$

    (a) $2 \cos \dfrac{\pi}{3} - 1 = 2\left(\dfrac{1}{2}\right) - 1 = 0$
                                                  (b) $2 \cos \dfrac{5\pi}{3} - 1 = 2\left(\dfrac{1}{2}\right) - 1 = 0$

3. $3 \tan^2 2x - 1 = 0$
                                                         5. $2 \sin^2 x - \sin x - 1 = 0$

    (a) $3\left[\tan 2\left(\dfrac{\pi}{12}\right)\right]^2 - 1 = 3 \tan^2 \dfrac{\pi}{6} - 1$
                                       (a) $2 \sin^2 \dfrac{\pi}{2} - \sin \dfrac{\pi}{2} - 1 = 2(1)^2 - 1 - 1$

                $= 3\left(\dfrac{1}{\sqrt{3}}\right)^2 - 1$
                                                             $= 0$

                $= 0$
                                (b) $2 \sin^2 \dfrac{7\pi}{6} - \sin \dfrac{7\pi}{6} - 1 = 2\left(-\dfrac{1}{2}\right)^2 - \left(-\dfrac{1}{2}\right) - 1$

    (b) $3\left[\tan 2\left(\dfrac{5\pi}{12}\right)\right]^2 - 1 = 3 \tan^2 \dfrac{5\pi}{6} - 1$
                                                          $= \dfrac{1}{2} + \dfrac{1}{2} - 1$

                $= 3\left(-\dfrac{1}{\sqrt{3}}\right)^2 - 1$
                                                             $= 0$

                $= 0$

7. $2 \cos x + 1 = 0$
                        9. $\sqrt{3} \csc x - 2 = 0$
                        11. $3 \sec^2 x - 4 = 0$

      $2 \cos x = -1$
                              $\sqrt{3} \csc x = 2$
                               $\sec^2 x = \dfrac{4}{3}$

        $\cos x = -\dfrac{1}{2}$
                              $\csc x = \dfrac{2}{\sqrt{3}}$
                               $\sec x = \pm\dfrac{2}{\sqrt{3}}$

        $x = \dfrac{2\pi}{3} + 2n\pi$
                              $x = \dfrac{\pi}{3} + 2n\pi$
                               $x = \dfrac{\pi}{6} + n\pi$

     or $x = \dfrac{4\pi}{3} + 2n\pi$
                          or $x = \dfrac{2\pi}{3} + 2n\pi$
                           or $x = \dfrac{5\pi}{6} + n\pi$

13. $\sin x(\sin x + 1) = 0$
                                    15. $4 \cos^2 x - 1 = 0$

    $\sin x = 0$ or $\sin x = -1$
                                       $\cos^2 x = \dfrac{1}{4}$

      $x = n\pi$ $x = \dfrac{3\pi}{2} + 2n\pi$
                                       $\cos^2 x = \pm\dfrac{1}{2}$

                                                    $x = \dfrac{\pi}{3} + n\pi$ or $x = \dfrac{2\pi}{3} + n\pi$

17. $2 \sin^2 2x = 1$

$$\sin 2x = \pm \frac{1}{\sqrt{2}} = \pm \frac{\sqrt{2}}{2}$$

$$2x = \frac{\pi}{4} + 2n\pi, \ 2x = \frac{3\pi}{4} + 2n\pi,$$

$$2x = \frac{5\pi}{4} + 2n\pi, \ 2x = \frac{7\pi}{4} + 2n\pi.$$

Thus, $x = \frac{\pi}{8} + n\pi, \ \frac{3\pi}{8} + n\pi, \ \frac{5\pi}{8} + n\pi, \ \frac{7\pi}{8} + n\pi.$

We can combine these as follows:

$$x = \frac{\pi}{8} + \frac{n\pi}{2}, x = \frac{3\pi}{8} + \frac{n\pi}{2}$$

19. $\tan 3x(\tan x - 1) = 0$

$\tan 3x = 0$ or $\tan x - 1 = 0$

$3x = n\pi$ $\tan x = 1$

$$x = \frac{n\pi}{3} \qquad\qquad x = \frac{\pi}{4} + n\pi$$

21. $\cos^3 x = \cos x$

$\cos^3 x - \cos x = 0$

$\cos x(\cos^2 x - 1) = 0$

$\cos x = 0$ or $\cos^2 x - 1 = 0$

$x = \frac{\pi}{2}, \frac{3\pi}{2}$ $\cos x = \pm 1$

$x = 0, \pi$

23. $3 \tan^3 x - \tan x = 0$

$\tan x(3 \tan^2 x - 1) = 0$

$\tan x = 0$ or $3 \tan^2 x - 1 = 0$

$x = 0, \pi$ $\tan x = \pm \frac{\sqrt{3}}{3}$

$$x = \frac{\pi}{6}, \frac{5\pi}{6}, \frac{7\pi}{6}, \frac{11\pi}{6}$$

25. $\sec^2 x - \sec x - 2 = 0$

$(\sec x - 2)(\sec x + 1) = 0$

$\sec x - 2 = 0$ or $\sec x + 1 = 0$

$\sec x = 2$ $\sec x = -1$

$x = \frac{\pi}{3}, \frac{5\pi}{3}$ $x = \pi$

27. $2 \sin x + \csc x = 0$

$2 \sin x + \frac{1}{\sin x} = 0$

$2 \sin^2 x + 1 = 0$

$\sin^2 x = -\frac{1}{2} \implies$ No solution

29. $2 \cos^2 x + \cos x - 1 = 0$

$(2 \cos x - 1)(\cos x + 1) = 0$

$2 \cos x - 1 = 0$ or $\cos x + 1 = 0$

$\cos x = \frac{1}{2}$ $\cos x = -1$

$x = \frac{\pi}{3}, \frac{5\pi}{3}$ $x = \pi$

31. $2 \sec^2 x + \tan^2 x - 3 = 0$

$2(\tan^2 x + 1) + \tan^2 x - 3 = 0$

$3 \tan^2 x - 1 = 0$

$\tan x = \pm \frac{\sqrt{3}}{3}$

$$x = \frac{\pi}{6}, \frac{5\pi}{6}, \frac{7\pi}{6}, \frac{11\pi}{6}$$

33.
$$\csc x + \cot x = 1$$
$$(\csc x + \cot x)^2 = 1^2$$
$$\csc^2 x + 2\csc x \cot x + \cot^2 x = 1$$
$$\cot^2 x + 1 + 2\csc x \cot x + \cot^2 x = 1$$
$$2\cot^2 x + 2\csc x \cot x = 0$$
$$2\cot x(\cot x + \csc x) = 0$$

$$2\cot x = 0 \qquad \text{or} \quad \cot x + \csc x = 0$$

$$x = \frac{\pi}{2}, \frac{3\pi}{2} \qquad \qquad \frac{\cos x}{\sin x} = -\frac{1}{\sin x}$$

$$\cos x = -1$$

$$x = \pi$$

By checking in the original equation, we find that $x = \pi$ and $x = 3\pi/2$ are extraneous. The only solution to the equation in the interval $[0, 2\pi)$ is $x = \pi/2$.

35. $\cos 2x = \dfrac{1}{2}$

$$2x = \frac{\pi}{3} + 2n\pi \quad \text{or} \quad 2x = \frac{5\pi}{3} + 2n\pi$$

$$x = \frac{\pi}{6} + n\pi \qquad \qquad x = \frac{5\pi}{6} + n\pi$$

37. $\tan 3x = 1$

$$3x = \frac{\pi}{4} + 2n\pi \qquad \text{or} \qquad 3x = \frac{5\pi}{4} + 2n\pi$$

$$x = \frac{\pi}{12} + \frac{2n\pi}{3} \qquad \qquad x = \frac{5\pi}{12} + \frac{2n\pi}{3}$$

These can be combined as $x = \dfrac{\pi}{12} + \dfrac{n\pi}{3}$.

39. $\cos\left(\dfrac{x}{2}\right) = \dfrac{\sqrt{2}}{2}$

$$\frac{x}{2} = \frac{\pi}{4} + 2n\pi \quad \text{or} \quad \frac{x}{2} = \frac{7\pi}{4} + 2n\pi$$

$$x = \frac{\pi}{2} + 4n\pi \qquad \qquad x = \frac{7\pi}{2} + 4n\pi$$

41. $y = \sin\dfrac{\pi x}{2} + 1$

From the graph in the textbook we see that the curve has x-intercepts at $x = -1$ and at $x = 3$.

In general, we have: $\sin\left(\dfrac{\pi x}{2}\right) = -1$

$$\frac{\pi x}{2} = \frac{3\pi}{2} + 2n\pi$$

$$x = 3 + 4n$$

43. $y = \tan^2\left(\dfrac{\pi x}{6}\right) - 3$

From the graph in the textbook we see that the curve has x-intercepts at $x = \pm 2$.

In general, we have: $\tan^2\left(\dfrac{\pi x}{6}\right) = 3$

$$\tan\left(\frac{\pi x}{6}\right) = \pm\sqrt{3}$$

$$\frac{\pi x}{6} = \pm\frac{\pi}{3} + n\pi$$

$$x = \pm 2 + 6n$$

45. Graph $y_1 = 2\sin x + \cos x$.

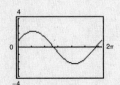

The x-intercepts occur at $x \approx 2.678$ and $x \approx 5.820$.

47. Graph $y_1 = \dfrac{1 + \sin x}{\cos x} + \dfrac{\cos x}{1 + \sin x} - 4$.

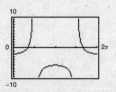

The x-intercepts occur at

$$x = \frac{\pi}{3} \approx 1.047 \text{ and } x = \frac{5\pi}{3} \approx 5.236.$$

49. $x \tan x - 1 = 0$

Graph $y_1 = x \tan x - 1$.

The x-intercepts occur at
$x \approx 0.860$ and $x \approx 3.426$.

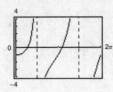

51. $\sec^2 x + 0.5 \tan x - 1 = 0$

Graph $y_1 = \dfrac{1}{(\cos x)^2} + 0.5 \tan x - 1$.

The x-intercepts occur at

$x = 0$, $x \approx 2.678$,

$x = \pi \approx 3.142$, and

$x \approx 5.820$.

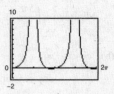

53. Graph $y_1 = 2 \tan^2 x + 7 \tan x - 15$.

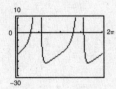

The x-intercepts occur at
$x \approx 0.983$, $x \approx 1.768$,
$x \approx 4.124$ and $x \approx 4.910$.

55. $12 \sin^2 x - 13 \sin x + 3 = 0$

$$\sin x = \frac{-(-13) \pm \sqrt{(-13)^2 - 4(12)(3)}}{2(12)}$$

$$= \frac{13 \pm 5}{24}$$

$\sin x = \dfrac{1}{3}$ or $\sin x = \dfrac{3}{4}$

$x \approx 0.3398,\ 2.8018$ $x \approx 0.8481,\ 2.2935$

Graph $y_1 = 12 \sin^2 x - 13 \sin x + 3$.

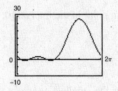

The x-intercepts occur at
$x \approx 0.3398$, $x \approx 0.8481$, $x \approx 2.2935$, and $x \approx 2.8018$.

57. $\tan^2 x + 3 \tan x + 1 = 0$

$$\tan x = \frac{-3 \pm \sqrt{3^2 - 4(1)(1)}}{2(1)} = \frac{-3 \pm \sqrt{5}}{2}$$

$\tan x = \dfrac{-3 - \sqrt{5}}{2}$ or $\tan x = \dfrac{-3 + \sqrt{5}}{2}$

$x \approx 1.9357,\ 5.0773$ $x \approx 2.7767,\ 5.9183$

Graph $y_1 = \tan^2 x + 3 \tan x + 1$.

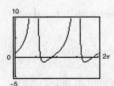

The x-intercepts occur at $x \approx 1.9357$, $x \approx 2.7767$,
$x \approx 5.0773$, and $x \approx 5.9183$.

59. $\tan^2 x - 6 \tan x + 5 = 0$

$(\tan x - 1)(\tan x - 5) = 0$

$\tan x - 1 = 0$ or $\tan x - 5 = 0$

$\tan x = 1$ $\tan x = 5$

$x = \dfrac{\pi}{4}, \dfrac{5\pi}{4}$ $x = \arctan 5,\ \arctan 5 + \pi$

61. $2 \cos^2 x - 5 \cos x + 2 = 0$

$(2 \cos x - 1)(\cos x - 2) = 0$

$2 \cos x - 1 = 0$ or $\cos x - 2 = 0$

$\cos x = \dfrac{1}{2}$ $\cos x = 2$

$x = \dfrac{\pi}{3}, \dfrac{5\pi}{3}$ No solution

63. (a) $f(x) = \sin x + \cos x$

Maximum: $\left(\dfrac{\pi}{4}, \sqrt{2}\right)$

Minimum: $\left(\dfrac{5\pi}{4}, -\sqrt{2}\right)$

(b) $\cos x - \sin x = 0$

$\qquad \cos x = \sin x$

$\qquad 1 = \dfrac{\sin x}{\cos x}$

$\qquad \tan x = 1$

$\qquad x = \dfrac{\pi}{4}, \dfrac{5\pi}{4}$

$f\left(\dfrac{\pi}{4}\right) = \sin \dfrac{\pi}{4} + \cos \dfrac{\pi}{4} = \dfrac{\sqrt{2}}{2} + \dfrac{\sqrt{2}}{2} = \sqrt{2}$

$f\left(\dfrac{5\pi}{4}\right) = \sin \dfrac{5\pi}{4} + \cos \dfrac{5\pi}{4} = -\sin \dfrac{\pi}{4} + \left(-\cos \dfrac{\pi}{4}\right) = -\dfrac{\sqrt{2}}{2} - \dfrac{\sqrt{2}}{2} = -\sqrt{2}$

Therefore, the maximum point in the interval $[0, 2\pi)$ is $\left(\pi/4, \sqrt{2}\right)$ and the minimum point is $\left(5\pi/4, -\sqrt{2}\right)$.

65. $f(x) = \tan \dfrac{\pi x}{4}$

Since $\tan \pi/4 = 1$, $x = 1$ is the smallest nonnegative fixed point.

67. $f(x) = \cos \dfrac{1}{x}$

(a) The domain of $f(x)$ is all real numbers x except $x = 0$.

(b) The graph has y-axis symmetry and a horizontal asymptote at $y = 1$.

(c) As $x \to 0$, $f(x)$ oscillates between 21 and 1.

(d) There are infinitely many solutions in the interval $[-1, 1]$. They occur at $x = \dfrac{2}{(2n+1)\pi}$ where n is any integer.

(e) The greatest solution appears to occur at $x \approx 0.6366$.

69.

$y = \dfrac{1}{12}(\cos 8t - 3 \sin 8t)$

$\dfrac{1}{12}(\cos 8t - 3 \sin 8t) = 0$

$\qquad \cos 8t = 3 \sin 8t$

$\qquad \dfrac{1}{3} = \tan 8t$

$\qquad 8t \approx 0.32175 + n\pi$

$\qquad t \approx 0.04 + \dfrac{n\pi}{8}$

In the interval $0 \le t \le 1$, $t \approx 0.04$, 0.43, and 0.83.

71. $S = 74.50 + 43.75 \sin \dfrac{\pi t}{6}$

t	1	2	3	4	5	6	7	8	9	10	11	12
S	96.4	112.4	118.3	112.4	96.4	74.5	52.6	36.6	30.8	36.6	52.6	74.5

Sales exceed 100,000 units during February, March, and April.

73.   Range $= 300$ feet

$v_0 = 100$ feet per second

$r = \frac{1}{32}v_0{}^2 \sin 2\theta$

$\frac{1}{32}(100)^2 \sin 2\theta = 300$

$\sin 2\theta = 0.96$

$2\theta = \arcsin(0.96) \approx 73.74°$

$\theta \approx 36.9°$

or

$2\theta = 180° - \arcsin(0.96) \approx 106.26°$

$\theta \approx 53.1°$

75. $h(t) = 53 + 50 \sin\left(\dfrac{\pi}{16}t - \dfrac{\pi}{2}\right)$

(a) $h(t) = 53$ when $50 \sin\left(\dfrac{\pi}{16}t - \dfrac{\pi}{2}\right) = 0.$

$\dfrac{\pi}{16}t - \dfrac{\pi}{2} = 0$ or $\dfrac{\pi}{16}t - \dfrac{\pi}{2} = \pi$

$\dfrac{\pi}{16}t = \dfrac{\pi}{2}$ $\dfrac{\pi}{16}t = \dfrac{3\pi}{2}$

$t = 8$ $t = 24$

The Ferris wheel will be 53 feet above ground at 8 seconds and at 24 seconds.

(b) The person will be at the top of the Ferris wheel when

$\sin\left(\dfrac{\pi}{16}t - \dfrac{\pi}{2}\right) = 1.$

$\dfrac{\pi}{16}t - \dfrac{\pi}{2} = \dfrac{\pi}{2}$

$\dfrac{\pi}{16}t = \pi$

$t = 16$

The first time this occurs is after 16 seconds. The period of this function is $\dfrac{2\pi}{\pi/16} = 32.$ During 160 seconds, 5 cycles will take place and the person will be at the top of the ride 5 times, spaced 32 seconds apart. The times are: 16 seconds, 48 seconds, 80 seconds, 112 seconds, and 144 seconds.

77. $A = 2x \cos x,\ 0 < x < \dfrac{\pi}{2}$

(a)

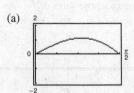

The maximum area of $A \approx 1.12$ occurs when $x \approx 0.86.$

(b) $A \ge 1$ for $0.6 < x < 1.1$

79. True. The period of $2 \sin 4t - 1$ is $\pi/2$ and the period of $2 \sin t - 1$ is 2π. In the interval $[0, 2\pi)$ the first equation has four cycles whereas the second equation has only one cycle, thus the first equation has four times the x-intercepts (solutions) as the second equation.

81. $y_1 = 2 \sin x$

$y_2 = 3x + 1$

From the graph we see that there is only one point of intersection.

83.
$$C = 90° - 66° = 24°$$

$$\cos 66° = \frac{22.3}{a}$$

$$a \cos 66° = 22.3$$

$$a = \frac{22.3}{\cos 66°} \approx 54.8$$

$$\tan 66° = \frac{b}{22.3}$$

$$b = 22.3 \tan 66° \approx 50.1$$

85. $\theta = 390°$, $\theta' = 390° - 360° = 30°$, θ is in Quadrant I.

$$\sin 390° = \sin 30° = \frac{1}{2}$$

$$\cos 390° = \cos 30° = \frac{\sqrt{3}}{2}$$

$$\tan 390° = \tan 30° = \frac{1}{\sqrt{3}} = \frac{\sqrt{3}}{3}$$

87. $\theta = -1845°$, $\theta' = 45°$, θ is in Quadrant IV.

$$\sin(-1845°) = -\sin 45° = -\frac{\sqrt{2}}{2}$$

$$\cos(-1845°) = \cos 45° = \frac{\sqrt{2}}{2}$$

$$\tan(-1845°) = -\tan 45° = -1$$

89. $\tan \theta = \dfrac{250 \text{ feet}}{2 \text{ miles}} \times \dfrac{1 \text{ mile}}{5280 \text{ feet}} \approx 0.02367$

$$\theta \approx 1.36°$$

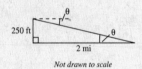

Not drawn to scale

91. Answers will vary.

Section 5.4   Sum and Difference Formulas

- ■ You should know the sum and difference formulas.

  $$\sin(u \pm v) = \sin u \cos v \pm \cos u \sin v$$

  $$\cos(u \pm v) = \cos u \cos v \mp \sin u \sin v$$

  $$\tan(u \pm v) = \frac{\tan u \pm \tan v}{1 \mp \tan u \tan v}$$

- ■ You should be able to use these formulas to find the values of the trigonometric functions of angles whose sums or differences are special angles.

- ■ You should be able to use these formulas to solve trigonometric equations.

Vocabulary Check

1. $\sin u \cos v - \cos u \sin v$

2. $\cos u \cos v - \sin u \sin v$

3. $\dfrac{\tan u + \tan v}{1 - \tan u \tan v}$

4. $\sin u \cos v + \cos u \sin v$

5. $\cos u \cos v + \sin u \sin v$

6. $\dfrac{\tan u - \tan v}{1 + \tan u \tan v}$

1. (a) $\cos(120° + 45°) = \cos 120° \cos 45° - \sin 120° \sin 45°$

$$= \left(-\frac{1}{2}\right)\left(\frac{\sqrt{2}}{2}\right) - \left(\frac{\sqrt{3}}{2}\right)\left(\frac{\sqrt{2}}{2}\right)$$

$$= \frac{-\sqrt{2} - \sqrt{6}}{4}$$

(b) $\cos 120° + \cos 45° = -\frac{1}{2} + \frac{\sqrt{2}}{2} = \frac{-1 + \sqrt{2}}{2}$

3. (a) $\cos\left(\frac{\pi}{4} + \frac{\pi}{3}\right) = \cos\frac{\pi}{4}\cos\frac{\pi}{3} - \sin\frac{\pi}{4}\sin\frac{\pi}{3}$

$$= \frac{\sqrt{2}}{2} \cdot \frac{1}{2} - \frac{\sqrt{2}}{2} \cdot \frac{\sqrt{3}}{2}$$

$$= \frac{\sqrt{2} - \sqrt{6}}{4}$$

(b) $\cos\frac{\pi}{4} + \cos\frac{\pi}{3} = \frac{\sqrt{2}}{2} + \frac{1}{2} = \frac{\sqrt{2} + 1}{2}$

5. (a) $\sin\left(\frac{7\pi}{6} - \frac{\pi}{3}\right) = \sin\frac{5\pi}{6} = \sin\frac{\pi}{6} = \frac{1}{2}$

(b) $\sin\frac{7\pi}{6} - \sin\frac{\pi}{3} = -\frac{1}{2} - \frac{\sqrt{3}}{2} = \frac{-1 - \sqrt{3}}{2}$

7. $\sin 105° = \sin(60° + 45°)$

$\qquad = \sin 60° \cos 45° + \cos 60° \sin 45°$

$$= \frac{\sqrt{3}}{2} \cdot \frac{\sqrt{2}}{2} + \frac{1}{2} \cdot \frac{\sqrt{2}}{2}$$

$$= \frac{\sqrt{2}}{4}\left(\sqrt{3} + 1\right)$$

$\cos 105° = \cos(60° + 45°)$

$\qquad = \cos 60° \cos 45° - \sin 60° \sin 45°$

$$= \frac{1}{2} \cdot \frac{\sqrt{2}}{2} - \frac{\sqrt{3}}{2} \cdot \frac{\sqrt{2}}{2}$$

$$= \frac{\sqrt{2}}{4}\left(1 - \sqrt{3}\right)$$

$\tan 105° = \tan(60° + 45°)$

$$= \frac{\tan 60° + \tan 45°}{1 - \tan 60° \tan 45°}$$

$$= \frac{\sqrt{3} + 1}{1 - \sqrt{3}} = \frac{\sqrt{3} + 1}{1 - \sqrt{3}} \cdot \frac{1 + \sqrt{3}}{1 + \sqrt{3}}$$

$$= \frac{4 + 2\sqrt{3}}{-2} = -2 - \sqrt{3}$$

9. $\sin 195° = \sin(225° - 30°)$

$\qquad = \sin 225° \cos 30° - \cos 225° \sin 30°$

$\qquad = -\sin 45° \cos 30° + \cos 45° \sin 30°$

$$= -\frac{\sqrt{2}}{2} \cdot \frac{\sqrt{3}}{2} + \frac{\sqrt{2}}{2} \cdot \frac{1}{2}$$

$$= \frac{\sqrt{2}}{4}\left(1 - \sqrt{3}\right)$$

$\cos 195° = \cos(225° - 30°)$

$\qquad = \cos 225° \cos 30° + \sin 225° \sin 30°$

$\qquad = -\cos 45° \cos 30° - \sin 45° \sin 30°$

$$= -\frac{\sqrt{2}}{2} \cdot \frac{\sqrt{3}}{2} - \frac{\sqrt{2}}{2} \cdot \frac{1}{2}$$

$$= -\frac{\sqrt{2}}{4}\left(\sqrt{3} + 1\right)$$

$\tan 195° = \tan(225° - 30°)$

$$= \frac{\tan 225° - \tan 30°}{1 + \tan 225° \tan 30°}$$

$$= \frac{\tan 45° - \tan 30°}{1 + \tan 45° \tan 30°}$$

$$= \frac{1 - \left(\frac{\sqrt{3}}{3}\right)}{1 + \left(\frac{\sqrt{3}}{3}\right)} = \frac{3 - \sqrt{3}}{3 + \sqrt{3}} \cdot \frac{3 - \sqrt{3}}{3 - \sqrt{3}}$$

$$= \frac{12 - 6\sqrt{3}}{6} = 2 - \sqrt{3}$$

11. $\sin\dfrac{11\pi}{12} = \sin\left(\dfrac{3\pi}{4} + \dfrac{\pi}{6}\right)$

$\qquad = \sin\dfrac{3\pi}{4}\cos\dfrac{\pi}{6} + \cos\dfrac{3\pi}{4}\sin\dfrac{\pi}{6}$

$\qquad = \dfrac{\sqrt{2}}{2}\cdot\dfrac{\sqrt{3}}{2} + \left(-\dfrac{\sqrt{2}}{2}\right)\dfrac{1}{2}$

$\qquad = \dfrac{\sqrt{2}}{4}\left(\sqrt{3} - 1\right)$

$\cos\dfrac{11\pi}{12} = \cos\left(\dfrac{3\pi}{4} + \dfrac{\pi}{6}\right)$

$\qquad = \cos\dfrac{3\pi}{4}\cos\dfrac{\pi}{6} - \sin\dfrac{3\pi}{4}\sin\dfrac{\pi}{6}$

$\qquad = -\dfrac{\sqrt{2}}{2}\cdot\dfrac{\sqrt{3}}{2} - \dfrac{\sqrt{2}}{2}\cdot\dfrac{1}{2} = -\dfrac{\sqrt{2}}{4}\left(\sqrt{3} + 1\right)$

13. $\sin\dfrac{17\pi}{12} = \sin\left(\dfrac{9\pi}{4} - \dfrac{5\pi}{6}\right)$

$\qquad = \sin\dfrac{9\pi}{4}\cos\dfrac{5\pi}{6} - \cos\dfrac{9\pi}{4}\sin\dfrac{5\pi}{6}$

$\qquad = \dfrac{\sqrt{2}}{2}\left(-\dfrac{\sqrt{3}}{2}\right) - \left(\dfrac{\sqrt{2}}{2}\right)\left(\dfrac{1}{2}\right)$

$\qquad = -\dfrac{\sqrt{2}}{4}\left(\sqrt{3} + 1\right)$

$\cos\dfrac{17\pi}{12} = \cos\left(\dfrac{9\pi}{4} - \dfrac{5\pi}{6}\right)$

$\qquad = \cos\dfrac{9\pi}{4}\cos\dfrac{5\pi}{6} + \sin\dfrac{9\pi}{4}\sin\dfrac{5\pi}{6}$

$\qquad = \dfrac{\sqrt{2}}{2}\left(-\dfrac{\sqrt{3}}{2}\right) + \dfrac{\sqrt{2}}{2}\left(\dfrac{1}{2}\right)$

$\qquad = \dfrac{\sqrt{2}}{4}\left(1 - \sqrt{3}\right)$

$\tan\dfrac{17\pi}{12} = \tan\left(\dfrac{9\pi}{4} - \dfrac{5\pi}{6}\right)$

$\qquad = \dfrac{\tan(9\pi/4) - \tan(5\pi/6)}{1 + \tan(9\pi/4)\tan(5\pi/6)}$

$\qquad = \dfrac{1 - \left(-\sqrt{3}/3\right)}{1 + \left(-\sqrt{3}/3\right)}$

$\qquad = \dfrac{3 + \sqrt{3}}{3 - \sqrt{3}}\cdot\dfrac{3 + \sqrt{3}}{3 + \sqrt{3}}$

$\qquad = \dfrac{12 + 6\sqrt{3}}{6} = 2 + \sqrt{3}$

$\tan 195° = \tan(225° - 30°)$

$\qquad = \dfrac{\tan 225° - \tan 30°}{1 + \tan 225°\tan 30°}$

$\qquad = \dfrac{\tan 45° - \tan 30°}{1 + \tan 45°\tan 30°}$

$\qquad = \dfrac{1 - \left(\dfrac{\sqrt{3}}{3}\right)}{1 + \left(\dfrac{\sqrt{3}}{3}\right)} = \dfrac{3 - \sqrt{3}}{3 + \sqrt{3}}\cdot\dfrac{3 - \sqrt{3}}{3 - \sqrt{3}}$

$\qquad = \dfrac{12 - 6\sqrt{3}}{6} = 2 - \sqrt{3}$

15. $285 = 225 + 60$

$\sin 285° = \sin(225° + 60°)$

$\qquad = \sin 225°\cos 60° + \cos 225°\sin 60°$

$\qquad = -\dfrac{\sqrt{2}}{2}\left(\dfrac{1}{2}\right) - \dfrac{\sqrt{2}}{2}\left(\dfrac{\sqrt{3}}{2}\right) = -\dfrac{\sqrt{2}}{4}\left(\sqrt{3} + 1\right)$

$\cos 285° = \cos(225° + 60°)$

$\qquad = \cos 225°\cos 60° - \sin 225°\sin 60°$

$\qquad = -\dfrac{\sqrt{2}}{2}\left(\dfrac{1}{2}\right) - \left(-\dfrac{\sqrt{2}}{2}\right)\left(\dfrac{\sqrt{3}}{2}\right) = \dfrac{\sqrt{2}}{4}\left(\sqrt{3} - 1\right)$

$\tan 285° = \tan(225° + 60°)$

$\qquad = \dfrac{\tan 225° + \tan 60°}{1 - \tan 225°\tan 60°} = \dfrac{1 + \sqrt{3}}{1 - \sqrt{3}}\cdot\dfrac{1 + \sqrt{3}}{1 + \sqrt{3}}$

$\qquad = \dfrac{4 + 2\sqrt{3}}{-2} = -2 - \sqrt{3} = -\left(2 + \sqrt{3}\right)$

17. $-165° = -(120° + 45°)$

$\sin(-165°) = \sin[-(120° + 45°)]$

$= -\sin(120° + 45°)$

$= -[\sin 120° \cos 45° + \cos 120° \sin 45°]$

$= -\left[\dfrac{\sqrt{3}}{2} \cdot \dfrac{\sqrt{2}}{2} - \dfrac{1}{2} \cdot \dfrac{\sqrt{2}}{2}\right]$

$= -\dfrac{\sqrt{2}}{4}(\sqrt{3} - 1)$

$\cos(-165°) = \cos[-(120° + 45°)]$

$= \cos(120° + 45°)$

$= \cos 120° \cos 45° - \sin 120° \sin 45°$

$= -\dfrac{1}{2} \cdot \dfrac{\sqrt{2}}{2} - \dfrac{\sqrt{3}}{2} \cdot \dfrac{\sqrt{2}}{2}$

$= -\dfrac{\sqrt{2}}{4}(1 + \sqrt{3})$

$\tan(-165°) = \tan[-(120° + 45°)]$

$= -\tan(120° + \tan 45°)$

$= -\dfrac{\tan 120° + \tan 45°}{1 - \tan 120° \tan 45°}$

$= -\dfrac{-\sqrt{3} + 1}{1 - (-\sqrt{3})(1)}$

$= -\dfrac{1 - \sqrt{3}}{1 + \sqrt{3}} \cdot \dfrac{1 - \sqrt{3}}{1 - \sqrt{3}}$

$= -\dfrac{4 - 2\sqrt{3}}{-2}$

$= 2 - \sqrt{3}$

19. $\dfrac{13\pi}{12} = \dfrac{3\pi}{4} + \dfrac{\pi}{3}$

$\sin \dfrac{13\pi}{12} = \sin\left(\dfrac{3\pi}{4} + \dfrac{\pi}{3}\right)$

$= \sin \dfrac{3\pi}{4} \cos \dfrac{\pi}{3} + \cos \dfrac{3\pi}{4} \sin \dfrac{\pi}{3}$

$= \dfrac{\sqrt{2}}{2} \cdot \dfrac{1}{2} + \left(-\dfrac{\sqrt{2}}{2}\right)\left(\dfrac{\sqrt{3}}{2}\right)$

$= \dfrac{\sqrt{2}}{4}(1 - \sqrt{3})$

$\cos \dfrac{13\pi}{12} = \cos\left(\dfrac{3\pi}{4} + \dfrac{\pi}{3}\right)$

$= \cos \dfrac{3\pi}{4} \cos \dfrac{\pi}{3} - \sin \dfrac{3\pi}{4} \sin \dfrac{\pi}{3}$

$= -\dfrac{\sqrt{2}}{2} \cdot \dfrac{1}{2} - \dfrac{\sqrt{2}}{2} \cdot \dfrac{\sqrt{3}}{2} = -\dfrac{\sqrt{2}}{4}(1 + \sqrt{3})$

$\tan \dfrac{13\pi}{12} = \tan\left(\dfrac{3\pi}{4} + \dfrac{\pi}{3}\right)$

$= \dfrac{\tan\left(\dfrac{3\pi}{4}\right) + \tan\left(\dfrac{\pi}{3}\right)}{1 - \tan\left(\dfrac{3\pi}{4}\right)\tan\left(\dfrac{\pi}{3}\right)}$

$= \dfrac{-1 + \sqrt{3}}{1 - (-1)(\sqrt{3})}$

$= -\dfrac{1 - \sqrt{3}}{1 + \sqrt{3}} \cdot \dfrac{1 - \sqrt{3}}{1 - \sqrt{3}}$

$= -\dfrac{4 - 2\sqrt{3}}{-2}$

$= 2 - \sqrt{3}$

21. $-\dfrac{13\pi}{12} = -\left(\dfrac{3\pi}{4} + \dfrac{\pi}{3}\right)$

$\sin\left[-\left(\dfrac{3\pi}{4} + \dfrac{\pi}{3}\right)\right] = -\sin\left(\dfrac{3\pi}{4} + \dfrac{\pi}{3}\right)$

$= -\left[\sin \dfrac{3\pi}{4} \cos \dfrac{\pi}{3} + \cos \dfrac{3\pi}{4} \sin \dfrac{\pi}{3}\right]$

$= -\left[\dfrac{\sqrt{2}}{2}\left(\dfrac{1}{2}\right) + \left(-\dfrac{\sqrt{2}}{2}\right)\left(\dfrac{\sqrt{3}}{2}\right)\right]$

$= -\dfrac{\sqrt{2}}{4}(1 - \sqrt{3}) = \dfrac{\sqrt{2}}{4}(\sqrt{3} - 1)$

$\cos\left[-\left(\dfrac{3\pi}{4} + \dfrac{\pi}{3}\right)\right] = \cos\left(\dfrac{3\pi}{4} + \dfrac{\pi}{3}\right)$

$= \cos \dfrac{3\pi}{4} \cos \dfrac{\pi}{3} - \sin \dfrac{3\pi}{4} \sin \dfrac{\pi}{3}$

$= -\dfrac{\sqrt{2}}{2}\left(\dfrac{1}{2}\right) - \dfrac{\sqrt{2}}{2}\left(\dfrac{\sqrt{3}}{2}\right) = -\dfrac{\sqrt{2}}{4}(\sqrt{3} + 1)$

$\tan\left[-\left(\dfrac{3\pi}{4} + \dfrac{\pi}{3}\right)\right] = -\tan\left(\dfrac{3\pi}{4} + \dfrac{\pi}{3}\right)$

$= -\dfrac{\tan \dfrac{3\pi}{4} + \tan \dfrac{\pi}{3}}{1 - \tan \dfrac{3\pi}{4} \tan \dfrac{\pi}{3}} = -\dfrac{-1 + \sqrt{3}}{1 - (-\sqrt{3})}$

$= \dfrac{1 - \sqrt{3}}{1 + \sqrt{3}} \cdot \dfrac{1 - \sqrt{3}}{1 - \sqrt{3}} = \dfrac{4 - 2\sqrt{3}}{-2}$

$= -2 + \sqrt{3}$

23. $\cos 25° \cos 15° - \sin 25° \sin 15° = \cos(25° + 15°) = \cos 40°$

25. $\dfrac{\tan 325° - \tan 86°}{1 + \tan 325° \tan 86°} = \tan(325° - 86°) = \tan 239°$

27. $\sin 3 \cos 1.2 - \cos 3 \sin 1.2 = \sin(3 - 1.2) = \sin 1.8$

29. $\dfrac{\tan 2x + \tan x}{1 - \tan 2x \tan x} = \tan(2x + x) = \tan 3x$

31. $\sin 330° \cos 30° - \cos 330° \sin 30° = \sin(330° - 30°)$
$$= \sin 300°$$
$$= -\frac{\sqrt{3}}{2}$$

33. $\sin \dfrac{\pi}{12} \cos \dfrac{\pi}{4} + \cos \dfrac{\pi}{12} \sin \dfrac{\pi}{4} = \sin\left(\dfrac{\pi}{12} + \dfrac{\pi}{4}\right)$
$$= \sin \frac{\pi}{3}$$
$$= \frac{\sqrt{3}}{2}$$

35. $\dfrac{\tan 25° + \tan 110°}{1 - \tan 25° \tan 110°} = \tan(25° + 110°)$
$$= \tan 135°$$
$$= -1$$

For Exercises 37–43, we have:

$$\sin u = \tfrac{5}{13}, \; u \text{ in Quadrant II} \implies \cos u = -\tfrac{12}{13}, \; \tan u = -\tfrac{5}{12}$$
$$\cos v = -\tfrac{3}{5}, \; v \text{ in Quadrant II} \implies \sin v = \tfrac{4}{5}, \; \tan v = -\tfrac{4}{3},$$

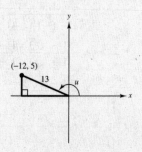
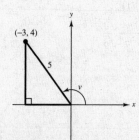

Figures for Exercises 37–43

37. $\sin(u + v) = \sin u \cos v + \cos u \sin v$
$$= \left(\frac{5}{13}\right)\left(-\frac{3}{5}\right) + \left(-\frac{12}{13}\right)\left(\frac{4}{5}\right)$$
$$= -\frac{63}{65}$$

39. $\cos(u + v) = \cos u \cos v - \sin u \sin v$
$$= \left(-\frac{12}{13}\right)\left(-\frac{3}{5}\right) - \left(\frac{5}{13}\right)\left(\frac{4}{5}\right)$$
$$= \frac{16}{65}$$

41. $\tan(u + v) = \dfrac{\tan u + \tan v}{1 - \tan u \tan v} = \dfrac{-\frac{5}{12} + \left(-\frac{4}{3}\right)}{1 - \left(-\frac{5}{12}\right)\left(-\frac{4}{3}\right)} = \dfrac{-\frac{21}{12}}{1 - \frac{5}{9}}$
$$= \left(-\frac{7}{4}\right)\left(\frac{9}{4}\right) = -\frac{63}{16}$$

43. $\sec(v - u) = \dfrac{1}{\cos(v - u)} = \dfrac{1}{\cos v \cos u + \sin v \sin u}$
$$= \dfrac{1}{\left(-\frac{3}{5}\right)\left(-\frac{12}{13}\right) + \left(\frac{4}{5}\right)\left(\frac{5}{13}\right)} = \dfrac{1}{\left(\frac{36}{65}\right) + \left(\frac{20}{65}\right)} = \dfrac{1}{\frac{56}{65}}$$
$$= \frac{65}{56}$$

For Exercises 45–49, we have:

$\sin u = -\frac{7}{25}$, u in Quadrant III $\Longrightarrow \cos u = -\frac{24}{25}$, $\tan u = \frac{7}{24}$

$\cos v = -\frac{4}{5}$, v in Quadrant III $\Longrightarrow \sin v = -\frac{3}{5}$, $\tan v = \frac{3}{4}$

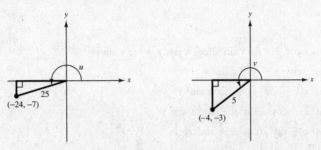

Figures for Exercises 45–49

45. $\cos(u + v) = \cos u \cos v - \sin u \sin v$

$\qquad = \left(-\frac{24}{25}\right)\left(-\frac{4}{5}\right) - \left(-\frac{7}{25}\right)\left(-\frac{3}{5}\right)$

$\qquad = \frac{3}{5}$

47. $\tan(u - v) = \dfrac{\tan u - \tan v}{1 + \tan u \tan v}$

$\qquad = \dfrac{\frac{7}{24} - \frac{3}{4}}{1 + \left(\frac{7}{24}\right)\left(\frac{3}{4}\right)} = \dfrac{-\frac{11}{24}}{\frac{39}{32}} = -\dfrac{44}{117}$

49. $\sec(u + v) = \dfrac{1}{\cos(u + v)} = \dfrac{1}{\frac{3}{5}} = \dfrac{5}{3}$

Use Exercise 45 for $\cos(u + v)$.

51. $\sin(\arcsin x + \arccos x) = \sin(\arcsin x)\cos(\arccos x) + \sin(\arccos x)\cos(\arcsin x)$

$\qquad\qquad = x \cdot x + \sqrt{1 - x^2} \cdot \sqrt{1 - x^2}$

$\qquad\qquad = x^2 + 1 - x^2$

$\qquad\qquad = 1$

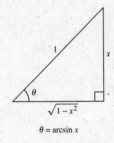

$\theta = \arcsin x$

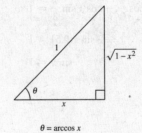

$\theta = \arccos x$

53. $\cos(\arccos x + \arcsin x) = \cos(\arccos x)\cos(\arcsin x) - \sin(\arccos x)\sin(\arcsin x)$

$\qquad\qquad = x \cdot \sqrt{1 - x^2} - \sqrt{1 - x^2} \cdot x$

$\qquad\qquad = 0$

(Use the triangles in Exercise 51.)

55. $\sin(3\pi - x) = \sin 3\pi \cos x - \sin x \cos 3\pi$

$\qquad\qquad = (0)(\cos x) - (-1)(\sin x)$

$\qquad\qquad = \sin x$

57. $\sin\left(\dfrac{\pi}{6} + x\right) = \sin\dfrac{\pi}{6}\cos x + \cos\dfrac{\pi}{6}\sin x$

$\qquad\qquad = \dfrac{1}{2}\left(\cos x + \sqrt{3}\sin x\right)$

59. $\cos(\pi - \theta) + \sin\left(\dfrac{\pi}{2} + \theta\right) = \cos\pi\cos\theta + \sin\pi\sin\theta + \sin\dfrac{\pi}{2}\cos\theta + \cos\dfrac{\pi}{2}\sin\theta$

$$= (-1)(\cos\theta) + (0)(\sin\theta) + (1)(\cos\theta) + (\sin\theta)(0)$$

$$= -\cos\theta + \cos\theta$$

$$= 0$$

61. $\cos(x + y)\cos(x - y) = (\cos x\cos y - \sin x\sin y)(\cos x\cos y + \sin x\sin y)$

$$= \cos^2 x\cos^2 y - \sin^2 x\sin^2 y$$

$$= \cos^2 x(1 - \sin^2 y) - \sin^2 x\sin^2 y$$

$$= \cos^2 x - \cos^2 x\sin^2 y - \sin^2 x\sin^2 y$$

$$= \cos^2 x - \sin^2 y(\cos^2 x + \sin^2 x)$$

$$= \cos^2 x - \sin^2 y$$

63. $\sin(x + y) + \sin(x - y) = \sin x\cos y + \cos x\sin y + \sin x\cos y - \cos x\sin y$

$$= 2\sin x\cos y$$

65. $\cos\left(\dfrac{3\pi}{2} - x\right) = \cos\dfrac{3\pi}{2}\cos x + \sin\dfrac{3\pi}{2}\sin x$

$$= (0)(\cos x) + (-1)(\sin x)$$

$$= -\sin x$$

67. $\sin\left(\dfrac{3\pi}{2} + \theta\right) = \sin\dfrac{3\pi}{2}\cos\theta + \cos\dfrac{3\pi}{2}\sin\theta$

$$= (-1)(\cos\theta) + (0)(\sin\theta)$$

$$= -\cos\theta$$

69.

$$\sin\left(x + \dfrac{\pi}{3}\right) + \sin\left(x - \dfrac{\pi}{3}\right) = 1$$

$$\sin x\cos\dfrac{\pi}{3} + \cos x\sin\dfrac{\pi}{3} + \sin x\cos\dfrac{\pi}{3} - \cos x\sin\dfrac{\pi}{3} = 1$$

$$2\sin x(0.5) = 1$$

$$\sin x = 1$$

$$x = \dfrac{\pi}{2}$$

71.

$$\cos\left(x + \dfrac{\pi}{4}\right) - \cos\left(x - \dfrac{\pi}{4}\right) = 1$$

$$\cos x\cos\dfrac{\pi}{4} - \sin x\sin\dfrac{\pi}{4} - \left(\cos x\cos\dfrac{\pi}{4} + \sin x\sin\dfrac{\pi}{4}\right) = 1$$

$$-2\sin x\left(\dfrac{\sqrt{2}}{2}\right) = 1$$

$$-\sqrt{2}\sin x = 1$$

$$\sin x = -\dfrac{1}{\sqrt{2}}$$

$$\sin x = -\dfrac{\sqrt{2}}{2}$$

$$x = \dfrac{5\pi}{4}, \dfrac{7\pi}{4}$$

73.   Analytically: $\cos\left(x + \dfrac{\pi}{4}\right) + \cos\left(x - \dfrac{\pi}{4}\right) = 1$

$$\cos x \cos \frac{\pi}{4} - \sin x \sin \frac{\pi}{4} + \cos x \cos \frac{\pi}{4} + \sin x \sin \frac{\pi}{4} = 1$$

$$2 \cos x\left(\frac{\sqrt{2}}{2}\right) = 1$$

$$\sqrt{2} \cos x = 1$$

$$\cos x = \frac{1}{\sqrt{2}}$$

$$\cos x = \frac{\sqrt{2}}{2}$$

$$x = \frac{\pi}{4}, \frac{7\pi}{4}$$

Graphically: Graph $y_1 = \cos\left(x + \dfrac{\pi}{4}\right) + \cos\left(x - \dfrac{\pi}{4}\right)$ and $y_2 = 1$.

The points of intersection occur at $x = \dfrac{\pi}{4}$ and $x = \dfrac{7\pi}{4}$.

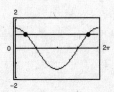

75. $y = \dfrac{1}{3} \sin 2t + \dfrac{1}{4} \cos 2t$

(a) $a = \dfrac{1}{3},\ b = \dfrac{1}{4},\ B = 2$

$C = \arctan \dfrac{b}{a} = \arctan \dfrac{3}{4} \approx 0.6435$

$y \approx \sqrt{\left(\dfrac{1}{3}\right)^2 + \left(\dfrac{1}{4}\right)^2}\ \sin(2t + 0.6435)$

$= \dfrac{5}{12} \sin(2t + 0.6435)$

(b) Amplitude: $\dfrac{5}{12}$ feet

(c) Frequency: $\dfrac{1}{\text{period}} = \dfrac{B}{2\pi} = \dfrac{2}{2\pi} = \dfrac{1}{\pi}$ cycle per second

77. False.

$\sin(u \pm v) = \sin u \cos v \pm \cos u \sin v$

79. False. $\cos\left(x - \dfrac{\pi}{2}\right) = \cos x \cos \dfrac{\pi}{2} + \sin x \sin \dfrac{\pi}{2}$

$= (\cos x)(0) + (\sin x)(1)$

$= \sin x$

81. $\cos(n\pi + \theta) = \cos n\pi \cos \theta - \sin n\pi \sin \theta$

$= (-1)^n(\cos \theta) - (0)(\sin \theta)$

$= (-1)^n(\cos \theta)$, where n is an integer.

83. $C = \arctan \dfrac{b}{a} \implies \sin C = \dfrac{b}{\sqrt{a^2 + b^2}},\ \cos C = \dfrac{a}{\sqrt{a^2 + b^2}}$

$\sqrt{a^2 + b^2}\ \sin(B\theta + C) = \sqrt{a^2 + b^2}\left(\sin B\theta \cdot \dfrac{a}{\sqrt{a^2 + b^2}} + \dfrac{b}{\sqrt{a^2 + b^2}} \cdot \cos B\theta\right) = a \sin B\theta + b \cos B\theta$

85. $\sin\theta + \cos\theta$

$a = 1, \ b = 1, \ B = 1$

(a) $C = \arctan\dfrac{b}{a} = \arctan 1 = \dfrac{\pi}{4}$

$\quad \sin\theta + \cos\theta = \sqrt{a^2 + b^2}\,\sin(B\theta + C)$

$\qquad\qquad\qquad = \sqrt{2}\,\sin\left(\theta + \dfrac{\pi}{4}\right)$

(b) $C = \arctan\dfrac{a}{b} = \arctan 1 = \dfrac{\pi}{4}$

$\quad \sin\theta + \cos\theta = \sqrt{a^2 + b^2}\,\cos(B\theta - C)$

$\qquad\qquad\qquad = \sqrt{2}\,\cos\left(\theta - \dfrac{\pi}{4}\right)$

87. $12\sin 3\theta + 5\cos 3\theta$

$a = 12, \ b = 5, \ B = 3$

(a) $C = \arctan\dfrac{b}{a} = \arctan\dfrac{5}{12} \approx 0.3948$

$\quad 12\sin 3\theta + 5\cos 3\theta = \sqrt{a^2 + b^2}\,\sin(B\theta + C)$

$\qquad\qquad\qquad\qquad \approx 13\sin(3\theta + 0.3948)$

(b) $C = \arctan\dfrac{a}{b} = \arctan\dfrac{12}{5} \approx 1.1760$

$\quad 12\sin 3\theta + 5\cos 3\theta = \sqrt{a^2 + b^2}\,\cos(B\theta - C)$

$\qquad\qquad\qquad\qquad \approx 13\cos(3\theta - 1.1760)$

89. $C = \arctan\dfrac{b}{a} = \dfrac{\pi}{2} \ \Rightarrow \ a = 0$

$\sqrt{a^2 + b^2} = 2 \ \Rightarrow \ b = 2$

$B = 1$

$2\sin\left(\theta + \dfrac{\pi}{2}\right) = (0)(\sin\theta) + (2)(\cos\theta) = 2\cos\theta$

91. $\dfrac{\cos(x + h) - \cos x}{h} = \dfrac{\cos x \cos h - \sin x \sin h - \cos x}{h}$

$\qquad\qquad\qquad\qquad = \dfrac{\cos x \cos h - \cos x - \sin x \sin h}{h}$

$\qquad\qquad\qquad\qquad = \dfrac{\cos x(\cos h - 1) - \sin x \sin h}{h}$

$\qquad\qquad\qquad\qquad = \dfrac{\cos x(\cos h - 1)}{h} - \dfrac{\sin x \sin h}{h}$

93.

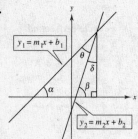

$m_1 = \tan\alpha$ and $m_2 = \tan\beta$

$\beta + \delta = 90° \ \Rightarrow \ \delta = 90° - \beta$

$\alpha + \theta + \delta = 90° \ \Rightarrow \ \alpha + \theta + (90° - \beta) = 90° \ \Rightarrow \ \theta = \beta - \alpha$

Therefore, $\theta = \arctan m_2 - \arctan m_1$.

For $y = x$ and $y = \sqrt{3}x$ we have $m_1 = 1$ and $m_2 = \sqrt{3}$.

$\theta = \arctan\sqrt{3} - \arctan 1$

$\quad = 60° - 45°$

$\quad = 15°$

95.

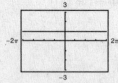

Conjecture: $\sin^2\left(\theta + \dfrac{\pi}{4}\right) + \sin^2\left(\theta - \dfrac{\pi}{4}\right) = 1$

$\sin^2\left(\theta + \dfrac{\pi}{4}\right) + \sin^2\left(\theta - \dfrac{\pi}{4}\right) = \left[\sin\theta\cos\dfrac{\pi}{4} + \cos\theta\sin\dfrac{\pi}{4}\right]^2 + \left[\sin\theta\cos\dfrac{\pi}{4} - \cos\theta\sin\dfrac{\pi}{4}\right]^2$

$\qquad\qquad\qquad\qquad\qquad\qquad = \left[\dfrac{\sin\theta}{\sqrt{2}} + \dfrac{\cos\theta}{\sqrt{2}}\right]^2 + \left[\dfrac{\sin\theta}{\sqrt{2}} - \dfrac{\cos\theta}{\sqrt{2}}\right]^2$

$\qquad\qquad\qquad\qquad\qquad\qquad = \dfrac{\sin^2\theta}{2} + \sin\theta\cos\theta + \dfrac{\cos^2\theta}{2} + \dfrac{\sin^2\theta}{2} - \sin\theta\cos\theta + \dfrac{\cos^2\theta}{2}$

$\qquad\qquad\qquad\qquad\qquad\qquad = \sin^2\theta + \cos^2\theta$

$\qquad\qquad\qquad\qquad\qquad\qquad = 1$

97. $f(x) = 5(x - 3)$

$y = 5(x - 3)$

$\dfrac{y}{5} = x - 3$

$\dfrac{y}{5} + 3 = x$

$\dfrac{x}{5} + 3 = y$

$f^{-1}(x) = \dfrac{x + 15}{5}$

$f(f^{-1}(x)) = f\left(\dfrac{x + 15}{5}\right) = 5\left[\dfrac{x + 15}{5} - 3\right]$

$\qquad\qquad = 5\left(\dfrac{x + 15}{5}\right) - 5(3)$

$\qquad\qquad = x + 15 - 15$

$\qquad\qquad = x$

$f^{-1}(f(x)) = f^{-1}(5(x - 3)) = \dfrac{5(x - 3) + 15}{5}$

$\qquad\qquad = \dfrac{5x - 15 + 15}{5}$

$\qquad\qquad = \dfrac{5x}{5}$

$\qquad\qquad = x$

99. $f(x) = x^2 - 8$

f is not one-to-one so f^{-1} does not exist.

101. $\log_3 3^{4x - 3} = 4x - 3$

103. $e^{\ln(6x - 3)} = 6x - 3$

Section 5.5   Multiple-Angle and Product-to-Sum Formulas

■ You should know the following double-angle formulas.

(a) $\sin 2u = 2 \sin u \cos u$

(b) $\cos 2u = \cos^2 u - \sin^2 u$

$\qquad = 2 \cos^2 u - 1$

$\qquad = 1 - 2 \sin^2 u$

(c) $\tan 2u = \dfrac{2 \tan u}{1 - \tan^2 u}$

■ You should be able to reduce the power of a trigonometric function.

(a) $\sin^2 u = \dfrac{1 - \cos 2u}{2}$

(b) $\cos^2 u = \dfrac{1 + \cos 2u}{2}$

(c) $\tan^2 u = \dfrac{1 - \cos 2u}{1 + \cos 2u}$

■ You should be able to use the half-angle formulas. The signs of $\sin \dfrac{u}{2}$ and $\cos \dfrac{u}{2}$ depend on the quadrant in which $\dfrac{u}{2}$ lies.

(a) $\sin \dfrac{u}{2} = \pm \sqrt{\dfrac{1 - \cos u}{2}}$

(b) $\cos \dfrac{u}{2} = \pm \sqrt{\dfrac{1 + \cos u}{2}}$

(c) $\tan \dfrac{u}{2} = \dfrac{1 - \cos u}{\sin u} = \dfrac{\sin u}{1 + \cos u}$

—CONTINUED—

—CONTINUED—

■ You should be able to use the product-sum formulas.

(a) $\sin u \sin v = \frac{1}{2}[\cos(u - v) - \cos(u + v)]$

(b) $\cos u \cos v = \frac{1}{2}[\cos(u - v) + \cos(u + v)]$

(c) $\sin u \cos v = \frac{1}{2}[\sin(u + v) + \sin(u - v)]$

(d) $\cos u \sin v = \frac{1}{2}[\sin(u + v) - \sin(u - v)]$

■ You should be able to use the sum-product formulas.

(a) $\sin x + \sin y = 2 \sin\left(\dfrac{x + y}{2}\right) \cos\left(\dfrac{x - y}{2}\right)$

(b) $\sin x - \sin y = 2 \cos\left(\dfrac{x + y}{2}\right) \sin\left(\dfrac{x - y}{2}\right)$

(c) $\cos x + \cos y = 2 \cos\left(\dfrac{x + y}{2}\right) \cos\left(\dfrac{x - y}{2}\right)$

(d) $\cos x - \cos y = -2 \sin\left(\dfrac{x + y}{2}\right) \sin\left(\dfrac{x - y}{2}\right)$

Vocabulary Check

1. $2 \sin u \cos u$

2. $\cos^2 u$

3. $\cos^2 u - \sin^2 u = 2\cos^2 u - 1 = 1 - 2\sin^2 u$

4. $\tan^2 u$

5. $\pm\sqrt{\dfrac{1 - \cos u}{2}}$

6. $\dfrac{1 - \cos u}{\sin u} = \dfrac{\sin u}{1 + \cos u}$

7. $\frac{1}{2}[\cos(u - v) + \cos(u + v)]$

8. $\frac{1}{2}[\sin(u + v) + \sin(u - v)]$

9. $2 \sin\left(\dfrac{u + v}{2}\right) \cos\left(\dfrac{u - v}{2}\right)$

10. $-2 \sin\left(\dfrac{u + v}{2}\right) \sin\left(\dfrac{u - v}{2}\right)$

Figure for Exercises 1–7

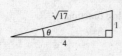

$\sin \theta = \dfrac{\sqrt{17}}{17}$

$\cos \theta = \dfrac{4\sqrt{17}}{17}$

$\tan \theta = \dfrac{1}{4}$

1. $\sin \theta = \dfrac{\sqrt{17}}{17}$

3. $\cos 2\theta = 2\cos^2 \theta - 1$

$= 2\left(\dfrac{4\sqrt{17}}{17}\right)^2 - 1$

$= \dfrac{32}{17} - 1$

$= \dfrac{15}{17}$

5. $\tan 2\theta = \dfrac{2 \tan \theta}{1 - \tan^2 \theta}$

$= \dfrac{2\left(\frac{1}{4}\right)}{1 - \left(\frac{1}{4}\right)^2}$

$= \dfrac{\frac{1}{2}}{1 - \frac{1}{16}}$

$= \dfrac{1}{2} \cdot \dfrac{16}{15}$

$= \dfrac{8}{15}$

7. $\csc 2\theta = \dfrac{1}{\sin 2\theta} = \dfrac{1}{2\sin\theta\cos\theta} = \dfrac{1}{2\left(\dfrac{\sqrt{17}}{17}\right)\left(\dfrac{4\sqrt{17}}{17}\right)} = \dfrac{17}{8}$

9.
$$\sin 2x - \sin x = 0$$
$$2\sin x\cos x - \sin x = 0$$
$$\sin x(2\cos x - 1) = 0$$
$$\sin x = 0 \quad\text{or}\quad 2\cos x - 1 = 0$$
$$x = 0, \pi \qquad\qquad \cos x = \frac{1}{2}$$
$$x = \frac{\pi}{3}, \frac{5\pi}{3}$$
$$x = 0, \frac{\pi}{3}, \pi, \frac{5\pi}{3}$$

11. $4\sin x\cos x = 1$
$$2\sin 2x = 1$$
$$\sin 2x = \frac{1}{2}$$
$$2x = \frac{\pi}{6} + 2n\pi \quad\text{or}\quad 2x = \frac{5\pi}{6} + 2n\pi$$
$$x = \frac{\pi}{12} + n\pi \qquad\qquad x = \frac{5\pi}{12} + n\pi$$
$$x = \frac{\pi}{12}, \frac{13\pi}{12} \qquad\qquad x = \frac{5\pi}{12}, \frac{17\pi}{12}$$

13.
$$\cos 2x - \cos x = 0$$
$$\cos 2x = \cos x$$
$$\cos^2 x - \sin^2 x = \cos x$$
$$\cos^2 x - (1 - \cos^2 x) - \cos x = 0$$
$$2\cos^2 x - \cos x - 1 = 0$$
$$(2\cos x + 1)(\cos x - 1) = 0$$
$$2\cos x + 1 = 0 \qquad\text{or}\qquad \cos x - 1 = 0$$
$$\cos x = -\frac{1}{2} \qquad\qquad \cos x = 1$$
$$x = \frac{2\pi}{3}, \frac{4\pi}{3} \qquad\qquad x = 0$$

15.
$$\tan 2x - \cot x = 0$$
$$\frac{2\tan x}{1 - \tan^2 x} = \cot x$$
$$2\tan x = \cot x(1 - \tan^2 x)$$
$$2\tan x = \cot x - \cot x\tan^2 x$$
$$2\tan x = \cot x - \tan x$$
$$3\tan x = \cot x$$
$$3\tan x - \cot x = 0$$
$$3\tan x - \frac{1}{\tan x} = 0$$
$$\frac{3\tan^2 x - 1}{\tan x} = 0$$
$$\frac{1}{\tan x}(3\tan^2 x - 1) = 0$$
$$\cot x(3\tan^2 x - 1) = 0$$

$$\cot x = 0 \qquad\text{or}\qquad 3\tan^2 x - 1 = 0$$
$$x = \frac{\pi}{2}, \frac{3\pi}{2} \qquad\qquad \tan^2 x = \frac{1}{3}$$
$$\tan x = \pm\frac{\sqrt{3}}{3}$$
$$x = \frac{\pi}{6}, \frac{5\pi}{6}, \frac{7\pi}{6}, \frac{11\pi}{6}$$

$$x = \frac{\pi}{6}, \frac{\pi}{2}, \frac{5\pi}{6}, \frac{7\pi}{6}, \frac{3\pi}{2}, \frac{11\pi}{6}$$

17.
$$\sin 4x = -2 \sin 2x$$
$$\sin 4x + 2 \sin 2x = 0$$
$$2 \sin 2x \cos 2x + 2 \sin 2x = 0$$
$$2 \sin 2x(\cos 2x + 1) = 0$$

$$2 \sin 2x = 0 \qquad \text{or} \qquad \cos 2x + 1 = 0$$
$$\sin 2x = 0 \qquad\qquad\qquad \cos 2x = -1$$
$$2x = n\pi \qquad\qquad\qquad 2x = \pi + 2n\pi$$
$$x = \frac{n}{2}\pi \qquad\qquad\qquad x = \frac{\pi}{2} + n\pi$$
$$x = 0,\ \frac{\pi}{2},\ \pi,\ \frac{3\pi}{2} \qquad\qquad x = \frac{\pi}{2},\ \frac{3\pi}{2}$$

19. $6 \sin x \cos x = 3(2 \sin x \cos x)$
$$= 3 \sin 2x$$

21. $4 - 8 \sin^2 x = 4(1 - 2 \sin^2 x)$
$$= 4 \cos 2x$$

23. $\sin u = -\dfrac{4}{5},\ \pi < u < \dfrac{3\pi}{2} \Rightarrow \cos u = -\dfrac{3}{5}$

$$\sin 2u = 2 \sin u \cos u = 2\left(-\frac{4}{5}\right)\left(-\frac{3}{5}\right) = \frac{24}{25}$$

$$\cos 2u = \cos^2 u - \sin^2 u = \frac{9}{25} - \frac{16}{25} = -\frac{7}{25}$$

$$\tan 2u = \frac{2 \tan u}{1 - \tan^2 u} = \frac{2\left(\frac{4}{3}\right)}{1 - \frac{16}{9}} = \frac{8}{3}\left(-\frac{9}{7}\right) = -\frac{24}{7}$$

25. $\tan u = \dfrac{3}{4},\ 0 < u < \dfrac{\pi}{2} \Rightarrow \sin u = \dfrac{3}{5}$ and $\cos u = \dfrac{4}{5}$

$$\sin 2u = 2 \sin u \cos u = 2\left(\frac{3}{5}\right)\left(\frac{4}{5}\right) = \frac{24}{25}$$

$$\cos 2u = \cos^2 u - \sin^2 u = \frac{16}{25} - \frac{9}{25} = \frac{7}{25}$$

$$\tan 2u = \frac{2 \tan u}{1 - \tan^2 u} = \frac{2\left(\frac{3}{4}\right)}{1 - \frac{9}{16}} = \frac{3}{2}\left(\frac{16}{7}\right) = \frac{24}{7}$$

27. $\sec u = -\dfrac{5}{2},\ \dfrac{\pi}{2} < u < \pi \Rightarrow \sin u = \dfrac{\sqrt{21}}{5}$ and $\cos u = -\dfrac{2}{5}$

$$\sin 2u = 2 \sin u \cos u = 2\left(\frac{\sqrt{21}}{5}\right)\left(-\frac{2}{5}\right) = -\frac{4\sqrt{21}}{25}$$

$$\cos 2u = \cos^2 u - \sin^2 u = \left(-\frac{2}{5}\right)^2 - \left(\frac{\sqrt{21}}{5}\right)^2 = -\frac{17}{25}$$

$$\tan 2u = \frac{2 \tan u}{1 - \tan^2 u} = \frac{2\left(-\frac{\sqrt{21}}{2}\right)}{1 - \left(-\frac{\sqrt{21}}{2}\right)^2}$$

$$= \frac{-\sqrt{21}}{1 - \frac{21}{4}} = \frac{4\sqrt{21}}{17}$$

29. $\cos^4 x = (\cos^2 x)(\cos^2 x) = \left(\dfrac{1 + \cos 2x}{2}\right)\left(\dfrac{1 + \cos 2x}{2}\right) = \dfrac{1 + 2\cos 2x + \cos^2 2x}{4}$

$$= \dfrac{1 + 2\cos 2x + \dfrac{1 + \cos 4x}{2}}{4}$$

$$= \dfrac{2 + 4\cos 2x + 1 + \cos 4x}{8}$$

$$= \dfrac{3 + 4\cos 2x + \cos 4x}{8}$$

$$= \dfrac{1}{8}(3 + 4\cos 2x + \cos 4x)$$

31. $(\sin^2 x)(\cos^2 x) = \left(\dfrac{1 - \cos 2x}{2}\right)\left(\dfrac{1 + \cos 2x}{2}\right)$

$$= \dfrac{1 - \cos^2 2x}{4}$$

$$= \dfrac{1}{4}\left(1 - \dfrac{1 + \cos 4x}{2}\right)$$

$$= \dfrac{1}{8}(2 - 1 - \cos 4x)$$

$$= \dfrac{1}{8}(1 - \cos 4x)$$

33. $\sin^2 x \cos^4 x = \sin^2 x \cos^2 x \cos^2 x = \left(\dfrac{1 - \cos 2x}{2}\right)\left(\dfrac{1 + \cos 2x}{2}\right)\left(\dfrac{1 + \cos 2x}{2}\right)$

$$= \dfrac{1}{8}(1 - \cos 2x)(1 + \cos 2x)(1 + \cos 2x)$$

$$= \dfrac{1}{8}(1 - \cos^2 2x)(1 + \cos 2x)$$

$$= \dfrac{1}{8}(1 + \cos 2x - \cos^2 2x - \cos^3 2x)$$

$$= \dfrac{1}{8}\left[1 + \cos 2x - \left(\dfrac{1 + \cos 4x}{2}\right) - \cos 2x\left(\dfrac{1 + \cos 4x}{2}\right)\right]$$

$$= \dfrac{1}{16}[2 + 2\cos 2x - 1 - \cos 4x - \cos 2x - \cos 2x \cos 4x]$$

$$= \dfrac{1}{16}(1 + \cos 2x - \cos 4x - \cos 2x \cos 4x)$$

Figure for Exercises 35–39

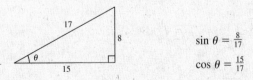

$\sin \theta = \frac{8}{17}$

$\cos \theta = \frac{15}{17}$

35. $\cos \dfrac{\theta}{2} = \sqrt{\dfrac{1 + \cos \theta}{2}} = \sqrt{\dfrac{1 + \frac{15}{17}}{2}} = \sqrt{\dfrac{32}{34}} = \sqrt{\dfrac{16}{17}} = \dfrac{4\sqrt{17}}{17}$

37. $\tan \dfrac{\theta}{2} = \dfrac{\sin \theta}{1 + \cos \theta} = \dfrac{\frac{8}{17}}{1 + \frac{15}{17}} = \dfrac{8}{17} \cdot \dfrac{17}{32} = \dfrac{1}{4}$

39. $\csc \dfrac{\theta}{2} = \dfrac{1}{\sin \dfrac{\theta}{2}} = \dfrac{1}{\sqrt{\dfrac{(1 - \cos \theta)}{2}}} = \dfrac{1}{\sqrt{\dfrac{1 - \frac{15}{17}}{2}}} = \dfrac{1}{\sqrt{\dfrac{1}{17}}} = \sqrt{17}$

41. $\sin 75° = \sin\left(\dfrac{1}{2} \cdot 150°\right) = \sqrt{\dfrac{1 - \cos 150°}{2}} = \sqrt{\dfrac{1 + \frac{\sqrt{3}}{2}}{2}}$

$\qquad = \dfrac{1}{2}\sqrt{2 + \sqrt{3}}$

$\quad\cos 75° = \cos\left(\dfrac{1}{2} \cdot 150°\right) = \sqrt{\dfrac{1 + \cos 150°}{2}} = \sqrt{\dfrac{1 - \frac{\sqrt{3}}{2}}{2}}$

$\qquad = \dfrac{1}{2}\sqrt{2 - \sqrt{3}}$

$\quad\tan 75° = \tan\left(\dfrac{1}{2} \cdot 150°\right) = \dfrac{\sin 150°}{1 + \cos 150°} = \dfrac{\dfrac{1}{2}}{1 - \dfrac{\sqrt{3}}{2}}$

$\qquad = \dfrac{1}{2 - \sqrt{3}} \cdot \dfrac{2 + \sqrt{3}}{2 + \sqrt{3}} = \dfrac{2 + \sqrt{3}}{4 - 3} = 2 + \sqrt{3}$

43. $\sin 112° \, 30' = \sin\left(\dfrac{1}{2} \cdot 225°\right) = \sqrt{\dfrac{1 - \cos 225°}{2}} = \sqrt{\dfrac{1 + \frac{\sqrt{2}}{2}}{2}} = \dfrac{1}{2}\sqrt{2 + \sqrt{2}}$

$\quad\cos 112° \, 30' = \cos\left(\dfrac{1}{2} \cdot 225°\right) = -\sqrt{\dfrac{1 + \cos 225°}{2}} = -\sqrt{\dfrac{1 - \frac{\sqrt{2}}{2}}{2}} = -\dfrac{1}{2}\sqrt{2 - \sqrt{2}}$

$\quad\tan 112° \, 30' = \tan\left(\dfrac{1}{2} \cdot 225°\right) = \dfrac{\sin 225°}{1 + \cos 225°} = \dfrac{-\dfrac{\sqrt{2}}{2}}{1 - \dfrac{\sqrt{2}}{2}} = -1 - \sqrt{2}$

45. $\sin \dfrac{\pi}{8} = \sin\left[\dfrac{1}{2}\left(\dfrac{\pi}{4}\right)\right] = \sqrt{\dfrac{1 - \cos \dfrac{\pi}{4}}{2}} = \dfrac{1}{2}\sqrt{2 - \sqrt{2}}$

$\quad\cos \dfrac{\pi}{8} = \cos\left[\dfrac{1}{2}\left(\dfrac{\pi}{4}\right)\right] = \sqrt{\dfrac{1 + \cos \dfrac{\pi}{4}}{2}} = \dfrac{1}{2}\sqrt{2 + \sqrt{2}}$

$\quad\tan \dfrac{\pi}{8} = \tan\left[\dfrac{1}{2}\left(\dfrac{\pi}{4}\right)\right] = \dfrac{\sin \dfrac{\pi}{4}}{1 + \cos \dfrac{\pi}{4}} = \dfrac{\dfrac{\sqrt{2}}{2}}{1 + \dfrac{\sqrt{2}}{2}} = \sqrt{2} - 1$

47. $\sin\dfrac{3\pi}{8} = \sin\left(\dfrac{1}{2}\cdot\dfrac{3\pi}{4}\right) = \sqrt{\dfrac{1-\cos\dfrac{3\pi}{4}}{2}} = \sqrt{\dfrac{1+\dfrac{\sqrt{2}}{2}}{2}} = \dfrac{1}{2}\sqrt{2+\sqrt{2}}$

$\cos\dfrac{3\pi}{8} = \cos\left(\dfrac{1}{2}\cdot\dfrac{3\pi}{4}\right) = \sqrt{\dfrac{1+\cos\dfrac{3\pi}{4}}{2}} = \sqrt{\dfrac{1-\dfrac{\sqrt{2}}{2}}{2}} = \dfrac{1}{2}\sqrt{2-\sqrt{2}}$

$\tan\dfrac{3\pi}{8} = \tan\left(\dfrac{1}{2}\cdot\dfrac{3\pi}{4}\right) = \dfrac{\sin\dfrac{3\pi}{4}}{1+\cos\dfrac{3\pi}{4}} = \dfrac{\dfrac{\sqrt{2}}{2}}{1-\dfrac{\sqrt{2}}{2}} = \dfrac{\dfrac{\sqrt{2}}{2}}{\dfrac{(2-\sqrt{2})}{2}} = \dfrac{\sqrt{2}}{2-\sqrt{2}} = \sqrt{2}+1$

49. $\sin u = \dfrac{5}{13},\ \dfrac{\pi}{2} < u < \pi \Rightarrow \cos u = -\dfrac{12}{13}$

$\sin\left(\dfrac{u}{2}\right) = \sqrt{\dfrac{1-\cos u}{2}} = \sqrt{\dfrac{1+\dfrac{12}{13}}{2}} = \dfrac{5\sqrt{26}}{26}$

$\cos\left(\dfrac{u}{2}\right) = \sqrt{\dfrac{1+\cos u}{2}} = \sqrt{\dfrac{1-\dfrac{12}{13}}{2}} = \dfrac{\sqrt{26}}{26}$

$\tan\left(\dfrac{u}{2}\right) = \dfrac{\sin u}{1+\cos u} = \dfrac{\dfrac{5}{13}}{1-\dfrac{12}{13}} = 5$

51. $\tan u = -\dfrac{5}{8},\ \dfrac{3\pi}{2} < u < 2\pi \Rightarrow \sin u = -\dfrac{5}{\sqrt{89}}$ and $\cos u = \dfrac{8}{\sqrt{89}}$

$\sin\left(\dfrac{u}{2}\right) = \sqrt{\dfrac{1-\cos u}{2}} = \sqrt{\dfrac{1-\dfrac{8}{\sqrt{89}}}{2}}\sqrt{\dfrac{\sqrt{89}-8}{2\sqrt{89}}} = \sqrt{\dfrac{89-8\sqrt{89}}{178}}$

$\cos\left(\dfrac{u}{2}\right) = -\sqrt{\dfrac{1+\cos u}{2}} = -\sqrt{\dfrac{1+\dfrac{8}{\sqrt{89}}}{2}} = -\sqrt{\dfrac{\sqrt{89}+8}{2\sqrt{89}}} = -\sqrt{\dfrac{89+8\sqrt{89}}{178}}$

$\tan\left(\dfrac{u}{2}\right) = \dfrac{1-\cos u}{\sin u} = \dfrac{1-\dfrac{8}{\sqrt{89}}}{-\dfrac{5}{\sqrt{89}}} = \dfrac{8-\sqrt{89}}{5}$

53. $\csc u = -\dfrac{5}{3},\ \pi < u < \dfrac{3\pi}{2} \Rightarrow \sin u = -\dfrac{3}{5}$ and $\cos u = -\dfrac{4}{5}$

$\sin\left(\dfrac{u}{2}\right) = \sqrt{\dfrac{1-\cos u}{2}} = \sqrt{\dfrac{1+\dfrac{4}{5}}{2}} = \dfrac{3\sqrt{10}}{10}$

$\cos\left(\dfrac{u}{2}\right) = -\sqrt{\dfrac{1+\cos u}{2}} = -\sqrt{\dfrac{1-\dfrac{4}{5}}{2}} = -\dfrac{\sqrt{10}}{10}$

$\tan\left(\dfrac{u}{2}\right) = \dfrac{1-\cos u}{\sin u} = \dfrac{1+\dfrac{4}{5}}{-\dfrac{3}{5}} = -3$

55. $\sqrt{\dfrac{1-\cos 6x}{2}} = |\sin 3x|$

57. $-\sqrt{\dfrac{1-\cos 8x}{1+\cos 8x}} = -\dfrac{\sqrt{\dfrac{1-\cos 8x}{2}}}{\sqrt{\dfrac{1+\cos 8x}{2}}}$

$= -\left|\dfrac{\sin 4x}{\cos 4x}\right| = -|\tan 4x|$

59. $\sin \dfrac{x}{2} + \cos x = 0$

$$\pm\sqrt{\dfrac{1 - \cos x}{2}} = -\cos x$$

$$\dfrac{1 - \cos x}{2} = \cos^2 x$$

$$0 = 2\cos^2 x + \cos x - 1$$

$$= (2\cos x - 1)(\cos x + 1)$$

$$\cos x = \dfrac{1}{2} \quad \text{or} \quad \cos x = -1$$

$$x = \dfrac{\pi}{3}, \dfrac{5\pi}{3} \qquad x = \pi$$

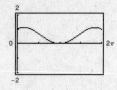

By checking these values in the original equation, we see that $x = \pi/3$ and $x = 5\pi/3$ are extraneous, and $x = \pi$ is the only solution.

61. $\cos \dfrac{x}{2} - \sin x = 0$

$$\pm\sqrt{\dfrac{1 + \cos x}{2}} = \sin x$$

$$\dfrac{1 + \cos x}{2} = \sin^2 x$$

$$1 + \cos x = 2\sin^2 x$$

$$1 + \cos x = 2 - 2\cos^2 x$$

$$2\cos^2 x + \cos x - 1 = 0$$

$$(2\cos x - 1)(\cos x + 1) = 0$$

$$2\cos x - 1 = 0 \quad \text{or} \quad \cos x + 1 = 0$$

$$\cos x = \dfrac{1}{2} \qquad\qquad \cos x = -1$$

$$x = \dfrac{\pi}{3}, \dfrac{5\pi}{3} \qquad\qquad x = \pi$$

$$x = \dfrac{\pi}{3}, \pi, \dfrac{5\pi}{3}$$

$\pi/3$, π, and $5\pi/3$ are all solutions to the equation.

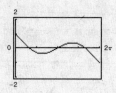

63. $6\sin\dfrac{\pi}{4}\cos\dfrac{\pi}{4} = 6 \cdot \dfrac{1}{2}\left[\sin\left(\dfrac{\pi}{4} + \dfrac{\pi}{4}\right) + \sin\left(\dfrac{\pi}{4} - \dfrac{\pi}{4}\right)\right] = 3\left(\sin\dfrac{\pi}{2} + \sin 0\right)$

65. $10\cos 75° \cos 15° = 10\left(\dfrac{1}{2}\right)[\cos(75° - 15°) + \cos(75° + 15°)] = 5[\cos 60° + \cos 90°]$

67. $\cos 4\theta \sin 6\theta = \dfrac{1}{2}[\sin(4\theta + 6\theta) - \sin(4\theta - 6\theta)] = \dfrac{1}{2}[\sin 10\theta - \sin(-2\theta)] = \dfrac{1}{2}(\sin 10\theta + \sin 2\theta)$

69. $5\cos(-5\beta)\cos 3\beta = 5 \cdot \dfrac{1}{2}[\cos(-5\beta - 3\beta) + \cos(-5\beta + 3\beta)] = \dfrac{5}{2}[\cos(-8\beta) + \cos(-2\beta)]$

$$= \dfrac{5}{2}(\cos 8\beta + \cos 2\beta)$$

71. $\sin(x + y)\sin(x - y) = \dfrac{1}{2}(\cos 2y - \cos 2x)$

73. $\cos(\theta - \pi)\sin(\theta + \pi) = \dfrac{1}{2}[\sin 2\theta - \sin(-2\pi)]$

$$= \dfrac{1}{2}(\sin 2\theta + \sin 2\pi)$$

75. $\sin 5\theta - \sin 3\theta = 2\cos\left(\dfrac{5\theta + 3\theta}{2}\right)\sin\left(\dfrac{5\theta - 3\theta}{2}\right)$

$$= 2\cos 4\theta \sin \theta$$

77. $\cos 6x + \cos 2x = 2\cos\left(\dfrac{6x + 2x}{2}\right)\cos\left(\dfrac{6x - 2x}{2}\right)$

$$= 2\cos 4x \cos 2x$$

79. $\sin(\alpha + \beta) - \sin(\alpha - \beta) = 2\cos\left(\dfrac{\alpha + \beta + \alpha - \beta}{2}\right)\sin\left(\dfrac{\alpha + \beta - \alpha + \beta}{2}\right) = 2\cos \alpha \sin \beta$

81. $\cos\left(\theta + \dfrac{\pi}{2}\right) - \cos\left(\theta - \dfrac{\pi}{2}\right) = -2\sin\left[\dfrac{\left(\theta + \frac{\pi}{2}\right) + \left(\theta - \frac{\pi}{2}\right)}{2}\right]\sin\left[\dfrac{\left(\theta + \frac{\pi}{2}\right) - \left(\theta - \frac{\pi}{2}\right)}{2}\right] = -2\sin\theta\sin\dfrac{\pi}{2}$

83. $\sin 60° + \sin 30° = 2\sin\left(\dfrac{60° + 30°}{2}\right)\cos\left(\dfrac{60° - 30°}{2}\right) = 2\sin 45°\cos 15°$

$\sin 60° + \sin 30° = \dfrac{\sqrt{3}}{2} + \dfrac{1}{2} = \dfrac{\sqrt{3} + 1}{2}$

85. $\cos\dfrac{3\pi}{4} - \cos\dfrac{\pi}{4} = -2\sin\left(\dfrac{\frac{3\pi}{4} + \frac{\pi}{4}}{2}\right)\sin\left(\dfrac{\frac{3\pi}{4} - \frac{\pi}{4}}{2}\right) = -2\sin\dfrac{\pi}{2}\sin\dfrac{\pi}{4}$

$\cos\dfrac{3\pi}{4} - \cos\dfrac{\pi}{4} = -\dfrac{\sqrt{2}}{2} - \dfrac{\sqrt{2}}{2} = -\sqrt{2}$

87. $\sin 6x + \sin 2x = 0$

$2\sin\left(\dfrac{6x + 2x}{2}\right)\cos\left(\dfrac{6x - 2x}{2}\right) = 0$

$2(\sin 4x)\cos 2x = 0$

$\sin 4x = 0 \quad \text{or} \quad \cos 2x = 0$

$4x = n\pi \qquad\qquad 2x = \dfrac{\pi}{2} + n\pi$

$x = \dfrac{n\pi}{4} \qquad\qquad x = \dfrac{\pi}{4} + \dfrac{n\pi}{2}$

In the interval $[0, 2\pi)$ we have

$x = 0, \dfrac{\pi}{4}, \dfrac{\pi}{2}, \dfrac{3\pi}{4}, \pi, \dfrac{5\pi}{4}, \dfrac{3\pi}{2}, \dfrac{7\pi}{4}.$

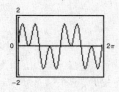

89. $\dfrac{\cos 2x}{\sin 3x - \sin x} - 1 = 0$

$\dfrac{\cos 2x}{\sin 3x - \sin x} = 1$

$\dfrac{\cos 2x}{2\cos 2x \sin x} = 1$

$2\sin x = 1$

$\sin x = \dfrac{1}{2}$

$x = \dfrac{\pi}{6}, \dfrac{5\pi}{6}$

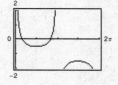

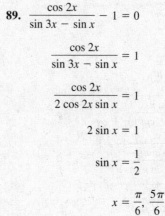

Figure for Exercises 91–93

91. $\sin^2\alpha = \left(\dfrac{5}{13}\right)^2 = \dfrac{25}{169}$

$\sin^2\alpha = 1 - \cos^2\alpha = 1 - \left(\dfrac{12}{13}\right)^2$

$= 1 - \dfrac{144}{169} = \dfrac{25}{169}$

93. $\sin\alpha\cos\beta = \left(\dfrac{5}{13}\right)\left(\dfrac{4}{5}\right) = \dfrac{4}{13}$

$\sin\alpha\cos\beta = \cos\left(\dfrac{\pi}{2} - \alpha\right)\sin\left(\dfrac{\pi}{2} - \beta\right)$

$= \left(\dfrac{5}{13}\right)\left(\dfrac{4}{5}\right) = \dfrac{4}{13}$

95. $\csc 2\theta = \dfrac{1}{\sin 2\theta}$

$\qquad = \dfrac{1}{2 \sin\theta \cos\theta}$

$\qquad = \dfrac{1}{\sin\theta} \cdot \dfrac{1}{2 \cos\theta}$

$\qquad = \dfrac{\csc\theta}{2 \cos\theta}$

97. $\cos^2 2\alpha - \sin^2 2\alpha = \cos[2(2\alpha)]$

$\qquad\qquad\qquad\qquad = \cos 4\alpha$

99. $(\sin x + \cos x)^2 = \sin^2 x + 2 \sin x \cos x + \cos^2 x$

$\qquad\qquad\qquad = (\sin^2 x + \cos^2 x) + 2 \sin x \cos x$

$\qquad\qquad\qquad = 1 + \sin 2x$

101. $1 + \cos 10y = 1 + \cos^2 5y - \sin^2 5y$

$\qquad\qquad\quad = 1 + \cos^2 5y - (1 - \cos^2 5y)$

$\qquad\qquad\quad = 2 \cos^2 5y$

103. $\sec \dfrac{u}{2} = \dfrac{1}{\cos \dfrac{u}{2}}$

$\qquad = \pm\sqrt{\dfrac{2}{1 + \cos u}}$

$\qquad = \pm\sqrt{\dfrac{2 \sin u}{\sin u(1 + \cos u)}}$

$\qquad = \pm\sqrt{\dfrac{2 \sin u}{\sin u + \sin u \cos u}}$

$\qquad = \pm\sqrt{\dfrac{\dfrac{2 \sin u}{\cos u}}{\dfrac{\sin u}{\cos u} + \dfrac{\sin u \cos u}{\cos u}}}$

$\qquad = \pm\sqrt{\dfrac{2 \tan u}{\tan u + \sin u}}$

105. $\dfrac{\sin x \pm \sin y}{\cos x + \cos y} = \dfrac{2 \sin\left(\dfrac{x \pm y}{2}\right) \cos\left(\dfrac{x \mp y}{2}\right)}{2 \cos\left(\dfrac{x + y}{2}\right) \cos\left(\dfrac{x - y}{2}\right)}$

$\qquad\qquad\qquad = \tan\left(\dfrac{x \pm y}{2}\right)$

107. $\dfrac{\cos 4x + \cos 2x}{\sin 4x + \sin 2x} = \dfrac{2 \cos\left(\dfrac{4x + 2x}{2}\right) \cos\left(\dfrac{4x - 2x}{2}\right)}{2 \sin\left(\dfrac{4x + 2x}{2}\right) \cos\left(\dfrac{4x - 2x}{2}\right)}$

$\qquad\qquad\qquad = \dfrac{2 \cos 3x \cos x}{2 \sin 3x \cos x} = \cot 3x$

109. $\sin\left(\dfrac{\pi}{6} + x\right) + \sin\left(\dfrac{\pi}{6} - x\right) = 2 \sin \dfrac{\pi}{6} \cos x$

$\qquad\qquad\qquad\qquad\qquad = 2 \cdot \dfrac{1}{2} \cos x$

$\qquad\qquad\qquad\qquad\qquad = \cos x$

111.

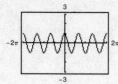

Let $y_1 = \cos(3x)$ and
$$y_2 = (\cos x)^3 - 3(\sin x)^2 \cos x.$$

$\cos 3\beta = \cos(2\beta + \beta)$

$\qquad = \cos 2\beta \cos\beta - \sin 2\beta \sin \beta$

$\qquad = (\cos^2 \beta - \sin^2 \beta) \cos \beta - 2 \sin \beta \cos \beta \sin \beta$

$\qquad = \cos^3 \beta - \sin^2 \beta \cos \beta - 2 \sin^2 \beta \cos \beta$

$\qquad = \cos^3 \beta - 3 \sin^2 \beta \cos \beta$

113.

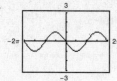

Let $y_1 = \dfrac{(\cos 4x - \cos 2x)}{(2 \sin 3x)}$

and $y_2 = -\sin x.$

$$\frac{\cos 4x - \cos 2x}{2 \sin 3x} = \frac{-2 \sin\left(\dfrac{4x + 2x}{2}\right) \sin\left(\dfrac{4x - 2x}{2}\right)}{2 \sin 3x}$$

$$= \frac{-2 \sin 3x \sin x}{2 \sin 3x} = -\sin x$$

115. $\sin^2 x = \dfrac{1 - \cos 2x}{2} = \dfrac{1}{2} - \dfrac{\cos 2x}{2}$

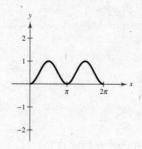

117. $\sin(2 \arcsin x) = 2 \sin(\arcsin x) \cos(\arcsin x)$

$\qquad\qquad\qquad = 2x\sqrt{1 - x^2}$

119. $\dfrac{1}{32}(75)^2 \sin 2\theta = 130$

$\qquad \sin 2\theta = \dfrac{130(32)}{75^2}$

$\qquad\qquad \theta = \dfrac{1}{2} \sin^{-1}\left(\dfrac{130(32)}{75^2}\right)$

$\qquad\qquad \theta \approx 23.85°$

121. $\sin \dfrac{\theta}{2} = \dfrac{1}{M}$

(a) $\sin \dfrac{\theta}{2} = 1$

$\quad \dfrac{\theta}{2} = \arcsin 1$

$\quad \dfrac{\theta}{2} = \dfrac{\pi}{2}$

$\quad \theta = \pi$

(c) $\dfrac{S}{760} = 1$

$\quad S = 760$ miles per hour

$\quad \dfrac{S}{760} = 4.5$

$\quad S = 3420$ miles per hour

(b) $\sin \dfrac{\theta}{2} = \dfrac{1}{4.5}$

$\quad \dfrac{\theta}{2} = \arcsin\left(\dfrac{1}{4.5}\right)$

$\quad \theta = 2 \arcsin\left(\dfrac{1}{4.5}\right)$

$\quad \theta \approx 0.4482$

(d) $\sin \dfrac{\theta}{2} = \dfrac{1}{M}$

$\quad \dfrac{\theta}{2} = \arcsin\left(\dfrac{1}{M}\right)$

$\quad \theta = 2 \arcsin\left(\dfrac{1}{M}\right)$

123. False. For $u < 0$,

$$\sin 2u = -\sin(-2u)$$
$$= -2\sin(-u)\cos(-u)$$
$$= -2(-\sin u)\cos u$$
$$= 2\sin u \cos u.$$

125. (a) $y = 4\sin\dfrac{x}{2} + \cos x$

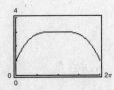

Maximum: $(\pi, 3)$

(b) $\qquad 2\cos\dfrac{x}{2} - \sin x = 0$

$$2\left(\pm\sqrt{\dfrac{1 + \cos x}{2}}\right) = \sin x$$

$$4\left(\dfrac{1 + \cos x}{2}\right) = \sin^2 x$$

$$2(1 + \cos x) = 1 - \cos^2 x$$

$$\cos^2 x + 2\cos x + 1 = 0$$

$$(\cos x + 1)^2 = 0$$

$$\cos x = -1$$

$$x = \pi$$

127. $f(x) = \sin^4 x + \cos^4 x$

(a) $\sin^4 x + \cos^4 x = (\sin^2 x)^2 + (\cos^2 x)^2$

$$= \left(\dfrac{1 - \cos 2x}{2}\right)^2 + \left(\dfrac{1 + \cos 2x}{2}\right)^2$$

$$= \dfrac{1}{4}[(1 - \cos 2x)^2 + (1 + \cos 2x)^2]$$

$$= \dfrac{1}{4}(1 - 2\cos 2x + \cos^2 2x + 1 + 2\cos 2x + \cos^2 2x)$$

$$= \dfrac{1}{4}(2 + 2\cos^2 2x)$$

$$= \dfrac{1}{4}\left[2 + 2\left(\dfrac{1 + \cos 2(2x)}{2}\right)\right]$$

$$= \dfrac{1}{4}(3 + \cos 4x)$$

(b) $\sin^4 x + \cos^4 x = (\sin^2 x)^2 + \cos^4 x$

$$= (1 - \cos^2 x)^2 + \cos^4 x$$

$$= 1 - 2\cos^2 x + \cos^4 x + \cos^4 x$$

$$= 2\cos^4 x - 2\cos^2 x + 1$$

(c) $\sin^4 x + \cos^4 x = \sin^4 x + 2\sin^2 x \cos^2 x + \cos^4 x - 2\sin^2 x \cos^2 x$

$$= (\sin^2 x + \cos^2 x)^2 - 2\sin^2 x \cos^2 x$$

$$= 1 - 2\sin^2 x \cos^2 x$$

(d) $1 - 2\sin^2 x \cos^2 x = 1 - (2\sin x \cos x)(\sin x \cos x)$

$$= 1 - (\sin 2x)\left(\dfrac{1}{2}\sin 2x\right)$$

$$= 1 - \dfrac{1}{2}\sin^2 2x$$

(e) No, it does not mean that one of you is wrong. There is often more than one way to rewrite a trigonometric expression.

129. (a)

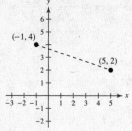

(b) $d = \sqrt{(-1-5)^2 + (4-2)^2} = \sqrt{(-6)^2 + (2)^2}$

$\quad = \sqrt{40} = 2\sqrt{10}$

(c) Midpoint: $\left(\dfrac{5+(-1)}{2}, \dfrac{2+4}{2} \right) = (2, 3)$

131. (a)

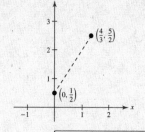

(b) $d = \sqrt{\left(\dfrac{4}{3} - 0 \right)^2 + \left(\dfrac{5}{2} - \dfrac{1}{2} \right)^2} = \sqrt{\dfrac{16}{9} + 4}$

$\quad = \sqrt{\dfrac{52}{9}} = \dfrac{2\sqrt{13}}{3}$

(c) Midpoint: $\left(\dfrac{0 + \frac{4}{3}}{2}, \dfrac{\frac{1}{2} + \frac{5}{2}}{2} \right) = \left(\dfrac{2}{3}, \dfrac{3}{2} \right)$

133. (a) Complement: $90° - 55° = 35°$

Supplement: $180° - 55° = 125°$

(b) Complement: Not possible. $162° > 90°$

Supplement: $180° - 162° = 18°$

135. (a) Complement: $\dfrac{\pi}{2} - \dfrac{\pi}{18} = \dfrac{4\pi}{9}$

Supplement: $\pi - \dfrac{\pi}{18} = \dfrac{17\pi}{18}$

(b) Complement: $\dfrac{\pi}{2} - \dfrac{9\pi}{20} = \dfrac{\pi}{20}$

Supplement: $\pi - \dfrac{9\pi}{20} = \dfrac{11\pi}{20}$

137. Let x = profit for September,
then $x + 0.16x$ = profit for October.

$x + (x + 0.16x) = 507{,}600$

$2.16x = 507{,}600$

$x = 235{,}000$

$x + 0.16x = 272{,}600$

Profit for September: \$235,000

Profit for October: \$272,600

139.

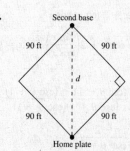

$d^2 = 90^2 + 90^2$

$\quad = 16{,}200$

$d = \sqrt{16{,}200}$

$\quad = 90\sqrt{2}$

$\quad \approx 127 \text{ feet}$

Review Exercises for Chapter 5

1. $\dfrac{1}{\cos x} = \sec x$

3. $\dfrac{1}{\sec x} = \cos x$

5. $\dfrac{\cos x}{\sin x} = \cot x$

7. $\sin x = \dfrac{3}{5}, \ \cos x = \dfrac{4}{5}$

$\tan x = \dfrac{\sin x}{\cos x} = \dfrac{\frac{3}{5}}{\frac{4}{5}} = \dfrac{3}{4}$

$\cot x = \dfrac{1}{\tan x} = \dfrac{4}{3}$

$\sec x = \dfrac{1}{\cos x} = \dfrac{5}{4}$

$\csc x = \dfrac{1}{\sin x} = \dfrac{5}{3}$

9. $\sin\left(\dfrac{\pi}{2} - x\right) = \dfrac{\sqrt{2}}{2} \implies \cos x = \dfrac{1}{\sqrt{2}} = \dfrac{\sqrt{2}}{2}$

$\sin x = -\dfrac{\sqrt{2}}{2}$

$\tan x = \dfrac{\sin x}{\cos x} = \dfrac{-\dfrac{1}{\sqrt{2}}}{\dfrac{1}{\sqrt{2}}} = -1$

$\cot x = \dfrac{1}{\tan x} = -1$

$\sec x = \dfrac{1}{\cos x} = \sqrt{2}$

$\csc x = \dfrac{1}{\sin x} = -\sqrt{2}$

11. $\dfrac{1}{\cot^2 x + 1} = \dfrac{1}{\csc^2 x} = \sin^2 x$

13. $\tan^2 x(\csc^2 x - 1) = \tan^2 x(\cot^2 x) = \tan^2 x\left(\dfrac{1}{\tan^2 x}\right) = 1$

15. $\dfrac{\sin\left(\dfrac{\pi}{2} - \theta\right)}{\sin \theta} = \dfrac{\cos \theta}{\sin \theta} = \cot \theta$

17. $\cos^2 x + \cos^2 x \cot^2 x = \cos^2 x(1 + \cot^2 x) = \cos^2 x(\csc^2 x)$

$= \cos^2 x\left(\dfrac{1}{\sin^2 x}\right) = \dfrac{\cos^2 x}{\sin^2 x} = \cot^2 x$

19. $(\tan x + 1)^2 \cos x = (\tan^2 x + 2\tan x + 1)\cos x$

$= (\sec^2 x + 2\tan x)\cos x$

$= \sec^2 x \cos x + 2\left(\dfrac{\sin x}{\cos x}\right)\cos x = \sec x + 2\sin x$

21. $\dfrac{1}{\csc \theta + 1} - \dfrac{1}{\csc \theta - 1} = \dfrac{(\csc \theta - 1) - (\csc \theta + 1)}{(\csc \theta + 1)(\csc \theta - 1)}$

$= \dfrac{-2}{\csc^2 \theta - 1}$

$= \dfrac{-2}{\cot^2 \theta}$

$= -2\tan^2 \theta$

23. $\csc^2 x - \csc x \cot x = \dfrac{1}{\sin^2 x} - \left(\dfrac{1}{\sin x}\right)\left(\dfrac{\cos x}{\sin x}\right)$

$= \dfrac{1 - \cos x}{\sin^2 x}$

25. $\cos x(\tan^2 x + 1) = \cos x \sec^2 x$

$= \dfrac{1}{\sec x}\sec^2 x$

$= \sec x$

27. $\cos\left(x + \dfrac{\pi}{2}\right) = \cos x \cos \dfrac{\pi}{2} - \sin x \sin \dfrac{\pi}{2}$

$= (\cos x)(0) - (\sin x)(1)$

$= -\sin x$

29. $\dfrac{1}{\tan \theta \csc \theta} = \dfrac{1}{\dfrac{\sin \theta}{\cos \theta} \cdot \dfrac{1}{\sin \theta}} = \cos \theta$

31. $\sin^5 x \cos^2 x = \sin^4 x \cos^2 x \sin x$

$= (1 - \cos^2 x)^2 \cos^2 x \sin x$

$= (1 - 2\cos^2 x + \cos^4 x)\cos^2 x \sin x$

$= (\cos^2 x - 2\cos^4 x + \cos^6 x)\sin x$

33. $\sin x = \sqrt{3} - \sin x$

$$\sin x = \frac{\sqrt{3}}{2}$$

$$x = \frac{\pi}{3} + 2\pi n, \frac{2\pi}{3} + 2\pi n$$

35. $3\sqrt{3} \tan u = 3$

$$\tan u = \frac{1}{\sqrt{3}}$$

$$u = \frac{\pi}{6} + n\pi$$

37. $3 \csc^2 x = 4$

$$\csc^2 x = \frac{4}{3}$$

$$\sin x = \pm\frac{\sqrt{3}}{2}$$

$$x = \frac{\pi}{3} + 2\pi n, \frac{2\pi}{3} + 2\pi n, \frac{4\pi}{3} + 2\pi n, \frac{5\pi}{3} + 2\pi n$$

These can be combined as:

$$x = \frac{\pi}{3} + n\pi \ \text{ or } \ x = \frac{2\pi}{3} + n\pi$$

39. $2 \cos^2 x - \cos x = 1$

$$2 \cos^2 x - \cos x - 1 = 0$$

$$(2 \cos x + 1)(\cos x - 1) = 0$$

$2 \cos x + 1 = 0 \qquad\qquad \cos x - 1 = 0$

$$\cos x = -\frac{1}{2} \qquad\qquad \cos x = 1$$

$$x = \frac{2\pi}{3}, \frac{4\pi}{3} \qquad\qquad x = 0$$

41. $\cos^2 x + \sin x = 1$

$$1 - \sin^2 x + \sin x - 1 = 0$$

$$-\sin x(\sin x - 1) = 0$$

$\sin x = 0 \qquad \sin x - 1 = 0$

$x = 0, \pi \qquad\quad \sin x = 1$

$$x = \frac{\pi}{2}$$

43. $2 \sin 2x - \sqrt{2} = 0$

$$\sin 2x = \frac{\sqrt{2}}{2}$$

$$2x = \frac{\pi}{4} + 2\pi n, \frac{3\pi}{4} + 2\pi n$$

$$x = \frac{\pi}{8} + \pi n, \frac{3\pi}{8} + \pi n$$

$$x = \frac{\pi}{8}, \frac{3\pi}{8}, \frac{9\pi}{8}, \frac{11\pi}{8}$$

45. $\cos 4x(\cos x - 1) = 0$

$\cos 4x = 0 \qquad\qquad\qquad\qquad \cos x - 1 = 0$

$$4x = \frac{\pi}{2} + 2\pi n, \frac{3\pi}{2} + 2\pi n \qquad \cos x = 1$$

$$x = \frac{\pi}{8} + \frac{\pi}{2}n, \frac{3\pi}{8} + \frac{\pi}{2}n \qquad\qquad x = 0$$

$$x = 0, \frac{\pi}{8}, \frac{3\pi}{8}, \frac{5\pi}{8}, \frac{7\pi}{8}, \frac{9\pi}{8}, \frac{11\pi}{8}, \frac{13\pi}{8}, \frac{15\pi}{8}$$

47. $\sin^2 x - 2 \sin x = 0$

$$\sin x(\sin x - 2) = 0$$

$\sin x = 0 \qquad \sin x - 2 = 0$

$x = 0, \pi \qquad$ No solution

49. $\tan^2 \theta + \tan \theta - 12 = 0$

$$(\tan \theta + 4)(\tan \theta - 3) = 0$$

$\tan \theta + 4 = 0 \qquad\qquad\qquad \tan \theta - 3 = 0$

$\theta = \arctan(-4) + n\pi \qquad\qquad \theta = \arctan 3 + n\pi$

$\theta = \arctan(-4) + \pi, \arctan(-4) + 2\pi, \arctan 3, \arctan 3 + \pi$

51. $\sin 285° = \sin(315° - 30°)$

$= \sin 315° \cos 30° - \cos 315° \sin 30°$

$= \left(-\dfrac{\sqrt{2}}{2}\right)\left(\dfrac{\sqrt{3}}{2}\right) - \left(\dfrac{\sqrt{2}}{2}\right)\left(\dfrac{1}{2}\right)$

$= -\dfrac{\sqrt{2}}{4}(\sqrt{3} + 1)$

$\cos 285° = \cos(315° - 30°)$

$= \cos 315° \cos 30° + \sin 315° \sin 30°$

$= \left(\dfrac{\sqrt{2}}{2}\right)\left(\dfrac{\sqrt{3}}{2}\right) + \left(-\dfrac{\sqrt{2}}{2}\right)\left(\dfrac{1}{2}\right)$

$= \dfrac{\sqrt{2}}{4}(\sqrt{3} - 1)$

$\tan 285° = \tan(315° - 30°) = \dfrac{\tan 315° - \tan 30°}{1 + \tan 315° \tan 30°}$

$= \dfrac{(-1) - \left(\dfrac{\sqrt{3}}{3}\right)}{1 + (-1)\left(\dfrac{\sqrt{3}}{3}\right)} = -2 - \sqrt{3}$

53. $\sin \dfrac{25\pi}{12} = \sin\left(\dfrac{11\pi}{6} + \dfrac{\pi}{4}\right) = \sin \dfrac{11\pi}{6} \cos \dfrac{\pi}{4} + \cos \dfrac{11\pi}{6} \sin \dfrac{\pi}{4}$

$= \left(-\dfrac{1}{2}\right)\left(\dfrac{\sqrt{2}}{2}\right) + \left(\dfrac{\sqrt{3}}{2}\right)\left(\dfrac{\sqrt{2}}{2}\right) = \dfrac{\sqrt{2}}{4}(\sqrt{3} - 1)$

$\cos \dfrac{25\pi}{12} = \cos\left(\dfrac{11\pi}{6} + \dfrac{\pi}{4}\right) = \cos \dfrac{11\pi}{6} \cos \dfrac{\pi}{4} - \sin \dfrac{11\pi}{6} \sin \dfrac{\pi}{4}$

$= \left(\dfrac{\sqrt{3}}{2}\right)\left(\dfrac{\sqrt{2}}{2}\right) - \left(-\dfrac{1}{2}\right)\left(\dfrac{\sqrt{2}}{2}\right) = \dfrac{\sqrt{2}}{4}(\sqrt{3} + 1)$

$\tan \dfrac{25\pi}{12} = \tan\left(\dfrac{11\pi}{6} + \dfrac{\pi}{4}\right) = \dfrac{\tan \dfrac{11\pi}{6} + \tan \dfrac{\pi}{4}}{1 - \tan \dfrac{11\pi}{6} \tan \dfrac{\pi}{4}}$

$= \dfrac{\left(-\dfrac{\sqrt{3}}{3}\right) + 1}{1 - \left(-\dfrac{\sqrt{3}}{3}\right)(1)} = 2 - \sqrt{3}$

55. $\sin 60° \cos 45° - \cos 60° \sin 45° = \sin(60° - 45°) = \sin 15°$

57. $\dfrac{\tan 25° + \tan 10°}{1 - \tan 25° \tan 10°} = \tan(25° + 10°) = \tan 35°$

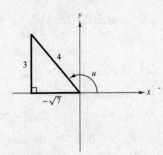

Figures for Exercises 59–63

59. $\sin(u + v) = \sin u \cos v + \cos u \sin v$

$= \left(\dfrac{3}{4}\right)\left(-\dfrac{5}{13}\right) + \left(-\dfrac{\sqrt{7}}{4}\right)\left(\dfrac{12}{13}\right)$

$= -\dfrac{3}{52}(5 + 4\sqrt{7})$

61. $\cos(u - v) = \cos u \cos v + \sin u \sin v$

$= \left(-\dfrac{\sqrt{7}}{4}\right)\left(-\dfrac{5}{13}\right) + \left(\dfrac{3}{4}\right)\left(\dfrac{12}{13}\right)$

$= \dfrac{1}{52}(5\sqrt{7} + 36)$

63. $\cos(u + v) = \cos u \cos v - \sin u \sin v$

$= \left(-\dfrac{\sqrt{7}}{4}\right)\left(-\dfrac{5}{13}\right) - \left(\dfrac{3}{4}\right)\left(\dfrac{12}{13}\right)$

$= \dfrac{1}{52}(5\sqrt{7} - 36)$

65. $\cos\left(x + \dfrac{\pi}{2}\right) = \cos x \cos \dfrac{\pi}{2} - \sin x \sin \dfrac{\pi}{2}$

$= \cos x(0) - \sin x(1)$

$= -\sin x$

67. $\cot\left(\dfrac{\pi}{2} - x\right) = \tan x$ by the cofunction identity.

69. $\cos 3x = \cos(2x + x)$

$= \cos 2x \cos x - \sin 2x \sin x$

$= (\cos^2 x - \sin^2 x)\cos x - (2 \sin x \cos x)\sin x$

$= \cos^3 x - \sin^2 x \cos x - 2 \sin^2 x \cos x$

$= \cos^3 x - 3 \sin^2 x \cos x$

$= \cos^3 x - 3(1 - \cos^2 x)\cos x$

$= \cos^3 x - 3 \cos x + 3 \cos^3 x$

$= 4 \cos^3 x - 3 \cos x$

71. $\sin\left(x + \dfrac{\pi}{4}\right) - \sin\left(x - \dfrac{\pi}{4}\right) = 1$

$2 \cos x \sin \dfrac{\pi}{4} = 1$

$\cos x = \dfrac{\sqrt{2}}{2}$

$x = \dfrac{\pi}{4}, \dfrac{7\pi}{4}$

73. $\sin\left(x + \dfrac{\pi}{2}\right) - \sin\left(x - \dfrac{\pi}{2}\right) = \sqrt{3}$

$2 \cos x \sin \dfrac{\pi}{2} = \sqrt{3}$

$\cos x = \dfrac{\sqrt{3}}{2}$

$x = \dfrac{\pi}{6}, \dfrac{11\pi}{6}$

75. $\sin u = -\dfrac{4}{5}, \ \pi < u < \dfrac{3\pi}{2}$

$\cos u = -\sqrt{1 - \sin^2 u} = \dfrac{-3}{5}$

$\tan u = \dfrac{\sin u}{\cos u} = \dfrac{4}{3}$

$\sin 2u = 2 \sin u \cos u = 2\left(-\dfrac{4}{5}\right)\left(-\dfrac{3}{5}\right) = \dfrac{24}{25}$

$\cos 2u = \cos^2 u - \sin^2 u = \left(-\dfrac{3}{5}\right)^2 - \left(-\dfrac{4}{5}\right)^2 = -\dfrac{7}{25}$

$\tan 2u = \dfrac{2 \tan u}{1 - \tan^2 u} = \dfrac{2\left(\dfrac{4}{3}\right)}{1 - \left(\dfrac{4}{3}\right)^2} = -\dfrac{24}{7}$

77. $\sin 4x = 2 \sin 2x \cos 2x$

$= 2[2 \sin x \cos x(\cos^2 x - \sin^2 x)]$

$= 4 \sin x \cos x(2 \cos^2 x - 1)$

$= 8 \cos^3 x \sin x - 4 \cos x \sin x$

79. $\tan^2 2x = \dfrac{\sin^2 2x}{\cos^2 2x} = \dfrac{\dfrac{1 - \cos 4x}{2}}{\dfrac{1 + \cos 4x}{2}} = \dfrac{1 - \cos 4x}{1 + \cos 4x}$

81. $\sin^2 x \tan^2 x = \sin^2 x\left(\dfrac{\sin^2 x}{\cos^2 x}\right) = \dfrac{\sin^4 x}{\cos^2 x}$

$= \dfrac{\left(\dfrac{1 - \cos 2x}{2}\right)^2}{\dfrac{1 + \cos 2x}{2}} = \dfrac{\dfrac{1 - 2\cos 2x + \cos^2 2x}{4}}{\dfrac{1 + \cos 2x}{2}}$

$= \dfrac{1 - 2 \cos 2x + \dfrac{1 + \cos 4x}{2}}{2(1 + \cos 2x)}$

$= \dfrac{2 - 4 \cos 2x + 1 + \cos 4x}{4(1 + \cos 2x)}$

$= \dfrac{3 - 4 \cos 2x + \cos 4x}{4(1 + \cos 2x)}$

83. $\sin(-75°) = -\sqrt{\dfrac{1 - \cos 150°}{2}} = -\sqrt{\dfrac{1 - \left(-\dfrac{\sqrt{3}}{2}\right)}{2}} = -\dfrac{\sqrt{2 + \sqrt{3}}}{2}$

$\qquad\qquad = -\dfrac{1}{2}\sqrt{2 + \sqrt{3}}$

$\cos(-75°) = \sqrt{\dfrac{1 + \cos 150°}{2}} = \sqrt{\dfrac{1 + \left(-\dfrac{\sqrt{3}}{2}\right)}{2}} = \dfrac{\sqrt{2 - \sqrt{3}}}{2}$

$\qquad\qquad = \dfrac{1}{2}\sqrt{2 - \sqrt{3}}$

$\tan(-75°) = -\left(\dfrac{1 - \cos 150°}{\sin 150°}\right) = -\left(\dfrac{1 - \left(-\dfrac{\sqrt{3}}{2}\right)}{\dfrac{1}{2}}\right) = -\left(2 + \sqrt{3}\right)$

$\qquad\qquad = -2 - \sqrt{3}$

85. $\sin\left(\dfrac{19\pi}{12}\right) = -\sqrt{\dfrac{1 - \cos\dfrac{19\pi}{6}}{2}} = -\sqrt{\dfrac{1 - \left(-\dfrac{\sqrt{3}}{2}\right)}{2}} = -\dfrac{\sqrt{2 + \sqrt{3}}}{2}$

$\qquad\qquad = -\dfrac{1}{2}\sqrt{2 + \sqrt{3}}$

$\cos\left(\dfrac{19\pi}{12}\right) = \sqrt{\dfrac{1 + \cos\dfrac{19\pi}{6}}{2}} = \sqrt{\dfrac{1 + \left(-\dfrac{\sqrt{3}}{2}\right)}{2}} = \dfrac{\sqrt{2 - \sqrt{3}}}{2}$

$\qquad\qquad = \dfrac{1}{2}\sqrt{2 - \sqrt{3}}$

$\tan\left(\dfrac{19\pi}{12}\right) = \dfrac{1 - \cos\dfrac{19\pi}{6}}{\sin\dfrac{19\pi}{6}} = \dfrac{1 - \left(-\dfrac{\sqrt{3}}{2}\right)}{-\dfrac{1}{2}} = -2 - \sqrt{3}$

87. Given $\sin u = \dfrac{3}{5}, 0 < u < \dfrac{\pi}{2} \implies \cos u = \dfrac{4}{5}$ and $\dfrac{u}{2}$ is in Quadrant I.

$\sin\left(\dfrac{u}{2}\right) = \sqrt{\dfrac{1 - \cos u}{2}} = \sqrt{\dfrac{1 - 4/5}{2}} = \sqrt{\dfrac{1}{10}} = \dfrac{\sqrt{10}}{10}$

$\cos\left(\dfrac{u}{2}\right) = \sqrt{\dfrac{1 + \cos u}{2}} = \sqrt{\dfrac{1 + 4/5}{2}} = \sqrt{\dfrac{9}{10}} = \dfrac{3\sqrt{10}}{10}$

$\tan\left(\dfrac{u}{2}\right) = \dfrac{1 - \cos u}{\sin u} = \dfrac{1 - 4/5}{3/5} = \dfrac{1}{3}$

89. Given $\cos u = -\dfrac{2}{7}, \dfrac{\pi}{2} < u < \pi \implies \sin u = \dfrac{3\sqrt{5}}{7}$ and $\dfrac{u}{2}$ is in Quadrant I.

$\sin\left(\dfrac{u}{2}\right) = \sqrt{\dfrac{1 - \cos u}{2}} = \sqrt{\dfrac{1 - (-2/7)}{2}} = \sqrt{\dfrac{9}{14}} = \dfrac{3}{\sqrt{14}} = \dfrac{3\sqrt{14}}{14}$

$\cos\left(\dfrac{u}{2}\right) = \sqrt{\dfrac{1 + \cos u}{2}} = \sqrt{\dfrac{1 + (-2/7)}{2}} = \sqrt{\dfrac{5}{14}} = \dfrac{\sqrt{70}}{14}$

$\tan\left(\dfrac{u}{2}\right) = \dfrac{1 - \cos u}{\sin u} = \dfrac{1 - (-2/7)}{3\sqrt{5}/7} = \dfrac{9/7}{3\sqrt{5}/7} = \dfrac{3}{\sqrt{5}} = \dfrac{3\sqrt{5}}{5}$

91. $-\sqrt{\dfrac{1 + \cos 10x}{2}} = -\left|\cos \dfrac{10x}{2}\right| = -|\cos 5x|$

93. $\cos \dfrac{\pi}{6} \sin \dfrac{\pi}{6} = \dfrac{1}{2}\left[\sin \dfrac{\pi}{3} - \sin 0\right] = \dfrac{1}{2}\sin \dfrac{\pi}{3}$

95. $\cos 5\theta \cos 3\theta = \dfrac{1}{2}[\cos 2\theta + \cos 8\theta]$

97. $\sin 4\theta - \sin 2\theta = 2\cos\left(\dfrac{4\theta + 2\theta}{2}\right)\sin\left(\dfrac{4\theta - 2\theta}{2}\right)$

$$= 2\cos 3\theta \sin \theta$$

99. $\cos\left(x + \dfrac{\pi}{6}\right) - \cos\left(x - \dfrac{\pi}{6}\right) = -2\sin x\, \sin \dfrac{\pi}{6}$

101. $r = \dfrac{1}{32}v_0^2 \sin 2\theta$

range $= 100$ feet

$v_0 = 80$ feet per second

$r = \dfrac{1}{32}(80)^2 \sin 2\theta = 100$

$\sin 2\theta = 0.5$

$2\theta = 30°$

$\theta = 15°$ or $\dfrac{\pi}{12}$

103. $y = 1.5 \sin 8t - 0.5 \cos 8t$

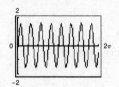

105. Amplitude $= \dfrac{\sqrt{10}}{2}$ feet

107. False. If $\dfrac{\pi}{2} < \theta < \pi$, then $\dfrac{\pi}{4} < \dfrac{\theta}{2} < \dfrac{\pi}{2}$ and $\dfrac{\theta}{2}$ is in Quadrant I.

$\cos \dfrac{\theta}{2} > 0$

109. True. $4\sin(-x)\cos(-x) = 4(-\sin x)\cos x$

$$= -4\sin x \cos x = -2(2\sin x \cos x)$$

$$= -2\sin 2x$$

111. Reciprocal Identities: $\sin \theta = \dfrac{1}{\csc \theta}$ $\csc \theta = \dfrac{1}{\sin \theta}$

$\cos \theta = \dfrac{1}{\sec \theta}$ $\sec \theta = \dfrac{1}{\cos \theta}$

$\tan \theta = \dfrac{1}{\cot \theta}$ $\cot \theta = \dfrac{1}{\tan \theta}$

Quotient Identities: $\tan \theta = \dfrac{\sin \theta}{\cos \theta}$ $\cot \theta = \dfrac{\cos \theta}{\sin \theta}$

Pythagorean Identities: $\sin^2 \theta + \cos^2 \theta = 1$

$$1 + \tan^2 \theta = \sec^2 \theta$$

$$1 + \cot^2 \theta = \csc^2 \theta$$

113. $a \sin x - b = 0$

$\sin x = \dfrac{b}{a}$

If $|b| > |a|$, then $\left|\dfrac{b}{a}\right| > 1$ and there is no solution

since $|\sin x| \le 1$ for all x.

115. The graph of y_1 is a vertical shift of the graph of y_2 one unit upward so $y_1 = y_2 + 1$.

117. $y = \sqrt{x + 3} + 4 \cos x$

Zeros: $x \approx -1.8431, 2.1758,$
3.9903, 8.8935, 9.8820

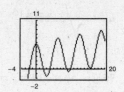

Problem Solving for Chapter 5

1. (a) Since $\sin^2 \theta + \cos^2 \theta = 1$ and $\cos^2 \theta = 1 - \sin^2 \theta$:

$$\cos \theta = \pm \sqrt{1 - \sin^2 \theta}$$

$$\tan \theta = \frac{\sin \theta}{\cos \theta} = \pm \frac{\sin \theta}{\sqrt{1 - \sin^2 \theta}}$$

$$\cot \theta = \frac{1}{\tan \theta} = \pm \frac{\sqrt{1 - \sin^2 \theta}}{\sin \theta}$$

$$\sec \theta = \frac{1}{\cos \theta} = \pm \frac{1}{\sqrt{1 - \sin^2 \theta}}$$

$$\cos \theta = \frac{1}{\sin \theta}$$

We also have the following relationships:

$$\cos \theta = \sin\left(\frac{\pi}{2} - \theta\right)$$

$$\tan \theta = \frac{\sin \theta}{\sin\left(\frac{\pi}{2} - \theta\right)}$$

$$\cot \theta = \frac{\sin\left(\frac{\pi}{2} - \theta\right)}{\sin \theta}$$

$$\sec \theta = \frac{1}{\sin\left(\frac{\pi}{2} - \theta\right)}$$

$$\csc \theta = \frac{1}{\sin \theta}$$

(b) $\sin \theta = \pm \sqrt{1 - \cos^2 \theta}$

$$\tan \theta = \frac{\sin \theta}{\cos \theta} = \pm \frac{\sqrt{1 - \cos^2 \theta}}{\cos \theta}$$

$$\csc \theta = \frac{1}{\sin \theta} = \pm \frac{1}{\sqrt{1 - \cos^2 \theta}}$$

$$\sec \theta = \frac{1}{\cos \theta}$$

$$\cot \theta = \frac{1}{\tan \theta} = \pm \frac{\cos \theta}{\sqrt{1 - \cos^2 \theta}}$$

We also have the following relationships:

$$\sin \theta = \cos\left(\frac{\pi}{2} - \theta\right)$$

$$\tan \theta = \frac{\cos\left(\frac{\pi}{2} - \theta\right)}{\cos \theta}$$

$$\csc \theta = \frac{1}{\cos\left(\frac{\pi}{2} - \theta\right)}$$

$$\sec \theta = \frac{1}{\cos \theta}$$

$$\cot \theta = \frac{\cos \theta}{\cos\left(\frac{\pi}{2} - \theta\right)}$$

3. $\sin\left[\dfrac{(12n + 1)\pi}{6}\right] = \sin\left[\dfrac{1}{6}(12n\pi + \pi)\right]$

$$= \sin\left(2n\pi + \frac{\pi}{6}\right)$$

$$= \sin \frac{\pi}{6} = \frac{1}{2}$$

Thus, $\sin\left[\dfrac{(12n + 1)\pi}{6}\right] = \dfrac{1}{2}$ for all integers n.

5. From the figure, it appears that $u + v = w$. Assume that u, v, and w are all in Quadrant I. From the figure:

$$\tan u = \frac{s}{3s} = \frac{1}{3}$$

$$\tan v = \frac{s}{2s} = \frac{1}{2}$$

$$\tan w = \frac{s}{s} = 1$$

$$\tan(u + v) = \frac{\tan u + \tan v}{1 - \tan u \tan v} = \frac{1/3 + 1/2}{1 - (1/3)(1/2)} = \frac{5/6}{1 - (1/6)} = 1 = \tan w.$$

Thus, $\tan(u + v) = \tan w$. Because u, v, and w are all in Quadrant I, we have

$$\arctan[\tan(u + v)] = \arctan[\tan w]$$

$$u + v = w.$$

7.

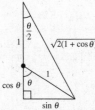

The hypotenuse of the larger right triangle is:

$$\sqrt{\sin^2 \theta + (1 + \cos \theta)^2} = \sqrt{\sin^2 \theta + 1 + 2 \cos \theta + \cos^2 \theta}$$

$$= \sqrt{2 + 2 \cos \theta}$$

$$= \sqrt{2(1 + \cos \theta)}$$

$$\sin\left(\frac{\theta}{2}\right) = \frac{\sin \theta}{\sqrt{2(1 + \cos \theta)}} = \frac{\sin \theta}{\sqrt{2(1 + \cos \theta)}} \cdot \frac{\sqrt{1 - \cos \theta}}{\sqrt{1 - \cos \theta}}$$

$$= \frac{\sin \theta \sqrt{1 - \cos \theta}}{\sqrt{2(1 - \cos^2 \theta)}} = \frac{\sin \theta \sqrt{1 - \cos \theta}}{\sqrt{2} \sin \theta}$$

$$= \sqrt{\frac{1 - \cos \theta}{2}}$$

$$\cos\left(\frac{\theta}{2}\right) = \frac{1 + \cos \theta}{\sqrt{2(1 + \cos \theta)}} = \frac{\sqrt{(1 + \cos \theta)^2}}{\sqrt{2(1 + \cos \theta)}} = \sqrt{\frac{1 + \cos \theta}{2}}$$

$$\tan\left(\frac{\theta}{2}\right) = \frac{\sin \theta}{1 + \cos \theta}$$

9. Seward: $D = 12.2 - 6.4 \cos\left[\dfrac{\pi (t + 0.2)}{182.6}\right]$

New Orleans: $D = 12.2 - 1.9 \cos\left[\dfrac{\pi (t + 0.2)}{182.6}\right]$

(a)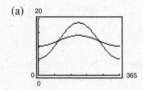

(b) The graphs intersect when $t \approx 91$ and when $t \approx 274$. These values correspond to April 1 and October 1, the spring equinox and the fall equinox.

(c) Seward has the greater variation in the number of daylight hours. This is determined by the amplitudes, 6.4 and 1.9.

(d) Period: $\dfrac{2\pi}{\dfrac{\pi}{182.6}} = 365.2$ days

11. (a) Let $y_1 = \sin x$ and $y_2 = 0.5$.

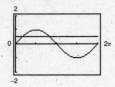

$\sin x \geq 0.5$ on the interval $\left[\dfrac{\pi}{6}, \dfrac{5\pi}{6}\right]$.

(b) Let $y_1 = \cos x$ and $y_2 = -0.5$.

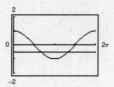

$\cos x \leq -0.5$ on the interval $\left[\dfrac{2\pi}{3}, \dfrac{4\pi}{3}\right]$.

(c) Let $y_1 = \tan x$ and $y_2 = \sin x$.

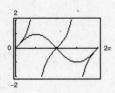

$\tan x < \sin x$ on the intervals $\left(\dfrac{\pi}{2}, \pi\right)$ and $\left(\dfrac{3\pi}{2}, 2\pi\right)$.

(d) Let $y_1 = \cos x$ and $y_2 = \sin x$.

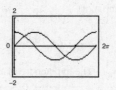

$\cos x \geq \sin x$ on the intervals $\left[0, \dfrac{\pi}{4}\right]$ and $\left[\dfrac{5\pi}{4}, 2\pi\right]$.

13. (a) $\sin(u + v + w) = \sin[(u + v) + w]$

$= \sin(u + v)\cos w + \cos(u + v)\sin w$

$= [\sin u \cos v + \cos u \sin v]\cos w + [\cos u \cos v - \sin u \sin v]\sin w$

$= \sin u \cos v \cos w + \cos u \sin v \cos w + \cos u \cos v \sin w - \sin u \sin v \sin w$

(b) $\tan(u + v + w) = \tan[(u + v) + w]$

$= \dfrac{\tan(u + v) + \tan w}{1 - \tan(u + v)\tan w}$

$= \dfrac{\left[\dfrac{\tan u + \tan v}{1 - \tan u \tan v}\right] + \tan w}{1 - \left[\dfrac{\tan u + \tan v}{1 - \tan u \tan v}\right]\tan w} \cdot \dfrac{(1 - \tan u \tan v)}{(1 - \tan u \tan v)}$

$= \dfrac{\tan u + \tan v + (1 - \tan u \tan v)\tan w}{(1 - \tan u \tan v) - (\tan u + \tan v)\tan w}$

$= \dfrac{\tan u + \tan v + \tan w - \tan u \tan v \tan w}{1 - \tan u \tan v - \tan u \tan w - \tan v \tan w}$

15. $h_1 = 3.75 \sin 733t + 7.5$

$h_2 = 3.75 \sin 733\left(t + \dfrac{4\pi}{3}\right) + 7.5$

(a)

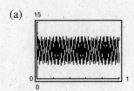

(b) The period for h_1 and h_2 is $\dfrac{2\pi}{733} \approx 0.0086$.

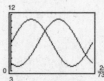

The graphs intersect twice per cycle.

There are $\dfrac{1}{2\pi/733} \approx 116.66$ cycles in the interval $[0, 1]$, so the graphs intersect approximately 233.3 times.

Chapter 5   Practice Test

1. Find the value of the other five trigonometric functions, given $\tan x = \frac{4}{11}$, $\sec x < 0$.

2. Simplify $\dfrac{\sec^2 x + \csc^2 x}{\csc^2 x(1 + \tan^2 x)}$.

3. Rewrite as a single logarithm and simplify $\ln|\tan \theta| - \ln|\cot \theta|$.

4. True or false:
$$\cos\left(\frac{\pi}{2} - x\right) = \frac{1}{\csc x}$$

5. Factor and simplify: $\sin^4 x + (\sin^2 x)\cos^2 x$

6. Multiply and simplify: $(\csc x + 1)(\csc x - 1)$

7. Rationalize the denominator and simplify:
$$\frac{\cos^2 x}{1 - \sin x}$$

8. Verify:
$$\frac{1 + \cos \theta}{\sin \theta} + \frac{\sin \theta}{1 + \cos \theta} = 2\csc \theta$$

9. Verify:
$$\tan^4 x + 2\tan^2 x + 1 = \sec^4 x$$

10. Use the sum or difference formulas to determine:

(a) $\sin 105°$ (b) $\tan 15°$

11. Simplify: $(\sin 42°)\cos 38° - (\cos 42°)\sin 38°$

12. Verify $\tan\left(\theta + \dfrac{\pi}{4}\right) = \dfrac{1 + \tan \theta}{1 - \tan \theta}$.

13. Write $\sin(\arcsin x - \arccos x)$ as an algebraic expression in x.

14. Use the double-angle formulas to determine:

(a) $\cos 120°$ (b) $\tan 300°$

15. Use the half-angle formulas to determine:

(a) $\sin 22.5°$ (d) $\tan\dfrac{\pi}{12}$

16. Given $\sin = 4/5$, θ lies in Quadrant II, find $\cos(\theta/2)$.

17. Use the power-reducing identities to write $(\sin^2 x)\cos^2 x$ in terms of the first power of cosine.

18. Rewrite as a sum: $6(\sin 5\theta)\cos 2\theta$.

19. Rewrite as a product:
$\sin(x + \pi) + \sin(x - \pi)$.

20. Verify $\dfrac{\sin 9x + \sin 5x}{\cos 9x - \cos 5x} = -\cot 2x$.

21. Verify:
$(\cos u)\sin v = \frac{1}{2}[\sin(u + v) - \sin(u - v)]$.

22. Find all solutions in the interval $[0, 2\pi)$:
$4\sin^2 x = 1$

23. Find all solutions in the interval $[0, 2\pi)$:
$\tan^2 \theta + \left(\sqrt{3} - 1\right)\tan\theta - \sqrt{3} = 0$

24. Find all solutions in the interval $[0, 2\pi)$:
$\sin 2x = \cos x$

25. Use the quadratic formula to find all solutions in the interval $[0, 2\pi)$:
$\tan^2 x - 6\tan x + 4 = 0$

CHAPTER 6
Additional Topics in Trigonometry

Section 6.1   Law of Sines **291**

Section 6.2   Law of Cosines **295**

Section 6.3   Vectors in the Plane **300**

Section 6.4   Vectors and Dot Products **306**

Section 6.5   Trigonometric Form of a Complex Number **311**

Review Exercises **321**

Problem Solving **328**

Practice Test **330**

CHAPTER 6
Additional Topics in Trigonometry

Section 6.1   Law of Sines

■   If ABC is any oblique triangle with sides a, b, and c, then

$$\frac{a}{\sin A} = \frac{b}{\sin B} = \frac{c}{\sin C}.$$

■   You should be able to use the Law of Sines to solve an oblique triangle for the remaining three parts, given:

(a) Two angles and any side (AAS or ASA)

(b) Two sides and an angle opposite one of them (SSA)

1. If A is acute and $h = b \sin A$:

(a) $a < h$, no triangle is possible.

(b) $a = h$ or $a > b$, one triangle is possible.

(c) $h < a < b$, two triangles are possible.

2. If A is obtuse and $h = b \sin A$:

(a) $a \leq b$, no triangle is possible.

(b) $a > b$, one triangle is possible.

■   The area of any triangle equals one-half the product of the lengths of two sides and the sine of their included angle.

$$A = \tfrac{1}{2}ab \sin C = \tfrac{1}{2}ac \sin B = \tfrac{1}{2}bc \sin A$$

Vocabulary Check

1. oblique

2. $\dfrac{b}{\sin B}$

3. $\dfrac{1}{2} ac \sin B$

1.

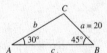

Given: $A = 30°$, $B = 45°$, $a = 20$

$C = 180° - A - B = 105°$

$b = \dfrac{a}{\sin A}(\sin B) = \dfrac{20 \sin 45°}{\sin 30°} = 20\sqrt{2} \approx 28.28$

$c = \dfrac{a}{\sin A}(\sin C) = \dfrac{20 \sin 105°}{\sin 30°} \approx 38.64$

3.

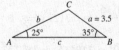

Given: $A = 25°$, $B = 35°$, $a = 3.5$

$C = 180° - A - B = 120°$

$b = \dfrac{a}{\sin A}(\sin B) = \dfrac{3.5}{\sin 25°}(\sin 35°) \approx 4.75$

$c = \dfrac{a}{\sin A}(\sin C) = \dfrac{3.5}{\sin 25°}(\sin 120°) \approx 7.17$

5. Given: $A = 36°$, $a = 8$, $b = 5$

$$\sin B = \frac{b \sin A}{a} = \frac{5 \sin 36°}{8} \approx 0.36737 \implies B \approx 21.55°$$

$$C = 180° - A - B \approx 180° - 36° - 21.55 = 122.45°$$

$$c = \frac{a}{\sin A}(\sin C) = \frac{8}{\sin 36°}(\sin 122.45°) \approx 11.49$$

7. Given: $A = 102.4°$, $C = 16.7°$, $a = 21.6$

$$B = 180° - A - C = 60.9°$$

$$b = \frac{a}{\sin A}(\sin B) = \frac{21.6}{\sin 102.4°}(\sin 60.9°) \approx 19.32$$

$$c = \frac{a}{\sin A}(\sin C) = \frac{21.6}{\sin 102.4°}(\sin 16.7°) \approx 6.36$$

9. Given: $A = 83° 20'$, $C = 54.6°$, $c = 18.1$

$$B = 180° - A - C = 180° - 83° 20' - 54° 36' = 42° 4'$$

$$a = \frac{c}{\sin C}(\sin A) = \frac{18.1}{\sin 54.6°}(\sin 83° 20') \approx 22.05$$

$$b = \frac{c}{\sin C}(\sin B) = \frac{18.1}{\sin 54.6°}(\sin 42° 4') \approx 14.88$$

11. Given: $B = 15° 30'$, $a = 4.5$, $b = 6.8$

$$\sin A = \frac{a \sin B}{b} = \frac{4.5 \sin 15° 30'}{6.8} \approx 0.17685 \implies A \approx 10° 11'$$

$$C = 180° - A - B \approx 180° - 10° 11' - 15° 30' = 154° 19'$$

$$c = \frac{b}{\sin B}(\sin C) = \frac{6.8}{\sin 15° 30'}(\sin 154° 19') \approx 11.03$$

13. Given: $C = 145°$, $b = 4$, $c = 14$

$$\sin B = \frac{b \sin C}{c} = \frac{4 \sin 145°}{14} \approx 0.16388 \implies B \approx 9.43°$$

$$A = 180° - B - C \approx 180° - 9.43° - 145° = 25.57°$$

$$a = \frac{c}{\sin C}(\sin A) \approx \frac{14}{\sin 145°}(\sin 25.57°) \approx 10.53$$

15. Given: $A = 110° 15'$, $a = 48$, $b = 16$

$$\sin B = \frac{b \sin A}{a} = \frac{16 \sin 110° 15'}{48} \approx 0.31273 \implies B \approx 18° 13'$$

$$C = 180° - A - B \approx 180° - 110° 15' - 18° 13' = 51° 32'$$

$$c = \frac{a}{\sin A}(\sin C) = \frac{48}{\sin 110° 15'}(\sin 51° 32') \approx 40.06$$

17. Given: $A = 55°$, $B = 42°$, $c = \dfrac{3}{4}$

$$C = 180° - A - B = 83°$$

$$a = \frac{c}{\sin C}(\sin A) = \frac{0.75}{\sin 83°}(\sin 55°) \approx 0.62$$

$$b = \frac{c}{\sin C}(\sin B) = \frac{0.75}{\sin 83°}(\sin 42°) \approx 0.51$$

19. Given: $A = 110°$, $a = 125$, $b = 100$

$$\sin B = \frac{b \sin A}{a} = \frac{100 \sin 110°}{125} \approx 0.75175 \implies B \approx 48.74°$$

$$C = 180° - A - B \approx 21.26°$$

$$c = \frac{a \sin C}{\sin A} \approx \frac{125 \sin 21.26°}{\sin 110°} \approx 48.23$$

21. Given: $a = 18, b = 20, A = 76°$

$h = 20 \sin 76° \approx 19.41$

Since $a < h$, no triangle is formed.

23. Given: $A = 58°, a = 11.4, c = 12.8$

$\sin B = \dfrac{b \sin A}{a} = \dfrac{12.8 \sin 58°}{11.4} \approx 0.9522 \implies B \approx 72.21°$ or $B \approx 107.79°$

<u>Case 1</u>

$B \approx 72.21°$

$C = 180° - A - B \approx 49.79°$

$c = \dfrac{a}{\sin A}(\sin C) \approx \dfrac{11.4 \sin 49.79°}{\sin 58°} \approx 10.27$

<u>Case 2</u>

$B \approx 107.79°$

$C = 180° - A - B \approx 14.21°$

$c = \dfrac{a}{\sin A}(\sin C) \approx \dfrac{11.4 \sin 14.21°}{\sin 58°} \approx 3.30$

25. Given: $A = 36°, a = 5$

(a) One solution if $b \leq 5$ or $b = \dfrac{5}{\sin 36°}$

(b) Two solutions if $5 < b < \dfrac{5}{\sin 36°}$

(c) No solution if $b > \dfrac{5}{\sin 36°}$

27. Given: $A = 10°, a = 10.8$

(a) One solution if $b \leq 10.8$ or $b = \dfrac{10.8}{\sin 10°}$

(b) Two solutions if $10.8 < b < \dfrac{10.8}{\sin 10°}$

(c) No solution if $b > \dfrac{10.8}{\sin 10°}$

29. Area $= \frac{1}{2}ab \sin C = \frac{1}{2}(4)(6) \sin 120° \approx 10.4$

31. Area $= \frac{1}{2}bc \sin A = \frac{1}{2}(57)(85) \sin 43° \, 45' \approx 1675.2$

33. Area $= \dfrac{1}{2}ac \sin B = \dfrac{1}{2}(105)(64)\sin(72°30') \approx 3204.5$

35. $C = 180° - 23° - 94° = 63°$

$h = \dfrac{35}{\sin 63°}(\sin 23°) \approx 15.3$ meters

37. $\dfrac{\sin(42° - \theta)}{10} = \dfrac{\sin 48°}{17}$

$\sin(42° - \theta) \approx 0.43714$

$42° - \theta \approx 25.9°$

$\theta \approx 16.1°$

39. Given: $c = 100$

$A = 74° - 28° = 46°,$

$B = 180° - 41° - 74° = 65°,$

$C = 180° - 46° - 65° = 69°$

$a = \dfrac{c}{\sin C}(\sin A) = \dfrac{100}{\sin 69°}(\sin 46°) \approx 77$ meters

(c) $\dfrac{y}{\sin 71.2°} = \dfrac{x}{\sin 90°}$

$y = x \sin 71.2° \approx 119{,}289.1261 \sin 71.2°$

$\approx 112{,}924.963$ feet ≈ 21.4 miles

(d) $z = x \sin 18.8° \approx 119{,}289.1261 \sin 18.8°$

$\approx 38{,}443$ feet ≈ 7.3 miles

41. (a)

Not drawn to scale

(b) $\dfrac{x}{\sin 17.5°} = \dfrac{9000}{\sin 1.3°}$

$x \approx 119{,}289.1261$ feet ≈ 22.6 miles

43.

In 15 minutes the boat has traveled

$$(10 \text{ mph})\left(\tfrac{1}{4} \text{ hr}\right) = \tfrac{10}{4} \text{ miles.}$$

$$\theta = 180° - 20° - (90° + 63°)$$

$$\theta = 7°$$

$$\frac{10/4}{\sin 7°} = \frac{y}{\sin 20°}$$

$$y \approx 7.0161$$

$$\sin 27° = \frac{d}{7.0161}$$

$$d \approx 3.2 \text{ miles}$$

45. True. If one angle of a triangle is obtuse, then there is less than 90° left for the other two angles, so it cannot contain a right angle. It must be oblique.

47. (a) $\dfrac{\sin \alpha}{9} = \dfrac{\sin \beta}{18}$

$\sin \alpha = 0.5 \sin \beta$

$\alpha = \arcsin(0.5 \sin \beta)$

(b)

Domain: $0 < \beta < \pi$

Range: $0 < \alpha \leq \dfrac{\pi}{6}$

(c) $\gamma = \pi - \alpha - \beta = \pi - \beta - \arcsin(0.5 \sin \beta)$

$$\frac{c}{\sin \gamma} = \frac{18}{\sin \beta}$$

$$c = \frac{18 \sin \gamma}{\sin \beta} = \frac{18 \sin[\pi - \beta - \arcsin(0.5 \sin \beta)]}{\sin \beta}$$

(d)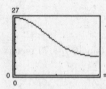

Domain: $0 < \beta < \pi$

Range: $9 < c < 27$

(e)

β	0.4	0.8	1.2	1.6	2.0	2.4	2.8
α	0.1960	0.3669	0.4848	0.5234	0.4720	0.3445	0.1683
c	25.95	23.07	19.19	15.33	12.29	10.31	9.27

As $\beta \to 0$, $c \to 27$

As $\beta \to \pi$, $c \to 9$

49. $\sin x \cot x = \sin x \dfrac{\cos x}{\sin x} = \cos x$

51. $1 - \sin^2\left(\dfrac{\pi}{2} - x\right) = 1 - \cos^2 x = \sin^2 x$

53. $6 \sin 8\theta \cos 3\theta = (6)\left(\tfrac{1}{2}\right)[\sin(8\theta + 3\theta) + \sin(8\theta - 3\theta)] = 3(\sin 11\theta + \sin 5\theta)$

Section 6.2   Law of Cosines

- If ABC is any oblique triangle with sides a, b, and c, the following equations are valid.

    (a) $a^2 = b^2 + c^2 - 2bc \cos A$ or $\cos A = \dfrac{b^2 + c^2 - a^2}{2bc}$

    (b) $b^2 = a^2 + c^2 - 2ac \cos B$ or $\cos B = \dfrac{a^2 + c^2 - b^2}{2ac}$

    (c) $c^2 = a^2 + b^2 - 2ab \cos C$ or $\cos C = \dfrac{a^2 + b^2 - c^2}{2ab}$

- You should be able to use the Law of Cosines to solve an oblique triangle for the remaining three parts, given:

    (a) Three sides (SSS)

    (b) Two sides and their included angle (SAS)

- Given any triangle with sides of length a, b, and c, the area of the triangle is

    $$\text{Area} = \sqrt{s(s-a)(s-b)(s-c)}, \text{ where } s = \frac{a+b+c}{2}. \qquad \text{(Heron's Formula)}$$

Vocabulary Check

1. Cosines

2. $b^2 = a^2 + c^2 - 2ac \cos B$

3. Heron's Area

---

1. Given: $a = 7, b = 10, c = 15$

$$\cos C = \frac{a^2 + b^2 - c^2}{2ab} = \frac{49 + 100 - 225}{2(7)(10)} \approx -0.5429 \implies C \approx 122.88°$$

$$\sin B = \frac{b \sin C}{c} = \frac{10 \sin 122.88°}{15} \approx 0.5599 \implies B \approx 34.05°$$

$$A \approx 180° - 34.05° - 122.88° \approx 23.07°$$

3. Given: $A = 30°, b = 15, c = 30$

$$a^2 = b^2 + c^2 - 2bc \cos A$$

$$= 225 + 900 - 2(15)(30) \cos 30° \approx 345.5771$$

$$a \approx 18.59$$

$$\cos C = \frac{a^2 + b^2 - c^2}{2ab} \approx \frac{(18.59)^2 + 15^2 - 30^2}{2(18.59)(15)} \approx -0.5907 \implies C \approx 126.21°$$

$$B \approx 180° - 30° - 126.21° = 13.79°$$

5. $a = 11, b = 14, c = 20$

$$\cos C = \frac{a^2 + b^2 - c^2}{2ab} = \frac{121 + 196 - 400}{2(11)(14)} \approx -0.2695 \implies C \approx 105.63°$$

$$\sin B = \frac{b \sin C}{c} = \frac{14 \sin 105.63°}{20} \approx 0.6741 \implies B \approx 42.38°$$

$$A \approx 180° - 42.38° - 105.63° \approx 31.99°$$

7. Given: $a = 75.4$, $b = 52$, $c = 52$

   $$\cos A = \frac{b^2 + c^2 - a^2}{2bc} = \frac{52^2 + 52^2 - 75.4^2}{2(52)(52)} = -0.05125 \implies A \approx 92.94°$$

   $$\sin B = \frac{b \sin A}{a} \approx \frac{52(0.9987)}{75.4} \approx 0.68876 \implies B \approx 43.53°$$

   $$C = B \approx 43.53°$$

9. Given: $A = 135°$, $b = 4$, $c = 9$

   $$a^2 = b^2 + c^2 - 2bc \cos A = 16 + 81 - 2(4)(9)\cos 135° \approx 147.9117 \implies a \approx 12.16$$

   $$\sin B = \frac{b \sin A}{a} = \frac{4 \sin 135°}{12.16} \approx 0.2326 \implies B \approx 13.45°$$

   $$C \approx 180° - 135° - 13.45° \approx 31.55°$$

11. Given: $B = 10° \, 35'$, $a = 40$, $c = 30$

    $$b^2 = a^2 + c^2 - 2ac \cos B = 1600 + 900 - 2(40)(30)\cos 10° \, 35' \approx 140.8268 \implies b \approx 11.87$$

    $$\sin C = \frac{c \sin B}{b} = \frac{30 \sin 10° \, 35'}{11.87} \approx 0.4642 \implies C \approx 27.66° \approx 27° \, 40'$$

    $$A \approx 180° - 10° \, 35' - 27° \, 40' = 141° \, 45'$$

13. Given: $B = 125° \, 40'$, $a = 32$, $c = 32$

    $$b^2 = a^2 + c^2 - 2ac \cos B \approx 32^2 + 32^2 - 2(32)(32) \cos 125° \, 40' \approx 3242.1888 \implies b \approx 56.94$$

    $$A = C \implies 2A = 180° - 125° \, 40' = 54° \, 20' \implies A = C = 27° \, 10'$$

15. $C = 43°$, $a = \dfrac{4}{9}$, $b = \dfrac{7}{9}$

    $$c^2 = a^2 + b^2 - 2ab \cos C = \left(\frac{4}{9}\right)^2 + \left(\frac{7}{9}\right)^2 - 2\left(\frac{4}{9}\right)\left(\frac{7}{9}\right)\cos 43° \approx 0.2968 \implies c \approx 0.54$$

    $$\sin A = \frac{a \sin C}{c} = \frac{(4/9) \sin 43°}{0.5448} \approx 0.5564 \implies A \approx 33.80°$$

    $$B \approx 180° - 43° - 33.8° \approx 103.20°$$

17. $d^2 = 5^2 + 8^2 - 2(5)(8)\cos 45° \approx 32.4315 \implies d \approx 5.69$

    $2\phi = 360° - 2(45°) = 270° \implies \phi = 135°$

    $c^2 = 5^2 + 8^2 - 2(5)(8)\cos 135° \approx 145.5685 \implies c \approx 12.07$

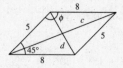

19. $\cos \phi = \dfrac{10^2 + 14^2 - 20^2}{2(10)(14)}$

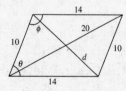

    $\phi \approx 111.8°$

    $2\theta \approx 360° - 2(111.8°)$

    $\theta = 68.2°$

    $d^2 = 10^2 + 14^2 - 2(10)(14) \cos 68.2°$

    $d \approx 13.86$

21. $\cos\alpha = \dfrac{(12.5)^2 + (15)^2 - 10^2}{2(12.5)(15)} = 0.75 \implies \alpha \approx 41.41°$

$\cos\beta = \dfrac{10^2 + 15^2 - (12.5)^2}{2(10)(15)} = 0.5625 \implies \beta \approx 55.77°$

$z = 180° - \alpha - \beta = 82.82°$

$u = 180° - z = 97.18°$

$b^2 = 12.5^2 + 10^2 - 2(12.5)(10)\cos 97.18° \approx 287.4967 \implies b \approx 16.96$

$\cos\delta = \dfrac{12.5^2 + 16.96^2 - 10^2}{2(12.5)(16.96)} \approx 0.8111 \implies \delta \approx 35.80°$

$\theta = \alpha + \delta = 41.41° + 35.80° \approx 77.2°$

$2\phi = 360° - 2\theta \implies \phi = \dfrac{360° - 2(77.21°)}{2} = 102.8°$

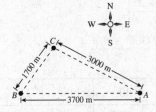

23. $a = 5, \ b = 7, \ c = 10 \implies s = \dfrac{a + b + c}{2} = 11$

Area $= \sqrt{s(s - a)(s - b)(s - c)} = \sqrt{11(6)(4)(1)} \approx 16.25$

25. $a = 2.5, b = 10.2, c = 9 \implies s = \dfrac{a + b + c}{2} = 10.85$

Area $= \sqrt{s(s - a)(s - b)(s - c)} = \sqrt{10.85(8.35)(0.65)(1.85)} \approx 10.4$

27. $a = 12.32, b = 8.46, c = 15.05 \implies s = \dfrac{a + b + c}{2} = 17.915$

Area $= \sqrt{s(s - a)(s - b)(s - c)} = \sqrt{17.915(5.595)(9.455)(2.865)} \approx 52.11$

29. $\cos B = \dfrac{1700^2 + 3700^2 - 3000^2}{2(1700)(3700)} \implies B \approx 52.9°$

Bearing: $90° - 52.9° = $ N $37.1°$ E

$\cos C = \dfrac{1700^2 + 3000^2 - 3700^2}{2(1700)(3000)} \implies C \approx 100.2°$

Bearing: $90° - 26.9° = $ S $63.1°$ E

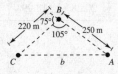

31. $b^2 = 220^2 + 250^2 - 2(220)(250)\cos 105° \implies b \approx 373.3$ meters

33.

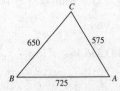

The largest angle is across from the largest side.

$\cos C = \dfrac{650^2 + 575^2 - 725^2}{2(650)(575)}$

$C \approx 72.3°$

35. $C = 180° - 53° - 67° = 60°$

$c^2 = a^2 + b^2 - 2ab\cos C$

$= 36^2 + 48^2 - 2(36)(48)(0.5)$

$= 1872$

$c \approx 43.3$ mi

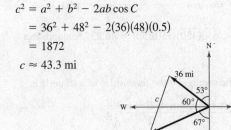

37. (a) $\cos\theta = \dfrac{273^2 + 178^2 - 235^2}{2(273)(178)}$

$\theta \approx 58.4°$

Bearing: N 58.4° W

(b) $\cos\phi = \dfrac{235^2 + 178^2 - 273^2}{2(235)(178)}$

$\phi \approx 81.5°$

Bearing: S 81.5° W

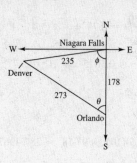

39. $d^2 = 60.5^2 + 90^2 - 2(60.5)(90)\cos 45° \approx 4059.8572 \implies d \approx 63.7$ ft

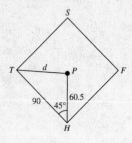

41. $a^2 = 35^2 + 20^2 - 2(35)(20)\cos 42° \implies a \approx 24.2$ miles

43. $\overline{RS} = \sqrt{8^2 + 10^2} = \sqrt{164} = 2\sqrt{41} \approx 12.8$ ft

$\overline{PQ} = \frac{1}{2}\sqrt{16^2 + 10^2} = \frac{1}{2}\sqrt{356} = \sqrt{89} \approx 9.4$ ft

$\tan P = \dfrac{10}{16} = \dfrac{\overline{QS}}{\overline{PS}} = \dfrac{\overline{QS}}{8} \implies \overline{QS} = 5$

45. $d^2 = 10^2 + 7^2 - 2(10)(7)\cos\theta$

$\theta = \arccos\left[\dfrac{10^2 + 7^2 - d^2}{2(10)(7)}\right]$

$s = \dfrac{360° - \theta}{360°}(2\pi r) = \dfrac{(360° - \theta)\pi}{45°}$

d (inches)	9	10	12	13	14	15	16
θ (degrees)	60.9°	69.5°	88.0°	98.2°	109.6°	122.9°	139.8°
s (inches)	20.88	20.28	18.99	18.28	17.48	16.55	15.37

47. $a = 200$

$b = 500$

$c = 600 \implies s = \dfrac{200 + 500 + 600}{2} = 650$

Area $= \sqrt{650(450)(150)(50)} \approx 46{,}837.5$ square feet

49. $s = \dfrac{510 + 840 + 1120}{2} = 1235$

Area $= \sqrt{1235(1235 - 510)(1235 - 840)(1235 - 1120)}$

$\approx 201{,}674$ square yards

Cost $\approx \left(\dfrac{201{,}674}{4840}\right)(2000) \approx \$83{,}336.36$

51. False. The average of the three sides of a triangle is

$\dfrac{a + b + c}{3}$, not $\dfrac{a + b + c}{2} = s$.

53. False. If $a = 10$, $b = 16$, and $c = 5$, then by the Law of Cosines, we would have:

$$\cos A = \frac{16^2 + 5^2 - 10^2}{2(16)(5)} = 1.13125 > 1$$

This is not possible. In general, if the sum of any two sides is less than the third side, then they cannot form a triangle. Here $10 + 5$ is less than 16.

55. $a = 25$, $b = 55$, $c = 72$

(a) area of triangle: $s = \frac{1}{2}(25 + 55 + 72) = 76$

area $= \sqrt{76(51)(21)(4)} \approx 570.60$

(b) area of circumscribed circle:

$$\cos C = \frac{25^2 + 55^2 - 72^2}{2(25)(55)} \approx -0.5578 \implies C \approx 123.9°$$

$$R = \frac{1}{2}\left(\frac{c}{\sin C}\right) \approx 43.37 \ \text{(see \#54)}$$

area $= \pi R^2 \approx 5909.2$

(c) area of inscribed circle:

$$r = \sqrt{\frac{(s-a)(s-b)(s-c)}{s}} = \sqrt{\frac{(51)(21)(4)}{76}} \approx 7.51 \ \text{(see \#54)}$$

area $= \pi r^2 \approx 177.09$

57. $\dfrac{1}{2}bc(1 + \cos A) = \dfrac{1}{2}bc\left[1 + \dfrac{b^2 + c^2 - a^2}{2bc}\right]$

$\qquad = \dfrac{1}{2}bc\left[\dfrac{2bc + b^2 + c^2 - a^2}{2bc}\right]$

$\qquad = \dfrac{1}{4}[(b + c)^2 - a^2]$

$\qquad = \dfrac{1}{4}[(b + c) + a][(b + c) - a]$

$\qquad = \dfrac{b + c + a}{2} \cdot \dfrac{b + c - a}{2}$

$\qquad = \dfrac{a + b + c}{2} \cdot \dfrac{-a + b + c}{2}$

59. $\arcsin(-1) = -\dfrac{\pi}{2}$

61. $\arctan\sqrt{3} = \dfrac{\pi}{3}$

63. $\arcsin\left(-\dfrac{\sqrt{3}}{2}\right) = -\dfrac{\pi}{3}$

65. Let $\theta = \arcsin 2x$, then

$\sin \theta = 2x = \dfrac{2x}{1}$ and

$\sec \theta = \dfrac{1}{\sqrt{1 - 4x^2}}$.

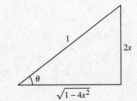

67. Let $\theta = \arctan(x - 2)$, then

$\tan \theta = x - 2 = \dfrac{x - 2}{1}$ and

$\cot \theta = \dfrac{1}{x - 2}$.

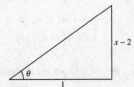

69. $5 = \sqrt{25 - x^2}, x = 5 \sin \theta$

$\quad 5 = \sqrt{25 - (5 \sin \theta)^2}$

$\quad 5 = \sqrt{25(1 - \sin^2 \theta)}$

$\quad 5 = 5 \cos \theta$

$\quad \cos \theta = 1$

$\quad \sec \theta = \dfrac{1}{\cos \theta} = 1$

$\quad \csc \theta$ is undefined.

71. $-\sqrt{3} = \sqrt{x^2 - 9}, x = 3 \sec \theta$

$\quad -\sqrt{3} = \sqrt{(3 \sec \theta)^2 - 9}$

$\quad -\sqrt{3} = \sqrt{9(\sec^2 \theta - 1)}$

$\quad -\sqrt{3} = 3 \tan \theta$

$\quad \tan \theta = -\dfrac{\sqrt{3}}{3}$

$\quad \sec \theta = \sqrt{1 + \tan^2 \theta} = \sqrt{1 + \left(-\dfrac{\sqrt{3}}{3}\right)^2} = \dfrac{2\sqrt{3}}{3}$

$\quad \cot \theta = \dfrac{1}{\tan \theta} = -\sqrt{3}$

$\quad \csc \theta = -\sqrt{1 + \cot^2 \theta} = -\sqrt{1 + (-\sqrt{3})^2} = -2$

73. $\cos \dfrac{5\pi}{6} - \cos \dfrac{\pi}{3} = -2 \sin\left(\dfrac{\dfrac{5\pi}{6} + \dfrac{\pi}{3}}{2}\right) \sin\left(\dfrac{\dfrac{5\pi}{6} - \dfrac{\pi}{3}}{2}\right)$

$\qquad\qquad\qquad\quad = -2 \sin \dfrac{7\pi}{12} \sin \dfrac{\pi}{4}$

Section 6.3   Vectors in the Plane

- A vector **v** is the collection of all directed line segments that are equivalent to a given directed line segment $\overrightarrow{PQ}$.

- You should be able to *geometrically* perform the operations of vector addition and scalar multiplication.

- The component form of the vector with initial point $P = (p_1, p_2)$ and terminal point $Q = (q_1, q_2)$ is

$$\overrightarrow{PQ} = \langle q_1 - p_1, q_2 - p_2 \rangle = \langle v_1, v_2 \rangle = \mathbf{v}.$$

- The magnitude of $\mathbf{v} = \langle v_1, v_2 \rangle$ is given by $\|\mathbf{v}\| = \sqrt{v_1^2 + v_2^2}$.

- If $\|\mathbf{v}\| = 1$, **v** is a unit vector.

- You should be able to perform the operations of scalar multiplication and vector addition in component form.

  (a) $\mathbf{u} + \mathbf{v} = \langle u_1 + v_1, u_2 + v_2 \rangle$ (b) $k\mathbf{u} = \langle ku_1, ku_2 \rangle$

- You should know the following properties of vector addition and scalar multiplication.

  (a) $\mathbf{u} + \mathbf{v} = \mathbf{v} + \mathbf{u}$

  (b) $(\mathbf{u} + \mathbf{v}) + \mathbf{w} = \mathbf{u} + (\mathbf{v} + \mathbf{w})$

  (c) $\mathbf{u} + \mathbf{0} = \mathbf{u}$

  (d) $\mathbf{u} + (-\mathbf{u}) = \mathbf{0}$

  (e) $c(d\mathbf{u}) = (cd)\mathbf{u}$

  (f) $(c + d)\mathbf{u} = c\mathbf{u} + d\mathbf{u}$

  (g) $c(\mathbf{u} + \mathbf{v}) = c\mathbf{u} + c\mathbf{v}$

  (h) $1(\mathbf{u}) = \mathbf{u}, 0\mathbf{u} = \mathbf{0}$

  (i) $\|c\mathbf{v}\| = |c| \, \|\mathbf{v}\|$

- A unit vector in the direction of **v** is $\mathbf{u} = \dfrac{\mathbf{v}}{\|\mathbf{v}\|}$.

- The standard unit vectors are $\mathbf{i} = \langle 1, 0 \rangle$ and $\mathbf{j} = \langle 0, 1 \rangle$. $\mathbf{v} = \langle v_1, v_2 \rangle$ can be written as $\mathbf{v} = v_1\mathbf{i} + v_2\mathbf{j}$.

- A vector **v** with magnitude $\|\mathbf{v}\|$ and direction θ can be written as $\mathbf{v} = a\mathbf{i} + b\mathbf{j} = \|\mathbf{v}\|(\cos \theta)\mathbf{i} + \|\mathbf{v}\|(\sin \theta)\mathbf{j}$, where $\tan \theta = b/a$.

Vocabulary Check

1. directed line segment

2. initial; terminal

3. magnitude

4. vector

5. standard position

6. unit vector

7. multiplication; addition

8. resultant

9. linear combination; horizontal; vertical

1. $\mathbf{v} = \langle 4 - 0, 1 - 0 \rangle = \langle 4, 1 \rangle$

$\mathbf{u} = \mathbf{v}$

3. Initial point: $(0, 0)$

Terminal point: $(3, 2)$

$\mathbf{v} = \langle 3 - 0, 2 - 0 \rangle = \langle 3, 2 \rangle$

$\|\mathbf{v}\| = \sqrt{3^2 + 2^2} = \sqrt{13}$

5. Initial point: $(2, 2)$

Terminal point: $(-1, 4)$

$\mathbf{v} = \langle -1 - 2, 4 - 2 \rangle = \langle -3, 2 \rangle$

$\|\mathbf{v}\| = \sqrt{(-3)^2 + 2^2} = \sqrt{13}$

7. Initial point: $(3, -2)$

Terminal point: $(3, 3)$

$\mathbf{v} = \langle 3 - 3, 3 - (-2) \rangle = \langle 0, 5 \rangle$

$\|\mathbf{v}\| = \sqrt{0^2 + 5^2} = \sqrt{25} = 5$

9. Initial point: $(-1, 5)$

Terminal point: $(15, 12)$

$\mathbf{v} = \langle 15 - (-1), 12 - 5 \rangle = \langle 16, 7 \rangle$

$\|\mathbf{v}\| = \sqrt{16^2 + 7^2} = \sqrt{305}$

11. Initial point: $(-3, -5)$

Terminal point: $(5, 1)$

$\mathbf{v} = \langle 5 - (-3), 1 - (-5) \rangle = \langle 8, 6 \rangle$

$\|\mathbf{v}\| = \sqrt{8^2 + 6^2} = \sqrt{100} = 10$

13. Initial point: $(1, 3)$

Terminal point: $(-8, -9)$

$\mathbf{v} = \langle -8 - 1, -9 - 3 \rangle = \langle -9, -12 \rangle$

$\|\mathbf{v}\| = \sqrt{(-9)^2 + (-12)^2} = \sqrt{225} = 15$

15.

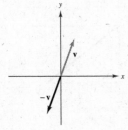

17.

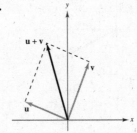

19. $\mathbf{u} + 2\mathbf{v}$

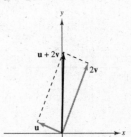

21. $\mathbf{u} = \langle 2, 1 \rangle, \mathbf{v} = \langle 1, 3 \rangle$

(a) $\mathbf{u} + \mathbf{v} = \langle 3, 4 \rangle$

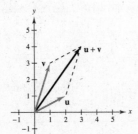

(b) $\mathbf{u} - \mathbf{v} = \langle 1, -2 \rangle$

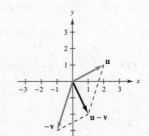

(c) $2\mathbf{u} - 3\mathbf{v} = \langle 4, 2 \rangle - \langle 3, 9 \rangle = \langle 1, -7 \rangle$

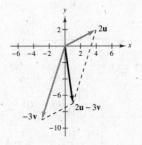

23. $\mathbf{u} = \langle -5, 3 \rangle$, $\mathbf{v} = \langle 0, 0 \rangle$

    (a) $\mathbf{u} + \mathbf{v} = \langle -5, 3 \rangle = \mathbf{u}$ (b) $\mathbf{u} - \mathbf{v} = \langle -5, 3 \rangle = \mathbf{u}$ (c) $2\mathbf{u} - 3\mathbf{v} = 2\mathbf{u} = \langle -10, 6 \rangle$

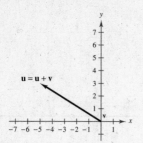 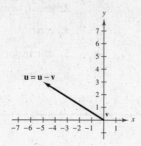

25. $\mathbf{u} = \mathbf{i} + \mathbf{j}$, $\mathbf{v} = 2\mathbf{i} - 3\mathbf{j}$

    (a) $\mathbf{u} + \mathbf{v} = 3\mathbf{i} - 2\mathbf{j}$ (b) $\mathbf{u} - \mathbf{v} = -\mathbf{i} + 4\mathbf{j}$ (c) $2\mathbf{u} - 3\mathbf{v} = (2\mathbf{i} + 2\mathbf{j}) - (6\mathbf{i} - 9\mathbf{j})$

$$= -4\mathbf{i} + 11\mathbf{j}$$

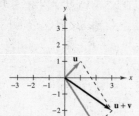 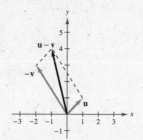 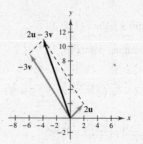

27. $\mathbf{u} = 2\mathbf{i}$, $\mathbf{v} = \mathbf{j}$

    (a) $\mathbf{u} + \mathbf{v} = 2\mathbf{i} + \mathbf{j}$ (b) $\mathbf{u} - \mathbf{v} = 2\mathbf{i} - \mathbf{j}$ (c) $2\mathbf{u} - 3\mathbf{v} = 4\mathbf{i} - 3\mathbf{j}$

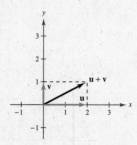 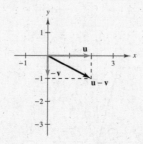 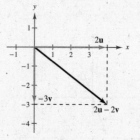

29. $\mathbf{v} = \dfrac{1}{\|\mathbf{u}\|}\mathbf{u} = \dfrac{1}{\sqrt{3^2 + 0^2}}\langle 3, 0 \rangle = \dfrac{1}{3}\langle 3, 0 \rangle = \langle 1, 0 \rangle$ **31.** $\mathbf{u} = \dfrac{1}{\|\mathbf{v}\|}\mathbf{v} = \dfrac{1}{\sqrt{(-2)^2 + 2^2}}\langle -2, 2 \rangle = \dfrac{1}{2\sqrt{2}}\langle -2, 2 \rangle$

$$= \left\langle -\dfrac{1}{\sqrt{2}}, \dfrac{1}{\sqrt{2}} \right\rangle$$

$$= \left\langle -\dfrac{\sqrt{2}}{2}, \dfrac{\sqrt{2}}{2} \right\rangle$$

33. $\mathbf{u} = \dfrac{1}{\|\mathbf{v}\|}\mathbf{v} = \dfrac{1}{\sqrt{6^2 + (-2)^2}}(6\mathbf{i} - 2\mathbf{j}) = \dfrac{1}{\sqrt{40}}(6\mathbf{i} - 2\mathbf{j})$ **35.** $\mathbf{u} = \dfrac{1}{\|\mathbf{w}\|}\mathbf{w} = \dfrac{1}{4}(4\mathbf{j}) = \mathbf{j}$

$$= \dfrac{1}{2\sqrt{10}}(6\mathbf{i} - 2\mathbf{j}) = \dfrac{3}{\sqrt{10}}\mathbf{i} - \dfrac{1}{\sqrt{10}}\mathbf{j}$$

$$= \dfrac{3\sqrt{10}}{10}\mathbf{i} - \dfrac{\sqrt{10}}{10}\mathbf{j}$$

37. $\mathbf{u} = \dfrac{1}{\|\mathbf{w}\|}\mathbf{w} = \dfrac{1}{\sqrt{1^2 + (-2)^2}}(\mathbf{i} - 2\mathbf{j}) = \dfrac{1}{\sqrt{5}}(\mathbf{i} - 2\mathbf{j})$

$\qquad = \dfrac{1}{\sqrt{5}}\mathbf{i} - \dfrac{2}{\sqrt{5}}\mathbf{j} = \dfrac{\sqrt{5}}{5}\mathbf{i} - \dfrac{2\sqrt{5}}{5}\mathbf{j}$

39. $5\left(\dfrac{1}{\|\mathbf{u}\|}\mathbf{u}\right) = 5\left(\dfrac{1}{\sqrt{3^2 + 3^2}}\langle 3, 3\rangle\right) = \dfrac{5}{3\sqrt{2}}\langle 3, 3\rangle$

$\qquad = \left\langle \dfrac{5}{\sqrt{2}}, \dfrac{5}{\sqrt{2}}\right\rangle = \left\langle \dfrac{5\sqrt{2}}{2}, \dfrac{5\sqrt{2}}{2}\right\rangle$

41. $9\left(\dfrac{1}{\|\mathbf{u}\|}\mathbf{u}\right) = 9\left(\dfrac{1}{\sqrt{2^2 + 5^2}}\langle 2, 5\rangle\right) = \dfrac{9}{\sqrt{29}}\langle 2, 5\rangle$

$\qquad = \left\langle \dfrac{18}{\sqrt{29}}, \dfrac{45}{\sqrt{29}}\right\rangle = \left\langle \dfrac{18\sqrt{29}}{29}, \dfrac{45\sqrt{29}}{29}\right\rangle$

43. $\mathbf{u} = \langle 4 - (-3), 5 - 1\rangle$
$\qquad = \langle 7, 4\rangle$
$\qquad = 7\mathbf{i} + 4\mathbf{j}$

45. $\mathbf{u} = \langle 2 - (-1), 3 - (-5)\rangle$
$\qquad = \langle 3, 8\rangle$
$\qquad = 3\mathbf{i} + 8\mathbf{j}$

47. $\mathbf{v} = \tfrac{3}{2}\mathbf{u}$
$\qquad = \tfrac{3}{2}(2\mathbf{i} - \mathbf{j})$
$\qquad = 3\mathbf{i} - \tfrac{3}{2}\mathbf{j} = \langle 3, -\tfrac{3}{2}\rangle$

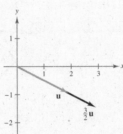

49. $\mathbf{v} = \mathbf{u} + 2\mathbf{w}$
$\qquad = (2\mathbf{i} - \mathbf{j}) + 2(\mathbf{i} + 2\mathbf{j})$
$\qquad = 4\mathbf{i} + 3\mathbf{j} = \langle 4, 3\rangle$

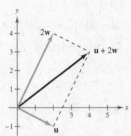

51. $\mathbf{v} = \tfrac{1}{2}(3\mathbf{u} + \mathbf{w})$
$\qquad = \tfrac{1}{2}(6\mathbf{i} - 3\mathbf{j} + \mathbf{i} + 2\mathbf{j})$
$\qquad = \tfrac{7}{2}\mathbf{i} - \tfrac{1}{2}\mathbf{j} = \langle \tfrac{7}{2}, -\tfrac{1}{2}\rangle$

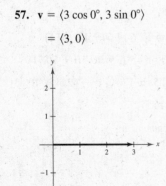

53. $\mathbf{v} = 3(\cos 60°\mathbf{i} + \sin 60°\mathbf{j})$
$\qquad \|\mathbf{v}\| = 3, \ \theta = 60°$

55. $\mathbf{v} = 6\mathbf{i} - 6\mathbf{j}$
$\qquad \|\mathbf{v}\| = \sqrt{6^2 + (-6)^2} = \sqrt{72}$
$\qquad = 6\sqrt{2}$
$\qquad \tan\theta = \dfrac{-6}{6} = -1$
Since $\mathbf{v}$ lies in Quadrant IV, $\theta = 315°$.

57. $\mathbf{v} = \langle 3\cos 0°, 3\sin 0°\rangle$
$\qquad = \langle 3, 0\rangle$

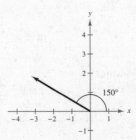

59. $\mathbf{v} = \left\langle \dfrac{7}{2}\cos 150°, \dfrac{7}{2}\sin 150°\right\rangle$
$\qquad = \left\langle -\dfrac{7\sqrt{3}}{4}, \dfrac{7}{4}\right\rangle$

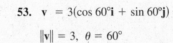

61. $\mathbf{v} = \langle 3\sqrt{2}\cos 150°, \ 3\sqrt{2}\sin 150°\rangle$
$\qquad = \left\langle -\dfrac{3\sqrt{6}}{2}, \dfrac{3\sqrt{2}}{2}\right\rangle$

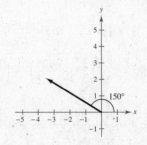
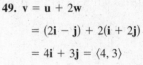

63. $\mathbf{v} = 2\left(\dfrac{1}{\sqrt{1^2 + 3^2}}\right)(\mathbf{i} + 3\mathbf{j})$

$\quad = \dfrac{2}{\sqrt{10}}(\mathbf{i} + 3\mathbf{j})$

$\quad = \dfrac{\sqrt{10}}{5}\mathbf{i} + \dfrac{3\sqrt{10}}{5}\mathbf{j} = \left\langle \dfrac{\sqrt{10}}{5}, \dfrac{3\sqrt{10}}{5} \right\rangle$

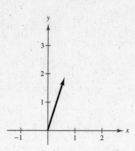

65. $\mathbf{u} = \langle 5\cos 0°, 5\sin 0° \rangle = \langle 5, 0 \rangle$

$\quad \mathbf{v} = \langle 5\cos 90°, 5\sin 90° \rangle = \langle 0, 5 \rangle$

$\quad \mathbf{u} + \mathbf{v} = \langle 5, 5 \rangle$

67. $\mathbf{u} = \langle 20\cos 45°, 20\sin 45° \rangle = \left\langle 10\sqrt{2}, 10\sqrt{2} \right\rangle$

$\quad \mathbf{v} = \langle 50\cos 180°, 50\sin 180° \rangle = \langle -50, 0 \rangle$

$\quad \mathbf{u} + \mathbf{v} = \left\langle 10\sqrt{2} - 50, 10\sqrt{2} \right\rangle$

69. $\mathbf{v} = \mathbf{i} + \mathbf{j}$

$\quad \mathbf{w} = 2\mathbf{i} - 2\mathbf{j}$

$\quad \mathbf{u} = \mathbf{v} - \mathbf{w} = -\mathbf{i} + 3\mathbf{j}$

$\quad \|\mathbf{v}\| = \sqrt{2}$

$\quad \|\mathbf{w}\| = 2\sqrt{2}$

$\quad \|\mathbf{v} - \mathbf{w}\| = \sqrt{10}$

$\quad \cos \alpha = \dfrac{\|\mathbf{v}\|^2 + \|\mathbf{w}\|^2 - \|\mathbf{v} - \mathbf{w}\|^2}{2\|\mathbf{v}\|\,\|\mathbf{w}\|} = \dfrac{2 + 8 - 10}{2\sqrt{2} \cdot 2\sqrt{2}} = 0$

$\quad\quad \alpha = 90°$

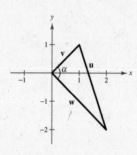

71. Force One: $\mathbf{u} = 45\mathbf{i}$

Force Two: $\mathbf{v} = 60\cos\theta\mathbf{i} + 60\sin\theta\mathbf{j}$

Resultant Force: $\mathbf{u} + \mathbf{v} = (45 + 60\cos\theta)\mathbf{i} + 60\sin\theta\mathbf{j}$

$\|\mathbf{u} + \mathbf{v}\| = \sqrt{(45 + 60\cos\theta)^2 + (60\sin\theta)^2} = 90$

$\quad\quad 2025 + 5400\cos\theta + 3600 = 8100$

$\quad\quad\quad\quad 5400\cos\theta = 2475$

$\quad\quad\quad\quad\quad \cos\theta = \dfrac{2475}{5400} \approx 0.4583$

$\quad\quad\quad\quad\quad\quad \theta \approx 62.7°$

73. $\quad \mathbf{u} = 300\mathbf{i}$

$\quad \mathbf{v} = (125\cos 45°)\mathbf{i} + (125\sin 45°)\mathbf{j} = \dfrac{125}{\sqrt{2}}\mathbf{i} + \dfrac{125}{\sqrt{2}}\mathbf{j}$

$\quad \mathbf{R} = \mathbf{u} + \mathbf{v} = \left(300 + \dfrac{125}{\sqrt{2}}\right)\mathbf{i} + \dfrac{125}{\sqrt{2}}\mathbf{j}$

$\quad \|\mathbf{R}\| = \sqrt{\left(300 + \dfrac{125}{\sqrt{2}}\right)^2 + \left(\dfrac{125}{\sqrt{2}}\right)^2} \approx 398.32$ newtons

$\quad \tan\theta = \dfrac{\dfrac{125}{\sqrt{2}}}{300 + \left(\dfrac{125}{\sqrt{2}}\right)} \Rightarrow \theta \approx 12.8°$

75. $\mathbf{u} = (75\cos 30°)\mathbf{i} + (75\sin 30°)\mathbf{j} \approx 64.95\mathbf{i} + 37.5\mathbf{j}$

$\quad \mathbf{v} = (100\cos 45°)\mathbf{i} + (100\sin 45°)\mathbf{j} \approx 70.71\mathbf{i} + 70.71\mathbf{j}$

$\quad \mathbf{w} = (125\cos 120°)\mathbf{i} + (125\sin 120°)\mathbf{j} \approx -62.5\mathbf{i} + 108.3\mathbf{j}$

$\quad \mathbf{u} + \mathbf{v} + \mathbf{w} \approx 73.16\mathbf{i} + 216.5\mathbf{j}$

$\quad \|\mathbf{u} + \mathbf{v} + \mathbf{w}\| \approx 228.5$ pounds

$\quad \tan\theta \approx \dfrac{216.5}{73.16} \approx 2.9593$

$\quad\quad \theta \approx 71.3°$

77. Horizontal component of velocity: $70 \cos 35° \approx 57.34$ feet per second

Vertical component of velocity: $70 \sin 35° \approx 40.15$ feet per second

79. Cable $\overrightarrow{AC}$: $\mathbf{u} = \|\mathbf{u}\|(\cos 50°\mathbf{i} - \sin 50°\mathbf{j})$

Cable $\overrightarrow{BC}$: $\mathbf{v} = \|\mathbf{v}\|(-\cos 30°\mathbf{i} - \sin 30°\mathbf{j})$

Resultant: $\mathbf{u} + \mathbf{v} = -2000\mathbf{j}$

$\|\mathbf{u}\| \cos 50° - \|\mathbf{v}\| \cos 30° = 0$

$-\|\mathbf{u}\| \sin 50° - \|\mathbf{v}\| \sin 30° = -2000$

Solving this system of equations yields:

$T_{AC} = \|\mathbf{u}\| \approx 1758.8$ pounds

$T_{BC} = \|\mathbf{v}\| \approx 1305.4$ pounds

81. Towline 1: $\mathbf{u} = \|\mathbf{u}\|(\cos 18°\mathbf{i} + \sin 18°\mathbf{j})$

Towline 2: $\mathbf{v} = \|\mathbf{u}\|(\cos 18°\mathbf{i} - \sin 18°\mathbf{j})$

Resultant: $\mathbf{u} + \mathbf{v} = 6000\mathbf{i}$

$\|\mathbf{u}\| \cos 18° + \|\mathbf{u}\| \cos 18° = 6000$

$\|\mathbf{u}\| \approx 3154.4$

Therefore, the tension on each towline is $\|\mathbf{u}\| \approx 3154.4$ pounds.

83. Airspeed: $\mathbf{u} = (875 \cos 58°)\mathbf{i} - (875 \sin 58°)\mathbf{j}$

Groundspeed: $\mathbf{v} = (800 \cos 50°)\mathbf{i} - (800 \sin 50°)\mathbf{j}$

Wind: $\mathbf{w} = \mathbf{v} - \mathbf{u} = (800 \cos 50° - 875 \cos 58°)\mathbf{i} + (-800 \sin 50° + 875 \sin 58°)\mathbf{j}$

$\approx 50.5507\mathbf{i} + 129.2065\mathbf{j}$

Wind speed: $\|\mathbf{w}\| \approx \sqrt{(50.5507)^2 + (129.2065)^2}$

≈ 138.7 kilometers per hour

Wind direction: $\tan \theta \approx \dfrac{129.2065}{50.5507}$

$\theta \approx 68.6°; \ 90° - \theta = 21.4°$

Bearing: N 21.4° E

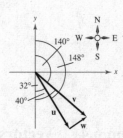

85. $W = FD = (100 \ \cos 50°)(30) = 1928.4$ foot–pounds

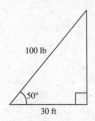

87. True. See Example 1.

89. (a) The angle between them is 0°.

(b) The angle between them is 180°.

(c) No. At most it can be equal to the sum when the angle between them is 0°.

91. Let $\mathbf{v} = (\cos \theta)\mathbf{i} + (\sin \theta)\mathbf{j}$.

$\|\mathbf{v}\| = \sqrt{\cos^2 \theta + \sin^2 \theta} = \sqrt{1} = 1$

Therefore, $\mathbf{v}$ is a unit vector for any value of θ.

93. $\mathbf{u} = \langle 5 - 1, 2 - 6 \rangle = \langle 4, -4 \rangle$

$\mathbf{v} = \langle 9 - 4, 4 - 5 \rangle = \langle 5, -1 \rangle$

$\mathbf{u} - \mathbf{v} = \langle -1, -3 \rangle$ or $\mathbf{v} - \mathbf{u} = \langle 1, 3 \rangle$

95. $\sqrt{x^2 - 64} = \sqrt{(8 \sec \theta)^2 - 64}$

$= \sqrt{64(\sec^2 \theta - 1)}$

$= 8\sqrt{\tan^2 \theta}$

$= 8 \tan \theta \ \text{ for } \ 0 < \theta < \dfrac{\pi}{2}$

97. $\sqrt{x^2 + 36} = \sqrt{(6 \tan \theta)^2 + 36}$

$$= \sqrt{36(\tan^2 \theta + 1)}$$

$$= 6\sqrt{\sec^2 \theta}$$

$$= 6 \sec \theta \ \text{ for } \ 0 < \theta < \frac{\pi}{2}$$

99. $\cos x(\cos x + 1) = 0$

$$\cos x = 0 \quad \text{or} \quad \cos x + 1 = 0$$

$$x = \frac{\pi}{2} + n\pi \quad \cos x = -1$$

$$x = \pi + 2n\pi$$

101. $3 \sec x \sin x - 2\sqrt{3} \sin x = 0$

$$\sin x\left(3 \sec x - 2\sqrt{3}\right) = 0$$

$$\sin x = 0 \quad \text{or} \quad 3 \sec x - 2\sqrt{3} = 0$$

$$x = n\pi \qquad\qquad \sec x = \frac{2\sqrt{3}}{3}$$

$$\cos x = \frac{3}{2\sqrt{3}} = \frac{\sqrt{3}}{2}$$

$$x = \frac{\pi}{6} + 2n\pi$$

$$x = \frac{11\pi}{6} + 2n\pi$$

Section 6.4   Vectors and Dot Products

■   Know the definition of the dot product of $\mathbf{u} = \langle u_1, u_2 \rangle$ and $\mathbf{v} = \langle v_1, v_2 \rangle$.

$\mathbf{u} \cdot \mathbf{v} = u_1 v_1 + u_2 v_2$

■   Know the following properties of the dot product:

1. $\mathbf{u} \cdot \mathbf{v} = \mathbf{v} \cdot \mathbf{u}$

2. $\mathbf{0} \cdot \mathbf{v} = 0$

3. $\mathbf{u} \cdot (\mathbf{v} + \mathbf{w}) = \mathbf{u} \cdot \mathbf{v} + \mathbf{u} \cdot \mathbf{w}$

4. $\mathbf{v} \cdot \mathbf{v} = \|\mathbf{v}\|^2$

5. $c(\mathbf{u} \cdot \mathbf{v}) = c\mathbf{u} \cdot \mathbf{v} = \mathbf{u} \cdot c\mathbf{v}$

■   If θ is the angle between two nonzero vectors $\mathbf{u}$ and $\mathbf{v}$, then

$$\cos \theta = \frac{\mathbf{u} \cdot \mathbf{v}}{\|\mathbf{u}\|\,\|\mathbf{v}\|}.$$

■   The vectors $\mathbf{u}$ and $\mathbf{v}$ are orthogonal if $\mathbf{u} \cdot \mathbf{v} = 0$.

■   Know the definition of vector components.

$\mathbf{u} = \mathbf{w}_1 + \mathbf{w}_2$ where $\mathbf{w}_1$ and $\mathbf{w}_2$ are orthogonal, and $\mathbf{w}_1$ is parallel to $\mathbf{v}$. $\mathbf{w}_1$ is called the projection of $\mathbf{u}$ onto $\mathbf{v}$

and is denoted by $\mathbf{w}_1 = \text{proj}_{\mathbf{v}}\mathbf{u} = \left(\dfrac{\mathbf{u} \cdot \mathbf{v}}{\|\mathbf{v}\|^2}\right)\mathbf{v}$. Then we have $\mathbf{w}_2 = \mathbf{u} - \mathbf{w}_1$.

■   Know the definition of work.

1.   Projection form: $w = \|\text{proj}_{\overrightarrow{PQ}}\,\mathbf{F}\|\,\|\overrightarrow{PQ}\|$

2.   Dot product form: $w = \mathbf{F} \cdot \overrightarrow{PQ}$

Vocabulary Check

1. dot product

2. $\dfrac{\mathbf{u} \cdot \mathbf{v}}{\|\mathbf{u}\| \, \|\mathbf{v}\|}$

3. orthogonal

4. $\left(\dfrac{\mathbf{u} \cdot \mathbf{v}}{\|\mathbf{v}\|^2}\right)\mathbf{v}$

5. $\|\text{proj}_{\overrightarrow{PQ}}\,\mathbf{F}\| \, \|\overrightarrow{PQ}\|$; $\mathbf{F} \cdot \overrightarrow{PQ}$

---

1. $\mathbf{u} = \langle 6, 1 \rangle$, $\mathbf{v} = \langle -2, 3 \rangle$

$\mathbf{u} \cdot \mathbf{v} = 6(-2) + 1(3) = -9$

3. $\mathbf{u} = \langle -4, 1 \rangle$, $\mathbf{v} = \langle 2, -3 \rangle$

$\mathbf{u} \cdot \mathbf{v} = -4(2) + 1(-3) = -11$

5. $\mathbf{u} = 4\mathbf{i} - 2\mathbf{j}$, $\mathbf{v} = \mathbf{i} - \mathbf{j}$

$\mathbf{u} \cdot \mathbf{v} = 4(1) + (-2)(-1) = 6$

7. $\mathbf{u} = 3\mathbf{i} + 2\mathbf{j}$, $\mathbf{v} = -2\mathbf{i} - 3\mathbf{j}$

$\mathbf{u} \cdot \mathbf{v} = 3(-2) + 2(-3) = -12$

9. $\mathbf{u} = \langle 2, 2 \rangle$

$\mathbf{u} \cdot \mathbf{u} = 2(2) + 2(2) = 8$

The result is a scalar.

11. $\mathbf{u} = \langle 2, 2 \rangle$, $\mathbf{v} = \langle -3, 4 \rangle$

$(\mathbf{u} \cdot \mathbf{v})\mathbf{v} = [(2)(-3) + 2(4)]\langle -3, 4 \rangle$

$= 2\langle -3, 4 \rangle = \langle -6, 8 \rangle$

The result is a vector.

13. $\mathbf{u} = \langle 2, 2 \rangle$, $\mathbf{v} = \langle -3, 4 \rangle$, $\mathbf{w} = \langle 1, -2 \rangle$

$(3\mathbf{w} \cdot \mathbf{v})\mathbf{u} = [3(1)(-3) + 3(-2)(4)]\langle 2, 2 \rangle$

$= -33\langle 2, 2 \rangle$

$= \langle -66, -66 \rangle$

The result is a vector.

15. $\mathbf{w} = \langle 1, -2 \rangle$

$\|\mathbf{w}\| - 1 = \sqrt{(1)^2 + (-2)^2} - 1 = \sqrt{5} - 1$

The result is a scalar.

17. $\mathbf{u} = \langle 2, 2 \rangle$, $\mathbf{v} = \langle -3, 4 \rangle$, $\mathbf{w} = \langle 1, -2 \rangle$

$(\mathbf{u} \cdot \mathbf{v}) - (\mathbf{u} \cdot \mathbf{w}) = [2(-3) + 2(4)] - [2(1) + 2(-2)]$

$= 2 - (-2)$

$= 4$

The result is a scalar.

19. $\mathbf{u} = \langle -5, 12 \rangle$

$\|\mathbf{u}\| = \sqrt{\mathbf{u} \cdot \mathbf{u}} = \sqrt{(-5)^2 + 12^2} = 13$

21. $\mathbf{u} = 20\mathbf{i} + 25\mathbf{j}$

$\|\mathbf{u}\| = \sqrt{\mathbf{u} \cdot \mathbf{u}} = \sqrt{(20)^2 + (25)^2} = \sqrt{1025} = 5\sqrt{41}$

23. $\mathbf{u} = 6\mathbf{j}$

$\|\mathbf{u}\| = \sqrt{\mathbf{u} \cdot \mathbf{u}} = \sqrt{(0)^2 + (6)^2} = \sqrt{36} = 6$

25. $\mathbf{u} = \langle 1, 0 \rangle$, $\mathbf{v} = \langle 0, -2 \rangle$

$\cos \theta = \dfrac{\mathbf{u} \cdot \mathbf{v}}{\|\mathbf{u}\| \, \|\mathbf{v}\|} = \dfrac{0}{(1)(2)} = 0$

$\theta = 90°$

27. $\mathbf{u} = 3\mathbf{i} + 4\mathbf{j}$, $\mathbf{v} = -2\mathbf{j}$

$\cos \theta = \dfrac{\mathbf{u} \cdot \mathbf{v}}{\|\mathbf{u}\| \, \|\mathbf{v}\|} = -\dfrac{8}{(5)(2)}$

$\theta = \arccos\left(-\dfrac{4}{5}\right)$

$\theta \approx 143.13°$

29. $\mathbf{u} = 2\mathbf{i} - \mathbf{j}$, $\mathbf{v} = 6\mathbf{i} + 4\mathbf{j}$

$\cos \theta = \dfrac{\mathbf{u} \cdot \mathbf{v}}{\|\mathbf{u}\| \, \|\mathbf{v}\|} = \dfrac{8}{\sqrt{5}\sqrt{52}} = 0.4961$

$\theta = 60.26°$

31. $\mathbf{u} = 5\mathbf{i} + 5\mathbf{j}$, $\mathbf{v} = -6\mathbf{i} + 6\mathbf{j}$

$\cos \theta = \dfrac{\mathbf{u} \cdot \mathbf{v}}{\|\mathbf{u}\| \, \|\mathbf{v}\|} = 0$

$\theta = 90°$

33. $\mathbf{u} = \left(\cos\dfrac{\pi}{3}\right)\mathbf{i} + \left(\sin\dfrac{\pi}{3}\right)\mathbf{j} = \dfrac{1}{2}\mathbf{i} + \dfrac{\sqrt{3}}{2}\mathbf{j}$

$\mathbf{v} = \left(\cos\dfrac{3\pi}{4}\right)\mathbf{i} + \left(\sin\dfrac{3\pi}{4}\right)\mathbf{j} = -\dfrac{\sqrt{2}}{2}\mathbf{i} + \dfrac{\sqrt{2}}{2}\mathbf{j}$

$\|\mathbf{u}\| = \|\mathbf{v}\| = 1$

$\cos\theta = \dfrac{\mathbf{u}\cdot\mathbf{v}}{\|\mathbf{u}\|\,\|\mathbf{v}\|} = \mathbf{u}\cdot\mathbf{v}$

$= \left(\dfrac{1}{2}\right)\left(-\dfrac{\sqrt{2}}{2}\right) + \left(\dfrac{\sqrt{3}}{2}\right)\left(\dfrac{\sqrt{2}}{2}\right) = \dfrac{-\sqrt{2}+\sqrt{6}}{4}$

$\theta = \arccos\left(\dfrac{-\sqrt{2}+\sqrt{6}}{4}\right) = 75° = \dfrac{5\pi}{12}$

35. $\mathbf{u} = 3\mathbf{i} + 4\mathbf{j}$

$\mathbf{v} = -7\mathbf{i} + 5\mathbf{j}$

$\cos\theta = \dfrac{\mathbf{u}\cdot\mathbf{v}}{\|\mathbf{u}\|\,\|\mathbf{v}\|}$

$= \dfrac{3(-7)+4(5)}{3\sqrt{74}}$

$= \dfrac{-1}{5\sqrt{74}} \approx -0.0232$

$\theta \approx 91.3°$

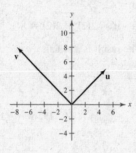

37. $\mathbf{u} = 5\mathbf{i} + 5\mathbf{j}$

$\mathbf{v} = -8\mathbf{i} + 8\mathbf{j}$

$\cos\theta = \dfrac{\mathbf{u}\cdot\mathbf{v}}{\|\mathbf{u}\|\,\|\mathbf{v}\|}$

$= \dfrac{5(-8)+5(8)}{\sqrt{50}\sqrt{128}}$

$= 0$

$\theta = 90°$

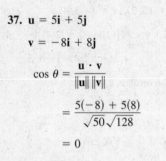

39. $P = (1, 2),\ Q = (3, 4),\ R = (2, 5)$

$\overrightarrow{PQ} = \langle 2, 2\rangle,\ \overrightarrow{PR} = \langle 1, 3\rangle,\ \overrightarrow{QR} = \langle -1, 1\rangle$

$\cos\alpha = \dfrac{\overrightarrow{PQ}\cdot\overrightarrow{PR}}{\|\overrightarrow{PQ}\|\,\|\overrightarrow{PR}\|} = \dfrac{8}{(2\sqrt{2})(\sqrt{10})} \Rightarrow \alpha = \arccos\dfrac{2}{\sqrt{5}} \approx 26.57°$

$\cos\beta = \dfrac{\overrightarrow{PQ}\cdot\overrightarrow{QR}}{\|\overrightarrow{PQ}\|\,\|\overrightarrow{QR}\|} = 0 \Rightarrow \beta = 90°.$ Thus, $\gamma = 180° - 26.57° - 90° = 63.43°.$

41. $P = (-3, 0),\ Q = (2, 2),\ R = (0, 6)$

$\overrightarrow{QP} = \langle -5, -2\rangle,\ \overrightarrow{PR} = \langle 3, 6\rangle,\ \overrightarrow{QR} = \langle -2, 4\rangle,\ \overrightarrow{PQ} = \langle 5, 2\rangle$

$\cos\alpha = \dfrac{\overrightarrow{PQ}\cdot\overrightarrow{PR}}{\|\overrightarrow{PQ}\|\,\|\overrightarrow{PR}\|} = \dfrac{27}{\sqrt{29}\sqrt{45}} \Rightarrow \alpha \approx 41.63°$

$\cos\beta = \dfrac{\overrightarrow{QP}\cdot\overrightarrow{QR}}{\|\overrightarrow{QP}\|\,\|\overrightarrow{PR}\|} = \dfrac{2}{\sqrt{29}\sqrt{20}} \Rightarrow \beta \approx 85.24°$

$\delta = 180° - 41.63° - 85.24° = 53.13°$

43. $\mathbf{u}\cdot\mathbf{v} = \|\mathbf{u}\|\,\|\mathbf{v}\|\cos\theta$

$= (4)(10)\cos\dfrac{2\pi}{3}$

$= 40\left(-\dfrac{1}{2}\right)$

$= -20$

45. $\mathbf{u}\cdot\mathbf{v} = \|\mathbf{u}\|\,\|\mathbf{v}\|\cos\theta$

$= (9)(36)\cos\dfrac{3\pi}{4}$

$= 324\left(-\dfrac{\sqrt{2}}{2}\right)$

$= -162\sqrt{2} \approx -229.1$

47. $\mathbf{u} = \langle -12, 30\rangle,\ \mathbf{v} = \left\langle \dfrac{1}{2}, -\dfrac{5}{4}\right\rangle$

$\mathbf{u} = -24\mathbf{v} \Rightarrow \mathbf{u}$ and $\mathbf{v}$ are parallel.

49. $\mathbf{u} = \frac{1}{4}(3\mathbf{i} - \mathbf{j})$, $\mathbf{v} = 5\mathbf{i} + 6\mathbf{j}$

$\mathbf{u} \neq k\mathbf{v} \implies$ Not parallel

$\mathbf{u} \cdot \mathbf{v} \neq 0 \implies$ Not orthogonal

Neither

51. $\mathbf{u} = 2\mathbf{i} - 2\mathbf{j}$, $\mathbf{v} = -\mathbf{i} - \mathbf{j}$

$\mathbf{u} \cdot \mathbf{v} = 0 \implies \mathbf{u}$ and $\mathbf{v}$ are orthogonal.

53. $\mathbf{u} = \langle 2, 2 \rangle$, $\mathbf{v} = \langle 6, 1 \rangle$

$\mathbf{w}_1 = \text{proj}_{\mathbf{v}}\mathbf{u} = \left(\dfrac{\mathbf{u} \cdot \mathbf{v}}{\|\mathbf{v}\|^2} \right)\mathbf{v} = \dfrac{14}{37}\langle 6, 1 \rangle = \dfrac{1}{37}\langle 84, 14 \rangle$

$\mathbf{w}_2 = \mathbf{u} - \mathbf{w}_1 = \langle 2, 2 \rangle - \dfrac{14}{37}\langle 6, 1 \rangle = \left\langle -\dfrac{10}{37}, \dfrac{60}{37} \right\rangle = \dfrac{10}{37}\langle -1, 6 \rangle = \dfrac{1}{37}\langle -10, 60 \rangle$

$\mathbf{u} = \dfrac{1}{37}\langle 84, 14 \rangle + \dfrac{1}{37}\langle -10, 60 \rangle = \langle 2, 2 \rangle$

55. $\mathbf{u} = \langle 0, 3 \rangle$, $\mathbf{v} = \langle 2, 15 \rangle$

$\mathbf{w}_1 = \text{proj}_{\mathbf{v}}\mathbf{u} = \left(\dfrac{\mathbf{u} \cdot \mathbf{v}}{\|\mathbf{v}\|^2} \right)\mathbf{v} = \dfrac{45}{229}\langle 2, 15 \rangle$

$\mathbf{w}_2 = \mathbf{u} - \mathbf{w}_1 = \langle 0, 3 \rangle - \dfrac{45}{229}\langle 2, 15 \rangle = \left\langle -\dfrac{90}{229}, \dfrac{12}{229} \right\rangle$

$\qquad = \dfrac{6}{229}\langle -15, 2 \rangle$

$\mathbf{u} = \dfrac{45}{229}\langle 2, 15 \rangle + \dfrac{6}{229}\langle -15, 2 \rangle = \langle 0, 3 \rangle$

57. $\text{proj}_{\mathbf{v}}\mathbf{u} = \mathbf{0}$ since they are perpendicular.

Since $\mathbf{u}$ and $\mathbf{v}$ are orthogonal, $\mathbf{u} \cdot \mathbf{v} = 0$ and $\text{proj}_{\mathbf{v}}\mathbf{u} = \mathbf{0}$.

$\text{proj}_{\mathbf{v}}\mathbf{u} = \dfrac{\mathbf{u} \cdot \mathbf{v}}{\|\mathbf{v}\|^2}\mathbf{v} = \mathbf{0}$, since $\mathbf{u} \cdot \mathbf{v} = \mathbf{0}$.

59. $\mathbf{u} = \langle 3, 5 \rangle$

For $\mathbf{v}$ to be orthogonal to $\mathbf{u}$, $\mathbf{u} \cdot \mathbf{v}$ must equal 0.

Two possibilities: $\langle -5, 3 \rangle$ and $\langle 5, -3 \rangle$

61. $\mathbf{u} = \frac{1}{2}\mathbf{i} - \frac{2}{3}\mathbf{j}$

For $\mathbf{u}$ and $\mathbf{v}$ to be orthogonal, $\mathbf{u} \cdot \mathbf{v}$ must equal 0.

Two possibilities: $\mathbf{v} = \frac{2}{3}\mathbf{i} + \frac{1}{2}\mathbf{j}$ and $\mathbf{v} = -\frac{2}{3}\mathbf{i} - \frac{1}{2}\mathbf{j}$

63. $w = \| \text{proj}_{\overrightarrow{PQ}}\, \mathbf{v} \| \|\overrightarrow{PQ}\|$ where $\overrightarrow{PQ} = \langle 4, 7 \rangle$ and $\mathbf{v} = \langle 1, 4 \rangle$.

$\text{proj}_{\overrightarrow{PQ}}\, \mathbf{v} = \left(\dfrac{\mathbf{v} \cdot \overrightarrow{PQ}}{\|\overrightarrow{PQ}\|^2} \right)\overrightarrow{PQ} = \left(\dfrac{32}{65} \right)\langle 4, 7 \rangle$

$w = \| \text{proj}_{\overrightarrow{PQ}}\, \mathbf{v} \| \|\overrightarrow{PQ}\| = \left(\dfrac{32\sqrt{65}}{65} \right)\left(\sqrt{65} \right) = 32$

65. (a) $\mathbf{u} = \langle 1650, 3200 \rangle$, $\mathbf{v} = \langle 15.25, 10.50 \rangle$

$\mathbf{u} \cdot \mathbf{v} = 1650(15.25) + 3200(10.50) = \$58{,}762.50$

This gives the total revenue that can be earned by selling all of the pans.

(b) Increase prices by 5%: $1.05\mathbf{v}$ The operation is scalar multiplication.

$\mathbf{u} \cdot 1.05\mathbf{v} = 1.05\mathbf{u} \cdot \mathbf{v}$

$\qquad = 1.05[1650(15.25) + 3200(10.50)]$

$\qquad = 1.05(58{,}762.50)$

$\qquad = 61{,}700.63$

67. (a) Force due to gravity:

$$\mathbf{F} = -30,000\mathbf{j}$$

Unit vector along hill:

$$\mathbf{v} = (\cos d)\mathbf{i} + (\sin d)\mathbf{j}$$

Projection of $\mathbf{F}$ onto $\mathbf{v}$:

$$\mathbf{w}_1 = \text{proj}_\mathbf{v}\,\mathbf{F} = \left(\frac{\mathbf{F} \cdot \mathbf{v}}{\|\mathbf{v}\|^2}\right)\mathbf{v} = (\mathbf{F} \cdot \mathbf{v})\mathbf{v} = -30,000 \sin d\,\mathbf{v}$$

The magnitude of the force is $30,000 \sin d$.

(b)

d	0°	1°	2°	3°	4°	5°	6°	7°	8°	9°	10°
Force	0	523.6	1047.0	1570.1	2092.7	2614.7	3135.9	3656.1	4175.2	4693.0	5209.4

(c) Force perpendicular to the hill when $d = 5°$:

$$\text{Force} = \sqrt{(30,000)^2 - (2614.7)^2} \approx 29,885.8 \text{ pounds}$$

69. $\mathbf{w} = (245)(3) = 735$ newton-meters

71. $\mathbf{w} = (\cos 30°)(45)(20) \approx 779.4$ foot-pounds

73. $\mathbf{w} = (\cos 30°)(250)(100) \approx 21,650.64$ foot-pounds

75. False. Work is represented by a scalar.

77. (a) $\mathbf{u} \cdot \mathbf{v} = 0 \implies \mathbf{u}$ and $\mathbf{v}$ are orthogonal and $\theta = \dfrac{\pi}{2}$.

(b) $\mathbf{u} \cdot \mathbf{v} > 0 \implies \cos\theta > 0 \implies 0 \le \theta < \dfrac{\pi}{2}$

(c) $\mathbf{u} \cdot \mathbf{v} < 0 \implies \cos\theta < 0 \implies \dfrac{\pi}{2} < \theta \le \pi$

79. In a rhombus, $\|\mathbf{u}\| = \|\mathbf{v}\|$. The diagonals are $\mathbf{u} + \mathbf{v}$ and $\mathbf{u} - \mathbf{v}$.

$$(\mathbf{u} + \mathbf{v}) \cdot (\mathbf{u} - \mathbf{v}) = (\mathbf{u} + \mathbf{v}) \cdot \mathbf{u} - (\mathbf{u} + \mathbf{v}) \cdot \mathbf{v}$$
$$= \mathbf{u} \cdot \mathbf{u} + \mathbf{v} \cdot \mathbf{u} - \mathbf{u} \cdot \mathbf{v} - \mathbf{v} \cdot \mathbf{v}$$
$$= \|\mathbf{u}\|^2 - \|\mathbf{v}\|^2 = 0$$

Therefore, the diagonals are orthogonal.

81. $\sqrt{42} \cdot \sqrt{24} = \sqrt{1008}$
$= \sqrt{144 \cdot 7}$
$= 12\sqrt{7}$

83. $\sqrt{-3}\,\sqrt{-8} = (\sqrt{3}i)(2\sqrt{2}i)$
$= 2\sqrt{6}i^2$
$= -2\sqrt{6}$

85. $\sin 2x - \sqrt{3}\sin x = 0$
$2\sin x \cos x - \sqrt{3}\sin x = 0$
$\sin x(2\cos x - \sqrt{3}) = 0$
$\sin x = 0 \quad \text{or} \quad 2\cos x - \sqrt{3} = 0$
$x = 0, \pi \qquad \cos x = \dfrac{\sqrt{3}}{2}$
$x = \dfrac{\pi}{6}, \dfrac{11\pi}{6}$

87. $2\tan x = \tan 2x$
$2\tan x = \dfrac{2\tan x}{1 - \tan^2 x}$
$2\tan x(1 - \tan^2 x) = 2\tan x$
$2\tan x(1 - \tan^2 x) - 2\tan x = 0$
$2\tan x[(1 - \tan^2 x) - 1] = 0$
$2\tan x(-\tan^2 x) = 0$
$-2\tan^3 x = 0$
$\tan x = 0$
$x = 0, \pi$

For Exercises 89–91:

$\sin u = -\frac{12}{13}$, u in Quadrant IV $\Rightarrow$ $\cos u = \frac{5}{13}$ $\cos v = \frac{24}{25}$, v in Quadrant IV $\Rightarrow$ $\sin v = -\frac{7}{25}$

89. $\sin(u - v) = \sin u \cos v - \cos u \sin v$

$= \left(-\frac{12}{13}\right)\left(\frac{24}{25}\right) - \left(\frac{5}{13}\right)\left(-\frac{7}{25}\right)$

$= -\frac{253}{325}$

91. $\cos(v - u) = \cos v \cos u + \sin v \sin u$

$= \left(\frac{24}{25}\right)\left(\frac{5}{13}\right) + \left(-\frac{7}{25}\right)\left(-\frac{12}{13}\right)$

$= \frac{204}{325}$

Section 6.5   Trigonometric Form of a Complex Number

- ■ You should be able to graphically represent complex numbers and know the following facts about them.

- ■ The absolute value of the complex number $z = a + bi$ is $|z| = \sqrt{a^2 + b^2}$.

- ■ The trigonometric form of the complex number $z = a + bi$ is $z = r(\cos\theta + i\sin\theta)$ where

    (a) $a = r\cos\theta$

    (b) $b = r\sin\theta$

    (c) $r = \sqrt{a^2 + b^2}$; r is called the modulus of z.

    (d) $\tan\theta = \dfrac{b}{a}$; θ is called the argument of z.

- ■ Given $z_1 = r_1(\cos\theta_1 + i\sin\theta_1)$ and $z_2 = r_2(\cos\theta_2 + i\sin\theta_2)$:

    (a) $z_1 z_2 = r_1 r_2 [\cos(\theta_1 + \theta_2) + i\sin(\theta_1 + \theta_2)]$

    (b) $\dfrac{z_1}{z_2} = \dfrac{r_1}{r_2} [\cos(\theta_1 - \theta_2) + i\sin(\theta_1 - \theta_2)]$, $z_2 \neq 0$

- ■ You should know DeMoivre's Theorem: If $z = r(\cos\theta + i\sin\theta)$, then for any positive integer n,

    $z^n = r^n(\cos n\theta + i\sin n\theta)$.

- ■ You should know that for any positive integer n, $z = r(\cos\theta + i\sin\theta)$ has n distinct nth roots given by

    $$\sqrt[n]{r}\left[\cos\left(\frac{\theta + 2\pi k}{n}\right) + i\sin\left(\frac{\theta + 2\pi k}{n}\right)\right]$$

    where $k = 0, 1, 2, \ldots, n - 1$.

Vocabulary Check

1. absolute value

2. trigonometric form; modulus; argument

3. DeMoivre's

4. n^{th} root

1. $|-7i| = \sqrt{0^2 + (-7)^2}$

$= \sqrt{49} = 7$

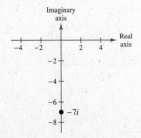

3. $|-4 + 4i| = \sqrt{(-4)^2 + (4)^2}$

$= \sqrt{32} = 4\sqrt{2}$

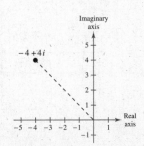

5. $|6 - 7i| = \sqrt{6^2 + (-7)^2}$

$= \sqrt{85}$

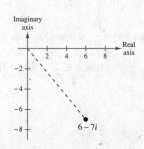

7. $z = 3i$

$r = \sqrt{0^2 + 3^2} = \sqrt{9} = 3$

$\tan \theta = \dfrac{3}{0}$, undefined $\Rightarrow \theta = \dfrac{\pi}{2}$

$z = 3\left(\cos \dfrac{\pi}{2} + i \sin \dfrac{\pi}{2} \right)$

9. $z = 3 - i$

$r = \sqrt{(3)^2 + (-1)^2} = \sqrt{10}$

$\tan \theta = -\dfrac{1}{3}$, θ is in Quadrant IV.

$\theta \approx 5.96$ radians

$z \approx \sqrt{10}(\cos 5.96 + i \sin 5.96)$

11. $z = 3 - 3i$

$r = \sqrt{3^2 + (-3)^2} = \sqrt{18} = 3\sqrt{2}$

$\tan \theta = \dfrac{-3}{3} = -1$, θ is in Quadrant IV $\Rightarrow \theta = \dfrac{7\pi}{4}$.

$z = 3\sqrt{2}\left(\cos \dfrac{7\pi}{4} + i \sin \dfrac{7\pi}{4} \right)$

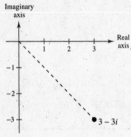

13. $z = \sqrt{3} + i$

$r = \sqrt{\left(\sqrt{3}\right)^2 + 1^2} = \sqrt{4} = 2$

$\tan \theta = \dfrac{1}{\sqrt{3}} = \dfrac{\sqrt{3}}{3} \Rightarrow \theta = \dfrac{\pi}{6}$

$z = 2\left(\cos \dfrac{\pi}{6} + i \sin \dfrac{\pi}{6} \right)$

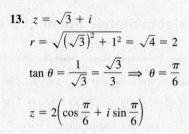

15. $z = -2\left(1 + \sqrt{3}i\right)$

$r = \sqrt{(-2)^2 + \left(-2\sqrt{3}\right)^2} = \sqrt{16} = 4$

$\tan \theta = \dfrac{\sqrt{3}}{1} = \sqrt{3}$, θ is in Quadrant III $\Rightarrow \theta = \dfrac{4\pi}{3}$.

$z = 4\left(\cos \dfrac{4\pi}{3} + i \sin \dfrac{4\pi}{3} \right)$

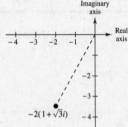

17. $z = -5i$

$r = \sqrt{0^2 + (-5)^2} = \sqrt{25} = 5$

$\tan \theta = \dfrac{-5}{0}$, undefined $\Rightarrow \theta = \dfrac{3\pi}{2}$

$z = 5\left(\cos \dfrac{3\pi}{2} + i \sin \dfrac{3\pi}{2} \right)$

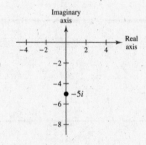

19. $z = -7 + 4i$

$r = \sqrt{(-7)^2 + (4)^2} = \sqrt{65}$

$\tan \theta = \dfrac{4}{-7}$, θ is in Quadrant II $\Rightarrow \theta \approx 2.62$.

$z \approx \sqrt{65}\,(\cos 2.62 + i \sin 2.62)$

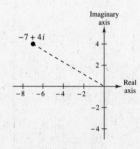

21. $z = 7 + 0i$

$r = \sqrt{(7)^2 + (0)^2} = \sqrt{49} = 7$

$\tan \theta = \dfrac{0}{7} = 0 \implies \theta = 0$

$z = 7(\cos 0 + i \sin 0)$

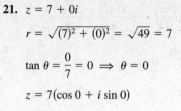

23. $z = 3 + \sqrt{3}i$

$r = \sqrt{(3)^2 + \left(\sqrt{3}\right)^2} = \sqrt{12}$

$= 2\sqrt{3}$

$\tan \theta = \dfrac{\sqrt{3}}{3} \implies \theta = \dfrac{\pi}{6}$

$z = 2\sqrt{3}\left(\cos \dfrac{\pi}{6} + i \sin \dfrac{\pi}{6}\right)$

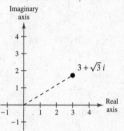

25. $z = -3 - i$

$r = \sqrt{(-3)^2 + (-1)^2} = \sqrt{10}$

$\tan \theta = \dfrac{-1}{-3} = \dfrac{1}{3}$, θ is in Quadrant III $\implies \theta \approx 3.46.$

$z \approx \sqrt{10}\,(\cos 3.46 + i \sin 3.46)$

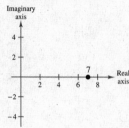

27. $z = 5 + 2i$

$r = \sqrt{5^2 + 2^2} = \sqrt{29}$

$\tan \theta = \frac{2}{5}$

$\theta \approx 0.38$

$z \approx \sqrt{29}(\cos 0.38 + i \sin 0.38)$

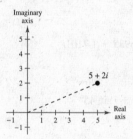

29. $z = -8 - 5\sqrt{3}i$

$r = \sqrt{(-8)^2 + \left(-5\sqrt{3}\right)^2} = \sqrt{139}$

$\tan \theta = \dfrac{5\sqrt{3}}{8}$

$\theta \approx 3.97$

$z \approx \sqrt{139}(\cos 3.97 + i \sin 3.97)$

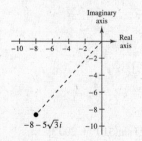

31. $3(\cos 120° + i \sin 120°) = 3\left(-\dfrac{1}{2} + \dfrac{\sqrt{3}}{2}i\right)$

$= -\dfrac{3}{2} + \dfrac{3\sqrt{3}}{2}i$

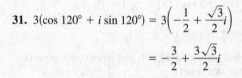

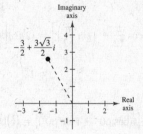

33. $\frac{3}{2}(\cos 300° + i \sin 300°) = \frac{3}{2}\left[\frac{1}{2} + i\left(-\frac{\sqrt{3}}{2}\right)\right]$

$$= \frac{3}{4} - \frac{3\sqrt{3}}{4}i$$

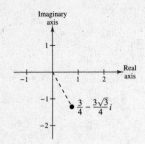

35. $3.75\left(\cos \frac{3\pi}{4} + i \sin \frac{3\pi}{4}\right) = -\frac{15\sqrt{2}}{8} + \frac{15\sqrt{2}}{8}i$

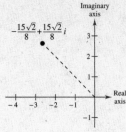

37. $8\left(\cos \frac{\pi}{2} + i \sin \frac{\pi}{2}\right) = 8(0 + i) = 8i$

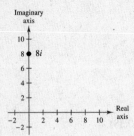

39. $3[\cos(18°45') + i \sin(18°45')] \approx 2.8408 + 0.9643i$

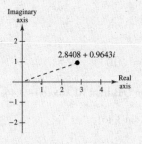

41. $5\left(\cos \frac{\pi}{9} + i \sin \frac{\pi}{9}\right) \approx 4.6985 + 1.7101i$

43. $3(\cos 165.5° + i \sin 165.5°) \approx -2.9044 + 0.7511i$

45. $z = \frac{\sqrt{2}}{2}(1 + i) = \cos 45° + i \sin 45°$

$z^2 = \cos 90° + i \sin 90° = i$

$z^3 = \cos 135° + i \sin 135° = \frac{\sqrt{2}}{2}(-1 + i)$

$z^4 = \cos 180° + i \sin 180° = -1$

The absolute value of each is 1, and consecutive powers of z are each 45° apart.

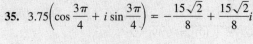

47. $\left[2\left(\cos \frac{\pi}{4} + i \sin \frac{\pi}{4}\right)\right]\left[6\left(\cos \frac{\pi}{12} + i \sin \frac{\pi}{12}\right)\right] = (2)(6)\left[\cos\left(\frac{\pi}{4} + \frac{\pi}{12}\right) + i \sin\left(\frac{\pi}{4} + \frac{\pi}{12}\right)\right]$

$$= 12\left(\cos \frac{\pi}{3} + i \sin \frac{\pi}{3}\right)$$

49. $\left[\frac{5}{3}(\cos 140° + i \sin 140°)\right]\left[\frac{2}{3}(\cos 60° + i \sin 60°)\right] = \left(\frac{5}{3}\right)\left(\frac{2}{3}\right)[\cos(140° + 60°) + i \sin(140° + 60°)]$

$$= \frac{10}{9}(\cos 200° + i \sin 200°)$$

51. $[0.45(\cos 310° + i \sin 310°)][0.60(\cos 200° + i \sin 200°)] = (0.45)(0.60)[\cos(310° + 200°) + i \sin(310° + 200°)]$

$$= 0.27(\cos 510° + i \sin 510°)$$

$$= 0.27(\cos 150° + i \sin 150°)$$

53. $\dfrac{\cos 50° + i \sin 50°}{\cos 20° + i \sin 20°} = \cos(50° - 20°) + i \sin(50° - 20°) = \cos 30° + i \sin 30°$

55. $\dfrac{\cos \dfrac{5\pi}{3} + i \sin \dfrac{5\pi}{3}}{\cos \pi + i \sin \pi} = \cos\left(\dfrac{5\pi}{3} - \pi\right) + i \sin\left(\dfrac{5\pi}{3} - \pi\right) = \cos\left(\dfrac{2\pi}{3}\right) + i \sin\left(\dfrac{2\pi}{3}\right)$

57. $\dfrac{12(\cos 52° + i \sin 52°)}{3(\cos 110° + i \sin 110°)} = 4[\cos(52° - 110°) + i \sin(52° - 110°)]$

$$= 4[\cos(-58°) + i \sin(-58°)]$$

$$= 4(\cos 302° + i \sin 302°)$$

59. (a) $2 + 2i = 2\sqrt{2}\left(\cos\dfrac{\pi}{4} + i \sin\dfrac{\pi}{4}\right)$

$1 - i = \sqrt{2}\left[\cos\left(-\dfrac{\pi}{4}\right) + i \sin\left(-\dfrac{\pi}{4}\right)\right] = \sqrt{2}\left(\cos\dfrac{7\pi}{4} + i \sin\dfrac{7\pi}{4}\right)$

(b) $(2 + 2i)(1 - i) = \left[2\sqrt{2}\left(\cos\dfrac{\pi}{4} + i \sin\dfrac{\pi}{4}\right)\right]\left[\sqrt{2}\left(\cos\left(\dfrac{7\pi}{4}\right) + i \sin\left(\dfrac{7\pi}{4}\right)\right)\right] = 4(\cos 2\pi + i \sin 2\pi)$

$$= 4(\cos 0 + i \sin 0) = 4$$

(c) $(2 + 2i)(1 - i) = 2 - 2i + 2i - 2i^2 = 2 + 2 = 4$

61. (a) $-2i = 2\left[\cos\left(-\dfrac{\pi}{2}\right) + i \sin\left(-\dfrac{\pi}{2}\right)\right] = 2\left(\cos\dfrac{3\pi}{2} + i \sin\dfrac{3\pi}{2}\right)$

$1 + i = \sqrt{2}\left(\cos\dfrac{\pi}{4} + i \sin\dfrac{\pi}{4}\right)$

(b) $-2i(1 + i) = 2\left[\cos\left(\dfrac{3\pi}{2}\right) + i \sin\left(\dfrac{3\pi}{2}\right)\right]\left[\sqrt{2}\left(\cos\dfrac{\pi}{4} + i \sin\dfrac{\pi}{4}\right)\right]$

$$= 2\sqrt{2}\left[\cos\left(\dfrac{7\pi}{4}\right) + i \sin\left(\dfrac{7\pi}{4}\right)\right]$$

$$= 2\sqrt{2}\left[\dfrac{1}{\sqrt{2}} - \dfrac{1}{\sqrt{2}}i\right] = 2 - 2i$$

(c) $-2i(1 + i) = -2i - 2i^2 = -2i + 2 = 2 - 2i$

63. (a) $3 + 4i \approx 5(\cos 0.93 + i \sin 0.93)$

$1 - \sqrt{3}i = 2\left(\cos\dfrac{5\pi}{3} + i \sin\dfrac{5\pi}{3}\right)$

(c) $\dfrac{3 + 4i}{1 - \sqrt{3}i} = \dfrac{3 + 4i}{1 - \sqrt{3}i} \cdot \dfrac{1 + \sqrt{3}i}{1 + \sqrt{3}i}$

$$= \dfrac{3 + \left(4 + 3\sqrt{3}\right)i + 4\sqrt{3}i^2}{1 + 3}$$

$$= \dfrac{3 - 4\sqrt{3}}{4} + \dfrac{4 + 3\sqrt{3}}{4}i$$

$$\approx -0.982 + 2.299i$$

(b) $\dfrac{3 + 4i}{1 - \sqrt{3}i} \approx \dfrac{5(\cos 0.93 + i \sin 0.93)}{2\left(\cos\dfrac{5\pi}{3} + i \sin\dfrac{5\pi}{3}\right)}$

$$\approx 2.5[\cos(-4.31) + i \sin(-4.31)]$$

$$= \dfrac{5}{2}(\cos 1.97 + i \sin 1.97)$$

$$\approx -0.982 + 2.299i$$

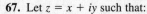

65. (a) $5 = 5(\cos 0 + i \sin 0)$

$2 + 3i \approx \sqrt{13}(\cos 0.98 + i \sin 0.98)$

(b) $\dfrac{5}{2 + 3i} \approx \dfrac{5(\cos 0 + i \sin 0)}{\sqrt{13}(\cos 0.98 + i \sin 0.98)} = \dfrac{5}{\sqrt{13}}[\cos(-0.98) + i \sin(-0.98)] = \dfrac{5}{\sqrt{13}}(\cos 5.30 + i \sin 5.30) \approx 0.769 - 1.154i$

(c) $\dfrac{5}{2 + 3i} = \dfrac{5}{2 + 3i} \cdot \dfrac{2 - 3i}{2 - 3i} = \dfrac{10 - 15i}{13} = \dfrac{10}{13} - \dfrac{15}{13}i \approx 0.769 - 1.154i$

67. Let $z = x + iy$ such that:

$|z| = 2 \implies 2 = \sqrt{x^2 + y^2}$

$\implies 4 = x^2 + y^2$:

circle with radius of 2

69. Let $\theta = \dfrac{\pi}{6}$.

Since $r \geq 0$, we have the portion of the line $\theta = \pi/6$ in Quadrant I.

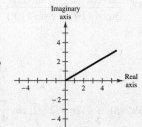

71. $(1 + i)^5 = \left[\sqrt{2}\left(\cos \dfrac{\pi}{4} + i \sin \dfrac{\pi}{4} \right) \right]^5$

$= (\sqrt{2})^5 \left(\cos \dfrac{5\pi}{4} + i \sin \dfrac{5\pi}{4} \right)$

$= 4\sqrt{2}\left(-\dfrac{\sqrt{2}}{2} - \dfrac{\sqrt{2}}{2}i \right)$

$= -4 - 4i$

73. $(-1 + i)^{10} = \left[\sqrt{2}\left(\cos \dfrac{3\pi}{4} + i \sin \dfrac{3\pi}{4} \right) \right]^{10}$

$= (\sqrt{2})^{10} \left(\cos \dfrac{30\pi}{4} + i \sin \dfrac{30\pi}{4} \right)$

$= 32\left[\cos\left(\dfrac{3\pi}{2} + 6\pi \right) + i \sin\left(\dfrac{3\pi}{2} + 6\pi \right) \right]$

$= 32\left(\cos \dfrac{3\pi}{2} + i \sin \dfrac{3\pi}{2} \right)$

$= 32[0 + i(-1)] = -32i$

75. $2(\sqrt{3} + i)^7 = 2\left[2\left(\cos \dfrac{\pi}{6} + i \sin \dfrac{\pi}{6} \right) \right]^7$

$= 2\left[2^7\left(\cos \dfrac{7\pi}{6} + i \sin \dfrac{7\pi}{6} \right) \right]$

$= 256\left(-\dfrac{\sqrt{3}}{2} - \dfrac{1}{2}i \right)$

$= -128\sqrt{3} - 128i$

77. $[5(\cos 20° + i \sin 20°)]^3 = 5^3(\cos 60° + i \sin 60°)$

$= \dfrac{125}{2} + \dfrac{125\sqrt{3}}{2}i$

79. $\left(\cos \dfrac{\pi}{4} + i \sin \dfrac{\pi}{4} \right)^{12} = \cos \dfrac{12\pi}{4} + i \sin \dfrac{12\pi}{4}$

$= \cos 3\pi + i \sin 3\pi$

$= -1$

81. $[5(\cos 3.2 + i \sin 3.2)]^4 = 5^4(\cos 12.8 + i \sin 12.8)$

$\approx 608.02 + 144.69i$

83. $(3 - 2i)^5 \approx [3.6056[\cos(-0.588) + i \sin(-0.588)]]^5$

$\approx (3.6056)^5[\cos(-2.94) + i \sin(-2.94)]$

$\approx -597 - 122i$

85. $[3(\cos 15° + i \sin 15°)]^4 = 81(\cos 60° + i \sin 60°)$

$= \dfrac{81}{2} + \dfrac{81\sqrt{3}}{2}i$

87. $\left[2\left(\cos \dfrac{\pi}{10} + i \sin \dfrac{\pi}{10} \right) \right]^5 = 2^5\left(\cos \dfrac{\pi}{2} + i \sin \dfrac{\pi}{2} \right) = 32i$

89. (a) Square roots of $5(\cos 120° + i \sin 120°)$:

$$\sqrt{5}\left[\cos\left(\frac{120° + 360°k}{2}\right) + i \sin\left(\frac{120° + 360°k}{2}\right)\right], \; k = 0, \; 1$$

$k = 0$: $\sqrt{5}(\cos 60° + i \sin 60°)$

$k = 1$: $\sqrt{5}(\cos 240° + i \sin 240°)$

(c) $\dfrac{\sqrt{5}}{2} + \dfrac{\sqrt{15}}{2}i, \; -\dfrac{\sqrt{5}}{2} - \dfrac{\sqrt{15}}{2}i$

(b)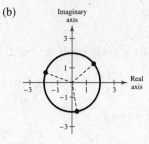

91. (a) Cube roots of $8\left(\cos\dfrac{2\pi}{3} + i \sin\dfrac{2\pi}{3}\right)$:

$$\sqrt[3]{8}\left[\cos\left(\frac{(2\pi/3) + 2\pi k}{3}\right) + i \sin\left(\frac{(2\pi/3) + 2\pi k}{3}\right)\right], \; k = 0, 1, 2$$

$k = 0$: $2\left(\cos\dfrac{2\pi}{9} + i \sin\dfrac{2\pi}{9}\right)$

$k = 1$: $2\left(\cos\dfrac{8\pi}{9} + i \sin\dfrac{8\pi}{9}\right)$

$k = 2$: $2\left(\cos\dfrac{14\pi}{9} + i \sin\dfrac{14\pi}{9}\right)$

(c) $1.5321 + 1.2856i, \; -1.8794 + 0.6840i, \; 0.3473 - 1.9696i$

(b)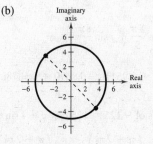

93. (a) Square roots of $-25i = 25\left(\cos\dfrac{3\pi}{2} + i \sin\dfrac{3\pi}{2}\right)$:

$$\sqrt{25}\left[\cos\left(\frac{\frac{3\pi}{2} + 2k\pi}{2}\right) + i \sin\left(\frac{\frac{3\pi}{2} + 2k\pi}{2}\right)\right], \; k = 0, 1$$

$k = 0$: $5\left(\cos\dfrac{3\pi}{4} + i \sin\dfrac{3\pi}{4}\right)$

$k = 1$: $5\left(\cos\dfrac{7\pi}{4} + i \sin\dfrac{7\pi}{4}\right)$

(c) $-\dfrac{5\sqrt{2}}{2} + \dfrac{5\sqrt{2}}{2}i, \; \dfrac{5\sqrt{2}}{2} - \dfrac{5\sqrt{2}}{2}i$

(b)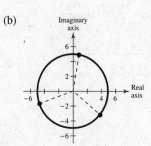

95. (a) Cube roots of $-\dfrac{125}{2}(1 + \sqrt{3}i) = 125\left(\cos\dfrac{4\pi}{3} + i \sin\dfrac{4\pi}{3}\right)$:

$$\sqrt[3]{125}\left[\cos\left(\frac{\frac{4\pi}{3} + 2k\pi}{3}\right) + i \sin\left(\frac{\frac{4\pi}{3} + 2k\pi}{3}\right)\right], \; k = 0, 1, 2$$

$k = 0$: $5\left(\cos\dfrac{4\pi}{9} + i \sin\dfrac{4\pi}{9}\right)$

$k = 1$: $5\left(\cos\dfrac{10\pi}{9} + i \sin\dfrac{10\pi}{9}\right)$

$k = 2$: $5\left(\cos\dfrac{16\pi}{9} + i \sin\dfrac{16\pi}{9}\right)$

(c) $0.8682 + 4.9240i, \; -4.6985 - 1.7101i, \; 3.8302 - 3.2140i$

(b)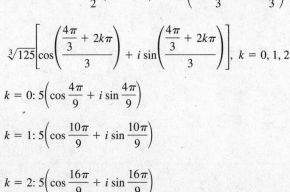

97. (a) Fourth roots of $16 = 16(\cos 0 + i \sin 0)$:

$$\sqrt[4]{16}\left[\cos\frac{0 + 2\pi k}{4} + i \sin\frac{0 + 2\pi k}{4}\right], k = 0, 1, 2, 3$$

$k = 0$: $2(\cos 0 + i \sin 0)$

$k = 1$: $2\left(\cos\frac{\pi}{2} + i \sin\frac{\pi}{2}\right)$

$k = 2$: $2(\cos \pi + i \sin \pi)$

$k = 3$: $2\left(\cos\frac{3\pi}{2} + i \sin\frac{3\pi}{2}\right)$

(c) $2, 2i, -2, -2i$

(b)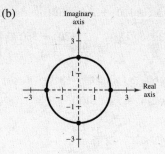

99. (a) Fifth roots of $1 = \cos 0 + i \sin 0$:

$$\cos\left(\frac{2k\pi}{5}\right) + i \sin\left(\frac{2k\pi}{5}\right), k = 0, 1, 2, 3, 4$$

$k = 0$: $\cos 0 + i \sin 0$

$k = 1$: $\cos\frac{2\pi}{5} + i \sin\frac{2\pi}{5}$

$k = 2$: $\cos\frac{4\pi}{5} + i \sin\frac{4\pi}{5}$

$k = 3$: $\cos\frac{6\pi}{5} + i \sin\frac{6\pi}{5}$

$k = 4$: $\cos\frac{8\pi}{5} + i \sin\frac{8\pi}{5}$

(c) $1, 0.3090 + 0.9511i, -0.8090 + 0.5878i, -0.8090 - 0.5878i, 0.3090 - 0.9511i$

(b)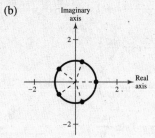

101. (a) Cube roots of $-125 = 125(\cos \pi + i \sin \pi)$:

$$\sqrt[3]{125}\left[\cos\left(\frac{\pi + 2\pi k}{3}\right) + i \sin\left(\frac{\pi + 2\pi k}{3}\right)\right], k = 0, 1, 2$$

$k = 0$: $5\left(\cos\frac{\pi}{3} + i \sin\frac{\pi}{3}\right)$

$k = 1$: $5(\cos \pi + i \sin \pi)$

$k = 2$: $5\left(\cos\frac{5\pi}{3} + i \sin\frac{5\pi}{3}\right)$

(c) $\frac{5}{2} + \frac{5\sqrt{3}}{2}i, -5, \frac{5}{2} - \frac{5\sqrt{3}}{2}i$

(b)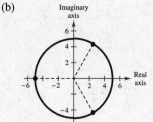

103. (a) Fifth roots of $128(-1 + i) = 128\sqrt{2}\left(\cos\dfrac{3\pi}{4} + i\sin\dfrac{3\pi}{4}\right) = 2^{15/2}\left(\cos\dfrac{3\pi}{4} + i\sin\dfrac{3\pi}{4}\right)$

$$2^{3/2}\left[\cos\left(\dfrac{\dfrac{3\pi}{4} + 2\pi k}{5}\right) + i\sin\left(\dfrac{\dfrac{3\pi}{4} + 2\pi k}{5}\right)\right],\ k = 0, 1, 2, 3, 4$$

$k = 0:\ 2\sqrt{2}\left(\cos\dfrac{3\pi}{20} + i\sin\dfrac{3\pi}{20}\right)$

$k = 1:\ 2\sqrt{2}\left(\cos\dfrac{11\pi}{20} + i\sin\dfrac{11\pi}{20}\right)$

$k = 2:\ 2\sqrt{2}\left(\cos\dfrac{19\pi}{20} + i\sin\dfrac{19\pi}{20}\right)$

$k = 3:\ 2\sqrt{2}\left(\cos\dfrac{27\pi}{20} + i\sin\dfrac{27\pi}{20}\right)$

$k = 4:\ 2\sqrt{2}\left(\cos\dfrac{7\pi}{4} + i\sin\dfrac{7\pi}{4}\right)$

(b)

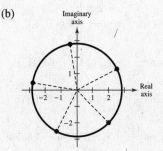

(c) $2.5201 + 1.2841i,\ -0.4425 + 2.7936i,$
$-2.7936 + 0.4425i,\ -1.2841 - 2.5201i,\ 2 - 2i$

105. $x^4 + i = 0$

$\qquad x^4 = -i$

The solutions are the fourth roots of $i = \cos\dfrac{3\pi}{2} + i\sin\dfrac{3\pi}{2}$:

$$\sqrt[4]{1}\left[\cos\left(\dfrac{\dfrac{3\pi}{2} + 2k\pi}{4}\right) + i\sin\left(\dfrac{\dfrac{3\pi}{2} + 2k\pi}{4}\right)\right],\ k = 0, 1, 2, 3$$

$k = 0:\ \cos\dfrac{3\pi}{8} + i\sin\dfrac{3\pi}{8} \approx 0.3827 + 0.9239i$

$k = 1:\ \cos\dfrac{7\pi}{8} + i\sin\dfrac{7\pi}{8} \approx -0.9239 + 0.3827i$

$k = 2:\ \cos\dfrac{11\pi}{8} + i\sin\dfrac{11\pi}{8} \approx -0.3827 - 0.9239i$

$k = 3:\ \cos\dfrac{15\pi}{8} + i\sin\dfrac{15\pi}{8} \approx 0.9239 - 0.3827i$

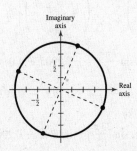

107. $x^5 + 243 = 0$

$\qquad x^5 = -243$

The solutions are the fifth roots of $-243 = 243(\cos\pi + i\sin\pi)$:

$$\sqrt[5]{243}\left[\cos\left(\dfrac{\pi + 2k\pi}{5}\right) + i\sin\left(\dfrac{\pi + 2k\pi}{5}\right)\right],\ k = 0, 1, 2, 3, 4$$

$k = 0:\ 3\left(\cos\dfrac{\pi}{5} + i\sin\dfrac{\pi}{5}\right) \approx 2.4271 + 1.7634i$

$k = 1:\ 3\left(\cos\dfrac{3\pi}{5} + i\sin\dfrac{3\pi}{5}\right) \approx -0.9271 + 2.8532i$

$k = 2:\ 3(\cos\pi + i\sin\pi) = -3$

$k = 3:\ 3\left(\cos\dfrac{7\pi}{5} + i\sin\dfrac{7\pi}{5}\right) \approx -0.9271 - 2.8532i$

$k = 4:\ 3\left(\cos\dfrac{9\pi}{5} + i\sin\dfrac{9\pi}{5}\right) \approx 2.4271 - 1.7634i$

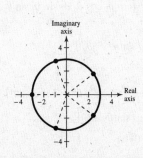

109. $x^4 + 16i = 0$

$$x^4 = -16i$$

The solutions are the fourth roots of $-16i = 16\left(\cos\dfrac{3\pi}{2} + i \sin\dfrac{3\pi}{2}\right)$:

$$\sqrt[4]{16}\left[\cos\dfrac{\dfrac{3\pi}{2} + 2\pi k}{4} + i \sin\dfrac{\dfrac{3\pi}{2} + 2\pi k}{4}\right], k = 0, 1, 2, 3$$

$k = 0$: $2\left(\cos\dfrac{3\pi}{8} + i \sin\dfrac{3\pi}{8}\right) \approx 0.7654 + 1.8478i$

$k = 1$: $2\left(\cos\dfrac{7\pi}{8} + i \sin\dfrac{7\pi}{8}\right) \approx -1.8478 + 0.7654i$

$k = 2$: $2\left(\cos\dfrac{11\pi}{8} + i \sin\dfrac{11\pi}{8}\right) \approx -0.7654 - 1.8478i$

$k = 3$: $2\left(\cos\dfrac{15\pi}{8} + i \sin\dfrac{15\pi}{8}\right) \approx 1.8478 - 0.7654i$

111. $x^3 - (1 - i) = 0$

$$x^3 = 1 - i = \sqrt{2}\left(\cos\dfrac{7\pi}{4} + i \sin\dfrac{7\pi}{4}\right)$$

The solutions are the cube roots of $1 - i$:

$$\sqrt[3]{\sqrt{2}}\left[\cos\left(\dfrac{\dfrac{7\pi}{4} + 2\pi k}{3}\right) + i \sin\left(\dfrac{\dfrac{7\pi}{4} + 2\pi k}{3}\right)\right], k = 0, 1, 2$$

$k = 0$: $\sqrt[6]{2}\left(\cos\dfrac{7\pi}{12} + i \sin\dfrac{7\pi}{12}\right) \approx -0.2905 + 1.0842i$

$k = 1$: $\sqrt[6]{2}\left(\cos\dfrac{5\pi}{4} + i \sin\dfrac{5\pi}{4}\right) \approx -0.7937 - 0.7937i$

$k = 2$: $\sqrt[6]{2}\left(\cos\dfrac{23\pi}{12} + i \sin\dfrac{23\pi}{12}\right) \approx 1.0842 - 0.2905i$

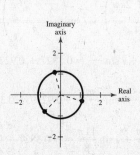

113. True, by the definition of the absolute value of a complex number.

115. True. $z_1 z_2 = r_1 r_2[\cos(\theta_1 + \theta_2) + i \sin(\theta_1 + \theta_2)]$ and $z_1 z_2 = 0$ if and only if $r_1 = 0$ and/or $r_2 = 0$.

117. $\dfrac{z_1}{z_2} = \dfrac{r_1(\cos\theta_1 + i \sin\theta_1)}{r_2(\cos\theta_2 + i \sin\theta_2)} \cdot \dfrac{\cos\theta_2 - i \sin\theta_2}{\cos\theta_2 - i \sin\theta_2}$

$$= \dfrac{r_1}{r_2(\cos^2\theta_2 + \sin^2\theta_2)}[\cos\theta_1 \cos\theta_2 + \sin\theta_1 \sin\theta_2 + i(\sin\theta_1 \cos\theta_2 - \sin\theta_2 \cos\theta_1)]$$

$$= \dfrac{r_1}{r_2}[\cos(\theta_1 - \theta_2) + i \sin(\theta_1 - \theta_2)]$$

119. (a) $z\bar{z} = [r(\cos\theta + i \sin\theta)][r(\cos(-\theta) + i \sin(-\theta))]$

$$= r^2[\cos(\theta - \theta) + i \sin(\theta - \theta)]$$

$$= r^2[\cos 0 + i \sin 0]$$

$$= r^2$$

(b) $\dfrac{z}{\bar{z}} = \dfrac{r(\cos\theta + i \sin\theta)}{r[\cos(-\theta) + i \sin(-\theta)]}$

$$= \dfrac{r}{r}[\cos(\theta - (-\theta)) + i \sin(\theta - (-\theta))]$$

$$= \cos 2\theta + i \sin 2\theta$$

121. $-\frac{1}{2}(1 + \sqrt{3}i) = -\left(\cos\frac{4\pi}{3} + i\sin\frac{4\pi}{3}\right)$

$$\left[-\frac{1}{2}(1 + \sqrt{3}i)\right]^6 = \left[-\left(\cos\frac{4\pi}{3} + i\sin\frac{4\pi}{3}\right)\right]^6$$

$$= \cos 8\pi + i\sin 8\pi$$

$$= 1$$

123. (a) $2(\cos 30° + i\sin 30°)$

$2(\cos 150° + i\sin 150°)$

$2(\cos 270° + i\sin 270°)$

(b) These are the cube roots of $8i$.

125. $A = 22°, a = 8$

$B = 90° - A = 68°$

$\tan 22° = \dfrac{8}{b} \implies b = \dfrac{8}{\tan 22°} \approx 19.80$

$\sin 22° = \dfrac{8}{c} \implies c = \dfrac{8}{\sin 22°} \approx 21.36$

127. $A = 30°, b = 112.6$

$B = 90° - A = 60°$

$\tan 30° = \dfrac{a}{112.6} \implies a = 112.6 \tan 30° \approx 65.01$

$\cos 30° = \dfrac{112.6}{c} \implies c = \dfrac{112.6}{\cos 30°} \approx 130.02$

129. $A = 42°15' = 42.25°, c = 11.2$

$B = 90° - A = 47°45'$

$\sin 42.25° = \dfrac{a}{11.2} \implies a = 11.2\sin 42.25° \approx 7.53$

$\cos 42.25° = \dfrac{b}{11.2} \implies b = 11.2\cos 42.25° \approx 8.29$

131. $d = 16\cos\dfrac{\pi}{4}t$

Maximum displacement: $|16| = 16$

$16\cos\dfrac{\pi}{4}t = 0 \implies \dfrac{\pi}{4}t = \dfrac{\pi}{2} \implies t = 2$

133. $d = \frac{1}{16}\sin\left(\frac{5}{4}\pi t\right)$

Maximum displacement: $\left|\frac{1}{16}\right| = \frac{1}{16}$

$\frac{1}{16}\sin\left(\frac{5}{4}\pi t\right) = 0$

$\frac{5}{4}\pi t = \pi$

$t = \frac{4}{5}$

135. $6\sin 8\theta\cos 3\theta = (6)\left(\frac{1}{2}\right)[\sin(8\theta + 3\theta) + \sin(8\theta - 3\theta)]$

$$= 3(\sin 11\theta + \sin 5\theta)$$

Review Exercises for Chapter 6

1. Given: $A = 35°, B = 71°, a = 8$

$C = 180° - 35° - 71° = 74°$

$b = \dfrac{a\sin B}{\sin A} = \dfrac{8\sin 71°}{\sin 35°} \approx 13.19$

$c = \dfrac{a\sin C}{\sin A} = \dfrac{8\sin 74°}{\sin 35°} \approx 13.41$

3. Given: $B = 72°, C = 82°, b = 54$

$A = 180° - 72° - 82° = 26°$

$a = \dfrac{b\sin A}{\sin B} = \dfrac{54\sin 26°}{\sin 72°} \approx 24.89$

$c = \dfrac{b\sin C}{\sin B} = \dfrac{54\sin 82°}{\sin 72°} \approx 56.23$

5. Given: $A = 16°, B = 98°, c = 8.4$

$C = 180° - 16° - 98° = 66°$

$a = \dfrac{c\sin A}{\sin C} = \dfrac{8.4\sin 16°}{\sin 66°} \approx 2.53$

$b = \dfrac{c\sin B}{\sin C} = \dfrac{8.4\sin 98°}{\sin 66°} \approx 9.11$

7. Given: $A = 24°, C = 48°, b = 27.5$

$B = 180° - 24° - 48° = 108°$

$a = \dfrac{b\sin A}{\sin B} = \dfrac{27.5\sin 24°}{\sin 108°} \approx 11.76$

$c = \dfrac{b\sin C}{\sin B} = \dfrac{27.5\sin 48°}{\sin 108°} \approx 21.49$

9. Given: $B = 150°$, $b = 30$, $c = 10$

$$\sin C = \frac{c \sin B}{b} = \frac{10 \sin 150°}{30} \approx 0.1667 \implies C \approx 9.59°$$

$$A \approx 180° - 150° - 9.59° = 20.41°$$

$$a = \frac{b \sin A}{\sin B} = \frac{30 \sin 20.41°}{\sin 150°} \approx 20.92$$

11. $A = 75°$, $a = 51.2$, $b = 33.7$

$$\sin B = \frac{b \sin A}{a} = \frac{33.7 \sin 75°}{51.2} \approx 0.6358 \implies B \approx 39.48°$$

$$C \approx 180° - 75° - 39.48° = 65.52°$$

$$c = \frac{a \sin C}{\sin A} = \frac{51.2 \sin 65.52°}{\sin 75°} \approx 48.24$$

13. Area $= \frac{1}{2}bc \sin A = \frac{1}{2}(5)(7)\sin 27° \approx 7.9$

15. Area $= \frac{1}{2}ab \sin C = \frac{1}{2}(16)(5)\sin 123° \approx 33.5$

17. $\tan 17° = \dfrac{h}{x + 50} \implies h = (x + 50) \tan 17°$

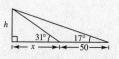

$$h = x \tan 17° + 50 \tan 17°$$

$$\tan 31° = \frac{h}{x} \implies h = x \tan 31°$$

$$x \tan 17° + 50 \tan 17° = x \tan 31°$$

$$50 \tan 17° = x(\tan 31° - \tan 17°)$$

$$\frac{50 \tan 17°}{\tan 31° - \tan 17°} = x$$

$$x \approx 51.7959$$

$$h = x \tan 31° \approx 51.7959 \tan 31° \approx 31.1 \text{ meters}$$

The height of the building is approximately 31.1 meters.

19. $\dfrac{h}{\sin 17°} = \dfrac{75}{\sin 45°}$

$$h = \frac{75 \sin 17°}{\sin 45°}$$

$$h \approx 31.01 \text{ feet}$$

21. Given: $a = 5$, $b = 8$, $c = 10$

$$\cos C = \frac{a^2 + b^2 - c^2}{2ab} = -0.1375 \implies C \approx 97.90°$$

$$\cos B = \frac{a^2 + c^2 - b^2}{2ac} = 0.61 \implies B \approx 52.41°$$

$$A = 180° - B - C \approx 29.69°$$

23. Given: $a = 2.5$, $b = 5.0$, $c = 4.5$

$$\cos B = \frac{a^2 + c^2 - b^2}{2ac} = 0.0667 \implies B \approx 86.18°$$

$$\cos C = \frac{a^2 + b^2 - c^2}{2ab} = 0.44 \implies C \approx 63.90°$$

$$A = 180° - B - C \approx 29.92°$$

25. Given: $B = 110°$, $a = 4$, $c = 4$

$$b = \sqrt{a^2 + c^2 - 2ac \cos B} \approx 6.55$$

$$A = C = \frac{1}{2}(180° - 110°) = 35°$$

27. Given: $C = 43°$, $a = 22.5$, $b = 31.4$

$$c = \sqrt{a^2 + b^2 - 2ab \cos C} \approx 21.42$$

$$\cos B = \frac{a^2 + c^2 - b^2}{2ac} \approx -0.02169 \implies B \approx 91.24°$$

$$A = 180° - B - C \approx 45.76°$$

29.

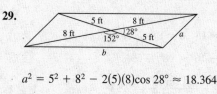

$$a^2 = 5^2 + 8^2 - 2(5)(8)\cos 28° \approx 18.364$$

$$a \approx 4.3 \text{ feet}$$

$$b^2 = 8^2 + 5^2 - 2(8)(5)\cos 152° \approx 159.636$$

$$b \approx 12.6 \text{ feet}$$

31. Length of AC = $\sqrt{300^2 + 425^2 - 2(300)(425)\cos 115°}$

≈ 615.1 meters

33. $a = 4$, $b = 5$, $c = 7$

$$s = \frac{a+b+c}{2} = \frac{4+5+7}{2} = 8$$

Area $= \sqrt{s(s-a)(s-b)(s-c)}$

$= \sqrt{8(4)(3)(1)} \approx 9.80$

35. $a = 12.3$, $b = 15.8$, $c = 3.7$

$$s = \frac{a+b+c}{2} = \frac{12.3+15.8+3.7}{2} = 15.9$$

Area $= \sqrt{s(s-a)(s-b)(s-c)}$

$= \sqrt{15.9(3.6)(0.1)(12.2)} = 8.36$

37. $\|\mathbf{u}\| = \sqrt{(4-(-2))^2 + (6-1)^2} = \sqrt{61}$

$\|\mathbf{v}\| = \sqrt{(6-0)^2 + (3-(-2))^2} = \sqrt{61}$

$\mathbf{u}$ is directed along a line with a slope of $\dfrac{6-1}{4-(-2)} = \dfrac{5}{6}$.

$\mathbf{v}$ is directed along a line with a slope of $\dfrac{3-(-2)}{6-0} = \dfrac{5}{6}$.

Since $\mathbf{u}$ and $\mathbf{v}$ have identical magnitudes and directions, $\mathbf{u} = \mathbf{v}$.

39. Initial point: $(-5, 4)$

Terminal point: $(2, -1)$

$\mathbf{v} = \langle 2-(-5), -1-4 \rangle = \langle 7, -5 \rangle$

41. Initial point: $(0, 10)$

Terminal point: $(7, 3)$

$\mathbf{v} = \langle 7-0, 3-10 \rangle = \langle 7, -7 \rangle$

43. $\|\mathbf{v}\| = 8$, $\theta = 120°$

$\langle 8\cos 120°, 8\sin 120° \rangle = \langle -4, 4\sqrt{3} \rangle$

45. $\mathbf{u} = \langle -1, -3 \rangle$, $\mathbf{v} = \langle -3, 6 \rangle$

(a) $\mathbf{u} + \mathbf{v} = \langle -1, -3 \rangle + \langle -3, 6 \rangle = \langle -4, 3 \rangle$

(b) $\mathbf{u} - \mathbf{v} = \langle -1, -3 \rangle - \langle -3, 6 \rangle = \langle 2, -9 \rangle$

(c) $3\mathbf{u} = 3\langle -1, -3 \rangle = \langle -3, -9 \rangle$

(d) $2\mathbf{v} + 5\mathbf{u} = 2\langle -3, 6 \rangle + 5\langle -1, -3 \rangle$

$= \langle -6, 12 \rangle + \langle -5, -15 \rangle = \langle -11, -3 \rangle$

47. $\mathbf{u} = \langle -5, 2 \rangle$, $\mathbf{v} = \langle 4, 4 \rangle$

(a) $\mathbf{u} + \mathbf{v} = \langle -5, 2 \rangle + \langle 4, 4 \rangle = \langle -1, 6 \rangle$

(b) $\mathbf{u} - \mathbf{v} = \langle -5, 2 \rangle - \langle 4, 4 \rangle = \langle -9, -2 \rangle$

(c) $3\mathbf{u} = 3\langle -5, 2 \rangle = \langle -15, 6 \rangle$

(d) $2\mathbf{v} + 5\mathbf{u} = 2\langle 4, 4 \rangle + 5\langle -5, 2 \rangle$

$= \langle 8, 8 \rangle + \langle -25, 10 \rangle = \langle -17, 18 \rangle$

49. $\mathbf{u} = 2\mathbf{i} - \mathbf{j}$, $\mathbf{v} = 5\mathbf{i} + 3\mathbf{j}$

(a) $\mathbf{u} + \mathbf{v} = (2\mathbf{i} - \mathbf{j}) + (5\mathbf{i} + 3\mathbf{j}) = 7\mathbf{i} + 2\mathbf{j}$

(b) $\mathbf{u} - \mathbf{v} = (2\mathbf{i} - \mathbf{j}) - (5\mathbf{i} + 3\mathbf{j}) = -3\mathbf{i} - 4\mathbf{j}$

(c) $3\mathbf{u} = 3(2\mathbf{i} - \mathbf{j}) = 6\mathbf{i} - 3\mathbf{j}$

(d) $2\mathbf{v} + 5\mathbf{u} = 2(5\mathbf{i} + 3\mathbf{j}) + 5(2\mathbf{i} - \mathbf{j})$

$= (10\mathbf{i} + 6\mathbf{j}) + (10\mathbf{i} - 5\mathbf{j}) = 20\mathbf{i} + \mathbf{j}$

51. $\mathbf{u} = 4\mathbf{i}$, $\mathbf{v} = -\mathbf{i} + 6\mathbf{j}$

(a) $\mathbf{u} + \mathbf{v} = 4\mathbf{i} + (-\mathbf{i} + 6\mathbf{j}) = 3\mathbf{i} + 6\mathbf{j}$

(b) $\mathbf{u} - \mathbf{v} = 4\mathbf{i} - (-\mathbf{i} + 6\mathbf{j}) = 5\mathbf{i} - 6\mathbf{j}$

(c) $3\mathbf{u} = 3(4\mathbf{i}) = 12\mathbf{i}$

(d) $2\mathbf{v} + 5\mathbf{u} = 2(-\mathbf{i} + 6\mathbf{j}) + 5(4\mathbf{i})$

$= (-2\mathbf{i} + 12\mathbf{j}) + 20\mathbf{i} = 18\mathbf{i} + 12\mathbf{j}$

53. $\mathbf{u} = 6\mathbf{i} - 5\mathbf{j}$, $\mathbf{v} = 10\mathbf{i} + 3\mathbf{j}$

$2\mathbf{u} + \mathbf{v} = 2(6\mathbf{i} - 5\mathbf{j}) + (10\mathbf{i} + 3\mathbf{j})$

$= 22\mathbf{i} - 7\mathbf{j}$

$= \langle 22, -7 \rangle$

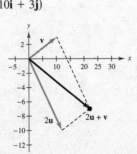

55. $\mathbf{v} = 10\mathbf{i} + 3\mathbf{j}$

$3\mathbf{v} = 3(10\mathbf{i} + 3\mathbf{j})$

$= 30\mathbf{i} + 9\mathbf{j}$

$= \langle 30, 9 \rangle$

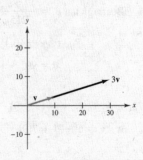

57. $\mathbf{u} = \langle -3, 4 \rangle = -3\mathbf{i} + 4\mathbf{j}$

59. Initial point: $(3, 4)$

Terminal point: $(9, 8)$

$\mathbf{u} = (9 - 3)\mathbf{i} + (8 - 4)\mathbf{j} = 6\mathbf{i} + 4\mathbf{j}$

61. $\mathbf{v} = -10\mathbf{i} + 10\mathbf{j}$

$\|\mathbf{v}\| = \sqrt{(-10)^2 + (10)^2} = \sqrt{200} = 10\sqrt{2}$

$\tan \theta = \dfrac{10}{-10} = -1 \implies \theta = 135°$ since

$\mathbf{v}$ is in Quadrant II.

$\mathbf{v} = 10\sqrt{2}(\mathbf{i} \cos 135° + \mathbf{j} \sin 135°)$

63. $\mathbf{v} = 7(\cos 60° \, \mathbf{i} + \sin 60° \, \mathbf{j})$

$\|\mathbf{v}\| = 7$

$\theta = 60°$

65. $\mathbf{v} = 5\mathbf{i} + 4\mathbf{j}$

$\|\mathbf{v}\| = \sqrt{5^2 + 4^2} = \sqrt{41}$

$\tan \theta = \dfrac{4}{5} \implies \theta \approx 38.7°$

67. $\mathbf{v} = -3\mathbf{i} - 3\mathbf{j}$

$\|\mathbf{v}\| = \sqrt{(-3)^2 + (-3)^2} = 3\sqrt{2}$

$\tan \theta = \dfrac{-3}{-3} = 1 \implies \theta = 225°$

69. Magnitude of resultant:

$c = \sqrt{85^2 + 50^2 - 2(85)(50) \cos 165°}$

≈ 133.92 pounds

Let θ be the angle between the resultant and the 85-pound force.

$\cos \theta \approx \dfrac{(133.92)^2 + 85^2 - 50^2}{2(133.92)(85)}$

≈ 0.9953

$\implies \theta \approx 5.6°$

71. Airspeed: $\mathbf{u} = 430(\cos 45°\mathbf{i} - \sin 45°\mathbf{j}) = 215\sqrt{2}(\mathbf{i} - \mathbf{j})$

Wind: $\mathbf{w} = 35(\cos 60° + \sin 60° \, \mathbf{j}) = \dfrac{35}{2}(\mathbf{i} + \sqrt{3}\mathbf{j})$

Groundspeed: $\mathbf{u} + \mathbf{w} = \left(215\sqrt{2} + \dfrac{35}{2}\right)\mathbf{i} + \left(\dfrac{35\sqrt{3}}{2} - 215\sqrt{2}\right)$

$\|\mathbf{u} + \mathbf{w}\| = \sqrt{\left(215\sqrt{2} + \dfrac{35}{2}\right)^2 + \left(\dfrac{35\sqrt{3}}{2} - 215\sqrt{2}\right)^2}$

≈ 422.30 miles per hour

Bearing: $\tan \theta' = \dfrac{17.5\sqrt{3} - 215\sqrt{2}}{215\sqrt{2} + 17.5}$

$\theta' \approx -40.4°$

$\theta = 90° + |\theta'| = 130.4°$

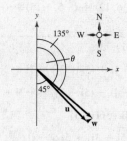

73. $\mathbf{u} = \langle 6, 7 \rangle$, $\mathbf{v} = \langle -3, 9 \rangle$

$\mathbf{u} \cdot \mathbf{v} = 6(-3) + 7(9) = 45$

75. $\mathbf{u} = 3\mathbf{i} + 7\mathbf{j}$, $\mathbf{v} = 11\mathbf{i} - 5\mathbf{j}$

$\mathbf{u} \cdot \mathbf{v} = 3(11) + 7(-5) = -2$

77. $\mathbf{u} = \langle -3, 4 \rangle$

$2\mathbf{u} = \langle -6, 8 \rangle$

$2\mathbf{u} \cdot \mathbf{u} = (-6)(-3) + 8(4) = 50$

The result is a scalar.

79. $\mathbf{u} = \langle -3, 4 \rangle, \mathbf{v} = \langle 2, 1 \rangle$

$\mathbf{u} \cdot \mathbf{v} = (-3)(2) + 4(1) = -2$

$\mathbf{u}(\mathbf{u} \cdot \mathbf{v}) = \mathbf{u}(-2) = -2\mathbf{u} = \langle 6, -8 \rangle$

The result is a vector.

81. $\mathbf{u} = \cos \dfrac{7\pi}{4}\mathbf{i} + \sin \dfrac{7\pi}{4}\mathbf{j} = \left\langle \dfrac{1}{\sqrt{2}}, -\dfrac{1}{\sqrt{2}} \right\rangle$

$\mathbf{v} = \cos \dfrac{5\pi}{6}\mathbf{i} + \sin \dfrac{5\pi}{6}\mathbf{j} = \left\langle -\dfrac{\sqrt{3}}{2}, \dfrac{1}{2} \right\rangle$

$\cos \theta = \dfrac{\mathbf{u} \cdot \mathbf{v}}{\|\mathbf{u}\| \, \|\mathbf{v}\|} = \dfrac{-\sqrt{3} - 1}{2\sqrt{2}} \implies \theta = \dfrac{11\pi}{12}$

83. $\mathbf{u} = \langle 2\sqrt{2}, -4 \rangle, \mathbf{v} = \langle -\sqrt{2}, 1 \rangle$

$\cos \theta = \dfrac{\mathbf{u} \cdot \mathbf{v}}{\|\mathbf{u}\| \, \|\mathbf{v}\|} = \dfrac{-8}{(\sqrt{24})(\sqrt{3})} \implies \theta \approx 160.5°$

85. $\mathbf{u} = \langle -3, 8 \rangle$

$\mathbf{v} = \langle 8, 3 \rangle$

$\mathbf{u} \cdot \mathbf{v} = -3(8) + 8(3) = 0$

$\mathbf{u}$ and $\mathbf{v}$ are orthogonal.

87. $\mathbf{u} = -\mathbf{i}$

$\mathbf{v} = \mathbf{i} + 2\mathbf{j}$

$\mathbf{u} \cdot \mathbf{v} \neq 0 \implies$ Not orthogonal

$\mathbf{v} \neq k\mathbf{u} \implies$ Not parallel

Neither

89. $\mathbf{u} = \langle -4, 3 \rangle, \mathbf{v} = \langle -8, -2 \rangle$

$\mathbf{w}_1 = \text{proj}_\mathbf{v}\mathbf{u} = \left(\dfrac{\mathbf{u} \cdot \mathbf{v}}{\|\mathbf{v}\|^2} \right)\mathbf{v} = \left(\dfrac{26}{68} \right)\langle -8, -2 \rangle = -\dfrac{13}{17}\langle 4, 1 \rangle$

$\mathbf{w}_2 = \mathbf{u} - \mathbf{w}_1 = \langle -4, 3 \rangle - \left(-\dfrac{13}{17} \right)\langle 4, 1 \rangle = \dfrac{16}{17}\langle -1, 4 \rangle$

$\mathbf{u} = \mathbf{w}_1 + \mathbf{w}_2 = -\dfrac{13}{17}\langle 4, 1 \rangle + \dfrac{16}{17}\langle -1, 4 \rangle$

91. $\mathbf{u} = \langle 2, 7 \rangle, \mathbf{v} = \langle 1, -1 \rangle$

$\mathbf{w}_1 = \text{proj}_\mathbf{v}\mathbf{u} = \left(\dfrac{\mathbf{u} \cdot \mathbf{v}}{\|\mathbf{v}\|^2} \right)\mathbf{v} = -\dfrac{5}{2}\langle 1, -1 \rangle = \dfrac{5}{2}\langle -1, 1 \rangle$

$\mathbf{w}_2 = \mathbf{u} - \mathbf{w}_1 = \langle 2, 7 \rangle - \left(\dfrac{5}{2} \right)\langle -1, 1 \rangle = \dfrac{9}{2}\langle 1, 1 \rangle$

$\mathbf{u} = \mathbf{w}_1 + \mathbf{w}_2 = \dfrac{5}{2}\langle -1, 1 \rangle + \dfrac{9}{2}\langle 1, 1 \rangle$

93. $P = (5, 3), Q = (8, 9) \implies \overrightarrow{PQ} = \langle 3, 6 \rangle$

$W = \mathbf{v} \cdot \overrightarrow{PQ} = \langle 2, 7 \rangle \cdot \langle 3, 6 \rangle = 48$

95. $w = (18,000)\left(\frac{48}{12} \right) = 72,000$ foot-pounds

97. $|7i| = \sqrt{0^2 + 7^2} = 7$

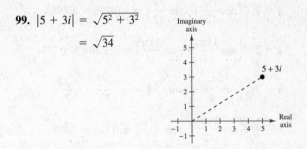

99. $|5 + 3i| = \sqrt{5^2 + 3^2}$

$= \sqrt{34}$

101. $5 - 5i$

$r = \sqrt{5^2 + (-5)^2} = \sqrt{50} = 5\sqrt{2}$

$\tan \theta = \dfrac{-5}{5} = -1 \implies \theta = \dfrac{7\pi}{4}$ since the

complex number is in Quadrant IV.

$5 - 5i = 5\sqrt{2}\left(\cos \dfrac{7\pi}{4} + i \sin \dfrac{7\pi}{4} \right)$

103. $-3\sqrt{3} + 3i$

$r = \sqrt{\left(-3\sqrt{3} \right)^2 + 3^2} = \sqrt{36} = 6$

$\tan \theta = \dfrac{3}{-3\sqrt{3}} = -\dfrac{1}{\sqrt{3}} \implies \theta = \dfrac{5\pi}{6}$

since the complex number is in Quadrant II.

$-3\sqrt{3} + 3i = 6\left(\cos \dfrac{5\pi}{6} + i \sin \dfrac{5\pi}{6} \right)$

105. (a) $z_1 = 2\sqrt{3} - 2i = 4\left(\cos\dfrac{11\pi}{6} + i\sin\dfrac{11\pi}{6}\right)$

$z_2 = -10i = 10\left(\cos\dfrac{3\pi}{2} + i\sin\dfrac{3\pi}{2}\right)$

(b) $z_1 z_2 = \left[4\left(\cos\dfrac{11\pi}{6} + i\sin\dfrac{11\pi}{6}\right)\right]\left[10\left(\cos\dfrac{3\pi}{2} + i\sin\dfrac{3\pi}{2}\right)\right]$

$= 40\left(\cos\dfrac{10\pi}{3} + i\sin\dfrac{10\pi}{3}\right)$

$\dfrac{z_1}{z_2} = \dfrac{4\left(\cos\dfrac{11\pi}{6} + i\sin\dfrac{11\pi}{6}\right)}{10\left(\cos\dfrac{3\pi}{2} + i\sin\dfrac{3\pi}{2}\right)} = \dfrac{2}{5}\left(\cos\dfrac{\pi}{3} + i\sin\dfrac{\pi}{3}\right)$

107. $\left[5\left(\cos\dfrac{\pi}{12} + i\sin\dfrac{\pi}{12}\right)\right]^4 = 5^4\left(\cos\dfrac{4\pi}{12} + i\sin\dfrac{4\pi}{12}\right)$

$= 625\left(\cos\dfrac{\pi}{3} + i\sin\dfrac{\pi}{3}\right)$

$= 625\left(\dfrac{1}{2} + \dfrac{\sqrt{3}}{2}i\right)$

$= \dfrac{625}{2} + \dfrac{625\sqrt{3}}{2}i$

109. $(2 + 3i)^6 \approx \left[\sqrt{13}(\cos 56.3° + i\sin 56.3°)\right]^6$

$= 13^3(\cos 337.9° + i\sin 337.9°)$

$\approx 13^3(0.9263 - 0.3769i)$

$\approx 2035 - 828i$

111. Sixth roots of $-729i = 729\left(\cos\dfrac{3\pi}{2} + i\sin\dfrac{3\pi}{2}\right)$:

(a) and (c)

$\sqrt[6]{729}\left[\cos\left(\dfrac{\dfrac{3\pi}{2} + 2k\pi}{6}\right) + i\sin\left(\dfrac{\dfrac{3\pi}{2} + 2k\pi}{6}\right)\right], k = 0, 1, 2, 3, 4, 5$

$k = 0: 3\left(\cos\dfrac{\pi}{4} + i\sin\dfrac{\pi}{4}\right) = \dfrac{3\sqrt{2}}{2} + \dfrac{3\sqrt{2}}{2}i$

$k = 1: 3\left(\cos\dfrac{7\pi}{12} + i\sin\dfrac{7\pi}{12}\right) \approx -0.776 + 2.898i$

$k = 2: 3\left(\cos\dfrac{11\pi}{12} + i\sin\dfrac{11\pi}{12}\right) \approx -2.898 + 0.776i$

$k = 3: 3\left(\cos\dfrac{5\pi}{4} + i\sin\dfrac{5\pi}{4}\right) = -\dfrac{3\sqrt{2}}{2} - \dfrac{3\sqrt{2}}{2}i$

$k = 4: 3\left(\cos\dfrac{19\pi}{12} + i\sin\dfrac{19\pi}{12}\right) \approx 0.776 - 2.898i$

$k = 5: 3\left(\cos\dfrac{23\pi}{12} + i\sin\dfrac{23\pi}{12}\right) \approx 2.898 - 0.776i$

(b)

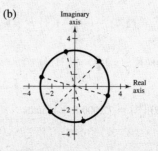

113. Cube roots of $8 = 8(\cos 0 + i \sin 0)$, $k = 0, 1, 2$

(a) and (c)

$$\sqrt[3]{8}\left[\cos\left(\frac{0 + 2\pi k}{3}\right) + i \sin\left(\frac{0 + 2\pi k}{3}\right)\right]$$

$k = 0;\ 2(\cos 0 + i \sin 0) = 2$

$k = 1:\ 2\left(\cos\frac{2\pi}{3} + i \sin\frac{2\pi}{3}\right) = -1 + \sqrt{3}i$

$k = 2:\ 2\left(\cos\frac{4\pi}{3} + i \sin\frac{4\pi}{3}\right) = -1 - \sqrt{3}i$

(b)

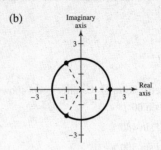

115. $x^4 + 81 = 0$

$x^4 = -81$ Solve by finding the fourth roots of -81.

$-81 = 81(\cos \pi + i \sin \pi)$

$$\sqrt[4]{-81} = \sqrt[4]{81}\left[\cos\left(\frac{\pi + 2\pi k}{4}\right) + i \sin\left(\frac{\pi + 2\pi k}{4}\right)\right],\ k = 0, 1, 2, 3$$

$k = 0:\ 3\left(\cos\frac{\pi}{4} + i \sin\frac{\pi}{4}\right) = \frac{3\sqrt{2}}{2} + \frac{3\sqrt{2}}{2}i$

$k = 1:\ 3\left(\cos\frac{3\pi}{4} + i \sin\frac{3\pi}{4}\right) = -\frac{3\sqrt{2}}{2} + \frac{3\sqrt{2}}{2}i$

$k = 2:\ 3\left(\cos\frac{5\pi}{4} + i \sin\frac{5\pi}{4}\right) = -\frac{3\sqrt{2}}{2} - \frac{3\sqrt{2}}{2}i$

$k = 3:\ 3\left(\cos\frac{7\pi}{4} + i \sin\frac{7\pi}{4}\right) = \frac{3\sqrt{2}}{2} - \frac{3\sqrt{2}}{2}i$

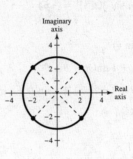

117. $x^3 + 8i = 0$

$x^3 = -8i$ Solve by finding the cube roots of $-8i$.

$$-8i = 8\left(\cos\frac{3\pi}{2} + i \sin\frac{3\pi}{2}\right)$$

$$\sqrt[3]{-8i} = \sqrt[3]{8}\left[\cos\left(\frac{\frac{3\pi}{2} + 2\pi k}{3}\right) + i \sin\left(\frac{\frac{3\pi}{2} + 2\pi k}{3}\right)\right],\ k = 0, 1, 2$$

$k = 0:\ 2\left(\cos\frac{\pi}{2} + i \sin\frac{\pi}{2}\right) = 2i$

$k = 1:\ 2\left(\cos\frac{7\pi}{6} + i \sin\frac{7\pi}{6}\right) = -\sqrt{3} - i$

$k = 2:\ 2\left(\cos\frac{11\pi}{6} + i \sin\frac{11\pi}{6}\right) = \sqrt{3} - i$

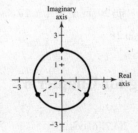

119. True. $\sin 90°$ is defined in the Law of Sines.

121. True, by the definition of a unit vector.

$$\mathbf{u} = \frac{\mathbf{v}}{\|\mathbf{v}\|} \text{ so } \mathbf{v} = \|\mathbf{v}\|\mathbf{u}$$

123. False. $x = \sqrt{3} + i$ is a solution to $x^3 - 8i = 0$, not $x^2 - 8i = 0$.

Also, $\left(\sqrt{3} + i\right)^2 - 8i = 2 + \left(2\sqrt{3} - 8\right)i \neq 0$.

125. $a^2 = b^2 + c^2 - 2bc \cos A$

$b^2 = a^2 + c^2 - 2ac \cos B$

$c^2 = a^2 + b^2 - 2ab \cos C$

127. *A* and *C* appear to have the same magnitude and direction.

129. If $k > 0$, the direction of $k\mathbf{u}$ is the same, and the magnitude is $k\|\mathbf{u}\|$.

If $k < 0$, the direction of $k\mathbf{u}$ is the opposite direction of $\mathbf{u}$, and the magnitude is $|k|\,\|\mathbf{u}\|$.

131. (a) The trigonometric form of the three roots shown is:

$4(\cos 60° + i \sin 60°)$

$4(\cos 180° + i \sin 180°)$

$4(\cos 300° + i \sin 300°)$

(b) Since there are three evenly spaced roots on the circle of radius 4, they are cube roots of a complex number of modulus $4^3 = 64$.

Cubing them yields -64.

$[4(\cos 60° + i \sin 60°)]^3 = -64$

$[4(\cos 180° + i \sin 180°)]^3 = -64$

$[4(\cos 300° + i \sin 300°)]^3 = -64$

133. $z_1 = 2(\cos \theta + i \sin \theta)$

$z_2 = 2(\cos(\pi - \theta) + i \sin(\pi - \theta))$

$z_1 z_2 = (2)(2)[\cos(\theta + (\pi - \theta)) + i \sin(\theta + (\pi - \theta))]$

$\qquad = 4(\cos \pi + i \sin \pi)$

$\qquad = -4$

$\dfrac{z_1}{z_2} = \dfrac{2(\cos \theta + i \sin \theta)}{2(\cos(\pi - \theta) + i \sin(\pi - \theta))}$

$\qquad = 1[\cos(\theta - (\pi - \theta)) + i \sin(\theta - (\pi - \theta))]$

$\qquad = \cos(2\theta - \pi) + i \sin(2\theta - \pi)$

$\qquad = \cos 2\theta \cos \pi + \sin 2\theta \sin \pi + i(\sin 2\theta \cos \pi - \cos 2\theta \sin \pi)$

$\qquad = -\cos 2\theta - i \sin 2\theta$

Problem Solving for Chapter 6

1. $\overrightarrow{PQ}^2 = 4.7^2 + 6^2 - 2(4.7)(6) \cos 25°$

$\overrightarrow{PQ} \approx 2.6409$ feet

$\dfrac{\sin \alpha}{4.7} = \dfrac{\sin 25°}{2.6409} \Rightarrow \alpha \approx 48.78°$

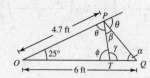

$\theta + \beta = 180° - 25° - 48.78° = 106.22°$

$(\theta + \beta) + \theta = 180° \Rightarrow \theta = 180° - 106.22° = 73.78°$

$\beta = 106.22° - 73.78° = 32.44°$

$\gamma = 180° - \alpha - \beta = 180° - 48.78° - 32.44° = 98.78°$

$\phi = 180° - \gamma = 180° - 98.78° = 81.22°$

$\dfrac{\overrightarrow{PT}}{\sin 25°} = \dfrac{4.7}{\sin 81.22°}$

$\overrightarrow{PT} \approx 2.01$ feet

3. (a)
A 75 mi B
30° 135° 15°
x y 75°
60° Lost party

(b) $\dfrac{x}{\sin 15°} = \dfrac{75}{\sin 135°}$ and $\dfrac{y}{\sin 30°} = \dfrac{75}{\sin 135°}$

$x \approx 27.45$ miles $y \approx 53.03$ miles

(c) $z^2 = (27.45)^2 + (20)^2 - 2(27.45)(20) \cos 20°$

$z \approx 11.03$ miles

$\dfrac{\sin \theta}{27.45} = \dfrac{\sin 20°}{11.03}$

$\sin \theta \approx 0.8511$

$\theta = 180° - \sin^{-1}(0.8511)$

$\theta \approx 121.7°$

A 10°
20 mi Rescue party
80° 20°
60° z
27.452 mi
Lost party

To find the bearing, we have $\theta - 10° - 90° \approx 21.7°$.

Bearing: S 21.7° E

5. If $\mathbf{u} \neq 0$, $\mathbf{v} \neq 0$, and $\mathbf{u} + \mathbf{v} \neq 0$, then $\left\| \dfrac{\mathbf{u}}{\|\mathbf{u}\|} \right\| = \left\| \dfrac{\mathbf{v}}{\|\mathbf{v}\|} \right\| = \left\| \dfrac{\mathbf{u} + \mathbf{v}}{\|\mathbf{u} + \mathbf{v}\|} \right\| = 1$ since all of these are magnitudes of **unit** vectors.

(a) $\mathbf{u} = \langle 1, -1 \rangle$, $\mathbf{v} = \langle -1, 2 \rangle$, $\mathbf{u} + \mathbf{v} = \langle 0, 1 \rangle$

$\|\mathbf{u}\| = \sqrt{2}$, $\|\mathbf{v}\| = \sqrt{5}$, $\|\mathbf{u} + \mathbf{v}\| = 1$

(b) $\mathbf{u} = \langle 0, 1 \rangle$, $\mathbf{v} = \langle 3, -3 \rangle$, $\mathbf{u} + \mathbf{v} = \langle 3, -2 \rangle$

$\|\mathbf{u}\| = 1$, $\|\mathbf{v}\| = \sqrt{18} = 3\sqrt{2}$, $\|\mathbf{u} + \mathbf{v}\| = \sqrt{13}$

(c) $\mathbf{u} = \left\langle 1, \dfrac{1}{2} \right\rangle$, $\mathbf{v} = \langle 2, 3 \rangle$, $\mathbf{u} + \mathbf{v} = \left\langle 3, \dfrac{7}{2} \right\rangle$

$\|\mathbf{u}\| = \dfrac{\sqrt{5}}{2}$, $\|\mathbf{v}\| = \sqrt{13}$, $\|\mathbf{u} + \mathbf{v}\| = \sqrt{9 + \dfrac{49}{4}} = \dfrac{\sqrt{85}}{2}$

(d) $\mathbf{u} = \langle 2, -4 \rangle$, $\mathbf{v} = \langle 5, 5 \rangle$, $\mathbf{u} + \mathbf{v} = \langle 7, 1 \rangle$

$\|\mathbf{u}\| = \sqrt{20} = 2\sqrt{5}$, $\|\mathbf{v}\| = \sqrt{50} = 5\sqrt{2}$, $\|\mathbf{u} + \mathbf{v}\| = \sqrt{50} = 5\sqrt{2}$

7. Initial point: $(0, 0)$

Terminal point: $\left(\dfrac{u_1 + v_1}{2}, \dfrac{u_2 + v_2}{2} \right)$

$\mathbf{w} = \left\langle \dfrac{u_1 + v_1}{2}, \dfrac{u_2 + v_2}{2} \right\rangle = \dfrac{1}{2}(\mathbf{u} + \mathbf{v})$

Initial point: (u_1, u_2)

Terminal point: $\dfrac{1}{2}(u_1 + v_1, u_2 + v_2)$

$\mathbf{w} = \left\langle \dfrac{u_1 + v_1}{2} - u_1, \dfrac{u_2 + v_2}{2} - u_2 \right\rangle = \left\langle \dfrac{v_1 - u_1}{2}, \dfrac{v_2 - u_2}{2} \right\rangle = \dfrac{1}{2}(\mathbf{v} - \mathbf{u})$

9. $W = (\cos \theta)\|F\| \, \|\overrightarrow{PQ}\|$ and $\|F_1\| = \|F_2\|$

(a)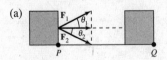
If $\theta_1 = -\theta_2$ then the work is the same since $\cos(-\theta) = \cos \theta$.

(b)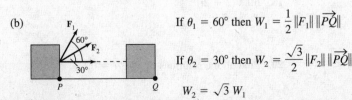
If $\theta_1 = 60°$ then $W_1 = \dfrac{1}{2} \|F_1\| \, \|\overrightarrow{PQ}\|$

If $\theta_2 = 30°$ then $W_2 = \dfrac{\sqrt{3}}{2} \|F_2\| \, \|\overrightarrow{PQ}\|$

$W_2 = \sqrt{3} \, W_1$

The amount of work done by F_2 is $\sqrt{3}$ times as great as the amount of work done by F_1.

Chapter 6   Practice Test

For Exercises 1 and 2, use the Law of Sines to find the remaining sides and angles of the triangle.

1. $A = 40°$, $B = 12°$, $b = 100$ **2.** $C = 150°$, $a = 5$, $c = 20$

3. Find the area of the triangle: $a = 3$, $b = 6$, $C = 130°$.

4. Determine the number of solutions to the triangle: $a = 10$, $b = 35$, $A = 22.5°$.

For Exercises 5 and 6, use the Law of Cosines to find the remaining sides and angles of the triangle.

5. $a = 49$, $b = 53$, $c = 38$ **6.** $C = 29°$, $a = 100$, $b = 300$

7. Use Heron's Formula to find the area of the triangle: $a = 4.1$, $b = 6.8$, $c = 5.5$.

8. A ship travels 40 miles due east, then adjusts its course $12°$ southward. After traveling 70 miles in that direction, how far is the ship from its point of departure?

9. $\mathbf{w} = 4\mathbf{u} - 7\mathbf{v}$ where $\mathbf{u} = 3\mathbf{i} + \mathbf{j}$ and $\mathbf{v} = -\mathbf{i} + 2\mathbf{j}$. Find $\mathbf{w}$.

10. Find a unit vector in the direction of $\mathbf{v} = 5\mathbf{i} - 3\mathbf{j}$.

11. Find the dot product and the angle between $\mathbf{u} = 6\mathbf{i} + 5\mathbf{j}$ and $\mathbf{v} = 2\mathbf{i} - 3\mathbf{j}$.

12. $\mathbf{v}$ is a vector of magnitude 4 making an angle of $30°$ with the positive x-axis. Find $\mathbf{v}$ in component form.

13. Find the projection of $\mathbf{u}$ onto $\mathbf{v}$ given $\mathbf{u} = \langle 3, -1 \rangle$ and $\mathbf{v} = \langle -2, 4 \rangle$.

14. Give the trigonometric form of $z = 5 - 5i$.

15. Give the standard form of $z = 6(\cos 225° + i \sin 225°)$.

16. Multiply $[7(\cos 23° + i \sin 23°)][4(\cos 7° + i \sin 7°)]$.

17. Divide $\dfrac{9\left(\cos \dfrac{5\pi}{4} + i \sin \dfrac{5\pi}{4}\right)}{3(\cos \pi + i \sin \pi)}$.   **18.** Find $(2 + 2i)^8$.

19. Find the cube roots of $8\left(\cos \dfrac{\pi}{3} + i \sin \dfrac{\pi}{3}\right)$.   **20.** Find all the solutions to $x^4 + i = 0$.

C H A P T E R   7
Systems of Equations and Inequalities

Section 7.1   Linear and Nonlinear Systems of Equations 332

Section 7.2   Two-Variable Linear Systems 339

Section 7.3   Multivariable Linear Systems 346

Section 7.4   Partial Fractions 358

Section 7.5   Systems of Inequalities 365

Section 7.6   Linear Programming 371

Review Exercises 377

Problem Solving 385

Practice Test 389

CHAPTER 7
Systems of Equations and Inequalities

Section 7.1   Linear and Nonlinear Systems of Equations

- ■ You should be able to solve systems of equations by the method of substitution.
    1. Solve one of the equations for one of the variables.
    2. Substitute this expression into the other equation and solve.
    3. Back-substitute into the first equation to find the value of the other variable.
    4. Check your answer in each of the original equations.
- ■ You should be able to find solutions graphically. (See Example 5 in textbook.)

Vocabulary Check

1. system of equations

2. solution

3. solving

4. substitution

5. point of intersection

6. break-even

---

1. $\begin{cases} 4x - y = 1 \\ 6x + y = -6 \end{cases}$

    (a) $4(0) - (-3) \neq 1$

    $(0, -3)$ *is not* a solution.

    (b) $4(-1) - (-4) \neq 1$

    $(-1, -4)$ *is not* a solution.

    (c) $4\left(-\frac{3}{2}\right) - (-2) \neq 1$

    $\left(-\frac{3}{2}, -2\right)$ *is not* a solution.

    (d) $4\left(-\frac{1}{2}\right) - (-3) = 1$

    $6\left(-\frac{1}{2}\right) + (-3) = -6$

    $\left(-\frac{1}{2}, -3\right)$ *is* a solution.

3. $\begin{cases} y = -2e^x \\ 3x - y = 2 \end{cases}$

    (a) $0 \neq -2e^{-2}$

    $(-2, 0)$ *is not* a solution.

    (b) $-2 = -2e^0$

    $3(0) - (-2) = 2$

    $(0, -2)$ *is* a solution.

    (c) $-3 \neq -2e^0$

    $(0, -3)$ *is not* a solution.

    (d) $2 \neq -2e^{-1}$

    $(-1, 2)$ *is not* a solution.

5. $\begin{cases} 2x + y = 6 \qquad \text{Equation 1} \\ -x + y = 0 \qquad \text{Equation 2} \end{cases}$

Solve for y in Equation 1: $y = 6 - 2x$

Substitute for y in Equation 2: $-x + (6 - 2x) = 0$

Solve for x: $-3x + 6 = 0 \implies x = 2$

Back-substitute $x = 2$: $y = 6 - 2(2) = 2$

Solution: $(2, 2)$

7. $\begin{cases} x - y = -4 & \text{Equation 1} \\ x^2 - y = -2 & \text{Equation 2} \end{cases}$

Solve for y in Equation 1: $y = x + 4$

Substitute for y in Equation 2: $x^2 - (x + 4) = -2$

Solve for x: $x^2 - x - 2 = 0 \implies (x + 1)(x - 2) = 0 \implies x = -1, 2$

Back-substitute $x = -1$: $y = -1 + 4 = 3$

Back-substitute $x = 2$: $y = 2 + 4 = 6$

Solutions: $(-1, 3), (2, 6)$

9. $\begin{cases} -2x + y = -5 & \text{Equation 1} \\ x^2 + y^2 = 25 & \text{Equation 2} \end{cases}$

Solve for y in Equation 1: $y = 2x - 5$

Substitute for y in Equation 2: $x^2 + (2x - 5)^2 = 25$

Solve for x:

$\qquad 5x^2 - 20x = 0 \implies 5x(x - 4) = 0 \implies x = 0, 4$

Back-substitute $x = 0$: $y = 2(0) - 5 = -5$

Back-substitute $x = 4$: $y = 2(4) - 5 = 3$

Solutions: $(0, -5), (4, 3)$

11. $\begin{cases} x^2 \qquad + y = 0 & \text{Equation 1} \\ x^2 - 4x - y = 0 & \text{Equation 2} \end{cases}$

Solve for y in Equation 1: $y = -x^2$

Substitute for y in Equation 2: $x^2 - 4x - (-x^2) = 0$

Solve for x: $2x^2 - 4x = 0 \implies 2x(x - 2) = 0 \implies x = 0, 2$

Back-substitute $x = 0$: $y = -0^2 = 0$

Back-substitute $x = 2$: $y = -2^2 = -4$

Solutions: $(0, 0), (2, -4)$

13. $\begin{cases} y = x^3 - 3x^2 + 1 & \text{Equation 1} \\ y = x^2 - 3x + 1 & \text{Equation 2} \end{cases}$

Substitute for y in Equation 2:

$\qquad x^3 - 3x^2 + 1 = x^2 - 3x + 1$

$\qquad x^3 - 4x^2 + 3x = 0$

$\qquad x(x - 1)(x - 3) = 0 \implies x = 0, 1, 3$

Back-substitute $x = 0$: $y = 0^3 - 3(0)^2 + 1 = 1$

Back-substitute $x = 1$: $y = 1^3 - 3(1)^2 + 1 = -1$

Back-substitute $x = 3$: $y = 3^3 - 3(3)^2 + 1 = 1$

Solutions: $(0, 1), (1, -1), (3, 1)$

15. $\begin{cases} x - y = 0 & \text{Equation 1} \\ 5x - 3y = 10 & \text{Equation 2} \end{cases}$

Solve for y in Equation 1: $y = x$

Substitute for y in Equation 2: $5x - 3x = 10$

Solve for x: $2x = 10 \implies x = 5$

Back-substitute in Equation 1: $y = x = 5$

Solution: $(5, 5)$

17. $\begin{cases} 2x - y + 2 = 0 & \text{Equation 1} \\ 4x + y - 5 = 0 & \text{Equation 2} \end{cases}$

Solve for y in Equation 1: $y = 2x + 2$

Substitute for y in Equation 2: $4x + (2x + 2) - 5 = 0$

Solve for x: $6x - 3 = 0 \implies x = \frac{1}{2}$

Back-substitute $x = \frac{1}{2}$: $y = 2x + 2 = 2\left(\frac{1}{2}\right) + 2 = 3$

Solution: $\left(\frac{1}{2}, 3\right)$

19. $\begin{cases} 1.5x + 0.8y = 2.3 & \text{Equation 1} \\ 0.3x - 0.2y = 0.1 & \text{Equation 2} \end{cases}$

Multiply the equations by 10.

$\qquad 15x + 8y = 23 \qquad$ Revised Equation 1

$\qquad 3x - 2y = 1 \qquad$ Revised Equation 2

Solve for y in revised Equation 2: $y = \frac{3}{2}x - \frac{1}{2}$

Substitute for y in revised Equation 1:
$15x + 8\left(\frac{3}{2}x - \frac{1}{2}\right) = 23$

Solve for x:
$15x + 12x - 4 = 23 \implies 27x = 27 \implies x = 1$

Back-substitute $x = 1$: $y = \frac{3}{2}(1) - \frac{1}{2} = 1$

Solution: $(1, 1)$

21. $\begin{cases} \frac{1}{5}x + \frac{1}{2}y = 8 & \text{Equation 1} \\ x + y = 20 & \text{Equation 2} \end{cases}$

Solve for x in Equation 2: $x = 20 - y$

Substitute for x in Equation 1: $\frac{1}{5}(20 - y) + \frac{1}{2}y = 8$

Solve for y: $4 + \frac{3}{10}y = 8 \implies y = \frac{40}{3}$

Back-substitute $y = \frac{40}{3}$: $x = 20 - y = 20 - \frac{40}{3} = \frac{20}{3}$

Solution: $\left(\frac{20}{3}, \frac{40}{3}\right)$

23. $\begin{cases} 6x + 5y = -3 & \text{Equation 1} \\ -x - \frac{5}{6}y = -7 & \text{Equation 2} \end{cases}$

Solve for x in Equation 2: $x = 7 - \frac{5}{6}y$

Substitute for x in Equation 1: $6\left(7 - \frac{5}{6}y\right) + 5y = -3$

Solve for y: $42 - 5y + 5y = -3 \implies 42 = -3$ (False)

No solution

25. $\begin{cases} x^2 - y = 0 & \text{Equation 1} \\ 2x + y = 0 & \text{Equation 2} \end{cases}$

Solve for y in Equation 2: $y = -2x$

Substitute for y in Equation 1: $x^2 - (-2x) = 0$

Solve for x:
$x^2 + 2x = 0 \implies x(x + 2) = 0 \implies x = 0, -2$

Back-substitute $x = 0$: $y = -2(0) = 0$

Back-substitute $x = -2$: $y = -2(-2) = 4$

Solutions: $(0, 0), (-2, 4)$

27. $\begin{cases} x - y = -1 & \text{Equation 1} \\ x^2 - y = -4 & \text{Equation 2} \end{cases}$

Solve for y in Equation 1: $y = x + 1$

Substitute for y in Equation 2: $x^2 - (x + 1) = -4$

Solve for x: $x^2 - x - 1 = -4 \implies x^2 - x + 3 = 0$

The Quadratic Formula yields no real solutions.

29. $\begin{cases} -x + 2y = 2 \implies y = \dfrac{x + 2}{2} \\ 3x + y = 15 \implies y = -3x + 15 \end{cases}$

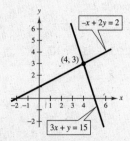

Point of intersection: $(4, 3)$

31. $\begin{cases} x - 3y = -2 \implies y = \frac{1}{3}(x + 2) \\ \end{cases}$

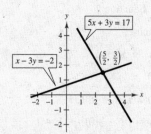

Point of intersection: $\left(\frac{5}{2}, \frac{3}{2}\right)$

33. $\begin{cases} x + y = 4 \implies y = -x + 4 \\ x^2 + y^2 - 4x = 0 \implies (x - 2)^2 + y^2 = 4 \end{cases}$

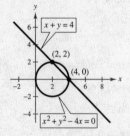

Points of intersection: $(2, 2), (4, 0)$

35. $\begin{cases} x - y + 3 = 0 \implies y = x + 3 \\ y = x^2 - 4x + 7 \implies y = (x - 2)^2 + 3 \end{cases}$

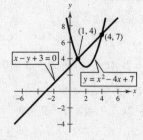

Points of intersection: $(1, 4), (4, 7)$

37. $\begin{cases} 7x + 8y = 24 \implies y = -\dfrac{7}{8}x + 3 \\[2mm] x - 8y = 8 \implies y = \dfrac{1}{8}x - 1 \end{cases}$

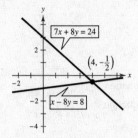

Point of intersection: $\left(4, -\dfrac{1}{2}\right)$

39. $\begin{cases} 3x - 2y = 0 \implies y = \dfrac{3}{2}x \\[2mm] x^2 - y^2 = 4 \implies \dfrac{x^2}{4} - \dfrac{y^2}{4} = 1 \end{cases}$

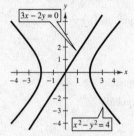

No points of intersection $\implies$ No solution

41. $\begin{cases} x^2 + y^2 = 25 \\ 3x^2 - 16y = 0 \implies y = \dfrac{3}{16}x^2 \end{cases}$

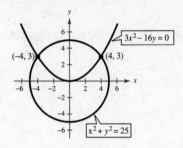

Points of intersection: $(-4, 3)$ and $(4, 3)$

Algebraically we have:

$$x^2 = 25 - y^2$$
$$\frac{16}{3}y = 25 - y^2$$
$$16y = 75 - 3y^2$$
$$3y^2 + 16y - 75 = 0$$
$$(3y + 25)(y - 3) = 0$$
$$y = -\frac{25}{3} \implies x^2 = -\frac{400}{9}, \quad \text{No real solution}$$
$$y = 3 \implies x^2 = 16$$

Solutions: $(\pm 4, 3)$

43. $\begin{cases} y = e^x \\ x - y + 1 = 0 \implies y = x + 1 \end{cases}$

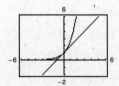

Point of intersection: $(0, 1)$

45. $\begin{cases} x + 2y = 8 \implies y = -\dfrac{1}{2}x + 4 \\[2mm] y = \log_2 x \implies y = \dfrac{\ln x}{\ln 2} \end{cases}$

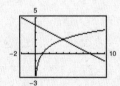

Point of intersection: $(4, 2)$

47. $\begin{cases} x^2 + y^2 = 169 \implies y_1 = \sqrt{169 - x^2} \text{ and } y_2 = -\sqrt{169 - x^2} \\ x^2 - 8y = 104 \implies y_3 = \dfrac{1}{8}x^2 - 13 \end{cases}$

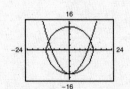

Points of intersection: $(0, -13)$, $(\pm 12, 5)$

49. $\begin{cases} y = 2x & \text{Equation 1} \\ y = x^2 + 1 & \text{Equation 2} \end{cases}$

Substitute for y in Equation 2: $2x = x^2 + 1$

Solve for x: $x^2 - 2x + 1 = (x - 1)^2 = 0 \implies x = 1$

Back-substitute $x = 1$ in Equation 1: $y = 2x = 2$

Solution: $(1, 2)$

51. $\begin{cases} 3x - 7y + 6 = 0 & \text{Equation 1} \\ x^2 - y^2 = 4 & \text{Equation 2} \end{cases}$

Solve for y in Equation 1: $y = \dfrac{3x + 6}{7}$

Substitute for y in Equation 2: $x^2 - \left(\dfrac{3x + 6}{7}\right)^2 = 4$

Solve for x: $x^2 - \left(\dfrac{9x^2 + 36x + 36}{49}\right) = 4$

$$49x^2 - (9x^2 + 36x + 36) = 196$$

$$40x^2 - 36x - 232 = 0$$

$$4(10x - 29)(x + 2) = 0 \implies x = \dfrac{29}{10}, -2$$

Back-substitute $x = \dfrac{29}{10}$: $y = \dfrac{3x + 6}{7} = \dfrac{3(29/10) + 6}{7} = \dfrac{21}{10}$

Back-substitute $x = -2$: $y = \dfrac{3x + 6}{7} = \dfrac{3(-2) + 6}{7} = 0$

Solutions: $\left(\dfrac{29}{10}, \dfrac{21}{10}\right)$, $(-2, 0)$

53. $\begin{cases} x - 2y = 4 & \text{Equation 1} \\ x^2 - y = 0 & \text{Equation 2} \end{cases}$

Solve for y in Equation 2: $y = x^2$

Substitute for y in Equation 1: $x - 2x^2 = 4$

Solve for x: $0 = 2x^2 - x + 4 \implies x = \dfrac{1 \pm \sqrt{1 - 4(2)(4)}}{2(2)}$

$$\implies x = \dfrac{1 \pm \sqrt{-31}}{4}$$

The discriminant in the Quadratic Formula is negative.

No real solution

55. $\begin{cases} y - e^{-x} = 1 \implies y = e^{-x} + 1 \\ y - \ln x = 3 \implies y = \ln x + 3 \end{cases}$

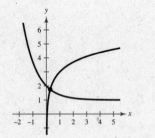

Point of intersection: approximately $(0.287, 1.751)$

57. $\begin{cases} y = x^4 - 2x^2 + 1 & \text{Equation 1} \\ y = 1 - x^2 & \text{Equation 2} \end{cases}$

Substitute for y in Equation 1: $1 - x^2 = x^4 - 2x^2 + 1$

Solve for x: $x^4 - x^2 = 0 \implies x^2(x^2 - 1) = 0$

$$\implies x = 0, \pm 1$$

Back-substitute $x = 0$: $1 - x^2 = 1 - 0^2 = 1$

Back-substitute $x = 1$: $1 - x^2 = 1 - 1^2 = 0$

Back-substitute $x = -1$: $1 - x^2 = 1 - (-1)^2 = 0$

Solutions: $(0, 1)$, $(\pm 1, 0)$

59. $\begin{cases} xy - 1 = 0 & \text{Equation 1} \\ 2x - 4y + 7 = 0 & \text{Equation 2} \end{cases}$

Solve for y in Equation 1: $y = \dfrac{1}{x}$

Substitute for y in Equation 2: $2x - 4\left(\dfrac{1}{x}\right) + 7 = 0$

Solve for x:

$$2x^2 - 4 + 7x = 0 \implies (2x - 1)(x + 4) = 0$$

$$\implies x = \frac{1}{2}, -4$$

Back-substitute $x = \dfrac{1}{2}$: $y = \dfrac{1}{1/2} = 2$

Back-substitute $x = -4$: $y = \dfrac{1}{-4} = -\dfrac{1}{4}$

Solutions: $\left(\dfrac{1}{2}, 2\right), \left(-4, -\dfrac{1}{4}\right)$

61. $C = 8650x + 250,000, \ R = 9950x$

$R = C$

$9950x = 8650x + 250,000$

$1300x = 250,000$

$x \approx 192$ units

63. $C = 35.45x + 16,000, \ R = 55.95x$

(a) $R = C$

$55.95x = 35.45x + 16,000$

$20.50x = 16,000$

$x \approx 781$ units

(b) $P = R - C$

$60,000 = 55.95x - (35.45x + 16,000)$

$60,000 = 20.50x - 16,000$

$76,000 = 20.50x$

$x \approx 3708$ units

65. $\begin{cases} R = 360 - 24x & \text{Equation 1} \\ R = 24 + 18x & \text{Equation 2} \end{cases}$

(a) Substitute for R in Equation 2: $360 - 24x = 24 + 18x$

Solve for x: $336 = 42x \implies x = 8$ weeks

(b)

Weeks	1	2	3	4	5	6	7	8	9	10
$R = 360 - 24x$	336	312	288	264	240	216	192	168	144	120
$R = 24 + 18x$	42	60	78	96	114	132	150	168	186	204

The rentals are equal when $x = 8$ weeks.

67. $0.06x = 0.03x + 350$

$0.03x = 350$

$x \approx \$11,666.67$

To make the straight commission offer the better offer, you would have to sell more than $11,666.67 per week.

69. (a) $\begin{cases} x + y = 25,000 \\ 0.06x + 0.085y = 2000 \end{cases}$

(b) $y_1 = 25,000 - x$

$y_2 = \dfrac{2000 - 0.06x}{0.085}$

As the amount at 6% increases, the amount at 8.5% decreases. The amount of interest is fixed at $2000.

(c) The point of intersection occurs when $x = 5000$, so the most that can be invested at 6% and still earn $2000 per year in interest is $5000.

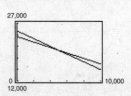

71.

t	8	9	10	11	12	13
Solar	70	69	66	65	64	63
Wind	31	46	57	68	105	108

(a) Solar: $C \approx 0.1429t^2 - 4.46t + 96.8$

   Wind: $C \approx 16.371t - 102.7$

(b)
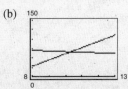

(c) Point of intersection: $(10.3, 66.01)$

   During the year 2000, the consumption of solar energy will equal the consumption of wind energy.

(d) $0.1429t^2 - 4.46t + 96.8 = 16.371t - 102.7$

   $0.1429t^2 - 20.831t + 199.5 = 0$

   By the Quadratic Formula we obtain $t \approx 10.3$ and $t \approx 135.47$.

(e) The results are the same for $t \approx 10.3$. The other "solution", $t \approx 135.47$, is too large to consider as a reasonable answer.

(f) Answers will vary.

73. $2l + 2w = 30 \implies l + w = 15$

$l = w + 3 \implies (w + 3) + w = 15$

$2w = 12$

$w = 6$

$l = w + 3 = 9$

Dimensions: 6×9 meters

75. $2l + 2w = 42 \implies l + w = 21$

$w = \frac{3}{4}l \implies l + \frac{3}{4}l = 21$

$\frac{7}{4}l = 21$

$l = 12$

$w = \frac{3}{4}l = 9$

Dimensions: 9×12 inches

77. $2l + 2w = 40 \implies l + w = 20 \implies w = 20 - l$

$lw = 96 \implies l(20 - l) = 96$

$20l - l^2 = 96$

$0 = l^2 - 20l + 96$

$0 = (l - 8)(l - 12)$

$l = 8 \text{ or } l = 12$

If $l = 8$, then $w = 12$.

If $l = 12$, then $w = 8$.

Since the length is supposed to be greater than the width, we have $l = 12$ kilometers and $w = 8$ kilometers.
Dimensions: 8×12 kilometers

79. False. To solve a system of equations by substitution, you can solve for either variable in one of the two equations and then back-substitute.

81. To solve a system of equations by substitution, use the following steps.

1. Solve one of the equations for one variable in terms of the other.

2. Substitute this expression into the other equation to obtain an equation in one variable.

3. Solve this equation.

4. Back-substitute the value(s) found in Step 3 into the expression found in Step 1 to find the value(s) of the other variable.

5. Check your solution(s) in each of the original equations.

83. $y = x^2$

  (a) Line with two points of intersection   (b) Line with one point of intersection   (c) Line with no points of intersection

     $y = 2x$ $y = 0$ $y = x - 2$

     $(0, 0)$ and $(2, 4)$ $(0, 0)$

85. $(-2, 7), (5, 5)$

$$m = \frac{5 - 7}{5 - (-2)} = -\frac{2}{7}$$

$$y - 7 = -\frac{2}{7}(x - (-2))$$

$$7y - 49 = -2x - 4$$

$$2x + 7y - 45 = 0$$

87. $(6, 3), (10, 3)$

$$m = \frac{3 - 3}{10 - 6} = 0 \implies \text{The line is horizontal.}$$

$$y = 3$$

$$y - 3 = 0$$

89. $\left(\frac{3}{5}, 0\right), (4, 6)$

$$m = \frac{6 - 0}{4 - (3/5)} = \frac{6}{17/5} = \frac{30}{17}$$

$$y - 6 = \frac{30}{17}(x - 4)$$

$$17y - 102 = 30x - 120$$

$$0 = 30x - 17y - 18$$

$$30x - 17y - 18 = 0$$

91. $f(x) = \dfrac{5}{x - 6}$

Domain: All real numbers except $x = 6$

Horizontal asymptote: $y = 0$

Vertical asymptote: $x = 6$

93. $f(x) = \dfrac{x^2 + 2}{x^2 - 16}$

Domain: All real numbers except $x = \pm 4$.

Horizontal asymptote: $y = 1$

Vertical asymptotes: $x = \pm 4$

Section 7.2   Two-Variable Linear Systems

- You should be able to solve a linear system by the method of elimination.
  1. Obtain coefficients for either x or y that differ only in sign. This is done by multiplying all the terms of one or both equations by appropriate constants.
  2. Add the equations to eliminate one of the variables and then solve for the remaining variable.
  3. Use back-substitution into either original equation and solve for the other variable.
  4. Check your answer.
- You should know that for a system of two linear equations, one of the following is true.
  1. There are infinitely many solutions; the lines are identical. The system is consistent. The slopes are equal.
  2. There is no solution; the lines are parallel. The system is inconsistent. The slopes are equal.
  3. There is one solution; the lines intersect at one point. The system is consistent. The slopes are not equal.

Vocabulary Check

  1. elimination   **2.** equivalent

  3. consistent; inconsistent   **4.** equilibrium price

1. $\begin{cases} 2x + y = 5 & \text{Equation 1} \\ x - y = 1 & \text{Equation 2} \end{cases}$

Add to eliminate y: $3x = 6 \implies x = 2$

Substitute $x = 2$ in Equation 2: $2 - y = 1 \implies y = 1$

Solution: $(2, 1)$

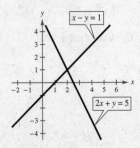

3. $\begin{cases} x + y = 0 & \text{Equation 1} \\ 3x + 2y = 1 & \text{Equation 2} \end{cases}$

Multiply Equation 1 by -2: $-2x - 2y = 0$

Add this to Equation 2 to eliminate y: $x = 1$

Substitute $x = 1$ in Equation 1: $1 + y = 0 \implies y = -1$

Solution: $(1, -1)$

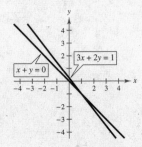

5. $\begin{cases} x - y = 2 & \text{Equation 1} \\ -2x + 2y = 5 & \text{Equation 2} \end{cases}$

Multiply Equation 1 by 2: $2x - 2y = 4$

Add this to Equation 2: $0 = 9$

There are no solutions.

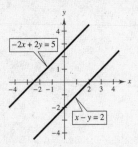

7. $\begin{cases} 3x - 2y = 5 & \text{Equation 1} \\ -6x + 4y = -10 & \text{Equation 2} \end{cases}$

Multiply Equation 1 by 2 and add to Equation 2: $0 = 0$

The equations are dependent. There are infinitely many solutions.

Let $x = a$, then $y = \dfrac{3a - 5}{2} = \dfrac{3}{2}a - \dfrac{5}{2}$.

Solution: $\left(a, \dfrac{3}{2}a - \dfrac{5}{2} \right)$ where a is any real number.

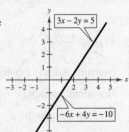

9. $\begin{cases} 9x + 3y = 1 & \text{Equation 1} \\ 3x - 6y = 5 & \text{Equation 2} \end{cases}$

Multiply Equation 2 by (-3): $\quad 9x + 3y = 1$

$\qquad\qquad\qquad\qquad\qquad -9x + 18y = -15$

Add to eliminate x: $21y = -14 \implies y = -\dfrac{2}{3}$

Substitute $y = -\dfrac{2}{3}$ in Equation 1: $9x + 3\left(-\dfrac{2}{3}\right) = 1$

$\qquad\qquad\qquad\qquad\qquad\qquad\qquad x = \dfrac{1}{3}$

Solution: $\left(\dfrac{1}{3}, -\dfrac{2}{3} \right)$

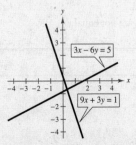

11. $\begin{cases} x + 2y = 4 & \text{Equation 1} \\ x - 2y = 1 & \text{Equation 2} \end{cases}$

Add to eliminate y:

$\qquad 2x = 5$

$\qquad x = \dfrac{5}{2}$

Substitute $x = \dfrac{5}{2}$ in Equation 1:

$\qquad \dfrac{5}{2} + 2y = 4 \implies y = \dfrac{3}{4}$

Solution: $\left(\dfrac{5}{2}, \dfrac{3}{4} \right)$

13. $\begin{cases} 2x + 3y = 18 & \text{Equation 1} \\ 5x - y = 11 & \text{Equation 2} \end{cases}$

Multiply Equation 2 by 3: $15x - 3y = 33$

Add this to Equation 1 to eliminate y:

$$17x = 51 \implies x = 3$$

Substitute $x = 3$ in Equation 1:

$$6 + 3y = 18 \implies y = 4$$

Solution: $(3, 4)$

17. $\begin{cases} 5u + 6v = 24 & \text{Equation 1} \\ 3u + 5v = 18 & \text{Equation 2} \end{cases}$

Multiply Equation 1 by 5 and Equation 2 by -6:

$$\begin{cases} 25u + 30v = 120 \\ -18u - 30v = -108 \end{cases}$$

Add to eliminate v: $7u = 12 \implies u = \frac{12}{7}$

Substitute $u = \frac{12}{7}$ in Equation 1:

$$5\left(\frac{12}{7}\right) + 6v = 24 \implies 6v = \frac{108}{7} \implies v = \frac{18}{7}$$

Solution: $\left(\frac{12}{7}, \frac{18}{7}\right)$

21. $\begin{cases} \dfrac{x}{4} + \dfrac{y}{6} = 1 & \text{Equation 1} \\ x - y = 3 & \text{Equation 2} \end{cases}$

Multiply Equation 1 by 6: $\dfrac{3}{2}x + y = 6$

Add this to Equation 2 to eliminate y:

$$\frac{5}{2}x = 9 \implies x = \frac{18}{5}$$

Substitute $x = \dfrac{18}{5}$ in Equation 2:

$$\frac{18}{5} - y = 3$$

$$y = \frac{3}{5}$$

Solution: $\left(\dfrac{18}{5}, \dfrac{3}{5}\right)$

15. $\begin{cases} 3x + 2y = 10 & \text{Equation 1} \\ 2x + 5y = 3 & \text{Equation 2} \end{cases}$

Multiply Equation 1 by 2 and
Equation 2 by (-3):

$$\begin{cases} 6x + 4y = 20 \\ -6x - 15y = -9 \end{cases}$$

Add to eliminate x: $-11y = 11 \implies y = -1$

Substitute $y = -1$ in Equation 1:

$$3x - 2 = 10 \implies x = 4$$

Solution: $(4, -1)$

19. $\begin{cases} \dfrac{9}{5}x + \dfrac{6}{5}y = 4 & \text{Equation 1} \\ 9x + 6y = 3 & \text{Equation 2} \end{cases}$

Multiply Equation 1 by 10 and Equation 2 by -2:

$$\begin{cases} 18x + 12y = 40 \\ -18x - 12y = -6 \end{cases}$$

Add to eliminate x and y: $0 = 34$

Inconsistent

No solution

23. $\begin{cases} -5x + 6y = -3 & \text{Equation 1} \\ 20x - 24y = 12 & \text{Equation 2} \end{cases}$

Multiply Equation 1 by 4:

$$\begin{cases} -20x + 24y = -12 \\ 20x - 24y = 12 \end{cases}$$

Add to eliminate x and y: $0 = 0$

The equations are dependent. There are infinitely many solutions.

Let $x = a$, then

$$-5a + 6y = -3 \implies y = \frac{5a - 3}{6} = \frac{5}{6}a - \frac{1}{2}.$$

Solution: $\left(a, \dfrac{5}{6}a - \dfrac{1}{2}\right)$ where a is any real number

25. $\begin{cases} 0.05x - 0.03y = 0.21 & \text{Equation 1} \\ 0.07x + 0.02y = 0.16 & \text{Equation 2} \end{cases}$

Multiply Equation 1 by 200 and Equation 2 by 300:

$\begin{cases} 10x - 6y = 42 \\ 21x + 6y = 48 \end{cases}$

Add to eliminate y: $31x = 90$

$$x = \tfrac{90}{31}$$

Substitute $x = \tfrac{90}{31}$ in Equation 2:

$$0.07\left(\tfrac{90}{31}\right) + 0.02y = 0.16$$

$$y = -\tfrac{67}{31}$$

Solution: $\left(\tfrac{90}{31}, -\tfrac{67}{31}\right)$

27. $\begin{cases} 4b + 3m = 3 & \text{Equation 1} \\ 3b + 11m = 13 & \text{Equation 2} \end{cases}$

Multiply Equation 1 by 3 and Equation 2 by (-4):

$\begin{cases} 12b + 9m = 9 \\ -12b - 44m = -52 \end{cases}$

Add to eliminate b: $-35m = -43$

$$m = \tfrac{43}{35}$$

Substitute $m = \tfrac{43}{35}$ in Equation 1:

$$4b + 3\left(\tfrac{43}{35}\right) = 3 \implies b = -\tfrac{6}{35}$$

Solution: $\left(-\tfrac{6}{35}, \tfrac{43}{35}\right)$

29. $\begin{cases} \dfrac{x+3}{4} + \dfrac{y-1}{3} = 1 & \text{Equation 1} \\ 2x - y = 12 & \text{Equation 2} \end{cases}$

Multiply Equation 1 by 12 and Equation 2 by 4:

$\begin{cases} 3x + 4y = 7 \\ 8x - 4y = 48 \end{cases}$

Add to eliminate y: $11x = 55 \implies x = 5$

Substitute $x = 5$ into Equation 2:

$$2(5) - y = 12 \implies y = -2$$

Solution: $(5, -2)$

31. $\begin{cases} 2x - 5y = 0 \\ x - y = 3 \end{cases}$

Multiply Equation 2 by -5:

$\begin{cases} 2x - 5y = 0 \\ -5x + 5y = -15 \end{cases}$

Add to eliminate y: $-3x = -15 \implies x = 5$

Matches graph (b).

Number of solutions: One

Consistent

33. $\begin{cases} 2x - 5y = 0 \\ 2x - 3y = -4 \end{cases}$

Multiply Equation 1 by -1:

$\begin{cases} -2x + 5y = 0 \\ 2x - 3y = -4 \end{cases}$

Add to eliminate x: $2y = -4 \implies y = -2$

Matches graph (c).

Number of solutions: One

Consistent

35. $\begin{cases} 3x - 5y = 7 & \text{Equation 1} \\ 2x + y = 9 & \text{Equation 2} \end{cases}$

Multiply Equation 2 by 5:

$$10x + 5y = 45$$

Add this to Equation 1:

$$13x = 52 \implies x = 4$$

Back-substitute $x = 4$ into Equation 2:

$$2(4) + y = 9 \implies y = 1$$

Solution: $(4, 1)$

37. $\begin{cases} y = 2x - 5 & \text{Equation 1} \\ y = 5x - 11 & \text{Equation 2} \end{cases}$

Since both equations are solved for y, set them equal to one another and solve for x.

$$2x - 5 = 5x - 11$$

$$6 = 3x$$

$$2 = x$$

Back-substitute $x = 2$ into Equation 1:

$$y = 2(2) - 5 = -1$$

Solution: $(2, -1)$

39. $\begin{cases} x - 5y = 21 & \text{Equation 1} \\ 6x + 5y = 21 & \text{Equation 2} \end{cases}$

Add the equations: $7x = 42 \implies x = 6$

Back-substitute $x = 6$ into Equation 1:

$$6 - 5y = 21 \implies -5y = 15 \implies y = -3$$

Solution: $(6, -3)$

41. $\begin{cases} -2x + 8y = 19 & \text{Equation 1} \\ y = x - 3 & \text{Equation 2} \end{cases}$

Substitute the expression for y from Equation 2 into Equation 1.

$$-2x + 8(x - 3) = 19 \implies -2x + 8x - 24 = 19$$

$$6x = 43$$

$$x = \frac{43}{6}$$

Back-substitute $x = \frac{43}{6}$ into Equation 2: $y = \frac{43}{6} - 3 \implies y = \frac{25}{6}$

Solution: $\left(\frac{43}{6}, \frac{25}{6}\right)$

43. Let $r_1 = $ the air speed of the plane
and $r_2 = $ the wind air speed.

$$\begin{array}{llll} 3.6(r_1 - r_2) = 1800 & \text{Equation 1} & \implies & r_1 - r_2 = 500 \\ 3(r_1 + r_2) = 1800 & \text{Equation 2} & \implies & \underline{r_1 + r_2 = 600} \\ & & & 2r_1 = 1100 \qquad \text{Add the equations.} \\ & & & r_1 = 550 \\ & & & 550 + r_2 = 600 \\ & & & r_2 = 50 \end{array}$$

The air speed of the plane is 550 mph and the speed of the wind is 50 mph.

45. $50 - 0.5x = 0.125x$

$50 = 0.625x$

$x = 80$ units

$p = \$10$

Solution: $(80, 10)$

47. $140 - 0.00002x = 80 + 0.00001x$

$60 = 0.00003x$

$x = 2{,}000{,}000$ units

$p = \$100.00$

Solution: $(2{,}000{,}000, 100)$

49. Let $x = $ number of calories in a cheeseburger

$y = $ number of calories in a small order of french fries

$$\begin{cases} 2x + y = 850 & \text{Equation 1} \\ 3x + 2y = 1390 & \text{Equation 2} \end{cases}$$

Multiply Equation 1 by -2:

$$\begin{cases} -4x - 2y = -1700 \\ 3x + 2y = 1390 \end{cases}$$

$$-x = -310 \qquad \text{Add the equations.}$$

$$x = 310$$

$$y = 230$$

Solution: The cheeseburger contains 310 calories and the fries contain 230 calories.

51. Let $x = $ the number of liters at 20%

Let $y = $ the number of liters at 50%.

(a) $\begin{cases} x + y = 10 \\ 0.2x + 0.5y = 0.3(10) \end{cases}$

$-2 \cdot$ Equation 1: $\quad -2x - 2y = -20$

$10 \cdot$ Equation 2: $\quad \underline{2x + 5y = 30}$

$$3y = 10$$

$$y = \frac{10}{3}$$

$$x + \frac{10}{3} = 10$$

$$x = \frac{20}{3}$$

(b)

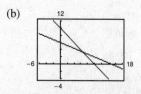

As x increases, y decreases.

(c) In order to obtain the specified concentration of the final mixture, $6\frac{2}{3}$ liters of the 20% solution and $3\frac{1}{3}$ liters of the 50% solution are required.

53. Let x = amount invested at 7.5%

y = amount invested at 9%.

$$\begin{cases} x + \quad y = 12{,}000 & \text{Equation 1} \\ 0.075x + 0.09y = \quad 990 & \text{Equation 2} \end{cases}$$

Multiply Equation 1 by 9 and Equation 2 by -100.

$$\begin{cases} 9x + 9y = \quad 108{,}000 \\ -7.5x - 9y = -99{,}000 \end{cases}$$
$$\quad 1.5x \quad = \quad 9{,}000 \qquad \text{Add the equations.}$$
$$x = \quad \$6000$$
$$y = \quad \$6000$$

The most that can be invested at 7.5% is $6000.

57. $\begin{cases} 5b + 10a = 20.2 \implies -10b - 20a = -40.4 \\ 10b + 30a = 50.1 \implies \quad 10b + 30a = \quad 50.1 \end{cases}$

$$10a = \quad 9.7$$
$$a = \quad 0.97$$
$$b = \quad 2.10$$

Least squares regression line: $y = 0.97x + 2.10$

61. $(0, 4), (1, 3), (1, 1), (2, 0)$

$$n = 4, \ \sum_{i=1}^{4} x_i = 4, \ \sum_{i=1}^{4} y_i = 8, \ \sum_{i=1}^{4} x_i^2 = 6, \ \sum_{i=1}^{4} x_i y_i = 4$$

$$\begin{cases} 4b + 4a = 8 \implies \quad 4b + 4a = \quad 8 \\ 4b + 6a = 4 \implies -4b - 6a = -4 \end{cases}$$
$$-2a = \quad 4$$
$$a = -2$$
$$b = \quad 4$$

Least squares regression line: $y = -2x + 4$

63. $(5, 66.65), (6, 70.93), (7, 75.31), (8, 78.62), (9, 81.33), (10, 85.89), (11, 88.27)$

(a) $n = 7, \ \sum_{i=1}^{7} x_i = 56, \ \sum_{i=1}^{7} x_i^2 = 476, \ \sum_{i=1}^{7} y_i = 547, \ \sum_{i=1}^{7} x_i y_i = 4476.8$

$$\begin{cases} 7b + \quad 56a = \quad 547 \\ 56b + 476a = 4476.8 \end{cases}$$

Multiply Equation 1 by -8.

$$\begin{cases} -56b - 448a = -4376 \\ 56b + 476a = \quad 4476.8 \end{cases}$$
$$28a = \quad 100.8$$
$$a = 3.6$$
$$b \approx 49.343$$

Least squares regression line: $y \approx 3.6t + 49.343$

(b) $y \approx 3.6t + 49.343$, This agrees with part (a).

55. Let x = number of student tickets

y = number of adult tickets.

$$\begin{cases} x + \quad y = 1435 & \text{Equation 1} \\ 1.50x + 5.00y = 3552.50 & \text{Equation 2} \end{cases}$$

Multiply Equation 1 by -1.50.

$$\begin{cases} -1.50x - 1.50y = -2152.50 \\ 1.50x + 5.00y = \quad 3552.50 \end{cases}$$
$$\quad 3.50y = \quad 1400.00 \qquad \text{Add the equations.}$$
$$y = \quad 400$$
$$x = \quad 1035$$

Solution: 1035 student tickets and 400 adult tickets were sold.

59. $\begin{cases} 7b + 21a = \quad 35.1 \implies -21b - 63a = -105.3 \\ 21b + 91a = 114.2 \implies \quad 21b + 91a = \quad 114.2 \end{cases}$

$$28a = \quad 8.9$$
$$a = \tfrac{89}{280}$$
$$b = \tfrac{1137}{280}$$

Least squares regression line: $y = \tfrac{1}{280}(89x + 1137)$

$$y \approx 0.32x + 4.1$$

—CONTINUED—

63. —CONTINUED—

(c)

t	Actual room rate	Model approximation
5	\$66.65	\$67.34
6	\$70.93	\$70.94
7	\$75.31	\$74.54
8	\$78.62	\$78.14
9	\$81.33	\$81.74
10	\$85.89	\$85.34
11	\$88.27	\$88.94

The model is a good fit to the data.

(d) When $t = 12$: $y \approx \$92.54$

This is a little off from the actual rate.

(e) $3.6t + 49.343 = 100$

$$3.6t = 50.657$$

$$t \approx 14.1$$

According to the model, room rates will average \$100.00 during the year 2004.

65. False. Two lines that coincide have infinitely many points of intersection.

67. No, it is not possible for a consistent system of linear equations to have exactly two solutions. Either the lines will intersect once or they will coincide and then the system would have infinite solutions.

69. $\begin{cases} 100y - x = 200 & \text{Equation 1} \\ 99y - x = -198 & \text{Equation 2} \end{cases}$

Subtract Equation 2 from Equation 1 to eliminate x:

$100y - x = 200$

$\underline{-99y + x = 198}$

$y = 398$

Substitute $y = 398$ into Equation 1:

$100(398) - x = 200 \implies x = 39{,}600$

Solution: $(39{,}600, \ 398)$

The lines are not parallel. The scale on the axes must be changed to see the point of intersection.

71. $\begin{cases} 4x - 8y = -3 & \text{Equation 1} \\ 2x + ky = 16 & \text{Equation 2} \end{cases}$

Multiply Equation 2 by -2: $-4x - 2ky = -32$

Add this to Equation 1: $\quad 4x - 8y = -3$

$ -4x - 2ky = -32$

$ -8y - 2ky = -35$

The system is inconsistent if $-8y - 2ky = 0$. This occurs when $k = -4$.

73. $-11 - 6x \geq 33$

$ -6x \geq 44$

$ x \leq -\tfrac{22}{3}$

75. $8x - 15 \le -4(2x - 1)$

$8x - 15 \le -8x + 4$

$16x \le 19$

$x \le \frac{19}{16}$

77. $|x - 8| < 10$

$-10 < x - 8 < 10$

$-2 < x < 18$

79. $2x^2 + 3x - 35 < 0$

$(2x - 7)(x + 5) < 0$

Critical numbers: $x = -5, \frac{7}{2}$

Test intervals: $(-\infty, -5), \left(-5, \frac{7}{2}\right), \left(\frac{7}{2}, \infty\right)$

Test: Is $(2x - 7)(x + 5) < 0$?

Solution: $-5 < x < \frac{7}{2}$

81. $\ln x + \ln 6 = \ln(6x)$

83. $\log_9 12 - \log_9 x = \log_9\left(\dfrac{12}{x}\right)$

85. $\begin{cases} 2x - y = 4 \implies y = 2x - 4 \\ -4x + 2y = -12 \end{cases}$

$-4x + 2(2x - 4) = -12$

$-4x + 4x - 8 = -12$

$-8 = -12$

There are no solutions.

87. Answers will vary.

Section 7.3   Multivariable Linear Systems

- You should know the operations that lead to equivalent systems of equations:
  - (a) Interchange any two equations.
  - (b) Multiply all terms of an equation by a nonzero constant.
  - (c) Replace an equation by the sum of itself and a constant multiple of any other equation in the system.
- You should be able to use the method of Gaussian elimination with back-substitution.

Vocabulary Check

1. row-echelon

2. ordered triple

3. Gaussian

4. row operation

5. nonsquare

6. position

1. $\begin{cases} 3x - y + z = 1 \\ 2x \quad - 3z = -14 \\ \quad 5y + 2z = 8 \end{cases}$

(a) $3(2) - (0) + (-3) \neq 1$

$(2, 0, -3)$ *is not* a solution.

(b) $3(-2) - (0) + 8 \neq 1$

$(-2, 0, 8)$ *is not* a solution.

(c) $3(0) - (-1) + 3 \neq 1$

$(0, -1, 3)$ *is not* a solution.

(d) $3(-1) - (0) + 4 = 1$

$2(-1) \quad - 3(4) = -14$

$5(0) + 2(4) = 8$

$(-1, 0, 4)$ *is* a solution.

3. $\begin{cases} 4x + y - z = 0 \\ -8x - 6y + z = -\frac{7}{4} \\ 3x - y \quad = -\frac{9}{4} \end{cases}$

(a) $4\left(\frac{1}{2}\right) + \left(-\frac{3}{4}\right) - \left(-\frac{7}{4}\right) \neq 0$

$\left(\frac{1}{2}, -\frac{3}{4}, -\frac{7}{4}\right)$ *is not* a solution.

(b) $4\left(-\frac{3}{2}\right) + \left(\frac{5}{4}\right) - \left(-\frac{5}{4}\right) \neq 0$

$\left(-\frac{3}{2}, \frac{5}{4}, -\frac{5}{4}\right)$ *is not* a solution.

(c) $4\left(-\frac{1}{2}\right) + \left(\frac{3}{4}\right) - \left(-\frac{5}{4}\right) = 0$

$-8\left(-\frac{1}{2}\right) - 6\left(\frac{3}{4}\right) + \left(-\frac{5}{4}\right) = -\frac{7}{4}$

$3\left(-\frac{1}{2}\right) - \left(\frac{3}{4}\right) \quad = -\frac{9}{4}$

$\left(-\frac{1}{2}, \frac{3}{4}, -\frac{5}{4}\right)$ *is* a solution.

(d) $4\left(-\frac{1}{2}\right) + \left(\frac{1}{6}\right) - \left(-\frac{3}{4}\right) \neq 0$

$\left(-\frac{1}{2}, \frac{1}{6}, -\frac{3}{4}\right)$ *is not* a solution.

5. $\begin{cases} 2x - y + 5z = 24 & \text{Equation 1} \\ y + 2z = 6 & \text{Equation 2} \\ z = 4 & \text{Equation 3} \end{cases}$

Back-substitute $z = 4$ into Equation 2:

$y + 2(4) = 6$

$y = -2$

Back-substitute $y = -2$ and $z = 4$ into Equation 1:

$2x - (-2) + 5(4) = 24$

$2x + 22 = 24$

$x = 1$

Solution: $(1, -2, 4)$

7. $\begin{cases} 2x + y - 3z = 10 & \text{Equation 1} \\ y + z = 12 & \text{Equation 2} \\ z = 2 & \text{Equation 3} \end{cases}$

Substitute $z = 2$ into Equation 2: $y + (2) = 12 \implies y = 10$

Substitute $y = 10$ and $z = 2$ into Equation 1:

$2x + (10) - 3(2) = 10$

$2x + 4 = 10$

$2x = 6$

$x = 3$

Solution: $(3, 10, 2)$

9. $\begin{cases} 4x - 2y + z = 8 & \text{Equation 1} \\ -y + z = 4 & \text{Equation 2} \\ z = 2 & \text{Equation 3} \end{cases}$

Substitute $z = 2$ into Equation 2:

$-y + (2) = 4 \implies y = -2$

Substitute $y = -2$ and $z = 2$ into Equation 1:

$4x - 2(-2) + (2) = 8$

$4x + 6 = 8$

$4x = 2$

$x = \frac{1}{2}$

Solution: $\left(\frac{1}{2}, -2, 2\right)$

11. $\begin{cases} x - 2y + 3z = 5 & \text{Equation 1} \\ -x + 3y - 5z = 4 & \text{Equation 2} \\ 2x \quad - 3z = 0 & \text{Equation 3} \end{cases}$

Add Equation 1 to Equation 2:

$\begin{cases} x - 2y + 3z = 5 \\ y - 2z = 9 \\ 2x \quad - 3z = 0 \end{cases}$

This is the first step in putting the system in row-echelon form.

13.
$$\begin{cases} x + y + z = 6 & \text{Equation 1} \\ 2x - y + z = 3 & \text{Equation 2} \\ 3x - z = 0 & \text{Equation 3} \end{cases}$$

$$\begin{cases} x + y + z = 6 \\ -3y - z = -9 & -2\text{Eq.1} + \text{Eq.2} \\ -3y - 4z = -18 & -3\text{Eq.1} + \text{Eq.3} \end{cases}$$

$$\begin{cases} x + y + z = 6 \\ -3y - z = -9 \\ -3z = -9 & -\text{Eq.2} + \text{Eq.3} \end{cases}$$

$$\begin{cases} x + y + z = 6 \\ -3y - z = -9 \\ z = 3 & -\frac{1}{3}\text{Eq.3} \end{cases}$$

$-3y - 3 = -9 \implies y = 2$

$x + 2 + 3 = 6 \implies x = 1$

Solution: $(1, 2, 3)$

15.
$$\begin{cases} 2x + 2z = 2 \\ 5x + 3y = 4 \\ 3y - 4z = 4 \end{cases}$$

$$\begin{cases} x + z = 1 & \frac{1}{2}\text{Eq.1} \\ 5x + 3y = 4 \\ 3y - 4z = 4 \end{cases}$$

$$\begin{cases} x + z = 1 \\ 3y - 5z = -1 & -5\text{Eq.1} + \text{Eq.2} \\ 3y - 4z = 4 \end{cases}$$

$$\begin{cases} x + z = 1 \\ 3y - 5z = -1 \\ z = 5 & -\text{Eq.2} + \text{Eq.3} \end{cases}$$

$3y - 5(5) = -1 \implies y = 8$

$x + 5 = 1 \implies x = -4$

Solution: $(-4, 8, 5)$

17.
$$\begin{cases} 3x + 3y = 9 & \text{Interchange equations.} \\ 2x - 3z = 10 \\ 6y + 4z = -12 \end{cases}$$

$$\begin{cases} x + y = 3 & \frac{1}{3}\text{Eq.1} \\ 2x - 3z = 10 \\ 6y + 4z = -12 \end{cases}$$

$$\begin{cases} x + y = 3 \\ -2y - 3z = 4 & -2\text{Eq.1} + \text{Eq.2} \\ 6y + 4z = -12 \end{cases}$$

$$\begin{cases} x + y = 3 \\ -2y - 3z = 4 \\ -5z = 0 & 3\text{Eq.2} + \text{Eq.3} \end{cases}$$

$$\begin{cases} x + y = 3 \\ -2y - 3z = 4 \\ z = 0 & -\frac{1}{5}\text{Eq.3} \end{cases}$$

$-2y - 3(0) = 4 \implies y = -2$

$x - 2 = 3 \implies x = 5$

Solution: $(5, -2, 0)$

19.
$$\begin{cases} x - 2y + 2z = -9 & \text{Interchange equations.} \\ 2x + y - z = 7 \\ 3x - y + z = 5 \end{cases}$$

$$\begin{cases} x - 2y + 2z = -9 \\ 5y - 5z = 25 & -2\text{Eq.1} + \text{Eq.2} \\ 5y - 5z = 32 & -3\text{Eq.1} + \text{Eq.3} \end{cases}$$

$$\begin{cases} x - 2y + 2z = -9 \\ 5y - 5z = 25 \\ 0 = 7 & -\text{Eq.2} + \text{Eq.3} \end{cases}$$

Inconsistent, no solution

21.
$$\begin{cases} 3x - 5y + 5z = 1 \\ 5x - 2y + 3z = 0 \\ 7x - y + 3z = 0 \end{cases}$$

$$\begin{cases} 6x - 10y + 10z = 2 & 2\text{Eq.1} \\ 5x - 2y + 3z = 0 \\ 7x - y + 3z = 0 \end{cases}$$

$$\begin{cases} x - 8y + 7z = 2 & -\text{Eq.2} + \text{Eq.1} \\ 5x - 2y + 3z = 0 \\ 7x - y + 3z = 0 \end{cases}$$

$$\begin{cases} x - 8y + 7z = 2 \\ 38y - 32z = -10 & -5\text{Eq.1} + \text{Eq.2} \\ 55y - 46z = -14 & -7\text{Eq.1} + \text{Eq.3} \end{cases}$$

$$\begin{cases} x - 8y + 7z = 2 \\ 2090y - 1760z = -550 & 55\text{Eq.2} \\ -2090y + 1748z = 532 & -38\text{Eq.3} \end{cases}$$

$$\begin{cases} x - 8y + 7z = 2 \\ 2090y - 1760z = -550 \\ -12z = -18 & \text{Eq.2} + \text{Eq.3} \end{cases}$$

$-12z = -18 \implies z = \frac{3}{2}$

$38y - 32\left(\frac{3}{2}\right) = -10 \implies y = 1$

$x - 8(1) + 7\left(\frac{3}{2}\right) = 2 \implies x = -\frac{1}{2}$

Solution: $\left(-\frac{1}{2}, 1, \frac{3}{2}\right)$

23. $\begin{cases} x + 2y - 7z = -4 \\ 2x + y + z = 13 \\ 3x + 9y - 36z = -33 \end{cases}$

$\begin{cases} x + 2y - 7z = -4 \\ \quad -3y + 15z = 21 \qquad -2\text{Eq.1} + \text{Eq.2} \\ \quad\quad 3y - 15z = -21 \qquad -3\text{Eq.1} + \text{Eq.3} \end{cases}$

$\begin{cases} x + 2y - 7z = -4 \\ \quad -3y + 15z = 21 \\ \quad\quad 0 = 0 \qquad\qquad \text{Eq.2} + \text{Eq.3} \end{cases}$

$\begin{cases} x + 2y - 7z = -4 \\ \quad y - 5z = -7 \qquad -\frac{1}{3}\text{Eq.2} \end{cases}$

$\begin{cases} x \quad\quad + 3z = 10 \qquad -2\text{Eq.2} + \text{Eq.1} \\ \quad y - 5z = -7 \end{cases}$

Let $z = a$, then:

$y = 5a - 7$

$x = -3a + 10$

Solution: $(-3a + 10, 5a - 7, a)$

27. $\begin{cases} x - 2y + 5z = 2 \\ 4x \quad\quad - z = 0 \end{cases}$

Let $z = a$, then $x = \frac{1}{4}a$.

$\frac{1}{4}a - 2y + 5a = 2$

$a - 8y + 20a = 8$

$\quad\quad -8y = -21a + 8$

$\quad\quad\quad y = \frac{21}{8}a - 1$

Answer: $\left(\frac{1}{4}a, \frac{21}{8}a - 1, a\right)$

To avoid fractions, we could go back and let $z = 8a$, then $4x - 8a = 0 \implies x = 2a$.

$2a - 2y + 5(8a) = 2$

$\quad -2y + 42a = 2$

$\quad\quad\quad y = 21a - 1$

Solution: $(2a, 21a - 1, 8a)$

25. $\begin{cases} 3x - 3y + 6z = 6 \\ x + 2y - z = 5 \\ 5x - 8y + 13z = 7 \end{cases}$

$\begin{cases} x - y + 2z = 2 \qquad \frac{1}{3}\text{Eq.1} \\ x + 2y - z = 5 \\ 5x - 8y + 13z = 7 \end{cases}$

$\begin{cases} x - y + 2z = 2 \\ \quad 3y - 3z = 3 \qquad -\text{Eq.1} + \text{Eq.2} \\ \quad -3y + 3z = -3 \qquad -5\text{Eq.1} + \text{Eq.3} \end{cases}$

$\begin{cases} x - y + 2z = 2 \\ \quad y - z = 1 \qquad \frac{1}{3}\text{Eq.2} \\ \quad\quad 0 = 0 \qquad \text{Eq.2} + \text{Eq.3} \end{cases}$

$\begin{cases} x \quad\quad + z = 3 \qquad \text{Eq.2} + \text{Eq.1} \\ \quad y - z = 1 \end{cases}$

Let $z = a$, then:

$y = a + 1$

$x = -a + 3$

Solution: $(-a + 3, a + 1, a)$

29. $\begin{cases} 2x - 3y + z = -2 \\ -4x + 9y = 7 \end{cases}$

$\begin{cases} 2x - 3y + z = -2 \\ \quad 3y + 2z = 3 \qquad 2\text{Eq.1} + \text{Eq.2} \end{cases}$

$\begin{cases} 2x \quad\quad + 3z = 1 \qquad \text{Eq.2} + \text{Eq.1} \\ \quad 3y + 2z = 3 \end{cases}$

Let $z = a$, then:

$y = -\frac{2}{3}a + 1$

$x = -\frac{3}{2}a + \frac{1}{2}$

Solution: $\left(-\frac{3}{2}a + \frac{1}{2}, -\frac{2}{3}a + 1, a\right)$

31.
$$\begin{cases} x & + 3w = 4 \\ 2y - z - w = 0 \\ 3y & - 2w = 1 \\ 2x - y + 4z & = 5 \end{cases}$$

$$\begin{cases} x & + 3w = 4 \\ 2y - z - w = 0 \\ 3y & - 2w = 1 \\ -y + 4z - 6w = -3 \end{cases} \quad -2\text{Eq.1} + \text{Eq.4}$$

$$\begin{cases} x & + 3w = 4 \\ y - 4z + 6w = 3 \\ 2y - z - w = 0 \\ 3y & - 2w = 1 \end{cases} \quad \begin{array}{l} -\text{Eq.4 and interchange} \\ \text{the equations.} \end{array}$$

$$\begin{cases} x & + 3w = 4 \\ y - 4z + 6w = 3 \\ 7z - 13w = -6 \quad -\text{Eq.2} + \text{Eq.3} \\ 12z - 20w = -8 \quad -3\text{Eq.2} + \text{Eq.4} \end{cases}$$

$$\begin{cases} x & + 3w = 4 \\ y - 4z + 6w = 3 \\ z - 3w = -2 \quad -\tfrac{1}{2}\text{Eq.4} + \text{Eq.3} \\ 12z - 20w = -8 \end{cases}$$

$$\begin{cases} x & + 3w = 4 \\ y - 4z + 6w = 3 \\ z - 3w = -2 \\ 16w = 16 \quad -12\text{Eq.3} + \text{Eq.4} \end{cases}$$

$$16w = 16 \implies w = 1$$
$$z - 3(1) = -2 \implies z = 1$$
$$y - 4(1) + 6(1) = 3 \implies y = 1$$
$$x + 3(1) = 4 \implies x = 1$$

Solution: $(1, 1, 1, 1)$

33.
$$\begin{cases} x & + 4z = 1 \\ x + y + 10z = 10 \\ 2x - y + 2z = -5 \end{cases}$$

$$\begin{cases} x & + 4z = 1 \\ y + 6z = 9 \quad -\text{Eq.1} + \text{Eq.2} \\ -y - 6z = -7 \quad -2\text{Eq.1} + \text{Eq.3} \end{cases}$$

$$\begin{cases} x & + 4z = 1 \\ y + 6z = 9 \\ 0 = 2 \quad \text{Eq.2} + \text{Eq.3} \end{cases}$$

No solution, inconsistent

35.
$$\begin{cases} 2x + 3y & = 0 \\ 4x + 3y - z = 0 \\ 8x + 3y + 3z = 0 \end{cases}$$

$$\begin{cases} 2x + 3y & = 0 \\ -3y - z = 0 \quad -2\text{Eq.1} + \text{Eq.2} \\ -9y + 3z = 0 \quad -4\text{Eq.1} + \text{Eq.3} \end{cases}$$

$$\begin{cases} 2x + 3y & = 0 \\ -3y - z = 0 \\ 6z = 0 \quad -3\text{Eq.2} + \text{Eq.3} \end{cases}$$

$$6z = 0 \implies z = 0$$
$$-3y - 0 = 0 \implies y = 0$$
$$2x + 3(0) = 0 \implies x = 0$$

Solution: $(0, 0, 0)$

37.
$$\begin{cases} 12x + 5y + z = 0 \\ 23x + 4y - z = 0 \end{cases}$$

$$\begin{cases} 24x + 10y + 2z = 0 \quad 2\text{Eq.1} \\ 23x + 4y - z = 0 \end{cases}$$

$$\begin{cases} x + 6y + 3z = 0 \quad -\text{Eq.2} + \text{Eq.1} \\ 23x + 4y - z = 0 \end{cases}$$

$$\begin{cases} x + 6y + 3z = 0 \\ -134y - 70z = 0 \quad -23\text{Eq.1} + \text{Eq.2} \end{cases}$$

$$\begin{cases} x + 6y + 3z = 0 \\ -67y - 35z = 0 \quad \tfrac{1}{2}\text{Eq.2} \end{cases}$$

To avoid fractions, let $z = 67a$, then:

$$-67y - 35(67a) = 0$$
$$y = -35a$$
$$x + 6(-35a) + 3(67a) = 0$$
$$x = 9a$$

Solution: $(9a, -35a, 67a)$

39. $s = \tfrac{1}{2}at^2 + v_0t + s_0$

$(1, 128), (2, 80), (3, 0)$

$$128 = \tfrac{1}{2}a + v_0 + s_0 \implies a + 2v_0 + 2s_0 = 256$$
$$80 = 2a + 2v_0 + s_0 \implies 2a + 2v_0 + s_0 = 80$$
$$0 = \tfrac{9}{2}a + 3v_0 + s_0 \implies 9a + 6v_0 + 2s_0 = 0$$

Solving this system yields $a = -32$, $v_0 = 0$, $s_0 = 144$.

Thus, $s = \tfrac{1}{2}(-32)t^2 + (0)t + 144 = -16t^2 + 144$.

41. $s = \frac{1}{2}at^2 + v_0t + s_0$

$(1, 452), (2, 372), (3, 260)$

$452 = \frac{1}{2}a + v_0 + s_0 \implies a + 2v_0 + 2s_0 = 904$

$372 = 2a + 2v_0 + s_0 \implies 2a + 2v_0 + s_0 = 372$

$260 = \frac{9}{2}a + 3v_0 + s_0 \implies 9a + 6v_0 + 2s_0 = 520$

Solving this system yields $a = -32, v_0 = -32, s_0 = 500$.

Thus, $s = \frac{1}{2}(-32)t^2 + (-32)t + 500$

$= -16t^2 - 32t + 500$.

43. $y = ax^2 + bx + c$ passing through $(0, 0), (2, -2), (4, 0)$

$(0, 0)$: $0 = c$

$(2, -2)$: $-2 = 4a + 2b + c \implies -1 = 2a + b$

$(4, 0)$: $0 = 16a + 4b + c \implies 0 = 4a + b$

Solution: $a = \frac{1}{2}, b = -2, c = 0$

The equation of the parabola is $y = \frac{1}{2}x^2 - 2x$.

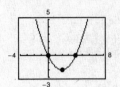

45. $y = ax^2 + bx + c$ passing through $(2, 0), (3, -1), (4, 0)$

$(2, 0)$: $0 = 4a + 2b + c$

$(3, -1)$: $-1 = 9a + 3b + c$

$(4, 0)$: $0 = 16a + 4b + c$

$$\begin{cases} 0 = 4a + 2b + c \\ -1 = 5a + b \qquad -\text{Eq.1} + \text{Eq.2} \\ 0 = 12a + 2b \qquad -\text{Eq.1} + \text{Eq.3} \end{cases}$$

$$\begin{cases} 0 = 4a + 2b + c \\ -1 = 5a + b \\ 2 = 2a \qquad -2\text{Eq.2} + \text{Eq.3} \end{cases}$$

Solution: $a = 1, b = -6, c = 8$

The equation of the parabola is $y = x^2 - 6x + 8$.

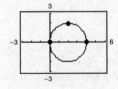

47. $x^2 + y^2 + Dx + Ey + F = 0$ passing through $(0, 0), (2, 2), (4, 0)$

$(0, 0)$: $F = 0$

$(2, 2)$: $8 + 2D + 2E + F = 0 \implies D + E = -4$

$(4, 0)$: $16 + 4D + F = 0 \implies D = -4$ and $E = 0$

The equation of the circle is $x^2 + y^2 - 4x = 0$.

To graph, let $y_1 = \sqrt{4x - x^2}$ and $y_2 = -\sqrt{4x - x^2}$.

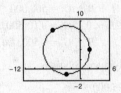

49. $x^2 + y^2 + Dx + Ey + F = 0$ passing through $(-3, -1), (2, 4), (-6, 8)$

$(-3, -1)$: $10 - 3D - E + F = 0 \implies 10 = 3D + E - F$

$(2, 4)$: $20 + 2D + 4E + F = 0 \implies 20 = -2D - 4E - F$

$(-6, 8)$: $100 - 6D + 8E + F = 0 \implies 100 = 6D - 8E - F$

Solution: $D = 6, E = -8, F = 0$

The equation of the circle is $x^2 + y^2 + 6x - 8y = 0$. To graph, complete the squares first, then solve for y.

$(x^2 + 6x + 9) + (y^2 - 8y + 16) = 0 + 9 + 16$

$(x + 3)^2 + (y - 4)^2 = 25$

$(y - 4)^2 = 25 - (x + 3)^2$

$y - 4 = \pm\sqrt{25 - (x + 3)^2}$

$y = 4 \pm \sqrt{25 - (x + 3)^2}$

Let $y_1 = 4 + \sqrt{25 - (x + 3)^2}$ and $y_2 = 4 - \sqrt{25 - (x + 3)^2}$.

51. Let x = number of touchdowns.

Let y = number of extra-point kicks.

Let z = number of field goals.

$$\begin{cases} x + y + z = 13 \\ 6x + y + 3z = 45 \\ x - y = 0 \\ x - 6z = 0 \end{cases}$$

$$\begin{cases} x + y + z = 13 \\ - 5y - 3z = -33 \quad -6\text{Eq.1} + \text{Eq.2} \\ - 2y - z = -13 \quad -\text{Eq.1} + \text{Eq.3} \\ - y - 7z = -13 \quad -\text{Eq.1} + \text{Eq.4} \end{cases}$$

$$\begin{cases} x + y + z = 13 \\ - y - 7z = -13 \quad \text{Interchange Eq.2 and Eq.4.} \\ - 2y - z = -13 \\ - 5y - 3z = -33 \end{cases}$$

$$\begin{cases} x + y + z = 13 \\ y + 7z = 13 \quad -\text{Eq.2} \\ - 2y - z = -13 \\ - 5y - 3z = -33 \end{cases}$$

$$\begin{cases} x + y + z = 13 \\ y + 7z = 13 \\ 13z = 13 \quad 2\text{Eq.2} + \text{Eq.3} \\ 32z = 32 \quad 5\text{Eq.2} + \text{Eq.4} \end{cases}$$

$z = 1$

$y + 7(1) = 13 \implies y = 6$

$x + 6 + 1 = 13 \implies x = 6$

Thus, 6 touchdowns, 6 extra-point kicks, and 1 field goal were scored.

55. Let C = amount in certificates of deposit.

Let M = amount in municipal bonds.

Let B = amount in blue-chip stocks.

Let G = amount in growth or speculative stocks.

$$\begin{cases} C + M + B + G = 500,000 \\ 0.10C + 0.08M + 0.12B + 0.13G = 0.10(500,000) \\ B + G = \tfrac{1}{4}(500,000) \end{cases}$$

This system has infinitely many solutions.

Let $G = s$, then $B = 125,000 - s$

$M = 125,000 + \tfrac{1}{2}s$

$C = 250,000 - \tfrac{1}{2}s$

One possible solution is to let $s = 50,000$.

Certificates of deposit: $225,000

Municipal bonds: $150,000

Blue-chip stocks: $75,000

Growth or speculative stocks: $50,000

53. Let x = amount at 8%.

Let y = amount at 9%.

Let z = amount at 10%.

$$\begin{cases} x + y + z = 775,000 \\ 0.08x + 0.09y + 0.10z = 67,500 \\ x = 4z \end{cases}$$

$y + 5z = 775,000$

$0.09y + 0.42z = 67,500$

$z = 75,000$

$y = 775,000 - 5z = 400,000$

$x = 4z = 300,000$

$300,000 was borrowed at 8%.

$400,000 was borrowed at 9%.

$75,000 was borrowed at 10%.

57. Let x = pounds of brand X.

Let y = pounds of brand Y.

Let z = pounds of brand Z.

Fertilizer A: $\phantom{\tfrac{1}{2}x +} \tfrac{1}{3}y + \tfrac{2}{9}z = 5$

Fertilizer B: $\tfrac{1}{2}x + \tfrac{2}{3}y + \tfrac{5}{9}z = 13$

Fertilizer C: $\tfrac{1}{2}x \phantom{+ \tfrac{2}{3}y} + \tfrac{2}{9}z = 4$

$$\begin{cases} \tfrac{1}{2}x + \tfrac{2}{3}y + \tfrac{5}{9}z = 13 \quad \text{Interchange Eq.1 and Eq.2.} \\ \phantom{\tfrac{1}{2}x +} \tfrac{1}{3}y + \tfrac{2}{9}z = 5 \\ \tfrac{1}{2}x \phantom{+ \tfrac{2}{3}y} + \tfrac{2}{9}z = 4 \end{cases}$$

$$\begin{cases} \tfrac{1}{2}x + \tfrac{2}{3}y + \tfrac{5}{9}z = 13 \\ \phantom{\tfrac{1}{2}x +} \tfrac{1}{3}y + \tfrac{2}{9}z = 5 \\ \phantom{\tfrac{1}{2}x} - \tfrac{2}{3}y - \tfrac{1}{3}z = -9 \quad -\text{Eq.1} + \text{Eq.3} \end{cases}$$

$$\begin{cases} \tfrac{1}{2}x + \tfrac{2}{3}y + \tfrac{5}{9}z = 13 \\ \phantom{\tfrac{1}{2}x +} \tfrac{1}{3}y + \tfrac{2}{9}z = 5 \\ \phantom{\tfrac{1}{2}x + \tfrac{2}{3}y} \tfrac{1}{9}z = 1 \quad 2\text{Eq.2} + \text{Eq.3} \end{cases}$$

$z = 9$

$\tfrac{1}{3}y + \tfrac{2}{9}(9) = 5 \implies y = 9$

$\tfrac{1}{2}x + \tfrac{2}{3}(9) + \tfrac{5}{9}(9) = 13 \implies x = 4$

4 pounds of brand X, 9 pounds of brand Y, and 9 pounds of brand Z are needed to obtain the desired mixture.

59. Let x = pounds of Vanilla coffee.

Let y = pounds of Hazelnut coffee.

Let z = pounds of French Roast coffee.

$$\begin{cases} x + y + z = 10 \\ 2x + 2.50y + 3z = 26 \\ y - z = 0 \end{cases}$$

$$\begin{cases} x + y + z = 10 \\ 0.5y + z = 6 \qquad -2\text{Eq.1} + \text{Eq.2} \\ y - z = 0 \end{cases}$$

$$\begin{cases} x + y + z = 10 \\ 0.5y + z = 6 \\ - 3z = -12 \qquad -2\text{Eq.2} + \text{Eq.3} \end{cases}$$

$z = 4$

$0.5y + 4 = 6 \implies y = 4$

$x + 4 + 4 = 10 \implies x = 2$

2 pounds of Vanilla coffee, 4 pounds of Hazelnut coffee, and 4 pounds of French Roast coffee are needed to obtain the desired mixture.

61. Let x = number of television ads.

Let y = number of radio ads.

Let z = number of local newspaper ads.

$$\begin{cases} x + y + z = 60 \\ 1000x + 200y + 500z = 42{,}000 \\ x - y - z = 0 \end{cases}$$

$$\begin{cases} x + y + z = 60 \\ - 800y - 500z = -18{,}000 \qquad -1000\text{Eq.1} + \text{Eq.2} \\ -2y - 2z = -60 \qquad -\text{Eq.1} + \text{Eq.3} \end{cases}$$

$$\begin{cases} x + y + z = 60 \\ -2y - 2z = -60 \qquad \text{Interchange} \\ - 800y - 500z = -18{,}000 \qquad \text{Eq.2 and Eq.3.} \end{cases}$$

$$\begin{cases} x + y + z = 60 \\ -2y - 2z = -60 \\ 300z = 6000 \qquad -400\text{Eq.2} + \text{Eq.3} \end{cases}$$

$z = 20$

$-2y - 2(20) = -60 \implies y = 10$

$x + 10 + 20 = 60 \implies x = 30$

30 television ads, 10 radio ads, and 20 newspaper ads can be run each month.

63. (a) To use 2 liters of the 50% solution:

Let x = amount of 10% solution.

Let y = amount of 20% solution.

$x + y = 8 \implies y = 8 - x$

$x(0.10) + y(0.20) + 2(0.50) = 10(0.25)$

$0.10x + 0.20(8 - x) + 1 = 2.5$

$0.10x + 1.6 - 0.20x + 1 = 2.5$

$-0.10x = -0.1$

$x = 1$ liter of 10% solution

$y = 7$ liters of 20% solution

Given: 2 liters of 50% solution

(b) To use as little of the 50% solution as possible, the chemist should use no 10% solution.

Let x = amount of 20% solution.

Let y = amount of 50% solution.

$x + y = 10 \implies y = 10 - x$

$x(0.20) + y(0.50) = 10(0.25)$

$x(0.20) + (10 - x)(0.50) = 10(0.25)$

$x(0.20) + 5 - 0.50x = 2.5$

$-0.30x = -2.5$

$x = 8\frac{1}{3}$ liters of 20% solution

$y = 1\frac{2}{3}$ liters of 50% solution

(c) To use as much of the 50% solution as possible, the chemist should use no 20% solution.

Let x = amount of 10% solution.

Let y = amount of 50% solution.

$x + y = 10 \implies y = 10 - x$

$x(0.10) + y(0.50) = 10(0.25)$

$0.10x + 0.50(10 - x) = 2.5$

$0.10x + 5 - 0.50x = 2.5$

$-0.40x = -2.5$

$x = 6\frac{1}{4}$ liters of 10% solution

$y = 3\frac{3}{4}$ liters of 50% solution

65. $\begin{cases} I_1 - I_2 + I_3 = 0 & \text{Equation 1} \\ 3I_1 + 2I_2 = 7 & \text{Equation 2} \\ 2I_2 + 4I_3 = 8 & \text{Equation 3} \end{cases}$

$\begin{cases} I_1 - I_2 + I_3 = 0 & \\ 5I_2 - 3I_3 = 7 & (-3)\text{Eq.1} + \text{Eq.2} \\ 2I_2 + 4I_3 = 8 & \end{cases}$

$\begin{cases} I_1 - I_2 + I_3 = 0 & \\ 10I_2 - 6I_3 = 14 & 2\text{Eq.2} \\ 10I_2 + 20I_3 = 40 & 5\text{Eq.3} \end{cases}$

$\begin{cases} I_1 - I_2 + I_3 = 0 & \\ 10I_2 - 6I_3 = 14 & \\ 26I_3 = 26 & (-1)\text{Eq.2} + \text{Eq.3} \end{cases}$

$26I_3 = 26 \implies I_3 = 1$

$10I_2 - 6(1) = 14 \implies I_2 = 2$

$I_1 - 2 + 1 = 0 \implies I_1 = 1$

Solution: $I_1 = 1, I_2 = 2, I_3 = 1$

67. $(-4, 5), (-2, 6), (2, 6), (4, 2)$

$n = 4, \displaystyle\sum_{i=1}^{4} x_i = 0, \sum_{i=1}^{4} x_i^2 = 40, \sum_{i=1}^{4} x_i^3 = 0, \sum_{i=1}^{4} x_i^4 = 544, \sum_{i=1}^{4} y_i = 19, \sum_{i=1}^{4} x_i y_i = -12, \sum_{i=1}^{4} x_i^2 y_i = 160$

$\begin{cases} 4c + 40a = 19 \\ 40b = -12 \\ 40c + 544a = 160 \end{cases}$

$\begin{cases} 4c + 40a = 19 \\ 40b = -12 \\ 144a = -30 & -10\text{Eq.1} + \text{Eq.3} \end{cases}$

$144a = -30 \implies a = -\frac{5}{24}$

$40b = -12 \implies b = -\frac{3}{10}$

$4c + 40\left(-\frac{5}{24}\right) = 19 \implies c = \frac{41}{6}$

Least squares regression parabola: $y = -\frac{5}{24}x^2 - \frac{3}{10}x + \frac{41}{6}$

69. $(0, 0), (2, 2), (3, 6), (4, 12)$

$n = 4, \displaystyle\sum_{i=1}^{4} x_i = 9, \sum_{i=1}^{4} x_i^2 = 29, \sum_{i=1}^{4} x_i^3 = 99, \sum_{i=1}^{4} x_i^4 = 353, \sum_{i=1}^{4} y_i = 20, \sum_{i=1}^{4} x_i y_i = 70, \sum_{i=1}^{4} x_i^2 y_i = 254$

$\begin{cases} 4c + 9b + 29a = 20 \\ 9c + 29b + 99a = 70 \\ 29c + 99b + 353a = 254 \end{cases}$

$\begin{cases} 9c + 29b + 99a = 70 & \text{Interchange equations.} \\ 4c + 9b + 29a = 20 \\ 29c + 99b + 353a = 254 \end{cases}$

$\begin{cases} c + 11b + 41a = 30 & -2\text{Eq.2} + \text{Eq.1} \\ -35b - 135a = -100 & -4\text{Eq.1} + \text{Eq.2} \\ -220b - 836a = -616 & -29\text{Eq.1} + \text{Eq.3} \end{cases}$

$\begin{cases} c + 11b + 41a = 30 \\ 1540b + 5940a = 4400 & -44\text{Eq.2} \\ -1540b - 5852a = -4312 & 7\text{Eq.3} \end{cases}$

—CONTINUED—

69. —CONTINUED—

$$\begin{cases} c + \quad 11b + \quad 41a = \quad 30 \\ \qquad 1540b + 5940a = \quad 4400 \\ \qquad\qquad\qquad 88a = \quad 88 \quad \text{Eq.2 + Eq.3} \end{cases}$$

$$88a = \quad 88 \implies a = \quad 1$$

$$1540b + 5940(1) = 4400 \implies b = -1$$

$$c + 11(-1) + 41(1) = \quad 30 \implies c = \quad 0$$

Least squares regression parabola: $y = x^2 - x$

71. (a) $(100, 75), (120, 68), (140, 55)$

$$n = 3, \sum_{i=1}^{3} x_i = 360, \sum_{i=1}^{3} x_i^2 = 44{,}000, \sum_{i=1}^{3} x_i^3 = 5{,}472{,}000$$

$$\sum_{i=1}^{3} x_i^4 = 691{,}520{,}000, \sum_{i=1}^{3} y_i = 198, \sum_{i=1}^{3} x_i y_i = 23{,}360,$$

$$\sum_{i=1}^{3} x_i^2 y_i = 2{,}807{,}200$$

$$\begin{cases} 3c + \qquad 360b + \qquad 44{,}000a = \qquad 198 \\ 360c + \qquad 44{,}000b + \qquad 5{,}472{,}000a = \qquad 23{,}360 \\ 44{,}000c + 5{,}472{,}000b + 691{,}520{,}000a = 2{,}807{,}200 \end{cases}$$

Solving this system yields $a = -0.0075$, $b = 1.3$ and $c = 20$.

Least squares regression parabola:
$y = -0.0075x^2 + 1.3x + 20$

(b)

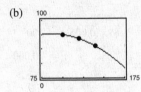

(c)

x	Actual Percent y	Model Approximation y
100	75	75
120	68	68
140	55	55

The model is a good fit to the actual data.
The values are the same.

(d) For $x = 170$:

$$y = -0.0075(170)^2 + 1.3(170) + 20$$
$$= 24.25\%$$

(e) For $y = 40$:

$$40 = -0.0075x^2 + 1.3x + 20$$

$$0.0075x^2 - 1.3x + 20 = 0$$

By the Quadratic Formula we have $x \approx 17$ or $x \approx 156$.

Choosing the value that fits with our data, we have 156 females.

73. Let x = number of touchdowns.

Let y = number of extra-point kicks.

Let z = number of two-point conversions.

Let w = number of field goals.

$$\begin{cases} x + y + \quad z + \quad w = 16 \\ 6x + y + 2z + 3w = 32 + 29 \\ x \qquad\qquad - 4w = 0 \quad \implies x = 4w \\ \qquad\quad 2z - \quad w = 0 \quad \implies z = \tfrac{1}{2}w \end{cases}$$

$$\begin{cases} 4w + y + \quad \tfrac{1}{2}w + \quad w = 16 \implies 5.5w + y = 16 \\ 6(4w) + y + 2\left(\tfrac{1}{2}\right)w + 3w = 61 \implies 28w + y = 61 \end{cases}$$

$$28w + y = 61$$
$$\underline{-5.5w - y = -16}$$
$$22.5w \quad = 45$$
$$w = 2$$
$$y = 5$$
$$x = 4w = 8$$
$$z = \tfrac{1}{2}w = 1$$

Thus, 8 touchdowns, 5 extra-point kicks, 1 two-point conversion, and 2 field goals were scored.

75. $\begin{cases} \quad y + \lambda = 0 \\ x \qquad + \lambda = 0 \\ x + y - 10 = 0 \end{cases} \implies x = y = -\lambda$

$\implies 2x - 10 = \quad 0$

$$x = \quad 5$$
$$y = \quad 5$$
$$\lambda = -5$$

77. $\begin{cases} 2x - 2x\lambda = 0 \implies 2x(1 - \lambda) = 0 \implies \lambda = 1 \text{ or } x = 0 \\ -2y + \quad \lambda = 0 \\ \quad y - \quad x^2 = 0 \end{cases}$

If $\lambda = 1$:

$$2y = \lambda \implies y = \frac{1}{2}$$

$$x^2 = y \implies x = \pm\sqrt{\frac{1}{2}} = \pm\frac{\sqrt{2}}{2}$$

If $x = 0$:

$$x^2 = y \implies y = 0$$

$$2y = \lambda \implies \lambda = 0$$

Solution: $x = \pm\dfrac{\sqrt{2}}{2}$ or $x = 0$

$$y = \frac{1}{2} \qquad\qquad y = 0$$

$$\lambda = 1 \qquad\qquad \lambda = 0$$

79. False. Equation 2 does not have a leading coefficient of 1.

81. No, they are not equivalent. There are two arithmetic errors. The constant in the second equation should be -11 and the coefficient of z in the third equation should be 2.

83. There are an infinite number of linear systems that have $(4, -1, 2)$ as their solution. Two such systems are as follows:

$$\begin{cases} 3x + \ y - \ z = 9 \\ \ x + 2y - \ z = 0 \\ -x + \ y + 3z = 1 \end{cases} \qquad \begin{cases} x + \ y + \ z = 5 \\ x \qquad\ - 2z = 0 \\ \quad 2y + \ z = 0 \end{cases}$$

85. There are an infinite number of linear systems that have $\left(3, -\frac{1}{2}, \frac{7}{4}\right)$ as their solution. Two such systems are as follows:

$$\begin{cases} \ x + 2y - 4z = -5 \\ -x - 4y + 8z = 13 \\ \ x + 6y + 4z = \ 7 \end{cases} \qquad \begin{cases} x + 2y + 4z = \ 9 \\ \quad y + 2z = \ 3 \\ x \qquad\ - 4z = -4 \end{cases}$$

87. $(0.075)(85) = 6.375$

89. $(0.005)n = 400$

$$n = 80{,}000$$

91. $(7 - i) + (4 + 2i) = (7 + 4) + (-i + 2i) = 11 + i$

93. $(4 - i)(5 + 2i) = 20 + 8i - 5i - 2i^2 = 20 + 3i + 2$

$$= 22 + 3i$$

95. $\dfrac{i}{1 + i} + \dfrac{6}{1 - i} = \dfrac{i(1 - i) + 6(1 + i)}{(1 + i)(1 - i)}$

$$= \frac{i - i^2 + 6 + 6i}{1 - i^2}$$

$$= \frac{7 + 7i}{2}$$

$$= \frac{7}{2} + \frac{7}{2}i$$

97. $f(x) = x^3 + x^2 - 12x$

(a) $x^3 + x^2 - 12x = 0$

$$x(x^2 + x - 12) = 0$$

$$x(x + 4)(x - 3) = 0$$

Zeros: $x = -4, 0, 3$

(b)

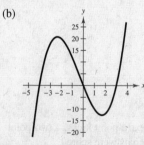

99. $f(x) = 2x^3 + 5x^2 - 21x - 36$

(a) $2x^3 + 5x^2 - 21x - 36 = 0$

$$
\begin{array}{r|rrrr}
3 & 2 & 5 & -21 & -36 \\
  &   & 6 & 33 & 36 \\
\hline
  & 2 & 11 & 12 & 0
\end{array}
$$

$f(x) = (x - 3)(2x^2 + 11x + 12)$

$\quad\quad = (x - 3)(x + 4)(2x + 3)$

Zeros: $x = -4, -\frac{3}{2}, 3$

(b)

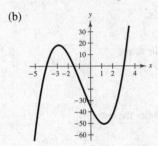

101. $y = 4^{x-4} - 5$

x	-2	0	2	4	5
y	-4.9998	-4.996	-4.938	-4	-1

Horizontal asymptote: $y = -5$

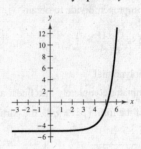

103. $y = 1.9^{-0.8x} + 3$

x	-2	-1	0	1	2
y	5.793	4.671	4	3.598	3.358

Horizontal asymptote: $y = 3$

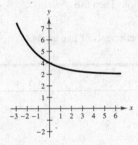

105. $\begin{cases} 2x + y = 120 & \text{Equation 1} \\ x + 2y = 120 & \text{Equation 2} \end{cases}$

$$
\begin{array}{r}
2x + y = 120 \\
-2x - 4y = -240 \quad (-2)\text{Eq.2} \\
\hline
-3y = -120 \\
y = 40
\end{array}
$$

$x + 2(40) = 120 \implies x = 40$

Solution: $(40, 40)$

107. Answers will vary.

Section 7.4   Partial Fractions

■ You should know how to decompose a rational function $\dfrac{N(x)}{D(x)}$ into partial fractions.

(a) If the fraction is improper, divide to obtain

$$\frac{N(x)}{D(x)} = p(x) + \frac{N_1(x)}{D(x)}$$

where $p(x)$ is a polynomial.

(b) Factor the denominator completely into linear and irreducible quadratic factors.

(c) For each factor of the form $(px + q)^m$, the partial fraction decomposition includes the terms

$$\frac{A_1}{(px + q)} + \frac{A_2}{(px + q)^2} + \cdots + \frac{A_m}{(px + q)^m}.$$

(d) For each factor of the form $(ax^2 + bx + c)^n$, the partial fraction decomposition includes the terms

$$\frac{B_1x + C_1}{ax^2 + bx + c} + \frac{B_2x + C_2}{(ax^2 + bx + c)^2} + \cdots + \frac{B_nx + C_n}{(ax^2 + bx + c)^n}.$$

■ You should know how to determine the values of the constants in the numerators.

(a) Set $\dfrac{N_1(x)}{D(x)}$ = partial fraction decomposition.

(b) Multiply both sides by $D(x)$ to obtain the basic equation.

(c) For distinct linear factors, substitute the zeros of the distinct linear factors into the basic equation.

(d) For repeated linear factors, use the coefficients found in part (c) to rewrite the basic equation. Then use other values of x to solve for the remaining coefficients.

(e) For quadratic factors, expand the basic equation, collect like terms, and then equate the coefficients of like terms.

Vocabulary Check

1. partial fraction decomposition

2. improper

3. m; n; irreducible

4. basic equation

---

1. $\dfrac{3x - 1}{x(x - 4)} = \dfrac{A}{x} + \dfrac{B}{x - 4}$

    Matches (b).

3. $\dfrac{3x - 1}{x(x^2 + 4)} = \dfrac{A}{x} + \dfrac{Bx + C}{x^2 + 4}$

    Matches (d).

5. $\dfrac{7}{x^2 - 14x} = \dfrac{7}{x(x - 14)} = \dfrac{A}{x} + \dfrac{B}{x - 14}$

7. $\dfrac{12}{x^3 - 10x^2} = \dfrac{12}{x^2(x - 10)} = \dfrac{A}{x} + \dfrac{B}{x^2} + \dfrac{C}{x - 10}$

9. $\dfrac{4x^2 + 3}{(x - 5)^3} = \dfrac{A}{x - 5} + \dfrac{B}{(x - 5)^2} + \dfrac{C}{(x - 5)^3}$

11. $\dfrac{2x - 3}{x^3 + 10x} = \dfrac{2x - 3}{x(x^2 + 10)} = \dfrac{A}{x} + \dfrac{Bx + C}{x^2 + 10}$

13. $\dfrac{x - 1}{x(x^2 + 1)^2} = \dfrac{A}{x} + \dfrac{Bx + C}{x^2 + 1} + \dfrac{Dx + E}{(x^2 + 1)^2}$

15. $\dfrac{1}{x^2 - 1} = \dfrac{A}{x + 1} + \dfrac{B}{x - 1}$

$1 = A(x - 1) + B(x + 1)$

Let $x = -1$: $1 = -2A \implies A = -\dfrac{1}{2}$

Let $x = 1$: $1 = 2B \implies B = \dfrac{1}{2}$

$\dfrac{1}{x^2 - 1} = \dfrac{1/2}{x - 1} - \dfrac{1/2}{x + 1} = \dfrac{1}{2}\left(\dfrac{1}{x - 1} - \dfrac{1}{x + 1}\right)$

17. $\dfrac{1}{x^2 + x} = \dfrac{A}{x} + \dfrac{B}{x + 1}$

$1 = A(x + 1) + Bx$

Let $x = 0$: $1 = A$

Let $x = -1$: $1 = -B \implies B = -1$

$\dfrac{1}{x^2 + x} = \dfrac{1}{x} - \dfrac{1}{x + 1}$

19. $\dfrac{1}{2x^2 + x} = \dfrac{A}{2x + 1} + \dfrac{B}{x}$

$1 = Ax + B(2x + 1)$

Let $x = -\dfrac{1}{2}$: $1 = -\dfrac{1}{2}A \implies A = -2$

Let $x = 0$: $1 = B$

$\dfrac{1}{2x^2 + x} = \dfrac{1}{x} - \dfrac{2}{2x + 1}$

21. $\dfrac{3}{x^2 + x - 2} = \dfrac{A}{x - 1} + \dfrac{B}{x + 2}$

$3 = A(x + 2) + B(x - 1)$

Let $x = 1$: $3 = 3A \implies A = 1$

Let $x = -2$: $3 = -3B \implies B = -1$

$\dfrac{3}{x^2 + x - 2} = \dfrac{1}{x - 1} - \dfrac{1}{x + 2}$

23. $\dfrac{x^2 + 12x + 12}{x^3 - 4x} = \dfrac{A}{x} + \dfrac{B}{x + 2} + \dfrac{C}{x - 2}$

$x^2 + 12x + 12 = A(x + 2)(x - 2) + Bx(x - 2) + Cx(x + 2)$

Let $x = 0$: $12 = -4A \implies A = -3$

Let $x = -2$: $-8 = 8B \implies B = -1$

Let $x = 2$: $40 = 8C \implies C = 5$

$\dfrac{x^2 + 12x + 12}{x^3 - 4x} = -\dfrac{3}{x} - \dfrac{1}{x + 2} + \dfrac{5}{x - 2}$

25. $\dfrac{4x^2 + 2x - 1}{x^2(x + 1)} = \dfrac{A}{x} + \dfrac{B}{x^2} + \dfrac{C}{x + 1}$

$4x^2 + 2x - 1 = Ax(x + 1) + B(x + 1) + Cx^2$

Let $x = 0$: $-1 = B$

Let $x = -1$: $1 = C$

Let $x = 1$: $5 = 2A + 2B + C$

$\qquad\qquad 5 = 2A - 2 + 1$

$\qquad\qquad 6 = 2A$

$\qquad\qquad 3 = A$

$\dfrac{4x^2 + 2x - 1}{x^2(x + 1)} = \dfrac{3}{x} - \dfrac{1}{x^2} + \dfrac{1}{x + 1}$

27. $\dfrac{3x}{(x - 3)^2} = \dfrac{A}{x - 3} + \dfrac{B}{(x - 3)^2}$

$3x = A(x - 3) + B$

Let $x = 3$: $9 = B$

Let $x = 0$: $0 = -3A + B$

$\qquad\qquad 0 = -3A + 9$

$\qquad\qquad 3 = A$

$\dfrac{3x}{(x - 3)^2} = \dfrac{3}{x - 3} + \dfrac{9}{(x - 3)^2}$

29. $\dfrac{x^2 - 1}{x(x^2 + 1)} = \dfrac{A}{x} + \dfrac{Bx + C}{x^2 + 1}$

$$x^2 - 1 = A(x^2 + 1) + (Bx + C)x$$

$$= Ax^2 + A + Bx^2 + Cx$$

$$= (A + B)x^2 + Cx + A$$

Equating coefficients of like terms gives
$1 = A + B, 0 = C$, and $-1 = A$.

Therefore, $A = -1, B = 2$, and $C = 0$.

$$\dfrac{x^2 - 1}{x(x^2 + 1)} = -\dfrac{1}{x} + \dfrac{2x}{x^2 + 1}$$

31. $\dfrac{x}{x^3 - x^2 - 2x + 2} = \dfrac{x}{(x - 1)(x^2 - 2)} = \dfrac{A}{x - 1} + \dfrac{Bx + C}{x^2 - 2}$

$$x = A(x^2 - 2) + (Bx + C)(x - 1)$$

$$= Ax^2 - 2A + Bx^2 - Bx + Cx - C$$

$$= (A + B)x^2 + (C - B)x - (2A + C)$$

Equating coefficients of like terms gives
$0 = A + B, 1 = C - B$, and $0 = 2A + C$.

Therefore, $A = -1, B = 1$, and $C = 2$.

$$\dfrac{x}{x^3 - x^2 - 2x + 2} = \dfrac{-1}{x - 1} + \dfrac{x + 2}{x^2 - 2}$$

33. $\dfrac{x^2}{x^4 - 2x^2 - 8} = \dfrac{x^2}{(x^2 - 4)(x^2 + 2)} = \dfrac{x^2}{(x + 2)(x - 2)(x^2 + 2)}$

$$= \dfrac{A}{x + 2} + \dfrac{B}{x - 2} + \dfrac{Cx + D}{x^2 + 2}$$

$$x^2 = A(x - 2)(x^2 + 2) + B(x + 2)(x^2 + 2) + (Cx + D)(x + 2)(x - 2)$$

$$= A(x^3 - 2x^2 + 2x - 4) + B(x^3 + 2x^2 + 2x + 4) + (Cx + D)(x^2 - 4)$$

$$= Ax^3 - 2Ax^2 + 2Ax - 4A + Bx^3 + 2Bx^2 + 2Bx + 4B + Cx^3 + Dx^2 - 4Cx - 4D$$

$$= (A + B + C)x^3 + (-2A + 2B + D)x^2 + (2A + 2B - 4C)x + (-4A + 4B - 4D)$$

Equating coefficients of like terms gives

$0 = A + B + C, 1 = -2A + 2B + D, 0 = 2A + 2B - 4C$, and $0 = -4A + 4B - 4D$.

Using the first and third equation, we have $A + B + C = 0$ and $A + B - 2C = 0$;
by subtraction, $C = 0$. Using the second and fourth equation, we have $-2A + 2B + D = 1$
and $-2A + 2B - 2D = 0$; by subtraction, $3D = 1$, so $D = \frac{1}{3}$. Substituting 0 for C and $\frac{1}{3}$ for
D in the first and second equations, we have

$A + B = 0$ and $-2A + 2B = \frac{2}{3}$, so $A = -\frac{1}{6}$ and $B = \frac{1}{6}$.

$$\dfrac{x^2}{x^4 - 2x^2 - 8} = \dfrac{-\frac{1}{6}}{x + 2} + \dfrac{\frac{1}{6}}{x - 2} + \dfrac{\frac{1}{3}}{x^2 + 2}$$

$$= \dfrac{1}{3(x^2 + 2)} - \dfrac{1}{6(x + 2)} + \dfrac{1}{6(x - 2)}$$

$$= \dfrac{1}{6}\left(\dfrac{2}{x^2 + 2} - \dfrac{1}{x + 2} + \dfrac{1}{x - 2}\right)$$

35. $\dfrac{x}{16x^4 - 1} = \dfrac{x}{(4x^2 - 1)(4x^2 + 1)} = \dfrac{x}{(2x + 1)(2x - 1)(4x^2 + 1)}$

$$= \dfrac{A}{2x + 1} + \dfrac{B}{2x - 1} + \dfrac{Cx + D}{4x^2 + 1}$$

$$x = A(2x - 1)(4x^2 + 1) + B(2x + 1)(4x^2 + 1) + (Cx + D)(2x + 1)(2x - 1)$$

$$= A(8x^3 - 4x^2 + 2x - 1) + B(8x^3 + 4x^2 + 2x + 1) + (Cx + D)(4x^2 - 1)$$

$$= 8Ax^3 - 4Ax^2 + 2Ax - A + 8Bx^3 + 4Bx^2 + 2Bx + B + 4Cx^3 + 4Dx^2 - Cx - D$$

$$= (8A + 8B + 4C)x^3 + (-4A + 4B + 4D)x^2 + (2A + 2B - C)x + (-A + B - D)$$

—CONTINUED—

35. **—CONTINUED—**

Equating coefficients of like terms gives $0 = 8A + 8B + 4C$, $0 = -4A + 4B + 4D$, $1 = 2A + 2B - C$, and $0 = -A + B - D$.

Using the first and third equations, we have $2A + 2B + C = 0$ and $2A + 2B - C = 1$; by subtraction, $2C = -1$, so $C = -\frac{1}{2}$.

Using the second and fourth equations, we have $-A + B + D = 0$ and $-A + B - D = 0$; by subtraction $2D = 0$, so $D = 0$.

Substituting $-\frac{1}{2}$ for C and 0 for D in the first and second equations, we have $8A + 8B = 2$ and $-4A + 4B = 0$, so $A = \frac{1}{8}$ and $B = \frac{1}{8}$.

$$\frac{x}{16x^4 - 1} = \frac{\frac{1}{8}}{2x + 1} + \frac{\frac{1}{8}}{2x - 1} + \frac{\left(-\frac{1}{2}\right)x}{4x^2 + 1}$$

$$= \frac{1}{8(2x + 1)} + \frac{1}{8(2x - 1)} - \frac{x}{2(4x^2 + 1)}$$

$$= \frac{1}{8}\left(\frac{1}{2x + 1} + \frac{1}{2x - 1} - \frac{4x}{4x^2 + 1}\right)$$

37. $\dfrac{x^2 + 5}{(x + 1)(x^2 - 2x + 3)} = \dfrac{A}{x + 1} + \dfrac{Bx + C}{x^2 - 2x + 3}$

$$x^2 + 5 = A(x^2 - 2x + 3) + (Bx + C)(x + 1)$$

$$= Ax^2 - 2Ax + 3A + Bx^2 + Bx + Cx + C$$

$$= (A + B)x^2 + (-2A + B + C)x + (3A + C)$$

Equating coefficients of like terms gives $1 = A + B$, $0 = -2A + B + C$, and $5 = 3A + C$.

Subtracting both sides of the second equation from the first gives $1 = 3A - C$; combining this with the third equation gives $A = 1$ and $C = 2$. Since $A + B = 1$, we also have $B = 0$.

$$\frac{x^2 + 5}{(x + 1)(x^2 - 2x + 3)} = \frac{1}{x + 1} + \frac{2}{x^2 - 2x + 3}$$

39. $\dfrac{x^2 - x}{x^2 + x + 1} = 1 + \dfrac{-2x - 1}{x^2 + x + 1} = 1 - \dfrac{2x + 1}{x^2 + x + 1}$

41. $\dfrac{2x^3 - x^2 + x + 5}{x^2 + 3x + 2} = 2x - 7 + \dfrac{18x + 19}{(x + 1)(x + 2)}$

$$\frac{18x + 19}{(x + 1)(x + 2)} = \frac{A}{x + 1} + \frac{B}{x + 2}$$

$$18x + 19 = A(x + 2) + B(x + 1)$$

Let $x = -1$: $1 = A$

Let $x = -2$: $-17 = -B \implies B = 17$

$$\frac{2x^3 - x^2 + x + 5}{x^2 + 3x + 2} = 2x - 7 + \frac{1}{x + 1} + \frac{17}{x + 2}$$

43. $\dfrac{x^4}{(x-1)^3} = \dfrac{x^4}{x^3 - 3x^2 + 3x - 1} = x + 3 + \dfrac{6x^2 - 8x + 3}{(x-1)^3}$

$\dfrac{6x^2 - 8x + 3}{(x-1)^3} = \dfrac{A}{x-1} + \dfrac{B}{(x-1)^2} + \dfrac{C}{(x-1)^3}$

$6x^2 - 8x + 3 = A(x-1)^2 + B(x-1) + C$

Let $x = 1$: $1 = C$

Let $x = 0$: $3 = A - B + 1$ $\left.\begin{array}{l} \\ \\ \end{array}\right\}$ $A - B = 2$

Let $x = 2$: $11 = A + B + 1$ $A + B = 10$

So, $A = 6$ and $B = 4$.

$\dfrac{x^4}{(x-1)^3} = x + 3 + \dfrac{6}{x-1} + \dfrac{4}{(x-1)^2} + \dfrac{1}{(x-1)^3}$

45. $\dfrac{5-x}{2x^2 + x - 1} = \dfrac{A}{2x-1} + \dfrac{B}{x+1}$

$\qquad -x + 5 = A(x+1) + B(2x-1)$

Let $x = \dfrac{1}{2}$: $\dfrac{9}{2} = \dfrac{3}{2}A \implies A = 3$

Let $x = -1$: $6 = -3B \implies B = -2$

$\dfrac{5-x}{2x^2 + x - 1} = \dfrac{3}{2x-1} - \dfrac{2}{x+1}$

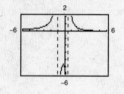

47. $\dfrac{x-1}{x^3 + x^2} = \dfrac{A}{x} + \dfrac{B}{x^2} + \dfrac{C}{x+1}$

$\qquad x - 1 = Ax(x+1) + B(x+1) + Cx^2$

Let $x = -1$: $-2 = C$

Let $x = 0$: $-1 = B$

Let $x = 1$: $0 = 2A + 2B + C$

$\qquad\qquad 0 = 2A - 2 - 2$

$\qquad\qquad 2 = A$

$\dfrac{x-1}{x^3 + x^2} = \dfrac{2}{x} - \dfrac{1}{x^2} - \dfrac{2}{x+1}$

49. $\dfrac{x^2 + x + 2}{(x^2 + 2)^2} = \dfrac{Ax + B}{x^2 + 2} + \dfrac{Cx + D}{(x^2 + 2)^2}$

$x^2 + x + 2 = (Ax + B)(x^2 + 2) + Cx + D$

$x^2 + x + 2 = Ax^3 + Bx^2 + (2A + C)x + (2B + D)$

Equating coefficients of like powers:

$\quad 0 = A$

$\quad 1 = B$

$\quad 1 = 2A + C \implies C = 1$

$\quad 2 = 2B + D \implies D = 0$

$\dfrac{x^2 + x + 2}{(x^2 + 2)^2} = \dfrac{1}{x^2 + 2} + \dfrac{x}{(x^2 + 2)^2}$

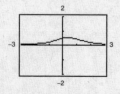

51. $\dfrac{2x^3 - 4x^2 - 15x + 5}{x^2 - 2x - 8} = 2x + \dfrac{x + 5}{(x+2)(x-4)}$

$\qquad \dfrac{x+5}{(x+2)(x-4)} = \dfrac{A}{x+2} + \dfrac{B}{x-4}$

$\qquad\qquad x + 5 = A(x-4) + B(x+2)$

Let $x = -2$: $3 = -6A \implies A = -\dfrac{1}{2}$

Let $x = 4$: $9 = 6B \implies B = \dfrac{3}{2}$

$\dfrac{2x^3 - 4x^2 - 15x + 5}{x^2 - 2x - 8} = 2x + \dfrac{1}{2}\left(\dfrac{3}{x-4} - \dfrac{1}{x+2}\right)$

53. (a) $\dfrac{x - 12}{x(x - 4)} = \dfrac{A}{x} + \dfrac{B}{x - 4}$

   $x - 12 = A(x - 4) + Bx$

   Let $x = 0$: $-12 = -4A \implies A = 3$

   Let $x = 4$: $-8 = 4B \implies B = -2$

   $\dfrac{x - 12}{x(x - 4)} = \dfrac{3}{x} - \dfrac{2}{x - 4}$

   (b) $y = \dfrac{x - 12}{x(x - 4)}$ $y = \dfrac{3}{x}$ $y = -\dfrac{2}{x - 4}$

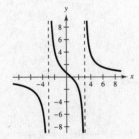

Vertical asymptotes: $x = 0$
and $x = 4$

 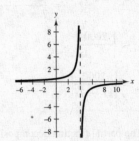

Vertical asymptote: $x = 0$

Vertical asymptote: $x = 4$

   (c) The combination of the vertical asymptotes of the terms of the decomposition are the same as the vertical asymptotes of the rational function.

55. (a) $\dfrac{2(4x - 3)}{x^2 - 9} = \dfrac{A}{x - 3} + \dfrac{B}{x + 3}$

   $2(4x - 3) = A(x + 3) + B(x - 3)$

   Let $x = 3$: $18 = 6A \implies A = 3$

   Let $x = -3$: $-30 = -6B \implies B = 5$

   $\dfrac{2(4x - 3)}{x^2 - 9} = \dfrac{3}{x - 3} + \dfrac{5}{x + 3}$

   (b) $y = \dfrac{2(4x - 3)}{x^2 - 9}$ $y = \dfrac{3}{x - 3}$ $y = \dfrac{5}{x + 3}$

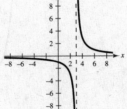 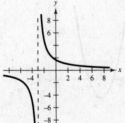

Vertical asymptotes: $x = \pm 3$ Vertical asymptote: $x = 3$ Vertical asymptote: $x = -3$

   (c) The combination of the vertical asymptotes of the terms of the decomposition are the same as the vertical asymptotes of the rational function.

57. (a) $\dfrac{2000(4 - 3x)}{(11 - 7x)(7 - 4x)} = \dfrac{A}{11 - 7x} + \dfrac{B}{7 - 4x}, \ 0 < x \le 1$

(c)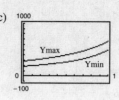

$$2000(4 - 3x) = A(7 - 4x) + B(11 - 7x)$$

Let $x = \dfrac{11}{7}$: $-\dfrac{10{,}000}{7} = \dfrac{5}{7}A \implies A = -2000$

Let $x = \dfrac{7}{4}$: $-2500 = -\dfrac{5}{4}B \implies B = 2000$

$$\dfrac{2000(4 - 3x)}{(11 - 7x)(7 - 4x)} = \dfrac{-2000}{11 - 7x} + \dfrac{2000}{7 - 4x}$$

$$= \dfrac{2000}{7 - 4x} - \dfrac{2000}{11 - 7x}, \ 0 < x \le 1$$

(b) $Y_{\max} = \left| \dfrac{2000}{7 - 4x} \right|$

$ Y_{\min} = \left| \dfrac{2000}{11 - 7x} \right|$

59. False. The partial fraction decomposition is $\dfrac{A}{x + 10} + \dfrac{B}{x - 10} + \dfrac{C}{(x - 10)^2}$.

61. $\dfrac{1}{a^2 - x^2} = \dfrac{A}{a + x} + \dfrac{B}{a - x}$, a is a constant.

$$1 = A(a - x) + B(a + x)$$

Let $x = -a$: $1 = 2aA \implies A = \dfrac{1}{2a}$

Let $x = a$: $1 = 2aB \implies B = \dfrac{1}{2a}$

$$\dfrac{1}{a^2 - x^2} = \dfrac{1}{2a}\left(\dfrac{1}{a + x} + \dfrac{1}{a - x} \right)$$

63. $\dfrac{1}{y(a - y)} = \dfrac{A}{y} + \dfrac{B}{a - y}$

$$1 = A(a - y) + By$$

Let $y = 0$: $1 = aA \implies A = \dfrac{1}{a}$

Let $y = a$: $1 = aB \implies B = \dfrac{1}{a}$

$$\dfrac{1}{y(a - y)} = \dfrac{1}{a}\left(\dfrac{1}{y} + \dfrac{1}{a - y} \right)$$

65. $f(x) = x^2 - 9x + 18 = (x - 6)(x - 3)$

Intercepts: $(0, 18), (3, 0), (6, 0)$

Graph rises to the left and rises to the right.

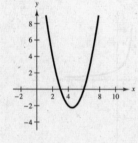

67. $f(x) = -x^2(x - 3)$

Intercepts: $(0, 0), (3, 0)$

Graph rises to the left and falls to the right.

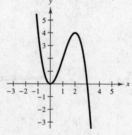

69. $f(x) = \dfrac{x^2 + x - 6}{x + 5}$

x-intercepts: $(-3, 0), (2, 0)$

y-intercept: $\left(0, -\dfrac{6}{5}\right)$

Vertical asymptote: $x = -5$

Slant asymptote: $y = x - 4$

No horizontal asymptote.

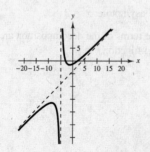

Section 7.5   Systems of Inequalities

■ You should be able to sketch the graph of an inequality in two variables.

    (a) Replace the inequality with an equal sign and graph the equation. Use a dashed line for < or >, a solid line for ≤ or ≥.

    (b) Test a point in each region formed by the graph. If the point satisfies the inequality, shade the whole region.

■ You should be able to sketch systems of inequalities.

Vocabulary Check

1. solution

2. graph

3. linear

4. solution

5. consumer surplus

---

1. $y < 2 - x^2$

Using a dashed line, graph $y = 2 - x^2$ and shade inside the parabola.

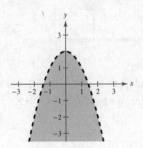

3. $x \geq 2$

Using a solid line, graph the vertical line $x = 2$ and shade to the right of this line.

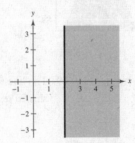

5. $y \geq -1$

Using a solid line, graph the horizontal line $y = -1$ and shade above this line.

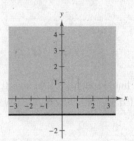

7. $y < 2 - x$

Using a dashed line, graph $y = 2 - x$, and then shade below the line. (Use $(0, 0)$ as a test point.)

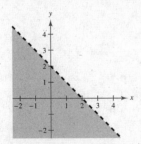

9. $2y - x \geq 4$

Using a solid line, graph $2y - x = 4$, and then shade above the line. (Use $(0, 0)$ as a test point.)

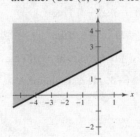

11. $(x + 1)^2 + (y - 2)^2 < 9$

Using a dashed line, sketch the circle $(x + 1)^2 + (y - 2)^2 = 9$.

Center: $(-1, 2)$

Radius: 3

Test point: $(0, 0)$

Shade the inside of the circle.

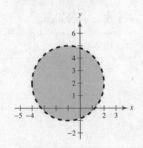

13. $y \leq \dfrac{1}{1 + x^2}$

Using a solid line, graph

$y = \dfrac{1}{1 + x^2}$, and then shade below the curve. (Use $(0, 0)$ as a test point.)

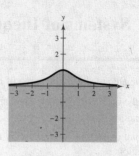

15. $y < \ln x$

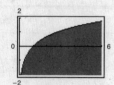

17. $y < 3^{-x-4}$

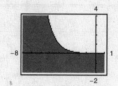

19. $y \geq \frac{2}{3}x - 1$

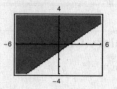

21. $y < -3.8x + 1.1$

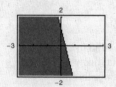

23. $x^2 + 5y - 10 \leq 0$

$$y \leq 2 - \dfrac{x^2}{5}$$

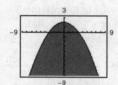

25. $\dfrac{5}{2}y - 3x^2 - 6 \geq 0$

$$y \geq \dfrac{2}{5}(3x^2 + 6)$$

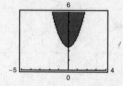

27. The line through $(-4, 0)$ and $(0, 2)$ is $y = \frac{1}{2}x + 2$. For the shaded region below the line, we have $y \leq \frac{1}{2}x + 2$.

29. The line through $(0, 2)$ and $(3, 0)$ is $y = -\frac{2}{3}x + 2$. For the shaded region above the line, we have $y \geq -\frac{2}{3}x + 2$.

31. $\begin{cases} x \geq -4 \\ y > -3 \\ y \leq -8x - 3 \end{cases}$

(a) $0 \leq -8(0) - 3$, False

    $(0, 0)$ *is not* a solution.

(b) $-3 > -3$, False

    $(-1, -3)$ *is not* a solution.

(c) $-4 \geq -4$, True

    $0 > -3$, True

    $0 \leq -8(-4) - 3$, True

    $(-4, 0)$ *is* a solution.

(d) $-3 \geq -4$, True

    $11 > -3$, True

    $11 < -8(-3) - 3$, True

    $(-3, 11)$ *is* a solution.

33. $\begin{cases} 3x + y > 1 \\ -y - \frac{1}{2}x^2 \le -4 \\ -15x + 4y > 0 \end{cases}$

(a) $3(0) + (10) > 1$, True

$-10 - \frac{1}{2}(0)^2 \le -4$, True

$-15(0) + 4(10) > 0$, True

$(0, 10)$ *is* a solution.

(b) $3(0) + (-1) > 1$, False $\implies (0, -1)$ *is not* a solution.

(c) $3(2) + (9) > 1$, True

$-9 - \frac{1}{2}(2)^2 \le -4$, True

$-15(2) + 4(9) > 0$, True

$(2, 9)$ *is* a solution.

(d) $3(-1) + 6 > 1$, True

$-6 - \frac{1}{2}(-1)^2 \le -4$, True

$-15(-1) + 4(6) > 0$, True

$(-1, 6)$ *is* a solution.

35. $\begin{cases} x + y \le 1 \\ -x + y \le 1 \\ y \ge 0 \end{cases}$

First, find the points of intersection of each pair of equations.

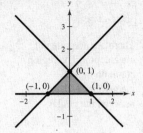

Vertex A	Vertex B	Vertex C
$x + y = 1$	$x + y = 1$	$-x + y = 1$
$-x + y = 1$	$y = 0$	$y = 0$
$(0, 1)$	$(1, 0)$	$(-1, 0)$

37. $\begin{cases} x^2 + y \le 5 \\ x \ge -1 \\ y \ge 0 \end{cases}$

First, find the points of intersection of each pair of equations.

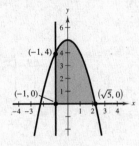

Vertex A	Vertex B	Vertex C
$x^2 + y = 5$	$x^2 + y = 5$	$x = -1$
$x = -1$	$y = 0$	$y = 0$
$(-1, 4)$	$(\pm\sqrt{5}, 0)$	$(-1, 0)$

39. $\begin{cases} 2x + y > 2 \\ 6x + 3y < 2 \end{cases}$

The graphs of $2x + y = 2$ and $6x + 3y = 2$ are parallel lines. The first inequality has the region above the line shaded. The second inequality has the region below the line shaded. There are no points that satisfy both inequalities.

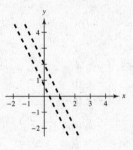

No solution

41. $\begin{cases} -3x + 2y < 6 \\ x - 4y > -2 \\ 2x + y < 3 \end{cases}$

First, find the points of intersection of each pair of equations.

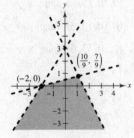

Vertex A	Point B	Vertex C
$-3x + 2y = 6$	$-3x + 2y = 6$	$x - 4y = -2$
$x - 4y = -2$	$2x + y = 3$	$2x + y = 3$
$(-2, 0)$	$(0, 3)$	$\left(\frac{10}{9}, \frac{7}{9}\right)$

Note that B is not a vertex of the solution region.

43. $\begin{cases} x > y^2 \\ x < y + 2 \end{cases}$

Points of intersection:

$$y^2 = y + 2$$
$$y^2 - y - 2 = 0$$
$$(y + 1)(y - 2) = 0$$
$$y = -1, 2$$

$(1, -1), (4, 2)$

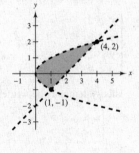

45. $\begin{cases} x^2 + y^2 \le 9 \\ x^2 + y^2 \ge 1 \end{cases}$

There are no points of intersection. The region common to both inequalities is the region between the circles.

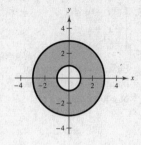

47. $3x + 4 \ge y^2$

$\quad x - y < 0$

Points of intersection:

$x - y = 0 \Longrightarrow y = x$

$3y + 4 = y^2$

$\quad 0 = y^2 - 3y - 4$

$\quad 0 = (y - 4)(y + 1)$

$\quad y = 4 \text{ or } y = -1$

$\quad x = 4 \qquad x = -1$

$(4, 4)$ and $(-1, -1)$

49. $\begin{cases} y \le \sqrt{3x} + 1 \\ y \ge x^2 + 1 \end{cases}$

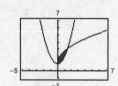

51. $\begin{cases} y < x^3 - 2x + 1 \\ y > -2x \\ x \le 1 \end{cases}$

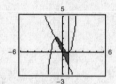

53. $\begin{cases} x^2 y \ge 1 \Longrightarrow y \ge \dfrac{1}{x^2} \\ 0 < x \le 4 \\ \quad y \le 4 \end{cases}$

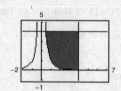

55. $\begin{cases} y \le 4 - x \\ x \ge 0 \\ y \ge 0 \end{cases}$

57. Line through points $(0, 4)$ and $(4, 0)$: $y = 4 - x$

Line through points $(0, 2)$ and $(8, 0)$: $y = 2 - \frac{1}{4}x$

$\begin{cases} y \ge 4 - x \\ y \ge 2 - \frac{1}{4}x \\ x \ge 0 \\ y \ge 0 \end{cases}$

59. $\begin{cases} x^2 + y^2 \le 16 \\ x \ge 0 \\ y \ge 0 \end{cases}$

61. Rectangular region with vertices at $(2, 1)$, $(5, 1)$, $(5, 7)$, and $(2, 7)$

$\begin{cases} x \ge 2 \\ x \le 5 \\ y \ge 1 \\ y \le 7 \end{cases}$

This system may be written as:

$\begin{cases} 2 \le x \le 5 \\ 1 \le y \le 7 \end{cases}$

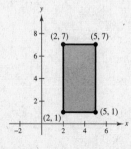

63. Triangle with vertices at $(0, 0)$, $(5, 0)$, $(2, 3)$

$(0, 0), (5, 0)$ Line: $y = 0$

$(0, 0), (2, 3)$ Line: $y = \frac{3}{2}x$

$(2, 3), (5, 0)$ Line: $y = -x + 5$

$\begin{cases} y \le \frac{3}{2}x \\ y \le -x + 5 \\ y \ge 0 \end{cases}$

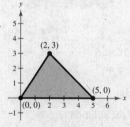

65. (a)   Demand = Supply

$$50 - 0.5x = 0.125x$$

$$50 = 0.625x$$

$$80 = x$$

$$10 = p$$

Point of equilibrium: (80, 10)

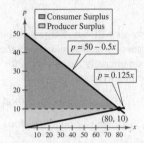

(b) The consumer surplus is the area of the triangular region defined by

$$\begin{cases} p \le 50 - 0.5x \\ p \ge 10 \\ x \ge \quad 0. \end{cases}$$

Consumer surplus = $\frac{1}{2}$(base)(height)

$$= \tfrac{1}{2}(80)(40)$$

$$= \$1600$$

The producer surplus is the area of the triangular region defined by

$$\begin{cases} p \ge 0.125x \\ p \le 10 \\ x \ge 0. \end{cases}$$

Producer surplus = $\frac{1}{2}$(base)(height)

$$= \tfrac{1}{2}(80)(10)$$

$$= \$400$$

67. (a)   Demand = Supply

$$140 - 0.00002x = 80 + 0.00001x$$

$$60 = 0.00003x$$

$$2{,}000{,}000 = x$$

$$100 = p$$

Point of equilibrium: (2,000,000, 100)

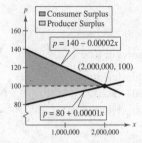

(b) The consumer surplus is the area of the triangular region defined by

$$\begin{cases} p \le 140 - 0.00002x \\ p \ge 100 \\ x \ge \quad 0. \end{cases}$$

Consumer surplus = $\frac{1}{2}$(base)(height)

$$= \tfrac{1}{2}(2{,}000{,}000)(40)$$

$$= \$40{,}000{,}000 \text{ or } \$40 \text{ million}$$

The producer surplus is the area of the triangular region defined by

$$\begin{cases} p \ge \quad 80 + 0.00001x \\ p \le 100 \\ x \ge \quad 0. \end{cases}$$

Producer surplus = $\frac{1}{2}$(base)(height)

$$= \tfrac{1}{2}(2{,}000{,}000)(20)$$

$$= \$20{,}000{,}000 \text{ or } \$20 \text{ million}$$

69. x = number of tables

y = number of chairs

$$\begin{cases} x + \tfrac{3}{2}y \le 12 & \text{Assembly center} \\ \tfrac{4}{3}x + \tfrac{3}{2}y \le 15 & \text{Finishing center} \\ \quad x \ge \ 0 \\ \quad y \ge \ 0 \end{cases}$$

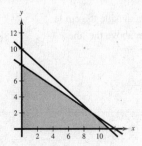

71. x = amount in smaller account

y = amount in larger account

Account constraints:

$$\begin{cases} x + y \le 20{,}000 \\ \quad\ y \ge \quad 2x \\ \quad\ x \ge \ 5{,}000 \\ \quad\ y \ge \ 5{,}000 \end{cases}$$

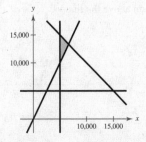

73. x = number of packages of gravel

  y = number of bags of stone

$$\begin{cases} 55x + 70y \le 7500 \quad \text{Weight} \\ \qquad x \ge \quad 50 \\ \qquad y \ge \quad 40 \end{cases}$$

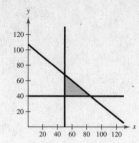

75. (a) x = number of ounces of food X

  y = number of ounces of food Y

$$\begin{cases} 20x + 10y \ge 300 \quad \text{(calcium)} \\ 15x + 10y \ge 150 \quad \text{(iron)} \\ 10x + 20y \ge 200 \quad \text{(vitamin B)} \\ \qquad x \ge 0 \\ \qquad y \ge 0 \end{cases}$$

(b)

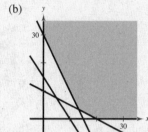

(c) Answers will vary. Some possible solutions which would satisfy the minimum daily requirements for calcium, iron, and vitamin B:

  $(0, 30) \implies 30$ ounces of food Y

  $(20, 0) \implies 20$ ounces of food X

  $\left(13\frac{1}{3}, 3\frac{1}{3}\right) \implies 13\frac{1}{3}$ ounces of food X and $3\frac{1}{3}$ ounces of food Y

77. (a) $(9, 125.8)$, $(10, 145.6)$, $(11, 164.1)$, $(12, 182.7)$, $(13, 203.1)$

  Linear model:
  $y = 19.17t - 46.61$

(b) $$\begin{cases} y \le 19.17t - 46.61 \\ t \ge 8.5 \\ t \le 13.5 \\ y \ge 0 \end{cases}$$

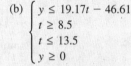

(c) Area of a trapezoid: $A = \dfrac{h}{2}(a + b)$

  $h = 13.5 - 8.5 = 5$

  $a = 19.17(8.5) - 46.61 = 116.335$

  $b = 19.17(13.5) - 46.61 = 212.185$

  $A = \dfrac{5}{2}(116.335 + 212.185)$

  $= \$821.3$ billion

79. True. The figure is a rectangle with length of 9 units and width of 11 units.

81. The graph is a half-line on the real number line; on the rectangular coordinate system, the graph is a half-plane.

83. x = radius of smaller circle

  y = radius of larger circle

(a) Constraints on circles:

$$\begin{aligned} \pi y^2 - \pi x^2 &\ge 10 \\ y &> x \\ x &> 0 \end{aligned}$$

(b)

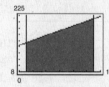

(c) The line is an asymptote to the boundary. The larger the circles, the closer the radii can be and the constraint still be satisfied.

85. $\begin{cases} x^2 + y^2 \le 16 \implies \text{region inside the circle} \\ x + y \ge 4 \implies \text{region above the line} \end{cases}$

  Matches graph (d).

87. $\begin{cases} x^2 + y^2 \ge 16 \implies \text{region outside the circle} \\ x + y \ge 4 \implies \text{region above the line} \end{cases}$

  Matches graph (c).

89. $(-2, 6), (4, -4)$

$$m = \frac{-4 - 6}{4 - (-2)} = \frac{-10}{6} = -\frac{5}{3}$$

$$y - (-4) = -\frac{5}{3}(x - 4)$$

$$3y + 12 = -5x + 20$$

$$5x + 3y - 8 = 0$$

91. $\left(\frac{3}{4}, -2\right), \left(-\frac{7}{2}, 5\right)$

$$m = \frac{5 - (-2)}{-\frac{7}{2} - \frac{3}{4}} = \frac{7}{-\frac{17}{4}} = -\frac{28}{17}$$

$$y - (-2) = -\frac{28}{17}\left(x - \frac{3}{4}\right)$$

$$17y + 34 = -28x + 21$$

$$28x + 17y + 13 = 0$$

93. $(3.4, -5.2), (-2.6, 0.8)$

$$m = \frac{0.8 - (-5.2)}{-2.6 - 3.4} = \frac{6}{-6} = -1$$

$$y - 0.8 = -1(x - (-2.6))$$

$$y - 0.8 = -x - 2.6$$

$$x + y + 1.8 = 0$$

95. (a) $(8, 39.43), (9, 41.24), (10, 45.27), (11, 47.37), (12, 48.40), (13, 49.91)$

Linear model: $y \approx 2.17t + 22.5$

Quadratic model: $y \approx -0.241t^2 + 7.23t - 3.4$

Exponential model: $y \approx 27(1.05^t)$

(b)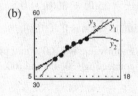

(c) The quadratic model is the best fit for the actual data.

(d) For 2008, use $t = 18$: $y \approx -0.241(18)^2 + 7.23(18) - 3.4 \approx \$48.66$

Section 7.6   Linear Programming

■ To solve a linear programming problem:
1. Sketch the solution set for the system of constraints.
2. Find the vertices of the region.
3. Test the objective function at each of the vertices.

Vocabulary Check

1. optimization

3. objective

5. vertex

2. linear programming

4. constraints; feasible solutions

1. $z = 4x + 3y$

At $(0, 5)$: $z = 4(0) + 3(5) = 15$

At $(0, 0)$: $z = 4(0) + 3(0) = 0$

At $(5, 0)$: $z = 4(5) + 3(0) = 20$

The minimum value is 0 at $(0, 0)$.

The maximum value is 20 at $(5, 0)$.

3. $z = 3x + 8y$

At $(0, 5)$: $z = 3(0) + 8(5) = 40$

At $(0, 0)$: $z = 3(0) + 8(0) = 0$

At $(5, 0)$: $z = 3(5) + 8(0) = 15$

The minimum value is 0 at $(0, 0)$.

The maximum value is 40 at $(0, 5)$.

5. $z = 3x + 2y$

At $(0, 5)$: $z = 3(0) + 2(5) = 10$

At $(4, 0)$: $z = 3(4) + 2(0) = 12$

At $(3, 4)$: $z = 3(3) + 2(4) = 17$

At $(0, 0)$: $z = 3(0) + 2(0) = 0$

The minimum value is 0 at $(0, 0)$.

The maximum value is 17 at $(3, 4)$.

7. $z = 5x + 0.5y$

At $(0, 5)$: $z = 5(0) + \frac{5}{2} = \frac{5}{2}$

At $(4, 0)$: $z = 5(4) + \frac{0}{2} = 20$

At $(3, 4)$: $z = 5(3) + \frac{4}{2} = 17$

At $(0, 0)$: $z = 5(0) + \frac{0}{2} = 0$

The minimum value is 0 at $(0, 0)$.

The maximum value is 20 at $(4, 0)$.

9. $z = 10x + 7y$

At $(0, 45)$: $z = 10(0) + 7(45) = 315$

At $(30, 45)$: $z = 10(30) + 7(45) = 615$

At $(60, 20)$: $z = 10(60) + 7(20) = 740$

At $(60, 0)$: $z = 10(60) + 7(0) = 600$

At $(0, 0)$: $z = 10(0) + 7(0) = 0$

The minimum value is 0 at $(0, 0)$.

The maximum value is 740 at $(60, 20)$.

11. $z = 25x + 30y$

At $(0, 45)$: $z = 25(0) + 30(45) = 1350$

At $(30, 45)$: $z = 25(30) + 30(45) = 2100$

At $(60, 20)$: $z = 25(60) + 30(20) = 2100$

At $(60, 0)$: $z = 25(60) + 30(0) = 1500$

At $(0, 0)$: $z = 25(0) + 30(0) = 0$

The minimum value is 0 at $(0, 0)$.

The maximum value is 2100 at any point along the line segment connecting $(30, 45)$ and $(60, 20)$.

13. $z = 6x + 10y$

At $(0, 2)$: $z = 6(0) + 10(2) = 20$

At $(5, 0)$: $z = 6(5) + 10(0) = 30$

At $(0, 0)$: $z = 6(0) + 10(0) = 0$

The minimum value is 0 at $(0, 0)$.

The maximum value is 30 at $(5, 0)$.

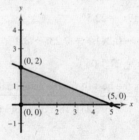

15. $z = 9x + 24y$

At $(0, 2)$: $z = 9(0) + 24(2) = 48$

At $(5, 0)$: $z = 9(5) + 24(0) = 45$

At $(0, 0)$: $z = 9(0) + 24(0) = 0$

The minimum value is 0 at $(0, 0)$.

The maximum value is 48 at $(0, 2)$.

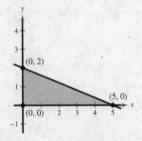

17. $z = 4x + 5y$

At $(10, 0)$: $z = 4(10) + 5(0) = 40$

At $(5, 3)$: $z = 4(5) + 5(3) = 35$

At $(0, 8)$: $z = 4(0) + 5(8) = 40$

The minimum value is 35 at $(5, 3)$.

The region is unbounded. There is no maximum.

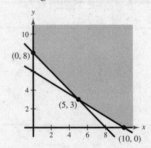

19. $z = 2x + 7y$

At $(10, 0)$: $z = 2(10) + 7(0) = 20$

At $(5, 3)$: $z = 2(5) + 7(3) = 31$

At $(0, 8)$: $z = 2(0) + 7(8) = 56$

The minimum value is 20 at $(10, 0)$.

The region is unbounded. There is no maximum.

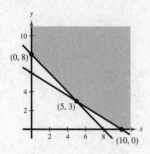

21. $z = 4x + y$

At $(36, 0)$: $z = 4(36) + 0 = 144$

At $(40, 0)$: $z = 4(40) + 0 = 160$

At $(24, 8)$: $z = 4(24) + 8 = 104$

The minimum value is 104 at $(24, 8)$.

The maximum value is 160 at $(40, 0)$.

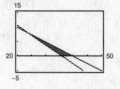

23. $z = x + 4y$

At $(36, 0)$: $z = 36 + 4(0) = 36$

At $(40, 0)$: $z = 40 + 4(0) = 40$

At $(24, 8)$: $z = 24 + 4(8) = 56$

The minimum value is 36 at $(36, 0)$.

The maximum value is 56 at $(24, 8)$.

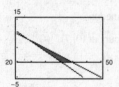

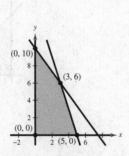

Figure for Exercises 25 and 27

25. $z = 2x + y$

At $(0, 10)$: $z = 2(0) + (10) = 10$

At $(3, 6)$: $z = 2(3) + (6) = 12$

At $(5, 0)$: $z = 2(5) + (0) = 10$

At $(0, 0)$: $z = 2(0) + (0) = 0$

The maximum value is 12 at $(3, 6)$.

27. $z = x + y$

At $(0, 10)$: $z = (0) + (10) = 10$

At $(3, 6)$: $z = (3) + (6) = 9$

At $(5, 0)$: $z = (5) + (0) = 5$

At $(0, 0)$: $z = (0) + (0) = 0$

The maximum value is 10 at $(0, 10)$.

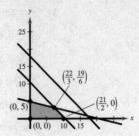

Figure for Exercises 29 and 31

29. $z = x + 5y$

At $(0, 5)$: $z = 0 + 5(5) = 25$

At $\left(\frac{22}{3}, \frac{19}{6}\right)$: $z = \frac{22}{3} + 5\left(\frac{19}{6}\right) = \frac{139}{6}$

At $\left(\frac{21}{2}, 0\right)$: $z = \frac{21}{2} + 5(0) = \frac{21}{2}$

At $(0, 0)$: $z = 0 + 5(0) = 0$

The maximum value is 25 at $(0, 5)$.

31. $z = 4x + 5y$

At $(0, 5)$: $z = 4(0) + 5(5) = 25$

At $\left(\frac{22}{3}, \frac{19}{6}\right)$: $z = 4\left(\frac{22}{3}\right) + 5\left(\frac{19}{6}\right) = \frac{271}{6}$

At $\left(\frac{21}{2}, 0\right)$: $z = 4\left(\frac{21}{2}\right) + 5(0) = 42$

At $(0, 0)$: $z = 4(0) + 5(0) = 0$

The maximum value is $\frac{271}{6}$ at $\left(\frac{22}{3}, \frac{19}{6}\right)$.

33. Objective function: $z = 2.5x + y$

Constraints: $x \geq 0, y \geq 0, 3x + 5y \leq 15, 5x + 2y \leq 10$

At $(0, 0)$: $z = 0$

At $(2, 0)$: $z = 5$

At $\left(\frac{20}{19}, \frac{45}{19}\right)$: $z = \frac{95}{19} = 5$

At $(0, 3)$: $z = 3$

The maximum value of 5 occurs at any point on the line segment connecting $(2, 0)$ and $\left(\frac{20}{19}, \frac{45}{19}\right)$.

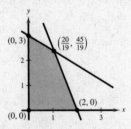

35. Objective function: $z = -x + 2y$

Constraints: $x \geq 0, y \geq 0, x \leq 10, x + y \leq 7$

At $(0, 0)$: $z = -0 + 2(0) = 0$

At $(0, 7)$: $z = -0 + 2(7) = 14$

At $(7, 0)$: $z = -7 + 2(0) = -7$

The constraint $x \leq 10$ is extraneous.

The maximum value of 14 occurs at $(0, 7)$.

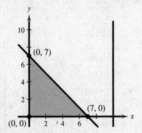

37. Objective function: $z = 3x + 4y$

Constraints: $x \geq 0, y \geq 0, x + y \leq 1, 2x + y \leq 4$

At $(0, 0)$: $z = 3(0) + 4(0) = 0$

At $(0, 1)$: $z = 3(0) + 4(1) = 4$

At $(1, 0)$: $z = 3(1) + 4(0) = 3$

The constraint $2x + y \leq 4$ is extraneous.

The maximum value of 4 occurs at $(0, 1)$.

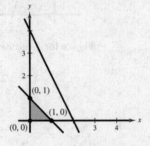

39. x = number of Model A

y = number of Model B

Constraints: $2x + 2.5y \leq 4000$
$4x + y \leq 4800$
$x + 0.75y \leq 1500$
$x \geq 0$
$y \geq 0$

Objective function: $P = 45x + 50y$

Vertices:

$(0, 0), (0, 1600), (750, 1000), (1050, 600), (1200, 0)$

At $(0, 0)$: $P = 45(0) + 50(0) = 0$

At $(0, 1600)$: $P = 45(0) + 50(1600) = 80{,}000$

At $(750, 1000)$: $P = 45(750) + 50(1000) = 83{,}750$

At $(1050, 600)$: $P = 45(1050) + 50(600) = 77{,}250$

At $(1200, 0)$: $P = 45(1200) + 50(0) = 54{,}000$

The optimal profit of \$83,750 occurs when 750 units of Model A and 1000 units of Model B are produced.

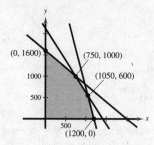

41. x = number of \$250 models

y = number of \$300 models

Constraints:
$$250x + 300y \leq 65{,}000$$
$$x + y \leq 250$$
$$x \geq 0$$
$$y \geq 0$$

Objective function: $P = 25x + 40y$

Vertices: $(0, 0), (250, 0), (200, 50), \left(0, 216\frac{2}{3}\right)$

At $(0, 0)$: $P = 25(0) + 40(0) = 0$

At $(250, 0)$: $P = 25(250) + 40(0) = 6250$

At $(200, 50)$: $P = 25(200) + 40(50) = 7000$

At $\left(0, 216\frac{2}{3}\right)$: $P = 25(0) + 40\left(216\frac{2}{3}\right) \approx 8666.67$

An optimal profit of \$8640 occurs when 0 units of the \$250 model and 216 units of the \$300 model are stocked in inventory. $\left(\text{Note: A merchant cannot sell } \frac{2}{3} \text{ of a unit.}\right)$

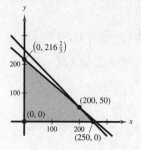

43. x = number of bags of Brand X

y = number of bags of Brand Y

Constraints:

$2x + y \geq 12$
$2x + 9y \geq 36$
$2x + 3y \geq 24$
$x \geq 0$
$y \geq 0$

Objective function:
$C = 25x + 20y$

Vertices: $(0, 12), (3, 6), (9, 2), (18, 0)$

At $(0, 12)$: $C = 25(0) + 20(12) = 240$

At $(3, 6)$: $C = 25(3) + 20(6) = 195$

At $(9, 2)$: $C = 25(9) + 20(2) = 265$

At $(18, 0)$: $C = 25(18) + 20(0) = 450$

To optimize cost, use three bags of Brand X and six bags of Brand Y for an optimal cost of \$195.

45. x = number of audits

y = number of tax returns

Constraints:
$75x + 12.5y \leq 900$
$10x + 2.5y \leq 155$
$x \geq 0$
$y \geq 0$

Objective function:
$R = 2500x + 350y$

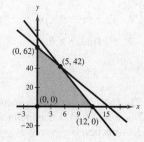

Vertices: $(0, 0), (12, 0), (5, 42), (0, 62)$

At $(0, 0)$: $R = 2500(0) + 350(0) = 0$

At $(12, 0)$: $R = 2500(12) + 350(0) = 30{,}000$

At $(5, 42)$: $R = 2500(5) + 350(42) = 27{,}200$

At $(0, 62)$: $R = 2500(0) + 350(62) = 21{,}700$

The revenue will be optimal if 12 audits and 0 tax returns are done each week. The optimal revenue is \$30,000.

47. x = amount of Type A

y = amount of Type B

Constraints: $x + y \leq 250{,}000$

$\qquad\qquad x \geq \frac{1}{4}(250{,}000)$

$\qquad\qquad y \geq \frac{1}{4}(250{,}000)$

Objective Function: $P = 0.08x + 0.10y$

Vertices: $(62{,}500, 62{,}500)$, $(62{,}500, 187{,}500)$, $(187{,}500, 62{,}500)$

At $(62{,}500, 62{,}500)$: $P = 0.08(62{,}500) + 0.10(62{,}500) = \$11{,}250$

At $(62{,}500, 187{,}500)$: $P = 0.08(62{,}500) + 0.10(187{,}500) = \$23{,}750$

At $(187{,}500, 62{,}500)$: $P = 0.08(187{,}500) + 0.10(62{,}500) = \$21{,}250$

To obtain an optimal return the investor should allocate \$62,500 to Type A and \$187,500 to Type B. The optimal return is \$23,750.

49. True. The objective function has a maximum value at any point on the line segment connecting the two vertices. Both of these points are on the line $y = -x + 11$ and lie between $(4, 7)$ and $(8, 3)$.

51. Constraints: $x \geq 0$, $y \geq 0$, $x + 3y \leq 15$, $4x + y \leq 16$

Vertex	Value of $z = 3x + ty$
$(0, 0)$	$z = 0$
$(0, 5)$	$z = 5t$
$(3, 4)$	$z = 9 + 4t$
$(4, 0)$	$z = 12$

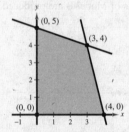

(a) For the maximum value to be at $(0, 5)$, $z = 5t$ must be greater than or equal to $z = 9 + 4t$ and $z = 12$.

$\qquad 5t \geq 9 + 4t$ and $5t \geq 12$

$\qquad\quad t \geq 9 \qquad\qquad t \geq \frac{12}{5}$

Thus, $t \geq 9$.

(b) For the maximum value to be at $(3, 4)$, $z = 9 + 4t$ must be greater than or equal to $z = 5t$ and $z = 12$.

$\qquad 9 + 4t \geq 5t$ and $9 + 4t \geq 12$

$\qquad\quad 9 \geq t \qquad\qquad 4t \geq 3$

$\qquad\qquad\qquad\qquad\qquad t \geq \frac{3}{4}$

Thus, $\frac{3}{4} \leq t \leq 9$.

53. There are an infinite number of objective functions that would have a maximum at $(0, 4)$. One such objective function is $z = x + 5y$.

55. There are an infinite number of objective functions that would have a maximum at $(5, 0)$. One such objective function is $z = 4x + y$.

57. $\dfrac{\dfrac{9}{x}}{\left(\dfrac{6}{x} + 2\right)} = \dfrac{\dfrac{9}{x}}{\dfrac{6 + 2x}{x}} = \dfrac{9}{x} \cdot \dfrac{x}{2(3 + x)} = \dfrac{9}{2(3 + x)} = \dfrac{9}{2(x + 3)}$, $x \neq 0$

59. $\dfrac{\left(\dfrac{4}{x^2-9}+\dfrac{2}{x-2}\right)}{\left(\dfrac{1}{x+3}+\dfrac{1}{x-3}\right)} = \dfrac{\dfrac{4(x-2)+2(x^2-9)}{(x-2)(x^2-9)}}{\dfrac{(x-3)+(x+3)}{x^2-9}}$

$\qquad = \dfrac{2x^2+4x-26}{(x-2)(x^2-9)} \cdot \dfrac{x^2-9}{2x}$

$\qquad = \dfrac{2(x^2+2x-13)}{(x-2)(2x)}$

$\qquad = \dfrac{x^2+2x-13}{x(x-2)}, \quad x \neq \pm 3$

61. $e^{2x} + 2e^x - 15 = 0$

$\quad (e^x+5)(e^x-3) = 0$

$\qquad e^x = -5 \ \text{or} \ e^x = 3$

$\begin{array}{ll} \text{No real} & x = \ln 3 \\ \text{solution.} & x \approx 1.099 \end{array}$

63. $8(62 - e^{x/4}) = 192$

$\qquad 62 - e^{x/4} = 24$

$\qquad -e^{x/4} = -38$

$\qquad e^{x/4} = 38$

$\qquad \dfrac{x}{4} = \ln 38$

$\qquad x = 4 \ln 38$

$\qquad x \approx 14.550$

65. $7 \ln 3x = 12$

$\qquad \ln 3x = \dfrac{12}{7}$

$\qquad 3x = e^{12/7}$

$\qquad x = \dfrac{e^{12/7}}{3}$

$\qquad x \approx 1.851$

67. $\begin{cases} -x - 2y + 3z = -23 \\ 2x + 6y - z = 17 \\ 5y + z = 8 \end{cases}$

$\begin{cases} -x - 2y + 3z = -23 \\ 2y + 5z = -29 \\ 5y + z = 8 \end{cases} \qquad 2\text{Eq.1} + \text{Eq.2}$

$\begin{cases} -x - 2y + 3z = -23 \\ 2y + 5z = -29 \\ -\frac{23}{2}z = \frac{161}{2} \end{cases} \qquad -\frac{5}{2}\text{Eq.2} + \text{Eq.3}$

$-\frac{23}{2}z = \frac{161}{2} \implies z = -7$

$2y + 5(-7) = -29 \implies y = 3$

$-x - 2(3) + 3(-7) = -23 \implies -x - 27 = -23$

$\qquad\qquad\qquad\qquad\qquad \implies x = -4$

Solution: $(-4, 3, -7)$

Review Exercises for Chapter 7

1. $\begin{cases} x + y = 2 \\ x - y = 0 \implies x = y \end{cases}$

$\quad x + x = 2$

$\quad\ \ 2x = 2$

$\quad\ \ \ x = 1$

$\quad\ \ \ y = 1$

Solution: $(1, 1)$

3. $\begin{cases} 0.5x + y = 0.75 \implies y = 0.75 - 0.5x \\ 1.25x - 4.5y = -2.5 \end{cases}$

$1.25x - 4.5(0.75 - 0.5x) = -2.5$

$1.25x - 3.375 + 2.25x = -2.5$

$\qquad\qquad\qquad 3.50x = 0.875$

$\qquad\qquad\qquad\quad x = 0.25$

$\qquad\qquad\qquad\quad y = 0.625$

Solution: $(0.25, 0.625)$

5. $\begin{cases} x^2 - y^2 = 9 \\ x - y = 1 \implies x = y + 1 \end{cases}$

$(y + 1)^2 - y^2 = 9$

$2y + 1 = 9$

$y = 4$

$x = 5$

Solution: $(5, 4)$

7. $\begin{cases} y = 2x^2 \\ y = x^4 - 2x^2 \implies 2x^2 = x^4 - 2x^2 \end{cases}$

$0 = x^4 - 4x^2$

$0 = x^2(x^2 - 4)$

$0 = x^2(x + 2)(x - 2)$

$x = 0, x = -2, x = 2$

$y = 0, y = 8, y = 8$

Solutions: $(0, 0), (-2, 8), (2, 8)$

9. $\begin{cases} 2x - y = 10 \\ x + 5y = -6 \end{cases}$

Point of intersection: $(4, -2)$

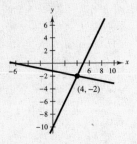

11. $\begin{cases} y = 2x^2 - 4x + 1 \\ y = x^2 - 4x + 3 \end{cases}$

Point of intersection: $(1.41, -0.66), (-1.41, 10.66)$

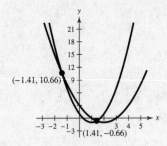

13. $\begin{cases} y = -2e^{-x} \\ 2e^x + y = 0 \implies y = -2e^x \end{cases}$

Point of intersection: $(0, -2)$

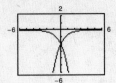

15. Let x = number of kits.

$C = 12x + 50,000$

$R = 25x$

Break-even: $R = C$

$25x = 12x + 50,000$

$13x = 50,000$

$x \approx 3846.15$

You would need to sell 3847 kits to cover your costs.

17. $2l + 2w = 480$

$l = 1.50w$

$2(1.50w) + 2w = 480$

$5w = 480$

$w = 96$

$l = 144$

The dimensions are 96×144 meters.

19. $\begin{cases} 2x - y = 2 \implies 16x - 8y = 16 \\ 6x + 8y = 39 \implies \underline{6x + 8y = 39} \end{cases}$

$ 22x = 55$

$ x = \frac{55}{22} = \frac{5}{2}$

Back-substitute $x = \frac{5}{2}$ into Equation 1.

$2\left(\frac{5}{2}\right) - y = 2$

$y = 3$

Solution: $\left(\frac{5}{2}, 3\right)$

21. $\begin{cases} 0.2x + 0.3y = 0.14 \implies 20x + 30y = 14 \implies \quad 20x + 30y = \quad 14 \\ 0.4x + 0.5y = 0.20 \implies 4x + 5y = 2 \implies \underline{-20x - 25y = -10} \end{cases}$

$$5y = 4$$

$$y = \tfrac{4}{5}$$

Back-substitute $y = \tfrac{4}{5}$ into Equation 2.

$$4x + 5\left(\tfrac{4}{5}\right) = 2$$

$$4x = -2$$

$$x = -\tfrac{1}{2}$$

Solution: $\left(-\tfrac{1}{2}, \tfrac{4}{5}\right) = (-0.5, 0.8)$

23. $\begin{cases} 3x - 2y = 0 \implies 3x - 2y = 0 \\ 3x + 2(y + 5) = 10 \implies \underline{3x + 2y = 0} \end{cases}$

$$6x \quad = 0$$

$$x \quad = 0$$

Back-substitute $x = 0$ into Equation 1.

$$3(0) - 2y = 0$$

$$2y = 0$$

$$y = 0$$

Solution: $(0, 0)$

25. $\begin{cases} 1.25x - 2y = 3.5 \implies 5x - 8y = 14 \\ 5x - 8y = 14 \implies \underline{-5x + 8y = -14} \end{cases}$

$$0 = 0$$

There are infinitely many solutions.

Let $y = a$, then $5x - 8a = 14 \implies x = \tfrac{8}{5}a + \tfrac{14}{5}$.

Solution: $\left(\tfrac{8}{5}a + \tfrac{14}{5}, a\right)$ where a is any real number.

27. $\begin{cases} x + 5y = 4 \implies x + 5y = 4 \\ x - 3y = 6 \implies \underline{-x + 3y = -6} \end{cases}$

$$8y = -2 \implies y = -\tfrac{1}{4}$$

Matches graph (d). The system has one solution and is consistent.

29. $\begin{cases} 3x - y = 7 \implies 6x - 2y = 14 \\ -6x + 2y = 8 \implies \underline{-6x + 2y = 8} \end{cases}$

$$0 \neq 22$$

Matches graph (b). The system has no solution and is inconsistent.

31. $37 - 0.0002x = 22 + 0.00001x$

$$15 = 0.00021x$$

$$x = \frac{500,000}{7}, p = \frac{159}{7}$$

Point of equilibrium: $\left(\dfrac{500,000}{7}, \dfrac{159}{7}\right)$

33. $\begin{cases} x - 4y + 3z = 3 \\ \quad -y + z = -1 \\ \quad\quad z = -5 \end{cases}$

$$-y + (-5) = -1 \implies y = -4$$

$$x - 4(-4) + 3(-5) = 3 \implies x = 2$$

Solution: $(2, -4, -5)$

35.
$$\begin{cases} x + 2y + 6z = 4 \\ -3x + 2y - z = -4 \\ 4x + 2z = 16 \end{cases}$$

$$\begin{cases} x + 2y + 6z = 4 \\ 8y + 17z = 8 \qquad \text{3Eq.1 + Eq.2} \\ -8y - 22z = 0 \qquad \text{−4Eq.1 + Eq.3} \end{cases}$$

$$\begin{cases} x + 2y + 6z = 4 \\ 8y + 17z = 8 \\ -5z = 8 \qquad \text{Eq.2 + Eq.3} \end{cases}$$

$$\begin{cases} x + 2y + 6z = 4 \\ 8y + 17z = 8 \\ z = -\frac{8}{5} \qquad -\frac{1}{5}\text{Eq.3} \end{cases}$$

$$8y + 17\left(-\frac{8}{5}\right) = 8 \implies y = \frac{22}{5}$$
$$x + 2\left(\frac{22}{5}\right) + 6\left(-\frac{8}{5}\right) = 4 \implies x = \frac{24}{5}$$

Solution: $\left(\frac{24}{5}, \frac{22}{5}, -\frac{8}{5}\right)$

37.
$$\begin{cases} x - 2y + z = -6 \\ 2x - 3y = -7 \\ -x + 3y - 3z = 11 \end{cases}$$

$$\begin{cases} x - 2y + z = -6 \\ y - 2z = 5 \qquad -2\text{Eq.1 + Eq.2} \\ y - 2z = 5 \qquad \text{Eq.1 + Eq.3} \end{cases}$$

$$\begin{cases} x - 2y + z = -6 \\ y - 2z = 5 \\ 0 = 0 \qquad -\text{Eq.2 + Eq.3} \end{cases}$$

Let $z = a$, then:

$$y = 2a + 5$$
$$x - 2(2a + 5) + a = -6$$
$$x - 3a - 10 = -6$$
$$x = 3a + 4$$

Solution: $(3a + 4, 2a + 5, a)$ where a is any real number.

39.
$$\begin{cases} 5x - 12y + 7z = 16 \\ 3x - 7y + 4z = 9 \end{cases} \implies \begin{cases} 15x - 36y + 21z = 48 \\ -15x + 35y - 20z = -45 \end{cases}$$
$$\overline{\qquad\qquad -y + z = 3}$$

Let $z = a$. Then $y = a - 3$ and

$$5x - 12(a - 3) + 7a = 16 \implies x = a - 4.$$

Solution: $(a - 4, a - 3, a)$ where a is any real number.

41. $y = ax^2 + bx + c$ through $(0, -5)$, $(1, -2)$, and $(2, 5)$.

$(0, -5)$: $-5 = \qquad + c \implies c = -5$
$(1, -2)$: $-2 = a + b + c \implies \begin{cases} a + b = 3 \\ 2a + b = 5 \end{cases}$
$(2, 5)$: $5 = 4a + 2b + c \implies$
$$\begin{cases} 2a + b = 5 \\ -a - b = -3 \end{cases}$$
$$a = 2$$
$$b = 1$$

The equation of the parabola is $y = 2x^2 + x - 5$.

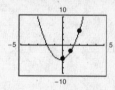

43. $x^2 + y^2 + Dx + Ey + F = 0$ through $(-1, -2)$, $(5, -2)$ and $(2, 1)$.

$(-1, -2)$: $5 - D - 2E + F = 0 \implies \begin{cases} D + 2E - F = 5 \\ 5D - 2E + F = -29 \\ 2D + E + F = -5 \end{cases}$
$(5, -2)$: $29 + 5D - 2E + F = 0 \implies$
$(2, 1)$: $5 + 2D + E + F = 0 \implies$

From the first two equations we have

$$6D = -24$$
$$D = -4.$$

Substituting $D = -4$ into the second and third equations yields:

$$-20 - 2E + F = -29 \implies \begin{cases} -2E + F = -9 \\ -E - F = -3 \end{cases}$$
$$-8 + E + F = -5 \implies$$
$$\overline{\qquad -3E = -12}$$
$$E = 4$$
$$F = -1$$

—CONTINUED—

43. —CONTINUED—

The equation of the circle is $x^2 + y^2 - 4x + 4y - 1 = 0$.

To verify the result using a graphing utility, solve the equation for y.

$$(x^2 - 4x + 4) + (y^2 + 4y + 4) = 1 + 4 + 4$$

$$(x - 2)^2 + (y + 2)^2 = 9$$

$$(y + 2)^2 = 9 - (x - 2)^2$$

$$y = -2 \pm \sqrt{9 - (x - 2)^2}$$

Let $y_1 = -2 + \sqrt{9 - (x - 2)^2}$ and $y_2 = -2 - \sqrt{9 - (x - 2)^2}$.

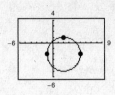

45. $(3, 101.7), (4, 108.4), (5, 121.1)$

(a) $n = 3, \sum_{i=1}^{3} x_i = 12, \sum_{i=1}^{3} x_i^2 = 50, \sum_{i=1}^{3} x_i^3 = 216, \sum_{i=1}^{3} x_i^4 = 962, \sum_{i=1}^{3} y_i = 331.2, \sum_{i=1}^{3} x_i y_i = 1344.2, \sum_{i=1}^{3} x_i^2 y_i = 5677.2$

$$\begin{cases} 3c + 12b + 50a = 331.2 \\ 12c + 50b + 216a = 1344.2 \\ 50c + 216b + 962a = 5677.2 \end{cases}$$

Solving this system yields $c = 117.6, b = -14.3, a = 3$.

Quadratic model: $y = 3x^2 - 14.3x + 117.6$

(b)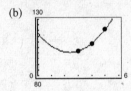

The model is a good fit to the data. The actual points lie on the parabola.

(c) For 2008, use $x = 8$:

$y = 3(8)^2 - 14.3(8) + 117.6$

$= 195.2$ million online shoppers

This answer seems reasonable.

47. Let x = amount invested at 7%

y = amount invested at 9%

z = amount invested at 11%.

$y = x - 3000$ and

$z = x - 5000 \implies y + z = 2x - 8000$

$$\begin{cases} x + y + z = 40{,}000 \\ 0.07x + 0.09y + 0.11z = 3500 \\ y + z = 2x - 8000 \end{cases}$$

$x + (2x - 8000) = 40{,}000 \implies x = 16{,}000$

$y = 16{,}000 - 3000 \implies y = 13{,}000$

$z = 16{,}000 - 5000 \implies z = 11{,}000$

Thus, \$16,000 was invested at 7%, \$13,000 at 9% and \$11,000 at 11%.

49. $\dfrac{3}{x^2 + 20x} = \dfrac{3}{x(x + 20)} = \dfrac{A}{x} + \dfrac{B}{x + 20}$

51. $\dfrac{3x - 4}{x^3 - 5x^2} = \dfrac{3x - 4}{x^2(x - 5)} = \dfrac{A}{x} + \dfrac{B}{x^2} + \dfrac{C}{x - 5}$

53. $\dfrac{4 - x}{x^2 + 6x + 8} = \dfrac{A}{x + 2} + \dfrac{B}{x + 4}$

$4 - x = A(x + 4) + B(x + 2)$

Let $x = -2$: $6 = 2A \implies A = 3$

Let $x = -4$: $8 = -2B \implies B = -4$

$\dfrac{4 - x}{x^2 + 6x + 8} = \dfrac{3}{x + 2} - \dfrac{4}{x + 4}$

55. $\dfrac{x^2}{x^2 + 2x - 15} = 1 - \dfrac{2x - 15}{x^2 + 2x - 15}$

$\dfrac{-2x + 15}{(x + 5)(x - 3)} = \dfrac{A}{x + 5} + \dfrac{B}{x - 3}$

$-2x + 15 = A(x - 3) + B(x + 5)$

Let $x = -5$: $25 = -8A \implies A = -\dfrac{25}{8}$

Let $x = 3$: $9 = 8B \implies B = \dfrac{9}{8}$

$\dfrac{x^2}{x^2 + 2x - 15} = 1 - \dfrac{25}{8(x + 5)} + \dfrac{9}{8(x - 3)}$

57. $\dfrac{x^2 + 2x}{x^3 - x^2 + x - 1} = \dfrac{x^2 + 2x}{(x - 1)(x^2 + 1)} = \dfrac{A}{x - 1} + \dfrac{Bx + C}{x^2 + 1}$

$x^2 + 2x = A(x^2 + 1) + (Bx + C)(x - 1)$

$= Ax^2 + A + Bx^2 - Bx + Cx - C$

$= (A + B)x^2 + (-B + C)x + (A - C)$

Equating coefficients of like terms gives $1 = A + B$, $2 = -B + C$, and $0 = A - C$. Adding both sides of all three equations gives $3 = 2A$. Therefore, $A = \frac{3}{2}$, $B = -\frac{1}{2}$, and $C = \frac{3}{2}$.

$\dfrac{x^2 + 2x}{x^3 - x^2 + x - 1} = \dfrac{\frac{3}{2}}{x - 1} + \dfrac{-\frac{1}{2}x + \frac{3}{2}}{x^2 + 1}$

$= \dfrac{1}{2}\left(\dfrac{3}{x - 1} - \dfrac{x - 3}{x^2 + 1}\right)$

59. $\dfrac{3x^3 + 4x}{(x^2 + 1)^2} = \dfrac{Ax + B}{x^2 + 1} + \dfrac{Cx + D}{(x^2 + 1)^2}$

$3x^3 + 4x = (Ax + B)(x^2 + 1) + Cx + D = Ax^3 + Bx^2 + (A + C)x + (B + D)$

Equating coefficients of like powers:

$3 = A$

$0 = B$

$4 = 3 + C \implies C = 1$

$0 = B + D \implies D = 0$

$\dfrac{3x^3 + 4x}{(x^2 + 1)^2} = \dfrac{3x}{x^2 + 1} + \dfrac{x}{(x^2 + 1)^2}$

61. $y \le 5 - \frac{1}{2}x$

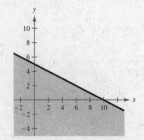

63. $y - 4x^2 > -1$

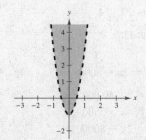

65. $\begin{cases} x + 2y \le 160 \\ 3x + y \le 180 \\ \quad x \ge 0 \\ \quad y \ge 0 \end{cases}$

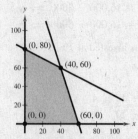

Vertex A	Vertex B	Vertex C	Vertex D
$x + 2y = 160$	$x + 2y = 160$	$3x + y = 180$	$x = 0$
$3x + y = 180$	$x = 0$	$y = 0$	$y = 0$
$(40, 60)$	$(0, 80)$	$(60, 0)$	$(0, 0)$

67. $\begin{cases} 3x + 2y \geq 24 \\ x + 2y \geq 12 \\ 2 \leq x \leq 15 \\ y \leq 15 \end{cases}$

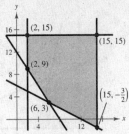

Vertex A	Vertex B	Vertex C	Vertex D	Vertex E
$3x + 2y = 24$	$3x + 2y = 24$	$x = 2$	$x = 15$	$x + 2y = 12$
$x + 2y = 12$	$x = 2$	$y = 15$	$y = 15$	$x = 15$
$(6, 3)$	$(2, 9)$	$(2, 15)$	$(15, 15)$	$\left(15, -\frac{3}{2}\right)$

69. $\begin{cases} y < x + 1 \\ y > x^2 - 1 \end{cases}$

Vertices:

$x + 1 = x^2 - 1$

$0 = x^2 - x - 2 = (x + 1)(x - 2)$

$x = -1$ or $x = 2$

$y = 0 \qquad y = 3$

$(-1, 0) \qquad (2, 3)$

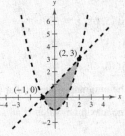

71. $\begin{cases} 2x - 3y \geq 0 \\ 2x - y \leq 8 \\ y \geq 0 \end{cases}$

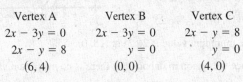

Vertex A	Vertex B	Vertex C
$2x - 3y = 0$	$2x - 3y = 0$	$2x - y = 8$
$2x - y = 8$	$y = 0$	$y = 0$
$(6, 4)$	$(0, 0)$	$(4, 0)$

73. $x = $ number of units of Product I

$y = $ number of units of Product II

$\begin{cases} 20x + 30y \leq 24{,}000 \\ 12x + 8y \leq 12{,}400 \\ x \geq 0 \\ y \geq 0 \end{cases}$

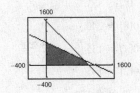

75. (a)

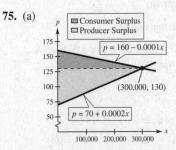

$160 - 0.0001x = 70 + 0.0002x$

$90 = 0.0003x$

$x = 300{,}000$ units

$p = \$130$

Point of equilibrium: $(300{,}000, 130)$

(b) Consumer surplus: $\frac{1}{2}(300{,}000)(30) = \$4{,}500{,}000$

Producer surplus: $\frac{1}{2}(300{,}000)(60) = \$9{,}000{,}000$

77. Objective function: $z = 3x + 4y$

Constraints:
$$\begin{cases} x \geq 0 \\ y \geq 0 \\ 2x + 5y \leq 50 \\ 4x + y \leq 28 \end{cases}$$

At $(0, 0)$: $z = 0$

At $(0, 10)$: $z = 40$

At $(5, 8)$: $z = 47$

At $(7, 0)$: $z = 21$

The minimum value is 0 at $(0, 0)$.

The maximum value is 47 at $(5, 8)$.

79. Objective function: $z = 1.75x + 2.25y$

Constraints:
$$\begin{cases} x \geq 0 \\ y \geq 0 \\ 2x + y \geq 25 \\ 3x + 2y \geq 45 \end{cases}$$

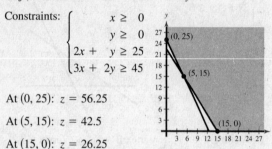

At $(0, 25)$: $z = 56.25$

At $(5, 15)$: $z = 42.5$

At $(15, 0)$: $z = 26.25$

The minimum value is 26.25 at $(15, 0)$.

Since the region in unbounded, there is no maximum value.

81. Objective function: $z = 5x + 11y$

Constraints:
$$\begin{cases} x \geq 0 \\ y \geq 0 \\ x + 3y \leq 12 \\ 3x + 2y \leq 15 \end{cases}$$

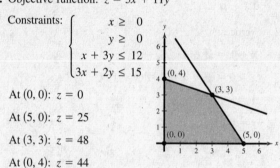

At $(0, 0)$: $z = 0$

At $(5, 0)$: $z = 25$

At $(3, 3)$: $z = 48$

At $(0, 4)$: $z = 44$

The minimum value is 0 at $(0, 0)$.

The maximum value is 48 at $(3, 3)$.

83. Let x = number of haircuts

y = number of permanents.

Objective function: Optimize $R = 25x + 70y$ subject to the following constraints:

$$\begin{cases} x \geq 0 \\ y \geq 0 \\ \left(\frac{20}{60}\right)x + \left(\frac{70}{60}\right)y \leq 24 \implies 2x + 7y \leq 144 \end{cases}$$

At $(0, 0)$: $R = 0$

At $(72, 0)$: $R = 1800$

At $\left(0, \frac{144}{7}\right)$: $R = 1440$

The revenue is optimal if the student does 72 haircuts and no permanents. The maximum revenue is $1800.

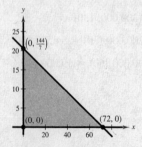

85. Let x = the number of bags of Brand X, and

y = the number of bags of Brand Y.

Objective function: Optimize $C = 15x + 30y$.

Constraints:
$$\begin{cases} 8x + 2y \geq 16 \\ x + y \geq 5 \\ 2x + 7y \geq 20 \\ x \geq 0 \\ y \geq 0 \end{cases}$$

At $(0, 8)$: $C = 15(0) + 30(8) = 240$

At $(1, 4)$: $C = 15(1) + 30(4) = 135$

At $(3, 2)$: $C = 15(3) + 30(2) = 105$

At $(10, 0)$: $C = 15(10) + 30(0) = 150$

To optimize cost, use three bags of Brand X and two bags of Brand Y. The minimum cost is $105.

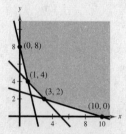

87. False. The system $y \le 5$, $y \ge -2$, $y \le \frac{7}{2}x - 9$, and $y \le -\frac{7}{2}x + 26$ represents the region covered by an isosceles trapezoid.

89. There are an infinite number of linear systems with the solution $(-6, 8)$. One possible solution is:
$$\begin{cases} x + y = 2 \\ x - y = -14 \end{cases}$$

91. There are infinite linear systems with the solution $\left(\frac{4}{3}, 3\right)$. One possible solution is:
$$\begin{cases} 3x + y = 7 \\ -6x + 3y = 1 \end{cases}$$

93. There are an infinite number of linear systems with the solution $(4, -1, 3)$. One possible system is as follows:
$$\begin{cases} x + y + z = 6 \\ x + y - z = 0 \\ x - y - z = 2 \end{cases}$$

95. There are an infinite number of linear systems with the solution $\left(5, \frac{3}{2}, 2\right)$. One possible solution is:
$$\begin{cases} 2x + 2y - 3z = 7 \\ x - 2y + z = 4 \\ -x + 4y - z = -1 \end{cases}$$

97. A system of linear equations is inconsistent if it has no solution.

99. If the solution to a system of equations is at fractional or irrational values, then the substitution method may yield an exact answer. The graphical method works well when the solution is at integer values, otherwise we can usually only approximate the solution.

Problem Solving for Chapter 7

1. The longest side of the triangle is a diameter of the circle and has a length of 20.

The lines $y = \frac{1}{2}x + 5$ and $y = -2x + 20$ intersect at the point $(6, 8)$.

The distance between $(-10, 0)$ and $(6, 8)$ is:
$$d_1 = \sqrt{(6 - (-10))^2 + (8 - 0)^2} = \sqrt{320} = 8\sqrt{5}$$

The distance between $(6, 8)$ and $(10, 0)$ is:
$$d_2 = \sqrt{(10 - 6)^2 + (0 - 8)^2} = \sqrt{80} = 4\sqrt{5}$$

Since $\left(\sqrt{320}\right)^2 + \left(\sqrt{80}\right)^2 = (20)^2$
$$400 = 400$$

the sides of the triangle satisfy the Pythagorean Theorem. Thus, the triangle is a right triangle.

3. The system will have exactly one solution when the slopes of the line are *not* equal.
$$\begin{cases} ax + by = e \implies y = -\frac{a}{b}x + \frac{e}{b} \\ cx + dy = f \implies y = -\frac{c}{d}x + \frac{f}{d} \end{cases}$$

$$-\frac{a}{b} \ne -\frac{c}{d}$$

$$\frac{a}{b} \ne \frac{c}{d}$$

$$ad \ne bc$$

5. There are a finite number of solutions.

(a) If both equations are linear, then the maximum number of solutions to a finite system is *one*.

(b) If one equation is linear and the other is quadratic, then the maximum number of solutions is *two*.

(c) If both equations are quadratic, then the maximum number of solutions to a finite system is *four*.

7. The point where the two sections meet is at a depth of 10.1 feet. The distance between $(0, -10.1)$ and $(252.5, 0)$ is:

$$d = \sqrt{(252.5 - 0)^2 + (0 - (-10.1))^2} = \sqrt{63858.26}$$

$$d \approx 252.7$$

Each section is approximately 252.7 feet long.

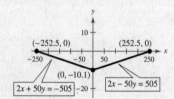

9. Let $x =$ cost of the cable, per foot.

Let $y =$ cost of a connector.

$$\begin{cases} 6x + 2y = 15.50 \implies \\ 3x + 2y = 10.25 \implies \end{cases} \begin{array}{r} 6x + 2y = 15.50 \\ -3x - 2y = -10.25 \\ \hline 3x = 5.25 \\ x = 1.75 \\ y = 2.50 \end{array}$$

For a four-foot cable with a connector on each end the cost should be $4(1.75) + 2(2.50) = \$12.00$

11. Let $X = \dfrac{1}{x}$, $Y = \dfrac{1}{y}$, and $Z = \dfrac{1}{z}$.

(a) $$\begin{cases} \dfrac{12}{x} - \dfrac{12}{y} = 7 \implies 12X - 12Y = 7 \implies 12X - 12Y = 7 \\[2mm] \dfrac{3}{x} + \dfrac{4}{y} = 0 \implies 3X + 4Y = 0 \implies \dfrac{9X + 12Y = 0}{} \end{cases}$$

$$21X = 7$$

$$X = \frac{1}{3}$$

$$Y = -\frac{1}{4}$$

Thus, $\dfrac{1}{x} = \dfrac{1}{3} \implies x = 3$ and $\dfrac{1}{y} = -\dfrac{1}{4} \implies y = -4$.

Solution: $(3, -4)$

—CONTINUED—

11. —CONTINUED—

(b)
$$\begin{cases} \dfrac{2}{x} + \dfrac{1}{y} - \dfrac{3}{z} = 4 \\[2mm] \dfrac{4}{x} + \dfrac{2}{z} = 10 \\[2mm] -\dfrac{2}{x} + \dfrac{3}{y} - \dfrac{13}{z} = -8 \end{cases} \Rightarrow \begin{matrix} 2X + Y - 3Z = 4 & \text{Eq.1} \\[2mm] 4X + 2Z = 10 & \text{Eq.2} \\[2mm] -2X + 3Y - 13Z = -8 & \text{Eq.3} \end{matrix}$$

$$\begin{cases} 2X + Y - 3Z = 4 \\ -2Y + 8Z = 2 \\ 4Y - 16Z = -4 \end{cases} \quad \begin{matrix} \\ -2\text{Eq.1} + \text{Eq.2} \\ \text{Eq.1} + \text{Eq.3} \end{matrix}$$

$$\begin{cases} 2X + Y - 3Z = 4 \\ -2Y + 8Z = 2 \\ 0 = 0 \end{cases} \quad \begin{matrix} \\ \\ 2\text{Eq.2} + \text{Eq.3} \end{matrix}$$

The system has infinite solutions.

Let $Z = a$, then $Y = 4a - 1$ and $X = \dfrac{-a + 5}{2}$.

Then $\dfrac{1}{z} = a \Rightarrow z = \dfrac{1}{a}, \dfrac{1}{y} = 4a - 1 \Rightarrow y = \dfrac{1}{4a - 1}$

$x = \dfrac{-a + 5}{2} \Rightarrow x = \dfrac{2}{-a + 5}.$

Solution: $\left(\dfrac{2}{-a + 5}, \dfrac{1}{4a - 1}, \dfrac{1}{a} \right),\ a \neq 5, \dfrac{1}{4}, 0$

13. Solution: $(1, -1, 2)$

$$\begin{cases} 4x - 2y + 5z = 16 & \text{Equation 1} \\ x + y = 0 & \text{Equation 2} \\ -x - 3y + 2z = 6 & \text{Equation 3} \end{cases}$$

(a) $\begin{cases} 4x - 2y + 5z = 16 \\ x + y = 0 \end{cases}$

$\begin{cases} x + y = 0 \\ 4x - 2y + 5z = 16 \end{cases}$ Interchange the equations.

$\begin{cases} x + y = 0 \\ -6y + 5z = 16 \end{cases}$ $-4\text{Eq.1} + \text{Eq.2}$

Let $z = a$, then $y = \dfrac{5a - 16}{6}$ and $x = \dfrac{-5a + 16}{6}$.

Solution: $\left(\dfrac{-5a + 16}{6}, \dfrac{5a - 16}{6}, a \right)$

When $a = 2$ we have the original solution.

(c) $\begin{cases} x + y = 0 \\ -x - 3y + 2z = 6 \end{cases}$

$\begin{cases} x + y = 0 \\ -2y + 2z = 6 \end{cases}$ $\text{Eq.1} + \text{Eq.2}$

Let $z = c$, then $y = c - 3$ and $x = -c + 3$

Solution: $(-c + 3, c - 3, c)$

When $c = 2$ we have the original solution.

(b) $\begin{cases} 4x - 2y + 5z = 16 \\ -x - 3y + 2z = 6 \end{cases}$

$\begin{cases} -x - 3y + 2z = 6 \\ 4x - 2y + 5z = 16 \end{cases}$ Interchange the equations.

$\begin{cases} -x - 3y + 2z = 6 \\ -14y + 13z = 40 \end{cases}$ $4\text{Eq.1} + \text{Eq.2}$

Let $z = b$, then $y = \dfrac{13b - 40}{14}$ and $x = \dfrac{-11b + 36}{14}$

Solution: $\left(\dfrac{-11b + 36}{14}, \dfrac{13b - 40}{14}, b \right)$

When $b = 2$ we have the original solution.

(d) Each of these systems has infinite solutions.

15. t = amount of terrestrial vegetation in kilograms

a = amount of aquatic vegetation in kilograms

$$a + t \leq 32$$

$$0.15a \geq 1.9$$

$$193a + 4(193)t \geq 11,000$$

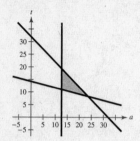

17. (a) x = HDL cholesterol (good)

y = LDL cholesterol (bad)

$$\begin{cases} 0 < y < 130 \\ x \geq 35 \\ x + y \leq 200 \end{cases}$$

(b)

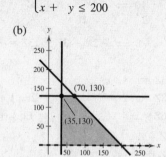

(c) $y = 120$ *is* in the region since $0 < y < 130$.

$x = 90$ *is* in the region since $35 < x < 200$.

$x + y = 210$ *is not* in the region since $x + y < 200$.

(d) If the LDL reading is 150 and the HDL reading is 40, then $x \geq 35$ and $x + y \leq 200$ but $y \not< 130$.

(e) $\dfrac{x + y}{x} < 4$

$x + y < 4x$

$y < 3x$

The point (50, 120) is in the region and $120 < 3(50)$.

Chapter 7   Practice Test

For Exercises 1–3, solve the given system by the method of substitution.

1. $\begin{cases} x + y = 1 \\ 3x - y = 15 \end{cases}$

2. $\begin{cases} x - 3y = -3 \\ x^2 + 6y = 5 \end{cases}$

3. $\begin{cases} x + y + z = 6 \\ 2x - y + 3z = 0 \\ 5x + 2y - z = -3 \end{cases}$

4. Find the two numbers whose sum is 110 and product is 2800.

5. Find the dimensions of a rectangle if its perimeter is 170 feet and its area is 1500 square feet.

For Exercises 6–8, solve the linear system by elimination.

6. $\begin{cases} 2x + 15y = 4 \\ x - 3y = 23 \end{cases}$

7. $\begin{cases} x + y = 2 \\ 38x - 19y = 7 \end{cases}$

8. $\begin{cases} 0.4x + 0.5y = 0.112 \\ 0.3x - 0.7y = -0.131 \end{cases}$

9. Herbert invests \$17,000 in two funds that pay 11% and 13% simple interest, respectively. If he receives \$2080 in yearly interest, how much is invested in each fund?

10. Find the least squares regression line for the points $(4, 3)$, $(1, 1)$, $(-1, -2)$, and $(-2, -1)$.

For Exercises 11–12, solve the system of equations.

11. $\begin{cases} x + y = -2 \\ 2x - y + z = 11 \\ 4y - 3z = -20 \end{cases}$

12. $\begin{cases} 3x + 2y - z = 5 \\ 6x - y + 5z = 2 \end{cases}$

13. Find the equation of the parabola $y = ax^2 + bx + c$ passing through the points $(0, -1)$, $(1, 4)$ and $(2, 13)$.

For Exercises 14–15, write the partial fraction decomposition of the rational functions.

14. $\dfrac{10x - 17}{x^2 - 7x - 8}$

15. $\dfrac{x^2 + 4}{x^4 + x^2}$

16. Graph $x^2 + y^2 \geq 9$.

17. Graph the solution of the system.

$$\begin{cases} x + y \leq 6 \\ \quad x \geq 2 \\ \quad y \geq 0 \end{cases}$$

18. Derive a set of inequalities to describe the triangle with vertices $(0, 0)$, $(0, 7)$, and $(2, 3)$.

19. Find the maximum value of the objective function, $z = 30x + 26y$, subject to the following constraints.

$$\begin{cases} \quad x \geq 0 \\ \quad y \geq 0 \\ 2x + 3y \leq 21 \\ 5x + 3y \leq 30 \end{cases}$$

20. Graph the system of inequalities.

$$\begin{cases} \quad x^2 + y^2 \leq 4 \\ (x - 2)^2 + y^2 \geq 4 \end{cases}$$

For Exercises 21–22, write the partial fraction decomposition for the rational expression.

21. $\dfrac{1 - 2x}{x^2 + x}$

22. $\dfrac{6x - 17}{(x - 3)^2}$

C H A P T E R 8
Matrices and Determinants

Section 8.1   Matrices and Systems of Equations **392**

Section 8.2   Operations with Matrices **400**

Section 8.3   The Inverse of a Square Matrix **407**

Section 8.4   The Determinant of a Square Matrix **415**

Section 8.5   Applications of Matrices and Determinants **421**

Review Exercises **427**

Problem Solving **438**

Practice Test **441**

CHAPTER 8
Matrices and Determinants

Section 8.1   Matrices and Systems of Equations

- You should be able to use elementary row operations to produce a row-echelon form (or reduced row-echelon form) of a matrix.
  1. Interchange two rows.
  2. Multiply a row by a nonzero constant.
  3. Add a multiple of one row to another row.
- You should be able to use either Gaussian elimination with back-substitution or Gauss-Jordan elimination to solve a system of linear equations.

Vocabulary Check

1. matrix
2. square
3. main diagonal
4. row; column
5. augmented
6. coefficient
7. row-equivalent
8. reduced row-echelon form
9. Gauss-Jordan elimination

---

1. Since the matrix has one row and two columns, its order is 1×2.

3. Since the matrix has three rows and one column, its order is 3×1.

5. Since the matrix has two rows and two columns, its order is 2×2.

7. $\begin{cases} 4x - 3y = -5 \\ -x + 3y = 12 \end{cases}$

$\begin{bmatrix} 4 & -3 & \vdots & -5 \\ -1 & 3 & \vdots & 12 \end{bmatrix}$

9. $\begin{cases} x + 10y - 2z = 2 \\ 5x - 3y + 4z = 0 \\ 2x + y = 6 \end{cases}$

$\begin{bmatrix} 1 & 10 & -2 & \vdots & 2 \\ 5 & -3 & 4 & \vdots & 0 \\ 2 & 1 & 0 & \vdots & 6 \end{bmatrix}$

11. $\begin{cases} 7x - 5y + z = 13 \\ 19x - 8z = 10 \end{cases}$

$\begin{bmatrix} 7 & -5 & 1 & \vdots & 13 \\ 19 & 0 & -8 & \vdots & 10 \end{bmatrix}$

13. $\begin{bmatrix} 1 & 2 & \vdots & 7 \\ 2 & -3 & \vdots & 4 \end{bmatrix}$

$\begin{cases} x + 2y = 7 \\ 2x - 3y = 4 \end{cases}$

15. $\begin{bmatrix} 2 & 0 & 5 & \vdots & -12 \\ 0 & 1 & -2 & \vdots & 7 \\ 6 & 3 & 0 & \vdots & 2 \end{bmatrix}$

$\begin{cases} 2x + 5z = -12 \\ y - 2z = 7 \\ 6x + 3y = 2 \end{cases}$

17. $\begin{bmatrix} 9 & 12 & 3 & 0 & \vdots & 0 \\ -2 & 18 & 5 & 2 & \vdots & 10 \\ 1 & 7 & -8 & 0 & \vdots & -4 \\ 3 & 0 & 2 & 0 & \vdots & -10 \end{bmatrix}$

$\begin{cases} 9x + 12y + 3z = 0 \\ -2x + 18y + 5z + 2w = 10 \\ x + 7y - 8z = -4 \\ 3x + 2z = -10 \end{cases}$

19. $\begin{bmatrix} 1 & 4 & 3 \\ 2 & 10 & 5 \end{bmatrix}$

$-2R_1 + R_2 \rightarrow \begin{bmatrix} 1 & 4 & 3 \\ 0 & \boxed{2} & -1 \end{bmatrix}$

21.
$$\begin{bmatrix} 1 & 1 & 4 & -1 \\ 3 & 8 & 10 & 3 \\ -2 & 1 & 12 & 6 \end{bmatrix}$$

$$\begin{matrix} \\ -3R_1 + R_2 \to \\ 2R_1 + R_3 \to \end{matrix} \begin{bmatrix} 1 & 1 & 4 & -1 \\ 0 & 5 & \boxed{-2} & \boxed{6} \\ 0 & 3 & \boxed{20} & \boxed{4} \end{bmatrix}$$

$$\tfrac{1}{5}R_2 \to \begin{bmatrix} 1 & 1 & 4 & -1 \\ 0 & 1 & -\tfrac{2}{5} & \tfrac{6}{5} \\ 0 & 3 & \boxed{20} & \boxed{4} \end{bmatrix}$$

23. $\begin{bmatrix} -2 & 5 & 1 \\ 3 & -1 & -8 \end{bmatrix} \to \begin{bmatrix} 13 & 0 & -39 \\ 3 & -1 & -8 \end{bmatrix}$

Add 5 times Row 2 to Row 1.

25. $\begin{bmatrix} 0 & -1 & -5 & 5 \\ -1 & 3 & -7 & 6 \\ 4 & -5 & 1 & 3 \end{bmatrix} \to \begin{bmatrix} -1 & 3 & -7 & 6 \\ 0 & -1 & -5 & 5 \\ 0 & 7 & -27 & 27 \end{bmatrix}$

Interchange Row 1 and Row 2. Then add 4 times the new Row 1 to Row 3.

27. $\begin{bmatrix} 1 & 2 & 3 \\ 2 & -1 & -4 \\ 3 & 1 & -1 \end{bmatrix}$

(a) $\begin{bmatrix} 1 & 2 & 3 \\ 0 & -5 & -10 \\ 3 & 1 & -1 \end{bmatrix}$ (b) $\begin{bmatrix} 1 & 2 & 3 \\ 0 & -5 & -10 \\ 0 & -5 & -10 \end{bmatrix}$ (c) $\begin{bmatrix} 1 & 2 & 3 \\ 0 & -5 & -10 \\ 0 & 0 & 0 \end{bmatrix}$

(d) $\begin{bmatrix} 1 & 2 & 3 \\ 0 & 1 & 2 \\ 0 & 0 & 0 \end{bmatrix}$ (e) $\begin{bmatrix} 1 & 0 & -1 \\ 0 & 1 & 2 \\ 0 & 0 & 0 \end{bmatrix}$ This matrix is in reduced row-echelon form.

29. $\begin{bmatrix} 1 & 0 & 0 & 0 \\ 0 & 1 & 1 & 5 \\ 0 & 0 & 0 & 0 \end{bmatrix}$

This matrix is in reduced row-echelon form.

31. $\begin{bmatrix} 2 & 0 & 4 & 0 \\ 0 & -1 & 3 & 6 \\ 0 & 0 & 1 & 5 \end{bmatrix}$

The first nonzero entries in Rows 1 and 2 are not 1. The matrix is not in row-echelon form.

33. $\begin{bmatrix} 1 & 1 & 0 & 5 \\ -2 & -1 & 2 & -10 \\ 3 & 6 & 7 & 14 \end{bmatrix}$

$$\begin{matrix} \\ 2R_1 + R_2 \to \\ -3R_1 + R_3 \to \end{matrix} \begin{bmatrix} 1 & 1 & 0 & 5 \\ 0 & 1 & 2 & 0 \\ 0 & 3 & 7 & -1 \end{bmatrix}$$

$$-3R_2 + R_3 \to \begin{bmatrix} 1 & 1 & 0 & 5 \\ 0 & 1 & 2 & 0 \\ 0 & 0 & 1 & -1 \end{bmatrix}$$

35. $\begin{bmatrix} 1 & -1 & -1 & 1 \\ 5 & -4 & 1 & 8 \\ -6 & 8 & 18 & 0 \end{bmatrix}$

$$\begin{matrix} -5R_1 + R_2 \to \\ 6R_1 + R_3 \to \end{matrix} \begin{bmatrix} 1 & -1 & -1 & 1 \\ 0 & 1 & 6 & 3 \\ 0 & 2 & 12 & 6 \end{bmatrix}$$

$$-2R_2 + R_3 \to \begin{bmatrix} 1 & -1 & -1 & 1 \\ 0 & 1 & 6 & 3 \\ 0 & 0 & 0 & 0 \end{bmatrix}$$

37. Use the reduced row-echelon form feature of a graphing utility.

$$\begin{bmatrix} 3 & 3 & 3 \\ -1 & 0 & -4 \\ 2 & 4 & -2 \end{bmatrix} \Rightarrow \begin{bmatrix} 1 & 0 & 0 \\ 0 & 1 & 0 \\ 0 & 0 & 1 \end{bmatrix}$$

39. Use the reduced row-echelon form feature of a graphing utility.

$$\begin{bmatrix} 1 & 2 & 3 & -5 \\ 1 & 2 & 4 & -9 \\ -2 & -4 & -4 & 3 \\ 4 & 8 & 11 & -14 \end{bmatrix} \Rightarrow \begin{bmatrix} 1 & 2 & 0 & 0 \\ 0 & 0 & 1 & 0 \\ 0 & 0 & 0 & 1 \\ 0 & 0 & 0 & 0 \end{bmatrix}$$

41. Use the reduced row-echelon form feature of a graphing utility.

$$\begin{bmatrix} -3 & 5 & 1 & 12 \\ 1 & -1 & 1 & 4 \end{bmatrix} \Rightarrow \begin{bmatrix} 1 & 0 & 3 & 16 \\ 0 & 1 & 2 & 12 \end{bmatrix}$$

43. $\begin{cases} x - 2y = 4 \\ \quad\ y = -3 \end{cases}$

$x - 2(-3) = 4$

$x = -2$

Solution: $(-2, -3)$

45. $\begin{cases} x - y + 2z = 4 \\ \quad y - z = 2 \\ \quad\quad\ z = -2 \end{cases}$

$y - (-2) = 2$

$y = 0$

$x - 0 + 2(-2) = 4$

$x = 8$

Solution: $(8, 0, -2)$

47. $\begin{bmatrix} 1 & 0 & \vdots & 3 \\ 0 & 1 & \vdots & -4 \end{bmatrix}$

$x = 3$

$y = -4$

Solution: $(3, -4)$

49. $\begin{bmatrix} 1 & 0 & 0 & \vdots & -4 \\ 0 & 1 & 0 & \vdots & -10 \\ 0 & 0 & 1 & \vdots & 4 \end{bmatrix}$

$x = -4$

$y = -10$

$z = 4$

Solution: $(-4, -10, 4)$

51. $\begin{cases} x + 2y = 7 \\ 2x + y = 8 \end{cases}$

$$\begin{bmatrix} 1 & 2 & \vdots & 7 \\ 2 & 1 & \vdots & 8 \end{bmatrix}$$

$-2R_1 + R_2 \rightarrow \begin{bmatrix} 1 & 2 & \vdots & 7 \\ 0 & -3 & \vdots & -6 \end{bmatrix}$

$-\frac{1}{3}R_2 \rightarrow \begin{bmatrix} 1 & 2 & \vdots & 7 \\ 0 & 1 & \vdots & 2 \end{bmatrix}$

$\begin{cases} x + 2y = 7 \\ \quad\ y = 2 \end{cases}$

$y = 2$

$x + 2(2) = 7 \implies x = 3$

Solution: $(3, 2)$

53. $\begin{cases} 3x - 2y = -27 \\ \ x + 3y = 13 \end{cases}$

$$\begin{bmatrix} 3 & -2 & \vdots & -27 \\ 1 & 3 & \vdots & 13 \end{bmatrix}$$

$\begin{matrix} R_1 \\ R_2 \end{matrix} \begin{bmatrix} 1 & 3 & \vdots & 13 \\ 3 & -2 & \vdots & -27 \end{bmatrix}$

$-3R_1 + R_2 \rightarrow \begin{bmatrix} 1 & 3 & \vdots & 13 \\ 0 & -11 & \vdots & -66 \end{bmatrix}$

$-\frac{1}{11}R_2 \rightarrow \begin{bmatrix} 1 & 3 & \vdots & 13 \\ 0 & 1 & \vdots & 6 \end{bmatrix}$

$\begin{cases} x + 3y = 13 \\ \quad\ y = 6 \end{cases}$

$y = 6$

$x + 3(6) = 13 \implies x = -5$

Solution: $(-5, 6)$

55. $\begin{cases} -2x + 6y = -22 \\ \ x + 2y = -9 \end{cases}$

$$\begin{bmatrix} -2 & 6 & \vdots & -22 \\ 1 & 2 & \vdots & -9 \end{bmatrix}$$

$\begin{matrix} R_1 \\ R_2 \end{matrix} \begin{bmatrix} 1 & 2 & \vdots & -9 \\ -2 & 6 & \vdots & -22 \end{bmatrix}$

$2R_1 + R_2 \rightarrow \begin{bmatrix} 1 & 2 & \vdots & -9 \\ 0 & 10 & \vdots & -40 \end{bmatrix}$

$\frac{1}{10}R_2 \rightarrow \begin{bmatrix} 1 & 2 & \vdots & -9 \\ 0 & 1 & \vdots & -4 \end{bmatrix}$

$\begin{cases} x + 2y = -9 \\ \quad\ y = -4 \end{cases}$

$y = -4$

$x + 2(-4) = -9 \implies x = -1$

Solution: $(-1, -4)$

57. $\begin{cases} -x + 2y = 1.5 \\ 2x - 4y = 3.0 \end{cases}$

$$\begin{bmatrix} -1 & 2 & \vdots & 1.5 \\ 2 & -4 & \vdots & 3.0 \end{bmatrix}$$

$2R_1 + R_2 \rightarrow \begin{bmatrix} -1 & 2 & \vdots & 1.5 \\ 0 & 0 & \vdots & 6.0 \end{bmatrix}$

The system is inconsistent and there is no solution.

59. $\begin{cases} x \quad\quad - 3z = -2 \\ 3x + y - 2z = 5 \\ 2x + 2y + z = 4 \end{cases}$

$$\begin{bmatrix} 1 & 0 & -3 & \vdots & -2 \\ 3 & 1 & -2 & \vdots & 5 \\ 2 & 2 & 1 & \vdots & 4 \end{bmatrix}$$

$\begin{matrix} -3R_1 + R_2 \to \\ -2R_1 + R_3 \to \end{matrix} \begin{bmatrix} 1 & 0 & -3 & \vdots & -2 \\ 0 & 1 & 7 & \vdots & 11 \\ 0 & 2 & 7 & \vdots & 8 \end{bmatrix}$

$-2R_2 + R_3 \to \begin{bmatrix} 1 & 0 & -3 & \vdots & -2 \\ 0 & 1 & 7 & \vdots & 11 \\ 0 & 0 & -7 & \vdots & -14 \end{bmatrix}$

$-\frac{1}{7}R_3 \to \begin{bmatrix} 1 & 0 & -3 & \vdots & -2 \\ 0 & 1 & 7 & \vdots & 11 \\ 0 & 0 & 1 & \vdots & 2 \end{bmatrix}$

$\begin{cases} x - 3z = -2 \\ y + 7z = 11 \\ \quad z = 2 \end{cases}$

$z = 2$

$y + 7(2) = 11 \implies y = -3$

$x - 3(2) = -2 \implies x = 4$

Solution: $(4, -3, 2)$

61. $\begin{cases} -x + y - z = -14 \\ 2x - y + z = 21 \\ 3x + 2y + z = 19 \end{cases}$

$$\begin{bmatrix} -1 & 1 & -1 & \vdots & -14 \\ 2 & -1 & 1 & \vdots & 21 \\ 3 & 2 & 1 & \vdots & 19 \end{bmatrix}$$

$-R_1 \to \begin{bmatrix} 1 & -1 & 1 & \vdots & 14 \\ 2 & -1 & 1 & \vdots & 21 \\ 3 & 2 & 1 & \vdots & 19 \end{bmatrix}$

$\begin{matrix} -2R_1 + R_2 \to \\ -3R_1 + R_3 \to \end{matrix} \begin{bmatrix} 1 & -1 & 1 & \vdots & 14 \\ 0 & 1 & -1 & \vdots & -7 \\ 0 & 5 & -2 & \vdots & -23 \end{bmatrix}$

$-5R_2 + R_3 \to \begin{bmatrix} 1 & -1 & 1 & \vdots & 14 \\ 0 & 1 & -1 & \vdots & -7 \\ 0 & 0 & 3 & \vdots & 12 \end{bmatrix}$

$\frac{1}{3}R_3 \to \begin{bmatrix} 1 & -1 & 1 & \vdots & 14 \\ 0 & 1 & -1 & \vdots & -7 \\ 0 & 0 & 1 & \vdots & 4 \end{bmatrix}$

$\begin{cases} x - y + z = 14 \\ y - z = -7 \\ \quad z = 4 \end{cases}$

$z = 4$

$y - 4 = -7 \implies y = -3$

$x - (-3) + 4 = 14 \implies x = 7$

Solution: $(7, -3, 4)$

63. $\begin{cases} x + 2y - 3z = -28 \\ 4y + 2z = 0 \\ -x + y - z = -5 \end{cases}$

$$\begin{bmatrix} 1 & 2 & -3 & \vdots & -28 \\ 0 & 4 & 2 & \vdots & 0 \\ -1 & 1 & -1 & \vdots & -5 \end{bmatrix}$$

$\begin{matrix} \frac{1}{4}R_2 \to \\ R_1 + R_3 \to \end{matrix} \begin{bmatrix} 1 & 2 & -3 & \vdots & -28 \\ 0 & 1 & \frac{1}{2} & \vdots & 0 \\ 0 & 3 & -4 & \vdots & -33 \end{bmatrix}$

$-3R_2 + R_3 \to \begin{bmatrix} 1 & 2 & -3 & \vdots & -28 \\ 0 & 1 & \frac{1}{2} & \vdots & 0 \\ 0 & 0 & -\frac{11}{2} & \vdots & -33 \end{bmatrix}$

$-\frac{2}{11}R_3 \to \begin{bmatrix} 1 & 2 & -3 & \vdots & -28 \\ 0 & 1 & \frac{1}{2} & \vdots & 0 \\ 0 & 0 & 1 & \vdots & 6 \end{bmatrix}$

$\begin{cases} x + 2y - 3z = -28 \\ y + \frac{1}{2}z = 0 \\ \quad z = 6 \end{cases}$

$z = 6$

$y + \frac{1}{2}(6) = 0 \implies y = -3$

$x + 2(-3) - 3(6) = -28 \implies x = -4$

Solution: $(-4, -3, 6)$

65. $\begin{cases} x + y - 5z = 3 \\ x \quad - 2z = 1 \\ 2x - y - z = 0 \end{cases}$

$$\begin{bmatrix} 1 & 1 & -5 & \vdots & 3 \\ 1 & 0 & -2 & \vdots & 1 \\ 2 & -1 & -1 & \vdots & 0 \end{bmatrix}$$

$\begin{matrix} -R_1 + R_2 \to \\ -2R_1 + R_3 \to \end{matrix} \begin{bmatrix} 1 & 1 & -5 & \vdots & 3 \\ 0 & -1 & 3 & \vdots & -2 \\ 0 & -3 & 9 & \vdots & -6 \end{bmatrix}$

$-R_2 \to \begin{bmatrix} 1 & 1 & -5 & \vdots & 3 \\ 0 & 1 & -3 & \vdots & 2 \\ 0 & -3 & 9 & \vdots & -6 \end{bmatrix}$

$\begin{matrix} -R_2 + R_1 \to \\ 3R_2 + R_3 \to \end{matrix} \begin{bmatrix} 1 & 0 & -2 & \vdots & 1 \\ 0 & 1 & -3 & \vdots & 2 \\ 0 & 0 & 0 & \vdots & 0 \end{bmatrix}$

$\begin{cases} x - 2z = 1 \\ y - 3z = 2 \end{cases}$

Let $z = a$.

$y - 3a = 2 \implies y = 3a + 2$

$x - 2a = 1 \implies x = 2a + 1$

Solution: $(2a + 1, 3a + 2, a)$ where a is any real number.

67. $\begin{cases} x + 2y + z + 2w = 8 \\ 3x + 7y + 6z + 9w = 26 \end{cases}$

$$\begin{bmatrix} 1 & 2 & 1 & 2 & \vdots & 8 \\ 3 & 7 & 6 & 9 & \vdots & 26 \end{bmatrix}$$

$-3R_1 + R_2 \rightarrow \begin{bmatrix} 1 & 2 & 1 & 2 & \vdots & 8 \\ 0 & 1 & 3 & 3 & \vdots & 2 \end{bmatrix}$

$-2R_2 + R_1 \rightarrow \begin{bmatrix} 1 & 0 & -5 & -4 & \vdots & 4 \\ 0 & 1 & 3 & 3 & \vdots & 2 \end{bmatrix}$

$\begin{cases} x - 5z - 4w = 4 \\ y + 3z + 3w = 2 \end{cases}$

Let $w = a$ and $z = b$.

$y + 3b + 3a = 2 \implies y = 2 - 3b - 3a$

$x - 5b - 4a = 4 \implies x = 4 + 5b + 4a$

Solution: $(4 + 5b + 4a, 2 - 3b - 3a, b, a)$
  where a and b are real numbers

69. $\begin{cases} -x + y = -22 \\ 3x + 4y = 4 \\ 4x - 8y = 32 \end{cases}$

$$\begin{bmatrix} -1 & 1 & \vdots & -22 \\ 3 & 4 & \vdots & 4 \\ 4 & -8 & \vdots & 32 \end{bmatrix}$$

$-R_1 \rightarrow \begin{bmatrix} 1 & -1 & \vdots & 22 \\ 3 & 4 & \vdots & 4 \\ 4 & -8 & \vdots & 32 \end{bmatrix}$

$\begin{matrix} \\ -3R_1 + R_2 \rightarrow \\ -4R_1 + R_3 \rightarrow \end{matrix} \begin{bmatrix} 1 & -1 & \vdots & 22 \\ 0 & 7 & \vdots & -62 \\ 0 & -4 & \vdots & -56 \end{bmatrix}$

$\begin{matrix} \\ \frac{1}{7}R_2 \rightarrow \\ -\frac{1}{4}R_3 \rightarrow \end{matrix} \begin{bmatrix} 1 & -1 & \vdots & 22 \\ 0 & 1 & \vdots & -\frac{62}{7} \\ 0 & 1 & \vdots & 14 \end{bmatrix}$

$\begin{matrix} \\ \\ -R_2 + R_3 \rightarrow \end{matrix} \begin{bmatrix} 1 & -1 & \vdots & 22 \\ 0 & 1 & \vdots & -\frac{62}{7} \\ 0 & 0 & \vdots & \frac{160}{7} \end{bmatrix}$

The system is inconsistent and there is no solution.

71. Use the reduced row-echelon form feature of a graphing utility.

$\begin{cases} 3x + 3y + 12z = 6 \\ x + y + 4z = 2 \\ 2x + 5y + 20z = 10 \\ -x + 2y + 8z = 4 \end{cases}$ $\begin{bmatrix} 3 & 3 & 12 & \vdots & 6 \\ 1 & 1 & 4 & \vdots & 2 \\ 2 & 5 & 20 & \vdots & 10 \\ -1 & 2 & 8 & \vdots & 4 \end{bmatrix} \Rightarrow \begin{bmatrix} 1 & 0 & 0 & \vdots & 0 \\ 0 & 1 & 4 & \vdots & 2 \\ 0 & 0 & 0 & \vdots & 0 \\ 0 & 0 & 0 & \vdots & 0 \end{bmatrix} \Rightarrow \begin{cases} x = 0 \\ y + 4z = 2 \end{cases}$

Let $z = a$.

$y = 2 - 4a$

$x = 0$

Solution: $(0, 2 - 4a, a)$ where a is any real number

73. Use the reduced row-echelon form feature of a graphing utility.

$\begin{cases} 2x + y - z + 2w = -6 \\ 3x + 4y + w = 1 \\ x + 5y + 2z + 6w = -3 \\ 5x + 2y - z - w = 3 \end{cases}$ $\begin{bmatrix} 2 & 1 & -1 & 2 & \vdots & -6 \\ 3 & 4 & 0 & 1 & \vdots & 1 \\ 1 & 5 & 2 & 6 & \vdots & -3 \\ 5 & 2 & -1 & -1 & \vdots & 3 \end{bmatrix} \Rightarrow \begin{bmatrix} 1 & 0 & 0 & 0 & \vdots & 1 \\ 0 & 1 & 0 & 0 & \vdots & 0 \\ 0 & 0 & 1 & 0 & \vdots & 4 \\ 0 & 0 & 0 & 1 & \vdots & -2 \end{bmatrix}$

$x = 1$

$y = 0$

$z = 4$

$w = -2$

Solution: $(1, 0, 4, -2)$

75. Use the reduced row-echelon form feature of a graphing utility.

$$\begin{cases} x + y + z + w = 0 \\ 2x + 3y + z - 2w = 0 \\ 3x + 5y + z \quad\quad = 0 \end{cases}$$
$$\begin{bmatrix} 1 & 1 & 1 & 1 & \vdots & 0 \\ 2 & 3 & 1 & -2 & \vdots & 0 \\ 3 & 5 & 1 & 0 & \vdots & 0 \end{bmatrix} \Rightarrow \begin{bmatrix} 1 & 0 & 2 & 0 & \vdots & 0 \\ 0 & 1 & -1 & 0 & \vdots & 0 \\ 0 & 0 & 0 & 1 & \vdots & 0 \end{bmatrix}$$

$$\begin{cases} x + 2z = 0 \\ y - z = 0 \\ \quad\quad w = 0 \end{cases}$$

Let $z = a$. Then $x = -2a$ and $y = a$.

Solution: $(-2a, a, a, 0)$ where a is a real number

77. (a)
$$\begin{cases} x - 2y + z = -6 \\ y - 5z = 16 \\ z = -3 \end{cases}$$

$$y - 5(-3) = 16$$
$$y = 1$$
$$x - 2(1) + (-3) = -6$$
$$x = -1$$

Solution: $(-1, 1, -3)$

(b)
$$\begin{cases} x + y - 2z = 6 \\ y + 3z = -8 \\ z = -3 \end{cases}$$

$$y + 3(-3) = -8$$
$$y = 1$$
$$x + (1) - 2(-3) = 6$$
$$x = -1$$

Solution: $(-1, 1, -3)$

Both systems yield the same solution, namely $(-1, 1, -3)$.

79. (a)
$$\begin{cases} x - 4y + 5z = 27 \\ y - 7z = -54 \\ z = 8 \end{cases}$$

$$y - 7(8) = -54$$
$$y = 2$$
$$x - 4(2) + 5(8) = 27$$
$$x = -5$$

Solution: $(-5, 2, 8)$

(b)
$$\begin{cases} x - 6y + z = 15 \\ y + 5z = 42 \\ z = 8 \end{cases}$$

$$y + 5(8) = 42$$
$$y = 2$$
$$x - 6(2) + (8) = 15$$
$$x = 19$$

Solution: $(19, 2, 8)$

The systems do *not* yield the same solution.

81.
$$\begin{cases} x + 3y + z = 3 \\ x + 5y + 5z = 1 \\ 2x + 6y + 3z = 8 \end{cases}$$

$$\begin{bmatrix} 1 & 3 & 1 & \vdots & 3 \\ 1 & 5 & 5 & \vdots & 1 \\ 2 & 6 & 3 & \vdots & 8 \end{bmatrix}$$

$$\begin{matrix} \\ -R_1 + R_2 \rightarrow \\ -2R_1 + R_3 \rightarrow \end{matrix} \begin{bmatrix} 1 & 3 & 1 & \vdots & 3 \\ 0 & 2 & 4 & \vdots & -2 \\ 0 & 0 & 1 & \vdots & 2 \end{bmatrix}$$

$$\tfrac{1}{2}R_2 \rightarrow \begin{bmatrix} 1 & 3 & 1 & \vdots & 3 \\ 0 & 1 & 2 & \vdots & -1 \\ 0 & 0 & 1 & \vdots & 2 \end{bmatrix}$$ This is a matrix in row-echelon form.

$$\begin{bmatrix} 1 & 3 & \tfrac{3}{2} & \vdots & 4 \\ 0 & 1 & \tfrac{7}{4} & \vdots & -\tfrac{3}{2} \\ 0 & 0 & 1 & \vdots & 2 \end{bmatrix}$$ The row-echelon form feature of a graphing utility yields this form.

There are infinitely many matrices in row-echelon form that correspond to the original system of equations. All such matrices will yield the same solution, namely $(16, -5, 2)$.

83. $\dfrac{4x^2}{(x+1)^2(x-1)} = \dfrac{A}{x-1} + \dfrac{B}{x+1} + \dfrac{C}{(x+1)^2}$

$$4x^2 = A(x+1)^2 + B(x+1)(x-1) + C(x-1)$$
$$4x^2 = A(x^2 + 2x + 1) + B(x^2 - 1) + C(x-1)$$
$$4x^2 = (A + B)x^2 + (2A + C)x + (A - B - C)$$

System of equations:
$$\begin{aligned} A + B \quad\quad &= 4 \\ 2A \quad + C &= 0 \\ A - B - C &= 0 \end{aligned}$$

$$\begin{bmatrix} 1 & 1 & 0 & \vdots & 4 \\ 2 & 0 & 1 & \vdots & 0 \\ 1 & -1 & -1 & \vdots & 0 \end{bmatrix} \xrightarrow{\text{rref}} \begin{bmatrix} 1 & 0 & 0 & \vdots & 1 \\ 0 & 1 & 0 & \vdots & 3 \\ 0 & 0 & 1 & \vdots & -2 \end{bmatrix}$$

Thus, $A = 1, B = 3, C = -2$.

So, $\dfrac{4x^2}{(x+1)^2(x-1)} = \dfrac{1}{x-1} + \dfrac{3}{x+1} - \dfrac{2}{(x+1)^2}$.

85. x = amount at 7%

y = amount at 8%,

z = amount at 10%

$z = 4x \implies -4x + z = 0$

$$\begin{cases} x + \quad y + \quad z = 1{,}500{,}000 \\ 0.07x + 0.08y + 0.10z = \quad 130{,}500 \\ -4x + \qquad\qquad z = \qquad\quad 0 \end{cases}$$

$$\begin{bmatrix} 1 & 1 & 1 & 1{,}500{,}000 \\ 0.07 & 0.08 & 0.10 & 130{,}500 \\ -4 & 0 & 1 & 0 \end{bmatrix}$$

$$\begin{matrix} -0.07R_1 + R_2 \to \\ 4R_1 + R_3 \to \end{matrix} \begin{bmatrix} 1 & 1 & 1 & : & 1{,}500{,}000 \\ 0 & 0.01 & 0.03 & : & 25{,}500 \\ 0 & 4 & 5 & : & 6{,}000{,}000 \end{bmatrix}$$

$$100R_2 \to \begin{bmatrix} 1 & 1 & 1 & : & 1{,}500{,}000 \\ 0 & 1 & 3 & : & 2{,}550{,}000 \\ 0 & 4 & 5 & : & 6{,}000{,}000 \end{bmatrix}$$

$$-4R_2 + R_3 \to \begin{bmatrix} 1 & 1 & 1 & : & 1{,}500{,}000 \\ 0 & 1 & 3 & : & 2{,}550{,}000 \\ 0 & 0 & -7 & : & -4{,}200{,}000 \end{bmatrix}$$

$$-\tfrac{1}{7}R_3 \to \begin{bmatrix} 1 & 1 & 1 & : & 1{,}500{,}000 \\ 0 & 1 & 3 & : & 2{,}550{,}000 \\ 0 & 0 & 1 & : & 600{,}000 \end{bmatrix}$$

$$\begin{cases} x + y + z = 1{,}500{,}000 \\ y + 3z = 2{,}550{,}000 \\ z = 600{,}000 \end{cases}$$

$y + 3(600{,}000) = 2{,}550{,}000 \implies y = 750{,}000$

$x + 750{,}000 + 600{,}000 = 1{,}500{,}000 \implies x = 150{,}000$

Solution: \$150,000 at 7%, \$750,000 at 8%,
and \$600,000 at 10%

89. (a) $(0, 5.0)$, $(15, 9.6)$, $(30, 12.4)$

$y = ax^2 + bx + c$

$$\begin{cases} c = 5 \\ 225a + 15b + c = 9.6 \implies 225a + 15b = 4.6 \\ 900a + 30b + c = 12.4 \implies 900a + 30b = 7.4 \end{cases}$$

$$\begin{bmatrix} 225 & 15 & : & 4.6 \\ 900 & 30 & : & 7.4 \end{bmatrix}$$

$$-4R_1 + R_2 \to \begin{bmatrix} 225 & 15 & : & 4.6 \\ 0 & -30 & : & -11 \end{bmatrix}$$

$$\begin{matrix} \tfrac{1}{225}R_1 \to \\ \left(-\tfrac{1}{30}\right)R_2 \to \end{matrix} \begin{bmatrix} 1 & \tfrac{1}{15} & : & \tfrac{23}{1125} \\ 0 & 1 & : & \tfrac{11}{30} \end{bmatrix}$$

$$\begin{cases} a + \tfrac{1}{15}b = \tfrac{23}{1125} \\ b = \tfrac{11}{30} \end{cases}$$

$a + \tfrac{1}{15}\left(\tfrac{11}{30}\right) = \tfrac{23}{1125} \implies a = -\tfrac{1}{250} = -0.004$

Equation of parabola: $y = -0.004x^2 + 0.367x + 5$.

87. $y = ax^2 + bx + c$

$$\begin{cases} a + \quad b + c = 8 \\ 4a + 2b + c = 13 \\ 9a + 3b + c = 20 \end{cases}$$

$$\begin{bmatrix} 1 & 1 & 1 & : & 8 \\ 4 & 2 & 1 & : & 13 \\ 9 & 3 & 1 & : & 20 \end{bmatrix}$$

$$\begin{matrix} -4R_1 + R_2 \to \\ -9R_1 + R_3 \to \end{matrix} \begin{bmatrix} 1 & 1 & 1 & : & 8 \\ 0 & -2 & -3 & : & -19 \\ 0 & -6 & -8 & : & -52 \end{bmatrix}$$

$$\begin{matrix} -\tfrac{1}{2}R_2 \to \\ -3R_2 + R_3 \to \end{matrix} \begin{bmatrix} 1 & 1 & 1 & : & 8 \\ 0 & 1 & \tfrac{3}{2} & : & \tfrac{19}{2} \\ 0 & 0 & 1 & : & 5 \end{bmatrix}$$

$$\begin{cases} a + b + c = 8 \\ b + \tfrac{3}{2}c = \tfrac{19}{2} \\ c = 5 \end{cases}$$

$c = 5$

$b + \tfrac{3}{2}(5) = \tfrac{19}{2} \implies b = 2$

$a + 2 + 5 = 8 \implies a = 1$

Equation of parabola: $y = x^2 + 2x + 5$

(b)

(c) The maximum height is approximately 13 feet and the ball strikes the ground at approximately 104 feet.

(d) The maximum occurs at the vertex.

$x = -\dfrac{b}{2a} = \dfrac{-0.367}{2(-0.004)} = 45.875$

$y = -0.004(45.875)^2 + 0.367(45.875) + 5$

$\quad = 13.418$ feet

The ball strikes the ground when $y = 0$.

$-0.004x^2 + 0.367x + 5 = 0$

By the Quadratic Formula and using the positive value for x we have $x \approx 103.793$ feet.

(e) The values found in part (d) are more accurate, but still very close to the estimates found in part (c).

91. (a) $x_1 + x_3 = 600$

$x_1 = x_2 + x_4 \implies x_1 - x_2 - x_4 = 0$

$x_2 + x_5 = 500$

$x_3 + x_6 = 600$

$x_4 + x_7 = x_6 \implies x_4 - x_6 + x_7 = 0$

$x_5 + x_7 = 500$

$$\left[\begin{array}{ccccccc:c} 1 & 0 & 1 & 0 & 0 & 0 & 0 & 600 \\ 1 & -1 & 0 & -1 & 0 & 0 & 0 & 0 \\ 0 & 1 & 0 & 0 & 1 & 0 & 0 & 500 \\ 0 & 0 & 1 & 0 & 0 & 1 & 0 & 600 \\ 0 & 0 & 0 & 1 & 0 & -1 & 1 & 0 \\ 0 & 0 & 0 & 0 & 1 & 0 & 1 & 500 \end{array}\right]$$

$$\begin{array}{l} \\ -R_1 + R_2 \to \\ R_2 + R_3 \to \\ R_3 + R_4 \to \\ R_4 + R_5 \to \\ -R_5 + R_6 \to \end{array} \left[\begin{array}{ccccccc:c} 1 & 0 & 1 & 0 & 0 & 0 & 0 & 600 \\ 0 & -1 & -1 & -1 & 0 & 0 & 0 & -600 \\ 0 & 0 & -1 & -1 & 1 & 0 & 0 & -100 \\ 0 & 0 & 0 & -1 & 1 & 1 & 0 & 500 \\ 0 & 0 & 0 & 0 & 1 & 0 & 1 & 500 \\ 0 & 0 & 0 & 0 & 0 & 0 & 0 & 0 \end{array}\right]$$

$$\begin{array}{l} \\ -R_3 + R_2 \to \\ -R_4 + R_3 \to \\ -R_4 \to \\ \\ \end{array} \left[\begin{array}{ccccccc:c} 1 & 0 & 1 & 0 & 0 & 0 & 0 & 600 \\ 0 & -1 & 0 & 0 & -1 & 0 & 0 & -500 \\ 0 & 0 & -1 & 0 & 0 & -1 & 0 & -600 \\ 0 & 0 & 0 & 1 & -1 & -1 & 0 & -500 \\ 0 & 0 & 0 & 0 & 1 & 0 & 1 & 500 \\ 0 & 0 & 0 & 0 & 0 & 0 & 0 & 0 \end{array}\right]$$

$$\begin{array}{l} \\ -R_2 \to \\ -R_3 \to \\ \\ \\ \end{array} \left[\begin{array}{ccccccc:c} 1 & 0 & 1 & 0 & 0 & 0 & 0 & 600 \\ 0 & 1 & 0 & 0 & 1 & 0 & 0 & 500 \\ 0 & 0 & 1 & 0 & 0 & 1 & 0 & 600 \\ 0 & 0 & 0 & 1 & -1 & -1 & 0 & -500 \\ 0 & 0 & 0 & 0 & 1 & 0 & 1 & 500 \\ 0 & 0 & 0 & 0 & 0 & 0 & 0 & 0 \end{array}\right]$$

$$\begin{cases} x_1 + x_3 = 600 \\ x_2 + x_5 = 500 \\ x_3 + x_6 = 600 \\ x_4 - x_5 - x_6 = -500 \\ x_5 + x_7 = 500 \end{cases}$$

Let $x_7 = t$ and $x_6 = s$, then $x_5 = 500 - t$,

$x_4 = -500 + s + (500 - t) = s - t$,

$x_3 = 600 - s$, $x_2 = 500 - (500 - t) = t$,

$x_1 = 600 - (600 - s) = s$.

Solution: $(s, t, 600 - s, s - t, 500 - t, s, t)$

(b) $s = 0, t = 0$: $x_1 = 0, x_2 = 0, x_3 = 600, x_4 = 0, x_5 = 500, x_6 = 0, x_7 = 0$

(c) $s = 0, t = -500$: $x_1 = 0, x_2 = -500, x_3 = 600, x_4 = 500, x_5 = 1000, x_6 = 0, x_7 = -500$

93. False. It is a 2×4 matrix.

95. False. Gaussian elimination reduces a matrix until a row-echelon form is obtained and Gauss-Jordan elimination reduces a matrix until a reduced row-echelon form is obtained.

97. (a) In the row-echelon form of an augmented matrix that corresponds to an inconsistent system of linear equations, there exists a row consisting of all zeros except for the entry in the last column.

(b) In the row-echelon form of an augmented matrix that corresponds to a system with an infinite number of solutions, there are fewer rows with nonzero entries than there are variables and no row has the first non-zero value in the last column.

99. They are the same.

101. $f(x) = \dfrac{2x^2 - 4x}{3x - x^2} = \dfrac{2x - 4}{3 - x}, x \neq 0$

x	-2	-1	0	1	2	3	4	5
$f(x)$	-1.6	-1.5	undef.	-1	0	undef.	-4	-3

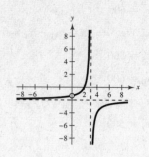

Vertical asymptote: $x = 3$

Horizontal asymptote: $y = -2$

Intercept: $(2, 0)$

103. $f(x) = 2^{x-1}$

x	-1	0	1	2	3
$f(x)$	$\frac{1}{4}$	$\frac{1}{2}$	1	2	4

Horizontal asymptote: $y = 0$

Intercept: $\left(0, \frac{1}{2}\right)$

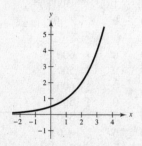

105. $h(x) = \ln(x - 1)$

x	1.5	2	3	4	5
$h(x)$	-0.693	0	0.693	1.099	1.386

Vertical asymptote: $x = 1$

Intercept: $(2, 0)$

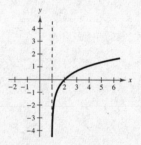

Section 8.2   Operations with Matrices

- $A = B$ if and only if they have the same order and $a_{ij} = b_{ij}$.
- You should be able to perform the operations of matrix addition, scalar multiplication, and matrix multiplication.
- Some properties of matrix addition and scalar multiplication are:
    - (a) $A + B = B + A$
    - (b) $A + (B + C) = (A + B) + C$
    - (c) $(cd)A = c(dA)$
    - (d) $1A = A$
    - (e) $c(A + B) = cA + cB$
    - (f) $(c + d)A = cA + dA$
- You should remember that $AB \neq BA$ in general.
- Some properties of matrix multiplication are:
    - (a) $A(BC) = (AB)C$
    - (b) $A(B + C) = AB + AC$
    - (c) $(A + B)C = AC + BC$
    - (d) $c(AB) = (cA)B = A(cB)$
- You should know that I_n, the identity matrix of order n, is an $n \times n$ matrix consisting of ones on its main diagonal and zeros elsewhere. If A is an $n \times n$ matrix, then $AI_n = I_nA = A$.

Vocabulary Check

1. equal

2. scalars

3. zero; O

4. identity

5. (a) (iii)   (b) (iv)   (c) (i)
(d) (v)   (e) (ii)

6. (a) (ii)   (b) (iv)   (c) (i)   (d) (iii)

---

1. $x = -4,\ y = 22$

3. $2x + 1 = 5,\ 3x = 6,\ 3y - 5 = 4$

$x = 2,\ y = 3$

5. (a) $A + B = \begin{bmatrix} 1 & -1 \\ 2 & -1 \end{bmatrix} + \begin{bmatrix} 2 & -1 \\ -1 & 8 \end{bmatrix} = \begin{bmatrix} 1+2 & -1-1 \\ 2-1 & -1+8 \end{bmatrix} = \begin{bmatrix} 3 & -2 \\ 1 & 7 \end{bmatrix}$

(b) $A - B = \begin{bmatrix} 1 & -1 \\ 2 & -1 \end{bmatrix} - \begin{bmatrix} 2 & -1 \\ -1 & 8 \end{bmatrix} = \begin{bmatrix} 1-2 & -1+1 \\ 2+1 & -1-8 \end{bmatrix} = \begin{bmatrix} -1 & 0 \\ 3 & -9 \end{bmatrix}$

(c) $3A = 3\begin{bmatrix} 1 & -1 \\ 2 & -1 \end{bmatrix} = \begin{bmatrix} 3(1) & 3(-1) \\ 3(2) & 3(-1) \end{bmatrix} = \begin{bmatrix} 3 & -3 \\ 6 & -3 \end{bmatrix}$

(d) $3A - 2B = \begin{bmatrix} 3 & -3 \\ 6 & -3 \end{bmatrix} - 2\begin{bmatrix} 2 & -1 \\ -1 & 8 \end{bmatrix} = \begin{bmatrix} 3 & -3 \\ 6 & -3 \end{bmatrix} + \begin{bmatrix} -4 & 2 \\ 2 & -16 \end{bmatrix} = \begin{bmatrix} -1 & -1 \\ 8 & -19 \end{bmatrix}$

7. $A = \begin{bmatrix} 6 & -1 \\ 2 & 4 \\ -3 & 5 \end{bmatrix},\ B = \begin{bmatrix} 1 & 4 \\ -1 & 5 \\ 1 & 10 \end{bmatrix}$

(a) $A + B = \begin{bmatrix} 7 & 3 \\ 1 & 9 \\ -2 & 15 \end{bmatrix}$

(b) $A - B = \begin{bmatrix} 5 & -5 \\ 3 & -1 \\ -4 & -5 \end{bmatrix}$

(c) $3A = \begin{bmatrix} 18 & -3 \\ 6 & 12 \\ -9 & 15 \end{bmatrix}$

(d) $3A - 2B = \begin{bmatrix} 18 & -3 \\ 6 & 12 \\ -9 & 15 \end{bmatrix} - \begin{bmatrix} 2 & 8 \\ -2 & 10 \\ 2 & 20 \end{bmatrix} = \begin{bmatrix} 16 & -11 \\ 8 & 2 \\ -11 & -5 \end{bmatrix}$

9. $A = \begin{bmatrix} 2 & 2 & -1 & 0 & 1 \\ 1 & 1 & -2 & 0 & -1 \end{bmatrix},\ B = \begin{bmatrix} 1 & 1 & -1 & 1 & 0 \\ -3 & 4 & 9 & -6 & -7 \end{bmatrix}$

(a) $A + B = \begin{bmatrix} 3 & 3 & -2 & 1 & 1 \\ -2 & 5 & 7 & -6 & -8 \end{bmatrix}$

(b) $A - B = \begin{bmatrix} 1 & 1 & 0 & -1 & 1 \\ 4 & -3 & -11 & 6 & 6 \end{bmatrix}$

(c) $3A = \begin{bmatrix} 6 & 6 & -3 & 0 & 3 \\ 3 & 3 & -6 & 0 & -3 \end{bmatrix}$

(d) $3A - 2B = \begin{bmatrix} 6 & 6 & -3 & 0 & 3 \\ 3 & 3 & -6 & 0 & -3 \end{bmatrix} - \begin{bmatrix} 2 & 2 & -2 & 2 & 0 \\ -6 & 8 & 18 & -12 & -14 \end{bmatrix} = \begin{bmatrix} 4 & 4 & -1 & -2 & 3 \\ 9 & -5 & -24 & 12 & 11 \end{bmatrix}$

11. $A = \begin{bmatrix} 6 & 0 & 3 \\ -1 & -4 & 0 \end{bmatrix},\ B = \begin{bmatrix} 8 & -1 \\ 4 & -3 \end{bmatrix}$

(a) $A + B$ is not possible. A and B do not have the same order.

(b) $A - B$ is not possible. A and B do not have the same order.

(c) $3A = \begin{bmatrix} 18 & 0 & 9 \\ -3 & -12 & 0 \end{bmatrix}$

(d) $3A - 2B$ is not possible. A and B do not have the same order.

13. $\begin{bmatrix} -5 & 0 \\ 3 & -6 \end{bmatrix} + \begin{bmatrix} 7 & 1 \\ -2 & -1 \end{bmatrix} + \begin{bmatrix} -10 & -8 \\ 14 & 6 \end{bmatrix} = \begin{bmatrix} -5+7+(-10) & 0+1+(-8) \\ 3+(-2)+14 & -6+(-1)+6 \end{bmatrix} = \begin{bmatrix} -8 & -7 \\ 15 & -1 \end{bmatrix}$

15. $4\left(\begin{bmatrix} -4 & 0 & 1 \\ 0 & 2 & 3 \end{bmatrix} - \begin{bmatrix} 2 & 1 & -2 \\ 3 & -6 & 0 \end{bmatrix}\right) = 4\begin{bmatrix} -6 & -1 & 3 \\ -3 & 8 & 3 \end{bmatrix} = \begin{bmatrix} -24 & -4 & 12 \\ -12 & 32 & 12 \end{bmatrix}$

17. $-3\left(\begin{bmatrix} 0 & -3 \\ 7 & 2 \end{bmatrix} + \begin{bmatrix} -6 & 3 \\ 8 & 1 \end{bmatrix}\right) - 2\begin{bmatrix} 4 & -4 \\ 7 & -9 \end{bmatrix} = -3\begin{bmatrix} -6 & 0 \\ 15 & 3 \end{bmatrix} - \begin{bmatrix} 8 & -8 \\ 14 & -18 \end{bmatrix} = \begin{bmatrix} 18 & 0 \\ -45 & -9 \end{bmatrix} - \begin{bmatrix} 8 & -8 \\ 14 & -18 \end{bmatrix} = \begin{bmatrix} 10 & 8 \\ -59 & 9 \end{bmatrix}$

19. $\frac{3}{7}\begin{bmatrix} 2 & 5 \\ -1 & -4 \end{bmatrix} + 6\begin{bmatrix} -3 & 0 \\ 2 & 2 \end{bmatrix} \approx \begin{bmatrix} -17.143 & 2.143 \\ 11.571 & 10.286 \end{bmatrix}$

21. $-\begin{bmatrix} 3.211 & 6.829 \\ -1.004 & 4.914 \\ 0.055 & -3.889 \end{bmatrix} - \begin{bmatrix} -1.630 & -3.090 \\ 5.256 & 8.335 \\ -9.768 & 4.251 \end{bmatrix} = \begin{bmatrix} -1.581 & -3.739 \\ -4.252 & -13.249 \\ 9.713 & -0.362 \end{bmatrix}$

23. $X = 3\begin{bmatrix} -2 & -1 \\ 1 & 0 \\ 3 & -4 \end{bmatrix} - 2\begin{bmatrix} 0 & 3 \\ 2 & 0 \\ -4 & -1 \end{bmatrix} = \begin{bmatrix} -6 & -3 \\ 3 & 0 \\ 9 & -12 \end{bmatrix} - \begin{bmatrix} 0 & 6 \\ 4 & 0 \\ -8 & -2 \end{bmatrix} = \begin{bmatrix} -6 & -9 \\ -1 & 0 \\ 17 & -10 \end{bmatrix}$

25. $X = -\frac{3}{2}A + \frac{1}{2}B = -\frac{3}{2}\begin{bmatrix} -2 & -1 \\ 1 & 0 \\ 3 & -4 \end{bmatrix} + \frac{1}{2}\begin{bmatrix} 0 & 3 \\ 2 & 0 \\ -4 & -1 \end{bmatrix} = \begin{bmatrix} 3 & \frac{3}{2} \\ -\frac{3}{2} & 0 \\ -\frac{9}{2} & 6 \end{bmatrix} + \begin{bmatrix} 0 & \frac{3}{2} \\ 1 & 0 \\ -2 & -\frac{1}{2} \end{bmatrix} = \begin{bmatrix} 3 & 3 \\ -\frac{1}{2} & 0 \\ -\frac{13}{2} & \frac{11}{2} \end{bmatrix}$

27. A is 3×2 and B is 3×3. AB is not possible.

29. A is 3×3, B is $3 \times 2 \implies AB$ is 3×2.

$\begin{bmatrix} 0 & -1 & 0 \\ 4 & 0 & 2 \\ 8 & -1 & 7 \end{bmatrix}\begin{bmatrix} 2 & 1 \\ -3 & 4 \\ 1 & 6 \end{bmatrix} = \begin{bmatrix} (0)(2) + (-1)(-3) + (0)(1) & (0)(1) + (-1)(4) + (0)(6) \\ (4)(2) + (0)(-3) + (2)(1) & (4)(1) + (0)(4) + (2)(6) \\ (8)(2) + (-1)(-3) + (7)(1) & (8)(1) + (-1)(4) + (7)(6) \end{bmatrix} = \begin{bmatrix} 3 & -4 \\ 10 & 16 \\ 26 & 46 \end{bmatrix}$

31. A is 3×3, B is $3 \times 3 \implies AB$ is 3×3.

$\begin{bmatrix} 1 & 0 & 0 \\ 0 & 4 & 0 \\ 0 & 0 & -2 \end{bmatrix}\begin{bmatrix} 3 & 0 & 0 \\ 0 & -1 & 0 \\ 0 & 0 & 5 \end{bmatrix} = \begin{bmatrix} (1)(3) + (0)(0) + (0)(0) & (1)(0) + (0)(-1) + (0)(0) & (1)(0) + (0)(0) + (0)(5) \\ (0)(3) + (4)(0) + (0)(0) & (0)(0) + (4)(-1) + (0)(0) & (0)(0) + (4)(0) + (0)(5) \\ (0)(3) + (0)(0) + (-2)(0) & (0)(0) + (0)(-1) + (-2)(0) & (0)(0) + (0)(0) + (-2)(5) \end{bmatrix}$

$= \begin{bmatrix} 3 & 0 & 0 \\ 0 & -4 & 0 \\ 0 & 0 & -10 \end{bmatrix}$

33. A is 3×3, B is $3 \times 3 \implies AB$ is 3×3.

$\begin{bmatrix} 0 & 0 & 5 \\ 0 & 0 & -3 \\ 0 & 0 & 4 \end{bmatrix}\begin{bmatrix} 6 & -11 & 4 \\ 8 & 16 & 4 \\ 0 & 0 & 0 \end{bmatrix} = \begin{bmatrix} (0)(6) + (0)(8) + (5)(0) & (0)(-11) + (0)(16) + (5)(0) & (0)(4) + (0)(4) + (5)(0) \\ (0)(6) + (0)(8) + (-3)(0) & (0)(-11) + (0)(16) + (-3)(0) & (0)(4) + (0)(4) + (-3)(0) \\ (0)(6) + (0)(8) + (4)(0) & (0)(-11) + (0)(16) + (4)(0) & (0)(4) + (0)(4) + (4)(0) \end{bmatrix}$

$= \begin{bmatrix} 0 & 0 & 0 \\ 0 & 0 & 0 \\ 0 & 0 & 0 \end{bmatrix}$

35. $\begin{bmatrix} 5 & 6 & -3 \\ -2 & 5 & 1 \\ 10 & -5 & 5 \end{bmatrix}\begin{bmatrix} 1 & -1 & 2 \\ 8 & 1 & 4 \\ 4 & -2 & 9 \end{bmatrix} = \begin{bmatrix} 41 & 7 & 7 \\ 42 & 5 & 25 \\ -10 & -25 & 45 \end{bmatrix}$

37. $\begin{bmatrix} -3 & 8 & -6 & 8 \\ -12 & 15 & 9 & 6 \\ 5 & -1 & 1 & 5 \end{bmatrix} \begin{bmatrix} 3 & 1 & 6 \\ 24 & 15 & 14 \\ 16 & 10 & 21 \\ 8 & -4 & 10 \end{bmatrix} = \begin{bmatrix} 151 & 25 & 48 \\ 516 & 279 & 387 \\ 47 & -20 & 87 \end{bmatrix}$

39. A is 2×4 and B is $2 \times 4 \implies AB$ is not possible.

41. (a) $AB = \begin{bmatrix} 1 & 2 \\ 4 & 2 \end{bmatrix} \begin{bmatrix} 2 & -1 \\ -1 & 8 \end{bmatrix} = \begin{bmatrix} (1)(2) + (2)(-1) & (1)(-1) + (2)(8) \\ (4)(2) + (2)(-1) & (4)(-1) + (2)(8) \end{bmatrix} = \begin{bmatrix} 0 & 15 \\ 6 & 12 \end{bmatrix}$

(b) $BA = \begin{bmatrix} 2 & -1 \\ -1 & 8 \end{bmatrix} \begin{bmatrix} 1 & 2 \\ 4 & 2 \end{bmatrix} = \begin{bmatrix} (2)(1) + (-1)(4) & (2)(2) + (-1)(2) \\ (-1)(1) + (8)(4) & (-1)(2) + (8)(2) \end{bmatrix} = \begin{bmatrix} -2 & 2 \\ 31 & 14 \end{bmatrix}$

(c) $A^2 = \begin{bmatrix} 1 & 2 \\ 4 & 2 \end{bmatrix} \begin{bmatrix} 1 & 2 \\ 4 & 2 \end{bmatrix} = \begin{bmatrix} (1)(1) + (2)(4) & (1)(2) + (2)(2) \\ (4)(1) + (2)(4) & (4)(2) + (2)(2) \end{bmatrix} = \begin{bmatrix} 9 & 6 \\ 12 & 12 \end{bmatrix}$

43. (a) $AB = \begin{bmatrix} 3 & -1 \\ 1 & 3 \end{bmatrix} \begin{bmatrix} 1 & -3 \\ 3 & 1 \end{bmatrix} = \begin{bmatrix} (3)(1) + (-1)(3) & (3)(-3) + (-1)(1) \\ (1)(1) + (3)(3) & (1)(-3) + (3)(1) \end{bmatrix} = \begin{bmatrix} 0 & -10 \\ 10 & 0 \end{bmatrix}$

(b) $BA = \begin{bmatrix} 1 & -3 \\ 3 & 1 \end{bmatrix} \begin{bmatrix} 3 & -1 \\ 1 & 3 \end{bmatrix} = \begin{bmatrix} (1)(3) + (-3)(1) & (1)(-1) + (-3)(3) \\ (3)(3) + (1)(1) & (3)(-1) + (1)(3) \end{bmatrix} = \begin{bmatrix} 0 & -10 \\ 10 & 0 \end{bmatrix}$

(c) $A^2 = \begin{bmatrix} 3 & -1 \\ 1 & 3 \end{bmatrix} \begin{bmatrix} 3 & -1 \\ 1 & 3 \end{bmatrix} = \begin{bmatrix} (3)(3) + (-1)(1) & (3)(-1) + (-1)(3) \\ (1)(3) + (3)(1) & (1)(-1) + (3)(3) \end{bmatrix} = \begin{bmatrix} 8 & -6 \\ 6 & 8 \end{bmatrix}$

45. (a) $AB = \begin{bmatrix} 7 \\ 8 \\ -1 \end{bmatrix} \begin{bmatrix} 1 & 1 & 2 \end{bmatrix} = \begin{bmatrix} 7(1) & 7(1) & 7(2) \\ 8(1) & 8(1) & 8(2) \\ -1(1) & -1(1) & -1(2) \end{bmatrix} = \begin{bmatrix} 7 & 7 & 14 \\ 8 & 8 & 16 \\ -1 & -1 & -2 \end{bmatrix}$

(b) $BA = \begin{bmatrix} 1 & 1 & 2 \end{bmatrix} \begin{bmatrix} 7 \\ 8 \\ -1 \end{bmatrix} = [(1)(7) + (1)(8) + (2)(-1)] = [13]$

(c) A^2 is not possible.

47. $\begin{bmatrix} 3 & 1 \\ 0 & -2 \end{bmatrix} \begin{bmatrix} 1 & 0 \\ -2 & 2 \end{bmatrix} \begin{bmatrix} 1 & 0 \\ 2 & 4 \end{bmatrix} = \begin{bmatrix} 1 & 2 \\ 4 & -4 \end{bmatrix} \begin{bmatrix} 1 & 0 \\ 2 & 4 \end{bmatrix} = \begin{bmatrix} 5 & 8 \\ -4 & -16 \end{bmatrix}$

49. $\begin{bmatrix} 0 & 2 & -2 \\ 4 & 1 & 2 \end{bmatrix} \left(\begin{bmatrix} 4 & 0 \\ 0 & -1 \\ -1 & 2 \end{bmatrix} + \begin{bmatrix} -2 & 3 \\ -3 & 5 \\ 0 & -3 \end{bmatrix} \right) = \begin{bmatrix} 0 & 2 & -2 \\ 4 & 1 & 2 \end{bmatrix} \begin{bmatrix} 2 & 3 \\ -3 & 4 \\ -1 & -1 \end{bmatrix} = \begin{bmatrix} -4 & 10 \\ 3 & 14 \end{bmatrix}$

51. (a) $\begin{bmatrix} -1 & 1 \\ -2 & 1 \end{bmatrix} \begin{bmatrix} x_1 \\ x_2 \end{bmatrix} = \begin{bmatrix} 4 \\ 0 \end{bmatrix}$

(b) $\begin{bmatrix} -1 & 1 & \vdots & 4 \\ -2 & 1 & \vdots & 0 \end{bmatrix}$

$-R_2 + R_1 \to \begin{bmatrix} 1 & 0 & \vdots & 4 \\ -2 & 1 & \vdots & 0 \end{bmatrix}$

$2R_1 + R_2 \to \begin{bmatrix} 1 & 0 & \vdots & 4 \\ 0 & 1 & \vdots & 8 \end{bmatrix}$

$X = \begin{bmatrix} 4 \\ 8 \end{bmatrix}$

53. (a) $\begin{bmatrix} -2 & -3 \\ 6 & 1 \end{bmatrix} \begin{bmatrix} x_1 \\ x_2 \end{bmatrix} = \begin{bmatrix} -4 \\ -36 \end{bmatrix}$

(b) $\begin{bmatrix} -2 & -3 & \vdots & -4 \\ 6 & 1 & \vdots & -36 \end{bmatrix}$

$3R_1 + R_2 \to \begin{bmatrix} -2 & -3 & \vdots & -4 \\ 0 & -8 & \vdots & -48 \end{bmatrix}$

$\begin{matrix} -\frac{1}{2}R_1 \to \\ -\frac{1}{8}R_2 \to \end{matrix} \begin{bmatrix} 1 & \frac{3}{2} & \vdots & 2 \\ 0 & 1 & \vdots & 6 \end{bmatrix}$

$-\frac{3}{2}R_2 + R_1 \to \begin{bmatrix} 1 & 0 & \vdots & -7 \\ 0 & 1 & \vdots & 6 \end{bmatrix}$

$X = \begin{bmatrix} -7 \\ 6 \end{bmatrix}$

55. (a) $A = \begin{bmatrix} 1 & -2 & 3 \\ -1 & 3 & -1 \\ 2 & -5 & 5 \end{bmatrix}\begin{bmatrix} x_1 \\ x_2 \\ x_3 \end{bmatrix} = \begin{bmatrix} 9 \\ -6 \\ 17 \end{bmatrix}$

(b) $\begin{bmatrix} 1 & -2 & 3 & \vdots & 9 \\ -1 & 3 & -1 & \vdots & -6 \\ 2 & -5 & 5 & \vdots & 17 \end{bmatrix}$

$\begin{array}{c} R_1 + R_2 \rightarrow \\ -2R_2 + R_3 \rightarrow \end{array}\begin{bmatrix} 1 & -2 & 3 & \vdots & 9 \\ 0 & 1 & 2 & \vdots & 3 \\ 0 & -1 & -1 & \vdots & -1 \end{bmatrix}$

$\begin{array}{c} 2R_2 + R_1 \rightarrow \\ \\ R_2 + R_3 \rightarrow \end{array}\begin{bmatrix} 1 & 0 & 7 & \vdots & 15 \\ 0 & 1 & 2 & \vdots & 3 \\ 0 & 0 & 1 & \vdots & 2 \end{bmatrix}$

$\begin{array}{c} -7R_3 + R_1 \rightarrow \\ -2R_3 + R_2 \rightarrow \end{array}\begin{bmatrix} 1 & 0 & 0 & \vdots & 1 \\ 0 & 1 & 0 & \vdots & -1 \\ 0 & 0 & 1 & \vdots & 2 \end{bmatrix}$

$X = \begin{bmatrix} 1 \\ -1 \\ 2 \end{bmatrix}$

57. (a) $\begin{bmatrix} 1 & -5 & 2 \\ -3 & 1 & -1 \\ 0 & -2 & 5 \end{bmatrix}\begin{bmatrix} x_1 \\ x_2 \\ x_3 \end{bmatrix} = \begin{bmatrix} -20 \\ 8 \\ -16 \end{bmatrix}$

(b) $\begin{bmatrix} 1 & -5 & 2 & \vdots & -20 \\ -3 & 1 & -1 & \vdots & 8 \\ 0 & -2 & 5 & \vdots & -16 \end{bmatrix}$

$3R_1 + R_2 \rightarrow \begin{bmatrix} 1 & -5 & 2 & \vdots & -20 \\ 0 & -14 & 5 & \vdots & -52 \\ 0 & -2 & 5 & \vdots & -16 \end{bmatrix}$

$-R_3 + R_2 \rightarrow \begin{bmatrix} 1 & -5 & 2 & \vdots & -20 \\ 0 & -12 & 0 & \vdots & -36 \\ 0 & -2 & 5 & \vdots & -16 \end{bmatrix}$

$-\tfrac{1}{12}R_2 \rightarrow \begin{bmatrix} 1 & -5 & 2 & \vdots & -20 \\ 0 & 1 & 0 & \vdots & 3 \\ 0 & -2 & 5 & \vdots & -16 \end{bmatrix}$

$\begin{array}{c} 5R_2 + R_1 \rightarrow \\ \\ 2R_2 + R_3 \rightarrow \end{array}\begin{bmatrix} 1 & 0 & 2 & \vdots & -5 \\ 0 & 1 & 0 & \vdots & 3 \\ 0 & 0 & 5 & \vdots & -10 \end{bmatrix}$

$\tfrac{1}{5}R_3 \rightarrow \begin{bmatrix} 1 & 0 & 2 & \vdots & -5 \\ 0 & 1 & 0 & \vdots & 3 \\ 0 & 0 & 1 & \vdots & -2 \end{bmatrix}$

$-2R_3 + R_1 \rightarrow \begin{bmatrix} 1 & 0 & 0 & \vdots & -1 \\ 0 & 1 & 0 & \vdots & 3 \\ 0 & 0 & 1 & \vdots & -2 \end{bmatrix}$

$X = \begin{bmatrix} -1 \\ 3 \\ -2 \end{bmatrix}$

59. $1.2\begin{bmatrix} 70 & 50 & 25 \\ 35 & 100 & 70 \end{bmatrix} = \begin{bmatrix} 84 & 60 & 30 \\ 42 & 120 & 84 \end{bmatrix}$

61. (a)

	Farmer's Market	Fruit Stand	Fruit Farm	
$A = $	125	100	75	Apples
	100	175	125	Peaches

Each entry represents the number of bushels of each type of crop that are shipped to each outlet.

(b) $B = [3.50 \quad 6.00]$

Each entry represents the profit per bushel for each type of crop.

(c) $BA = [3.50 \quad 6.00]\begin{bmatrix} 125 & 100 & 75 \\ 100 & 175 & 125 \end{bmatrix}$

$= [\$1037.50 \quad \$1400.00 \quad \$1012.50]$

The entries in the matrix represent the profits for both crops at each of the three outlets.

63. $ST = \begin{bmatrix} 3 & 2 & 2 & 3 & 0 \\ 0 & 2 & 3 & 4 & 3 \\ 4 & 2 & 1 & 3 & 2 \end{bmatrix}\begin{bmatrix} 840 & 1100 \\ 1200 & 1350 \\ 1450 & 1650 \\ 2650 & 3000 \\ 3050 & 3200 \end{bmatrix} = \begin{bmatrix} \$15,770 & \$18,300 \\ \$26,500 & \$29,250 \\ \$21,260 & \$24,150 \end{bmatrix}$

The entries represent the wholesale and retail inventory values of the inventories at the three outlets.

65. $P^3 = P^2P = \begin{bmatrix} 0.40 & 0.15 & 0.15 \\ 0.28 & 0.53 & 0.17 \\ 0.32 & 0.32 & 0.68 \end{bmatrix} \begin{bmatrix} 0.6 & 0.1 & 0.1 \\ 0.2 & 0.7 & 0.1 \\ 0.2 & 0.2 & 0.8 \end{bmatrix} = \begin{bmatrix} 0.300 & 0.175 & 0.175 \\ 0.308 & 0.433 & 0.217 \\ 0.392 & 0.392 & 0.608 \end{bmatrix}$

$P^4 = P^3P = \begin{bmatrix} 0.300 & 0.175 & 0.175 \\ 0.308 & 0.433 & 0.217 \\ 0.392 & 0.392 & 0.608 \end{bmatrix} \begin{bmatrix} 0.6 & 0.1 & 0.1 \\ 0.2 & 0.7 & 0.1 \\ 0.2 & 0.2 & 0.8 \end{bmatrix} = \begin{bmatrix} 0.250 & 0.188 & 0.188 \\ 0.315 & 0.377 & 0.248 \\ 0.435 & 0.435 & 0.565 \end{bmatrix}$

$P^5 = P^4P = \begin{bmatrix} 0.250 & 0.188 & 0.188 \\ 0.315 & 0.377 & 0.248 \\ 0.435 & 0.435 & 0.565 \end{bmatrix} \begin{bmatrix} 0.6 & 0.1 & 0.1 \\ 0.2 & 0.7 & 0.1 \\ 0.2 & 0.2 & 0.8 \end{bmatrix} = \begin{bmatrix} 0.225 & 0.194 & 0.194 \\ 0.314 & 0.345 & 0.267 \\ 0.461 & 0.461 & 0.539 \end{bmatrix}$

$P^6 = \begin{bmatrix} 0.213 & 0.197 & 0.197 \\ 0.311 & 0.326 & 0.280 \\ 0.477 & 0.477 & 0.523 \end{bmatrix}$

$P^7 = \begin{bmatrix} 0.206 & 0.198 & 0.198 \\ 0.308 & 0.316 & 0.288 \\ 0.486 & 0.486 & 0.514 \end{bmatrix}$

$P^8 = \begin{bmatrix} 0.203 & 0.199 & 0.199 \\ 0.305 & 0.309 & 0.292 \\ 0.492 & 0.492 & 0.508 \end{bmatrix}$

As P is raised to higher and higher powers, the resulting matrices appear to be approaching the matrix

$\begin{bmatrix} 0.2 & 0.2 & 0.2 \\ 0.3 & 0.3 & 0.3 \\ 0.5 & 0.5 & 0.5 \end{bmatrix}$.

67. (a) $AB = \begin{bmatrix} 40 & 64 & 52 \\ 60 & 82 & 76 \\ 76 & 96 & 84 \end{bmatrix} \begin{bmatrix} 2.65 & 0.65 \\ 2.85 & 0.70 \\ 3.05 & 0.85 \end{bmatrix} = \begin{matrix} & \text{Sales} & \text{Profit} \\ & \begin{bmatrix} 447 & 115 \\ 624.50 & 161 \\ 731.20 & 188 \end{bmatrix} & \begin{matrix} \text{Friday} \\ \text{Saturday} \\ \text{Sunday} \end{matrix} \end{matrix}$

The entries in Column 1 represent the total sales of the three kinds of milk for Friday, Saturday, and Sunday. The entries in Column 2 represent each days' total profit.

(b) Total profit for the weekend: $115 + 161 + 188 = \$464$

69. (a) $\begin{matrix} & \text{Bicycled} & \text{Jogged} & \text{Walked} \\ B = & [2 & 0.5 & 3] \end{matrix}$ 20-minute time periods

(b) $BA = \begin{matrix} \\ [2 \quad 0.5 \quad 3] \end{matrix} \begin{matrix} \\ \begin{bmatrix} 109 & 136 \\ 127 & 159 \\ 64 & 79 \end{bmatrix} \end{matrix} = \begin{matrix} \text{120-pound} & \text{150-pound} \\ \text{person} & \text{person} \\ [473.5 & 588.5] \end{matrix}$ Calories burned

The first entry represents the total calories burned by the 120-pound person and the second entry represents the total calories burned by the 150-pound person.

71. True. The sum of two matrices of different orders is undefined.

For 73–79, A is of order 2×3, B is of order 2×3, C is of order 3×2 and D is of order 2×2.

73. $A + 2C$ is not possible. A and C are not of the same order.

75. AB is not possible. The number of columns of A does not equal the number of rows of B.

77. $BC - D$ is possible. The resulting order is 2×2.

79. $D(A - 3B)$ is possible. The resulting order is 2×3.

81. $AC = \begin{bmatrix} 0 & 1 \\ 0 & 1 \end{bmatrix}\begin{bmatrix} 2 & 3 \\ 2 & 3 \end{bmatrix} = \begin{bmatrix} 2 & 3 \\ 2 & 3 \end{bmatrix}$

$BC = \begin{bmatrix} 1 & 0 \\ 1 & 0 \end{bmatrix}\begin{bmatrix} 2 & 3 \\ 2 & 3 \end{bmatrix} = \begin{bmatrix} 2 & 3 \\ 2 & 3 \end{bmatrix}$

Thus, $AC = BC$ even though $A \neq B$.

83. The product of two diagonal matrices of the same order is a diagonal matrix whose entries are the products of the corresponding diagonal entries of A and B.

85. $3x^2 + 20x - 32 = 0$

$(3x - 4)(x + 8) = 0$

$3x - 4 = 0$ or $x + 8 = 0$

$x = \frac{4}{3}$ or $x = -8$

Solutions: $\frac{4}{3}, -8$

87. $4x^3 + 10x^2 - 3x = 0$

$x(4x^2 + 10x - 3) = 0$

$x = 0$ or $4x^2 + 10x - 3 = 0$

$x = \frac{-10 \pm \sqrt{10^2 - 4(4)(-3)}}{2(4)} = \frac{-10 \pm \sqrt{148}}{8}$

$= \frac{-5 \pm \sqrt{37}}{4}$ by the Quadratic Formula

Solutions: $0, \dfrac{-5 \pm \sqrt{37}}{4}$

89. $3x^3 - 12x^2 + 5x - 20 = 0$

$3x^2(x - 4) + 5(x - 4) = 0$

$(x - 4)(3x^2 + 5) = 0$

$x - 4 = 0$ or $3x^2 + 5 = 0$

$x = 4$ $\qquad x^2 = -\frac{5}{3}$

$x = \pm\sqrt{-\frac{5}{3}} = \pm\frac{\sqrt{15}}{3}i$

Solutions: $4, \pm\dfrac{\sqrt{15}}{3}i$

91. $\begin{cases} -x + 4y = -9 & \text{Eq.1} \\ 5x - 8y = 39 & \text{Eq.2} \end{cases}$

$\begin{aligned} -5x + 20y &= -45 \quad \text{(5)Eq.1} \\ 5x - 8y &= 39 \\ \hline 12y &= -6 \quad \text{Add equations.} \\ y &= -\tfrac{1}{2} \end{aligned}$

$-x + 4\left(-\tfrac{1}{2}\right) = -9 \Rightarrow x = 7$

Solution: $\left(7, -\tfrac{1}{2}\right)$

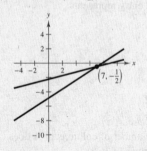

93. $\begin{cases} -x + 2y = -5 & \text{Equation 1} \\ -3x - y = -8 & \text{Equation 2} \end{cases}$

$\begin{aligned} -x + 2y &= -5 \\ -6x - 2y &= -16 \quad \text{(2)Eq.2} \\ \hline -7x &= -21 \quad \text{Add equations.} \\ x &= 3 \end{aligned}$

$-3 + 2y = -5 \Rightarrow y = -1$

Solution: $(3, -1)$

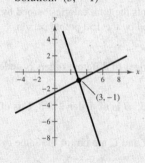

Section 8.3   The Inverse of a Square Matrix

- You should know that the inverse of an $n \times n$ matrix A is the $n \times n$ matrix A^{-1}, if is exists, such that $AA^{-1} = A^{-1}A = I$, where I is the $n \times n$ identity matrix.

- You should be able to find the inverse, if it exists, of a square matrix.

  (a) Write the $n \times 2n$ matrix that consists of the given matrix A on the left and the $n \times n$ identity matrix I on the right to obtain $[A \;\vdots\; I]$. Note that we separate the matrices A and I by a dotted line. We call this process **adjoining** the matrices A and I.

  (b) If possible, row reduce A to I using elementary row operations on the *entire* matrix $[A \;\vdots\; I]$. The result will be the matrix $[I \;\vdots\; A^{-1}]$. If this is not possible, then A is not invertible.

  (c) Check your work by multiplying to see that $AA^{-1} = I = A^{-1}A$.

- The inverse of $A = \begin{bmatrix} a & b \\ c & d \end{bmatrix}$ is $A^{-1} = \dfrac{1}{ad - bc}\begin{bmatrix} d & -b \\ -c & a \end{bmatrix}$ if $ad - cb \neq 0$.

- You should be able to use inverse matrices to solve systems of linear equations if the coefficient matrix is square and invertible.

Vocabulary Check

1. square   **2.** inverse   **3.** nonsingular; singular   **4.** $A^{-1}B$

---

1. $AB = \begin{bmatrix} 2 & 1 \\ 5 & 3 \end{bmatrix}\begin{bmatrix} 3 & -1 \\ -5 & 2 \end{bmatrix} = \begin{bmatrix} 6-5 & -2+2 \\ 15-15 & -5+6 \end{bmatrix} = \begin{bmatrix} 1 & 0 \\ 0 & 1 \end{bmatrix}$

$BA = \begin{bmatrix} 3 & -1 \\ -5 & 2 \end{bmatrix}\begin{bmatrix} 2 & 1 \\ 5 & 3 \end{bmatrix} = \begin{bmatrix} 6-5 & 3-3 \\ -10+10 & -5+6 \end{bmatrix} = \begin{bmatrix} 1 & 0 \\ 0 & 1 \end{bmatrix}$

3. $AB = \begin{bmatrix} 1 & 2 \\ 3 & 4 \end{bmatrix}\begin{bmatrix} -2 & 1 \\ \frac{3}{2} & -\frac{1}{2} \end{bmatrix} = \begin{bmatrix} -2+3 & 1-1 \\ -6+6 & 3-2 \end{bmatrix} = \begin{bmatrix} 1 & 0 \\ 0 & 1 \end{bmatrix}$

$BA = \begin{bmatrix} -2 & 1 \\ \frac{3}{2} & -\frac{1}{2} \end{bmatrix}\begin{bmatrix} 1 & 2 \\ 3 & 4 \end{bmatrix} = \begin{bmatrix} -2+3 & -4+4 \\ \frac{3}{2}-\frac{3}{2} & 3-2 \end{bmatrix} = \begin{bmatrix} 1 & 0 \\ 0 & 1 \end{bmatrix}$

5. $AB = \begin{bmatrix} 2 & -17 & 11 \\ -1 & 11 & -7 \\ 0 & 3 & -2 \end{bmatrix}\begin{bmatrix} 1 & 1 & 2 \\ 2 & 4 & -3 \\ 3 & 6 & -5 \end{bmatrix} = \begin{bmatrix} 2-34+33 & 2-68+66 & 4+51-55 \\ -1+22-21 & -1+44-42 & -2-33+35 \\ 6-6 & 12-12 & -9+10 \end{bmatrix} = \begin{bmatrix} 1 & 0 & 0 \\ 0 & 1 & 0 \\ 0 & 0 & 1 \end{bmatrix}$

$BA = \begin{bmatrix} 1 & 1 & 2 \\ 2 & 4 & -3 \\ 3 & 6 & -5 \end{bmatrix}\begin{bmatrix} 2 & -17 & 11 \\ -1 & 11 & -7 \\ 0 & 3 & -2 \end{bmatrix} = \begin{bmatrix} 2-1 & -17+11+6 & 11-7-4 \\ 4-4 & -34+44-9 & 22-28+6 \\ 6-6 & -51+66-15 & 33-42+10 \end{bmatrix} = \begin{bmatrix} 1 & 0 & 0 \\ 0 & 1 & 0 \\ 0 & 0 & 1 \end{bmatrix}$

7. $AB = \begin{bmatrix} 2 & 0 & 1 & 1 \\ 3 & 0 & 0 & 1 \\ -1 & 1 & -2 & 1 \\ 4 & -1 & 1 & 0 \end{bmatrix}\begin{bmatrix} -1 & 2 & -1 & -1 \\ -4 & 9 & -5 & -6 \\ 0 & 1 & -1 & -1 \\ 3 & -5 & 3 & 3 \end{bmatrix}$

$= \begin{bmatrix} -2+3 & 4+1-5 & -2-1+3 & -2-1+3 \\ 0 & 6-5 & 0 & 0 \\ 1-4+3 & -2+9-2-5 & 1-5+2+3 & 1-6+2+3 \\ 0 & 8-9+1 & -4+5-1 & -4+6-1 \end{bmatrix} = \begin{bmatrix} 1 & 0 & 0 & 0 \\ 0 & 1 & 0 & 0 \\ 0 & 0 & 1 & 0 \\ 0 & 0 & 0 & 1 \end{bmatrix}$

—CONTINUED—

7. —CONTINUED—

$$BA = \begin{bmatrix} -1 & 2 & -1 & -1 \\ -4 & 9 & -5 & -6 \\ 0 & 1 & -1 & -1 \\ 3 & -5 & 3 & 3 \end{bmatrix} \begin{bmatrix} 2 & 0 & 1 & 1 \\ 3 & 0 & 0 & 1 \\ -1 & 1 & -2 & 1 \\ 4 & -1 & 1 & 0 \end{bmatrix}$$

$$= \begin{bmatrix} -2+6+1-4 & 0 & -1+2-1 & -1+2-1 \\ -8+27+5-24 & -5+6 & -4+10-6 & -4+9-5 \\ 3+1-4 & 0 & 2-1 & 0 \\ 6-15-3+12 & 0 & 3-6+3 & 3-5+3 \end{bmatrix} = \begin{bmatrix} 1 & 0 & 0 & 0 \\ 0 & 1 & 0 & 0 \\ 0 & 0 & 1 & 0 \\ 0 & 0 & 0 & 1 \end{bmatrix}$$

9. $AB = \dfrac{1}{3}\begin{bmatrix} -2 & 2 & 3 \\ 1 & -1 & 0 \\ 0 & 1 & 4 \end{bmatrix} \begin{bmatrix} -4 & -5 & 3 \\ -4 & -8 & 3 \\ 1 & 2 & 0 \end{bmatrix} = \dfrac{1}{3}\begin{bmatrix} 8-8+3 & 10-16+6 & -6+6 \\ -4+4 & -5+8 & 3-3 \\ -4+4 & -8+8 & 3 \end{bmatrix}$

$$= \dfrac{1}{3}\begin{bmatrix} 3 & 0 & 0 \\ 0 & 3 & 0 \\ 0 & 0 & 3 \end{bmatrix} = \begin{bmatrix} 1 & 0 & 0 \\ 0 & 1 & 0 \\ 0 & 0 & 1 \end{bmatrix}$$

$$BA = \dfrac{1}{3}\begin{bmatrix} -4 & -5 & 3 \\ -4 & -8 & 3 \\ 1 & 2 & 0 \end{bmatrix} \begin{bmatrix} -2 & 2 & 3 \\ 1 & -1 & 0 \\ 0 & 1 & 4 \end{bmatrix} = \dfrac{1}{3}\begin{bmatrix} 8-5 & -8+5+3 & -12+12 \\ 8-8 & -8+8+3 & -12+12 \\ -2+2 & 2-2 & 3 \end{bmatrix} = \begin{bmatrix} 1 & 0 & 0 \\ 0 & 1 & 0 \\ 0 & 0 & 1 \end{bmatrix}$$

11. $[A \;\vdots\; I] = \begin{bmatrix} 2 & 0 & \vdots & 1 & 0 \\ 0 & 3 & \vdots & 0 & 1 \end{bmatrix}$

$$\begin{array}{c} \frac{1}{2}R_1 \to \\ \frac{1}{3}R_2 \to \end{array} \begin{bmatrix} 1 & 0 & \vdots & \frac{1}{2} & 0 \\ 0 & 1 & \vdots & 0 & \frac{1}{3} \end{bmatrix} = [I \;\vdots\; A^{-1}]$$

$$A^{-1} = \begin{bmatrix} \frac{1}{2} & 0 \\ 0 & \frac{1}{3} \end{bmatrix}$$

13. $[A \;\vdots\; I] = \begin{bmatrix} 1 & -2 & \vdots & 1 & 0 \\ 2 & -3 & \vdots & 0 & 1 \end{bmatrix}$

$$-2R_1 + R_2 \to \begin{bmatrix} 1 & -2 & \vdots & 1 & 0 \\ 0 & 1 & \vdots & -2 & 1 \end{bmatrix}$$

$$2R_2 + R_1 \to \begin{bmatrix} 1 & 0 & \vdots & -3 & 2 \\ 0 & 1 & \vdots & -2 & 1 \end{bmatrix} = [I \;\vdots\; A^{-1}]$$

$$A^{-1} = \begin{bmatrix} -3 & 2 \\ -2 & 1 \end{bmatrix}$$

15. $[A \;\vdots\; I] = \begin{bmatrix} -1 & 1 & \vdots & 1 & 0 \\ -2 & 1 & \vdots & 0 & 1 \end{bmatrix}$

$$-R_2 + R_1 \to \begin{bmatrix} 1 & 0 & \vdots & 1 & -1 \\ -2 & 1 & \vdots & 0 & 1 \end{bmatrix}$$

$$2R_1 + R_2 \to \begin{bmatrix} 1 & 0 & \vdots & 1 & -1 \\ 0 & 1 & \vdots & 2 & -1 \end{bmatrix} = [I \;\vdots\; A^{-1}]$$

$$A^{-1} = \begin{bmatrix} 1 & -1 \\ 2 & -1 \end{bmatrix}$$

17. $[A \;\vdots\; I] = \begin{bmatrix} 2 & 4 & \vdots & 1 & 0 \\ 4 & 8 & \vdots & 0 & 1 \end{bmatrix}$

$$-2R_1 + R_2 \to \begin{bmatrix} 2 & 4 & \vdots & 1 & 0 \\ 0 & 0 & \vdots & -2 & 1 \end{bmatrix}$$

The two zeros in the second row imply that the inverse does not exist.

19. $A = \begin{bmatrix} 2 & 7 & 1 \\ -3 & -9 & 2 \end{bmatrix}$

A has no inverse because it is not square.

21. $[A \vdots I] = \begin{bmatrix} 1 & 1 & 1 & \vdots & 1 & 0 & 0 \\ 3 & 5 & 4 & \vdots & 0 & 1 & 0 \\ 3 & 6 & 5 & \vdots & 0 & 0 & 1 \end{bmatrix}$

$\begin{matrix} -3R_1 + R_2 \to \\ -3R_1 + R_3 \to \end{matrix} \begin{bmatrix} 1 & 1 & 1 & \vdots & 1 & 0 & 0 \\ 0 & 2 & 1 & \vdots & -3 & 1 & 0 \\ 0 & 3 & 2 & \vdots & -3 & 0 & 1 \end{bmatrix}$

$\begin{matrix} \\ \frac{1}{2}R_2 \to \\ \\ \end{matrix} \begin{bmatrix} 1 & 1 & 1 & \vdots & 1 & 0 & 0 \\ 0 & 1 & \frac{1}{2} & \vdots & -\frac{3}{2} & \frac{1}{2} & 0 \\ 0 & 3 & 2 & \vdots & -3 & 0 & 1 \end{bmatrix}$

$\begin{matrix} -R_2 + R_1 \to \\ \\ -3R_2 + R_3 \to \end{matrix} \begin{bmatrix} 1 & 0 & \frac{1}{2} & \vdots & \frac{5}{2} & -\frac{1}{2} & 0 \\ 0 & 1 & \frac{1}{2} & \vdots & -\frac{3}{2} & \frac{1}{2} & 0 \\ 0 & 0 & \frac{1}{2} & \vdots & \frac{3}{2} & -\frac{3}{2} & 1 \end{bmatrix}$

$\begin{matrix} -R_3 + R_1 \to \\ -R_3 + R_2 \to \\ \\ \end{matrix} \begin{bmatrix} 1 & 0 & 0 & \vdots & 1 & 1 & -1 \\ 0 & 1 & 0 & \vdots & -3 & 2 & -1 \\ 0 & 0 & \frac{1}{2} & \vdots & \frac{3}{2} & -\frac{3}{2} & 1 \end{bmatrix}$

$\begin{matrix} \\ \\ 2R_3 \to \end{matrix} \begin{bmatrix} 1 & 0 & 0 & \vdots & 1 & 1 & -1 \\ 0 & 1 & 0 & \vdots & -3 & 2 & -1 \\ 0 & 0 & 1 & \vdots & 3 & -3 & 2 \end{bmatrix} = [I \vdots A^{-1}]$

$A^{-1} = \begin{bmatrix} 1 & 1 & -1 \\ -3 & 2 & -1 \\ 3 & -3 & 2 \end{bmatrix}$

23. $[A \vdots I] = \begin{bmatrix} 1 & 0 & 0 & \vdots & 1 & 0 & 0 \\ 3 & 4 & 0 & \vdots & 0 & 1 & 0 \\ 2 & 5 & 5 & \vdots & 0 & 0 & 1 \end{bmatrix}$

$\begin{matrix} -3R_1 + R_2 \to \\ -2R_1 + R_3 \to \end{matrix} \begin{bmatrix} 1 & 0 & 0 & \vdots & 1 & 0 & 0 \\ 0 & 4 & 0 & \vdots & -3 & 1 & 0 \\ 0 & 5 & 5 & \vdots & -2 & 0 & 1 \end{bmatrix}$

$\begin{matrix} \\ \\ -\frac{5}{4}R_2 + R_3 \to \end{matrix} \begin{bmatrix} 1 & 0 & 0 & \vdots & 1 & 0 & 0 \\ 0 & 4 & 0 & \vdots & -3 & 1 & 0 \\ 0 & 0 & 5 & \vdots & \frac{7}{4} & -\frac{5}{4} & 1 \end{bmatrix}$

$\begin{matrix} \\ \frac{1}{4}R_2 \to \\ \frac{1}{5}R_3 \to \end{matrix} \begin{bmatrix} 1 & 0 & 0 & \vdots & 1 & 0 & 0 \\ 0 & 1 & 0 & \vdots & -\frac{3}{4} & \frac{1}{4} & 0 \\ 0 & 0 & 1 & \vdots & \frac{7}{20} & -\frac{1}{4} & \frac{1}{5} \end{bmatrix} = [I \vdots A^{-1}]$

$A^{-1} = \begin{bmatrix} 1 & 0 & 0 \\ -\frac{3}{4} & \frac{1}{4} & 0 \\ \frac{7}{20} & -\frac{1}{4} & \frac{1}{5} \end{bmatrix}$

25. $[A \vdots I] = \begin{bmatrix} -8 & 0 & 0 & 0 & \vdots & 1 & 0 & 0 & 0 \\ 0 & 1 & 0 & 0 & \vdots & 0 & 1 & 0 & 0 \\ 0 & 0 & 4 & 0 & \vdots & 0 & 0 & 1 & 0 \\ 0 & 0 & 0 & -5 & \vdots & 0 & 0 & 0 & 1 \end{bmatrix}$

$\begin{matrix} -\frac{1}{8}R_1 \to \\ \\ \frac{1}{4}R_3 \to \\ -\frac{1}{5}R_4 \to \end{matrix} \begin{bmatrix} 1 & 0 & 0 & 0 & \vdots & -\frac{1}{8} & 0 & 0 & 0 \\ 0 & 1 & 0 & 0 & \vdots & 0 & 1 & 0 & 0 \\ 0 & 0 & 1 & 0 & \vdots & 0 & 0 & \frac{1}{4} & 0 \\ 0 & 0 & 0 & 1 & \vdots & 0 & 0 & 0 & -\frac{1}{5} \end{bmatrix} = [I \vdots A^{-1}]$

$A^{-1} = \begin{bmatrix} -\frac{1}{8} & 0 & 0 & 0 \\ 0 & 1 & 0 & 0 \\ 0 & 0 & \frac{1}{4} & 0 \\ 0 & 0 & 0 & -\frac{1}{5} \end{bmatrix}$

27. $A = \begin{bmatrix} 1 & 2 & -1 \\ 3 & 7 & -10 \\ -5 & -7 & -15 \end{bmatrix}$

$A^{-1} = \begin{bmatrix} -175 & 37 & -13 \\ 95 & -20 & 7 \\ 14 & -3 & 1 \end{bmatrix}$

29. $A = \begin{bmatrix} 1 & 1 & 2 \\ 3 & 1 & 0 \\ -2 & 0 & 3 \end{bmatrix}$

$A^{-1} = \frac{1}{2}\begin{bmatrix} -3 & 3 & 2 \\ 9 & -7 & -6 \\ -2 & 2 & 2 \end{bmatrix} = \begin{bmatrix} -1.5 & 1.5 & 1 \\ 4.5 & -3.5 & -3 \\ -1 & 1 & 1 \end{bmatrix}$

31. $A = \begin{bmatrix} -\frac{1}{2} & \frac{3}{4} & \frac{1}{4} \\ 1 & 0 & -\frac{3}{2} \\ 0 & -1 & \frac{1}{2} \end{bmatrix}$

$A^{-1} = \begin{bmatrix} -12 & -5 & -9 \\ -4 & -2 & -4 \\ -8 & -4 & -6 \end{bmatrix}$

33. $A = \begin{bmatrix} 0.1 & 0.2 & 0.3 \\ -0.3 & 0.2 & 0.2 \\ 0.5 & 0.4 & 0.4 \end{bmatrix}$

$A^{-1} = \frac{5}{11}\begin{bmatrix} 0 & -4 & 2 \\ -22 & 11 & 11 \\ 22 & -6 & -8 \end{bmatrix} = \begin{bmatrix} 0 & -1.\overline{81} & 0.\overline{90} \\ -10 & 5 & 5 \\ 10 & -2.\overline{72} & -3.\overline{63} \end{bmatrix}$

35. $A = \begin{bmatrix} 1 & 0 & 3 & 0 \\ 0 & 2 & 0 & 4 \\ 1 & 0 & 3 & 0 \\ 0 & 2 & 0 & 4 \end{bmatrix}$

A^{-1} does not exist.

37. $A = \begin{bmatrix} -1 & 0 & 1 & 0 \\ 0 & 2 & 0 & -1 \\ 2 & 0 & -1 & 0 \\ 0 & -1 & 0 & 1 \end{bmatrix}$

$A^{-1} = \begin{bmatrix} 1 & 0 & 1 & 0 \\ 0 & 1 & 0 & 1 \\ 2 & 0 & 1 & 0 \\ 0 & 1 & 0 & 2 \end{bmatrix}$

39. $A = \begin{bmatrix} a & b \\ c & d \end{bmatrix}, A^{-1} = \frac{1}{ad - bc}\begin{bmatrix} d & -b \\ -c & a \end{bmatrix}$

$A = \begin{bmatrix} 5 & -2 \\ 2 & 3 \end{bmatrix}$

$ad - bc = (5)(3) - (-2)(2) = 19$

$A^{-1} = \frac{1}{19}\begin{bmatrix} 3 & 2 \\ -2 & 5 \end{bmatrix} = \begin{bmatrix} \frac{3}{19} & \frac{2}{19} \\ -\frac{2}{19} & \frac{5}{19} \end{bmatrix}$

41. $A = \begin{bmatrix} -4 & -6 \\ 2 & 3 \end{bmatrix}$

$ad - bc = (-4)(3) - (-2)(-6) = 0$

Since $ad - bc = 0$, A^{-1} does not exist.

43. $A = \begin{bmatrix} \frac{7}{2} & -\frac{3}{4} \\ \frac{1}{5} & \frac{4}{5} \end{bmatrix}$

$ad - bc = \left(\frac{7}{2}\right)\left(\frac{4}{5}\right) - \left(-\frac{3}{4}\right)\left(\frac{1}{5}\right) = \frac{28}{10} + \frac{3}{20} = \frac{59}{20}$

$A^{-1} = \frac{1}{59/20}\begin{bmatrix} \frac{4}{5} & \frac{3}{4} \\ -\frac{1}{5} & \frac{7}{2} \end{bmatrix} = \frac{20}{59}\begin{bmatrix} \frac{4}{5} & \frac{3}{4} \\ -\frac{1}{5} & \frac{7}{2} \end{bmatrix} = \begin{bmatrix} \frac{16}{59} & \frac{15}{59} \\ -\frac{4}{59} & \frac{70}{59} \end{bmatrix}$

45. $\begin{bmatrix} x \\ y \end{bmatrix} = \begin{bmatrix} -3 & 2 \\ -2 & 1 \end{bmatrix}\begin{bmatrix} 5 \\ 10 \end{bmatrix} = \begin{bmatrix} 5 \\ 0 \end{bmatrix}$

Solution: $(5, 0)$

47. $\begin{bmatrix} x \\ y \end{bmatrix} = \begin{bmatrix} -3 & 2 \\ -2 & 1 \end{bmatrix}\begin{bmatrix} 4 \\ 2 \end{bmatrix} = \begin{bmatrix} -8 \\ -6 \end{bmatrix}$

Solution: $(-8, -6)$

49. $\begin{bmatrix} x \\ y \\ z \end{bmatrix} = \begin{bmatrix} 1 & 1 & -1 \\ -3 & 2 & -1 \\ 3 & -3 & 2 \end{bmatrix}\begin{bmatrix} 0 \\ 5 \\ 2 \end{bmatrix} = \begin{bmatrix} 3 \\ 8 \\ -11 \end{bmatrix}$

Solution: $(3, 8, -11)$

51. $\begin{bmatrix} x_1 \\ x_2 \\ x_3 \\ x_4 \end{bmatrix} = \begin{bmatrix} -24 & 7 & 1 & -2 \\ -10 & 3 & 0 & -1 \\ -29 & 7 & 3 & -2 \\ 12 & -3 & -1 & 1 \end{bmatrix} \begin{bmatrix} 0 \\ 1 \\ -1 \\ 2 \end{bmatrix} = \begin{bmatrix} 2 \\ 1 \\ 0 \\ 0 \end{bmatrix}$

Solution: $(2, 1, 0, 0)$

53. $A = \begin{bmatrix} 3 & 4 \\ 5 & 3 \end{bmatrix}$

$A^{-1} = \dfrac{1}{9 - 20} \begin{bmatrix} 3 & -4 \\ -5 & 3 \end{bmatrix}$

$\begin{bmatrix} x \\ y \end{bmatrix} = -\dfrac{1}{11} \begin{bmatrix} 3 & -4 \\ -5 & 3 \end{bmatrix} \begin{bmatrix} -2 \\ 4 \end{bmatrix} = -\dfrac{1}{11} \begin{bmatrix} -22 \\ 22 \end{bmatrix} = \begin{bmatrix} 2 \\ -2 \end{bmatrix}$

Solution: $(2, -2)$

55. $A = \begin{bmatrix} -0.4 & 0.8 \\ 2 & -4 \end{bmatrix}$

$A^{-1} = \dfrac{1}{1.6 - 1.6} \begin{bmatrix} -4 & -0.8 \\ -2 & -0.4 \end{bmatrix}$

A^{-1} does not exist.

This implies that there is no unique solution; that is, either the system is inconsistent *or* there are infinitely many solutions.

Find the reduced row-echelon form of the matrix corresponding to the system.

$\begin{bmatrix} -0.4 & 0.8 & \vdots & 1.6 \\ 2 & -4 & \vdots & 5 \end{bmatrix}$

$\begin{array}{c} -2.5R_1 \rightarrow \end{array} \begin{bmatrix} 1 & -2 & \vdots & -4 \\ 2 & -4 & \vdots & 5 \end{bmatrix}$

$\begin{array}{c} -2R_1 + R_2 \rightarrow \end{array} \begin{bmatrix} 1 & -2 & \vdots & -4 \\ 0 & 0 & \vdots & 13 \end{bmatrix}$

The given system is inconsistent and there is no solution.

57. $A = \begin{bmatrix} -\frac{1}{4} & \frac{3}{8} \\ \frac{3}{2} & \frac{3}{4} \end{bmatrix}$

$A^{-1} = \dfrac{1}{-\frac{3}{16} - \frac{9}{16}} \begin{bmatrix} \frac{3}{4} & -\frac{3}{8} \\ -\frac{3}{2} & -\frac{1}{4} \end{bmatrix} = -\dfrac{4}{3} \begin{bmatrix} \frac{3}{4} & -\frac{3}{8} \\ -\frac{3}{2} & -\frac{1}{4} \end{bmatrix} = \begin{bmatrix} -1 & \frac{1}{2} \\ 2 & \frac{1}{3} \end{bmatrix}$

$\begin{bmatrix} x \\ y \end{bmatrix} = \begin{bmatrix} -1 & \frac{1}{2} \\ 2 & \frac{1}{3} \end{bmatrix} \begin{bmatrix} -2 \\ -12 \end{bmatrix} = \begin{bmatrix} -4 \\ -8 \end{bmatrix}$

Solution: $(-4, -8)$

59. $A = \begin{bmatrix} 4 & -1 & 1 \\ 2 & 2 & 3 \\ 5 & -2 & 6 \end{bmatrix}$

Find A^{-1}.

$[A \vdots I] = \begin{bmatrix} 4 & -1 & 1 & \vdots & 1 & 0 & 0 \\ 2 & 2 & 3 & \vdots & 0 & 1 & 0 \\ 5 & -2 & 6 & \vdots & 0 & 0 & 1 \end{bmatrix}$

$\begin{array}{c} R_1 \\ \\ R_3 \end{array} \begin{bmatrix} 5 & -2 & 6 & \vdots & 0 & 0 & 1 \\ 2 & 2 & 3 & \vdots & 0 & 1 & 0 \\ 4 & -1 & 1 & \vdots & 1 & 0 & 0 \end{bmatrix}$

$\begin{array}{c} -R_3 + R_1 \rightarrow \end{array} \begin{bmatrix} 1 & -1 & 5 & \vdots & -1 & 0 & 1 \\ 2 & 2 & 3 & \vdots & 0 & 1 & 0 \\ 4 & -1 & 1 & \vdots & 1 & 0 & 0 \end{bmatrix}$

$\begin{array}{c} -2R_1 + R_2 \rightarrow \\ -4R_1 + R_3 \rightarrow \end{array} \begin{bmatrix} 1 & -1 & 5 & \vdots & -1 & 0 & 1 \\ 0 & 4 & -7 & \vdots & 2 & 1 & -2 \\ 0 & 3 & -19 & \vdots & 5 & 0 & -4 \end{bmatrix}$

$\begin{array}{c} -R_3 + R_2 \rightarrow \end{array} \begin{bmatrix} 1 & -1 & 5 & \vdots & -1 & 0 & 1 \\ 0 & 1 & 12 & \vdots & -3 & 1 & 2 \\ 0 & 3 & -19 & \vdots & 5 & 0 & -4 \end{bmatrix}$

—CONTINUED—

59. —CONTINUED—

$$
\begin{array}{c} R_2 + R_1 \to \\ \\ -3R_2 + R_3 \to \end{array}
\left[\begin{array}{ccc:ccc}
1 & 0 & 17 & -4 & 1 & 3 \\
0 & 1 & 12 & -3 & 1 & 2 \\
0 & 0 & -55 & 14 & -3 & -10
\end{array}\right]
$$

$$
\begin{array}{c} \\ \\ -\frac{1}{55}R_3 \to \end{array}
\left[\begin{array}{ccc:ccc}
1 & 0 & 17 & -4 & 1 & 3 \\
0 & 1 & 12 & -3 & 1 & 2 \\
0 & 0 & 1 & -\frac{14}{55} & \frac{3}{55} & \frac{2}{11}
\end{array}\right]
$$

$$
\begin{array}{c} -17R_3 + R_1 \to \\ -12R_3 + R_2 \to \\ \\ \end{array}
\left[\begin{array}{ccc:ccc}
1 & 0 & 0 & \frac{18}{55} & \frac{4}{55} & -\frac{1}{11} \\
0 & 1 & 0 & \frac{3}{55} & \frac{19}{55} & -\frac{2}{11} \\
0 & 0 & 1 & -\frac{14}{55} & \frac{3}{55} & \frac{2}{11}
\end{array}\right] = [I \ : \ A^{-1}]
$$

$$
A^{-1} = \frac{1}{55}\begin{bmatrix} 18 & 4 & -5 \\ 3 & 19 & -10 \\ -14 & 3 & 10 \end{bmatrix}
$$

$$
\begin{bmatrix} x \\ y \\ z \end{bmatrix} = \frac{1}{55}\begin{bmatrix} 18 & 4 & -5 \\ 3 & 19 & -10 \\ -14 & 3 & 10 \end{bmatrix}\begin{bmatrix} -5 \\ 10 \\ 1 \end{bmatrix} = \frac{1}{55}\begin{bmatrix} -55 \\ 165 \\ 110 \end{bmatrix} = \begin{bmatrix} -1 \\ 3 \\ 2 \end{bmatrix}
$$

Solution: $(-1, 3, 2)$

61. $A = \begin{bmatrix} 5 & -3 & 2 \\ 2 & 2 & -3 \\ 1 & -7 & 8 \end{bmatrix}$

A^{-1} does not exist. This implies that there is no unique solution; that is, either the system is inconsistent *or* the system has infinitely many solutions. Use a graphing utility to find the reduced row-echelon form of the matrix corresponding to the system.

$$
\left[\begin{array}{ccc:c}
5 & -3 & 2 & 2 \\
2 & 2 & -3 & 3 \\
1 & -7 & 8 & -4
\end{array}\right]
$$

$$
\left[\begin{array}{ccc:c}
1 & 0 & -\frac{5}{16} & \frac{13}{16} \\
0 & 1 & -\frac{19}{16} & \frac{11}{16} \\
0 & 0 & 0 & 0
\end{array}\right]
$$

$$
\begin{cases} x - \frac{5}{16}z = \frac{13}{16} \\ y - \frac{19}{16}z = \frac{11}{16} \end{cases}
$$

Let $z = a$. Then $x = \frac{5}{16}a + \frac{13}{16}$ and $y = \frac{19}{16}a + \frac{11}{16}$.

Solution: $\left(\frac{5}{16}a + \frac{13}{16}, \frac{19}{16}a + \frac{11}{16}, a\right)$ where a is a real number

63. $A = \begin{bmatrix} 3 & -2 & 1 \\ -4 & 1 & -3 \\ 1 & -5 & 1 \end{bmatrix}$

$$
A^{-1} = \begin{bmatrix} 0.56 & 0.12 & -0.2 \\ -0.04 & -0.08 & -0.2 \\ -0.76 & -0.52 & 0.2 \end{bmatrix}
$$

$$
\begin{bmatrix} x \\ y \\ z \end{bmatrix} = \begin{bmatrix} 0.56 & 0.12 & -0.2 \\ -0.04 & -0.08 & -0.2 \\ -0.76 & -0.52 & 0.2 \end{bmatrix}\begin{bmatrix} -29 \\ 37 \\ -24 \end{bmatrix} = \begin{bmatrix} -7 \\ 3 \\ -2 \end{bmatrix}
$$

Solution: $(-7, 3, -2)$

65. $A = \begin{bmatrix} 7 & -3 & 0 & 2 \\ -2 & 1 & 0 & -1 \\ 4 & 0 & 1 & -2 \\ -1 & 1 & 0 & -1 \end{bmatrix}$

$$
A^{-1} = \begin{bmatrix} 0 & -1 & 0 & 1 \\ -1 & -5 & 0 & 3 \\ -2 & -4 & 1 & -2 \\ -1 & -4 & 0 & 1 \end{bmatrix}
$$

$$
\begin{bmatrix} x \\ y \\ z \\ w \end{bmatrix} = \begin{bmatrix} 0 & -1 & 0 & 1 \\ -1 & -5 & 0 & 3 \\ -2 & -4 & 1 & -2 \\ -1 & -4 & 0 & 1 \end{bmatrix}\begin{bmatrix} 41 \\ -13 \\ 12 \\ -8 \end{bmatrix} = \begin{bmatrix} 5 \\ 0 \\ -2 \\ 3 \end{bmatrix}
$$

Solution: $(5, 0, -2, 3)$

67. $A = \begin{bmatrix} 1 & 1 & 1 \\ 0.065 & 0.07 & 0.09 \\ 0 & 2 & -1 \end{bmatrix}$

$$[A \;\vdots\; I] = \begin{bmatrix} 1 & 1 & 1 & \vdots & 1 & 0 & 0 \\ 0.065 & 0.07 & 0.09 & \vdots & 0 & 1 & 0 \\ 0 & 2 & -1 & \vdots & 0 & 0 & 1 \end{bmatrix}$$

$$200R_2 \rightarrow \begin{bmatrix} 1 & 1 & 1 & \vdots & 1 & 0 & 0 \\ 13 & 14 & 18 & \vdots & 0 & 200 & 0 \\ 0 & 2 & -1 & \vdots & 0 & 0 & 1 \end{bmatrix}$$

$$-13R_1 + R_2 \rightarrow \begin{bmatrix} 1 & 1 & 1 & \vdots & 1 & 0 & 0 \\ 0 & 1 & 5 & \vdots & -13 & 200 & 0 \\ 0 & 2 & -1 & \vdots & 0 & 0 & 1 \end{bmatrix}$$

$$\begin{matrix} -R_2 + R_1 \rightarrow \\ \\ -2R_2 + R_3 \rightarrow \end{matrix} \begin{bmatrix} 1 & 0 & -4 & \vdots & 14 & -200 & 0 \\ 0 & 1 & 5 & \vdots & -13 & 200 & 0 \\ 0 & 0 & -11 & \vdots & 26 & -400 & 1 \end{bmatrix}$$

$$-\tfrac{1}{11}R_3 \rightarrow \begin{bmatrix} 1 & 0 & -4 & \vdots & 14 & -200 & 0 \\ 0 & 1 & 5 & \vdots & -13 & 200 & 0 \\ 0 & 0 & 1 & \vdots & -\tfrac{26}{11} & \tfrac{400}{11} & -\tfrac{1}{11} \end{bmatrix}$$

$$\begin{matrix} 4R_3 + R_1 \rightarrow \\ \\ -5R_3 + R_2 \rightarrow \end{matrix} \begin{bmatrix} 1 & 0 & 0 & \vdots & \tfrac{50}{11} & -\tfrac{600}{11} & -\tfrac{4}{11} \\ 0 & 1 & 0 & \vdots & -\tfrac{13}{11} & \tfrac{200}{11} & \tfrac{5}{11} \\ 0 & 0 & 1 & \vdots & -\tfrac{26}{11} & \tfrac{400}{11} & -\tfrac{1}{11} \end{bmatrix} = [I \;\vdots\; A^{-1}]$$

$$X = A^{-1}B = \tfrac{1}{11}\begin{bmatrix} 50 & -600 & -4 \\ -13 & 200 & 5 \\ -26 & 400 & -1 \end{bmatrix}\begin{bmatrix} 10{,}000 \\ 705 \\ 0 \end{bmatrix} = \begin{bmatrix} 7000 \\ 1000 \\ 2000 \end{bmatrix}$$

Solution: $7000 in AAA-rated bonds, $1000 in A-rated bonds, $2000 in B-rated bonds

69. Use the inverse matrix A^{-1} from Exercise 67.

$$X = A^{-1}B = \tfrac{1}{11}\begin{bmatrix} 50 & -600 & -4 \\ -13 & 200 & 5 \\ -26 & 400 & -1 \end{bmatrix}\begin{bmatrix} 12{,}000 \\ 835 \\ 0 \end{bmatrix} = \begin{bmatrix} 9000 \\ 1000 \\ 2000 \end{bmatrix}$$

Solution: $9000 in AAA-rated bonds, $1000 in A-rated bonds, $2000 in B-rated bonds

71. (a) $A = \begin{bmatrix} 2 & 0 & 4 \\ 0 & 1 & 4 \\ 1 & 1 & -1 \end{bmatrix}$

$[A \,\vdots\, I] = \begin{bmatrix} 2 & 0 & 4 & \vdots & 1 & 0 & 0 \\ 0 & 1 & 4 & \vdots & 0 & 1 & 0 \\ 1 & 1 & -1 & \vdots & 0 & 0 & 1 \end{bmatrix}$

$\begin{matrix} R_1 \\ \\ R_3 \end{matrix} \begin{bmatrix} 1 & 1 & -1 & \vdots & 0 & 0 & 1 \\ 0 & 1 & 4 & \vdots & 0 & 1 & 0 \\ 2 & 0 & 4 & \vdots & 1 & 0 & 0 \end{bmatrix}$

$-2R_1 + R_3 \rightarrow \begin{bmatrix} 1 & 1 & -1 & \vdots & 0 & 0 & 1 \\ 0 & 1 & 4 & \vdots & 0 & 1 & 0 \\ 0 & -2 & 6 & \vdots & 1 & 0 & -2 \end{bmatrix}$

$\begin{matrix} -R_2 + R_1 \rightarrow \\ \\ 2R_2 + R_3 \rightarrow \end{matrix} \begin{bmatrix} 1 & 0 & -5 & \vdots & 0 & -1 & 1 \\ 0 & 1 & 4 & \vdots & 0 & 1 & 0 \\ 0 & 0 & 14 & \vdots & 1 & 2 & -2 \end{bmatrix}$

$\frac{1}{14}R_3 \rightarrow \begin{bmatrix} 1 & 0 & -5 & \vdots & 0 & -1 & 1 \\ 0 & 1 & 4 & \vdots & 0 & 1 & 0 \\ 0 & 0 & 1 & \vdots & \frac{1}{14} & \frac{1}{7} & -\frac{1}{7} \end{bmatrix}$

$\begin{matrix} 5R_3 + R_1 \rightarrow \\ -4R_3 + R_2 \rightarrow \\ \\ \end{matrix} \begin{bmatrix} 1 & 0 & 0 & \vdots & \frac{5}{14} & -\frac{2}{7} & \frac{2}{7} \\ 0 & 1 & 0 & \vdots & -\frac{2}{7} & \frac{3}{7} & \frac{4}{7} \\ 0 & 0 & 1 & \vdots & \frac{1}{14} & \frac{1}{7} & -\frac{1}{7} \end{bmatrix} = [I \,\vdots\, A^{-1}]$

$A^{-1} = \frac{1}{14}\begin{bmatrix} 5 & -4 & 4 \\ -4 & 6 & 8 \\ 1 & 2 & -2 \end{bmatrix}$

$\begin{bmatrix} I_1 \\ I_2 \\ I_3 \end{bmatrix} = \frac{1}{14}\begin{bmatrix} 5 & -4 & 4 \\ -4 & 6 & 8 \\ 1 & 2 & -2 \end{bmatrix}\begin{bmatrix} 14 \\ 28 \\ 0 \end{bmatrix} = \begin{bmatrix} -3 \\ 8 \\ 5 \end{bmatrix}$

Solution: $I_1 = -3$ amperes, $I_2 = 8$ amperes, $I_3 = 5$ amperes

(b) $\begin{bmatrix} I_1 \\ I_2 \\ I_3 \end{bmatrix} = \frac{1}{14}\begin{bmatrix} 5 & -4 & 4 \\ -4 & 6 & 8 \\ 1 & 2 & -2 \end{bmatrix}\begin{bmatrix} 24 \\ 23 \\ 0 \end{bmatrix} = \begin{bmatrix} 2 \\ 3 \\ 5 \end{bmatrix}$

Solution:

$I_1 = 2$ amperes, $I_2 = 3$ amperes, $I_3 = 5$ amperes

73. True. If B is the inverse of A, then $AB = I = BA$.

75. $AA^{-1} = \begin{bmatrix} a & b \\ c & d \end{bmatrix}\left(\dfrac{1}{ad - bc}\right)\begin{bmatrix} d & -b \\ -c & a \end{bmatrix} = \dfrac{1}{ad - bc}\begin{bmatrix} a & b \\ c & d \end{bmatrix}\begin{bmatrix} d & -b \\ -c & a \end{bmatrix}$

$= \dfrac{1}{ad - bc}\begin{bmatrix} ad - bc & 0 \\ 0 & ad - bc \end{bmatrix} = \begin{bmatrix} 1 & 0 \\ 0 & 1 \end{bmatrix}$

$A^{-1}A = \dfrac{1}{ad - bc}\begin{bmatrix} d & -b \\ -c & a \end{bmatrix}\begin{bmatrix} a & b \\ c & d \end{bmatrix} = \dfrac{1}{ad - bc}\begin{bmatrix} ad - bc & 0 \\ 0 & ad - bc \end{bmatrix} = \begin{bmatrix} 1 & 0 \\ 0 & 1 \end{bmatrix}$

77. $|x + 7| \geq 2$

$x + 7 \leq -2 \text{ or } x + 7 \geq 2$

$x \leq -9 \text{ or } \quad x \geq -5$

79. $3^{x/2} = 315$

$\ln 3^{x/2} = \ln 315$

$\dfrac{x}{2}\ln 3 = \ln 315$

$x = \dfrac{2\ln 315}{\ln 3} \approx 10.472$

81. $\log_2 x - 2 = 4.5$

$\log_2 x = 6.5$

$x = 2^{6.5} \approx 90.510$

83. Answers will vary.

Section 8.4   The Determinant of a Square Matrix

- ■ You should be able to determine the determinant of a matrix of order 2×2 by using the difference of the products of the diagonals.
- ■ You should be able to use expansion by cofactors to find the determinant of a matrix of order 3×3 or greater.
- ■ The determinant of a triangular matrix equals the product of the entries on the main diagonal.

Vocabulary Check

1. determinant   **2.** minor   **3.** cofactor   **4.** expanding by cofactors

---

1. 5

3. $\begin{vmatrix} 2 & 1 \\ 3 & 4 \end{vmatrix} = 2(4) - 1(3) = 8 - 3 = 5$

5. $\begin{vmatrix} 5 & 2 \\ -6 & 3 \end{vmatrix} = 5(3) - 2(-6) = 15 + 12 = 27$

7. $\begin{vmatrix} -7 & 0 \\ 3 & 0 \end{vmatrix} = -7(0) - 0(3) = 0$

9. $\begin{vmatrix} 2 & 6 \\ 0 & 3 \end{vmatrix} = 2(3) - 6(0) = 6$

11. $\begin{vmatrix} -3 & -2 \\ -6 & -1 \end{vmatrix} = (-3)(-1) - (-2)(-6) = 3 - 12 = -9$

13. $\begin{vmatrix} 9 & 0 \\ 7 & 8 \end{vmatrix} = 9(8) - 0(7) = 72 - 0 = 72$

15. $\begin{vmatrix} -\frac{1}{2} & \frac{1}{3} \\ -6 & \frac{1}{3} \end{vmatrix} = -\frac{1}{2}\left(\frac{1}{3}\right) - \frac{1}{3}(-6) = -\frac{1}{6} + 2 = \frac{11}{6}$

17. $\begin{vmatrix} 0.3 & 0.2 & 0.2 \\ 0.2 & 0.2 & 0.2 \\ -0.4 & 0.4 & 0.3 \end{vmatrix} = -0.002$

19. $\begin{vmatrix} 0.9 & 0.7 & 0 \\ -0.1 & 0.3 & 1.3 \\ -2.2 & 4.2 & 6.1 \end{vmatrix} = -4.842$

21. $\begin{vmatrix} 1 & 4 & -2 \\ 3 & 6 & -6 \\ -2 & 1 & 4 \end{vmatrix} = 0$

23. $\begin{bmatrix} 3 & 4 \\ 2 & -5 \end{bmatrix}$

    (a) $M_{11} = -5$ (b) $C_{11} = M_{11} = -5$

        $M_{12} = 2$ $C_{12} = -M_{12} = -2$

        $M_{21} = 4$ $C_{21} = -M_{21} = -4$

        $M_{22} = 3$ $C_{22} = M_{22} = 3$

25. $\begin{bmatrix} 3 & 1 \\ -2 & -4 \end{bmatrix}$

    (a) $M_{11} = -4$ (b) $C_{11} = M_{11} = -4$

        $M_{12} = -2$ $C_{12} = -M_{12} = 2$

        $M_{21} = 1$ $C_{21} = -M_{21} = -1$

        $M_{22} = 3$ $C_{22} = M_{22} = 3$

27. $\begin{bmatrix} 4 & 0 & 2 \\ -3 & 2 & 1 \\ 1 & -1 & 1 \end{bmatrix}$

(a) $M_{11} = \begin{vmatrix} 2 & 1 \\ -1 & 1 \end{vmatrix} = 2 - (-1) = 3$

$M_{12} = \begin{vmatrix} -3 & 1 \\ 1 & 1 \end{vmatrix} = -3 - 1 = -4$

$M_{13} = \begin{vmatrix} -3 & 2 \\ 1 & -1 \end{vmatrix} = 3 - 2 = 1$

$M_{21} = \begin{vmatrix} 0 & 2 \\ -1 & 1 \end{vmatrix} = 0 - (-2) = 2$

$M_{22} = \begin{vmatrix} 4 & 2 \\ 1 & 1 \end{vmatrix} = 4 - 2 = 2$

$M_{23} = \begin{vmatrix} 4 & 0 \\ 1 & -1 \end{vmatrix} = -4 - 0 = -4$

$M_{31} = \begin{vmatrix} 0 & 2 \\ 2 & 1 \end{vmatrix} = 0 - 4 = -4$

$M_{32} = \begin{vmatrix} 4 & 2 \\ -3 & 1 \end{vmatrix} = 4 - (-6) = 10$

$M_{33} = \begin{vmatrix} 4 & 0 \\ -3 & 2 \end{vmatrix} = 8 - 0 = 8$

(b) $C_{11} = (-1)^2 M_{11} = 3$

$C_{12} = (-1)^3 M_{12} = 4$

$C_{13} = (-1)^4 M_{13} = 1$

$C_{21} = (-1)^3 M_{21} = -2$

$C_{22} = (-1)^4 M_{22} = 2$

$C_{23} = (-1)^5 M_{23} = 4$

$C_{31} = (-1)^4 M_{31} = -4$

$C_{32} = (-1)^5 M_{32} = -10$

$C_{33} = (-1)^6 M_{33} = 8$

29. $\begin{bmatrix} 3 & -2 & 8 \\ 3 & 2 & -6 \\ -1 & 3 & 6 \end{bmatrix}$

(a) $M_{11} = \begin{vmatrix} 2 & -6 \\ 3 & 6 \end{vmatrix} = 12 + 18 = 30$

$M_{12} = \begin{vmatrix} 3 & -6 \\ -1 & 6 \end{vmatrix} = 18 - 6 = 12$

$M_{13} = \begin{vmatrix} 3 & 2 \\ -1 & 3 \end{vmatrix} = 9 + 2 = 11$

$M_{21} = \begin{vmatrix} -2 & 8 \\ 3 & 6 \end{vmatrix} = -12 - 24 = -36$

$M_{22} = \begin{vmatrix} 3 & 8 \\ -1 & 6 \end{vmatrix} = 18 + 8 = 26$

$M_{23} = \begin{vmatrix} 3 & -2 \\ -1 & 3 \end{vmatrix} = 9 - 2 = 7$

$M_{31} = \begin{vmatrix} -2 & 8 \\ 2 & -6 \end{vmatrix} = 12 - 16 = -4$

$M_{32} = \begin{vmatrix} 3 & 8 \\ 3 & -6 \end{vmatrix} = -18 - 24 = -42$

$M_{33} = \begin{vmatrix} 3 & -2 \\ 3 & 2 \end{vmatrix} = 6 + 6 = 12$

(b) $C_{11} = (-1)^2 M_{11} = 30$

$C_{12} = (-1)^3 M_{12} = -12$

$C_{13} = (-1)^4 M_{13} = 11$

$C_{21} = (-1)^3 M_{21} = 36$

$C_{22} = (-1)^4 M_{22} = 26$

$C_{23} = (-1)^5 M_{23} = -7$

$C_{31} = (-1)^4 M_{31} = -4$

$C_{32} = (-1)^5 M_{32} = 42$

$C_{33} = (-1)^6 M_{33} = 12$

31. (a) $\begin{vmatrix} -3 & 2 & 1 \\ 4 & 5 & 6 \\ 2 & -3 & 1 \end{vmatrix} = -3 \begin{vmatrix} 5 & 6 \\ -3 & 1 \end{vmatrix} - 2 \begin{vmatrix} 4 & 6 \\ 2 & 1 \end{vmatrix} + \begin{vmatrix} 4 & 5 \\ 2 & -3 \end{vmatrix} = -3(23) - 2(-8) - 22 = -75$

(b) $\begin{vmatrix} -3 & 2 & 1 \\ 4 & 5 & 6 \\ 2 & -3 & 1 \end{vmatrix} = -2 \begin{vmatrix} 4 & 6 \\ 2 & 1 \end{vmatrix} + 5 \begin{vmatrix} -3 & 1 \\ 2 & 1 \end{vmatrix} + 3 \begin{vmatrix} -3 & 1 \\ 4 & 6 \end{vmatrix} = -2(-8) + 5(-5) + 3(-22) = -75$

33. (a) $\begin{vmatrix} 5 & 0 & -3 \\ 0 & 12 & 4 \\ 1 & 6 & 3 \end{vmatrix} = 0 \begin{vmatrix} 0 & -3 \\ 6 & 3 \end{vmatrix} + 12 \begin{vmatrix} 5 & -3 \\ 1 & 3 \end{vmatrix} - 4 \begin{vmatrix} 5 & 0 \\ 1 & 6 \end{vmatrix} = 0(18) + 12(18) - 4(30) = 96$

(b) $\begin{vmatrix} 5 & 0 & -3 \\ 0 & 12 & 4 \\ 1 & 6 & 3 \end{vmatrix} = 0 \begin{vmatrix} 0 & 4 \\ 1 & 3 \end{vmatrix} + 12 \begin{vmatrix} 5 & -3 \\ 1 & 3 \end{vmatrix} - 6 \begin{vmatrix} 5 & -3 \\ 0 & 4 \end{vmatrix} = 0(-4) + 12(18) - 6(20) = 96$

35. (a) $\begin{vmatrix} 6 & 0 & -3 & 5 \\ 4 & 13 & 6 & -8 \\ -1 & 0 & 7 & 4 \\ 8 & 6 & 0 & 2 \end{vmatrix} = -4 \begin{vmatrix} 0 & -3 & 5 \\ 0 & 7 & 4 \\ 6 & 0 & 2 \end{vmatrix} + 13 \begin{vmatrix} 6 & -3 & 5 \\ -1 & 7 & 4 \\ 8 & 0 & 2 \end{vmatrix} - 6 \begin{vmatrix} 6 & 0 & 5 \\ -1 & 0 & 4 \\ 8 & 6 & 2 \end{vmatrix} - 8 \begin{vmatrix} 6 & 0 & -3 \\ -1 & 0 & 7 \\ 8 & 6 & 0 \end{vmatrix}$

$= -4(-282) + 13(-298) - 6(-174) - 8(-234) = 170$

(b) $\begin{vmatrix} 6 & 0 & -3 & 5 \\ 4 & 13 & 6 & -8 \\ -1 & 0 & 7 & 4 \\ 8 & 6 & 0 & 2 \end{vmatrix} = 0 \begin{vmatrix} 4 & 6 & -8 \\ -1 & 7 & 4 \\ 8 & 0 & 2 \end{vmatrix} + 13 \begin{vmatrix} 6 & -3 & 5 \\ -1 & 7 & 4 \\ 8 & 0 & 2 \end{vmatrix} + 0 \begin{vmatrix} 6 & -3 & 5 \\ 4 & 6 & -8 \\ 8 & 0 & 2 \end{vmatrix} + 6 \begin{vmatrix} 6 & -3 & 5 \\ 4 & 6 & -8 \\ -1 & 7 & 4 \end{vmatrix}$

$= 0 + 13(-298) + 0 + 6(674) = 170$

37. Expand along Column 1.

$\begin{vmatrix} 2 & -1 & 0 \\ 4 & 2 & 1 \\ 4 & 2 & 1 \end{vmatrix} = 2 \begin{vmatrix} 2 & 1 \\ 2 & 1 \end{vmatrix} - 4 \begin{vmatrix} -1 & 0 \\ 2 & 1 \end{vmatrix} + 4 \begin{vmatrix} -1 & 0 \\ 2 & 1 \end{vmatrix} = 2(0) - 4(-1) + 4(-1) = 0$

39. Expand along Row 2.

$\begin{vmatrix} 6 & 3 & -7 \\ 0 & 0 & 0 \\ 4 & -6 & 3 \end{vmatrix} = 0 \begin{vmatrix} 3 & -7 \\ -6 & 3 \end{vmatrix} - 0 \begin{vmatrix} 6 & -7 \\ 4 & 3 \end{vmatrix} + 0 \begin{vmatrix} 6 & 3 \\ 4 & -6 \end{vmatrix} = 0$

41. $\begin{vmatrix} -1 & 2 & -5 \\ 0 & 3 & 4 \\ 0 & 0 & 3 \end{vmatrix} = (-1)(3)(3) = -9$ (Upper triangular)

43. Expand along Column 3.

$\begin{vmatrix} 1 & 4 & -2 \\ 3 & 2 & 0 \\ -1 & 4 & 3 \end{vmatrix} = -2 \begin{vmatrix} 3 & 2 \\ -1 & 4 \end{vmatrix} + 3 \begin{vmatrix} 1 & 4 \\ 3 & 2 \end{vmatrix}$

$= -2(14) + 3(-10) = -58$

45. $\begin{vmatrix} 2 & 4 & 6 \\ 0 & 3 & 1 \\ 0 & 0 & -5 \end{vmatrix} = (2)(3)(-5) = -30$ (Upper triangular)

47. Expand along Column 3.

$\begin{vmatrix} 2 & 6 & 6 & 2 \\ 2 & 7 & 3 & 6 \\ 1 & 5 & 0 & 1 \\ 3 & 7 & 0 & 7 \end{vmatrix} = 6 \begin{vmatrix} 2 & 7 & 6 \\ 1 & 5 & 1 \\ 3 & 7 & 7 \end{vmatrix} - 3 \begin{vmatrix} 2 & 6 & 2 \\ 1 & 5 & 1 \\ 3 & 7 & 7 \end{vmatrix} = 6(-20) - 3(16) = -168$

49. Expand along Column 1.

$\begin{vmatrix} 5 & 3 & 0 & 6 \\ 4 & 6 & 4 & 12 \\ 0 & 2 & -3 & 4 \\ 0 & 1 & -2 & 2 \end{vmatrix} = 5 \begin{vmatrix} 6 & 4 & 12 \\ 2 & -3 & 4 \\ 1 & -2 & 2 \end{vmatrix} - 4 \begin{vmatrix} 3 & 0 & 6 \\ 2 & -3 & 4 \\ 1 & -2 & 2 \end{vmatrix} = 5(0) - 4(0) = 0$

51. Expand along Column 2, then along Column 4.

$$\begin{vmatrix} 3 & 2 & 4 & -1 & 5 \\ -2 & 0 & 1 & 3 & 2 \\ 1 & 0 & 0 & 4 & 0 \\ 6 & 0 & 2 & -1 & 0 \\ 3 & 0 & 5 & 1 & 0 \end{vmatrix} = -2 \begin{vmatrix} -2 & 1 & 3 & 2 \\ 1 & 0 & 4 & 0 \\ 6 & 2 & -1 & 0 \\ 3 & 5 & 1 & 0 \end{vmatrix} = (-2)(-2)\begin{vmatrix} 1 & 0 & 4 \\ 6 & 2 & -1 \\ 3 & 5 & 1 \end{vmatrix} = 4(103) = 412$$

53. $\begin{vmatrix} 3 & 8 & -7 \\ 0 & -5 & 4 \\ 8 & 1 & 6 \end{vmatrix} = -126$

55. $\begin{vmatrix} 7 & 0 & -14 \\ -2 & 5 & 4 \\ -6 & 2 & 12 \end{vmatrix} = 0$

57. $\begin{vmatrix} 1 & -1 & 8 & 4 \\ 2 & 6 & 0 & -4 \\ 2 & 0 & 2 & 6 \\ 0 & 2 & 8 & 0 \end{vmatrix} = -336$

59. $\begin{vmatrix} 3 & -2 & 4 & 3 & 1 \\ -1 & 0 & 2 & 1 & 0 \\ 5 & -1 & 0 & 3 & 2 \\ 4 & 7 & -8 & 0 & 0 \\ 1 & 2 & 3 & 0 & 2 \end{vmatrix} = 410$

61. (a) $\begin{vmatrix} -1 & 0 \\ 0 & 3 \end{vmatrix} = -3$

(b) $\begin{vmatrix} 2 & 0 \\ 0 & -1 \end{vmatrix} = -2$

(c) $\begin{bmatrix} -1 & 0 \\ 0 & 3 \end{bmatrix}\begin{bmatrix} 2 & 0 \\ 0 & -1 \end{bmatrix} = \begin{bmatrix} -2 & 0 \\ 0 & -3 \end{bmatrix}$

(d) $\begin{vmatrix} -2 & 0 \\ 0 & -3 \end{vmatrix} = 6$

63. (a) $\begin{vmatrix} 4 & 0 \\ 3 & -2 \end{vmatrix} = -8$

(b) $\begin{vmatrix} -1 & 1 \\ -2 & 2 \end{vmatrix} = 0$

(c) $\begin{bmatrix} 4 & 0 \\ 3 & -2 \end{bmatrix}\begin{bmatrix} -1 & 1 \\ -2 & 2 \end{bmatrix} = \begin{bmatrix} -4 & 4 \\ 1 & -1 \end{bmatrix}$

(d) $\begin{vmatrix} -4 & 4 \\ 1 & -1 \end{vmatrix} = 0$

65. (a) $\begin{vmatrix} 0 & 1 & 2 \\ -3 & -2 & 1 \\ 0 & 4 & 1 \end{vmatrix} = -21$

(b) $\begin{vmatrix} 3 & -2 & 0 \\ 1 & -1 & 2 \\ 3 & 1 & 1 \end{vmatrix} = -19$

(c) $\begin{bmatrix} 0 & 1 & 2 \\ -3 & -2 & 1 \\ 0 & 4 & 1 \end{bmatrix}\begin{bmatrix} 3 & -2 & 0 \\ 1 & -1 & 2 \\ 3 & 1 & 1 \end{bmatrix} = \begin{bmatrix} 7 & 1 & 4 \\ -8 & 9 & -3 \\ 7 & -3 & 9 \end{bmatrix}$

(d) $\begin{vmatrix} 7 & 1 & 4 \\ -8 & 9 & -3 \\ 7 & -3 & 9 \end{vmatrix} = 399$

67. (a) $\begin{vmatrix} -1 & 2 & 1 \\ 1 & 0 & 1 \\ 0 & 1 & 0 \end{vmatrix} = 2$

(b) $\begin{vmatrix} -1 & 0 & 0 \\ 0 & 2 & 0 \\ 0 & 0 & 3 \end{vmatrix} = -6$

(c) $\begin{bmatrix} -1 & 2 & 1 \\ 1 & 0 & 1 \\ 0 & 1 & 0 \end{bmatrix}\begin{bmatrix} -1 & 0 & 0 \\ 0 & 2 & 0 \\ 0 & 0 & 3 \end{bmatrix} = \begin{bmatrix} 1 & 4 & 3 \\ -1 & 0 & 3 \\ 0 & 2 & 0 \end{bmatrix}$

(d) $\begin{vmatrix} 1 & 4 & 3 \\ -1 & 0 & 3 \\ 0 & 2 & 0 \end{vmatrix} = -12$

69. $\begin{vmatrix} w & x \\ y & z \end{vmatrix} = wz - xy$

$-\begin{vmatrix} y & z \\ w & x \end{vmatrix} = -(xy - wz) = wz - xy$

Thus, $\begin{vmatrix} w & x \\ y & z \end{vmatrix} = -\begin{vmatrix} y & z \\ w & x \end{vmatrix}$.

71. $\begin{vmatrix} w & x \\ y & z \end{vmatrix} = wz - xy$

$\begin{vmatrix} w & x + cw \\ y & z + cy \end{vmatrix} = w(z + cy) - y(x + cw) = wz - xy$

Thus, $\begin{vmatrix} w & x \\ y & z \end{vmatrix} = \begin{vmatrix} w & x + cw \\ y & z + cy \end{vmatrix}$.

73. $\begin{vmatrix} 1 & x & x^2 \\ 1 & y & y^2 \\ 1 & z & z^2 \end{vmatrix} = \begin{vmatrix} y & y^2 \\ z & z^2 \end{vmatrix} - \begin{vmatrix} x & x^2 \\ z & z^2 \end{vmatrix} + \begin{vmatrix} x & x^2 \\ y & y^2 \end{vmatrix}$

$= (yz^2 - y^2z) - (xz^2 - x^2z) + (xy^2 - x^2y)$

$= yz^2 - xz^2 - y^2z + x^2z + xy(y - x)$

$= z^2(y - x) - z(y^2 - x^2) + xy(y - x)$

$= z^2(y - x) - z(y - x)(y + x) + xy(y - x)$

$= (y - x)[z^2 - z(y + x) + xy]$

$= (y - x)[z^2 - zy - zx + xy]$

$= (y - x)[z^2 - zx - zy + xy]$

$= (y - x)[z(z - x) - y(z - x)]$

$= (y - x)(z - x)(z - y)$

75. $\begin{vmatrix} x - 1 & 2 \\ 3 & x - 2 \end{vmatrix} = 0$

$(x - 1)(x - 2) - 6 = 0$

$x^2 - 3x - 4 = 0$

$(x + 1)(x - 4) = 0$

$x = -1 \text{ or } x = 4$

77. $\begin{vmatrix} x + 3 & 2 \\ 1 & x + 2 \end{vmatrix} = 0$

$(x + 3)(x + 2) - 2 = 0$

$x^2 + 5x + 4 = 0$

$(x + 1)(x + 4) = 0$

$x = -1 \text{ or } x = -4$

79. $\begin{vmatrix} 4u & -1 \\ -1 & 2v \end{vmatrix} = 8uv - 1$

81. $\begin{vmatrix} e^{2x} & e^{3x} \\ 2e^{2x} & 3e^{3x} \end{vmatrix} = 3e^{5x} - 2e^{5x} = e^{5x}$

83. $\begin{vmatrix} x & \ln x \\ 1 & \dfrac{1}{x} \end{vmatrix} = 1 - \ln x$

85. True. If an entire row is zero, then each cofactor in the expansion is multiplied by zero.

87. Let $A = \begin{bmatrix} 1 & 3 \\ -2 & 4 \end{bmatrix}$ and $B = \begin{bmatrix} -4 & 0 \\ 3 & 5 \end{bmatrix}$.

$|A| = \begin{vmatrix} 1 & 3 \\ -2 & 4 \end{vmatrix} = 10, \ |B| = \begin{vmatrix} -4 & 0 \\ 3 & 5 \end{vmatrix} = -20, \ |A| + |B| = -10$

$A + B = \begin{bmatrix} -3 & 3 \\ 1 & 9 \end{bmatrix}, \ |A + B| = \begin{vmatrix} -3 & 3 \\ 1 & 9 \end{vmatrix} = -30$

Thus, $|A + B| \ne |A| + |B|$. Your answer may differ, depending on how you choose A and B.

89. A square matrix is a square array of numbers. The determinant of a square matrix is a real number.

91. (a) $\begin{vmatrix} 1 & 3 & 4 \\ -7 & 2 & -5 \\ 6 & 1 & 2 \end{vmatrix} = -115$

$-\begin{vmatrix} 1 & 4 & 3 \\ -7 & -5 & 2 \\ 6 & 2 & 1 \end{vmatrix} = -115$

Column 2 and Column 3 were interchanged.

(b) $\begin{vmatrix} 1 & 3 & 4 \\ -2 & 2 & 0 \\ 1 & 6 & 2 \end{vmatrix} = -40$

$-\begin{vmatrix} 1 & 6 & 2 \\ -2 & 2 & 0 \\ 1 & 3 & 4 \end{vmatrix} = -40$

Row 1 and Row 3 were interchanged.

93. (a) $A = \begin{bmatrix} 1 & 2 \\ 2 & -3 \end{bmatrix}, B = \begin{bmatrix} 5 & 10 \\ 2 & -3 \end{bmatrix}$

$|B| = \begin{vmatrix} 5 & 10 \\ 2 & -3 \end{vmatrix} = -35$

$5|A| = 5\begin{vmatrix} 1 & 2 \\ 2 & -3 \end{vmatrix} = -35$

Row 1 was multiplied by 5.

$|B| = 5|A|$

(b) $A = \begin{bmatrix} 1 & 2 & -1 \\ 3 & -3 & 2 \\ 7 & 1 & 3 \end{bmatrix}, B = \begin{bmatrix} 1 & 8 & -3 \\ 3 & -12 & 6 \\ 7 & 4 & 9 \end{bmatrix}$

$|B| = \begin{vmatrix} 1 & 8 & -3 \\ 3 & -12 & 6 \\ 7 & 4 & 9 \end{vmatrix} = -300$

$12|A| = 12\begin{vmatrix} 1 & 2 & -1 \\ 3 & -3 & 2 \\ 7 & 1 & 3 \end{vmatrix} = -300$

Column 2 was multiplied by 4 and Column 3 was multiplied by 3.

$|B| = (4)(3)|A| = 12|A|$

95. $f(x) = x^3 - 2x$

Since f is a polynomial, the domain is all real numbers x.

97. $h(x) = \sqrt{16 - x^2}$

$16 - x^2 \geq 0$

$(4 + x)(4 - x) \geq 0$

Critical numbers: $x = \pm 4$

Test intervals: $(-\infty, -4), (-4, 4), (4, \infty)$

Test: Is $16 - x^2 \geq 0$?

Solution: $[-4, 4]$

Domain of h: $-4 \leq x \leq 4$

99. $g(t) = \ln(t - 1)$

$t - 1 > 0$

$t > 1$

Domain: all real numbers $t > 1$

101. $\begin{cases} x + y \leq 8 \\ \quad\quad x \geq -3 \\ 2x - y < 5 \end{cases}$

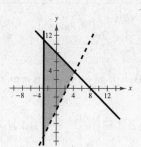

103. $[A \;\vdots\; I] = \begin{bmatrix} -4 & 1 & \vdots & 1 & 0 \\ 8 & -1 & \vdots & 0 & 1 \end{bmatrix}$

$2R_1 + R_2 \rightarrow \begin{bmatrix} -4 & 1 & \vdots & 1 & 0 \\ 0 & 1 & \vdots & 2 & 1 \end{bmatrix}$

$-R_2 + R_1 \rightarrow \begin{bmatrix} -4 & 0 & \vdots & -1 & -1 \\ 0 & 1 & \vdots & 2 & 1 \end{bmatrix}$

$-\frac{1}{4}R_1 \rightarrow \begin{bmatrix} 1 & 0 & \vdots & \frac{1}{4} & \frac{1}{4} \\ 0 & 1 & \vdots & 2 & 1 \end{bmatrix} = [I \;\vdots\; A^{-1}]$

$A^{-1} = \begin{bmatrix} \frac{1}{4} & \frac{1}{4} \\ 2 & 1 \end{bmatrix}$

105. $[A \;\vdots\; I] = \begin{bmatrix} -7 & 2 & 9 & \vdots & 1 & 0 & 0 \\ 2 & -4 & -6 & \vdots & 0 & 1 & 0 \\ 3 & 5 & 2 & \vdots & 0 & 0 & 1 \end{bmatrix}$

$4R_2 + R_1 \rightarrow \begin{bmatrix} 1 & -14 & -15 & \vdots & 1 & 4 & 0 \\ 2 & -4 & -6 & \vdots & 0 & 1 & 0 \\ 3 & 5 & 2 & \vdots & 0 & 0 & 1 \end{bmatrix}$

$\begin{matrix} \\ -2R_1 + R_2 \rightarrow \\ -3R_1 + R_3 \rightarrow \end{matrix} \begin{bmatrix} 1 & -14 & -15 & \vdots & 1 & 4 & 0 \\ 0 & 24 & 24 & \vdots & -2 & -7 & 0 \\ 0 & 47 & 47 & \vdots & -3 & -12 & 1 \end{bmatrix}$

$-\frac{47}{24}R_2 + R_3 \rightarrow \begin{bmatrix} 1 & -14 & -15 & \vdots & 1 & 4 & 0 \\ 0 & 24 & 24 & \vdots & -2 & -7 & 0 \\ 0 & 0 & 0 & \vdots & \frac{11}{12} & \frac{41}{24} & 1 \end{bmatrix}$

The zeros in Row 3 imply that the inverse does not exist.

Section 8.5   Applications of Matrices and Determinants

- You should be able to use Cramer's Rule to solve a system of linear equations.

- Now you should be able to solve a system of linear equations by graphing, substitution, elimination, elementary row operations on an augmented matrix, using the inverse matrix, or Cramer's Rule.

- You should be able to find the area of a triangle with vertices (x_1, y_1), (x_2, y_2), and (x_3, y_3).

$$\text{Area} = \pm\frac{1}{2}\begin{vmatrix} x_1 & y_1 & 1 \\ x_2 & y_2 & 1 \\ x_3 & y_3 & 1 \end{vmatrix}$$

  The $\pm$ symbol indicates that the appropriate sign should be chosen so that the area is positive.

- You should be able to test to see if three points, (x_1, y_1), (x_2, y_2), and (x_3, y_3), are collinear.

$$\begin{vmatrix} x_1 & y_1 & 1 \\ x_2 & y_2 & 1 \\ x_3 & y_3 & 1 \end{vmatrix} = 0, \text{ if and only if they are collinear.}$$

- You should be able to find the equation of the line through (x_1, y_1) and (x_2, y_2) by evaluating.

$$\begin{vmatrix} x & y & 1 \\ x_1 & y_1 & 1 \\ x_2 & y_2 & 1 \end{vmatrix} = 0$$

- You should be able to encode and decode messages by using an invertible $n \times n$ matrix.

Vocabulary Check

1. Cramer's Rule   **2.** colinear   **3.** $A = \pm\frac{1}{2}\begin{vmatrix} x_1 & y_1 & 1 \\ x_2 & y_2 & 1 \\ x_3 & y_3 & 1 \end{vmatrix}$ **4.** cryptogram   **5.** uncoded; coded

---

1. $\begin{cases} 3x + 4y = -2 \\ 5x + 3y = 4 \end{cases}$

$$x = \frac{\begin{vmatrix} -2 & 4 \\ 4 & 3 \end{vmatrix}}{\begin{vmatrix} 3 & 4 \\ 5 & 3 \end{vmatrix}} = \frac{-22}{-11} = 2$$

$$y = \frac{\begin{vmatrix} 3 & -2 \\ 5 & 4 \end{vmatrix}}{\begin{vmatrix} 3 & 4 \\ 5 & 3 \end{vmatrix}} = \frac{22}{-11} = -2$$

Solution: $(2, -2)$

3. $\begin{cases} 3x + 2y = -2 \\ 6x + 4y = 4 \end{cases}$

Since $\begin{vmatrix} 3 & 2 \\ 6 & 4 \end{vmatrix} = 0$, Cramer's Rule does not apply.

The system is inconsistent in this case and has no solution.

5. $\begin{cases} -0.4x + 0.8y = 1.6 \\ 0.2x + 0.3y = 2.2 \end{cases}$

$$x = \frac{\begin{vmatrix} 1.6 & 0.8 \\ 2.2 & 0.3 \end{vmatrix}}{\begin{vmatrix} -0.4 & 0.8 \\ 0.2 & 0.3 \end{vmatrix}} = \frac{-1.28}{-0.28} = \frac{32}{7}$$

$$y = \frac{\begin{vmatrix} -0.4 & 1.6 \\ 0.2 & 2.2 \end{vmatrix}}{\begin{vmatrix} -0.4 & 0.8 \\ 0.2 & 0.3 \end{vmatrix}} = \frac{-1.20}{-0.28} = \frac{30}{7}$$

Solution: $\left(\dfrac{32}{7}, \dfrac{30}{7} \right)$

7. $\begin{cases} 4x - y + z = -5 \\ 2x + 2y + 3z = 10, \\ 5x - 2y + 6z = 1 \end{cases}$ $\quad D = \begin{vmatrix} 4 & -1 & 1 \\ 2 & 2 & 3 \\ 5 & -2 & 6 \end{vmatrix} = 55$

$$x = \frac{\begin{vmatrix} -5 & -1 & 1 \\ 10 & 2 & 3 \\ 1 & -2 & 6 \end{vmatrix}}{55} = \frac{-55}{55} = -1$$

$$y = \frac{\begin{vmatrix} 4 & -5 & 1 \\ 2 & 10 & 3 \\ 5 & 1 & 6 \end{vmatrix}}{55} = \frac{165}{55} = 3$$

$$z = \frac{\begin{vmatrix} 4 & -1 & -5 \\ 2 & 2 & 10 \\ 5 & -2 & 1 \end{vmatrix}}{55} = \frac{110}{55} = 2$$

Solution: $(-1, 3, 2)$

9. $\begin{cases} x + 2y + 3z = -3 \\ -2x + y - z = 6, \\ 3x - 3y + 2z = -11 \end{cases}$ $\quad D = \begin{vmatrix} 1 & 2 & 3 \\ -2 & 1 & -1 \\ 3 & -3 & 2 \end{vmatrix} = 10$

$$x = \frac{\begin{vmatrix} -3 & 2 & 3 \\ 6 & 1 & -1 \\ -11 & -3 & 2 \end{vmatrix}}{10} = \frac{-20}{10} = -2$$

$$y = \frac{\begin{vmatrix} 1 & -3 & 3 \\ -2 & 6 & -1 \\ 3 & -11 & 2 \end{vmatrix}}{10} = \frac{10}{10} = 1$$

$$z = \frac{\begin{vmatrix} 1 & 2 & -3 \\ -2 & 1 & 6 \\ 3 & -3 & -11 \end{vmatrix}}{10} = \frac{-10}{10} = -1$$

Solution: $(-2, 1, -1)$

11. $\begin{cases} 3x + 3y + 5z = 1 \\ 3x + 5y + 9z = 2, \\ 5x + 9y + 17z = 4 \end{cases}$ $\quad D = \begin{vmatrix} 3 & 3 & 5 \\ 3 & 5 & 9 \\ 5 & 9 & 17 \end{vmatrix} = 4$

$$x = \frac{\begin{vmatrix} 1 & 3 & 5 \\ 2 & 5 & 9 \\ 4 & 9 & 17 \end{vmatrix}}{4} = 0$$

$$y = \frac{\begin{vmatrix} 3 & 1 & 5 \\ 3 & 2 & 9 \\ 5 & 4 & 17 \end{vmatrix}}{4} = -\frac{1}{2}$$

$$z = \frac{\begin{vmatrix} 3 & 3 & 1 \\ 3 & 5 & 2 \\ 5 & 9 & 4 \end{vmatrix}}{4} = \frac{1}{2}$$

Solution: $\left(0, -\dfrac{1}{2}, \dfrac{1}{2} \right)$

13. $\begin{cases} 2x + y + 2z = 6 \\ -x + 2y - 3z = 0 \\ 3x + 2y - z = 6 \end{cases}$ $\quad D = \begin{vmatrix} 2 & 1 & 2 \\ -1 & 2 & -3 \\ 3 & 2 & -1 \end{vmatrix} = -18$

$$x = \frac{\begin{vmatrix} 6 & 1 & 2 \\ 0 & 2 & -3 \\ 6 & 2 & -1 \end{vmatrix}}{-18} = 1, \quad y = \frac{\begin{vmatrix} 2 & 6 & 2 \\ -1 & 0 & -3 \\ 3 & 6 & -1 \end{vmatrix}}{-18} = 2, \quad z = \frac{\begin{vmatrix} 2 & 1 & 6 \\ -1 & 2 & 0 \\ 3 & 2 & 6 \end{vmatrix}}{-18} = 1$$

Solution: $(1, 2, 1)$

15. Vertices: $(0, 0), (3, 1), (1, 5)$

$$\text{Area} = \frac{1}{2} \begin{vmatrix} 0 & 0 & 1 \\ 3 & 1 & 1 \\ 1 & 5 & 1 \end{vmatrix} = \frac{1}{2} \begin{vmatrix} 3 & 1 \\ 1 & 5 \end{vmatrix} = 7 \text{ square units}$$

17. Vertices: $(-2, -3), (2, -3), (0, 4)$

$$\text{Area} = \frac{1}{2} \begin{vmatrix} -2 & -3 & 1 \\ 2 & -3 & 1 \\ 0 & 4 & 1 \end{vmatrix} = \frac{1}{2}\left(-2\begin{vmatrix} -3 & 1 \\ 4 & 1 \end{vmatrix} - 2\begin{vmatrix} -3 & 1 \\ 4 & 1 \end{vmatrix}\right) = \frac{1}{2}(14 + 14) = 14 \text{ square units}$$

19. Vertices: $\left(0, \frac{1}{2}\right), \left(\frac{5}{2}, 0\right), (4, 3)$

$$\text{Area} = \frac{1}{2} \begin{vmatrix} 0 & \frac{1}{2} & 1 \\ \frac{5}{2} & 0 & 1 \\ 4 & 3 & 1 \end{vmatrix} = \frac{1}{2}\left(-\frac{1}{2}\begin{vmatrix} \frac{5}{2} & 1 \\ 4 & 1 \end{vmatrix} + 1\begin{vmatrix} \frac{5}{2} & 0 \\ 4 & 3 \end{vmatrix}\right) = \frac{1}{2}\left(\frac{3}{4} + \frac{15}{2}\right) = \frac{33}{8} \text{ square units}$$

21. Vertices: $(-2, 4), (2, 3), (-1, 5)$

$$\text{Area} = \frac{1}{2} \begin{vmatrix} -2 & 4 & 1 \\ 2 & 3 & 1 \\ -1 & 5 & 1 \end{vmatrix} = \frac{1}{2}\left[\begin{vmatrix} 2 & 3 \\ -1 & 5 \end{vmatrix} - \begin{vmatrix} -2 & 4 \\ -1 & 5 \end{vmatrix} + \begin{vmatrix} -2 & 4 \\ 2 & 3 \end{vmatrix}\right] = \frac{1}{2}(13 + 6 - 14) = \frac{5}{2} \text{ square units}$$

23. Vertices: $(-3, 5), (2, 6), (3, -5)$

$$\text{Area} = -\frac{1}{2} \begin{vmatrix} -3 & 5 & 1 \\ 2 & 6 & 1 \\ 3 & -5 & 1 \end{vmatrix} = -\frac{1}{2}\left[\begin{vmatrix} 2 & 6 \\ 3 & -5 \end{vmatrix} - \begin{vmatrix} -3 & 5 \\ 3 & -5 \end{vmatrix} + \begin{vmatrix} -3 & 5 \\ 2 & 6 \end{vmatrix}\right] = -\frac{1}{2}(-28 + 0 - 28) = 28 \text{ square units}$$

25. $4 = \pm\frac{1}{2} \begin{vmatrix} -5 & 1 & 1 \\ 0 & 2 & 1 \\ -2 & y & 1 \end{vmatrix}$

$\pm 8 = -5\begin{vmatrix} 2 & 1 \\ y & 1 \end{vmatrix} - 2\begin{vmatrix} 1 & 1 \\ 2 & 1 \end{vmatrix}$

$\pm 8 = -5(2 - y) - 2(-1)$

$\pm 8 = 5y - 8$

$y = \frac{8 \pm 8}{5}$

$y = \frac{16}{5} \text{ or } y = 0$

27. $6 = \pm\frac{1}{2} \begin{vmatrix} -2 & -3 & 1 \\ 1 & -1 & 1 \\ -8 & y & 1 \end{vmatrix}$

$\pm 12 = \begin{vmatrix} 1 & -1 \\ -8 & y \end{vmatrix} - \begin{vmatrix} -2 & -3 \\ -8 & y \end{vmatrix} + \begin{vmatrix} -2 & -3 \\ 1 & -1 \end{vmatrix}$

$\pm 12 = (y - 8) - (-2y - 24) + 5$

$\pm 12 = 3y + 21$

$y = \frac{-21 \pm 12}{3} = -7 \pm 4$

$y = -3 \text{ or } y = -11$

29. Vertices: $(0, 25), (10, 0), (28, 5)$

$$\text{Area} = \frac{1}{2} \begin{vmatrix} 0 & 25 & 1 \\ 10 & 0 & 1 \\ 28 & 5 & 1 \end{vmatrix} = 250 \text{ square miles}$$

31. Points: $(3, -1), (0, -3), (12, 5)$

$$\begin{vmatrix} 3 & -1 & 1 \\ 0 & -3 & 1 \\ 12 & 5 & 1 \end{vmatrix} = 3\begin{vmatrix} -3 & 1 \\ 5 & 1 \end{vmatrix} + 12\begin{vmatrix} -1 & 1 \\ -3 & 1 \end{vmatrix} = 3(-8) + 12(2) = 0$$

The points are collinear.

33. Points: $\left(2, -\frac{1}{2}\right), (-4, 4), (6, -3)$

$$\begin{vmatrix} 2 & -\frac{1}{2} & 1 \\ -4 & 4 & 1 \\ 6 & -3 & 1 \end{vmatrix} = \begin{vmatrix} -4 & 4 \\ 6 & -3 \end{vmatrix} - \begin{vmatrix} 2 & -\frac{1}{2} \\ 6 & -3 \end{vmatrix} + \begin{vmatrix} 2 & -\frac{1}{2} \\ -4 & 4 \end{vmatrix} = -12 + 3 + 6 = -3 \neq 0$$

The points are not collinear.

35. Points: $(0, 2), (1, 2.4), (-1, 1.6)$

$$\begin{vmatrix} 0 & 2 & 1 \\ 1 & 2.4 & 1 \\ -1 & 1.6 & 1 \end{vmatrix} = -2\begin{vmatrix} 1 & 1 \\ -1 & 1 \end{vmatrix} + \begin{vmatrix} 1 & 2.4 \\ -1 & 1.6 \end{vmatrix} = -2(2) + 4 = 0$$

The points are collinear.

37.

$$\begin{vmatrix} 2 & -5 & 1 \\ 4 & y & 1 \\ 5 & -2 & 1 \end{vmatrix} = 0$$

$$2\begin{vmatrix} y & 1 \\ -2 & 1 \end{vmatrix} + 5\begin{vmatrix} 4 & 1 \\ 5 & 1 \end{vmatrix} + \begin{vmatrix} 4 & y \\ 5 & -2 \end{vmatrix} = 0$$

$$2(y + 2) + 5(-1) + (-8 - 5y) = 0$$

$$-3y - 9 = 0$$

$$y = -3$$

39. Points: $(0, 0), (5, 3)$

Equation: $\begin{vmatrix} x & y & 1 \\ 0 & 0 & 1 \\ 5 & 3 & 1 \end{vmatrix} = -\begin{vmatrix} x & y \\ 5 & 3 \end{vmatrix} = 5y - 3x = 0 \implies 3x - 5y = 0$

41. Points: $(-4, 3), (2, 1)$

Equation: $\begin{vmatrix} x & y & 1 \\ -4 & 3 & 1 \\ 2 & 1 & 1 \end{vmatrix} = x\begin{vmatrix} 3 & 1 \\ 1 & 1 \end{vmatrix} - y\begin{vmatrix} -4 & 1 \\ 2 & 1 \end{vmatrix} + \begin{vmatrix} -4 & 3 \\ 2 & 1 \end{vmatrix} = 2x + 6y - 10 = 0 \implies x + 3y - 5 = 0$

43. Points: $\left(-\frac{1}{2}, 3\right), \left(\frac{5}{2}, 1\right)$

Equation: $\begin{vmatrix} x & y & 1 \\ -\frac{1}{2} & 3 & 1 \\ \frac{5}{2} & 1 & 1 \end{vmatrix} = x\begin{vmatrix} 3 & 1 \\ 1 & 1 \end{vmatrix} - y\begin{vmatrix} -\frac{1}{2} & 1 \\ \frac{5}{2} & 1 \end{vmatrix} + \begin{vmatrix} -\frac{1}{2} & 3 \\ \frac{5}{2} & 1 \end{vmatrix} = 2x + 3y - 8 = 0$

45. The uncoded row matrices are the rows of the 7×3 matrix on the left.

$$\begin{matrix} \text{T} & \text{R} & \text{O} \\ \text{U} & \text{B} & \text{L} \\ \text{E} & & \text{I} \\ \text{N} & & \text{R} \\ \text{I} & \text{V} & \text{E} \\ \text{R} & & \text{C} \\ \text{I} & \text{T} & \text{Y} \end{matrix} \begin{bmatrix} 20 & 18 & 15 \\ 21 & 2 & 12 \\ 5 & 0 & 9 \\ 14 & 0 & 18 \\ 9 & 22 & 5 \\ 18 & 0 & 3 \\ 9 & 20 & 25 \end{bmatrix} \begin{bmatrix} 1 & -1 & 0 \\ 1 & 0 & -1 \\ -6 & 2 & 3 \end{bmatrix} = \begin{bmatrix} -52 & 10 & 27 \\ -49 & 3 & 34 \\ -49 & 13 & 27 \\ -94 & 22 & 54 \\ 1 & 1 & -7 \\ 0 & -12 & 9 \\ -121 & 41 & 55 \end{bmatrix}$$

Solution: -52 10   27 -49 3   34 -49 13   27 -94 22   54   1   1 -7 0 -12 9 -121 41   55

In Exercises 47 and 49, use the matrix $A = \begin{bmatrix} 1 & 2 & 2 \\ 3 & 7 & 9 \\ -1 & -4 & -7 \end{bmatrix}$.

47. C A L L _ A T _ N O O N
$[3 \quad 1 \quad 12] [12 \quad 0 \quad 1] [20 \quad 0 \quad 14] [15 \quad 15 \quad 14]$

$[3 \quad 1 \quad 12] A = [-6 \; -35 \; -69]$

$[12 \quad 0 \quad 1] A = [11 \quad 20 \quad 17]$

$[20 \quad 0 \quad 14] A = [6 \; -16 \; -58]$

$[15 \quad 15 \quad 14] A = [46 \quad 79 \quad 67]$

Cryptogram: $-6 \; -35 \; -69 \; 11 \; 20 \; 17 \; 6 \; -16 \; -58 \; 46 \; 79 \; 67$

49. H A P P Y _ B I R T H D A Y _
$[8 \quad 1 \quad 16] [16 \quad 25 \quad 0] [2 \quad 9 \quad 18] [20 \quad 8 \quad 4] [1 \quad 25 \quad 0]$

$[8 \quad 1 \quad 16] A = [-5 \quad -41 \quad -87]$

$[16 \quad 25 \quad 0] A = [91 \quad 207 \quad 257]$

$[2 \quad 9 \quad 18] A = [11 \quad -5 \quad -41]$

$[20 \quad 8 \quad 4] A = [40 \quad 80 \quad 84]$

$[1 \quad 25 \quad 0] A = [76 \quad 177 \quad 227]$

Cryptogram: $-5 \; -41 \; -87 \; 91 \; 207 \; 257 \; 11 \; -5 \; -41 \; 40 \; 80 \; 84 \; 76 \; 177 \; 227$

51. $A^{-1} = \begin{bmatrix} 1 & 2 \\ 3 & 5 \end{bmatrix}^{-1} = \begin{bmatrix} -5 & 2 \\ 3 & -1 \end{bmatrix}$

$\begin{bmatrix} 11 & 21 \\ 64 & 112 \\ 25 & 50 \\ 29 & 53 \\ 23 & 46 \\ 40 & 75 \\ 55 & 92 \end{bmatrix} \begin{bmatrix} -5 & 2 \\ 3 & -1 \end{bmatrix} = \begin{bmatrix} 8 & 1 \\ 16 & 16 \\ 25 & 0 \\ 14 & 5 \\ 23 & 0 \\ 25 & 5 \\ 1 & 18 \end{bmatrix} \begin{matrix} H & A \\ P & P \\ Y & \_ \\ N & E \\ W & \_ \\ Y & E \\ A & R \end{matrix}$ Message: HAPPY NEW YEAR

53. $A^{-1} = \begin{bmatrix} 1 & -1 & 0 \\ 1 & 0 & -1 \\ -6 & 2 & 3 \end{bmatrix}^{-1} = \begin{bmatrix} -2 & -3 & -1 \\ -3 & -3 & -1 \\ -2 & -4 & -1 \end{bmatrix}$

$\begin{bmatrix} 9 & -1 & -9 \\ 38 & -19 & -19 \\ 28 & -9 & -19 \\ -80 & 25 & 41 \\ -64 & 21 & 31 \\ 9 & -5 & -4 \end{bmatrix} \begin{bmatrix} -2 & -3 & -1 \\ -3 & -3 & -1 \\ -2 & -4 & -1 \end{bmatrix} = \begin{bmatrix} 3 & 12 & 1 \\ 19 & 19 & 0 \\ 9 & 19 & 0 \\ 3 & 1 & 14 \\ 3 & 5 & 12 \\ 5 & 4 & 0 \end{bmatrix} \begin{matrix} C & L & A \\ S & S & \_ \\ I & S & \_ \\ C & A & N \\ C & E & L \\ E & D & \_ \end{matrix}$ Message: CLASS IS CANCELED

55. $A^{-1} = \begin{bmatrix} 1 & 2 & 2 \\ 3 & 7 & 9 \\ -1 & -4 & -7 \end{bmatrix}^{-1} = \begin{bmatrix} -13 & 6 & 4 \\ 12 & -5 & -3 \\ -5 & 2 & 1 \end{bmatrix}$

$\begin{bmatrix} 20 & 17 & -15 \\ -12 & -56 & -104 \\ 1 & -25 & -65 \\ 62 & 143 & 181 \end{bmatrix} \begin{bmatrix} -13 & 6 & 4 \\ 12 & -5 & -3 \\ -5 & 2 & 1 \end{bmatrix} = \begin{bmatrix} 19 & 5 & 14 \\ 4 & 0 & 16 \\ 12 & 1 & 14 \\ 5 & 19 & 0 \end{bmatrix} \begin{matrix} S & E & N \\ D & \_ & P \\ L & A & N \\ E & S & \_ \end{matrix}$ Message: SEND PLANES

57. Let A be the 2×2 matrix needed to decode the message.

$$\begin{bmatrix} -18 & -18 \\ 1 & 16 \end{bmatrix} A = \begin{bmatrix} 0 & 18 \\ 15 & 14 \end{bmatrix} \begin{matrix} \text{R} \\ \text{N} \end{matrix}$$

$$A = \begin{bmatrix} -18 & -18 \\ 1 & 16 \end{bmatrix}^{-1} \begin{bmatrix} 0 & 18 \\ 15 & 14 \end{bmatrix} = \begin{bmatrix} -\frac{8}{135} & -\frac{1}{15} \\ \frac{1}{270} & \frac{1}{15} \end{bmatrix} \begin{bmatrix} 0 & 18 \\ 15 & 14 \end{bmatrix} = \begin{bmatrix} -1 & -2 \\ 1 & 1 \end{bmatrix}$$

8	21		13	5	M	E
−15	−10		5	20	E	T
−13	−13		0	13	_	M
5	10		5	0	E	_
5	25	$\begin{bmatrix} -1 & -2 \\ 1 & 1 \end{bmatrix} =$	20	15	T	O
5	19		14	9	N	I
−1	6		7	8	G	H
20	40		20	0	T	_
−18	−18		0	18	_	R
1	16		15	14	O	N

Message: MEET ME TONIGHT RON

59. False. In Cramer's Rule, the **denominator** is the determinant of the coefficient matrix.

61. False. If the determinant of the coefficient matrix is zero, the system has either no solution or infinitely many solutions.

63.
$$\begin{cases} -x - 7y = -22 \\ 5x + y = -26 \end{cases}$$
Equation 1
Equation 2

$$-5x - 35y = -110 \quad \text{(5)Eq.1}$$
$$5x + y = -26$$
$$-34y = -136 \quad \text{Add equations.}$$
$$y = 4$$
$$-x - 7(4) = -22$$
$$x = -6$$

Solution: $(-6, 4)$

65.
$$\begin{cases} -x - 3y + 5z = -14 \\ 4x + 2y - z = -1 \\ 5x - 3y + 2z = -11 \end{cases}$$

$$A^{-1} = \begin{bmatrix} -1 & -3 & 5 \\ 4 & 2 & -1 \\ 5 & -3 & 2 \end{bmatrix}^{-1}$$

$$= \frac{1}{72} \begin{bmatrix} -1 & 9 & 7 \\ 13 & 27 & -19 \\ 22 & 18 & -10 \end{bmatrix}$$

$$\begin{bmatrix} x \\ y \\ z \end{bmatrix} = A^{-1} \begin{bmatrix} -14 \\ -1 \\ -11 \end{bmatrix} = \begin{bmatrix} -1 \\ 0 \\ -3 \end{bmatrix}$$

Solution: $(-1, 0, -3)$

67. Objective function: $z = 6x + 4y$

Constraints:
$$x \geq 0$$
$$y \geq 0$$
$$x + 6y \leq 30$$
$$6x + y \leq 40$$

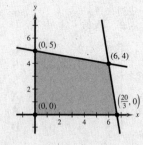

At $(0, 0)$: $z = 6(0) + 4(0) = 0$

At $(0, 5)$: $z = 6(0) + 4(5) = 20$

At $(6, 4)$: $z = 6(6) + 4(4) = 52$

At $\left(\frac{20}{3}, 0\right)$: $z = 6\left(\frac{20}{3}\right) + 4(0) = 40$

The minimum value of 0 occurs at $(0, 0)$.

The maximum value of 52 occurs at $(6, 4)$.

Review Exercises for Chapter 8

1. $\begin{bmatrix} -4 \\ 0 \\ 5 \end{bmatrix}$

Order: 3×1

3. $[3]$

Order: 1×1

5. $\begin{cases} 3x - 10y = 15 \\ 5x + 4y = 22 \end{cases}$

$\begin{bmatrix} 3 & -10 & \vdots & 15 \\ 5 & 4 & \vdots & 22 \end{bmatrix}$

7. $\begin{bmatrix} 5 & 1 & 7 & \vdots & -9 \\ 4 & 2 & 0 & \vdots & 10 \\ 9 & 4 & 2 & \vdots & 3 \end{bmatrix}$

$\begin{cases} 5x + y + 7z = -9 \\ 4x + 2y = 10 \\ 9x + 4y + 2z = 3 \end{cases}$

9. $\begin{bmatrix} 0 & 1 & 1 \\ 1 & 2 & 3 \\ 2 & 2 & 2 \end{bmatrix}$

$\begin{matrix} R_1 \\ R_2 \end{matrix} \begin{bmatrix} 1 & 2 & 3 \\ 0 & 1 & 1 \\ 2 & 2 & 2 \end{bmatrix}$

$-2R_1 + R_3 \rightarrow \begin{bmatrix} 1 & 2 & 3 \\ 0 & 1 & 1 \\ 0 & -2 & -4 \end{bmatrix}$

$2R_2 + R_3 \rightarrow \begin{bmatrix} 1 & 2 & 3 \\ 0 & 1 & 1 \\ 0 & 0 & -2 \end{bmatrix}$

$-\frac{1}{2}R_3 \rightarrow \begin{bmatrix} 1 & 2 & 3 \\ 0 & 1 & 1 \\ 0 & 0 & 1 \end{bmatrix}$

11. $\begin{bmatrix} 1 & 2 & 3 & \vdots & 9 \\ 0 & 1 & -2 & \vdots & 2 \\ 0 & 0 & 1 & \vdots & 0 \end{bmatrix} \Rightarrow \begin{cases} x + 2y + 3z = 9 \\ y - 2z = 2 \\ z = 0 \end{cases}$

$y - 2(0) = 2 \Rightarrow y = 2$

$x + 2(2) + 3(0) = 9 \Rightarrow x = 5$

Solution: $(5, 2, 0)$

13. $\begin{bmatrix} 1 & -5 & 4 & \vdots & 1 \\ 0 & 1 & 2 & \vdots & 3 \\ 0 & 0 & 1 & \vdots & 4 \end{bmatrix} \Rightarrow \begin{cases} x - 5y + 4z = 1 \\ y + 2z = 3 \\ z = 4 \end{cases}$

$y + 2(4) = 3 \Rightarrow y = -5$

$x - 5(-5) + 4(4) = 1 \Rightarrow x = -40$

Solution: $(-40, -5, 4)$

15. $\begin{bmatrix} 5 & 4 & \vdots & 2 \\ -1 & 1 & \vdots & -22 \end{bmatrix}$

$4R_2 + R_1 \rightarrow \begin{bmatrix} 1 & 8 & \vdots & -86 \\ -1 & 1 & \vdots & -22 \end{bmatrix}$

$R_1 + R_2 \rightarrow \begin{bmatrix} 1 & 8 & \vdots & -86 \\ 0 & 9 & \vdots & -108 \end{bmatrix}$

$\frac{1}{9}R_2 \rightarrow \begin{bmatrix} 1 & 8 & \vdots & -86 \\ 0 & 1 & \vdots & -12 \end{bmatrix}$

$\begin{cases} x + 8y = -86 \\ y = -12 \end{cases}$

$y = -12$

$x + 8(-12) = -86 \Rightarrow x = 10$

Solution: $(10, -12)$

17. $\begin{bmatrix} 0.3 & -0.1 & \vdots & -0.13 \\ 0.2 & -0.3 & \vdots & -0.25 \end{bmatrix}$

$\begin{matrix} 10R_1 \rightarrow \\ 10R_2 \rightarrow \end{matrix} \begin{bmatrix} 3 & -1 & \vdots & -1.3 \\ 2 & -3 & \vdots & -2.5 \end{bmatrix}$

$-R_2 + R_1 \rightarrow \begin{bmatrix} 1 & 2 & \vdots & 1.2 \\ 2 & -3 & \vdots & -2.5 \end{bmatrix}$

$-2R_1 + R_2 \rightarrow \begin{bmatrix} 1 & 2 & \vdots & 1.2 \\ 0 & -7 & \vdots & -4.9 \end{bmatrix}$

$-\frac{1}{7}R_2 \rightarrow \begin{bmatrix} 1 & 2 & \vdots & 1.2 \\ 0 & 1 & \vdots & 0.7 \end{bmatrix}$

$\begin{cases} x + 2y = 1.2 \\ y = 0.7 \end{cases}$

$y = 0.7$

$x + 2(0.7) = 1.2 \Rightarrow x = -0.2$

Solution: $(-0.2, 0.7) = \left(-\frac{1}{5}, \frac{7}{10}\right)$

19.
$$\begin{bmatrix} 2 & 3 & 1 & \vdots & 10 \\ 2 & -3 & -3 & \vdots & 22 \\ 4 & -2 & 3 & \vdots & -2 \end{bmatrix}$$

$$\begin{matrix} -R_1 + R_2 \to \\ -2R_1 + R_3 \to \end{matrix} \begin{bmatrix} 2 & 3 & 1 & \vdots & 10 \\ 0 & -6 & -4 & \vdots & 12 \\ 0 & -8 & 1 & \vdots & -22 \end{bmatrix}$$

$$\begin{matrix} \tfrac{1}{2}R_1 \to \\ -\tfrac{1}{6}R_2 \to \end{matrix} \begin{bmatrix} 1 & \tfrac{3}{2} & \tfrac{1}{2} & \vdots & 5 \\ 0 & 1 & \tfrac{2}{3} & \vdots & -2 \\ 0 & -8 & 1 & \vdots & -22 \end{bmatrix}$$

$$8R_2 + R_3 \to \begin{bmatrix} 1 & \tfrac{3}{2} & \tfrac{1}{2} & \vdots & 5 \\ 0 & 1 & \tfrac{2}{3} & \vdots & -2 \\ 0 & 0 & \tfrac{19}{3} & \vdots & -38 \end{bmatrix}$$

$$\tfrac{3}{19}R_3 \to \begin{bmatrix} 1 & \tfrac{3}{2} & \tfrac{1}{2} & \vdots & 5 \\ 0 & 1 & \tfrac{2}{3} & \vdots & -2 \\ 0 & 0 & 1 & \vdots & -6 \end{bmatrix}$$

$$z = -6$$

$$y + \tfrac{2}{3}(-6) = -2 \implies y = 2$$

$$x + \tfrac{3}{2}(2) + \tfrac{1}{2}(-6) = 5 \implies x = 5$$

Solution: $(5, 2, -6)$

21.
$$\begin{bmatrix} 2 & 1 & 2 & \vdots & 4 \\ 2 & 2 & 0 & \vdots & 5 \\ 2 & -1 & 6 & \vdots & 2 \end{bmatrix}$$

$$\begin{matrix} -R_1 + R_2 \to \\ -R_1 + R_3 \to \end{matrix} \begin{bmatrix} 2 & 1 & 2 & \vdots & 4 \\ 0 & 1 & -2 & \vdots & 1 \\ 0 & -2 & 4 & \vdots & -2 \end{bmatrix}$$

$$\begin{matrix} -R_2 + R_1 \to \\ \\ 2R_2 + R_3 \to \end{matrix} \begin{bmatrix} 2 & 0 & 4 & \vdots & 3 \\ 0 & 1 & -2 & \vdots & 1 \\ 0 & 0 & 0 & \vdots & 0 \end{bmatrix}$$

$$\tfrac{1}{2}R_1 \to \begin{bmatrix} 1 & 0 & 2 & \vdots & \tfrac{3}{2} \\ 0 & 1 & -2 & \vdots & 1 \\ 0 & 0 & 0 & \vdots & 0 \end{bmatrix}$$

Let $z = a$, then:

$$y - 2a = 1 \implies y = 2a + 1$$

$$x + 2a = \tfrac{3}{2} \implies x = -2a + \tfrac{3}{2}$$

Solution: $\left(-2a + \tfrac{3}{2}, 2a + 1, a\right)$ where a is any real number

23.
$$\begin{bmatrix} 2 & 1 & 1 & 0 & \vdots & 6 \\ 0 & -2 & 3 & -1 & \vdots & 9 \\ 3 & 3 & -2 & -2 & \vdots & -11 \\ 1 & 0 & 1 & 3 & \vdots & 14 \end{bmatrix}$$

$$-R_4 + R_1 \begin{bmatrix} 1 & 1 & 0 & -3 & \vdots & -8 \\ 0 & -2 & 3 & -1 & \vdots & 9 \\ 3 & 3 & -2 & -2 & \vdots & -11 \\ 1 & 0 & 1 & 3 & \vdots & 14 \end{bmatrix}$$

$$\begin{matrix} \\ \\ -3R_1 + R_3 \to \\ -R_1 + R_4 \to \end{matrix} \begin{bmatrix} 1 & 1 & 0 & -3 & \vdots & -8 \\ 0 & -2 & 3 & -1 & \vdots & 9 \\ 0 & 0 & -2 & 7 & \vdots & 13 \\ 0 & -1 & 1 & 6 & \vdots & 22 \end{bmatrix}$$

$$-3R_4 + R_2 \to \begin{bmatrix} 1 & 1 & 0 & -3 & \vdots & -8 \\ 0 & 1 & 0 & -19 & \vdots & -57 \\ 0 & 0 & -2 & 7 & \vdots & 13 \\ 0 & -1 & 1 & 6 & \vdots & 22 \end{bmatrix}$$

$$R_2 + R_4 \to \begin{bmatrix} 1 & 1 & 0 & -3 & \vdots & -8 \\ 0 & 1 & 0 & -19 & \vdots & -57 \\ 0 & 0 & -2 & 7 & \vdots & 13 \\ 0 & 0 & 1 & -13 & \vdots & -35 \end{bmatrix}$$

$$\begin{matrix} \\ \\ R_4 \\ R_3 \end{matrix} \begin{bmatrix} 1 & 1 & 0 & -3 & \vdots & -8 \\ 0 & 1 & 0 & -19 & \vdots & -57 \\ 0 & 0 & 1 & -13 & \vdots & -35 \\ 0 & 0 & -2 & 7 & \vdots & 13 \end{bmatrix}$$

$$2R_3 + R_4 \to \begin{bmatrix} 1 & 1 & 0 & -3 & \vdots & -8 \\ 0 & 1 & 0 & -19 & \vdots & -57 \\ 0 & 0 & 1 & -13 & \vdots & -35 \\ 0 & 0 & 0 & -19 & \vdots & -57 \end{bmatrix}$$

$$\tfrac{1}{19}R_4 \to \begin{bmatrix} 1 & 1 & 0 & -3 & \vdots & -8 \\ 0 & 1 & 0 & -19 & \vdots & -57 \\ 0 & 0 & 1 & -13 & \vdots & -35 \\ 0 & 0 & 0 & 1 & \vdots & 3 \end{bmatrix}$$

$$w = 3$$

$$z - 13(3) = -35 \implies z = 4$$

$$y - 19(3) = -57 \implies y = 0$$

$$x + 0 - 3(3) = -8 \implies x = 1$$

Solution: $(1, 0, 4, 3)$

25.
$$\begin{bmatrix} -1 & 1 & 2 & \vdots & 1 \\ 2 & 3 & 1 & \vdots & -2 \\ 5 & 4 & 2 & \vdots & 4 \end{bmatrix}$$

$$-R_1 \rightarrow \begin{bmatrix} 1 & -1 & -2 & \vdots & -1 \\ 2 & 3 & 1 & \vdots & -2 \\ 5 & 4 & 2 & \vdots & 4 \end{bmatrix}$$

$$\begin{matrix} -2R_1+R_2 \rightarrow \\ -5R_1+R_3 \rightarrow \end{matrix} \begin{bmatrix} 1 & -1 & -2 & \vdots & -1 \\ 0 & 5 & 5 & \vdots & 0 \\ 0 & 9 & 12 & \vdots & 9 \end{bmatrix}$$

$$\tfrac{1}{5}R_2 \rightarrow \begin{bmatrix} 1 & -1 & -2 & \vdots & -1 \\ 0 & 1 & 1 & \vdots & 0 \\ 0 & 9 & 12 & \vdots & 9 \end{bmatrix}$$

$$\begin{matrix} R_2+R_1 \rightarrow \\ \\ -9R_2+R_3 \rightarrow \end{matrix} \begin{bmatrix} 1 & 0 & -1 & \vdots & -1 \\ 0 & 1 & 1 & \vdots & 0 \\ 0 & 0 & 3 & \vdots & 9 \end{bmatrix}$$

$$\tfrac{1}{3}R_3 \rightarrow \begin{bmatrix} 1 & 0 & -1 & \vdots & -1 \\ 0 & 1 & 1 & \vdots & 0 \\ 0 & 0 & 1 & \vdots & 3 \end{bmatrix}$$

$$\begin{matrix} R_3+R_1 \rightarrow \\ -R_3+R_2 \rightarrow \end{matrix} \begin{bmatrix} 1 & 0 & 0 & \vdots & 2 \\ 0 & 1 & 0 & \vdots & -3 \\ 0 & 0 & 1 & \vdots & 3 \end{bmatrix}$$

$x = 2, y = -3, z = 3$

Solution: $(2, -3, 3)$

27.
$$\begin{bmatrix} 2 & -1 & 9 & \vdots & -8 \\ -1 & -3 & 4 & \vdots & -15 \\ 5 & 2 & -1 & \vdots & 17 \end{bmatrix}$$

$$R_2+R_1 \rightarrow \begin{bmatrix} 1 & -4 & 13 & \vdots & -23 \\ -1 & -3 & 4 & \vdots & -15 \\ 5 & 2 & -1 & \vdots & 17 \end{bmatrix}$$

$$\begin{matrix} R_1+R_2 \rightarrow \\ -5R_1+R_3 \rightarrow \end{matrix} \begin{bmatrix} 1 & -4 & 13 & \vdots & -23 \\ 0 & -7 & 17 & \vdots & -38 \\ 0 & 22 & -66 & \vdots & 132 \end{bmatrix}$$

$$\begin{matrix} R_3 \\ R_2 \end{matrix} \begin{bmatrix} 1 & -4 & 13 & \vdots & -23 \\ 0 & 22 & -66 & \vdots & 132 \\ 0 & -7 & 17 & \vdots & -38 \end{bmatrix}$$

$$\tfrac{1}{22}R_2 \rightarrow \begin{bmatrix} 1 & -4 & 13 & \vdots & -23 \\ 0 & 1 & -3 & \vdots & 6 \\ 0 & -7 & 17 & \vdots & -38 \end{bmatrix}$$

$$7R_2+R_3 \rightarrow \begin{bmatrix} 1 & -4 & 13 & \vdots & -23 \\ 0 & 1 & -3 & \vdots & 6 \\ 0 & 0 & -4 & \vdots & 4 \end{bmatrix}$$

$$-\tfrac{1}{4}R_3 \rightarrow \begin{bmatrix} 1 & -4 & 13 & \vdots & -23 \\ 0 & 1 & -3 & \vdots & 6 \\ 0 & 0 & 1 & \vdots & -1 \end{bmatrix}$$

$$4R_2+R_1 \rightarrow \begin{bmatrix} 1 & 0 & 1 & \vdots & 1 \\ 0 & 1 & -3 & \vdots & 6 \\ 0 & 0 & 1 & \vdots & -1 \end{bmatrix}$$

$$\begin{matrix} -R_3+R_1 \rightarrow \\ 3R_3+R_2 \rightarrow \end{matrix} \begin{bmatrix} 1 & 0 & 0 & \vdots & 2 \\ 0 & 1 & 0 & \vdots & 3 \\ 0 & 0 & 1 & \vdots & -1 \end{bmatrix}$$

$x = 2, y = 3, z = -1$

Solution: $(2, 3, -1)$

29. Use the reduced row-echelon form feature of a graphing utility.
$$\begin{bmatrix} 3 & -1 & 5 & -2 & \vdots & -44 \\ 1 & 6 & 4 & -1 & \vdots & 1 \\ 5 & -1 & 1 & 3 & \vdots & -15 \\ 0 & 4 & -1 & -8 & \vdots & 58 \end{bmatrix} \Rightarrow \begin{bmatrix} 1 & 0 & 0 & 0 & \vdots & 2 \\ 0 & 1 & 0 & 0 & \vdots & 6 \\ 0 & 0 & 1 & 0 & \vdots & -10 \\ 0 & 0 & 0 & 1 & \vdots & -3 \end{bmatrix}$$

$x = 2, y = 6, z = -10, w = -3$

Solution: $(2, 6, -10, -3)$

31. $\begin{bmatrix} -1 & x \\ y & 9 \end{bmatrix} = \begin{bmatrix} -1 & 12 \\ -7 & 9 \end{bmatrix} \Rightarrow x = 12$ and $y = -7$

33. $\begin{bmatrix} x+3 & -4 & 4y \\ 0 & -3 & 2 \\ -2 & y+5 & 6x \end{bmatrix} = \begin{bmatrix} 5x-1 & -4 & 44 \\ 0 & -3 & 2 \\ -2 & 16 & 6 \end{bmatrix}$

$\left.\begin{matrix} x+3 = 5x-1 \\ 4y = 44 \\ y+5 = 16 \\ 6x = 6 \end{matrix}\right\} \; x = 1$ and $y = 11$

35. (a) $A + B = \begin{bmatrix} 2 & -2 \\ 3 & 5 \end{bmatrix} + \begin{bmatrix} -3 & 10 \\ 12 & 8 \end{bmatrix} = \begin{bmatrix} -1 & 8 \\ 15 & 13 \end{bmatrix}$

(b) $A - B = \begin{bmatrix} 2 & -2 \\ 3 & 5 \end{bmatrix} - \begin{bmatrix} -3 & 10 \\ 12 & 8 \end{bmatrix} = \begin{bmatrix} 5 & -12 \\ -9 & -3 \end{bmatrix}$

(c) $4A = 4\begin{bmatrix} 2 & -2 \\ 3 & 5 \end{bmatrix} = \begin{bmatrix} 8 & -8 \\ 12 & 20 \end{bmatrix}$

(d) $A + 3B = \begin{bmatrix} 2 & -2 \\ 3 & 5 \end{bmatrix} + 3\begin{bmatrix} -3 & 10 \\ 12 & 8 \end{bmatrix} = \begin{bmatrix} 2 & -2 \\ 3 & 5 \end{bmatrix} + \begin{bmatrix} -9 & 30 \\ 36 & 24 \end{bmatrix} = \begin{bmatrix} -7 & 28 \\ 39 & 29 \end{bmatrix}$

37. (a) $A + B = \begin{bmatrix} 5 & 4 \\ -7 & 2 \\ 11 & 2 \end{bmatrix} + \begin{bmatrix} 0 & 3 \\ 4 & 12 \\ 20 & 40 \end{bmatrix} = \begin{bmatrix} 5 & 7 \\ -3 & 14 \\ 31 & 42 \end{bmatrix}$

(b) $A - B = \begin{bmatrix} 5 & 4 \\ -7 & 2 \\ 11 & 2 \end{bmatrix} - \begin{bmatrix} 0 & 3 \\ 4 & 12 \\ 20 & 40 \end{bmatrix} = \begin{bmatrix} 5 & 1 \\ -11 & -10 \\ -9 & -38 \end{bmatrix}$

(c) $4A = 4\begin{bmatrix} 5 & 4 \\ -7 & 2 \\ 11 & 2 \end{bmatrix} = \begin{bmatrix} 20 & 16 \\ -28 & 8 \\ 44 & 8 \end{bmatrix}$

(d) $A + 3B = \begin{bmatrix} 5 & 4 \\ -7 & 2 \\ 11 & 2 \end{bmatrix} + 3\begin{bmatrix} 0 & 3 \\ 4 & 12 \\ 20 & 40 \end{bmatrix} = \begin{bmatrix} 5 & 4 \\ -7 & 2 \\ 11 & 2 \end{bmatrix} + \begin{bmatrix} 0 & 9 \\ 12 & 36 \\ 60 & 120 \end{bmatrix} = \begin{bmatrix} 5 & 13 \\ 5 & 38 \\ 71 & 122 \end{bmatrix}$

39. $\begin{bmatrix} 7 & 3 \\ -1 & 5 \end{bmatrix} + \begin{bmatrix} 10 & -20 \\ 14 & -3 \end{bmatrix} = \begin{bmatrix} 7+10 & 3-20 \\ -1+14 & 5-3 \end{bmatrix} = \begin{bmatrix} 17 & -17 \\ 13 & 2 \end{bmatrix}$

41. $-2\begin{bmatrix} 1 & 2 \\ 5 & -4 \\ 6 & 0 \end{bmatrix} + 8\begin{bmatrix} 7 & 1 \\ 1 & 2 \\ 1 & 4 \end{bmatrix} = \begin{bmatrix} -2 & -4 \\ -10 & 8 \\ -12 & 0 \end{bmatrix} + \begin{bmatrix} 56 & 8 \\ 8 & 16 \\ 8 & 32 \end{bmatrix} = \begin{bmatrix} 54 & 4 \\ -2 & 24 \\ -4 & 32 \end{bmatrix}$

43. $3\begin{bmatrix} 8 & -2 & 5 \\ 1 & 3 & -1 \end{bmatrix} + 6\begin{bmatrix} 4 & -2 & -3 \\ 2 & 7 & 6 \end{bmatrix} = \begin{bmatrix} 24 & -6 & 15 \\ 3 & 9 & -3 \end{bmatrix} + \begin{bmatrix} 24 & -12 & -18 \\ 12 & 42 & 36 \end{bmatrix} = \begin{bmatrix} 48 & -18 & -3 \\ 15 & 51 & 33 \end{bmatrix}$

45. $X = 3A - 2B = 3\begin{bmatrix} -4 & 0 \\ 1 & -5 \\ -3 & 2 \end{bmatrix} - 2\begin{bmatrix} 1 & 2 \\ -2 & 1 \\ 4 & 4 \end{bmatrix}$

$= \begin{bmatrix} -14 & -4 \\ 7 & -17 \\ -17 & -2 \end{bmatrix}$

47. $X = \frac{1}{3}[B - 2A] = \frac{1}{3}\left(\begin{bmatrix} 1 & 2 \\ -2 & 1 \\ 4 & 4 \end{bmatrix} - 2\begin{bmatrix} -4 & 0 \\ 1 & -5 \\ -3 & 2 \end{bmatrix} \right)$

$= \frac{1}{3}\begin{bmatrix} 9 & 2 \\ -4 & 11 \\ 10 & 0 \end{bmatrix} = \begin{bmatrix} 3 & \frac{2}{3} \\ -\frac{4}{3} & \frac{11}{3} \\ \frac{10}{3} & 0 \end{bmatrix}$

49. A and B are both 2×2 so AB exists.

$AB = \begin{bmatrix} 2 & -2 \\ 3 & 5 \end{bmatrix}\begin{bmatrix} -3 & 10 \\ 12 & 8 \end{bmatrix} = \begin{bmatrix} 2(-3)+(-2)(12) & 2(10)+(-2)(8) \\ 3(-3)+5(12) & 3(10)+5(8) \end{bmatrix} = \begin{bmatrix} -30 & 4 \\ 51 & 70 \end{bmatrix}$

51. Since A is 3×2 and B is 2×2, AB exists.

$AB = \begin{bmatrix} 5 & 4 \\ -7 & 2 \\ 11 & 2 \end{bmatrix}\begin{bmatrix} 4 & 12 \\ 20 & 40 \end{bmatrix} = \begin{bmatrix} 5(4)+4(20) & 5(12)+4(40) \\ -7(4)+2(20) & -7(12)+2(40) \\ 11(4)+2(20) & 11(12)+2(40) \end{bmatrix} = \begin{bmatrix} 100 & 220 \\ 12 & -4 \\ 84 & 212 \end{bmatrix}$

53. $\begin{bmatrix} 1 & 2 \\ 5 & -4 \\ 6 & 0 \end{bmatrix} \begin{bmatrix} 6 & -2 & 8 \\ 4 & 0 & 0 \end{bmatrix} = \begin{bmatrix} 1(6) + 2(4) & 1(-2) + 2(0) & 1(8) + 2(0) \\ 5(6) + (-4)(4) & 5(-2) + (-4)(0) & 5(8) + (-4)(0) \\ 6(6) + (0)(4) & 6(-2) + (0)(0) & 6(8) + (0)(0) \end{bmatrix}$

$$= \begin{bmatrix} 14 & -2 & 8 \\ 14 & -10 & 40 \\ 36 & -12 & 48 \end{bmatrix}$$

55. $\begin{bmatrix} 1 & 5 & 6 \\ 2 & -4 & 0 \end{bmatrix} \begin{bmatrix} 6 & 4 \\ -2 & 0 \\ 8 & 0 \end{bmatrix} = \begin{bmatrix} 1(6) + 5(-2) + 6(8) & 1(4) + 5(0) + 6(0) \\ 2(6) - 4(-2) + 0(8) & 2(4) - 4(0) + 0(0) \end{bmatrix}$

$$= \begin{bmatrix} 44 & 4 \\ 20 & 8 \end{bmatrix}$$

57. $\begin{bmatrix} 4 \\ 6 \end{bmatrix} \begin{bmatrix} 6 & -2 \end{bmatrix} = \begin{bmatrix} 4(6) & 4(-2) \\ 6(6) & 6(-2) \end{bmatrix} = \begin{bmatrix} 24 & -8 \\ 36 & -12 \end{bmatrix}$

59. $\begin{bmatrix} 2 & 1 \\ 6 & 0 \end{bmatrix} \left(\begin{bmatrix} 4 & 2 \\ -3 & 1 \end{bmatrix} + \begin{bmatrix} -2 & 4 \\ 0 & 4 \end{bmatrix} \right) = \begin{bmatrix} 2 & 1 \\ 6 & 0 \end{bmatrix} \begin{bmatrix} 2 & 6 \\ -3 & 5 \end{bmatrix}$

$$= \begin{bmatrix} 2(2) + 1(-3) & 2(6) + 1(5) \\ 6(2) + 0 & 6(6) + 0 \end{bmatrix}$$

$$= \begin{bmatrix} 1 & 17 \\ 12 & 36 \end{bmatrix}$$

61. $\begin{bmatrix} 4 & 1 \\ 11 & -7 \\ 12 & 3 \end{bmatrix} \begin{bmatrix} 3 & -5 & 6 \\ 2 & -2 & -2 \end{bmatrix} = \begin{bmatrix} 14 & -22 & 22 \\ 19 & -41 & 80 \\ 42 & -66 & 66 \end{bmatrix}$

63. $0.95A = 0.95 \begin{bmatrix} 80 & 120 & 140 \\ 40 & 100 & 80 \end{bmatrix} = \begin{bmatrix} 76 & 114 & 133 \\ 38 & 95 & 76 \end{bmatrix}$

65. $BA = \begin{bmatrix} 10.25 & 14.50 & 17.75 \end{bmatrix} \begin{bmatrix} 8200 & 7400 \\ 6500 & 9800 \\ 5400 & 4800 \end{bmatrix} = \begin{bmatrix} \$274,150 & \$303,150 \end{bmatrix}$

The merchandise shipped to warehouse 1 is worth \$274,150, and the merchandise shipped to warehouse 2 is worth \$303,150.

67. $AB = \begin{bmatrix} -4 & -1 \\ 7 & 2 \end{bmatrix} \begin{bmatrix} -2 & -1 \\ 7 & 4 \end{bmatrix} = \begin{bmatrix} -4(-2) + (-1)(7) & -4(-1) + (-1)(4) \\ 7(-2) + 2(7) & 7(-1) + 2(4) \end{bmatrix}$

$$= \begin{bmatrix} 1 & 0 \\ 0 & 1 \end{bmatrix} = I$$

$BA = \begin{bmatrix} -2 & -1 \\ 7 & 4 \end{bmatrix} \begin{bmatrix} -4 & -1 \\ 7 & 2 \end{bmatrix} = \begin{bmatrix} -2(-4) + (-1)(7) & -2(-1) + (-1)(2) \\ 7(-4) + 4(7) & 7(-1) + 4(2) \end{bmatrix}$

$$= \begin{bmatrix} 1 & 0 \\ 0 & 1 \end{bmatrix} = I$$

69. $AB = \begin{bmatrix} 1 & 1 & 0 \\ 1 & 0 & 1 \\ 6 & 2 & 3 \end{bmatrix} \begin{bmatrix} -2 & -3 & 1 \\ 3 & 3 & -1 \\ 2 & 4 & -1 \end{bmatrix}$

$= \begin{bmatrix} 1(-2) + 1(3) + 0(2) & 1(-3) + 1(3) + 0(4) & 1(1) + 1(-1) + 0(-1) \\ 1(-2) + 0(3) + 1(2) & 1(-3) + 0(3) + 1(4) & 1(1) + 0(-1) + 1(-1) \\ 6(-2) + 2(3) + 3(2) & 6(-3) + 2(3) + 3(4) & 6(1) + 2(-1) + 3(-1) \end{bmatrix}$

$= \begin{bmatrix} 1 & 0 & 0 \\ 0 & 1 & 0 \\ 0 & 0 & 1 \end{bmatrix} = I$

$BA = \begin{bmatrix} -2 & -3 & 1 \\ 3 & 3 & -1 \\ 2 & 4 & -1 \end{bmatrix} \begin{bmatrix} 1 & 1 & 0 \\ 1 & 0 & 1 \\ 6 & 2 & 3 \end{bmatrix}$

$= \begin{bmatrix} -2(1) + (-3)(1) + 1(6) & -2(1) + (-3)(0) + 1(2) & -2(0) + (-3)(1) + 1(3) \\ 3(1) + 3(1) + (-1)(6) & 3(1) + 3(0) + (-1)(2) & 3(0) + 3(1) + (-1)(3) \\ 2(1) + 4(1) + (-1)(6) & 2(1) + 4(0) + (-1)(2) & 2(0) + 4(1) + (-1)(3) \end{bmatrix}$

$= \begin{bmatrix} 1 & 0 & 0 \\ 0 & 1 & 0 \\ 0 & 0 & 1 \end{bmatrix} = I$

71. $[A \;\vdots\; I] = \begin{bmatrix} -6 & 5 & \vdots & 1 & 0 \\ -5 & 4 & \vdots & 0 & 1 \end{bmatrix}$

$-\frac{1}{6}R_1 \rightarrow \begin{bmatrix} 1 & -\frac{5}{6} & \vdots & -\frac{1}{6} & 0 \\ -5 & 4 & \vdots & 0 & 1 \end{bmatrix}$

$5R_1 + R_2 \rightarrow \begin{bmatrix} 1 & -\frac{5}{6} & \vdots & -\frac{1}{6} & 0 \\ 0 & -\frac{1}{6} & \vdots & -\frac{5}{6} & 1 \end{bmatrix}$

$-6R_2 \rightarrow \begin{bmatrix} 1 & -\frac{5}{6} & \vdots & -\frac{1}{6} & 0 \\ 0 & 1 & \vdots & 5 & -6 \end{bmatrix}$

$\frac{5}{6}R_2 + R_1 \rightarrow \begin{bmatrix} 1 & 0 & \vdots & 4 & -5 \\ 0 & 1 & \vdots & 5 & -6 \end{bmatrix} = [I \;\vdots\; A^{-1}]$

$A^{-1} = \begin{bmatrix} 4 & -5 \\ 5 & -6 \end{bmatrix}$

73. $[A \;\vdots\; I] = \begin{bmatrix} -1 & -2 & -2 & \vdots & 1 & 0 & 0 \\ 3 & 7 & 9 & \vdots & 0 & 1 & 0 \\ 1 & 4 & 7 & \vdots & 0 & 0 & 1 \end{bmatrix}$

$-R_1 \rightarrow \begin{bmatrix} 1 & 2 & 2 & \vdots & -1 & 0 & 0 \\ 3 & 7 & 9 & \vdots & 0 & 1 & 0 \\ 1 & 4 & 7 & \vdots & 0 & 0 & 1 \end{bmatrix}$

$\begin{matrix} -3R_1 + R_2 \rightarrow \\ -R_1 + R_3 \rightarrow \end{matrix} \begin{bmatrix} 1 & 2 & 2 & \vdots & -1 & 0 & 0 \\ 0 & 1 & 3 & \vdots & 3 & 1 & 0 \\ 0 & 2 & 5 & \vdots & 1 & 0 & 1 \end{bmatrix}$

$\begin{matrix} -2R_2 + R_1 \rightarrow \\ \\ -2R_2 + R_3 \rightarrow \end{matrix} \begin{bmatrix} 1 & 0 & -4 & \vdots & -7 & -2 & 0 \\ 0 & 1 & 3 & \vdots & 3 & 1 & 0 \\ 0 & 0 & -1 & \vdots & -5 & -2 & 1 \end{bmatrix}$

$\begin{matrix} -4R_3 + R_1 \rightarrow \\ 3R_3 + R_2 \rightarrow \\ -R_3 \rightarrow \end{matrix} \begin{bmatrix} 1 & 0 & 0 & \vdots & 13 & 6 & -4 \\ 0 & 1 & 0 & \vdots & -12 & -5 & 3 \\ 0 & 0 & 1 & \vdots & 5 & 2 & -1 \end{bmatrix} = [I \;\vdots\; A^{-1}]$

$A^{-1} = \begin{bmatrix} 13 & 6 & -4 \\ -12 & -5 & 3 \\ 5 & 2 & -1 \end{bmatrix}$

75. $\begin{bmatrix} 2 & 0 & 3 \\ -1 & 1 & 1 \\ 2 & -2 & 1 \end{bmatrix}^{-1} = \begin{bmatrix} \frac{1}{2} & -1 & -\frac{1}{2} \\ \frac{1}{2} & -\frac{2}{3} & -\frac{5}{6} \\ 0 & \frac{2}{3} & \frac{1}{3} \end{bmatrix}$

77. $\begin{bmatrix} 1 & 3 & 1 & 6 \\ 4 & 4 & 2 & 6 \\ 3 & 4 & 1 & 2 \\ -1 & 2 & -1 & -2 \end{bmatrix}^{-1} = \begin{bmatrix} -3 & 6 & -\frac{11}{2} & \frac{7}{2} \\ 1 & -2 & 2 & -1 \\ 7 & -15 & \frac{29}{2} & -\frac{19}{2} \\ -1 & \frac{5}{2} & -\frac{5}{2} & \frac{3}{2} \end{bmatrix}$

$= \begin{bmatrix} -3 & 6 & -5.5 & 3.5 \\ 1 & -2 & 2 & -1 \\ 7 & -15 & 14.5 & -9.5 \\ -1 & 2.5 & -2.5 & 1.5 \end{bmatrix}$

79. $A = \begin{bmatrix} -7 & 2 \\ -8 & 2 \end{bmatrix}$

$A^{-1} = \dfrac{1}{-7(2) - 2(-8)} \begin{bmatrix} 2 & -2 \\ 8 & -7 \end{bmatrix} = \dfrac{1}{2} \begin{bmatrix} 2 & -2 \\ 8 & -7 \end{bmatrix} = \begin{bmatrix} 1 & -1 \\ 4 & -\frac{7}{2} \end{bmatrix}$

81. $A = \begin{bmatrix} -\frac{1}{2} & 20 \\ \frac{3}{10} & -6 \end{bmatrix}$

$A^{-1} = \dfrac{1}{-\frac{1}{2}(-6) - 20\left(\frac{3}{10}\right)} \begin{bmatrix} -6 & -20 \\ -\frac{3}{10} & -\frac{1}{2} \end{bmatrix} = -\dfrac{1}{3} \begin{bmatrix} -6 & -20 \\ -\frac{3}{10} & -\frac{1}{2} \end{bmatrix}$

$= \begin{bmatrix} 2 & \frac{20}{3} \\ \frac{1}{10} & \frac{1}{6} \end{bmatrix}$

83. $\begin{cases} -x + 4y = 8 \\ 2x - 7y = -5 \end{cases}$

$\begin{bmatrix} x \\ y \end{bmatrix} = \begin{bmatrix} -1 & 4 \\ 2 & -7 \end{bmatrix}^{-1} \begin{bmatrix} 8 \\ -5 \end{bmatrix} = \begin{bmatrix} 7 & 4 \\ 2 & 1 \end{bmatrix} \begin{bmatrix} 8 \\ -5 \end{bmatrix}$

$= \begin{bmatrix} 7(8) + 4(-5) \\ 2(8) + 1(-5) \end{bmatrix} = \begin{bmatrix} 36 \\ 11 \end{bmatrix}$

Solution: $(36, 11)$

85. $\begin{cases} -3x + 10y = 8 \\ 5x - 17y = -13 \end{cases}$

$\begin{bmatrix} x \\ y \end{bmatrix} = \begin{bmatrix} -3 & 10 \\ 5 & -17 \end{bmatrix}^{-1} \begin{bmatrix} 8 \\ -13 \end{bmatrix} = \begin{bmatrix} -17 & -10 \\ -5 & -3 \end{bmatrix} \begin{bmatrix} 8 \\ -13 \end{bmatrix}$

$= \begin{bmatrix} -17(8) + (-10)(-13) \\ -5(8) + (-3)(-13) \end{bmatrix} = \begin{bmatrix} -6 \\ -1 \end{bmatrix}$

Solution: $(-6, -1)$

87. $\begin{cases} 3x + 2y - z = 6 \\ x - y + 2z = -1 \\ 5x + y + z = 7 \end{cases}$

$\begin{bmatrix} x \\ y \\ z \end{bmatrix} = \begin{bmatrix} 3 & 2 & -1 \\ 1 & -1 & 2 \\ 5 & 1 & 1 \end{bmatrix}^{-1} \begin{bmatrix} 6 \\ -1 \\ 7 \end{bmatrix} = \begin{bmatrix} -1 & -1 & 1 \\ 3 & \frac{8}{3} & -\frac{7}{3} \\ 2 & \frac{7}{3} & -\frac{5}{3} \end{bmatrix} \begin{bmatrix} 6 \\ -1 \\ 7 \end{bmatrix}$

$= \begin{bmatrix} -1(6) - 1(-1) + 1(7) \\ 3(6) + \frac{8}{3}(-1) - \frac{7}{3}(7) \\ 2(6) + \frac{7}{3}(-1) - \frac{5}{3}(7) \end{bmatrix} = \begin{bmatrix} 2 \\ -1 \\ -2 \end{bmatrix}$

Solution: $(2, -1, -2)$

89. $\begin{cases} -2x + y + 2z = -13 \\ -x - 4y + z = -11 \\ -y - z = 0 \end{cases}$

$\begin{bmatrix} x \\ y \\ z \end{bmatrix} = \begin{bmatrix} -2 & 1 & 2 \\ -1 & -4 & 1 \\ 0 & -1 & -1 \end{bmatrix}^{-1} \begin{bmatrix} -13 \\ -11 \\ 0 \end{bmatrix} = \begin{bmatrix} -\frac{5}{9} & \frac{1}{9} & -1 \\ \frac{1}{9} & -\frac{2}{9} & 0 \\ -\frac{1}{9} & \frac{2}{9} & -1 \end{bmatrix} \begin{bmatrix} -13 \\ -11 \\ 0 \end{bmatrix}$

$= \begin{bmatrix} -\frac{5}{9}(-13) + \frac{1}{9}(-11) - 1(0) \\ \frac{1}{9}(-13) - \frac{2}{9}(-11) + 0(0) \\ -\frac{1}{9}(-13) + \frac{2}{9}(-11) - 1(0) \end{bmatrix} = \begin{bmatrix} 6 \\ 1 \\ -1 \end{bmatrix}$

Solution: $(6, 1, -1)$

91. $\begin{cases} x + 2y = -1 \\ 3x + 4y = -5 \end{cases}$

$\begin{bmatrix} x \\ y \end{bmatrix} = \begin{bmatrix} 1 & 2 \\ 3 & 4 \end{bmatrix}^{-1} \begin{bmatrix} -1 \\ -5 \end{bmatrix} = \begin{bmatrix} -2 & 1 \\ \frac{3}{2} & -\frac{1}{2} \end{bmatrix} \begin{bmatrix} -1 \\ -5 \end{bmatrix} = \begin{bmatrix} -3 \\ 1 \end{bmatrix}$

Solution: $(-3, 1)$

93. $\begin{cases} -3x - 3y - 4z = 2 \\ y + z = -1 \\ 4x + 3y + 4z = -1 \end{cases}$

$\begin{bmatrix} x \\ y \\ z \end{bmatrix} = \begin{bmatrix} -3 & -3 & -4 \\ 0 & 1 & 1 \\ 4 & 3 & 4 \end{bmatrix}^{-1} \begin{bmatrix} 2 \\ -1 \\ -1 \end{bmatrix} = \begin{bmatrix} 1 & 0 & 1 \\ 4 & 4 & 3 \\ -4 & -3 & -3 \end{bmatrix} \begin{bmatrix} 2 \\ -1 \\ -1 \end{bmatrix} = \begin{bmatrix} 1 \\ 1 \\ -2 \end{bmatrix}$

Solution: $(1, 1, -2)$

95. $\begin{vmatrix} 8 & 5 \\ 2 & -4 \end{vmatrix} = 8(-4) - 5(2) = -42$

97. $\begin{vmatrix} 50 & -30 \\ 10 & 5 \end{vmatrix} = 50(5) - (-30)(10) = 550$

99. $\begin{bmatrix} 2 & -1 \\ 7 & 4 \end{bmatrix}$

(a) $M_{11} = 4$

$M_{12} = 7$

$M_{21} = -1$

$M_{22} = 2$

(b) $C_{11} = M_{11} = 4$

$C_{12} = -M_{12} = -7$

$C_{21} = -M_{21} = 1$

$C_{22} = M_{22} = 2$

101. $\begin{bmatrix} 3 & 2 & -1 \\ -2 & 5 & 0 \\ 1 & 8 & 6 \end{bmatrix}$

(a) $M_{11} = \begin{vmatrix} 5 & 0 \\ 8 & 6 \end{vmatrix} = 30$

$M_{12} = \begin{vmatrix} -2 & 0 \\ 1 & 6 \end{vmatrix} = -12$

$M_{13} = \begin{vmatrix} -2 & 5 \\ 1 & 8 \end{vmatrix} = -21$

$M_{21} = \begin{vmatrix} 2 & -1 \\ 8 & 6 \end{vmatrix} = 20$

$M_{22} = \begin{vmatrix} 3 & -1 \\ 1 & 6 \end{vmatrix} = 19$

$M_{23} = \begin{vmatrix} 3 & 2 \\ 1 & 8 \end{vmatrix} = 22$

$M_{31} = \begin{vmatrix} 2 & -1 \\ 5 & 0 \end{vmatrix} = 5$

$M_{32} = \begin{vmatrix} 3 & -1 \\ -2 & 0 \end{vmatrix} = -2$

$M_{33} = \begin{vmatrix} 3 & 2 \\ -2 & 5 \end{vmatrix} = 19$

(b) $C_{11} = M_{11} = 30$

$C_{12} = -M_{12} = 12$

$C_{13} = M_{13} = -21$

$C_{21} = -M_{21} = -20$

$C_{22} = M_{22} = 19$

$C_{23} = -M_{23} = -22$

$C_{31} = M_{31} = 5$

$C_{32} = -M_{32} = 2$

$C_{33} = M_{33} = 19$

103. Expand using Column 2.

$\begin{vmatrix} -2 & 4 & 1 \\ -6 & 0 & 2 \\ 5 & 3 & 4 \end{vmatrix} = -4 \begin{vmatrix} -6 & 2 \\ 5 & 4 \end{vmatrix} - 3 \begin{vmatrix} -2 & 1 \\ -6 & 2 \end{vmatrix}$

$= -4(-34) - 3(2) = 130$

105. Expand along Row 1.

$$\begin{vmatrix} 3 & 0 & -4 & 0 \\ 0 & 8 & 1 & 2 \\ 6 & 1 & 8 & 2 \\ 0 & 3 & -4 & 1 \end{vmatrix} = 3\begin{vmatrix} 8 & 1 & 2 \\ 1 & 8 & 2 \\ 3 & -4 & 1 \end{vmatrix} + (-4)\begin{vmatrix} 0 & 8 & 2 \\ 6 & 1 & 2 \\ 0 & 3 & 1 \end{vmatrix}$$

$$= 3[8(8 - (-8)) - 1(1 - 6) + 2(-4 - 24)] - 4[0 - 6(8 - 6) + 0]$$

$$= 3[128 + 5 - 56] - 4[-12]$$

$$= 279$$

107. $\begin{cases} 5x - 2y = 6 \\ -11x + 3y = -23 \end{cases}$

$$x = \frac{\begin{vmatrix} 6 & -2 \\ -23 & 3 \end{vmatrix}}{\begin{vmatrix} 5 & -2 \\ -11 & 3 \end{vmatrix}} = \frac{-28}{-7} = 4, \qquad y = \frac{\begin{vmatrix} 5 & 6 \\ -11 & -23 \end{vmatrix}}{\begin{vmatrix} 5 & -2 \\ -11 & 3 \end{vmatrix}} = \frac{-49}{-7} = 7$$

Solution: $(4, 7)$

109. $\begin{cases} -2x + 3y - 5z = -11 \\ 4x - y + z = -3 \\ -x - 4y + 6z = 15 \end{cases}$

$$D = \begin{vmatrix} -2 & 3 & -5 \\ 4 & -1 & 1 \\ -1 & -4 & 6 \end{vmatrix} = -2(-1)^2\begin{vmatrix} -1 & 1 \\ -4 & 6 \end{vmatrix} + 4(-1)^3\begin{vmatrix} 3 & -5 \\ -4 & 6 \end{vmatrix} - 1(-1)^4\begin{vmatrix} 3 & -5 \\ -1 & 1 \end{vmatrix}$$

$$= -2(-2) - 4(-2) - (-2) = 14$$

$$x = \frac{\begin{vmatrix} -11 & 3 & -5 \\ -3 & -1 & 1 \\ 15 & -4 & 6 \end{vmatrix}}{14} = \frac{-11(-1)^2\begin{vmatrix} -1 & 1 \\ -4 & 6 \end{vmatrix} - 3(-1)^3\begin{vmatrix} 3 & -5 \\ -4 & 6 \end{vmatrix} + 15(-1)^4\begin{vmatrix} 3 & -5 \\ -1 & 1 \end{vmatrix}}{14}$$

$$= \frac{-11(-2) + 3(-2) + 15(-2)}{14} = \frac{-14}{14} = -1$$

$$y = \frac{\begin{vmatrix} -2 & -11 & -5 \\ 4 & -3 & 1 \\ -1 & 15 & 6 \end{vmatrix}}{14} = \frac{-2(-1)^2\begin{vmatrix} -3 & 1 \\ 15 & 6 \end{vmatrix} + 4(-1)^3\begin{vmatrix} -11 & -5 \\ 15 & 6 \end{vmatrix} - 1(-1)^4\begin{vmatrix} -11 & -5 \\ -3 & 1 \end{vmatrix}}{14}$$

$$= \frac{-2(-33) - 4(9) - 1(-26)}{14} = \frac{56}{14} = 4$$

$$z = \frac{\begin{vmatrix} -2 & 3 & -11 \\ 4 & -1 & -3 \\ -1 & -4 & 15 \end{vmatrix}}{14} = \frac{-2(-1)^2\begin{vmatrix} -1 & -3 \\ -4 & 15 \end{vmatrix} + 4(-1)^3\begin{vmatrix} 3 & -11 \\ -4 & 15 \end{vmatrix} - 1(-1)^4\begin{vmatrix} 3 & -11 \\ -1 & -3 \end{vmatrix}}{14}$$

$$= \frac{-2(-27) - 4(1) - 1(-20)}{14} = \frac{70}{14} = 5$$

Solution: $(-1, 4, 5)$

111. $(1, 0), (5, 0), (5, 8)$

$$\text{Area} = \frac{1}{2}\begin{vmatrix} 1 & 0 & 1 \\ 5 & 0 & 1 \\ 5 & 8 & 1 \end{vmatrix} = \frac{1}{2}\left(1\begin{vmatrix} 0 & 1 \\ 8 & 1 \end{vmatrix} + 1\begin{vmatrix} 5 & 0 \\ 5 & 8 \end{vmatrix}\right) = \frac{1}{2}(-8 + 40) = \frac{1}{2}(32) = 16 \text{ square units}$$

113. $(1, -4), (-2, 3), (0, 5)$

$$\text{Area} = -\frac{1}{2}\begin{vmatrix} 1 & -4 & 1 \\ -2 & 3 & 1 \\ 0 & 5 & 1 \end{vmatrix}$$

$$= -\frac{1}{2}\left(-5\begin{vmatrix} 1 & 1 \\ -2 & 1 \end{vmatrix} + \begin{vmatrix} 1 & -4 \\ -2 & 3 \end{vmatrix}\right)$$

$$= -\frac{1}{2}(-5(3) + (-5)) = 10 \text{ square units}$$

115. $(-1, 7), (3, -9), (-3, 15)$

$$\begin{vmatrix} -1 & 7 & 1 \\ 3 & -9 & 1 \\ -3 & 15 & 1 \end{vmatrix} = 0$$

The points are collinear.

117. $(-4, 0), (4, 4)$

$$\begin{vmatrix} x & y & 1 \\ -4 & 0 & 1 \\ 4 & 4 & 1 \end{vmatrix} = 0$$

$$1\begin{vmatrix} -4 & 0 \\ 4 & 4 \end{vmatrix} - 1\begin{vmatrix} x & y \\ 4 & 4 \end{vmatrix} + 1\begin{vmatrix} x & y \\ -4 & 0 \end{vmatrix} = 0$$

$$-16 - (4x - 4y) + 4y = 0$$

$$-4x + 8y - 16 = 0$$

$$x - 2y + 4 = 0$$

119. $\left(-\frac{5}{2}, 3\right), \left(\frac{7}{2}, 1\right)$

$$\begin{vmatrix} x & y & 1 \\ -\frac{5}{2} & 3 & 1 \\ \frac{7}{2} & 1 & 1 \end{vmatrix} = 0$$

$$1\begin{vmatrix} -\frac{5}{2} & 3 \\ \frac{7}{2} & 1 \end{vmatrix} - 1\begin{vmatrix} x & y \\ \frac{7}{2} & 1 \end{vmatrix} + 1\begin{vmatrix} x & y \\ -\frac{5}{2} & 3 \end{vmatrix} = 0$$

$$-13 - \left(x - \tfrac{7}{2}y\right) + \left(3x + \tfrac{5}{2}y\right) = 0$$

$$2x + 6y - 13 = 0$$

121. L   O   O   K   __   O   U   T   __   B   E   L   O   W   __

[12   15   15]   [11   0   15]   [21   20   0]   [2   5   12]   [15   23   0]

$$A = \begin{bmatrix} 2 & -2 & 0 \\ 3 & 0 & -3 \\ -6 & 2 & 3 \end{bmatrix}$$

$$\begin{bmatrix} 12 & 15 & 15 \end{bmatrix}\begin{bmatrix} 2 & -2 & 0 \\ 3 & 0 & -3 \\ -6 & 2 & 3 \end{bmatrix} = \begin{bmatrix} -21 & 6 & 0 \end{bmatrix}$$

$$\begin{bmatrix} 11 & 0 & 15 \end{bmatrix}\begin{bmatrix} 2 & -2 & 0 \\ 3 & 0 & -3 \\ -6 & 2 & 3 \end{bmatrix} = \begin{bmatrix} -68 & 8 & 45 \end{bmatrix}$$

$$\begin{bmatrix} 21 & 20 & 0 \end{bmatrix}\begin{bmatrix} 2 & -2 & 0 \\ 3 & 0 & -3 \\ -6 & 2 & 3 \end{bmatrix} = \begin{bmatrix} 102 & -42 & -60 \end{bmatrix}$$

$$\begin{bmatrix} 2 & 5 & 12 \end{bmatrix}\begin{bmatrix} 2 & -2 & 0 \\ 3 & 0 & -3 \\ -6 & 2 & 3 \end{bmatrix} = \begin{bmatrix} -53 & 20 & 21 \end{bmatrix}$$

$$\begin{bmatrix} 15 & 23 & 0 \end{bmatrix}\begin{bmatrix} 2 & -2 & 0 \\ 3 & 0 & -3 \\ -6 & 2 & 3 \end{bmatrix} = \begin{bmatrix} 99 & -30 & -69 \end{bmatrix}$$

Cryptogram: -21 6   0 -68 8   45   102
           -42 -60 -53 20   21   99 -30 -69

123. $A^{-1} = \begin{bmatrix} -1 & 2 & -3 \\ 2 & 1 & 0 \\ 4 & -2 & 5 \end{bmatrix}$

$\begin{bmatrix} -5 & 11 & -2 \end{bmatrix} \begin{bmatrix} -1 & 2 & -3 \\ 2 & 1 & 0 \\ 4 & -2 & 5 \end{bmatrix} = \begin{bmatrix} 19 & 5 & 5 \end{bmatrix}$ S   E   E

$\begin{bmatrix} 370 & -265 & 225 \end{bmatrix} \begin{bmatrix} -1 & 2 & -3 \\ 2 & 1 & 0 \\ 4 & -2 & 5 \end{bmatrix} = \begin{bmatrix} 0 & 25 & 15 \end{bmatrix}$ _   Y   O

$\begin{bmatrix} -57 & 48 & -33 \end{bmatrix} \begin{bmatrix} -1 & 2 & -3 \\ 2 & 1 & 0 \\ 4 & -2 & 5 \end{bmatrix} = \begin{bmatrix} 21 & 0 & 6 \end{bmatrix}$ U   _   F

$\begin{bmatrix} 32 & -15 & 20 \end{bmatrix} \begin{bmatrix} -1 & 2 & -3 \\ 2 & 1 & 0 \\ 4 & -2 & 5 \end{bmatrix} = \begin{bmatrix} 18 & 9 & 4 \end{bmatrix}$ R   I   D

$\begin{bmatrix} 245 & -171 & 147 \end{bmatrix} \begin{bmatrix} -1 & 2 & -3 \\ 2 & 1 & 0 \\ 4 & -2 & 5 \end{bmatrix} = \begin{bmatrix} 1 & 25 & 0 \end{bmatrix}$ A   Y   _

Message: SEE YOU FRIDAY

125. False. The matrix must be square.

127. The matrix must be square and its determinant nonzero to have an inverse.

129. No. Each matrix is in row-echelon form, but the third matrix cannot be achieved from the first or second matrix with elementary row operations. Also, the first two matrices describe a system of equations with one solution. The third matrix describes a system with infinitely many solutions.

131.
$$\begin{vmatrix} 2 - \lambda & 5 \\ 3 & -8 - \lambda \end{vmatrix} = 0$$

$$(2 - \lambda)(-8 - \lambda) - 15 = 0$$

$$-16 + 6\lambda + \lambda^2 - 15 = 0$$

$$\lambda^2 + 6\lambda - 31 = 0$$

$$\lambda = \frac{-6 \pm \sqrt{36 - 4(-31)}}{2}$$

$$\lambda = -3 \pm 2\sqrt{10}$$

Problem Solving for Chapter 8

1. $A = \begin{bmatrix} 0 & -1 \\ 1 & 0 \end{bmatrix}$ $T = \begin{bmatrix} 1 & 2 & 3 \\ 1 & 4 & 2 \end{bmatrix}$

(a) $AT = \begin{bmatrix} -1 & -4 & -2 \\ 1 & 2 & 3 \end{bmatrix}$ $AAT = \begin{bmatrix} -1 & -2 & -3 \\ -1 & -4 & -2 \end{bmatrix}$

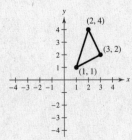

Original Triangle

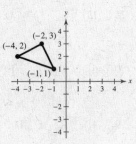

AT Triangle

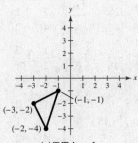

AAT Triangle

The transformation A interchanges the x and y coordinates and then takes the negative of the x coordinate. A represents a counterclockwise rotation by $90°$.

(b) $A^{-1}(AAT) = (A^{-1}A)(AT) = (I)(AT) = AT$

$A^{-1}(AT) = (A^{-1}A)T = IT = T$

$A^{-1} = \begin{bmatrix} 0 & 1 \\ -1 & 0 \end{bmatrix}$

A^{-1} represents a clockwise rotation by $90°$.

3. (a) $A^2 = \begin{bmatrix} 1 & 0 \\ 0 & 0 \end{bmatrix}\begin{bmatrix} 1 & 0 \\ 0 & 0 \end{bmatrix} = \begin{bmatrix} 1 & 0 \\ 0 & 0 \end{bmatrix} = A$

A *is* idempotent.

(b) $A^2 = \begin{bmatrix} 0 & 1 \\ 1 & 0 \end{bmatrix}\begin{bmatrix} 0 & 1 \\ 1 & 0 \end{bmatrix} = \begin{bmatrix} 1 & 0 \\ 0 & 1 \end{bmatrix} \neq A$

A is *not* idempotent.

(c) $A^2 = \begin{bmatrix} 2 & 3 \\ -1 & -2 \end{bmatrix}\begin{bmatrix} 2 & 3 \\ -1 & -2 \end{bmatrix} = \begin{bmatrix} 1 & 0 \\ 0 & 1 \end{bmatrix} \neq A$

A is *not* idempotent.

(d) $A^2 = \begin{bmatrix} 2 & 3 \\ 1 & 2 \end{bmatrix}\begin{bmatrix} 2 & 3 \\ 1 & 2 \end{bmatrix} = \begin{bmatrix} 7 & 12 \\ 4 & 7 \end{bmatrix} \neq A$

A is *not* idempotent.

5. (a) $\begin{bmatrix} 0.70 & 0.15 & 0.15 \\ 0.20 & 0.80 & 0.15 \\ 0.10 & 0.05 & 0.70 \end{bmatrix}\begin{bmatrix} 25,000 \\ 30,000 \\ 45,000 \end{bmatrix} = \begin{bmatrix} 28,750 \\ 35,750 \\ 35,500 \end{bmatrix}$

Gold Cable Company: 28,750 households

Galaxy Cable Company: 35,750 households

Nonsubscribers: 35,500 households

(c) $\begin{bmatrix} 0.70 & 0.15 & 0.15 \\ 0.20 & 0.80 & 0.15 \\ 0.10 & 0.05 & 0.70 \end{bmatrix}\begin{bmatrix} 30,812.5 \\ 39,675 \\ 29,512.5 \end{bmatrix} \approx \begin{bmatrix} 31,947 \\ 42,329 \\ 25,724 \end{bmatrix}$

Gold Cable Company: 31,947 households

Galaxy Cable Company: 42,329 households

Nonsubscribers: 25,724 households

(b) $\begin{bmatrix} 0.70 & 0.15 & 0.15 \\ 0.20 & 0.80 & 0.15 \\ 0.10 & 0.05 & 0.70 \end{bmatrix}\begin{bmatrix} 28,750 \\ 35,750 \\ 35,500 \end{bmatrix} \approx \begin{bmatrix} 30,813 \\ 39,675 \\ 29,513 \end{bmatrix}$

Gold Cable Company: 30,813 households

Galaxy Cable Company: 39,675 households

Nonsubscribers: 29,513 households

(d) Both cable companies are increasing the number of subscribers, while the number of nonsubscribers is decreasing each year.

7. If $A = \begin{bmatrix} 4 & x \\ -2 & -3 \end{bmatrix}$ is singular then

$$ad - bc = -12 + 2x = 0.$$

Thus, $x = 6$.

9. $(a - b)(b - c)(c - a) = -a^2b + a^2c + ab^2 - ac^2 - b^2c + bc^2$

$$\begin{vmatrix} 1 & 1 & 1 \\ a & b & c \\ a^2 & b^2 & c^2 \end{vmatrix} = \begin{vmatrix} b & c \\ b^2 & c^2 \end{vmatrix} - \begin{vmatrix} a & c \\ a^2 & c^2 \end{vmatrix} + \begin{vmatrix} a & b \\ a^2 & b^2 \end{vmatrix} = bc^2 - b^2c - ac^2 + a^2c + ab^2 - a^2b$$

Thus, $\begin{vmatrix} 1 & 1 & 1 \\ a & b & c \\ a^2 & b^2 & c^2 \end{vmatrix} = (a - b)(b - c)(c - a)$.

11. $\begin{vmatrix} x & 0 & c \\ -1 & x & b \\ 0 & -1 & a \end{vmatrix} = x \begin{vmatrix} x & b \\ -1 & a \end{vmatrix} + c \begin{vmatrix} -1 & x \\ 0 & -1 \end{vmatrix} = x(ax + b) + c(1 - 0) = ax^2 + bx + c$

13. $4S + 4N \quad\quad = 184$

$\quad S \quad\quad + 6F = 146$

$\quad\quad 2N + 4F = 104$

$D = \begin{vmatrix} 4 & 4 & 0 \\ 1 & 0 & 6 \\ 0 & 2 & 4 \end{vmatrix} = -64$

$N = \dfrac{\begin{vmatrix} 4 & 184 & 0 \\ 1 & 146 & 6 \\ 0 & 104 & 4 \end{vmatrix}}{-64} = \dfrac{-896}{-64} = 14$

$S = \dfrac{\begin{vmatrix} 184 & 4 & 0 \\ 146 & 0 & 6 \\ 104 & 2 & 4 \end{vmatrix}}{-64} = \dfrac{-2048}{-64} = 32$

$F = \dfrac{\begin{vmatrix} 4 & 4 & 184 \\ 1 & 0 & 146 \\ 0 & 2 & 104 \end{vmatrix}}{-64} = \dfrac{-1216}{-64} = 19$

Element	Atomic mass
Sulfur	32
Nitrogen	14
Fluoride	19

15. $A = \begin{bmatrix} -1 & 1 & -2 \\ 2 & 0 & 1 \end{bmatrix}, \quad\quad B = \begin{bmatrix} -3 & 0 \\ 1 & 2 \\ 1 & -1 \end{bmatrix}$

$A^T = \begin{bmatrix} -1 & 2 \\ 1 & 0 \\ -2 & 1 \end{bmatrix}, \quad\quad B^T = \begin{bmatrix} -3 & 1 & 1 \\ 0 & 2 & -1 \end{bmatrix}$

$AB = \begin{bmatrix} 2 & 4 \\ -5 & -1 \end{bmatrix}, \quad\quad (AB)^T = \begin{bmatrix} 2 & -5 \\ 4 & -1 \end{bmatrix}$

$B^T A^T = \begin{bmatrix} -3 & 1 & 1 \\ 0 & 2 & -1 \end{bmatrix} \begin{bmatrix} -1 & 2 \\ 1 & 0 \\ -2 & 1 \end{bmatrix} = \begin{bmatrix} 2 & -5 \\ 4 & -1 \end{bmatrix}$

Thus, $(AB)^T = B^T A^T$.

17. (a) $\begin{bmatrix} 45 & -35 \end{bmatrix}\begin{bmatrix} w & x \\ y & z \end{bmatrix} = \begin{bmatrix} 10 & 15 \end{bmatrix}$

$\begin{bmatrix} 38 & -30 \end{bmatrix}\begin{bmatrix} w & x \\ y & z \end{bmatrix} = \begin{bmatrix} 8 & 14 \end{bmatrix}$

$45w - 35y = 10$

$45x - 35z = 15$

$38w - 30y = 8$

$38x - 30z = 14$

$\left.\begin{array}{r} 45w - 35y = 10 \\ 38w - 30y = 8 \end{array}\right\} \Rightarrow w = 1, y = 1$

$\left.\begin{array}{r} 45x - 35z = 15 \\ 38x - 30z = 14 \end{array}\right\} \Rightarrow x = -2, z = -3$

$A^{-1} = \begin{bmatrix} 1 & -2 \\ 1 & -3 \end{bmatrix}$

(b) $\begin{bmatrix} 45 & -35 \\ 38 & -30 \\ 18 & -18 \\ 35 & -30 \\ 81 & -60 \\ 42 & -28 \\ 75 & -55 \\ 2 & -2 \\ 22 & -21 \\ 15 & -10 \end{bmatrix}\begin{bmatrix} 1 & -2 \\ 1 & -3 \end{bmatrix} = \begin{bmatrix} 10 & 15 \\ 8 & 14 \\ 0 & 18 \\ 5 & 20 \\ 21 & 18 \\ 14 & 0 \\ 20 & 15 \\ 0 & 2 \\ 1 & 19 \\ 5 & 0 \end{bmatrix}$

J	O
H	N
—	R
E	T
U	R
N	—
T	O
—	B
A	S
E	—

JOHN RETURN TO BASE

19. Let $A = \begin{bmatrix} 3 & -3 \\ 5 & -5 \end{bmatrix}$, then $|A| = 0$.

Let $A = \begin{bmatrix} 2 & 4 & -6 \\ -3 & 1 & 2 \\ 5 & -8 & 3 \end{bmatrix}$, then $|A| = 0$.

Let $A = \begin{bmatrix} 3 & -7 & 5 & -1 \\ -6 & 4 & 0 & 2 \\ 5 & 8 & -6 & -7 \\ 9 & 11 & -4 & -16 \end{bmatrix}$, then $|A| = 0$.

Conjecture: If A is an $n \times n$ matrix, each of whose rows add up to zero, then $|A| = 0$.

Chapter 8   Practice Test

1. Put the matrix in reduced row-echelon form.

$$\begin{bmatrix} 1 & -2 & 4 \\ 3 & -5 & 9 \end{bmatrix}$$

For Exercises 2–4, use matrices to solve the system of equations.

2. $\begin{cases} 3x + 5y = 3 \\ 2x - y = -11 \end{cases}$

3. $\begin{cases} 2x + 3y = -3 \\ 3x + 2y = 8 \\ x + y = 1 \end{cases}$

4. $\begin{cases} x + 3z = -5 \\ 2x + y = 0 \\ 3x + y - z = 3 \end{cases}$

5. Multiply $\begin{bmatrix} 1 & 4 & 5 \\ 2 & 0 & -3 \end{bmatrix} \begin{bmatrix} 1 & 6 \\ 0 & -7 \\ -1 & 2 \end{bmatrix}$.

6. Given $A = \begin{bmatrix} 9 & 1 \\ -4 & 8 \end{bmatrix}$ and $B = \begin{bmatrix} 6 & -2 \\ 3 & 5 \end{bmatrix}$, find $3A - 5B$.

7. Find $f(A)$.

$$f(x) = x^2 - 7x + 8, \ A = \begin{bmatrix} 3 & 0 \\ 7 & 1 \end{bmatrix}$$

8. True or false:

$(A + B)(A + 3B) = A^2 + 4AB + 3B^2$ where A and B are matrices.

(Assume that A^2, AB, and B^2 exist.)

For Exercises 9–10, find the inverse of the matrix, if it exists.

9. $\begin{bmatrix} 1 & 2 \\ 3 & 5 \end{bmatrix}$

10. $\begin{bmatrix} 1 & 1 & 1 \\ 3 & 6 & 5 \\ 6 & 10 & 8 \end{bmatrix}$

11. Use an inverse matrix to solve the systems.

(a) $x + 2y = 4$
    $3x + 5y = 1$

(b) $x + 2y = 3$
    $3x + 5y = -2$

For Exercises 12–14, find the determinant of the matrix.

12. $\begin{bmatrix} 6 & -1 \\ 3 & 4 \end{bmatrix}$

13. $\begin{bmatrix} 1 & 3 & -1 \\ 5 & 9 & 0 \\ 6 & 2 & -5 \end{bmatrix}$

14. $\begin{bmatrix} 1 & 4 & 2 & 3 \\ 0 & 1 & -2 & 0 \\ 3 & 5 & -1 & 1 \\ 2 & 0 & 6 & 1 \end{bmatrix}$

15. Evaluate $\begin{vmatrix} 6 & 4 & 3 & 0 & 6 \\ 0 & 5 & 1 & 4 & 8 \\ 0 & 0 & 2 & 7 & 3 \\ 0 & 0 & 0 & 9 & 2 \\ 0 & 0 & 0 & 0 & 1 \end{vmatrix}$.

16. Use a determinant to find the area of the triangle with vertices $(0, 7)$, $(5, 0)$, and $(3, 9)$.

17. Find the equation of the line through $(2, 7)$ and $(-1, 4)$.

For Exercises 18–20, use Cramer's Rule to find the indicated value.

18. Find x.

$$\begin{cases} 6x - 7y = 4 \\ 2x + 5y = 11 \end{cases}$$

19. Find z.

$$\begin{cases} 3x \quad\;\; + z = 1 \\ \quad\;\; y + 4z = 3 \\ x - y \quad\;\; = 2 \end{cases}$$

20. Find y.

$$\begin{cases} 721.4x - 29.1y = 33.77 \\ 45.9x + 105.6y = 19.85 \end{cases}$$

C H A P T E R 9
Sequences, Series, and Probability

Section 9.1   Sequences and Series **444**

Section 9.2   Arithmetic Sequences and Partial Sums **450**

Section 9.3   Geometric Sequences and Series **454**

Section 9.4   Mathematical Induction **461**

Section 9.5   The Binomial Theorem **469**

Section 9.6   Counting Principles **474**

Section 9.7   Probability **477**

Review Exercises **480**

Problem Solving **485**

Practice Test **488**

CHAPTER 9
Sequences, Series, and Probability

Section 9.1   Sequences and Series

- Given the general nth term in a sequence, you should be able to find, or list, some of the terms.
- You should be able to find an expression for the apparent nth term of a sequence.
- You should be able to use and evaluate factorials.
- You should be able to use summation notation for a sum.
- You should know that the sum of the terms of a sequence is a series.

Vocabulary Check

1. infinite sequence
2. terms
3. finite
4. recursively
5. factorial
6. summation notation
7. index; upper; lower
8. series
9. nth partial sum

---

1. $a_n = 3n + 1$

$a_1 = 3(1) + 1 = 4$

$a_2 = 3(2) + 1 = 7$

$a_3 = 3(3) + 1 = 10$

$a_4 = 3(4) + 1 = 13$

$a_5 = 3(5) + 1 = 16$

3. $a_n = 2^n$

$a_1 = 2^1 = 2$

$a_2 = 2^2 = 4$

$a_3 = 2^3 = 8$

$a_4 = 2^4 = 16$

$a_5 = 2^5 = 32$

5. $a_n = (-2)^n$

$a_1 = (-2)^1 = -2$

$a_2 = (-2)^2 = 4$

$a_3 = (-2)^3 = -8$

$a_4 = (-2)^4 = 16$

$a_5 = (-2)^5 = -32$

7. $a_n = \dfrac{n+2}{n}$

$a_1 = \dfrac{1+2}{1} = 3$

$a_2 = \dfrac{4}{2} = 2$

$a_3 = \dfrac{5}{3}$

$a_4 = \dfrac{6}{4} = \dfrac{3}{2}$

$a_5 = \dfrac{7}{5}$

9. $a_n = \dfrac{6n}{3n^2 - 1}$

$a_1 = \dfrac{6(1)}{3(1)^2 - 1} = 3$

$a_2 = \dfrac{6(2)}{3(2)^2 - 1} = \dfrac{12}{11}$

$a_3 = \dfrac{6(3)}{3(3)^2 - 1} = \dfrac{9}{13}$

$a_4 = \dfrac{6(4)}{3(4)^2 - 1} = \dfrac{24}{47}$

$a_5 = \dfrac{6(5)}{3(5)^2 - 1} = \dfrac{15}{37}$

11. $a_n = \dfrac{1 + (-1)^n}{n}$

$a_1 = 0$

$a_2 = \dfrac{2}{2} = 1$

$a_3 = 0$

$a_4 = \dfrac{2}{4} = \dfrac{1}{2}$

$a_5 = 0$

13. $a_n = 2 - \dfrac{1}{3^n}$

$a_1 = 2 - \dfrac{1}{3} = \dfrac{5}{3}$

$a_2 = 2 - \dfrac{1}{9} = \dfrac{17}{9}$

$a_3 = 2 - \dfrac{1}{27} = \dfrac{53}{27}$

$a_4 = 2 - \dfrac{1}{81} = \dfrac{161}{81}$

$a_5 = 2 - \dfrac{1}{243} = \dfrac{485}{243}$

15. $a_n = \dfrac{1}{n^{3/2}}$

$a_1 = \dfrac{1}{1} = 1$

$a_2 = \dfrac{1}{2^{3/2}}$

$a_3 = \dfrac{1}{3^{3/2}}$

$a_4 = \dfrac{1}{4^{3/2}} = \dfrac{1}{8}$

$a_5 = \dfrac{1}{5^{3/2}}$

17. $a_n = \dfrac{(-1)^n}{n^2}$

$a_1 = -\dfrac{1}{1} = -1$

$a_2 = \dfrac{1}{4}$

$a_3 = -\dfrac{1}{9}$

$a_4 = \dfrac{1}{16}$

$a_5 = -\dfrac{1}{25}$

19. $a_n = \dfrac{2}{3}$

$a_1 = \dfrac{2}{3}$

$a_2 = \dfrac{2}{3}$

$a_3 = \dfrac{2}{3}$

$a_4 = \dfrac{2}{3}$

$a_5 = \dfrac{2}{3}$

21. $a_n = n(n-1)(n-2)$

$a_1 = (1)(0)(-1) = 0$

$a_2 = (2)(1)(0) = 0$

$a_3 = (3)(2)(1) = 6$

$a_4 = (4)(3)(2) = 24$

$a_5 = (5)(4)(3) = 60$

23. $a_{25} = (-1)^{25}(3(25) - 2) = -73$

25. $a_{11} = \dfrac{4(11)}{2(11)^2 - 3} = \dfrac{44}{239}$

27. $a_n = \dfrac{3}{4}n$

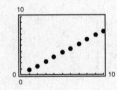

29. $a_n = 16(-0.5)^{n-1}$

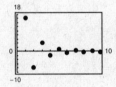

31. $a_n = \dfrac{2n}{n+1}$

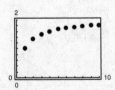

33. $a_n = \dfrac{8}{n+1}$

$a_1 = 4, \; a_{10} = \dfrac{8}{11}$

The sequence decreases.

Matches graph (c).

35. $a_n = 4(0.5)^{n-1}$

$a_1 = 4, \; a_{10} = \dfrac{1}{128}$

The sequence decreases.

Matches graph (d).

37. $1, 4, 7, 10, 13, \ldots$

$a_n = 1 + (n-1)3 = 3n - 2$

39. $0, 3, 8, 15, 24, \ldots$

$a_n = n^2 - 1$

41. $-\dfrac{2}{3}, \dfrac{3}{4}, -\dfrac{4}{5}, \dfrac{5}{6}, -\dfrac{6}{7}, \ldots$

$a_n = (-1)^n \left(\dfrac{n+1}{n+2} \right)$

43. $\dfrac{2}{1}, \dfrac{3}{3}, \dfrac{4}{5}, \dfrac{5}{7}, \dfrac{6}{9}, \ldots$

$a_n = \dfrac{n+1}{2n-1}$

45. $1, \dfrac{1}{4}, \dfrac{1}{9}, \dfrac{1}{16}, \dfrac{1}{25}, \ldots$

$a_n = \dfrac{1}{n^2}$

47. $1, -1, 1, -1, 1, \ldots$

$a_n = (-1)^{n+1}$

49. $1 + \dfrac{1}{1}, 1 + \dfrac{1}{2}, 1 + \dfrac{1}{3}, 1 + \dfrac{1}{4}, 1 + \dfrac{1}{5}, \ldots$

$a_n = 1 + \dfrac{1}{n}$

51. $a_1 = 28$ and $a_{k+1} = a_k - 4$

$a_1 = 28$

$a_2 = a_1 - 4 = 28 - 4 = 24$

$a_3 = a_2 - 4 = 24 - 4 = 20$

$a_4 = a_3 - 4 = 20 - 4 = 16$

$a_5 = a_4 - 4 = 16 - 4 = 12$

53. $a_1 = 3$ and $a_{k+1} = 2(a_k - 1)$

$a_1 = 3$

$a_2 = 2(a_1 - 1) = 2(3 - 1) = 4$

$a_3 = 2(a_2 - 1) = 2(4 - 1) = 6$

$a_4 = 2(a_3 - 1) = 2(6 - 1) = 10$

$a_5 = 2(a_4 - 1) = 2(10 - 1) = 18$

55. $a_1 = 6$ and $a_{k+1} = a_k + 2$

$a_1 = 6$

$a_2 = a_1 + 2 = 6 + 2 = 8$

$a_3 = a_2 + 2 = 8 + 2 = 10$

$a_4 = a_3 + 2 = 10 + 2 = 12$

$a_5 = a_4 + 2 = 12 + 2 = 14$

In general, $a_n = 2n + 4$.

57. $a_1 = 81$ and $a_{k+1} = \dfrac{1}{3} a_k$

$a_1 = 81$

$a_2 = \dfrac{1}{3} a_1 = \dfrac{1}{3}(81) = 27$

$a_3 = \dfrac{1}{3} a_2 = \dfrac{1}{3}(27) = 9$

$a_4 = \dfrac{1}{3} a_3 = \dfrac{1}{3}(9) = 3$

$a_5 = \dfrac{1}{3} a_4 = \dfrac{1}{3}(3) = 1$

In general,

$a_n = 81 \left(\dfrac{1}{3}\right)^{n-1} = 81(3)\left(\dfrac{1}{3}\right)^n = \dfrac{243}{3^n}.$

59. $a_n = \dfrac{3^n}{n!}$

$a_0 = \dfrac{3^0}{0!} = 1$

$a_1 = \dfrac{3^1}{1!} = 3$

$a_2 = \dfrac{3^2}{2!} = \dfrac{9}{2}$

$a_3 = \dfrac{3^3}{3!} = \dfrac{27}{6} = \dfrac{9}{2}$

$a_4 = \dfrac{3^4}{4!} = \dfrac{81}{24} = \dfrac{27}{8}$

61. $a_n = \dfrac{1}{(n+1)!}$

$a_0 = \dfrac{1}{1!} = 1$

$a_1 = \dfrac{1}{2!} = \dfrac{1}{2}$

$a_2 = \dfrac{1}{3!} = \dfrac{1}{6}$

$a_3 = \dfrac{1}{4!} = \dfrac{1}{24}$

$a_4 = \dfrac{1}{5!} = \dfrac{1}{120}$

63. $a_n = \dfrac{(-1)^{2n}}{(2n)!} = \dfrac{1}{(2n)!}$

$a_0 = \dfrac{1}{0!} = 1$

$a_1 = \dfrac{1}{2!} = \dfrac{1}{2}$

$a_2 = \dfrac{1}{4!} = \dfrac{1}{24}$

$a_3 = \dfrac{1}{6!} = \dfrac{1}{720}$

$a_4 = \dfrac{1}{8!} = \dfrac{1}{40,320}$

65. $\dfrac{4!}{6!} = \dfrac{1 \cdot 2 \cdot 3 \cdot 4}{1 \cdot 2 \cdot 3 \cdot 4 \cdot 5 \cdot 6} = \dfrac{1}{5 \cdot 6} = \dfrac{1}{30}$

67. $\dfrac{10!}{8!} = \dfrac{1 \cdot 2 \cdot 3 \cdot 4 \cdot 5 \cdot 6 \cdot 7 \cdot 8 \cdot 9 \cdot 10}{1 \cdot 2 \cdot 3 \cdot 4 \cdot 5 \cdot 6 \cdot 7 \cdot 8} = \dfrac{9 \cdot 10}{1} = 90$

69. $\dfrac{(n+1)!}{n!} = \dfrac{1 \cdot 2 \cdot 3 \cdots n \cdot (n+1)}{1 \cdot 2 \cdot 3 \cdots n} = \dfrac{n+1}{1}$

$= n + 1$

71. $\dfrac{(2n-1)!}{(2n+1)!} = \dfrac{1 \cdot 2 \cdot 3 \cdots (2n-1)}{1 \cdot 2 \cdot 3 \cdots (2n-1) \cdot (2n) \cdot (2n+1)}$

$= \dfrac{1}{2n(2n+1)}$

73. $\displaystyle\sum_{i=1}^{5} (2i + 1) = (2 + 1) + (4 + 1) + (6 + 1) + (8 + 1) + (10 + 1) = 35$

75. $\displaystyle\sum_{k=1}^{4} 10 = 10 + 10 + 10 + 10 = 40$

77. $\displaystyle\sum_{i=0}^{4} i^2 = 0^2 + 1^2 + 2^2 + 3^2 + 4^2 = 30$

79. $\displaystyle\sum_{k=0}^{3} \frac{1}{k^2 + 1} = \frac{1}{1} + \frac{1}{1+1} + \frac{1}{4+1} + \frac{1}{9+1} = \frac{9}{5}$

81. $\displaystyle\sum_{k=2}^{5} (k+1)^2(k-3) = (3)^2(-1) + (4)^2(0) + (5)^2(1) + (6)^2(2) = 88$

83. $\displaystyle\sum_{i=1}^{4} 2^i = 2^1 + 2^2 + 2^3 + 2^4 = 30$ **85.** $\displaystyle\sum_{j=1}^{6} (24 - 3j) = 81$ **87.** $\displaystyle\sum_{k=0}^{4} \frac{(-1)^k}{k+1} = \frac{47}{60}$

89. $\displaystyle\frac{1}{3(1)} + \frac{1}{3(2)} + \frac{1}{3(3)} + \cdots + \frac{1}{3(9)} = \sum_{i=1}^{9} \frac{1}{3i}$

91. $\displaystyle\left[2\left(\frac{1}{8}\right) + 3\right] + \left[2\left(\frac{2}{8}\right) + 3\right] + \left[2\left(\frac{3}{8}\right) + 3\right] + \cdots + \left[2\left(\frac{8}{8}\right) + 3\right] = \sum_{i=1}^{8}\left[2\left(\frac{i}{8}\right) + 3\right]$

93. $3 - 9 + 27 - 81 + 243 - 729 = \displaystyle\sum_{i=1}^{6} (-1)^{i+1}3^i$ **95.** $\displaystyle\frac{1}{1^2} - \frac{1}{2^2} + \frac{1}{3^2} - \frac{1}{4^2} + \cdots - \frac{1}{20^2} = \sum_{i=1}^{20} \frac{(-1)^{i+1}}{i^2}$

97. $\displaystyle\frac{1}{4} + \frac{3}{8} + \frac{7}{16} + \frac{15}{32} + \frac{31}{64} = \sum_{i=1}^{5} \frac{2^i - 1}{2^{i+1}}$ **99.** $\displaystyle\sum_{i=1}^{4} 5\left(\frac{1}{2}\right)^i = 5\left(\frac{1}{2}\right) + 5\left(\frac{1}{2}\right)^2 + 5\left(\frac{1}{2}\right)^3 + 5\left(\frac{1}{2}\right)^4 = \frac{75}{16}$

101. $\displaystyle\sum_{n=1}^{3} 4\left(-\frac{1}{2}\right)^n = 4\left(-\frac{1}{2}\right) + 4\left(-\frac{1}{2}\right)^2 + 4\left(-\frac{1}{2}\right)^3 = -\frac{3}{2}$ **103.** $\displaystyle\sum_{i=1}^{\infty} 6\left(\frac{1}{10}\right)^i = 0.6 + 0.06 + 0.006 + 0.0006 + \cdots = \frac{2}{3}$

105. By using a calculator, we have

$\displaystyle\sum_{k=1}^{10} 7\left(\frac{1}{10}\right)^k \approx 0.7777777777$

$\displaystyle\sum_{k=1}^{50} 7\left(\frac{1}{10}\right)^k \approx 0.7777777778$

$\displaystyle\sum_{k=1}^{100} 7\left(\frac{1}{10}\right)^k \approx \frac{7}{9}.$

The terms approach zero as $n \rightarrow \infty$.

Thus, we conclude that $\displaystyle\sum_{k=1}^{\infty} 7\left(\frac{1}{10}\right)^k = \frac{7}{9}.$

107. $A_n = 5000\left(1 + \dfrac{0.08}{4}\right)^n, n = 1, 2, 3, \ldots$

(a) $A_1 = \$5100.00$

$A_2 = \$5202.00$

$A_3 = \$5306.04$

$A_4 = \$5412.16$

$A_5 = \$5520.40$

$A_6 = \$5630.81$

$A_7 = \$5743.43$

$A_8 = \$5858.30$

(b) $A_{40} = \$11,040.20$

109. (a) Linear model: $a_n \approx 60.57n - 182$

(b) Quadratic model: $a_n \approx 1.61n^2 + 26.8n - 9.5$

(c)

Year	n	Actual Data	Linear Model	Quadratic Model
1998	8	311	303	308
1999	9	357	363	362
2000	10	419	424	420
2001	11	481	484	480
2002	12	548	545	544
2003	13	608	605	611

The quadratic model is a better fit.

(d) For the year 2008 we have the following predictions:

Linear model: 908 stores

Quadratic model: 995 stores

Since the quadratic model is a better fit, the predicted number of stores in 2008 is 995.

111. (a) $a_n = 2.7698n^3 - 61.372n^2 + 600.00n + 3102.9$

$a_0 = \$3102.9$ billion	$a_7 \approx \$5245.7$ billion
$a_1 \approx \$3644.3$ billion	$a_8 \approx \$5393.2$ billion
$a_2 \approx \$4079.6$ billion	$a_9 \approx \$5551.0$ billion
$a_3 \approx \$4425.3$ billion	$a_{10} = \$5735.5$ billion
$a_4 \approx \$4698.2$ billion	$a_{11} \approx \$5963.5$ billion
$a_5 \approx \$4914.8$ billion	$a_{12} \approx \$6251.5$ billion
$a_6 \approx \$5091.8$ billion	$a_{13} \approx \$6616.3$ billion

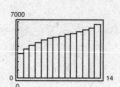

(b) The federal debt is increasing.

113. True, $\displaystyle\sum_{i=1}^{4} (i^2 + 2i) = \sum_{i=1}^{4} i^2 + 2\sum_{i=1}^{4} i$ by the Properties of Sums.

115. $a_1 = 1, a_2 = 1, \ a_{k+2} = a_{k+1} + a_k, k \geq 1$

$a_1 = 1$	$b_1 = \frac{1}{1} = 1$
$a_2 = 1$	$b_2 = \frac{2}{1} = 2$
$a_3 = 1 + 1 = 2$	$b_3 = \frac{3}{2}$
$a_4 = 2 + 1 = 3$	$b_4 = \frac{5}{3}$
$a_5 = 3 + 2 = 5$	$b_5 = \frac{8}{5}$
$a_6 = 5 + 3 = 8$	$b_6 = \frac{13}{8}$
$a_7 = 8 + 5 = 13$	$b_7 = \frac{21}{13}$
$a_8 = 13 + 8 = 21$	$b_8 = \frac{34}{21}$
$a_9 = 21 + 13 = 34$	$b_9 = \frac{55}{34}$
$a_{10} = 34 + 21 = 55$	$b_{10} = \frac{89}{55}$
$a_{11} = 55 + 34 = 89$	
$a_{12} = 89 + 55 = 144$	

117. $\dfrac{327.15 + 785.69 + 433.04 + 265.38 + 604.12 + 590.30}{6} \approx \$500.95$

119. $\displaystyle\sum_{i=1}^{n}(x_i - \bar{x}) = \sum_{i=1}^{n} x_i - \sum_{i=1}^{n} \bar{x}$

$\displaystyle = \left(\sum_{i=1}^{n} x_i\right) - n\bar{x}$

$\displaystyle = \left(\sum_{i=1}^{n} x_i\right) - n\left(\frac{1}{n}\sum_{i=1}^{n} x_i\right)$

$= 0$

121. $a_n = \dfrac{x^n}{n!}$

$a_1 = \dfrac{x^1}{1!} = x$

$a_2 = \dfrac{x^2}{2!} = \dfrac{x^2}{2}$

$a_3 = \dfrac{x^3}{3!} = \dfrac{x^3}{6}$

$a_4 = \dfrac{x^4}{4!} = \dfrac{x^4}{24}$

$a_5 = \dfrac{x^5}{5!} = \dfrac{x^5}{120}$

123. $a_n = \dfrac{(-1)^n x^{2n}}{(2n)!}$

$a_1 = \dfrac{-x^2}{2!} = -\dfrac{x^2}{2}$

$a_2 = \dfrac{x^4}{4!} = \dfrac{x^4}{24}$

$a_3 = \dfrac{-x^6}{6!} = -\dfrac{x^6}{720}$

$a_4 = \dfrac{x^8}{8!} = \dfrac{x^8}{40,320}$

$a_5 = \dfrac{-x^{10}}{10!} = -\dfrac{x^{10}}{3,628,800}$

125. $f(x) = 4x - 3$ is one-to-one, so it has an inverse.

$y = 4x - 3$

$x = 4y - 3$

$\dfrac{x + 3}{4} = y$

$f^{-1}(x) = \dfrac{x + 3}{4}$

127. $h(x) = \sqrt{5x + 1}$ is one-to-one, so it has in inverse.

Domain: $x \geq -\dfrac{1}{5}$

Range: $y \geq 0$

$y = \sqrt{5x + 1},\, x \geq -\dfrac{1}{5},\, y \geq 0$

$x = \sqrt{5y + 1},\, x \geq 0,\, y \geq -\dfrac{1}{5}$

$x^2 = 5y + 1,\, x \geq 0$

$\dfrac{x^2 - 1}{5} = y,\, x \geq 0$

$h^{-1}(x) = \dfrac{x^2 - 1}{5} = \dfrac{1}{5}(x^2 - 1),\, x \geq 0$

129. (a) $A - B = \begin{bmatrix} 6 & 5 \\ 3 & 4 \end{bmatrix} - \begin{bmatrix} -2 & 4 \\ 6 & -3 \end{bmatrix} = \begin{bmatrix} 6-(-2) & 5-4 \\ 3-6 & 4-(-3) \end{bmatrix} = \begin{bmatrix} 8 & 1 \\ -3 & 7 \end{bmatrix}$

(b) $4B - 3A = 4\begin{bmatrix} -2 & 4 \\ 6 & -3 \end{bmatrix} - 3\begin{bmatrix} 6 & 5 \\ 3 & 4 \end{bmatrix} = \begin{bmatrix} -8-18 & 16-15 \\ 24-9 & -12-12 \end{bmatrix} = \begin{bmatrix} -26 & 1 \\ 15 & -24 \end{bmatrix}$

(c) $AB = \begin{bmatrix} 6 & 5 \\ 3 & 4 \end{bmatrix}\begin{bmatrix} -2 & 4 \\ 6 & -3 \end{bmatrix} = \begin{bmatrix} -12+30 & 24-15 \\ -6+24 & 12-12 \end{bmatrix} = \begin{bmatrix} 18 & 9 \\ 18 & 0 \end{bmatrix}$

(d) $BA = \begin{bmatrix} -2 & 4 \\ 6 & -3 \end{bmatrix}\begin{bmatrix} 6 & 5 \\ 3 & 4 \end{bmatrix} = \begin{bmatrix} -12+16 & -10+12 \\ 36-9 & 30-12 \end{bmatrix} = \begin{bmatrix} 0 & 6 \\ 27 & 18 \end{bmatrix}$

131. (a) $A - B = \begin{bmatrix} -2 & -3 & 6 \\ 4 & 5 & 7 \\ 1 & 7 & 4 \end{bmatrix} - \begin{bmatrix} 1 & 4 & 2 \\ 0 & 1 & 6 \\ 0 & 3 & 1 \end{bmatrix} = \begin{bmatrix} -2-1 & -3-4 & 6-2 \\ 4-0 & 5-1 & 7-6 \\ 1-0 & 7-3 & 4-1 \end{bmatrix} = \begin{bmatrix} -3 & -7 & 4 \\ 4 & 4 & 1 \\ 1 & 4 & 3 \end{bmatrix}$

(b) $4B - 3A = 4\begin{bmatrix} 1 & 4 & 2 \\ 0 & 1 & 6 \\ 0 & 3 & 1 \end{bmatrix} - 3\begin{bmatrix} -2 & -3 & 6 \\ 4 & 5 & 7 \\ 1 & 7 & 4 \end{bmatrix} = \begin{bmatrix} 4-(-6) & 16-(-9) & 8-18 \\ 0-12 & 4-15 & 24-21 \\ 0-3 & 12-21 & 4-12 \end{bmatrix} = \begin{bmatrix} 10 & 25 & -10 \\ -12 & -11 & 3 \\ -3 & -9 & -8 \end{bmatrix}$

(c) $AB = \begin{bmatrix} -2 & -3 & 6 \\ 4 & 5 & 7 \\ 1 & 7 & 4 \end{bmatrix}\begin{bmatrix} 1 & 4 & 2 \\ 0 & 1 & 6 \\ 0 & 3 & 1 \end{bmatrix} = \begin{bmatrix} -2+0+0 & -8-3+18 & -4-18+6 \\ 4+0+0 & 16+5+21 & 8+30+7 \\ 1+0+0 & 4+7+12 & 2+42+4 \end{bmatrix} = \begin{bmatrix} -2 & 7 & -16 \\ 4 & 42 & 45 \\ 1 & 23 & 48 \end{bmatrix}$

(d) $BA = \begin{bmatrix} 1 & 4 & 2 \\ 0 & 1 & 6 \\ 0 & 3 & 1 \end{bmatrix}\begin{bmatrix} -2 & -3 & 6 \\ 4 & 5 & 7 \\ 1 & 7 & 4 \end{bmatrix} = \begin{bmatrix} -2+16+2 & -3+20+14 & 6+28+8 \\ 0+4+6 & 0+5+42 & 0+7+24 \\ 0+12+1 & 0+15+7 & 0+21+4 \end{bmatrix} = \begin{bmatrix} 16 & 31 & 42 \\ 10 & 47 & 31 \\ 13 & 22 & 25 \end{bmatrix}$

133. $|A| = \begin{vmatrix} 3 & 5 \\ -1 & 7 \end{vmatrix} = 3(7) - 5(-1) = 26$

135. $|A| = \begin{vmatrix} 3 & 4 & 5 \\ 0 & 7 & 3 \\ 4 & 9 & -1 \end{vmatrix} = 3\begin{vmatrix} 7 & 3 \\ 9 & -1 \end{vmatrix} + 4\begin{vmatrix} 4 & 5 \\ 7 & 3 \end{vmatrix}$

$= 3[7(-1) - 3(9)] + 4[4(3) - 5(7)] = -194$

Section 9.2   Arithmetic Sequences and Partial Sums

■ You should be able to recognize an arithmetic sequence, find its common difference, and find its nth term.

■ You should be able to find the nth partial sum of an arithmetic sequence by using the formula

$$S_n = \frac{n}{2}(a_1 + a_n).$$

Vocabulary Check

1. arithmetic; common

2. $a_n = dn + c$

3. sum of a finite arithmetic sequence

---

1. $10, 8, 6, 4, 2, \ldots$

Arithmetic sequence, $d = -2$

3. $1, 2, 4, 8, 16, \ldots$

Not an arithmetic sequence

5. $\frac{9}{4}, 2, \frac{7}{4}, \frac{3}{2}, \frac{5}{4}, \ldots$

Arithmetic sequence, $d = -\frac{1}{4}$

7. $\frac{1}{3}, \frac{2}{3}, 1, \frac{4}{3}, \frac{5}{6}, \ldots$

Not an arithmetic sequence

9. $\ln 1, \ln 2, \ln 3, \ln 4, \ln 5, \ldots$

Not an arithmetic sequence

11. $a_n = 5 + 3n$

8, 11, 14, 17, 20

Arithmetic sequence, $d = 3$

13. $a_n = 3 - 4(n - 2)$

$7, 3, -1, -5, -9$

Arithmetic sequence, $d = -4$

15. $a_n = (-1)^n$

$-1, 1, -1, 1, -1$

Not an arithmetic sequence

17. $a_n = \frac{(-1)^n 3}{n}$

$-3, \frac{3}{2}, -1, \frac{3}{4}, -\frac{3}{5}$

Not an arithmetic sequence

19. $a_1 = 1, \; d = 3$

$a_n = a_1 + (n - 1)d = 1 + (n - 1)(3) = 3n - 2$

21. $a_1 = 100, \; d = -8$

$a_n = a_1 + (n - 1)d = 100 + (n - 1)(-8)$

$= -8n + 108$

23. $a_1 = x, \; d = 2x$

$a_n = a_1 + (n - 1)d = x + (n - 1)(2x) = 2xn - x$

25. $4, \frac{3}{2}, -1, -\frac{7}{2}, \ldots$

$d = -\frac{5}{2}$

$a_n = a_1 + (n - 1)d = 4 + (n - 1)\left(-\frac{5}{2}\right) = -\frac{5}{2}n + \frac{13}{2}$

27. $a_1 = 5, \; a_4 = 15$

$a_4 = a_1 + 3d \implies 15 = 5 + 3d \implies d = \frac{10}{3}$

$a_n = a_1 + (n - 1)d = 5 + (n - 1)\left(\frac{10}{3}\right) = \frac{10}{3}n + \frac{5}{3}$

29. $a_3 = 94, \; a_6 = 85$

$a_6 = a_3 + 3d \implies 85 = 94 + 3d \implies d = -3$

$a_1 = a_3 - 2d \implies a_1 = 94 - 2(-3) = 100$

$a_n = a_1 + (n - 1)d = 100 + (n - 1)(-3)$

$= -3n + 103$

31. $a_1 = 5,\ d = 6$

$a_1 = 5$

$a_2 = 5 + 6 = 11$

$a_3 = 11 + 6 = 17$

$a_4 = 17 + 6 = 23$

$a_5 = 23 + 6 = 29$

33. $a_1 = -2.6,\ d = -0.4$

$a_1 = -2.6$

$a_2 = -2.6 + (-0.4) = -3.0$

$a_3 = -3.0 + (-0.4) = -3.4$

$a_4 = -3.4 + (-0.4) = -3.8$

$a_5 = -3.8 + (-0.4) = -4.2$

35. $a_1 = 2,\ a_{12} = 46$

$46 = 2 + (12 - 1)d$

$44 = 11d$

$4 = d$

$a_1 = 2$

$a_2 = 2 + 4 = 6$

$a_3 = 6 + 4 = 10$

$a_4 = 10 + 4 = 14$

$a_5 = 14 + 4 = 18$

37. $a_8 = 26,\ a_{12} = 42$

$a_{12} = a_8 + 4d$

$42 = 26 + 4d \Rightarrow d = 4$

$a_8 = a_1 + 7d$

$26 = a_1 + 28 \Rightarrow a_1 = -2$

$a_1 = -2$

$a_2 = -2 + 4 = 2$

$a_3 = 2 + 4 = 6$

$a_4 = 6 + 4 = 10$

$a_5 = 10 + 4 = 14$

39. $a_1 = 15,\ a_{k+1} = a_k + 4$

$a_2 = 15 + 4 = 19$

$a_3 = 19 + 4 = 23$

$a_4 = 23 + 4 = 27$

$a_5 = 27 + 4 = 31$

$d = 4$

$c = a_1 - d = 15 - 4 = 11$

$a_n = 4n + 11$

41. $a_1 = 200,\ a_{k+1} = a_k - 10$

$a_2 = 200 - 10 = 190$

$a_3 = 190 - 10 = 180$

$a_4 = 180 - 10 = 170$

$a_5 = 170 - 10 = 160$

$d = -10$

$c = a_1 - d = 200 - (-10) = 210$

$a_n = -10n + 210$

43. $a_1 = \frac{5}{8},\ a_{k+1} = a_k - \frac{1}{8}$

$a_1 = \frac{5}{8}$

$a_2 = \frac{5}{8} - \frac{1}{8} = \frac{1}{2}$

$a_3 = \frac{1}{2} - \frac{1}{8} = \frac{3}{8}$

$a_4 = \frac{3}{8} - \frac{1}{8} = \frac{1}{4}$

$a_5 = \frac{1}{4} - \frac{1}{8} = \frac{1}{8}$

$d = -\frac{1}{8}$

$c = a_1 - d = \frac{5}{8} - \left(-\frac{1}{8}\right) = \frac{3}{4}$

$a_n = -\frac{1}{8}n + \frac{3}{4}$

45. $a_1 = 5,\ a_2 = 11 \Rightarrow d = 11 - 5 = 6$

$a_n = a_1 + (n - 1)d \Rightarrow a_{10} = 5 + 9(6) = 59$

47. $a_1 = 4.2,\ a_2 = 6.6 \Rightarrow d = 6.6 - 4.2 = 2.4$

$a_n = a_1 + (n - 1)d \Rightarrow a_7 = 4.2 + 6(2.4) = 18.6$

49. $a_n = -\frac{3}{4}n + 8$

$d = -\frac{3}{4}$ so the sequence is decreasing and $a_1 = 7\frac{1}{4}$.

Matches (b).

51. $a_n = 2 + \frac{3}{4}n$

$d = \frac{3}{4}$ so the sequence is increasing and $a_1 = 2\frac{3}{4}$.

Matches (c).

53. $a_n = 15 - \frac{3}{2}n$

55. $a_n = 0.2n + 3$

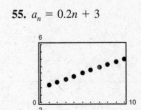

57. 8, 20, 32, 44, . . .

$a_1 = 8, d = 12, n = 10$

$a_{10} = 8 + 9(12) = 116$

$S_{10} = \frac{10}{2}(8 + 116) = 620$

59. 4.2, 3.7, 3.2, 2.7, . . .

$a_1 = 4.2, d = -0.5, n = 12$

$a_{12} = 4.2 + 11(-0.5) = -1.3$

$S_{12} = \frac{12}{2}[4.2 + (-1.3)] = 17.4$

61. 40, 37, 34, 31, . . .

$a_1 = 40, d = -3, n = 10$

$a_{10} = 40 + 9(-3) = 13$

$S_{10} = \frac{10}{2}(40 + 13) = 265$

63. $a_1 = 100, a_{25} = 220, n = 25$

$S_n = \frac{n}{2}[a_1 + a_n]$

$S_{25} = \frac{25}{2}(100 + 220) = 4000$

65. $a_n = 2n - 1$

$a_1 = 1, a_{100} = 199$

$\sum_{n=1}^{100}(2n - 1) = \frac{100}{2}(1 + 199)$

$= 10,000$

67. $a_1 = 1, a_{50} = 50, n = 50$

$\sum_{n=1}^{50} n = \frac{50}{2}(1 + 50) = 1275$

69. $a_{10} = 60, a_{100} = 600, n = 91$

$\sum_{n=10}^{100} 6n = \frac{91}{2}(60 + 600) = 30,030$

71. $\sum_{n=11}^{30} n - \sum_{n=1}^{10} n = \frac{20}{2}(11 + 30) - \frac{10}{2}(1 + 10) = 355$

73. $a_1 = 1, a_{400} = 799, n = 400$

$\sum_{n=1}^{400}(2n - 1) = \frac{400}{2}(1 + 799) = 160,000$

75. $\sum_{n=1}^{20}(2n + 5) = 520$

77. $\sum_{n=1}^{100} \frac{n+4}{2} = 2725$

79. $\sum_{i=1}^{60}\left(250 - \frac{8}{3}i\right) = 10,120$

81. (a) $a_1 = 32,500, d = 1500$

$a_6 = a_1 + 5d = 32,500 + 5(1500) = \$40,000$

(b) $S_6 = \frac{6}{2}[32,500 + 40,000] = \$217,500$

83. $a_1 = 20, d = 4, n = 30$

$a_{30} = 20 + 29(4) = 136$

$S_{30} = \frac{30}{2}(20 + 136) = 2340$ seats

85. $a_1 = 14, a_{18} = 31$

$S_{18} = \frac{18}{2}(14 + 31) = 405$ bricks

87. 4.9, 14.7, 24.5, 34.3, . . .

$d = 9.8$

$a_{10} = 4.9 + 9(9.8) = 93.1$ meters

$S_{10} = \frac{10}{2}(4.9 + 93.1) = 490$ meters

89. (a) $a_1 = 200, a_2 = 175 \Rightarrow d = -25$

$c = 200 - (-25) = 225$

$a_n = -25n + 225$

(b) $a_8 = -25(8) + 225 = 25$

$S_8 = \frac{8}{2}(200 + 25) = \$900$

91. $a_n = 1500n + 6500$

$a_1 = 8000, a_6 = 15,500$

$S_6 = \frac{6}{2}(8000 + 15,500) = \$70,500$

The cost of gasoline, labor, equipment, insurance, and maintenance are a few economic factors that could prevent the company from meeting its goals, but the biggest unknown variable is the amount of annual snowfall.

93. (a)

Monthly Payment	Unpaid Balance
$a_1 = 200 + 0.01(2000) = \$220$	$1800
$a_2 = 200 + 0.01(1800) = \$218$	$1600
$a_3 = 200 + 0.01(1600) = \$216$	$1400
$a_4 = 200 + 0.01(1400) = \$214$	$1200
$a_5 = 200 + 0.01(1200) = \$212$	$1000
$a_6 = 200 + 0.01(1000) = \$210$	$800

(b) $a_n = -2n + 222 \Rightarrow a_{10} = 202$

$S_{10} = \frac{10}{2}(220 + 202) = \$2110$

Interest paid: $110

95. (a) Using $(5, 23{,}078)$ and $(6, 24{,}176)$ we have $d = 1098$
and $c = 23{,}078 - 5(1098) = 17{,}588$.

$a_n \approx 1098n + 17{,}588$

(b) $a_n \approx 1114.95n + 17{,}795.07$

The models are similar.

(c)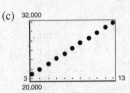

(d) For 2004 use $n = 14$: $32,960$

For 2005 use $n = 15$: $34,058$

(e) Answers will vary.

97. True; given a_1 and a_2 then $d = a_2 - a_1$ and
$a_n = a_1 + (n - 1)d$.

99. A sequence is arithmetic if the differences between
consecutive terms are the same.

$a_{n+1} - a_n = d$ for $n \geq 1$

101. (a) $a_n = 2 + 3n$

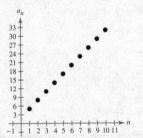

(b) $y = 3x + 2$

(c) The graph of $a_n = 2 + 3n$ contains only points at
the positive integers. The graph of $y = 3x + 2$ is a
solid line which contains these points.

(d) The slope $m = 3$ is equal to the common difference
$d = 3$. In general, these should be equal.

103.
$$S_{20} = \frac{20}{2}\{a_1 + [a_1 + (20 - 1)(3)]\} = 650$$

$10(2a_1 + 57) = 650$

$2a_1 + 57 = 65$

$2a_1 = 8$

$a_1 = 4$

105. $2x - 4y = 3$

$y = \frac{1}{2}x - \frac{3}{4}$

Slope: $m = \frac{1}{2}$

y-intercept: $\left(0, -\frac{3}{4}\right)$

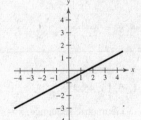

107. $x - 7 = 0$

$x = 7$

Vertical line

No slope

No y-intercept

109.
$$\begin{cases} 2x - y + 7z = -10 \\ 3x + 2y - 4z = 17 \\ 6x - 5y + z = -20 \end{cases}$$
Equation 1
Equation 2
Equation 3

$$\begin{cases} x - \frac{1}{2}y + \frac{7}{2}z = -5 \\ 3x + 2y - 4z = 17 \\ 6x - 5y + z = -20 \end{cases}$$
$\frac{1}{2}$Eq.1

$$\begin{cases} x - \frac{1}{2}y + \frac{7}{2}z = -5 \\ \frac{7}{2}y - \frac{29}{2}z = 32 \\ -2y - 20z = 10 \end{cases}$$
(-3)Eq.1 + Eq.2
(-6)Eq.1 + Eq.3

$$\begin{cases} x - \frac{1}{2}y + \frac{7}{2}z = -5 \\ -2y - 20z = 10 \\ \frac{7}{2}y - \frac{29}{2}z = 32 \end{cases}$$

$$\begin{cases} x - \frac{1}{2}y + \frac{7}{2}z = -5 \\ y + 10z = -5 \\ 7y - 29z = 64 \end{cases}$$
$(-\frac{1}{2})$Eq.2
2 Eq.3

$$\begin{cases} x + \frac{17}{2}z = -\frac{15}{2} \\ y + 10z = -5 \\ -99z = 99 \end{cases}$$
$(\frac{1}{2})$Eq.2 + Eq.1
(-7)Eq.2 + Eq.3

$$\begin{cases} x + \frac{17}{2}z = -\frac{15}{2} \\ y + 10z = -5 \\ z = -1 \end{cases}$$
$(-\frac{1}{99})$Eq.3

$$\begin{cases} x = 1 \\ y = 5 \\ z = -1 \end{cases}$$
$(-\frac{17}{2})$Eq.3 + Eq.1
(-10)Eq.3 + Eq.2

Answer: $x = 1, y = 5, z = -1$

111. Answers will vary.

Section 9.3 Geometric Sequences and Series

- You should be able to identify a geometric sequence, find its common ratio, and find the nth term.
- You should know that the nth term of a geometric sequence with common ratio r is given by $a_n = a_1 r^{n-1}$.
- You should know that the nth partial sum of a geometric sequence with common ratio $r \neq 1$ is given by

$$S_n = a_1\left(\frac{1 - r^n}{1 - r}\right).$$

- You should know that if $|r| < 1$, then

$$\sum_{n=1}^{\infty} a_1 r^{n-1} = \sum_{n=0}^{\infty} a_1 r^n = \frac{a_1}{1 - r}.$$

Vocabulary Check

1. geometric; common

2. $a_n = a_1 r^{n-1}$

3. $S_n = a_1\left(\dfrac{1 - r^n}{1 - r}\right)$

4. geometric series

5. $S = \dfrac{a_1}{1 - r}$

---

1. 5, 15, 45, 135, . . .

Geometric sequence, $r = 3$

3. 3, 12, 21, 30, . . .

Not a geometric sequence

Note: It is an arithmetic sequence with $d = 9$.

5. $1, -\frac{1}{2}, \frac{1}{4}, -\frac{1}{8}, \dots$

Geometric sequence, $r = -\frac{1}{2}$

7. $\frac{1}{8}, \frac{1}{4}, \frac{1}{2}, 1, \dots$

Geometric sequence, $r = 2$

9. $1, \frac{1}{2}, \frac{1}{3}, \frac{1}{4}, \dots$

Not a geometric sequence

11. $a_1 = 2, r = 3$

$a_1 = 2$

$a_2 = 2(3) = 6$

$a_3 = 6(3) = 18$

$a_4 = 18(3) = 54$

$a_5 = 54(3) = 162$

13. $a_1 = 1,\ r = \frac{1}{2}$

$a_1 = 1$

$a_2 = 1\left(\frac{1}{2}\right) = \frac{1}{2}$

$a_3 = \frac{1}{2}\left(\frac{1}{2}\right) = \frac{1}{4}$

$a_4 = \frac{1}{4}\left(\frac{1}{2}\right) = \frac{1}{8}$

$a_5 = \frac{1}{8}\left(\frac{1}{2}\right) = \frac{1}{16}$

15. $a_1 = 5,\ r = -\frac{1}{10}$

$a_1 = 5$

$a_2 = 5\left(-\frac{1}{10}\right) = -\frac{1}{2}$

$a_3 = \left(-\frac{1}{2}\right)\left(-\frac{1}{10}\right) = \frac{1}{20}$

$a_4 = \frac{1}{20}\left(-\frac{1}{10}\right) = -\frac{1}{200}$

$a_5 = \left(-\frac{1}{200}\right)\left(-\frac{1}{10}\right) = \frac{1}{2000}$

17. $a_1 = 1,\ r = e$

$a_1 = 1$

$a_2 = 1(e) = e$

$a_3 = (e)(e) = e^2$

$a_4 = (e^2)(e) = e^3$

$a_5 = (e^3)(e) = e^4$

19. $a_1 = 2,\ r = \frac{x}{4}$

$a_1 = 2$

$a_2 = 2\left(\frac{x}{4}\right) = \frac{x}{2}$

$a_3 = \left(\frac{x}{2}\right)\left(\frac{x}{4}\right) = \frac{x^2}{8}$

$a_4 = \left(\frac{x^2}{8}\right)\left(\frac{x}{4}\right) = \frac{x^3}{32}$

$a_5 = \left(\frac{x^3}{32}\right)\left(\frac{x}{4}\right) = \frac{x^4}{128}$

21. $a_1 = 64,\ a_{k+1} = \frac{1}{2}a_k$

$a_1 = 64$

$a_2 = \frac{1}{2}(64) = 32$

$a_3 = \frac{1}{2}(32) = 16$

$a_4 = \frac{1}{2}(16) = 8$

$a_5 = \frac{1}{2}(8) = 4$

$r = \frac{1}{2}$

$a_n = 64\left(\frac{1}{2}\right)^{n-1} = 128\left(\frac{1}{2}\right)^n$

23. $a_1 = 7,\ a_{k+1} = 2a_k$

$a_1 = 7$

$a_2 = 2(7) = 14$

$a_3 = 2(14) = 28$

$a_4 = 2(28) = 56$

$a_5 = 2(56) = 112$

$r = 2$

$a_n = 7(2)^{n-1} = \frac{7}{2}(2)^n$

25. $a_1 = 6,\ a_{k+1} = -\frac{3}{2}a_k$

$a_1 = 6$

$a_2 = -\frac{3}{2}(6) = -9$

$a_3 = -\frac{3}{2}(-9) = \frac{27}{2}$

$a_4 = -\frac{3}{2}\left(\frac{27}{2}\right) = -\frac{81}{4}$

$a_5 = -\frac{3}{2}\left(-\frac{81}{4}\right) = \frac{243}{8}$

$r = -\frac{3}{2}$

$a_n = 6\left(-\frac{3}{2}\right)^{n-1}$ or $a_n = -4\left(-\frac{3}{2}\right)^n$

27. $a_1 = 4,\ r = \frac{1}{2},\ n = 10$

$a_n = a_1 r^{n-1} = 4\left(\frac{1}{2}\right)^{n-1}$

$a_{10} = 4\left(\frac{1}{2}\right)^9 = \left(\frac{1}{2}\right)^7 = \frac{1}{128}$

29. $a_1 = 6,\ r = -\frac{1}{3},\ n = 12$

$a_n = a_1 r^{n-1} = 6\left(-\frac{1}{3}\right)^{n-1}$

$a_{12} = 6\left(-\frac{1}{3}\right)^{11} = -\frac{2}{3^{10}}$

31. $a_1 = 100,\ r = e^x,\ n = 9$

$a_n = a_1 r^{n-1} = 100(e^x)^{n-1}$

$a_9 = 100(e^x)^8 = 100e^{8x}$

33. $a_1 = 500,\ r = 1.02,\ n = 40$

$a_n = a_1 r^{n-1} = 500(1.02)^{n-1}$

$a_{40} = 500(1.02)^{39} \approx 1082.372$

35. $7, 21, 63, \ldots \Rightarrow r = 3$

$a_n = 7(3)^{n-1}$

$a_9 = 7(3)^8 = 45{,}927$

37. $5, 30, 180, \ldots \implies r = 6$

$a_n = 5(6)^{n-1}$

$a_{10} = 5(6)^9 = 50,388,480$

39. $a_1 = 16, a_4 = \dfrac{27}{4}$

$a_4 = a_1 r^3$

$\dfrac{27}{4} = 16r^3$

$\dfrac{27}{64} = r^3$

$\dfrac{3}{4} = r$

$a_n = 16\left(\dfrac{3}{4}\right)^{n-1}$

$a_3 = 16\left(\dfrac{3}{4}\right)^2 = 9$

41. $a_4 = -18, a_7 = \dfrac{2}{3}$

$a_7 = a_4 r^3$

$\dfrac{2}{3} = -18r^3$

$-\dfrac{1}{27} = r^3$

$-\dfrac{1}{3} = r$

$a_6 = \dfrac{a_7}{r} = \dfrac{2/3}{-1/3} = -2$

43. $a_n = 18\left(\frac{2}{3}\right)^{n-1}$

$a_1 = 18$ and $r = \dfrac{2}{3}$

Since $0 < r < 1$, the sequence is decreasing.

Matches (a).

45. $a_n = 18\left(\frac{3}{2}\right)^{n-1}$

$a_1 = 18$ and $r = \dfrac{3}{2} > 1$, so the sequence is increasing.

Matches (b).

47. $a_n = 12(-0.75)^{n-1}$

49. $a_n = 12(-0.4)^{n-1}$

51. $a_n = 2(1.3)^{n-1}$

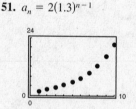

53. $\displaystyle\sum_{n=1}^{9} 2^{n-1} = 1 + 2^1 + 2^2 + \cdots + 2^8 \implies a_1 = 1, r = 2$

$S_9 = \dfrac{1(1 - 2^9)}{1 - 2} = 511$

55. $\displaystyle\sum_{n=1}^{9} (-2)^{n-1} \implies a_1 = 1, r = -2, n = 9$

$S_9 = 1\left(\dfrac{1 - (-2)^9}{1 - (-2)}\right) = 171$

57. $\displaystyle\sum_{i=1}^{7} 64\left(-\dfrac{1}{2}\right)^{i-1} = 64 + 64\left(-\dfrac{1}{2}\right)^1 + 64\left(-\dfrac{1}{2}\right)^2 + \cdots + 64\left(-\dfrac{1}{2}\right)^6 \implies a_1 = 64, r = -\dfrac{1}{2}$

$S_7 = 64\left[\dfrac{1 - \left(-\frac{1}{2}\right)^7}{1 - \left(-\frac{1}{2}\right)}\right] = \dfrac{128}{3}\left[1 - \left(-\dfrac{1}{2}\right)^7\right] = 43$

59. $\displaystyle\sum_{i=1}^{6} 32\left(\dfrac{1}{4}\right)^{i-1} = 32 + 32\left(\dfrac{1}{4}\right)^1 + 32\left(\dfrac{1}{4}\right)^2 + 32\left(\dfrac{1}{4}\right)^3 + 32\left(\dfrac{1}{4}\right)^4 + 32\left(\dfrac{1}{4}\right)^5 \implies a_1 = 32, r = \dfrac{1}{4}, n = 6$

$S_6 = 32\left(\dfrac{1 - \left(\frac{1}{4}\right)^6}{1 - \frac{1}{4}}\right) = \dfrac{1365}{32}$

61. $\displaystyle\sum_{n=0}^{20} 3\left(\dfrac{3}{2}\right)^n = \sum_{n=1}^{21} 3\left(\dfrac{3}{2}\right)^{n-1} = 3 + 3\left(\dfrac{3}{2}\right)^1 + 3\left(\dfrac{3}{2}\right)^2 + \cdots + 3\left(\dfrac{3}{2}\right)^{20} \implies a_1 = 3, r = \dfrac{3}{2}$

$S_{21} = 3\left[\dfrac{1 - \left(\frac{3}{2}\right)^{21}}{1 - \frac{3}{2}}\right] = -6\left[1 - \left(\dfrac{3}{2}\right)^{21}\right] \approx 29,921.311$

63. $\sum_{n=0}^{15} 2\left(\frac{4}{3}\right)^n = \sum_{n=1}^{16} 2\left(\frac{4}{3}\right)^{n-1} = 2 + 2\left(\frac{4}{3}\right)^1 + 2\left(\frac{4}{3}\right)^2 + \cdots + 2\left(\frac{4}{3}\right)^{15} \Rightarrow a_1 = 2, r = \frac{4}{3}, n = 16$

$S_{16} = 2\left(\frac{1 - \left(\frac{4}{3}\right)^{16}}{1 - \frac{4}{3}}\right) \approx 592.647$

65. $\sum_{n=0}^{5} 300(1.06)^n = \sum_{n=1}^{6} 300(1.06)^{n-1}$

$\qquad = 300 + 300(1.06)^1 + 300(1.06)^2 + 300(1.06)^3 + 300(1.06)^4 + 300(1.06)^5 \Rightarrow a_1 = 300, \ r = 1.06$

$S_6 = 300\left[\frac{1 - (1.06)^6}{1 - 1.06}\right] \approx 2092.596$

67. $\sum_{n=0}^{40} 2\left(-\frac{1}{4}\right)^n = 2 + 2\left(-\frac{1}{4}\right) + 2\left(-\frac{1}{4}\right)^2 + \cdots + 2\left(-\frac{1}{4}\right)^{40} \Rightarrow a_1 = 2, \ r = -\frac{1}{4}, n = 41$

$S_{41} = 2\left[\frac{1 - \left(-\frac{1}{4}\right)^{41}}{1 - \left(-\frac{1}{4}\right)}\right] = \frac{8}{5}\left[1 - \left(-\frac{1}{4}\right)^{41}\right] \approx 1.6 = \frac{8}{5}$

69. $\sum_{i=1}^{10} 8\left(-\frac{1}{4}\right)^{i-1} = 8 + 8\left(-\frac{1}{4}\right)^1 + 8\left(-\frac{1}{4}\right)^2 + \cdots + 8\left(-\frac{1}{4}\right)^9 \Rightarrow a_1 = 8, \ r = -\frac{1}{4}$

$S_{10} = 8\left[\frac{1 - \left(-\frac{1}{4}\right)^{10}}{1 - \left(-\frac{1}{4}\right)}\right] = \frac{32}{5}\left[1 - \left(-\frac{1}{4}\right)^{10}\right] \approx 6.400$

71. $\sum_{i=1}^{10} 5\left(-\frac{1}{3}\right)^{i-1} = 5 + 5\left(-\frac{1}{3}\right)^1 + 5\left(-\frac{1}{3}\right)^2 + \cdots + 5\left(-\frac{1}{3}\right)^9 \Rightarrow a_1 = 5, r = -\frac{1}{3}, n = 10$

$S_{10} = 5\left(\frac{1 - \left(-\frac{1}{3}\right)^{10}}{1 - \left(-\frac{1}{3}\right)}\right) \approx 3.750$

73. $5 + 15 + 45 + \cdots + 3645$

$r = 3$ and $3645 = 5(3)^{n-1}$

$729 = 3^{n-1} \Rightarrow 6 = n - 1 \Rightarrow n = 7$

Thus, the sum can be written as $\sum_{n=1}^{7} 5(3)^{n-1}$.

75. $2 - \frac{1}{2} + \frac{1}{8} - \cdots + \frac{1}{2048}$

$r = -\frac{1}{4}$ and $\frac{1}{2048} = 2\left(-\frac{1}{4}\right)^{n-1}$

By trial and error, we find that $n = 7$.

Thus, the sum can be written as $\sum_{n=1}^{7} 2\left(-\frac{1}{4}\right)^{n-1}$.

77. $0.1 + 0.4 + 1.6 + \cdots + 102.4$

$r = 4$ and $102.4 = 0.1(4)^{n-1}$

$1024 = 4^{n-1} \Rightarrow 5 = n - 1 \Rightarrow n = 6$

Thus, the sum can be written as $\sum_{n=1}^{6} 0.1(4)^{n-1}$.

79. $\sum_{n=0}^{\infty} \left(\frac{1}{2}\right)^n = 1 + \left(\frac{1}{2}\right)^1 + \left(\frac{1}{2}\right)^2 + \cdots$

$a_1 = 1, \ r = \frac{1}{2}$

$\sum_{n=0}^{\infty} \left(\frac{1}{2}\right)^n = \frac{a_1}{1 - r} = \frac{1}{1 - \left(\frac{1}{2}\right)} = 2$

81. $\displaystyle\sum_{n=0}^{\infty}\left(-\frac{1}{2}\right)^n = 1 + \left(-\frac{1}{2}\right)^1 + \left(-\frac{1}{2}\right)^2 + \cdots$

$a_1 = 1, \; r = -\dfrac{1}{2}$

$\displaystyle\sum_{n=0}^{\infty}\left(-\frac{1}{2}\right)^n = \frac{a_1}{1-r} = \frac{1}{1-\left(-\frac{1}{2}\right)} = \frac{2}{3}$

83. $\displaystyle\sum_{n=0}^{\infty} 4\left(\frac{1}{4}\right)^n = 4 + 4\left(\frac{1}{4}\right)^1 + 4\left(\frac{1}{4}\right)^2 + \cdots$

$a_1 = 4, \; r = \dfrac{1}{4}$

$\displaystyle\sum_{n=0}^{\infty} 4\left(\frac{1}{4}\right)^n = \frac{a_1}{1-r} = \frac{4}{1-\left(\frac{1}{4}\right)} = \frac{16}{3}$

85. $\displaystyle\sum_{n=0}^{\infty} (0.4)^n = 1 + (0.4)^1 + (0.4)^2 + \cdots$

$a_1 = 1, \; r = 0.4$

$\displaystyle\sum_{n=0}^{\infty} (0.4)^n = \frac{1}{1-0.4} = \frac{5}{3}$

87. $\displaystyle\sum_{n=0}^{\infty} -3(0.9)^n = -3 - 3(0.9)^1 - 3(0.9)^2 - \cdots$

$a_1 = -3, \; r = 0.9$

$\displaystyle\sum_{n=0}^{\infty} -3(0.9)^n = \frac{-3}{1-0.9} = -30$

89. $8 + 6 + \dfrac{9}{2} + \dfrac{27}{8} + \cdots = \displaystyle\sum_{n=0}^{\infty} 8\left(\frac{3}{4}\right)^n = \dfrac{8}{1-\frac{3}{4}} = 32$

91. $\dfrac{1}{9} - \dfrac{1}{3} + 1 - 3 + \cdots = \displaystyle\sum_{n=0}^{\infty} \frac{1}{9}(-3)^n$

The sum is undefined because

$|r| = |-3| = 3 > 1.$

93. $0.\overline{36} = \displaystyle\sum_{n=0}^{\infty} 0.36(0.01)^n = \dfrac{0.36}{1-0.01} = \dfrac{0.36}{0.99} = \dfrac{36}{99} = \dfrac{4}{11}$

95. $0.3\overline{18} = 0.3 + \displaystyle\sum_{n=0}^{\infty} 0.018(0.01)^n = \dfrac{3}{10} + \dfrac{0.018}{1-0.01}$

$= \dfrac{3}{10} + \dfrac{0.018}{0.99} = \dfrac{3}{10} + \dfrac{18}{990} = \dfrac{3}{10} + \dfrac{2}{110}$

$= \dfrac{35}{110} = \dfrac{7}{22}$

97. $f(x) = 6\left[\dfrac{1 - (0.5)^x}{1 - (0.5)}\right], \; \displaystyle\sum_{n=0}^{\infty} 6\left(\frac{1}{2}\right)^n = \dfrac{6}{1-\frac{1}{2}} = 12$

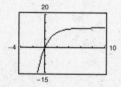

The horizontal asymptote of $f(x)$ is $y = 12$.
This corresponds to the sum of the series.

99. (a) $a_n \approx 1190.88(1.006)^n$

(b) The population is growing at a rate of 0.6% per year.

(c) For 2010, let $n = 20$: $a_n = 1190.88(1.006)^{20}$

≈ 1342.2 million

(d) $1190.88(1.006)^n = 1320$

$1.006^n = \dfrac{1320}{1190.88}$

$\ln 1.006^n = \ln\left(\dfrac{1320}{1190.88}\right)$

$n \ln 1.006 = \ln\left(\dfrac{1320}{1190.88}\right)$

$n = \dfrac{\ln\left(\dfrac{1320}{1190.88}\right)}{\ln 1.006} \approx 17.21$

This corresponds with the year 2008.

101. $A = P\left(1 + \dfrac{r}{n}\right)^{nt} = 2500\left(1 + \dfrac{0.02}{n}\right)^{n(20)}$

(a) $n = 1$: $A = 2500\left(1 + \dfrac{0.02}{1}\right)^{(1)(20)} \approx \$3714.87$

(b) $n = 2$: $A = 2500\left(1 + \dfrac{0.02}{2}\right)^{(2)(20)} \approx \$3722.16$

(c) $n = 4$: $A = 2500\left(1 + \dfrac{0.02}{4}\right)^{(4)(20)} \approx \$3725.85$

(d) $n = 12$: $A = 2500\left(1 + \dfrac{0.02}{12}\right)^{(12)(20)} \approx \$3728.32$

(e) $n = 365$: $A = 2500\left(1 + \dfrac{0.02}{365}\right)^{(365)(20)} \approx \$3729.52$

103. $A = \displaystyle\sum_{n=1}^{60} 100\left(1 + \dfrac{0.06}{12}\right)^n$

$= \displaystyle\sum_{n=1}^{60} 100(1.005)^n$

$= 100(1.005) \cdot \dfrac{[1 - 1.005^{60}]}{[1 - 1.005]}$

$\approx \$7011.89$

105. Let $N = 12t$ be the total number of deposits.

$A = P\left(1 + \dfrac{r}{12}\right) + P\left(1 + \dfrac{r}{12}\right)^2 + \cdots + P\left(1 + \dfrac{r}{12}\right)^N$

$= \left(1 + \dfrac{r}{12}\right)\left[P + P\left(1 + \dfrac{r}{12}\right) + \cdots + P\left(1 + \dfrac{r}{12}\right)^{N-1}\right]$

$= P\left(1 + \dfrac{r}{12}\right)\displaystyle\sum_{n=1}^{N}\left(1 + \dfrac{r}{12}\right)^{n-1}$

$= P\left(1 + \dfrac{r}{12}\right)\left[\dfrac{1 - \left(1 + \dfrac{r}{12}\right)^N}{1 - \left(1 + \dfrac{r}{12}\right)}\right]$

$= P\left(1 + \dfrac{r}{12}\right)\left(-\dfrac{12}{r}\right)\left[1 - \left(1 + \dfrac{r}{12}\right)^N\right]$

$= P\left(\dfrac{12}{r} + 1\right)\left[-1 + \left(1 + \dfrac{r}{12}\right)^N\right]$

$= P\left[\left(1 + \dfrac{r}{12}\right)^N - 1\right]\left(1 + \dfrac{12}{r}\right)$

$= P\left[\left(1 + \dfrac{r}{12}\right)^{12t} - 1\right]\left(1 + \dfrac{12}{r}\right)$

107. $P = \$50$, $r = 7\%$, $t = 20$ years

(a) Compounded monthly:

$A = 50\left[\left(1 + \dfrac{0.07}{12}\right)^{12(20)} - 1\right]\left(1 + \dfrac{12}{0.07}\right)$

$\approx \$26,198.27$

(b) Compounded continuously:

$A = \dfrac{50e^{0.07/12}(e^{0.07(20)} - 1)}{e^{0.07/12} - 1} \approx \$26,263.88$

109. $P = \$100$, $r = 10\%$, $t = 40$ years

(a) Compounded monthly: $A = 100\left[\left(1 + \dfrac{0.10}{12}\right)^{12(40)} - 1\right]\left(1 + \dfrac{12}{0.10}\right) \approx \$637,678.02$

(b) Compounded continuously: $A = \dfrac{100e^{0.10/12}(e^{(0.10)(40)} - 1)}{e^{0.10/12} - 1} \approx \$645,861.43$

111. $P = W \sum\limits_{n=1}^{12t} \left[\left(1 + \dfrac{r}{12} \right)^{-1} \right]^n$

$= W \left(1 + \dfrac{r}{12} \right)^{-1} \left[\dfrac{1 - \left(1 + \dfrac{r}{12} \right)^{-12t}}{1 - \left(1 + \dfrac{r}{12} \right)^{-1}} \right]$

$= W \left(\dfrac{1}{1 + \dfrac{r}{12}} \right) \left[\dfrac{1 - \left(1 + \dfrac{r}{12} \right)^{-12t}}{1 - \dfrac{1}{\left(1 + \dfrac{r}{12} \right)}} \right]$

$= W \dfrac{\left[1 - \left(1 + \dfrac{r}{12} \right)^{-12t} \right]}{\left(1 + \dfrac{r}{12} \right) - 1}$

$= W \left(\dfrac{12}{r} \right) \left[1 - \left(1 + \dfrac{r}{12} \right)^{-12t} \right]$

113. $\sum\limits_{n=1}^{\infty} 400(0.75)^n = \dfrac{300}{1 - 0.75} = \$1200$

115. $\sum\limits_{n=1}^{\infty} 600(0.725)^n = \dfrac{435}{1 - 0.725} \approx \$1581.82$

117. $64 + 32 + 16 + 8 + 4 + 2 = 126$

Total area of shaded region is approximately 126 square inches.

119. $a_n = 30{,}000(1.05)^{n-1}$

$T = \sum\limits_{n=1}^{40} 30{,}000(1.05)^{n-1} = 30{,}000 \dfrac{(1 - 1.05^{40})}{(1 - 1.05)} \approx \$3{,}623{,}993.23$

121. False. A sequence is geometric if the ratios of consecutive terms are the same.

123. Given a real number r between -1 and 1, as the exponent n increases, r^n approaches zero.

125. $g(x) = x^2 - 1$

$g(x + 1) = (x + 1)^2 - 1$

$\qquad = x^2 + 2x + 1 - 1 = x^2 + 2x$

127. $f(x) = 3x + 1, g(x) = x^2 - 1$

$f(g(x + 1)) = f(x^2 + 2x)$

$\qquad = 3(x^2 + 2x) + 1$

$\qquad = 3x^2 + 6x + 1$

129. $9x^3 - 64x = x(9x^2 - 64) = x(3x + 8)(3x - 8)$

131. $6x^2 - 13x - 5 = (3x + 1)(2x - 5)$

133. $\dfrac{3}{x + 3} \cdot \dfrac{x(x + 3)}{x - 3} = \dfrac{3x}{x - 3}, x \neq -3$

135. $\dfrac{x}{3} \div \dfrac{3x}{6x + 3} = \dfrac{x}{3} \cdot \dfrac{3(2x + 1)}{3x} = \dfrac{2x + 1}{3}, x \neq 0, -\dfrac{1}{2}$

137. $5 + \dfrac{7}{x + 2} + \dfrac{2}{x - 2} = \dfrac{5(x + 2)(x - 2) + 7(x - 2) + 2(x + 2)}{(x + 2)(x - 2)}$

$= \dfrac{5(x^2 - 4) + 7(x - 2) + 2(x + 2)}{(x + 2)(x - 2)}$

$= \dfrac{5x^2 - 20 + 7x - 14 + 2x + 4}{(x + 2)(x - 2)} = \dfrac{5x^2 + 9x - 30}{(x + 2)(x - 2)}$

139. Answers will vary.

Section 9.4   Mathematical Induction

- You should be sure that you understand the principle of mathematical induction. If P_n is a statement involving the positive integer n, where P_1 is true and the truth of P_k implies the truth of P_{k+1} for every positive k, then P_n is true for all positive integers n.

- You should be able to verify (by induction) the formulas for the sums of powers of integers and be able to use these formulas.

- You should be able to calculate the first and second differences of a sequence.

- You should be able to find the quadratic model for a sequence, when it exists.

Vocabulary Check

1. mathematical induction

2. first

3. arithmetic

4. second

---

1. $P_k = \dfrac{5}{k(k+1)}$

$$P_{k+1} = \frac{5}{(k+1)[(k+1)+1]} = \frac{5}{(k+1)(k+2)}$$

3. $P_k = \dfrac{k^2(k+1)^2}{4}$

$$P_{k+1} = \frac{(k+1)^2[(k+1)+1]^2}{4} = \frac{(k+1)^2(k+2)^2}{4}$$

5. 1. When $n = 1$, $S_1 = 2 = 1(1+1)$.

2. Assume that

$$S_k = 2 + 4 + 6 + 8 + \cdots + 2k = k(k+1).$$

Then,

$$S_{k+1} = 2 + 4 + 6 + 8 + \cdots + 2k + 2(k+1)$$
$$= S_k + 2(k+1) = k(k+1) + 2(k+1) = (k+1)(k+2).$$

Therefore, we conclude that the formula is valid for all positive integer values of n.

7. 1. When $n = 1$, $S_1 = 2 = \dfrac{1}{2}(5(1) - 1)$.

2. Assume that

$$S_k = 2 + 7 + 12 + 17 + \cdots + (5k - 3) = \frac{k}{2}(5k - 1).$$

Then,

$$S_{k+1} = 2 + 7 + 12 + 17 + \cdots + (5k - 3) + [5(k+1) - 3]$$

$$= S_k + (5k + 5 - 3) = \frac{k}{2}(5k - 1) + 5k + 2$$

$$= \frac{5k^2 - k + 10k + 4}{2} = \frac{5k^2 + 9k + 4}{2}$$

$$= \frac{(k+1)(5k+4)}{2} = \frac{(k+1)}{2}[5(k+1) - 1].$$

Therefore, we conclude that this formula is valid for all positive integer values of n.

9. 1. When $n = 1$, $S_1 = 1 = 2^1 - 1$.

2. Assume that

$$S_k = 1 + 2 + 2^2 + 2^3 + \cdots + 2^{k-1} = 2^k - 1.$$

Then,

$$S_{k+1} = 1 + 2 + 2^2 + 2^3 + \cdots + 2^{k-1} + 2^k$$
$$= S_k + 2^k = 2^k - 1 + 2^k = 2(2^k) - 1 = 2^{k+1} - 1.$$

Therefore, we conclude that this formula is valid for all positive integer values of n.

11. 1. When $n = 1$, $S_1 = 1 = \dfrac{1(1 + 1)}{2}$.

2. Assume that

$$S_k = 1 + 2 + 3 + 4 + \cdots + k = \frac{k(k + 1)}{2}.$$

Then,

$$S_{k+1} = 1 + 2 + 3 + 4 + \cdots + k + (k + 1)$$

$$= S_k + (k + 1) = \frac{k(k + 1)}{2} + \frac{2(k + 1)}{2} = \frac{(k + 1)(k + 2)}{2}.$$

Therefore, we conclude that this formula is valid for all positive integer values of n.

13. 1. When $n = 1$, $S_1 = 1 = \dfrac{(1)^2(1 + 1)^2(2(1)^2 + 2(1) - 1)}{12}$.

2. Assume that

$$S_k = \sum_{i=1}^{k} i^5 = \frac{k^2(k + 1)^2(2k^2 + 2k - 1)}{12}.$$

Then,

$$S_{k+1} = \sum_{i=1}^{k+1} i^5 = \left(\sum_{i=1}^{k} i^5 \right) + (k + 1)^5$$

$$= \frac{k^2(k + 1)^2(2k^2 + 2k - 1)}{12} + \frac{12(k + 1)^5}{12}$$

$$= \frac{(k + 1)^2[k^2(2k^2 + 2k - 1) + 12(k + 1)^3]}{12}$$

$$= \frac{(k + 1)^2[2k^4 + 2k^3 - k^2 + 12(k^3 + 3k^2 + 3k + 1)]}{12}$$

$$= \frac{(k + 1)^2[2k^4 + 14k^3 + 35k^2 + 36k + 12]}{12}$$

$$= \frac{(k + 1)^2(k^2 + 4k + 4)(2k^2 + 6k + 3)}{12}$$

$$= \frac{(k + 1)^2(k + 2)^2[2(k + 1)^2 + 2(k + 1) - 1]}{12}.$$

Therefore, we conclude that this formula is valid for all positive integer values of n.

Note: The easiest way to complete the last two steps is to "work backwards." Start with the desired expression for S_{k+1} and multiply out to show that it is equal to the expression you found for $S_k + (k + 1)^5$.

15. 1. When $n = 1, S_1 = 2 = \dfrac{1(2)(3)}{3}$.

2. Assume that

$$S_k = 1(2) + 2(3) + 3(4) + \cdots + k(k + 1) = \dfrac{k(k + 1)(k + 2)}{3}.$$

Then,

$$S_{k+1} = 1(2) + 2(3) + 3(4) + \cdots + k(k + 1) + (k + 1)(k + 2)$$

$$= S_k + (k + 1)(k + 2) = \dfrac{k(k + 1)(k + 2)}{3} + \dfrac{3(k + 1)(k + 2)}{3}$$

$$= \dfrac{(k + 1)(k + 2)(k + 3)}{3}.$$

Therefore, we conclude that this formula is valid for all positive integer values of n.

17. 1. When $n = 4, 4! = 24$ and $2^4 = 16$, thus $4! > 2^4$.

2. Assume

$k! > 2^k, \ k > 4$.

Then,

$(k + 1)! = k!(k + 1) > 2^k(2)$ since $k! > 2^k$ and $k + 1 > 2$.

Thus, $(k + 1)! > 2^{k+1}$.

Therefore, by extended mathematical induction, the inequality is valid for all integers n such that $n \geq 4$.

19. 1. When $n = 2, \dfrac{1}{\sqrt{1}} + \dfrac{1}{\sqrt{2}} \approx 1.707$ and $\sqrt{2} \approx 1.414$, thus $\dfrac{1}{\sqrt{1}} + \dfrac{1}{\sqrt{2}} > \sqrt{2}$.

2. Assume that

$$\dfrac{1}{\sqrt{1}} + \dfrac{1}{\sqrt{2}} + \dfrac{1}{\sqrt{3}} + \cdots + \dfrac{1}{\sqrt{k}} > \sqrt{k}, k > 2.$$

Then,

$$\dfrac{1}{\sqrt{1}} + \dfrac{1}{\sqrt{2}} + \dfrac{1}{\sqrt{3}} + \cdots + \dfrac{1}{\sqrt{k}} + \dfrac{1}{\sqrt{k + 1}} > \sqrt{k} + \dfrac{1}{\sqrt{k + 1}}.$$

Now it is sufficient to show that

$$\sqrt{k} + \dfrac{1}{\sqrt{k + 1}} > \sqrt{k + 1}, \ k > 2,$$

or equivalently $\left(\text{multiplying by } \sqrt{k + 1}\right)$,

$\sqrt{k}\sqrt{k + 1} + 1 > k + 1$.

This is true because

$\sqrt{k}\sqrt{k + 1} + 1 > \sqrt{k}\sqrt{k} + 1 = k + 1$.

Therefore,

$$\dfrac{1}{\sqrt{1}} + \dfrac{1}{\sqrt{2}} + \dfrac{1}{\sqrt{3}} + \ldots + \dfrac{1}{\sqrt{k}} + \dfrac{1}{\sqrt{k + 1}} > \sqrt{k + 1}.$$

Therefore, by extended mathematical induction, the inequality is valid for all integers n such that $n \geq 2$.

21. $(1 + a)^n \geq na, n \geq 1$ and $a > 0$

Since a is positive, then all of the terms in the binomial expansion are positive.

$(1 + a)^n = 1 + na + \cdots + na^{n-1} + a^n > na$

23. 1. When $n = 1$, $(ab)^1 = a^1b^1 = ab$.

2. Assume that $(ab)^k = a^kb^k$.

Then, $(ab)^{k+1} = (ab)^k(ab)$

$$= a^kb^kab$$

$$= a^{k+1}b^{k+1}.$$

Thus, $(ab)^n = a^nb^n$.

25. 1. When $n = 2$, $(x_1x_2)^{-1} = \dfrac{1}{x_1x_2} = \dfrac{1}{x_1} \cdot \dfrac{1}{x_2} = x_1^{-1}x_2^{-1}$.

2. Assume that

$$(x_1x_2x_3 \cdots x_k)^{-1} = x_1^{-1}x_2^{-1}x_3^{-1} \cdots x_k^{-1}.$$

Then,

$$(x_1x_2x_3 \cdots x_kx_{k+1})^{-1} = [(x_1x_2x_3 \cdots x_k)x_{k+1}]^{-1}$$

$$= (x_1x_2x_3 \ldots x_k)^{-1}x_{k+1}^{-1}$$

$$= x_1^{-1}x_2^{-1}x_3^{-1} \cdots x_k^{-1}x_{k+1}^{-1}.$$

Thus, the formula is valid.

27. 1. When $n = 1$, $x(y_1) = xy_1$.

2. Assume that

$$x(y_1 + y_2 + \cdots + y_k) = xy_1 + xy_2 + \cdots + xy_k.$$

Then,

$$xy_1 + xy_2 + \cdots + xy_k + xy_{k+1} = x(y_1 + y_2 + \cdots + y_k) + xy_{k+1}$$

$$= x[(y_1 + y_2 + \cdots + y_k) + y_{k+1}]$$

$$= x(y_1 + y_2 + \cdots + y_k + y_{k+1}).$$

Hence, the formula holds.

29. 1. When $n = 1$, $[1^3 + 3(1)^2 + 2(1)] = 6$ and 3 is a factor.

2. Assume that 3 is a factor of $k^3 + 3k^2 + 2k$.

Then,

$$(k + 1)^3 + 3(k + 1)^2 + 2(k + 1) = k^3 + 3k^2 + 3k + 1 + 3k^2 + 6k + 3 + 2k + 2$$

$$= (k^3 + 3k^2 + 2k) + (3k^2 + 9k + 6)$$

$$= (k^3 + 3k^2 + 2k) + 3(k^2 + 3k + 2).$$

Since 3 is a factor of $(k^3 + 3k^2 + 2k)$, our assumption, and 3 is a factor of $3(k^2 + 3k + 2)$, we conclude that 3 is a factor of the whole sum.

Thus, 3 is a factor of $(n^3 + 3n^2 + 2n)$ for every positive integer n.

31. A factor of $n^4 - n + 4$ is 2.

1. When $n = 1$, $1^4 - 1 + 4 = 4$ and 2 is a factor.

2. Assume that 2 is a factor of $k^4 - k + 4$.

Then,

$$(k + 1)^4 - (k + 1) + 4 = k^4 + 4k^3 + 6k^2 + 4k + 1 - k - 1 + 4$$

$$= (k^4 - k + 4) + (4k^3 + 6k^2 + 4k)$$

$$= (k^4 - k + 4) + 2(2k^3 + 3k^2 + 2k).$$

Since 2 is a factor of $k^4 - k + 4$, our assumption, and 2 is a factor of $2(2k^3 + 3k^2 + 2k)$, we conclude that 2 is a factor of the entire expression.

Thus, 2 is a factor of $n^4 - n + 4$ for every positive integer n.

33. A factor of $2^{4n-2} + 1$ is 5.

1. When $n = 1$,

$2^{4(1)-2} + 1 = 5$ and 5 is a factor.

2. Assume that 5 is a factor of $2^{4k-2} + 1$.

Then,

$$2^{4(k+1)-2} + 1 = 2^{4k+4-2} + 1$$
$$= 2^{4k-2} \cdot 2^4 + 1$$
$$= 2^{4k-2} \cdot 16 + 1$$
$$= (2^{4k-2} + 1) + 15 \cdot 2^{4k-2}.$$

Since 5 is a factor of $2^{4k-2} + 1$, our assumption, and 5 is a factor of $15 \cdot 2^{4k-2}$, we conclude that 5 is a factor of the entire expression.

Thus, 5 is a factor of $2^{4n-2} + 1$ for every positive integer n.

35. $S_n = 1 + 5 + 9 + 13 + \cdots + (4n - 3)$

$S_1 = 1 = 1 \cdot 1$

$S_2 = 1 + 5 = 6 = 2 \cdot 3$

$S_3 = 1 + 5 + 9 = 15 = 3 \cdot 5$

$S_4 = 1 + 5 + 9 + 13 = 28 = 4 \cdot 7$

From this sequence, it appears that $S_n = n(2n - 1)$. This can be verified by mathematical induction. The formula has already been verified for $n = 1$. Assume that the formula is valid for $n = k$. Then,

$$S_{k+1} = [1 + 5 + 9 + 13 + \cdots + (4k - 3)] + [4(k + 1) - 3]$$
$$= k(2k - 1) + (4k + 1)$$
$$= 2k^2 + 3k + 1$$
$$= (k + 1)(2k + 1)$$
$$= (k + 1)[2(k + 1) - 1].$$

Thus, the formula is valid.

37. $S_n = 1 + \dfrac{9}{10} + \dfrac{81}{100} + \dfrac{729}{1000} + \cdots + \left(\dfrac{9}{10}\right)^{n-1}$

Since this series is geometric, we have

$$S_n = \sum_{i=1}^{n} \left(\frac{9}{10}\right)^{i-1} = \frac{1 - \left(\dfrac{9}{10}\right)^n}{1 - \dfrac{9}{10}} = 10\left[1 - \left(\frac{9}{10}\right)^n\right]$$

$$= 10 - 10\left(\frac{9}{10}\right)^n.$$

39. $S_n = \dfrac{1}{4} + \dfrac{1}{12} + \dfrac{1}{24} + \dfrac{1}{40} + \cdots + \dfrac{1}{2n(n+1)}$

$S_1 = \dfrac{1}{4} = \dfrac{1}{2(2)}$

$S_2 = \dfrac{1}{4} + \dfrac{1}{12} = \dfrac{4}{12} = \dfrac{2}{6} = \dfrac{2}{2(3)}$

$S_3 = \dfrac{1}{4} + \dfrac{1}{12} + \dfrac{1}{24} = \dfrac{9}{24} = \dfrac{3}{8} = \dfrac{3}{2(4)}$

$S_4 = \dfrac{1}{4} + \dfrac{1}{12} + \dfrac{1}{24} + \dfrac{1}{40} = \dfrac{16}{40} = \dfrac{4}{10} = \dfrac{4}{2(5)}$

From this sequence, it appears that

$S_n = \dfrac{n}{2(n+1)}.$

This can be verified by mathematical induction. The formula has already been verified for $n = 1$. Assume that the formula is valid for $n = k$. Then,

$S_{k+1} = \left[\dfrac{1}{4} + \dfrac{1}{12} + \dfrac{1}{40} + \cdots + \dfrac{1}{2k(k+1)} \right] + \dfrac{1}{2(k+1)(k+2)}$

$= \dfrac{k}{2(k+1)} + \dfrac{1}{2(k+1)(k+2)}$

$= \dfrac{k(k+2) + 1}{2(k+1)(k+2)}$

$= \dfrac{k^2 + 2k + 1}{2(k+1)(k+2)}$

$= \dfrac{(k+1)^2}{2(k+1)(k+2)}$

$= \dfrac{k+1}{2(k+2)}.$

Thus, the formula is valid.

41. $\displaystyle\sum_{n=1}^{15} n = \dfrac{15(15+1)}{2} = 120$

43. $\displaystyle\sum_{n=1}^{6} n^2 = \dfrac{6(6+1)[2(6)+1]}{6} = 91$

45. $\displaystyle\sum_{n=1}^{5} n^4 = \dfrac{5(5+1)[2(5)+1][3(5)^2 + 3(5) - 1]}{30} = 979$

47. $\displaystyle\sum_{n=1}^{6} (n^2 - n) = \sum_{n=1}^{6} n^2 - \sum_{n=1}^{6} n$

$= \dfrac{6(6+1)[2(6)+1]}{6} - \dfrac{6(6+1)}{2}$

$= 91 - 21 = 70$

49. $\displaystyle\sum_{i=1}^{6} (6i - 8i^3) = 6\sum_{i=1}^{6} i - 8\sum_{i=1}^{6} i^3 = 6\left[\dfrac{6(6+1)}{2} \right] - 8\left[\dfrac{(6)^2(6+1)^2}{4} \right] = 6(21) - 8(441) = -3402$

51. $a_1 = 0, a_n = a_{n-1} + 3$

$a_1 = a_1 = 0$

$a_2 = a_1 + 3 = 0 + 3 = 3$

$a_3 = a_2 + 3 = 3 + 3 = 6$

$a_4 = a_3 + 3 = 6 + 3 = 9$

$a_5 = a_4 + 3 = 9 + 3 = 12$

$a_6 = a_5 + 3 = 12 + 3 = 15$

a_n: 0   3   6   9   12   15

First differences: 3   3   3   3   3

Second differences: 0   0   0   0

Since the first differences are equal, the sequence has a linear model.

53. $a_1 = 3, a_n = a_{n-1} - n$

$a_1 = a_1 = 3$

$a_2 = a_1 - 2 = 3 - 2 = 1$

$a_3 = a_2 - 3 = 1 - 3 = -2$

$a_4 = a_3 - 4 = -2 - 4 = -6$

$a_5 = a_4 - 5 = -6 - 5 = -11$

$a_6 = a_5 - 6 = -11 - 6 = -17$

a_n: 3   1 -2 -6 -11 -17

First differences: -2 -3 -4 -5 -6

Second differences: -1 -1 -1 -1

Since the second differences are all the same, the sequence has a quadratic model.

55. $a_0 = 2, a_n = (a_{n-1})^2$

$a_0 = 2$

$a_1 = a_0^2 = 2^2 = 4$

$a_2 = a_1^2 = 4^2 = 16$

$a_3 = a_2^2 = 16^2 = 256$

$a_4 = a_3^2 = 256^2 = 65{,}536$

$a_5 = a_4^2 = 65{,}536^2 = 4{,}294{,}967{,}296$

a_n: 2   4   16   256   65,536   4,294,967,296

First differences: 2   12   240   65,280   4,294,901,760

Second differences: 10   228   65,040   4,294,836,480

Since neither the first differences nor the second differences are equal, the sequence does not have a linear or quadratic model.

57. $a_0 = 3, a_1 = 3, a_4 = 15$

Let $a_n = an^2 + bn + c$.

Thus: $a_0 = a(0)^2 + b(0) + c = 3 \Rightarrow c = 3$

$a_1 = a(1)^2 + b(1) + c = 3 \Rightarrow a + b + c = 3$

$a + b = 0$

$a_4 = a(4)^2 + b(4) + c = 15 \Rightarrow 16a + 4b + c = 15$

$16a + 4b = 12$

$4a + b = 3$

By elimination: $-a - b = 0$

$\underline{4a + b = 3}$

$3a \quad = 3$

$a = 1 \Rightarrow b = -1$

Thus, $a_n = n^2 - n + 3$.

59. $a_0 = -3, a_2 = 1, a_4 = 9$

Let $a_n = an^2 + bn + c$.

Then: $a_0 = a(0)^2 + b(0) + c = -3 \Rightarrow c = -3$

$a_2 = a(2)^2 + b(2) + c = 1 \Rightarrow 4a + 2b + c = 1$

$4a + 2b = 4$

$2a + b = 2$

$a_4 = a(4)^2 + b(4) + c = 9 \Rightarrow 16a + 4b + c = 9$

$16a + 4b = 12$

$4a + b = 3$

By elimination: $-2a - b = -2$

$\underline{4a + b = \quad 3}$

$2a \quad = 1$

$a = \frac{1}{2} \Rightarrow b = 1$

Thus, $a_n = \frac{1}{2}n^2 + n - 3$.

61. (a)

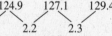

120.3 122.5 124.9 127.1 129.4 130.3

First differences: 2.2   2.4   2.2   2.3   0.9

(b) The first differences are not equal, but are fairly close to each other, so a linear model can be used.
If we let $m = 2.2$, then $b = 120.3 - 2.2(8) = 102.7$

$a_n \approx 2.2n + 102.7$

(c) $a_n \approx 2.08n + 103.9$ is obtained by using the regression feature of a graphing utility.

(d) For 2008, let $n = 18$.

$a_n \approx 2.2(18) + 102.7 = 142.3$

$a_n = 2.08(18) + 103.9 = 141.34$

These are very similar.

63. True. P_7 may be false.

65. True. If the second differences are all zero, then the first differences are all the same, so the sequence is arithmetic.

67. $(2x^2 - 1)^2 = (2x^2 - 1)(2x^2 - 1) = 4x^4 - 4x^2 + 1$ **69.** $(5 - 4x)^3 = -64x^3 + 240x^2 - 300x + 125$

71. $f(x) = \dfrac{x}{x + 3}$

(a) Domain: All real numbers x except $x = -3$

(b) Intercept: $(0, 0)$

(c) Vertical asymptote: $x = -3$

   Horizontal asymptote: $y = 1$

(d)

x	-5	-4	-2	-1	1
$f(x)$	$\frac{5}{2}$	4	-2	$-\frac{1}{2}$	$\frac{1}{4}$

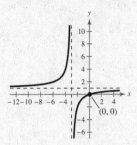

73. $h(t) = \dfrac{t - 7}{t}$

(a) Domain: All real numbers t except $t = 0$

(b) Intercept: $(7, 0)$

(c) Vertical asymptote: $t = 0$

   Horizontal asymptote: $y = 1$

(d)

t	-2	-1	1	2	3
$h(t)$	$\frac{9}{2}$	8	-6	$-\frac{5}{2}$	$-\frac{4}{3}$

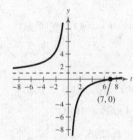

Section 9.5   The Binomial Theorem

■ You should be able to use the formula

$$(x + y)^n = x^n + nx^{n-1}y + \frac{n(n-1)}{2!}x^{n-2}y^2 + \cdots + {}_nC_r x^{n-r}y^r + \cdots + y^n$$

where ${}_nC_r = \dfrac{n!}{(n-r)!\,r!}$, to expand $(x + y)^n$. Also, ${}_nC_r = \dbinom{n}{r}$.

■ You should be able to use Pascal's Triangle in binomial expansion.

Vocabulary Check

1. binomial coefficients

2. Binomial Theorem/Pascal's Triangle

3. $\dbinom{n}{r}$ or ${}_nC_r$

4. expanding a binomial

1. ${}_5C_3 = \dfrac{5!}{3!2!} = \dfrac{5 \cdot 4}{2 \cdot 1} = 10$

3. ${}_{12}C_0 = \dfrac{12!}{0!12!} = 1$

5. ${}_{20}C_{15} = \dfrac{20!}{15!5!} = \dfrac{20 \cdot 19 \cdot 18 \cdot 17 \cdot 16}{5 \cdot 4 \cdot 3 \cdot 2 \cdot 1} = 15{,}504$

7. $\dbinom{10}{4} = \dfrac{10!}{6!4!} = \dfrac{10 \cdot 9 \cdot 8 \cdot 7 \cdot 6!}{6!(24)} = 210$

9. $\binom{100}{98} = \dfrac{100!}{2!98!} = \dfrac{100 \cdot 99}{2 \cdot 1} = 4950$

11.
```
              1
            1   1
          1   2   1
        1   3   3   1
      1   4   6   4   1
    1   5  10  10   5   1
  1   6  15  20  15   6   1
1 · 7  21  35  35  21   7   1
1   8  28  56  70 (56) 28   8   1
```

$\binom{8}{5} = 56$, the 6th entry in the 8th row.

13.
```
              1
            1   1
          1   2   1
        1   3   3   1
      1   4   6   4   1
    1   5  10  10   5   1
  1   6  15  20  15   6   1
1   7  21  35 (35) 21   7   1
```

$_7C_4 = 35$, the 5th entry in the 7th row.

15. $(x + 1)^4 = {}_4C_0x^4 + {}_4C_1x^3(1) + {}_4C_2x^2(1)^2 + {}_4C_3x(1)^3 + {}_4C_4(1)^4$

$\qquad = x^4 + 4x^3 + 6x^2 + 4x + 1$

17. $(a + 6)^4 = {}_4C_0a^4 + {}_4C_1a^3(6) + {}_4C_2a^2(6)^2 + {}_4C_3a(6)^3 + {}_4C_4(6)^4$

$\qquad = 1a^4 + 4a^3(6) + 6a^2(6)^2 + 4a(6)^3 + 1(6)^4$

$\qquad = a^4 + 24a^3 + 216a^2 + 864a + 1296$

19. $(y - 4)^3 = {}_3C_0y^3 - {}_3C_1y^2(4) + {}_3C_2y(4)^2 - {}_3C_3(4)^3$

$\qquad = 1y^3 - 3y^2(4) + 3y(4)^2 - 1(4)^3$

$\qquad = y^3 - 12y^2 + 48y - 64$

21. $(x + y)^5 = {}_5C_0x^5 + {}_5C_1x^4y + {}_5C_2x^3y^2 + {}_5C_3x^2y^3 + {}_5C_4xy^4 + {}_5C_5y^5$

$\qquad = x^5 + 5x^4y + 10x^3y^2 + 10x^2y^3 + 5xy^4 + y^5$

23. $(r + 3s)^6 = {}_6C_0r^6 + {}_6C_1r^5(3s) + {}_6C_2r^4(3s)^2 + {}_6C_3r^3(3s)^3 + {}_6C_4r^2(3s)^4 + {}_6C_5r(3s)^5 + {}_6C_6(3s)^6$

$\qquad = 1r^6 + 6r^5(3s) + 15r^4(3s)^2 + 20r^3(3s)^3 + 15r^2(3s)^4 + 6r(3s)^5 + 1(3s)^6$

$\qquad = r^6 + 18r^5s + 135r^4s^2 + 540r^3s^3 + 1215r^2s^4 + 1458rs^5 + 729s^6$

25. $(3a - 4b)^5 = {}_5C_0(3a)^5 - {}_5C_1(3a)^4(4b) + {}_5C_2(3a)^3(4b)^2 - {}_5C_3(3a)^2(4b)^3 + {}_5C_4(3a)(4b)^4 - {}_5C_5(4b)^5$

$\qquad = (1)(243a^5) - 5(81a^4)(4b) + 10(27a^3)(16b^2) - 10(9a^2)(64b^3) + 5(3a)(256b^4) - (1)(1024b^5)$

$\qquad = 243a^5 - 1620a^4b + 4320a^3b^2 - 5760a^2b^3 + 3840ab^4 - 1024b^5$

27. $(2x + y)^3 = {}_3C_0(2x)^3 + {}_3C_1(2x)^2(y) + {}_3C_2(2x)(y^2) + {}_3C_3(y^3)$

$\qquad = (1)(8x^3) + (3)(4x^2)(y) + (3)(2x)(y^2) + (1)(y^3)$

$\qquad = 8x^3 + 12x^2y + 6xy^2 + y^3$

29. $(x^2 + y^2)^4 = {}_4C_0(x^2)^4 + {}_4C_1(x^2)^3(y^2) + {}_4C_2(x^2)^2(y^2)^2 + {}_4C_3(x^2)(y^2)^3 + {}_4C_4(y^2)^4$

$\qquad = (1)(x^8) + (4)(x^6y^2) + (6)(x^4y^4) + (4)(x^2y^6) + (1)(y^8)$

$\qquad = x^8 + 4x^6y^2 + 6x^4y^4 + 4x^2y^6 + y^8$

31. $\left(\dfrac{1}{x} + y\right)^5 = {_5C_0}\left(\dfrac{1}{x}\right)^5 + {_5C_1}\left(\dfrac{1}{x}\right)^4 y + {_5C_2}\left(\dfrac{1}{x}\right)^3 y^2 + {_5C_3}\left(\dfrac{1}{x}\right)^2 y^3 + {_5C_4}\left(\dfrac{1}{x}\right)y^4 + {_5C_5}y^5$

$\qquad = \dfrac{1}{x^5} + \dfrac{5y}{x^4} + \dfrac{10y^2}{x^3} + \dfrac{10y^3}{x^2} + \dfrac{5y^4}{x} + y^5$

33. $2(x - 3)^4 + 5(x - 3)^2 = 2[x^4 - 4(x^3)(3) + 6(x^2)(3^2) - 4(x)(3^3) + 3^4] + 5[x^2 - 2(x)(3) + 3^2]$

$\qquad = 2(x^4 - 12x^3 + 54x^2 - 108x + 81) + 5(x^2 - 6x + 9)$

$\qquad = 2x^4 - 24x^3 + 113x^2 - 246x + 207$

35. 5th Row of Pascal's Triangle: 1   5   10   10   5   1

$\qquad (2t - s)^5 = 1(2t)^5 - 5(2t)^4(s) + 10(2t)^3(s)^2 - 10(2t)^2(s)^3 + 5(2t)(s)^4 - 1(s)^5$

$\qquad\qquad = 32t^5 - 80t^4s + 80t^3s^2 - 40t^2s^3 + 10ts^4 - s^5$

37. 5th Row of Pascal's Triangle: 1   5   10   10   5   1

$\qquad (x + 2y)^5 = 1x^5 + 5x^4(2y) + 10x^3(2y)^2 + 10x^2(2y)^3 + 5x(2y)^4 + 1(2y)^5$

$\qquad\qquad = x^5 + 10x^4y + 40x^3y^2 + 80x^2y^3 + 80xy^4 + 32y^5$

39. The 4th term in the expansion of $(x + y)^{10}$ is

$\qquad {_{10}C_3}x^{10-3}y^3 = 120x^7y^3.$

41. The 3rd term in the expansion of $(x - 6y)^5$ is

$\qquad {_5C_2}x^{5-2}(-6y)^2 = 10x^3(36y^2) = 360x^3y^2.$

43. The 8th term in the expansion of $(4x + 3y)^9$ is

$\qquad {_9C_7}(4x)^{9-7}(3y)^7 = 36(16x^2)(2187y^7)$

$\qquad\qquad = 1{,}259{,}712x^2y^7.$

45. The 9th term in the expansion of $(10x - 3y)^{12}$ is

$\qquad {_{12}C_8}(10x)^{12-8}(-3y)^8 = 495(10{,}000x^4)(6561y^8)$

$\qquad\qquad = 32{,}476{,}950{,}000x^4y^8.$

47. The term involving x^5 in the expansion of $(x + 3)^{12}$ is

$\qquad {_{12}C_7}x^5(3)^7 = \dfrac{12!}{7!5!} \cdot 3^7 x^5 = 1{,}732{,}104x^5.$

The coefficient is $1{,}732{,}104$.

49. The term involving x^8y^2 in the expansion of $(x - 2y)^{10}$ is

$\qquad {_{10}C_2}x^8(-2y)^2 = \dfrac{10!}{2!8!} \cdot 4x^8y^2 = 180x^8y^2.$

The coefficient is 180.

51. The term involving x^4y^5 in the expansion of $(3x - 2y)^9$ is

$\qquad {_9C_5}(3x)^4(-2y)^5 = \dfrac{9!}{5!4!}(81x^4)(-32y^5) = -326{,}592x^4y^5.$

The coefficient is $-326{,}592$.

53. The term involving $x^8y^6 = (x^2)^4y^6$ in the expansion of

$\qquad (x^2 + y)^{10}$ is ${_{10}C_6}(x^2)^4y^6 = \dfrac{10!}{4!6!}(x^2)^4y^6 = 210x^8y^6.$

The coefficient is 210.

55. $\left(\sqrt{x} + 3\right)^4 = \left(\sqrt{x}\right)^4 + 4\left(\sqrt{x}\right)^3(3) + 6\left(\sqrt{x}\right)^2(3)^2 + 4\left(\sqrt{x}\right)(3)^3 + (3)^4$

$\qquad = x^2 + 12x\sqrt{x} + 54x + 108\sqrt{x} + 81$

$\qquad = x^2 + 12x^{3/2} + 54x + 108x^{1/2} + 81$

57. $(x^{2/3} - y^{1/3})^3 = (x^{2/3})^3 - 3(x^{2/3})^2\,(y^{1/3}) + 3(x^{2/3})\,(y^{1/3})^2 - (y^{1/3})^3$

$\qquad = x^2 - 3x^{4/3}y^{1/3} + 3x^{2/3}y^{2/3} - y$

59. $\dfrac{f(x+h)-f(x)}{h} = \dfrac{(x+h)^3 - x^3}{h}$

$= \dfrac{x^3 + 3x^2h + 3xh^2 + h^3 - x^3}{h}$

$= \dfrac{h(3x^2 + 3xh + h^2)}{h}$

$= 3x^2 + 3xh + h^2,\ h \neq 0$

61. $\dfrac{f(x+h)-f(x)}{h} = \dfrac{\sqrt{x+h} - \sqrt{x}}{h}$

$= \dfrac{\sqrt{x+h} - \sqrt{x}}{h} \cdot \dfrac{\sqrt{x+h} + \sqrt{x}}{\sqrt{x+h} + \sqrt{x}}$

$= \dfrac{(x+h) - x}{h\left(\sqrt{x+h} + \sqrt{x}\right)}$

$= \dfrac{1}{\sqrt{x+h} + \sqrt{x}},\ h \neq 0$

63. $(1+i)^4 = {}_4C_0(1)^4 + {}_4C_1(1)^3 i + {}_4C_2(1)^2 i^2 + {}_4C_3(1)i^3 + {}_4C_4 i^4$

$= 1 + 4i - 6 - 4i + 1$

$= -4$

65. $(2-3i)^6 = {}_6C_0 2^6 - {}_6C_1 2^5(3i) + {}_6C_2 2^4(3i)^2 - {}_6C_3 2^3(3i)^3 + {}_6C_4 2^2(3i)^4 - {}_6C_5 2(3i)^5 + {}_6C_6(3i)^6$

$= (1)(64) - (6)(32)(3i) + 15(16)(-9) - 20(8)(-27i) + 15(4)(81) - 6(2)(243i) + (1)(-729)$

$= 64 - 576i - 2160 + 4320i + 4860 - 2916i - 729$

$= 2035 + 828i$

67. $\left(-\dfrac{1}{2} + \dfrac{\sqrt{3}}{2}i\right)^3 = \dfrac{1}{8}\left[(-1)^3 + 3(-1)^2\left(\sqrt{3}i\right) + 3(-1)\left(\sqrt{3}i\right)^2 + \left(\sqrt{3}i\right)^3\right]$

$= \dfrac{1}{8}\left[-1 + 3\sqrt{3}i + 9 - 3\sqrt{3}i\right]$

$= 1$

69. $(1.02)^8 = (1 + 0.02)^8$

$= 1 + 8(0.02) + 28(0.02)^2 + 56(0.02)^3 + 70(0.02)^4 + 56(0.02)^5 + 28(0.02)^6 + 8(0.02)^7 + (0.02)^8$

$= 1 + 0.16 + 0.0112 + 0.000448 + \cdots \approx 1.172$

71. $(2.99)^{12} = (3 - 0.01)^{12}$

$= 3^{12} - 12(3)^{11}(0.01) + 66(3)^{10}(0.01)^2 - 220(3)^9(0.01)^3 + 495(3)^8(0.01)^4$

$\quad - 792(3)^7(0.01)^5 + 924(3)^6(0.01)^6 - 792(3)^5(0.01)^7 + 495(3)^4(0.01)^8$

$\quad - 220(3)^3(0.01)^9 + 66(3)^2(0.01)^{10} - 12(3)(0.01)^{11} + (0.01)^{12}$

$\approx 531{,}441 - 21{,}257.64 + 389.7234 - 4.3303 + 0.0325 - 0.0002 + \cdots \approx 510{,}568.785$

73. $f(x) = x^3 - 4x$

$g(x) = f(x+4)$

$= (x+4)^3 - 4(x+4)$

$= x^3 + 3x^2(4) + 3x(4)^2 + (4)^3 - 4x - 16$

$= x^3 + 12x^2 + 48x + 64 - 4x - 16$

$= x^3 + 12x^2 + 44x + 48$

The graph of g is the same as the graph of f shifted four units to the left.

75. $_7C_4\left(\dfrac{1}{2}\right)^4\left(\dfrac{1}{2}\right)^3 = \dfrac{7!}{3!4!}\left(\dfrac{1}{16}\right)\left(\dfrac{1}{8}\right) = 35\left(\dfrac{1}{16}\right)\left(\dfrac{1}{8}\right) \approx 0.273$

77. $_8C_4\left(\dfrac{1}{3}\right)^4\left(\dfrac{2}{3}\right)^4 = \dfrac{8!}{4!4!}\left(\dfrac{1}{81}\right)\left(\dfrac{16}{81}\right) = 70\left(\dfrac{1}{81}\right)\left(\dfrac{16}{81}\right) \approx 0.171$

79. (a) $f(t) \approx 0.0025t^3 - 0.015t^2 + 0.88t + 7.7$

(b)

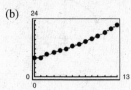

(c) $g(t) = f(t + 10) = 0.0025(t + 10)^3 - 0.015(t + 10)^2$

$$+ 0.88(t + 10) + 7.7$$

$$= 0.0025(t^3 + 30t^2 + 300t + 1000)$$

$$- 0.015(t^2 + 20t + 100) + 0.88(t + 10) + 7.7$$

$$= 0.0025t^3 + 0.06t^2 + 1.33t + 17.5$$

(d)

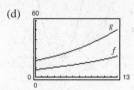

(e) For 2008 use $t = 18$ in $f(t)$ and $t = 8$ in $g(t)$.

$f(18) = 33.26$ gallons

$g(8) = 33.26$ gallons

Both models yield the same answer.

(f) The trend is for the per capita consumption of bottled water to increase. This may be due to the increasing concern with contaminants in tap water.

81. True. The coefficients from the Binomial Theorem can be used to find the numbers in Pascal's Triangle.

83. False.

The coefficient of the x^{10}-term is $_{12}C_7(3)^7 = 1,732,104$.

The coefficient of the x^{14}-term is $_{12}C_5(3)^5 = 192,456$.

85.

```
                        1
                     1     1
                  1     2     1
               1     3     3     1
            1     4     6     4     1
         1     5    10    10     5     1
      1     6    15    20    15     6     1
   1     7    21    35    35    21     7     1
1     8    28    56    70    56    28     8     1
1   9   36   84   126   126   84   36   9   1
1  10  45  120  210  252  210  120  45  10   1
```

87. The signs of the terms in the expansion of $(x - y)^n$ alternate from positive to negative.

89. $_nC_{n-r} = \dfrac{n!}{(n - (n - r))!(n - r)!}$

$$= \dfrac{n!}{r!(n - r)!}$$

$$= \dfrac{n!}{(n - r)!r!}$$

$$= \,_nC_r$$

91. $_nC_r + \,_nC_{r-1} = \dfrac{n!}{(n - r)!r!} + \dfrac{n!}{(n - r + 1)!(r - 1)!}$

$$= \dfrac{n!(n - r + 1)!(r - 1)! + n!(n - r)!r!}{(n - r)!r!(n - r + 1)!(r - 1)!}$$

$$= \dfrac{n![(n - r + 1)!(r - 1)! + r!(n - r)!]}{(n - r)!r!(n - r + 1)!(r - 1)!}$$

$$= \dfrac{n!(r - 1)![(n - r + 1)! + r(n - r)!]}{(n - r)!r!(n - r + 1)!(r - 1)!}$$

$$= \dfrac{n!(n - r)![(n - r + 1) + r]}{(n - r)!r!(n - r + 1)!}$$

$$= \dfrac{n![n + 1]}{r!(n - r + 1)!}$$

$$= \dfrac{(n + 1)!}{[(n + 1) - r]!r!}$$

$$= \,_{n+1}C_r$$

93. The graph of $f(x) = x^2$ is shifted three units to the right. Thus, $g(x) = (x - 3)^2$.

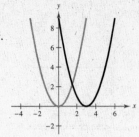

95. The graph of $f(x) = \sqrt{x}$ is shifted two units to the left and shifted one unit upward. Thus, $g(x) = \sqrt{x + 2} + 1$.

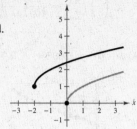

97. $A^{-1} = \dfrac{1}{(-6)(4) - (5)(-5)} \begin{bmatrix} 4 & -5 \\ 5 & -6 \end{bmatrix} = \begin{bmatrix} 4 & -5 \\ 5 & -6 \end{bmatrix}$

Section 9.6 Counting Principles

- You should know The Fundamental Counting Principle.

- $_nP_r = \dfrac{n!}{(n - r)!}$ is the number of permutations of n elements taken r at a time.

- Given a set of n objects that has n_1 of one kind, n_2 of a second kind, and so on, the number of distinguishable permutations is

$$\frac{n!}{n_1! n_2! \ldots n_k!}$$

- $_nC_r = \dfrac{n!}{(n - r)! r!}$ is the number of combinations of n elements taken r at a time.

Vocabulary Check

1. Fundamental Counting Principle

2. permutation

3. $_nP_r = \dfrac{n!}{(n - r)!}$

4. distinguishable permutations

5. combinations

1. Odd integers: 1, 3, 5, 7, 9, 11

6 ways

3. Prime integers: 2, 3, 5, 7, 11

5 ways

5. Divisible by 4: 4, 8, 12

3 ways

7. Sum is 9: $1 + 8, 2 + 7, 3 + 6, 4 + 5, 5 + 4,$
$\qquad\qquad 6 + 3, 7 + 2, 8 + 1$

8 ways

9. Amplifiers: 3 choices

Compact disc players: 2 choices

Speakers: 5 choices

Total: $3 \cdot 2 \cdot 5 = 30$ ways

11. Math courses: 2

Science courses: 3

Social sciences and humanities courses: 5

Total: $2 \cdot 3 \cdot 5 = 30$ schedules

13. $2^6 = 64$

15. $26 \cdot 26 \cdot 26 \cdot 10 \cdot 10 \cdot 10 \cdot 10 = 175,760,000$
distinct license plate numbers

17. (a) $9 \cdot 10 \cdot 10 = 900$

  (b) $9 \cdot 9 \cdot 8 = 648$

  (c) $9 \cdot 10 \cdot 2 = 180$

  (d) $6 \cdot 10 \cdot 10 = 600$

19. $40^3 = 64,000$

21. (a) $8 \cdot 7 \cdot 6 \cdot 5 \cdot 4 \cdot 3 \cdot 2 \cdot 1 = 40,320$

  (b) $8 \cdot 1 \cdot 6 \cdot 1 \cdot 4 \cdot 1 \cdot 2 \cdot 1 = 384$

23. $_nP_r = \dfrac{n!}{(n-r)!}$

So, $_4P_4 = \dfrac{4!}{0!} = 4! = 24.$

25. $_8P_3 = \dfrac{8!}{5!} = 8 \cdot 7 \cdot 6 = 336$

27. $_5P_4 = \dfrac{5!}{1!} = 120$

29. $14 \cdot {}_nP_3 = {}_{n+2}P_4$ **Note:** $n \geq 3$ for this to be defined.

$$14\left(\frac{n!}{(n-3)!}\right) = \frac{(n+2)!}{(n-2)!}$$

$14n(n-1)(n-2) = (n+2)(n+1)n(n-1)$ (We can divide here by $n(n-1)$ since $n \neq 0, n \neq 1.$)

$$14(n-2) = (n+2)(n+1)$$

$$14n - 28 = n^2 + 3n + 2$$

$$0 = n^2 - 11n + 30$$

$$0 = (n-5)(n-6)$$

$$n = 5 \text{ or } n = 6$$

31. $_{20}P_5 = 1,860,480$

33. $_{100}P_3 = 970,200$

35. $_{20}C_5 = 15,504$

37. $5! = 120$ ways

39. $_{12}P_4 = \dfrac{12!}{8!} = 12 \cdot 11 \cdot 10 \cdot 9 = 11,880$ ways

41. $\dfrac{7!}{2!1!3!1!} = \dfrac{7!}{2!3!} = 420$

43. $\dfrac{7!}{2!1!1!1!1!1!1!} = \dfrac{7!}{2!} = 7 \cdot 6 \cdot 5 \cdot 4 \cdot 3 = 2520$

45.

ABCD	BACD	CABD	DABC
ABDC	BADC	CADB	DACB
ACBD	BCAD	CBAD	DBAC
ACDB	BCDA	CBDA	DBCA
ADBC	BDAC	CDAB	DCAB
ADCB	BDCA	CDBA	DCBA

47. $_{15}P_9 = \dfrac{15!}{6!} = 1,816,214,400$

different batting orders

49. $_{40}C_{12} = \dfrac{40!}{28!12!} = 5,586,853,480$ ways

51. $_6C_2 = 15$

The 15 ways are listed below.

AB, AC, AD, AE, AF, BC, BD, BE,

BF, CD, CE, CF, DE, DF, EF

53. $_{35}C_5 = \dfrac{35!}{30!5!} = 324{,}632$ ways

55. There are 7 good units and 3 defective units.

 (a) $_7C_4 = \dfrac{7!}{3!4!} = 35$ ways

 (b) $_7C_2 \cdot {_3}C_2 = \dfrac{7!}{5!2!} \cdot \dfrac{3!}{1!2!} = 21 \cdot 3 = 63$ ways

 (c) $_7C_4 + {_7}C_3 \cdot {_3}C_1 + {_7}C_2 \cdot {_3}C_2 = \dfrac{7!}{3!4!} + \dfrac{7!}{4!3!} \cdot \dfrac{3!}{2!1!} + \dfrac{7!}{5!2!} \cdot \dfrac{3!}{1!2!}$

$$= 35 + 35 \cdot 3 + 21 \cdot 3$$
$$= 203 \text{ ways}$$

57. (a) Select type of card for three of a kind: $_{13}C_1$

 Select three of four cards for three of a kind: $_4C_3$

 Select type of card for pair: $_{12}C_1$

 Select two of four cards for pair: $_4C_2$

$$_{13}C_1 \cdot {_4}C_3 \cdot {_{12}}C_1 \cdot {_4}C_2 = \frac{13!}{(13-1)!1!} \cdot \frac{4!}{(4-3)!3!} \cdot \frac{12!}{(12-1)!1!} \cdot \frac{4!}{(4-2)!2!} = 3744$$

 (b) Select two jacks: $_4C_2$

 Select three aces: $_4C_3$

$$_4C_2 \cdot {_4}C_3 = \frac{4!}{(4-2)!2!} \cdot \frac{4!}{(4-3)!3!} = 24$$

59. $_7C_1 \cdot {_{12}}C_3 \cdot {_{20}}C_2 = \dfrac{7!}{(7-1)!1!} \cdot \dfrac{12!}{(12-3)!3!} \cdot \dfrac{20!}{(20-2)!2!} = 292{,}600$

61. $_5C_2 - 5 = 10 - 5 = 5$ diagonals

63. $_8C_2 - 8 = 28 - 8 = 20$ diagonals

65. (a) $_{53}C_5 \cdot (42) = 120{,}526{,}770$

 (b) 1. If the jackpot is won, then there is only one winning number.

 (c) There are 22,957,480 possible winning numbers in the state lottery, which is less than the possible number of winning Powerball numbers.

67. False.

 It is an example of a combination.

69. $_nC_r = {_n}C_{n-r}$ They are the same.

71. $_nP_{n-1} = \dfrac{n!}{(n-(n-1))!} = \dfrac{n!}{1!} = \dfrac{n!}{0!} = {_n}P_n$

73. $_nC_{n-1} = \dfrac{n!}{(n-(n-1))!(n-1)!} = \dfrac{n!}{(1)!(n-1)!}$

$$= \frac{n!}{(n-1)!1!} = {_n}C_1$$

75. $_{100}P_{80} \approx 3.836 \times 10^{139}$

 This number is too large for some calculators to evaluate.

77. $f(x) = 3x^2 + 8$

  (a) $f(3) = 3(3)^2 + 8 = 35$

  (b) $f(0) = 3(0)^2 + 8 = 8$

  (c) $f(-5) = 3(-5)^2 + 8 = 83$

79. $f(x) = -|x - 5| + 6$

  (a) $f(-5) = -|-5 - 5| + 6 = -10 + 6 = -4$

  (b) $f(-1) = -|-1 - 5| + 6 = -6 + 6 = 0$

  (c) $f(11) = -|11 - 5| + 6 = -6 + 6 = 0$

81. $\sqrt{x - 3} = x - 6$

$\left(\sqrt{x - 3}\right)^2 = (x - 6)^2$

$x - 3 = x^2 - 12x + 36$

$0 = x^2 - 13x + 39$

By the Quadratic Formula we have: $x = \dfrac{13 \pm \sqrt{13}}{2}$

$x = \dfrac{13 - \sqrt{13}}{2}$ is extraneous.

The only valid solution is $x = \dfrac{13 + \sqrt{13}}{2} \approx 8.30$.

83. $\log_2(x - 3) = 5$

$x - 3 = 2^5$

$x - 3 = 32$

$x = 35$

Section 9.7   Probability

You should know the following basic principles of probability.

■ If an event E has $n(E)$ equally likely outcomes and its sample space has $n(S)$ equally likely outcomes, then the probability of event E is

$$P(E) = \frac{n(E)}{n(S)}, \text{ where } 0 \le P(E) \le 1.$$

■ If A and B are mutually exclusive events, then $P(A \cup B) = P(A) + P(B)$.

  If A and B are not mutually exclusive events, then $P(A \cup B) = P(A) + P(B) - P(A \cap B)$.

■ If A and B are independent events, then the probability that both A and B will occur is $P(A)P(B)$.

■ The complement of an event A is denoted by A' and its probability is $P(A') = 1 - P(A)$.

Vocabulary Check

  1. experiment; outcomes

  3. probability

  5. mutually exclusive

  7. complement

  2. sample space

  4. impossible; certain

  6. independent

  8. (a) iii   (b) i   (c) iv   (d) ii

1. $\{(H, 1), (H, 2), (H, 3), (H, 4), (H, 5), (H, 6),$
  $(T, 1), (T, 2), (T, 3), (T, 4), (T, 5), (T, 6)\}$

3. $\{ABC, ACB, BAC, BCA, CAB, CBA\}$

5. $\{AB, AC, AD, AE, BC, BD, BE, CD, CE, DE\}$

7. $E = \{HHT, HTH, THH\}$

$$P(E) = \frac{n(E)}{n(S)} = \frac{3}{8}$$

9. $E = \{HHH, HHT, HTH, HTT, THH, THT, TTH\}$

$$P(E) = \frac{n(E)}{n(S)} = \frac{7}{8}$$

11. $E = \{K\clubsuit, K\diamondsuit, K\heartsuit, K\spadesuit, Q\clubsuit, Q\diamondsuit, Q\heartsuit, Q\spadesuit, J\clubsuit, J\diamondsuit, J\heartsuit, J\spadesuit\}$

$$P(E) = \frac{n(E)}{n(S)} = \frac{12}{52} = \frac{3}{13}$$

13. $E = \{K\diamondsuit, K\heartsuit, Q\diamondsuit, Q\heartsuit, J\diamondsuit, J\heartsuit\}$

$$P(E) = \frac{n(E)}{n(S)} = \frac{6}{52} = \frac{3}{26}$$

15. $E = \{(1, 3), (2, 2), (3, 1)\}$

$$P(E) = \frac{n(E)}{n(S)} = \frac{3}{36} = \frac{1}{12}$$

17. Use the complement.

$$E' = \{(5, 6), (6, 5), (6, 6)\}$$

$$P(E') = \frac{n(E')}{n(S)} = \frac{3}{36} = \frac{1}{12}$$

$$P(E) = 1 - P(E') = 1 - \frac{1}{12} = \frac{11}{12}$$

19. $E_3 = \{(1, 2), (2, 1)\}, \ n(E_3) = 2$

$E_5 = \{(1, 4), (2, 3), (3, 2), (4, 1)\}, \ n(E_5) = 4$

$E_7 = \{(1, 6), (2, 5), (3, 4), (4, 3), (5, 2), (6, 1)\}, \ n(E_7) = 6$

$E = E_3 \cup E_5 \cup E_7$

$n(E) = 2 + 4 + 6 = 12$

$$P(E) = \frac{n(E)}{n(S)} = \frac{12}{36} = \frac{1}{3}$$

21. $P(E) = \dfrac{_3C_2}{_6C_2} = \dfrac{3}{15} = \dfrac{1}{5}$

23. $P(E) = \dfrac{_4C_2}{_6C_2} = \dfrac{6}{15} = \dfrac{2}{5}$

25. $P(E') = 1 - P(E) = 1 - 0.7 = 0.3$

27. $P(E') = 1 - P(E) = 1 - \frac{1}{4} = \frac{3}{4}$

29. $P(E) = 1 - P(E')$

$= 1 - 0.14 = 0.86$

31. $P(E) = 1 - P(E') = 1 - \frac{17}{35} = \frac{18}{35}$

33. (a) $\dfrac{290}{500} = 0.58 = 58\%$

(b) $\dfrac{478}{500} = 0.956 = 95.6\%$

(c) $\dfrac{2}{500} = 0.004 = 0.4\%$

35. (a) $0.24(1011) \approx 243$ adults

(b) $2\% = \dfrac{1}{50}$

(c) $52\% + 12\% = 64\% = \dfrac{16}{25}$

37. (a) $\dfrac{672}{1254} = \dfrac{112}{209}$

(b) $\dfrac{582}{1254} = \dfrac{97}{209}$

(c) $\dfrac{672 - 124}{1254} = \dfrac{548}{1254} = \dfrac{274}{627}$

39. $p + p + 2p = 1$

$p = 0.25$

Taylor: $0.50 = \frac{1}{2}$, Moore: $0.25 = \frac{1}{4}$, Jenkins: $0.25 = \frac{1}{4}$

41. (a) $\dfrac{_{15}C_{10}}{_{20}C_{10}} = \dfrac{3003}{184,756} = \dfrac{21}{1292} \approx 0.016$

(b) $\dfrac{_{15}C_8 \cdot {_5C_2}}{_{20}C_{10}} = \dfrac{64,350}{184,756} = \dfrac{225}{646} \approx 0.348$

(c) $\dfrac{_{15}C_9 \cdot {_5C_1}}{_{20}C_{10}} + \dfrac{_{15}C_{10}}{_{20}C_{10}} = \dfrac{25,025 + 3003}{184,756} = \dfrac{28,028}{184,756} = \dfrac{49}{323} \approx 0.152$

43. (a) $\dfrac{1}{_5P_5} = \dfrac{1}{120}$

(b) $\dfrac{1}{_4P_4} = \dfrac{1}{24}$

45. (a) $\dfrac{20}{52} = \dfrac{5}{13}$

(b) $\dfrac{26}{52} = \dfrac{1}{2}$

(c) $\dfrac{16}{52} = \dfrac{4}{13}$

47. (a) $\dfrac{{}_9C_4}{{}_{12}C_4} = \dfrac{126}{495} = \dfrac{14}{55}$ (4 good units)

  (b) $\dfrac{{}_9C_2 \cdot {}_3C_2}{{}_{12}C_4} = \dfrac{108}{495} = \dfrac{12}{55}$ (2 good units) ·

  (c) $\dfrac{{}_9C_3 \cdot {}_3C_1}{{}_{12}C_4} = \dfrac{252}{495} = \dfrac{28}{55}$ (3 good units)

  At least 2 good units: $\dfrac{12}{55} + \dfrac{28}{55} + \dfrac{14}{55} = \dfrac{54}{55}$

49. $(0.78)^3 \approx 0.4746$

51. (a) $P(SS) = (0.985)^2 \approx 0.9702$

  (b) $P(S) = 1 - P(FF) = 1 - (0.015)^2 \approx 0.9998$

  (c) $P(FF) = (0.015)^2 \approx 0.0002$

53. (a) $P(BBBB) = \left(\tfrac{1}{2}\right)^4 = \tfrac{1}{16}$

  (b) $P(BBBB) + P(GGGG) = \left(\tfrac{1}{2}\right)^4 + \left(\tfrac{1}{2}\right)^4 = \tfrac{1}{8}$

  (c) $P(\text{at least one boy}) = 1 - P(\text{no boys})$

   $= 1 - P(GGGG) = 1 - \tfrac{1}{16} = \tfrac{15}{16}$

55. $1 - \dfrac{(45)^2}{(60)^2} = 1 - \left(\dfrac{45}{60}\right)^2 = 1 - \left(\dfrac{3}{4}\right)^2 = 1 - \dfrac{9}{16} = \dfrac{7}{16}$

57. True. Two events are independent if the occurance of one has no effect on the occurance of the other.

59. (a) As you consider successive people with distinct birthdays, the probabilities must decrease to take into account the birth dates already used. Because the birth dates of people are independent events, multiply the respective probabilities of distinct birthdays.

  (b) $\dfrac{365}{365} \cdot \dfrac{364}{365} \cdot \dfrac{363}{365} \cdot \dfrac{362}{365}$

  (c) $P_1 = \dfrac{365}{365} = 1$

   $P_2 = \dfrac{365}{365} \cdot \dfrac{364}{365} = \dfrac{364}{365} P_1 = \dfrac{365 - (2-1)}{365} P_1$

   $P_3 = \dfrac{365}{365} \cdot \dfrac{364}{365} \cdot \dfrac{363}{365} = \dfrac{363}{365} P_2 = \dfrac{365 - (3-1)}{365} P_2$

   $P_n = \dfrac{365}{365} \cdot \dfrac{364}{365} \cdot \dfrac{363}{365} \cdot \ldots \cdot \dfrac{365 - (n-1)}{365} = \dfrac{365 - (n-1)}{365} P_{n-1}$

  (d) Q_n is the probability that the birthdays are not distinct which is equivalent to at least two people having the same birthday.

  (e)

n	10	15	20	23	30	40	50
P_n	0.88	0.75	0.59	0.49	0.29	0.11	0.03
Q_n	0.12	0.25	0.41	(0.51)	0.71	0.89	0.97

  (f) 23, see the chart above.

61. $6x^2 + 8 = 0$

   $6x^2 = -8$

   $x^2 = -\tfrac{4}{3}$

  No real solution

63. $x^3 - x^2 - 3x = 0$

   $x(x^2 - x - 3) = 0$

   $x = 0$ or $x^2 - x - 3 = 0$

   $x = \dfrac{1 \pm \sqrt{1 - 4(1)(-3)}}{2(1)} = \dfrac{1 \pm \sqrt{13}}{2}$

65. $\dfrac{12}{x} = -3$

$12 = -3x$

$-4 = x$

67. $\dfrac{2}{x - 5} = 4$

$2 = 4(x - 5)$

$2 = 4x - 20$

$22 = 4x$

$\dfrac{11}{2} = x$

69. $\dfrac{3}{x - 2} + \dfrac{x}{x + 2} = 1$

$3(x + 2) + x(x - 2) = 1(x - 2)(x + 2)$

$3x + 6 + x^2 - 2x = x^2 - 4$

$x^2 + x + 6 = x^2 - 4$

$x + 6 = -4$

$x = -10$

71. $\begin{cases} y \geq -3 \\ x \geq -1 \\ -x - y \geq -8 \end{cases}$

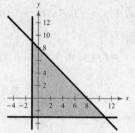

73. $\begin{cases} x^2 + y \geq -2 \\ y \geq x - 4 \end{cases}$

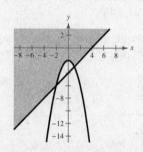

Review Exercises for Chapter 9

1. $a_n = 2 + \dfrac{6}{n}$

$a_1 = 2 + \dfrac{6}{1} = 8$

$a_2 = 2 + \dfrac{6}{2} = 5$

$a_3 = 2 + \dfrac{6}{3} = 4$

$a_4 = 2 + \dfrac{6}{4} = \dfrac{7}{2}$

$a_5 = 2 + \dfrac{6}{5} = \dfrac{16}{5}$

3. $a_n = \dfrac{72}{n!}$

$a_1 = \dfrac{72}{1!} = 72$

$a_2 = \dfrac{72}{2!} = 36$

$a_3 = \dfrac{72}{3!} = 12$

$a_4 = \dfrac{72}{4!} = 3$

$a_5 = \dfrac{72}{5!} = \dfrac{3}{5}$

5. $-2, 2, -2, 2, -2, \ldots$

$a_n = 2(-1)^n$

7. $4, 2, \dfrac{4}{3}, 1, \dfrac{4}{5}, \ldots$

$a_n = \dfrac{4}{n}$

9. $5! = 5 \cdot 4 \cdot 3 \cdot 2 \cdot 1 = 120$

11. $\dfrac{3! \, 5!}{6!} = \dfrac{(3 \cdot 2 \cdot 1)5!}{6 \cdot 5!} = 1$

13. $\displaystyle\sum_{i=1}^{6} 5 = 6(5) = 30$

15. $\displaystyle\sum_{j=1}^{4} \frac{6}{j^2} = \frac{6}{1^2} + \frac{6}{2^2} + \frac{6}{3^2} + \frac{6}{4^2} = 6 + \frac{3}{2} + \frac{2}{3} + \frac{3}{8} = \frac{205}{24}$

17. $\displaystyle\sum_{k=1}^{10} 2k^3 = 2(1)^3 + 2(2)^3 + 2(3)^3 + \cdots + 2(10)^3 = 6050$

19. $\dfrac{1}{2(1)} + \dfrac{1}{2(2)} + \dfrac{1}{2(3)} + \cdots + \dfrac{1}{2(20)} = \displaystyle\sum_{k=1}^{20} \frac{1}{2k}$

21. $\displaystyle\sum_{i=1}^{\infty} \frac{5}{10^i} = 0.5 + 0.05 + 0.005 + 0.0005 + \cdots = 0.5555\cdots = \frac{5}{9}$

23. $\displaystyle\sum_{k=1}^{\infty} \frac{2}{100^k} = 0.02 + 0.0002 + 0.000002 + \cdots = 0.020202\cdots = \frac{2}{99}$

25. $A_n = 10{,}000\left(1 + \dfrac{0.08}{12}\right)^n$

    (a) $A_1 \approx \$10{,}066.67$ $A_6 \approx \$10{,}406.73$

         $A_2 \approx \$10{,}133.78$ $A_7 \approx \$10{,}476.10$

         $A_3 \approx \$10{,}201.34$ $A_8 \approx \$10{,}545.95$

         $A_4 \approx \$10{,}269.35$ $A_9 \approx \$10{,}616.25$

         $A_5 \approx \$10{,}337.81$ $A_{10} \approx \$10{,}687.03$

    (b) $A_{120} \approx \$22{,}196.40$

27. $5, 3, 1, -1, -3, \ldots$

Arithmetic sequence, $d = -2$

29. $\frac{1}{2}, 1, \frac{3}{2}, 2, \frac{5}{2}, \ldots$

Arithmetic sequence, $d = \frac{1}{2}$

31. $a_1 = 4,\ d = 3$

$a_1 = 4$

$a_2 = 4 + 3 = 7$

$a_3 = 7 + 3 = 10$

$a_4 = 10 + 3 = 13$

$a_5 = 13 + 3 = 16$

33. $a_1 = 25,\ a_{k+1} = a_k + 3$

$a_1 = 25$

$a_2 = 25 + 3 = 28$

$a_3 = 28 + 3 = 31$

$a_4 = 31 + 3 = 34$

$a_5 = 34 + 3 = 37$

35. $a_1 = 7,\ d = 12$

$a_n = 7 + (n-1)12$

$= 7 + 12n - 12$

$= 12n - 5$

37. $a_1 = y,\ d = 3y$

$a_n = y + (n-1)3y$

$= y + 3ny - 3y$

$= 3ny - 2y$

39. $a_2 = 93,\ a_6 = 65$

$a_6 = a_2 + 4d \Rightarrow 65 = 93 + 4d \Rightarrow -28 = 4d \Rightarrow d = -7$

$a_1 = a_2 - d \Rightarrow a_1 = 93 - (-7) = 100$

$a_n = a_1 + (n-1)d = 100 + (n-1)(-7) = -7n + 107$

41. $\displaystyle\sum_{j=1}^{10} (2j - 3)$ is arithmetic. Therefore, $a_1 = -1,\ a_{10} = 17,\ S_{10} = \frac{10}{2}[-1 + 17] = 80$.

43. $\displaystyle\sum_{k=1}^{11} \left(\frac{2}{3}k + 4\right)$ is arithmetic. Therefore, $a_1 = \frac{14}{3},\ a_{11} = \frac{34}{3},\ S_{11} = \frac{11}{2}\left[\frac{14}{3} + \frac{34}{3}\right] = 88$.

45. $\sum_{k=1}^{100} 5k$ is arithmetic. Therefore, $a_1 = 5$, $a_{100} = 500$, $S_{500} = \frac{100}{2}(5 + 500) = 25{,}250$.

47. $a_n = 34{,}000 + (n-1)(2250)$

(a) $a_5 = 34{,}000 + 4(2250) = \$43{,}000$

(b) $S_5 = \frac{5}{2}(34{,}000 + 43{,}000) = \$192{,}500$

49. $5, 10, 20, 40, \ldots$

The sequence *is* geometric, $r = 2$

51. $\frac{1}{3}, -\frac{2}{3}, \frac{4}{3}, -\frac{8}{3}, \ldots$

The sequence *is* geometric, $r = -2$

53. $a_1 = 4$, $r = -\frac{1}{4}$

$a_1 = 4$

$a_2 = 4\left(-\frac{1}{4}\right) = -1$

$a_3 = -1\left(-\frac{1}{4}\right) = \frac{1}{4}$

$a_4 = \frac{1}{4}\left(-\frac{1}{4}\right) = -\frac{1}{16}$

$a_5 = -\frac{1}{16}\left(-\frac{1}{4}\right) = \frac{1}{64}$

55. $a_1 = 9$, $a_3 = 4$

$a_3 = a_1 r^2$

$4 = 9r^2$

$\frac{4}{9} = r^2 \implies r = \pm\frac{2}{3}$

$a_1 = 9$	$a_1 = 9$
$a_2 = 9\left(\frac{2}{3}\right) = 6$	$a_2 = 9\left(-\frac{2}{3}\right) = -6$
$a_3 = 6\left(\frac{2}{3}\right) = 4$ or	$a_3 = -6\left(-\frac{2}{3}\right) = 4$
$a_4 = 4\left(\frac{2}{3}\right) = \frac{8}{3}$	$a_4 = 4\left(-\frac{2}{3}\right) = -\frac{8}{3}$
$a_5 = \frac{8}{3}\left(\frac{2}{3}\right) = \frac{16}{9}$	$a_5 = -\frac{8}{3}\left(-\frac{2}{3}\right) = \frac{16}{9}$

57. $a_1 = 16$, $a_2 = -8$

$a_2 = a_1 r \implies -8 = 16r \implies r = -\frac{1}{2}$

$a_n = 16\left(-\frac{1}{2}\right)^{n-1}$

$a_{20} = 16\left(-\frac{1}{2}\right)^{19} \approx -3.052 \times 10^{-5}$

59. $a_1 = 100$, $r = 1.05$

$a_n = 100(1.05)^{n-1}$

$a_{20} = 100(1.05)^{19} \approx 252.695$

61. $\sum_{i=1}^{7} 2^{i-1} = \frac{1-2^7}{1-2} = 127$

63. $\sum_{i=1}^{4} \left(\frac{1}{2}\right)^i = \frac{1}{2} + \frac{1}{4} + \frac{1}{8} + \frac{1}{16} = \frac{15}{16}$

65. $\sum_{i=1}^{5} (2)^{i-1} = 1 + 2 + 4 + 8 + 16 = 31$

67. $\sum_{i=1}^{10} 10\left(\frac{3}{5}\right)^{i-1} \approx 24.85$

69. $\sum_{i=1}^{25} 100(1.06)^{i-1} \approx 5486.45$

71. $\sum_{i=1}^{\infty} \left(\frac{7}{8}\right)^{i-1} = \frac{1}{1-\frac{7}{8}} = 8$

73. $\sum_{i=1}^{\infty} (0.1)^{i-1} = \frac{1}{1-0.1} = \frac{10}{9}$

75. $\sum_{k=1}^{\infty} 4\left(\frac{2}{3}\right)^{k-1} = \frac{4}{1-\frac{2}{3}} = 12$

77. (a) $a_t = 120{,}000(0.7)^t$

(b) $a_5 = 120{,}000(0.7)^5$

$= \$20{,}168.40$

79. 1. When $n = 1, 3 = 1(1 + 2)$.

   2. Assume that $S_k = 3 + 5 + 7 + \cdots + (2k + 1) = k(k + 2)$.

   Then, $S_{k+1} = 3 + 5 + 7 + \cdots + (2k + 1) + [2(k + 1) + 1] = S_k + (2k + 3)$

   $$= k(k + 2) + 2k + 3$$

   $$= k^2 + 4k + 3$$

   $$= (k + 1)(k + 3)$$

   $$= (k + 1)[(k + 1) + 2].$$

Therefore, by mathematical induction, the formula is valid for all positive integer values of n.

81. 1. When $n = 1, a = a\left(\dfrac{1 - r}{1 - r}\right)$.

   2. Assume that $S_k = \displaystyle\sum_{i=0}^{k-1} ar^i = \dfrac{a(1 - r^k)}{1 - r}$.

   Then, $S_{k+1} = \displaystyle\sum_{i=0}^{k} ar^i = \left(\displaystyle\sum_{i=0}^{k-1} ar^i\right) + ar^k = \dfrac{a(1 - r^k)}{1 - r} + ar^k$

   $$= \dfrac{a(1 - r^k + r^k - r^{k+1})}{1 - r} = \dfrac{a(1 - r^{k+1})}{1 - r}.$$

Therefore, by mathematical induction, the formula is valid for all positive integer values of n.

83. $S_1 = 9 = 1(9) = 1[2(1) + 7]$

$S_2 = 9 + 13 = 22 = 2(11) = 2[2(2) + 7]$

$S_3 = 9 + 13 + 17 = 39 = 3(13) = 3[2(3) + 7]$

$S_4 = 9 + 13 + 17 + 21 = 60 = 4(15) = 4[2(4) + 7]$

$S_n = n(2n + 7)$

85. $S_1 = 1$

$S_2 = 1 + \dfrac{3}{5} = \dfrac{8}{5}$

$S_3 = 1 + \dfrac{3}{5} + \dfrac{9}{25} = \dfrac{49}{25}$

$S_4 = 1 + \dfrac{3}{5} + \dfrac{9}{25} + \dfrac{27}{125} = \dfrac{272}{125}$

Since the series is geometric,

$$S_n = \dfrac{1 - \left(\frac{3}{5}\right)^n}{1 - \frac{3}{5}} = \dfrac{5}{2}\left[1 - \left(\dfrac{3}{5}\right)^n\right].$$

87. $\displaystyle\sum_{n=1}^{30} n = \dfrac{30(31)}{2} = 465$

89. $\displaystyle\sum_{n=1}^{7} (n^4 - n) = \sum_{n=1}^{7} n^4 - \sum_{n=1}^{7} n = \dfrac{(7)(8)(15)[(3)(49) + 21 - 1]}{30} - \dfrac{(7)(8)}{2}$

   $$= \dfrac{(7)(8)(15)(167)}{30} - \dfrac{(7)(8)}{2}$$

   $$= 4676 - 28 = 4648$$

91. $a_1 = f(1) = 5, \quad a_n = a_{n-1} + 5$

$a_1 = 5$

$a_2 = 5 + 5 = 10$

$a_3 = 10 + 5 = 15$

$a_4 = 15 + 5 = 20$

$a_5 = 20 + 5 = 25$

n:	1	2	3	4	5
a_n:	5	10	15	20	25

First differences: 5 5 5 5

Second differences: 0 0 0

The sequence has a linear model.

93. $a_1 = f(1) = 16, \quad a_n = a_{n-1} - 1$

$a_1 = 16$

$a_2 = 16 - 1 = 15$

$a_3 = 15 - 1 = 14$

$a_4 = 14 - 1 = 13$

$a_5 = 13 - 1 = 12$

n:	1	2	3	4	5
a_n:	16	15	14	13	12

First differences: -1 -1 -1 -1

Second differences: 0 0 0

The sequence has a linear model.

95. $_6C_4 = \dfrac{6!}{2!4!} = 15$

97. $_8C_5 = \dfrac{8!}{3!5!} = 56$

99. $\dbinom{7}{3} = 35$

```
            1
          1   1
        1   2   1
      1   3   3   1
    1   4   6   4   1
  1   5  10  10   5   1
1   6  15  20  15   6   1
1   7  21  35  (35)  21   7   1
```

$\dbinom{7}{3} = 35$, the 5th entry in the 7th row

101. $\dbinom{8}{6} = 28$

```
            1
          1   1
        1   2   1
      1   3   3   1
    1   4   6   4   1
  1   5  10  10   5   1
1   6  15  20  15   6   1
1   7  21  35  35  21   7   1
1   8  28  56  70  56 (28)  8   1
```

$\dbinom{8}{6} = 28$, the 7th entry in the 8th row

103. $(x + 4)^4 = x^4 + 4x^3(4) + 6x^2(4)^2 + 4x(4)^3 + 4^4$

$\quad\quad\quad\quad = x^4 + 16x^3 + 96x^2 + 256x + 256$

105. $(a - 3b)^5 = a^5 - 5a^4(3b) + 10a^3(3b)^2 - 10a^2(3b)^3 + 5a(3b)^4 - (3b)^5$

$\quad\quad\quad\quad\quad = a^5 - 15a^4b + 90a^3b^2 - 270a^2b^3 + 405ab^4 - 243b^5$

107. $(5 + 2i)^4 = (5)^4 + 4(5)^3(2i) + 6(5)^2(2i)^2 + 4(5)(2i)^3 + (2i)^4$

$\quad\quad\quad\quad\quad = 625 + 1000i + 600i^2 + 160i^3 + 16i^4$

$\quad\quad\quad\quad\quad = 625 + 1000i - 600 - 160i + 16 = 41 + 840i$

109.

First number:	1	2	3	4	5	6	7	8	9	10	11
Second number:	11	10	9	8	7	6	5	4	3	2	1

From this list, you can see that a total of 12 occurs 11 different ways.

111. $(10)(10)(10)(10) = 10,000$ different telephone numbers

113. $_{10}P_3 = \dfrac{10!}{7!} = \dfrac{10 \cdot 9 \cdot 8 \cdot 7!}{7!}$

$\quad\quad\quad = 10 \cdot 9 \cdot 8 = 720$ different ways

115. $_8C_3 = \dfrac{8!}{5!3!} = 56$ **117.** $(1)\left(\dfrac{1}{9}\right) = \dfrac{1}{9}$ **119.** (a) $25\% + 18\% = 43\%$

(b) $100\% - 18\% = 82\%$

121. $\left(\dfrac{1}{6}\right)\left(\dfrac{1}{6}\right)\left(\dfrac{1}{6}\right) = \dfrac{1}{216}$ **123.** $1 - \dfrac{13}{52} = 1 - \dfrac{1}{4} = \dfrac{3}{4}$

125. True. $\dfrac{(n+2)!}{n!} = \dfrac{(n+2)(n+1)n!}{n!} = (n+2)(n+1)$ **127.** True. $\displaystyle\sum_{k=1}^{8} 3k = 3\sum_{k=1}^{8} k$ by the Properties of Sums.

129. False. If $r = 0$ or $r = 1$, then $_nP_r = {}_nC_r$.   **131.** (a) Odd-numbered terms are negative.

(b) Even-numbered terms are negative.

133. Each term of the sequence is defined in terms of preceding terms.   **135.** $a_n = 4\left(\dfrac{1}{2}\right)^{n-1}$

$a_1 = 4, a_2 = 2, a_{10} = \dfrac{1}{128}$

The sequence is geometric and is decreasing.

Matches graph (d).

137. $a_n = \displaystyle\sum_{k=1}^{n} 4\left(\dfrac{1}{2}\right)^{k-1}$

$a_1 = 4$ and $a_n \rightarrow 8$ as $n \rightarrow \infty$

Matches graph (b).

139. $S_6 = S_5 + S_4 + S_3 = 130 + 70 + 40 = 240$

$S_7 = S_6 + S_5 + S_4 = 240 + 130 + 70 = 440$

$S_8 = S_7 + S_6 + S_5 = 440 + 240 + 130 = 810$

$S_9 = S_8 + S_7 + S_6 = 810 + 440 + 240 = 1490$

$S_{10} = S_9 + S_8 + S_7 = 1490 + 810 + 440 = 2740$

Problem Solving for Chapter 9

1. $x_0 = 1$ and $x_n = \dfrac{1}{2}x_{n-1} + \dfrac{1}{x_{n-1}}, n = 1, 2, \ldots$

$x_0 = 1$

$x_1 = \dfrac{1}{2}(1) + \dfrac{1}{1} = \dfrac{3}{2} = 1.5$

$x_2 = \dfrac{1}{2}\left(\dfrac{3}{2}\right) + \dfrac{1}{3/2} = \dfrac{17}{12} = 1.41\overline{6}$

$x_3 = \dfrac{1}{2}\left(\dfrac{17}{12}\right) + \dfrac{1}{17/12} = \dfrac{577}{408} \approx 1.414215686$

$x_4 = \dfrac{1}{2}\left(\dfrac{577}{408}\right) + \dfrac{1}{577/408} \approx 1.414213562$

$x_5 = \dfrac{1}{2}x_4 + \dfrac{1}{x_4} \approx 1.414213562$

$x_6 \approx x_7 \approx x_8 \approx x_9 \approx 1.414213562$

Conjecture: $x_n \rightarrow \sqrt{2}$ as $n \rightarrow \infty$

3. $a_n = 3 + (-1)^n$

(a)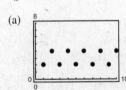

(b) $a_n = \begin{cases} 2, & \text{if } n \text{ is odd} \\ 4, & \text{if } n \text{ is even} \end{cases}$

(c)

n	1	10	101	1000	10,001
a_n	2	4	2	4	2

(d) As $n \rightarrow \infty$, a_n oscillates between 2 and 4 and does not approach a fixed value.

5. (a)

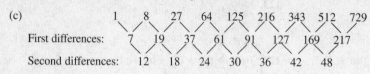

First differences: 3   5   7   9   11   13   15   17

In general, $b_n = 2n + 1$ for the first differences.

(b) Find the second differences of the perfect cubes.

(c)

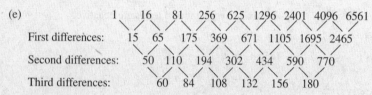

First differences: 7   19   37   61   91   127   169   217

Second differences: 12   18   24   30   36   42   48

In general, $c_n = 6(n + 1) = 6n + 6$ for the second differences.

(d) Find the third differences of the perfect fourth powers.

(e)   1   16   81   256   625   1296   2401   4096   6561

First differences: 15   65   175   369   671   1105   1695   2465

Second differences: 50   110   194   302   434   590   770

Third differences: 60   84   108   132   156   180

In general, $d_n = 24n + 36$ for the third differences.

7. Side lengths: $1, \dfrac{1}{2}, \dfrac{1}{4}, \dfrac{1}{8}, \ldots$

$$S_n = \left(\frac{1}{2}\right)^{n-1} \text{ for } n \geq 1$$

Areas: $\dfrac{\sqrt{3}}{4}, \dfrac{\sqrt{3}}{4}\left(\dfrac{1}{2}\right)^2, \dfrac{\sqrt{3}}{4}\left(\dfrac{1}{4}\right)^2, \dfrac{\sqrt{3}}{4}\left(\dfrac{1}{8}\right)^2, \ldots$

$$A_n = \frac{\sqrt{3}}{4}\left[\left(\frac{1}{2}\right)^{n-1}\right]^2 = \frac{\sqrt{3}}{4}\left(\frac{1}{2}\right)^{2n-2} = \frac{\sqrt{3}}{4}S_n^{\,2}$$

9. The numbers 1, 5, 12, 22, 35, 51, . . . can be written recursively as $P_n = P_{n-1} + (3n - 2)$. Show that $P_n = n(3n - 1)/2$.

1. For $n = 1$: $1 = \dfrac{1(3 - 1)}{2}$

2. Assume $P_k = \dfrac{k(3k - 1)}{2}$.

Then, $P_{k+1} = P_k + [3(k + 1) - 2]$

$$= \frac{k(3k - 1)}{2} + (3k + 1) = \frac{k(3k - 1) + 2(3k + 1)}{2}$$

$$= \frac{3k^2 + 5k + 2}{2} = \frac{(k + 1)(3k + 2)}{2}$$

$$= \frac{(k + 1)[3(k + 1) - 1]}{2}.$$

Therefore, by mathematical induction, the formula is valid for all integers $n \geq 1$.

11. (a) The Fibonacci sequence is defined as follows: $f_1 = 1, f_2 = 1, f_n = f_{n-2} + f_{n-1}$ for $n \geq 3$.

By this definition $f_3 = f_1 + f_2 = 2, f_4 = f_2 + f_3 = 3, f_5 = f_4 + f_3 = 5, f_6 = f_5 + f_4 = 8, \ldots$

1. For $n = 2$: $f_1 + f_2 = 2$ and $f_4 - 1 = 2$

2. Assume $f_1 + f_2 + \ldots + f_k = f_{k+2} - 1$.

Then, $f_1 + f_2 + f_3 + \ldots + f_k + f_{k+1} = f_{k+2} - 1 + f_{k+1} = (f_{k+2} + f_{k+1}) - 1 = f_{k+3} - 1 = f_{(k+1)+2} - 1$.

Therefore, by mathematical induction, the formula is valid for all integers $n \geq 2$.

(b) $S_{20} = f_{22} - 1 = 17,711 - 1 = 17,710$

13. $\dfrac{1}{3}$

15. (a) $V = \left(\dfrac{1}{_{47}C_5(27)}\right)(12,000,000) + \left(1 - \dfrac{1}{_{47}C_5(27)}\right)(-1)$

$\approx -\$0.71$

(b) $V = \dfrac{1}{36}(1) + \dfrac{1}{36}(4) + \dfrac{1}{36}(9) + \dfrac{1}{36}(16) + \dfrac{1}{36}(25) + \dfrac{1}{36}(36) + \dfrac{30}{36}(0) \approx 2.53$

$\dfrac{60}{2.53} \approx 24$ turns

Chapter 9   Practice Test

1. Write out the first five terms of the sequence $a_n = \dfrac{2n}{(n+2)!}$.

2. Write an expression for the nth term of the sequence $\frac{4}{3}, \frac{5}{9}, \frac{6}{27}, \frac{7}{81}, \frac{8}{243}, \ldots$.

3. Find the sum $\displaystyle\sum_{i=1}^{6}(2i-1)$.

4. Write out the first five terms of the arithmetic sequence where $a_1 = 23$ and $d = -2$.

5. Find a_n for the arithmetic sequence with $a_1 = 12$, $d = 3$, and $n = 50$.

6. Find the sum of the first 200 positive integers.

7. Write out the first five terms of the geometric sequence with $a_1 = 7$ and $r = 2$.

8. Evaluate $\displaystyle\sum_{n=1}^{10} 6\left(\dfrac{2}{3}\right)^{n-1}$.

9. Evaluate $\displaystyle\sum_{n=0}^{\infty}(0.03)^n$.

10. Use mathematical induction to prove that $1 + 2 + 3 + 4 + \cdots + n = \dfrac{n(n+1)}{2}$.

11. Use mathematical induction to prove that $n! > 2^n$, $n \geq 4$.

12. Evaluate $_{13}C_4$.

13. Expand $(x+3)^5$.

14. Find the term involving x^7 in $(x-2)^{12}$.

15. Evaluate $_{30}P_4$.

16. How many ways can six people sit at a table with six chairs?

17. Twelve cars run in a race. How many different ways can they come in first, second, and third place? (Assume that there are no ties.)

18. Two six-sided dice are tossed. Find the probability that the total of the two dice is less than 5.

19. Two cards are selected at random form a deck of 52 playing cards without replacement. Find the probability that the first card is a King and the second card is a black ten.

20. A manufacturer has determined that for every 1000 units it produces, 3 will be faulty. What is the probability that an order of 50 units will have one or more faulty units?

C H A P T E R 1 0
Topics in Analytic Geometry

Section 10.1 Lines **490**

Section 10.2 Introduction to Conics: Parabolas **494**

Section 10.3 Ellipses **499**

Section 10.4 Hyperbolas **505**

Section 10.5 Rotation of Conics **511**

Section 10.6 Parametric Equations **520**

Section 10.7 Polar Coordinates **527**

Section 10.8 Graphs of Polar Equations **532**

Section 10.9 Polar Equations of Conics **538**

Review Exercises **544**

Problem Solving **552**

Practice Test **555**

CHAPTER 10
Topics in Analytic Geometry

Section 10.1   Lines

- The **inclination** of a nonhorizontal line is the positive angle θ, $(\theta < 180°)$ measured counterclockwise from the x-axis to the line. A horizontal line has an inclination of zero.
- If a nonvertical line has inclination of θ and slope m, then $m = \tan\theta$.
- If two nonperpendicular lines have slopes m_1 and m_2, then the angle between the lines is given by

$$\tan\theta = \left|\frac{m_2 - m_1}{1 + m_1 m_2}\right|.$$

- The distance between a point (x_1, y_1) and a line $Ax + By + C = 0$ is given by

$$d = \frac{|Ax_1 + By_1 + C|}{\sqrt{A^2 + B^2}}.$$

Vocabulary Check

1. inclination

2. $\tan\theta$

3. $\left|\dfrac{m_2 - m_1}{1 + m_1 m_2}\right|$

4. $\dfrac{|Ax_1 + By_1 + C|}{\sqrt{A^2 + B^2}}$

---

1. $m = \tan\dfrac{\pi}{6} = \dfrac{\sqrt{3}}{3}$

3. $m = \tan\dfrac{3\pi}{4} = -1$

5. $m = \tan\dfrac{\pi}{3} = \sqrt{3}$

7. $m = \tan 1.27 \approx 3.2236$

9. $m = -1$

$-1 = \tan\theta$

$\theta = 180° + \arctan(-1)$

$\quad = \dfrac{3\pi}{4}$ radians $= 135°$

11. $m = 1$

$1 = \tan\theta$

$\theta = \dfrac{\pi}{4}$ radian $= 45°$

13. $m = \dfrac{3}{4}$

$\dfrac{3}{4} = \tan\theta$

$\theta = \arctan\left(\dfrac{3}{4}\right) \approx 0.6435$ radian $\approx 36.9°$

15. $(6, 1), (10, 8)$

$m = \dfrac{8 - 1}{10 - 6} = \dfrac{7}{4}$

$\dfrac{7}{4} = \tan\theta$

$\theta = \arctan\left(\dfrac{7}{4}\right) \approx 1.0517$ radians $\approx 60.3°$

17. $(-2, 20), (10, 0)$

$m = \dfrac{0 - 20}{10 - (-2)} = -\dfrac{20}{12} = -\dfrac{5}{3}$

$-\dfrac{5}{3} = \tan\theta$

$\theta = \pi + \arctan\left(-\dfrac{5}{3}\right) \approx 2.1112$ radians $\approx 121.0°$

19. $6x - 2y + 8 = 0$

$y = 3x + 4 \Rightarrow m = 3$

$3 = \tan\theta$

$\theta = \arctan 3 \approx 1.2490$ radians $\approx 71.6°$

21. $5x + 3y = 0$

$$y = -\frac{5}{3}x \Rightarrow m = -\frac{5}{3}$$

$$-\frac{5}{3} = \tan\theta$$

$$\theta = \pi + \arctan\left(-\frac{5}{3}\right) \approx 2.1112 \text{ radians} \approx 121.0°$$

23. $3x + y = 3 \Rightarrow y = -3x + 3 \Rightarrow m_1 = -3$

$x - y = 2 \Rightarrow y = x - 2 \quad \Rightarrow m_2 = 1$

$$\tan\theta = \left|\frac{1 - (-3)}{1 + (-3)(1)}\right| = 2$$

$$\theta = \arctan 2 \approx 1.1071 \text{ radians} \approx 63.4°$$

25. $x - y = 0 \Rightarrow y = x \quad \Rightarrow m_1 = 1$

$3x - 2y = -1 \Rightarrow y = \frac{3}{2}x + \frac{1}{2} \Rightarrow m_2 = \frac{3}{2}$

$$\tan\theta = \left|\frac{\frac{3}{2} - 1}{1 + \left(\frac{3}{2}\right)(1)}\right| = \frac{1}{5}$$

$$\theta = \arctan\frac{1}{5} \approx 0.1974 \text{ radian} \approx 11.3°$$

27. $x - 2y = 7 \Rightarrow y = \frac{1}{2}x - \frac{7}{2} \Rightarrow m_1 = \frac{1}{2}$

$6x + 2y = 5 \Rightarrow y = -3x + \frac{5}{2} \Rightarrow m_2 = -3$

$$\tan\theta = \left|\frac{-3 - \frac{1}{2}}{1 + \left(\frac{1}{2}\right)(-3)}\right| = 7$$

$$\theta = \arctan 7 \approx 1.4289 \text{ radians} \approx 81.9°$$

29. $x + 2y = 8 \Rightarrow y = -\frac{1}{2}x + 4 \Rightarrow m_1 = -\frac{1}{2}$

$x - 2y = 2 \Rightarrow y = \frac{1}{2}x - 1 \quad \Rightarrow m_2 = \frac{1}{2}$

$$\tan\theta = \left|\frac{\frac{1}{2} - \left(-\frac{1}{2}\right)}{1 + \left(-\frac{1}{2}\right)\left(\frac{1}{2}\right)}\right| = \frac{4}{3}$$

$$\theta = \arctan\left(\frac{4}{3}\right) \approx 0.9273 \text{ radian} \approx 53.1°$$

31. $0.05x - 0.03y = 0.21 \Rightarrow y = \frac{5}{3}x - 7 \Rightarrow m_1 = \frac{5}{3}$

$0.07x + 0.02y = 0.16 \Rightarrow y = -\frac{7}{2}x + 8 \Rightarrow m_2 = -\frac{7}{2}$

$$\tan\theta = \left|\frac{\left(-\frac{7}{2}\right) - \left(\frac{5}{3}\right)}{1 + \left(\frac{5}{3}\right)\left(-\frac{7}{2}\right)}\right| = \frac{31}{29}$$

$$\theta = \arctan\left(\frac{31}{29}\right) \approx 0.8187 \text{ radian} \approx 46.9°$$

33. Let $A = (2, 1)$, $B = (4, 4)$, and $C = (6, 2)$.

Slope of AB: $m_1 = \dfrac{1 - 4}{2 - 4} = \dfrac{3}{2}$

Slope of BC: $m_2 = \dfrac{4 - 2}{4 - 6} = -1$

Slope of AC: $m_3 = \dfrac{1 - 2}{2 - 6} = \dfrac{1}{4}$

$$\tan A = \left|\frac{\frac{1}{4} - \frac{3}{2}}{1 + \left(\frac{3}{2}\right)\left(\frac{1}{4}\right)}\right| = \frac{\frac{5}{4}}{\frac{11}{8}} = \frac{10}{11}$$

$$A = \arctan\left(\frac{10}{11}\right) \approx 42.3°$$

$$\tan B = \left|\frac{\frac{3}{2} - (-1)}{1 + (-1)\left(\frac{3}{2}\right)}\right| = \frac{\frac{5}{2}}{\frac{1}{2}} = 5$$

$$B = \arctan 5 \approx 78.7°$$

$$\tan C = \left|\frac{-1 - \frac{1}{4}}{1 + \left(\frac{1}{4}\right)(-1)}\right| = \frac{\frac{5}{4}}{\frac{3}{4}} = \frac{5}{3}$$

$$C = \arctan\left(\frac{5}{3}\right) \approx 59.0°$$

35. Let $A = (-4, -1)$, $B = (3, 2)$, and $C = (1, 0)$.

Slope of AB: $m_1 = \dfrac{-1 - 2}{-4 - 3} = \dfrac{3}{7}$

Slope of BC: $m_2 = \dfrac{2 - 0}{3 - 1} = 1$

Slope of AC: $m_3 = \dfrac{-1 - 0}{-4 - 1} = \dfrac{1}{5}$

$$\tan A = \left|\frac{\frac{1}{5} - \frac{3}{7}}{1 + \left(\frac{3}{7}\right)\left(\frac{1}{5}\right)}\right| = \frac{\frac{8}{35}}{\frac{38}{35}} = \frac{4}{19}$$

$$A = \arctan\left(\frac{4}{19}\right) \approx 11.9°$$

$$\tan B = \left|\frac{1 - \frac{3}{7}}{1 + \left(\frac{3}{7}\right)(1)}\right| = \frac{\frac{4}{7}}{\frac{10}{7}} = \frac{2}{5}$$

$$B = \arctan\left(\frac{2}{5}\right) \approx 21.8°$$

$$C = 180° - A - B$$

$$\approx 180° - 11.9° - 21.8° = 146.3°$$

37. $(0, 0) \implies x_1 = 0$ and $y_1 = 0$

$4x + 3y = 0 \implies A = 4, B = 3$, and $C = 0$

$$d = \frac{|4(0) + 3(0) + 0|}{\sqrt{4^2 + 3^2}} = \frac{0}{5} = 0$$

Note: The point is *on* the line.

39. $(2, 3) \implies x_1 = 2$ and $y_1 = 3$

$4x + 3y - 10 = 0 \implies A = 4, B = 3$, and $C = -10$

$$d = \frac{|4(2) + 3(3) + (-10)|}{\sqrt{4^2 + 3^2}} = \frac{7}{5}$$

41. $(6, 2) \implies x_1 = 6$ and $y_1 = 2$

$x + 1 = 0 \implies A = 1, B = 0$, and $C = 1$

$$d = \frac{|1(6) + 0(2) + 1|}{\sqrt{1^2 + 0^2}} = 7$$

43. $(0, 8) \implies x_1 = 0$ and $y_1 = 8$

$6x - y = 0 \implies A = 6, B = -1$, and $C = 0$

$$d = \frac{|6(0) + (-1)(8) + 0|}{\sqrt{6^2 + (-1)^2}}$$

$$= \frac{8}{\sqrt{37}} = \frac{8\sqrt{37}}{37} \approx 1.3152$$

45. $A = (0, 0), B = (1, 4), C = (4, 0)$

(a)

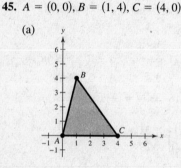

(b) The slope the line through AC is $m = \dfrac{0 - 0}{4 - 0} = 0$.

The equation of the line through AC is $y = 0$.

The distance between the line and $B = (1, 4)$ is

$$d = \frac{|0(1) + (1)(4) + 0|}{\sqrt{0^2 + 1^2}} = 4.$$

(c) The distance between A and C is 4.

$$A = \frac{1}{2}(4)(4) = 8 \text{ square units}$$

47. $A = \left(-\dfrac{1}{2}, \dfrac{1}{2}\right), B = (2, 3), C = \left(\dfrac{5}{2}, 0\right)$

(a)

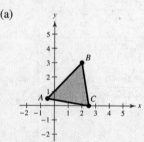

(b) The slope of the line through AC is $m = \dfrac{\frac{1}{2} - 0}{\left(-\frac{1}{2}\right) - \frac{5}{2}} = -\dfrac{1}{6}$.

The equation of the line through AC is $y - 0 = -\dfrac{1}{6}\left(x - \dfrac{5}{2}\right) \implies 2x + 12y - 5 = 0.$

The distance between the line and $B = (2, 3)$ is $d = \dfrac{|2(2) + 12(3) + (-5)|}{\sqrt{2^2 + 12^2}} = \dfrac{35}{\sqrt{148}} = \dfrac{35\sqrt{37}}{74}.$

(c) The distance between A and C is $d = \sqrt{\left[\left(-\dfrac{1}{2}\right) - \left(\dfrac{5}{2}\right)\right]^2 + \left[\left(\dfrac{1}{2}\right) - 0\right]^2} = \dfrac{\sqrt{37}}{2}.$

$$A = \frac{1}{2}\left(\frac{\sqrt{37}}{2}\right)\left(\frac{35\sqrt{37}}{74}\right) = \frac{35}{8} \text{ square units}$$

49. $x + y = 1 \Rightarrow (0, 1)$ is a point on the line $\Rightarrow x_1 = 0$
and $y_1 = 1$

$x + y = 5 \Rightarrow A = 1, B = 1,$ and $C = -5$

$d = \dfrac{|1(0) + 1(1) + (-5)|}{\sqrt{1^2 + 1^2}} = \dfrac{4}{\sqrt{2}} = 2\sqrt{2}$

51. Slope: $m = \tan 0.1 \approx 0.1003$

Change in elevation: $\sin 0.1 = \dfrac{x}{2(5280)}$

$x \approx 1054$ feet

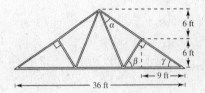

Not drawn to scale

53. Slope $= \frac{3}{5}$

Inclination $= \tan^{-1}\frac{3}{5} \approx 31.0°$

55. $\tan \gamma = \frac{6}{9}$

$\gamma = \arctan\left(\frac{2}{3}\right) \approx 33.69°$

$\beta = 90 - \gamma \approx 56.31°$

Also, since the right triangles containing α and β are equal, $\alpha = \gamma \approx 33.69°$.

57. True. The inclination of a line is related to its slope by $m = \tan \theta$. If the angle is greater than $\pi/2$ but less than π, then the angle is in the second quadrant where the tangent function is negative.

59. (a) $(0, 0) \Rightarrow x_1 = 0$ and $y_1 = 0$

$y = mx + 4 \Rightarrow 0 = mx - y + 4$

$d = \dfrac{|m(0) + (-1)(0) + 4|}{\sqrt{m^2 + (-1)^2}} = \dfrac{4}{\sqrt{m^2 + 1}}$

(b)

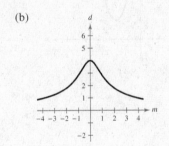

(c) The maximum distance of 4 occurs when the slope m is 0 and the line through $(0, 4)$ is horizontal.

(d) The graph has a horizontal asymptote at $d = 0$. As the slope becomes larger, the distance between the origin and the line, $y = mx + 4$, becomes smaller and approaches 0.

61. $f(x) = (x - 7)^2$

x-intercept: $0 = (x - 7)^2 \Rightarrow x = 7$

$(7, 0)$

y-intercept: $y = (0 - 7)^2 = 49$

$(0, 49)$

63. $f(x) = (x - 5)^2 - 5$

x-intercepts: $0 = (x - 5)^2 - 5$

$5 = (x - 5)^2$

$\pm\sqrt{5} = x - 5$

$5 \pm \sqrt{5} = x$

$\left(5 \pm \sqrt{5}, 0\right)$

y-intercept: $y = (0 - 5)^2 - 5 = 20$

$(0, 20)$

65. $f(x) = x^2 - 7x - 1$

x-intercepts: $0 = x^2 - 7x - 1$

$x = \dfrac{7 \pm \sqrt{53}}{2}$ by the Quadratic Formula

$\left(\dfrac{7 \pm \sqrt{53}}{2}, 0\right)$

y-intercept: $y = 0^2 - 7(0) - 1 = -1$

$(0, -1)$

67. $f(x) = 3x^2 + 2x - 16$

$= 3\left(x^2 + \dfrac{2}{3}x\right) - 16$

$= 3\left(x^2 + \dfrac{2}{3}x + \dfrac{1}{9}\right) - \dfrac{1}{3} - 16$

$= 3\left(x + \dfrac{1}{3}\right)^2 - \dfrac{49}{3}$

Vertex: $\left(-\dfrac{1}{3}, -\dfrac{49}{3}\right)$

69. $f(x) = 5x^2 + 34x - 7$

$= 5\left(x^2 + \dfrac{34}{5}x\right) - 7$

$= 5\left(x^2 + \dfrac{34}{5}x + \dfrac{289}{25}\right) - \dfrac{289}{5} - 7$

$= 5\left(x + \dfrac{17}{5}\right)^2 - \dfrac{324}{5}$

Vertex: $\left(-\dfrac{17}{5}, -\dfrac{324}{5}\right)$

71. $f(x) = 6x^2 - x - 12$

$= 6\left(x^2 - \dfrac{1}{6}x\right) - 12$

$= 6\left(x^2 - \dfrac{1}{6}x + \dfrac{1}{144}\right) - \dfrac{1}{24} - 12$

$= 6\left(x - \dfrac{1}{12}\right)^2 - \dfrac{289}{24}$

Vertex: $\left(\dfrac{1}{12}, -\dfrac{289}{24}\right)$

73. $f(x) = (x - 4)^2 + 3$

Vertex: $(4, 3)$

y-intercept: $(0, 19)$

x-intercept: None

75. $g(x) = 2x^2 - 3x + 1$

$= 2\left(x^2 - \dfrac{3}{2}x + \dfrac{9}{16}\right) - \dfrac{9}{8} + 1$

$= 2\left(x - \dfrac{3}{4}\right)^2 - \dfrac{1}{8}$

Vertex: $\left(\dfrac{3}{4}, -\dfrac{1}{8}\right)$

y-intercept: $(0, 1)$

x-intercept: $\left(\dfrac{1}{2}, 0\right), (1, 0)$

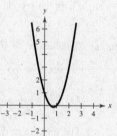

Section 10.2   Introduction to Conics: Parabolas

- A **parabola** is the set of all points (x, y) that are equidistant from a fixed line (**directrix**) and a fixed point (**focus**) not on the line.

- The standard equation of a parabola with vertex (h, k) and:
  - (a) Vertical axis $x = h$ and directrix $y = k - p$ is: $(x - h)^2 = 4p(y - k), p \neq 0$
  - (b) Horizontal axis $y = k$ and directrix $x = h - p$ is: $(y - k)^2 = 4p(x - h), p \neq 0$

- The tangent line to a parabola at a point P makes **equal angles** with:
  - (a) the line through P and the focus.
  - (b) the axis of the parabola.

Vocabulary Check

1. conic

2. locus

3. parabola; directrix; focus

4. axis

5. vertex

6. focal chord

7. tangent

1. A circle is formed when a plane intersects the top or bottom half of a double-napped cone and is perpendicular to the axis of the cone.

3. A parabola is formed when a plane intersects the top or bottom half of a double-napped cone, is parallel to the side of the cone, and does not intersect the vertex.

5. $y^2 = -4x$

Vertex: $(0, 0)$

Opens to the left since p is negative; matches graph (e).

7. $x^2 = -8y$

Vertex: $(0, 0)$

Opens downward since p is negative; matches graph (d).

9. $(y - 1)^2 = 4(x - 3)$

Vertex: $(3, 1)$

Opens to the right since p is positive; matches graph (a).

11. $y = \frac{1}{2}x^2$

$x^2 = 2y$

$x^2 = 4\left(\frac{1}{2}\right)y \implies h = 0, k = 0, p = \frac{1}{2}$

Vertex: $(0, 0)$

Focus: $\left(0, \frac{1}{2}\right)$

Directrix: $y = -\frac{1}{2}$

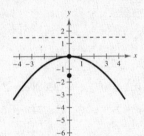

13. $y^2 = -6x$

$y^2 = 4\left(-\frac{3}{2}\right)x \implies h = 0, k = 0, p = -\frac{3}{2}$

Vertex: $(0, 0)$

Focus: $\left(-\frac{3}{2}, 0\right)$

Directrix: $x = \frac{3}{2}$

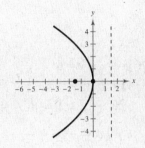

15. $x^2 + 6y = 0$

$x^2 = -6y = 4\left(-\frac{3}{2}\right)y \implies h = 0, k = 0, p = -\frac{3}{2}$

Vertex: $(0, 0)$

Focus: $\left(0, -\frac{3}{2}\right)$

Directrix: $y = \frac{3}{2}$

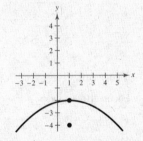

17. $(x - 1)^2 + 8(y + 2) = 0$

$(x - 1)^2 = 4(-2)(y + 2)$

$h = 1, k = -2, p = -2$

Vertex: $(1, -2)$

Focus: $(1, -4)$

Directrix: $y = 0$

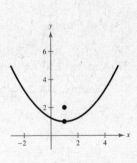

19. $\left(x + \frac{3}{2}\right)^2 = 4(y - 2)$

$\left(x + \frac{3}{2}\right)^2 = 4(1)(y - 2)$

$h = -\frac{3}{2}, k = 2, p = 1$

Vertex: $\left(-\frac{3}{2}, 2\right)$

Focus: $\left(-\frac{3}{2}, 3\right)$

Directrix: $y = 1$

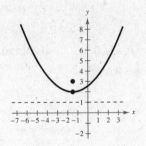

21.

$y = \frac{1}{4}(x^2 - 2x + 5)$

$4y = x^2 - 2x + 5$

$4y - 5 + 1 = x^2 - 2x + 1$

$4y - 4 = (x - 1)^2$

$(x - 1)^2 = 4(1)(y - 1)$

$h = 1, k = 1, p = 1$

Vertex: $(1, 1)$

Focus: $(1, 2)$

Directrix: $y = 0$

23. $y^2 + 6y + 8x + 25 = 0$

$$y^2 + 6y + 9 = -8x - 25 + 9$$

$$(y + 3)^2 = 4(-2)(x + 2)$$

$h = -2, k = -3, p = -2$

Vertex: $(-2, -3)$

Focus: $(-4, -3)$

Directrix: $x = 0$

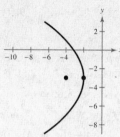

25. $x^2 + 4x + 6y - 2 = 0$

$$x^2 + 4x = -6y + 2$$

$$x^2 + 4x + 4 = -6y + 2 + 4$$

$$(x + 2)^2 = -6(y - 1)$$

$$(x + 2)^2 = 4\left(-\tfrac{3}{2}\right)(y - 1)$$

$h = -2, k = 1, p = -\tfrac{3}{2}$

Vertex: $(-2, 1)$

Focus: $\left(-2, -\tfrac{1}{2}\right)$

Directrix: $y = \tfrac{5}{2}$

On a graphing calculator, enter:

$$y_1 = -\tfrac{1}{6}(x^2 + 4x - 2)$$

27. $y^2 + x + y = 0$

$$y^2 + y + \tfrac{1}{4} = -x + \tfrac{1}{4}$$

$$\left(y + \tfrac{1}{2}\right)^2 = 4\left(-\tfrac{1}{4}\right)\left(x - \tfrac{1}{4}\right)$$

$h = \tfrac{1}{4}, k = -\tfrac{1}{2}, p = -\tfrac{1}{4}$

Vertex: $\left(\tfrac{1}{4}, -\tfrac{1}{2}\right)$

Focus: $\left(0, -\tfrac{1}{2}\right)$

Directrix: $x = \tfrac{1}{2}$

To use a graphing calculator, enter:

$$y_1 = -\tfrac{1}{2} + \sqrt{\tfrac{1}{4} - x}$$

$$y_2 = -\tfrac{1}{2} - \sqrt{\tfrac{1}{4} - x}$$

29. Vertex: $(0, 0) \Rightarrow h = 0, k = 0$

Graph opens upward.

$x^2 = 4py$

Point on graph: $(3, 6)$

$3^2 = 4p(6)$

$9 = 24p$

$\tfrac{3}{8} = p$

Thus, $x^2 = 4\left(\tfrac{3}{8}\right)y \Rightarrow x^2 = \tfrac{3}{2}y$.

31. Vertex: $(0, 0) \Rightarrow h = 0, k = 0$

Focus: $\left(0, -\tfrac{3}{2}\right) \Rightarrow p = -\tfrac{3}{2}$

$x^2 = 4py$

$x^2 = 4\left(-\tfrac{3}{2}\right)y$

$x^2 = -6y$

33. Vertex: $(0, 0) \Rightarrow h = 0, k = 0$

Focus: $(-2, 0) \Rightarrow p = -2$

$y^2 = 4px$

$y^2 = 4(-2)x$

$y^2 = -8x$

35. Vertex: $(0, 0) \Rightarrow h = 0, k = 0$

Directrix: $y = -1 \Rightarrow p = 1$

$x^2 = 4py$

$x^2 = 4(1)y$

$x^2 = 4y$

37. Vertex: $(0, 0) \Rightarrow h = 0, k = 0$

Directrix: $x = 2 \Rightarrow p = -2$

$y^2 = 4px$

$y^2 = 4(-2)x$

$y^2 = -8x$

39. Vertex: $(0, 0) \Rightarrow h = 0, k = 0$

Horizontal axis and passes through the point $(4, 6)$

$y^2 = 4px$

$6^2 = 4p(4)$

$36 = 16p \Rightarrow p = \tfrac{9}{4}$

$y^2 = 4\left(\tfrac{9}{4}\right)x$

$y^2 = 9x$

41. Vertex: $(3, 1)$ and opens downward. Passes through $(2, 0)$ and $(4, 0)$.

$$y = -(x - 2)(x - 4)$$

$$= -x^2 + 6x - 8$$

$$= -(x - 3)^2 + 1$$

$$(x - 3)^2 = -(y - 1)$$

43. Vertex: $(-4, 0)$ and opens to the right.

Passes through $(0, 4)$.

$$(y - 0)^2 = 4p(x + 4)$$
$$4^2 = 4p(0 + 4)$$
$$16 = 16p$$
$$1 = p$$
$$y^2 = 4(x + 4)$$

45. Vertex: $(5, 2)$

Focus: $(3, 2)$

Horizontal axis

$$p = 3 - 5 = -2$$
$$(y - 2)^2 = 4(-2)(x - 5)$$
$$(y - 2)^2 = -8(x - 5)$$

47. Vertex: $(0, 4)$

Directrix: $y = 2$

Vertical axis

$$p = 4 - 2 = 2$$
$$(x - 0)^2 = 4(2)(y - 4)$$
$$x^2 = 8(y - 4)$$

49. Focus: $(2, 2)$

Directrix: $x = -2$

Horizontal axis

Vertex: $(0, 2)$

$$p = 2 - 0 = 2$$
$$(y - 2)^2 = 4(2)(x - 0)$$
$$(y - 2)^2 = 8x$$

51. $(y - 3)^2 = 6(x + 1)$

For the upper half of the parabola:

$$y - 3 = \sqrt{6(x + 1)}$$
$$y = \sqrt{6(x + 1)} + 3$$

53. $y^2 - 8x = 0 \implies y = \pm\sqrt{8x}$

$$x - y + 2 = 0 \implies y = x + 2$$

The point of tangency is $(2, 4)$.

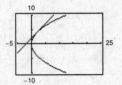

55. $x^2 = 2y \implies p = \dfrac{1}{2}$

Point: $(4, 8)$

Focus: $\left(0, \dfrac{1}{2}\right)$

$$d_1 = \dfrac{1}{2} - b$$
$$d_2 = \sqrt{(4 - 0)^2 + \left(8 - \dfrac{1}{2}\right)^2}$$
$$= \dfrac{17}{2}$$
$$d_1 = d_2 \implies b = -8$$

Slope: $m = \dfrac{8 - (-8)}{4 - 0} = 4$

$$y = 4x - 8 \implies 0 = 4x - y - 8$$

x-intercept: $(2, 0)$

57. $y = -2x^2 \implies x^2 = -\dfrac{1}{2}y \implies p = -\dfrac{1}{8}$

Point: $(-1, -2)$

Focus: $\left(0, -\dfrac{1}{8}\right)$

$$d_1 = b - \left(-\dfrac{1}{8}\right) = b + \dfrac{1}{8}$$
$$d_2 = \sqrt{(-1 - 0)^2 + \left(-2 - \left(-\dfrac{1}{8}\right)\right)^2}$$
$$= \dfrac{17}{8}$$
$$d_1 = d_2 \implies b = 2$$

Slope: $m = \dfrac{-2 - 2}{-1 - 0} = 4$

$$y = 4x + 2 \implies 0 = 4x - y + 2$$

x-intercept: $\left(-\dfrac{1}{2}, 0\right)$

59.
$$(x - 106)^2 = -\tfrac{4}{5}(R - 14{,}045)$$
$$x^2 - 212x + 11{,}236 = -\tfrac{4}{5}R + 11{,}236$$
$$R = 265x - \tfrac{5}{4}x^2$$

The revenue is maximum when $x = 106$ units.

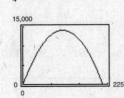

61. Vertex: $(0, 0) \implies h = 0, k = 0$

Focus: $(0, 4.5) \implies p = 4.5$

$$(x - h)^2 = 4p(y - k)$$
$$(x - 0)^2 = 4(4.5)(y - 0)$$
$$x^2 = 18y \text{ or } y = \tfrac{1}{18}x^2$$

63. (a) Vertex: $(0, 0) \implies h = 0, k = 0$

Points on the parabola: $(\pm 16, -0.4)$

$$x^2 = 4py$$

$$(\pm 16)^2 = 4p(-0.4)$$

$$256 = -1.6p$$

$$-160 = p$$

$$x^2 = 4(-160y)$$

$$x^2 = -640y$$

$$y = -\tfrac{1}{640}x^2$$

(b) When $y = -0.1$ we have

$$-0.1 = -\tfrac{1}{640}x^2$$

$$64 = x^2$$

$$\pm 8 = x.$$

Thus, 8 feet away from the center of the road, the road surface is 0.1 foot lower than in the middle.

65. (a) $V = 17,500\sqrt{2}$ mi/hr

$\approx 24,750$ mi/hr

(b) $p = -4100, (h, k) = (0, 4100)$

$$(x - 0)^2 = 4(-4100)(y - 4100)$$

$$x^2 = -16,400(y - 4100)$$

67. (a) $x^2 = -\dfrac{(32)^2}{16}(y - 75)$

$$x^2 = -64(y - 75)$$

(b) When $y = 0, x^2 = -64(-75) = 4800$.

Thus, $x = \sqrt{4800} = 40\sqrt{3} \approx 69.3$ feet.

69. False. It is not possible for a parabola to intersect its directrix. If the graph crossed the directrix there would exist points closer to the directrix than the focus.

71. (a)

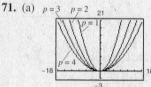

As p increases, the graph becomes wider.

(b) $(0, 1), (0, 2), (0, 3), (0, 4)$

(c) 4, 8, 12, 16. The chord passing through the focus and parallel to the directrix has length $|4p|$.

(d) This provides an easy way to determine two additional points on the graph, each of which is $|2p|$ units away from the focus on the chord.

73. $y - y_1 = \dfrac{x_1}{2p}(x - x_1)$

Slope: $m = \dfrac{x_1}{2p}$

75. $f(x) = x^3 - 2x^2 + 2x - 4$

Possible rational zeros: $\pm 1, \pm 2, \pm 4$

77. $f(x) = 2x^5 + x^2 + 16$

Possible rational zeros: $\pm 1, \pm 2, \pm 4, \pm 8, \pm 16, \pm \tfrac{1}{2}$

79. $f(x) = (x - 3)[x - (2 + i)][x - (2 - i)]$

$= (x - 3)[(x - 2) - i][(x - 2) + i]$

$= (x - 3)(x^2 - 4x + 5)$

$= x^3 - 7x^2 + 17x - 15$

81. $g(x) = 6x^4 + 7x^3 - 29x^2 - 28x + 20$

Possible rational roots: $\pm 1, \pm 2, \pm 4, \pm 5, \pm 10, \pm 20,$

$\pm \tfrac{1}{2}, \pm \tfrac{5}{2}, \pm \tfrac{1}{3}, \pm \tfrac{2}{3}, \pm \tfrac{4}{3}, \pm \tfrac{5}{3}, \pm \tfrac{10}{3}, \pm \tfrac{20}{3}, \pm \tfrac{1}{6}, \pm \tfrac{5}{6}$

$x = \pm 2$ are both solutions.

$$
\begin{array}{r|rrrrr}
2 & 6 & 7 & -29 & -28 & 20 \\
  &   & 12 & 38 & 18 & -20 \\
\hline
  & 6 & 19 & 9 & -10 & 0 \\
-2 & 6 & 19 & 9 & -10 & \\
   &   & -12 & -14 & 10 & \\
\hline
   & 6 & 7 & -5 & 0 &
\end{array}
$$

$g(x) = (x - 2)(x + 2)(6x^2 + 7x - 5)$

$= (x - 2)(x + 2)(2x - 1)(3x + 5)$

The zeros of $g(x)$ are $x = \pm 2, x = \tfrac{1}{2}, x = -\tfrac{5}{3}$.

83. $A = 35°, a = 10, b = 7$

$$\frac{\sin B}{7} = \frac{\sin 35°}{10} \implies \sin B \approx 0.4015 \implies B \approx 23.67°$$

$$C \approx 180° - 35° - 23.67° = 121.33°$$

$$\frac{c}{\sin 121.33°} = \frac{10}{\sin 35°} \implies c \approx 14.89$$

85. $A = 40°, B = 51°, c = 3$

$$C = 180° - 40° - 51° = 89°$$

$$\frac{a}{\sin 40°} = \frac{3}{\sin 89°} \implies a \approx 1.93$$

$$\frac{b}{\sin 51°} = \frac{3}{\sin 89°} \implies b \approx 2.33$$

87. $a = 7, b = 10, c = 16$

$$\cos C = \frac{7^2 + 10^2 - 16^2}{2(7)(10)} \approx -0.7643 \implies C \approx 139.84°$$

$$\frac{\sin B}{10} = \frac{\sin 139.84°}{16} \implies \sin B \approx 0.4031 \implies B \approx 23.77°$$

$$A = 180° - B - C \implies A \approx 16.39°$$

89. $A = 65°, b = 5, c = 12$

$$a^2 = 5^2 + 12^2 - 2(5)(12) \cos 65° \implies a \approx 10.8759 \approx 10.88$$

$$\frac{\sin B}{5} = \frac{\sin 65°}{10.8759} \implies \sin B \approx 0.4167 \implies B \approx 24.62°$$

$$C = 180° - A - B \implies C \approx 90.38°$$

Section 10.3   Ellipses

- An **ellipse** is the set of all points (x, y) the sum of whose distances from two distinct fixed points (**foci**) is constant.

- The standard equation of an ellipse with center (h, k) and major and minor axes of lengths $2a$ and $2b$ is:

  (a) $\dfrac{(x - h)^2}{a^2} + \dfrac{(y - k)^2}{b^2} = 1$ if the major axis is horizontal.

  (b) $\dfrac{(x - h)^2}{b^2} + \dfrac{(y - k)^2}{a^2} = 1$ is the major axis is vertical.

- $c^2 = a^2 - b^2$ where c is the distance from the center to a focus.

- The eccentricity of an ellipse is $e = \dfrac{c}{a}$.

Vocabulary Check

1. ellipse; foci

2. major axis, center

3. minor axis

4. eccentricity

1. $\dfrac{x^2}{4} + \dfrac{y^2}{9} = 1$

Center: $(0, 0)$

$a = 3, b = 2$

Vertical major axis

Matches graph (b).

3. $\dfrac{x^2}{4} + \dfrac{y^2}{25} = 1$

Center: $(0, 0)$

$a = 5, b = 2$

Vertical major axis

Matches graph (d).

5. $\dfrac{(x - 2)^2}{16} + (y + 1)^2 = 1$

Center: $(2, -1)$

$a = 4, b = 1$

Horizontal major axis

Matches graph (a).

7. $\dfrac{x^2}{25} + \dfrac{y^2}{16} = 1$

Ellipse

Center: $(0, 0)$

$a = 5, b = 4, c = 3$

Vertices: $(\pm 5, 0)$

Foci: $(\pm 3, 0)$

$e = \dfrac{3}{5}$

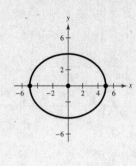

9. $\dfrac{x^2}{25} + \dfrac{y^2}{25} = 1 \Rightarrow x^2 + y^2 = 25$

Circle

Center: $(0, 0)$

Radius: 5

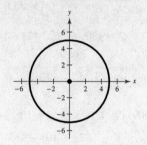

11. $\dfrac{x^2}{5} + \dfrac{y^2}{9} = 1$

Ellipse

$a = 3, b = \sqrt{5}, c = 2$

Center: $(0, 0)$

Vertices: $(0, \pm 3)$

Foci: $(0, \pm 2)$

$e = \dfrac{2}{3}$

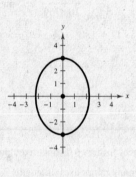

13. $\dfrac{(x + 3)^2}{16} + \dfrac{(y - 5)^2}{25} = 1$

Ellipse

$a = 5, b = 4, c = 3$

Center: $(-3, 5)$

Vertices: $(-3, 10)(-3, 0)$

Foci: $(-3, 8)(-3, 2)$

$e = \dfrac{3}{5}$

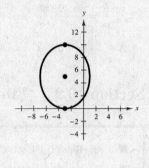

15. $\dfrac{x^2}{4/9} + \dfrac{(y + 1)^2}{4/9} = 1 \Rightarrow x^2 + (y + 1)^2 = \dfrac{4}{9}$

Circle

Center: $(0, -1)$

Radius: $\dfrac{2}{3}$

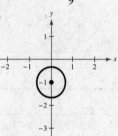

17. $\dfrac{(x + 2)^2}{1} + \dfrac{(y + 4)^2}{1/4} = 1$

Ellipse

$a = 1, b = \dfrac{1}{2}, c = \dfrac{\sqrt{3}}{2}$

Center: $(-2, -4)$

Vertices: $(-1, -4), (-3, -4)$

Foci: $\left(-2 \pm \dfrac{\sqrt{3}}{2}, -4\right) = \left(\dfrac{-4 \pm \sqrt{3}}{2}, -4\right)$

$e = \dfrac{\sqrt{3}}{2}$

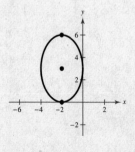

19. $\quad 9x^2 + 4y^2 + 36x - 24y + 36 = 0$

$9(x^2 + 4x + 4) + 4(y^2 - 6y + 9) = -36 + 36 + 36$

$9(x + 2)^2 + 4(y - 3)^2 = 36$

$$\dfrac{(x + 2)^2}{4} + \dfrac{(y - 3)^2}{9} = 1$$

Ellipse

$a = 3, b = 2, c = \sqrt{5}$

Center: $(-2, 3)$

Vertices: $(-2, 6), (-2, 0)$

Foci: $\left(-2, 3 \pm \sqrt{5}\right)$

$e = \dfrac{\sqrt{5}}{3}$

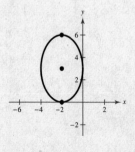

21. $\quad x^2 + y^2 - 2x + 4y - 31 = 0$

$(x^2 - 2x + 1) + (y^2 + 4y + 4) = 31 + 1 + 4$

$(x - 1)^2 + (y + 2)^2 = 36$

$$\dfrac{(x - 1)^2}{36} + \dfrac{(y + 2)^2}{36} = 1$$

Circle

Center: $(1, -2)$

Radius: 6

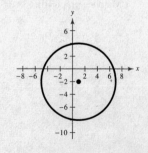

23. $\qquad 3x^2 + y^2 + 18x - 2y - 8 = 0$

$$3(x^2 + 6x + 9) + (y^2 - 2y + 1) = 8 + 27 + 1$$

$$3(x + 3)^2 + (y - 1)^2 = 36$$

$$\frac{(x + 3)^2}{12} + \frac{(y - 1)^2}{36} = 1$$

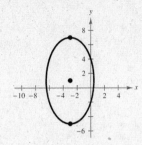

Ellipse

$a = 6, b = \sqrt{12} = 2\sqrt{3}, c = \sqrt{24} = 2\sqrt{6}$

Center: $(-3, 1)$

Vertices: $(-3, 7), (-3, -5)$

Foci: $\left(-3, 1 \pm 2\sqrt{6}\right)$

Eccentricity: $e = \dfrac{\sqrt{6}}{3}$

25. $\qquad x^2 + 4y^2 - 6x + 20y - 2 = 0$

$$(x^2 - 6x + 9) + 4\left(y^2 + 5y + \frac{25}{4}\right) = 2 + 9 + 25$$

$$(x - 3)^2 + 4\left(y + \frac{5}{2}\right)^2 = 36$$

$$\frac{(x - 3)^2}{36} + \frac{\left(y + \frac{5}{2}\right)^2}{9} = 1$$

Ellipse

$a = 6, b = 3, c = \sqrt{27} = 3\sqrt{3}$

Center: $\left(3, -\dfrac{5}{2}\right)$

Vertices: $\left(9, -\dfrac{5}{2}\right), \left(-3, -\dfrac{5}{2}\right)$

Foci: $\left(3 \pm 3\sqrt{3}, -\dfrac{5}{2}\right)$

Eccentricity: $e = \dfrac{\sqrt{3}}{2}$

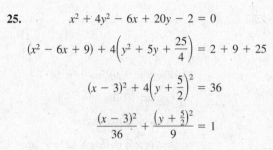

27. $\qquad 9x^2 + 9y^2 + 18x - 18y + 14 = 0$

$$9(x^2 + 2x + 1) + 9(y^2 - 2y + 1) = -14 + 9 + 9$$

$$9(x + 1)^2 + 9(y - 1)^2 = 4$$

$$(x + 1)^2 + (y - 1)^2 = \frac{4}{9}$$

$$\frac{(x + 1)^2}{4/9} + \frac{(y - 1)^2}{4/9} = 1$$

Circle

Center: $(-1, 1)$

Radius: $\dfrac{2}{3}$

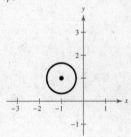

29. $\qquad 9x^2 + 25y^2 - 36x - 50y + 60 = 0$

$$9(x^2 - 4x + 4) + 25(y^2 - 2y + 1) = -60 + 36 + 25$$

$$9(x - 2)^2 + 25(y - 1)^2 = 1$$

$$\frac{(x - 2)^2}{1/9} + \frac{(y - 1)^2}{1/25} = 1$$

Ellipse

$a = \dfrac{1}{3}, b = \dfrac{1}{5}, c = \dfrac{4}{15}$

Center: $(2, 1)$

Vertices: $\left(\dfrac{5}{3}, 1\right), \left(\dfrac{7}{3}, 1\right)$

Foci: $\left(\dfrac{34}{15}, 1\right), \left(\dfrac{26}{15}, 1\right)$

Eccentricity: $e = \dfrac{4}{5}$

31. $5x^2 + 3y^2 = 15$

$$\frac{x^2}{3} + \frac{y^2}{5} = 1$$

Center: $(0, 0)$

$a = \sqrt{5}, b = \sqrt{3}, c = \sqrt{2}$

Foci: $\left(0, \pm\sqrt{2}\right)$

Vertices: $\left(0, \pm\sqrt{5}\right)$

$e = \dfrac{\sqrt{10}}{5}$

To graph, solve for y.

$$y^2 = \frac{15 - 5x^2}{3}$$

$$y_1 = \sqrt{\frac{15 - 5x^2}{3}}$$

$$y_2 = -\sqrt{\frac{15 - 5x^2}{3}}$$

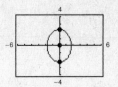

33. $12x^2 + 20y^2 - 12x + 40y - 37 = 0$

$$12\left(x^2 - x + \frac{1}{4}\right) + 20(y^2 + 2y + 1) = 37 + 3 + 20$$

$$12\left(x - \frac{1}{2}\right)^2 + 20(y + 1)^2 = 60$$

$$\frac{\left(x - \frac{1}{2}\right)^2}{5} + \frac{(y + 1)^2}{3} = 1$$

$a = \sqrt{5}, b = \sqrt{3}, c = \sqrt{2}$

Center: $\left(\dfrac{1}{2}, -1\right)$

Foci: $\left(\dfrac{1}{2} \pm \sqrt{2}, -1\right)$

Vertices: $\left(\dfrac{1}{2} \pm \sqrt{5}, -1\right)$

$e = \dfrac{\sqrt{10}}{5}$

To graph, solve for y.

$$(y + 1)^2 = 3\left[1 - \frac{(x - 0.5)^2}{5}\right]$$

$$y_1 = -1 + \sqrt{3\left[1 - \frac{(x - 0.5)^2}{5}\right]}$$

$$y_2 = -1 - \sqrt{3\left[1 - \frac{(x - 0.5)^2}{5}\right]}$$

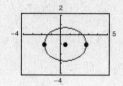

35. Center: $(0, 0)$

$a = 4, b = 2$

Vertical major axis

$$\frac{(x - h)^2}{b^2} + \frac{(y - k)^2}{a^2} = 1$$

$$\frac{x^2}{4} + \frac{y^2}{16} = 1$$

37. Vertices: $(\pm 6, 0)$

$a = 6, c = 2 \implies b = \sqrt{32} = 4\sqrt{2}$

Foci: $(\pm 2, 0)$

Horizontal major axis

Center: $(0, 0)$

$$\frac{(x - h)^2}{a^2} + \frac{(y - k)^2}{b^2} = 1$$

$$\frac{x^2}{36} + \frac{y^2}{32} = 1$$

39. Foci: $(\pm 5, 0) \implies c = 5$

Center: $(0, 0)$

Horizontal major axis

Major axis of length 12 $\implies 2a = 12$

$$a = 6$$

$6^2 - b^2 = 5^2 \implies b^2 = 11$

$$\frac{(x - h)^2}{a^2} + \frac{(y - k)^2}{b^2} = 1$$

$$\frac{x^2}{36} + \frac{y^2}{11} = 1$$

41. Vertices: $(0, \pm 5) \implies a = 5$

Center: $(0, 0)$

Vertical major axis

$$\frac{(x - h)^2}{b^2} + \frac{(y - k)^2}{a^2} = 1$$

$$\frac{x^2}{b^2} + \frac{y^2}{25} = 1$$

Point: $(4, 2)$

$$\frac{4^2}{b^2} + \frac{2^2}{25} = 1$$

$$\frac{16}{b^2} = 1 - \frac{4}{25} = \frac{21}{25}$$

$$400 = 21b^2$$

$$\frac{400}{21} = b^2$$

$$\frac{x^2}{400/21} + \frac{y^2}{25} = 1$$

$$\frac{21x^2}{400} + \frac{y^2}{25} = 1$$

43. Center: $(2, 3)$

$a = 3, b = 1$

Vertical major axis

$$\dfrac{(x - h)^2}{b^2} + \dfrac{(y - k)^2}{a^2} = 1$$

$$\dfrac{(x - 2)^2}{1} + \dfrac{(y - 3)^2}{9} = 1$$

45. Center: $(-2, 3)$

$a = 4, b = 3$

Horizontal major axis

$$\dfrac{(x - h)^2}{a^2} + \dfrac{(y - k)^2}{b^2} = 1$$

$$\dfrac{(x + 2)^2}{16} + \dfrac{(y - 3)^2}{9} = 1$$

47. Vertices: $(0, 4), (4, 4) \implies a = 2$

Minor axis of length $2 \implies b = 1$

Center: $(2, 4) = (h, k)$

$$\dfrac{(x - h)^2}{a^2} + \dfrac{(y - k)^2}{b^2} = 1$$

$$\dfrac{(x - 2)^2}{4} + \dfrac{(y - 4)^2}{1} = 1$$

49. Foci: $(0, 0), (0, 8) \implies c = 4$

Major axis of length $16 \implies a = 8$

$b^2 = a^2 - c^2 = 64 - 16 = 48$

Center: $(0, 4) = (h, k)$

$$\dfrac{(x - h)^2}{b^2} + \dfrac{(y - k)^2}{a^2} = 1$$

$$\dfrac{x^2}{48} + \dfrac{(y - 4)^2}{64} = 1$$

51. Center: $(0, 4)$

Vertices: $(-4, 4), (4, 4) \implies a = 4$

$a = 2c \implies 4 = 2c \implies c = 2$

$2^2 = 4^2 - b^2 \implies b^2 = 12$

Horizontal major axis

$$\dfrac{(x - h)^2}{a^2} + \dfrac{(y - k)^2}{b^2} = 1$$

$$\dfrac{x^2}{16} + \dfrac{(y - 4)^2}{12} = 1$$

53. Vertices: $(0, 2), (4, 2) \implies a = 2$

Center: $(2, 2)$

Endpoints of the minor axis: $(2, 3), (2, 1) \implies b = 1$

Horizontal major axis

$$\dfrac{(x - h)^2}{a^2} + \dfrac{(y - k)^2}{b^2} = 1$$

$$\dfrac{(x - 2)^2}{4} + \dfrac{(y - 2)^2}{1} = 1$$

55. Vertices: $(\pm 5, 0) \implies a = 5$

Eccentricity: $\dfrac{3}{5} \implies c = \dfrac{3}{5} a = 3$

$b^2 = a^2 - c^2 = 25 - 9 = 16$

Center: $(0, 0) = (h, k)$

$$\dfrac{(x - h)^2}{a^2} + \dfrac{(y - k)^2}{b^2} = 1$$

$$\dfrac{x^2}{25} + \dfrac{y^2}{16} = 1$$

57. (a)

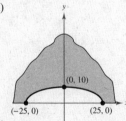

(b) $a = 25, b = 10$

$$\dfrac{x^2}{a^2} + \dfrac{y^2}{b^2} = 1$$

$$\dfrac{x^2}{625} + \dfrac{y^2}{100} = 1$$

(c) When $x = \pm 4$:

$$\dfrac{4^2}{625} + \dfrac{y^2}{100} = 1$$

$$y^2 = 100\left(1 - \dfrac{16}{625}\right) = \dfrac{2436}{25}$$

$$y = \sqrt{\dfrac{2436}{25}} \approx 9.87 \text{ feet} > 9 \text{ feet}$$

Yes. If the truck travels down the center of the tunnel, it will clear the opening of the arch.

59. (a) $a = \dfrac{35.88}{2} = 17.94$

$e = \dfrac{c}{a} = 0.967$

$c = ea \approx 17.35$

$b^2 = a^2 - c^2 \approx 20.82$

$$\dfrac{x^2}{a^2} + \dfrac{y^2}{b^2} = 1$$

$$\dfrac{x^2}{321.84} + \dfrac{y^2}{20.82} = 1$$

(b)

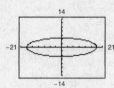

(c) The sun's center is at a focus of the orbit, 17.35 astronomical units from the center of the orbit.

Apogee $\approx 17.35 + \frac{1}{2}(35.88) = 35.29$ astronomical units

Perigee $\approx \frac{1}{2}(35.88) - 17.35 + \ = 0.59$ astronomical units

61. (a) The equation is the bottom half of the ellipse.

$$\dfrac{\theta^2}{(0.2)^2} + \dfrac{y^2}{(1.6)^2} = 1$$

$$y = -1.6\sqrt{1 - \dfrac{\theta^2}{0.04}}$$

$$= -8\sqrt{0.04 - \theta^2}$$

(b)

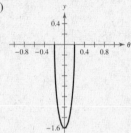

(c) The bottom half models the motion of the pendulum.

63. $\dfrac{x^2}{9} + \dfrac{y^2}{16} = 1$

$a = 4, b = 3, c = \sqrt{7}$

Points on the ellipse:

$(\pm 3, 0), (0, \pm 4)$

Length of latus recta:

$\dfrac{2b^2}{a} = \dfrac{2(3)^2}{4} = \dfrac{9}{2}$

Additional points: $\left(\pm\dfrac{9}{4}, -\sqrt{7}\right), \left(\pm\dfrac{9}{4}, \sqrt{7}\right)$

65. $5x^2 + 3y^2 = 15$

$$\dfrac{x^2}{3} + \dfrac{y^2}{5} = 1$$

$a = \sqrt{5}, b = \sqrt{3}, c = \sqrt{2}$

Points on the ellipse:

$\left(\pm\sqrt{3}, 0\right), \left(0, \pm\sqrt{5}\right)$

Length of latus recta:

$\dfrac{2b^2}{a} = \dfrac{2 \cdot 3}{\sqrt{5}} = \dfrac{6\sqrt{5}}{5}$

Additional points: $\left(\pm\dfrac{3\sqrt{5}}{5}, \pm\sqrt{2}\right)$

67. False. The graph of $\dfrac{x^2}{4} + y^4 = 1$ is not an ellipse.

The degree on y is 4, not 2.

69. $\dfrac{x^2}{a^2} + \dfrac{y^2}{b^2} = 1$

(a) $a + b = 20 \implies b = 20 - a$

$A = \pi a b = \pi a(20 - a)$

(b) $264 = \pi a(20 - a)$

$0 = -\pi a^2 + 20\pi a - 264$

$0 = \pi a^2 - 20\pi a + 264$

By the Quadratic Formula: $a \approx 14$ or $a \approx 6$. Choosing the larger value of a, we have $a \approx 14$ and $b \approx 6$. The equation of an ellipse with an area of 264 is

$$\dfrac{x^2}{196} + \dfrac{y^2}{36} = 1.$$

—**CONTINUED**—

69. **—CONTINUED—**

(c)

a	8	9	10	11	12	13
A	301.6	311.0	314.2	311.0	301.6	285.9

The area is maximum when $a = 10$ and the ellipse is a circle.

(d)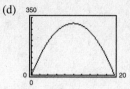

The area is maximum (314.16) when $a = b = 10$ and the ellipse is a circle.

71. $80, 40, 20, 10, 5, \ldots$

Geometric, $r = \frac{1}{2}$

73. $-\frac{1}{2}, \frac{1}{2}, \frac{3}{2}, \frac{5}{2}, \frac{7}{2}, \ldots$

Arithmetic, $d = 1$

75. $\displaystyle\sum_{n=0}^{6} (-3)^n = 1 - 3 + 9 - 27 + 81 - 243 + 729$

$= 547$

77. $\displaystyle\sum_{n=0}^{10} 5\left(\frac{4}{3}\right)^n = 5 \dfrac{\left(1 - \left(\frac{4}{3}\right)^{11}\right)}{1 - \frac{4}{3}}$

≈ 340.15

Section 10.4   Hyperbolas

- A **hyperbola** is the set of all points (x, y) the difference of whose distances from two distinct fixed points (**foci**) is constant.

- The standard equation of a hyperbola with center (h, k) and transverse and conjugate axes of lengths $2a$ and $2b$ is:

   (a) $\dfrac{(x - h)^2}{a^2} - \dfrac{(y - k)^2}{b^2} = 1$ if the traverse axis is horizontal.

   (b) $\dfrac{(y - k)^2}{a^2} - \dfrac{(x - h)^2}{b^2} = 1$ if the traverse axis is vertical.

- $c^2 = a^2 + b^2$ where c is the distance from the center to a focus.

- The asymptotes of a hyperbola are:

   (a) $y = k \pm \dfrac{b}{a}(x - h)$ if the transverse axis is horizontal.

   (b) $y = k \pm \dfrac{a}{b}(x - h)$ if the transverse axis is vertical.

- The eccentricity of a hyperbola is $e = \dfrac{c}{a}$.

- To classify a nondegenerate conic from its general equation $Ax^2 + Cy^2 + Dx + Ey + F = 0$:

   (a) If $A = C$ $(A \neq 0, C \neq 0)$, then it is a circle.

   (b) If $AC = 0$ $(A = 0$ or $C = 0$, but not both), then it is a parabola.

   (c) If $AC > 0$, then it is an ellipse.

   (d) If $AC < 0$, then it is a hyperbola.

Vocabulary Check

1. hyperbola

2. branches

3. transverse axis; center

4. asymptotes

5. $Ax^2 + Cy^2 + Dx + Ey + F = 0$

1. $\dfrac{y^2}{9} - \dfrac{x^2}{25} = 1$

Center: $(0, 0)$

$a = 3, b = 5$

Vertical transverse axis

Matches graph (b).

3. $\dfrac{(x - 1)^2}{16} - \dfrac{y^2}{4} = 1$

Center: $(1, 0)$

$a = 4, b = 2$

Horizontal transverse axis

Matches graph (a).

5. $x^2 - y^2 = 1$

$a = 1, b = 1, c = \sqrt{2}$

Center: $(0, 0)$

Vertices: $(\pm 1, 0)$

Foci: $\left(\pm \sqrt{2}, 0\right)$

Asymptotes: $y = \pm x$

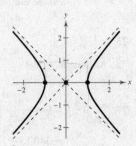

7. $\dfrac{y^2}{25} - \dfrac{x^2}{81} = 1$

$a = 5, b = 9, c = \sqrt{106}$

Center: $(0, 0)$

Vertices: $(0, \pm 5)$

Foci: $\left(0, \pm \sqrt{106}\right)$

Asymptotes: $y = \pm \dfrac{5}{9} x$

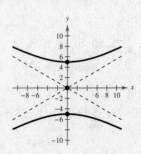

9. $\dfrac{(x - 1)^2}{4} - \dfrac{(y + 2)^2}{1} = 1$

$a = 2, b = 1, c = \sqrt{5}$

Center: $(1, -2)$

Vertices: $(-1, -2), (3, -2)$

Foci: $\left(1 \pm \sqrt{5}, -2\right)$

Asymptotes:

$y = -2 \pm \dfrac{1}{2}(x - 1)$

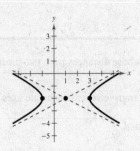

11. $\dfrac{(y + 6)^2}{1/9} - \dfrac{(x - 2)^2}{1/4} = 1$

$a = \dfrac{1}{3}, b = \dfrac{1}{2}, c = \dfrac{\sqrt{13}}{6}$

Center: $(2, -6)$

Vertices: $\left(2, -\dfrac{17}{3}\right), \left(2, -\dfrac{19}{3}\right)$

Foci: $\left(2, -6 \pm \dfrac{\sqrt{13}}{6}\right)$

Asymptotes: $y = -6 \pm \dfrac{2}{3}(x - 2)$

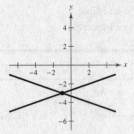

13. $\qquad 9x^2 - y^2 - 36x - 6y + 18 = 0$

$9(x^2 - 4x + 4) - (y^2 + 6y + 9) = -18 + 36 - 9$

$\qquad 9(x - 2)^2 - (y + 3)^2 = 9$

$\qquad \dfrac{(x - 2)^2}{1} - \dfrac{(y + 3)^2}{9} = 1$

$a = 1, b = 3, c = \sqrt{10}$

Center: $(2, -3)$

Vertices: $(1, -3), (3, -3)$

Foci: $\left(2 \pm \sqrt{10}, -3\right)$

Asymptotes:

$y = -3 \pm 3(x - 2)$

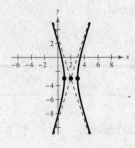

15. $\qquad x^2 - 9y^2 + 2x - 54y - 80 = 0$

$(x^2 + 2x + 1) - 9(y^2 + 6y + 9) = 80 + 1 - 81$

$\qquad (x + 1)^2 - 9(y + 3)^2 = 0$

$\qquad\qquad y + 3 = \pm \tfrac{1}{3}(x + 1)$

Degenerate hyperbola is two lines intersecting at $(-1, -3)$.

17. $2x^2 - 3y^2 = 6$

$$\frac{x^2}{3} - \frac{y^2}{2} = 1$$

$a = \sqrt{3}, b = \sqrt{2}, c = \sqrt{5}$

Center: $(0, 0)$

Vertices: $\left(\pm\sqrt{3}, 0\right)$

Foci: $\left(\pm\sqrt{5}, 0\right)$

Asymptotes: $y = \pm\sqrt{\dfrac{2}{3}}x = \pm\dfrac{\sqrt{6}}{3}x$

To use a graphing calculator, solve for y first.

$$y^2 = \frac{2x^2 - 6}{3}$$

$\left. \begin{array}{l} y_1 = \sqrt{\dfrac{2x^2 - 6}{3}} \\[3ex] y_2 = -\sqrt{\dfrac{2x^2 - 6}{3}} \end{array} \right\}$ Hyperbola

$\left. \begin{array}{l} y_3 = \dfrac{\sqrt{6}}{3}x \\[3ex] y_4 = -\dfrac{\sqrt{6}}{3}x \end{array} \right\}$ Asymptotes

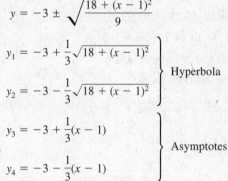

19. $9y^2 - x^2 + 2x + 54y + 62 = 0$

$9(y^2 + 6y + 9) - (x^2 - 2x + 1) = -62 - 1 + 81$

$9(y + 3)^2 - (x - 1)^2 = 18$

$$\frac{(y + 3)^2}{2} - \frac{(x - 1)^2}{18} = 1$$

$a = \sqrt{2}, b = 3\sqrt{2}, c = 2\sqrt{5}$

Center: $(1, -3)$

Vertices: $\left(1, -3 \pm \sqrt{2}\right)$

Foci: $\left(1, -3 \pm 2\sqrt{5}\right)$

Asymptotes: $y = -3 \pm \dfrac{1}{3}(x - 1)$

To use a graphing calculator, solve for y first.

$9(y + 3)^2 = 18 + (x - 1)^2$

$$y = -3 \pm \sqrt{\frac{18 + (x - 1)^2}{9}}$$

$\left. \begin{array}{l} y_1 = -3 + \dfrac{1}{3}\sqrt{18 + (x - 1)^2} \\[3ex] y_2 = -3 - \dfrac{1}{3}\sqrt{18 + (x - 1)^2} \end{array} \right\}$ Hyperbola

$\left. \begin{array}{l} y_3 = -3 + \dfrac{1}{3}(x - 1) \\[3ex] y_4 = -3 - \dfrac{1}{3}(x - 1) \end{array} \right\}$ Asymptotes

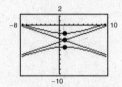

21. Vertices: $(0, \pm 2) \Rightarrow a = 2$

Foci: $(0, \pm 4) \Rightarrow c = 4$

$b^2 = c^2 - a^2 = 16 - 4 = 12$

Center: $(0, 0) = (h, k)$

$$\frac{(y - k)^2}{a^2} - \frac{(x - h)^2}{b^2} = 1$$

$$\frac{y^2}{4} - \frac{x^2}{12} = 1$$

23. Vertices: $(\pm 1, 0) \Rightarrow a = 1$

Asymptotes: $y = \pm 5x \Rightarrow \dfrac{b}{a} = 5, b = 5$

Center: $(0, 0) = (h, k)$

$$\frac{(x - h)^2}{a^2} - \frac{(y - k)^2}{b^2} = 1$$

$$\frac{x^2}{1} - \frac{y^2}{25} = 1$$

25. Foci: $(0, \pm 8) \implies c = 8$

Asymptotes: $y = \pm 4x \implies \dfrac{a}{b} = 4 \implies a = 4b$

Center: $(0, 0) = (h, k)$

$c^2 = a^2 + b^2 \implies 64 = 16b^2 + b^2$

$$\dfrac{64}{17} = b^2 \implies a^2 = \dfrac{1024}{17}$$

$$\dfrac{(y - k)^2}{a^2} - \dfrac{(x - h)^2}{b^2} = 1$$

$$\dfrac{y^2}{1024/17} - \dfrac{x^2}{64/17} = 1$$

$$\dfrac{17y^2}{1024} - \dfrac{17x^2}{64} = 1$$

27. Vertices: $(2, 0), (6, 0) \implies a = 2$

Foci: $(0, 0), (8, 0) \implies c = 4$

$b^2 = c^2 - a^2 = 16 - 4 = 12$

Center: $(4, 0) = (h, k)$

$$\dfrac{(x - h)^2}{a^2} - \dfrac{(y - k)^2}{b^2} = 1$$

$$\dfrac{(x - 4)^2}{4} - \dfrac{y^2}{12} = 1$$

29. Vertices: $(4, 1), (4, 9) \implies a = 4$

Foci: $(4, 0), (4, 10) \implies c = 5$

$b^2 = c^2 - a^2 = 25 - 16 = 9$

Center: $(4, 5) = (h, k)$

$$\dfrac{(y - k)^2}{a^2} - \dfrac{(x - h)^2}{b^2} = 1$$

$$\dfrac{(y - 5)^2}{16} - \dfrac{(x - 4)^2}{9} = 1$$

31. Vertices: $(2, 3), (2, -3) \implies a = 3$

Passes through the point: $(0, 5)$

Center: $(2, 0) = (h, k)$

$$\dfrac{(y - k)^2}{a^2} - \dfrac{(x - h)^2}{b^2} = 1$$

$$\dfrac{y^2}{9} - \dfrac{(x - 2)^2}{b^2} = 1 \implies$$

$$\dfrac{(x - 2)^2}{b^2} = \dfrac{y^2}{9} - 1 = \dfrac{y^2 - 9}{9} \implies$$

$$b^2 = \dfrac{9(x - 2)^2}{y^2 - 9} = \dfrac{9(-2)^2}{25 - 9} = \dfrac{36}{16} = \dfrac{9}{4}$$

$$\dfrac{y^2}{9} - \dfrac{(x - 2)^2}{9/4} = 1$$

$$\dfrac{y^2}{9} - \dfrac{4(x - 2)^2}{9} = 1$$

33. Vertices: $(0, 4), (0, 0) \implies a = 2$

Passes through the point $\left(\sqrt{5}, -1 \right)$

Center: $(0, 2) = (h, k)$

$$\dfrac{(y - k)^2}{a^2} - \dfrac{(x - h)^2}{b^2} = 1$$

$$\dfrac{(y - 2)^2}{4} - \dfrac{x^2}{b^2} = 1 \implies \dfrac{x^2}{b^2} = \dfrac{(y - 2)^2}{4} - 1 = \dfrac{(y - 2)^2 - 4}{4}$$

$$\implies b^2 = \dfrac{4x^2}{(y - 2)^2 - 4} = \dfrac{4\left(\sqrt{5} \right)^2}{(-1 - 2)^2 - 4} = \dfrac{20}{5} = 4$$

$$\dfrac{(y - 2)^2}{4} - \dfrac{x^2}{4} = 1$$

35. Vertices: $(1, 2), (3, 2) \implies a = 1$

Asymptotes: $y = x, y = 4 - x$

$\dfrac{b}{a} = 1 \implies \dfrac{b}{1} = 1 \implies b = 1$

Center: $(2, 2) = (h, k)$

$\dfrac{(x - h)^2}{a^2} - \dfrac{(y - k)^2}{b^2} = 1$

$\dfrac{(x - 2)^2}{1} - \dfrac{(y - 2)^2}{1} = 1$

37. Vertices: $(0, 2), (6, 2) \implies a = 3$

Asymptotes: $y = \dfrac{2}{3}x, y = 4 - \dfrac{2}{3}x$

$\dfrac{b}{a} = \dfrac{2}{3} \implies b = 2$

Center: $(3, 2) = (h, k)$

$\dfrac{(x - h)^2}{a^2} - \dfrac{(y - k)^2}{b^2} = 1$

$\dfrac{(x - 3)^2}{9} - \dfrac{(y - 2)^2}{4} = 1$

39. (a) Vertices: $(\pm 1, 0) \implies a = 1$

Horizontal transverse axis

Center: $(0, 0)$

$\dfrac{x^2}{a^2} - \dfrac{y^2}{b^2} = 1$

Point on the graph: $(2, 13)$

$\dfrac{2^2}{1^2} - \dfrac{13^2}{b^2} = 1$

$4 - \dfrac{169}{b^2} = 1$

$3b^2 = 169$

$b^2 = \dfrac{169}{3} \approx 56.33$

Thus we have $\dfrac{x^2}{1} - \dfrac{y^2}{56.33} = 1.$

(b) When $y = 5$: $x^2 = 1 + \dfrac{5^2}{56.33}$

$x = \sqrt{1 + \dfrac{25}{56.33}} \approx 1.2016$

Width: $2x \approx 2.403$ feet

41. Since listening station C heard the explosion 4 seconds after listening station A, and since listening station B heard the explosion one second after listening station A, and sound travels 1100 feet per second, the explosion is located in Quadrant IV on the line $x = 3300$. The locus of all points 4400 feet closer to A than C is one branch of the hyperbola.

$\dfrac{x^2}{a^2} - \dfrac{y^2}{b^2} = 1$ where $c = 3300$ feet and $a = \dfrac{4400}{2} = 2200$ feet, $b^2 = c^2 - a^2 = 6{,}050{,}000.$

When $x = 3300$ we have $\dfrac{3300^2}{2200^2} - \dfrac{y^2}{6{,}050{,}000} = 1.$

Solving for y: $y^2 = 6{,}050{,}000 \left(\dfrac{3300^2}{2200^2} - 1 \right)$

$= 7{,}562{,}500$

$y = \pm 2750$

Since the explosion is in Quadrant IV, its coordinates are $(3300, -2750).$

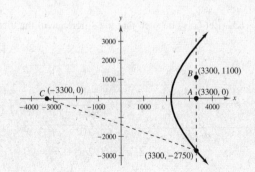

43. Center: $(0, 0) = (h, k)$

Focus: $(24, 0) \implies c = 24$

Solution point: $(24, 24)$

$24^2 = a^2 + b^2 \implies b^2 = 24^2 - a^2$

$\dfrac{(x - h)^2}{a^2} - \dfrac{(y - k)^2}{b^2} = 1$

$\dfrac{x^2}{a^2} - \dfrac{y^2}{24^2 - a^2} = 1 \implies \dfrac{24^2}{a^2} - \dfrac{24^2}{24^2 - a^2} = 1$

Solving yields $a = 12\sqrt{2\left(3 - \sqrt{5}\right)}$ OR

$12\left(\sqrt{5} - 1\right) \approx 14.83$ and $b^2 \approx 355.9876$.

Thus, we have $\dfrac{x^2}{220.0124} - \dfrac{y^2}{355.9876} = 1$.

The right vertex is at $(a, 0) \approx (14.83, 0)$.

45. $x^2 + y^2 - 6x + 4y + 9 = 0$

$A = 1, C = 1$

$A = C \implies$ Circle

47. $4x^2 - y^2 - 4x - 3 = 0$

$A = 4, C = -1$

$AC = (4)(-1) = -4 < 0 \implies$ Hyperbola

49. $y^2 - 4x^2 + 4x - 2y - 4 = 0$

$A = -4, C = 1$

$AC = (-4)(1) = -4 < 0 \implies$ Hyperbola

51. $x^2 - 4x - 8y + 2 = 0$

$A = 1, C = 0$

$AC = (1)(0) = 0 \implies$ Parabola

53. $4x^2 + 3y^2 + 8x - 24y + 51 = 0$

$A = 4, C = 3$

$AC = 4(3) = 12 > 0$ and $A \ne C \implies$ Ellipse

55. $25x^2 - 10x - 200y - 119 = 0$

$A = 25, C = 0$

$AC = 25(0) = 0 \implies$ Parabola

57. $4x^2 + 16y^2 - 4x - 32y + 1 = 0$

$A = 4, C = 16$

$AC = (4)(16) = 64 > 0$ and $A \ne C \implies$ Ellipse

59. $100x^2 + 100y^2 - 100x + 400y + 409 = 0$

$A = 100, C = 100$

$A = C \implies$ Circle

61. True. For a hyperbola, $c^2 = a^2 + b^2$ or

$e^2 = \dfrac{c^2}{a^2} = 1 + \dfrac{b^2}{a^2}$.

The larger the ratio of b to a, the larger the eccentricity $e = c/a$ of the hyperbola.

63. Let (x, y) be such that the difference of the distances from $(c, 0)$ and $(-c, 0)$ is $2a$ (again only deriving one of the forms).

$$2a = \left| \sqrt{(x + c)^2 + y^2} - \sqrt{(x - c)^2 + y^2} \right|$$

$$2a + \sqrt{(x - c)^2 + y^2} = \sqrt{(x + c)^2 + y^2}$$

$$4a^2 + 4a\sqrt{(x - c)^2 + y^2} + (x - c)^2 + y^2 = (x + c)^2 + y^2$$

$$4a\sqrt{(x - c)^2 + y^2} = 4cx - 4a^2$$

$$a\sqrt{(x - c)^2 + y^2} = cx - a^2$$

$$a^2(x^2 - 2cx + c^2 + y^2) = c^2x^2 - 2a^2cx + a^4$$

$$a^2(c^2 - a^2) = (c^2 - a^2)x^2 - a^2y^2$$

Let $b^2 = c^2 - a^2$. Then $a^2b^2 = b^2x^2 - a^2y^2 \implies 1 = \dfrac{x^2}{a^2} - \dfrac{y^2}{b^2}$.

65. $9x^2 - 54x - 4y^2 + 8y + 41 = 0$

$9(x^2 - 6x + 9) - 4(y^2 - 2y + 1) = -41 + 81 - 4$

$9(x - 3)^2 - 4(y - 1)^2 = 36$

$$\frac{(x - 3)^2}{4} - \frac{(y - 1)^2}{9} = 1$$

$$\frac{(y - 1)^2}{9} = \frac{(x - 3)^2}{4} - 1$$

$$(y - 1)^2 = 9\left[\frac{(x - 3)^2}{4} - 1\right]$$

The bottom half of the hyperbola is:

$$y - 1 = -\sqrt{9\left[\frac{(x - 3)^2}{4} - 1\right]}$$

$$y = 1 - 3\sqrt{\frac{(x - 3)^2}{4} - 1}$$

67. $x^3 - 16x = x(x^2 - 16) = x(x + 4)(x - 4)$

69. $2x^3 - 24x^2 + 72x = 2x(x^2 - 12x + 36) = 2x(x - 6)^2$

71. $16x^3 + 54 = 2(8x^3 + 27)$

$= 2[(2x)^3 + (3)^3]$

$= 2(2x + 3)(4x^2 - 6x + 9)$

73. $y = 2\cos x + 1$

Amplitude: 2

Period: 2π

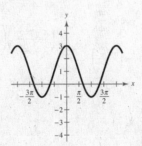

75. $y = \tan 2x$

Period: $\dfrac{\pi}{2}$

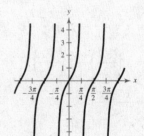

Section 10.5   Rotation of Conics

■ The general second-degree equation $Ax^2 + Bxy + Cy^2 + Dx + Ey + F = 0$ can be rewritten as $A'(x')^2 + C'(y')^2 + D'x' + E'y' + F' = 0$ by rotating the coordinate axes through the angle θ, where $\cot 2\theta = (A - C)/B$ and the following quantities are invariant under rotation:

   1. $F = F'$

   2. $A + C = A' + C'$

   3. $B^2 - 4AC = (B')^2 - 4A'C'$

■ $x = x'\cos\theta - y'\sin\theta$
  $y = x'\sin\theta + y'\cos\theta$

■ The graph of the nondegenerate equation $Ax^2 + Bxy + Cy^2 + Dx + Ey + F = 0$ is:

   (a) An ellipse or circle if $B^2 - 4AC < 0$.

   (b) A parabola if $B^2 - 4AC = 0$.

   (c) A hyperbola if $B^2 - 4AC > 0$.

Vocabulary Check

  1. rotation of axes

  3. invariant under rotation

  2. $A'(x')^2 + C'(y')^2 + D'x' + E'y' + F' = 0$

  4. discriminant

1. $\theta = 90°$; Point: $(0, 3)$

$x = x' \cos\theta - y' \sin\theta \qquad\qquad y = x' \sin\theta + y' \cos\theta$

$0 = x' \cos 90° - y' \sin 90° \qquad 3 = x' \sin 90° - y' \cos 90°$

$0 = y' \qquad\qquad\qquad\qquad\quad 3 = x'$

So, $(x', y') = (3, 0)$.

3. $\theta = 30°$; Point: $(1, 3)$

$\begin{aligned} x &= x' \cos\theta - y' \sin\theta \\ y &= x' \sin\theta + y' \cos\theta \end{aligned} \Rightarrow \begin{cases} 1 = x' \cos 30° - y' \sin 30° \\ 3 = x' \sin 30° + y' \cos 30° \end{cases}$

Solving the system yields $(x', y') = \left(\dfrac{3+\sqrt{3}}{2}, \dfrac{3\sqrt{3}-1}{2}\right)$.

5. $\theta = 45°$; Point $(2, 1)$

$\begin{aligned} x &= x' \cos\theta - y' \sin\theta \\ y &= x' \sin\theta + y' \cos\theta \end{aligned} \Rightarrow \begin{cases} 2 = x' \cos 45° - y' \sin 45° \\ 1 = x' \sin 45° + y' \cos 45° \end{cases}$

Solving the system yields $(x', y') = \left(\dfrac{3\sqrt{2}}{2}, -\dfrac{\sqrt{2}}{2}\right)$.

7. $xy + 1 = 0, A = 0, B = 1, C = 0$

$\cot 2\theta = \dfrac{A-C}{B} = 0 \Rightarrow 2\theta = \dfrac{\pi}{2} \Rightarrow \theta = \dfrac{\pi}{4}$

$x = x' \cos\dfrac{\pi}{4} - y' \sin\dfrac{\pi}{4} \qquad y = x' \sin\dfrac{\pi}{4} + y' \cos\dfrac{\pi}{4}$

$\quad = x'\left(\dfrac{\sqrt{2}}{2}\right) - y'\left(\dfrac{\sqrt{2}}{2}\right) \qquad = x'\left(\dfrac{\sqrt{2}}{2}\right) + y'\left(\dfrac{\sqrt{2}}{2}\right)$

$\quad = \dfrac{x'-y'}{\sqrt{2}} \qquad\qquad\qquad\quad = \dfrac{x'+y'}{\sqrt{2}}$

$xy + 1 = 0$

$\left(\dfrac{x'-y'}{\sqrt{2}}\right)\left(\dfrac{x'+y'}{\sqrt{2}}\right) + 1 = 0$

$\dfrac{(y')^2}{2} - \dfrac{(x')^2}{2} = 1$

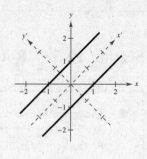

9. $x^2 - 2xy + y^2 - 1 = 0, A = 1, B = -2, C = 1$

$\cot 2\theta = \dfrac{A-C}{B} = 0 \Rightarrow 2\theta = \dfrac{\pi}{2} \Rightarrow \theta = \dfrac{\pi}{4}$

$x = x' \cos\dfrac{\pi}{4} - y' \sin\dfrac{\pi}{4} \qquad y = x' \sin\dfrac{\pi}{4} + y' \cos\dfrac{\pi}{4}$

$\quad = x'\left(\dfrac{\sqrt{2}}{2}\right) - y'\left(\dfrac{\sqrt{2}}{2}\right) \qquad = x'\left(\dfrac{\sqrt{2}}{2}\right) + y'\left(\dfrac{\sqrt{2}}{2}\right)$

$\quad = \dfrac{x'-y'}{\sqrt{2}} \qquad\qquad\qquad\quad = \dfrac{x'+y'}{\sqrt{2}}$

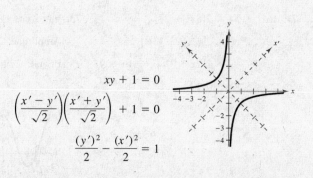

$x^2 - 2xy + y^2 - 1 = 0$

$\left(\dfrac{x'-y'}{\sqrt{2}}\right)^2 - 2\left(\dfrac{x'-y'}{\sqrt{2}}\right)\left(\dfrac{x'+y'}{\sqrt{2}}\right) + \left(\dfrac{x'+y'}{\sqrt{2}}\right)^2 - 1 = 0$

$\dfrac{(x')^2 - 2(x')(y') + (y')^2}{2} - \dfrac{2((x')^2 - (y')^2)}{2} + \dfrac{(x')^2 + 2(x')(y') + (y')^2}{2} - 1 = 0$

$2(y')^2 - 1 = 0$

$(y')^2 = \dfrac{1}{2}$

$y' = \pm\sqrt{\dfrac{1}{2}} = \pm\dfrac{\sqrt{2}}{2}$

The graph is two parallel lines.

Alternate solution:

$x^2 - 2xy + y^2 - 1 = 0$

$(x - y)^2 = 1$

$x - y = \pm 1$

$y = x \pm 1$

11. $xy - 2y - 4x = 0$

$A = 0, B = 1, C = 0$

$$\cot 2\theta = \frac{A - C}{B} = 0 \implies 2\theta = \frac{\pi}{2} \implies \theta = \frac{\pi}{4}$$

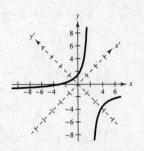

$$x = x'\cos\frac{\pi}{4} - y'\sin\frac{\pi}{4} \qquad\qquad y = x'\sin\frac{\pi}{4} + y'\cos\frac{\pi}{4}$$

$$= x'\left(\frac{\sqrt{2}}{2}\right) - y'\left(\frac{\sqrt{2}}{2}\right) \qquad\qquad = x'\left(\frac{\sqrt{2}}{2}\right) + y'\left(\frac{\sqrt{2}}{2}\right)$$

$$= \frac{x' - y'}{\sqrt{2}} \qquad\qquad\qquad\qquad = \frac{x' + y'}{\sqrt{2}}$$

$$xy - 2y - 4x = 0$$

$$\left(\frac{x' - y'}{\sqrt{2}}\right)\left(\frac{x' + y'}{\sqrt{2}}\right) - 2\left(\frac{x' + y'}{\sqrt{2}}\right) - 4\left(\frac{x' - y'}{\sqrt{2}}\right) = 0$$

$$\frac{(x')^2}{2} - \frac{(y')^2}{2} - \sqrt{2}x' - \sqrt{2}y' - 2\sqrt{2}x' + 2\sqrt{2}y' = 0$$

$$\left[(x')^2 - 6\sqrt{2}x' + \left(3\sqrt{2}\right)^2\right] - \left[(y')^2 - 2\sqrt{2}y' + \left(\sqrt{2}\right)^2\right] = 0 + \left(3\sqrt{2}\right)^2 - \left(\sqrt{2}\right)^2$$

$$\left(x' - 3\sqrt{2}\right)^2 - \left(y' - \sqrt{2}\right)^2 = 16$$

$$\frac{\left(x' - 3\sqrt{2}\right)^2}{16} - \frac{\left(y' - \sqrt{2}\right)^2}{16} = 1$$

13. $5x^2 - 6xy + 5y^2 - 12 = 0$

$A = 5, B = -6, C = 5$

$$\cot 2\theta = \frac{A - C}{B} = 0 \implies 2\theta = \frac{\pi}{2} \implies \theta = \frac{\pi}{4}$$

$$x = x'\cos\frac{\pi}{4} - y'\sin\frac{\pi}{4} \qquad\qquad y = x'\sin\frac{\pi}{4} + y'\cos\frac{\pi}{4}$$

$$= x'\left(\frac{\sqrt{2}}{2}\right) - y'\left(\frac{\sqrt{2}}{2}\right) \qquad\qquad = x'\left(\frac{\sqrt{2}}{2}\right) + y'\left(\frac{\sqrt{2}}{2}\right)$$

$$= \frac{x' - y'}{\sqrt{2}} \qquad\qquad\qquad\qquad = \frac{x' + y'}{\sqrt{2}}$$

$$5x^2 - 6xy + 5y^2 - 12 = 0$$

$$5\left(\frac{x' - y'}{\sqrt{2}}\right)^2 - 6\left(\frac{x' - y'}{\sqrt{2}}\right)\left(\frac{x' + y'}{\sqrt{2}}\right) + 5\left(\frac{x' + y'}{\sqrt{2}}\right)^2 - 12 = 0$$

$$\frac{5(x')^2}{2} - 5x'y' + \frac{5(y')^2}{2} - 3(x')^2 + 3(y')^2 + \frac{5(x')^2}{2} + 5x'y' + \frac{5(y')^2}{2} - 12 = 0$$

$$2(x')^2 + 8(y')^2 = 12$$

$$\frac{(x')^2}{6} + \frac{(y')^2}{3/2} = 1$$

15. $3x^2 - 2\sqrt{3}xy + y^2 + 2x + 2\sqrt{3}y = 0$

$A = 3, B = -2\sqrt{3}, C = 1$

$\cot 2\theta = \dfrac{A - C}{B} = -\dfrac{1}{\sqrt{3}} \implies \theta = 60°$

$x = x'\cos 60° - y'\sin 60°$

$= x'\left(\dfrac{1}{2}\right) - y'\left(\dfrac{\sqrt{3}}{2}\right) = \dfrac{x' - \sqrt{3}y'}{2}$

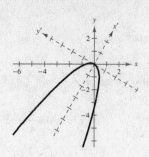

$$3x^2 - 2\sqrt{3}xy + y^2 + 2x + 2\sqrt{3}y = 0$$

$$3\left(\dfrac{x' - \sqrt{3}y'}{2}\right)^2 - 2\sqrt{3}\left(\dfrac{x' - \sqrt{3}y'}{2}\right)\left(\dfrac{\sqrt{3}x' + y'}{2}\right) + \left(\dfrac{\sqrt{3}x' + y'}{2}\right)^2 + 2\left(\dfrac{x' - \sqrt{3}y'}{2}\right) + 2\sqrt{3}\left(\dfrac{\sqrt{3}x' + y'}{2}\right) = 0$$

$$\dfrac{3(x')^2}{4} - \dfrac{6\sqrt{3}x'y'}{4} + \dfrac{9(y')^2}{4} - \dfrac{6(x')^2}{4} + \dfrac{4\sqrt{3}x'y'}{4} + \dfrac{6(y')^2}{4} + \dfrac{3(x')^2}{4} + \dfrac{2\sqrt{3}x'y'}{4} + \dfrac{(y')^2}{4}$$

$$+ x' - \sqrt{3}y' + 3x' + \sqrt{3}y' = 0$$

$$4(y')^2 + 4x' = 0$$

$$(y')^2 = -x'$$

17. $9x^2 + 24xy + 16y^2 + 90x - 130y = 0$

$A = 9, B = 24, C = 16$

$\cot 2\theta = \dfrac{A - C}{B} = -\dfrac{7}{24} \implies \theta \approx 53.13°$

$\cos 2\theta = -\dfrac{7}{25}$

$\sin \theta = \sqrt{\dfrac{1 - \cos 2\theta}{2}} = \sqrt{\dfrac{1 - \left(-\frac{7}{25}\right)}{2}} = \dfrac{4}{5}$

$\cos \theta = \sqrt{\dfrac{1 + \cos 2\theta}{2}} = \sqrt{\dfrac{1 + \left(-\frac{7}{25}\right)}{2}} = \dfrac{3}{5}$

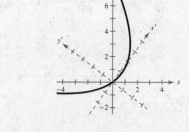

$x = x'\cos \theta - y'\sin \theta$

$\qquad = x'\left(\dfrac{3}{5}\right) - y'\left(\dfrac{4}{5}\right) = \dfrac{3x' - 4y'}{5}$

$y = x'\sin \theta + y'\cos \theta$

$\qquad = x'\left(\dfrac{4}{5}\right) + y'\left(\dfrac{3}{5}\right)$

$\qquad = \dfrac{4x' + 3y'}{5}$

$$9x^2 + 24xy + 16y^2 + 90x - 130y = 0$$

$$9\left(\dfrac{3x' - 4y'}{5}\right)^2 + 24\left(\dfrac{3x' - 4y'}{5}\right)\left(\dfrac{4x' + 3y'}{5}\right) + 16\left(\dfrac{4x' + 3y'}{5}\right)^2 + 90\left(\dfrac{3x' - 4y'}{5}\right) - 130\left(\dfrac{4x' + 3y'}{5}\right) = 0$$

$$\dfrac{81(x')^2}{25} - \dfrac{216x'y'}{25} + \dfrac{144(y')^2}{25} + \dfrac{288(x')^2}{25} - \dfrac{168x'y'}{25} - \dfrac{288(y')^2}{25} + \dfrac{256(x')^2}{25} + \dfrac{384x'y'}{25} + \dfrac{144(y')^2}{25}$$

$$+ 54x' - 72y' - 104x' - 78y' = 0$$

$$25(x')^2 - 50x' - 150y' = 0$$

$$(x')^2 - 2x' = 6y'$$

$$(x')^2 - 2x' + 1 = 6y' + 1$$

$$(x' - 1)^2 = 6\left(y' + \dfrac{1}{6}\right)$$

19. $x^2 + 2xy + y^2 = 20$

$A = 1, B = 2, C = 1$

$\cot 2\theta = \dfrac{A - C}{B} = \dfrac{1 - 1}{2} = 0 \implies \theta = \dfrac{\pi}{4}$ or $45°$

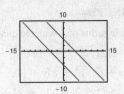

To graph the conic using a graphing calculator, we need to solve for y in terms of x.

$(x + y)^2 = 20$

$x + y = \pm\sqrt{20}$

$y = -x \pm \sqrt{20}$

Use $y_1 = -x + \sqrt{20}$ and $y_2 = -x - \sqrt{20}$.

21. $17x^2 + 32xy - 7y^2 = 75$

$\cot 2\theta = \dfrac{A - C}{B} = \dfrac{17 + 7}{32} = \dfrac{24}{32} = \dfrac{3}{4} \implies \theta \approx 26.57°$

Solve for y in terms of x by completing the square.

$-7y^2 + 32xy = -17x^2 + 75$

$y^2 - \dfrac{32}{7}xy = \dfrac{17}{7}x^2 - \dfrac{75}{7}$

$y^2 - \dfrac{32}{7}xy + \dfrac{256}{49}x^2 = \dfrac{119}{49}x^2 - \dfrac{525}{49} + \dfrac{256}{49}x^2$

$\left(y - \dfrac{16}{7}x\right)^2 = \dfrac{375x^2 - 525}{49}$

$y = \dfrac{16}{7}x \pm \sqrt{\dfrac{375x^2 - 525}{49}}$

$y = \dfrac{16x \pm 5\sqrt{15x^2 - 21}}{7}$

Use $y_1 = \dfrac{16x + 5\sqrt{15x^2 - 21}}{7}$

and $y_2 = \dfrac{16x - 5\sqrt{15x^2 - 21}}{7}$.

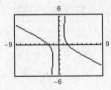

23. $32x^2 + 48xy + 8y^2 = 50$

$\cot 2\theta = \dfrac{A - C}{B} = \dfrac{24}{48} = \dfrac{1}{2} \implies \theta \approx 31.72°$

Solve for y in terms of x by completing the square.

$8y^2 + 48xy = -32x^2 + 50$

$y^2 + 6xy = -4x^2 + \dfrac{25}{4}$

$y^2 + 6xy + 9x^2 = -4x^2 + \dfrac{25}{4} + 9x^2$

$(y + 3x)^2 = 5x^2 + \dfrac{25}{4}$

$y + 3x = \pm\sqrt{5x^2 + \dfrac{25}{4}}$

$y = -3x \pm \sqrt{5x^2 + \dfrac{25}{4}}$

Use $y_1 = -3x + \sqrt{5x^2 + \dfrac{25}{4}}$ and

$y_2 = -3x - \sqrt{5x^2 + \dfrac{25}{4}}$.

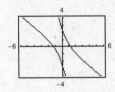

25. $4x^2 - 12xy + 9y^2 + \left(4\sqrt{13} - 12\right)x - \left(6\sqrt{13} + 8\right)y = 91$

$A = 4, B = -12, C = 9$

$\cot 2\theta = \dfrac{A - C}{B} = \dfrac{4 - 9}{-12} = \dfrac{5}{12}$

$\dfrac{1}{\tan 2\theta} = \dfrac{5}{12}$

$\tan 2\theta = \dfrac{12}{5}$

$2\theta \approx 67.38°$

—CONTINUED—

25. **—CONTINUED—**

Solve for y in terms of x with the quadratic formula:

$$4x^2 - 12xy + 9y^2 + \left(4\sqrt{13} - 12\right)x - \left(6\sqrt{13} + 8\right)y = 91$$

$$9y^2 - \left(12x + 6\sqrt{13} + 8\right)y + \left(4x^2 + 4\sqrt{13}x - 12x - 91\right) = 0$$

$$a = 9, \, b = -\left(12x + 6\sqrt{13} + 8\right), \, c = 4x^2 + 4\sqrt{13}x - 12x - 91$$

$$y = \frac{-b \pm \sqrt{b^2 - 4ac}}{2a}$$

$$y = \frac{\left(12x + 6\sqrt{13} + 8\right) \pm \sqrt{\left(12x + 6\sqrt{13} + 8\right)^2 - 4(9)(4x^2 + 4\sqrt{13}x - 12x - 91)}}{18}$$

$$= \frac{\left(12x + 6\sqrt{13} + 8\right) \pm \sqrt{624x + 3808 + 96\sqrt{13}}}{18}$$

Enter $y_1 = \dfrac{12x + 6\sqrt{13} + 8 + \sqrt{624x + 3808 + 96\sqrt{13}}}{18}$

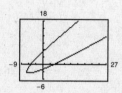

and $y_2 = \dfrac{12x + 6\sqrt{13} + 8 - \sqrt{624x + 3808 + 96\sqrt{13}}}{18}$.

27. $xy + 2 = 0$

$B^2 - 4AC = 1 \implies$ The graph is a hyperbola.

$\cot 2\theta = \dfrac{A - C}{B} = 0 \implies \theta = 45°$

Matches graph (e).

29. $-2x^2 + 3xy + 2y^2 + 3 = 0$

$B^2 - 4AC = (3)^2 - 4(-2)(2) = 25 \implies$

The graph is a hyperbola.

$\cot 2\theta = \dfrac{A - C}{B} = -\dfrac{4}{3} \implies \theta \approx -18.43°$

Matches graph (b).

31. $3x^2 + 2xy + y^2 - 10 = 0$

$B^2 - 4AC = (2)^2 - 4(3)(1) = -8 \implies$

The graph is an ellipse or circle.

$\cot 2\theta = \dfrac{A - C}{B} = 1 \implies \theta = 22.5°$

Matches graph (d).

33. (a) $16x^2 - 8xy + y^2 - 10x + 5y = 0$

$B^2 - 4AC = (-8)^2 - 4(16)(1) = 0$

The graph is a parabola.

(b) $y^2 + (-8x + 5)y + (16x^2 - 10x) = 0$

$$y = \frac{-(-8x + 5) \pm \sqrt{(-8x + 5)^2 - 4(1)(16x^2 - 10x)}}{2(1)}$$

$$= \frac{(8x - 5) \pm \sqrt{(8x - 5)^2 - 4(16x^2 - 10x)}}{2}$$

(c)

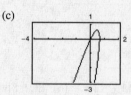

35. (a) $12x^2 - 6xy + 7y^2 - 45 = 0$

$B^2 - 4AC = (-6)^2 - 4(12)(7) = -300 < 0$

The graph is an ellipse.

(b) $7y^2 + (-6x)y + (12x^2 - 45) = 0$

$$y = \frac{-(-6x) \pm \sqrt{(-6x)^2 - 4(7)(12x^2 - 45)}}{2(7)}$$

$$= \frac{6x \pm \sqrt{36x^2 - 28(12x^2 - 45)}}{14}$$

(c)

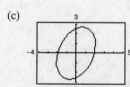

37. (a) $x^2 - 6xy - 5y^2 + 4x - 22 = 0$

$B^2 - 4AC = (-6)^2 - 4(1)(-5) = 56 > 0$

The graph is a hyperbola.

(b) $-5y^2 + (-6x)y + (x^2 + 4x - 22) = 0$

$$y = \frac{-(-6x) \pm \sqrt{(-6x)^2 - 4(-5)(x^2 + 4x - 22)}}{2(-5)}$$

$$= \frac{6x \pm \sqrt{36x^2 + 20(x^2 + 4x - 22)}}{-10}$$

$$= \frac{-6x \pm \sqrt{36x^2 + 20(x^2 + 4x - 22)}}{10}$$

(c)

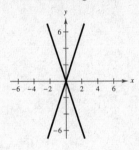

39. (a) $x^2 + 4xy + 4y^2 - 5x - y - 3 = 0$

$B^2 - 4AC = (4)^2 - 4(1)(4) = 0$

The graph is a parabola.

(b) $4y^2 + (4x - 1)y + (x^2 - 5x - 3) = 0$

$$y = \frac{-(4x - 1) \pm \sqrt{(4x - 1)^2 - 4(4)(x^2 - 5x - 3)}}{2(4)}$$

$$= \frac{-(4x - 1) \pm \sqrt{(4x - 1)^2 - 16(x^2 - 5x - 3)}}{8}$$

(c)

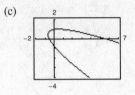

41. $y^2 - 9x^2 = 0$

$y^2 = 9x^2$

$y = \pm 3x$

Two intersecting lines

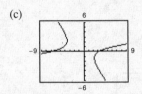

43. $x^2 + 2xy + y^2 - 1 = 0$

$(x + y)^2 - 1 = 0$

$(x + y)^2 = 1$

$x + y = \pm 1$

$y = -x \pm 1$

Two parallel lines

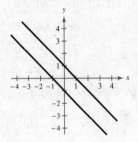

45.

$-x^2 + y^2 + 4x - 6y + 4 = 0 \implies (y - 3)^2 - (x - 2)^2 = 1$

$\underline{x^2 + y^2 - 4x - 6y + 12 = 0} \implies (x - 2)^2 + (y - 3)^2 = 1$

$2y^2 - 12y + 16 = 0$

$2(y - 2)(y - 4) = 0$

$y = 2 \text{ or } y = 4$

For $y = 2$: $x^2 + 2^2 - 4x - 6(2) + 12 = 0$

$x^2 - 4x + 4 = 0$

$(x - 2)^2 = 0$

$x = 2$

For $y = 4$: $x^2 + 4^2 - 4x - 6(4) + 12 = 0$

$x^2 - 4x + 4 = 0$

$(x - 2)^2 = 0$

$x = 2$

The points of intersection are $(2, 2)$ and $(2, 4)$.

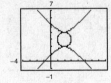

47. $-4x^2 - y^2 - 16x + 24y - 16 = 0$

$\underline{4x^2 + y^2 + 40x - 24y + 208 = 0}$

$24x + 192 = 0$

$x = -8$

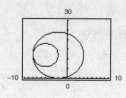

When $x = -8$: $4(-8)^2 + y^2 + 40(-8) - 24y + 208 = 0$

$y^2 - 24y + 144 = 0$

$(y - 12)^2 = 0$

$y = 12$

The point of intersection is $(-8, 12)$. In standard form the equations are:

$\dfrac{(x+2)^2}{36} + \dfrac{(y-12)^2}{144} = 1$ and $\dfrac{(x+5)^2}{9} + \dfrac{(y-12)^2}{36} = 1$

49. $x^2 - y^2 - 12x + 16y - 64 = 0$

$\underline{x^2 + y^2 - 12x - 16y + 64 = 0}$

$2x^2 - 24x = 0$

$2x(x - 12) = 0$

$x = 0$ or $x = 12$

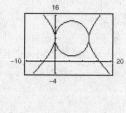

When $x = 0$: $0^2 + y^2 - 12(0) - 16y + 64 = 0$

$y^2 - 16y + 64 = 0$

$(y - 8)^2 = 0$

$y = 8$

When $x = 12$: $12^2 + y^2 - 12(12) - 16y + 64 = 0$

$y^2 - 16y + 64 = 0$

$(y - 8)^2 = 0$

$y = 8$

The points of intersection are $(0, 8)$ and $(12, 8)$. The standard forms of the equations are:

$\dfrac{(x-6)^2}{36} - \dfrac{(y-8)^2}{36} = 1$ and $(x - 6)^2 + (y - 8)^2 = 36$

51. $-16x^2 - y^2 + 24y - 80 = 0$

$\underline{16x^2 + 25y^2 - 400 = 0}$

$24y^2 + 24y - 480 = 0$

$24(y + 5)(y - 4) = 0$

$y = -5$ or $y = 4$

When $y = -5$: $16x^2 + 25(-5)^2 - 400 = 0$

$16x^2 = -225$

No real solution

When $y = 4$: $16x^2 + 25(4)^2 - 400 = 0$

$16x^2 = 0$

$x = 0$

The point of intersection is $(0, 4)$.
In standard form the equations are:

$\dfrac{x^2}{4} + \dfrac{(y-12)^2}{64} = 1$

$\dfrac{x^2}{25} + \dfrac{y^2}{16} = 1$

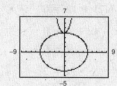

53. $x^2 \qquad + y^2 - 4 = 0$

$$\underline{3x - y^2 \qquad = 0}$$

$x^2 + 3x \qquad - 4 = 0$

$(x + 4)(x - 1) = 0$

$x = -4 \quad \text{or} \quad x = 1$

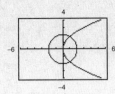

When $x = -4$: $3(-4) - y^2 = 0$

$$y^2 = -12$$

No real solution

When $x = 1$: $3(1) - y^2 = 0$

$$y^2 = 3$$

$$y = \pm\sqrt{3}$$

The points of intersection are $\left(1, \sqrt{3}\right)$ and $\left(1, -\sqrt{3}\right)$.

The standard forms of the equations are:

$x^2 + y^2 = 4$

$ y^2 = 3x$

55. $x^2 + 2y^2 - 4x + 6y - 5 = 0$

$-x + y - 4 = 0 \implies y = x + 4$

$x^2 + 2(x + 4)^2 - 4x + 6(x + 4) - 5 = 0$

$x^2 + 2(x^2 + 8x + 16) - 4x + 6x + 24 - 5 = 0$

$3x^2 + 18x + 51 = 0$

$3(x^2 + 6x + 17) = 0$

$x^2 + 6x + 17 = 0$

$x^2 + 6x + 9 = -17 + 9$

$(x + 3)^2 = -8$

No real solution

No points of intersection

The standard forms of the equations are:

$$\frac{(x - 2)^2}{\frac{27}{2}} + \frac{\left(y + \frac{3}{2}\right)^2}{\frac{27}{4}} = 1$$

$x - y = -4$

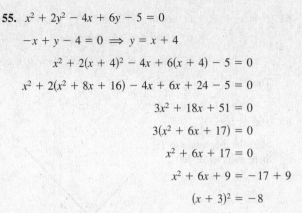

57.
$$xy + x - 2y + 3 = 0 \implies y = \frac{-x - 3}{x - 2}$$

$$x^2 + 4y^2 - 9 = 0$$

$$x^2 + 4\left(\frac{-x - 3}{x - 2}\right)^2 = 9$$

$$x^2(x - 2)^2 + 4(-x - 3)^2 = 9(x - 2)^2$$

$$x^2(x^2 - 4x + 4) + 4(x^2 + 6x + 9) = 9(x^2 - 4x + 4)$$

$$x^4 - 4x^3 + 4x^2 + 4x^2 + 24x + 36 = 9x^2 - 36x + 36$$

$$x^4 - 4x^3 - x^2 + 60x = 0$$

$$x(x + 3)(x^2 - 7x + 20) = 0$$

$$x = 0 \quad \text{or} \quad x = -3$$

Note: $x^2 - 7x + 20 = 0$ has no real solution.

When $x = 0$; $\quad y = \dfrac{-0 - 3}{0 - 2} = \dfrac{3}{2}$

When $x = -3$: $y = \dfrac{-(-3) - 3}{-3 - 2} = 0$

The points of intersection are $\left(0, \dfrac{3}{2}\right), (-3, 0)$.

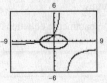

59. $x^2 + xy + ky^2 + 6x + 10 = 0$

$B^2 - 4AC = 1^2 - 4(1)(k) = 1 - 4k > 0 \implies -4k > -1 \implies k < \frac{1}{4}$

True. For the graph to be a hyperbola, the discriminant must be greater than zero.

61. $r^2 = x^2 + y^2 = (x' \cos \theta - y' \sin \theta)^2 + (y' \cos \theta + x' \sin \theta)^2$

$\qquad = (x')^2 \cos^2 \theta - 2x'y' \cos \theta \sin \theta + (y')^2 \sin^2 \theta + (y')^2 \cos^2 \theta + 2x'y' \cos \theta \sin \theta + (x')^2 \sin^2 \theta$

$\qquad = (x')^2(\cos^2 \theta + \sin^2 \theta) + (y')^2(\sin^2 \theta + \cos^2 \theta) = (x')^2 + (y')^2$

Thus, $(x')^2 + (y')^2 = r^2$.

63. $f(x) = |x + 3|$

Shift the graph of $y = |x|$ three units to the left.

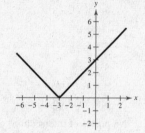

65. $g(x) = \sqrt{4 - x^2}$

$\qquad y^2 = 4 - x^2$

$\qquad x^2 + y^2 = 4$

$g(x)$ is the top half of this circle since $y \geq 0$.

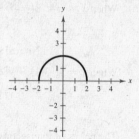

67. $h(t) = -(t - 2)^3 + 3$

Reflect the graph of $y = x^3$ about the x-axis, shift it to the right two units, and upward three units.

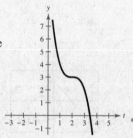

69. $f(t) = [\![t - 5]\!] + 1$

Shift the graph of $y = [\![x]\!]$ five units to the right and upward one unit.

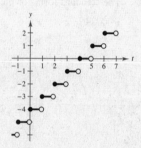

71. Area $= \dfrac{1}{2}ab \sin C$

$\qquad = \dfrac{1}{2}(8)(12) \sin 110°$

$\qquad \approx 45.11$ square units

73. $s = \dfrac{a + b + c}{2} = \dfrac{11 + 18 + 10}{2} = 19.5$

$\qquad$ Area $= \sqrt{s(s - a)(s - b)(s - c)}$

$\qquad = \sqrt{(19.5)(8.5)(1.5)(9.5)}$

$\qquad \approx 48.60$ square units

Section 10.6   Parametric Equations

- If f and g are continuous functions of t on an interval I, then the set of ordered pairs $(f(t), g(t))$ is a *plane curve C*. The equations $x = f(t)$ and $y = g(t)$ are *parametric equations* for C and t is the *parameter*.

- To eliminate the parameter:
  (a) Solve for t in one equation and substitute into the second equation.
  (b) Use trigonometric identities.

- You should be able to find the parametric equations for a graph.

Vocabulary Check

1. plane curve; parametric; parameter

2. orientation

3. eliminating the parameter

1. $x = \sqrt{t}, y = 3 - t$

(a)

t	0	1	2	3	4
x	0	1	$\sqrt{2}$	$\sqrt{3}$	2
y	3	2	1	0	-1

(b)

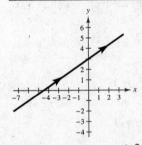

(c) $x = \sqrt{t} \quad \Rightarrow x^2 = t$

$y = 3 - t \quad \Rightarrow \quad y = 3 - x^2$

The graph of the parametric equations only shows the right half of the parabola, whereas the rectangular equation yields the entire parabola.

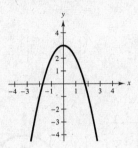

3. (a) $x = 3t - 3, y = 2t + 1$

t	-2	-1	0	1	2
x	-9	-6	-3	0	3
y	-3	-1	1	3	5

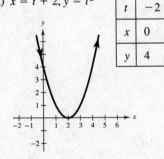

(b) $x = 3t - 3 \Rightarrow t = \dfrac{x + 3}{3}$

$y = 2t + 1 \Rightarrow y = \dfrac{2}{3}(x + 3) + 1 = \dfrac{2}{3}x + 3$

5. (a) $x = \dfrac{1}{4}t, y = t^2$

t	-2	-1	0	1	2
x	$-\frac{1}{2}$	$-\frac{1}{4}$	0	$\frac{1}{4}$	$\frac{1}{2}$
y	4	1	0	1	4

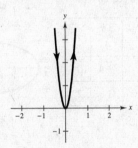

(b) $x = \dfrac{1}{4}t \Rightarrow t = 4x$

$y = t^2 \Rightarrow y = 16x^2$

7. (a) $x = t + 2, y = t^2$

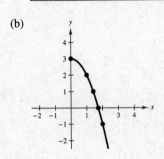

t	-2	-1	0	1	2
x	0	1	2	3	4
y	4	1	0	1	4

(b) $x = t + 2 \Rightarrow t = x - 2$

$y = t^2 \quad \Rightarrow y = (x - 2)^2 = x^2 - 4x + 4$

9. (a) $x = t + 1, y = \dfrac{t}{t + 1}$

t	-3	-2	0	1	2
x	-2	-1	1	2	3
y	$\frac{3}{2}$	2	0	$\frac{2}{3}$	$\frac{3}{4}$

(b) $x = t + 1 \Rightarrow t = x - 1$

$y = \dfrac{t}{t + 1} \Rightarrow y = \dfrac{x - 1}{x}$

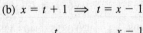

11. (a) $x = 2(t + 1), y = |t - 2|$

t	0	2	4	6	8	10
x	2	6	10	14	18	22
y	2	0	2	4	6	8

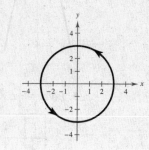

(b) $x = 2(t + 1) \implies \dfrac{x}{2} - 1 = t$ or $t = \dfrac{x - 2}{2}$

$y = |t - 2| \implies y = \left|\dfrac{x}{2} - 1 - 2\right| = \left|\dfrac{x}{2} - 3\right|$

13. (a) $x = 3 \cos \theta, y = 3 \sin \theta$

θ	0	$\dfrac{\pi}{2}$	π	$\dfrac{3\pi}{2}$	2π
x	3	0	-3	0	3
y	0	3	0	-3	0

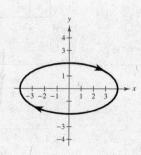

(b) $x = 3 \cos \theta \implies \left(\dfrac{x}{3}\right)^2 = \cos^2 \theta$

$y = 3 \sin \theta \implies \left(\dfrac{y}{3}\right)^2 = \sin^2 \theta$

$\left(\dfrac{x}{3}\right)^2 + \left(\dfrac{y}{3}\right)^2 = 1$

$\dfrac{x^2}{9} + \dfrac{y^2}{9} = 1$

15. (a) $x = 4 \sin 2\theta, y = 2 \cos 2\theta$

θ	0	$\dfrac{\pi}{4}$	$\dfrac{\pi}{2}$	$\dfrac{3\pi}{4}$	π
x	0	4	0	-4	0
y	2	0	-2	0	2

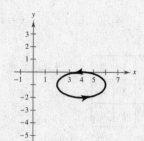

(b) $x = 4 \sin 2\theta \implies \left(\dfrac{x}{4}\right)^2 = \sin^2 2\theta$

$y = 2 \cos 2\theta \implies \left(\dfrac{y}{2}\right)^2 = \cos^2 2\theta$

$\left(\dfrac{x}{4}\right)^2 + \left(\dfrac{y}{2}\right)^2 = 1$

$\dfrac{x^2}{16} + \dfrac{y^2}{4} = 1$

17. (a) $x = 4 + 2 \cos \theta, y = -1 + \sin \theta$

θ	0	$\dfrac{\pi}{2}$	π	$\dfrac{3\pi}{2}$	2π
x	6	4	2	4	6
y	-1	0	-1	-2	-1

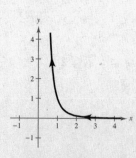

(b) $x = 4 + 2 \cos \theta \implies \left(\dfrac{x - 4}{2}\right)^2 = \cos^2 \theta$

$y = -1 + \sin \theta \implies (y + 1)^2 = \sin^2 \theta$

$\dfrac{(x - 4)^2}{4} + \dfrac{(y + 1)^2}{1} = 1$

19. (a) $x = e^{-t}, y = e^{3t}$

t	-2	-1	0	1	2
x	7.3891	2.7183	1	0.3679	0.1353
y	0.0025	0.0498	1	20.0855	403.4288

(b) $x = e^{-t} \implies \dfrac{1}{x} = e^t$

$y = e^{3t} \implies y = (e^t)^3$

$y = \left(\dfrac{1}{x}\right)^3$

$y = \dfrac{1}{x^3}, x > 0, y > 0$

21. (a) $x = t^3, y = 3 \ln t$

t	$\frac{1}{2}$	1	2	3	4
x	$\frac{1}{8}$	1	8	27	64
y	-2.0794	0	2.0794	3.2958	4.1589

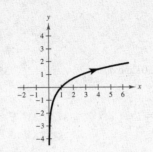

(b) $x = t^3 \quad \Rightarrow x^{1/3} = t$

$y = 3 \ln t \Rightarrow y = \ln t^3$

$y = \ln(x^{1/3})^3$

$y = \ln x$

23. By eliminating the parameter, each curve becomes $y = 2x + 1$.

(a) $x = t$

$y = 2t + 1$

There are no restrictions on x and y.

Domain: $(-\infty, \infty)$

Orientation: Left to right

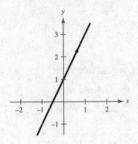

(b) $x = \cos \theta \qquad \Rightarrow -1 \le x \le 1$

$y = 2 \cos \theta + 1 \Rightarrow -1 \le y \le 3$

The graph oscillates.

Domain: $[-1, 1]$

Orientation: Depends on θ

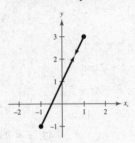

(c) $x = e^{-t} \qquad \Rightarrow x > 0$

$y = 2e^{-t} + 1 \Rightarrow y > 1$

Domain: $(0, \infty)$

Orientation: Downward or right to left

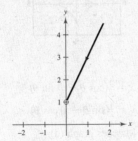

(d) $x = e^t \qquad \Rightarrow x > 0$

$y = 2e^t + 1 \Rightarrow y > 1$

Domain: $(0, \infty)$

Orientation: Upward or left to right

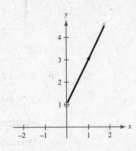

25. $x = x_1 + t(x_2 - x_1), y = y_1 + t(y_2 - y_1)$

$$\frac{x - x_1}{x_2 - x_1} = t$$

$$y = y_1 + \left(\frac{x - x_1}{x_2 - x_1}\right)(y_2 - y_1)$$

$$y - y_1 = \frac{y_2 - y_1}{x_2 - x_1}(x - x_1) = m(x - x_1)$$

27. $x = h + a \cos \theta, y = k + b \sin \theta$

$$\frac{x - h}{a} = \cos \theta, \frac{y - k}{b} = \sin \theta$$

$$\frac{(x - h)^2}{a^2} + \frac{(y - k)^2}{b^2} = 1$$

29. From Exercise 25 we have:

$x = 0 + t(6 - 0) = 6t$

$y = 0 + t(-3 - 0) = -3t$

31. From Exercise 26 we have:

$x = 3 + 4 \cos \theta$

$y = 2 + 4 \sin \theta$

33. Vertices: $(\pm 4, 0) \implies (h, k) = (0, 0)$ and $a = 4$

Foci: $(\pm 3, 0) \implies c = 3$

$c^2 = a^2 - b^2 \implies 9 = 16 - b^2 \implies b = \sqrt{7}$

From Exercise 27 we have:

$x = 4 \cos \theta$

$y = \sqrt{7} \sin \theta$

35. Vertices: $(\pm 4, 0) \implies (h, k) = (0, 0)$ and $a = 4$

Foci: $(\pm 5, 0) \implies c = 5$

$c^2 = a^2 + b^2 \implies 25 = 16 + b^2 \implies b = 3$

From Exercise 28 we have:

$x = 4 \sec \theta$

$y = 3 \tan \theta$

37. $y = 3x - 2$

(a) $t = x \implies x = t$ and $y = 3t - 2$

(b) $t = 2 - x \implies x = -t + 2$ and

$y = 3(-t + 2) - 2 = -3t + 4$

39. $y = x^2$

(a) $t = x \implies x = t$ and $y = t^2$

(b) $t = 2 - x \implies x = -t + 2$ and

$y = (-t + 2)^2 = t^2 - 4t + 4$

41. $y = x^2 + 1$

(a) $t = x \implies x = t$ and $y = t^2 + 1$

(b) $t = 2 - x \implies x = -t + 2$ and

$y = (-t + 2)^2 + 1 = t^2 - 4t + 5$

43. $y = \dfrac{1}{x}$

(a) $t = x \implies x = t$ and $y = \dfrac{1}{t}$

(b) $t = 2 - x \implies x = -t + 2$ and $y = \dfrac{1}{-t + 2} = \dfrac{-1}{t - 2}$

45. $x = 4(\theta - \sin \theta)$

$y = 4(1 - \cos \theta)$

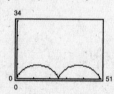

47. $x = \theta - \dfrac{3}{2} \sin \theta$

$y = 1 - \dfrac{3}{2} \cos \theta$

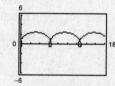

49. $x = 3 \cos^3 \theta$

$y = 3 \sin^3 \theta$

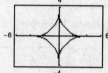

51. $x = 2 \cot \theta$

$y = 2 \sin^2 \theta$

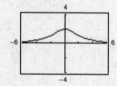

53. $x = 2 \cos \theta \implies -2 \leq x \leq 2$

$y = \sin 2\theta \implies -1 \leq y \leq 1$

Matches graph (b).

Domain: $[-2, 2]$

Range: $[-1, 1]$

55. $x = \dfrac{1}{2}(\cos \theta + \theta \sin \theta)$

$y = \dfrac{1}{2}(\sin \theta - \theta \cos \theta)$

Matches graph (d).

Domain: $(-\infty, \infty)$

Range: $(-\infty, \infty)$

57. $x = (v_0 \cos \theta)t$ and $y = h + (v_0 \sin \theta)t - 16t^2$

    (a) $\theta = 60°$, $v_0 = 88$ ft/sec

        $x = (88 \cos 60°)t$ and $y = (88 \sin 60°)t - 16t^2$

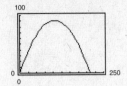

    Maximum height: 90.7 feet

    Range: 209.6 feet

    (c) $\theta = 45°$, $v_0 = 88$ ft/sec

        $x = (88 \cos 45°)t$ and $y = (88 \sin 45°)t - 16t^2$

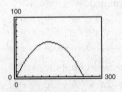

    Maximum height: 60.5 ft

    Range: 242.0 ft

    (b) $\theta = 60°$, $v_0 = 132$ ft/sec

        $x = (132 \cos 60°)t$ and $y = (132 \sin 60°)t - 16t^2$

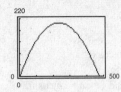

    Maximum height: 204.2 feet

    Range: 471.6 feet

    (d) $\theta = 45°$, $v_0 = 132$ ft/sec

        $x = (132 \cos 45°)t$ and $y = (132 \sin 45°)t - 16t^2$

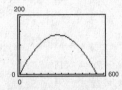

    Maximum height: 136.1 ft

    Range: 544.5 ft

59. (a) 100 miles per hour $= 100\left(\frac{5280}{3600}\right)$ ft/sec $= \frac{440}{3}$ ft/sec

        $x = \left(\frac{440}{3} \cos \theta\right)t \approx (146.67 \cos \theta)t$

        $y = 3 + \left(\frac{440}{3} \sin \theta\right)t - 16t^2 \approx 3 + (146.67 \sin \theta)t - 16t^2$

    (b) For $\theta = 15°$, we have:

        $x = \left(\frac{440}{3} \cos 15°\right)t \approx 141.7t$

        $y = 3 + \left(\frac{440}{3} \sin 15°\right)t - 16t^2 \approx 3 + 38.0t - 16t^2$

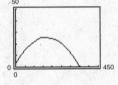

        The ball hits the ground inside the ballpark, so it is not a home run.

    (c) For $\theta = 23°$, we have:

        $x = \left(\frac{440}{3} \cos 23°\right)t \approx 135.0t$

        $y = 3 + \left(\frac{440}{3} \sin 23°\right)t - 16t^2 \approx 3 + 57.3t - 16t^2$

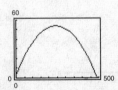

        The ball easily clears the 7-foot fence at 408 feet so it is a home run.

    (d) Find θ so that $y = 7$ when $x = 408$ by graphing the parametric
        equations for θ values between 15° and 23°. This occurs when $\theta \approx 19.3°$.

61. $x = (v_0 \cos \theta)t \implies t = \dfrac{x}{v_0 \cos \theta}$

    $y = h + (v_0 \sin \theta)t - 16t^2$

      $= h + (v_0 \sin \theta)\left(\dfrac{x}{v_0 \cos \theta}\right) - 16\left(\dfrac{x}{v_0 \cos \theta}\right)^2$

      $= h + (\tan \theta)x - \dfrac{16x^2}{v_0^2 \cos^2 \theta}$

      $= -\dfrac{16 \sec^2 \theta}{v_0^2}x^2 + (\tan \theta)x + h$

63. When the circle has rolled θ radians, the center is at $(a\theta, a)$.

$$\sin \theta = \sin(180° - \theta)$$

$$= \frac{|AC|}{b} = \frac{|BD|}{b} \implies |BD| = b \sin \theta$$

$$\cos \theta = -\cos(180° - \theta)$$

$$= \frac{|AP|}{-b} \implies |AP| = -b \cos \theta$$

Therefore, $x = a\theta - b \sin \theta$ and $y = a - b \cos \theta$.

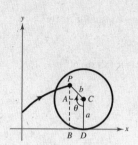

65. True

$$x = t$$

$$y = t^2 + 1 \implies y = x^2 + 1$$

$$x = 3t$$

$$y = 9t^2 + 1 \implies y = x^2 + 1$$

67. The use of parametric equations is useful when graphing two functions simultaneously on the same coordinate system. For example, this is useful when tracking the path of an object so the position and the time associated with that position can be determined.

69.

$$\begin{aligned} 5x - 7y &= 11 \implies & 5x - 7y &= 11 \\ -3x + y &= -13 \implies & -21x + 7y &= -91 \\ & & -16x &= -80 \\ & & x &= 5 \end{aligned}$$

$$5(5) - 7y = 11 \implies y = 2$$

Solution: $(5, 2)$

71.

$$\begin{aligned} 3a - 2b + c &= 8 \implies & 9a - 6b + 3c &= 24 \\ 2a + b - 3c &= -3 \implies & 2a + b - 3c &= -3 \\ & & 11a - 5b &= 21 \\ 2a + b - 3c &= -3 \implies & 6a + 3b - 9c &= -9 \\ a - 3b + 9c &= 16 \implies & a - 3b + 9c &= 16 \\ & & 7a &= 7 \\ & & a &= 1 \end{aligned}$$

$$11(1) - 5b = 21 \implies b = -2$$

$$3(1) - 2(-2) + c = 8 \implies c = 1$$

Solution: $(1, -2, 1)$

73. $\theta = 105°$

$$\theta' = 180° - 105° = 75°$$

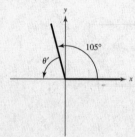

75. $\theta = -\dfrac{2\pi}{3}$

$$\theta' = -\frac{2\pi}{3} + \pi = \frac{\pi}{3}$$

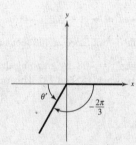

Section 10.7   Polar Coordinates

■ In polar coordinates you do not have unique representation of points. The point (r, θ) can be represented by $(r, \theta \pm 2n\pi)$ or by $(-r, \theta \pm (2n + 1)\pi)$ where n is any integer. The pole is represented by $(0, \theta)$ where θ is any angle.

■ To convert from polar coordinates to rectangular coordinates, use the following relationships.

$x = r \cos \theta$
$y = r \sin \theta$

■ To convert from rectangular coordinates to polar coordinates, use the following relationships.

$r = \pm\sqrt{x^2 + y^2}$
$\tan \theta = y/x$

If θ is in the same quadrant as the point (x, y), then r is positive. If θ is in the opposite quadrant as the point (x, y), then r is negative.

■ You should be able to convert rectangular equations to polar form and vice versa.

Vocabulary Check

1. pole

2. directed distance; directed angle

3. polar

4. $x = r \cos \theta, \quad \tan \theta = \dfrac{y}{x}$

$y = r \sin \theta, \quad r^2 = x^2 + y^2$

---

1. Polar coordinates: $\left(4, -\dfrac{\pi}{3}\right)$

Additional representations:

$\left(4, -\dfrac{\pi}{3} + 2\pi\right) = \left(4, \dfrac{5\pi}{3}\right)$

$\left(-4, -\dfrac{\pi}{3} - \pi\right) = \left(-4, -\dfrac{4\pi}{3}\right)$

3. Polar coordinates: $\left(0, -\dfrac{7\pi}{6}\right)$

Additional representations:

$\left(0, -\dfrac{7\pi}{6} + 2\pi\right) = \left(0, \dfrac{5\pi}{6}\right)$

$\left(0, -\dfrac{7\pi}{6} + \pi\right) = \left(0, -\dfrac{\pi}{6}\right)$ or $(0, \theta)$ for any θ, $-2\pi < \theta < 2\pi$

5. Polar coordinates: $\left(\sqrt{2}, 2.36\right)$

Additional representations:

$\left(\sqrt{2}, 2.36 - 2\pi\right) \approx \left(\sqrt{2}, -3.92\right)$

$\left(-\sqrt{2}, 2.36 - \pi\right) \approx \left(-\sqrt{2}, -0.78\right)$

7. Polar coordinates: $\left(2\sqrt{2}, 4.71\right)$

Additional representations:

$\left(2\sqrt{2}, 4.71 - 2\pi\right) \approx \left(2\sqrt{2}, -1.57\right)$

$\left(-2\sqrt{2}, 2\pi - 4.71\right) \approx \left(-2\sqrt{2}, 1.57\right)$

9. Polar coordinates: $\left(3, \dfrac{\pi}{2}\right)$

$x = 3 \cos \dfrac{\pi}{2} = 0$

$y = 3 \sin \dfrac{\pi}{2} = 3$

Rectangular coordinates: $(0, 3)$

11. Polar coordinates: $\left(-1, \dfrac{5\pi}{4}\right)$

$x = -1 \cos\left(\dfrac{5\pi}{4}\right) = \dfrac{\sqrt{2}}{2}, y = -1 \sin\left(\dfrac{5\pi}{4}\right) = \dfrac{\sqrt{2}}{2}$

Rectangular coordinates: $\left(\dfrac{\sqrt{2}}{2}, \dfrac{\sqrt{2}}{2}\right)$

13. Polar coordinates: $\left(2, \dfrac{3\pi}{4}\right)$

$x = 2 \cos \dfrac{3\pi}{4} = -\sqrt{2}$

$y = 2 \sin \dfrac{3\pi}{4} = \sqrt{2}$

Rectangular coordinates: $\left(-\sqrt{2}, \sqrt{2}\right)$

15. Polar coordinates: $(-2.5, 1.1)$

$x = -2.5 \cos 1.1 \approx -1.1340$

$y = -2.5 \sin 1.1 \approx -2.2280$

Rectangular coordinates: $(-1.1340, -2.2280)$

17. Rectangular coordinates: $(1, 1)$

$r = \pm\sqrt{2}, \tan \theta = 1, \theta = \dfrac{\pi}{4} \text{ or } \dfrac{5\pi}{4}$

Polar coordinates: $\left(\sqrt{2}, \dfrac{\pi}{4}\right), \left(-\sqrt{2}, \dfrac{5\pi}{4}\right)$

19. Rectangular coordinates: $(-6, 0)$

$r = \pm 6, \tan \theta = 0, \theta = 0 \text{ or } \pi$

Polar coordinates: $(6, \pi), (-6, 0)$

21. Rectangular coordinates: $(-3, 4)$

$r = \pm\sqrt{9 + 16} = \pm 5, \tan \theta = -\dfrac{4}{3}, \theta \approx 2.2143, 5.3559$

Polar coordinates: $(5, 2.2143), (-5, 5.3559)$

23. Rectangular coordinates: $\left(-\sqrt{3}, -\sqrt{3}\right)$

$r = \pm\sqrt{3 + 3} = \pm\sqrt{6}, \tan \theta = 1, \theta = \dfrac{\pi}{4} \text{ or } \dfrac{5\pi}{4}$

Polar coordinates: $\left(\sqrt{6}, \dfrac{5\pi}{4}\right), \left(-\sqrt{6}, \dfrac{\pi}{4}\right)$

25. Rectangular coordinates: $(6, 9)$

$r = \pm\sqrt{6^2 + 9^2} = \pm\sqrt{117} = \pm 3\sqrt{13}$

$\tan \theta = \dfrac{9}{6}, \theta \approx 0.9828, 4.1244$

Polar coordinates: $\left(3\sqrt{13}, 0.9828\right), \left(-3\sqrt{13}, 4.1244\right)$

27. Rectangular: $(3, -2)$

$(3, -2) \blacktriangleright \text{Pol}$

$\approx (3.606, -0.5880)$

or $\left(\sqrt{13}, -0.5880\right)$

or $\left(\sqrt{13}, 5.6952\right)$

29. Rectangular: $\left(\sqrt{3}, 2\right)$

$\left(\sqrt{3}, 2\right) \blacktriangleright \text{Pol}$

$\approx (2.646, 0.8571)$

or $\left(\sqrt{7}, 0.8571\right)$

31. Rectangular: $\left(\frac{5}{2}, \frac{4}{3}\right)$

$\left(\frac{5}{2}, \frac{4}{3}\right) \blacktriangleright \text{Pol}$

$\approx (2.833, 0.4900)$

or $\left(\frac{17}{6}, 0.4900\right)$

33. $x^2 + y^2 = 9$

$r = 3$

35. $y = 4$

$r \sin \theta = 4$

$r = 4 \csc \theta$

37. $x = 10$

$r \cos \theta = 10$

$r = 10 \sec \theta$

39. $3x - y + 2 = 0$

$3r \cos \theta - r \sin \theta + 2 = 0$

$r(3 \cos \theta - \sin \theta) = -2$

$$r = \frac{-2}{3 \cos \theta - \sin \theta}$$

41. $xy = 16$

$(r \cos \theta)(r \sin \theta) = 16$

$r^2 = 16 \sec \theta \csc \theta = 32 \csc 2\theta$

43. $y^2 - 8x - 16 = 0$

$r^2 \sin^2 \theta - 8r \cos \theta - 16 = 0$

By the Quadratic Formula, we have:

$$r = \frac{-(-8 \cos \theta) \pm \sqrt{(-8 \cos \theta)^2 - 4(\sin^2 \theta)(-16)}}{2 \sin^2 \theta}$$

$$= \frac{8 \cos \theta \pm \sqrt{64 \cos^2 \theta + 64 \sin^2 \theta}}{2 \sin^2 \theta}$$

$$= \frac{8 \cos \theta \pm \sqrt{64 (\cos^2 \theta + \sin^2 \theta)}}{2 \sin^2 \theta}$$

$$= \frac{8 \cos \theta \pm 8}{2 \sin^2 \theta}$$

$$= \frac{4(\cos \theta \pm 1)}{1 - \cos^2 \theta}$$

$$r = \frac{4(\cos \theta + 1)}{(1 + \cos \theta)(1 - \cos \theta)} = \frac{4}{1 - \cos \theta}$$

or

$$r = \frac{4(\cos \theta - 1)}{(1 + \cos \theta)(1 - \cos \theta)} = \frac{-4}{1 + \cos \theta}$$

45. $x^2 + y^2 = a^2$

$r^2 = a^2$

$r = a$

47. $x^2 + y^2 - 2ax = 0$

$r^2 - 2a \, r \cos \theta = 0$

$r(r - 2a \cos \theta) = 0$

$r - 2a \cos \theta = 0$

$r = 2a \cos \theta$

49. $r = 4 \sin \theta$

$r^2 = 4r \sin \theta$

$x^2 + y^2 = 4y$

$x^2 + y^2 - 4y = 0$

51. $\theta = \dfrac{2\pi}{3}$

$\tan \theta = \tan \dfrac{2\pi}{3}$

$\dfrac{y}{x} = -\sqrt{3}$

$y = -\sqrt{3}x$

$\sqrt{3}x + y = 0$

53. $r = 4$

$r^2 = 16$

$x^2 + y^2 = 16$

55. $r = 4 \csc \theta$

$r \sin \theta = 4$

$y = 4$

57. $r^2 = \cos \theta$

$r^3 = r \cos \theta$

$\left(\pm \sqrt{x^2 + y^2}\right)^3 = x$

$\pm \left(x^2 + y^2\right)^{3/2} = x$

$\left(x^2 + y^2\right)^3 = x^2$

$x^2 + y^2 = x^{2/3}$

$x^2 + y^2 - x^{2/3} = 0$

59.
$$r = 2 \sin 3\theta$$
$$r = 2 \sin(\theta + 2\theta)$$
$$r = 2[\sin \theta \cos 2\theta + \cos \theta \sin 2\theta]$$
$$r = 2[\sin \theta(1 - 2 \sin^2 \theta) + \cos \theta(2 \sin \theta \cos \theta)]$$
$$r = 2[\sin \theta - 2 \sin^3 \theta + 2 \sin \theta \cos^2 \theta]$$
$$r = 2[\sin \theta - 2 \sin^3 \theta + 2 \sin \theta(1 - \sin^2 \theta)]$$
$$r = 2(3 \sin \theta - 4 \sin^3 \theta)$$
$$r^4 = 6r^3 \sin \theta - 8r^3 \sin^3 \theta$$
$$(x^2 + y^2)^2 = 6(x^2 + y^2)y - 8y^3$$
$$(x^2 + y^2)^2 = 6x^2y - 2y^3$$

61.
$$r = \frac{2}{1 + \sin \theta}$$
$$r(1 + \sin \theta) = 2$$
$$r + r \sin \theta = 2$$
$$r = 2 - r \sin \theta$$
$$\pm\sqrt{x^2 + y^2} = 2 - y$$
$$x^2 + y^2 = (2 - y)^2$$
$$x^2 + y^2 = 4 - 4y + y^2$$
$$x^2 + 4y - 4 = 0$$

63.
$$r = \frac{6}{2 - 3 \sin \theta}$$
$$r(2 - 3 \sin \theta) = 6$$
$$2r = 6 + 3r \sin \theta$$
$$2(\pm\sqrt{x^2 + y^2}) = 6 + 3y$$
$$4(x^2 + y^2) = (6 + 3y)^2$$
$$4x^2 + 4y^2 = 36 + 36y + 9y^2$$
$$4x^2 - 5y^2 - 36y - 36 = 0$$

65. The graph of the polar equation consists of all points that are six units from the pole.
$$r = 6$$
$$r^2 = 36$$
$$x^2 + y^2 = 36$$

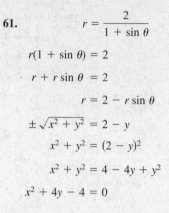

67. The graph of the polar equation consists of all points that make an angle of $\pi/6$ with the polar axis.
$$\theta = \frac{\pi}{6}$$
$$\tan \theta = \tan \frac{\pi}{6}$$
$$\frac{y}{x} = \frac{\sqrt{3}}{3}$$
$$y = \frac{\sqrt{3}}{3}x$$
$$3y = \sqrt{3}x$$
$$-\sqrt{3}x + 3y = 0$$

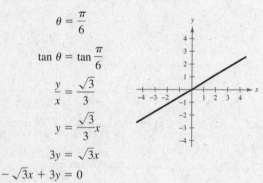

69. The graph of the polar equation is not evident by simple inspection. Convert to rectangular form first.
$$r = 3 \sec \theta$$
$$r \cos\theta = 3$$
$$x = 3$$
$$x - 3 = 0$$

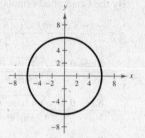

71. True. Because r is a directed distance, then the point (r, θ) can be represented as $(r, \theta \pm 2n\pi)$.

73.
$$r = 2(h \cos \theta + k \sin \theta)$$
$$r = 2\left(h\left(\frac{x}{r}\right) + k\left(\frac{y}{r}\right)\right)$$
$$r = \frac{2hx + 2ky}{r}$$
$$r^2 = 2hx + 2ky$$
$$x^2 + y^2 = 2hx + 2ky$$
$$x^2 - 2hx + y^2 - 2ky = 0$$
$$(x^2 - 2hx + h^2) + (y^2 - 2ky + k^2) = h^2 + k^2$$
$$(x - h)^2 + (y - k)^2 = h^2 + k^2$$

Center: (h, k)

Radius: $\sqrt{h^2 + k^2}$

75. (a) $(r_1, \theta_1) = (x_1, y_1)$ where $x_1 = r_1 \cos \theta_1$ and $y_1 = r_1 \sin \theta_1$.

$(r_2, \theta_2) = (x_2, y_2)$ where $x_2 = r_2 \cos \theta_2$ and $y_2 = r_2 \sin \theta_2$.

$$d = \sqrt{(x_1 - x_2)^2 + (y_1 - y_2)^2}$$
$$= \sqrt{x_1^2 - 2x_1x_2 + x_2^2 + y_1^2 - 2y_1y_2 + y_2^2}$$
$$= \sqrt{(x_1^2 + y_1^2) + (x_2^2 + y_2^2) - 2(x_1x_2 + y_1y_2)}$$
$$= \sqrt{r_1^2 + r_2^2 - 2(r_1r_2 \cos \theta_1 \cos \theta_2 + r_1r_2 \sin \theta_1 \sin \theta_2)}$$
$$= \sqrt{r_1^2 + r_2^2 - 2r_1r_2 \cos(\theta_1 - \theta_2)}$$

(b) If $\theta_1 = \theta_2$, then

$$d = \sqrt{r_1^2 + r_2^2 - 2r_1r_2}$$
$$= \sqrt{(r_1 - r_2)^2}$$
$$= |r_1 - r_2|.$$

This represents the distance between two points on the line $\theta = \theta_1 = \theta_2$.

(c) If $\theta_1 - \theta_2 = 90°$, then

$$d = \sqrt{r_1^2 + r_2^2}.$$

This is the result of the Pythagorean Theorem.

(d) The results should be the same. For example, use the points

$$\left(3, \frac{\pi}{6}\right) \text{ and } \left(4, \frac{\pi}{3}\right).$$

The distance is $d \approx 2.053$.
Now use the representations

$$\left(-3, \frac{7\pi}{6}\right) \text{ and } \left(-4, \frac{4\pi}{3}\right).$$

The distance is still $d \approx 2.053$.

77. $\log_6 \dfrac{x^2z}{3y} = \log_6 x^2z - \log_6 3y$

$$= \log_6 x^2 + \log_6 z - (\log_6 3 + \log_6 y)$$

$$= 2\log_6 x + \log_6 z - \log_6 3 - \log_6 y$$

79. $\ln x(x+4)^2 = \ln x + \ln(x+4)^2$

$$= \ln x + 2\ln(x+4)$$

81. $\log_7 x - \log_7 3y = \log_7 \dfrac{x}{3y}$

83. $\dfrac{1}{2}\ln x + \ln(x-2) = \ln \sqrt{x} + \ln(x-2)$

$$= \ln \sqrt{x}(x-2)$$

85. $\begin{cases} 5x - 7y = -11 \\ -3x + y = -3 \end{cases}$

By Cramer's Rule we have:

$$x = \frac{\begin{vmatrix} -11 & -7 \\ -3 & 1 \end{vmatrix}}{\begin{vmatrix} 5 & -7 \\ -3 & 1 \end{vmatrix}} = \frac{-32}{-16} = 2$$

$$y = \frac{\begin{vmatrix} 5 & -11 \\ -3 & -3 \end{vmatrix}}{\begin{vmatrix} 5 & -7 \\ -3 & 1 \end{vmatrix}} = \frac{-48}{-16} = 3$$

Solution: $(2, 3)$

87. $\begin{cases} 3a - 2b + c = 0 \\ 2a + b - 3c = 0 \\ a - 3b + 9c = 8 \end{cases}$

$$\begin{vmatrix} 3 & -2 & 1 \\ 2 & 1 & -3 \\ 1 & -3 & 9 \end{vmatrix} = 35$$

By Cramer's Rule we have:

$$a = \frac{\begin{vmatrix} 0 & -2 & 1 \\ 0 & 1 & -3 \\ 8 & -3 & 9 \end{vmatrix}}{35} = \frac{40}{35} = \frac{8}{7}$$

$$b = \frac{\begin{vmatrix} 3 & 0 & 1 \\ 2 & 0 & -3 \\ 1 & 8 & 9 \end{vmatrix}}{35} = \frac{88}{35}$$

$$c = \frac{\begin{vmatrix} 3 & -2 & 0 \\ 2 & 1 & 0 \\ 1 & -3 & 8 \end{vmatrix}}{35} = \frac{56}{35} = \frac{8}{5}$$

Solution: $\left(\dfrac{8}{7}, \dfrac{88}{35}, \dfrac{8}{5}\right)$

89. $\begin{cases} -x + y + 2z = 1 \\ 2x + 3y + z = -2 \\ 5x + 4y + 2z = 4 \end{cases}$

$$\begin{vmatrix} -1 & 1 & 2 \\ 2 & 3 & 1 \\ 5 & 4 & 2 \end{vmatrix} = -15$$

By Cramer's Rule we have:

$$x = \frac{\begin{vmatrix} 1 & 1 & 2 \\ -2 & 3 & 1 \\ 4 & 4 & 2 \end{vmatrix}}{-15} = \frac{-30}{-15} = 2$$

$$y = \frac{\begin{vmatrix} -1 & 1 & 2 \\ 2 & -2 & 1 \\ 5 & 4 & 2 \end{vmatrix}}{-15} = \frac{45}{-15} = -3$$

$$z = \frac{\begin{vmatrix} -1 & 1 & 1 \\ 2 & 3 & -2 \\ 5 & 4 & 4 \end{vmatrix}}{-15} = \frac{-45}{-15} = 3$$

Solution: $(2, -3, 3)$

91. Points: $(4, -3), (6, -7), (-2, -1)$

$$\begin{vmatrix} 4 & -3 & 1 \\ 6 & -7 & 1 \\ -2 & -1 & 1 \end{vmatrix} = -20 \neq 0$$

The points are not collinear.

93. Points: $(-6, -4), (-1, -3), (1.5, -2.5)$

$$\begin{vmatrix} -6 & -4 & 1 \\ -1 & -3 & 1 \\ 1.5 & -2.5 & 1 \end{vmatrix} = 0$$

The points are collinear.

Section 10.8   Graphs of Polar Equations

■   When graphing polar equations:

1. Test for symmetry.
   (a) $\theta = \pi/2$: Replace (r, θ) by $(r, \pi - \theta)$ or $(-r, -\theta)$.
   (b) Polar axis: Replace (r, θ) by $(r, -\theta)$ or $(-r, \pi - \theta)$.
   (c) Pole: Replace (r, θ) by $(r, \pi + \theta)$ or $(-r, \theta)$.
   (d) $r = f(\sin \theta)$ is symmetric with respect to the line $\theta = \pi/2$.
   (e) $r = f(\cos \theta)$ is symmetric with respect to the polar axis.

2. Find the θ values for which $|r|$ is maximum.

3. Find the θ values for which $r = 0$.

4. Know the different types of polar graphs.
   (a) Limaçons $(0 < a, 0 < b)$
   $r = a \pm b \cos \theta$
   $r = a \pm b \sin \theta$

   (b) Rose curves, $n \geq 2$
   $r = a \cos n\theta$
   $r = a \sin n\theta$

   (c) Circles
   $r = a \cos \theta$
   $r = a \sin \theta$
   $r = a$

   (d) Lemniscates
   $r^2 = a^2 \cos 2\theta$
   $r^2 = a^2 \sin 2\theta$

5. Plot additional points.

Vocabulary Check

1. $\theta = \dfrac{\pi}{2}$

4. circle

2. polar axis

5. lemniscate

3. convex limaçon

6. cardioid

---

1. $r = 3 \cos 2\theta$
Rose curve with 4 petals

3. $r = 3(1 - 2 \cos \theta)$
Limaçon with inner loop

5. $r = 6 \sin 2\theta$
Rose curve with 4 petals

7. $r = 5 + 4 \cos \theta$

$\theta = \dfrac{\pi}{2}$: $-r = 5 + 4 \cos(-\theta)$

$-r = 5 + 4 \cos \theta$

Not an equivalent equation

Polar axis: $r = 5 + 4 \cos(-\theta)$

$r = 5 + 4 \cos \theta$

Equivalent equation

Pole: $-r = 5 + 4 \cos \theta$

Not an equivalent equation

Answer: Symmetric with respect to polar axis

9. $r = \dfrac{2}{1 + \sin \theta}$

$\theta = \dfrac{\pi}{2}:\quad r = \dfrac{2}{1 + \sin(\pi - \theta)}$

$r = \dfrac{2}{1 + \sin \pi \cos \theta - \cos \pi \sin \theta}$

$r = \dfrac{2}{1 + \sin \theta}$

Equivalent equation

Polar axis: $r = \dfrac{2}{1 + \sin(-\theta)}$

$r = \dfrac{2}{1 - \sin \theta}$

Not an equivalent equation

Pole: $-r = \dfrac{2}{1 + \sin \theta}$

Answer: Symmetric with respect to $\theta = \pi/2$

11. $r^2 = 16 \cos 2\theta$

$\theta = \dfrac{\pi}{2}:\quad (-r)^2 = 16 \cos 2(-\theta)$

$r^2 = 16 \cos 2\theta$

Equivalent equation

Polar axis: $r^2 = 16 \cos 2(-\theta)$

$r^2 = 16 \cos 2\theta$

Equivalent equation

Pole: $(-r)^2 = 16 \cos 2\theta$

$r^2 = 16 \cos 2\theta$

Equivalent equation

Answer: Symmetric with respect to $\theta = \dfrac{\pi}{2}$, the polar axis, and the pole

13. $|r| = |10(1 - \sin \theta)| = 10|1 - \sin \theta| \le 10(2) = 20$

$|1 - \sin \theta| = 2$

$1 - \sin \theta = 2$

$\sin \theta = -1$

$\theta = \dfrac{3\pi}{2}$

Maximum: $|r| = 20$ when $\theta = \dfrac{3\pi}{2}$

$0 = 10(1 - \sin \theta)$

$\sin \theta = 1$

$\theta = \dfrac{\pi}{2}$

Zero: $r = 0$ when $\theta = \dfrac{\pi}{2}$

15. $|r| = |4 \cos 3\theta| = 4|\cos 3\theta| \le 4$

$|\cos 3\theta| = 1$

$\cos 3\theta = \pm 1$

$\theta = 0, \dfrac{\pi}{3}, \dfrac{2\pi}{3}$

Maximum: $|r| = 4$ when $\theta = 0, \dfrac{\pi}{3}, \dfrac{2\pi}{3}$

$0 = 4 \cos 3\theta$

$\cos 3\theta = 0$

$\theta = \dfrac{\pi}{6}, \dfrac{\pi}{2}, \dfrac{5\pi}{6}$

Zero: $r = 0$ when $\theta = \dfrac{\pi}{6}, \dfrac{\pi}{2}, \dfrac{5\pi}{6}$

17. Circle: $r = 5$

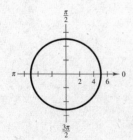

19. Circle: $r = \dfrac{\pi}{6}$

21. $r = 3 \sin \theta$

Symmetric with respect to $\theta = \pi/2$

Circle with a radius of $3/2$

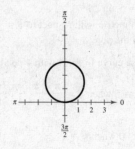

23. $r = 3(1 - \cos \theta)$

Symmetric with respect to the polar axis

$\frac{a}{b} = \frac{3}{3} = 1 \Rightarrow$ Cardioid

$|r| = 6$ when $\theta = \pi$

$r = 0$ when $\pi = 0$

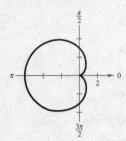

25. $r = 4(1 + \sin \theta)$

Symmetric with respect to $\theta = \frac{\pi}{2}$

$\frac{a}{b} = \frac{4}{4} = 1 \Rightarrow$ Cardioid

$|r| = 8$ when $\theta = \frac{\pi}{2}$

$r = 0$ when $\theta = \frac{3\pi}{2}$

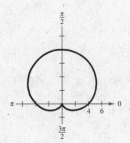

27. $r = 3 + 6 \sin \theta$

Symmetric with respect to $\theta = \frac{\pi}{2}$

$\frac{a}{b} = \frac{3}{6} < 1 \Rightarrow$ Limaçon with inner loop

$|r| = 9$ when $\theta = \frac{\pi}{2}$

$r = 0$ when $\theta = \frac{7\pi}{6}, \frac{11\pi}{6}$

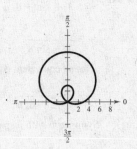

29. $r = 1 - 2 \sin \theta$

Symmetric with respect to $\theta = \frac{\pi}{2}$

$\frac{a}{b} = \frac{1}{2} < 1 \Rightarrow$ Limaçon with inner loop

$|r| = 3$ when $\theta = \frac{3\pi}{2}$

$r = 0$ when $\theta = \frac{\pi}{6}, \frac{5\pi}{6}$

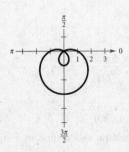

31. $r = 3 - 4 \cos \theta$

Symmetric with respect to the polar axis

$\frac{a}{b} = \frac{3}{4} < 1 \Rightarrow$ Limaçon with inner loop

$|r| = 7$ when $\theta = \pi$

$r = 0$ when $\cos \theta = \frac{3}{4}$ or $\theta \approx 0.723, 5.560$

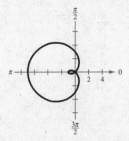

33. $r = 5 \sin 2\theta$

Symmetric with respect to $\theta = \pi/2$, the polar axis, and the pole

Rose curve ($n = 2$) with 4 petals

$|r| = 5$ when $\theta = \frac{\pi}{4}, \frac{3\pi}{4}, \frac{5\pi}{4}, \frac{7\pi}{4}$

$r = 0$ when $\theta = 0, \frac{\pi}{2}, \pi$

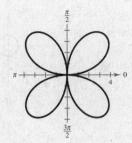

35. $r = 2 \sec \theta$

$r = \dfrac{2}{\cos \theta}$

$r \cos \theta = 2$

$x = 2 \implies$ Line

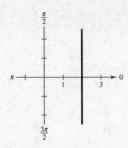

37. $r = \dfrac{3}{\sin \theta - 2 \cos \theta}$

$r(\sin \theta - 2 \cos \theta) = 3$

$y - 2x = 3$

$y = 2x + 3 \implies$ Line

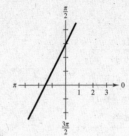

39. $r^2 = 9 \cos 2\theta$

Symmetric with respect to the polar axis, $\theta = \pi/2$, and the pole

Lemniscate

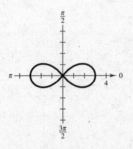

41. $r = 8 \cos \theta$

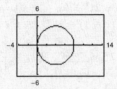

$0 \le \theta \le 2\pi$

θmin = 0
θmax = 2π
θstep = $\pi/24$
Xmin = -4
Xmax = 14
Xscl = 2
Ymin = -6
Ymax = 6
Yscl = 2

43. $r = 3(2 - \sin \theta)$

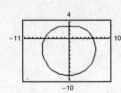

$0 \le \theta \le 2\pi$

θmin = 0
θmax = 2π
θstep = $\pi/24$
Xmin = -10
Xmax = 10
Xscl = 1
Ymin = -10
Ymax = 4
Yscl = 1

45. $r = 8 \sin \theta \cos^2 \theta$

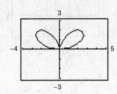

$0 \le \theta \le 2\pi$

θmin = 0
θmax = 2π
θstep = $\pi/24$
Xmin = -4
Xmax = 4
Xscl = 1
Ymin = -3
Ymax = 3
Yscl = 1

47. $r = 3 - 4 \cos \theta$

$0 \le \theta < 2\pi$

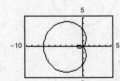

49. $r = 2 \cos\left(\dfrac{3\theta}{2}\right)$

$0 \le \theta < 4\pi$

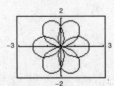

51. $r^2 = 9 \sin 2\theta$

$0 \le \theta < \pi$

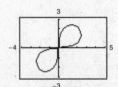

53.
$$r = 2 - \sec\theta = 2 - \frac{1}{\cos\theta}$$

$$r\cos\theta = 2\cos\theta - 1$$

$$r(r\cos\theta) = 2r\cos\theta - r$$

$$\left(\pm\sqrt{x^2+y^2}\right)x = 2x - \left(\pm\sqrt{x^2+y^2}\right)$$

$$\left(\pm\sqrt{x^2+y^2}\right)(x+1) = 2x$$

$$\left(\pm\sqrt{x^2+y^2}\right) = \frac{2x}{x+1}$$

$$x^2+y^2 = \frac{4x^2}{(x+1)^2}$$

$$y^2 = \frac{4x^2}{(x+1)^2} - x^2$$

$$= \frac{4x^2 - x^2(x+1)^2}{(x+1)^2} = \frac{4x^2 - x^2(x^2+2x+1)}{(x+1)^2}$$

$$= \frac{-x^4 - 2x^3 + 3x^2}{(x+1)^2} = \frac{-x^2(x^2+2x-3)}{(x+1)^2}$$

$$y = \pm\sqrt{\frac{x^2(3-2x-x^2)}{(x+1)^2}} = \pm\left|\frac{x}{x+1}\right|\sqrt{3-2x-x^2}$$

The graph has an asymptote at $x = -1$.

55. $r = \dfrac{3}{\theta}$

$$\theta = \frac{3}{r} = \frac{3\sin\theta}{r\sin\theta} = \frac{3\sin\theta}{y}$$

$$y = \frac{3\sin\theta}{\theta}$$

As $\theta \to 0, y \to 3$

57. True. For a graph to have polar axis symmetry, replace (r, θ) by $(r, -\theta)$ or $(-r, \pi - \theta)$.

59. $r = 6\cos\theta$

(a) $0 \le \theta \le \dfrac{\pi}{2}$

Upper half of circle

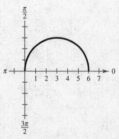

(b) $\dfrac{\pi}{2} \le \theta \le \pi$

Lower half of circle

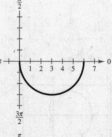

(c) $-\dfrac{\pi}{2} \le \theta \le \dfrac{\pi}{2}$

Entire circle

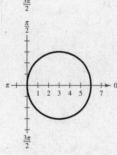

(d) $\dfrac{\pi}{4} \le \theta \le \dfrac{3\pi}{4}$

Left half of circle

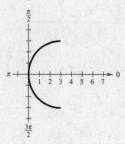

61. Let the curve $r = f(\theta)$ be rotated by ϕ to form the curve $r = g(\theta)$. If $\left(r_1, \theta_1\right)$ is a point on $r = f(\theta)$, then $\left(r_1, \theta_1 + \phi\right)$ is on $r = g(\theta)$. That is, $g\left(\theta_1 + \phi\right) = r_1 = f\left(\theta_1\right)$. Letting $\theta = \theta_1 + \phi$, or $\theta_1 = \theta - \phi$, we see that $g(\theta) = g\left(\theta_1 + \phi\right) = f\left(\theta_1\right) = f\left(\theta - \phi\right)$.

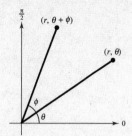

63. (a) $r = 2 - \sin\left(\theta - \dfrac{\pi}{4}\right)$

$= 2 - \left[\sin\theta\cos\dfrac{\pi}{4} - \cos\theta\sin\dfrac{\pi}{4}\right]$

$= 2 - \dfrac{\sqrt{2}}{2}(\sin\theta - \cos\theta)$

(c) $r = 2 - \sin(\theta - \pi)$

$= 2 - [\sin\theta\cos\pi - \cos\theta\sin\pi]$

$= 2 + \sin\theta$

(b) $r = 2 - \sin\left(\theta - \dfrac{\pi}{2}\right)$

$= 2 - \left[\sin\theta\cos\dfrac{\pi}{2} - \cos\theta\sin\dfrac{\pi}{2}\right]$

$= 2 + \cos\theta$

(d) $r = 2 - \sin\left(\theta - \dfrac{3\pi}{2}\right)$

$= 2 - \left[\sin\theta\cos\dfrac{3\pi}{2} - \cos\theta\sin\dfrac{3\pi}{2}\right]$

$= 2 - \cos\theta$

65. (a) $r = 1 - \sin\theta$

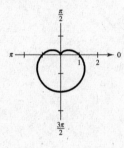

(b) $r = 1 - \sin\left(\theta - \dfrac{\pi}{4}\right)$

Rotate the graph in part (a) through the angle $\dfrac{\pi}{4}$.

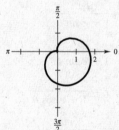

67. $r = 2 + k\sin\theta$

$k = 0$: $r = 2$

    Circle

$k = 1$: $r = 2 + \sin\theta$

    Convex limaçon

$k = 2$: $r = 2 + 2\sin\theta$

    Cardioid

$k = 3$: $r = 2 + 3\sin\theta$

    Limaçon with inner loop

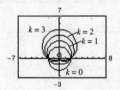

69. $y = \dfrac{x^2 - 9}{x + 1}$

$\dfrac{x^2 - 9}{x + 1} = 0$

$x^2 - 9 = 0$

$x^2 = 9$

$x = \pm 3$

71.
$$y = 5 - \frac{3}{x - 2}$$

$$5 - \frac{3}{x - 2} = 0$$

$$5 = \frac{3}{x - 2}$$

$$5(x - 2) = 3$$

$$5x - 10 = 3$$

$$5x = 13$$

$$x = \frac{13}{5}$$

73. Vertices: $(-4, 2), (2, 2) \implies$ Center at $(-1, 2)$ and $a = 3$

Minor axis of length 4: $2b = 4 \implies b = 2$

Horizontal major axis

$$\frac{(x - h)^2}{a^2} + \frac{(y - k)^2}{b^2} = 1$$

$$\frac{(x + 1)^2}{9} + \frac{(y - 2)^2}{4} = 1$$

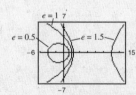

Section 10.9   Polar Equations of Conics

■ The graph of a polar equation of the form

$$r = \frac{ep}{1 \pm e \cos \theta} \quad \text{or} \quad r = \frac{ep}{1 \pm e \sin \theta}$$

is a conic, where $e > 0$ is the eccentricity and $|p|$ is the distance between the focus (pole) and the directrix.

  (a) If $e < 1$, the graph is an ellipse.

  (b) If $e = 1$, the graph is a parabola.

  (c) If $e > 1$, the graph is a hyperbola.

■ Guidelines for finding polar equations of conics:

  (a) Horizontal directrix above the pole: $r = \dfrac{ep}{1 + e \sin \theta}$

  (b) Horizontal directrix below the pole: $r = \dfrac{ep}{1 - e \sin \theta}$

  (c) Vertical directrix to the right of the pole: $r = \dfrac{ep}{1 + e \cos \theta}$

  (d) Vertical directrix to the left of the pole: $r = \dfrac{ep}{1 - e \cos \theta}$

Vocabulary Check

1. conic

2. eccentricity; e

3. vertical; right

4. (a) iii   (b) i   (c) ii

1. $r = \dfrac{4e}{1 + e \cos \theta}$

  (a) $e = 1, r = \dfrac{4}{1 + \cos \theta}$, parabola

  (b) $e = 0.5, r = \dfrac{2}{1 + 0.5 \cos \theta} = \dfrac{4}{2 + \cos \theta}$, ellipse

  (c) $e = 1.5, r = \dfrac{6}{1 + 1.5 \cos \theta} = \dfrac{12}{2 + 3 \cos \theta}$, hyperbola

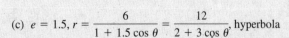

3. $r = \dfrac{4e}{1 - e \sin \theta}$

(a) $e = 1$, $r = \dfrac{4}{1 - \sin \theta}$, parabola

(b) $e = 0.5$, $r = \dfrac{2}{1 - 0.5 \sin \theta} = \dfrac{4}{2 - \sin \theta}$, ellipse

(c) $e = 1.5$, $r = \dfrac{6}{1 - 1.5 \sin \theta} = \dfrac{12}{2 - 3 \sin \theta}$, hyperbola

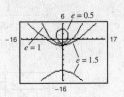

5. $r = \dfrac{2}{1 + \cos \theta}$

$e = 1 \Rightarrow$ Parabola

Vertical directrix to the right
of the pole
Matches graph (f).

7. $r = \dfrac{3}{1 + 2 \sin \theta}$

$e = 2 \Rightarrow$ Hyperbola

Matches graph (d).

9. $r = \dfrac{4}{2 + \cos \theta}$

$= \dfrac{2}{1 + 0.5 \cos \theta}$

$e = 0.5 \Rightarrow$ Ellipse
Matches graph (a).

11. $r = \dfrac{2}{1 - \cos \theta}$

$e = 1$, the graph is a parabola.

Vertex: $(1, \pi)$

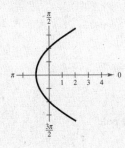

13. $r = \dfrac{5}{1 + \sin \theta}$

$e = 1$, the graph is a parabola.

Vertex: $\left(\dfrac{5}{2}, \dfrac{\pi}{2}\right)$

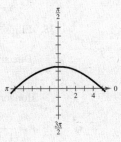

15. $r = \dfrac{2}{2 - \cos \theta} = \dfrac{1}{1 - (1/2) \cos \theta}$

$e = \dfrac{1}{2} < 1$, the graph is an ellipse.

Vertices: $(2, 0), \left(\dfrac{2}{3}, \pi\right)$

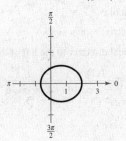

17. $r = \dfrac{6}{2 + \sin \theta} = \dfrac{3}{1 + (1/2) \sin \theta}$

$e = \dfrac{1}{2} < 1$, the graph is an ellipse.

Vertices: $\left(2, \dfrac{\pi}{2}\right), \left(6, \dfrac{3\pi}{2}\right)$

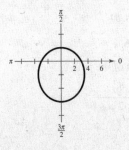

19. $r = \dfrac{3}{2 + 4 \sin \theta} = \dfrac{3/2}{1 + 2 \sin \theta}$

$e = 2 > 1$, the graph is a hyperbola.

Vertices: $\left(\dfrac{1}{2}, \dfrac{\pi}{2}\right), \left(-\dfrac{3}{2}, \dfrac{3\pi}{2}\right)$

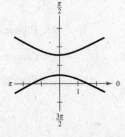

21. $r = \dfrac{3}{2 - 6 \cos \theta} = \dfrac{3/2}{1 - 3 \cos \theta}$

$e = 3 > 1$, the graph is a hyperbola.

Vertices: $\left(-\dfrac{3}{4}, 0\right), \left(\dfrac{3}{8}, \pi\right)$

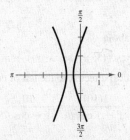

23. $r = \dfrac{4}{2 - \cos\theta} = \dfrac{2}{1 - (1/2)\cos\theta}$

$e = \dfrac{1}{2} < 1$, the graph is an ellipse.

Vertices: $(4, 0), \left(\dfrac{4}{3}, \pi\right)$

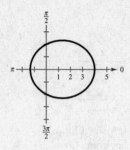

25. $r = \dfrac{-1}{1 - \sin\theta}$

$e = 1 \implies$ Parabola

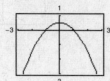

27. $r = \dfrac{3}{-4 + 2\cos\theta}$

$e = \dfrac{1}{2} \implies$ Ellipse

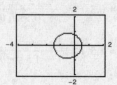

29. $r = \dfrac{2}{1 - \cos\left(\theta - \dfrac{\pi}{4}\right)}$

Rotate the graph in Exercise 11 through the angle $\pi/4$.

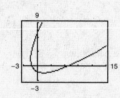

31. $r = \dfrac{6}{2 + \sin\left(\theta + \dfrac{\pi}{6}\right)}$

Rotate the graph in Exercise 17 through the angle $-\pi/6$.

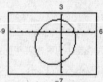
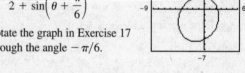

33. Parabola: $e = 1$

Directrix: $x = -1$

Vertical directrix to the left of the pole

$r = \dfrac{1(1)}{1 - 1\cos\theta} = \dfrac{1}{1 - \cos\theta}$

35. Ellipse: $e = \dfrac{1}{2}$

Directrix: $y = 1$

$p = 1$

Horizontal directrix above the pole

$r = \dfrac{(1/2)(1)}{1 + (1/2)\sin\theta} = \dfrac{1}{2 + \sin\theta}$

37. Hyperbola: $e = 2$

Directrix: $x = 1$

$p = 1$

Vertical directrix to the right of the pole

$r = \dfrac{2(1)}{1 + 2\cos\theta} = \dfrac{2}{1 + 2\cos\theta}$

39. Parabola

Vertex: $\left(1, -\dfrac{\pi}{2}\right) \implies e = 1, p = 2$

Horizontal directrix below the pole

$r = \dfrac{1(2)}{1 - 1\sin\theta} = \dfrac{2}{1 - \sin\theta}$

41. Parabola

Vertex: $(5, \pi) \implies e = 1, p = 10$

Vertical directrix to the left of the pole

$r = \dfrac{1(10)}{1 - 1\cos\theta} = \dfrac{10}{1 - \cos\theta}$

43. Ellipse: Vertices $(2, 0), (10, \pi)$

Center: $(4, \pi); c = 4, a = 6, e = \dfrac{2}{3}$

Vertical directrix to the right of the pole

$r = \dfrac{(2/3)p}{1 + (2/3)\cos\theta} = \dfrac{2p}{3 + 2\cos\theta}$

$2 = \dfrac{2p}{3 + 2\cos 0}$

$p = 5$

$r = \dfrac{2(5)}{3 + 2\cos\theta} = \dfrac{10}{3 + 2\cos\theta}$

45. Ellipse: Vertices $(20, 0)$, $(4, \pi)$

Center: $(8, 0)$; $c = 8$, $a = 12$, $e = \dfrac{2}{3}$

Vertical directrix to the left of the pole

$$r = \frac{(2/3)p}{1 - (2/3)\cos\theta} = \frac{2p}{3 - 2\cos\theta}$$

$$20 = \frac{2p}{3 - 2\cos 0}$$

$$p = 10$$

$$r = \frac{2(10)}{3 - 2\cos\theta} = \frac{20}{3 - 2\cos\theta}$$

47. Hyperbola: Vertices $\left(1, \dfrac{3\pi}{2}\right)$, $\left(9, \dfrac{3\pi}{2}\right)$

Center: $\left(5, \dfrac{3\pi}{2}\right)$; $c = 5$, $a = 4$, $e = \dfrac{5}{4}$

Horizontal directrix below the pole

$$r = \frac{(5/4)p}{1 - (5/4)\sin\theta} = \frac{5p}{4 - 5\sin\theta}$$

$$1 = \frac{5p}{4 - 5\sin(3\pi/2)}$$

$$p = \frac{9}{5}$$

$$r = \frac{5(9/5)}{4 - 5\sin\theta} = \frac{9}{4 - 5\sin\theta}$$

49. When $\theta = 0$, $r = c + a = ea + a = a(1 + e)$.

Therefore,

$$a(1 + e) = \frac{ep}{1 - e\cos 0}$$

$$a(1 + e)(1 - e) = ep$$

$$a(1 - e^2) = ep.$$

Thus, $r = \dfrac{ep}{1 - e\cos\theta} = \dfrac{(1 - e^2)a}{1 - e\cos\theta}$.

51. $r = \dfrac{[1 - (0.0167)^2](95.956 \times 10^6)}{1 - 0.0167\cos\theta} \approx \dfrac{9.5929 \times 10^7}{1 - 0.0167\cos\theta}$

Perihelion distance:
$r = 95.956 \times 10^6(1 - 0.0167) \approx 9.4354 \times 10^7$ miles

Aphelion distance:
$r = 95.956 \times 10^6(1 + 0.0167) \approx 9.7558 \times 10^7$ miles

53. $r = \dfrac{[1 - (0.0068)^2](108.209 \times 10^6)}{1 - 0.0068\cos\theta} \approx \dfrac{1.0820 \times 10^8}{1 - 0.0068\cos\theta}$

Perihelion distance: $r = 108.209 \times 10^6(1 - 0.0068) \approx 1.0747 \times 10^8$ kilometers

Aphelion distance: $r = 108.209 \times 10^6(1 + 0.0068) \approx 1.0894 \times 10^8$ kilometers

55. $r = \dfrac{[1 - (0.0934)^2](141.63 \times 10^6)}{1 - 0.0934\cos\theta} \approx \dfrac{1.4039 \times 10^8}{1 - 0.0934\cos\theta}$

Perihelion distance: $r = 141.63 \times 10^6(1 - 0.0934) \approx 1.2840 \times 10^8$ miles

Aphelion distance: $r = 141.63 \times 10^6(1 + 0.0934) \approx 1.5486 \times 10^8$ miles

57. $e \approx 0.847$, $a \approx \dfrac{4.42}{2} = 2.21$

$$2a = \frac{0.847p}{1 + 0.847} + \frac{0.847p}{1 - 0.847} \approx 5.9945p \approx 4.42$$

$$p \approx 0.737, \; ep \approx 0.624$$

$$r = \frac{0.624}{1 + 0.847\sin\theta}$$

To find the closest point to the sun, let $\theta = \dfrac{\pi}{2}$.

$$r = \frac{0.624}{1 + 0.847\sin(\pi/2)} \approx 0.338 \text{ astronomical units}$$

59. True. The graphs represent the same hyperbola, although the graphs are not traced out in the same order as θ goes from 0 to 2π.

61. True. See Exercise 63.

$$e = \frac{2}{3} < 1$$

63.
$$\frac{x^2}{a^2} + \frac{y^2}{b^2} = 1$$

$$\frac{r^2 \cos^2 \theta}{a^2} + \frac{r^2 \sin^2 \theta}{b^2} = 1$$

$$\frac{r^2 \cos^2 \theta}{a^2} + \frac{r^2(1 - \cos^2 \theta)}{b^2} = 1$$

$$r^2 b^2 \cos^2 \theta + r^2 a^2 - r^2 a^2 \cos^2 \theta = a^2 b^2$$

$$r^2(b^2 - a^2)\cos^2 \theta + r^2 a^2 = a^2 b^2$$

Since $b^2 - a^2 = -c^2$, we have:

$$-r^2 c^2 \cos^2 \theta + r^2 a^2 = a^2 b^2$$

$$-r^2 \left(\frac{c}{a}\right)^2 \cos^2 \theta + r^2 = b^2, e = \frac{c}{a}$$

$$-r^2 e^2 \cos^2 \theta + r^2 = b^2$$

$$r^2(1 - e^2 \cos^2 \theta) = b^2$$

$$r^2 = \frac{b^2}{1 - e^2 \cos^2 \theta}$$

65. $\dfrac{x^2}{169} + \dfrac{y^2}{144} = 1$

$a = 13, b = 12, c = 5, e = \dfrac{5}{13}$

$$r^2 = \frac{144}{1 - (25/169) \cos^2 \theta} = \frac{24{,}336}{169 - 25 \cos^2 \theta}$$

67. $\dfrac{x^2}{9} - \dfrac{y^2}{16} = 1$

$a = 3, b = 4, c = 5, e = \dfrac{5}{3}$

$$r^2 = \frac{-16}{1 - (25/9) \cos^2 \theta} = \frac{144}{25 \cos^2 \theta - 9}$$

69. One focus: $(5, 0)$

Vertices: $(4, 0), (4, 0)$

$a = 4, c = 5 \implies b = 3$ and $e = \dfrac{5}{4}$

$$\frac{x^2}{16} - \frac{y^2}{9} = 1$$

$$r^2 = \frac{-9}{1 - (25/16) \cos^2 \theta} = \frac{-144}{16 - 25 \cos^2 \theta}$$

71. $r = \dfrac{4}{1 - 0.4 \cos \theta}$

(a) Since $e < 1$, the conic is an ellipse.

(b) $r = \dfrac{4}{1 + 0.4 \cos \theta}$ has a vertical directrix to the right

of the pole and $r = \dfrac{4}{1 - 0.4 \sin \theta}$ has a horizontal directrix below

the pole. The given polar equation, $r = \dfrac{4}{1 - 0.4 \cos \theta}$, has a verti-

cal directrix to the left of the pole.

(c)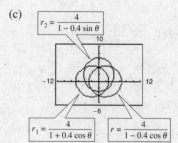

73. $4\sqrt{3}\tan\theta - 3 = 1$

$4\sqrt{3}\tan\theta = 4$

$\tan\theta = \dfrac{1}{\sqrt{3}}$

$\theta = \dfrac{\pi}{6} + n\pi$

75. $12\sin^2\theta = 9$

$\sin^2\theta = \dfrac{3}{4}$

$\sin\theta = \pm\dfrac{\sqrt{3}}{2}$

$\theta = \dfrac{\pi}{3} + n\pi, \dfrac{2\pi}{3} + n\pi$

77. $2\cot x = 5\cos\dfrac{\pi}{2}$

$2\cot x = 0$

$\cot x = 0$

$x = \dfrac{\pi}{2} + n\pi$

For 79 and 81 use the following:

    u and v are in Quadrant IV; $\sin u = -\dfrac{3}{5} \Rightarrow \cos u = \dfrac{4}{5}$; $\cos v = \dfrac{1}{\sqrt{2}} \Rightarrow \sin v = -\dfrac{1}{\sqrt{2}}$

79. $\cos(u + v) = \cos u \cos v - \sin u \sin v$

$= \left(\dfrac{4}{5}\right)\left(\dfrac{1}{\sqrt{2}}\right) - \left(-\dfrac{3}{5}\right)\left(-\dfrac{1}{\sqrt{2}}\right)$

$= \dfrac{4}{5\sqrt{2}} - \dfrac{3}{5\sqrt{2}}$

$= \dfrac{1}{5\sqrt{2}}$

$= \dfrac{\sqrt{2}}{10}$

81. $\cos(u - v) = \cos u \cos v + \sin u \sin v$

$= \left(\dfrac{4}{5}\right)\left(\dfrac{1}{\sqrt{2}}\right) + \left(-\dfrac{3}{5}\right)\left(-\dfrac{1}{\sqrt{2}}\right)$

$= \dfrac{4}{5\sqrt{2}} + \dfrac{3}{5\sqrt{2}}$

$= \dfrac{7}{5\sqrt{2}}$

$= \dfrac{7\sqrt{2}}{10}$

83. $\sin u = \dfrac{4}{5}, \dfrac{\pi}{2} < u < \pi \Rightarrow \cos u = -\dfrac{3}{5}$

$\sin 2u = 2\sin u \cos u$

$= 2\left(\dfrac{4}{5}\right)\left(-\dfrac{3}{5}\right)$

$= -\dfrac{24}{25}$

$\cos 2u = \cos^2 u - \sin^2 u$

$= \left(-\dfrac{3}{5}\right)^2 - \left(\dfrac{4}{5}\right)^2$

$= \dfrac{9}{25} - \dfrac{16}{25} = -\dfrac{7}{25}$

$\tan 2u = \dfrac{\sin 2u}{\cos 2u}$

$= \dfrac{-24/25}{-7/25}$

$= \dfrac{24}{7}$

85. $a_1 = 0, d = -\dfrac{1}{4}$

$a_n = a_1 + (n - 1)d$

$= 0 + (n - 1)\left(-\dfrac{1}{4}\right)$

$= -\dfrac{1}{4}n + \dfrac{1}{4}$

87. $a_3 = 27, a_8 = 72$

$a_8 = a_3 + 5d$

$72 = 27 + 5d \Rightarrow d = 9$

$a_1 = 27 - 2(9) = 9$

$a_n = a_1 + (n - 1)d$

$= 9 + (n - 1)(9)$

$= 9n$

89. $_{12}C_9 = \dfrac{12!}{(12 - 9)!9!} = \dfrac{12 \cdot 11 \cdot 10}{3!} = 220$

91. $_{10}P_3 = \dfrac{10!}{(10 - 3)!} = \dfrac{10!}{7!} = 10 \cdot 9 \cdot 8 = 720$

Review Exercises for Chapter 10

1. Points: $(-1, 2)$ and $(2, 5)$

$$m = \frac{5 - 2}{2 - (-1)} = \frac{3}{3} = 1$$

$$\tan \theta = 1 \implies \theta = \frac{\pi}{4} \text{ radian } = 45°$$

3. $y = 2x + 4 \implies m = 2$

$$\tan \theta = 2 \implies \theta = \arctan 2 \approx 1.1071 \text{ radians} \approx 63.43°$$

5. $4x + y = 2 \implies y = -4x + 2 \implies m_1 = -4$

$-5x + y = -1 \implies y = 5x - 1 \implies m_2 = 5$

$$\tan \theta = \left| \frac{5 - (-4)}{1 + (-4)(5)} \right| = \frac{9}{19}$$

$$\theta = \arctan \frac{9}{19} \approx 0.4424 \text{ radian} \approx 25.35°$$

7. $2x - 7y = 8 \implies y = \frac{2}{7}x - \frac{8}{7} \implies m_1 = \frac{2}{7}$

$0.4x + y = 0 \implies y = -0.4x \implies m_2 = -0.4$

$$\tan \theta = \left| \frac{-0.4 - (2/7)}{1 + (2/7)(-0.4)} \right| = \frac{24}{31}$$

$$\theta = \arctan \left(\frac{24}{31} \right) \approx 0.6588 \text{ radian} \approx 37.75°$$

9. $(1, 2) \implies x_1 = 1, y_1 = 2$

$x - y - 3 = 0 \implies A = 1, B = -1, C = -3$

$$d = \frac{|1(1) + (-1)(2) + (-3)|}{\sqrt{1^2 + (-1)^2}} = \frac{4}{\sqrt{2}} = 2\sqrt{2}$$

11. Hyperbola

13. Vertex: $(0, 0) = (h, k)$

Focus: $(4, 0) \implies p = 4$

$(y - k)^2 = 4p(x - h)$

$(y - 0)^2 = 4(4)(x - 0)$

$$y^2 = 16x$$

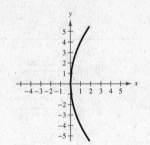

15. Vertex: $(0, 2) = (h, k)$

Directrix: $x = -3 \implies p = 3$

$(y - k)^2 = 4p(x - h)$

$(y - 2)^2 = 12x$

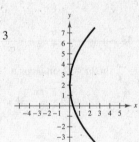

17. $x^2 = -2y \implies p = -\frac{1}{2}$

Focus: $\left(0, -\frac{1}{2} \right)$

$$d_1 = b + \frac{1}{2}$$

$$d_2 = \sqrt{(2 - 0)^2 + \left(-2 + \frac{1}{2} \right)^2}$$

$$= \sqrt{4 + \frac{9}{4}} = \frac{5}{2}$$

$d_1 = d_2$

$$b + \frac{1}{2} = \frac{5}{2}$$

$$b = 2$$

The slope of the line is

$$m = \frac{-2 - 2}{2 - 0} = -2.$$

Tangent line: $y = -2x + 2$

x-intercept: $(1, 0)$

19. Parabola

Opens downward

Vertex: $(0, 12)$

$(x - h)^2 = 4p(y - k)$

$x^2 = 4p(y - 12)$

Solution points: $(\pm 4, 10)$

$16 = 4p(10 - 12)$

$16 = -8p$

$-2 = p$

$x^2 = -8(y - 12)$

To find the x-intercepts, let $y = 0$.

$x^2 = 96$

$x = \pm\sqrt{96} = \pm 4\sqrt{6}$

At the base, the archway is $2(4\sqrt{6}) = 8\sqrt{6}$ meters wide.

21. Vertices: $(-3, 0), (7, 0) \implies a = 5$

$\qquad (h, k) = (2, 0)$

Foci: $(0, 0), (4, 0) \implies c = 2$

$b^2 = a^2 - c^2 = 25 - 4 = 21$

$$\frac{(x-h)^2}{a^2} + \frac{(y-k)^2}{b^2} = 1$$

$$\frac{(x-2)^2}{25} + \frac{y^2}{21} = 1$$

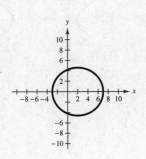

23. Vertices: $(0, 1), (4, 1) \implies a = 2, (h, k) = (2, 1)$

Endpoints of minor axis: $(2, 0), (2, 2) \implies b = 1$

$$\frac{(x-h)^2}{a^2} + \frac{(y-k)^2}{b^2} = 1$$

$$\frac{(x-2)^2}{4} + (y-1)^2 = 1$$

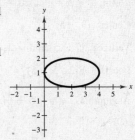

25. $2a = 10 \implies a = 5$

$b = 4$

$c^2 = a^2 - b^2 = 25 - 16 = 9 \implies c = 3$

The foci occur 3 feet from the center of the arch on a line connecting the tops of the pillars.

27. $\dfrac{(x+2)^2}{81} + \dfrac{(y-1)^2}{100} = 1$

$a = 10, b = 9, c = \sqrt{19}$

Center: $(-2, 1)$

Vertices: $(-2, 11)$ and $(-2, -9)$

Foci: $\left(-2, 1 \pm \sqrt{19}\right)$

Eccentricity: $e = \dfrac{\sqrt{19}}{10}$

29. $\qquad 16x^2 + 9y^2 - 32x + 72y + 16 = 0$

$16(x^2 - 2x + 1) + 9(y^2 + 8y + 16) = -16 + 16 + 144$

$\qquad 16(x-1)^2 + 9(y+4)^2 = 144$

$$\frac{(x-1)^2}{9} + \frac{(y+4)^2}{16} = 1$$

$a = 4, b = 3, c = \sqrt{7}$

Center: $(1, -4)$

Vertices: $(1, 0)$ and $(1, -8)$

Foci: $\left(1, -4 \pm \sqrt{7}\right)$

Eccentricity: $e = \dfrac{\sqrt{7}}{4}$

31. Vertices: $(0, \pm 1) \implies a = 1, (h, k) = (0, 0)$

Foci: $(0, \pm 3) \implies c = 3$

$b^2 = c^2 - a^2 = 9 - 1 = 8$

$$\frac{(y-k)^2}{a^2} - \frac{(x-h)^2}{b^2} = 1$$

$$y^2 - \frac{x^2}{8} = 1$$

33. Foci: $(0, 0), (8, 0) \implies c = 4, (h, k) = (4, 0)$

Asymptotes: $y = \pm 2(x - 4) \implies \dfrac{b}{a} = 2, b = 2a$

$b^2 = c^2 - a^2 \implies 4a^2 = 16 - a^2 \implies$

$a^2 = \dfrac{16}{5}, b^2 = \dfrac{64}{5}$

$$\frac{(x-h)^2}{a^2} - \frac{(y-k)^2}{b^2} = 1$$

$$\frac{(x-4)^2}{16/5} - \frac{y^2}{64/5} = 1$$

$$\frac{5(x-4)^2}{16} - \frac{5y^2}{64} = 1$$

35. $\dfrac{(x-3)^2}{16} - \dfrac{(y+5)^2}{4} = 1$

$a = 4, b = 2, c = \sqrt{20} = 2\sqrt{5}$

Center: $(3, -5)$

Vertices: $(7, -5)$ and $(-1, -5)$

Foci: $\left(3 \pm 2\sqrt{5}, -5\right)$

Asymptotes: $y = -5 \pm \dfrac{1}{2}(x-3)$

$$y = \dfrac{1}{2}x - \dfrac{13}{2} \quad \text{or} \quad y = -\dfrac{1}{2}x - \dfrac{7}{2}$$

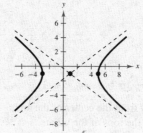

37. $\quad 9x^2 - 16y^2 - 18x - 32y - 151 = 0$

$9(x^2 - 2x + 1) - 16(y^2 + 2y + 1) = 151 + 9 - 16$

$9(x-1)^2 - 16(y+1)^2 = 144$

$\dfrac{(x-1)^2}{16} - \dfrac{(y+1)^2}{9} = 1$

$a = 4, b = 3, c = 5$

Center: $(1, -1)$

Vertices: $(5, -1)$ and $(-3, -1)$

Foci: $(6, -1)$ and $(-4, -1)$

Asymptotes: $y = -1 \pm \dfrac{3}{4}(x-1)$

$$y = \dfrac{3}{4}x - \dfrac{7}{4} \quad \text{or} \quad y = -\dfrac{3}{4}x - \dfrac{1}{4}$$

39. Foci: $(\pm 100, 0) \implies c = 100$

Center: $(0, 0)$

$\dfrac{d_2}{186{,}000} - \dfrac{d_1}{186{,}000} = 0.0005 \implies d_2 - d_1 = 93 = 2a \implies a = 46.5$

$b^2 = c^2 - a^2 = 100^2 - 46.5^2 = 7837.75$

$\dfrac{x^2}{2162.25} - \dfrac{y^2}{7837.75} = 1$

$y^2 = 7837.75\left(\dfrac{60^2}{2162.25} - 1\right) \approx 5211.5736$

$y \approx 72$ miles

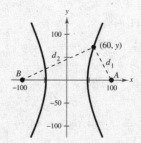

41. $5x^2 - 2y^2 + 10x - 4y + 17 = 0$

$AC = 5(-2) = -10 < 0$

The graph is a hyperbola.

43. $3x^2 + 2y^2 - 12x + 12y + 29 = 0$

$A = 3, C = 2$

$AC = 3(2) = 6 > 0$

The graph is an ellipse.

45. $xy - 4 = 0$

$A = C = 0, B = 1$

$B^2 - 4AC = 1^2 - 4(0)(0) = 1 > 0$

The graph is a hyperbola.

$\cot 2\theta = 0 \implies 2\theta = \dfrac{\pi}{2} \implies \theta = \dfrac{\pi}{4}$

$x = x' \cos \dfrac{\pi}{4} - y' \sin \dfrac{\pi}{4} = \dfrac{x' - y'}{\sqrt{2}}$

$y = x' \sin \dfrac{\pi}{4} + y' \cos \dfrac{\pi}{4} = \dfrac{x' + y'}{\sqrt{2}}$

$\left(\dfrac{x' - y'}{\sqrt{2}}\right)\left(\dfrac{x' + y'}{\sqrt{2}}\right) - 4 = 0$

$\dfrac{(x')^2 - (y')^2}{2} = 4$

$\dfrac{(x')^2}{8} - \dfrac{(y')^2}{8} = 1$

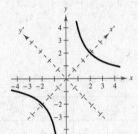

47. $5x^2 - 2xy + 5y^2 - 12 = 0$

$A = C = 5, B = -2$

$B^2 - 4AC = (-2)^2 - 4(5)(5) = -96 < 0$

The graph is an ellipse.

$\cot 2\theta = 0 \implies 2\theta = \dfrac{\pi}{2} \implies \theta = \dfrac{\pi}{4}$

$x = x' \cos \dfrac{\pi}{4} - y' \sin \dfrac{\pi}{4} = \dfrac{x' - y'}{\sqrt{2}}$

$y = x' \sin \dfrac{\pi}{4} + y' \cos \dfrac{\pi}{4} = \dfrac{x' + y'}{\sqrt{2}}$

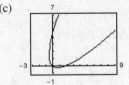

$5\left(\dfrac{x' - y'}{\sqrt{2}}\right)^2 - 2\left(\dfrac{x' - y'}{\sqrt{2}}\right)\left(\dfrac{x' + y'}{\sqrt{2}}\right) + 5\left(\dfrac{x' + y'}{\sqrt{2}}\right)^2 - 12 = 0$

$\dfrac{5}{2}[(x')^2 - 2(x'y') + (y')^2] - [(x')^2 - (y')^2] + \dfrac{5}{2}[(x')^2 + 2(x'y') + (y')^2] = 12$

$4(x')^2 + 6(y')^2 = 12$

$\dfrac{(x')^2}{3} + \dfrac{(y')^2}{2} = 1$

49. (a) $16x^2 - 24xy + 9y^2 - 30x - 40y = 0$

$B^2 - 4AC = (-24)^2 - 4(16)(9) = 0$

The graph is a parabola.

(b) To use a graphing utility, we need to solve for y in terms of x.

$9y^2 + (-24x - 40)y + (16x^2 - 30x) = 0$

$y = \dfrac{-(-24x - 40) \pm \sqrt{(-24x - 40)^2 - 4(9)(16x^2 - 30x)}}{2(9)}$

$= \dfrac{(24x + 40) \pm \sqrt{(24x + 40)^2 - 36(16x^2 - 30x)}}{18}$

(c)

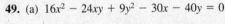

51. (a) $x^2 + y^2 + 2xy + 2\sqrt{2}x - 2\sqrt{2}y + 2 = 0$

$B^2 - 4AC = 2^2 - 4(1)(1) = 0$

The graph is a parabola.

(b) To use a graphing utility, we need to solve for y in terms of x.

$y^2 + \left(2x - 2\sqrt{2}\right)y + \left(x^2 + 2\sqrt{2}x + 2\right) = 0$

$$y = \frac{-\left(2x - 2\sqrt{2}\right) \pm \sqrt{\left(2x - 2\sqrt{2}\right)^2 - 4\left(x^2 + 2\sqrt{2}x + 2\right)}}{2}$$

(c)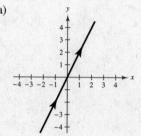

53. $x = 3t - 2, y = 7 - 4t$

t	-3	-2	0	1	2	3
x	-11	-8	-2	1	4	7
y	19	15	7	3	-1	-5

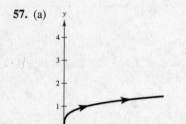

55. (a)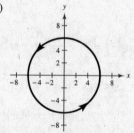

(b) $x = 2t \implies \dfrac{x}{2} = t$

$y = 4t \implies y = 4\left(\dfrac{x}{2}\right) = 2x$

57. (a)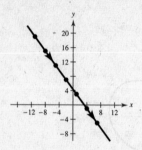

(b) $x = t^2, \ x \geq 0$

$y = \sqrt{t} \implies y^2 = t$

$x = (y^2)^2 \implies x$

$= y^4 \implies y = \sqrt[4]{x}$

59. (a)

(b) $x = 6\cos\theta, y = 6\sin\theta$

$\cos\theta = \dfrac{x}{6}, \sin\theta = \dfrac{y}{6}$

$\dfrac{x^2}{36} + \dfrac{y^2}{36} = 1$

$x^2 + y^2 = 36$

61. Center: $(5, 4)$

Radius: 6

$x = h + r\cos\theta = 5 + 6\cos\theta$

$y = k + r\sin\theta = 4 + 6\sin\theta$

63. Hyperbola

Vertices: $(0, \pm 4)$

Foci: $(0, \pm 5)$

Center: $(0, 0)$

$a = 4, c = 5, b = \sqrt{c^2 - a^2} = 3$

$x = 3\tan\theta, y = 4\sec\theta$

65. Polar coordinates: $\left(2, \dfrac{\pi}{4}\right)$

Additional polar representations:

$\left(2, -\dfrac{7\pi}{4}\right), \left(-2, \dfrac{5\pi}{4}\right)$

67. Polar coordinates: $(-7, 4.19)$

Additional polar representations: $(7, 1.05), (-7, -2.09)$

69. Polar coordinates: $\left(-1, \dfrac{\pi}{3}\right)$

$x = -1 \cos \dfrac{\pi}{3} = -\dfrac{1}{2}$

$y = -1 \sin \dfrac{\pi}{3} = -\dfrac{\sqrt{3}}{2}$

Rectangular coordinates: $\left(-\dfrac{1}{2}, -\dfrac{\sqrt{3}}{2}\right)$

71. Polar coordinates: $\left(3, \dfrac{3\pi}{4}\right)$

$x = 3 \cos \dfrac{3\pi}{4} = -\dfrac{3\sqrt{2}}{2}$

$y = 3 \sin \dfrac{3\pi}{4} = \dfrac{3\sqrt{2}}{2}$

Rectangular coordinates: $\left(-\dfrac{3\sqrt{2}}{2}, \dfrac{3\sqrt{2}}{2}\right)$

73. Rectangular coordinates: $(0, 2)$

$r = \pm \sqrt{0^2 + 2^2} = \pm 2$

$\tan \theta$ is undefined $\implies \theta = \dfrac{\pi}{2}, \dfrac{3\pi}{2}$

Polar coordinates: $\left(2, \dfrac{\pi}{2}\right)$ or $\left(-2, \dfrac{3\pi}{2}\right)$

75. Rectangular coordinates: $(4, 6)$

$r = \pm \sqrt{4^2 + 6^2} = \pm \sqrt{52} = \pm 2\sqrt{13}$

$\tan \theta = \dfrac{6}{4} \implies \theta \approx 0.9828, 4.1244$

Polar coordinates: $\left(2\sqrt{13}, 0.9828\right)$ or $\left(-2\sqrt{13}, 4.1244\right)$

77. $x^2 + y^2 = 49$

$r^2 = 49$

$r = 7$

79. $x^2 + y^2 - 6y = 0$

$r^2 - 6r \sin \theta = 0$

$r(r - 6 \sin \theta) = 0$

$\qquad r = 0 \text{ or } r = 6 \sin \theta$

Since $r = 6 \sin \theta$ contains $r = 0$, we just have $r = 6 \sin \theta$.

81. $xy = 5$

$(r \cos \theta)(r \sin \theta) = 5$

$r^2 = \dfrac{5}{\sin \theta \cos \theta}$

$\quad = \dfrac{10}{\sin 2\theta} = 10 \csc 2\theta$

83. $\quad r = 5$

$\quad r^2 = 25$

$x^2 + y^2 = 25$

85. $\qquad r = 3 \cos \theta$

$\quad r^2 = 3r \cos \theta$

$x^2 + y^2 = 3x$

87. $\qquad r^2 = \sin \theta$

$\qquad r^3 = r \sin \theta$

$\left(\pm \sqrt{x^2 + y^2}\right)^3 = y$

$\qquad (x^2 + y^2)^3 = y^2$

$\qquad x^2 + y^2 = y^{2/3}$

89. $r = 4$

Circle of radius 4 centered at the pole

Symmetric with respect to $\theta = \pi/2$, the polar axis and the pole

Maximum value of $|r| = 4$, for all values of θ

Zeros: None

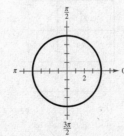

91. $r = 4 \sin 2\theta$

Rose curve ($n = 2$) with 4 petals

Symmetric with respect to $\theta = \pi/2$, the polar axis, and the pole

Maximum value of $|r| = 4$ when $\theta = \dfrac{\pi}{4}, \dfrac{3\pi}{4}, \dfrac{5\pi}{4}, \dfrac{7\pi}{4}$

Zeros: $r = 0$ when $\theta = 0, \dfrac{\pi}{2}, \pi, \dfrac{3\pi}{2}$

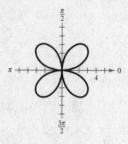

93. $r = -2(1 + \cos \theta)$

Symmetric with respect to the polar axis

Maximum value of $|r| = 4$ when $\theta = 0$

Zeros: $r = 0$ when $\theta = \pi$

$\dfrac{a}{b} = \dfrac{2}{2} = 1 \implies$ Cardioid

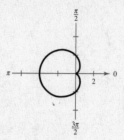

95. $r = 2 + 6 \sin \theta$

Limaçon with inner loop

$r = f(\sin \theta) \implies \theta = \dfrac{\pi}{2}$ symmetry

Maximum value: $|r| = 8$ when $\theta = \dfrac{\pi}{2}$

Zeros: $2 + 6 \sin \theta = 0 \implies \sin \theta = -\dfrac{1}{3} \implies \theta \approx 3.4814, 5.9433$

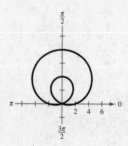

97. $r = -3 \cos 2\theta$

Rose curve with 4 petals

$r = f(\cos \theta) \implies$ polar axis symmetry

$\theta = \dfrac{\pi}{2}: \quad r = -3 \cos 2(\pi - \theta) = -3 \cos(2\pi - 2\theta) = -3 \cos 2\theta$

Equivalent equation $\implies \theta = \dfrac{\pi}{2}$ symmetry

Pole: $\quad r = -3 \cos 2(\pi + \theta) = -3 \cos(2\pi + 2\theta) = -3 \cos 2\theta$

Equivalent equation $\implies$ pole symmetry

Maximum value: $|r| = 3$ when $\theta = 0, \dfrac{\pi}{2}, \pi, \dfrac{3\pi}{2}$

Zeros: $-3 \cos 2\theta = 0$ when $\cos 2\theta = 0 \implies \theta = \dfrac{\pi}{4}, \dfrac{3\pi}{4}, \dfrac{5\pi}{4}, \dfrac{7\pi}{4}$

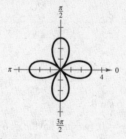

99. $r = 3(2 - \cos \theta)$

$= 6 - 3 \cos \theta$

$\dfrac{a}{b} = \dfrac{6}{3} = 2$

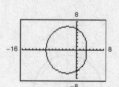

The graph is a convex limaçon.

101. $r = 4 \cos 3\theta$

The graph is a rose curve with 3 petals.

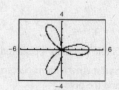

103. $r = \dfrac{1}{1 + 2 \sin \theta}$, $e = 2$

Hyperbola symmetric with respect to $\theta = \dfrac{\pi}{2}$ and having

vertices at $\left(\dfrac{1}{3}, \dfrac{\pi}{2}\right)$ and $\left(-1, \dfrac{3\pi}{2}\right)$.

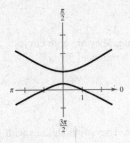

105. $r = \dfrac{4}{5 - 3 \cos \theta}$

$r = \dfrac{4/5}{1 - (3/5) \cos \theta}$, $e = \dfrac{3}{5}$

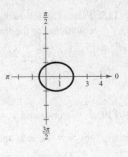

Ellipse symmetric with
respect to the polar axis and
having vertices at $(2, 0)$ and
$(1/2, \pi)$.

107. Parabola: $r = \dfrac{ep}{1 - e \cos \theta}$, $e = 1$

Vertex: $(2, \pi)$

Focus: $(0, 0) \implies p = 4$

$r = \dfrac{4}{1 - \cos \theta}$

109. Ellipse: $r = \dfrac{ep}{1 - e \cos \theta}$

Vertices: $(5, 0), (1, \pi) \implies a = 3$

One focus: $(0, 0) \implies c = 2$

$e = \dfrac{c}{a} = \dfrac{2}{3}, p = \dfrac{5}{2}$

$r = \dfrac{(2/3)(5/2)}{1 - (2/3) \cos \theta} = \dfrac{5/3}{1 - (2/3) \cos \theta}$

$= \dfrac{5}{3 - 2 \cos \theta}$

111.

$$a + c = 122{,}800 + 4000 \implies a + c = 126{,}800$$
$$a - c = 119 + 4000 \implies \underline{a - c = 4{,}119}$$
$$2a = 130{,}919$$
$$a = 65{,}459.5$$
$$c = 61{,}340.5$$

$e = \dfrac{c}{a} = \dfrac{61{,}340.5}{65{,}459.5} \approx 0.937$

$r = \dfrac{ep}{1 - e \cos \theta} \approx \dfrac{0.937p}{1 - 0.937 \cos \theta}$

$r = 126{,}800$ when $\theta = 0$

$126{,}800 = \dfrac{ep}{1 - e \cos 0}$

$ep = 126{,}800\left(1 - \dfrac{61{,}340.5}{65{,}459.5}\right) \approx 7978.81$

Thus, $r \approx \dfrac{7978.81}{1 - 0.937 \cos \theta}$.

When $\theta = \dfrac{\pi}{3}, r \approx \dfrac{7978.81}{1 - 0.937 \cos(\pi/3)} \approx 15{,}011.87$ miles.

The distance from the surface of Earth and the
satellite is $15{,}011.87 - 4000 \approx 11{,}011.87$ miles.

113. False. When classifying equations of the form
$Ax^2 + Bxy + Cy^2 + Dx + Ey + F = 0$,
its graph can be determined by its discriminant. For a
graph to be a parabola, its discriminant, $B^2 - 4AC$, must
equal zero. So, if $B = 0$, then A **or** C equals 0, but not
both.

115. False. The following are **two** sets of parametric equations for the line.

$x = t, y = 3 - 2t$

$x = 3t, y = 3 - 6t$

117. $2a = 10 \implies a = 5$

b must be less than 5; $0 < b < 5$.

As b approaches 5, the ellipse becomes more circular and approaches a circle of radius 5.

119. $x = 4 \cos t$ and $y = 3 \sin t$

(a) $x = 4 \cos 2t$ and $y = 3 \sin 2t$

The speed would double.

(b) $x = 5 \cos t$ and $y = 3 \sin t$

The elliptical orbit would be flatter. The length of the major axis is greater.

121. (a) $x^2 + y^2 = 25$

$r = 5$

The graphs are the same. They are both circles centered at $(0, 0)$ with a radius of 5.

(b) $x - y = 0 \implies y = x$

$$\theta = \frac{\pi}{4}$$

The graphs are the same. They are both lines with slope 1 and intercept $(0, 0)$.

Problem Solving for Chapter 10

1. (a) $\theta = \pi - 1.10 - 0.84 \approx 1.2016$ radians

(b) $\sin 0.84 = \dfrac{x}{3250} \implies x = 3250 \sin 0.84 \approx 2420$ feet

$\sin 1.10 = \dfrac{y}{6700} \implies y = 6700 \sin 1.10 \approx 5971$ feet

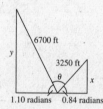

3. Since the axis of symmetry is the x-axis, the vertex is $(h, 0)$ and $y^2 = 4p(x - h)$. Also, since the focus is $(0, 0)$, $0 - h = p \implies h = -p$ and $y^2 = 4p(x + p)$.

5. (a)

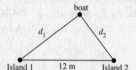

Since $d_1 + d_2 \le 20$, by definition, the outer bound that the boat can travel is an ellipse. The islands are the foci.

(c) $d_1 + d_2 = 2a = 20 \implies a = 10$

The boat traveled 20 miles. The vertex is $(10, 0)$.

(b)

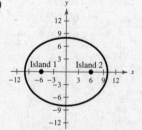

Island 1 is located at $(-6, 0)$ and Island 2 is located at $(6, 0)$.

(d) $c = 6, a = 10 \implies b^2 = a^2 - c^2 = 64$

$$\frac{x^2}{100} + \frac{y^2}{64} = 1$$

7. $Ax^2 + Cy^2 + Dx + Ey + F = 0$

Assume that the conic is *not* degenerate.

(a) $A = C, A \neq 0$

$$Ax^2 + Ay^2 + Dx + Ey + F = 0$$

$$x^2 + y^2 + \frac{D}{A}x + \frac{E}{A}y + \frac{F}{A} = 0$$

$$\left(x^2 + \frac{D}{A}x + \frac{D^2}{4A^2}\right) + \left(y^2 + \frac{E}{A}y + \frac{E^2}{4A^2}\right) = -\frac{F}{A} + \frac{D^2}{4A^2} + \frac{E^2}{4A^2}$$

$$\left(x + \frac{D}{2A}\right)^2 + \left(y + \frac{E}{2A}\right)^2 = \frac{D^2 + E^2 - 4AF}{4A^2}$$

This is a circle with center $\left(-\dfrac{D}{2A}, -\dfrac{E}{2A}\right)$ and radius $\dfrac{\sqrt{D^2 + E^2 - 4AF}}{2|A|}$.

(b) $A = 0$ or $C = 0$ (but not both). Let $C = 0$.

$$Ax^2 + Dx + Ey + F = 0$$

$$x^2 + \frac{D}{A}x = -\frac{E}{A}y - \frac{F}{A}$$

$$x^2 + \frac{D}{A}x + \frac{D^2}{4A^2} = -\frac{E}{A}y - \frac{F}{A} + \frac{D^2}{4A^2}$$

$$\left(x + \frac{D}{2A}\right)^2 = -\frac{E}{A}\left(y + \frac{F}{E} - \frac{D^2}{4AE}\right)$$

This is a parabola with vertex $\left(-\dfrac{D}{2A}, \dfrac{D^2 - 4AF}{4AE}\right)$.

$A = 0$ yields a similar result.

(c) $AC > 0 \implies A$ and C are either both positive or are both negative (if that is the case, move the terms to the other side of the equation so that they are both positive).

$$Ax^2 + Cy^2 + Dx + Ey + F = 0$$

$$A\left(x^2 + \frac{D}{A}x + \frac{D^2}{4A^2}\right) + C\left(y^2 + \frac{E}{C}y + \frac{E^2}{4C^2}\right) = -F + \frac{D^2}{4A} + \frac{E^2}{4C}$$

$$A\left(x + \frac{D}{2A}\right)^2 + C\left(y + \frac{E}{2C}\right)^2 = \frac{CD^2 + AE^2 - 4ACF}{4AC}$$

$$\frac{\left(x + \dfrac{D}{2A}\right)^2}{\dfrac{CD^2 + AE^2 - 4ACF}{4A^2C}} + \frac{\left(y + \dfrac{E}{2C}\right)^2}{\dfrac{CD^2 + AE^2 - 4ACF}{4AC^2}} = 1$$

Since A and C are both positive, $4A^2C$ and $4AC^2$ are both positive. $CD^2 + AE^2 - 4ACF$ must be positive or the conic is degenerate. Thus, we have an ellipse with center $\left(-\dfrac{D}{2A}, -\dfrac{E}{2C}\right)$.

(d) $AC < 0 \implies A$ and C have opposite signs. Let's assume that A is positive and C is negative. (If A is negative and C is positive, move the terms to the other side of the equation.) From part (c) we have

$$\frac{\left(x + \dfrac{D}{2A}\right)^2}{\dfrac{CD^2 + AE^2 - 4ACF}{4A^2C}} + \frac{\left(y + \dfrac{E}{2C}\right)^2}{\dfrac{CD^2 + AE^2 - 4ACF}{4AC^2}} = 1.$$

Since $A > 0$ and $C < 0$, the first denominator is positive if $CD^2 + AE^2 - 4ACF < 0$ and is negative if $CD^2 + AE^2 - 4ACF > 0$, since $4A^2C$ is negative. The second denominator would have the *opposite* sign since $4AC^2 > 0$. Thus, we have a hyperbola with center $\left(-\dfrac{D}{2A}, -\dfrac{E}{2C}\right)$.

9. To change the orientation, we can just replace t with $-t$.

$$x = \cos(-t) = \cos t$$

$$y = 2\sin(-t) = -2\sin t$$

11. (a) $y^2 = \dfrac{t^2(1-t^2)^2}{(1+t^2)^2}$, $x^2 = \dfrac{(1-t^2)^2}{(1+t^2)^2}$

(b)
$$r^2 \sin^2 \theta = r^2 \cos^2 \theta \left(\frac{1 - r\cos\theta}{1 + r\cos\theta} \right)$$

$$\frac{1-x}{1+x} = \frac{1 - \left(\dfrac{1-t^2}{1+t^2}\right)}{1 + \left(\dfrac{1-t^2}{1+t^2}\right)} = \frac{2t^2}{2} = t^2$$

$$\sin^2 \theta (1 + r\cos\theta) = \cos^2 \theta (1 - r\cos\theta)$$

$$r\cos\theta \sin^2 \theta + \sin^2 \theta = \cos^2 \theta - r\cos^3 \theta$$

$$r\cos\theta(\sin^2\theta + \cos^2\theta) = \cos^2\theta - \sin^2\theta$$

Thus, $y^2 = x^2 \left(\dfrac{1-x}{1+x} \right)$.

$$r\cos\theta = \cos 2\theta$$

$$r = \cos 2\theta \cdot \sec \theta$$

(c)

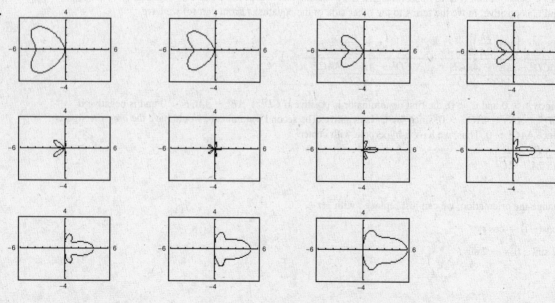

13. $r = a\sin\theta + b\cos\theta$

$r^2 = r(a\sin\theta + b\cos\theta)$

$r^2 = ar\,\sin\theta + br\cos\theta$

$$x^2 + y^2 = ay + bx$$

$$x^2 + y^2 - bx - ay = 0$$

$$\left(x^2 - bx + \frac{b^2}{4}\right) + \left(y^2 - ay + \frac{a^2}{4}\right) = \frac{a^2}{4} + \frac{b^2}{4}$$

$$\left(x - \frac{b}{2}\right)^2 + \left(y - \frac{a}{2}\right)^2 = \frac{a^2 + b^2}{4}$$

This represents a circle with center $\left(\dfrac{b}{2}, \dfrac{a}{2}\right)$ and radius $r = \dfrac{1}{2}\sqrt{a^2 + b^2}$.

15.

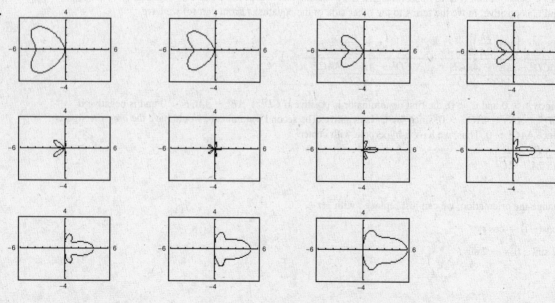

$n = 1, 2, 3, 4, 5$ produce "bells"; $n = -1, -2, -3, -4, -5$ produce "hearts".

Chapter 10   Practice Test

1. Find the angle, θ, between the lines $3x + 4y = 12$ and $4x - 3y = 12$.

2. Find the distance between the point $(5, -9)$ and the line $3x - 7y = 21$.

3. Find the vertex, focus and directrix of the parabola $x^2 - 6x - 4y + 1 = 0$.

4. Find an equation of the parabola with its vertex at $(2, -5)$ and focus at $(2, -6)$.

5. Find the center, foci, vertices, and eccentricity of the ellipse $x^2 + 4y^2 - 2x + 32y + 61 = 0$.

6. Find an equation of the ellipse with vertices $(0, \pm 6)$ and eccentricity $e = \frac{1}{2}$.

7. Find the center, vertices, foci, and asymptotes of the hyperbola $16y^2 - x^2 - 6x - 128y + 231 = 0$.

8. Find an equation of the hyperbola with vertices at $(\pm 3, 2)$ and foci at $(\pm 5, 2)$.

9. Rotate the axes to eliminate the xy-term. Sketch the graph of the resulting equation, showing both sets of axes.

   $5x^2 + 2xy + 5y^2 - 10 = 0$

10. Use the discriminant to determine whether the graph of the equation is a parabola, ellipse, or hyperbola.

    (a) $6x^2 - 2xy + y^2 = 0$ (b) $x^2 + 4xy + 4y^2 - x - y + 17 = 0$

11. Convert the polar point $\left(\sqrt{2}, \dfrac{3\pi}{4}\right)$ to rectangular coordinates.

12. Convert the rectangular point $\left(\sqrt{3}, -1\right)$ to polar coordinates.

13. Convert the rectangular equation $4x - 3y = 12$ to polar form.

14. Convert the polar equation $r = 5 \cos \theta$ to rectangular form.

15. Sketch the graph of $r = 1 - \cos \theta$.

16. Sketch the graph of $r = 5 \sin 2\theta$.

17. Sketch the graph of $r = \dfrac{3}{6 - \cos \theta}$.

18. Find a polar equation of the parabola with its vertex at $\left(6, \dfrac{\pi}{2}\right)$ and focus at $(0, 0)$.

For Exercises 19 and 20, eliminate the parameter and write the corresponding rectangular equation.

19. $x = 3 - 2 \sin \theta, y = 1 + 5 \cos \theta$ 20. $x = e^{2t}, y = e^{4t}$

C H A P T E R   1 1
Analytic Geometry in Three Dimensions

Section 11.1   The Three-Dimensional Coordinate System **557**

Section 11.2   Vectors in Space **561**

Section 11.3   The Cross Product of Two Vectors **564**

Section 11.4   Lines and Planes in Space **568**

Review Exercises **571**

Problem Solving **575**

Practice Test **578**

CHAPTER 11
Analytic Geometry in Three Dimensions

Section 11.1   The Three-Dimensional Coordinate System

■ You should be able to plot points in the three-dimensional coordinate system.
■ The distance between the points (x_1, y_1, z_1) and (x_2, y_2, z_2) is
$$d = \sqrt{(x_2 - x_1)^2 + (y_2 - y_1)^2 + (z_2 - z_1)^2}.$$
■ The midpoint of the line segment joining the points (x_1, y_1, z_1) and (x_2, y_2, z_2) is
$$\left(\frac{x_1 + x_2}{2}, \frac{y_1 + y_2}{2}, \frac{z_1 + z_2}{2}\right).$$
■ The equation of the sphere with center (h, k, j) and radius r is
$$(x - h)^2 + (y - k)^2 + (z - j)^2 = r^2.$$
■ You should be able to find the trace of a surface in space.

Vocabulary Check

1. three-dimensional

2. xy-plane, xz-plane, yz-plane

3. octants

4. Distance Formula

5. $\left(\dfrac{x_1 + x_2}{2}, \dfrac{y_1 + y_2}{2}, \dfrac{z_1 + z_2}{2}\right)$

6. sphere

7. surface, space

8. trace

---

1. $A(-1, 4, 3)$
$B(1, 3, -2)$
$C(-3, 0, -2)$

3.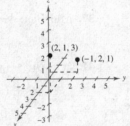

5.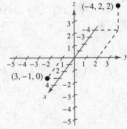

7. $x = -3, y = 3, z = 4$: $(-3, 3, 4)$

9. $y = z = 0, x = 10$: $(10, 0, 0)$

11. Octant IV

13. Octants I, II, III, IV
(above the xy-plane)

15. Octants II, IV, VI, VIII

17. $d = \sqrt{(5 - 0)^2 + (2 - 0)^2 + (6 - 0)^2}$
$= \sqrt{25 + 4 + 36}$
$= \sqrt{65}$ units

19. $d = \sqrt{(7 - 3)^2 + (4 - 2)^2 + (8 - 5)^2}$
$= \sqrt{4^2 + 2^2 + 3^2}$
$= \sqrt{16 + 4 + 9}$
$= \sqrt{29}$
≈ 5.385

21. $d = \sqrt{[6 - (-1)]^2 + [0 - 4]^2 + [-9 - (-2)]^2}$

$\quad = \sqrt{7^2 + 4^2 + 7^2}$

$\quad = \sqrt{49 + 16 + 49}$

$\quad = \sqrt{114}$

$\quad \approx 10.677$

23. $d = \sqrt{(1 - 0)^2 + [0 - (-3)]^2 + (-10 - 0)^2}$

$\quad = \sqrt{1 + 9 + 100}$

$\quad = \sqrt{110} \approx 10.488$

25. $d_1 = \sqrt{(-2 - 0)^2 + (5 - 0)^2 + (2 - 2)^2} = \sqrt{4 + 25} = \sqrt{29}$

$\quad d_2 = \sqrt{(0 - 0)^2 + (4 - 0)^2 + (0 - 2)^2} = \sqrt{16 + 4} = \sqrt{20} = 2\sqrt{5}$

$\quad d_3 = \sqrt{(0 + 2)^2 + (4 - 5)^2 + (0 - 2)^2} = \sqrt{4 + 1 + 4} = \sqrt{9} = 3$

$\quad d_1^2 = d_2^2 + d_3^2 = 29$

27. $d_1 = \sqrt{(2 - 0)^2 + (2 - 0)^2 + (1 - 0)^2} = \sqrt{4 + 4 + 1} = \sqrt{9} = 3$

$\quad d_2 = \sqrt{(2 - 0)^2 + (-4 - 0)^2 + (4 - 0)^2} = \sqrt{4 + 16 + 16} = \sqrt{36} = 6$

$\quad d_3 = \sqrt{(2 - 2)^2 + (-4 - 2)^2 + (4 - 1)^2} = \sqrt{36 + 9} = \sqrt{45} = 3\sqrt{5}$

$\quad d_1^2 + d_2^2 = 9 + 36 = 45 = d_3^2$

29. $d_1 = \sqrt{(5 - 1)^2 + (-1 + 3)^2 + (2 + 2)^2} = \sqrt{16 + 4 + 16} = \sqrt{36} = 6$

$\quad d_2 = \sqrt{(5 + 1)^2 + (-1 - 1)^2 + (2 - 2)^2} = \sqrt{36 + 4} = \sqrt{40} = 2\sqrt{10}$

$\quad d_3 = \sqrt{(-1 - 1)^2 + (1 + 3)^2 + (2 + 2)^2} = \sqrt{4 + 16 + 16} = \sqrt{36} = 6$

$\quad d_1 = d_3$ Isosceles triangle

31. $\left(\dfrac{3 + 0}{2}, \dfrac{-2 + 0}{2}, \dfrac{4 + 0}{2} \right) = \left(\dfrac{3}{2}, -1, 2 \right)$

33. Midpoint: $\left(\dfrac{3 - 3}{2}, \dfrac{-6 + 4}{2}, \dfrac{10 + 4}{2} \right) = (0, -1, 7)$

35. Midpoint: $\left(\dfrac{6 - 4}{2}, \dfrac{-2 + 2}{2}, \dfrac{5 + 6}{2} \right) = \left(1, 0, \dfrac{11}{2} \right)$

37. Midpoint: $\left(\dfrac{-2 + 7}{2}, \dfrac{8 - 4}{2}, \dfrac{10 + 2}{2} \right) = \left(\dfrac{5}{2}, 2, 6 \right)$

39. $(x - 3)^2 + (y - 2)^2 + (z - 4)^2 = 16$

41. $(x - 0)^2 + (y - 4)^2 + (z - 3)^2 = 3^2$

$\quad\quad x^2 + (y - 4)^2 + (z - 3)^2 = 9$

43. Radius $= \dfrac{\text{Diameter}}{2} = 5$

$\quad (x + 3)^2 + (y - 7)^2 + (z - 5)^2 = 5^2 = 25$

45. Center: $\left(\dfrac{3 + 0}{2}, \dfrac{0 + 0}{2}, \dfrac{0 + 6}{2} \right) = \left(\dfrac{3}{2}, 0, 3 \right)$

$\quad$ Radius: $\sqrt{\left(3 - \dfrac{3}{2} \right)^2 + (0 - 0)^2 + (0 - 3)^2} = \sqrt{\dfrac{9}{4} + 9} = \sqrt{\dfrac{45}{4}}$

$\quad$ Sphere: $\left(x - \dfrac{3}{2} \right)^2 + (y - 0)^2 + (z - 3)^2 = \dfrac{45}{4}$

47. $\left(x^2 - 5x + \frac{25}{4}\right) + y^2 + z^2 = \frac{25}{4}$

$\left(x - \frac{5}{2}\right)^2 + y^2 + z^2 = \frac{25}{4}$

Center: $\left(\frac{5}{2}, 0, 0\right)$

Radius: $\frac{5}{2}$

49. $(x^2 - 4x + 4) + (y^2 + 2y + 1) + (z^2 - 6z + 9) = -10 + 4 + 1 + 9$

$(x - 2)^2 + (y + 1)^2 + (z - 3)^2 = 4$

Center: $(2, -1, 3)$

Radius: 2

51. $(x^2 + 4x + 4) + y^2 + (z^2 - 8z + 16) = -19 + 4 + 16$

$(x + 2)^2 + y^2 + (z - 4)^2 = 1$

Center: $(-2, 0, 4)$

Radius: 1

53. $$x^2 + y^2 + z^2 - 2x - \frac{2}{3}y - 8z = -\frac{73}{9}$$

$(x^2 - 2x + 1) + \left(y^2 - \frac{2}{3}y + \frac{1}{9}\right) + (z^2 - 8z + 16) = -\frac{73}{9} + 1 + \frac{1}{9} + 16$

$(x - 1)^2 + \left(y - \frac{1}{3}\right)^2 + (z - 4)^2 = 9$

Center: $\left(1, \frac{1}{3}, 4\right)$

Radius: 3

55. $9x^2 - 6x + 9y^2 + 18y + 9z^2 = -1$

$x^2 - \frac{2}{3}x + \frac{1}{9} + y^2 + 2y + 1 + z^2 = -\frac{1}{9} + \frac{1}{9} + 1$

$\left(x - \frac{1}{3}\right)^2 + (y + 1)^2 + z^2 = 1$

Center: $\left(\frac{1}{3}, -1, 0\right)$

Radius: 1

57.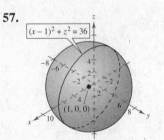

59.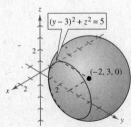

61.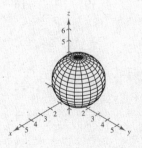

63. The length of each side is 3.
Thus, $(x, y, z) = (3, 3, 3)$.

65. $d = 165 \Rightarrow r = \frac{165}{2} = 82.5$

$x^2 + y^2 + z^2 = \left(\frac{165}{2}\right)^2$

67. False. x is the directed distance from the yz-plane to P.

69. In the xy-plane, the z-coordinate is 0.
In the xz-plane, the y-coordinate is 0.
In the yz-plane, the x-coordinate is 0.

71. The trace is a circle, or a single point.

73. $x_m = \dfrac{x_2 + x_1}{2} \implies x_2 = 2x_m - x_1$

Similarly for y_2 and z_2,

$(x_2, y_2, z_2) = (2x_m - x_1, 2y_m - y_1, 2z_m - z_1).$

75. $v^2 + 3v + \dfrac{9}{4} = 2 + \dfrac{9}{4}$

$$\left(v + \dfrac{3}{2}\right)^2 = \dfrac{17}{4}$$

$$v + \dfrac{3}{2} = \pm\dfrac{\sqrt{17}}{2}$$

$$v = -\dfrac{3}{2} \pm \dfrac{\sqrt{17}}{2}$$

77. $x^2 - 5x + \dfrac{25}{4} = -5 + \dfrac{25}{4}$

$$\left(x - \dfrac{5}{2}\right)^2 = \dfrac{5}{4}$$

$$x - \dfrac{5}{2} = \pm\dfrac{\sqrt{5}}{2}$$

$$x = \dfrac{5}{2} \pm \dfrac{\sqrt{5}}{2}$$

79. $4y^2 + 4y = 9$

$$y^2 + y + \dfrac{1}{4} = \dfrac{9}{4} + \dfrac{1}{4}$$

$$\left(y + \dfrac{1}{2}\right)^2 = \dfrac{10}{4}$$

$$y + \dfrac{1}{2} = \pm\dfrac{\sqrt{10}}{2}$$

$$y = -\dfrac{1}{2} \pm \dfrac{\sqrt{10}}{2}$$

81. $\mathbf{v} = 3\mathbf{i} - 3\mathbf{j}$, Quadrant IV

$\|\mathbf{v}\| = \sqrt{3^2 + (-3)^2}$

$\quad = \sqrt{18}$

$\quad = 3\sqrt{2}$

$\tan\theta = -\dfrac{3}{3} = -1 \implies$

$\quad \theta = -45°$ or $315°$

83. $\mathbf{v} = 4\mathbf{i} + 5\mathbf{j}$, Quadrant I

$\|\mathbf{v}\| = \sqrt{16 + 25} = \sqrt{41}$

$\tan\theta = \frac{5}{4} \implies \theta \approx 51.34°$

85. $\mathbf{u} \cdot \mathbf{v} = \langle -4, 1\rangle \cdot \langle 3, 5\rangle$

$\quad = -4(3) + 1(5)$

$\quad = -7$

87. $a_0 = 1, a_n = a_{n-1} + n^2$

$a_1 = 1 + 1^2 = 2$

$a_2 = 2 + 2^2 = 6$

$a_3 = 6 + 3^2 = 15$

$a_4 = 15 + 4^2 = 31$

	1		2		6		15		31
First differences:		1		4		9		16	
Second differences:			3		5		7		

Neither model

89. $a_1 = -1, a_n = a_{n-1} + 3$

$a_2 = -1 + 3 = 2$

$a_3 = 2 + 3 = 5$

$a_4 = 5 + 3 = 8$

$a_5 = 8 + 3 = 11$

	-1		2		5		8		11
First differences:		3		3		3		3	
Second differences:			0		0		0		

Linear model

91. $(x + 5)^2 + (y - 1)^2 = 49$

93. $(y - 1)^2 = 4p(x - 4)$, $p = -3$

$(y - 1)^2 = 4(-3)(x - 4)$

$(y - 1)^2 = -12(x - 4)$

95. $a = 3, b = 2$, center: $(3, 3)$, horizontal major axis

$$\frac{(x - 3)^2}{9} + \frac{(y - 3)^2}{4} = 1$$

97. Center: $(6, 0)$, horizontal transverse axis

$a = 2, c = 6, b^2 = c^2 - a^2 = 36 - 4 = 32$

$$\frac{(x - 6)^2}{4} - \frac{y^2}{32} = 1$$

Section 11.2   Vectors in Space

- Vectors in space $\mathbf{v} = \langle v_1, v_2, v_3 \rangle$ have many of the same properties as vectors in the plane.
- The dot product of two vectors $\mathbf{u} = \langle u_1, u_2, u_3 \rangle$ and $\mathbf{v} = \langle v_1, v_2, v_3 \rangle$ in space is $\mathbf{u} \cdot \mathbf{v} = u_1 v_1 + u_2 v_2 + u_3 v_3$.
- Two nonzero vectors $\mathbf{u}$ and $\mathbf{v}$ are said to be parallel if there is some scalar c such that $\mathbf{u} = c\mathbf{v}$.
- You should be able to use vectors to solve real life problems.

Vocabulary Check

1. zero

2. $\mathbf{v} = v_1 \mathbf{i} + v_2 \mathbf{j} + v_3 \mathbf{k}$

3. component form

4. orthogonal

5. parallel

1. $\mathbf{v} = \langle 0 - 2, 3 - 0, 2 - 1 \rangle = \langle -2, 3, 1 \rangle$

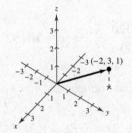

3. (a) $\mathbf{v} = \langle 1 - (-6), -1 - 4, 3 - (-2) \rangle$

$= \langle 7, -5, 5 \rangle$

(b) $\|\mathbf{v}\| = \sqrt{7^2 + (-5)^2 + 5^2}$

$= \sqrt{49 + 25 + 25}$

$= \sqrt{99}$

$= 3\sqrt{11}$

(c) $\dfrac{\mathbf{v}}{\|\mathbf{v}\|} = \dfrac{1}{3\sqrt{11}} \langle 7, -5, 5 \rangle = \dfrac{\sqrt{11}}{33} \langle 7, -5, 5 \rangle$

5. (a)

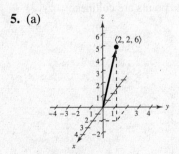

(b)

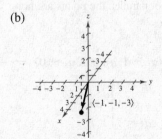

—CONTINUED—

5. —CONTINUED—

(c)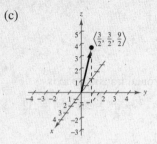

(d)

7. $\mathbf{z} = \mathbf{u} - 2\mathbf{v} = \langle -1, 3, 2 \rangle - 2\langle 1, -2, -2 \rangle = \langle -3, 7, 6 \rangle$

9. $2\mathbf{z} - 4\mathbf{u} = \mathbf{w} \implies \mathbf{z} = \frac{1}{2}(4\mathbf{u} + \mathbf{w}) = \frac{1}{2}(4\langle -1, 3, 2 \rangle + \langle 5, 0, -5 \rangle) = \langle \frac{1}{2}, 6, \frac{3}{2} \rangle$

11. $\|\mathbf{v}\| = \|\langle 7, 8, 7 \rangle\|$
$= \sqrt{49 + 64 + 49} = \sqrt{162} = 9\sqrt{2}$

13. $\|\mathbf{v}\| = \sqrt{4^2 + (-3)^2 + (-7)^2}$
$= \sqrt{16 + 9 + 49} = \sqrt{74}$

15. $\mathbf{v} = \langle 1 - 1, 0 - (-3), -1 - 4 \rangle = \langle 0, 3, -5 \rangle$
$\|\mathbf{v}\| = \sqrt{0 + 3^2 + (-5)^2} = \sqrt{34}$

17. (a) $\dfrac{\mathbf{u}}{\|\mathbf{u}\|} = \dfrac{\langle 8, 3, -1 \rangle}{\sqrt{74}}$

$= \dfrac{1}{\sqrt{74}}(8\mathbf{i} + 3\mathbf{j} - \mathbf{k}) = \dfrac{\sqrt{74}}{74}\langle 8, 3, -1 \rangle$

(b) $-\dfrac{1}{\sqrt{74}}(8\mathbf{i} + 3\mathbf{j} - \mathbf{k}) = -\dfrac{\sqrt{74}}{74}\langle 8, 3, -1 \rangle$

19. $\mathbf{u} \cdot \mathbf{v} = \langle 4, 4, -1 \rangle \cdot \langle 2, -5, -8 \rangle$
$= 8 - 20 + 8 = -4$

21. $\mathbf{u} \cdot \mathbf{v} = \langle 2, -5, 3 \rangle \cdot \langle 9, 3, -1 \rangle$
$= 18 - 15 - 3 = 0$

23. $\cos \theta = \dfrac{\mathbf{u} \cdot \mathbf{v}}{\|\mathbf{u}\| \, \|\mathbf{v}\|} = \dfrac{-8}{\sqrt{8}\sqrt{25}} \implies \theta \approx 124.45°$

25. $\cos \theta = \dfrac{\mathbf{u} \cdot \mathbf{v}}{\|\mathbf{u}\| \, \|\mathbf{v}\|} = \dfrac{-120}{\sqrt{1700}\sqrt{73}} \implies \theta \approx 109.92°$

27. $-\frac{3}{2}\langle 8, -4, -10 \rangle = \langle -12, 6, 15 \rangle \implies$ parallel

29. $\mathbf{u} \cdot \mathbf{v} = 3 - 5 + 2 = 0 \implies$ orthogonal

31. $\mathbf{v} = \langle 7 - 5, 3 - 4, -1 - 1 \rangle = \langle 2, -1, -2 \rangle$
$\mathbf{u} = \langle 4 - 7, 5 - 3, 3 - (-1) \rangle = \langle -3, 2, 4 \rangle$

Since $\mathbf{u}$ and $\mathbf{v}$ are not parallel, the points are not collinear.

33. $\mathbf{v} = \langle -1 - 1, 2 - 3, 5 - 2 \rangle = \langle -2, -1, 3 \rangle$
$\mathbf{u} = \langle 3 - (-1), 4 - 2, -1 - 5 \rangle = \langle 4, 2, -6 \rangle$

Since $\mathbf{u} = -2\mathbf{v}$, the points are collinear.

35. $\mathbf{v} = \langle 2, -4, 7 \rangle = \langle q_1 - 1, q_2 - 5, q_3 - 0 \rangle \implies$

$\left. \begin{array}{l} 2 = q_1 - 1 \\ -4 = q_2 - 5 \\ 7 = q_3 \end{array} \right\} \implies \left. \begin{array}{l} q_1 = 3 \\ q_2 = 1 \\ q_3 = 7 \end{array} \right\} \implies$

Terminal point is $(3, 1, 7)$.

37. $\mathbf{v} = \left\langle 4, \frac{3}{2}, -\frac{1}{4} \right\rangle = \left\langle q_1 - 2, q_2 - 1, q_3 + \frac{3}{2} \right\rangle$

$4 = q_1 - 2 \implies q_1 = 6$

$\frac{3}{2} = q_2 - 1 \implies q_2 = \frac{5}{2}$

$-\frac{1}{4} = q_3 + \frac{3}{2} \implies q_3 = -\frac{7}{4}$

Terminal point: $\left(6, \frac{5}{2}, -\frac{7}{4} \right)$

39. $c\mathbf{u} = c\mathbf{i} + 2c\mathbf{j} + 3c\mathbf{k}$

$\|c\mathbf{u}\| = \sqrt{c^2 + 4c^2 + 9c^2} = |c|\sqrt{14} = 3 \implies$

$c = \pm\dfrac{3}{\sqrt{14}} = \pm\dfrac{3\sqrt{14}}{14}$

41. $\mathbf{v} = \langle q_1, q_2, q_3 \rangle$

Since $\mathbf{v}$ lies in the yz-plane, $q_1 = 0$. Since $\mathbf{v}$ makes an angle of $45°$, $q_2 = q_3$. Finally, $\|\mathbf{v}\| = 4$ implies that $q_2{}^2 + q_3{}^2 = 16$. Thus, $q_2 = q_3 = 2\sqrt{2}$ and $\mathbf{v} = \left\langle 0, 2\sqrt{2}, 2\sqrt{2} \right\rangle$, or $q_2 = 2\sqrt{2}$ and $q_3 = -2\sqrt{2}$ and $\mathbf{v} = \left\langle 0, 2\sqrt{2}, -2\sqrt{2} \right\rangle$.

43. $\overrightarrow{AB} = \langle 0, 70, 115 \rangle$. $F_1 = C_1\langle 0, 70, 115 \rangle$

$\overrightarrow{AC} = \langle -60, 0, 115 \rangle$. $F_2 = C_2\langle -60, 0, 115 \rangle$

$\overrightarrow{AD} = \langle 45, -65, 115 \rangle$. $F_3 = C_3\langle 45, -65, 115 \rangle$

$F_1 + F_2 + F_3 = \langle 0, 0, -500 \rangle$. Thus

$$-60C_2 + 45C_3 = 0$$

$$70C_1 \quad\quad - \quad 65C_3 = 0$$

$$115C_1 + 115C_2 + 115C_3 = -500$$

Solving this system yields $C_1 = \dfrac{-104}{69}$, $C_2 = \dfrac{-28}{23}$, $C_3 = \dfrac{-112}{69}$.

Thus,

$\|\mathbf{F}_1\| \approx 202.919$ N

$\|\mathbf{F}_2\| \approx 157.909$ N

$\|\mathbf{F}_3\| \approx 226.521$ N

45. True. $\cos\theta = 0 \implies \theta = 90°$

47. If $\mathbf{u} \cdot \mathbf{v} < 0$, then $\cos\theta < 0$ and the angle between $\mathbf{u}$ and $\mathbf{v}$ is obtuse, $180° > \theta > 90°$.

49. (a) $x = t, y = 3t + 2$

(b) $x = t - 1, y = 3(t - 1) + 2 = 3t - 1$

51. (a) $x = t, y = t^2 - 8$

(b) $x = t - 1, y = (t - 1)^2 - 8 = t^2 - 2t - 7$

Section 11.3 The Cross Product of Two Vectors

■ The cross product of two vectors $\mathbf{u} = u_1\mathbf{i} + u_2\mathbf{j} + u_3\mathbf{k}$ and $\mathbf{v} = v_1\mathbf{i} + v_2\mathbf{j} + v_3\mathbf{k}$ is given by

$$\mathbf{u} \times \mathbf{v} = (u_2v_3 - u_3v_2)\mathbf{i} - (u_1v_3 - u_3v_1)\mathbf{j} + (u_1v_2 - u_2v_1)\mathbf{k}$$

$$= \begin{vmatrix} \mathbf{i} & \mathbf{j} & \mathbf{k} \\ u_1 & u_2 & u_3 \\ v_1 & v_2 & v_3 \end{vmatrix}.$$

■ The cross product satisfies the following algebraic properties.

(a) $\mathbf{u} \times \mathbf{v} = -(\mathbf{v} \times \mathbf{u})$

(b) $\mathbf{u} \times (\mathbf{v} + \mathbf{w}) = (\mathbf{u} \times \mathbf{v}) + (\mathbf{u} \times \mathbf{w})$

(c) $c(\mathbf{u} \times \mathbf{v}) = (c\mathbf{u}) \times \mathbf{v} = \mathbf{u} \times (c\mathbf{v})$

(d) $\mathbf{u} \times \mathbf{0} = \mathbf{0} \times \mathbf{u} = \mathbf{0}$

(e) $\mathbf{u} \times \mathbf{u} = \mathbf{0}$

(f) $\mathbf{u} \cdot (\mathbf{v} \times \mathbf{w}) = (\mathbf{u} \times \mathbf{v}) \cdot \mathbf{w}$

■ The following geometric properties of the cross product are valid, where θ is the angle between the vectors $\mathbf{u}$ and $\mathbf{v}$:

(a) $\mathbf{u} \times \mathbf{v}$ is orthogonal to both $\mathbf{u}$ and $\mathbf{v}$.

(b) $\|\mathbf{u} \times \mathbf{v}\| = \|\mathbf{u}\| \|\mathbf{v}\| \sin \theta$

(c) $\mathbf{u} \times \mathbf{v} = \mathbf{0}$ if and only if $\mathbf{u}$ and $\mathbf{v}$ are scalar multiples.

(d) $\|\mathbf{u} \times \mathbf{v}\|$ is the area of the parallelogram having $\mathbf{u}$ and $\mathbf{v}$ as sides.

■ The absolute value of the triple scalar product is the volume of the parallelepiped having $\mathbf{u}$, $\mathbf{v}$, and $\mathbf{w}$ as sides.

$$\mathbf{u} \cdot (\mathbf{v} \times \mathbf{w}) = \begin{vmatrix} u_1 & u_2 & u_3 \\ v_1 & v_2 & v_3 \\ w_1 & w_2 & w_3 \end{vmatrix}$$

Vocabulary Check

1. cross product

2. 0

3. $\|\mathbf{u}\| \|\mathbf{v}\| \sin \theta$

4. triple scalar product

1. $\mathbf{j} \times \mathbf{i} = \begin{vmatrix} \mathbf{i} & \mathbf{j} & \mathbf{k} \\ 0 & 1 & 0 \\ 1 & 0 & 0 \end{vmatrix} = -\mathbf{k}$

3. $\mathbf{i} \times \mathbf{k} = \begin{vmatrix} \mathbf{i} & \mathbf{j} & \mathbf{k} \\ 1 & 0 & 0 \\ 0 & 0 & 1 \end{vmatrix} = -\mathbf{j}$

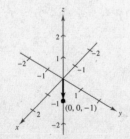

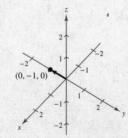

5. $\mathbf{u} \times \mathbf{v} = \begin{vmatrix} \mathbf{i} & \mathbf{j} & \mathbf{k} \\ 3 & -2 & 5 \\ 0 & -1 & 1 \end{vmatrix} = \langle 3, -3, -3 \rangle$

$(\mathbf{u} \times \mathbf{v}) \cdot \mathbf{u} = \langle 3, -3, -3 \rangle \cdot \langle 3, -2, 5 \rangle = 0$

$(\mathbf{u} \times \mathbf{v}) \cdot \mathbf{v} = \langle 3, -3, -3 \rangle \cdot \langle 0, -1, 1 \rangle = 0$

7. $\mathbf{u} \times \mathbf{v} = \begin{vmatrix} \mathbf{i} & \mathbf{j} & \mathbf{k} \\ -10 & 0 & 6 \\ 7 & 0 & 0 \end{vmatrix} = \langle 0, 42, 0 \rangle$

$(\mathbf{u} \times \mathbf{v}) \cdot \mathbf{u} = \langle 0, 42, 0 \rangle \cdot \langle -10, 0, 6 \rangle = 0$

$(\mathbf{u} \times \mathbf{v}) \cdot \mathbf{v} = \langle 0, 42, 0 \rangle \cdot \langle 7, 0, 0 \rangle = 0$

9. $\mathbf{u} \times \mathbf{v} = \begin{vmatrix} \mathbf{i} & \mathbf{j} & \mathbf{k} \\ 6 & 2 & 1 \\ 1 & 3 & -2 \end{vmatrix} = \langle -7, 13, 16 \rangle$

$\qquad\qquad = -7\mathbf{i} + 13\mathbf{j} + 16\mathbf{k}$

11. $\mathbf{u} \times \mathbf{v} = \begin{vmatrix} \mathbf{i} & \mathbf{j} & \mathbf{k} \\ 0 & 0 & 6 \\ -1 & 3 & 1 \end{vmatrix} = \langle -18, -6, 0 \rangle$

$\qquad\qquad = -18\mathbf{i} - 6\mathbf{j}$

13. $\mathbf{u} \times \mathbf{v} = \begin{vmatrix} \mathbf{i} & \mathbf{j} & \mathbf{k} \\ -1 & 0 & 1 \\ 0 & 1 & -2 \end{vmatrix} = \langle -1, -2, -1 \rangle$

$\qquad\qquad = -\mathbf{i} - 2\mathbf{j} - \mathbf{k}$

15. $\mathbf{u} \times \mathbf{v} = \begin{vmatrix} \mathbf{i} & \mathbf{j} & \mathbf{k} \\ 3 & 1 & 0 \\ 0 & 1 & 1 \end{vmatrix} = \mathbf{i} - 3\mathbf{j} + 3\mathbf{k}$

$\|\mathbf{u} \times \mathbf{v}\| = \sqrt{19}$

Unit vector $= \dfrac{\mathbf{u} \times \mathbf{v}}{\|\mathbf{u} \times \mathbf{v}\|} = \dfrac{1}{\sqrt{19}}(\mathbf{i} - 3\mathbf{j} + 3\mathbf{k})$

$\qquad\qquad = \dfrac{\sqrt{19}}{19} \langle 1, -3, 3 \rangle$

17. $\mathbf{u} \times \mathbf{v} = \begin{vmatrix} \mathbf{i} & \mathbf{j} & \mathbf{k} \\ -3 & 2 & -5 \\ \frac{1}{2} & -\frac{3}{4} & \frac{1}{10} \end{vmatrix} = \left\langle -\dfrac{71}{20}, -\dfrac{11}{5}, \dfrac{5}{4} \right\rangle$

Consider the parallel vector $\langle -71, -44, 25 \rangle = \mathbf{w}$.

$\|\mathbf{w}\| = \sqrt{71^2 + 44^2 + 25^2} = \sqrt{7602}$

Unit vector $= \dfrac{1}{\sqrt{7602}} \langle -71, -44, 25 \rangle$

$\qquad\qquad = \dfrac{\sqrt{7602}}{7602} \langle -71, -44, 25 \rangle$

19. $\mathbf{u} \times \mathbf{v} = \begin{vmatrix} \mathbf{i} & \mathbf{j} & \mathbf{k} \\ 1 & 1 & -1 \\ 1 & 1 & 1 \end{vmatrix} = 2\mathbf{i} - 2\mathbf{j}$

$\|\mathbf{u} \times \mathbf{v}\| = 2\sqrt{2}$

Unit vector $= \dfrac{\mathbf{u} \times \mathbf{v}}{\|\mathbf{u} \times \mathbf{v}\|} = \dfrac{1}{2\sqrt{2}}(2\mathbf{i} - 2\mathbf{j})$

$\qquad\qquad = \dfrac{1}{\sqrt{2}}\mathbf{i} - \dfrac{1}{\sqrt{2}}\mathbf{j}$

$\qquad\qquad = \dfrac{\sqrt{2}}{2}\mathbf{i} - \dfrac{\sqrt{2}}{2}\mathbf{j}$

21. $\mathbf{u} \times \mathbf{v} = \begin{vmatrix} \mathbf{i} & \mathbf{j} & \mathbf{k} \\ 0 & 0 & 1 \\ 1 & 0 & 1 \end{vmatrix} = \mathbf{j}$

Area $= \|\mathbf{u} \times \mathbf{v}\| = \|\mathbf{j}\| = 1$ square unit

23. $\mathbf{u} \times \mathbf{v} = \begin{vmatrix} \mathbf{i} & \mathbf{j} & \mathbf{k} \\ 3 & 4 & 6 \\ 2 & -1 & 5 \end{vmatrix} = 26\mathbf{i} - 3\mathbf{j} - 11\mathbf{k}$

Area $= \|\mathbf{u} \times \mathbf{v}\| = \sqrt{26^2 + (-3)^2 + (-11)^2}$

$\qquad\qquad = \sqrt{806}$ square units

25. $\mathbf{u} \times \mathbf{v} = \begin{vmatrix} \mathbf{i} & \mathbf{j} & \mathbf{k} \\ 2 & 2 & -3 \\ 0 & 2 & 3 \end{vmatrix} = \langle 12, -6, 4 \rangle$

Area $= \|\mathbf{u} \times \mathbf{v}\| = \sqrt{12^2 + (-6)^2 + 4^2}$

$\qquad = 14$ square units

27. (a) $\overrightarrow{AB} = \langle 3 - 2, 1 - (-1), 2 - 4 \rangle = \langle 1, 2, -2 \rangle$ is parallel to

$\overrightarrow{DC} = \langle 0 - (-1), 5 - 3, 6 - 8 \rangle = \langle 1, 2, -2 \rangle$.

$\overrightarrow{AD} = \langle -3, 4, 4 \rangle$ is parallel to $\overrightarrow{BC} = \langle -3, 4, 4 \rangle$.

(b) $\overrightarrow{AB} \times \overrightarrow{AD} = \begin{vmatrix} \mathbf{i} & \mathbf{j} & \mathbf{k} \\ 1 & 2 & -2 \\ -3 & 4 & 4 \end{vmatrix} = \langle 16, 2, 10 \rangle$

Area $= \|\overrightarrow{AB} \times \overrightarrow{AD}\| = \sqrt{16^2 + 2^2 + 10^2} = \sqrt{360} = 6\sqrt{10}$ square units

(c) $\overrightarrow{AB} \cdot \overrightarrow{AD} = \langle 1, 2, -2 \rangle \cdot \langle -3, 4, 4 \rangle \neq 0 \implies$ not a rectangle

29. $\mathbf{u} = \langle 1, 2, 3 \rangle, \mathbf{v} = \langle -3, 0, 0 \rangle$

$\mathbf{u} \times \mathbf{v} = \begin{vmatrix} \mathbf{i} & \mathbf{j} & \mathbf{k} \\ 1 & 2 & 3 \\ -3 & 0 & 0 \end{vmatrix} = \langle 0, -9, 6 \rangle$

Area $= \frac{1}{2}\|\mathbf{u} \times \mathbf{v}\| = \frac{1}{2}\sqrt{81 + 36} = \frac{3}{2}\sqrt{13}$ square units

31. $\mathbf{u} = \langle -2 - 2, -2 - 3, 0 - (-5) \rangle = \langle -4, -5, 5 \rangle$

$\mathbf{v} = \langle 3 - 2, 0 - 3, 6 - (-5) \rangle = \langle 1, -3, 11 \rangle$

$\mathbf{u} \times \mathbf{v} = \begin{vmatrix} \mathbf{i} & \mathbf{j} & \mathbf{k} \\ -4 & -5 & 5 \\ 1 & -3 & 11 \end{vmatrix} = \langle -40, 49, 17 \rangle$

Area $= \frac{1}{2}\|\mathbf{u} \times \mathbf{v}\| = \frac{1}{2}\sqrt{(-40)^2 + 49^2 + 17^2}$

$= \frac{1}{2}\sqrt{4290}$ square units

33. $\mathbf{u} \cdot (\mathbf{v} \times \mathbf{w}) = \begin{vmatrix} 2 & 3 & 3 \\ 4 & 4 & 0 \\ 0 & 0 & 4 \end{vmatrix}$

$= 2(16) - 3(16) + 3(0) = -16$

35. $\mathbf{u} \cdot (\mathbf{v} \times \mathbf{w}) = \begin{vmatrix} 2 & 3 & 1 \\ 1 & -1 & 0 \\ 4 & 3 & 1 \end{vmatrix}$

$= 2(-1) - 3(1) + 1(7) = 2$

37. $\mathbf{u} \cdot (\mathbf{v} \times \mathbf{w}) = \begin{vmatrix} 1 & 1 & 0 \\ 0 & 1 & 1 \\ 1 & 0 & 1 \end{vmatrix} = 1 + 1 = 2$

Volume $= |\mathbf{u} \cdot (\mathbf{v} \times \mathbf{w})| = 2$ cubic units

39. $\mathbf{u} \cdot (\mathbf{v} \times \mathbf{w}) = \begin{vmatrix} 0 & 2 & 2 \\ 0 & 0 & -2 \\ 3 & 0 & 2 \end{vmatrix}$

$= 0 - 2(6) + 2(0) = -12$

Volume $= |\mathbf{u} \cdot (\mathbf{v} \times \mathbf{w})| = 12$ cubic units

41. $\mathbf{u} = \langle 4, 0, 0 \rangle, \mathbf{v} = \langle 0, -2, 3 \rangle, \mathbf{w} = \langle 0, 5, 3 \rangle$

$\mathbf{u} \cdot (\mathbf{v} \times \mathbf{w}) = \begin{vmatrix} 4 & 0 & 0 \\ 0 & -2 & 3 \\ 0 & 5 & 3 \end{vmatrix} = 4(-21) = -84$

Volume $= |-84| = 84$ cubic units

43. $\mathbf{V} = \dfrac{1}{2}(-\cos 40° \,\mathbf{j} - \sin 40° \,\mathbf{k})$

$\mathbf{F} = -p\mathbf{k}$

(a) $\mathbf{V} \times \mathbf{F} = \begin{vmatrix} \mathbf{i} & \mathbf{j} & \mathbf{k} \\ 0 & -\dfrac{1}{2}\cos 40° & -\dfrac{1}{2}\sin 40° \\ 0 & 0 & -p \end{vmatrix}$

$= \dfrac{1}{2}p \cos 40° \,\mathbf{i}$

$T = \|\mathbf{V} \times \mathbf{F}\| = \dfrac{p}{2} \cos 40°$

(b)

p	T
15	5.75
20	7.66
25	9.58
30	11.49
35	13.41
40	15.32
45	17.24

45. True. The cross product is not defined for two-dimensional vectors.

47. If the magnitudes of two vectors are doubled, the magnitude of the cross product will be four times as large.

49. $\cos 480° = \cos 120° = -\dfrac{1}{2}$

51. $\sin 690° = \sin 330° = -\dfrac{1}{2}$

53. $\sin \dfrac{19\pi}{6} = \sin\left(\dfrac{7\pi}{6}\right) = -\dfrac{1}{2}$

55. $\tan \dfrac{15\pi}{4} = \tan \dfrac{7\pi}{4} = -1$

57. $z = 6x + 4y$

At $(0, 5)$: $z = 6(0) + 4(5) = 20$

At $(0, 0)$: $z = 6(0) + 4(0) = 0$

At $\left(\dfrac{20}{3}, 0\right)$: $z = 6\left(\dfrac{20}{3}\right) + 4(0) = 40$

At $(6, 4)$: $z = 6(6) + 4(4) = 52$

The maximum value of z, $z = 52$, is found at $(6, 4)$.

The minimum value of z, $z = 0$, is found at $(0, 0)$.

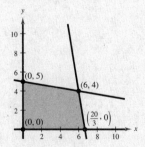

Section 11.4   Lines and Planes in Space

- The parametric equations of the line in space parallel to the vector $\langle a, b, c \rangle$ and passing through the point (x_1, y_1, z_1) are
$$x = x_1 + at, \quad y = y_1 + bt, \quad z = z_1 + ct.$$

- The standard equation of the plane in space containing the point (x_1, y_1, z_1) and having normal vector (a, b, c) is
$$a(x - x_1) + b(y - y_1) + c(z - z_1) = 0.$$

- You should be able to find the angle between two planes by calculating the angle between their normal vectors.

- You should be able to sketch a plane in space.

- The distance between a point Q and a plane having normal $\mathbf{n}$ is
$$D = \|\text{proj}_{\mathbf{n}} \overrightarrow{PQ}\| = \frac{|\overrightarrow{PQ} \cdot \mathbf{n}|}{\|\mathbf{n}\|}$$

where P is a point in the plane.

Vocabulary Check

1. direction, $\dfrac{\overrightarrow{PQ}}{t}$

2. parametric equations

3. symmetric equations

4. normal

5. $a(x - x_1) + b(y - y_1) + c(z - z_1) = 0$

---

1. $x = x_1 + at = 0 + t, y = y_1 + bt = 0 + 2t, z = z_1 + ct = 0 + 3t$

(a) Parametric equations: $x = t, y = 2t, z = 3t$

(b) Symmetric equations: $\dfrac{x}{1} = \dfrac{y}{2} = \dfrac{z}{3}$

3. $x = x_1 + at = -4 + \dfrac{1}{2}t, \ y = y_1 + bt = 1 + \dfrac{4}{3}t, \ z = z_1 + ct = 0 - t$

(a) Parametric equations: $x = -4 + \dfrac{1}{2}t, y = 1 + \dfrac{4}{3}t, z = -t$

Equivalently: $x = -4 + 3t, y = 1 + 8t, z = -6t$

(b) Symmetric equations: $\dfrac{x + 4}{3} = \dfrac{y - 1}{8} = \dfrac{z}{-6}$

5. $x = x_1 + at = 2 + 2t, \ y = y_1 + bt = -3 - 3t, \ z = z_1 + ct = 5 + t$

(a) Parametric equations: $x = 2 + 2t, y = -3 - 3t, z = 5 + t$

(b) Symmetric equations: $\dfrac{x - 2}{2} = \dfrac{y + 3}{-3} = z - 5$

7. (a) $\mathbf{v} = \langle 1 - 2, 4 - 0, -3 - 2 \rangle = \langle -1, 4, -5 \rangle$

   Point: $(2, 0, 2)$

   $x = 2 - t, y = 4t, z = 2 - 5t$

   (b) $\dfrac{x - 2}{-1} = \dfrac{y}{4} = \dfrac{z - 2}{-5}$

9. (a) $\mathbf{v} = \langle 1 - (-3), -2 - 8, 16 - 15 \rangle = \langle 4, -10, 1 \rangle$

   Point: $(-3, 8, 15)$

   $x = -3 + 4t, y = 8 - 10t, z = 15 + t$

   (b) $\dfrac{x + 3}{4} = \dfrac{y - 8}{-10} = \dfrac{z - 15}{1}$

11. $(3, 1, 2), (-1, 1, 5)$

   (a) $\mathbf{v} = \langle -1 - 3, 1 - 1, 5 - 2 \rangle = \langle -4, 0, 3 \rangle$

   Parametric: $x = 3 - 4t, y = 1, z = 2 + 3t$

   (b) Since $b = 0$, there are no symmetric equations.

13. $\left(-\dfrac{1}{2}, 2, \dfrac{1}{2} \right), \left(1, -\dfrac{1}{2}, 0 \right)$

   (a) $\mathbf{v} = \left\langle 1 - \left(-\dfrac{1}{2} \right), -\dfrac{1}{2} - 2, 0 - \dfrac{1}{2} \right\rangle = \left\langle \dfrac{3}{2}, -\dfrac{5}{2}, -\dfrac{1}{2} \right\rangle$

   Direction numbers: $3, -5, -1$

   Parametric: $x = -\dfrac{1}{2} + 3t, y = 2 - 5t, z = \dfrac{1}{2} - t$

   (b) Symmetric: $\dfrac{2x + 1}{6} = \dfrac{y - 2}{-5} = \dfrac{2z - 1}{-2}$

15.

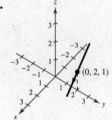

$(0, 2, 1)$

17. $a(x - x_1) + b(y - y_1) + c(z - z_1) = 0$

   $1(x - 2) + 0(y - 1) + 0(z - 2) = 0$

   $x - 2 = 0$

19. $-2(x - 5) + 1(y - 6) - 2(z - 3) = 0$

   $-2x + y - 2z + 10 = 0$

21. $\mathbf{n} = \langle -1, -2, 1 \rangle \implies -1(x - 2) - 2(y - 0) + 1(z - 0) = 0$

   $-x - 2y + z + 2 = 0$

23. $\mathbf{u} = \langle 1 - 0, 2 - 0, 3 - 0 \rangle = \langle 1, 2, 3 \rangle$

   $\mathbf{v} = \langle -2 - 0, 3 - 0, 3 - 0 \rangle = \langle -2, 3, 3 \rangle$

   $\mathbf{n} = \mathbf{u} \times \mathbf{v} = \begin{vmatrix} \mathbf{i} & \mathbf{j} & \mathbf{k} \\ 1 & 2 & 3 \\ -2 & 3 & 3 \end{vmatrix} = \langle -3, -9, 7 \rangle$

   $-3(x - 0) - 9(y - 0) + 7(z - 0) = 0$

   $-3x - 9y + 7z = 0$

   $3x + 9y - 7z = 0$

25. $\mathbf{u} = \langle 3 - 2, 4 - 3, 2 + 2 \rangle = \langle 1, 1, 4 \rangle$

   $\mathbf{v} = \langle 1 - 2, -1 - 3, 0 + 2 \rangle = \langle -1, -4, 2 \rangle$

   $\mathbf{n} = \mathbf{u} \times \mathbf{v} = \begin{vmatrix} \mathbf{i} & \mathbf{j} & \mathbf{k} \\ 1 & 1 & 4 \\ -1 & -4 & 2 \end{vmatrix} = \langle 18, -6, -3 \rangle$

   $18(x - 2) - 6(y - 3) - 3(z + 2) = 0$

   $18x - 6y - 3z - 24 = 0$

   $6x - 2y - z - 8 = 0$

27. $\mathbf{n} = \mathbf{j}$: $0(x - 2) + 1(y - 5) + 0(z - 3) = 0$

$$y - 5 = 0$$

29. $\mathbf{n}_1 = \langle 5, -3, 1 \rangle$, $\mathbf{n}_2 = \langle 1, 4, 7 \rangle$

$\mathbf{n}_1 \cdot \mathbf{n}_2 = 5 - 12 + 7 = 0$; orthogonal

31. $\mathbf{n}_1 = \langle 2, 0, -1 \rangle$, $\mathbf{n}_2 = \langle 4, 1, 8 \rangle$

$\mathbf{n}_1 \cdot \mathbf{n}_2 = 8 - 8 = 0$; orthogonal

33. (a) $\mathbf{n}_1 = \langle 3, -4, 5 \rangle$, $\mathbf{n}_2 = \langle 1, 1, -1 \rangle$; normal vectors to planes

$$\cos \theta = \frac{|\mathbf{n}_1 \cdot \mathbf{n}_2|}{\|\mathbf{n}_1\| \|\mathbf{n}_2\|} = \frac{|-6|}{\sqrt{50}\sqrt{3}} = \frac{6}{\sqrt{150}} \implies \theta \approx 60.67°$$

(b) $3x - 4y + 5z = 6$ Equation 1

$\quad x + y - z = 2$ Equation 2

(-3) times Equation 2 added to Equation 1 gives

$-7y + 8z = 0$

$$y = \frac{8}{7}z.$$

Substituting back into Equation 2, $x = 2 - y + z = 2 - \frac{8}{7}z + z = 2 - \frac{1}{7}z$.

Letting $t = z/7$, we obtain $x = 2 - t$, $y = 8t$, $z = 7t$.

35. (a) $\mathbf{n}_1 = \langle 1, 1, -1 \rangle$, $\mathbf{n}_2 = \langle 2, -5, -1 \rangle$; normal vectors to planes

$$\cos \theta = \frac{|\mathbf{n}_1 \cdot \mathbf{n}_2|}{\|\mathbf{n}_1\| \|\mathbf{n}_2\|} = \frac{|-2|}{\sqrt{3}\sqrt{30}} = \frac{2}{\sqrt{90}} \implies \theta \approx 77.83°$$

(b) $\quad x + y - z = 0$ Equation 1

$2x - 5y - z = 1$ Equation 2

(-2) times Equation 1 added to Equation 2 gives

$-7y + z = 1$

$$y = \frac{z - 1}{7}.$$

Substituting back into Equation 1, $x = z - y = z - \frac{z - 1}{7} = \frac{6z}{7} + \frac{1}{7} = \frac{1}{7}(6z + 1)$.

Letting $z = t$, $x = \frac{6t + 1}{7}$, $y = \frac{t - 1}{7}$.

Equivalently, let $y = t$, $z = 7t + 1$ and $x = 6t + 1$.

37. $x + 2y + 3z = 6$

39. $x + 2y = 4$

41. $3x + 2y - z = 6$

43. $D = \dfrac{|\overrightarrow{PQ} \cdot \mathbf{n}|}{\|\mathbf{n}\|}$

$P = (1, 0, 0)$ on plane, $Q = (0, 0, 0)$,

$\mathbf{n} = \langle 8, -4, 1 \rangle, \overrightarrow{PQ} = \langle -1, 0, 0 \rangle$

$D = \dfrac{|\langle -1, 0, 0 \rangle \cdot \langle 8, -4, 1 \rangle|}{\sqrt{64 + 16 + 1}} = \dfrac{|-8|}{\sqrt{81}} = \dfrac{8}{9}$

45. $D = \dfrac{|\overrightarrow{PQ} \cdot \mathbf{n}|}{\|\mathbf{n}\|}$

$P = (2, 0, 0)$ on plane, $Q = (4, -2, -2)$,

$\mathbf{n} = \langle 2, -1, 1 \rangle, \overrightarrow{PQ} = \langle 2, -2, -2 \rangle$

$D = \dfrac{|\langle 2, -2, -2 \rangle \cdot \langle 2, -1, 1 \rangle|}{\sqrt{6}} = \dfrac{4}{\sqrt{6}} = \dfrac{2\sqrt{6}}{3}$

47. (a) $z = 0.81x + 0.36y + 0.2$

Year	x	y	z (Actual)	z (Model)
1999	6.2	7.3	7.8	7.85
2000	6.1	7.1	7.7	7.70
2001	5.9	7.0	7.4	7.50
2002	5.7	7.0	7.3	7.34
2003	5.6	6.9	7.2	7.22

(b) The approximations are very similar to the actual values of z.

(c) If the consumption of the two types of milk increases (or decreases), so does the consumption of the third type of milk.

49. False. They might be skew lines, such as:

$L_1: x = t, y = 0, z = 0$ (x-axis)

and $L_2: x = 0, y = t, z = 1$

51. The lines are parallel:

$-\frac{3}{2}\langle 10, -18, 20 \rangle = \langle -15, 27, -30 \rangle$

53. $x^2 + y^2 = 10^2 = 100$

55. $r = 3\cos\theta$

$r^2 = 3r\cos\theta$

$x^2 + y^2 = 3x$

57. $r^2 = 49$

$r = 7$

59. $y = 5$

$r\sin\theta = 5$

$r = 5\csc\theta$

Review Exercises for Chapter 11

1. (a) and (b)

3. $(-5, 4, 0)$

5. $d = \sqrt{(5-4)^2 + (2-0)^2 + (1-7)^2}$

$= \sqrt{1 + 4 + 36}$

$= \sqrt{41}$

7. $d_1 = \sqrt{(3-0)^2 + (-2-3)^2 + (0-2)^2} = \sqrt{9 + 25 + 4} = \sqrt{38}$

$d_2 = \sqrt{(0-0)^2 + (5-3)^2 + (-3-2)^2} = \sqrt{4 + 25} = \sqrt{29}$

$d_3 = \sqrt{(0-3)^2 + (5-(-2))^2 + (-3-0)^2} = \sqrt{9 + 49 + 9} = \sqrt{67}$

$d_1^2 + d_2^2 = 38 + 29 = 67 = d_3^2$

9. Midpoint: $\left(\dfrac{8+5}{2}, \dfrac{-2+6}{2}, \dfrac{3+7}{2}\right) = \left(\dfrac{13}{2}, 2, 5\right)$

11. Midpoint: $\left(\dfrac{10-8}{2}, \dfrac{6-2}{2}, \dfrac{-12-6}{2}\right) = (1, 2, -9)$

13. $(x-2)^2 + (y-3)^2 + (z-5)^2 = 1$

15. Radius: 6

$(x-1)^2 + (y-5)^2 + (z-2)^2 = 36$

17. $(x^2 - 4x + 4) + (y^2 - 6y + 9) + z^2 = -4 + 4 + 9$

$(x-2)^2 + (y-3)^2 + z^2 = 9$

Center: $(2, 3, 0)$

Radius: 3

19. (a) *xz*-trace $(y = 0)$: $x^2 + z^2 = 7$, circle

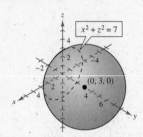

(b) *yz*-trace $(x = 0)$: $(y-3)^2 + z^2 = 16$, circle

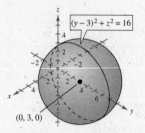

21. Initial point: $(2, -1, 4)$

Terminal point: $(3, 3, 0)$

(a) $\mathbf{v} = \langle 3-2, 3-(-1), 0-4 \rangle = \langle 1, 4, -4 \rangle$

(b) $\|\mathbf{v}\| = \sqrt{(1)^2 + (4)^2 + (-4)^2} = \sqrt{33}$

(c) $\mathbf{u} = \dfrac{\mathbf{v}}{\|\mathbf{v}\|} = \left\langle \dfrac{1}{\sqrt{33}}, \dfrac{4}{\sqrt{33}}, -\dfrac{4}{\sqrt{33}} \right\rangle$

23. Initial point: $(7, -4, 3)$

Terminal point: $(-3, 2, 10)$

(a) $\mathbf{v} = \langle -3-7, 2-(-4), 10-3 \rangle = \langle -10, 6, 7 \rangle$

(b) $\|\mathbf{v}\| = \sqrt{(-10)^2 + (6)^2 + (7)^2} = \sqrt{185}$

(c) $\mathbf{u} = \dfrac{\mathbf{v}}{\|\mathbf{v}\|} = \left\langle -\dfrac{10}{\sqrt{185}}, \dfrac{6}{\sqrt{185}}, \dfrac{7}{\sqrt{185}} \right\rangle$

25. $\mathbf{u} \cdot \mathbf{v} = -1(0) + 4(-6) + 3(5) = -9$

27. $\mathbf{u} \cdot \mathbf{v} = 2(1) - 1(0) + 1(-1) = 1$

29. Since $\mathbf{u} \cdot \mathbf{v} = 0$, the angle is $90°$.

31. Since $-\frac{2}{3}\langle 39, -12, 21 \rangle = \langle -26, 8, -14 \rangle$, the vectors are parallel.

33. First two points: $\mathbf{u} = \langle -3, 4, 1 \rangle$

Last two points: $\mathbf{v} = \langle 0, -2, 6 \rangle$

Since $\mathbf{u} \neq c\mathbf{v}$, the points are not collinear.

35. Let **a**, **b**, and **c** be the three force vectors determined by $A(0, 10, 10)$, $B(-4, -6, 10)$ and $C(4, -6, 10)$.

$$\mathbf{a} = \|\mathbf{a}\|\frac{\langle 0, 10, 10\rangle}{10\sqrt{2}} = \|\mathbf{a}\|\left\langle 0, \frac{1}{\sqrt{2}}, \frac{1}{\sqrt{2}}\right\rangle$$

$$\mathbf{b} = \|\mathbf{b}\|\frac{\langle -4, -6, 10\rangle}{\sqrt{152}} = \|\mathbf{b}\|\left\langle \frac{-2}{\sqrt{38}}, \frac{-3}{\sqrt{38}}, \frac{5}{\sqrt{38}}\right\rangle$$

$$\mathbf{c} = \|\mathbf{c}\|\frac{\langle 4, -6, 10\rangle}{\sqrt{152}} = \|\mathbf{c}\|\left\langle \frac{2}{\sqrt{38}}, \frac{-3}{\sqrt{38}}, \frac{5}{\sqrt{38}}\right\rangle$$

Must have $\mathbf{a} + \mathbf{b} + \mathbf{c} = 300\mathbf{k}$. Thus:

$$\frac{-2}{\sqrt{38}}\|\mathbf{b}\| + \frac{2}{\sqrt{38}}\|\mathbf{c}\| = 0$$

$$\frac{1}{\sqrt{2}}\|\mathbf{a}\| - \frac{3}{\sqrt{38}}\|\mathbf{b}\| - \frac{3}{\sqrt{38}}\|\mathbf{c}\| = 0$$

$$\frac{1}{\sqrt{2}}\|\mathbf{a}\| + \frac{5}{\sqrt{38}}\|\mathbf{b}\| + \frac{5}{\sqrt{38}}\|\mathbf{c}\| = 300.$$

From the first equation $\|\mathbf{b}\| = \|\mathbf{c}\|$. From the second equation, $\frac{1}{\sqrt{2}}\|\mathbf{a}\| = \frac{6}{\sqrt{38}}\|\mathbf{b}\|$.

From the third equation, $\frac{1}{\sqrt{2}}\|\mathbf{a}\| = 300 - \frac{10}{\sqrt{38}}\|\mathbf{b}\|$. Thus,

$$\frac{6}{\sqrt{38}}\|\mathbf{b}\| = 300 - \frac{10}{\sqrt{38}}\|\mathbf{b}\| \implies \frac{16}{\sqrt{38}}\|\mathbf{b}\| = 300 \text{ and } \|\mathbf{b}\| = \|\mathbf{c}\| = \frac{75\sqrt{38}}{4} \approx 115.58.$$

Finally, $\|\mathbf{a}\| = \sqrt{2}\left(\frac{6}{\sqrt{38}}\right)\left(\frac{75\sqrt{38}}{4}\right) = \frac{225\sqrt{2}}{2} \approx 159.10.$

37. $\mathbf{u} \times \mathbf{v} = \begin{vmatrix} \mathbf{i} & \mathbf{j} & \mathbf{k} \\ -2 & 8 & 2 \\ 1 & 1 & -1 \end{vmatrix} = \langle -10, 0, -10\rangle$

39. $\mathbf{u} \times \mathbf{v} = \begin{vmatrix} \mathbf{i} & \mathbf{j} & \mathbf{k} \\ -3 & 2 & -5 \\ 10 & -15 & 2 \end{vmatrix} = \langle -71, -44, 25\rangle$

$\|\mathbf{u} \times \mathbf{v}\| = \sqrt{7602}$

Unit vector: $\frac{1}{\sqrt{7602}}\langle -71, -44, 25\rangle$

41. First two points: $\langle 3, 2, 3\rangle$

Last two points: $\langle 3, 2, 3\rangle$

First and third points: $\langle -2, 2, 0\rangle$

$\begin{vmatrix} \mathbf{i} & \mathbf{j} & \mathbf{k} \\ 3 & 2 & 3 \\ -2 & 2 & 0 \end{vmatrix} = \langle -6, -6, 10\rangle$

Area $= |\langle -6, -6, 10\rangle| = \sqrt{36 + 36 + 100}$

$= \sqrt{172} = 2\sqrt{43}$ square units

43. The parallelogram is determined by the three vectors with initial point $(0, 0, 0)$.

$\mathbf{u} = \langle 3, 0, 0\rangle$, $\mathbf{v} = \langle 2, 0, 5\rangle$, $\mathbf{w} = \langle 0, 5, 1\rangle$

$\mathbf{u} \cdot (\mathbf{v} \times \mathbf{w}) = \begin{vmatrix} 3 & 0 & 0 \\ 2 & 0 & 5 \\ 0 & 5 & 1 \end{vmatrix} = -75$

Volume $= |-75| = 75$ cubic units

45. $\mathbf{v} = \langle 3 + 1, 6 - 3, -1 - 5\rangle = \langle 4, 3, -6\rangle$, point: $(-1, 3, 5)$

(a) Parametric equations: $x = -1 + 4t, y = 3 + 3t, z = 5 - 6t$

(b) Symmetric equations: $\frac{x + 1}{4} = \frac{y - 3}{3} = \frac{z - 5}{-6}$

47. Use $2\mathbf{v} = \langle -4, 5, 2 \rangle$, point: $(0, 0, 0)$.

    (a) Parametric equations: $x = -4t, y = 5t, z = 2t$

    (b) Symmetric equations: $\dfrac{x}{-4} = \dfrac{y}{5} = \dfrac{z}{2}$

49. $\mathbf{u} = \langle 5, 0, 2 \rangle$, $\mathbf{v} = \langle 2, 3, 8 \rangle$

$$\mathbf{u} \times \mathbf{v} = \begin{vmatrix} \mathbf{i} & \mathbf{j} & \mathbf{k} \\ 5 & 0 & 2 \\ 2 & 3 & 8 \end{vmatrix} = \langle -6, -36, 15 \rangle$$

$$\mathbf{n} = \langle 2, 12, -5 \rangle$$

$$a(x - x_0) + b(y - y_0) + c(z - z_0) = 0$$

$$2(x - 0) + 12(y - 0) - 5(z - 0) = 0$$

$$2x + 12y - 5z = 0$$

51. $\mathbf{n} = \mathbf{k}$, normal vector

    Plane: $0(x - 5) + 0(y - 3) + 1(z - 2) = 0$

$$z - 2 = 0$$

53. $3x - 2y + 3z = 6$

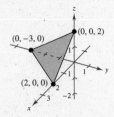

55. $2x - 3z = 6$

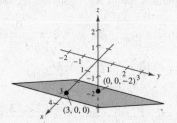

57. $\mathbf{n} = \langle 2, -20, 6 \rangle$, $P = (0, 0, 1)$ in plane, $Q = (2, 3, 10)$, $\overrightarrow{PQ} = \langle 2, 3, 9 \rangle$

$$D = \frac{|\overrightarrow{PQ} \cdot \mathbf{n}|}{\|\mathbf{n}\|} = \frac{|-2|}{\sqrt{440}} = \frac{1}{\sqrt{110}} = \frac{\sqrt{110}}{110} \approx 0.0953$$

59. $\mathbf{n} = \langle 1, -10, 3 \rangle$, $P = (2, 0, 0)$ in plane, $Q = (0, 0, 0)$, $\overrightarrow{PQ} = \langle -2, 0, 0 \rangle$

$$D = \frac{|\overrightarrow{PQ} \cdot \mathbf{n}|}{\|\mathbf{n}\|} = \frac{|-2|}{\sqrt{1 + 100 + 9}} = \frac{2}{\sqrt{110}} = \frac{2\sqrt{110}}{110} = \frac{\sqrt{110}}{55} \approx 0.191$$

61. False. $\mathbf{a} \times \mathbf{b} = -(\mathbf{b} \times \mathbf{a})$

63. $\mathbf{u} \cdot \mathbf{u} = \langle 3, -2, 1 \rangle \cdot \langle 3, -2, 1 \rangle$

$$= 9 + 4 + 1$$

$$= 14$$

$$= \|\mathbf{u}\|^2$$

65. $\mathbf{u} \cdot (\mathbf{v} + \mathbf{w}) = \langle 3, -2, 1 \rangle \cdot \langle 1, -2, -1 \rangle = 6$

    $\mathbf{u} \cdot \mathbf{v} + \mathbf{u} \cdot \mathbf{w} = 11 + (-5) = 6$

Problem Solving for Chapter 11

1. (a)

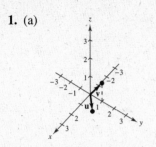

(b) $\mathbf{w} = a\mathbf{u} + b\mathbf{v} = a\langle 1, 1, 0\rangle + b\langle 0, 1, 1\rangle$

$$\mathbf{0} = \langle a, a + b, b\rangle \Longrightarrow a = b = 0$$

(c) $\mathbf{w} = \langle 1, 2, 1\rangle = a\langle 1, 1, 0\rangle + b\langle 0, 1, 1\rangle$

$$1 = a$$

$$2 = a + b$$

$$1 = b$$

Hence, $a = b = 1$.

(d) $\mathbf{w} = \langle 1, 2, 3\rangle = a\langle 1, 1, 0\rangle + b\langle 0, 1, 1\rangle$

$$1 = a$$

$$2 = a + b$$

$$3 = b$$

Impossible

3. Programs will vary. See online website.

5. The largest angle in a triangle is always opposite the longest side of the triangle. First, determine the lengths of the three sides. Then, once the largest angle has been identified, use the fact that $\cos \theta = \dfrac{\mathbf{u} \cdot \mathbf{v}}{\|\mathbf{u}\| \, \|\mathbf{v}\|}$, where $\mathbf{u}$ and $\mathbf{v}$ are defined to be the vectors that form θ. If $\mathbf{u} \cdot \mathbf{v} = 0$, the angle is a right angle. If $\mathbf{u} \cdot \mathbf{v} > 0$, the angle is acute. If $\mathbf{u} \cdot \mathbf{v} < 0$, the angle is obtuse.

(a) A: $(1, 2, 0)$

B: $(0, 0, 0)$

C: $(-2, 1, 0)$

$d(AB) = \sqrt{5}$, $d(AC) = \sqrt{10}$, $d(BC) = \sqrt{5}$

Angle B is largest.

$\overrightarrow{BA} = \langle 1, 2, 0\rangle$, $\overrightarrow{BC} = \langle -2, 1, 0\rangle$

$\overrightarrow{BA} \cdot \overrightarrow{BC} = 0 \Longrightarrow$ The triangle is a right triangle.

(b) A: $(-3, 0, 0)$

B: $(0, 0, 0)$

C: $(1, 2, 3)$

$d(AB) = 3$, $d(AC) = \sqrt{29}$, $d(BC) = \sqrt{14}$

Angle B is largest.

$\overrightarrow{BA} = \langle -3, 0, 0\rangle$, $\overrightarrow{BC} = \langle 1, 2, 3\rangle$

$\overrightarrow{BA} \cdot \overrightarrow{BC} = -3 < 0 \Longrightarrow$ The triangle is an obtuse triangle.

(c) A: $(2, -3, 4)$

B: $(0, 1, 2)$

C: $(-1, 2, 0)$

$d(AB) = \sqrt{24}$, $d(AC) = \sqrt{50}$, $d(BC) = \sqrt{6}$

Angle B is largest.

$\overrightarrow{BA} = \langle 2, -4, 2\rangle$, $\overrightarrow{BC} = \langle -1, 1, -2\rangle$

$\overrightarrow{BA} \cdot \overrightarrow{BC} = -10 < 0 \Longrightarrow$ The triangle is an obtuse triangle.

(d) A: $(2, -7, 3)$

B: $(-1, 5, 8)$

C: $(4, 6, -1)$

$d(AB) = \sqrt{178}$, $d(AC) = \sqrt{189}$, $d(BC) = \sqrt{107}$

Angle B is largest.

$\overrightarrow{BA} = \langle 3, -12, -5\rangle$, $\overrightarrow{BC} = \langle 5, 1, -9\rangle$

$\overrightarrow{BA} \cdot \overrightarrow{BC} = 48 \Longrightarrow$ The triangle is an acute triangle.

7. Let A lie on the y-axis and the wall on the x-axis. Then, $A = (0, 10, 0)$, $B = (8, 0, 6)$, $C = (-10, 0, 6)$ and $\overrightarrow{AB} = \langle 8, -10, 6 \rangle$, $\overrightarrow{AC} = \langle -10, -10, 6 \rangle$.

$$\|\overrightarrow{AB}\| = \sqrt{8^2 + (-10)^2 + 6^2} = 10\sqrt{2}$$

$$\|\overrightarrow{AC}\| = \sqrt{(-10)^2 + (-10)^2 + 6^2} = 2\sqrt{59}$$

$$\mathbf{F}_1 = 420 \frac{\overrightarrow{AB}}{\|\overrightarrow{AB}\|} = \frac{420}{10\sqrt{2}} \langle 8, -10, 6 \rangle = \frac{84}{\sqrt{2}} \langle 4, -5, 3 \rangle$$

$$\mathbf{F}_2 = 650 \frac{\overrightarrow{AC}}{\|\overrightarrow{AC}\|} = \frac{650}{2\sqrt{59}} \langle -10, -10, 6 \rangle = \frac{650}{\sqrt{59}} \langle -5, -5, 3 \rangle$$

$$\mathbf{F} = \mathbf{F}_1 + \mathbf{F}_2 = \left\langle \frac{(4)(84)}{\sqrt{2}} + \frac{(-5)(650)}{\sqrt{59}}, \frac{(-5)(84)}{\sqrt{2}} + \frac{(-5)(650)}{\sqrt{59}}, \frac{(3)(84)}{\sqrt{2}} + \frac{(3)(650)}{\sqrt{59}} \right\rangle$$

$$\approx \langle -185.526, -720.099, 432.059 \rangle$$

$$\|\mathbf{F}\| \approx 860.0 \text{ lb}$$

9. $\mathbf{u} = \langle a_1, b_1, c_1 \rangle$, $\mathbf{v} = \langle a_2, b_2, c_2 \rangle$, $\mathbf{w} = \langle a_3, b_3, c_3 \rangle$

$$\mathbf{v} \times \mathbf{w} = \begin{vmatrix} \mathbf{i} & \mathbf{j} & \mathbf{k} \\ a_2 & b_2 & c_2 \\ a_3 & b_3 & c_3 \end{vmatrix} = (b_2 c_3 - b_3 c_2)\mathbf{i} - (a_2 c_3 - a_3 c_2)\mathbf{j} + (a_2 b_3 - a_3 b_2)\mathbf{k}$$

$$\mathbf{u} \times (\mathbf{v} \times \mathbf{w}) = \begin{vmatrix} \mathbf{i} & \mathbf{j} & \mathbf{k} \\ a_1 & b_1 & c_1 \\ (b_2 c_3 - b_3 c_2) & (a_3 c_2 - a_2 c_3) & (a_2 b_3 - a_3 b_2) \end{vmatrix}$$

$$\mathbf{u} \times (\mathbf{v} \times \mathbf{w}) = [b_1(a_2 b_3 - a_3 b_2) - c_1(a_3 c_2 - a_2 c_3)]\mathbf{i} - [a_1(a_2 b_3 - a_3 b_2) - c_1(b_2 c_3 - b_3 c_2)]\mathbf{j}$$

$$+ [a_1(a_3 c_2 - a_2 c_3) - b_1(b_2 c_3 - b_3 c_2)]\mathbf{k}$$

$$= [a_2(a_1 a_3 + b_1 b_3 + c_1 c_3) - a_3(a_1 a_2 + b_1 b_2 + c_1 c_2)]\mathbf{i}$$

$$+ [b_2(a_1 a_3 + b_1 b_3 + c_1 c_3) - b_3(a_1 a_2 + b_1 b_2 + c_1 c_2)]\mathbf{j}$$

$$+ [c_2(a_1 a_3 + b_1 b_3 + c_1 c_3) - c_3(a_1 a_2 + b_1 b_2 + c_1 c_2)]\mathbf{k}$$

$$= (\mathbf{u} \cdot \mathbf{w})\mathbf{v} - (\mathbf{u} \cdot \mathbf{v})\mathbf{w}$$

11.

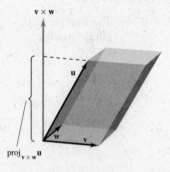

$\|\mathbf{v} \times \mathbf{w}\| =$ area of base and $\|\text{proj}_{\mathbf{v} \times \mathbf{w}} \mathbf{u}\| =$ height of parallelepiped

Therefore, the volume is

$$V = (\text{height})(\text{area of base}) = \|\text{proj}_{\mathbf{v} \times \mathbf{w}} \mathbf{u}\| \|\mathbf{v} \times \mathbf{w}\|$$

$$= \left| \frac{\mathbf{u} \cdot (\mathbf{v} \times \mathbf{w})}{\|\mathbf{v} \times \mathbf{w}\|} \right| \|\mathbf{v} \times \mathbf{w}\|$$

$$= |\mathbf{u} \cdot (\mathbf{v} \times \mathbf{w})|.$$

13. (a) In inches: $\overrightarrow{AB} = -15\mathbf{j} + 12\mathbf{k}$

In feet: $\overrightarrow{AB} = -\frac{5}{4}\mathbf{j} + \mathbf{k}$

$$\mathbf{F} = -200(\cos\theta\mathbf{j} + \sin\theta\mathbf{k})$$

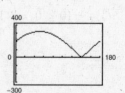

—CONTINUED—

13. —CONTINUED—

(b) $\overrightarrow{AB} \times \mathbf{F} = \begin{vmatrix} \mathbf{i} & \mathbf{j} & \mathbf{k} \\ 0 & & 1 \\ 0 & -200\cos\theta & -200\sin\theta \end{vmatrix} = (250\sin\theta + 200\cos\theta)\mathbf{i}$

$\|\overrightarrow{AB} \times \mathbf{F}\| = |250\sin\theta + 200\cos\theta| = 25|10\sin\theta + 8\cos\theta|$

(c) When $\theta = 30°$: $\|\overrightarrow{AB} \times \mathbf{F}\| = 25\left[10\left(\dfrac{1}{2}\right) + 8\left(\dfrac{\sqrt{3}}{2}\right)\right] = 25(5 + 4\sqrt{3}) \approx 298.2$

(d) From the graph we see that the maximum value occurs when $\theta \approx 51.34°$.

(e) From the graph we see that the zero occurs when $\theta \approx 141.34°$, the angle making $\overrightarrow{AB}$ parallel to $\mathbf{F}$.

15. First insect: $x = 6 + t, y = 8 - t, z = 3 + t$

Second insect: $x = 1 + t, y = 2 + t, z = 2t$

(a) When $t = 0$ the first insect is located at $(6, 8, 3)$ and the second insect is located at $(1, 2, 0)$.

$d = \sqrt{(1-6)^2 + (2-8)^2 + (0-3)^2} = \sqrt{70}$ inches

(b) $d = \sqrt{[(1+t) - (6+t)]^2 + [(2+t) - (8-t)]^2 + [2t - (3+t)]^2}$

$= \sqrt{(-5)^2 + (2t-6)^2 + (t-3)^2}$

$= \sqrt{5t^2 - 30t + 70}$

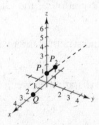

t	0	1	2	3	4	5	6	7	8	9	10
d	$\sqrt{70}$	$\sqrt{45}$	$\sqrt{30}$	5	$\sqrt{30}$	$\sqrt{45}$	$\sqrt{70}$	$\sqrt{105}$	$\sqrt{150}$	$\sqrt{205}$	$\sqrt{270}$

(c) The distance between the two insects appears to lessen in the first 3 seconds, but then begins to increase with time.

(d) When $t = 3$, the insects get within 5 inches of each other.

17. (a) $\mathbf{u} = \langle 0, 1, 1 \rangle$ direction vector of line determined by P_1 and P_2

$D = \dfrac{\|\overrightarrow{P_1Q} \times \mathbf{u}\|}{\|\mathbf{u}\|} = \dfrac{\|\langle 2, 0, -1 \rangle \times \langle 0, 1, 1 \rangle\|}{\sqrt{2}}$

$= \dfrac{\|\langle 1, -2, 2 \rangle\|}{\sqrt{2}} = \dfrac{3}{\sqrt{2}} = \dfrac{3\sqrt{2}}{2}$

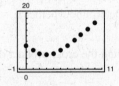

(b) The shortest distance to the line **segment** is $\|\overrightarrow{P_1Q}\| = \|\langle 2, 0, -1 \rangle\| = \sqrt{5}$.

Chapter 11   Practice Test

1. Find the lengths of the sides of the triangle with vertices $(0, 0, 0)$, $(1, 2, -4)$, and $(0, -2, -1)$. Show that the triangle is a right triangle.

2. Find the standard form of the equation of a sphere having center $(0, 4, 1)$ and radius 5.

3. Find the center and radius of the sphere $x^2 + y^2 + z^2 + 2x - 4z - 11 = 0$.

4. Find the vector $\mathbf{u} - 3\mathbf{v}$ given $\mathbf{u} = \langle 1, 0, -1 \rangle$ and $\mathbf{v} = \langle 4, 3, -6 \rangle$.

5. Find the length of $\frac{1}{2}\mathbf{v}$ if $\mathbf{v} = \langle 2, 4, -6 \rangle$.

6. Find the dot product of $\mathbf{u} = \langle 2, 1, -3 \rangle$ and $\mathbf{v} = \langle 1, 1, -2 \rangle$.

7. Determine whether $\mathbf{u} = \langle 1, 1, -1 \rangle$ and $\mathbf{v} = \langle -3, -3, 3 \rangle$ are orthogonal, parallel, or neither.

8. Find the cross product of $\mathbf{u} = \langle -1, 0, 2 \rangle$ and $\mathbf{v} = \langle 1, -1, 3 \rangle$. What is $\mathbf{v} \times \mathbf{u}$?

9. Use the triple scalar product to find the volume of the parallelepiped having adjacent edges $\mathbf{u} = \langle 1, 1, 1 \rangle$, $\mathbf{v} = \langle 0, -1, 1 \rangle$, and $\mathbf{w} = \langle 1, 0, 4 \rangle$.

10. Find a set of parametric equations for the line through the points $(0, -3, 3)$ and $(2, -3, 4)$.

11. Find an equation of the plane passing through $(1, 2, 3)$ and perpendicular to the vector $\mathbf{n} = \langle 1, -1, 0 \rangle$.

12. Find an equation of the plane passing through the three points $A = (0, 0, 0)$, $B = (1, 1, 1)$, and $C = (1, 2, 3)$.

13. Determine whether the planes $x + y - z = 12$ and $3x - 4y - z = 9$ are parallel, orthogonal or neither.

14. Find the distance between the point $(1, 1, 1)$ and the plane $x + 2y + z = 6$.

CHAPTER 12
Limits and an Introduction to Calculus

Section 12.1   Introduction to Limits **580**

Section 12.2   Techniques for Evaluating Limits **584**

Section 12.3   The Tangent Line Problem **590**

Section 12.4   Limits at Infinity and Limits of Sequences **596**

Section 12.5   The Area Problem **600**

Review Exercises **604**

Problem Solving **610**

Practice Test **613**

CHAPTER 12
Limits and an Introduction to Calculus

Section 12.1   Introduction to Limits

■ If $f(x)$ becomes arbitrarily close to a unique number L as x approaches c from either side, then the limit of $f(x)$ as x approaches c is L:

$$\lim_{x \to c} f(x) = L.$$

■ You should be able to use a calculator to find a limit.

■ You should be able to use a graph to find a limit.

■ You should understand how limits can fail to exist:

(a) $f(x)$ approaches a different number from the right of c than it approaches from the left of c.

(b) $f(x)$ increases or decreases without bound as x approaches c.

(c) $f(x)$ oscillates between two fixed values as x approaches c.

■ You should know and be able to use the elementary properties of limits.

Vocabulary Check

1. limit   **2.** oscillates   **3.** direct substitution

---

1. (a)
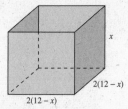

(b) $V = (\text{base})\text{height} = (24 - 2x)^2 x = 4x(12 - x)^2$

(d)
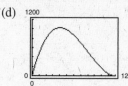

(c) $\lim_{x \to 4} V = 1024$ Maximum at $x = 4$

x	3	3.5	3.9	4	4.1	4.5	5
V	972.0	1011.5	1023.5	1024.0	1023.5	1012.5	980.0

3. $\lim_{x \to 2} (5x + 4) = 14$

x	1.9	1.99	1.999	2	2.001	2.01	2.1
$f(x)$	13.5	13.95	13.995	14	14.005	14.05	14.5

The limit is reached.

5. $\lim\limits_{x\to 3} \dfrac{x-3}{x^2-9} = \dfrac{1}{6}$

x	2.9	2.99	2.999	3	3.001	3.01	3.1
f(x)	0.1695	0.1669	0.16669	?	0.16664	0.1664	0.1639

The limit is not reached.

7. $f(x) = \dfrac{x-1}{x^2+2x-3}$

x	0.9	0.99	0.999	1	1.001	1.01	1.1
f(x)	0.2564	0.2506	0.2501	?	0.2499	0.2494	0.2439

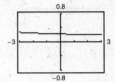

$\lim\limits_{x\to 1} \dfrac{x-1}{x^2+2x-3} = \dfrac{1}{4}$

9. $f(x) = \dfrac{\sqrt{x+5}-\sqrt{5}}{x}$

x	−0.1	−0.01	−0.001	0	0.001	0.01	0.1
f(x)	0.2247	0.2237	0.2236	?	0.2236	0.2235	0.2225

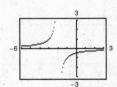

$\lim\limits_{x\to 0} \dfrac{\sqrt{x+5}-\sqrt{5}}{x} \approx 0.2236 \;\left(\text{Actual limit: } \dfrac{1}{2\sqrt{5}}\right)$

11. $f(x) = \dfrac{\dfrac{x}{x+2}-2}{x+4}$

x	−4.1	−4.01	−4.001	−4	−3.999	−3.99	−3.9
f(x)	0.4762	0.4975	0.4998	?	0.5003	0.5025	0.5263

$\lim\limits_{x\to -4} \dfrac{\dfrac{x}{x+2}-2}{x+4} = \dfrac{1}{2}$

13. $f(x) = \dfrac{\sin x}{x}$

Make sure your calculator is set in radian mode.

x	−0.1	−0.01	−0.001	0	0.001	0.01	0.1
f(x)	0.9983	0.99998	0.9999998	?	0.9999998	0.99998	0.9983

$\lim\limits_{x\to 0} \dfrac{\sin x}{x} = 1$

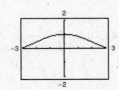

15. $f(x) = \begin{cases} 2x + 1, & x < 2 \\ x + 3, & x \geq 2 \end{cases}$

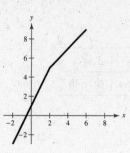

The limit exists as x approaches 2:

$\lim\limits_{x \to 2} f(x) = 5$

17. $\lim\limits_{x \to -4} (x^2 - 3) = 13$

19. $\lim\limits_{x \to -2} \dfrac{|x + 2|}{x + 2}$ does not exist. $f(x) = \dfrac{|x + 2|}{x + 2}$ equals -1 to the left of -2, and equals 1 to the right of -2.

21. The limit does not exist because $f(x)$ oscillates between 2 and -2.

23.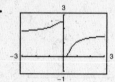

$\lim\limits_{x \to 0} \dfrac{5}{2 + e^{1/x}}$ does not exist.

25.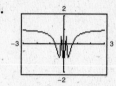

$\lim\limits_{x \to 0} \cos \dfrac{1}{x}$ does not exist.

The graph oscillates between -1 and 1.

27.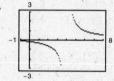

$\lim\limits_{x \to 4} \dfrac{\sqrt{x + 3} - 1}{x - 4}$ does not exist.

29.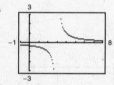

$\lim\limits_{x \to 1} \dfrac{x - 1}{x^2 - 4x + 3} = -\dfrac{1}{2}$

31.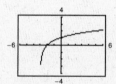

$\lim\limits_{x \to 4} \ln(x + 3) \approx 1.946$

(Exact limit is $\ln 7$.)

33. (a) $\lim\limits_{x \to c} [-2g(x)] = -2(6) = -12$

(b) $\lim\limits_{x \to c} [f(x) + g(x)] = 3 + 6 = 9$

(c) $\lim\limits_{x \to c} \dfrac{f(x)}{g(x)} = \dfrac{3}{6} = \dfrac{1}{2}$

(d) $\lim\limits_{x \to c} \sqrt{f(x)} = \sqrt{3}$

35. (a) $\lim\limits_{x \to 2} f(x) = 2^3 = 8$

(b) $\lim\limits_{x \to 2} g(x) = \dfrac{\sqrt{2^2 + 5}}{2(2^2)} = \dfrac{3}{8}$

(c) $\lim\limits_{x \to 2} [f(x)g(x)] = 8\left(\dfrac{3}{8}\right) = 3$

(d) $\lim\limits_{x \to 2} [g(x) - f(x)] = \dfrac{3}{8} - 8 = -\dfrac{61}{8}$

37. $\lim\limits_{x \to 5} (10 - x^2) = 10 - 5^2 = -15$

39. $\lim\limits_{x \to -3} (2x^2 + 4x + 1) = 2(-3)^2 + 4(-3) + 1 = 7$

41. $\lim\limits_{x \to 3} \left(-\dfrac{9}{x}\right) = -\dfrac{9}{3} = -3$

43. $\lim\limits_{x \to -3} \dfrac{3x}{x^2 + 1} = -\dfrac{9}{10}$

45. $\lim\limits_{x \to -2} \dfrac{5x + 3}{2x - 9} = \dfrac{5(-2) + 3}{2(-2) - 9} = \dfrac{-7}{-13} = \dfrac{7}{13}$

47. $\lim\limits_{x \to -1} \sqrt{x + 2} = \sqrt{-1 + 2} = 1$

49. $\lim\limits_{x \to 7} \dfrac{5x}{\sqrt{x + 2}} = \dfrac{5(7)}{\sqrt{7 + 2}} = \dfrac{35}{3}$

51. $\lim\limits_{x \to 3} e^x = e^3 \approx 20.0855$

53. $\lim\limits_{x \to \pi} \sin 2x = \sin 2\pi = 0$

55. $\lim\limits_{x \to 1/2} \arcsin x = \arcsin \dfrac{1}{2} = \dfrac{\pi}{6} \approx 0.5236$

57. True

59. Answers will vary.

61. (a) No. The limit may or may not exist. And if it does exist, it may not equal 4.

(b) No. $f(2)$ may or may not exist. And if $f(2)$ exists, it may not equal 4.

63. $f(x) = \dfrac{x - 9}{\sqrt{x} - 3}$

(a)

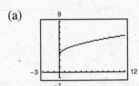

$\lim\limits_{x \to 9} = \dfrac{x - 9}{\sqrt{x} - 3} = 6$

(c) $\sqrt{x} - 3 \neq 0 \Rightarrow x \neq 9; \ \sqrt{x} \Rightarrow x \geq 0$

Domain: $x \geq 0, \ x \neq 9$

(b) Domain: $x \geq 0, \ x \neq 9$

(d) It may not be clear from a graph that a function is not defined at a single point. Examining a function graphically and algebraically ensures that you will find all points at which the function is not defined.

65. $\dfrac{5 - x}{3x - 15} = \dfrac{5 - x}{-3(5 - x)} = -\dfrac{1}{3}, \ x \neq 5$

67. $\dfrac{15x^2 + 7x - 4}{15x^2 + x - 2} = \dfrac{(3x - 1)(5x + 4)}{(3x - 1)(5x + 2)}$

$= \dfrac{5x + 4}{5x + 2}, \ x \neq \dfrac{1}{3}$

69. $\dfrac{x^3 + 27}{x^2 + x - 6} = \dfrac{(x + 3)(x^2 - 3x + 9)}{(x + 3)(x - 2)}$

$= \dfrac{x^2 - 3x + 9}{x - 2}, \ x \neq -3$

71. (a)

(b) $d = \sqrt{(3 - 3)^2 + (2 - 2)^2 + (8 - 7)^2} = 1$

(c) Midpoint: $\left(\dfrac{3 + 3}{2}, \dfrac{2 + 2}{2}, \dfrac{7 + 8}{2} \right) = \left(3, 2, \dfrac{15}{2} \right)$

73. (a)

(b) $d = \sqrt{(0 - 3)^2 + [5 - (-3)]^2 + (-5 - 0)^2}$

$= \sqrt{98} = 7\sqrt{2}$

(c) Midpoint:

$\left(\dfrac{3 + 0}{2}, \dfrac{-3 + 5}{2}, \dfrac{0 + (-5)}{2} \right) = \left(\dfrac{3}{2}, 1, -\dfrac{5}{2} \right)$

Section 12.2 Techniques For Evaluating Limits

- You can use direct substitution to find the limit of a polynomial function $p(x)$:
$$\lim_{x \to c} p(x) = p(c).$$

- You can use direct substitution to find the limit of a rational function $r(x) = \dfrac{p(x)}{q(x)}$, as long as $q(c) \neq 0$:
$$\lim_{x \to c} r(x) = r(c) = \frac{p(c)}{q(c)}, \, q(c) \neq 0.$$

- You should be able to use cancellation techniques to find a limit.
- You should know how to use rationalization techniques to find a limit.
- You should know how to use technology to find a limit.
- You should be able to calculate one-sided limits.

Vocabulary Check

1. dividing out technique

2. indeterminate form

3. one-sided limit

4. difference quotient

---

1. $g(x) = \dfrac{-2x^2 + x}{x}$, $g_2(x) = -2x + 1$

   (a) $\lim_{x \to 0} g(x) = 1$

   (b) $\lim_{x \to -1} g(x) = 3$

   (c) $\lim_{x \to -2} g(x) = 5$

3. $g(x) = \dfrac{x^3 - x}{x - 1}$, $g_2(x) = x^2 + x = x(x + 1)$

   (a) $\lim_{x \to 1} g(x) = 2$

   (b) $\lim_{x \to -1} g(x) = 0$

   (c) $\lim_{x \to 0} g(x) = 0$

5. $\displaystyle\lim_{x \to 6} \frac{x - 6}{x^2 - 36} = \lim_{x \to 6} \frac{x - 6}{(x - 6)(x + 6)}$

   $\displaystyle = \lim_{x \to 6} \frac{1}{x + 6} = \frac{1}{12}$

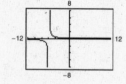

7. $\displaystyle\lim_{x \to -1} \frac{1 - 2x - 3x^2}{1 + x} = \lim_{x \to -1} \frac{(1 + x)(1 - 3x)}{1 + x}$

   $\displaystyle = \lim_{x \to -1} (1 - 3x) = 4$

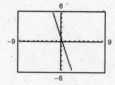

9. $\displaystyle\lim_{t \to 2} \frac{t^3 - 8}{t - 2} = \lim_{t \to 2} \frac{(t - 2)(t^2 + 2t + 4)}{t - 2}$

   $\displaystyle = \lim_{t \to 2} (t^2 + 2t + 4)$

   $= 4 + 4 + 4 = 12$

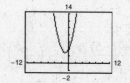

11. $\displaystyle\lim_{x \to 2} \frac{x^5 - 32}{x - 2} = 80$

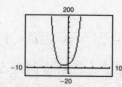

13. $\lim\limits_{y\to 0} \dfrac{\sqrt{5+y}-\sqrt{5}}{y} = \lim\limits_{y\to 0} \dfrac{\sqrt{5+y}-\sqrt{5}}{y}\cdot\dfrac{\sqrt{5+y}+\sqrt{5}}{\sqrt{5+y}+\sqrt{5}}$

$\qquad\qquad = \lim\limits_{y\to 0} \dfrac{(5+y)-5}{y\left(\sqrt{5+y}+\sqrt{5}\right)}$

$\qquad\qquad = \lim\limits_{y\to 0} \dfrac{1}{\sqrt{5+y}+\sqrt{5}}$

$\qquad\qquad = \dfrac{1}{2\sqrt{5}}$

$\qquad\qquad = \dfrac{\sqrt{5}}{10}$

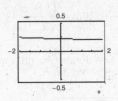

15. $\lim\limits_{x\to 0} = \dfrac{\sqrt{x+3}-\sqrt{3}}{x}$

$\qquad = \lim\limits_{x\to 0} \dfrac{\sqrt{x+3}-\sqrt{3}}{x}\cdot\dfrac{\sqrt{x+3}+\sqrt{3}}{\sqrt{x+3}+\sqrt{3}}$

$\qquad = \lim\limits_{x\to 0} \dfrac{1}{\sqrt{x+3}+\sqrt{3}} = \dfrac{1}{2\sqrt{3}} = \dfrac{\sqrt{3}}{6}$

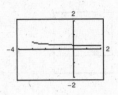

17. $\lim\limits_{x\to 0} \dfrac{\sqrt{2x+1}-1}{x} = \lim\limits_{x\to 0} \dfrac{\sqrt{2x+1}-1}{x}\cdot\dfrac{\sqrt{2x+1}+1}{\sqrt{2x+1}+1}$

$\qquad\qquad = \lim\limits_{x\to 0} \dfrac{2}{\sqrt{2x+1}+1} = \dfrac{2}{2} = 1$

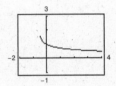

19. $\lim\limits_{x\to -3} \dfrac{\sqrt{x+7}-2}{x+3} = \lim\limits_{x\to -3} \dfrac{\sqrt{x+7}-2}{x+3}\cdot\dfrac{\sqrt{x+7}+2}{\sqrt{x+7}+2}$

$\qquad\qquad = \lim\limits_{x\to -3} \dfrac{(x+7)-4}{(x+3)\left(\sqrt{x+7}+2\right)}$

$\qquad\qquad = \lim\limits_{x\to -3} \dfrac{1}{\sqrt{x+7}+2}$

$\qquad\qquad = \dfrac{1}{4}$

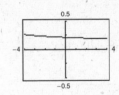

21. $\lim\limits_{x\to 0} \dfrac{1/(1+x)-1}{x} = \lim\limits_{x\to 0} \dfrac{1-(1+x)}{(1+x)x}$

$\qquad\qquad = \lim\limits_{x\to 0} \dfrac{-1}{1+x} = -1$

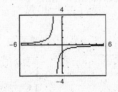

23. $f(x) = \dfrac{\dfrac{1}{x+4}-\dfrac{1}{4}}{x}$

$\lim\limits_{x\to 0} f(x) = -\dfrac{1}{16},\ (-0.0625)$

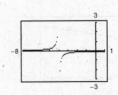

25. $\lim\limits_{x \to 0} \dfrac{\sec x}{\tan x} = \lim\limits_{x \to 0} \dfrac{1}{\cos x} \cdot \dfrac{\cos x}{\sin x}$

$= \lim\limits_{x \to 0} \dfrac{1}{\sin x}$, does not exist

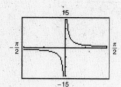

27. $f(x) = \dfrac{e^{2x} - 1}{x}$

$\lim\limits_{x \to 0} f(x) = 2$

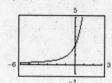

29. $\lim\limits_{x \to 0^{+}} x \ln x = 0$

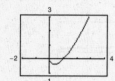

31. $\lim\limits_{x \to 0} \dfrac{\sin 2x}{x} = 2$

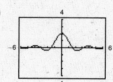

33. $\lim\limits_{x \to 0} \dfrac{\tan x}{x} = 1$

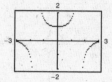

35. $\lim\limits_{x \to 1} \dfrac{1 - \sqrt[3]{x}}{1 - x} = \dfrac{1}{3} \approx 0.333$

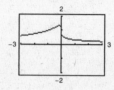

37. $f(x) = (1 - x)^{2/x}$

$\lim\limits_{x \to 0} f(x) \approx 0.135$

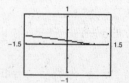

39. $f(x) = \dfrac{x - 1}{x^2 - 1}$

(a) Graphically, $\lim\limits_{x \to 1^{-}} \dfrac{x - 1}{x^2 - 1} = \dfrac{1}{2}$.

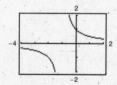

(b)

x	0.5	0.9	0.99	0.999	1
$f(x)$	0.6667	0.5263	0.5025	0.5003	0.5

Numerically, $\lim\limits_{x \to 1^{-}} \dfrac{x - 1}{x^2 - 1} = \dfrac{1}{2}$.

(c) Algebraically, $\lim\limits_{x \to 1^{-}} \dfrac{x - 1}{x^2 - 1} = \lim\limits_{x \to 1^{-}} \dfrac{x - 1}{(x - 1)(x + 1)} = \lim\limits_{x \to 1^{-}} \dfrac{1}{x + 1} = \dfrac{1}{2}$.

41. $f(x) = \dfrac{4 - \sqrt{x}}{x - 16}$

(a) Graphically, $\lim\limits_{x \to 16^{+}} \dfrac{4 - \sqrt{x}}{x - 16} = -\dfrac{1}{8}$.

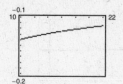

—CONTINUED—

41.—CONTINUED—

(b)

x	16	16.001	16.01	16.1	16.5
$f(x)$	?	-0.1250	-0.1250	-0.1248	-0.1240

(c) Algebraically, $\displaystyle\lim_{x\to 16^+} \frac{4-\sqrt{x}}{x-16} = \lim_{x\to 16^+} \frac{4-\sqrt{x}}{(\sqrt{x}-4)(\sqrt{x}+4)}$

$$= \lim_{x\to 16^+} \frac{-1}{\sqrt{x}+4}$$

$$= \frac{-1}{4+4} = -\frac{1}{8}.$$

43. $f(x) = \dfrac{|x-6|}{x-6}$

$\displaystyle\lim_{x\to 6^+} f(x) = 1$

$\displaystyle\lim_{x\to 6^-} f(x) = -1$

Limit does not exist.

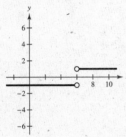

45. $f(x) = \dfrac{1}{x^2+1}$

$\displaystyle\lim_{x\to 1^-} \frac{1}{x^2+1} = \lim_{x\to 1^+} \frac{1}{x^2+1}$

$$= \lim_{x\to 1} \frac{1}{x^2+1} = \frac{1}{2}$$

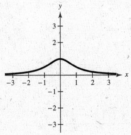

47. $\displaystyle\lim_{x\to 2^-} f(x) = 2 - 1 = 1$

$\displaystyle\lim_{x\to 2^+} f(x) = 2(2) - 3 = 1$

$\displaystyle\lim_{x\to 2} f(x) = 1$

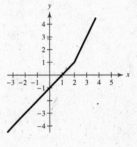

49. $f(x) = \begin{cases} 4 - x^2, & x \le 1 \\ 3 - x, & x > 1 \end{cases}$

$\displaystyle\lim_{x\to 1^-} f(x) = 4 - 1 = 3$

$\displaystyle\lim_{x\to 1^+} f(x) = 3 - 1 = 2$

$\displaystyle\lim_{x\to 1} f(x)$ does not exist.

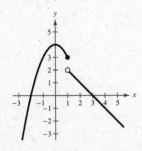

51. $\displaystyle\lim_{x\to 0} f(x) = 0$

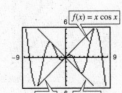

53. $\displaystyle\lim_{x\to 0} f(x) = 0$

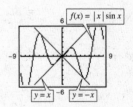

55. $\displaystyle\lim_{x\to 0} f(x) = 0$

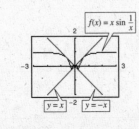

57. (a) Can be evaluated by direct substitution:

$$\lim_{x\to 0} x^2 \sin x^2 = 0^2 \sin 0^2 = 0$$

(b) Cannot be evaluated by direct substitution:

$$\lim_{x\to 0} \frac{\sin x^2}{x^2} = 1$$

59. $\displaystyle\lim_{h\to 0}\frac{f(x+h)-f(x)}{h}=\lim_{h\to 0}\frac{3(x+h)-1-(3x-1)}{h}$

$\displaystyle\qquad\qquad =\lim_{h\to 0}\frac{3x+3h-1-3x+1}{h}$

$\displaystyle\qquad\qquad =\lim_{h\to 0}\frac{3h}{h}=3$

61. $\displaystyle\lim_{h\to 0}\frac{f(x+h)-f(x)}{h}=\lim_{h\to 0}\frac{\sqrt{x+h}-\sqrt{x}}{h}\cdot\left(\frac{\sqrt{x+h}+\sqrt{x}}{\sqrt{x+h}+\sqrt{x}}\right)$

$\displaystyle\qquad\qquad =\lim_{h\to 0}\frac{(x+h)-x}{h\left(\sqrt{x+h}+\sqrt{x}\right)}$

$\displaystyle\qquad\qquad =\lim_{h\to 0}\frac{1}{\sqrt{x+h}+\sqrt{x}}=\frac{1}{2\sqrt{x}}$

63. $\displaystyle\lim_{h\to 0}\frac{f(x+h)-f(x)}{h}=\lim_{h\to 0}\frac{((x+h)^2-3(x+h))-(x^2-3x)}{h}$

$\displaystyle\qquad\qquad =\lim_{h\to 0}\frac{x^2+2xh+h^2-3x-3h-x^2+3x}{h}$

$\displaystyle\qquad\qquad =\lim_{h\to 0}\frac{2xh+h^2-3h}{h}$

$\displaystyle\qquad\qquad =\lim_{h\to 0}(2x+h-3)=2x-3$

65. $\displaystyle\lim_{h\to 0}\frac{f(x+h)-f(x)}{h}=\lim_{h\to 0}\frac{1/(x+h+2)-1/(x+2)}{h}$

$\displaystyle\qquad\qquad =\lim_{h\to 0}\frac{(x+2)-(x+h+2)}{h(x+h+2)(x+2)}$

$\displaystyle\qquad\qquad =\lim_{h\to 0}\frac{-h}{h(x+h+2)(x+2)}$

$\displaystyle\qquad\qquad =\lim_{h\to 0}\frac{-1}{(x+h+2)(x+2)}$

$\displaystyle\qquad\qquad =\frac{-1}{(x+2)^2}$

67. $\displaystyle\lim_{t\to 1}\frac{(-16(1)+128)-(-16t^2+128)}{1-t}=\lim_{t\to 1}\frac{16t^2-16}{1-t}$

$\displaystyle\qquad\qquad\qquad =\lim_{t\to 1}\frac{16(t-1)(t+1)}{1-t}$

$\displaystyle\qquad\qquad\qquad =\lim_{t\to 1}-16(t+1)$

$\displaystyle\qquad\qquad\qquad =-32\;\frac{\text{ft}}{\text{sec}}$

69. $\displaystyle\lim_{t\to 2^-}f(t)=30.80,\;\lim_{t\to 2^+}f(t)=33.88$

Thus, the limit of f as $t\to 2$ does not exist.

71. $C(t) = 0.75 - 0.50[\![-(t-1)]\!]$

(a)
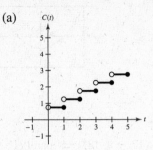

(b)

t	3	3.3	3.4	3.5	3.6	3.7	4
C	1.75	2.25	2.25	2.25	2.25	2.25	2.25

$\lim\limits_{t \to 3.5} C(t) = 2.25$

(c)

t	2	2.5	2.9	3	3.1	3.5	4
C	1.25	1.75	1.75	1.75	2.25	2.25	2.25

No, $\lim\limits_{t \to 3} C(t)$ does not exist.

$\lim\limits_{t \to 3^-} C(t) = 1.75$, $\lim\limits_{t \to 3^+} C(t) = 2.25$

73. True

75. Many answers possible

(a)
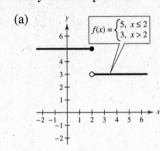

(b)
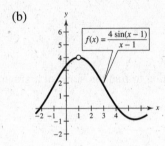

77. Slope of line through $(4, -6)$ and $(3, -4)$:

$\dfrac{-6+4}{4-3} = -2$

Slope of perpendicular line: $\dfrac{1}{2}$

Equation: $y + 10 = \dfrac{1}{2}(x - 6)$

$2y - x + 26 = 0$

79. $s = r\theta$

$= (8.5)\left[45\left(\dfrac{\pi}{180}\right)\right]$

$= 2.125\pi$ inches

≈ 6.676 inches

81. $A = \dfrac{\theta r^2}{2}$

$= \dfrac{1}{2}\left[135\left(\dfrac{\pi}{180}\right)\right](9)^2$

$= \dfrac{1}{2}\left(\dfrac{3\pi}{4}\right)(81)$

$= 30.375\pi$ square centimeters

≈ 95.426 square centimeters

83. $r = \dfrac{3}{1 + \cos\theta}$, $e = 1$, Parabola

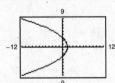

85. $r = \dfrac{9}{2 + 3\cos\theta} = \dfrac{9/2}{1 + (3/2)\cos\theta}, e = \dfrac{3}{2},$

   Hyperbola

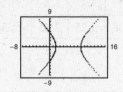

87. $r = \dfrac{5}{1 - \sin\theta}, e = 1,$

   Parabola

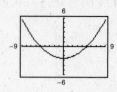

89. $\langle 7, -2, 3\rangle \cdot \langle -1, 4, 5\rangle = -7 - 8 + 15$

$$= 0 \Rightarrow \text{orthogonal}$$

91. $-3\langle -4, 3, -6\rangle = \langle 12, -9, 18\rangle \Rightarrow \text{parallel}$

Section 12.3   The Tangent Line Problem

- ■ You should be able to visually approximate the slope of a graph.
- ■ The slope m of the graph of f at the point $(x, f(x))$ is given by

$$m = \lim_{h \to 0} \frac{f(x + h) - f(x)}{h}$$

   provided this limit exists.

- ■ You should be able to use the limit definition to find the slope of a graph.
- ■ The derivative of f at x is given by

$$f'(x) = \lim_{h \to 0} \frac{f(x + h) - f(x)}{h}$$

   provided this limit exists. Notice that this is the same limit as that for the tangent line slope.

- ■ You should be able to use the limit definition to find the derivative of a function.

Vocabulary Check

1. calculus   **2.** tangent line   **3.** secant line

4. difference quotient   **5.** derivative

---

1. Slope is 0 at (x, y).   **3.** Slope is $\frac{1}{2}$ at (x, y).

5. $m_{\text{sec}} = \dfrac{g(3 + h) - g(3)}{h} = \dfrac{(3 + h)^2 - 4(3 + h) - (-3)}{h} = \dfrac{h^2 + 2h}{h}$

$m = \lim\limits_{h \to 0} \dfrac{h^2 + 2h}{h} = \lim\limits_{h \to 0} \dfrac{h(h + 2)}{h} = \lim\limits_{h \to 0} (h + 2) = 2$

7. $m_{\text{sec}} = \dfrac{g(1 + h) - g(1)}{h} = \dfrac{5 - 2(1 + h) - 3}{h} = \dfrac{-2h}{h}$

$m = \lim\limits_{h \to 0} \dfrac{-2h}{h} = -2$

9. $m_{\text{sec}} = \dfrac{g(2 + h) - g(2)}{h} = \dfrac{[4/(2 + h)] - 2}{h} = \dfrac{4 - 2(2 + h)}{(2 + h)h} = \dfrac{-2}{2 + h}, \ h \neq 0$

$m = \lim\limits_{h \to 0}\left(\dfrac{-2}{2 + h}\right) = -1$

11. $m_{\text{sec}} = \dfrac{h(9 + k) - h(9)}{k} = \dfrac{\sqrt{9 + k} - 3}{k} \cdot \dfrac{\sqrt{9 + k} + 3}{\sqrt{9 + k} + 3} = \dfrac{(9 + k) - 9}{k[\sqrt{9 + k} + 3]} = \dfrac{1}{\sqrt{9 + k} + 3}, \ k \neq 0$

$m = \lim\limits_{k \to 0} \dfrac{1}{\sqrt{9 + k} + 3} = \dfrac{1}{6}$

13. $f(x) = 4 - x^2$

$m_{\text{sec}} = \dfrac{f(x + h) - f(x)}{h} = \dfrac{4 - (x + h)^2 - (4 - x^2)}{h} = \dfrac{-2xh - h^2}{h} = -2x - h, \ h \neq 0$

$m = \lim\limits_{h \to 0}(-2x - h) = -2x$

(a) At $(0, 4)$, $m = -2(0) = 0$.   (b) At $(-1, 3)$, $m = -2(-1) = 2$.

15. $f(x) = \dfrac{1}{x + 4}$

$m_{\text{sec}} = \dfrac{f(x + h) - f(x)}{h} = \dfrac{\dfrac{1}{x + h + 4} - \dfrac{1}{x + 4}}{h} = \dfrac{(x + 4) - (x + 4 + h)}{(x + h + 4)(x + 4)(h)}$

$= \dfrac{-h}{(x + h + 4)(x + 4)h} = \dfrac{-1}{(x + h + 4)(x + 4)}, \ h \neq 0$

$m = \lim\limits_{h \to 0} \dfrac{-1}{(x + h + 4)(x + 4)} = \dfrac{-1}{(x + 4)^2}$

(a) At $\left(0, \dfrac{1}{4}\right)$, $m = \dfrac{-1}{(0 + 4)^2} = \dfrac{-1}{16}$.   (b) At $\left(-2, \dfrac{1}{2}\right)$, $m = \dfrac{-1}{(-2 + 4)^2} = \dfrac{-1}{4}$.

17. $f(x) = \sqrt{x - 1}$

$m_{\text{sec}} = \dfrac{f(x + h) - f(x)}{h} = \dfrac{\sqrt{x + h - 1} - \sqrt{x - 1}}{h} \cdot \dfrac{\sqrt{x + h - 1} + \sqrt{x - 1}}{\sqrt{x + h - 1} + \sqrt{x - 1}}$

$= \dfrac{(x + h - 1) - (x - 1)}{h\left(\sqrt{x + h - 1} + \sqrt{x - 1}\right)} = \dfrac{1}{\sqrt{x + h - 1} + \sqrt{x - 1}}, \ h \neq 0$

$m = \lim\limits_{h \to 0}\left(\dfrac{1}{\sqrt{x + h - 1} + \sqrt{x - 1}}\right) = \dfrac{1}{2\sqrt{x - 1}}$

(a) At $(5, 2)$, $m = \dfrac{1}{2\sqrt{5 - 1}} = \dfrac{1}{4}$.   (b) At $(10, 3)$, $m = \dfrac{1}{2\sqrt{10 - 1}} = \dfrac{1}{6}$.

19. $f(x) = x^2 - 2$

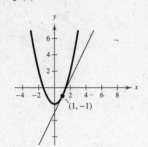

Slope at $(1, -1)$ is 2.

21. $f(x) = \sqrt{2 - x}$

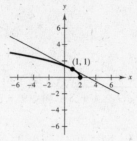

Slope at $(1, 1)$ is $-\dfrac{1}{2}$.

23. $f(x) = \dfrac{4}{x + 1}$

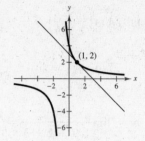

Slope at $(1, 2)$ is -1.

25. $f'(x) = \lim\limits_{h \to 0} \dfrac{f(x + h) - f(x)}{h} = \lim\limits_{h \to 0} \dfrac{5 - 5}{h} = 0$

27. $g'(x) = \lim\limits_{h \to 0} \dfrac{g(x + h) - g(x)}{h} = \lim\limits_{h \to 0} \dfrac{\left[9 - \frac{1}{3}(x + h)\right] - \left[9 - \frac{1}{3}x\right]}{h} = \lim\limits_{h \to 0} \dfrac{-\frac{1}{3}h}{h} = -\dfrac{1}{3}$

29. $f'(x) = \lim\limits_{h \to 0} \dfrac{f(x + h) - f(x)}{h} = \lim\limits_{h \to 0} \dfrac{\left[4 - 3(x + h)^2\right] - (4 - 3x^2)}{h}$

$= \lim\limits_{h \to 0} \dfrac{-3(x^2 + 2xh + h^2) + 3x^2}{h} = \lim\limits_{h \to 0} \dfrac{-6xh - 3h^2}{h} = \lim\limits_{h \to 0}(-6x - 3h) = -6x$

31. $f'(x) = \lim\limits_{h \to 0} \dfrac{f(x + h) - f(x)}{h} = \lim\limits_{h \to 0} \dfrac{\dfrac{1}{(x + h)^2} - \dfrac{1}{x^2}}{h}$

$= \lim\limits_{h \to 0} \dfrac{x^2 - (x^2 + 2xh + h^2)}{(x + h)^2 x^2 h} = \lim\limits_{h \to 0} \dfrac{-2x - h}{(x + h)^2 x^2} = -\dfrac{2x}{x^4} = -\dfrac{2}{x^3}$

33. $f'(x) = \lim\limits_{h \to 0} \dfrac{f(x + h) - f(x)}{h} = \lim\limits_{h \to 0} \dfrac{\dfrac{1}{\sqrt{x + h - 9}} - \dfrac{1}{\sqrt{x - 9}}}{h} \cdot \dfrac{\dfrac{1}{\sqrt{x + h - 9}} + \dfrac{1}{\sqrt{x - 9}}}{\dfrac{1}{\sqrt{x + h - 9}} + \dfrac{1}{\sqrt{x - 9}}}$

$= \lim\limits_{h \to 0} \dfrac{\dfrac{1}{(x + h - 9)} - \dfrac{1}{(x - 9)}}{h\left[\dfrac{1}{\sqrt{x + h - 9}} + \dfrac{1}{\sqrt{x - 9}}\right]} = \lim\limits_{h \to 0} \dfrac{(x - 9) - (x + h - 9)}{h(x + h - 9)(x - 9)\left[\dfrac{1}{\sqrt{x + h - 9}} + \dfrac{1}{\sqrt{x - 9}}\right]}$

$= \lim\limits_{h \to 0} \dfrac{-1}{(x + h - 9)(x - 9)\left[\dfrac{1}{\sqrt{x + h - 9}} + \dfrac{1}{\sqrt{x - 9}}\right]} = \dfrac{-1}{(x - 9)^2\left[\dfrac{2}{\sqrt{x - 9}}\right]} = \dfrac{-1}{2(x - 9)^{3/2}}$

35. $f(x) = x^2 - 1, (2, 3)$

(a) $m_{\sec} = \dfrac{f(2 + h) - f(2)}{h} = \dfrac{(2 + h)^2 - 1 - 3}{h} = \dfrac{4h + h^2}{h} = 4 + h, h \neq 0$

(c)

$m = \lim\limits_{h \to 0}(4 + h) = 4$

(b) Tangent line: $y - 3 = 4(x - 2)$

$$y = 4x - 5$$

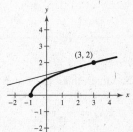

37. $f(x) = \sqrt{x + 1}, (3, 2)$

(a) $m_{\sec} = \dfrac{f(3 + h) - f(3)}{h} = \dfrac{\sqrt{3 + h + 1} - 2}{h} \cdot \dfrac{\sqrt{4 + h} + 2}{\sqrt{4 + h} + 2} = \dfrac{(4 + h) - 4}{h\left[\sqrt{4 + h} + 2\right]} = \dfrac{1}{\sqrt{4 + h} + 2}$

(c)

$m = \lim\limits_{h \to 0} \dfrac{1}{\sqrt{4 + h} + 2} = \dfrac{1}{4}$

(b) Tangent line: $y - 2 = \dfrac{1}{4}(x - 3)$

$$4y = x + 5$$

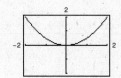

39.

x	-2	-1.5	-1	-0.5	0	0.5	1	1.5	2
$f(x)$	2	1.125	0.5	0.125	0	0.125	0.5	0.125	2
$f'(x)$	-2	-1.5	-1	-0.5	0	0.5	1	1.5	2

$f(x) = \frac{1}{2}x^2$

$f'(x) = x$

They appear to be the same.

41.

x	-2	-1.5	-1	-0.5	0	0.5	1	1.5	2
$f(x)$	1	1.225	1.414	1.581	1.732	1.871	2	2.121	2.236
$f'(x)$	0.5	0.408	0.354	0.316	0.289	0.267	0.25	0.236	0.224

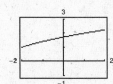

$f(x) = \sqrt{x + 3}$

$f'(x) = \dfrac{1}{2\sqrt{x + 3}}$

They appear to be the same.

43. Given line: $x + y = 0 \implies y = -x \implies m = -1 \implies m_{\tan} = -1$ since the lines are parallel.

$$f(x) = -\frac{1}{4}x^2$$

$$m_{\tan} = \lim_{h \to 0} \frac{f(x + h) - f(x)}{h} = \lim_{h \to 0} \frac{-\frac{1}{4}(x + h)^2 - (-\frac{1}{4}x^2)}{h}$$

$$= \lim_{h \to 0} \frac{-\frac{1}{4}(x^2 + 2xh + h^2) + \frac{1}{4}x^2}{h} = \lim_{h \to 0} \frac{-\frac{1}{4}h(2x + h)}{h}$$

$$= \lim_{h \to 0} -\frac{1}{4}(2x + h) = -\frac{1}{2}x$$

$$m_{\tan} = -\frac{1}{2}x = -1 \implies x = 2$$

Point: $(2, f(2)) = (2, -1)$

Tangent line: $y - (-1) = -1(x - 2)$

$$y + 1 = -x + 2$$

$$y = -x + 1$$

45. Given line: $6x + y + 4 = 0 \implies y = -6x - 4 \implies m = -6 \implies m_{\tan} = -6$ since the lines are parallel.

$$f(x) = -\frac{1}{2}x^3$$

$$m_{\tan} = \lim_{h \to 0} \frac{f(x + h) - f(x)}{h} = \lim_{h \to 0} \frac{-\frac{1}{2}(x + h)^3 - (-\frac{1}{2}x^3)}{h}$$

$$= \lim_{h \to 0} \frac{-\frac{1}{2}(x^3 + 3x^2h + 3xh^2 + h^3) + \frac{1}{2}x^3}{h} = \lim_{h \to 0} \frac{-\frac{1}{2}h(3x^2 + 3xh + h^2)}{h}$$

$$= \lim_{h \to 0} -\frac{1}{2}(3x^2 + 3xh + h^2) = -\frac{3x^2}{2}$$

$$m_{\tan} = -\frac{3x^2}{2} = -6 \implies x^2 = 4 \implies x = \pm 2$$

Points: $(2, f(2)) = (2, -4)$ and $(-2, f(-2)) = (-2, 4)$

Tangent lines: $y - (-4) = -6(x - 2)$ and

$$y + 4 = -6x + 12$$

$$y = -6x + 8$$

47. $f'(x) = \lim_{h \to 0} \frac{f(x + h) - f(x)}{h} = \lim_{h \to 0} \frac{[(x + h)^2 - 4(x + h) + 3] - [x^2 - 4x + 3]}{h}$

$$= \lim_{h \to 0} \frac{(x^2 + 2xh + h^2 - 4x - 4h + 3) - (x^2 - 4x + 3)}{h}$$

$$= \lim_{h \to 0} \frac{2xh + h^2 - 4h}{h} = \lim_{h \to 0} (2x + h - 4) = 2x - 4$$

$$f'(x) = 0 = 2x - 4 \implies x = 2$$

f has a horizontal tangent at $(2, -1)$.

49. $f'(x) = \lim\limits_{h \to 0} \dfrac{f(x + h) - f(x)}{h} = \lim\limits_{h \to 0} \dfrac{3(x + h)^3 - 9(x + h) - (3x^3 - 9x)}{h}$

$\quad = \lim\limits_{h \to 0} \dfrac{9x^2h + 9xh^2 + 3h^3 - 9h}{h} = 9x^2 - 9$

$f'(x) = 0 = 9x^2 - 9 \implies x = \pm 1$

f has horizontal tangents at $(1, -6)$ and $(-1, 6)$.

51.

Year	x	Revenue, y
1999	9	4463.5
2000	10	4960.1
2001	11	5156.7
2002	12	5565.9
2003	13	5911.7
2004	14	6053.2

(a) Quadratic model:

$\quad y \approx -21.048x^2 + 804.47x - 1054.5$

(b)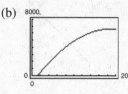

When $x = 12$ the slope is approximately 299.31. This represents a \$299.31 million rate of change of revenue in 2002.

(c) Tangent line: When $x = 12$, $y \approx 5568.16$ (Model value).

$\quad\quad y - 5568.16 = 299.31(x - 12)$

$\quad\quad\quad\quad y \approx 299.31x + 1976.44$

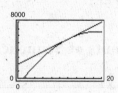

The slopes are the same.

53. $f(x) = -x^2 + 5x + 2$

Using the definition of slope, you obtain $f'(x) = -2x + 5$.

For $0 \le x \le 2$, $f'(x) > 0 \implies$ height increasing.

For $4 \le x \le 6$, $f'(x) < 0 \implies$ height decreasing.

55. True. The slope is $2x$, which is different for all x.

57. Matches (b). (Derivative is always positive, but decreasing.)

59. Matches (d). (Derivative is -1 for $x < 0$, 1 for $x > 0$.)

61. Answers will vary.

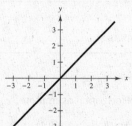

63. Answers will vary.

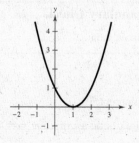

65. $f(x) = \dfrac{1}{x^2 - x - 2} = \dfrac{1}{(x-2)(x+1)}$

Vertical asymptotes: $x = 2, -1$

Horizontal asymptote: $y = 0$

Intercept: $\left(0, -\dfrac{1}{2}\right)$

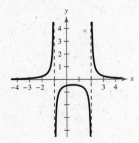

67. $f(x) = \dfrac{x^2 - x - 2}{x - 2}$

$= \dfrac{(x-2)(x+1)}{x-2} = x + 1, \; x \neq 2$

Line with hole at $(2, 3)$

Intercepts: $(0, 1), (-1, 0)$

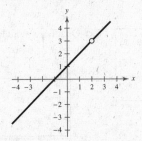

69. $\langle 1, 1, 1 \rangle \times \langle 2, 1, -1 \rangle = \begin{vmatrix} \mathbf{i} & \mathbf{j} & \mathbf{k} \\ 1 & 1 & 1 \\ 2 & 1 & -1 \end{vmatrix}$

$= \langle -2, 3, -1 \rangle$

71. $\langle -4, 10, 0 \rangle \times \langle 4, -1, 0 \rangle = \begin{vmatrix} \mathbf{i} & \mathbf{j} & \mathbf{k} \\ -4 & 10 & 0 \\ 4 & -1 & 0 \end{vmatrix}$

$= \langle 0, 0, -36 \rangle$

73. Answers will vary.

Section 12.4 Limits at Infinity and Limits of Sequences

- The limit at infinity

  $\lim\limits_{x \to \infty} f(x) = L$

  means that $f(x)$ gets arbitrarily close to L as x increases without bound.

- Similarly, the limit at infinity

  $\lim\limits_{x \to -\infty} f(x) = L$

  means that $F(x)$ gets arbitrarily close to L as x decreases without bound.

- You should be able to calculate limits at infinity, especially those arising from rational functions.

- Limits of functions can be used to evaluate limits of sequences. If f is a function such that $\lim\limits_{x \to \infty} f(x) = L$ and if a_n is a sequence such that $f(n) = a_n$, then $\lim\limits_{n \to \infty} a_n = L$.

Vocabulary Check

1. limit, infinity

2. converge

3. diverge

1. Intercept: $(0, 0)$

Horizontal asymptote: $y = 4$

Matches (c).

3. Horizontal asymptote: $y = 4$

Vertical asymptote: $x = 0$

Matches (d).

5. $\lim\limits_{x \to \infty} \dfrac{3}{x^2} = 0$

7. $\lim\limits_{x\to\infty} \dfrac{3 + x}{3 - x} = -1$

9. $\lim\limits_{x\to-\infty} \dfrac{4x - 3}{2x + 1} = 2$

11. $\lim\limits_{x\to-\infty} \dfrac{3x^2 - 4}{1 - x^2} = -3$

13. $\lim\limits_{t\to\infty} \dfrac{t^2}{t + 3}$ does not exist.

15. $\lim\limits_{t\to\infty} \dfrac{1 - 2t + 6t^2}{5 + 3t - 4t^2} = \dfrac{6}{-4} = -\dfrac{3}{2}$

17. $\lim\limits_{x\to-\infty} \dfrac{-(x^2 + 3)}{(2 - x)^2} = \lim\limits_{x\to-\infty} \dfrac{-x^2 - 3}{x^2 - 4x + 4} = -1$

19. $\lim\limits_{x\to-\infty} \left[\dfrac{x}{(x + 1)^2} - 4 \right] = 0 - 4 = -4$

21. $\lim\limits_{t\to\infty} \left(\dfrac{1}{3t^2} - \dfrac{5t}{t + 2} \right) = 0 - 5$

$$= -5$$

23. $y = \dfrac{3x}{1 - x}$

Horizontal asymptote:
$y = -3$

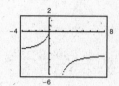

25. Horizontal asymptote: $y = 0$

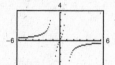

27. $y = 1 - \dfrac{3}{x^2}$

Horizontal asymptote: $y = 1$

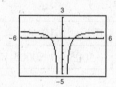

29. (a)

x	10^0	10^1	10^2	10^3	10^4	10^5	10^6
$f(x)$	-0.7321	-0.0995	-0.00999	-0.001	-1×10^{-4}	-1×10^{-5}	-1×10^{-6}

$$\lim\limits_{x\to\infty} \left(x - \sqrt{x^2 + 2}\right) = 0$$

(b) 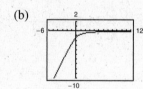 $\lim\limits_{x\to\infty} \left(x - \sqrt{x^2 + 2}\right) = 0$

31. (a)

x	10^0	10^1	10^2	10^3	10^4	10^5	10^6
$f(x)$	-0.7082	-0.7454	-0.7495	-0.74995	-0.749995	-0.75	-0.75

$$\lim\limits_{x\to\infty} 3\left(2x - \sqrt{4x^2 + x}\right) = -\tfrac{3}{4}$$

(b) 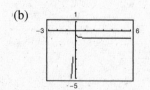 $\lim\limits_{x\to\infty} 3\left(2x - \sqrt{4x^2 + x}\right) = -\tfrac{3}{4}$

33. $a_n = \dfrac{n+1}{n^2+1}$

$a_1 = \dfrac{1+1}{1^2+1} = 1$ $\qquad$ $a_4 = \dfrac{5}{17}$

$a_2 = \dfrac{2+1}{2^2+1} = \dfrac{3}{5}$ $\qquad$ $a_5 = \dfrac{6}{26} = \dfrac{3}{13}$

$a_3 = \dfrac{4}{10} = \dfrac{2}{5}$

$\lim\limits_{n\to\infty} a_n = 0$

35. $a_n = \dfrac{n}{2n+1}$

$a_1 = \dfrac{1}{3}$ $\qquad$ $a_4 = \dfrac{4}{9}$

$a_2 = \dfrac{2}{5}$ $\qquad$ $a_5 = \dfrac{5}{11}$

$a_3 = \dfrac{3}{7}$

$\lim\limits_{n\to\infty} a_n = \dfrac{1}{2}$

37. $\dfrac{1}{5}, \dfrac{1}{2}, \dfrac{9}{11}, \dfrac{8}{7}, \dfrac{25}{17}$

$\lim\limits_{n\to\infty} \dfrac{n^2}{3n+2}$ does not exist.

39. 2, 3, 4, 5, 6

$\lim\limits_{n\to\infty} \dfrac{(n+1)!}{n!} = \lim\limits_{n\to\infty} (n+1)$
does not exist.

41. $-1, \dfrac{1}{2}, -\dfrac{1}{3}, \dfrac{1}{4}, -\dfrac{1}{5}$

$\lim\limits_{n\to\infty} \dfrac{(-1)^n}{n} = 0$

43. $\lim\limits_{n\to\infty} a_n = \lim\limits_{n\to\infty}\left[1 + \dfrac{n(n+1)}{2n^2}\right] = 1 + \dfrac{1}{2} = \dfrac{3}{2}$

n	10^0	10^1	10^2	10^3	10^4	10^5	10^6
a_n	2	1.55	1.505	1.5005	1.50005	1.500005	1.5000005

45. $\lim\limits_{n\to\infty} a_n = \dfrac{16}{1}\left[\dfrac{2}{6}\right] = \dfrac{16}{3}$

n	10^0	10^1	10^2	10^3	10^4	10^5	10^6
a_n	16	6.16	5.4136	5.341336	5.3341	5.33341	5.333341

47. $f(t) = \dfrac{t^2-t+1}{t^2+1}$

(a) $\lim\limits_{t\to\infty} \dfrac{t^2-t+1}{t^2+1} = 1$

(b)

(c) Over a long period of time, the level of oxygen in the pond returns to the normal level.

49. (a) Average cost $= \overline{C} = \dfrac{C}{x} = 13.50 + \dfrac{45{,}750}{x}$

(b) $\overline{C}(100) \doteq \$471$

$\overline{C}(1000) = \$59.25$

(c) $\lim\limits_{x\to\infty} C(x) = 13.50$

As more units are produced, the fixed costs (45,750) become less dominant.

51.

Year	t	Benefit, B
1997	7	765
1998	8	780
1999	9	804
2000	10	844
2001	11	874
2002	12	896
2003	13	922

(a)
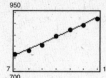

The model is a good fit to the data.

(b) For 2006: $B(16) \approx \$1032$

(c) The graph has a vertical asymptote and approaches infinity when t is slightly greater than 45.

53. False. $f(x) = \dfrac{x^2 + 1}{1}$ does not have a horizontal asymptote.

55. True

57. For example, let $f(x) = \dfrac{1}{x^2}$ and $g(x) = \dfrac{1}{x^2}$.

Then, $\displaystyle\lim_{x \to 0} \dfrac{1}{x^2}$ increases without bound, but $\displaystyle\lim_{x \to 0} [f(x) - g(x)] = 0$.

59. $a_n = 4\left(\dfrac{2}{3}\right)^n$

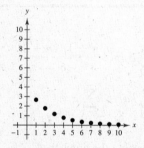

Converges to 0

61. $a_n = \dfrac{3[1 - (1.5)^n]}{1 - 1.5}$

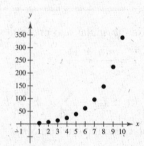

Diverges

63. $y = x^4$

(a) $f(x) = (x + 3)^4$

(b) $f(x) = x^4 - 1$

(c) $f(x) = -2 + x^4$

(d) $f(x) = \dfrac{1}{2}(x - 4)^4$

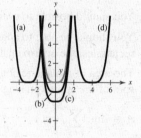

65.

$$
\begin{array}{r}
x^2 + 2x + 1 \\
x^2 - 4 \overline{\smash{\big)}\, x^4 + 2x^3 - 3x^2 - 8x - 4} \\
\underline{x^4 - 4x^2} \\
2x^3 + x^2 \\
\underline{2x^3 - 8x} \\
x^2 - 4
\end{array}
$$

$x^4 + 2x^3 - 3x^2 - 8x - 4 = (x^2 - 4)(x^2 + 2x + 1)$

67.

$$\begin{array}{r} x^3 + 5x^2 - 3 \\ 3x+2\overline{\smash{\big)}\,3x^4 + 17x^3 + 10x^2 - 9x - 8} \\ \underline{3x^4 + 2x^3} \\ 15x^3 + 10x^2 \\ \underline{15x^3 + 10x^2} \\ -9x - 8 \\ \underline{-9x - 6} \\ -2 \end{array}$$

$$\frac{3x^4 + 17x^3 + 10x^2 - 9x - 8}{3x + 2} = x^3 + 5x^2 - 3 + \frac{-2}{3x + 2}$$

69. $f(x) = x^4 - x^3 - 20x^2$

$ = x^2(x^2 - x - 20)$

$ = x^2(x - 5)(x + 4)$

Real zeros: $0, 0, 5, -4$

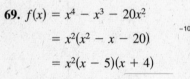

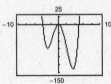

71. $f(x) = x^3 - 3x^2 + 2x - 6$

$ = x^2(x - 3) + 2(x - 3)$

$ = (x - 3)(x^2 + 2)$

Real zero: 3

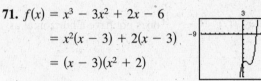

73. $\displaystyle\sum_{i=1}^{6} (2i + 3) = 5 + 7 + 9 + 11 + 13 + 15 = 60$

75. $\displaystyle\sum_{k=1}^{10} 15 = 10(15) = 150$

Section 12.5 The Area Problem

- You should know the following summation formulas and properties.

  (a) $\displaystyle\sum_{i=1}^{n} c = cn$

  (b) $\displaystyle\sum_{i=1}^{n} i = \frac{n(n + 1)}{2}$

  (c) $\displaystyle\sum_{i=1}^{n} i^2 = \frac{n(n + 1)(2n + 1)}{6}$

  (d) $\displaystyle\sum_{i=1}^{n} i^3 = \frac{n^2(n + 1)^2}{4}$

  (e) $\displaystyle\sum_{i=1}^{n} (a_i \pm b_i) = \sum_{i=1}^{n} a_i \pm \sum_{i=1}^{n} b_i$

  (f) $\displaystyle\sum_{i=1}^{n} ka_i = k\sum_{i=1}^{n} a_i$

- You should be able to evaluate a limit of a summation, $\displaystyle\lim_{n\to\infty} S(n)$.

- You should be able to approximate the area of a region using rectangles. By increasing the number of rectangles, the approximation improves.

- The area of a plane region above the x-axis bounded by f between $x = a$ and $x = b$ is the limit of the sum of the approximating rectangles:

  $$A = \lim_{n\to\infty} \sum_{i=1}^{n} f\left(a + \frac{(b - a)i}{n}\right)\left(\frac{b - a}{n}\right)$$

- You should be able to use the limit definition of area to find the area bounded by simple functions in the plane.

Vocabulary Check

1. $\dfrac{n(n + 1)}{2}$

2. $\dfrac{n^2(n + 1)^2}{4}$

3. area

1. $\displaystyle\sum_{i=1}^{60} 7 = 7(60) = 420$

3. $\displaystyle\sum_{k=1}^{20} (k^3 + 2) = \frac{20^2(21)^2}{4} + 2(20)$

$$= 44{,}100 + 40 = 44{,}140$$

5. $\displaystyle\sum_{j=1}^{25} (j^2 + j) = \frac{25(26)(51)}{6} + \frac{25(26)}{2} = 5850$

7. (a) $\displaystyle S(n) = \sum_{i=1}^{n} \frac{i^3}{n^4} = \frac{1}{n^4}\left[\frac{n^2(n+1)^2}{4}\right] = \frac{n^2 + 2n + 1}{4n^2}$

(b)

n	10^0	10^1	10^2	10^3	10^4
$S(n)$	1	0.3025	0.255025	0.25050025	0.25005

(c) $\displaystyle\lim_{n\to\infty} S(n) = \frac{1}{4}$

9. (a) $\displaystyle S(n) = \sum_{i=1}^{n} \frac{3}{n^3}(1 + i^2) = \frac{3}{n^3}\left[n + \frac{n(n+1)(2n+1)}{6}\right] = \frac{3}{n^2} + \frac{6n^2 + 9n + 3}{6n^2} = \frac{2n^2 + 3n + 7}{2n^2}$

(b)

n	10^0	10^1	10^2	10^3	10^4
$S(n)$	6	1.185	1.0154	1.0015	1.00015

(c) $\displaystyle\lim_{n\to\infty} S(n) = 1$

11. (a) $\displaystyle S(n) = \sum_{i=1}^{n} \left(\frac{i^2}{n^3} + \frac{2}{n}\right)\left(\frac{1}{n}\right) = \frac{1}{n}\left[\frac{n(n+1)(2n+1)}{6n^3} + \frac{2n}{n}\right] = \frac{1}{6n^3}(2n^2 + 3n + 1) + \frac{2}{n} = \frac{14n^2 + 3n + 1}{6n^3}$

(b)

n	10^0	10^1	10^2	10^3	10^4
$S(n)$	3	0.2385	0.02338	0.00233	0.0002333

(c) $\displaystyle\lim_{n\to\infty} S(n) = 0$

13. (a) $\displaystyle S(n) = \sum_{i=1}^{n} \left[1 - \left(\frac{i}{n}\right)^2\right]\left(\frac{1}{n}\right) = \frac{1}{n}\left[n - \frac{1}{n^2}\left(\frac{n(n+1)(2n+1)}{6}\right)\right] = 1 - \frac{2n^2 + 3n + 1}{6n^2} = \frac{4n^2 - 3n - 1}{6n^2}$

(b)

n	10^0	10^1	10^2	10^3	10^4
$S(n)$	0	0.615	0.66165	0.66617	0.666617

(c) $\displaystyle\lim_{n\to\infty} S(n) = \frac{2}{3}$

15. $f(x) = x + 4, [-1, 2], n = 6$, width $= \dfrac{1}{2}$

Area $\approx \dfrac{1}{2}[3.5 + 4 + 4.5 + 5 + 5.5 + 6]$

$= 14.25$ square units

17. The width of each rectangle is $\frac{1}{4}$. The height is obtained by evaluating f at the right-hand endpoint of each interval.

$$A \approx \sum_{i=1}^{8} f\left(\frac{i}{4}\right)\left(\frac{1}{4}\right) = \sum_{i=1}^{8} \frac{1}{4}\left(\frac{i}{4}\right)^3\left(\frac{1}{4}\right)$$

$$= 1.265625 \text{ square units}$$

19. Width of each rectangle is $\dfrac{12}{n}$. The height is $f\left(\dfrac{12}{n}i\right) = -\dfrac{1}{3}\left(\dfrac{12}{n}i\right) + 4$.

$$A = \sum_{i=1}^{n} \left[-\frac{1}{3}\left(\frac{12i}{n}\right) + 4\right]\left(\frac{12}{n}\right)$$

(**Note:** Exact area is 24.)

n	4	8	20	50
Approximate area	18	21	22.8	23.52

21. The width of each rectangle is $\dfrac{3}{n}$. The height is $\dfrac{1}{9}\left(\dfrac{3i}{n}\right)^3$.

$$A \approx \sum_{i=1}^{n} \frac{1}{9}\left(\frac{3i}{n}\right)^3\left(\frac{3}{n}\right)$$

n	4	8	20	50
Approximate area	3.52	2.85	2.48	2.34

23. $A \approx \displaystyle\sum_{i=1}^{n} f\left(\dfrac{i}{n}\right)\left(\dfrac{1}{n}\right)$

$= \displaystyle\sum_{i=1}^{n}\left[4\left(\dfrac{i}{n}\right)+1\right]\left(\dfrac{1}{n}\right)$

$= \dfrac{1}{n}\displaystyle\sum_{i=1}^{n}\left[\dfrac{4}{n}i+1\right]$

$= \dfrac{1}{n}\left[\dfrac{4}{n}\dfrac{n(n+1)}{2}+n\right]$

$= \dfrac{1}{n}\left[2(n+1)+n\right]$

$= \dfrac{3n+2}{n}$

$A = \displaystyle\lim_{n\to\infty}\dfrac{3n+2}{n}=3$ square units

25. $A \approx \displaystyle\sum_{i=1}^{n} f\left(\dfrac{i}{n}\right)\left(\dfrac{1}{n}\right)$

$= \displaystyle\sum_{i=1}^{n}\left[-2\left(\dfrac{i}{n}\right)+3\right]\left(\dfrac{1}{n}\right)$

$= \dfrac{1}{n}\displaystyle\sum_{i=1}^{n}\left[-\dfrac{2i}{n}+3\right]$

$= \dfrac{1}{n}\left[-\dfrac{2}{n}\dfrac{n(n+1)}{2}+3n\right]$

$= \dfrac{1}{n}\left[2n-1\right]$

$A = \displaystyle\lim_{n\to\infty}\dfrac{2n-1}{n}=2$ square units

27. $A \approx \displaystyle\sum_{i=1}^{n} f\left(-1+\dfrac{2i}{n}\right)\left(\dfrac{2}{n}\right)$

$= \displaystyle\sum_{i=1}^{n}\left[2-\left(-1+\dfrac{2i}{n}\right)^2\right]\dfrac{2}{n}$

$= \displaystyle\sum_{i=1}^{n}\left[2-1+\dfrac{4i}{n}-\dfrac{4i^2}{n^2}\right]\left(\dfrac{2}{n}\right)$

$= \dfrac{2}{n}\displaystyle\sum_{i=1}^{n}1+\dfrac{8}{n^2}\displaystyle\sum_{i=1}^{n}i-\dfrac{8}{n^3}\displaystyle\sum_{i=1}^{n}i^2$

$= \dfrac{2}{n}(n)+\dfrac{8}{n^2}\dfrac{n(n+1)}{2}-\dfrac{8}{n^3}\dfrac{n(n+1)(2n+1)}{6}$

$A = \displaystyle\lim_{n\to\infty}\left[2+4\dfrac{n(n+1)}{n^2}-\dfrac{4}{3}\dfrac{n(n+1)(2n+1)}{n^3}\right]=2+4-\dfrac{8}{3}=\dfrac{10}{3}$ square units

29. $A \approx \displaystyle\sum_{i=1}^{n} g\left(1+\dfrac{i}{n}\right)\left(\dfrac{1}{n}\right)$

$= \displaystyle\sum_{i=1}^{n}\left[8-\left(1+\dfrac{i}{n}\right)^3\right]\dfrac{1}{n}$

$= \displaystyle\sum_{i=1}^{n}\left[7-\dfrac{3i}{n}-\dfrac{3i^2}{n^2}-\dfrac{i^3}{n^3}\right]\dfrac{1}{n}$

$= \dfrac{7}{n}\displaystyle\sum_{i=1}^{n}1-\dfrac{3}{n^2}\displaystyle\sum_{i=1}^{n}i-\dfrac{3}{n^3}\displaystyle\sum_{i=1}^{n}i^2-\dfrac{1}{n^4}\displaystyle\sum_{i=1}^{n}i^3$

$= \dfrac{7}{n}(n)-\dfrac{3}{n^2}\dfrac{n(n+1)}{2}-\dfrac{3}{n^3}\dfrac{n(n+1)(2n+1)}{6}-\dfrac{1}{n^4}\dfrac{n^2(n+1)^2}{4}$

$A = \displaystyle\lim_{n\to\infty}\left[7-\dfrac{3}{2}\dfrac{n(n+1)}{n^2}-\dfrac{1}{2n^3}n(n+1)(2n+1)-\dfrac{1}{n^4}\dfrac{n^2(n+1)^2}{4}\right]=7-\dfrac{3}{2}-1-\dfrac{1}{4}=\dfrac{17}{4}$ square units

31. $A \approx \sum_{i=1}^{n} g\left(\dfrac{i}{n}\right)\left(\dfrac{1}{n}\right)$

$\quad = \sum_{i=1}^{n} \left[2\left(\dfrac{i}{n}\right) - \left(\dfrac{i}{n}\right)^3\right]\left(\dfrac{1}{n}\right)$

$\quad = \dfrac{1}{n}\sum_{i=1}^{n}\left[\dfrac{2}{n}i - \dfrac{1}{n^3}i^3\right]$

$\quad = \dfrac{1}{n}\left[\dfrac{2}{n}\dfrac{n(n+1)}{2} - \dfrac{1}{n^3}\dfrac{n^2(n+1)^2}{4}\right]$

$\quad = \dfrac{n+1}{n} - \dfrac{(n+1)^2}{4n^2}$

$A = \lim_{n\to\infty}\left[\dfrac{n+1}{n} - \dfrac{(n+1)^2}{4n^2}\right]$

$\quad = 1 - \dfrac{1}{4} = \dfrac{3}{4}$ square unit

33. $A \approx \sum_{i=1}^{n} f\left(1 + \dfrac{3i}{n}\right)\left(\dfrac{3}{n}\right)$

$\quad = \sum_{i=1}^{n}\left[\dfrac{1}{4}\left(1 + \dfrac{3i}{n}\right)^2 + \left(1 + \dfrac{3i}{n}\right)\right]\left(\dfrac{3}{n}\right)$

$\quad = \sum_{i=1}^{n}\left(\dfrac{1}{4} + \dfrac{3}{2}\dfrac{i}{n} + \dfrac{9}{4}\dfrac{i^2}{n^2} + 1 + \dfrac{3i}{n}\right)\left(\dfrac{3}{n}\right)$

$\quad = \dfrac{15}{4n}\sum_{i=1}^{n} 1 + \dfrac{27}{2n^2}\sum_{i=1}^{n} i + \dfrac{27}{4n^3}\sum_{i=1}^{n} i^2$

$\quad = \dfrac{15}{4n}(n) + \dfrac{27}{2n^2}\left(\dfrac{n(n+1)}{2}\right) + \dfrac{27}{4n^3}\dfrac{n(n+1)(2n+1)}{6}$

$A = \lim_{n\to\infty}\left[\dfrac{15}{4} + \dfrac{27}{4}\dfrac{n(n+1)}{n^2} + \dfrac{9}{8n^3}n(n+1)(2n+1)\right]$

$\quad = \dfrac{15}{4} + \dfrac{27}{4} + \dfrac{9}{4} = \dfrac{51}{4}$ square units

35. $y = (-3.0 \cdot 10^{-6})x^3 + 0.002x^2 - 1.05x + 400$

Note that $y = 0$ when $x = 500$.

Area $\approx 105{,}208.33$ square feet ≈ 2.4153 acres

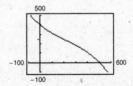

37. True. See Formula 2, page 892.

39. Answers will vary.

41. $\quad\quad \sin 2x - \sqrt{3}\sin x = 0$

$\quad 2\sin x \cos x - \sqrt{3}\sin x = 0$

$\quad\quad \sin x\left(2\cos x - \sqrt{3}\right) = 0$

$\quad\quad\quad\quad \sin x = 0 \implies x = n\pi$

$\cos x = \dfrac{\sqrt{3}}{2} \implies x = \dfrac{\pi}{6} + 2n\pi,\ x = \dfrac{11\pi}{6} + 2n\pi$

43. $2\tan x = \tan 2x = \dfrac{2\tan x}{1 - \tan^2 x}$

$\quad \tan x = 0 \implies x = n\pi$

45. $2 \cot x = 5 \cos \dfrac{\pi}{2} = 0$

$\cot x = 0 \implies x = \dfrac{\pi}{2} + n\pi$

47. $(\mathbf{u} \cdot \mathbf{v})\mathbf{u} = (\langle 4, -5 \rangle \cdot \langle -1, -2 \rangle)\langle 4, -5 \rangle$

$= 6\langle 4, -5 \rangle$

$= \langle 24, -30 \rangle$

49. $\|\mathbf{v}\| - 2 = \sqrt{5} - 2$

Review Exercises for Chapter 12

1. $\lim\limits_{x \to 3} (6x - 1)$

The limit (17) can be reached.

x	2.9	2.99	2.999	3	3.001	3.01	3.1
$f(x)$	16.4	16.94	16.994	17	17.006	17.06	17.6

3. $\lim\limits_{x \to 1} (3 - x) = 2$

5. $\lim\limits_{x \to 1} \dfrac{x^2 - 1}{x - 1} = 2$

7. (a) $\lim\limits_{x \to c} [f(x)]^3 = 4^3 = 64$

(b) $\lim\limits_{x \to c} [3f(x) - g(x)] = 3(4) - 5 = 7$

(c) $\lim\limits_{x \to c} [f(x)g(x)] = (4)(5) = 20$

(d) $\lim\limits_{x \to c} \dfrac{f(x)}{g(x)} = \dfrac{4}{5}$

9. $\lim\limits_{x \to 4} \left(\dfrac{1}{2}x + 3 \right) = \dfrac{1}{2}(4) + 3 = 5$

11. $\lim\limits_{x \to 2} \dfrac{x^2 - 1}{x^3 + 2} = \dfrac{2^2 - 1}{2^3 + 2} = \dfrac{3}{10}$

13. $\lim\limits_{x \to \pi} \sin 3x = \sin 3\pi = 0$

15. $\lim\limits_{x \to 3} (5x - 4) = 5(3) - 4 = 11$

17. $\lim\limits_{x \to 2} (5x - 3)(3x + 5) = (5(2) - 3)(3(2) + 5)$

$= (7)(11) = 77$

19. $\lim\limits_{t \to 3} \dfrac{t^2 + 1}{t} = \dfrac{9 + 1}{3} = \dfrac{10}{3}$

21. $\lim\limits_{t \to -2} \dfrac{t + 2}{t^2 - 4} = \lim\limits_{t \to -2} \dfrac{t + 2}{(t + 2)(t - 2)}$

$= \lim\limits_{t \to -2} \dfrac{1}{t - 2} = -\dfrac{1}{4}$

23. $\lim\limits_{x \to 5} \dfrac{x - 5}{x^2 + 5x - 50} = \lim\limits_{x \to 5} \dfrac{x - 5}{(x - 5)(x + 10)}$

$= \lim\limits_{x \to 5} \dfrac{1}{x + 10} = \dfrac{1}{15}$

25. $\lim\limits_{x \to -2} \dfrac{x^2 - 4}{x^3 + 8} = \lim\limits_{x \to -2} \dfrac{(x + 2)(x - 2)}{(x + 2)(x^2 - 2x + 4)}$

$= \lim\limits_{x \to -2} \dfrac{x - 2}{x^2 - 2x + 4}$

$= \dfrac{-4}{12} = \dfrac{-1}{3}$

27. $\lim\limits_{x \to -1} \dfrac{1/(x + 2) - 1}{x + 1} = \lim\limits_{x \to -1} \dfrac{1 - (x + 2)}{(x + 2)(x + 1)}$

$= \lim\limits_{x \to -1} \dfrac{-(x + 1)}{(x + 2)(x + 1)}$

$= \lim\limits_{x \to -1} \dfrac{-1}{(x + 2)} = -1$

29. $\lim\limits_{u \to 0} \dfrac{\sqrt{4+u} - 2}{u} = \lim\limits_{u \to 0} \dfrac{\sqrt{4+u} - 2}{u} \cdot \dfrac{\sqrt{4+u} + 2}{\sqrt{4+u} + 2}$

$\qquad = \lim\limits_{u \to 0} \dfrac{(4+u) - 4}{u\left(\sqrt{4+u} + 2\right)}$

$\qquad = \lim\limits_{u \to 0} \dfrac{1}{\sqrt{4+u} + 2} = \dfrac{1}{4}$

31. $\lim\limits_{x \to 5} \dfrac{\sqrt{x-1} - 2}{x - 5} = \lim\limits_{x \to 5} \dfrac{\sqrt{x-1} - 2}{x - 5} \cdot \dfrac{\sqrt{x-1} + 2}{\sqrt{x-1} + 2}$

$\qquad = \lim\limits_{x \to 5} \dfrac{(x-1) - 4}{(x-5)\left(\sqrt{x-1} + 2\right)}$

$\qquad = \lim\limits_{x \to 5} \dfrac{1}{\sqrt{x-1} + 2} = \dfrac{1}{2+2} = \dfrac{1}{4}$

33. (a)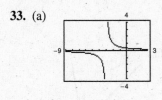

(b)

x	2.9	2.99	3	3.01	3.1
$f(x)$	0.1695	0.1669	Error	0.1664	0.1639

$\lim\limits_{x \to 3} \dfrac{x - 3}{x^2 - 9} = \dfrac{1}{6}$

35. (a)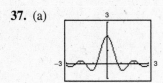

(b) (Answers will vary.)

x	-0.1	-0.01	-0.001	0	0.001	0.01	0.1
y_1	4.85 E 8	7.2 E 86	Error	Error	0	1 E -87	2.1 E -9

$\lim\limits_{x \to 0} e^{-2/x}$ does not exist.

37. (a)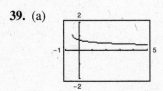

(b)

x	-0.1	-0.01	-0.001	0	0.001	0.01	0.1
y_1	1.9471	1.9995	1.999995	Error	1.999995	1.9995	1.9471

$\lim\limits_{x \to 0} \dfrac{\sin 4x}{2x} = 2$

39. (a)

(b)

x	1.1	1.01	1.001	1.0001
$f(x)$	0.5680	0.5764	0.5773	0.5773

$\lim\limits_{x \to 1^+} \dfrac{\sqrt{2x+1} - \sqrt{3}}{x - 1} \approx 0.577$

$\left(\text{Exact value: } \dfrac{\sqrt{3}}{3}\right)$

41. $f(x) = \dfrac{|x - 3|}{x - 3}$

Limit does not exist because

$\lim\limits_{x \to 3^{+}} f(x) = 1$ and

$\lim\limits_{x \to 3^{-}} f(x) = -1$.

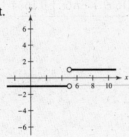

43. $f(x) = \dfrac{2}{x^2 - 4}$

Limit does not exist.

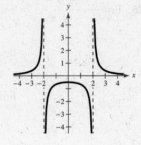

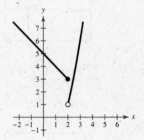

45. $\lim\limits_{x \to 5} \dfrac{|x - 5|}{x - 5}$ does not exist.

47. $\lim\limits_{x \to 2} f(x)$ does not exist.

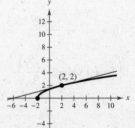

49. $f(x) = 4x + 3$

$$\lim_{h \to 0} \frac{f(x + h) - f(x)}{h} = \lim_{h \to 0} \frac{[4(x + h) + 3] - (4x + 3)}{h}$$

$$= \lim_{h \to 0} \frac{4x + 4h + 3 - 4x - 3}{h}$$

$$= \lim_{h \to 0} \frac{4h}{h}$$

$$= \lim_{h \to 0} 4 = 4$$

51. $\lim\limits_{h \to 0} \dfrac{f(x + h) - f(x)}{h} = \lim\limits_{h \to 0} \dfrac{3(x + h) - (x + h)^2 - (3x - x^2)}{h}$

$$= \lim_{h \to 0} \frac{3x + 3h - x^2 - 2xh - h^2 - 3x + x^2}{h} = \lim_{h \to 0} \frac{3h - 2xh - h^2}{h}$$

$$= \lim_{h \to 0} (3 - 2x - h) = 3 - 2x$$

53. Slope ≈ 2

(Answers will vary.)

55.

Slope at $(2, f(2))$ is approximately 2.

57.

Slope is $\frac{1}{4}$ at $(2, 2)$.

59. $f(x) = x^2 - 4x$

$$m = \lim_{h \to 0} \frac{f(x + h) - f(x)}{h}$$

$$= \lim_{h \to 0} \frac{(x + h)^2 - 4(x + h) - (x^2 - 4x)}{h}$$

$$= \lim_{h \to 0} \frac{x^2 + 2xh + h^2 - 4x - 4h - x^2 - 4x}{h}$$

$$= \lim_{h \to 0} \frac{2xh + h^2 - 4h}{h}$$

$$= \lim_{h \to 0} (2x + h - 4) = 2x - 4$$

(a) At $(0, 0)$, $m = 2(0) - 4 = -4$.

(b) At $(5, 5)$, $m = 2(5) - 4 = 6$.

61. $f(x) = \dfrac{4}{x - 6}$

$$m = \lim_{h \to 0} \frac{f(x + h) - f(x)}{h} = \lim_{h \to 0} \frac{\dfrac{4}{x + h - 6} - \dfrac{4}{x - 6}}{h}$$

$$= \lim_{h \to 0} \frac{4(x - 6) - 4(x + h - 6)}{(x + h - 6)(x - 6)h}$$

$$= \lim_{h \to 0} \frac{-4h}{(x + h - 6)(x - 6)h}$$

$$= \lim_{h \to 0} \frac{-4}{(x + h - 6)(x - 6)} = \frac{-4}{(x - 6)^2}$$

(a) At $(7, 4)$, $m = \dfrac{-4}{(7 - 6)^2} = -4$.

(b) At $(8, 2)$, $m = \dfrac{-4}{(8 - 6)^2} = -1$.

63. $f'(x) = \lim_{h \to 0} \dfrac{f(x + h) - f(x)}{h} = \lim_{h \to 0} \dfrac{5 - 5}{h} = 0$

65. $h'(x) = \lim_{k \to 0} \dfrac{h(x + k) - h(x)}{k}$

$$= \lim_{k \to 0} \frac{\left[5 - \frac{1}{2}(x + k)\right] - \left[5 - \frac{1}{2}x\right]}{k}$$

$$= \lim_{k \to 0} \frac{-\frac{1}{2}k}{k} = -\frac{1}{2}$$

67. $g'(x) = \lim_{h \to 0} \dfrac{g(x + h) - g(x)}{h}$

$$= \lim_{h \to 0} \frac{2(x + h)^2 - 1 - (2x^2 - 1)}{h}$$

$$= \lim_{h \to 0} \frac{2x^2 + 4xh + 2h^2 - 2x^2}{h}$$

$$= \lim_{h \to 0} (4x + 2h)$$

$$= 4x$$

69. $f'(t) = \lim_{h \to 0} \dfrac{f(t + h) - f(t)}{h}$

$$= \lim_{h \to 0} \frac{\sqrt{t + h + 5} - \sqrt{t + 5}}{h} \cdot \frac{\sqrt{t + h + 5} + \sqrt{t + 5}}{\sqrt{t + h + 5} + \sqrt{t + 5}}$$

$$= \lim_{h \to 0} \frac{(t + h + 5) - (t + 5)}{h\left(\sqrt{t + h + 5} + \sqrt{t + 5}\right)}$$

$$= \lim_{h \to 0} \frac{1}{\sqrt{t + h + 5} + \sqrt{t + 5}}$$

$$= \frac{1}{2\sqrt{t + 5}}$$

71. $g'(s) = \dfrac{g(s+h) - g(s)}{h} = \lim\limits_{h \to 0} \dfrac{\dfrac{4}{s+h+5} - \dfrac{4}{s+5}}{h}$

$$= \lim\limits_{h \to 0} \frac{4s + 20 - 4s - 4h - 20}{(s+h+5)(s+5)h}$$

$$= \lim\limits_{h \to 0} \frac{-4h}{(s+h+5)(s+5)h}$$

$$= \lim\limits_{h \to 0} \frac{-4}{(s+h+5)(s+5)} = \frac{-4}{(s+5)^2}$$

73. $f(x) = 2x^2 - 1, \ (0, -1)$

(a) $m_{\text{tan}} = \lim\limits_{h \to 0} \dfrac{f(x+h) - f(x)}{h}$

$\qquad = \lim\limits_{h \to 0} \dfrac{f(0+h) - f(0)}{h}$

$\qquad = \lim\limits_{h \to 0} \dfrac{2h^2 - 1 - (-1)}{h}$

$\qquad = \lim\limits_{h \to 0} 2h = 0$

(b) Tangent line: $\quad y - (-1) = 0(x - 0)$

$$y + 1 = 0$$

$$y = -1$$

(c)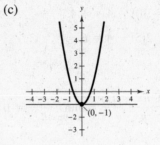

75. $\lim\limits_{x \to \infty} \dfrac{4x}{2x - 3} = \dfrac{4}{2} = 2$

77. $\lim\limits_{x \to -\infty} \dfrac{2x}{x^2 - 25} = 0$

79. $\lim\limits_{x \to \infty} \dfrac{x^2}{2x + 3}$ does not exist.

81. $\lim\limits_{x \to \infty} \left[\dfrac{x}{(x-2)^2} + 3 \right] = 0 + 3 = 3$

83. $\dfrac{2}{3}, \dfrac{5}{5} = 1, \dfrac{8}{7}, \dfrac{11}{9}, \dfrac{14}{11}$

$\qquad \lim\limits_{n \to \infty} a_n = \dfrac{3}{2}$

85. $a_n = \dfrac{1}{2n^2}[3 - 2n(n+1)] = \dfrac{3}{2n^2} - \dfrac{n+1}{n}$

$\quad -0.5, -1.125, -1.16\overline{6}, -1.15625, -1.14$

$\quad \lim\limits_{n \to \infty} a_n = 0 - 1 = -1$

87. (a) $\displaystyle\sum_{i=1}^{n}\left(\frac{4i^2}{n^2}-\frac{i}{n}\right)\frac{1}{n}=\frac{4}{n^3}\sum_{i=1}^{n}i^2-\frac{1}{n^2}\sum_{i=1}^{n}i$

$$=\frac{4}{n^3}\frac{n(n+1)(2n+1)}{6}-\frac{1}{n^2}\frac{n(n+1)}{2}$$

$$=\frac{4n(n+1)(2n+1)-3n^2(n+1)}{6n^3}$$

$$=\frac{n(n+1)(8n+4-3n)}{6n^3}$$

$$=\frac{(n+1)(5n+4)}{6n^2}$$

(b)

n	10^0	10^1	10^2	10^3	10^4
$S(n)$	3	0.99	0.8484	0.8348	0.8335

(c) $\displaystyle\lim_{n\to\infty}S(n)=\frac{5}{6}$

89. Area $\approx\frac{1}{2}\left[\frac{7}{2}+3+\frac{5}{2}+2+\frac{3}{2}+1\right]=\frac{1}{2}\frac{27}{2}=\frac{27}{4}=6.75$

91. $f(x)=\dfrac{1}{4}x^2,\ b-a=4-0=4$

$A\approx\displaystyle\sum_{i=1}^{n}f\left(\frac{4i}{n}\right)\left(\frac{4}{n}\right)$

$=\displaystyle\sum_{i=1}^{n}\frac{1}{4}\left(\frac{4i}{n}\right)^2\left(\frac{4}{n}\right)$

$=\dfrac{1}{n}\displaystyle\sum_{i=1}^{n}\frac{16}{n^2}i^2$

$=\dfrac{16}{n^3}\dfrac{n(n+1)(2n+1)}{6}$

$=\dfrac{8(n+1)(2n+1)}{3n^2}$

n	4	8	20	50
Approximate area	7.5	6.375	5.74	5.4944

$\left(\text{Exact area is }\dfrac{16}{3}\approx5.33.\right)$

93. $A=\displaystyle\lim_{n\to\infty}\sum_{i=1}^{n}\left(10-\frac{10i}{n}\right)\left(\frac{10}{n}\right)$

$=\displaystyle\lim_{n\to\infty}\left[\frac{100}{n}\sum_{i=1}^{n}1-\frac{100}{n^2}\sum_{i=1}^{n}i\right]$

$=\displaystyle\lim_{n\to\infty}\left[\frac{100}{n}(n)-\frac{100}{n^2}\left(\frac{n(n+1)}{2}\right)\right]$

$=\displaystyle\lim_{n\to\infty}\left[100-50\frac{n(n+1)}{n^2}\right]$

$=100-50=50,\ \text{exact area}$

95. $A=\displaystyle\lim_{n\to\infty}\sum_{i=1}^{n}\left[\left(-1+\frac{3i}{n}\right)^2+4\right]\left(\frac{3}{n}\right)$

$=\displaystyle\lim_{n\to\infty}\sum_{i=1}^{n}\left[5-\frac{6i}{n}+\frac{9i^2}{n^2}\right]\frac{3}{n}$

$=\displaystyle\lim_{n\to\infty}\left[\frac{15}{n}\sum_{i=1}^{n}1-\frac{18}{n^2}\sum_{i=1}^{n}i+\frac{27}{n^3}\sum_{i=1}^{n}i^2\right]$

$=\displaystyle\lim_{n\to\infty}\left[\frac{15}{n}(n)-\frac{18}{n^2}\frac{n(n+1)}{2}+\frac{27}{n^3}\frac{n(n+1)(2n+1)}{6}\right]$

$=15-9+9=15,\ \text{exact area}$

97. $A = \lim_{n \to \infty} \sum_{i=1}^{n} 2\left[\left(-1 + \frac{2i}{n}\right)^2 - \left(-1 + \frac{2i}{n}\right)^3\right]\left(\frac{2}{n}\right)$

$= \lim_{n \to \infty} \sum_{i=1}^{n} \frac{4}{n}\left(1 - \frac{4i}{n} + \frac{4i^2}{n^2} - \left(-1 + \frac{6i}{n} - \frac{12i^2}{n^2} + \frac{8i^3}{n^3}\right)\right)$

$= \lim_{n \to \infty} \sum_{i=1}^{n} \frac{4}{n}\left(2 - \frac{10i}{n} + \frac{16i^2}{n^2} - \frac{8i^3}{n^3}\right)$

$= \lim_{n \to \infty} \left[\frac{8}{n}\sum_{i=1}^{n} 1 - \frac{40}{n^2}\sum_{i=1}^{n} i + \frac{64}{n^3}\sum_{i=1}^{n} i^2 - \frac{32}{n^4}\sum_{i=1}^{n} i^3\right]$

$= \lim_{n \to \infty} \left[\frac{8}{n}(n) - \frac{40}{n^2}\frac{n(n+1)}{2} + \frac{64}{n^3}\frac{n(n+1)(2n+1)}{6} - \frac{32}{n^4}\frac{n^2(n+1)^2}{4}\right] = 8 - 20 + \frac{64}{3} - 8 = \frac{4}{3}$, exact area

99. (a) $y = (-3.376068 \times 10^{-7})x^3 + (3.7529 \times 10^{-4})x^2 - 0.17x + 132$ (b)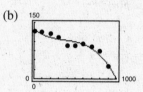

(c) Area $\approx$ 88,000 square feet, answers will vary.

101. False. The limit does not exist.

Problem Solving for Chapter 12

1. (a) $\lim_{x \to 2} f(x) = 3; g_1, g_4$ (b) $\lim_{x \to 2^-} f(x) = 3; g_1, g_3, g_4$ (c) $\lim_{x \to 2^+} f(x) = 3; g_1, g_4$

3.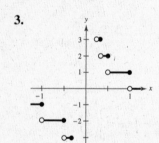

(a) $f\left(\frac{1}{4}\right) = [\![4]\!] = 4$

$f(3) = [\![\frac{1}{3}]\!] = 0$

$f(1) = [\![1]\!] = 1$

(b) $\lim_{x \to 1^-} f(x) = 1$

$\lim_{x \to 1^+} f(x) = 0$

$\lim_{x \to (1/2)^-} f(x) = 2$

$\lim_{x \to (1/2)^+} f(x) = 1$

5. Since $\lim_{x \to 0} \frac{\sqrt{a + bx} - \sqrt{3}}{x}$ exists $\Rightarrow \lim_{x \to 0}\left(\sqrt{a + bx} - \sqrt{3}\right) = 0 \Rightarrow \sqrt{a} - \sqrt{3} = 0 \Rightarrow a = 3$.

$\lim_{x \to 0}\left(\frac{\sqrt{3 + bx} - \sqrt{3}}{x} \cdot \frac{\sqrt{3 + bx} + \sqrt{3}}{\sqrt{3 + bx} + \sqrt{3}}\right) = \lim_{x \to 0} \frac{3 + bx - 3}{x\left(\sqrt{3 + bx} + \sqrt{3}\right)}$

$= \lim_{x \to 0} \frac{bx}{x\left(\sqrt{3 + bx} + \sqrt{3}\right)}$

$= \lim_{x \to 0} \frac{b}{\sqrt{3 + bx} + \sqrt{3}} = \frac{b}{2\sqrt{3}} = \sqrt{3} \Rightarrow b = 6$

Thus, $a = 3$ and $b = 6$.

7. $f(x) = \begin{cases} 0, & \text{if } x \text{ is rational} \\ 1, & \text{if } x \text{ is irrational} \end{cases}$

$\lim_{x \to 0} f(x)$ does not exist.

No matter how close to 0 x is, there are still an infinite number of rational and irrational numbers

$g(x) = \begin{cases} 0, & \text{if } x \text{ is rational} \\ x, & \text{if } x \text{ is irrational} \end{cases}$

$\lim_{x \to 0} g(x) = 0$

When x is close to 0, both parts of the function are close to zero.

9. $f(x) = a + b\sqrt{x}$

$f(1) = 4 \implies a + b = 4$

Tangent line: $2y - 3x = 5 \implies y = \frac{3}{2}x + \frac{5}{2} \implies m_{\text{tan}} = \frac{3}{2}$ at $(1, 4)$

$$m_{\text{tan}} = \lim_{h \to 0} \frac{f(x + h) - f(x)}{h} = \lim_{h \to 0} \frac{\left(a + b\sqrt{x + h}\right) - \left(a + b\sqrt{x}\right)}{h}$$

$$= \lim_{h \to 0} \left(\frac{b\left(\sqrt{x + h} - \sqrt{x}\right)}{h} \cdot \frac{\sqrt{x + h} + \sqrt{x}}{\sqrt{x + h} + \sqrt{x}} \right)$$

$$= \lim_{h \to 0} \frac{b(x + h - x)}{h\left(\sqrt{x + h} + \sqrt{x}\right)} = \lim_{h \to 0} \frac{b}{\sqrt{x + h} + \sqrt{x}} = \frac{b}{2\sqrt{x}}$$

At $(1, 4)$, $m_{\text{tan}} = \frac{b}{2} = \frac{3}{2} \implies b = 3$ and $a = 1$.

Thus, $f(x) = 1 + 3\sqrt{x}$.

11. Slope m through $(0, 4) \implies y = mx + 4$ or $mx - y + 4 = 0$

(a) $d(m) = \frac{|3m + (-1)(1) + 4|}{\sqrt{m^2 + (-1)^2}} = \frac{|3m + 3|}{\sqrt{m^2 + 1}}$

(b)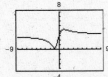

(c) $\lim_{m \to \infty} d(m) = 3$ and $\lim_{m \to -\infty} d(m) = 3$

This indicates that the distance between the point and the line approaches 3 as the slope approaches positive or negative infinity.

13. The error was probably due to the calculator being in degree mode rather than radian mode.

15. (a)

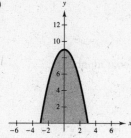

(b) If we find the area of the region in Quadrant I, bounded by $y = 9 - x^2$ and the x- and y-axes, we can find the area of the entire region by doubling this value.

Width: $\dfrac{b-a}{n} = \dfrac{3-0}{n} = \dfrac{3}{n}$

Height: $f\left(a + \dfrac{(b-a)i}{n}\right) = f\left(0 + \dfrac{3i}{n}\right) = f\left(\dfrac{3i}{n}\right) = 9 - \left(\dfrac{3i}{n}\right)^2 = 9 - \dfrac{9i^2}{n^2}$

$$\lim_{n\to\infty} \sum_{i=1}^{n} f\left(a + \dfrac{(b-a)i}{n}\right)\left(\dfrac{b-a}{n}\right) = \lim_{n\to\infty} \sum_{i=1}^{n} \left(9 - \dfrac{9i^2}{n^2}\right)\left(\dfrac{3}{n}\right)$$

$$= \lim_{n\to\infty} \sum_{i=1}^{n} \left(\dfrac{27}{n} - \dfrac{27i^2}{n^3}\right)$$

$$= \lim_{n\to\infty} \left[\left(\dfrac{27}{n}\right)(n) - \dfrac{27}{n^3}\sum_{i=1}^{n} i^2\right]$$

$$= \lim_{n\to\infty} \left[27 - \dfrac{27}{n^3}\left(\dfrac{n(n+1)(2n+1)}{6}\right)\right]$$

$$= 27 - \dfrac{27(2)}{6} = 18$$

Area $= 2(18) = 36$ square units

(c) Base $= 6$, Height $= 9$, Area $= \dfrac{2}{3}bh = \dfrac{2}{3}(6)(9) = 36$

Chapter 12   Practice Test

1. Use a graphing utility to complete the table and use the result to estimate the limit

$$\lim_{x \to 3} \frac{x - 3}{x^2 - 9}.$$

x	2.9	2.99	3	3.01	3.1
$f(x)$			?		

2. Graph the function $f(x) = \dfrac{\sqrt{x + 4} - 2}{x}$ and estimate the limit $\lim\limits_{x \to 0} \dfrac{\sqrt{x + 4} - 2}{x}$.

3. Find the limit $\lim\limits_{x \to 2} e^{x - 2}$ by direct substitution.

4. Find the limit $\lim\limits_{x \to 1} \dfrac{x^3 - 1}{x - 1}$ analytically.

5. Use a graphing utility to estimate the limit.

$$\lim_{x \to 0} \frac{\sin 5x}{2x}$$

6. Find the limit.

$$\lim_{x \to -2} \frac{|x + 2|}{x + 2}$$

7. Use the limit process to find the slope of the graph of $f(x) = \sqrt{x}$ at the point $(4, 2)$.

8. Find the derivative of the function $f(x) = 3x - 1$.

9. Find the limits.

(a) $\lim\limits_{x \to \infty} \dfrac{3}{x^4}$ (b) $\lim\limits_{x \to -\infty} \dfrac{x^2}{x^2 + 3}$ (c) $\lim\limits_{x \to \infty} \dfrac{|x|}{1 - x}$

10. Write the first four terms of the sequence $a_n = \dfrac{1 - n^2}{2n^2 + 1}$ and find the limit of the sequence.

11. Find the sum $\sum\limits_{i=1}^{25} (i^2 + i)$.

12. Write the sum $\sum\limits_{i=1}^{n} \dfrac{i^2}{n^3}$ as a rational function $S(n)$, and find $\lim\limits_{n \to \infty} S(n)$.

13. Find the area of the region bounded by $f(x) = 1 - x^2$ over the interval $0 \le x \le 1$.

APPENDIX A
Review of Fundamental Concepts of Algebra

Appendix A.1   Review of Real Numbers and Their Properties **615**

Appendix A.2   Exponents and Radicals **620**

Appendix A.3   Polynomials and Factoring **624**

Appendix A.4   Rational Expressions **631**

Appendix A.5   Solving Equations **635**

Appendix A.6   Linear Inequalities in One Variable **647**

Appendix A.7   Errors and the Algebra of Calculus **652**

APPENDIX A
Review of Fundamental Concepts of Algebra

Appendix A.1   Review of Real Numbers and Their Properties

■ You should know the following sets.

(a) The set of real numbers includes the rational numbers and the irrational numbers.

(b) The set of rational numbers includes all real numbers that can be written as the ratio p/q of two integers, where $q \neq 0$.

(c) The set of irrational numbers includes all real numbers which are not rational.

(d) The set of integers: $\{\ldots, -3, -2, -1, 0, 1, 2, 3, \ldots\}$

(e) The set of whole numbers: $\{0, 1, 2, 3, 4, \ldots\}$

(f) The set of natural numbers: $\{1, 2, 3, 4, \ldots\}$

■ The real number line is used to represent the real numbers.

■ Know the inequality symbols.

(a) $a < b$ means a is less than b.

(b) $a \leq b$ means a is less than or equal to b.

(c) $a > b$ means a is greater than b.

(d) $a \geq b$ means a is greater than or equal to b.

■

Interval Notation	Inequality Notation	Graph	Type
$[a, b]$	$a \leq x \leq b$		Bounded and Closed
(a, b)	$a < x < b$		Bounded and Open
$[a, b)$	$a \leq x < b$		Bounded
$(a, b]$	$a < x \leq b$		Bounded
$[a, \infty)$	$x \geq a$		Unbounded
(a, ∞)	$x > a$		Unbounded
$(-\infty, b]$	$x \leq b$		Unbounded
$(-\infty, b)$	$x < b$		Unbounded
$(-\infty, \infty)$	$-\infty < x < \infty$		Unbounded

■ You should know that $|a| = \begin{cases} a, & \text{if } a \geq 0 \\ -a, & \text{if } a < 0 \end{cases}$

■ Know the properties of absolute value.

(a) $|a| \geq 0$

(b) $|-a| = |a|$

(c) $|ab| = |a|\,|b|$

(d) $\left|\dfrac{a}{b}\right| = \dfrac{|a|}{|b|}, \; b \neq 0$

—CONTINUED—

- The distance between a and b on the real line is $d(a, b) = |b - a| = |a - b|$.

- You should be able to identify the terms in an algebraic expression.

- You should know and be able to use the basic rules of algebra.

- Commutative Property

  (a) Addition: $a + b = b + a$ (b) Multiplication: $a \cdot b = b \cdot a$

- Associative Property

  (a) Addition: $(a + b) + c = a + (b + c)$ (b) Multiplication: $(ab)c = a(bc)$

- Identity Property

  (a) Addition: 0 is the identity; $a + 0 = 0 + a = a$.   (b) Multiplication: 1 is the identity; $a \cdot 1 = 1 \cdot a = a$.

- Inverse Property

  (a) Addition: $-a$ is the additive inverse of a; $a + (-a) = -a + a = 0$.

  (b) Multiplication: $1/a$ is the multiplicative inverse of a, $a \neq 0$; $a(1/a) = (1/a)a = 1$.

- Distributive Property

  (a) $a(b + c) = ab + ac$ (b) $(a + b)c = ac + bc$

- Properties of Negation

  (a) $(-1)a = -a$ (b) $-(-a) = a$ (c) $(-a)b = a(-b) = -ab$

  (d) $(-a)(-b) = ab$ (e) $-(a + b) = (-a) + (-b) = -a - b$

- Properties of Equality

  (a) If $a = b$, then $a \pm c = b \pm c$.   (b) If $a = b$, then $ac = bc$.

  (c) If $a \pm c = b \pm c$, then $a = b$.   (d) If $ac = bc$ and $c \neq 0$, then $a = b$.

- Properties of Zero

  (a) $a \pm 0 = a$ (b) $a \cdot 0 = 0$ (c) $0 \div a = 0/a = 0, a \neq 0$

  (d) $a/0$ is undefined.   (e) If $ab = 0$, then $a = 0$ or $b = 0$.

- Properties of Fractions ($b \neq 0, d \neq 0$)

  (a) Equivalent Fractions: $a/b = c/d$ if and only if $ad = bc$.

  (b) Rule of Signs: $-a/b = a/-b = -(a/b)$ and $-a/-b = a/b$

  (c) Equivalent Fractions: $a/b = ac/bc, c \neq 0$

  (d) Addition and Subtraction

     1. Like Denominators: $(a/b) \pm (c/b) = (a \pm c)/b$ 2. Unlike Denominators: $(a/b) \pm (c/d) = (ad \pm bc)/bd$

  (e) Multiplication: $(a/b) \cdot (c/d) = (ac)/(bd)$

  (f) Division: $(a/b) \div (c/d) = (a/b) \cdot (d/c) = (ad)/(bc)$ if $c \neq 0$.

Vocabulary Check

1. rational   **2.** irrational   **3.** absolute value

4. composite   **5.** prime   **6.** variables; constants

7. terms   **8.** coefficient   **9.** zero-factor property

1. $-9, -\frac{7}{2}, 5, \frac{2}{3}, \sqrt{2}, 0, 1, -4, 2, -11$

   (a) Natural numbers: $5, 1, 2$

   (b) Whole numbers: $0, 5, 1, 2$

   (c) Integers: $-9, 5, 0, 1, -4, 2, -11$

   (d) Rational numbers: $-9, -\frac{7}{2}, 5, \frac{2}{3}, 0, 1, -4, 2, -11$

   (e) Irrational numbers: $\sqrt{2}$

3. $2.01, 0.666\ldots, -13, 0.010110111\ldots, 1, -6$

   (a) Natural numbers: 1

   (b) Whole numbers: 1

   (c) Integers: $-13, 1, -6$

   (d) Rational numbers: $2.01, 0.666\ldots, -13, 1, -6$

   (e) Irrational numbers: $0.010110111\ldots$

5. $-\pi, -\frac{1}{3}, \frac{6}{3}, \frac{1}{2}\sqrt{2}, -7.5, -1, 8, -22$

   (a) Natural numbers: $\frac{6}{3}$ (since it equals 2), 8

   (b) Whole numbers: $\frac{6}{3}, 8$

   (c) Integers: $\frac{6}{3}, -1, 8, -22$

   (d) Rational numbers: $-\frac{1}{3}, \frac{6}{3}, -7.5, -1, 8, -22$

   (e) Irrational numbers: $-\pi, \frac{1}{2}\sqrt{2}$

7. $\frac{5}{8} = 0.625$

9. $\frac{41}{333} = 0.\overline{123}$

11. $-1 < 2.5$

13. $-4 > -8$

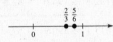

15. $\frac{3}{2} < 7$

17. $\frac{5}{6} > \frac{2}{3}$

19. (a) The inequality $x \le 5$ denotes the set of all real numbers less than or equal to 5.

   (b)

   (c) The interval is unbounded.

21. (a) The inequality $x < 0$ denotes the set of all negative real numbers.

   (b)

   (c) The interval is unbounded.

23. (a) The interval $[4, \infty)$ denotes the set of all real numbers greater than or equal to 4.

   (b)

   (c) The interval is unbounded.

25. (a) The inequality $-2 < x < 2$ denotes the set of all real numbers greater than -2 and less than 2.

   (b)

   (c) The interval is bounded.

27. (a) The inequality $-1 \le x < 0$ denotes the set of all negative real numbers greater than or equal to -1.

   (b)

   (c) The interval is bounded.

29. (a) The interval $[-2, 5)$ denotes the set of all real numbers greater than or equal to -2 and less than 5.

   (b)

   (c) The interval is bounded.

31. $-2 < x \le 4$

33. $y \ge 0$

35. $10 \le t \le 22$

37. $W > 65$

39. $|-10| = -(-10) = 10$

41. $|3 - 8| = |-5| = -(-5) = 5$

43. $|-1| - |-2| = 1 - 2 = -1$

45. $\dfrac{-5}{|-5|} = \dfrac{-5}{-(-5)} = \dfrac{-5}{5} = -1$

47. If $x < -2$, then $x + 2$ is negative.

Thus $\dfrac{|x+2|}{x+2} = \dfrac{-(x+2)}{x+2} = -1$.

49. $|-3| > -|-3|$ since $3 > -3$.

51. $-5 = -|5|$ since $-5 = -5$.

53. $-|-2| = -|2|$ since $-2 = -2$.

55. $d(126, 75) = |75 - 126| = 51$

57. $d\left(-\frac{5}{2}, 0\right) = \left|0 - \left(-\frac{5}{2}\right)\right| = \frac{5}{2}$

59. $d\left(\frac{16}{5}, \frac{112}{75}\right) = \left|\frac{112}{75} - \frac{16}{5}\right| = \frac{128}{75}$

61.

Budgeted Expense, b	Actual Expense, a	$\|a - b\|$	$0.05b$
$112,700	$113,356	$656	0.05(112,700) = $5635

Since $656 < $5635 but $656 > $500, the actual expense does not pass the "budget variance test."

63.

Budgeted Expense, b	Actual Expense, a	$\|a - b\|$	$0.05b$
$37,640	$37,335	$305	0.05(37,640) = $1882

Since $305 < $500 and $305 < $1882, the actual expense passes the "budget variance test."

65. (a)

Year	Expenditures (in billions)	Surplus or Deficit (in billions)
1960	$92.2	$|92.5 - 92.2| = \$0.3$ surplus
1970	$195.6	$|192.8 - 195.6| = \$2.8$ deficit
1980	$590.9	$|517.1 - 590.9| = \$73.8$ deficit
1990	$1253.2	$|1032.0 - 1253.2| = \$221.2$ deficit
2000	$1788.8	$|2025.2 - 1788.8| = \$236.4$ surplus

(b)

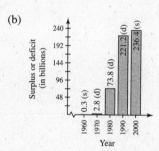

67. $d(x, 5) = |x - 5|$ and $d(x, 5) \le 3$, thus $|x - 5| \le 3$.

69. $d(y, 0) = |y - 0| = |y|$ and $d(y, 0) \ge 6$, thus $|y| \ge 6$.

71. $d(326, 351) = |351 - 326|$
$= 25$ miles

73. $7x + 4$

Terms: $7x, 4$

Coefficient: 7

75. $\sqrt{3}x^2 - 8x - 11$

Terms: $\sqrt{3}x^2, -8x, -11$

Coefficients: $\sqrt{3}, -8$

77. $4x^3 + \dfrac{x}{2} - 5$

Terms: $4x^3, \dfrac{x}{2}, -5$

Coefficients: $4, \dfrac{1}{2}$

79. $4x - 6$

(a) $4(-1) - 6 = -4 - 6 = -10$

(b) $4(0) - 6 = 0 - 6 = -6$

81. $x^2 - 3x + 4$

(a) $(-2)^2 - 3(-2) + 4 = 4 + 6 + 4 = 14$

(b) $(2)^2 - 3(2) + 4 = 4 - 6 + 4 = 2$

83. $\dfrac{x+1}{x-1}$

(a) $\dfrac{1+1}{1-1} = \dfrac{2}{0}$

Division by zero is undefined

(b) $\dfrac{-1+1}{-1-1} = \dfrac{0}{-2} = 0$

85. $x + 9 = 9 + x$

Commutative Property of Addition

87. $\dfrac{1}{(h+6)}(h+6) = 1, h \neq -6$

Multiplicative Inverse Property

89. $2(x+3) = 2x + 6$

Distributive Property

91. $1 \cdot (1+x) = 1 + x$

Multiplicative Identity Property

93. $x + (y+10) = (x+y) + 10$

Associative Property of Addition

95. $3(t-4) = 3 \cdot t - 3 \cdot 4$

Distributive Property

97. $\frac{3}{16} + \frac{5}{16} = \frac{8}{16} = \frac{1}{2}$

99. $\frac{5}{8} - \frac{5}{12} + \frac{1}{6} = \frac{15}{24} - \frac{10}{24} + \frac{4}{24} = \frac{9}{24} = \frac{3}{8}$

101. $12 \div \frac{1}{4} = 12 \cdot \frac{4}{1} = 12 \cdot 4 = 48$

103. $\dfrac{2x}{3} - \dfrac{x}{4} = \dfrac{8x}{12} - \dfrac{3x}{12} = \dfrac{5x}{12}$

105. (a)

n	1	0.5	0.01	0.0001	0.000001
$5/n$	5	10	500	50,000	5,000,000

(b) The value of $5/n$ approaches infinity as n approaches 0.

107. False. If $a < b$, then $\dfrac{1}{a} > \dfrac{1}{b}$, where $a \neq b \neq 0$.

109. (a) $|u+v| \neq |u| + |v|$ if u is positive and v is negative or vice versa.

(b) $|u+v| \leq |u| + |v|$

They are equal when u and v have the same sign. If they differ in sign, $|u+v|$ is less than $|u| + |v|$.

111. The only even prime number is 2, because its factors are itself and 1.

113. (a) Since $A > 0$, $-A < 0$. The expression is negative.

(b) Since $B < A$, $B - A < 0$. The expression is negative.

115. Yes, if a is a negative number, then $-a$ is positive. Thus, $|a| = -a$ if a is negative.

Appendix A.2   Exponents and Radicals

■ You should know the properties of exponents.

(a) $a^1 = a$

(b) $a^0 = 1, a \neq 0$

(c) $a^m a^n = a^{m+n}$

(d) $a^m / a^n = a^{m-n}, a \neq 0$

(e) $a^{-n} = 1/a^n = (1/a)^n, a \neq 0$

(f) $(a^m)^n = a^{mn}$

(g) $(ab)^n = a^n b^n$

(h) $(a/b)^n = a^n / b^n, b \neq 0$

(i) $(a/b)^{-n} = (b/a)^n, a \neq 0, b \neq 0$

(j) $|a^2| = |a|^2 = a^2$

■ You should be able to write numbers in scientific notation, $c \times 10^n$, where $1 \leq c < 10$ and n is an integer.

■ You should be able to use your calculator to evaluate expressions involving exponents.

■ You should know the properties of radicals.

(a) $\sqrt[n]{a^m} = \left(\sqrt[n]{a}\right)^m, a > 0$

(b) $\sqrt[n]{a} \cdot \sqrt[n]{b} = \sqrt[n]{ab}$

(c) $\dfrac{\sqrt[n]{a}}{\sqrt[n]{b}} = \sqrt[n]{\dfrac{a}{b}}, b \neq 0$

(d) $\sqrt[m]{\sqrt[n]{a}} = \sqrt[mn]{a}$

(e) $\left(\sqrt[n]{a}\right)^n = a$

(f) For n even, $\sqrt[n]{a^n} = |a|$.
    For n odd, $\sqrt[n]{a^n} = a$.

(g) $a^{1/n} = \sqrt[n]{a}$

(h) $a^{m/n} = \left(\sqrt[n]{a}\right)^m = \sqrt[n]{a^m}, a \geq 0$

■ You should be able to simplify radicals.

(a) All possible factors have been removed from the radical sign.

(b) All fractions have radical-free denominators.

(c) The index for the radical has been reduced as far as possible.

■ You should be able to use your calculator to evaluate radicals.

Vocabulary Check

1. exponent; base

2. scientific notation

3. square root

4. principal nth root

5. index; radicand

6. simplest form

7. conjugates

8. rationalizing

9. power; index

---

1. $8^5 = 8 \times 8 \times 8 \times 8 \times 8$

3. $(4.9)(4.9)(4.9)(4.9)(4.9)(4.9) = 4.9^6$

5. (a) $3^2 \cdot 3 = 3^3 = 27$

   (b) $3 \cdot 3^3 = 3^4 = 81$

7. (a) $(3^3)^0 = 1$

   (b) $-3^2 = -9$

9. (a) $\dfrac{3 \cdot 4^{-4}}{3^{-4} \cdot 4^{-1}} = 3^{1-(-4)} \cdot 4^{-4-(-1)} = 3^5 \cdot 4^{-3}$

            $= \dfrac{3^5}{4^3} = \dfrac{243}{64}$

   (b) $32(-2)^{-5} = \dfrac{32}{(-2)^5} = \dfrac{32}{-32} = -1$

11. (a) $2^{-1} + 3^{-1} = \dfrac{1}{2} + \dfrac{1}{3} = \dfrac{3}{6} + \dfrac{2}{6} = \dfrac{5}{6}$

    (b) $(2^{-1})^{-2} = 2^{(-1)(-2)} = 2^2 = 4$

13. $(-4)^3(5^2) = (-64)(25) = -1600$

15. $\dfrac{3^6}{7^3} = \dfrac{729}{343} \approx 2.125$

17. When $x = 2$,
$$-3x^3 = -3(2)^3 = -24.$$

19. When $x = 10$,
$$6x^0 = 6(10)^0 = 6(1) = 6.$$

21. When $x = -3$,
$$2x^3 = 2(-3)^3 = 2(-27) = -54.$$

23. When $x = -\frac{1}{2}$,
$$4x^2 = 4\left(-\frac{1}{2}\right)^2 = 4\left(\frac{1}{4}\right) = 1.$$

25. (a) $(-5z)^3 = (-5)^3 z^3 = -125z^3$

(b) $5x^4(x^2) = 5x^{4+2} = 5x^6$

27. (a) $6y^2(2y^0)^2 = 6y^2(2 \cdot 1)^2 = 6y^2(4) = 24y^2$

(b) $\dfrac{3x^5}{x^3} = 3x^{5-3} = 3x^2$

29. (a) $\dfrac{7x^2}{x^3} = 7x^{2-3} = 7x^{-1} = \dfrac{7}{x}$

(b) $\dfrac{12(x+y)^3}{9(x+y)} = \dfrac{4}{3}(x+y)^{3-1} = \dfrac{4}{3}(x+y)^2$

31. (a) $(x+5)^0 = 1,\ x \neq -5$

(b) $(2x^2)^{-2} = \dfrac{1}{(2x^2)^2} = \dfrac{1}{4x^4}$

33. (a) $(-2x^2)^3(4x^3)^{-1} = \dfrac{-8x^6}{4x^3} = -2x^3$

(b) $\left(\dfrac{x}{10}\right)^{-1} = \dfrac{10}{x}$

35. (a) $3^n \cdot 3^{2n} = 3^{n+2n} = 3^{3n}$

(b) $\left(\dfrac{a^{-2}}{b^{-2}}\right)\left(\dfrac{b}{a}\right)^3 = \left(\dfrac{b^2}{a^2}\right)\left(\dfrac{b^3}{a^3}\right) = \dfrac{b^5}{a^5}$

37. $57{,}300{,}000 = 5.73 \times 10^7$ square miles

39. $0.0000899 = 8.99 \times 10^{-5}$ gram per cubic centimeter

41. $4.568 \times 10^9 = 4{,}568{,}000{,}000$ ounces

43. $1.6022 \times 10^{-19} = 0.00000000000000000016022$ coulomb

45. (a) $\sqrt{25 \times 10^8} = 5 \times 10^4 = 50{,}000$

(b) $\sqrt[3]{8 \times 10^{15}} = 2 \times 10^5 = 200{,}000$

47. (a) $750\left(1 + \dfrac{0.11}{365}\right)^{800} \approx 954.448$

(b) $\dfrac{67{,}000{,}000 + 93{,}000{,}000}{0.0052} = 30{,}769{,}230{,}769.2$
$$\approx 3.077 \times 10^{10}$$

49. (a) $\sqrt{4.5 \times 10^9} \approx 67{,}082.039$

(b) $\sqrt[3]{6.3 \times 10^4} \approx 39.791$

51. (a) $\sqrt{9} = 3$

(b) $\sqrt[3]{\dfrac{27}{8}} = \dfrac{\sqrt[3]{27}}{\sqrt[3]{8}} = \dfrac{3}{2}$

53. (a) $32^{-3/5} = \dfrac{1}{32^{3/5}} = \dfrac{1}{\left(\sqrt[5]{32}\right)^3} = \dfrac{1}{(2)^3} = \dfrac{1}{8}$

(b) $\left(\dfrac{16}{81}\right)^{-3/4} = \left(\dfrac{81}{16}\right)^{3/4} = \left(\sqrt[4]{\dfrac{81}{16}}\right)^3 = \left(\dfrac{3}{2}\right)^3 = \dfrac{27}{8}$

55. (a) $\left(-\dfrac{1}{64}\right)^{-1/3} = (-64)^{1/3} = \sqrt[3]{-64} = -4$

(b) $\left(\dfrac{1}{\sqrt{32}}\right)^{-2/5} = \left(\sqrt{32}\right)^{2/5} = \sqrt[5]{\left(\sqrt{32}\right)^2} = \sqrt[5]{32} = 2$

57. (a) $\sqrt{57} \approx 7.550$

(b) $\sqrt[5]{-27^3} = (-27)^{3/5}$
$$\approx -7.225$$

59. (a) $(-12.4)^{-1.8} \approx -0.011$

(b) $\left(5\sqrt{3}\right)^{-2.5} \approx 0.005$

61. (a) $\left(\sqrt[3]{4}\right)^3 = 4^{3/3} = 4^1 = 4$

(b) $\sqrt[5]{96x^5} = \sqrt[5]{32x^5 \cdot 3}$
$$= 2x\sqrt[5]{3}$$
$$= 2 \cdot 3^{1/5} \cdot x$$

63. (a) $\sqrt{8} = \sqrt{4 \cdot 2}$
$= \sqrt{4}\sqrt{2} = 2\sqrt{2}$

(b) $\sqrt[3]{24} = \sqrt[3]{8 \cdot 3}$
$= \sqrt[3]{8}\sqrt[3]{3} = 2\sqrt[3]{3}$

65. (a) $\sqrt{72x^3} = \sqrt{36x^2 \cdot 2x}$
$= 6x\sqrt{2x}$

(b) $\sqrt{\dfrac{18^2}{z^3}} = \dfrac{\sqrt{18^2}}{\sqrt{z^2 \cdot z}} = \dfrac{18}{z\sqrt{z}}$

67. (a) $\sqrt[3]{16x^5} = \sqrt[3]{8x^3 \cdot 2x^2}$
$= 2x\sqrt[3]{2x^2}$

(b) $\sqrt{75x^2y^{-4}} = \sqrt{\dfrac{75x^2}{y^4}}$
$= \dfrac{\sqrt{25x^2 \cdot 3}}{\sqrt{y^4}}$
$= \dfrac{5|x|\sqrt{3}}{y^2}$

69. (a) $2\sqrt{50} + 12\sqrt{8} = 2\sqrt{25 \cdot 2} + 12\sqrt{4 \cdot 2} = 2(5\sqrt{2}) + 12(2\sqrt{2}) = 10\sqrt{2} + 24\sqrt{2} = 34\sqrt{2}$

(b) $10\sqrt{32} - 6\sqrt{18} = 10\sqrt{16 \cdot 2} - 6\sqrt{9 \cdot 2} = 10(4\sqrt{2}) - 6(3\sqrt{2}) = 40\sqrt{2} - 18\sqrt{2} = 22\sqrt{2}$

71. (a) $5\sqrt{x} - 3\sqrt{x} = 2\sqrt{x}$

(b) $-2\sqrt{9y} + 10\sqrt{y} = -2(3\sqrt{y}) + 10\sqrt{y}$
$= -6\sqrt{y} + 10\sqrt{y} = 4\sqrt{y}$

73. (a) $3\sqrt{x+1} + 10\sqrt{x+1} = 13\sqrt{x+1}$

(b) $7\sqrt{80x} - 2\sqrt{125x} = 7\sqrt{16 \cdot 5x} - 2\sqrt{25 \cdot 5x} = 7(4\sqrt{5x}) - 2(5\sqrt{5x}) = 28\sqrt{5x} - 10\sqrt{5x} = 18\sqrt{5x}$

75. $\sqrt{5} + \sqrt{3} \approx 3.968$ and
$\sqrt{5+3} = \sqrt{8} \approx 2.828$
Thus, $\sqrt{5} + \sqrt{3} > \sqrt{5+3}$.

77. $\sqrt{3^2 + 2^2} = \sqrt{9+4}$
$= \sqrt{13} \approx 3.606$
Thus, $5 > \sqrt{3^2 + 2^2}$.

79. $\dfrac{1}{\sqrt{3}} = \dfrac{1}{\sqrt{3}} \cdot \dfrac{\sqrt{3}}{\sqrt{3}} = \dfrac{\sqrt{3}}{3}$

81. $\dfrac{2}{5 - \sqrt{3}} = \dfrac{2}{5 - \sqrt{3}} \cdot \dfrac{5 + \sqrt{3}}{5 + \sqrt{3}} = \dfrac{2(5 + \sqrt{3})}{5^2 - (\sqrt{3})^2} = \dfrac{2(5 + \sqrt{3})}{25 - 3} = \dfrac{2(5 + \sqrt{3})}{22} = \dfrac{5 + \sqrt{3}}{11}$

83. $\dfrac{\sqrt{8}}{2} = \dfrac{\sqrt{4 \cdot 2}}{2} = \dfrac{2\sqrt{2}}{2} = \dfrac{\sqrt{2}}{1} \cdot \dfrac{\sqrt{2}}{\sqrt{2}} = \dfrac{2}{\sqrt{2}}$

85. $\dfrac{\sqrt{5} + \sqrt{3}}{3} = \dfrac{\sqrt{5} + \sqrt{3}}{3} \cdot \dfrac{\sqrt{5} - \sqrt{3}}{\sqrt{5} - \sqrt{3}} = \dfrac{5 - 3}{3(\sqrt{5} - \sqrt{3})} = \dfrac{2}{3(\sqrt{5} - \sqrt{3})}$

Radical Form	*Rational Exponent Form*
87. $\sqrt{9} = 3$, Given	$9^{1/2} = 3$, Answer
89. $\sqrt[5]{32} = 2$, Answer	$32^{1/5} = 2$, Given
91. $\sqrt[3]{-216} = -6$, Given	$(-216)^{1/3} = -6$, Answer
93. $\sqrt[4]{81^3} = 27$, Given	$81^{3/4} = 27$, Answer

95. $\dfrac{(2x^2)^{3/2}}{2^{1/2}x^4} = \dfrac{2^{3/2}(x^2)^{3/2}}{2^{1/2}x^4}$
$= \dfrac{2^{3/2}x^3}{2^{1/2}x^4} = 2^{3/2 - 1/2}x^{3-4} = 2^1 x^{-1} = \dfrac{2}{x}$

97. $\dfrac{x^{-3} \cdot x^{1/2}}{x^{3/2} \cdot x^{-1}} = \dfrac{x^{1/2} \cdot x^1}{x^{3/2} \cdot x^3}$
$= x^{1/2 + 1 - 3/2 - 3} = x^{-3} = \dfrac{1}{x^3}, \; x > 0$

99. (a) $\sqrt[4]{3^2} = 3^{2/4} = 3^{1/2} = \sqrt{3}$

(b) $\sqrt[6]{(x+1)^4} = (x+1)^{4/6} = (x+1)^{2/3} = \sqrt[3]{(x+1)^2}$

101. (a) $\sqrt{\sqrt{32}} = (32^{1/2})^{1/2}$

$= 32^{1/4} = \sqrt[4]{32} = \sqrt[4]{16 \cdot 2} = 2\sqrt[4]{2}$

(b) $\sqrt{\sqrt[4]{2x}} = ((2x)^{1/4})^{1/2} = (2x)^{1/8} = \sqrt[8]{2x}$

103. $T = 2\pi\sqrt{\dfrac{2}{32}}$

$= 2\pi\sqrt{\dfrac{1}{16}}$

$= 2\pi\left(\dfrac{1}{4}\right)$

$= \dfrac{\pi}{2} \approx 1.57$ seconds

105. $t = 0.03[12^{5/2} - (12-h)^{5/2}], 0 \le h \le 12$

(a)

h (in centimeters)	t (in seconds)
0	0
1	2.93
2	5.48
3	7.67
4	9.53
5	11.08
6	12.32
7	13.29
8	14.00
9	14.50
10	14.80
11	14.93
12	14.96

(b) As h approaches 12, t approaches

$0.03(12^{5/2}) = 8.64\sqrt{3} \approx 14.96$ seconds.

107. True. When dividing variables, you subtract exponents.

109. $1 = \dfrac{a^m}{a^m} = a^{m-m} = a^0, a \ne 0$

111. When any positive integer is squared, the units digit is 0, 1, 4, 5, 6, or 9. Therefore, $\sqrt{5233}$ is not an integer.

Appendix A.3   Polynomials and Factoring

■   Given a polynomial in x, $a_n x^n + a_{n-1} x^{n-1} + \ldots + a_1 x + a_0$, where $a_n \neq 0$, and n is a nonnegative integer, you should be able to identify the following.

  (a) Degree: n (b) Terms: $a_n x^n, a_{n-1} x^{n-1}, \ldots, a_1 x, a_0$

  (c) Coefficients: $a_n, a_{n-1}, \ldots, a_1, a_0$ (d) Leading coefficient: a_n

  (e) Constant term: a_0

■   You should be able to add and subtract polynomials.

■   You should be able to multiply polynomials by the Distributive Properties.

■   You should be able to multiply two binomials by the FOIL Method.

■   You should know the special binomial products.

  (a) $(u + v)(u - v) = u^2 - v^2$ (b) $(u \pm v)^2 = u^2 \pm 2uv + v^2$

  (c) $(u \pm v)^3 = u^3 \pm 3u^2 v + 3uv^2 \pm v^3$

■   You should be able to factor out all common factors, the first step in factoring.

■   You should be able to factor the following special polynomial forms.

  (a) $u^2 - v^2 = (u + v)(u - v)$ (b) $u^2 \pm 2uv + v^2 = (u \pm v)^2$

  (c) $u^3 \pm v^3 = (u \pm v)(u^2 \mp uv + v^2)$

■   You should be able to factor by grouping.

■   You should be able to factor some trinomials by grouping.

Vocabulary Check

1. n; a_n; a_0 2. descending

3. monomial; binomial; trinomial   4. like terms

5. First terms; Outer terms; Inner terms; Last terms   6. factoring

7. completely factored

---

1. (d) 12 is a polynomial of degree zero.

3. (b) $1 - 2x^3 = -2x^3 + 1$ is a binomial with leading coefficient -2.

5. (f) $\frac{2}{3} x^4 + x^2 + 10$ is a trinomial with leading coefficient $\frac{2}{3}$.

7. $-2x^3$; $-2x^3 + 5$; $-2x^3 + 4x^2 - 3x + 20$, etc. (Answers will vary.)

9. $-15x^4 + 1$; $-3x^4 + 7x^2$; $-5x^4 - 6x$, etc. (Answers will vary.)

11. (a) Standard form: $-\frac{1}{2} x^5 + 14x$

    (b) Degree: 5

       Leading coefficient: $-\frac{1}{2}$

    (c) Binomial

13. (a) Standard form: $-3x^4 + 2x^2 - 5$

    (b) Degree: 4

       Leading coefficient: -3

    (c) Trinomial

15. (a) Standard form: $x^5 - 1$

    (b) Degree: 5

       Leading coefficient: 1

    (c) Binomial

17. (a) Standard form: 3

    (b) Degree: 0

       Leading coefficient: 3

    (c) Monomial

19. (a) Standard form: $-4x^5 + 6x^4 + 1$

(b) Degree: 5

Leading coefficient: -4

(c) Trinomial

21. (a) Standard form: $4x^3y$

(b) Degree: 4 (add the exponents on x and y)

Leading coefficient: 4

(c) Monomial

23. $2x - 3x^3 + 8$ *is* a polynomial.

Standard form: $-3x^3 + 2x + 8$

25. $\dfrac{3x + 4}{x} = 3 + \dfrac{4}{x} = 3 + 4x^{-1}$ is *not* a polynomial because it includes a term with a negative exponent.

27. $y^2 - y^4 + y^3$ *is* a polynomial.

Standard form: $-y^4 + y^3 + y^2$

29.
$$
\begin{aligned}
(6x + 5) - (8x + 15) &= 6x + 5 - 8x - 15 \\
&= (6x - 8x) + (5 - 15) \\
&= -2x - 10
\end{aligned}
$$

31.
$$
\begin{aligned}
-(x^3 - 2) + (4x^3 - 2x) &= -x^3 + 2 + 4x^3 - 2x \\
&= (4x^3 - x^3) - 2x + 2 \\
&= 3x^3 - 2x + 2
\end{aligned}
$$

33.
$$
\begin{aligned}
(15x^2 - 6) - (-8.3x^3 - 14.7x^2 - 17) &= 15x^2 - 6 + 8.3x^3 + 14.7x^2 + 17 \\
&= 8.3x^3 + (15x^2 + 14.7x^2) + (-6 + 17) \\
&= 8.3x^3 + 29.7x^2 + 11
\end{aligned}
$$

35.
$$
\begin{aligned}
5z - [3z - (10z + 8)] &= 5z - (3z - 10z - 8) \\
&= 5z - 3z + 10z + 8 \\
&= (5z - 3z + 10z) + 8 \\
&= 12z + 8
\end{aligned}
$$

37.
$$
\begin{aligned}
3x(x^2 - 2x + 1) &= 3x(x^2) + 3x(-2x) + 3x(1) \\
&= 3x^3 - 6x^2 + 3x
\end{aligned}
$$

39.
$$
\begin{aligned}
-5z(3z - 1) &= -5z(3z) + (-5z)(-1) \\
&= -15z^2 + 5z
\end{aligned}
$$

41.
$$
\begin{aligned}
(1 - x^3)(4x) &= 1(4x) - x^3(4x) \\
&= 4x - 4x^4 \\
&= -4x^4 + 4x
\end{aligned}
$$

43.
$$
\begin{aligned}
(2.5x^2 + 3)(3x) &= (2.5x^2)(3x) + (3)(3x) \\
&= 7.5x^3 + 9x
\end{aligned}
$$

45.
$$
\begin{aligned}
-4x\left(\tfrac{1}{8}x + 3\right) &= (-4x)\left(\tfrac{1}{8}x\right) + (-4x)(3) \\
&= -\tfrac{1}{2}x^2 - 12x
\end{aligned}
$$

47.
$$
\begin{aligned}
(x + 3)(x + 4) &= x^2 + 4x + 3x + 12 \quad \text{FOIL} \\
&= x^2 + 7x + 12
\end{aligned}
$$

49.
$$
\begin{aligned}
(3x - 5)(2x + 1) &= 6x^2 + 3x - 10x - 5 \quad \text{FOIL} \\
&= 6x^2 - 7x - 5
\end{aligned}
$$

51. Multiply:
$$
\begin{array}{r}
x^2 - x + 1 \\
x^2 + x + 1 \\
\hline
x^4 - x^3 + x^2 \\
x^3 - x^2 + x \\
x^2 - x + 1 \\
\hline
x^4 - 0x^3 + x^2 + 0x + 1 = x^4 + x^2 + 1
\end{array}
$$

53. $(x + 10)(x - 10) = x^2 - 10^2 = x^2 - 100$

55. $(x + 2y)(x - 2y) = x^2 - (2y)^2 = x^2 - 4y^2$

57.
$$
\begin{aligned}
(2x + 3)^2 &= (2x)^2 + 2(2x)(3) + 3^2 \\
&= 4x^2 + 12x + 9
\end{aligned}
$$

59.
$$
\begin{aligned}
(2x - 5y)^2 &= (2x)^2 - 2(2x)(5y) + (5y)^2 \\
&= 4x^2 - 20xy + 25y^2
\end{aligned}
$$

61. $(x + 1)^3 = x^3 + 3x^2(1) + 3x(1^2) + 1^3$
$= x^3 + 3x^2 + 3x + 1$

63. $(2x - y)^3 = (2x)^3 - 3(2x)^2y + 3(2x)y^2 - y^3$
$= 8x^3 - 12x^2y + 6xy^2 - y^3$

65. $(4x^3 - 3)^2 = (4x^3)^2 - 2(4x^3)(3) + (3)^2$
$= 16x^6 - 24x^3 + 9$

67. $[(m - 3) + n][(m - 3) - n] = (m - 3)^2 - n^2$
$= m^2 - 6m + 9 - n^2$
$= m^2 - n^2 - 6m + 9$

69. $[(x - 3) + y]^2 = (x - 3)^2 + 2y(x - 3) + y^2$
$= x^2 - 6x + 9 + 2xy - 6y + y^2$
$= x^2 + 2xy + y^2 - 6x - 6y + 9$

71. $(2r^2 - 5)(2r^2 + 5) = (2r^2)^2 - 5^2 = 4r^4 - 25$

73. $\left(\frac{1}{2}x - 3\right)^2 = \left(\frac{1}{2}x\right)^2 - 2\left(\frac{1}{2}x\right)(3) + 3^2$
$= \frac{1}{4}x^2 - 3x + 9$

75. $\left(\frac{1}{3}x - 2\right)\left(\frac{1}{3}x + 2\right) = \left(\frac{1}{3}x\right)^2 - (2)^2$
$= \frac{1}{9}x^2 - 4$

77. $(1.2x + 3)^2 = (1.2x)^2 + 2(1.2x)(3) + 3^2$
$= 1.44x^2 + 7.2x + 9$

79. $(1.5x - 4)(1.5x + 4) = (1.5x)^2 - 4^2$
$= 2.25x^2 - 16$

81. $5x(x + 1) - 3x(x + 1) = 2x(x + 1)$
$= 2x^2 + 2x$

83. $(u + 2)(u - 2)(u^2 + 4) = (u^2 - 4)(u^2 + 4)$
$= u^4 - 16$

85. $\left(\sqrt{x} + \sqrt{y}\right)\left(\sqrt{x} - \sqrt{y}\right) = \left(\sqrt{x}\right)^2 - \left(\sqrt{y}\right)^2$
$= x - y$

87. $\left(x - \sqrt{5}\right)^2 = x^2 - 2(x)\left(\sqrt{5}\right) + \left(\sqrt{5}\right)^2$
$= x^2 - 2\sqrt{5}x + 5$

89. $3x + 6 = 3(x + 2)$

91. $2x^3 - 6x = 2x(x^2 - 3)$

93. $x(x - 1) + 6(x - 1) = (x - 1)(x + 6)$

95. $(x + 3)^2 - 4(x + 3) = (x + 3)[(x + 3) - 4]$
$= (x + 3)(x - 1)$

97. $\frac{1}{2}x + 4 = \frac{1}{2}x + \frac{8}{2}$
$= \frac{1}{2}(x + 8)$

99. $\frac{1}{2}x^3 + 2x^2 - 5x = \frac{1}{2}x^3 + \frac{4}{2}x^2 - \frac{10}{2}x$
$= \frac{1}{2}x(x^2 + 4x - 10)$

101. $\frac{2}{3}x(x - 3) - 4(x - 3) = \frac{2}{3}x(x - 3) - \frac{12}{3}(x - 3)$
$= \frac{2}{3}(x - 3)(x - 6)$

103. $x^2 - 81 = x^2 - 9^2$
$= (x + 9)(x - 9)$

105. $32y^2 - 18 = 2(16y^2 - 9)$
$= 2[(4y)^2 - 3^2]$
$= 2(4y + 3)(4y - 3)$

107. $16x^2 - \frac{1}{9} = (4x)^2 - \left(\frac{1}{3}\right)^2$
$= \left(4x + \frac{1}{3}\right)\left(4x - \frac{1}{3}\right)$

109. $(x - 1)^2 - 4 = (x - 1)^2 - (2)^2$
$= [(x - 1) + 2][(x - 1) - 2]$
$= (x + 1)(x - 3)$

111. $9u^2 - 4v^2 = (3u)^2 - (2v)^2$
$= (3u + 2v)(3u - 2v)$

113. $x^2 - 4x + 4 = x^2 - 2(2)x + 2^2$
$$= (x - 2)^2$$

115. $4t^2 + 4t + 1 = (2t)^2 + 2(2t)(1) + 1^2$
$$= (2t + 1)^2$$

117. $25y^2 - 10y + 1 = (5y)^2 - 2(5y)(1) + 1^2$
$$= (5y - 1)^2$$

119. $9u^2 + 24uv + 16v^2 = (3u)^2 + 2(3u)(4v) + (4v)^2$
$$= (3u + 4v)^2$$

121. $x^2 - \frac{4}{3}x + \frac{4}{9} = x^2 - 2(x)\left(\frac{2}{3}\right) + \left(\frac{2}{3}\right)^2$
$$= \left(x - \frac{2}{3}\right)^2$$

123. $x^3 - 8 = x^3 - 2^3$
$$= (x - 2)(x^2 + 2x + 4)$$

125. $y^3 + 64 = y^3 + 4^3$
$$= (y + 4)(y^2 - 4y + 16)$$

127. $8t^3 - 1 = (2t)^3 - 1^3$
$$= (2t - 1)(4t^2 + 2t + 1)$$

129. $u^3 + 27v^3 = u^3 + (3v)^3$
$$= (u + 3v)(u^2 - 3uv + 9v^2)$$

131. $x^2 + x - 2 = (x + 2)(x - 1)$

133. $s^2 - 5s + 6 = (s - 3)(s - 2)$

135. $20 - y - y^2 = -(y^2 + y - 20)$
$$= -(y + 5)(y - 4)$$

137. $x^2 - 30x + 200 = (x - 20)(x - 10)$

139. $3x^2 - 5x + 2 = (3x - 2)(x - 1)$

141. $5x^2 + 26x + 5 = (5x + 1)(x + 5)$

143. $-9z^2 + 3z + 2 = -(9z^2 - 3z - 2)$
$$= -(3z - 2)(3z + 1)$$

145. $x^3 - x^2 + 2x - 2 = x^2(x - 1) + 2(x - 1)$
$$= (x - 1)(x^2 + 2)$$

147. $2x^3 - x^2 - 6x + 3 = x^2(2x - 1) - 3(2x - 1)$
$$= (2x - 1)(x^2 - 3)$$

149. $6 + 2x - 3x^3 - x^4 = 2(3 + x) - x^3(3 + x)$
$$= (3 + x)(2 - x^3)$$

151. $6x^3 - 2x + 3x^2 - 1 = 2x(3x^2 - 1) + 1(3x^2 - 1)$
$$= (3x^2 - 1)(2x + 1)$$

153. $a \cdot c = (3)(8) = 24$. Rewrite the middle term, $10x = 6x + 4x$, since $(6)(4) = 24$ and $6 + 4 = 10$.

$3x^2 + 10x + 8 = 3x^2 + 6x + 4x + 8$
$$= 3x(x + 2) + 4(x + 2)$$
$$= (x + 2)(3x + 4)$$

155. $a \cdot c = (6)(-2) = -12$. Rewrite the middle term, $x = 4x - 3x$, since $4(-3) = -12$ and $4 + (-3) = 1$.

$6x^2 + x - 2 = 6x^2 + 4x - 3x - 2$
$$= 2x(3x + 2) - 1(3x + 2)$$
$$= (2x - 1)(3x + 2)$$

157. $a \cdot c = (15)(2) = 30$. Rewrite the middle term, $-11x = -6x - 5x$, since $(-6)(-5) = 30$ and $(-6) + (-5) = -11$.

$15x^2 - 11x + 2 = 15x^2 - 6x - 5x + 2$
$$= 3x(5x - 2) - 1(5x - 2)$$
$$= (3x - 1)(5x - 2)$$

159. $6x^2 - 54 = 6(x^2 - 9)$
$$= 6(x + 3)(x - 3)$$

161. $x^3 - 4x^2 = x^2(x - 4)$

163. $x^2 - 2x + 1 = (x - 1)^2$

165. $1 - 4x + 4x^2 = (1 - 2x)^2$

167.
$$2x^2 + 4x - 2x^3 = -2x(-x - 2 + x^2)$$
$$= -2x(x^2 - x - 2)$$
$$= -2x(x + 1)(x - 2)$$

169. $9x^2 + 10x + 1 = (9x + 1)(x + 1)$

171.
$$\tfrac{1}{81}x^2 + \tfrac{2}{9}x - 8 = \tfrac{1}{81}x^2 + \tfrac{18}{81}x - \tfrac{648}{81}$$
$$= \tfrac{1}{81}(x^2 + 18x - 648)$$
$$= \tfrac{1}{81}(x + 36)(x - 18)$$

173.
$$3x^3 + x^2 + 15x + 5 = x^2(3x + 1) + 5(3x + 1)$$
$$= (3x + 1)(x^2 + 5)$$

175.
$$x^4 - 4x^3 + x^2 - 4x = x(x^3 - 4x^2 + x - 4)$$
$$= x[x^2(x - 4) + (x - 4)]$$
$$= x(x - 4)(x^2 + 1)$$

177.
$$\tfrac{1}{4}x^3 + 3x^2 + \tfrac{3}{4}x + 9 = \tfrac{1}{4}x^3 + \tfrac{12}{4}x^2 + \tfrac{3}{4}x + \tfrac{36}{4}$$
$$= \tfrac{1}{4}(x^3 + 12x^2 + 3x + 36)$$
$$= \tfrac{1}{4}[x^2(x + 12) + 3(x + 12)]$$
$$= \tfrac{1}{4}(x + 12)(x^2 + 3)$$

179.
$$(t - 1)^2 - 49 = (t - 1)^2 - (7)^2$$
$$= [(t - 1) + 7][(t - 1) - 7]$$
$$= (t + 6)(t - 8)$$

181.
$$(x^2 + 8)^2 - 36x^2 = (x^2 + 8)^2 - (6x)^2$$
$$= [(x^2 + 8) - 6x][(x^2 + 8) + 6x]$$
$$= (x^2 - 6x + 8)(x^2 + 6x + 8)$$
$$= (x - 4)(x - 2)(x + 4)(x + 2)$$

183.
$$5x^3 + 40 = 5(x^3 + 8)$$
$$= 5(x^3 + 2^3)$$
$$= 5(x + 2)(x^2 - 2x + 4)$$

185.
$$5(3 - 4x)^2 - 8(3 - 4x)(5x - 1) = (3 - 4x)[5(3 - 4x) - 8(5x - 1)]$$
$$= (3 - 4x)[15 - 20x - 40x + 8]$$
$$= (3 - 4x)(23 - 60x)$$

187.
$$7(3x + 2)^2(1 - x)^2 + (3x + 2)(1 - x)^3 = (3x + 2)(1 - x)^2[7(3x + 2) + (1 - x)]$$
$$= (3x + 2)(1 - x)^2(21x + 14 + 1 - x)$$
$$= (3x + 2)(1 - x)^2(20x + 15)$$
$$= 5(3x + 2)(1 - x)^2(4x + 3)$$

189.
$$3(x - 2)^2(x + 1)^4 + (x - 2)^3(4)(x + 1)^3 = (x - 2)^2(x + 1)^3[3(x + 1) + 4(x - 2)]$$
$$= (x - 2)^2(x + 1)^3(3x + 3 + 4x - 8)$$
$$= (x - 2)^2(x + 1)^3(7x - 5)$$

191.
$$5(x^6 + 1)^4(6x^5)(3x + 2)^3 + 3(3x + 2)^2(3)(x^6 + 1)^5 = 3(x^6 + 1)^4(3x + 2)^2[10x^5(3x + 2) + 3(x^6 + 1)]$$
$$= 3(x^6 + 1)^4(3x + 2)^2(30x^6 + 20x^5 + 3x^6 + 3)$$
$$= 3(x^6 + 1)^4(3x + 2)^2(33x^6 + 20x^5 + 3)$$
$$= 3[(x^2)^3 + 1]^4(3x + 2)^2(33x^6 + 20x^5 + 3)$$
$$= 3[(x^2 + 1)(x^4 - x^2 + 1)]^4(3x + 2)^2(33x^6 + 20x^5 + 3)$$
$$= 3(x^2 + 1)^4(x^4 - x^2 + 1)^4(3x + 2)^2(33x^6 + 20x^5 + 3)$$

193. For $x^2 + bx - 15$ to be factorable, b must equal $m + n$ where $mn = -15$.

Factors of -15	Sum of factors
$(15)(-1)$	$15 + (-1) = 14$
$(-15)(1)$	$-15 + 1 = -14$
$(3)(-5)$	$3 + (-5) = -2$
$(-3)(5)$	$-3 + 5 = 2$

The possible b-values are $14, -14, -2$, or 2.

195. For $x^2 + bx - 12$ to be factorable, b must equal $m + n$ where $mn = -12$.

Factors of -12	Sum of factors
$(12)(-1)$	$12 + (-1) = 11$
$(-12)(1)$	$-12 + 1 = -11$
$(2)(-6)$	$2 + (-6) = -4$
$(-2)(6)$	$-2 + 6 = 4$
$(3)(-4)$	$3 + (-4) = -1$
$(-3)(4)$	$-3 + 4 = 1$

The possible b-values are $11, -11, -4, 4, -1, 1$.

197. For $2x^2 + 5x + c$ to be factorable, the factors of $2c$ must add up to 5.

Possible c-values	$2c$	Factors of $2c$ that add up to 5
2	4	$(1)(4) = 4$ and $1 + 4 = 5$
3	6	$(2)(3) = 6$ and $2 + 3 = 5$
-3	-6	$(6)(-1) = -6$ and $6 + (-1) = 5$
-7	-14	$(7)(-2) = -14$ and $7 + (-2) = 5$
-12	-24	$(8)(-3) = -24$ and $8 + (-3) = 5$

These are a few possible c-values. There are *many* correct answers.

If $c = 2$: $2x^2 + 5x + 2 = (2x + 1)(x + 2)$

If $c = 3$: $2x^2 + 5x + 3 = (2x + 3)(x + 1)$

If $c = -3$: $2x^2 + 5x - 3 = (2x - 1)(x + 3)$

If $c = -7$: $2x^2 + 5x - 7 = (2x + 7)(x - 1)$

If $c = -12$: $2x^2 + 5x - 12 = (2x - 3)(x + 4)$

199. For $3x^2 - x + c$ to be factorable, the factors of $3c$ must add up to -1.

Possible c-values	$3c$	Factors of $3c$ must add up to -1
-2	-6	$(2)(-3) = -6$ and $2 + (-3) = -1$
-4	-12	$(3)(-4) = -12$ and $3 + (-4) = -1$
-10	-30	$(5)(-6) = -30$ and $5 + (-6) = -1$

These are a few possible c-values. There are *many* correct answers.

If $c = -2$: $3x^2 - x - 2 = (3x + 2)(x - 1)$

If $c = -4$: $3x^2 - x - 4 = (3x - 4)(x + 1)$

If $c = -10$: $3x^2 - x - 10 = (3x + 5)(x - 2)$

201. (a) Profit = Revenue − Cost

Profit $= 95x - (73x + 25{,}000)$

$= 95x - 73x - 25{,}000 = 22x - 25{,}000$

(b) For $x = 5000$:

Profit $= 22(5000) - 25{,}000$

$= 110{,}000 - 25{,}000 = \$85{,}000$

203. (a) $500(1 + r)^2 = 500(r + 1)^2 = 500(r^2 + 2r + 1)$

$$= 500r^2 + 1000r + 500$$

(b)

r	$2\frac{1}{2}\%$	3%	4%	$4\frac{1}{2}\%$	5%
$500(1 + r)^2$	\$525.31	\$530.45	\$540.80	\$546.01	\$551.25

(c) As r increases, the amount increases.

205. (a) $V = l \cdot w \cdot h = (26 - 2x)(18 - 2x)(x)$

$$= 2(13 - x)(2)(9 - x)(x)$$

$$= 4x(-1)(x - 13)(-1)(x - 9)$$

$$= 4x(x - 13)(x - 9)$$

$$= 4x^3 - 88x^2 + 468x$$

(b)

x (cm)	1	2	3
V (cm^3)	384	616	720

207. Area = length $\times$ width

$$= (2x + 14)(22)$$

$$= (2x)(22) + (14)(22)$$

$$= 44x + 308$$

209. (a) Area of shaded region = Area of outer rectangle $-$ Area of inner rectangle

$$A = 2x(2x + 6) - x(x + 4)$$

$$= 4x^2 + 12x - x^2 - 4x$$

$$= 3x^2 + 8x$$

(b) Area of shaded region = Area of outer triangle $-$ Area of inner triangle

$$A = \tfrac{1}{2}(9x)(12x) - \tfrac{1}{2}(6x)(8x)$$

$$= 54x^2 - 24x^2$$

$$= 30x^2$$

211. $3x^2 + 7x + 2 = (3x + 1)(x + 2)$

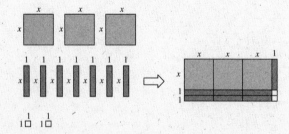

213. $2x^2 + 7x + 3 = (2x + 1)(x + 3)$

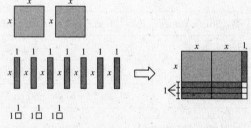

215. $A = \pi(r + 2)^2 - \pi r^2$

$$= \pi[(r + 2)^2 - r^2]$$

$$= \pi[r^2 + 4r + 4 - r^2]$$

$$= \pi(4r + 4)$$

$$= 4\pi(r + 1)$$

217. $A = 8(18) - 4x^2$

$$= 4(36 - x^2)$$

$$= 4(6 - x)(6 + x)$$

219. (a) $V = \pi R^2 h - \pi r^2 h$

$$= \pi h(R^2 - r^2)$$

$$= \pi h(R - r)(R + r)$$

(b) The average radius is $(R + r)/2$. The thickness of the tank is $R - r$.

$$V = \pi h(R - r)(R + r) = 2\pi\left(\frac{R + r}{2}\right)(R - r)h$$

$$= 2\pi(\text{average radius})(\text{thickness})h$$

221. False. $(4x^2 + 1)(3x + 1) = 12x^3 + 4x^2 + 3x + 1$

223. True. $a^2 - b^2 = (a + b)(a - b)$

225. Since $x^m x^n = x^{m+n}$, the degree of the product is $m + n$.

227. The unknown polynomial may be found by adding $-x^3 + 3x^2 + 2x - 1$ and $5x^2 + 8$:

$$(-x^3 + 3x^2 + 2x - 1) + (5x^2 + 8) = -x^3 + (3x^2 + 5x^2) + 2x + (-1 + 8)$$

$$= -x^3 + 8x^2 + 2x + 7$$

229. $x^{2n} - y^{2n} = (x^n)^2 - (y^n)^2$

$$= (x^n + y^n)(x^n - y^n)$$

This is not completely factored unless $n = 1$.

For $n = 2$: $(x^2 + y^2)(x^2 - y^2) = (x^2 + y^2)(x + y)(x - y)$

For $n = 3$: $(x^3 + y^3)(x^3 - y^3) = (x + y)(x^2 - xy + y^2)(x - y)(x^2 + xy + y^2)$

For $n = 4$: $(x^4 + y^4)(x^4 - y^4) = (x^4 + y^4)(x^2 + y^2)(x + y)(x - y)$

231. $x^{3n} - y^{2n} = (x^n)^3 - (y^n)^2 = x^{3n} - y^{2n}$ is completely factored. For integer values of n greater than 4, the factorizations become more complicated.

233. Answers will vary. Some examples:

$$x^2 - 3;\ x^2 + x + 1; x^2 + 16$$

Appendix A.4   Rational Expressions

- You should be able to find the domain of a rational expression.

- You should know that a rational expression is the quotient of two polynomials.

- You should be able to simplify rational expressions by reducing them to lowest terms. This may involve factoring both the numerator and the denominator.

- You should be able to add, subtract, multiply, and divide rational expressions.

- You should be able to simplify complex fractions.

- You should be able to simplify expressions with negative or fraction exponents.

Vocabulary Check

1. domain

2. rational expression

3. complex

4. smaller

5. equivalent

6. difference quotient

1. The domain of the polynomial $3x^2 - 4x + 7$ is the set of all real numbers.

3. The domain of the polynomial $4x^3 + 3$, $x \geq 0$ is the set of non-negative real numbers, since the polynomial is restricted to that set.

5. The domain of $1/(x - 2)$ is the set of all real numbers x such that $x \neq 2$.

7. The domain of $\sqrt{x + 1}$ is the set of all real numbers x such that $x \geq -1$.

9. $\dfrac{5}{2x} = \dfrac{5(3x)}{(2x)(3x)} = \dfrac{5(3x)}{6x^2}, \quad x \neq 0$

The missing factor is $3x, \; x \neq 0$.

11. $\dfrac{15x^2}{10x} = \dfrac{5x(3x)}{5x(2)} = \dfrac{3x}{2}, \quad x \neq 0$

13. $\dfrac{3xy}{xy + x} = \dfrac{x(3y)}{x(y + 1)} = \dfrac{3y}{y + 1}, \; x \neq 0$

15. $\dfrac{4y - 8y^2}{10y - 5} = \dfrac{-4y(2y - 1)}{5(2y - 1)}$

$\qquad = -\dfrac{4y}{5}, \; y \neq \dfrac{1}{2}$

17. $\dfrac{x - 5}{10 - 2x} = \dfrac{x - 5}{-2(x - 5)}$

$\qquad = -\dfrac{1}{2}, \; x \neq 5$

19. $\dfrac{y^2 - 16}{y + 4} = \dfrac{(y + 4)(y - 4)}{y + 4}$

$\qquad = y - 4, \; y \neq -4$

21. $\dfrac{x^3 + 5x^2 + 6x}{x^2 - 4} = \dfrac{x(x + 2)(x + 3)}{(x + 2)(x - 2)} = \dfrac{x(x + 3)}{x - 2}, \quad x \neq -2$

23. $\dfrac{y^2 - 7y + 12}{y^2 + 3y - 18} = \dfrac{(y - 3)(y - 4)}{(y + 6)(y - 3)} = \dfrac{y - 4}{y + 6}, \quad y \neq 3$

25. $\dfrac{2 - x + 2x^2 - x^3}{x^2 - 4} = \dfrac{(2 - x) + x^2(2 - x)}{(x + 2)(x - 2)}$

$\qquad = \dfrac{(2 - x)(1 + x^2)}{(x + 2)(x - 2)}$

$\qquad = \dfrac{-(x - 2)(x^2 + 1)}{(x + 2)(x - 2)}$

$\qquad = -\dfrac{x^2 + 1}{x + 2}, \; x \neq 2$

27. $\dfrac{z^3 - 8}{z^2 + 2z + 4} = \dfrac{(z - 2)(z^2 + 2z + 4)}{z^2 + 2z + 4} = z - 2$

29.

x	0	1	2	3	4	5	6
$\dfrac{x^2 - 2x - 3}{x - 3}$	1	2	3	Undef.	5	6	7
$x + 1$	1	2	3	4	5	6	7

The expressions are equivalent except at $x = 3$.

31. $\dfrac{5x^3}{2x^3 + 4} = \dfrac{5x^3}{2(x^3 + 2)}$

There are no common factors so this expression cannot be simplified. In this case factors of terms were incorrectly cancelled.

33. $\dfrac{\pi r^2}{(2r)^2} = \dfrac{\pi r^2}{4r^2} = \dfrac{\pi}{4}, \; r \neq 0$

35. $\dfrac{5}{x - 1} \cdot \dfrac{x - 1}{25(x - 2)} = \dfrac{1}{5(x - 2)}, \; x \neq 1$

37. $\dfrac{r}{r - 1} \cdot \dfrac{r^2 - 1}{r^2} = \dfrac{r(r + 1)(r - 1)}{r^2(r - 1)} = \dfrac{r + 1}{r}, \; r \neq 1, r \neq 0$

39. $\dfrac{t^2 - t - 6}{t^2 + 6t + 9} \cdot \dfrac{t + 3}{t^2 - 4} = \dfrac{(t - 3)(t + 2)(t + 3)}{(t + 3)^2(t + 2)(t - 2)} = \dfrac{t - 3}{(t + 3)(t - 2)}, \; t \neq -2$

41. $\dfrac{x^2 - 36}{x} \div \dfrac{x^3 - 6x^2}{x^2 + x} = \dfrac{x^2 - 36}{x} \cdot \dfrac{x^2 + x}{x^3 - 6x^2}$

$\qquad = \dfrac{(x + 6)(x - 6)}{x} \cdot \dfrac{x(x + 1)}{x^2(x - 6)}$

$\qquad = \dfrac{(x + 6)(x + 1)}{x^2}, \; x \neq 6$

43. $\dfrac{5}{x - 1} + \dfrac{x}{x - 1} = \dfrac{5 + x}{x - 1} = \dfrac{x + 5}{x - 1}$

45. $6 - \dfrac{5}{x+3} = \dfrac{6(x+3)}{(x+3)} - \dfrac{5}{x+3}$

$= \dfrac{6(x+3) - 5}{x+3}$

$= \dfrac{6x + 18 - 5}{x+3}$

$= \dfrac{6x + 13}{x+3}$

47. $\dfrac{3}{x-2} + \dfrac{5}{2-x} = \dfrac{3}{x-2} - \dfrac{5}{x-2} = -\dfrac{2}{x-2}$

49. $\dfrac{1}{x^2 - x - 2} - \dfrac{x}{x^2 - 5x + 6} = \dfrac{1}{(x-2)(x+1)} - \dfrac{x}{(x-2)(x-3)}$

$= \dfrac{(x-3) - x(x+1)}{(x+1)(x-2)(x-3)} = \dfrac{x - 3 - x^2 - x}{(x+1)(x-2)(x-3)}$

$= \dfrac{-x^2 - 3}{(x+1)(x-2)(x-3)} = -\dfrac{x^2 + 3}{(x+1)(x-2)(x-3)}$

51. $-\dfrac{1}{x} + \dfrac{2}{x^2 + 1} + \dfrac{1}{x^3 + x} = \dfrac{-(x^2 + 1)}{x(x^2 + 1)} + \dfrac{2x}{x(x^2 + 1)} + \dfrac{1}{x(x^2 + 1)}$

$= \dfrac{-x^2 - 1 + 2x + 1}{x(x^2 + 1)} = \dfrac{-x^2 + 2x}{x(x^2 + 1)} = \dfrac{-x(x - 2)}{x(x^2 + 1)}$

$= -\dfrac{x - 2}{x^2 + 1} = \dfrac{2 - x}{x^2 + 1}, \quad x \neq 0$

53. $\dfrac{x+4}{x+2} - \dfrac{3x-8}{x+2} = \dfrac{(x+4) - (3x-8)}{x+2}$

$= \dfrac{x + 4 - 3x + 8}{x+2} = \dfrac{-2x + 12}{x+2} = \dfrac{-2(x-6)}{x+2}$

The error was incorrect subtraction in the numerator.

55. $\dfrac{\left(\dfrac{x}{2} - 1\right)}{(x-2)} = \dfrac{\left(\dfrac{x}{2} - \dfrac{2}{2}\right)}{\left(\dfrac{x-2}{1}\right)}$

$= \dfrac{x-2}{2} \cdot \dfrac{1}{x-2}$

$= \dfrac{1}{2}, \quad x \neq 2$

57. $\dfrac{\left[\dfrac{x^2}{(x+1)^2}\right]}{\left[\dfrac{x}{(x+1)^3}\right]} = \dfrac{x^2}{(x+1)^2} \cdot \dfrac{(x+1)^3}{x}$

$= x(x + 1), \quad x \neq -1, 0$

59. $\dfrac{\left(\sqrt{x} - \dfrac{1}{2\sqrt{x}}\right)}{\sqrt{x}} = \dfrac{\left(\sqrt{x} - \dfrac{1}{2\sqrt{x}}\right)}{\sqrt{x}} \cdot \dfrac{2\sqrt{x}}{2\sqrt{x}} = \dfrac{2x - 1}{2x}, \quad x > 0$

61. $x^5 - 2x^{-2} = x^{-2}(x^7 - 2) = \dfrac{x^7 - 2}{x^2}$

63. $x^2(x^2 + 1)^{-5} - (x^2 + 1)^{-4} = (x^2 + 1)^{-5}[x^2 - (x^2 + 1)] = -\dfrac{1}{(x^2 + 1)^5}$

65. $2x^2(x-1)^{1/2} - 5(x-1)^{-1/2} = (x-1)^{-1/2}[2x^2(x-1)^1 - 5] = \dfrac{2x^3 - 2x^2 - 5}{(x-1)^{1/2}}$

67. $\dfrac{3x^{1/3} - x^{-2/3}}{3x^{-2/3}} = \dfrac{3x^{1/3} - x^{-2/3}}{3x^{-2/3}} \cdot \dfrac{x^{2/3}}{x^{2/3}} = \dfrac{3x^1 - x^0}{3x^0} = \dfrac{3x - 1}{3}, \ x \neq 0$

69. $\dfrac{\left(\dfrac{1}{x + h} - \dfrac{1}{x}\right)}{h} = \dfrac{\left(\dfrac{1}{x + h} - \dfrac{1}{x}\right)}{h} \cdot \dfrac{x(x + h)}{x(x + h)}$

$\qquad = \dfrac{x - (x + h)}{hx(x + h)}$

$\qquad = \dfrac{-h}{hx(x + h)}$

$\qquad = -\dfrac{1}{x(x + h)}, \ \ h \neq 0$

71. $\dfrac{\left(\dfrac{1}{x + h - 4} - \dfrac{1}{x - 4}\right)}{h} = \dfrac{\left(\dfrac{1}{x + h - 4} - \dfrac{1}{x - 4}\right)}{h} \cdot \dfrac{(x - 4)(x + h - 4)}{(x - 4)(x + h - 4)}$

$\qquad = \dfrac{(x - 4) - (x + h - 4)}{h(x - 4)(x + h - 4)}$

$\qquad = \dfrac{-h}{h(x - 4)(x + h - 4)}$

$\qquad = -\dfrac{1}{(x - 4)(x + h - 4)}, \ \ h \neq 0$

73. $\dfrac{\sqrt{x + 2} - \sqrt{x}}{2} = \dfrac{\sqrt{x + 2} - \sqrt{x}}{2} \cdot \dfrac{\sqrt{x + 2} + \sqrt{x}}{\sqrt{x + 2} + \sqrt{x}}$

$\qquad = \dfrac{(x + 2) - x}{2(\sqrt{x + 2} + \sqrt{x})}$

$\qquad = \dfrac{2}{2(\sqrt{x + 2} + \sqrt{x})}$

$\qquad = \dfrac{1}{\sqrt{x + 2} + \sqrt{x}}$

75. $\dfrac{\sqrt{x + h + 1} - \sqrt{x + 1}}{h} = \dfrac{\sqrt{x + h + 1} - \sqrt{x + 1}}{h} \cdot \dfrac{\sqrt{x + h + 1} + \sqrt{x + 1}}{\sqrt{x + h + 1} + \sqrt{x + 1}}$

$\qquad = \dfrac{(x + h + 1) - (x + 1)}{h(\sqrt{x + h + 1} + \sqrt{x + 1})}$

$\qquad = \dfrac{h}{h(\sqrt{x + h + 1} + \sqrt{x + 1})}$

$\qquad = \dfrac{1}{\sqrt{x + h + 1} + \sqrt{x + 1}}, \ \ h \neq 0$

77. Probability $= \dfrac{\text{Shaded area}}{\text{Total area}} = \dfrac{x(x/2)}{x(2x + 1)} = \dfrac{x/2}{2x + 1} \cdot \dfrac{2}{2} = \dfrac{x}{2(2x + 1)}$

79. (a) $\dfrac{1}{16}$ minute   (b) $x\left(\dfrac{1}{16}\right) = \dfrac{x}{16}$ minutes   (c) $\dfrac{60}{16} = \dfrac{15}{4}$ minutes

81. (a) $r = \dfrac{\left(\dfrac{24[48(400) - 16{,}000]}{48}\right)}{\left[16{,}000 + \dfrac{48(400)}{12}\right]} \approx 0.0909 = 9.09\%$

(b) $r = \dfrac{\left[\dfrac{24(NM - P)}{N}\right]}{\left(P + \dfrac{NM}{12}\right)} = \dfrac{24(NM - P)}{N} \cdot \dfrac{12}{12P + NM} = \dfrac{288(NM - P)}{N(12P + NM)}$

$r = \dfrac{288[48(400) - 16{,}000]}{48[12(16{,}000) + 48(400)]} \approx 0.0909 = 9.09\%$

83. $T = 10\left(\dfrac{4t^2 + 16t + 75}{t^2 + 4t + 10}\right)$

(a)

t	0	2	4	6	8	10	12	14	16	18	20	22
T	75°	55.9°	48.3°	45°	43.3°	42.3°	41.7°	41.3°	41.1°	40.9°	40.7°	40.6°

(b) T is approaching 40°.

85. False. In order for the simplified expression to be equivalent to the original expression, the domain of the simplified expression needs to be restricted. If n is even, $x \neq \pm 1$. If n is odd, $x \neq 1$.

87. Completely factor the numerator and the denominator. A rational expression is in **simplest** form if there are no common factors in the numerator and the denominator other than ± 1.

Appendix A.5   Solving Equations

- You should know how to solve linear equations.
  $ax + b = 0$
- An identity is an equation whose solution consists of every real number in its domain.
- To solve an equation you can:
  - (a) Add or subtract the same quantity from both sides.
  - (b) Multiply or divide both sides by the same nonzero quantity.
- To solve an equation that can be simplified to a linear equation:
  - (a) Remove all symbols of grouping and all fractions.
  - (b) Combine like terms.
  - (c) Solve by algebra.
  - (d) Check the answer.
- A "solution" that does not satisfy the original equation is called an extraneous solution.
- You should be able to solve a quadratic equation by factoring, if possible.
- You should be able to solve a quadratic equation of the form $u^2 = d$ by extracting square roots.
- You should be able to solve a quadratic equation by completing the square.
- You should know and be able to use the Quadratic Formula: For $ax^2 + bx + c = 0$, $a \neq 0$,

  $$x = \dfrac{-b \pm \sqrt{b^2 - 4ac}}{2a}.$$

- You should be able to solve polynomials of higher degree by factoring.
- For equations involving radicals or fractional powers, raise both sides to the same power.
- For equations with fractions, multiply both sides by the least common denominator to clear the fractions.
- For equations involving absolute value, remember that the expression inside the absolute value can be positive or negative.

Vocabulary Check

1. equation   **2.** solve   **3.** identities; conditional

4. $ax + b = 0$ **5.** extraneous   **6.** quadratic equation

7. factoring; extracting square roots; completing the square; Quadratic Formula

---

1. $2(x - 1) = 2x - 2$ is an *identity* by the Distributive Property. It is true for all real values of x.

3. $-6(x - 3) + 5 = -2x + 10$ is *conditional*. There are real values of x for which the equation is not true.

5. $4(x + 1) - 2x = 4x + 4 - 2x = 2x + 4 = 2(x + 2)$

This is an *identity* by simplification. It is true for all real values of x.

7. $(x - 4)^2 - 11 = x^2 - 8x + 16 - 11 = x^2 - 8x + 5$

Thus, $x^2 - 8x + 5 = (x - 4)^2 - 11$ is an *identity* by simplification. It is true for all real values of x.

9. $3 + \dfrac{1}{x + 1} = \dfrac{4x}{x + 1}$ is *conditional*. There are real values of x for which the equation is not true.

11.
$$x + 11 = 15$$
$$x + 11 - 11 = 15 - 11$$
$$x = 4$$

13.
$$7 - 2x = 25$$
$$7 - 7 - 2x = 25 - 7$$
$$-2x = 18$$
$$\frac{-2x}{-2} = \frac{18}{-2}$$
$$x = -9$$

15.
$$8x - 5 = 3x + 20$$
$$8x - 3x - 5 = 3x - 3x + 20$$
$$5x - 5 = 20$$
$$5x - 5 + 5 = 20 + 5$$
$$5x = 25$$
$$\frac{5x}{5} = \frac{25}{5}$$
$$x = 5$$

17. $2(x + 5) - 7 = 3(x - 2)$
$$2x + 10 - 7 = 3x - 6$$
$$2x + 3 = 3x - 6$$
$$2x - 3x + 3 = 3x - 3x - 6$$
$$-x + 3 = -6$$
$$-x + 3 - 3 = -6 - 3$$
$$-x = -9$$
$$x = 9$$

19. $x - 3(2x + 3) = 8 - 5x$
$$x - 6x - 9 = 8 - 5x$$
$$-5x - 9 = 8 - 5x$$
$$-5x + 5x - 9 = 8 - 5x + 5x$$
$$-9 \neq 8$$
No solution

21.
$$\frac{5x}{4} + \frac{1}{2} = x - \frac{1}{2}$$
$$4\left(\frac{5x}{4} + \frac{1}{2}\right) = 4\left(x - \frac{1}{2}\right)$$
$$4\left(\frac{5x}{4}\right) + 4\left(\frac{1}{2}\right) = 4(x) - 4\left(\frac{1}{2}\right)$$
$$5x + 2 = 4x - 2$$
$$x = -4$$

23.
$$\tfrac{3}{2}(z + 5) - \tfrac{1}{4}(z + 24) = 0$$
$$4\left[\tfrac{3}{2}(z + 5) - \tfrac{1}{4}(z + 24)\right] = 4(0)$$
$$4\left(\tfrac{3}{2}\right)(z + 5) - 4\left(\tfrac{1}{4}\right)(z + 24) = 4(0)$$
$$6(z + 5) - (z + 24) = 0$$
$$6z + 30 - z - 24 = 0$$
$$5z = -6$$
$$z = -\tfrac{6}{5}$$

25. $0.25x + 0.75(10 - x) = 3$
$$0.25x + 7.5 - 0.75x = 3$$
$$-0.50x + 7.5 = 3$$
$$-0.50x = -4.5$$
$$x = 9$$

27. $x + 8 = 2(x - 2) - x$

$x + 8 = 2x - 4 - x$

$x + 8 = x - 4$

$8 \neq -4$

Contradiction; no solution

29. $\dfrac{100 - 4x}{3} = \dfrac{5x + 6}{4} + 6$

$12\left(\dfrac{100 - 4x}{3}\right) = 12\left(\dfrac{5x + 6}{4}\right) + 12(6)$

$4(100 - 4x) = 3(5x + 6) + 72$

$400 - 16x = 15x + 18 + 72$

$-31x = -310$

$x = 10$

31. $\dfrac{5x - 4}{5x + 4} = \dfrac{2}{3}$

$3(5x - 4) = 2(5x + 4)$

$15x - 12 = 10x + 8$

$5x = 20$

$x = 4$

33. $10 - \dfrac{13}{x} = 4 + \dfrac{5}{x}$

$\dfrac{10x - 13}{x} = \dfrac{4x + 5}{x}$

$10x - 13 = 4x + 5$

$6x = 18$

$x = 3$

35. $3 = 2 + \dfrac{2}{z + 2}$

$3(z + 2) = \left(2 + \dfrac{2}{z + 2}\right)(z + 2)$

$3z + 6 = 2z + 4 + 2$

$z = 0$

37. $\dfrac{x}{x + 4} + \dfrac{4}{x + 4} + 2 = 0$

$\dfrac{x + 4}{x + 4} + 2 = 0$

$1 + 2 = 0$

$3 \neq 0$

Contradiction; no solution

39. $\dfrac{2}{(x - 4)(x - 2)} = \dfrac{1}{x - 4} + \dfrac{2}{x - 2}$ Multiply both sides by $(x - 4)(x - 2)$.

$2 = 1(x - 2) + 2(x - 4)$

$2 = x - 2 + 2x - 8$

$2 = 3x - 10$

$12 = 3x$

$4 = x$

A check reveals that $x = 4$ is an extraneous solution—it makes the denominator zero. There is no real solution.

41. $\dfrac{1}{x - 3} + \dfrac{1}{x + 3} = \dfrac{10}{x^2 - 9}$

$\dfrac{1}{x - 3} + \dfrac{1}{x + 3} = \dfrac{10}{(x + 3)(x - 3)}$ Multiply both sides by $(x + 3)(x - 3)$.

$1(x + 3) + 1(x - 3) = 10$

$2x = 10$

$x = 5$

43. $\dfrac{3}{x^2 - 3x} + \dfrac{4}{x} = \dfrac{1}{x - 3}$

$\dfrac{3}{x(x - 3)} + \dfrac{4}{x} = \dfrac{1}{x - 3}$ Multiply both sides by $x(x - 3)$.

$3 + 4(x - 3) = x$

$3 + 4x - 12 = x$

$\qquad\qquad 3x = 9$

$\qquad\qquad\ x = 3$

A check reveals that $x = 3$ is an extraneous solution since it makes the denominator zero, so there is no solution.

45. $(x + 2)^2 + 5 = (x + 3)^2$

$x^2 + 4x + 4 + 5 = x^2 + 6x + 9$

$\qquad\quad 4x + 9 = 6x + 9$

$\qquad\qquad\ -2x = 0$

$\qquad\qquad\quad\ x = 0$

47. $(x + 2)^2 - x^2 = 4(x + 1)$

$x^2 + 4x + 4 - x^2 = 4x + 4$

$\qquad\qquad\quad 4 = 4$

The equation is an identity; every real number is a solution.

49. $2x^2 = 3 - 8x$

General form: $2x^2 + 8x - 3 = 0$

51. $(x - 3)^2 = 3$

$x^2 - 6x + 9 = 3$

General form: $x^2 - 6x + 6 = 0$

53. $\frac{1}{5}(3x^2 - 10) = 18x$

$3x^2 - 10 = 90x$

General form:
$3x^2 - 90x - 10 = 0$

55. $6x^2 + 3x = 0$

$3x(2x + 1) = 0$

$3x = 0$ or $2x + 1 = 0$

$\ x = 0$ or $\qquad x = -\frac{1}{2}$

57. $x^2 - 2x - 8 = 0$

$(x - 4)(x + 2) = 0$

$x - 4 = 0$ or $x + 2 = 0$

$\quad x = 4$ or $\qquad x = -2$

59. $x^2 + 10x + 25 = 0$

$(x + 5)^2 = 0$

$x + 5 = 0$

$\quad x = -5$

61. $3 + 5x - 2x^2 = 0$

$(3 - x)(1 + 2x) = 0$

$3 - x = 0$ or $1 + 2x = 0$

$\quad x = 3$ or $\qquad x = -\frac{1}{2}$

63. $x^2 + 4x = 12$

$x^2 + 4x - 12 = 0$

$(x + 6)(x - 2) = 0$

$x + 6 = 0$ or $x - 2 = 0$

$\quad x = -6$ or $\quad x = 2$

65. $\frac{3}{4}x^2 + 8x + 20 = 0$

$4\left(\frac{3}{4}x^2 + 8x + 20\right) = 4(0)$

$3x^2 + 32x + 80 = 0$

$(3x + 20)(x + 4) = 0$

$3x + 20 = 0$ or $x + 4 = 0$

$\quad x = -\frac{20}{3}$ or $\quad x = -4$

67. $x^2 + 2ax + a^2 = 0$

$(x + a)^2 = 0$

$x + a = 0$

$\quad x = -a$

69. $x^2 = 49$

$x = \pm 7$

71. $x^2 = 11$

$x = \pm\sqrt{11}$

73. $3x^2 = 81$

$x^2 = 27$

$x = \pm 3\sqrt{3}$

75. $(x - 12)^2 = 16$

$x - 12 = \pm 4$

$x = 12 \pm 4$

$x = 16$ or $x = 8$

77. $(x + 2)^2 = 14$

$x + 2 = \pm\sqrt{14}$

$x = -2 \pm \sqrt{14}$

79. $(2x - 1)^2 = 18$

$2x - 1 = \pm\sqrt{18}$

$2x = 1 \pm 3\sqrt{2}$

$x = \dfrac{1 \pm 3\sqrt{2}}{2}$

81. $(x - 7)^2 = (x + 3)^2$

$x - 7 = \pm(x + 3)$

$x - 7 = x + 3 \quad \text{or} \quad x - 7 = -x - 3$

$-7 \neq 3 \quad \text{or} \quad 2x = 4$

$x = 2$

The only solution to the equation is $x = 2$.

83. $x^2 + 4x - 32 = 0$

$x^2 + 4x = 32$

$x^2 + 4x + 2^2 = 32 + 2^2$

$(x + 2)^2 = 36$

$x + 2 = \pm 6$

$x = -2 \pm 6$

$x = 4 \quad \text{or} \quad x = -8$

85. $x^2 + 12x + 25 = 0$

$x^2 + 12x = -25$

$x^2 + 12x + 6^2 = -25 + 6^2$

$(x + 6)^2 = 11$

$x + 6 = \pm\sqrt{11}$

$x = -6 \pm \sqrt{11}$

87. $9x^2 - 18x = -3$

$x^2 - 2x = -\dfrac{1}{3}$

$x^2 - 2x + 1^2 = -\dfrac{1}{3} + 1^2$

$(x - 1)^2 = \dfrac{2}{3}$

$x - 1 = \pm\sqrt{\dfrac{2}{3}}$

$x = 1 \pm \sqrt{\dfrac{2}{3}}$

$x = 1 \pm \dfrac{\sqrt{6}}{3}$

89. $8 + 4x - x^2 = 0$

$-x^2 + 4x + 8 = 0$

$x^2 - 4x - 8 = 0$

$x^2 - 4x = 8$

$x^2 - 4x + 2^2 = 8 + 2^2$

$(x - 2)^2 = 12$

$x - 2 = \pm\sqrt{12}$

$x = 2 \pm 2\sqrt{3}$

91. $2x^2 + 5x - 8 = 0$

$2x^2 + 5x = 8$

$x^2 + \dfrac{5}{2}x = 4$

$x^2 + \dfrac{5}{2}x + \left(\dfrac{5}{4}\right)^2 = 4 + \left(\dfrac{5}{4}\right)^2$

$\left(x + \dfrac{5}{4}\right)^2 = \dfrac{89}{16}$

$x + \dfrac{5}{4} = \pm\dfrac{\sqrt{89}}{4}$

$x = -\dfrac{5}{4} \pm \dfrac{\sqrt{89}}{4}$

$x = \dfrac{-5 \pm \sqrt{89}}{4}$

93. $2x^2 + x - 1 = 0$

$x = \dfrac{-b \pm \sqrt{b^2 - 4ac}}{2a}$

$= \dfrac{-1 \pm \sqrt{1^2 - 4(2)(-1)}}{2(2)}$

$= \dfrac{-1 \pm 3}{4} = \dfrac{1}{2}, -1$

95. $16x^2 + 8x - 3 = 0$

$x = \dfrac{-b \pm \sqrt{b^2 - 4ac}}{2a}$

$= \dfrac{-8 \pm \sqrt{8^2 - 4(16)(-3)}}{2(16)}$

$= \dfrac{-8 \pm 16}{32} = \dfrac{1}{4}, -\dfrac{3}{4}$

97. $2 + 2x - x^2 = 0$

$-x^2 + 2x + 2 = 0$

$x = \dfrac{-b \pm \sqrt{b^2 - 4ac}}{2a}$

$= \dfrac{-2 \pm \sqrt{2^2 - 4(-1)(2)}}{2(-1)}$

$= \dfrac{-2 \pm 2\sqrt{3}}{-2} = 1 \pm \sqrt{3}$

99. $x^2 + 14x + 44 = 0$

$x = \dfrac{-b \pm \sqrt{b^2 - 4ac}}{2a}$

$= \dfrac{-14 \pm \sqrt{14^2 - 4(1)(44)}}{2(1)}$

$= \dfrac{-14 \pm 2\sqrt{5}}{2} = -7 \pm \sqrt{5}$

101. $x^2 + 8x - 4 = 0$

$$x = \frac{-b \pm \sqrt{b^2 - 4ac}}{2a}$$

$$= \frac{-8 \pm \sqrt{8^2 - 4(1)(-4)}}{2(1)}$$

$$= \frac{-8 \pm 4\sqrt{5}}{2} = -4 \pm 2\sqrt{5}$$

103. $12x - 9x^2 = -3$

$$-9x^2 + 12x + 3 = 0$$

$$x = \frac{-b \pm \sqrt{b^2 - 4ac}}{2a}$$

$$= \frac{-12 \pm \sqrt{12^2 - 4(-9)(3)}}{2(-9)}$$

$$= \frac{-12 \pm 6\sqrt{7}}{-18} = \frac{2}{3} \pm \frac{\sqrt{7}}{3}$$

105. $9x^2 + 24x + 16 = 0$

$$x = \frac{-b \pm \sqrt{b^2 - 4ac}}{2a}$$

$$= \frac{-24 \pm \sqrt{24^2 - 4(9)(16)}}{2(9)}$$

$$= \frac{-24 \pm 0}{18}$$

$$= -\frac{4}{3}$$

107. $4x^2 + 4x = 7$

$$4x^2 + 4x - 7 = 0$$

$$x = \frac{-b \pm \sqrt{b^2 - 4ac}}{2a}$$

$$= \frac{-4 \pm \sqrt{4^2 - 4(4)(-7)}}{2(4)}$$

$$= \frac{-4 \pm 8\sqrt{2}}{8} = -\frac{1}{2} \pm \sqrt{2}$$

109. $28x - 49x^2 = 4$

$$-49x^2 + 28x - 4 = 0$$

$$x = \frac{-b \pm \sqrt{b^2 - 4ac}}{2a}$$

$$= \frac{-28 \pm \sqrt{28^2 - 4(-49)(-4)}}{2(-49)}$$

$$= \frac{-28 \pm 0}{-98} = \frac{2}{7}$$

111. $8t = 5 + 2t^2$

$$-2t^2 + 8t - 5 = 0$$

$$t = \frac{-b \pm \sqrt{b^2 - 4ac}}{2a}$$

$$= \frac{-8 \pm \sqrt{8^2 - 4(-2)(-5)}}{2(-2)}$$

$$= \frac{-8 \pm 2\sqrt{6}}{-4} = 2 \pm \frac{\sqrt{6}}{2}$$

113. $(y - 5)^2 = 2y$

$$y^2 - 12y + 25 = 0$$

$$y = \frac{-b \pm \sqrt{b^2 - 4ac}}{2a}$$

$$= \frac{-(-12) \pm \sqrt{(-12)^2 - 4(1)(25)}}{2(1)}$$

$$= \frac{12 \pm 2\sqrt{11}}{2} = 6 \pm \sqrt{11}$$

115. $\frac{1}{2}x^2 + \frac{3}{8}x = 2$

$$4x^2 + 3x = 16$$

$$4x^2 + 3x - 16 = 0$$

$$x = \frac{-b \pm \sqrt{b^2 - 4ac}}{2a}$$

$$= \frac{-3 \pm \sqrt{3^2 - 4(4)(-16)}}{2(4)}$$

$$= \frac{-3 \pm \sqrt{265}}{8}$$

$$= -\frac{3}{8} \pm \frac{\sqrt{265}}{8}$$

117. $5.1x^2 - 1.7x - 3.2 = 0$

$$x = \frac{1.7 \pm \sqrt{(-1.7)^2 - 4(5.1)(-3.2)}}{2(5.1)}$$

$$x \approx 0.976, -0.643$$

119. $-0.067x^2 - 0.852x + 1.277 = 0$

$$x = \frac{-(-0.852) \pm \sqrt{(-0.852)^2 - 4(-0.067)(1.277)}}{2(-0.067)}$$

$$x \approx -14.071, 1.355$$

121. $422x^2 - 506x - 347 = 0$

$$x = \frac{506 \pm \sqrt{(-506)^2 - 4(422)(-347)}}{2(422)}$$

$$x \approx 1.687, -0.488$$

123. $12.67x^2 + 31.55x + 8.09 = 0$

$$x = \frac{-31.55 \pm \sqrt{(31.55)^2 - 4(12.67)(8.09)}}{2(12.67)}$$

$$x \approx -2.200, -0.290$$

125. $x^2 - 2x - 1 = 0$ Complete the square.

$$x^2 - 2x = 1$$
$$x^2 - 2x + 1^2 = 1 + 1^2$$
$$(x - 1)^2 = 2$$
$$x - 1 = \pm\sqrt{2}$$
$$x = 1 \pm \sqrt{2}$$

127. $(x + 3)^2 = 81$ Extract square roots.

$$x + 3 = \pm 9$$
$$x + 3 = 9 \quad \text{or} \quad x + 3 = -9$$
$$x = 6 \quad \text{or} \qquad x = -12$$

129. $x^2 - x - \frac{11}{4} = 0$ Complete the square.

$$x^2 - x = \frac{11}{4}$$
$$x^2 - x + \left(\frac{1}{2}\right)^2 = \frac{11}{4} + \left(\frac{1}{2}\right)^2$$
$$\left(x - \frac{1}{2}\right)^2 = \frac{12}{4}$$
$$x - \frac{1}{2} = \pm\sqrt{\frac{12}{4}}$$
$$x = \frac{1}{2} \pm \sqrt{3}$$

131. $(x + 1)^2 = x^2$ Extract square roots.

$$x^2 = (x + 1)^2$$
$$x = \pm(x + 1)$$

For $x = +(x + 1)$:

$$0 \neq 1 \quad \text{No solution}$$

For $x = -(x + 1)$:

$$2x = -1$$
$$x = -\frac{1}{2}$$

133. $3x + 4 = 2x^2 - 7$ Quadratic Formula

$$0 = 2x^2 - 3x - 11$$
$$x = \frac{-(-3) \pm \sqrt{(-3)^2 - 4(2)(-11)}}{2(2)}$$
$$= \frac{3 \pm \sqrt{97}}{4}$$
$$= \frac{3}{4} \pm \frac{\sqrt{97}}{4}$$

135. $4x^4 - 18x^2 = 0$

$$2x^2(2x^2 - 9) = 0$$
$$2x^2 = 0 \Rightarrow x = 0$$
$$2x^2 - 9 = 0 \Rightarrow x = \pm\frac{3\sqrt{2}}{2}$$

137.
$$x^4 - 81 = 0$$
$$(x^2 + 9)(x + 3)(x - 3) = 0$$
$$x^2 + 9 = 0 \Rightarrow \text{No real solution}$$
$$x + 3 = 0 \Rightarrow x = -3$$
$$x - 3 = 0 \Rightarrow x = 3$$

139.
$$x^3 + 216 = 0$$
$$x^3 + 6^3 = 0$$
$$(x + 6)(x^2 - 6x + 36) = 0$$
$$x + 6 = 0 \Rightarrow x = -6$$
$$x^2 - 6x + 36 = 0 \Rightarrow \text{No real solution (by the Quadratic Formula)}$$

141. $5x^3 + 30x^2 + 45x = 0$

$\qquad 5x(x^2 + 6x + 9) = 0$

$\qquad 5x(x + 3)^2 = 0$

$\qquad\qquad 5x = 0 \implies x = 0$

$\qquad\qquad x + 3 = 0 \implies x = -3$

143. $\qquad x^3 - 3x^2 - x + 3 = 0$

$\qquad x^2(x - 3) - (x - 3) = 0$

$\qquad\qquad (x - 3)(x^2 - 1) = 0$

$\qquad (x - 3)(x + 1)(x - 1) = 0$

$\qquad\qquad x - 3 = 0 \implies x = 3$

$\qquad\qquad x + 1 = 0 \implies x = -1$

$\qquad\qquad x - 1 = 0 \implies x = 1$

145. $\qquad x^4 - x^3 + x - 1 = 0$

$\qquad x^3(x - 1) + (x - 1) = 0$

$\qquad\qquad (x - 1)(x^3 + 1) = 0$

$\qquad (x - 1)(x + 1)(x^2 - x + 1) = 0$

$\qquad\qquad x - 1 = 0 \implies x = 1$

$\qquad\qquad x + 1 = 0 \implies x = -1$

$\qquad x^2 - x + 1 = 0 \implies$ No real solution (by the Quadratic Formula)

147. $\qquad x^4 - 4x^2 + 3 = 0$

$\qquad (x^2 - 3)(x^2 - 1) = 0$

$\left(x + \sqrt{3}\right)\left(x - \sqrt{3}\right)(x + 1)(x - 1) = 0$

$\qquad x + \sqrt{3} = 0 \implies x = -\sqrt{3}$

$\qquad x - \sqrt{3} = 0 \implies x = \sqrt{3}$

$\qquad\qquad x + 1 = 0 \implies x = -1$

$\qquad\qquad x - 1 = 0 \implies x = 1$

149. $\qquad 4x^4 - 65x^2 + 16 = 0$

$\qquad (4x^2 - 1)(x^2 - 16) = 0$

$\qquad (2x + 1)(2x - 1)(x + 4)(x - 4) = 0$

$\qquad 2x + 1 = 0 \implies x = -\frac{1}{2}$

$\qquad 2x - 1 = 0 \implies x = \frac{1}{2}$

$\qquad\qquad x + 4 = 0 \implies x = -4$

$\qquad\qquad x - 4 = 0 \implies x = 4$

151. $\qquad x^6 + 7x^3 - 8 = 0$

$\qquad (x^3 + 8)(x^3 - 1) = 0$

$(x + 2)(x^2 - 2x + 4)(x - 1)(x^2 + x + 1) = 0$

$\qquad\qquad x + 2 = 0 \implies x = -2$

$\qquad x^2 - 2x + 4 = 0 \implies$ No real solution (by the Quadratic Formula)

$\qquad\qquad x - 1 = 0 \implies x = 1$

$\qquad x^2 + x + 1 = 0 \implies$ No real solution (by the Quadratic Formula)

153. $\sqrt{2x} - 10 = 0$

$\qquad \sqrt{2x} = 10$

$\qquad 2x = 100$

$\qquad x = 50$

155. $\sqrt{x - 10} - 4 = 0$

$\qquad \sqrt{x - 10} = 4$

$\qquad x - 10 = 16$

$\qquad x = 26$

157. $\sqrt[3]{2x + 5} + 3 = 0$

$\qquad \sqrt[3]{2x + 5} = -3$

$\qquad 2x + 5 = -27$

$\qquad 2x = -32$

$\qquad x = -16$

159. $-\sqrt{26 - 11x} + 4 = x$

$$4 - x = \sqrt{26 - 11x}$$
$$16 - 8x + x^2 = 26 - 11x$$
$$x^2 + 3x - 10 = 0$$
$$(x + 5)(x - 2) = 0$$
$$x + 5 = 0 \implies x = -5$$
$$x - 2 = 0 \implies x = 2$$

161. $\sqrt{x + 1} = \sqrt{3x + 1}$

$$x + 1 = 3x + 1$$
$$-2x = 0$$
$$x = 0$$

163. $(x - 5)^{3/2} = 8$

$$(x - 5)^3 = 8^2$$
$$x - 5 = \sqrt[3]{64}$$
$$x = 5 + 4 = 9$$

165. $(x + 3)^{2/3} = 8$

$$(x + 3)^2 = 8^3$$
$$x + 3 = \pm\sqrt{8^3}$$
$$x + 3 = \pm\sqrt{512}$$
$$x = -3 \pm 16\sqrt{2}$$

167. $(x^2 - 5)^{3/2} = 27$

$$(x^2 - 5)^3 = 27^2$$
$$x^2 - 5 = \sqrt[3]{27^2}$$
$$x^2 = 5 + 9$$
$$x^2 = 14$$
$$x = \pm\sqrt{14}$$

169. $3x(x - 1)^{1/2} + 2(x - 1)^{3/2} = 0$

$$(x - 1)^{1/2}[3x + 2(x - 1)] = 0$$
$$(x - 1)^{1/2}(5x - 2) = 0$$
$$(x - 1)^{1/2} = 0 \implies x - 1 = 0 \implies x = 1$$
$$5x - 2 = 0 \implies x = \tfrac{2}{5}, \text{ extraneous}$$

171.

$$x = \frac{3}{x} + \frac{1}{2}$$
$$(2x)(x) = (2x)\left(\frac{3}{x}\right) + (2x)\left(\frac{1}{2}\right)$$
$$2x^2 = 6 + x$$
$$2x^2 - x - 6 = 0$$
$$(2x + 3)(x - 2) = 0$$
$$2x + 3 = 0 \implies x = -\frac{3}{2}$$
$$x - 2 = 0 \implies x = 2$$

173.

$$\frac{1}{x} - \frac{1}{x + 1} = 3$$
$$x(x + 1)\frac{1}{x} - x(x + 1)\frac{1}{x + 1} = x(x + 1)(3)$$
$$x + 1 - x = 3x(x + 1)$$
$$1 = 3x^2 + 3x$$
$$0 = 3x^2 + 3x - 1$$
$$a = 3, \quad b = 3, \quad c = -1$$
$$x = \frac{-3 \pm \sqrt{(3)^2 - 4(3)(-1)}}{2(3)} = \frac{-3 \pm \sqrt{21}}{6}$$

175. $\dfrac{20 - x}{x} = x$

$$20 - x = x^2$$
$$0 = x^2 + x - 20$$
$$0 = (x + 5)(x - 4)$$
$$x + 5 = 0 \implies x = -5$$
$$x - 4 = 0 \implies x = 4$$

177.

$$\frac{x}{x^2 - 4} + \frac{1}{x + 2} = 3$$

$$(x + 2)(x - 2)\frac{x}{x^2 - 4} + (x + 2)(x - 2)\frac{1}{x + 2} = 3(x + 2)(x - 2)$$

$$x + x - 2 = 3x^2 - 12$$

$$3x^2 - 2x - 10 = 0$$

$a = 3, \ b = -2, \ c = -10$

$$x = \frac{-(-2) \pm \sqrt{(-2)^2 - 4(3)(-10)}}{2(3)}$$

$$= \frac{2 \pm \sqrt{124}}{6} = \frac{2 \pm 2\sqrt{31}}{6} = \frac{1 \pm \sqrt{31}}{3}$$

179. $|2x - 1| = 5$

$$2x - 1 = 5 \implies x = 3$$

$$-(2x - 1) = 5 \implies x = -2$$

181. $|x| = x^2 + x - 3$

First equation:

$$x = x^2 + x - 3$$

$$x^2 - 3 = 0$$

$$x = \pm\sqrt{3}$$

Second equation:

$$-x = x^2 + x - 3$$

$$x^2 + 2x - 3 = 0$$

$$(x - 1)(x + 3) = 0$$

$$x - 1 = 0 \implies x = 1$$

$$x + 3 = 0 \implies x = -3$$

Only $x = \sqrt{3}$ and $x = -3$ are solutions to the original equation. $x = -\sqrt{3}$ and $x = 1$ are extraneous.

183. $|x + 1| = x^2 - 5$

First equation:

$$x + 1 = x^2 - 5$$

$$x^2 - x - 6 = 0$$

$$(x - 3)(x + 2) = 0$$

$$x - 3 = 0 \implies x = 3$$

$$x + 2 = 0 \implies x = -2$$

Second equation:

$$-(x + 1) = x^2 - 5$$

$$-x - 1 = x^2 - 5$$

$$x^2 + x - 4 = 0$$

$$x = \frac{-1 \pm \sqrt{17}}{2}$$

Only $x = 3$ and $x = \dfrac{-1 - \sqrt{17}}{2}$ are solutions to the original equation. $x = -2$ and $x = \dfrac{-1 + \sqrt{17}}{2}$ are extraneous.

185. (a) Female: $y = 0.432x - 10.44$

For $y = 16$: $\quad 16 = 0.432x - 10.44$

$$26.44 = 0.432x$$

$$\frac{26.44}{0.432} = x$$

$$x \approx 61.2 \text{ inches}$$

(b) Male: $y = 0.449x - 12.15$

For $y = 19$: $\quad 19 = 0.449x - 12.15$

$$31.15 = 0.449x$$

$$69.4 \approx x$$

Yes, it is likely that both bones came from the same person because the estimated height of a male with a 19-inch thigh bone is 69.4 inches.

—CONTINUED—

185. —CONTINUED—

(c)

Height x	Female Femur Length	Male Femur Length
60	15.48	14.79
70	19.80	19.28
80	24.12	23.77
90	28.44	28.26
100	32.76	32.75
110	37.08	37.24

The lengths of the male and female femurs are approximately equal when the lengths are 100 inches.

187.
$$y = -0.25t + 8$$
$$1 = -0.25t + 8$$
$$0.25t = 7$$
$$t = 28 \text{ hours}$$

189.
$$S = x^2 + 4xh$$
$$84 = x^2 + 4x(2)$$
$$0 = x^2 + 8x - 84$$
$$0 = (x + 14)(x - 6)$$
$$x = -14 \quad \text{or} \quad x = 6$$

Since x must be positive, we have $x = 6$ inches. The dimensions of the box are 6 inches × 6 inches × 2 inches.

191. *Model*: $(\text{height})^2 + (\text{half of side})^2 = (\text{side})^2$

Labels: height = 10 inches, side = s, half of side = $\dfrac{s}{2}$

Equation: $10^2 + \left(\dfrac{s}{2}\right)^2 = s^2$

$$100 + \frac{s^2}{4} = s^2$$

$$\frac{3}{4}s^2 = 100$$

$$s^2 = \frac{400}{3}$$

$$s = \sqrt{\frac{400}{3}} = \frac{20\sqrt{3}}{3} \approx 11.55 \text{ inches}$$

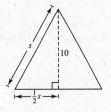

Each side of the equilateral triangle is approximately 11.55 inches long.

193. (a) $P = 200$ million when:

$$\frac{182.45 - 3.189t}{1.00 - 0.026t} = 200$$

$$182.45 - 3.189t = 200(1.00 - 0.026t)$$

$$182.45 - 3.189t = 200 - 5.2t$$

$$2.011t = 17.55$$

$$t = 8.7$$

So the total voting-age population reached 200 million during 1998.

(b) For $P = 230$:

$$\frac{182.45 - 3.189t}{1.00 - 0.026t} = 230$$

$$182.45 - 3.189t = 230(1.00 - 0.026t)$$

$$182.45 - 3.189t = 230 - 5.98t$$

$$\text{so} \quad 2.791t = 47.55$$

$$t = 17$$

The model predicts that the total voting-age population will reach 230 million during 2007. This value is reasonable but the model is reaching its limit since it soon begins to rise very fast due to its asymptotic behavior.

195.
$$37.55 = 40 - \sqrt{0.01x + 1}$$
$$\sqrt{0.01x + 1} = 2.45$$
$$0.01x + 1 = 6.0025$$
$$0.01x = 5.0025$$
$$x = 500.25$$

Rounding x to the nearest whole unit yields
$x \approx 500$ units.

197. False. $x(3 - x) = 10 \implies 3x - x^2 = 10$

This is a quadratic equation. The equation cannot be written in the form $ax + b = 0$.

199. False—See Example 14 on page A55.

201. Equivalent equations are derived from the substitution principle and simplification techniques. They have the same solution(s).

$2x + 3 = 8$ and $2x = 5$ are equivalent equations.

203. The student should have subtracted $15x$ from both sides so that the equation is equal to zero. By factoring out an x, there are two solutions, $x = 0$ and $x = 6$.

205. -3 and 6

One possible equation is:
$$(x - (-3))(x - 6) = 0$$
$$(x + 3)(x - 6) = 0$$
$$x^2 - 3x - 18 = 0$$

Any non-zero multiple of this equation would also have these solutions.

207. 8 and 14

One possible equation is:
$$(x - 8)(x - 14) = 0$$
$$x^2 - 22x + 112 = 0$$

Any non-zero multiple of this equation would also have these solutions.

209. $1 + \sqrt{2}$ and $1 - \sqrt{2}$

One possible equation is:
$$\left[x - \left(1 + \sqrt{2}\right)\right]\left[x - \left(1 - \sqrt{2}\right)\right] = 0$$
$$\left[(x - 1) - \sqrt{2}\right]\left[(x - 1) + \sqrt{2}\right] = 0$$
$$(x - 1)^2 - \left(\sqrt{2}\right)^2 = 0$$
$$x^2 - 2x + 1 - 2 = 0$$
$$x^2 - 2x - 1 = 0$$

Any non-zero multiple of this equation would also have these solutions.

211.
$$9 + |9 - a| = b$$
$$|9 - a| = b - 9$$
$$9 - a = b - 9 \quad \text{OR} \quad 9 - a = -(b - 9)$$
$$-a = b - 18 \qquad\qquad 9 - a = -b + 9$$
$$a = 18 - b \qquad\qquad -a = -b$$
$$\qquad\qquad\qquad a = b$$

Thus, $a = 18 - b$ or $a = b$. From the original equation we know that $b \geq 9$.

Some possibilities are: $b = 9, a = 9$

$b = 10, a = 8$ or $a = 10$

$b = 11, a = 7$ or $a = 11$

$b = 12, a = 6$ or $a = 12$

$b = 13, a = 5$ or $a = 13$

$b = 14, a = 4$ or $a = 14$

213. (a) $ax^2 + bx = 0$
$$x(ax + b) = 0$$
$$x = 0$$
$$ax + b = 0 \implies x = -\frac{b}{a}$$

(b) $ax^2 - ax = 0$
$$ax(x - 1) = 0$$
$$ax = 0 \implies x = 0$$
$$x - 1 = 0 \implies x = 1$$

Appendix A.6   Linear Inequalities in One Variable

- You should know the properties of inequalities.

  (a) Transitive: $a < b$ and $b < c$ implies $a < c$.

  (b) Addition: $a < b$ and $c < d$ implies $a + c < b + d$.

  (c) Adding or Subtracting a Constant: $a \pm c < b \pm c$ if $a < b$.

  (d) Multiplying or Dividing a Constant: For $a < b$,

      1. If $c > 0$, then $ac < bc$ and $\dfrac{a}{c} < \dfrac{b}{c}$.   2. If $c < 0$, then $ac > bc$ and $\dfrac{a}{c} > \dfrac{b}{c}$.

- You should be able to solve absolute value inequalities.

  (a) $|x| < a$ if and only if $-a < x < a$.   (b) $|x| > a$ if and only if $x < -a$ or $x > a$.

Vocabulary Check

1. solution set   **2.** graph   **3.** negative

4. equivalent   **5.** double   **6.** union

---

1. Interval: $[-1, 5]$

    (a) Inequality: $-1 \le x \le 5$

    (b) The interval is bounded.

3. Interval: $(11, \infty)$

    (a) Inequality: $x > 11$

    (b) The interval is unbounded.

5. Interval: $(-\infty, -2)$

    (a) Inequality: $x < -2$

    (b) The interval is unbounded.

7. $x < 3$

    Matches (b).

9. $-3 < x \le 4$

    Matches (d).

11. $|x| < 3 \implies -3 < x < 3$

    Matches (e).

13. $5x - 12 > 0$

(a) $x = 3$

$$5(3) - 12 \overset{?}{>} 0$$

$$3 > 0$$

Yes, $x = 3$ *is* a solution.

(b) $x = -3$

$$5(-3) - 12 \overset{?}{>} 0$$

$$-27 \not> 0$$

No, $x = -3$ *is not* a solution.

(c) $x = \frac{5}{2}$

$$5\left(\tfrac{5}{2}\right) - 12 \overset{?}{>} 0$$

$$\tfrac{1}{2} > 0$$

Yes, $x = \frac{5}{2}$ *is* a solution.

(d) $x = \frac{3}{2}$

$$5\left(\tfrac{3}{2}\right) - 12 \overset{?}{>} 0$$

$$-\tfrac{9}{2} \not> 0$$

No, $x = \frac{3}{2}$ *is not* a solution.

15. $0 < \dfrac{x - 2}{4} < 2$

(a) $x = 4$

$$0 \overset{?}{<} \frac{4 - 2}{4} \overset{?}{<} 2$$

$$0 < \frac{1}{2} < 2$$

Yes, $x = 4$ *is* a solution.

(b) $x = 10$

$$0 \overset{?}{<} \frac{10 - 2}{4} \overset{?}{<} 2$$

$$0 < 2 \not< 2$$

No, $x = 10$ *is not* a solution.

(c) $x = 0$

$$0 \overset{?}{<} \frac{0 - 2}{4} \overset{?}{<} 2$$

$$0 \not< -\frac{1}{2} < 2$$

No, $x = 0$ *is not* a solution.

(d) $x = \dfrac{7}{2}$

$$0 \overset{?}{<} \frac{(7/2) - 2}{4} \overset{?}{<} 2$$

$$0 < \frac{3}{8} < 2$$

Yes, $x = \frac{7}{2}$ *is* a solution.

17. $|x - 10| \geq 3$

(a) $x = 13$

$|13 - 10| \overset{?}{\geq} 3$

$3 \geq 3$

Yes, $x = 13$ *is* a solution.

(b) $x = -1$

$|-1 - 10| \overset{?}{\geq} 3$

$11 \geq 3$

Yes, $x = -1$ *is* a solution.

(c) $x = 14$

$|14 - 10| \overset{?}{\geq} 3$

$4 \geq 3$

Yes, $x = 14$ *is* a solution.

(d) $x = 9$

$|9 - 10| \overset{?}{\geq} 3$

$1 \ngeq 3$

No, $x = 9$ *is not* a solution.

19. $4x < 12$

$\frac{1}{4}(4x) < \frac{1}{4}(12)$

$x < 3$

21. $-2x > -3$

$-\frac{1}{2}(-2x) < \left(-\frac{1}{2}\right)(-3)$

$x < \frac{3}{2}$

23. $x - 5 \geq 7$

$x \geq 12$

25. $2x + 7 < 3 + 4x$

$-2x < -4$

$x > 2$

27. $2x - 1 \geq 1 - 5x$

$7x \geq 2$

$x \geq \frac{2}{7}$

29. $4 - 2x < 3(3 - x)$

$4 - 2x < 9 - 3x$

$x < 5$

31. $\frac{3}{4}x - 6 \leq x - 7$

$-\frac{1}{4}x \leq -1$

$x \geq 4$

33. $\frac{1}{2}(8x + 1) \geq 3x + \frac{5}{2}$

$4x + \frac{1}{2} \geq 3x + \frac{5}{2}$

$x \geq 2$

35. $3.6x + 11 \geq -3.4$

$3.6x \geq -14.4$

$x \geq -4$

37. $1 < 2x + 3 < 9$

$-2 < 2x < 6$

$-1 < x < 3$

39. $-4 < \dfrac{2x - 3}{3} < 4$

$-12 < 2x - 3 < 12$

$-9 < 2x < 15$

$-\frac{9}{2} < x < \frac{15}{2}$

41. $\dfrac{3}{4} > x + 1 > \dfrac{1}{4}$

$-\frac{1}{4} > x > -\frac{3}{4}$

$-\frac{3}{4} < x < -\frac{1}{4}$

43. $3.2 \leq 0.4x - 1 \leq 4.4$

$4.2 \leq 0.4x \leq 5.4$

$10.5 \leq x \leq 13.5$

45. $|x| < 6$

$-6 < x < 6$

47. $\left|\dfrac{x}{2}\right| > 1$

$\dfrac{x}{2} < -1$ or $\dfrac{x}{2} > 1$

$x < -2 \qquad x > 2$

49. $|x - 5| < -1$

No solution. The absolute value of a number cannot be less than a negative number.

51. $|x - 20| \leq 6$

$-6 \leq x - 20 \leq 6$

$14 \leq x \leq 26$

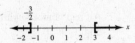

53. $|3 - 4x| \geq 9$

$3 - 4x \leq -9$ or $3 - 4x \geq 9$

$-4x \leq -12$ $\qquad$ $-4x \geq 6$

$x \geq 3$ $\qquad\qquad$ $x \leq -\dfrac{3}{2}$

55. $\left|\dfrac{x - 3}{2}\right| \geq 4$

$\dfrac{x - 3}{2} \leq -4$ or $\dfrac{x - 3}{2} \geq 4$

$x - 3 \leq -8$ $\qquad$ $x - 3 \geq 8$

$x \leq -5$ $\qquad\qquad$ $x \geq 11$

57. $|9 - 2x| - 2 < -1$

$|9 - 2x| < 1$

$-1 < 9 - 2x < 1$

$-10 < -2x < -8$

$5 > x > 4$

$4 < x < 5$

59. $2|x + 10| \geq 9$

$|x + 10| \geq \dfrac{9}{2}$

$x + 10 \leq -\dfrac{9}{2}$ or $x + 10 \geq \dfrac{9}{2}$

$x \leq -\dfrac{29}{2}$ $\qquad\qquad$ $x \geq -\dfrac{11}{2}$

61. $6x > 12$

$x > 2$

63. $5 - 2x \geq 1$

$-2x \geq -4$

$x \leq 2$

65. $|x - 8| \leq 14$

$-14 \leq x - 8 \leq 14$

$-6 \leq x \leq 22$

67. $2|x + 7| \geq 13$

$|x + 7| \geq \dfrac{13}{2}$

$x + 7 \leq -\dfrac{13}{2}$ or $x + 7 \geq \dfrac{13}{2}$

$x \leq -\dfrac{27}{2}$ $\qquad\qquad$ $x \geq -\dfrac{1}{2}$

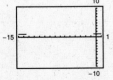

69. $y = 2x - 3$

(a) $\qquad$ $y \geq 1$

$\qquad$ $2x - 3 \geq 1$

$\qquad$ $2x \geq 4$

$\qquad$ $x \geq 2$

(b) $\qquad$ $y \leq 0$

$\qquad$ $2x - 3 \leq 0$

$\qquad$ $2x \leq 3$

$\qquad$ $x \leq \dfrac{3}{2}$

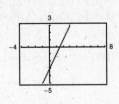

71. $y = -\frac{1}{2}x + 2$

   (a) $0 \le y \le 3$

         $0 \le -\frac{1}{2}x + 2 \le 3$

         $-2 \le -\frac{1}{2}x \le 1$

         $4 \ge x \ge -2$

   (b) $y \ge 0$

         $-\frac{1}{2}x + 2 \ge 0$

         $-\frac{1}{2}x \ge -2$

         $x \le 4$

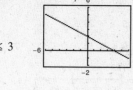

73. $y = |x - 3|$

   (a) $y \le 2$

         $|x - 3| \le 2$

         $-2 \le x - 3 \le 2$

         $1 \le x \le 5$

   (b) $y \ge 4$

         $|x - 3| \ge 4$

         $x - 3 \le -4$ or $x - 3 \ge 4$

         $x \le -1$ or $x \ge 7$

75. $x - 5 \ge 0$

     $x \ge 5$

     $[5, \infty)$

77. $x + 3 \ge 0$

     $x \ge -3$

     $[-3, \infty)$

79. $7 - 2x \ge 0$

     $-2x \ge -7$

     $x \le \frac{7}{2}$

     $\left(-\infty, \frac{7}{2}\right]$

81. $|x - 10| < 8$

     All real numbers within 8 units of 10.

83. The midpoint of the interval $[-3, 3]$ is 0. The interval represents all real numbers x no more than 3 units from 0.

     $|x - 0| \le 3$

       $|x| \le 3$

85. The graph shows all real numbers at least 3 units from 7.

     $|x - 7| \ge 3$

87. All real numbers within 10 units of 12

     $|x - 12| < 10$

89. All real numbers more than 4 units from -3

     $|x - (-3)| > 4$

         $|x + 3| > 4$

91. Let $x =$ the number of checks written in a month.

     Type A account charges: $6.00 + 0.25x$

     Type B account charges: $4.50 + 0.50x$

     $6.00 + 0.25x < 4.50 + 0.50x$

         $1.50 < 0.25x$

           $6 < x$

If you write more than six checks a month, then the charges for the type A account are less than the charges for the type B account.

93. $1000(1 + r(2)) > 1062.50$

       $1 + 2r > 1.0625$

         $2r > 0.0625$

           $r > 0.03125$

           $r > 3.125\%$

95. $R > C$

     $115.95x > 95x + 750$

     $20.95x > 750$

         $x > 35.7995$

         $x \ge 36$ units

97. Let x = daily sales level (in dozens) of doughnuts.

Revenue: $R = 2.95x$

Cost: $C = 150 + 1.45x$

Profit: $P = R - C$

$\qquad\qquad = 2.95x - (150 + 1.45x)$

$\qquad\qquad = 1.50x - 150$

$\qquad\quad 50 \le P \le 200$

$\qquad\quad 50 \le 1.50x - 150 \le 200$

$\qquad\quad 200 \le 1.50x \le 350$

$\qquad\quad 133\frac{1}{3} \le x \le 233\frac{1}{3}$

In whole dozens, $134 \le x \le 234$.

99. (a) $y = 0.067x - 5.638$

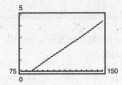

(b) From the graph we see that $y \ge 3$ when $x \ge 129$. Algebraically we have:

$\qquad 3 \le 0.067x - 5.638$

$\qquad 8.638 \le 0.067x$

$\qquad\quad x \ge 129$

IQ scores are not a good predictor of GPAs. Other factors include study habits, class attendance, and attitude.

101. $S = 1.05t + 31.0,\ 0 \le t \le 12$

(a) $32 \le 1.05t + 31 \le 42$

$\qquad 1 \le 1.05t \le 11$

$\qquad 0.95 \le t \le 10.48$

Rounding to the nearest year, $1 \le t \le 10$. The average salary was at least \$32,000 but not more than \$42,000 between 1991 and 2000.

(b) $1.05t + 31 > 48$

$\qquad 1.05t > 17$

$\qquad\quad t > 16$

According to the model, the average salary will exceed \$48,000 in 2006.

103. $|s - 10.4| \le \frac{1}{16}$

$\qquad -\frac{1}{16} \le s - 10.4 \le \frac{1}{16}$

$\qquad -0.0625 \le s - 10.4 \le 0.0625$

$\qquad 10.3375 \le s \le 10.4625$

Since $A = s^2$, we have

$\qquad (10.3375)^2 \le \text{area} \le (10.4625)^2$

$\qquad 106.864 \le \text{area} \le 109.464.$

105. $|x - 15| \le \dfrac{1}{10}$

$\qquad -\dfrac{1}{10} \le x - 15 \le \dfrac{1}{10}$

$\qquad 14.9 \le x \le 15.1$ gallons

$\dfrac{1}{10}(\$1.89) \approx \$0.19$

You might have been undercharged or overcharged by \$0.19.

107. $\left|\dfrac{t - 15.6}{1.9}\right| < 1$

$\qquad -1 < \dfrac{t - 15.6}{1.9} < 1$

$\qquad -1.9 < t - 15.6 < 1.9$

$\qquad\quad 13.7 < t < 17.5$

Two-thirds of the workers could perform the task in the time interval between 13.7 minutes and 17.5 minutes.

109. $|h - 50| \le 30$

$\qquad -30 \le h - 50 \le 30$

$\qquad\quad 20 \le h \le 80$

The minimum relative humidity is 20 and the maximum is 80.

111. False. If c is negative, then $ac \ge bc$.

113. $|x - a| \ge 2$ Matches (b).

$\qquad x - a \le -2$

$\qquad\quad x \le a - 2$ or

$\qquad x - a \ge 2$

$\qquad\quad x \ge a + 2$

Appendix A.7   Errors and the Algebra of Calculus

■ You should be able to recognize and avoid the common algebraic errors involving parentheses, fractions, exponents, radicals, and cancellation.

■ You should be able to "unsimplify" algebraic expressions by the following methods.

(a) Unusual Factoring   (b) Rewriting with Negative Exponents

(c) Writing a Fraction as a Sum of Terms   (d) Inserting Factors or Terms

---

Vocabulary Check

1. numerator   **2.** reciprocal

---

1. $2x - (3y + 4) \neq 2x - 3y + 4$

Change all signs when distributing the minus sign.

$2x - (3y + 4) = 2x - 3y - 4$

3. $\dfrac{4}{16x - (2x + 1)} \neq \dfrac{4}{14x + 1}$

Change all signs when distributing the minus sign.

$\dfrac{4}{16x - (2x + 1)} = \dfrac{4}{16x - 2x - 1} = \dfrac{4}{14x - 1}$

5. $(5z)(6z) \neq 30z$

z occurs twice as a factor.

$(5z)(6z) = 30z^2$

7. $a\left(\dfrac{x}{y}\right) \neq \dfrac{ax}{ay}$

The fraction as a whole is multiplied by a, not the numerator and denominator separately.

$a\left(\dfrac{x}{y}\right) = \dfrac{a}{1} \cdot \dfrac{x}{y} = \dfrac{ax}{y}$

9. $\sqrt{x + 9} \neq \sqrt{x} + 3$

Do not apply the radical to the terms.

$\sqrt{x + 9}$ does not simplify.

11. $\dfrac{2x^2 + 1}{5x} \neq \dfrac{2x + 1}{5}$

Divide out common factors not common terms.

$\dfrac{2x^2 + 1}{5x}$ cannot be simplified.

13. $\dfrac{1}{a^{-1} + b^{-1}} \neq \left(\dfrac{1}{a + b}\right)^{-1}$

To get rid of negative exponents:

$\dfrac{1}{a^{-1} + b^{-1}} = \dfrac{1}{a^{-1} + b^{-1}} \cdot \dfrac{ab}{ab} = \dfrac{ab}{b + a}$

15. $(x^2 + 5x)^{1/2} \neq x(x + 5)^{1/2}$

Factor within grouping symbols before applying the exponent to each factor.

$(x^2 + 5x)^{1/2} = [x(x + 5)]^{1/2} = x^{1/2}(x + 5)^{1/2}$

17. $\dfrac{3}{x} + \dfrac{4}{y} = \dfrac{3}{x} \cdot \dfrac{y}{y} + \dfrac{4}{y} \cdot \dfrac{x}{x} = \dfrac{3y + 4x}{xy}$

To add fractions, they must have a common denominator.

19. $\dfrac{3x + 2}{5} = \dfrac{1}{5}(3x + 2)$

The required factor is $3x + 2$.

21. $\frac{2}{3}x^2 + \frac{1}{3}x + 5 = \frac{2}{3}x^2 + \frac{1}{3}x + \frac{15}{3} = \frac{1}{3}(2x^2 + x + 15)$

The required factor is $2x^2 + x + 15$.

23. $x^2(x^3 - 1)^4 = \frac{1}{3}(x^3 - 1)^4(3x^2)$

The required factor is $\frac{1}{3}$.

25. $\dfrac{4x + 6}{(x^2 + 3x + 7)^3} = \dfrac{2(2x + 3)}{(x^2 + 3x + 7)^3} = \dfrac{2}{1} \cdot \dfrac{(2x + 3)}{1} \cdot \dfrac{1}{(x^2 + 3x + 7)^3} = (2)\dfrac{1}{(x^2 + 3x + 7)^3}(2x + 3)$

The required factor is 2.

27. $\dfrac{3}{x} + \dfrac{5}{2x^2} - \dfrac{3}{2}x = \dfrac{6x}{2x^2} + \dfrac{5}{2x^2} - \dfrac{3x^3}{2x^2}$

$\qquad = \left(\dfrac{1}{2x^2}\right)(6x + 5 - 3x^3)$

The required factor is $\dfrac{1}{2x^2}$.

29. $\dfrac{9x^2}{25} + \dfrac{16y^2}{49} = \dfrac{9}{25} \cdot \dfrac{x^2}{1} + \dfrac{16}{49} \cdot \dfrac{y^2}{1}$

$\qquad = \dfrac{1}{25/9} \cdot \dfrac{x^2}{1} + \dfrac{1}{49/16} \cdot \dfrac{y^2}{1}$

$\qquad = \dfrac{x^2}{(25/9)} + \dfrac{y^2}{(49/16)}$

The required factors are $\frac{25}{9}$ and $\frac{49}{16}$.

31. $\dfrac{x^2}{1/12} - \dfrac{y^2}{2/3} = x^2\left(\dfrac{12}{1}\right) - y^2\left(\dfrac{3}{2}\right) = \dfrac{12x^2}{1} - \dfrac{3y^2}{2}$

The required factors are 1 and 2.

33. $x^{1/3} - 5x^{4/3} = x^{1/3}(1 - 5x^{3/3}) = x^{1/3}(1 - 5x)$

The required factor is $1 - 5x$.

35. $(1 - 3x)^{4/3} - 4x(1 - 3x)^{1/3} = (1 - 3x)^{1/3}[(1 - 3x)^1 - 4x]$

$\qquad = (1 - 3x)^{1/3}(1 - 7x)$

The required factor is $1 - 7x$.

37. $\frac{1}{10}(2x + 1)^{5/2} - \frac{1}{6}(2x + 1)^{3/2} = \frac{3}{30}(2x + 1)^{3/2}(2x + 1)^1 - \frac{5}{30}(2x + 1)^{3/2}$

$\qquad = \frac{1}{30}(2x + 1)^{3/2}[3(2x + 1) - 5]$

$\qquad = \frac{1}{30}(2x + 1)^{3/2}(6x - 2)$

$\qquad = \frac{1}{30}(2x + 1)^{3/2}2(3x - 1)$

$\qquad = \frac{1}{15}(2x + 1)^{3/2}(3x - 1)$

The required factor is $3x - 1$.

39. $\dfrac{3x^2}{(2x - 1)^3} = 3x^2(2x - 1)^{-3}$

41. $\dfrac{4}{3x} + \dfrac{4}{x^4} - \dfrac{7x}{\sqrt[3]{2x}} = 4(3x)^{-1} + 4x^{-4} - 7x(2x)^{-1/3}$

43. $\dfrac{16 - 5x - x^2}{x} = \dfrac{16}{x} - \dfrac{5x}{x} - \dfrac{x^2}{x} = \dfrac{16}{x} - 5 - x$

45. $\dfrac{4x^3 - 7x^2 + 1}{x^{1/3}} = \dfrac{4x^3}{x^{1/3}} - \dfrac{7x^2}{x^{1/3}} + \dfrac{1}{x^{1/3}}$

$\qquad = 4x^{3-1/3} - 7x^{2-1/3} + \dfrac{1}{x^{1/3}}$

$\qquad = 4x^{8/3} - 7x^{5/3} + \dfrac{1}{x^{1/3}}$

47. $\dfrac{3 - 5x^2 - x^4}{\sqrt{x}} = \dfrac{3}{\sqrt{x}} - \dfrac{5x^2}{\sqrt{x}} - \dfrac{x^4}{\sqrt{x}}$

$\qquad = \dfrac{3}{\sqrt{x}} - 5x^{2-1/2} - x^{4-1/2}$

$\qquad = \dfrac{3}{x^{1/2}} - 5x^{3/2} - x^{7/2}$

49. $\dfrac{-2(x^2 - 3)^{-3}(2x)(x + 1)^3 - 3(x + 1)^2(x^2 - 3)^{-2}}{[(x + 1)^3]^2} = \dfrac{(x^2 - 3)^{-3}(x + 1)^2[-4x(x + 1) - 3(x^2 - 3)]}{(x + 1)^6}$

$\qquad = \dfrac{-4x^2 - 4x - 3x^2 + 9}{(x^2 - 3)^3(x + 1)^4}$

$\qquad = \dfrac{-7x^2 - 4x + 9}{(x^2 - 3)^3(x + 1)^4}$

51. $\dfrac{(6x + 1)^3(27x^2 + 2) - (9x^3 + 2x)(3)(6x + 1)^2(6)}{[(6x + 1)^3]^2} = \dfrac{(6x + 1)^2[(6x + 1)(27x^2 + 2) - 18(9x^3 + 2x)]}{(6x + 1)^6}$

$$= \dfrac{162x^3 + 12x + 27x^2 + 2 - 162x^3 - 36x}{(6x + 1)^4}$$

$$= \dfrac{27x^2 - 24x + 2}{(6x + 1)^4}$$

53. $\dfrac{(x + 2)^{3/4}(x + 3)^{-2/3} - (x + 3)^{1/3}(x + 2)^{-1/4}}{[(x + 2)^{3/4}]^2} = \dfrac{(x + 2)^{-1/4}(x + 3)^{-2/3}[(x + 2) - (x + 3)]}{(x + 2)^{6/4}}$

$$= \dfrac{x + 2 - x - 3}{(x + 2)^{1/4}(x + 3)^{2/3}(x + 2)^{6/4}}$$

$$= -\dfrac{1}{(x + 3)^{2/3}(x + 2)^{7/4}}$$

55. $\dfrac{2(3x - 1)^{1/3} - (2x + 1)(1/3)(3x - 1)^{-2/3}(3)}{(3x - 1)^{2/3}} = \dfrac{(3x - 1)^{-2/3}[2(3x - 1) - (2x + 1)]}{(3x - 1)^{2/3}}$

$$= \dfrac{6x - 2 - 2x - 1}{(3x - 1)^{2/3}(3x - 1)^{2/3}}$$

$$= \dfrac{4x - 3}{(3x - 1)^{4/3}}$$

57. $\dfrac{1}{(x^2 + 4)^{1/2}} \cdot \dfrac{1}{2}(x^2 + 4)^{-1/2}(2x) = \dfrac{1}{(x^2 + 4)^{1/2}} \cdot \dfrac{1}{(x^2 + 4)^{1/2}} \cdot \dfrac{1}{2}(2x)$

$$= \dfrac{1}{(x^2 + 4)^1}(x)$$

$$= \dfrac{x}{x^2 + 4}$$

59. $(x^2 + 5)^{1/2}\left(\dfrac{3}{2}\right)(3x - 2)^{1/2}(3) + (3x - 2)^{3/2}\left(\dfrac{1}{2}\right)(x^2 + 5)^{-1/2}(2x) = \dfrac{9}{2}(x^2 + 5)^{1/2}(3x - 2)^{1/2} + x(x^2 + 5)^{-1/2}(3x - 2)^{3/2}$

$$= \dfrac{9}{2}(x^2 + 5)^{1/2}(3x - 2)^{1/2} + \dfrac{2}{2}x(x^2 + 5)^{-1/2}(3x - 2)^{3/2}$$

$$= \dfrac{1}{2}(x^2 + 5)^{-1/2}(3x - 2)^{1/2}[9(x^2 + 5)^1 + 2x(3x - 2)^1]$$

$$= \dfrac{1}{2}(x^2 + 5)^{-1/2}(3x - 2)^{1/2}(9x^2 + 45 + 6x^2 - 4x)$$

$$= \dfrac{(3x - 2)^{1/2}(15x^2 - 4x + 45)}{2(x^2 + 5)^{1/2}}$$

61. $t = \dfrac{\sqrt{x^2 + 4}}{2} + \dfrac{\sqrt{(4 - x)^2 + 4}}{6}$

(a)

x	t
0.5	1.70
1.0	1.72
1.5	1.78
2.0	1.89
2.5	2.02
3.0	2.18
3.5	2.36
4.0	2.57

(b) She should swim to a point about $\frac{1}{2}$ mile down the coast to minimize the time required to reach the finish line.

(c) $\dfrac{1}{2}x(x^2 + 4)^{-1/2} + \dfrac{1}{6}(x - 4)(x^2 - 8x + 20)^{-1/2} = \dfrac{3}{6}x(x^2 + 4)^{-1/2} + \dfrac{1}{6}(x - 4)(x^2 - 8x + 20)^{-1/2}$

$$= \dfrac{1}{6}\left[3x(x^2 + 4)^{-1/2} + (x - 4)(x^2 - 8x + 20)^{-1/2}\right]$$

$$= \dfrac{1}{6}\left[\dfrac{3x}{(x^2 + 4)^{1/2}} + \dfrac{x - 4}{(x^2 - 8x + 20)^{1/2}}\right]$$

$$= \dfrac{3x\sqrt{x^2 - 8x + 20} + (x - 4)\sqrt{x^2 + 4}}{6\sqrt{x^2 + 4}\,\sqrt{x^2 - 8x + 20}}$$

63. True.

$$x^{-1} + y^{-2} = \dfrac{1}{x} + \dfrac{1}{y^2} = \dfrac{y^2 + x}{xy^2}$$

65. True.

$$\dfrac{1}{\sqrt{x} + 4} = \dfrac{1}{\sqrt{x} + 4} \cdot \dfrac{\sqrt{x} - 4}{\sqrt{x} - 4} = \dfrac{\sqrt{x} - 4}{x - 16}$$

67. $x^n \cdot x^{3n} \neq x^{3n^2}$

Add exponents when multiplying powers with like bases.

$x^n \cdot x^{3n} = x^{4n}$

69. $x^{2n} + y^{2n} \neq (x^n + y^n)^2$

When squaring binomials, there is also a middle term.

$(x^n + y^n)^2 = x^{2n} + 2x^n y^n + y^{2n}$

71. The two answers are equivalent and can be obtained by factoring.

$\dfrac{1}{10}(2x - 1)^{5/2} + \dfrac{1}{6}(2x - 1)^{3/2} = \dfrac{1}{60}(2x - 1)^{3/2}[6(2x - 1) + 10]$

$$= \dfrac{1}{60}(2x - 1)^{3/2}(12x + 4)$$

$$= \dfrac{4}{60}(2x - 1)^{3/2}(3x + 1)$$

$$= \dfrac{1}{15}(2x - 1)^{3/2}(3x + 1)$$

(a) $\dfrac{2}{3}x(2x - 3)^{3/2} - \dfrac{2}{15}(2x - 3)^{5/2} = \dfrac{2}{15}(2x - 3)^{3/2}[5x - (2x - 3)]$

$$= \dfrac{2}{15}(2x - 3)^{3/2}(3x + 3)$$

$$= \dfrac{2}{15}(2x - 3)^{3/2}3(x + 1)$$

$$= \dfrac{2}{5}(2x - 3)^{3/2}(x + 1)$$

(b) $\dfrac{2}{3}x(4 + x)^{3/2} - \dfrac{2}{15}(4 + x)^{5/2} = \dfrac{2}{15}(4 + x)^{3/2}[5x - (4 + x)]$

$$= \dfrac{2}{15}(4 + x)^{3/2}(4x - 4) = \dfrac{2}{15}(4 + x)^{3/2}4(x - 1) = \dfrac{8}{15}(4 + x)^{3/2}(x - 1)$$

Chapter 1   Practice Test Solutions

1. (a) Midpoint: $\left(\dfrac{-3+5}{2}, \dfrac{4+(-6)}{2}\right) = (1, -1)$

   (b) Distance: $d = \sqrt{[5-(-3)]^2 + (-6-4)^2}$

$$= \sqrt{(8)^2 + (-10)^2}$$

$$= \sqrt{164} = 2\sqrt{41}$$

2. $y = \sqrt{7-x}$

Domain: $x \le 7$

x	7	6	3	-2
y	0	1	2	3

3. $[x-(-3)]^2 + (y-5)^2 = 6^2$

   $(x+3)^2 + (y-5)^2 = 36$

4. $m = \dfrac{-1-4}{3-2} = -5$

$$y - 4 = -5(x-2)$$

$$y - 4 = -5x + 10$$

$$y = -5x + 14$$

5. $y = \dfrac{4}{3}x - 3$

6. $2x + 3y = 0$

$$y = -\frac{2}{3}x$$

$$m_1 = -\frac{2}{3}$$

$$\perp m_2 = \frac{3}{2} \text{ through } (4, 1)$$

$$y - 1 = \frac{3}{2}(x-4)$$

$$y - 1 = \frac{3}{2}x - 6$$

$$y = \frac{3}{2}x - 5$$

7. $(5, 32)$ and $(9, 44)$

$$m = \frac{44-32}{9-5} = \frac{12}{4} = 3$$

$$y - 32 = 3(x - 5)$$

$$y - 32 = 3x - 15$$

$$y = 3x + 17$$

When $x = 20$, $y = 3(20) + 17$

$$y = \$77.$$

8. $f(x-3) = (x-3)^2 - 2(x-3) + 1$

$$= x^2 - 6x + 9 - 2x + 6 + 1$$

$$= x^2 - 8x + 16$$

9. $f(3) = 12 - 11 = 1$

$$\frac{f(x) - f(3)}{x - 3} = \frac{(4x - 11) - 1}{x - 3}$$

$$= \frac{4x - 12}{x - 3}$$

$$= \frac{4(x-3)}{x-3} = 4, \ x \ne 3$$

10. $f(x) = \sqrt{36 - x^2} = \sqrt{(6+x)(6-x)}$

Domain: $[-6, 6]$, because $(6+x)(6-x) \ge 0$ on this interval.

Range: $[0, 6]$, because $0 \le (6+x)(6-x) \le 36$ on this interval.

11. (a) $6x - 5y + 4 = 0$

$$y = \frac{6x + 4}{5} \text{ is a function of } x.$$

(b) $x^2 + y^2 = 9$

$\quad$ $y = \pm\sqrt{9 - x^2}$ is not a function of x.

(c) $y^3 = x^2 + 6$

$\quad$ $y = \sqrt[3]{x^2 + 6}$ is a function of x.

12. Parabola

Vertex: $(0, -5)$

Intercepts: $(0, -5)$, $(\pm\sqrt{5}, 0)$

y-axis symmetry

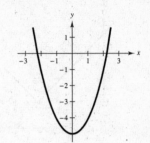

13. Intercepts: $(0, 3)$, $(-3, 0)$

x	0	1	-1	2	-2	-3	-4
y	3	4	2	5	1	0	1

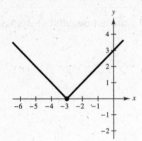

14.

x	0	1	2	3	-1	-2	-3
y	1	3	5	7	2	6	12

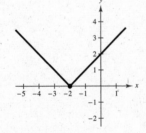

15. (a) $f(x + 2)$

Horizontal shift
two units to the left

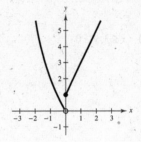

(b) $-f(x) + 2$

Reflection in the x-axis
and a vertical shift two
units upward

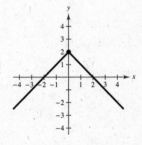

16. (a) $(g - f)(x) = g(x) - f(x)$

$\qquad = (2x^2 - 5) - (3x + 7)$

$\qquad = 2x^2 - 3x - 12$

(b) $(fg)(x) = f(x)g(x)$

$\qquad = (3x + 7)(2x^2 - 5)$

$\qquad = 6x^3 + 14x^2 - 15x - 35$

17. $f(g(x)) = f(2x + 3)$

$\qquad = (2x + 3)^2 - 2(2x + 3) + 16$

$\qquad = 4x^2 + 12x + 9 - 4x - 6 + 16$

$\qquad = 4x^2 + 8x + 19$

18. $\quad f(x) = x^3 + 7$

$\qquad y = x^3 + 7$

$\qquad x = y^3 + 7$

$\qquad x - 7 = y^3$

$\qquad \sqrt[3]{x - 7} = y$

$\qquad f^{-1}(x) = \sqrt[3]{x - 7}$

19. (a) $f(x) = |x - 6|$ does not have an inverse.

Its graph does not pass the horizontal line test.

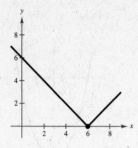

(b) $f(x) = ax + b, a \neq 0$ does have an inverse.

$$y = ax + b$$

$$x = ay + b$$

$$\frac{x - b}{a} = y$$

$$f^{-1}(x) = \frac{x - b}{a}$$

(c) $f(x) = x^3 - 19$ does have an inverse.

$$y = x^3 - 19$$

$$x = y^3 - 19$$

$$x + 19 = y^3$$

$$\sqrt[3]{x + 19} = y$$

$$f^{-1}(x) = \sqrt[3]{x + 19}$$

20.
$$f(x) = \sqrt{\frac{3 - x}{x}}, \; 0 < x \le 3, y \ge 0$$

$$y = \sqrt{\frac{3 - x}{x}}$$

$$x = \sqrt{\frac{3 - y}{y}}$$

$$x^2 = \frac{3 - y}{\bullet y}$$

$$x^2 y = 3 - y$$

$$x^2 y + y = 3$$

$$y(x^2 + 1) = 3$$

$$y = \frac{3}{x^2 + 1}$$

$$f^{-1}(x) = \frac{3}{x^2 + 1}, \; x \ge 0$$

21. False. The slopes of 3 and $\frac{1}{3}$ are not **negative** reciprocals.

22. True. Let $y = (f \circ g)(x)$. Then $x = (f \circ g)^{-1}(y)$.

Also,

$$(f \circ g)(x) = y$$

$$f(g(x)) = y$$

$$g(x) = f^{-1}(y)$$

$$x = g^{-1}(f^{-1}(y))$$

$$x = (g^{-1} \circ f^{-1})(y)$$

Since $x = x$, we have $(f \circ g)^{-1}(y) = (g^{-1} \circ f^{-1})(y)$.

23. True. It must pass the vertical line test to be a function and it must pass the horizontal line test to have an inverse.

24. $z = \dfrac{cx^3}{\sqrt{y}}$

$-1 = \dfrac{c(-1)^3}{\sqrt{25}}$

$-1 = \dfrac{-c}{5}$

$5 = c$

$z = \dfrac{5x^3}{\sqrt{y}}$

25. $y \approx 0.669x + 2.669$

Chapter 2   Practice Test Solutions

1. x-intercepts: $(1, 0), (5, 0)$

y-intercepts: $(0, 5)$

Vertex: $(3, -4)$

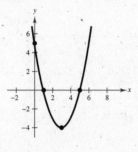

2. $a = 0.01, b = -90$

$\dfrac{-b}{2a} = \dfrac{90}{2(.01)} = 4500 \text{ units}$

3. Vertex: $(1, 7)$ opening downward through $(2, 5)$

$y = a(x - 1)^2 + 7$ Standard form

$5 = a(2 - 1)^2 + 7$

$5 = a + 7$

$a = -2$

$y = -2(x - 1)^2 + 7$

$\quad = -2(x^2 - 2x + 1) + 7$

$\quad = -2x^2 + 4x + 5$

4. $y = \pm a(x - 2)(3x - 4)$ where a is any real number

$y = \pm(3x^2 - 10x + 8)$

5. Leading coefficient: -3

Degree: 5

Moves down to the right and up to the left

6. $0 = x^5 - 5x^3 + 4x$

$\quad = x(x^4 - 5x^2 + 4)$

$\quad = x(x^2 - 1)(x^2 - 4)$

$\quad = x(x + 1)(x - 1)(x + 2)(x - 2)$

$x = 0, x = \pm 1, x = \pm 2$

7. $f(x) = x(x - 3)(x + 2)$

$\quad = x(x^2 - x - 6)$

$\quad = x^3 - x^2 - 6x$

8. Intercepts: $(0, 0), \left(\pm 2\sqrt{3}, 0\right)$

Moves up to the right

Moves down to the left

Origin symmetry

x	-2	-1	0	1	2
y	16	11	0	-11	-16

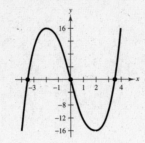

9.
$$
\begin{array}{r}
3x^3 + 9x^2 + 20x + 62 + \dfrac{176}{x-3} \\
x-3\overline{\smash{\big)}\,3x^4 + 0x^3 - 7x^2 + 2x - 10} \\
\underline{3x^4 - 9x^3} \\
9x^3 - 7x^2 \\
\underline{9x^3 - 27x^2} \\
20x^2 + 2x \\
\underline{20x^2 - 60x} \\
62x - 10 \\
\underline{62x - 186} \\
176
\end{array}
$$

10.
$$
\begin{array}{r}
x - 2 + \dfrac{5x - 13}{x^2 + 2x - 1} \\
x^2 + 2x - 1\overline{\smash{\big)}\,x^3 + 0x^2 + 0x - 11} \\
\underline{x^3 + 2x^2 - x} \\
-2x^2 + x - 11 \\
\underline{-2x^2 - 4x + 2} \\
5x - 13
\end{array}
$$

11.

$$
\begin{array}{r|rrrrrr}
-5 & 3 & 13 & 0 & 0 & 12 & -1 \\
   &   & -15 & 10 & -50 & 250 & -1310 \\
\hline
   & 3 & -2 & 10 & -50 & 262 & -1311
\end{array}
$$

$$
\frac{3x^5 + 13x^4 + 12x - 1}{x + 5} = 3x^4 - 2x^3 + 10x^2 - 50x + 262 - \frac{1311}{x + 5}
$$

12.

$$
\begin{array}{r|rrrr}
-6 & 7 & 40 & -12 & 15 \\
   &   & -42 & 12 & 0 \\
\hline
   & 7 & -2 & 0 & 15
\end{array}
$$

$f(-6) = 15$

13. $0 = x^3 - 19x - 30$

Possible rational roots: $\pm 1, \pm 2, \pm 3, \pm 5, \pm 6, \pm 10, \pm 15, \pm 30$

$$
\begin{array}{r|rrrr}
-2 & 1 & 0 & -19 & -30 \\
   &   & -2 & 4 & 30 \\
\hline
   & 1 & -2 & -15 & 0
\end{array}
\qquad x = -2 \text{ is a zero.}
$$

$0 = (x + 2)(x^2 - 2x - 15)$

$0 = (x + 2)(x + 3)(x - 5)$

Zeros: $x = -2, x = -3, x = 5$

14. $0 = x^4 + x^3 - 8x^2 - 9x - 9$

Possible rational roots: $\pm 1, \pm 3, \pm 9$

$$
\begin{array}{r|rrrr}
3 & 1 & 1 & -8 & -9 & -9 \\
  &   & 3 & 12 & 12 & 9 \\
\hline
  & 1 & 4 & 4 & 3 & 0
\end{array}
\qquad x = 3 \text{ is a zero.}
$$

$0 = (x - 3)(x^3 + 4x^2 + 4x + 3)$

The zeros of $x^2 + x + 1$ are $x = \dfrac{-1 \pm \sqrt{3}\,i}{2}$ (by the Quadratic Formula).

Zeros: $x = 3, x = -3, x = -\dfrac{1}{2} + \dfrac{\sqrt{3}}{2}i, x = -\dfrac{1}{2} - \dfrac{\sqrt{3}}{2}i$

Possible rational roots of $x^3 + 4x^2 + 4x + 3$: $\pm 1, \pm 3$

$$
\begin{array}{r|rrrr}
-3 & 1 & 4 & 4 & 3 \\
   &   & -3 & -3 & -3 \\
\hline
   & 1 & 1 & 1 & 0
\end{array}
\qquad x = -3 \text{ is a zero.}
$$

$0 = (x - 3)(x + 3)(x^2 + x + 1)$

15. $0 = 6x^3 - 5x^2 + 4x - 15$

Possible rational roots: $\pm 1, \pm 3, \pm 5, \pm 15, \pm \frac{1}{2}, \pm \frac{3}{2}, \pm \frac{5}{2}, \pm \frac{15}{2}, \pm \frac{1}{3}, \pm \frac{5}{3}, \pm \frac{1}{6}, \pm \frac{5}{6}$

16. $0 = x^3 - \frac{20}{3}x^2 + 9x - \frac{10}{3}$

$0 = 3x^3 - 20x^2 + 27x - 10$

Possible rational roots:
$\pm 1, \pm 2, \pm 5, \pm 10, \pm \frac{1}{3}, \pm \frac{2}{3}, \pm \frac{5}{3}, \pm \frac{10}{3}$

$$
\begin{array}{r|rrrr}
1 & 3 & -20 & 27 & -10 \\
  &   & 3 & -17 & 10 \\
\hline
  & 3 & -17 & 10 & 0
\end{array}
$$

$0 = (x - 1)(3x^2 - 17x + 10)$

$0 = (x - 1)(3x - 2)(x - 5)$

Zeros: $x = 1, x = \frac{2}{3}, x = 5$

17. Possible rational roots: $\pm 1, \pm 2, \pm 5, \pm 10$

$$
\begin{array}{r|rrrrr}
1 & 1 & 1 & 3 & 5 & -10 \\
  &   & 1 & 2 & 5 & 10 \\
\hline
  & 1 & 2 & 5 & 10 & 0
\end{array}
\qquad x = 1 \text{ is a zero.}
$$

$$
\begin{array}{r|rrrr}
-2 & 1 & 2 & 5 & 10 \\
   &   & -2 & 0 & -10 \\
\hline
   & 1 & 0 & 5 & 0
\end{array}
\qquad x = -2 \text{ is a zero.}
$$

$f(x) = (x - 1)(x + 2)(x^2 + 5)$
$\quad = (x - 1)(x + 2)(x + \sqrt{5}i)(x - \sqrt{5}i)$

18. $f(x) = (x - 2)[x - (3 + i)][x - (3 - i)]$

$\quad = (x - 2)[(x - 3) - i][(x - 3) + i]$

$\quad = (x - 2)[(x - 3)^2 - i^2]$

$\quad = (x - 2)[x^2 - 6x + 10]$

$\quad = x^3 - 8x^2 + 22x - 20$

19.
$$
\begin{array}{r|rrrr}
3i & 1 & 4 & 9 & 36 \\
   &   & 3i & 12i - 9 & -36 \\
\hline
   & 1 & 4 + 3i & 12i & 0
\end{array}
$$

20. Vertical asymptote: $x = 0$

Horizontal asymptote: $y = \frac{1}{2}$

x-intercept: $(1, 0)$

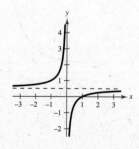

21. $y = 8$ is a horizontal asymptote since the degree on the numerator equals the degree of the denominator. There are no vertical asymptotes.

22. $x = 1$ is a vertical asymptote.

$$\frac{4x^2 - 2x + 7}{x - 1} = 4x + 2 + \frac{9}{x - 1}$$

Thus, $y = 4x + 2$ is a slant asymptote.

23. (a) $(4 - 3i) - (-2 + i) = 4 - 3i + 2 - i = 6 - 4i$

(b) $(4 - 3i)(-2 + i) = -8 + 4i + 6i - 3i^2 = -8 + 10i + 3 = -5 + 10i$

(c) $\dfrac{4 - 3i}{-2 + i} = \dfrac{4 - 3i}{-2 + i} \cdot \dfrac{-2 - i}{-2 - i} = \dfrac{-8 - 4i + 6i + 3i^2}{4 + 1}$

$\qquad = \dfrac{-11 + 2i}{5} = -\dfrac{11}{5} + \dfrac{2}{5}i$

24. $\qquad x^2 - 49 \le 0$

$(x + 7)(x - 7) \le 0$

Critical numbers: $x = -7$ and $x = 7$

Test intervals: $(-\infty, -7), (-7, 7), (7, \infty)$

Test: Is $x^2 - 49 \le 0$?

Solution set: $[-7, 7]$

25. $\dfrac{x + 3}{x - 7} \ge 0$

Critical numbers: $x = -3$ and $x = 7$

Test intervals: $(-\infty, -3), (-3, 7), (7, \infty)$

Test: Is $\dfrac{x + 3}{x - 7} \ge 0$?

Solution set: $(-\infty, -3] \cup [7, \infty)$

Chapter 3   Practice Test Solutions

1. $x^{3/5} = 8$

$x = 8^{5/3} = \left(\sqrt[3]{8}\right)^5 = 2^5 = 32$

2. $3^{x-1} = \frac{1}{81}$

$3^{x-1} = 3^{-4}$

$x - 1 = -4$

$x = -3$

3. $f(x) = 2^{-x} = \left(\frac{1}{2}\right)^x$

x	-2	-1	0	1	2
$f(x)$	4	2	1	$\frac{1}{2}$	$\frac{1}{4}$

4. $g(x) = e^x + 1$

x	-2	-1	0	1	2
$g(x)$	1.14	1.37	2	3.72	8.39

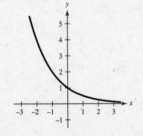

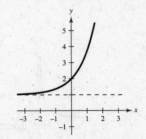

5. (a) $A = P\left(1 + \dfrac{r}{n}\right)^{nt}$

$A = 5000\left(1 + \dfrac{0.09}{12}\right)^{12(3)} \approx \$6543.23$

(b) $A = P\left(1 + \dfrac{r}{n}\right)^{nt}$

$A = 5000\left(1 + \dfrac{0.09}{4}\right)^{4(3)} \approx \$6530.25$

(c) $A = Pe^{rt}$

$A = 5000e^{(0.09)(3)} \approx \$6549.82$

6. $7^{-2} = \dfrac{1}{49}$

$\log_7 \dfrac{1}{49} = -2$

7. $x - 4 = \log_2 \dfrac{1}{64}$

$2^{x-4} = \dfrac{1}{64}$

$2^{x-4} = 2^{-6}$

$x - 4 = -6$

$x = -2$

8. $\log_b \sqrt[4]{\dfrac{8}{25}} = \dfrac{1}{4}\log_b \dfrac{8}{25}$

$= \dfrac{1}{4}[\log_b 8 - \log_b 25]$

$= \dfrac{1}{4}[\log_b 2^3 - \log_b 5^2]$

$= \dfrac{1}{4}[3\log_b 2 - 2\log_b 5]$

$= \dfrac{1}{4}[3(0.3562) - 2(0.8271)]$

$= -0.1464$

9. $5\ln x - \dfrac{1}{2}\ln y + 6\ln z = \ln x^5 - \ln \sqrt{y} + \ln z^6 = \ln\left(\dfrac{x^5 z^6}{\sqrt{y}}\right),\ z > 0$

10. $\log_9 28 = \dfrac{\log 28}{\log 9} \approx 1.5166$

11. $\log N = 0.6646$

$N = 10^{0.6646} \approx 4.62$

12.

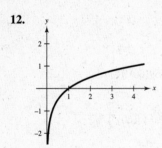

13. Domain:

$x^2 - 9 > 0$

$(x + 3)(x - 3) > 0$

$x < -3 \text{ or } x > 3$

14.

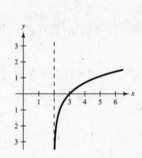

15. False. $\dfrac{\ln x}{\ln y} \neq \ln(x - y)$ since $\dfrac{\ln x}{\ln y} = \log_y x$.

16. $5^3 = 41$

$x = \log_5 41 = \dfrac{\ln 41}{\ln 5} \approx 2.3074$

17. $x - x^2 = \log_5 \frac{1}{25}$

$5^{x-x^2} = \frac{1}{25}$

$5^{x-x^2} = 5^{-2}$

$x - x^2 = -2$

$0 = x^2 - x - 2$

$0 = (x + 1)(x - 2)$

$x = -1$ or $x = 2$

18. $\log_2 x + \log_2(x - 3) = 2$

$\log_2[x(x - 3)] = 2$

$x(x - 3) = 2^2$

$x^2 - 3x = 4$

$x^2 - 3x - 4 = 0$

$(x + 1)(x - 4) = 0$

$x = 4$

$x = -1$ (extraneous)

$x = 4$ is the only solution.

19. $\dfrac{e^x + e^{-x}}{3} = 4$

$e^x(e^x + e^{-x}) = 12e^x$

$e^{2x} + 1 = 12e^x$

$e^{2x} - 12e^x + 1 = 0$

$e^x = \dfrac{12 \pm \sqrt{144 - 4}}{2}$

$e^x \approx 11.9161$ or $e^x \approx 0.0839$

$x = \ln 11.9161$ $x = \ln 0.0839$

$x \approx 2.478$ $x \approx -2.478$

20. $A = Pe^{et}$

$12,000 = 6000e^{0.13t}$

$2 = e^{0.13t}$

$0.13t = \ln 2$

$t = \dfrac{\ln 2}{0.13}$

$t \approx 5.3319$ years or 5 years 4 months

Chapter 4   Practice Test Solutions

1. $350° = 350\left(\dfrac{\pi}{180}\right) = \dfrac{35\pi}{18}$

2. $\dfrac{5\pi}{9} = \dfrac{5\pi}{9} \cdot \dfrac{180}{\pi} = 100°$

3. $135°\,14'\,12'' = \left(135 + \dfrac{14}{60} + \dfrac{12}{3600}\right)^{\circ}$

$\approx 135.2367°$

4. $-22.569° = -(22° + 0.569(60)')$

$= -22°\,34.14'$

$= -(22°\,34' + 0.14(60)'')$

$\approx -22°\,34'\,8''$

5. $\cos \theta = \dfrac{2}{3}$

$x = 2, r = 3, y = \pm\sqrt{9 - 4} = \pm\sqrt{5}$

$\tan \theta = \dfrac{y}{x} = \pm\dfrac{\sqrt{5}}{2}$

6. $\sin \theta = 0.9063$

$\theta = \arcsin(0.9063)$

$\theta = 65° = \dfrac{13\pi}{36}$ or $\theta = 180° - 65° = 115° = \dfrac{23\pi}{36}$

7. $\tan 20° = \dfrac{35}{x}$

$x = \dfrac{35}{\tan 20°} \approx 96.1617$

8. $\theta = \dfrac{6\pi}{5}$, θ is in Quadrant III.

Reference angle: $\dfrac{6\pi}{5} - \pi = \dfrac{\pi}{5}$ or $36°$

9. $\csc 3.92 = \dfrac{1}{\sin 3.92} \approx -1.4242$

10. $\tan \theta = 6 = \dfrac{6}{1}$, θ lies in Quandrant III.

$y = -6, x = -1, r = \sqrt{36 + 1} = \sqrt{37}$,

so $\sec \theta = \dfrac{\sqrt{37}}{-1} \approx -6.0828$.

11. Period: 4π

Amplitude: 3

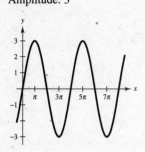

12. Period: 2π

Amplitude: 2

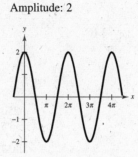

13. Period: $\dfrac{\pi}{2}$

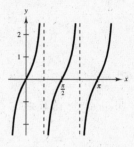

14. Period: 2π

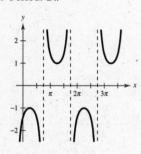

15.

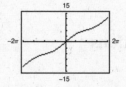

16.

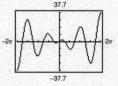

17. $\theta = \arcsin 1$

$\sin \theta = 1$

$\theta = \dfrac{\pi}{2} = 90°$

18. $\theta = \arctan(-3)$

$\tan \theta = -3$

$\theta \approx -1.249 \approx -71.565°$

19. $\sin\left(\arccos \dfrac{4}{\sqrt{35}}\right)$

$\sin \theta = \dfrac{\sqrt{19}}{\sqrt{35}} \approx 0.7368$

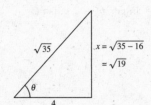

$x = \sqrt{35 - 16}$
$\quad = \sqrt{19}$

20. $\cos\left(\arcsin \dfrac{x}{4}\right)$

$\cos \theta = \dfrac{\sqrt{16 - x^2}}{4}$

21. Given $A = 40°$, $c = 12$

$B = 90° - 40° = 50°$

$\sin 40° = \dfrac{a}{12}$

$a = 12 \sin 40° \approx 7.713$

$\cos 40° = \dfrac{b}{12}$

$b = 12 \cos 40° \approx 9.193$

22. Given $B = 6.84°$, $a = 21.3$

$A = 90° - 6.84° = 83.16°$

$\sin 83.16° = \dfrac{21.3}{c}$

$c = \dfrac{21.3}{\sin 83.16°} \approx 21.453$

$\tan 83.16° = \dfrac{21.3}{b}$

$b = \dfrac{21.3}{\tan 83.16°} \approx 2.555$

23. Given $a = 5$, $b = 9$

$c = \sqrt{25 + 81} = \sqrt{106} \approx 10.296$

$\tan A = \dfrac{5}{9}$

$A = \arctan \dfrac{5}{9} \approx 29.055°$

$B \approx 90° - 29.055° = 60.945°$

24. $\sin 67° = \dfrac{x}{20}$

$x = 20 \sin 67° \approx 18.41$ feet

25. $\tan 5° = \dfrac{250}{x}$

$x = \dfrac{250}{\tan 5°}$

≈ 2857.513 feet

≈ 0.541 mi

Chapter 5   Practice Test Solutions

1. $\tan x = \dfrac{4}{11}$, $\sec x < 0 \implies x$ is in Quadrant III.

$y = -4$, $x = -11$, $r = \sqrt{16 + 121} = \sqrt{137}$

$\sin x = -\dfrac{4}{\sqrt{137}} = -\dfrac{4\sqrt{137}}{137}$ $\csc x = -\dfrac{\sqrt{137}}{4}$

$\cos x = -\dfrac{11}{\sqrt{137}} = -\dfrac{11\sqrt{137}}{137}$ $\sec x = -\dfrac{\sqrt{137}}{11}$

$\tan x = \dfrac{4}{11}$ $\cot x = \dfrac{11}{4}$

2.
$$\dfrac{\sec^2 x + \csc^2 x}{\csc^2 x(1 + \tan^2 x)} = \dfrac{\sec^2 x + \csc^2 x}{\csc^2 x + (\csc^2 x)\tan^2 x}$$

$$= \dfrac{\sec^2 x + \csc^2 x}{\csc^2 x + \dfrac{1}{\sin^2 x} \cdot \dfrac{\sin^2 x}{\cos^2 x}}$$

$$= \dfrac{\sec^2 x + \csc^2 x}{\csc^2 x + \dfrac{1}{\cos^2 x}}$$

$$= \dfrac{\sec^2 x + \csc^2 x}{\csc^2 x + \sec^2 x} = 1$$

3. $\ln|\tan \theta| - \ln|\cot \theta| = \ln\left|\dfrac{\tan \theta}{\cot \theta}\right| = \ln\left|\dfrac{\sin \theta/\cos \theta}{\cos \theta/\sin \theta}\right| = \ln\left|\dfrac{\sin^2 \theta}{\cos^2 \theta}\right| = \ln|\tan^2 \theta| = 2\ln|\tan \theta|$

4. $\cos\left(\dfrac{\pi}{2} - x\right) = \dfrac{1}{\csc x}$ is true since $\cos\left(\dfrac{\pi}{2} - x\right) = \sin x = \dfrac{1}{\csc x}$.

5. $\sin^4 x + (\sin^2 x)\cos^2 x = \sin^2 x(\sin^2 x + \cos^2 x)$

$= \sin^2 x(1) = \sin^2 x$

6. $(\csc x + 1)(\csc x - 1) = \csc^2 x - 1 = \cot^2 x$

7. $\dfrac{\cos^2 x}{1 - \sin x} \cdot \dfrac{1 + \sin x}{1 + \sin x} = \dfrac{\cos^2 x(1 + \sin x)}{1 - \sin^2 x} = \dfrac{\cos^2 x(1 + \sin x)}{\cos^2 x} = 1 + \sin x$

8. $\dfrac{1 + \cos \theta}{\sin \theta} + \dfrac{\sin \theta}{1 + \cos \theta} = \dfrac{(1 + \cos \theta)^2 + \sin^2 \theta}{\sin \theta (1 + \cos \theta)}$

$$= \frac{1 + 2\cos \theta + \cos^2 \theta + \sin^2 \theta}{\sin \theta (1 + \cos \theta)} = \frac{2 + 2\cos \theta}{\sin \theta (1 + \cos \theta)} = \frac{2}{\sin \theta} = 2 \csc \theta$$

9. $\tan^4 x + 2 \tan^2 x + 1 = (\tan^2 x + 1)^2 = (\sec^2 x)^2 = \sec^4 x$

10. (a) $\sin 105° = \sin(60° + 45°) = \sin 60° \cos 45° + \cos 60° \sin 45°$

$$= \frac{\sqrt{3}}{2} \cdot \frac{\sqrt{2}}{2} + \frac{1}{2} \cdot \frac{\sqrt{2}}{2} = \frac{\sqrt{2}}{4}\left(\sqrt{3} + 1\right)$$

(b) $\tan 15° = \tan(60° - 45°) = \dfrac{\tan 60° - \tan 45°}{1 + \tan 60° \tan 45°}$

$$= \frac{\sqrt{3} - 1}{1 + \sqrt{3}} \cdot \frac{1 - \sqrt{3}}{1 - \sqrt{3}} = \frac{2\sqrt{3} - 1 - 3}{1 - 3} = \frac{2\sqrt{3} - 4}{-2} = 2 - \sqrt{3}$$

11. $(\sin 42°) \cos 38° - (\cos 42°) \sin 38° = \sin(42° - 38°) = \sin 4°$

12. $\tan\left(\theta + \dfrac{\pi}{4}\right) = \dfrac{\tan \theta + \tan\left(\dfrac{\pi}{4}\right)}{1 - (\tan \theta) \tan\left(\dfrac{\pi}{4}\right)} = \dfrac{\tan \theta + 1}{1 - \tan \theta(1)} = \dfrac{1 + \tan \theta}{1 - \tan \theta}$

13. $\sin(\arcsin x - \arccos x) = \sin(\arcsin x) \cos(\arccos x) - \cos(\arcsin x) \sin(\arccos x)$

$$= (x)(x) - \left(\sqrt{1 - x^2}\right)\left(\sqrt{1 - x^2}\right) = x^2 - (1 - x^2) = 2x^2 - 1$$

14. (a) $\cos(120°) = \cos[2(60°)] = 2\cos^2 60° - 1 = 2\left(\dfrac{1}{2}\right)^2 - 1 = -\dfrac{1}{2}$

(b) $\tan(300°) = \tan[2(150°)] = \dfrac{2 \tan 150°}{1 - \tan^2 150°} = \dfrac{-\dfrac{2\sqrt{3}}{3}}{1 - \left(\dfrac{1}{3}\right)} = -\sqrt{3}$

15. (a) $\sin 22.5° = \sin \dfrac{45°}{2} = \sqrt{\dfrac{1 - \cos 45°}{2}} = \sqrt{\dfrac{1 - \dfrac{\sqrt{2}}{2}}{2}} = \dfrac{\sqrt{2 - \sqrt{2}}}{2}$

(b) $\tan \dfrac{\pi}{12} = \tan \dfrac{\dfrac{\pi}{6}}{2} = \dfrac{\sin \dfrac{\pi}{6}}{1 + \cos\left(\dfrac{\pi}{6}\right)} = \dfrac{\dfrac{1}{2}}{1 + \dfrac{\sqrt{3}}{2}} = \dfrac{1}{2 + \sqrt{3}} = 2 - \sqrt{3}$

16. $\sin \theta = \dfrac{4}{5}$, θ lies in Quadrant II $\Rightarrow$ $\cos \theta = -\dfrac{3}{5}$.

$$\cos \frac{\theta}{2} = \sqrt{\frac{1 + \cos \theta}{2}} = \sqrt{\frac{1 - \dfrac{3}{5}}{2}} = \sqrt{\frac{2}{10}} = \frac{1}{\sqrt{5}} = \frac{\sqrt{5}}{5}$$

17. $(\sin^2 x) \cos^2 x = \dfrac{1 - \cos 2x}{2} \cdot \dfrac{1 + \cos 2x}{2} = \dfrac{1}{4}[1 - \cos^2 2x] = \dfrac{1}{4}\left[1 - \dfrac{1 + \cos 4x}{2}\right]$

$$= \dfrac{1}{8}[2 - (1 + \cos 4x)] = \dfrac{1}{8}[1 - \cos 4x]$$

18. $6(\sin 5\theta) \cos 2\theta = 6\left\{\dfrac{1}{2}[\sin(5\theta + 2\theta) + \sin(5\theta - 2\theta)]\right\} = 3[\sin 7\theta + \sin 3\theta]$

19. $\sin(x + \pi) + \sin(x - \pi) = 2\left(\sin \dfrac{[(x + \pi) + (x - \pi)]}{2}\right) \cos \dfrac{[(x + \pi) - (x - \pi)]}{2}$

$$= 2 \sin x \cos \pi = -2 \sin x$$

20. $\dfrac{\sin 9x + \sin 5x}{\cos 9x - \cos 5x} = \dfrac{2 \sin 7x \cos 2x}{-2 \sin 7x \sin 2x} = -\dfrac{\cos 2x}{\sin 2x} = -\cot 2x$

21. $\dfrac{1}{2}[\sin(u + v) - \sin(u - v)] = \dfrac{1}{2}\{(\sin u) \cos v + (\cos u) \sin v - [(\sin u) \cos v - (\cos u) \sin v]\}$

$$= \dfrac{1}{2}[2(\cos u) \sin v] = (\cos u) \sin v$$

22. $4 \sin^2 x = 1$

$\sin^2 x = \dfrac{1}{4}$

$\sin x = \pm \dfrac{1}{2}$

$\sin x = \dfrac{1}{2}$ or $\sin x = -\dfrac{1}{2}$

$x = \dfrac{\pi}{6}$ or $\dfrac{5\pi}{6}$ $x = \dfrac{7\pi}{6}$ or $\dfrac{11\pi}{6}$

23. $\tan^2 \theta + \left(\sqrt{3} - 1\right) \tan \theta - \sqrt{3} = 0$

$(\tan\theta - 1)\left(\tan\theta + \sqrt{3}\right) = 0$

$\tan \theta = 1$ or $\tan \theta = -\sqrt{3}$

$\theta = \dfrac{\pi}{4}$ or $\dfrac{5\pi}{4}$ $\theta = \dfrac{2\pi}{3}$ or $\dfrac{5\pi}{3}$

24. $\sin 2x = \cos x$

$2(\sin x) \cos x - \cos x = 0$

$\cos x(2 \sin x - 1) = 0$

$\cos x = 0$ or $\sin x = \dfrac{1}{2}$

$x = \dfrac{\pi}{2}$ or $\dfrac{3\pi}{2}$ $x = \dfrac{\pi}{6}$ or $\dfrac{5\pi}{6}$

25. $\tan^2 x - 6 \tan x + 4 = 0$

$\tan x = \dfrac{-(-6) \pm \sqrt{(-6)^2 - 4(1)(4)}}{2(1)}$

$\tan x = \dfrac{6 \pm \sqrt{20}}{2} = 3 \pm \sqrt{5}$

$\tan x = 3 + \sqrt{5}$ or $\tan x = 3 - \sqrt{5}$

$x \approx 1.3821$ or 4.5237 $x = 0.6524$ or 3.7940

Chapter 6   Practice Test Solutions

1. $C = 180° - (40° + 12°) = 128°$

$a = \sin 40°\left(\dfrac{100}{\sin 12°}\right) \approx 309.164$

$c = \sin 128°\left(\dfrac{100}{\sin 12°}\right) \approx 379.012$

2. $\sin A = 5\left(\dfrac{\sin 150°}{20}\right) = 0.125$

$A \approx 7.181°$

$B \approx 180° - (150° + 7.181°) = 22.819°$

$b = \sin 22.819°\left(\dfrac{20}{\sin 150°}\right) \approx 15.513$

3. Area $= \frac{1}{2}ab \sin C = \frac{1}{2}(3)(6) \sin 130° \approx 6.894$ square units

4. $h = b \sin A = 35 \sin 22.5° \approx 13.394$

$a = 10$

Since $a < h$ and A is acute, the triangle has no solution.

5. $\cos A = \dfrac{(53)^2 + (38)^2 - (49)^2}{2(53)(38)} \approx 0.4598$

$A \approx 62.627°$

$\cos B = \dfrac{(49)^2 + (38)^2 - (53)^2}{2(49)(38)} \approx 0.2782$

$B \approx 73.847°$

$C \approx 180° - (62.627° + 73.847°)$

$= 43.526°$

6. $c^2 = (100)^2 + (300)^2 - 2(100)(300) \cos 29°$

≈ 47522.8176

$c \approx 218$

$\cos A = \dfrac{(300)^2 + (218)^2 - (100)^2}{2(300)(218)} \approx 0.97495$

$A \approx 12.85°$

$B \approx 180° - (12.85° + 29°) = 138.15°$

7. $s = \dfrac{a + b + c}{2} = \dfrac{4.1 + 6.8 + 5.5}{2} = 8.2$

Area $= \sqrt{s(s - a)(s - b)(s - c)}$

$= \sqrt{8.2(8.2 - 4.1)\,(8.2 - 6.8)(8.2 - 5.5)}$

≈ 11.273 square units

8. $x^2 = (40)^2 + (70)^2 - 2(40)(70)\cos 168°$

≈ 11977.6266

$x \approx 190.442$ miles

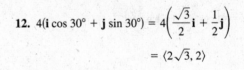

9. $\mathbf{w} = 4(3\mathbf{i} + \mathbf{j}) - 7(-\mathbf{i} + 2\mathbf{j})$

$= 19\mathbf{i} - 10\mathbf{j}$

10. $\dfrac{\mathbf{v}}{\|\mathbf{v}\|} = \dfrac{5\mathbf{i} - 3\mathbf{j}}{\sqrt{25 + 9}} = \dfrac{5}{\sqrt{34}}\mathbf{i} - \dfrac{3}{\sqrt{34}}\mathbf{j}$

$= \dfrac{5\sqrt{34}}{34}\mathbf{i} - \dfrac{3\sqrt{34}}{34}\mathbf{j}$

11. $\mathbf{u} = 6\mathbf{i} + 5\mathbf{j} \qquad \mathbf{v} = 2\mathbf{i} - 3\mathbf{j}$

$\mathbf{u} \cdot \mathbf{v} = 6(2) + 5(-3) = -3$

$\|\mathbf{u}\| = \sqrt{61}, \qquad \|\mathbf{v}\| = \sqrt{13}$

$\cos \theta = \dfrac{-3}{\sqrt{61}\sqrt{13}}$

$\theta \approx 96.116°$

12. $4(\mathbf{i} \cos 30° + \mathbf{j} \sin 30°) = 4\left(\dfrac{\sqrt{3}}{2}\mathbf{i} + \dfrac{1}{2}\mathbf{j}\right)$

$= \langle 2\sqrt{3}, 2\rangle$

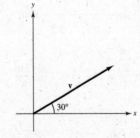

13. $\text{proj}_{\mathbf{v}}\mathbf{u} = \left(\dfrac{\mathbf{u} \cdot \mathbf{v}}{\|\mathbf{v}\|^2}\right)\mathbf{v} = \dfrac{-10}{20}\langle -2, 4\rangle = \langle 1, -2\rangle$

14. $r = \sqrt{25 + 25} = \sqrt{50} = 5\sqrt{2}$

$\tan \theta = \dfrac{-5}{5} = -1$

Since z is in Quadrant IV, $\theta = 315°$

$z = 5\sqrt{2}(\cos 315° + i \sin 315°)$.

15. $\cos 225° = -\dfrac{\sqrt{2}}{2}, \ \sin 225° = -\dfrac{\sqrt{2}}{2}$

$z = 6\left(-\dfrac{\sqrt{2}}{2} - i\dfrac{\sqrt{2}}{2}\right)$

$= -3\sqrt{2} - 3\sqrt{2}i$

16. $[7(\cos 23° + i \sin 23°)][4(\cos 7° + i \sin 7°)] = 7(4)[\cos(23° + 7°) + i \sin(23° + 7°)]$

$$= 28(\cos 30° + i \sin 30°)$$

17. $\dfrac{9\left(\cos \dfrac{5\pi}{4} + i \sin \dfrac{5\pi}{4}\right)}{3(\cos \pi + i \sin \pi)} = \dfrac{9}{3}\left[\cos\left(\dfrac{5\pi}{4} - \pi\right) + i \sin\left(\dfrac{5\pi}{4} - \pi\right)\right] = 3\left(\cos \dfrac{\pi}{4} + i \sin \dfrac{\pi}{4}\right)$

18. $(2 + 2i)^8 = [2\sqrt{2}(\cos 45° + i \sin 45°)]^8 = \left(2\sqrt{2}\right)^8[\cos(8)(45°) + i \sin(8)(45°)]$

$$= 4096[\cos 360° + i \sin 360°] = 4096$$

19. $z = 8\left(\cos \dfrac{\pi}{3} + i \sin \dfrac{\pi}{3}\right),\ n = 3$

The cube roots of z are: $\sqrt[3]{8}\left[\cos \dfrac{\dfrac{\pi}{3} + 2\pi k}{3} + i \sin \dfrac{\dfrac{\pi}{3} + 2\pi k}{3}\right], k = 0, 1, 2$

For $k = 0$, $\sqrt[3]{8}\left[\cos \dfrac{\dfrac{\pi}{3}}{3} + i \sin \dfrac{\dfrac{\pi}{3}}{3}\right] = 2\left(\cos \dfrac{\pi}{9} + i \sin \dfrac{\pi}{9}\right)$

For $k = 1$, $\sqrt[3]{8}\left[\cos \dfrac{\left(\dfrac{\pi}{3}\right) + 2\pi}{3} + i \sin \dfrac{\left(\dfrac{\pi}{3}\right) + 2\pi}{3}\right] = 2\left(\cos \dfrac{7\pi}{9} + i \sin \dfrac{7\pi}{9}\right)$

For $k = 2$, $\sqrt[3]{8}\left[\cos \dfrac{\dfrac{\pi}{3} + 4\pi}{3} + i \sin \dfrac{\dfrac{\pi}{3} + 4\pi}{3}\right] = 2\left(\cos \dfrac{13\pi}{9} + i \sin \dfrac{13\pi}{9}\right)$

20. $x^4 = -i = 1\left(\cos \dfrac{3\pi}{2} + i \sin \dfrac{3\pi}{2}\right)$

The fourth roots are: $\sqrt[4]{1}\left[\cos \dfrac{\left(\dfrac{3\pi}{2}\right) + 2\pi k}{4} + i \sin \dfrac{\left(\dfrac{3\pi}{2}\right) + 2\pi k}{4}\right], k = 0, 1, 2, 3$

For $k = 0$, $\cos \dfrac{\dfrac{3\pi}{2}}{4} + i \sin \dfrac{\dfrac{3\pi}{2}}{4} = \cos \dfrac{3\pi}{8} + i \sin \dfrac{3\pi}{8}$

For $k = 1$, $\cos \dfrac{\dfrac{3\pi}{2} + 2\pi}{4} + i \sin \dfrac{\dfrac{3\pi}{2} + 2\pi}{4} = \cos \dfrac{7\pi}{8} + i \sin \dfrac{7\pi}{8}$

For $k = 2$, $\cos \dfrac{\dfrac{3\pi}{2} + 4\pi}{4} + i \sin \dfrac{\dfrac{3\pi}{2} + 4\pi}{4} = \cos \dfrac{11\pi}{8} + i \sin \dfrac{11\pi}{8}$

For $k = 3$, $\cos \dfrac{\dfrac{3\pi}{2} + 6\pi}{4} + i \sin \dfrac{\dfrac{3\pi}{2} + 6\pi}{4} = \cos \dfrac{15\pi}{8} + i \sin \dfrac{15\pi}{8}$

Chapter 7   Practice Test Solutions

1. $\begin{cases} x + y = 1 \\ 3x - y = 15 \implies y = 3x - 15 \end{cases}$

$x + (3x - 15) = 1$

$4x = 16$

$x = 4$

$y = -3$

Solution: $(4, -3)$

2. $\begin{cases} x - 3y = -3 \implies x = 3y - 3 \\ x^2 + 6y = 5 \end{cases}$

$(3y - 3)^2 + 6y = 5$

$9y^2 - 18y + 9 + 6y = 5$

$9y^2 - 12y + 4 = 0$

$(3y - 2)^2 = 0$

$y = \frac{2}{3}$

$x = -1$

Solution: $\left(-1, \frac{2}{3}\right)$

3. $\begin{cases} x + y + z = 6 \implies z = 6 - x - y \\ 2x - y + 3z = 0 \implies 2x - y + 3(6 - x - y) = 0 \implies -x - 4y = -18 \implies x = 18 - 4y \\ 5x + 2y - z = -3 \implies 5x + 2y - (6 - x - y) = -3 \implies 6x + 3y = 3 \end{cases}$

$6(18 - 4y) + 3y = 3$

$-21y = -105$

$y = 5$

$x = 18 - 4y = -2$

$z = 6 - x - y = 3$

Solution: $(-2, 5, 3)$

4. $x + y = 110 \implies y = 110 - x$

$xy = 2800$

$x(110 - x) = 2800$

$0 = x^2 - 110x + 2800$

$0 = (x - 40)(x - 70)$

$x = 40$ or $x = 70$

$y = 70 \qquad y = 40$

Solution: The two numbers are 40 and 70.

5. $2x + 2y = 170 \implies y = \dfrac{170 - 2x}{2} = 85 - x$

$xy = 1500$

$x(85 - x) = 1500$

$0 = x^2 - 85x + 1500$

$0 = (x - 25)(x - 60)$

$x = 25$ or $x = 60$

$y = 60 \qquad y = 25$

Dimensions: 60 ft × 25 ft

6. $\begin{cases} 2x + 15y = 4 \implies 2x + 15y = 4 \\ x - 3y = 23 \implies \underline{5x - 15y = 115} \end{cases}$

$7x = 119$

$x = 17$

$y = \dfrac{x - 23}{3}$

$= -2$

Solution: $(17, -2)$

7. $\begin{cases} x + y = 2 \implies 19x + 19y = 38 \\ 38x - 19y = 7 \implies 38x - 19y = 7 \end{cases}$

$57x = 45$

$x = \dfrac{45}{57} = \dfrac{15}{19}$

$y = 2 - x = \dfrac{38}{19} - \dfrac{15}{19} = \dfrac{23}{19}$

Solution: $\left(\dfrac{15}{19}, \dfrac{23}{19}\right)$

8. $\begin{cases} 0.4x + 0.5y = 0.112 \\ 0.3x - 0.7y = -0.131 \end{cases}$ $\begin{array}{l} \Rightarrow 0.28x + 0.35y = 0.0784 \\ \Rightarrow 0.15x - 0.35y = -0.0655 \\ \phantom{\Rightarrow 0.28x +} 0.43x = 0.0129 \end{array}$

$$x = \frac{0.0129}{0.43} = 0.03$$

$$y = \frac{0.112 - 0.4x}{0.5} = 0.20$$

Solution: $(0.03, 0.20)$

9. Let x = amount in 11% fund and y = amount in 13% fund.

$$x + y = 17000 \implies y = 17000 - x$$

$$0.11x + 0.13y = 2080$$

$$0.11x + 0.13(17000 - x) = 2080$$

$$-0.02x = -130$$

$$x = \$6500 \quad \text{at } 11\%$$

$$y = \$10{,}500 \text{ at } 13\%$$

10. $(4, 3), (1, 1), (-1, -2), (-2, -1)$

Use a calculator.

$$y = ax + b = \tfrac{11}{14}x - \tfrac{1}{7}$$

11. $\begin{cases} x + y = -2 \\ 2x - y + z = 11 \\ 4y - 3z = -20 \end{cases}$

$\begin{cases} x + y = -2 \\ -3y + z = 15 \\ 4y - 3z = -20. \end{cases}$ $\quad -2\text{Eq.}1 + \text{Eq.}2$

$\begin{cases} x + y = -2 \\ y - 2z = -5 \\ 4y - 3z = -20 \end{cases}$ $\quad \text{Eq.}3 + \text{Eq.}2$

$\begin{cases} x + y = -2 \\ y - 2z = -5 \\ 5z = 0 \end{cases}$ $\quad -4\text{Eq.}2 + \text{Eq.}3$

$\begin{cases} x + y = -2 \\ y - 2z = -5 \\ z = 0 \end{cases}$

$$y - 2(0) = -5 \implies y = -5$$
$$x + (-5) = -2 \implies x = 3$$

Solution: $(3, -5, 0)$

12. $\begin{cases} 4x - y + 5z = 4 \\ 2x + y - z = 0 \\ 2x + 4y + 8z = 0 \end{cases}$

$\begin{cases} 2x + 4y + 8z = 0 \\ 2x + y - z = 0 \\ 4x - y + 5z = 4 \end{cases}$ $\quad$ Interchange equations.

$\begin{cases} 2x + 4y + 8z = 0 \\ -3y - 9z = 0 \\ -9y - 11z = 4 \end{cases}$ $\quad \begin{array}{l} -\text{Eq.}1 + \text{Eq.}2 \\ -2\text{Eq.}1 + \text{Eq.}3 \end{array}$

$\begin{cases} 2x + 4y + 8z = 0 \\ -3y - 9z = 0 \\ 16z = 4 \end{cases}$ $\quad -3\text{Eq.}2 + \text{Eq.}3$

$\begin{cases} x + 2y + 4z = 0 \\ y + 3z = 0 \\ z = \tfrac{1}{4} \end{cases}$ $\quad \begin{array}{l} \tfrac{1}{2}\text{Eq.}1 \\ -\tfrac{1}{3}\text{Eq.}2 \\ \tfrac{1}{16}\text{Eq.}3 \end{array}$

$$y + 3\left(\tfrac{1}{4}\right) = 0 \implies y = -\tfrac{3}{4}$$
$$x + 2\left(-\tfrac{3}{4}\right) + 4\left(\tfrac{1}{4}\right) = 0 \implies x = \tfrac{1}{2}$$

Solution: $\left(-\tfrac{1}{2}, -\tfrac{3}{4}, \tfrac{1}{4}\right)$

13. $\begin{cases} 3x + 2y - z = 5 \\ 6x - y + 5z = 2 \end{cases}$

$\begin{cases} 3x + 2y - z = 5 \\ -5y + 7z = -8 \end{cases}$ $\quad -2\text{Eq.}1 + \text{Eq.}2$

$\begin{cases} x + \tfrac{2}{3}y - \tfrac{1}{3}z = \tfrac{5}{3} \\ y - \tfrac{7}{5}z = \tfrac{8}{5} \end{cases}$ $\quad \begin{array}{l} \tfrac{1}{3}\text{Eq.}1 \\ -\tfrac{1}{5}\text{Eq.}2 \end{array}$

Let $a = z$.

Then $y = \tfrac{7}{5}a + \tfrac{8}{5}$, and

$$x + \tfrac{2}{3}\left(\tfrac{7}{5}a + \tfrac{8}{5}\right) - \tfrac{1}{3}a = \tfrac{5}{3}$$

$$x + \tfrac{3}{5}a = \tfrac{3}{5}$$

$$x = -\tfrac{3}{5}a + \tfrac{3}{5}.$$

Solution: $\left(-\tfrac{3}{5}a + \tfrac{3}{5}, \tfrac{7}{5}a + \tfrac{8}{5}, a\right)$, where a is any real number.

14. $y = ax^2 + bx + c$ passes through $(0, -1)$, $(1, 4)$, and $(2, 13)$.

At $(0, -1)$: $-1 = a(0)^2 + b(0) + c \implies c = -1$

At $(1, 4)$: $4 = a(1)^2 + b(1) - 1 \implies 5 = a + b \implies 5 = a + b$

At $(2, 13)$: $13 = a(2)^2 + b(2) - 1 \implies 14 = 4a + 2b \implies -7 = -2a - b$

$$-2 = -a$$
$$a = 2$$
$$b = 3$$

Thus, the equation of the parabola is $y = 2x^2 + 3x - 1$.

15. $s = \frac{1}{2}at^2 + v_0 t + s_0$ passes through $(1, 12)$, $(2, 5)$, and $(3, 4)$.

At $(1, 12)$: $12 = \frac{1}{2}a + v_0 + s_0$

At $(2, 5)$: $5 = 2a + 2v_0 + s_0$

At $(3, 4)$: $4 = \frac{9}{2}a + 3v_0 + s_0$

$$\begin{cases} a + & 2v_0 + & 2s_0 = & 24 \\ 2a + & 2v_0 + & s_0 = & 5 \\ 9a + & 6v_0 + & 2s_0 = & 8 \end{cases}$$

$$\begin{cases} a + & 2v_0 + & 2s_0 = & 24 \\ & -2v_0 - & 3s_0 = & -43 \quad -2\text{Eq.1} + \text{Eq.2} \\ & -12v_0 - & 16s_0 = & -208 \quad -9\text{Eq.1} + \text{Eq.3} \end{cases}$$

$$\begin{cases} a + & 2v_0 + 2s_0 = & 24 \\ & -2v_0 - 3s_0 = & -43 \\ & 2s_0 = & 50 \quad -6\text{Eq.2} + \text{Eq.3} \end{cases}$$

$$\begin{cases} a + & 2v_0 + & 2s_0 = & 24 \\ & v_0 + & \frac{3}{2}s_0 = & \frac{43}{2} \quad -\frac{1}{2}\text{Eq.2} \\ & & s_0 = & 25 \quad \frac{1}{2}\text{Eq.3} \end{cases}$$

$$v_0 + \tfrac{3}{2}(25) = \tfrac{43}{2} \implies v_0 = -16$$
$$a + 2(-16) + 2(25) = 24 \implies a = 6$$

Thus, $s = \frac{1}{2}(6)t^2 - 16t + 25 = 3t^2 - 16t + 25$.

16. $x^2 + y^2 \geq 9$

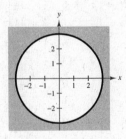

17. $\begin{cases} x + y \leq 6 \\ x \geq 2 \\ y \geq 0 \end{cases}$

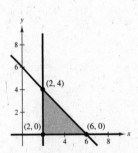

18. Line through $(0, 0)$ and $(0, 7)$:

$x = 0$

Line through $(0, 0)$ and $(2, 3)$:

$y = \frac{3}{2}x$ or $3x - 2y = 0$

Line through $(0, 7)$ and $(2, 3)$:

$y = -2x + 7$ or $2x + y = 7$

Inequalities: $\begin{cases} x \geq 0 \\ 3x - 2y \leq 0 \\ 2x + y \leq 7 \end{cases}$

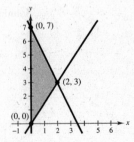

19. Vertices: $(0, 0), (0, 7), (6, 0), (3, 5)$

$z = 30x + 26y$

At $(0, 0)$: $z = 0$

At $(0, 7)$: $z = 182$

At $(6, 0)$: $z = 180$

At $(3, 5)$: $z = 220$

The maximum value of z occurs at $(3, 5)$ and is 220.

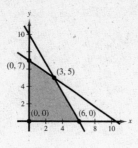

20. $x^2 + y^2 \le 4$

$(x - 2)^2 + y^2 \ge 4$

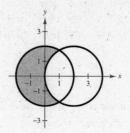

21. $\dfrac{1 - 2x}{x^2 + x} = \dfrac{1 - 2x}{x(x + 1)} = \dfrac{A}{x} + \dfrac{B}{x + 1}$

$1 - 2x = A(x + 1) + Bx$

When $x = 0, 1 = A.$

When $x = -1, 3 = -B \Rightarrow B = -3.$

$\dfrac{1 - 2x}{x^2 + x} = \dfrac{1}{x} - \dfrac{3}{x + 1}$

22. $\dfrac{6x - 17}{(x - 3)^2} = \dfrac{A}{x - 3} + \dfrac{B}{(x - 3)^2}$

$6x - 17 = A(x - 3) + B$

When $x = 3, 1 = B.$

When $x = 0, -17 = -3A + B \Rightarrow A = 6.$

$\dfrac{6x - 17}{(x - 3)^2} = \dfrac{6}{x - 3} + \dfrac{1}{(x - 3)^2}$

Chapter 8   Practice Test Solutions

1.
$$\begin{bmatrix} 1 & -2 & 4 \\ 3 & -5 & 9 \end{bmatrix}$$

$-3R_1 + R_2 \rightarrow \begin{bmatrix} 1 & -2 & 4 \\ 0 & 1 & -3 \end{bmatrix}$

$2R_2 + R_1 \rightarrow \begin{bmatrix} 1 & 0 & -2 \\ 0 & 1 & -3 \end{bmatrix}$

2. $\begin{cases} 3x + 5y = 3 \\ 2x - y = -11 \end{cases}$

$$\begin{bmatrix} 3 & 5 & \vdots & 3 \\ 2 & -1 & \vdots & -11 \end{bmatrix}$$

$-R_2 + R_1 \rightarrow \begin{bmatrix} 1 & 6 & \vdots & 14 \\ 2 & -1 & \vdots & -11 \end{bmatrix}$

$-2R_1 + R_2 \rightarrow \begin{bmatrix} 1 & 6 & \vdots & 14 \\ 0 & -13 & \vdots & -39 \end{bmatrix}$

$-\frac{1}{13}R_2 \rightarrow \begin{bmatrix} 1 & 6 & \vdots & 14 \\ 0 & 1 & \vdots & 3 \end{bmatrix}$

$-6R_2 + R_1 \rightarrow \begin{bmatrix} 1 & 0 & \vdots & -4 \\ 0 & 1 & \vdots & 3 \end{bmatrix}$

$x = -4, y = 3$

Solution: $(-4, 3)$

3. $\begin{cases} 2x + 3y = -3 \\ 3x - 2y = 8 \\ x + y = 1 \end{cases}$

$$\begin{bmatrix} 2 & 3 & \vdots & -3 \\ 3 & 2 & \vdots & 8 \\ 1 & 1 & \vdots & 1 \end{bmatrix}$$

$$\begin{matrix} R_3 \to \\ \\ R_1 \to \end{matrix} \begin{bmatrix} 1 & 1 & \vdots & 1 \\ 3 & 2 & \vdots & 8 \\ 2 & 3 & \vdots & -3 \end{bmatrix}$$

$$\begin{matrix} \\ -3R_1 + R_2 \to \\ -2R_1 + R_3 \to \end{matrix} \begin{bmatrix} 1 & 1 & \vdots & 1 \\ 0 & -1 & \vdots & 5 \\ 0 & 1 & \vdots & -5 \end{bmatrix}$$

$$\begin{matrix} \\ -R_2 \to \\ \\ \end{matrix} \begin{bmatrix} 1 & 1 & \vdots & 1 \\ 0 & 1 & \vdots & -5 \\ 0 & 1 & \vdots & -5 \end{bmatrix}$$

$$\begin{matrix} -R_2 + R_1 \to \\ \\ -R_2 + R_3 \to \end{matrix} \begin{bmatrix} 1 & 0 & \vdots & 6 \\ 0 & 1 & \vdots & -5 \\ 0 & 0 & \vdots & 0 \end{bmatrix}$$

$x = 6, y = -5$

Solution: $(6, -5)$

4. $\begin{cases} x \quad + 3z = -5 \\ 2x + y \quad = 0 \\ 3x + y - z = -3 \end{cases}$

$$\begin{bmatrix} 1 & 0 & 3 & \vdots & -5 \\ 2 & 1 & 0 & \vdots & 0 \\ 3 & 1 & -1 & \vdots & 3 \end{bmatrix}$$

$$\begin{matrix} \\ -2R_1 + R_2 \to \\ -3R_1 + R_3 \to \end{matrix} \begin{bmatrix} 1 & 0 & 3 & \vdots & -5 \\ 0 & 1 & -6 & \vdots & 10 \\ 0 & 1 & -10 & \vdots & 18 \end{bmatrix}$$

$$\begin{matrix} \\ \\ -R_2 + R_3 \to \end{matrix} \begin{bmatrix} 1 & 0 & 3 & \vdots & -5 \\ 0 & 1 & -6 & \vdots & 10 \\ 0 & 0 & -4 & \vdots & 8 \end{bmatrix}$$

$$\begin{matrix} \\ \\ -\frac{1}{4}R_3 \to \end{matrix} \begin{bmatrix} 1 & 0 & 3 & \vdots & -5 \\ 0 & 1 & -6 & \vdots & 10 \\ 0 & 0 & 1 & \vdots & -2 \end{bmatrix}$$

$$\begin{matrix} -3R_3 + R_1 \to \\ 6R_3 + R_2 \to \\ \end{matrix} \begin{bmatrix} 1 & 0 & 0 & \vdots & 1 \\ 0 & 1 & 0 & \vdots & -2 \\ 0 & 0 & 1 & \vdots & -2 \end{bmatrix}$$

$x = 1, y = -2, z = -2$

Solution: $(1, -2, -2)$

5. $\begin{bmatrix} 1 & 4 & 5 \\ 2 & 0 & -3 \end{bmatrix} \begin{bmatrix} 1 & 6 \\ 0 & -7 \\ -1 & 2 \end{bmatrix} = \begin{bmatrix} (1)(1) + (4)(0) + (5)(-1) & (1)(6) + (4)(-7) + (5)(2) \\ (2)(1) + (0)(0) + (-3)(-1) & (2)(6) + (0)(-7) + (-3)(2) \end{bmatrix} = \begin{bmatrix} -4 & -12 \\ 5 & 6 \end{bmatrix}$

6. $3A - 5B = 3\begin{bmatrix} 9 & 1 \\ -4 & 8 \end{bmatrix} - 5\begin{bmatrix} 6 & -2 \\ 3 & 5 \end{bmatrix}$

$= \begin{bmatrix} 27 & 3 \\ -12 & 24 \end{bmatrix} - \begin{bmatrix} 30 & -10 \\ 15 & 25 \end{bmatrix}$

$= \begin{bmatrix} -3 & 13 \\ -27 & -1 \end{bmatrix}$

7. $f(A) = \begin{bmatrix} 3 & 0 \\ 7 & 1 \end{bmatrix}^2 - 7\begin{bmatrix} 3 & 0 \\ 7 & 1 \end{bmatrix} + 8\begin{bmatrix} 1 & 0 \\ 0 & 1 \end{bmatrix}$

$= \begin{bmatrix} 3 & 0 \\ 7 & 1 \end{bmatrix}\begin{bmatrix} 3 & 0 \\ 7 & 1 \end{bmatrix} - \begin{bmatrix} 21 & 0 \\ 49 & 7 \end{bmatrix} + \begin{bmatrix} 8 & 0 \\ 0 & 8 \end{bmatrix}$

$= \begin{bmatrix} 9 & 0 \\ 28 & 1 \end{bmatrix} - \begin{bmatrix} 21 & 0 \\ 49 & 7 \end{bmatrix} + \begin{bmatrix} 8 & 0 \\ 0 & 8 \end{bmatrix}$

$= \begin{bmatrix} -4 & 0 \\ -21 & 2 \end{bmatrix}$

8. False since

$(A + B)(A + 3B) = A(A + 3B) + B(A + 3B)$

$= A^2 + 3AB + BA + 3B^2 \quad$ and, in general, $AB \neq BA$.

9.
$$\begin{bmatrix} 1 & 2 & \vdots & 1 & 0 \\ 3 & 5 & \vdots & 0 & 1 \end{bmatrix}$$

$$-3R_1 + R_2 \rightarrow \begin{bmatrix} 1 & 2 & \vdots & 1 & 0 \\ 0 & -1 & \vdots & -3 & 1 \end{bmatrix}$$

$$2R_2 + R_1 \rightarrow \begin{bmatrix} 1 & 0 & \vdots & -5 & 2 \\ 0 & -1 & \vdots & -3 & 1 \end{bmatrix}$$

$$-R_2 \rightarrow \begin{bmatrix} 1 & 0 & \vdots & -5 & 2 \\ 0 & 1 & \vdots & 3 & -1 \end{bmatrix}$$

$$A^{-1} = \begin{bmatrix} -5 & 2 \\ 3 & -1 \end{bmatrix}$$

10.
$$\begin{bmatrix} 1 & 1 & 1 & \vdots & 1 & 0 & 0 \\ 3 & 6 & 5 & \vdots & 0 & 1 & 0 \\ 6 & 10 & 8 & \vdots & 0 & 0 & 1 \end{bmatrix}$$

$$\begin{matrix} -3R_1 + R_2 \rightarrow \\ -6R_1 + R_3 \rightarrow \end{matrix} \begin{bmatrix} 1 & 1 & 1 & \vdots & 1 & 0 & 0 \\ 0 & 3 & 2 & \vdots & -3 & 1 & 0 \\ 0 & 4 & 2 & \vdots & -6 & 0 & 1 \end{bmatrix}$$

$$-R_3 + R_2 \rightarrow \begin{bmatrix} 1 & 1 & 1 & \vdots & 1 & 0 & 0 \\ 0 & -1 & 0 & \vdots & 3 & 1 & -1 \\ 0 & 4 & 2 & \vdots & -6 & 0 & 1 \end{bmatrix}$$

$$\begin{matrix} R_2 + R_1 \rightarrow \\ 4R_2 + R_3 \rightarrow \end{matrix} \begin{bmatrix} 1 & 0 & 1 & \vdots & 4 & 1 & -1 \\ 0 & -1 & 0 & \vdots & 3 & 1 & -1 \\ 0 & 0 & 2 & \vdots & 6 & 4 & -3 \end{bmatrix}$$

$$\begin{matrix} -R_2 \rightarrow \\ \frac{1}{2}R_3 \rightarrow \end{matrix} \begin{bmatrix} 1 & 0 & 1 & \vdots & 4 & 1 & -1 \\ 0 & 1 & 0 & \vdots & -3 & -1 & 1 \\ 0 & 0 & 1 & \vdots & 3 & 2 & -\frac{3}{2} \end{bmatrix}$$

$$-R_3 + R_1 \rightarrow \begin{bmatrix} 1 & 0 & 0 & \vdots & 1 & -1 & \frac{1}{2} \\ 0 & 1 & 0 & \vdots & -3 & -1 & 1 \\ 0 & 0 & 1 & \vdots & 3 & 2 & -\frac{3}{2} \end{bmatrix}$$

$$A^{-1} = \begin{bmatrix} 1 & -1 & \frac{1}{2} \\ -3 & -1 & 1 \\ 3 & 2 & -\frac{3}{2} \end{bmatrix}$$

11. (a) $\begin{cases} x + 2y = 4 \\ 3x + 5y = 1 \end{cases}$

$$A = \begin{bmatrix} 1 & 2 \\ 3 & 5 \end{bmatrix}$$

$$A^{-1} = \frac{1}{5-6}\begin{bmatrix} 5 & -2 \\ -3 & 1 \end{bmatrix} = \begin{bmatrix} -5 & 2 \\ 3 & -1 \end{bmatrix}$$

$$\begin{bmatrix} x \\ y \end{bmatrix} = A^{-1}B = \begin{bmatrix} -5 & 2 \\ 3 & -1 \end{bmatrix}\begin{bmatrix} 4 \\ 1 \end{bmatrix} = \begin{bmatrix} -18 \\ 11 \end{bmatrix}$$

$x = -18, y = 11$

Solution: $(-18, 11)$

(b) $\begin{cases} x + 2y = 3 \\ 3x + 5y = -2 \end{cases}$

Again, $A^{-1} = \begin{bmatrix} -5 & 2 \\ 3 & -1 \end{bmatrix}$.

$$\begin{bmatrix} x \\ y \end{bmatrix} = A^{-1}B = \begin{bmatrix} -5 & 2 \\ 3 & -1 \end{bmatrix}\begin{bmatrix} 3 \\ -2 \end{bmatrix} = \begin{bmatrix} -19 \\ 11 \end{bmatrix}$$

$x = -19, y = 11$

Solution: $(-19, 11)$

12. $\begin{vmatrix} 6 & -1 \\ 3 & 4 \end{vmatrix} = 24 - (-3) = 27$

13. $\begin{vmatrix} 1 & 3 & -1 \\ 5 & 9 & 0 \\ 6 & 2 & -5 \end{vmatrix} = -1\begin{vmatrix} 5 & 9 \\ 6 & 2 \end{vmatrix} - 5\begin{vmatrix} 1 & 3 \\ 5 & 9 \end{vmatrix} = -(-44) - 5(-6) = 74$

14. Expand along Row 2.

$$\begin{vmatrix} 1 & 4 & 2 & 3 \\ 0 & 1 & -2 & 0 \\ 3 & 5 & -1 & 1 \\ 2 & 0 & 6 & 1 \end{vmatrix} = \begin{vmatrix} 1 & 2 & 3 \\ 3 & -1 & 1 \\ 2 & 6 & 1 \end{vmatrix} + 2\begin{vmatrix} 1 & 4 & 3 \\ 3 & 5 & 1 \\ 2 & 0 & 1 \end{vmatrix}$$

$$= 51 + 2(-29) = -7$$

15. $\begin{vmatrix} 6 & 4 & 3 & 0 & 6 \\ 0 & 5 & 1 & 4 & 8 \\ 0 & 0 & 2 & 7 & 3 \\ 0 & 0 & 0 & 9 & 2 \\ 0 & 0 & 0 & 0 & 1 \end{vmatrix} = 6 \begin{vmatrix} 5 & 1 & 4 & 8 \\ 0 & 2 & 7 & 3 \\ 0 & 0 & 9 & 2 \\ 0 & 0 & 0 & 1 \end{vmatrix} = 6(5) \begin{vmatrix} 2 & 7 & 3 \\ 0 & 9 & 2 \\ 0 & 0 & 1 \end{vmatrix} = 6(5)(2) \begin{vmatrix} 9 & 2 \\ 0 & 1 \end{vmatrix} = 6(5)(2)(9) = 540$

16. Area $= \dfrac{1}{2} \begin{vmatrix} 0 & 7 & 1 \\ 5 & 0 & 1 \\ 3 & 9 & 1 \end{vmatrix} = \dfrac{1}{2}(31) = \dfrac{31}{2}$

17. $\begin{vmatrix} x & y & 1 \\ 2 & 7 & 1 \\ -1 & 4 & 1 \end{vmatrix} = 3x - 3y + 15 = 0$ or, equivalently, $x - y + 5 = 0$

18. $x = \dfrac{\begin{vmatrix} 4 & -7 \\ 11 & 5 \end{vmatrix}}{\begin{vmatrix} 6 & -7 \\ 2 & 5 \end{vmatrix}} = \dfrac{97}{44}$

19. $z = \dfrac{\begin{vmatrix} 3 & 0 & 1 \\ 0 & 1 & 3 \\ 1 & -1 & 2 \end{vmatrix}}{\begin{vmatrix} 3 & 0 & 1 \\ 0 & 1 & 4 \\ 1 & -1 & 0 \end{vmatrix}} = \dfrac{14}{11}$

20. $y = \dfrac{\begin{vmatrix} 721.4 & 33.77 \\ 45.9 & 19.85 \end{vmatrix}}{\begin{vmatrix} 721.4 & -29.1 \\ 45.9 & 105.6 \end{vmatrix}} = \dfrac{12,769.747}{77,515.530} \approx 0.1647$

Chapter 9   Practice Test Solutions

1. $a_n = \dfrac{2n}{(n+2)!}$

$a_1 = \dfrac{2(1)}{3!} = \dfrac{2}{6} = \dfrac{1}{3}$

$a_2 = \dfrac{2(2)}{4!} = \dfrac{4}{24} = \dfrac{1}{6}$

$a_3 = \dfrac{2(3)}{5!} = \dfrac{6}{120} = \dfrac{1}{20}$

$a_4 = \dfrac{2(4)}{6!} = \dfrac{8}{720} = \dfrac{1}{90}$

$a_5 = \dfrac{2(5)}{7!} = \dfrac{10}{5040} = \dfrac{1}{504}$

Terms: $\dfrac{1}{3}, \dfrac{1}{6}, \dfrac{1}{20}, \dfrac{1}{90}, \dfrac{1}{504}$

2. $a_n = \dfrac{n+3}{3^n}$

3. $\displaystyle\sum_{i=1}^{6} (2i - 1) = 1 + 3 + 5 + 7 + 9 + 11 = 36$

4. $a_1 = 23, \ d = -2$

$a_2 = 23 + (-2) = 21$

$a_3 = 21 + (-2) = 19$

$a_4 = 19 + (-2) = 17$

$a_5 = 17 + (-2) = 15$

Terms: 23, 21, 19, 17, 15

5. $a_1 = 12, d = 3, n = 50$

$a_n = a_1 + (n - 1)d$

$a_{50} = 12 + (50 - 1)3 = 159$

6. $a_1 = 1$

$a_{200} = 200$

$S_n = \dfrac{n}{2}(a_1 + a_n)$

$S_{200} = \dfrac{200}{2}(1 + 200) = 20{,}100$

7. $a_1 = 7, \ r = 2$

$a_2 = 7(2) = 14$

$a_3 = 7(2)^2 = 28$

$a_4 = 7(2)^3 = 56$

$a_5 = 7(2)^4 = 112$

Terms: 7, 14, 28, 56, 112

8. $\displaystyle\sum_{n=1}^{10} 6\left(\dfrac{2}{3}\right)^{n-1}, a_1 = 6, r = \dfrac{2}{3}, n = 10$

$S_n = \dfrac{a_1(1 - r^n)}{1 - r} = \dfrac{6\left[1 - \left(\frac{2}{3}\right)^{10}\right]}{1 - \frac{2}{3}} = 18\left(1 - \dfrac{1024}{59{,}049}\right) = \dfrac{116{,}050}{6561} \approx 17.6879$

9. $\displaystyle\sum_{n=0}^{\infty} (0.03)^n = \sum_{n=1}^{\infty} (0.03)^{n-1}, a_1 = 1, r = 0.03$

$S = \dfrac{a_1}{1 - r} = \dfrac{1}{1 - 0.03} = \dfrac{1}{0.97} = \dfrac{100}{97} \approx 1.0309$

10. For $n = 1, 1 = \dfrac{1(1 + 1)}{2}$.

Assume that $S_k = 1 + 2 + 3 + 4 + \cdots + k = \dfrac{k(k + 1)}{2}$.

Then $S_{k+1} = 1 + 2 + 3 + 4 + \cdots + k + (k + 1) = \dfrac{k(k + 1)}{2} + k + 1$

$= \dfrac{k(k + 1)}{2} + \dfrac{2(k + 1)}{2}$

$= \dfrac{(k + 1)(k + 2)}{2}$.

Thus, by the principle of mathematical induction, $1 + 2 + 3 + 4 + \cdots + n = \dfrac{n(n + 1)}{2}$ for all integers $n \geq 1$.

11. For $n = 4, 4! > 2^4$. Assume that $k! > 2^k$.

Then $(k + 1)! = (k + 1)(k!) > (k + 1)2^k > 2 \cdot 2^k = 2^{k+1}$.

Thus, by the extended principle of mathematical induction, $n! > 2^n$ for all integers $n \geq 4$.

12. $_{13}C_4 = \dfrac{13!}{(13-4)!4!} = 715$

13. $(x+3)^5 = x^5 + 5x^4(3) + 10x^3(3)^2 + 10x^2(3)^3 + 5x(3)^4 + (3)^5$

$\qquad = x^5 + 15x^4 + 90x^3 + 270x^2 + 405x + 243$

14. $-_{12}C_5 x^7(2)^5 = -25{,}344x^7$

15. $_{30}P_4 = \dfrac{30!}{(30-4)!} = 657{,}720$

16. $6! = 720$ ways

17. $_{12}P_3 = 1320$

18. $P(2) + P(3) + P(4) = \dfrac{1}{36} + \dfrac{2}{36} + \dfrac{3}{36}$

$\qquad\qquad\qquad\qquad = \dfrac{6}{36} = \dfrac{1}{6}$

19. $P(K, B10) = \dfrac{4}{52} \cdot \dfrac{2}{51} = \dfrac{2}{663}$

20. Let A = probability of no faulty units.

$P(A) = \left(\dfrac{997}{1000}\right)^{50} \approx 0.8605$

$P(A') = 1 - P(A) \approx 0.1395$

Chapter 10   Practice Test Solutions

1. $3x + 4y = 12 \implies y = -\dfrac{3}{4}x + 3 \implies m_1 = -\dfrac{3}{4}$

$4x - 3y = 12 \implies y = \dfrac{4}{3}x - 4 \implies m_2 = \dfrac{4}{3}$

$\tan\theta = \left|\dfrac{(4/3) - (-3/4)}{1 + (4/3)(-3/4)}\right| = \left|\dfrac{25/12}{0}\right|$

Since $\tan\theta$ is undefined, the lines are perpendicular
(note that $m_2 = -1/m_1$) and $\theta = 90°$.

2. $x_1 = 5, x_2 = -9, A = 3, B = -7, C = -21$

$d = \dfrac{|3(5) + (-7)(-9) + (-21)|}{\sqrt{3^2 + (-7)^2}} = \dfrac{57}{\sqrt{58}} \approx 7.484$

3. $x^2 - 6x - 4y + 1 = 0$

$\qquad x^2 - 6x + 9 = 4y - 1 + 9$

$\qquad\qquad (x-3)^2 = 4y + 8$

$\qquad\qquad (x-3)^2 = 4(1)(y+2) \implies p = 1$

Vertex: $(3, -2)$

Focus: $(3, -1)$

Directrix: $y = -3$

4. Vertex: $(2, -5)$

Focus: $(2, -6)$

Vertical axis; opens downward with $p = -1$

$\qquad\qquad (x-h)^2 = 4p(y-k)$

$\qquad\qquad (x-2)^2 = 4(-1)(y+5)$

$\qquad\qquad x^2 - 4x + 4 = -4y - 20$

$\qquad x^2 - 4x + 4y + 24 = 0$

5.
$$x^2 + 4y^2 - 2x + 32y + 61 = 0$$
$$(x^2 - 2x + 1) + 4(y^2 + 8y + 16) = -61 + 1 + 64$$
$$(x - 1)^2 + 4(y + 4)^2 = 4$$
$$\frac{(x - 1)^2}{4} + \frac{(y + 4)^2}{1} = 1$$

$a = 2, b = 1, c = \sqrt{3}$

Horizontal major axis

Center: $(1, -4)$

Foci: $\left(1 \pm \sqrt{3}, -4\right)$.

Vertices: $(3, -4), (-1, -4)$

Eccentricity: $e = \dfrac{\sqrt{3}}{2}$

6. Vertices: $(0, \pm 6)$

Eccentricity: $e = \dfrac{1}{2}$

Center: $(0, 0)$

Vertical major axis

$a = 6, e = \dfrac{c}{a} = \dfrac{c}{6} = \dfrac{1}{2} \Rightarrow c = 3$

$b^2 = (6)^2 - (3)^2 = 27$

$\dfrac{x^2}{27} + \dfrac{y^2}{36} = 1$

7.
$$16y^2 - x^2 - 6x - 128y + 231 = 0$$
$$16(y^2 - 8y + 16) - (x^2 + 6x + 9) = -231 + 256 - 9$$
$$16(y - 4)^2 - (x + 3)^2 = 16$$
$$\frac{(y - 4)^2}{1} - \frac{(x + 3)^2}{16} = 1$$

$a = 1, b = 4, c = \sqrt{17}$

Center: $(-3, 4)$

Vertical transverse axis

Vertices: $(-3, 5), (-3, 3)$

Foci: $\left(-3, 4 \pm \sqrt{17}\right)$

Asymptotes: $y = 4 \pm \dfrac{1}{4}(x + 3)$

8. Vertices: $(\pm 3, 2)$

Foci: $(\pm 5, 2)$

Center: $(0, 2)$

Horizontal transverse axis

$a = 3, c = 5, b = 4$

$\dfrac{(x - 0)^2}{9} - \dfrac{(y - 2)^2}{16} = 1$

$\dfrac{x^2}{9} - \dfrac{(y - 2)^2}{16} = 1$

9. $5x^2 + 2xy + 5y^2 - 10 = 0$

$A = 5, B = 2, C = 5$

$\cot 2\theta = \dfrac{5 - 5}{2} = 0$

$2\theta = \dfrac{\pi}{2} \Rightarrow \theta = \dfrac{\pi}{4}$

$x = x' \cos \dfrac{\pi}{4} - y' \sin \dfrac{\pi}{4}$ $x = x' \cos \dfrac{\pi}{4} + y' \sin \dfrac{\pi}{4}$

$= \dfrac{x' - y'}{\sqrt{2}}$ $= \dfrac{x' + y'}{\sqrt{2}}$

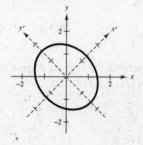

$$5\left(\frac{x' - y'}{\sqrt{2}}\right)^2 + 2\left(\frac{x' - y'}{\sqrt{2}}\right)\left(\frac{x' + y'}{\sqrt{2}}\right) + 5\left(\frac{x' + y'}{\sqrt{2}}\right)^2 - 10 = 0$$

$$\frac{5(x')^2}{2} - \frac{10x'y'}{2} + \frac{5(y')^2}{2} + (x')^2 - (y')^2 + \frac{5(x')^2}{2} + \frac{10x'y'}{2} + \frac{5(y')^2}{2} - 10 = 0$$

$$6(x')^2 + 4(y')^2 - 10 = 0$$

$$\frac{3(x')^2}{5} + \frac{2(y')^2}{5} = 1$$

$$\frac{(x')^2}{5/3} + \frac{(y')^2}{5/2} = 1$$

Ellipse centered at the origin

10. (a) $6x^2 - 2xy + y^2 = 0$

$A = 6, B = -2, C = 1$

$B^2 - 4AC = (-2)^2 - 4(6)(1) = -20 < 0$

Ellipse

(b) $x^2 + 4xy + 4y^2 - x - y + 17 = 0$

$A = 1, B = 4, C = 4$

$B^2 - 4AC = (4)^2 - 4(1)(4) = 0$

Parabola

11. Polar: $\left(\sqrt{2}, \dfrac{3\pi}{4}\right)$

$x = \sqrt{2} \cos \dfrac{3\pi}{4} = \sqrt{2}\left(-\dfrac{1}{\sqrt{2}}\right) = -1$

$y = \sqrt{2} \sin \dfrac{3\pi}{4} = \sqrt{2}\left(\dfrac{1}{\sqrt{2}}\right) = 1$

Rectangular: $(-1, 1)$

12. Rectangular: $\left(\sqrt{3}, -1\right)$

$r = \pm\sqrt{\left(\sqrt{3}\right)^2 + (-1)^2} = \pm 2$

$\tan \theta = \dfrac{\sqrt{3}}{-1} = -\sqrt{3}$

$\theta = \dfrac{2\pi}{3}$ or $\theta = \dfrac{5\pi}{3}$

Polar: $\left(-2, \dfrac{2\pi}{3}\right)$ or $\left(2, \dfrac{5\pi}{3}\right)$

13. Rectangular: $4x - 3y = 12$

Polar: $4r \cos \theta - 3r \sin \theta = 12$

$r(4 \cos \theta - 3 \sin \theta) = 12$

$r = \dfrac{12}{4 \cos \theta - 3 \sin \theta}$

14. Polar: $r = 5 \cos \theta$

$r^2 = 5r \cos \theta$

Rectangular: $x^2 + y^2 = 5x$

$x^2 + y^2 - 5x = 0$

15. $r = 1 - \cos \theta$

Cardioid

Symmetry: Polar axis

Maximum value of $|r|$: $r = 2$ when $\theta = \pi$

Zero of r: $r = 0$ when $\theta = 0$

θ	0	$\dfrac{\pi}{2}$	π	$\dfrac{3\pi}{2}$
r	0	1	2	1

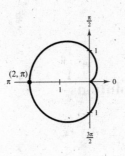

16. $r = 5 \sin 2\theta$

Rose curve with four petals

Symmetry: Polar axis, $\theta = \dfrac{\pi}{2}$, and pole

Maximum value of $|r|$: $|r| = 5$ when $\theta = \dfrac{\pi}{4}, \dfrac{3\pi}{4}, \dfrac{5\pi}{4}, \dfrac{7\pi}{4}$

Zeros of r: $r = 0$ when $\theta = 0, \dfrac{\pi}{2}, \pi, \dfrac{3\pi}{2}$

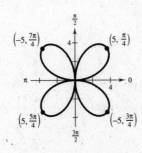

17. $r = \dfrac{3}{6 - \cos \theta}$

$r = \dfrac{1/2}{1 - (1/6)\cos\theta}$

$e = \dfrac{1}{6} < 1$, so the graph is an ellipse.

θ	0	$\dfrac{\pi}{2}$	π	$\dfrac{3\pi}{2}$
r	$\dfrac{3}{5}$	$\dfrac{1}{2}$	$\dfrac{3}{7}$	$\dfrac{1}{2}$

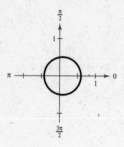

18. Parabola

Vertex: $\left(6, \dfrac{\pi}{2}\right)$

Focus: $(0, 0)$

$e = 1$

$r = \dfrac{ep}{1 + e\sin\theta}$

$r = \dfrac{p}{1 + \sin\theta}$

$6 = \dfrac{p}{1 + \sin(\pi/2)}$

$6 = \dfrac{p}{2}$

$12 = p$

$r = \dfrac{12}{1 + \sin\theta}$

19. $x = 3 - 2\sin\theta, \; y = 1 + 5\cos\theta$

$\dfrac{x - 3}{-2} = \sin\theta, \; \dfrac{y - 1}{5} = \cos\theta$

$\left(\dfrac{x - 3}{-2}\right)^2 + \left(\dfrac{y - 1}{5}\right)^2 = 1$

$\dfrac{(x - 3)^2}{4} + \dfrac{(y - 1)^2}{25} = 1$

20. $x = e^{2t}, \; y = e^{4t}$

$x > 0, \; y > 0$

$y = (e^{2t})^2 = (x)^2 = x^2, \; x > 0, \; y > 0$

Chapter 11   Practice Test Solutions

1. Let $A = (0, 0, 0)$, $B = (1, 2, -4)$, $C = (0, -2, -1)$.

Side AB: $\sqrt{1^2 + 2^2 + 4^2} = \sqrt{21}$

Side AC: $\sqrt{0^2 + 2^2 + 1^2} = \sqrt{5}$

Side BC: $\sqrt{(-1)^2 + (-2 - 2)^2 + (-1 + 4)^2} = \sqrt{1 + 16 + 9} = \sqrt{26}$

$BC^2 = AB^2 + AC^2$

$26 = 21 + 5$

2. $(x - 0)^2 + (y - 4)^2 + (z - 1)^2 = 5^2$

$ x^2 + (y - 4)^2 + (z - 1)^2 = 25$

3. $(x^2 + 2x + 1) + y^2 + (z^2 - 4z + 4) = 1 + 4 + 11$

$ (x + 1)^2 + y^2 + (z - 2)^2 = 16$

Center: $(-1, 0, 2)$

Radius: 4

4. $\mathbf{u} - 3\mathbf{v} = \langle 1, 0, -1 \rangle - 3\langle 4, 3, -6 \rangle$

$\qquad = \langle 1, 0, -1 \rangle - \langle 12, 9, -18 \rangle$

$\qquad = \langle -11, -9, 17 \rangle$

5. $\frac{1}{2}\mathbf{v} = \frac{1}{2}\langle 2, 4, -6 \rangle = \langle 1, 2, -3 \rangle$

$\qquad \left\| \frac{1}{2}\mathbf{v} \right\| = \sqrt{1^2 + 2^2 + (-3)^2} = \sqrt{14}$

6. $\mathbf{u} \cdot \mathbf{v} = \langle 2, 1, -3 \rangle \cdot \langle 1, 1, -2 \rangle$

$\qquad = 2 + 1 + 6 = 9$

7. Because $\mathbf{v} = \langle -3, -3, 3 \rangle = -3\langle 1, 1, -1 \rangle = -3\mathbf{u}$, $\mathbf{u}$ and $\mathbf{v}$ are parallel.

8. $\mathbf{u} \times \mathbf{v} = \begin{vmatrix} \mathbf{i} & \mathbf{j} & \mathbf{k} \\ -1 & 0 & 2 \\ 1 & -1 & 3 \end{vmatrix} = \langle 2, 5, 1 \rangle$

$\qquad \mathbf{v} \times \mathbf{u} = -(\mathbf{u} \times \mathbf{v}) = \langle -2, -5, -1 \rangle$

9. $\mathbf{u} \cdot (\mathbf{v} \times \mathbf{w}) = \begin{vmatrix} 1 & 1 & 1 \\ 0 & -1 & 1 \\ 1 & 0 & 4 \end{vmatrix}$

$\qquad = 1(-4) - 1(-1) + 1(1)$

$\qquad = -4 + 1 + 1 = -2$

Volume $= |\mathbf{u} \cdot (\mathbf{v} \times \mathbf{w})| = |-2| = 2$

10. $\mathbf{v} = \langle (2 - 0), -3 - (-3), 4 - 3 \rangle = \langle 2, 0, 1 \rangle$

$\qquad x = 2 + 2t, y = -3, z = 4 + t$

11. $1(x - 1) - 1(y - 2) + 0(z - 3) = 0$

$\qquad x - 1 - y + 2 = 0$

$\qquad x - y + 1 = 0$

12. $\overrightarrow{AB} = \langle 1, 1, 1 \rangle, \overrightarrow{AC} = \langle 1, 2, 3 \rangle$

$\qquad \mathbf{n} = \overrightarrow{AB} \times \overrightarrow{AC} = \begin{vmatrix} \mathbf{i} & \mathbf{j} & \mathbf{k} \\ 1 & 1 & 1 \\ 1 & 2 & 3 \end{vmatrix} = \langle 1, -2, 1 \rangle$

Plane: $1(x - 0) - 2(y - 0) + (z - 0) = 0$

$\qquad x - 2y + z = 0$

13. $\mathbf{n}_1 = \langle 1, 1, -1 \rangle, \mathbf{n}_2 = \langle 3, -4, -1 \rangle$

$\qquad \mathbf{n}_1 \cdot \mathbf{n}_2 = 3 - 4 + 1 = 0 \implies$ Orthogonal planes

14. $\mathbf{n} = \langle 1, 2, 1 \rangle, Q = (1, 1, 1), P = (0, 0, 6)$ on plane, $\overrightarrow{PQ} = \langle 1, 1, -5 \rangle$

$D = \dfrac{|\overrightarrow{PQ} \cdot \mathbf{n}|}{\|\mathbf{n}\|} = \dfrac{|1 + 2 - 5|}{\sqrt{1 + 4 + 1}} = \dfrac{2}{\sqrt{6}} = \dfrac{\sqrt{6}}{3}$

Chapter 12   Practice Test Solutions

1.

x	2.9	2.99	3	3.01	3.1
$f(x)$	0.1695	0.1669	?	0.1664	0.1639

$\displaystyle \lim_{x \to 3} \frac{x - 3}{x^2 - 9} \approx 0.1667$

2. $\displaystyle \lim_{x \to 0} \frac{\sqrt{x + 4} - 2}{x} \approx \frac{1}{4}$

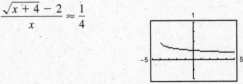

3. $\displaystyle \lim_{x \to 2} e^{x-2} = e^{2-2} = e^0 = 1$

4. $\displaystyle \lim_{x \to 1} \frac{x^3 - 1}{x - 1} = \lim_{x \to 1} \frac{(x - 1)(x^2 + x + 1)}{x - 1}$

$\qquad = \displaystyle \lim_{x \to 1} (x^2 + x + 1) = 3$

5. $\displaystyle\lim_{x\to 0}\frac{\sin 5x}{2x}\approx 2.5$

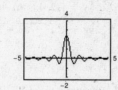

6. The limit does not exist. If

$$f(x)=\frac{|x+2|}{x+2},$$

then $f(x)=1$ for $x>-2$, and $f(x)=-1$ for $x<-2$.

7. $m_{\text{sec}}=\dfrac{f(4+h)-f(4)}{h}$

$\qquad = \dfrac{\sqrt{4+h}-2}{h}$

$\qquad = \dfrac{\sqrt{4+h}-2}{h}\cdot\dfrac{\sqrt{4+h}+2}{\sqrt{4+h}+2}$

$\qquad = \dfrac{(4+h)-4}{h\left[\sqrt{4+h}+2\right]}$

$\qquad = \dfrac{h}{h\left[\sqrt{4+h}+2\right]}$

$\qquad = \dfrac{1}{\sqrt{4+h}+2},\ h\neq 0$

$m=\displaystyle\lim_{h\to 0}\frac{1}{\sqrt{4+h}+2}=\frac{1}{\sqrt{4}+2}=\frac{1}{4}$

8. $f'(x)=\displaystyle\lim_{h\to 0}\frac{f(x+h)-f(x)}{h}$

$\qquad = \displaystyle\lim_{h\to 0}\frac{[3(x+h)-1]-[3x-1]}{h}$

$\qquad = \displaystyle\lim_{h\to 0}\frac{3x+3h-1-3x+1}{h}$

$\qquad = \displaystyle\lim_{h\to 0}\frac{3h}{h}=\lim_{h\to 0}3=3$

9. (a) $\displaystyle\lim_{x\to\infty}\frac{3}{x^4}=0$

(b) $\displaystyle\lim_{x\to-\infty}\frac{x^2}{x^2+3}=1$

(c) $\displaystyle\lim_{x\to\infty}\frac{|x|}{1-x}=-1$

10. $a_1=0,\ a_2=\dfrac{1-4}{8+1}=-\dfrac{1}{3},\ a_3=\dfrac{1-9}{18+1}=-\dfrac{8}{19},$

$a_4=\dfrac{1-16}{33}=-\dfrac{15}{33}$

$\displaystyle\lim_{n\to\infty}a_n=\lim_{n\to\infty}\frac{1-n^2}{2n^2+1}=-\frac{1}{2}$

11. $\displaystyle\sum_{i=1}^{25}i^2+\sum_{i=1}^{25}i=\frac{25(26)(51)}{6}+\frac{25(26)}{2}=\frac{25(26)}{6}[51+3]=\frac{25(26)(54)}{6}=5850$

12. $\displaystyle\sum_{i=1}^{n}\frac{i^2}{n^3}=\frac{1}{n^3}\sum_{i=1}^{n}i^2=\frac{1}{n^3}\left[\frac{n(n+1)(2n+1)}{6}\right]=\frac{2n^2+3n+1}{6n^2}=S(n)$

$\displaystyle\lim_{n\to\infty}S(n)=\frac{1}{3}$

13. Width of rectangles: $\dfrac{b-a}{n}=\dfrac{1}{n}$

Height: $f\left(a+\dfrac{(b-a)i}{n}\right)=f\left(\dfrac{i}{n}\right)=1-\left(\dfrac{i}{n}\right)^2$

$A\approx\displaystyle\sum_{i=1}^{n}\left[1-\frac{i^2}{n^2}\right]\frac{1}{n}=\sum_{i=1}^{n}\frac{1}{n}-\sum_{i=1}^{n}\frac{i^2}{n^3}=1-\frac{1}{n^3}\frac{n(n+1)(2n+1)}{6}$

$A=\displaystyle\lim_{n\to\infty}A_n=1-\frac{1}{3}=\frac{2}{3}$

PART II

Chapter 1   Chapter Test Solutions

1. Midpoint: $\left(\dfrac{-2+6}{2}, \dfrac{5+0}{2}\right) = \left(2, \dfrac{5}{2}\right)$

Distance: $d = \sqrt{(-2-6)^2 + (5-0)^2}$

$\quad\quad\quad\quad = \sqrt{64 + 25}$

$\quad\quad\quad\quad = \sqrt{89}$

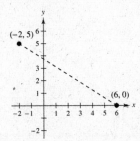

2. $\pi r^2 h = V$

$\quad \pi(4)^2 h = 600$

$\quad\quad\quad h = \dfrac{600}{16\pi}$

$\quad\quad\quad h \approx 11.937$ centimeters

3. $y = 3 - 5x$

x-intercept: $\left(\frac{3}{5}, 0\right)$

y-intercept: $(0, 3)$

No axis or origin symmetry

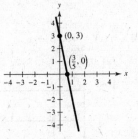

4. $y = 4 - |x|$

x-intercepts: $(\pm 4, 0)$

y-intercept: $(0, 4)$

y-axis symmetry

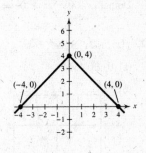

5. $y = x^2 - 1$

x-intercepts: $(\pm 1, 0)$

y-intercept: $(0, -1)$

y-axis symmetry

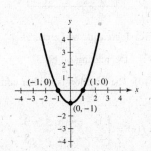

6. Center: $(1, 3)$

Radius: 4

Standard form: $(x - 1)^2 + (y - 3)^2 = 16$

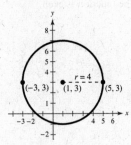

7. $(2, -3)$ and $(-4, 9)$

$m = \dfrac{9 - (-3)}{-4 - 2} = -2$

$y - (-3) = -2(x - 2)$

$y + 3 = -2x + 4$

$y = -2x + 1$

$2x + y - 1 = 0$

8. $(3, 0.8)$ and $(7, -6)$

$m = \dfrac{-6 - 0.8}{7 - 3} = -1.7$

$y - (-6) = -1.7(x - 7)$

$y + 6 = -1.7x + 11.9$

$y = -1.7x + 5.9$

$10y = -17x + 59$

$17x + 10y - 59 = 0$

9. $-4x + 7y = -5$

$$7y = 4x - 5$$

$$y = \tfrac{4}{7}x - \tfrac{5}{7} \implies m_1 = \tfrac{4}{7}$$

(a) Parallel line: $m_2 = \tfrac{4}{7}$

$$y - 8 = \tfrac{4}{7}(x - 3)$$

$$7y - 56 = 4x - 12$$

$$-4x + 7y = 44$$

$$4x - 7y + 44 = 0$$

(b) Perpendicular line: $m_2 = -\tfrac{7}{4}$

$$y - 8 = -\tfrac{7}{4}(x - 3)$$

$$4y - 32 = -7x + 21$$

$$7x + 4y - 53 = 0$$

10. $f(x) = \dfrac{\sqrt{x + 9}}{x^2 - 81}$

(a) $f(7) = \dfrac{4}{-32} = -\dfrac{1}{8}$

(b) $f(-5) = \dfrac{2}{-56} = -\dfrac{1}{28}$

(c) $f(x - 9) = \dfrac{\sqrt{x}}{(x - 9)^2 - 81} = \dfrac{\sqrt{x}}{x^2 - 18x}$

11. $f(x) = \sqrt{100 - x^2}$

Domain: $100 - x^2 \geq 0 \implies -10 \leq x \leq 10$ or $[-10, 10]$

12. $f(x) = 2x^6 + 5x^4 - x^2$

(a) $0, \pm 0.4314$

(b)

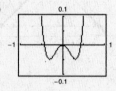

(c) Increasing on $(-0.31, 0), (0.31, \infty)$

Decreasing on $(-\infty, -0.31), (0, 0.31)$

(d) y-axis symmetry $\implies$ The function is even.

13. $f(x) = 4x\sqrt{3 - x}$

(a) $0, 3$

(b)

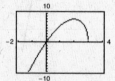

(c) Increasing on $(-\infty, 2)$

Decreasing on $(2, 3)$

(d) The function is neither odd nor even.

14. $f(x) = |x + 5|$

(a) -5

(b)

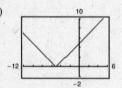

(c) Increasing on $(-5, \infty)$

Decreasing on $(-\infty, -5)$

(d) The function is neither odd nor even.

15. $f(x) = \begin{cases} 3x + 7, & x \leq -3 \\ 4x^2 - 1, & x > -3 \end{cases}$

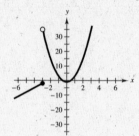

16. $h(x) = -[\![x]\!]$

Common function: $f(x) = [\![x]\!]$

Transformation: Reflection in the x-axis

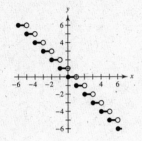

17. $h(x) = -\sqrt{x+5} + 8$

Common function: $f(x) = \sqrt{x}$

Transformation: Reflection in the x-axis, a horizontal shift 5 units to the left, and a vertical shift 8 units upward

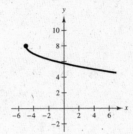

18. $f(x) = 3x^2 - 7,\; g(x) = -x^2 - 4x + 5$

   (a) $(f+g)(x) = (3x^2 - 7) + (-x^2 - 4x + 5) = 2x^2 - 4x - 2$

   (b) $(f-g)(x) = (3x^2 - 7) - (-x^2 - 4x + 5) = 4x^2 + 4x - 12$

   (c) $(fg)(x) = (3x^2 - 7)(-x^2 - 4x + 5) = -3x^4 - 12x^3 + 22x^2 + 28x - 35$

   (d) $\left(\dfrac{f}{g}\right)(x) = \dfrac{3x^2 - 7}{-x^2 - 4x + 5},\; x \neq -5, 1$

   (e) $(f \circ g)(x) = f(g(x)) = f(-x^2 - 4x + 5) = 3(-x^2 - 4x + 5)^2 - 7 = 3x^4 + 24x^3 + 18x^2 - 120x + 68$

   (f) $(g \circ f)(x) = g(f(x)) = g(3x^2 - 7) = -(3x^2 - 7)^2 - 4(3x^2 - 7) + 5 = -9x^4 + 30x^2 - 16$

19. $f(x) = \dfrac{1}{x},\; g(x) = 2\sqrt{x}$

   (a) $(f+g)(x) = \dfrac{1}{x} + 2\sqrt{x} = \dfrac{1 + 2x^{3/2}}{x},\; x > 0$

   (b) $(f-g)(x) = \dfrac{1}{x} - 2\sqrt{x} = \dfrac{1 - 2x^{3/2}}{x},\; x > 0$

   (c) $(fg)(x) = \left(\dfrac{1}{x}\right)(2\sqrt{x}) = \dfrac{2\sqrt{x}}{x},\; x > 0$

   (d) $\left(\dfrac{f}{g}\right)(x) = \dfrac{\dfrac{1}{x}}{2\sqrt{x}} = \dfrac{1}{2x\sqrt{x}} = \dfrac{1}{2x^{3/2}},\; x > 0$

   (e) $(f \circ g)(x) = f(g(x)) = f(2\sqrt{x}) = \dfrac{1}{2\sqrt{x}} = \dfrac{\sqrt{x}}{2x},\; x > 0$

   (f) $(g \circ f)(x) = g(f(x)) = g\left(\dfrac{1}{x}\right) = 2\sqrt{\dfrac{1}{x}} = \dfrac{2}{\sqrt{x}} = \dfrac{2\sqrt{x}}{x},\; x > 0$

20. $f(x) = x^3 + 8$

Since f is one-to-one, f has an inverse.

$$y = x^3 + 8$$

$$x = y^3 + 8$$

$$x - 8 = y^3$$

$$\sqrt[3]{x - 8} = y$$

$$f^{-1}(x) = \sqrt[3]{x - 8}$$

21. $f(x) = |x^2 - 3| + 6$

Since f is not one-to-one, f does not have an inverse.

22. $f(x) = 3x\sqrt{x} = 3x^{3/2}$

Domain: $[0, \infty)$

Range: $[0, \infty)$

The graph of $f(x)$ passes the Horizontal Line Test, so $f(x)$ is one-to-one and has an inverse.

$$f(x) = 3x^{3/2}$$

$$y = 3x^{3/2}$$

$$x = 3y^{3/2}$$

$$\frac{x}{3} = y^{3/2}$$

$$\left(\frac{x}{3}\right)^{2/3} = y$$

$$f^{-1}(x) = \left(\frac{x}{3}\right)^{2/3} = \sqrt[3]{\frac{x^2}{9}}, x \geq 0$$

23. $v = k\sqrt{s}$

$24 = k\sqrt{16}$

$6 = k$

$v = 6\sqrt{s}$

24. $A = kxy$

$500 = k(15)(8)$

$500 = k(120)$

$\dfrac{25}{6} = k$

$A = \dfrac{25}{6}xy$

25. $b = \dfrac{k}{a}$

$32 = \dfrac{k}{1.5}$

$48 = k$

$b = \dfrac{48}{a}$

Chapter 2   Chapter Test Solutions

1. $f(x) = x^2$

(a) $g(x) = 2 - x^2$

Reflection in the x-axis followed by a vertical translation two units upward

(b) $g(x) = \left(x - \frac{3}{2}\right)^2$

Horizontal translation $\frac{3}{2}$ units to the right

2. Vertex: $(3, -6)$

$y = a(x - 3)^2 - 6$

Point on the graph: $(0, 3)$

$3 = a(0 - 3)^2 - 6$

$9 = 9a \implies a = 1$

Thus, $y = (x - 3)^2 - 6$.

3. (a) $y = -\frac{1}{20}x^2 + 3x + 5$

$\quad\quad = -\frac{1}{20}(x^2 - 60x + 900 - 900) + 5$

$\quad\quad = -\frac{1}{20}[(x - 30)^2 - 900] + 5$

$\quad\quad = -\frac{1}{20}(x - 30)^2 + 50$

Vertex: $(30, 50)$

The maximum height is 50 feet.

(b) The constant term, $c = 5$, determines the height at which the ball was thrown. Changing this constant results in a vertical translation of the graph, and, therefore, changes the maximum height.

4. $h(t) = -\frac{3}{4}t^5 + 2t^2$

The degree is odd and the leading coefficient is negative. The graph rises to the left and falls to the right.

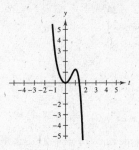

5.

$$x^2 + 0x + 1 \overline{\smash{\big)}\ 3x^3 + 0x^2 + 4x - 1} \quad\quad 3x + \frac{x - 1}{x^2 + 1}$$

$$\underline{3x^3 + 0x^2 + 3x}$$

$$x - 1$$

Thus, $\dfrac{3x^3 + 4x - 1}{x^2 + 1} = 3x + \dfrac{x - 1}{x^2 + 1}.$

6.

$$\begin{array}{r|rrrrr} 2 & 2 & 0 & -5 & 0 & -3 \\ & & 4 & 8 & 6 & 12 \\ \hline & 2 & 4 & 3 & 6 & 9 \end{array}$$

Thus, $\dfrac{2x^4 - 5x^2 - 3}{x - 2} = 2x^3 + 4x^2 + 3x + 6 + \dfrac{9}{x - 2}.$

7.

$$\begin{array}{r|rrrr} \sqrt{3} & 4 & -1 & -12 & 3 \\ & & 4\sqrt{3} & 12 - \sqrt{3} & -3 \\ \hline & 4 & 4\sqrt{3} - 1 & -\sqrt{3} & 0 \end{array}$$

$$\begin{array}{r|rrr} -\sqrt{3} & 4 & 4\sqrt{3} - 1 & -\sqrt{3} \\ & & -4\sqrt{3} & \sqrt{3} \\ \hline & 4 & -1 & 0 \end{array}$$

$4x^3 - x^2 - 12x + 3 = (x - \sqrt{3})(x + \sqrt{3})(4x - 1)$

The real solutions are $x = \pm\sqrt{3}$ and $x = \frac{1}{4}$.

8. (a) $10i - (3 + \sqrt{-25}) = 10i - 3 - 5i$

$\quad\quad\quad\quad\quad\quad\quad\quad\quad\quad = -3 + 5i$

(b) $(2 + \sqrt{3}\,i)(2 - \sqrt{3}\,i) = 4 - 3i^2$

$\quad\quad\quad\quad\quad\quad\quad\quad\quad = 4 + 3$

$\quad\quad\quad\quad\quad\quad\quad\quad\quad = 7$

9. $\dfrac{5}{2 + i} = \dfrac{5}{2 + i} \cdot \dfrac{2 - i}{2 - i}$

$\quad\quad = \dfrac{5(2 - i)}{4 + 1}$

$\quad\quad = 2 - i$

10. $f(x) = x(x - 3)[x - (3 + i)][x - (3 - i)]$

$\quad\quad = (x^2 - 3x)[(x - 3) - i][(x - 3) + i]$

$\quad\quad = (x^2 - 3x)[(x - 3)^2 - i^2]$

$\quad\quad = (x^2 - 3x)(x^2 - 6x + 10)$

$\quad\quad = x^4 - 9x^3 + 28x^2 - 30x$

11. $f(x) = [x - (1 + \sqrt{3}\,i)][x - (1 - \sqrt{3}\,i)](x - 2)(x - 2)$

$\quad\quad = [(x - 1) - \sqrt{3}\,i][(x - 1) + \sqrt{3}\,i](x^2 - 4x + 4)$

$\quad\quad = [(x - 1)^2 - 3i^2](x^2 - 4x + 4)$

$\quad\quad = (x^2 - 2x + 4)(x^2 - 4x + 4)$

$\quad\quad = x^4 - 6x^3 + 16x^2 - 24x + 16$

12. $f(x) = x^3 + 2x^2 + 5x + 10$

$\quad\quad = x^2(x + 2) + 5(x + 2)$

$\quad\quad = (x + 2)(x^2 + 5)$

$\quad\quad = (x + 2)(x + \sqrt{5}\,i)(x - \sqrt{5}\,i)$

Zeros: $x = -2, \pm\sqrt{5}\,i$

13. $f(x) = x^4 - 9x^2 - 22x - 24$

Possible rational zeros: $\pm1, \pm2, \pm3, \pm4, \pm6, \pm8, \pm12, \pm24$

$$
\begin{array}{r|rrrrr}
-2 & 1 & 0 & -9 & -22 & -24 \\
   &   & -2 & 4 & 10 & 24 \\
\hline
   & 1 & -2 & -5 & -12 & 0
\end{array}
$$

$$
\begin{array}{r|rrrr}
4 & 1 & -2 & -5 & -12 \\
  &   & 4 & 8 & 12 \\
\hline
  & 1 & 2 & 3 & 0
\end{array}
$$

$f(x) = (x + 2)(x - 4)(x^2 + 2x + 3)$

By the Quadratic Formula the zeros of $x^2 + 2x + 3$ are
$x = -1 \pm \sqrt{2}i$. The zeros of f are: $x = -2, 4, -1 \pm \sqrt{2}i$.

14. $h(x) = \dfrac{4}{x^2} - 1$

$$= \dfrac{4 - x^2}{x^2}$$

$$= \dfrac{(2 - x)(2 + x)}{x^2}$$

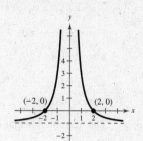

Vertical asymptote: $x = 0$

Horizontal asymptote:
$y = -1$

x-intercepts: $(\pm2, 0)$

15. $f(x) = \dfrac{2x^2 - 5x - 12}{x^2 - 16}$

$$= \dfrac{(2x + 3)(x - 4)}{(x + 4)(x - 4)}$$

$$= \dfrac{2x + 3}{x + 4}, \quad x \neq 4$$

x-intercept: $\left(-\dfrac{3}{2}, 0\right)$

y-intercept: $\left(0, \dfrac{3}{4}\right)$

Vertical asymptote: $x = -4$

Horizontal asymptote: $y = 2$

16. $g(x) = \dfrac{x^2 + 2}{x - 1} = x + 1 + \dfrac{3}{x - 1}$

Vertical asymptote: $x = 1$

Slant asymptote: $y = x + 1$

y-intercept: $(0, -2)$

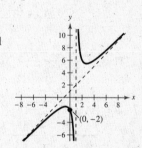

17. $2x^2 + 5x > 12$

$2x^2 + 5x - 12 > 0$

$(2x - 3)(x + 4) > 0$

Critical numbers: $x = \dfrac{3}{2}, x = -4$

Test intervals: $(-\infty, -4), \left(-4, \dfrac{3}{2}\right), \left(\dfrac{3}{2}, \infty\right)$

Test: Is $(2x - 3)(x + 4) > 0$?

Solution set: $(-\infty, -4) \cup \left(\dfrac{3}{2}, \infty\right)$

In inequality notation: $x < -4$ or $x > \dfrac{3}{2}$

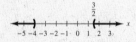

18. $\dfrac{2}{x} > \dfrac{5}{x + 6}$

$\dfrac{2}{x} - \dfrac{5}{x + 6} > 0$

$\dfrac{2(x + 6) - 5x}{x(x + 6)} > 0$

$\dfrac{-3x + 12}{x(x + 6)} > 0$

$\dfrac{-3(x - 4)}{x(x + 6)} > 0$

Critical numbers: $x = 4, x = 0, x = -6$

Test intervals: $(-\infty, -6), (-6, 0), (0, 4), (4, \infty)$

Test: Is $\dfrac{-3(x - 4)}{x(x + 6)} > 0$?

Solution set: $(-\infty, -6) \cup (0, 4)$

In inequality notation: $x < -6$ or $0 < x < 4$

Chapter 3   Chapter Test Solutions

1. $12.4^{2.79} \approx 1123.690$ **2.** $4^{3\pi/2} \approx 687.291$ **3.** $e^{-7/10} \approx 0.497$ **4.** $e^{3.1} \approx 22.198$

5. $f(x) = 10^{-x}$

x	-1	$-\frac{1}{2}$	0	$\frac{1}{2}$	1
$f(x)$	10	3.162	1	0.316	0.1

Horizontal asymptote: $y = 0$

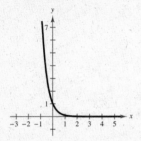

6. $f(x) = -6^{x-2}$

x	-1	0	1	2	3
$f(x)$	-0.005	-0.028	-0.167	-1	-6

Horizontal asymptote: $y = 0$

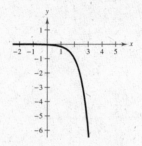

7. $f(x) = 1 - e^{2x}$

x	-1	$-\frac{1}{2}$	0	$\frac{1}{2}$	1
$f(x)$	0.865	0.632	0	-1.718	-6.389

Horizontal asymptote: $y = 1$

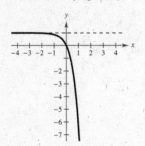

8. (a) $\log_7 7^{-0.89} = -0.89$

(b) $4.6 \ln e^2 = 4.6(2) = 9.2$

9. $f(x) = -\log x - 6$

x	$\frac{1}{2}$	1	$\frac{3}{2}$	2	4
$f(x)$	-5.699	-6	-6.176	-6.301	-6.602

Vertical asymptote: $x = 0$

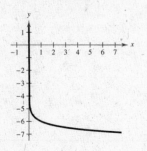

10. $f(x) = \ln(x - 4)$

x	5	7	9	11	13
$f(x)$	0	1.099	1.609	1.946	2.197

Vertical asymptote: $x = 4$

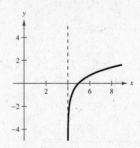

11. $f(x) = 1 + \ln(x + 6)$

x	-5	-3	-1	0	1
$f(x)$	1	2.099	2.609	2.792	2.946

Vertical asymptote: $x = -6$

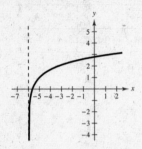

12. $\log_7 44 = \dfrac{\ln 44}{\ln 7} = \dfrac{\log 44}{\log 7} \approx 1.945$

13. $\log_{2/5} 0.9 = \dfrac{\ln 0.9}{\ln(2/5)} = \dfrac{\log 0.9}{\log(2/5)} \approx 0.115$

14. $\log_{24} 68 = \dfrac{\ln 68}{\ln 24} = \dfrac{\log 68}{\log 24} \approx 1.328$

15. $\log_2 3a^4 = \log_2 3 + \log_2 a^4 = \log_2 3 + 4 \log_2 |a|$

16. $\ln \dfrac{5\sqrt{x}}{6} = \ln(5\sqrt{x}) - \ln 6 = \ln 5 + \ln \sqrt{x} - \ln 6 = \ln 5 + \dfrac{1}{2}\ln x - \ln 6$

17. $\log\left(\dfrac{7x^2}{yz^3}\right) = \log 7x^2 - \log yz^3 = \log 7 + \log x^2 - (\log y + \log z^3) = \log 7 + 2 \log x - \log y - 3 \log z$

18. $\log_3 13 + \log_3 y = \log_3 13y$

19. $4 \ln x - 4 \ln y = \ln x^4 - \ln y^4 = \ln\left(\dfrac{x^4}{y^4}\right)$, $x > 0, y > 0$

20. $2 \ln x + \ln(x - 5) - 3 \ln y = \ln x^2 + \ln(x - 5) - \ln y^3$
$$= \ln x^2(x - 5) - \ln y^3$$
$$= \ln \dfrac{x^2(x - 5)}{y^3}$$

21. $5^x = \dfrac{1}{25}$
$$5^x = 5^{-2}$$
$$x = -2$$

22. $3e^{-5x} = 132$
$$e^{-5x} = 44$$
$$-5x = \ln 44$$
$$x = \dfrac{\ln 44}{-5} \approx -0.757$$

23. $\dfrac{1025}{8 + e^{4x}} = 5$
$$1025 = 5(8 + e^{4x})$$
$$205 = 8 + e^{4x}$$
$$197 = e^{4x}$$
$$\ln 197 = 4x$$
$$\dfrac{\ln 197}{4} = x$$
$$x \approx 1.321$$

24. $\ln x = \frac{1}{2}$

$x = e^{1/2} \approx 1.649$

25. $18 + 4 \ln x = 7$

$4 \ln x = -11$

$\ln x = -\frac{11}{4}$

$x = e^{-11/4} \approx 0.064$

26. $\log x - \log(8 - 5x) = 2$

$\log \dfrac{x}{8 - 5x} = 2$

$\dfrac{x}{8 - 5x} = 10^2$

$x = 100(8 - 5x)$

$x = 800 - 500x$

$501x = 800$

$x = \dfrac{800}{501} \approx 1.597$

27. $y = ae^{bt}$

$(0, 2745):$ $2745 = ae^{b(0)} \Rightarrow a = 2745$

$y = 2745e^{bt}$

$(9, 11{,}277):$ $11{,}277 = 2745e^{b(9)}$

$\dfrac{11{,}277}{2745} = e^{9b}$

$\ln\!\left(\dfrac{11{,}277}{2745}\right) = 9b$

$\dfrac{1}{9} \ln\!\left(\dfrac{11{,}277}{2745}\right) = b \Rightarrow b \approx 0.1570$

Thus, $y = 2745e^{0.1570t}$.

28. $y = ae^{bt}$

$\dfrac{1}{2}a = ae^{b(21.77)}$

$\dfrac{1}{2} = e^{21.77b}$

$\ln\!\left(\dfrac{1}{2}\right) = 21.77b$

$b = \dfrac{\ln(1/2)}{21.77} \approx -0.0318$

$y = ae^{-0.0318t}$

When $t = 19$: $y = ae^{-0.0318(19)} \approx 0.55a$

Thus, 55% will remain after 19 years.

29. $H = 70.228 + 5.104x + 9.222 \ln x,\ \frac{1}{4} \le x \le 6$

(a)

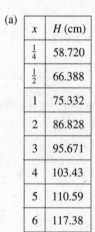

x	H (cm)
$\frac{1}{4}$	58.720
$\frac{1}{2}$	66.388
1	75.332
2	86.828
3	95.671
4	103.43
5	110.59
6	117.38

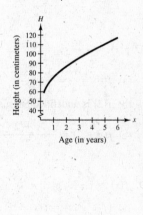

(b) When $x = 4$, $H \approx 103.43$ cm.

Chapters 1–3   Cumulative Test Solutions

1.

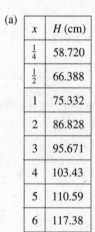

Midpoint: $\left(\dfrac{3 + (-1)}{2}, \dfrac{4 + (-1)}{2}\right) = \left(1, \dfrac{3}{2}\right)$

Distance: $d = \sqrt{(3 - (-1))^2 + (4 - (-1))^2} = \sqrt{(4)^2 + (5)^2} = \sqrt{16 + 25} = \sqrt{41}$

2. $x - 3y + 12 = 0$

Line

x-intercept: $(-12, 0)$

y-intercept: $(0, 4)$

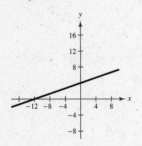

3. $y = x^2 - 9$

Parabola

x-intercepts: $(\pm 3, 0)$

y-intercept: $(0, -9)$

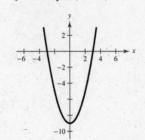

4. $y = \sqrt{4 - x}$

Domain: $x \le 4$

x-intercept: $(4, 0)$

y-intercept: $(0, 2)$

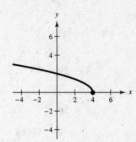

5. $\left(-\dfrac{1}{2}, 1\right)$ and $(3, 8)$

$$m = \frac{8 - 1}{3 - (-1/2)} = \frac{7}{7/2} = 2$$

$$y - 8 = 2(x - 3)$$

$$y - 8 = 2x - 6$$

$$0 = 2x - y + 2$$

6. It fails the Vertical Line Test. For some values of x there correspond two values of y.

7. $f(x) = \dfrac{x}{x - 2}$

(a) $f(6) = \dfrac{6}{4} = \dfrac{3}{2}$

(b) $f(2)$ is undefined because division by zero is undefined.

(c) $f(s + 2) = \dfrac{s + 2}{(s + 2) - 2} = \dfrac{s + 2}{s}$

8. $y = \sqrt[3]{x}$

(a) $r(x) = \dfrac{1}{2}\sqrt[3]{x}$ is a vertical shrink by a factor of $\dfrac{1}{2}$.

(b) $h(x) = \sqrt[3]{x} + 2$ is a vertical shift two units upward.

(c) $g(x) = \sqrt[3]{x + 2}$ is a horizontal shift two units to the left.

9. $f(x) = x - 3$, $g(x) = 4x + 1$

(a) $(f + g)(x) = f(x) + g(x)$

$$= (x - 3) + (4x + 1)$$

$$= 5x - 2$$

(b) $(f - g)(x) = f(x) - g(x)$

$$= (x - 3) - (4x + 1)$$

$$= -3x - 4$$

(c) $(fg)(x) = f(x)g(x)$

$$= (x - 3)(4x + 1)$$

$$= 4x^2 - 11x - 3$$

(d) $\left(\dfrac{f}{g}\right)(x) = \dfrac{f(x)}{g(x)} = \dfrac{x - 3}{4x + 1}$

Domain: all real numbers except $x = -\dfrac{1}{4}$

10. $f(x) = \sqrt{x - 1}$, $g(x) = x^2 + 1$

(a) $(f + g)(x) = f(x) + g(x)$

$$= \sqrt{x - 1} + x^2 + 1$$

(b) $(f - g)(x) = f(x) - g(x)$

$$= \sqrt{x - 1} - x^2 - 1$$

(c) $(fg)(x) = f(x)g(x)$

$$= \sqrt{x - 1}(x^2 + 1) = x^2\sqrt{x - 1} + \sqrt{x - 1}$$

(d) $\left(\dfrac{f}{g}\right)(x) = \dfrac{f(x)}{g(x)} = \dfrac{\sqrt{x - 1}}{x^2 + 1}$

Domain: $x \ge 1$

11. $f(x) = 2x^2, g(x) = \sqrt{x+6}$

    (a) $(f \circ g)(x) = f(g(x))$

$$= f\left(\sqrt{x+6}\right)$$

$$= 2\left(\sqrt{x+6}\right)^2$$

$$= 2(x+6)$$

$$= 2x + 12, \text{ Domain: } x \geq -6$$

    (b) $(g \circ f)(x) = g(f(x))$

$$= g(2x^2)$$

$$= \sqrt{2x^2 + 6}, \text{ Domain: all real numbers}$$

12. $f(x) = x - 2, g(x) = |x|$

    (a) $(f \circ g)(x) = f(g(x))$

$$= f(|x|)$$

$$= |x| - 2, \text{ Domain: all real numbers}$$

    (b) $(g \circ f)(x) = g(f(x))$

$$= g(x - 2)$$

$$= |x - 2|, \text{ Domain: all real numbers}$$

13. The graph of h is a one-to-one line so h has an inverse.

$$h(x) = 5x - 2$$

$$y = 5x - 2$$

$$x = 5y - 2$$

$$x + 2 = 5y$$

$$\tfrac{1}{5}(x + 2) = y$$

$$h^{-1}(x) = \tfrac{1}{5}(x + 2)$$

14. $P = kS^3$

$$750 = k(27)^3 \implies k = \frac{750}{(27)^3} = \frac{250}{6561}$$

$$P = \frac{250}{6561}S^3$$

When $S = 40$: $P = \left(\dfrac{250}{6561}\right)(40)^3 \approx 2438.65$ kilowatts

15. Vertex $(-8, 5)$

Point $(-4, -7)$

$$y - k = a(x - h)^2$$

$$y - 5 = a(x + 8)^2$$

$$-7 - 5 = a(-4 + 8)^2$$

$$-12 = 16a$$

$$-\tfrac{3}{4} = a$$

$$y = -\tfrac{3}{4}(x + 8)^2 + 5$$

16. $h(x) = -(x^2 + 4x)$

$$= -(x^2 + 4x + 4 - 4)$$

$$= -(x + 2)^2 + 4$$

Parabola

Vertex: $(-2, 4)$

Intercepts: $(-4, 0), (0, 0)$

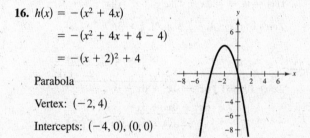

17. $f(t) = \tfrac{1}{4}t(t - 2)^2$

Cubic

Falls to the left

Rises to the right

Intercepts: $(0, 0), (2, 0)$

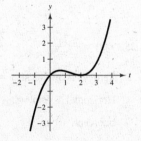

18. $g(s) = s^2 + 4s + 10$

$$= (s^2 + 4s + 4) - 4 + 10$$

$$= (s + 2)^2 + 6$$

Parabola

Vertex: $(-2, 6)$

Intercept: $(0, 10)$

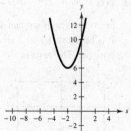

19. $f(x) = x^3 + 2x^2 + 4x + 8$

$\quad = x^2(x + 2) + 4(x + 2)$

$\quad = (x + 2)(x^2 + 4)$

$x + 2 = 0 \Longrightarrow x = -2$

$x^2 + 4 = 0 \Longrightarrow x = \pm 2i$

The zeros of $f(x)$ are -2 and $\pm 2i$.

20. $f(x) = x^4 + 4x^3 - 21x^2$

$\quad = x^2(x^2 + 4x - 21)$

$\quad = x^2(x + 7)(x - 3)$

The zeros of $f(x)$ are 0, -7, and 3.

21. $f(x) = 2x^4 - 11x^3 + 30x^2 - 62x - 40$

Possible Rational Zeros: $\pm 1, \pm 2, \pm 4, \pm 5, \pm 8, \pm 10, \pm 20, \pm 40, \pm\frac{1}{2}, \pm\frac{5}{2}$

By testing (or by looking at the graph of $f(x)$) we see that $x = 4$ and $x = -\frac{1}{2}$ are zeros.

$$
\begin{array}{r|rrrrr}
4 & 2 & -11 & 30 & -62 & -40 \\
  &   & 8 & -12 & 72 & 40 \\
\hline
  & 2 & -3 & 18 & 10 & 0
\end{array}
$$

$$
\begin{array}{r|rrrr}
-\frac{1}{2} & 2 & -3 & 18 & 10 \\
  &   & -1 & 2 & -10 \\
\hline
  & 2 & -4 & 20 & 0
\end{array}
$$

$f(x) = (x - 4)\left(x + \frac{1}{2}\right)(2x^2 - 4x + 20) = (x - 4)\left(x + \frac{1}{2}\right)(2)(x^2 - 2x + 10) = (x - 4)(2x + 1)(x^2 - 2x + 10)$

By Completing the Square (or by the Quadratic Formula) we find the zeros of $x^2 - 2x + 10$ to be $1 \pm 3i$.

$f(x) = (x - 4)(2x + 1)(x - 1 - 3i)(x - 1 + 3i)$

Zeros of $f(x)$: $4, -\frac{1}{2}, 1 + 3i, 1 - 3i$

22.

$$
\begin{array}{r}
3x - 2 + \dfrac{-3x + 2}{2x^2 + 1} \\
2x^2 + 0x + 1 \overline{)\, 6x^3 - 4x^2 + 0x + 0} \\
\underline{6x^3 + 0x^2 + 3x} \\
-4x^2 - 3x + 0 \\
\underline{-4x^2 + 0x - 2} \\
-3x + 2
\end{array}
$$

Thus, $\dfrac{6x^3 - 4x^2}{2x^2 + 1} = 3x - 2 - \dfrac{3x - 2}{2x^2 + 1}$.

23.

$$
\begin{array}{r|rrrrr}
-2 & 2 & 3 & 0 & -6 & 5 \\
  &   & -4 & 2 & -4 & 20 \\
\hline
  & 2 & -1 & 2 & -10 & 25
\end{array}
$$

Thus, $\dfrac{2x^4 + 3x^3 - 6x + 5}{x + 2} = 2x^3 - x^2 + 2x - 10 + \dfrac{25}{x + 2}$.

24. $g(x) = x^3 + 3x^2 - 6$

From the graph we can see that $g(x)$ has one real zero. It is between 1 and 2 since $g(1)$ is negative and $g(2)$ is positive. The zero is $x \approx 1.20$.

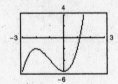

25. $f(x) = \dfrac{2x}{x^2 - 9}$

Vertical asymptotes: $x = \pm 3$

Horizontal asymptote: $y = 0$

Intercept: $(0, 0)$

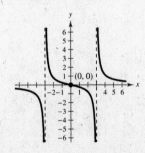

26. $f(x) = \dfrac{x^2 - 4x + 3}{x^2 - 2x - 3}$

$= \dfrac{(x - 1)(x - 3)}{(x + 1)(x - 3)}$

$= \dfrac{x - 1}{x + 1}, \quad x \neq 3$

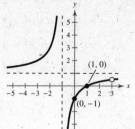

Vertical asymptote: $x = -1$

Horizontal asymptote: $y = 1$

x-intercept: $(1, 0)$

y-intercept: $(0, -1)$

27. $f(x) = \dfrac{x^3 + 3x^2 - 4x - 12}{x^2 - x - 2}$

$= \dfrac{(x + 2)(x - 2)(x + 3)}{(x + 1)(x - 2)}$

$= \dfrac{(x + 2)(x + 3)}{x + 1}$

$= \dfrac{x^2 + 5x + 6}{x + 1}$

$= x + 4 + \dfrac{2}{x + 1}, \quad x \neq 2$

Vertical asymptote: $x = -1$

Slant asymptote: $y = x + 4$

x-intercepts: $(-2, 0), (-3, 0)$

y-intercept: $(0, 6)$

28. $3x^3 - 12x \leq 0$

$3x(x - 2)(x + 2) \leq 0$

Critical numbers: $x = 0, x = \pm 2$

Test intervals: $(-\infty, -2), (-2, 0), (0, 2), (2, \infty)$

Test: Is $3x(x - 2)(x + 2) \leq 0$?

By testing a value in each interval, we have the following solution set: $(-\infty, -2] \cup [0, 2]$.

In inequality form, $x \leq -2$ or $0 \leq x \leq 2$.

29. $\dfrac{1}{x + 1} \geq \dfrac{1}{x + 5}$

$\dfrac{1}{x + 1} - \dfrac{1}{x + 5} \geq 0$

$\dfrac{4}{(x + 1)(x + 5)} \geq 0$

Critical numbers: $x = -1, x = -5$

Test intervals: $(-\infty, -5), (-5, -1), (-1, \infty)$

Test: Is $\dfrac{4}{(x + 1)(x + 5)} \geq 0$?

By testing a value in each interval, we have the following solution set: $(-\infty, -5) \cup (-1, \infty)$.

In inequality form, $x < -5$ or $x > -1$.

30. $f(x) = \left(\frac{2}{5}\right)^x$

$g(x) = -\left(\frac{2}{5}\right)^{-x + 3}$

g is a reflection in the x-axis, a reflection in the y-axis, and a horizontal shift three units to the right of the graph of f.

31. $f(x) = 2.2^x$

$g(x) = -2.2^x + 4$

g is a reflection in the x-axis, and a vertical shift four units upward of the graph of f.

32. $\log 98 \approx 1.991$

33. $\log\left(\frac{6}{7}\right) \approx -0.067$

34. $\ln\sqrt{31} \approx 1.717$

35. $\ln\left(\sqrt{40} - 5\right) \approx 0.281$

36. $\ln\left(\dfrac{x^2 - 16}{x^4}\right) = \ln(x^2 - 16) - \ln x^4$

$= \ln(x + 4)(x - 4) - 4\ln x$

$= \ln(x + 4) + \ln(x - 4) - 4\ln x, \quad x > 4$

37. $2\ln x - \dfrac{1}{2}\ln(x + 5) = \ln x^2 - \ln\sqrt{x + 5}$

$= \ln\dfrac{x^2}{\sqrt{x + 5}}, \quad x > 0$

38. $6e^{2x} = 72$

$e^{2x} = 12$

$2x = \ln 12$

$x = \dfrac{\ln 12}{2} \approx 1.242$

39. $e^{2x} - 11e^x + 24 = 0$

$(e^x - 3)(e^x - 8) = 0$

$e^x - 3 = 0 \Rightarrow e^x = 3 \Rightarrow x = \ln 3 \approx 1.099$

$e^x - 8 = 0 \Rightarrow e^x = 8 \Rightarrow x = \ln 8 \approx 2.079$

40. $\ln \sqrt{x + 2} = 3$

$\dfrac{1}{2} \ln(x + 2) = 3$

$\ln(x + 2) = 6$

$x + 2 = e^6$

$x = e^6 - 2 \approx 401.429$

41. (a)

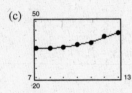

(b) $S \approx 0.274t^2 - 4.08t + 50.6$

(c)

The model is a good fit to the actual data.

(d) For 2008, use $t = 18$: $S(18) \approx \$65.9$ billion

Yes, this seems reasonable, but since the model goes to infinity as t increases, it cannot be used for predictions much beyond this value.

42. $N = 175e^{kt}$

$420 = 175e^{k(8)}$

$2.4 = e^{8k}$

$\ln 2.4 = 8k$

$\dfrac{\ln 2.4}{8} = k$

$k \approx 0.1094$

$N = 175e^{0.1094t}$

$350 = 175e^{0.1094t}$

$2 = e^{0.1094t}$

$\ln 2 = 0.1094t$

$t = \dfrac{\ln 2}{0.1094} \approx 6.3$ hours to double

Chapter 4   Chapter Test Solutions

1. $\theta = \dfrac{5\pi}{4}$ (a)

(b) $\dfrac{5\pi}{4} + 2\pi = \dfrac{13\pi}{4}$ (c) $\dfrac{5\pi}{4}\left(\dfrac{180°}{\pi}\right) = 225°$

$\dfrac{5\pi}{4} - 2\pi = -\dfrac{3\pi}{4}$

2. $90 \dfrac{\text{km}}{\text{hr}} \times \dfrac{1 \text{ hr}}{60 \text{ min}} \times \dfrac{1000 \text{ m}}{1 \text{ km}} = 1500$ meters per minute

$\dfrac{\text{Revolutions}}{\text{minute}} = \dfrac{1500}{\pi}$

Circumference $= 2\pi\left(\dfrac{1}{2}\right) = \pi = \pi$ meters

Angular speed $= \left(\dfrac{1500 \text{ revolutions}}{\pi \text{ minute}}\right)\left(\dfrac{2\pi \text{ radians}}{\text{revolution}}\right)$

$= 3000$ radians per minute

3. $130° = \dfrac{130\pi}{180} = \dfrac{13\pi}{18}$ radians

$A = \dfrac{1}{2}r^2\theta = \dfrac{1}{2}(25)^2\left(\dfrac{13\pi}{18}\right) \approx 709.04$ square feet

4. $x = -2, y = 6$

$r = \sqrt{(-2)^2 + (6)^2} = 2\sqrt{10}$

$\sin\theta = \dfrac{y}{r} = \dfrac{6}{2\sqrt{10}} = \dfrac{3}{\sqrt{10}} = \dfrac{3\sqrt{10}}{10}$

$\csc\theta = \dfrac{r}{y} = \dfrac{2\sqrt{10}}{6} = \dfrac{\sqrt{10}}{3}$

$\cos\theta = \dfrac{x}{r} = \dfrac{-2}{2\sqrt{10}} = -\dfrac{1}{\sqrt{10}} = -\dfrac{\sqrt{10}}{10}$

$\sec\theta = \dfrac{r}{x} = \dfrac{2\sqrt{10}}{-2} = -\sqrt{10}$

$\tan\theta = \dfrac{y}{x} = \dfrac{6}{-2} = -3$

$\cot\theta = \dfrac{x}{y} = \dfrac{-2}{6} = -\dfrac{1}{3}$

5.

For $0 \le \theta < \dfrac{\pi}{2}$, we have:

$\sin\theta = \dfrac{\text{opp}}{\text{hyp}} = \dfrac{3}{\sqrt{13}} = \dfrac{3\sqrt{13}}{13}$

$\cos\theta = \dfrac{\text{adj}}{\text{hyp}} = \dfrac{2}{\sqrt{13}} = \dfrac{2\sqrt{13}}{13}$

$\csc\theta = \dfrac{\text{hyp}}{\text{opp}} = \dfrac{\sqrt{13}}{3}$

$\sec\theta = \dfrac{\text{hyp}}{\text{adj}} = \dfrac{\sqrt{13}}{2}$

$\cot\theta = \dfrac{\text{adj}}{\text{opp}} = \dfrac{2}{3}$

For $\pi \le \theta < \dfrac{3\pi}{2}$, we have:

$\sin\theta = -\dfrac{3\sqrt{13}}{13}$

$\cos\theta = -\dfrac{2\sqrt{13}}{13}$

$\csc\theta = -\dfrac{\sqrt{13}}{3}$

$\sec\theta = -\dfrac{\sqrt{13}}{2}$

$\cot\theta = \dfrac{2}{3}$

6. $\theta = 290°$

$\theta' = 360° - 290° = 70°$

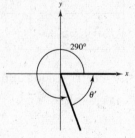

7. $\sec\theta < 0$ and $\tan\theta > 0$

$\dfrac{r}{x} < 0$ and $\dfrac{y}{x} > 0$

Quandrant III

8. $\cos\theta = -\dfrac{\sqrt{3}}{2}$

Reference angle is $30°$ and θ is in Quadrant II or III.

$\theta = 150°$ or $210°$

9. $\csc\theta = 1.030$

$\dfrac{1}{\sin\theta} = 1.030$

$\sin\theta = \dfrac{1}{1.030}$

$\theta = \arcsin\dfrac{1}{1.030}$

$\theta \approx 1.33$ and $\pi - 1.33 \approx 1.81$

10. $\cos\theta = \frac{3}{5}$, $\tan\theta < 0 \implies \theta$ lies in Quadrant IV.

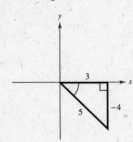

Let $x = 3$, $r = 5 \implies y = -4$.

$\sin\theta = -\frac{4}{5}$ $\csc\theta = -\frac{5}{4}$

$\cos\theta = \frac{3}{5}$ $\sec\theta = \frac{5}{3}$

$\tan\theta = -\frac{4}{3}$ $\cot\theta = -\frac{3}{4}$

11. $\sec\theta = -\frac{17}{8}$, $\sin\theta > 0 \implies \theta$ lies in Quadrant II.

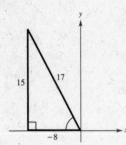

Let $r = 17$, $x = -8 \implies y = 15$.

$\sin\theta = \frac{15}{17}$ $\csc\theta = \frac{17}{15}$

$\cos\theta = -\frac{8}{17}$ $\sec\theta = -\frac{17}{8}$

$\tan\theta = -\frac{15}{8}$ $\cot\theta = -\frac{8}{15}$

12. $g(x) = -2\sin\left(x - \frac{\pi}{4}\right)$

Period: 2π

Amplitude: $|-2| = 2$

Shifted to the right by $\frac{\pi}{4}$ units
and reflected in the x-axis.

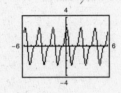

x	0	$\frac{\pi}{4}$	$\frac{3\pi}{4}$	$\frac{5\pi}{4}$	$\frac{7\pi}{4}$
y	$\sqrt{2}$	0	-2	0	2

13. $f(\alpha) = \frac{1}{2}\tan 2\alpha$

Period: $\frac{\pi}{2}$

Asymptotes:

$x = -\frac{\pi}{4}, x = \frac{\pi}{4}$

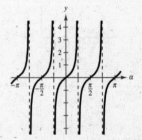

α	$-\frac{\pi}{8}$	0	$\frac{\pi}{8}$
$f(\alpha)$	$-\frac{1}{2}$	0	$\frac{1}{2}$

14. $y = \sin 2\pi x + 2\cos\pi x$

Periodic: period $= 2$

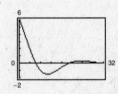

15. $y = 6e^{-0.12t}\cos(0.25t)$, $0 \le t \le 32$

Not periodic

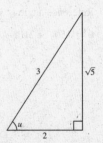

16. $f(x) = a\sin(bx + c)$

Amplitude: $2 \implies |a| = 2$

Reflected in the x-axis: $a = -2$

Period: $4\pi = \frac{2\pi}{b} \implies b = \frac{1}{2}$

Phase shift: $\frac{c}{b} = -\frac{\pi}{2} \implies c = -\frac{\pi}{4}$

$f(x) = -2\sin\left(\frac{x}{2} - \frac{\pi}{4}\right)$

17. Let $u = \arccos\frac{2}{3}$,

$\cos u = \frac{2}{3}$.

$\tan\left(\arccos\frac{2}{3}\right) = \tan u = \frac{\sqrt{5}}{2}$

18. $f(x) = 2 \arcsin\left(\frac{1}{2}x\right)$

Domain: $[-2, 2]$

Range: $[-\pi, \pi]$

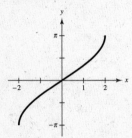

19.

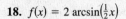

$\tan \theta = -\frac{80}{95} \implies \theta \approx -40.1°$

Bearing: $360° - (90° - 40.1°) = 310.1°$

20. $d = a \cos bt$

$a = -6$

$\dfrac{2\pi}{b} = 2 \implies b = \pi$

$d = -6 \cos \pi t$

Chapter 5   Chapter Test Solutions

1. $\tan \theta = \dfrac{3}{2}$ and $\cos \theta < 0$

θ is in Quadrant III.

$\sec \theta = -\sqrt{1 + \tan^2 \theta} = -\sqrt{1 + \left(\dfrac{3}{2}\right)^2} = -\dfrac{\sqrt{13}}{2}$

$\cos \theta = \dfrac{1}{\sec \theta} = -\dfrac{2}{\sqrt{13}} = -\dfrac{2\sqrt{13}}{13}$

$\sin \theta = \tan \theta \cos \theta = \left(\dfrac{3}{2}\right)\left(-\dfrac{2}{\sqrt{13}}\right) = -\dfrac{3}{\sqrt{13}} = -\dfrac{3\sqrt{13}}{13}$

$\csc \theta = \dfrac{1}{\sin \theta} = -\dfrac{\sqrt{13}}{3}$

$\cot \theta = \dfrac{1}{\tan \theta} = \dfrac{2}{3}$

2. $\csc^2 \beta (1 - \cos^2 \beta) = \dfrac{1}{\sin^2 \beta}(\sin^2 \beta) = 1$

3. $\dfrac{\sec^4 x - \tan^4 x}{\sec^2 x + \tan^2 x} = \dfrac{(\sec^2 x + \tan^2 x)(\sec^2 x - \tan^2 x)}{\sec^2 x + \tan^2 x}$

$\qquad = \sec^2 x - \tan^2 x = 1$

4. $\dfrac{\cos \theta}{\sin \theta} + \dfrac{\sin \theta}{\cos \theta} = \dfrac{\cos^2 \theta + \sin^2 \theta}{\sin \theta \cos \theta} = \dfrac{1}{\sin \theta \cos \theta}$

$\qquad = \csc \theta \sec \theta$

5. $y = \tan\theta,\ y = -\sqrt{\sec^2\theta - 1}$

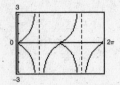

$\tan\theta = -\sqrt{\sec^2\theta - 1}$ on

$\theta = 0, \dfrac{\pi}{2} < \theta \le \pi, \dfrac{3\pi}{2} < \theta < 2\pi.$

6. $y_1 = \cos x + \sin x \tan x,\ y_2 = \sec x$

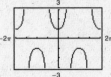

It appears that $y_1 = y_2$.

$\cos x + \sin x \tan x = \cos + \sin x\, \dfrac{\sin x}{\cos x}$

$= \cos + \dfrac{\sin^2 x}{\cos x}$

$= \dfrac{\cos^2 x + \sin^2 x}{\cos x}$

$= \dfrac{1}{\cos x} = \sec x$

7. $\sin\theta \sec\theta = \sin\theta\, \dfrac{1}{\cos\theta} = \dfrac{\sin\theta}{\cos\theta} = \tan\theta$

8. $\sec^2 x \tan^2 x + \sec^2 x = \sec^2 x(\sec^2 x - 1) + \sec^2 x$

$= \sec^4 x - \sec^2 x + \sec^2 x$

$= \sec^4 x$

9. $\dfrac{\csc\alpha + \sec\alpha}{\sin\alpha + \cos\alpha} = \dfrac{\dfrac{1}{\sin\alpha} + \dfrac{1}{\cos\alpha}}{\sin\alpha + \cos\alpha} = \dfrac{\dfrac{\cos\alpha + \sin\alpha}{\sin\alpha \cos\alpha}}{\sin\alpha + \cos\alpha} = \dfrac{1}{\sin\alpha \cos\alpha}$

$= \dfrac{\cos^2\alpha + \sin^2\alpha}{\sin\alpha \cos\alpha} = \dfrac{\cos^2\alpha}{\sin\alpha \cos\alpha} + \dfrac{\sin^2\alpha}{\sin\alpha \cos\alpha}$

$= \dfrac{\cos\alpha}{\sin\alpha} + \dfrac{\sin\alpha}{\cos\alpha} = \cot\alpha + \tan\alpha$

10. $\cos\left(x + \dfrac{\pi}{2}\right) = \cos\left(\dfrac{\pi}{2} - (-x)\right) = \sin(-x) = -\sin x$

11. $\sin(n\pi + \theta) = (-1)^n \sin\theta,\ n$ is an integer.

For n odd: $\sin(n\pi + \theta) = \sin n\pi \cos\theta + \cos n\pi \sin\theta$

$= (0)\cos\theta + (-1)\sin\theta = -\sin\theta$

For n even: $\sin(n\pi + \theta) = \sin n\pi \cos\theta + \cos n\pi \sin\theta$

$= (0)\cos\theta + (1)\sin\theta = \sin\theta$

When n is odd, $(-1)^n = -1$. When n is even $(-1)^n = 1$.

Thus, $\sin(n\pi + \theta) = (-1)^n \sin\theta$ for any integer n.

12. $(\sin x + \cos x)^2 = \sin^2 x + 2\sin x \cos x + \cos^2 x$

$= 1 + 2\sin x \cos x$

$= 1 + \sin 2x$

13. $\sin^4 x \tan^2 x = \sin^4 x \left(\dfrac{\sin^2 x}{\cos^2 x}\right) = \dfrac{\sin^6 x}{\cos^2 x} = \dfrac{(\sin^2 x)^3}{\cos^2 x}$

$= \dfrac{\left(\dfrac{1 - \cos 2x}{2}\right)^3}{\dfrac{1 + \cos 2x}{2}}$

$= \dfrac{\dfrac{1 - 3\cos 2x + 3\cos^2 2x - \cos^3 2x}{8}}{\dfrac{1 + \cos 2x}{2}}$

$= \dfrac{\dfrac{1}{4}\left[1 - 3\cos 2x + 3\left(\dfrac{1 + \cos 4x}{2}\right) - \cos 2x\left(\dfrac{1 + \cos 4x}{2}\right)\right]}{1 + \cos 2x}$

$= \dfrac{\dfrac{1}{8}[2 - 6\cos 2x + 3 + 3\cos 4x - \cos 2x - \cos 2x \cos 4x]}{1 + \cos 2x}$

$= \dfrac{1}{8}\left[\dfrac{5 - 7\cos 2x + 3\cos 4x - \dfrac{1}{2}(\cos(-2x) + \cos(6x))}{1 + \cos 2x}\right]$

$= \dfrac{1}{16}\left[\dfrac{10 - 14\cos 2x + 6\cos 4x - \cos 2x - \cos 6x}{1 + \cos 2x}\right]$

$= \dfrac{1}{16}\left[\dfrac{10 - 15\cos 2x + 6\cos 4x - \cos 6x}{1 + \cos 2x}\right]$

14. $\dfrac{\sin 4\theta}{1 + \cos 4\theta} = \tan \dfrac{4\theta}{2} = \tan 2\theta$

15. $4\cos 2\theta \sin 4\theta = 4\left(\dfrac{1}{2}\right)[\sin(2\theta + 4\theta) - \sin(2\theta - 4\theta)]$

$= 2[\sin 6\theta - \sin(-2\theta)]$

$= 2(\sin 6\theta + \sin 2\theta)$

16. $\sin 3\theta - \sin 4\theta = 2\cos\left(\dfrac{3\theta + 4\theta}{2}\right)\sin\left(\dfrac{3\theta - 4\theta}{2}\right)$

$= 2\cos\dfrac{7\theta}{2}\sin\left(\dfrac{-\theta}{2}\right)$

$= -2\cos\dfrac{7\theta}{2}\sin\dfrac{\theta}{2}$

17. $\tan^2 x + \tan x = 0$

$\tan x(\tan x + 1) = 0$

$\tan x = 0 \quad \text{or} \quad \tan x + 1 = 0$

$x = 0, \pi \qquad \tan x = -1$

$\qquad\qquad x = \dfrac{3\pi}{4}, \dfrac{7\pi}{4}$

18. $\sin 2\alpha - \cos \alpha = 0$

$2\sin \alpha \cos \alpha - \cos \alpha = 0$

$\cos \alpha(2\sin \alpha - 1) = 0$

$\cos \alpha = 0 \quad \text{or} \quad 2\sin \alpha - 1 = 0$

$\alpha = \dfrac{\pi}{2}, \dfrac{3\pi}{2} \qquad \sin \alpha = \dfrac{1}{2}$

$\qquad\qquad \alpha = \dfrac{\pi}{6}, \dfrac{5\pi}{6}$

19. $4\cos^2 x - 3 = 0$

$\cos^2 x = \dfrac{3}{4}$

$\cos x = \pm\sqrt{\dfrac{3}{4}} = \pm\dfrac{\sqrt{3}}{2}$

$x = \dfrac{\pi}{6}, \dfrac{5\pi}{6}, \dfrac{7\pi}{6}, \dfrac{11\pi}{6}$

20. $\csc^2 x - \csc x - 2 = 0$

$(\csc x - 2)(\csc x + 1) = 0$

$\csc x - 2 = 0$ or $\csc x + 1 = 0$

$\csc x = 2$ $\csc = -1$

$\dfrac{1}{\sin x} = 2$ $\dfrac{1}{\sin x} = -1$

$\sin x = \dfrac{1}{2}$ $\sin x = -1$

$x = \dfrac{\pi}{6}, \dfrac{5\pi}{6}$ $x = \dfrac{3\pi}{2}$

21. $3 \cos x - x = 0$

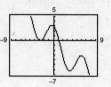

$x \approx -2.938, -2.663, 1.170$

22. $105° = 135° - 30°$

$\cos 105° = \cos(135° - 30°)$

$\quad = \cos 135° \cos 30° + \sin 135° \sin 30°$

$\quad = -\cos 45° \cos 30° + \sin 45° \sin 30°$

$\quad = \left(-\dfrac{\sqrt{2}}{2}\right)\left(\dfrac{\sqrt{3}}{2}\right) + \left(\dfrac{\sqrt{2}}{2}\right)\left(\dfrac{1}{2}\right)$

$\quad = \dfrac{-\sqrt{6} + \sqrt{2}}{4} = \dfrac{\sqrt{2} - \sqrt{6}}{4}$

23. $x = 1, y = 2, r = \sqrt{5}$

$\sin 2u = 2 \sin u \cos u$

$\quad = 2\left(\dfrac{2}{\sqrt{5}}\right)\left(\dfrac{1}{\sqrt{5}}\right) = \dfrac{4}{5}$

$\cos 2u = \cos^2 u - \sin^2 u$

$\quad = \left(\dfrac{1}{\sqrt{5}}\right)^2 - \left(\dfrac{2}{\sqrt{5}}\right)^2 = \dfrac{1}{5} - \dfrac{4}{5} = -\dfrac{3}{5}$

$\tan 2u = \dfrac{2 \tan u}{1 - \tan^2 u} = \dfrac{2(2)}{1 - (2)^2} = \dfrac{4}{-3} = -\dfrac{4}{3}$

24. Let $y_1 = 31 \sin\left(\dfrac{2\pi t}{365} - 1.4\right)$ and $y_2 = 20$.

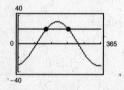

The points of intersection occur when $t \approx 123$ and $t \approx 223$. The number of days that $D > 20°$ is 100, from day 123 to day 223.

25. $28 \cos 10t + 38 = 28 \cos\left[10\left(t - \dfrac{\pi}{6}\right)\right] + 38$

$$\cos 10t = \cos\left[10\left(t - \dfrac{\pi}{6}\right)\right]$$

$$0 = \cos\left[10\left(t - \dfrac{\pi}{6}\right)\right] - \cos 10t$$

$$= -2 \sin\left(\dfrac{10(t - (\pi/6)) + 10t}{2}\right) \sin\left(\dfrac{10(t - (\pi/6)) - 10t}{2}\right)$$

$$= -2 \sin\left(10t - \dfrac{5\pi}{6}\right) \sin\left(-\dfrac{5\pi}{6}\right)$$

$$= -2 \sin\left(10t - \dfrac{5\pi}{6}\right)\left(-\dfrac{1}{2}\right)$$

$$= \sin\left(10t - \dfrac{5\pi}{6}\right)$$

$10t - \dfrac{5\pi}{6} = n\pi$ where n is any integer.

$t = \dfrac{n\pi}{10} + \dfrac{\pi}{12}$ where n is any integer.

The first six times the two people are at the same height are:

0.26 minutes, 0.58 minutes, 0.89 minutes, 1.20 minutes, 1.52 minutes, 1.83 minutes.

Chapter 6   Chapter Test Solutions

1. $A = 24°, B = 68°, a = 12.2$

$C = 180° - 24° - 68° = 88°$

$b = \dfrac{a \sin B}{\sin A} = \dfrac{12.2 \sin 68°}{\sin 24°} \approx 27.81$

$c = \dfrac{a \sin C}{\sin A} = \dfrac{12.2 \sin 88°}{\sin 24°} \approx 29.98$

2. $B = 104°, C = 33°, a = 18.1$

$A = 180° - 104° - 33° = 43°$

$b = \dfrac{a \sin B}{\sin A} = \dfrac{18.1 \sin 104°}{\sin 43°} \approx 25.75$

$c = \dfrac{a \sin C}{\sin A} = \dfrac{18.1 \sin 33°}{\sin 43°} \approx 14.45$

3. $A = 24°, a = 11.2, b = 13.4$

$\sin B = \dfrac{b \sin A}{a} = \dfrac{13.4 \sin 24°}{11.2} \approx 0.4866$

Two Solutions

$B \approx 29.12°$ or $B \approx 150.88°$

$C \approx 126.88°$ $C \approx 5.12°$

$c = \dfrac{a \sin C}{\sin A} = \dfrac{11.2 \sin 126.88°}{\sin 24°}$ $c = \dfrac{11.2 \sin 5.12°}{\sin 24°}$

$c \approx 22.03$ $c \approx 2.46$

4. $a = 4.0, b = 7.3, c = 12.4$

$\cos C = \dfrac{a^2 + b^2 - c^2}{2ab} = \dfrac{4^2 + 7.3^2 - 12.4^2}{2(4)(7.3)} \approx -1.4464 < -1$

No solution

5. $B = 100°, a = 15, b = 23$

$$\sin A = \frac{a \sin B}{b} = \frac{15 \sin 100°}{23} \Rightarrow A \approx 39.96°$$

$$C \approx 180° - 100° - 39.96° = 40.04°$$

$$c \approx \frac{b \sin C}{\sin B} = \frac{23 \sin 40.04°}{\sin 100°} \approx 15.02$$

6. $C = 123°, a = 41, b = 57$

$$c^2 = 41^2 + 57^2 - 2(41)(57)\cos 123° \Rightarrow c \approx 86.46$$

$$\sin A = \frac{a \sin C}{c} = \frac{41 \sin 123°}{86.46} \Rightarrow A \approx 23.43°$$

$$B \approx 180° - 23.43° - 123° = 33.57°$$

7. $a = 60, b = 70, c = 82$

$$s = \frac{60 + 70 + 82}{2} = 106$$

$$\text{Area} = \sqrt{106(46)(36)(24)} \approx 2052.5 \text{ square meters}$$

8.
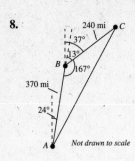
Not drawn to scale

$$b^2 = 370^2 + 240^2 - 2(370)(240)\cos 167°$$

$$b \approx 606.3 \text{ miles}$$

$$\sin A = \frac{a \sin B}{b} = \frac{240 \sin 167°}{606.3}$$

$$A \approx 5.1°$$

Bearing: $24° + 5.1° = 29.1°$

9. Initial point: $(-3, 7)$

Terminal point: $(11, -16)$

$$\mathbf{v} = \langle 11 - (-3), -16 - 7 \rangle = \langle 14, -23 \rangle$$

10. $\mathbf{v} = 12\left(\dfrac{\mathbf{u}}{\|\mathbf{u}\|}\right) = 12\left(\dfrac{\langle 3, -5 \rangle}{\sqrt{3^2 + (-5)^2}}\right) = \dfrac{12}{\sqrt{34}}\langle 3, -5 \rangle$

$$= \frac{6\sqrt{34}}{17}\langle 3, -5 \rangle = \left\langle \frac{18\sqrt{34}}{17}, -\frac{30\sqrt{34}}{17} \right\rangle$$

11. $\mathbf{u} + \mathbf{v} = \langle 3, 5 \rangle + \langle -7, 1 \rangle = \langle -4, 6 \rangle$

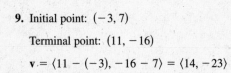

12. $\mathbf{u} - \mathbf{v} = \langle 3, 5 \rangle - \langle -7, 1 \rangle = \langle 10, 4 \rangle$

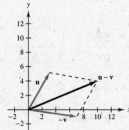

13. $5\mathbf{u} - 3\mathbf{v} = 5\langle 3, 5 \rangle - 3\langle -7, 1 \rangle = \langle 15, 25 \rangle + \langle 21, -3 \rangle$

$$= \langle 36, 22 \rangle$$

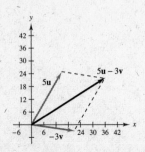

14. $\dfrac{\mathbf{u}}{\|\mathbf{u}\|} = \dfrac{\langle 4, -3 \rangle}{\sqrt{4^2 + (-3)^2}} = \dfrac{1}{5}\langle 4, -3 \rangle = \left\langle \dfrac{4}{5}, -\dfrac{3}{5} \right\rangle$

15. $\mathbf{u} = 250(\cos 45° \, \mathbf{i} + \sin 45° \, \mathbf{j})$

$\mathbf{v} = 130(\cos(-60°)\mathbf{i} + \sin(-60°)\mathbf{j})$

$\mathbf{R} = \mathbf{u} + \mathbf{v} \approx 241.7767 \, \mathbf{i} + 64.1934 \, \mathbf{j}$

$\|\mathbf{R}\| \approx \sqrt{241.7767^2 + 64.1934^2} \approx 250.15$ pounds

$\tan \theta \approx \dfrac{64.1934}{241.7767} \implies \theta \approx 14.9°$

16. $\mathbf{u} = \langle -1, 5 \rangle, \mathbf{v} = \langle 3, -2 \rangle$

$\cos \theta = \dfrac{\mathbf{u} \cdot \mathbf{v}}{\|\mathbf{u}\| \|\mathbf{v}\|} = \dfrac{-13}{\sqrt{26}\sqrt{13}} \implies \theta = 135°$

17. $\mathbf{u} = \langle 6, 10 \rangle, \mathbf{v} = \langle 2, 3 \rangle$

$\mathbf{u} \cdot \mathbf{v} = 42 \neq 0 \implies \mathbf{u}$ and $\mathbf{v}$ are not orthogonal.

18. $\mathbf{u} = \langle 6, 7 \rangle, \mathbf{v} = \langle -5, -1 \rangle$

$\mathbf{w}_1 = \text{proj}_{\mathbf{v}} \, \mathbf{u} = \left(\dfrac{\mathbf{u} \cdot \mathbf{v}}{\|\mathbf{v}\|^2} \right)\mathbf{v} = -\dfrac{37}{26}\langle -5, -1 \rangle = \dfrac{37}{26}\langle 5, 1 \rangle$

$\mathbf{w}_2 = \mathbf{u} - \mathbf{w}_1 = \langle 6, 7 \rangle - \dfrac{37}{26}\langle 5, 1 \rangle$

$\qquad = \left\langle -\dfrac{29}{26}, \dfrac{145}{26} \right\rangle$

$\qquad = \dfrac{29}{26}\langle -1, 5 \rangle$

$\mathbf{u} = \mathbf{w}_1 + \mathbf{w}_2 = \dfrac{37}{26}\langle 5, 1 \rangle + \dfrac{29}{26}\langle -1, 5 \rangle$

19. $\mathbf{F} = -500\mathbf{j}, \mathbf{v} = (\cos 12°)\mathbf{i} + (\sin 12°)\mathbf{j}$

$\mathbf{w}_1 = \text{proj}_{\mathbf{v}} \, \mathbf{F} = \left(\dfrac{\mathbf{F} \cdot \mathbf{v}}{\|\mathbf{v}\|^2} \right)\mathbf{v} = (\mathbf{F} \cdot \mathbf{v})\mathbf{v}$

$\qquad = (-500 \sin 12°)\mathbf{v}$

The magnitude of the force is $500 \sin 12° \approx 104$ pounds.

20. $z = 5 - 5i$

$|z| = \sqrt{5^2 + (-5)^2} = \sqrt{50} = 5\sqrt{2}$

$\tan \theta = \dfrac{-5}{5} = -1$ and θ is in Quadrant IV $\implies \theta = \dfrac{7\pi}{4}$

$z = 5\sqrt{2}\left(\cos \dfrac{7\pi}{4} + i \sin \dfrac{7\pi}{4} \right)$

21. $z = 6(\cos 120° + i \sin 120°)$

$\qquad = 6\left(-\dfrac{1}{2} + \dfrac{\sqrt{3}}{2}i \right) = -3 + 3\sqrt{3}i$

22. $\left[3\left(\cos \dfrac{7\pi}{6} + i \sin \dfrac{7\pi}{6} \right) \right]^8 = 3^8\left(\cos \dfrac{28\pi}{3} + i \sin \dfrac{28\pi}{3} \right)$

$\qquad\qquad = 6561\left(-\dfrac{1}{2} - \dfrac{\sqrt{3}}{2}i \right)$

$\qquad\qquad = -\dfrac{6561}{2} - \dfrac{6561\sqrt{3}}{2}i$

23. $(3 - 3i)^6 = \left[3\sqrt{2}\left(\cos \dfrac{7\pi}{4} + i \sin \dfrac{7\pi}{4} \right) \right]^6$

$\qquad\qquad = \left(3\sqrt{2} \right)^6\left(\cos \dfrac{21\pi}{2} + i \sin \dfrac{21\pi}{2} \right)$

$\qquad\qquad = 5832(0 + i)$

$\qquad\qquad = 5832i$

24. $z = 256\left(1 + \sqrt{3}i\right)$

$\quad |z| = 256\sqrt{1^2 + \left(\sqrt{3}\right)^2} = 256\sqrt{4} = 512$

$\tan \theta = \dfrac{\sqrt{3}}{1} \implies \theta = \dfrac{\pi}{3}$

$\quad z = 512\left(\cos \dfrac{\pi}{3} + i \sin \dfrac{\pi}{3}\right)$

Fourth roots of $z = \sqrt[4]{512}\left[\cos \dfrac{\dfrac{\pi}{3} + 2\pi k}{4} + i \sin \dfrac{\dfrac{\pi}{3} + 2\pi k}{4}\right]$, $k = 0, 1, 2, 3$

$k = 0$: $4\sqrt[4]{2}\left(\cos \dfrac{\pi}{12} + i \sin \dfrac{\pi}{12}\right)$

$k = 1$: $4\sqrt[4]{2}\left(\cos \dfrac{7\pi}{12} + i \sin \dfrac{7\pi}{12}\right)$

$k = 2$: $4\sqrt[4]{2}\left(\cos \dfrac{13\pi}{12} + i \sin \dfrac{13\pi}{12}\right)$

$k = 3$: $4\sqrt[4]{2}\left(\cos \dfrac{19\pi}{12} + i \sin \dfrac{19\pi}{12}\right)$

25. $x^3 - 27i = 0 \implies x = 27i$

The solutions to the equation are the cube roots of $27i = 27\left(\cos \dfrac{\pi}{2} + i \sin \dfrac{\pi}{2}\right)$.

Cube roots: $\sqrt[3]{27}\left[\cos \dfrac{\dfrac{\pi}{2} + 2\pi k}{3} + i \sin \dfrac{\dfrac{\pi}{2} + 2\pi k}{3}\right]$, $k = 0, 1, 2$

$k = 0$: $3\left(\cos \dfrac{\pi}{6} + i \sin \dfrac{\pi}{6}\right) = 3\left(\dfrac{\sqrt{3}}{2} + \dfrac{1}{2}i\right) = \dfrac{3\sqrt{3}}{2} + \dfrac{3}{2}i$

$k = 1$: $3\left(\cos \dfrac{5\pi}{6} + i \sin \dfrac{5\pi}{6}\right) = 3\left(-\dfrac{\sqrt{3}}{2} + \dfrac{1}{2}i\right) = -\dfrac{3\sqrt{3}}{2} + \dfrac{3}{2}i$

$k = 2$: $3\left(\cos \dfrac{3\pi}{2} + i \sin \dfrac{3\pi}{2}\right) = 3(0 - i) = -3i$

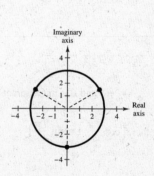

Chapters 4–6   Cumulative Test Solutions

1. (a)

(b) $-120° + 360° = 240°$

(c) $-120\left(\dfrac{\pi}{180°}\right) = -\dfrac{2\pi}{3}$

(d) $-120° + 360° = 240°$

$\quad \theta' = 240° - 180° = 60°$

(e) $\sin(-120°) = -\sin 60° = -\dfrac{\sqrt{3}}{2}$

$\cos(-120°) = -\cos 60° = -\dfrac{1}{2}$

$\tan(-120°) = \tan 60° = \sqrt{3}$

$\csc(-120°) = \dfrac{1}{-\sin 60°} = -\dfrac{2\sqrt{3}}{3}$

$\sec(-120°) = \dfrac{1}{-\cos 60°} = -2$

$\cot(-120°) = \dfrac{1}{\tan 60°} = \dfrac{\sqrt{3}}{3}$

2. $2.35\left(\dfrac{180°}{\pi}\right) \approx 134.6°$

3. $\tan\theta = \dfrac{y}{x} = -\dfrac{4}{3} \implies r = 5$

Since $\sin\theta < 0$ θ is in Quadrant IV, $\implies x = 3$.

$\cos\theta = \dfrac{x}{r} = \dfrac{3}{5}$

4. $f(x) = 3 - 2\sin\pi x$

Period: $\dfrac{2\pi}{\pi} = 2$

Amplitude: $|a| = |-2| = 2$

Upward shift of 3 units (reflected in x-axis prior to shift)

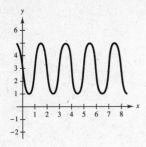

5. $g(x) = \dfrac{1}{2}\tan\left(x - \dfrac{\pi}{2}\right)$

Period: π

Asymptotes: $x = 0, x = \pi$

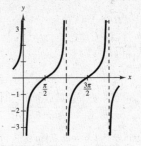

6. $h(x) = -\sec(x + \pi)$

Graph $y = -\cos(x + \pi)$ first.

Period: 2π

Amplitude: 1

Set $x + \pi = 0$ and $x + \pi = 2\pi$ for one cycle

$\qquad x = -\pi \qquad\qquad x = \pi$

The asymptotes of $h(x)$ corresponds to the x-intercepts of

$y = -\cos(x + \pi)$

$x + \pi = \dfrac{(2n + 1)\pi}{2}$

$\quad x = \dfrac{(2n - 1)\pi}{2}$ where n is any integer

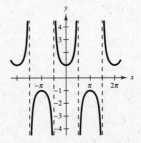

7. $h(x) = a\cos(bx + c)$

Graph is reflected in x-axis.

Amplitude: $a = -3$

Period: $2 = \dfrac{2\pi}{\pi} \implies b = \pi$

No phase shift: $c = 0$

$h(x) = -3\cos(\pi x)$

8. $f(x) = \dfrac{x}{2}\sin x, \ -3\pi \le x \le 3\pi$

$-\dfrac{x}{2} \le f(x) \le \dfrac{x}{2}$

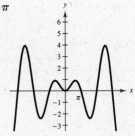

9. $\tan(\arctan 6.7) = 6.7$

10. $\tan\left(\arcsin\dfrac{3}{5}\right) = \dfrac{3}{4}$

11.
$$y = \arccos(2x)$$
$$\sin y = \sin(\arccos(2x)) = \sqrt{1 - 4x^2}$$

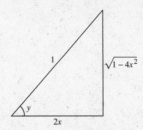

12. $\cos\left(\dfrac{\pi}{2} - x\right)\csc x = \sin x\left(\dfrac{1}{\sin x}\right) = 1$

13. $\dfrac{\sin\theta - 1}{\cos\theta} - \dfrac{\cos\theta}{\sin\theta - 1} = \dfrac{\sin\theta - 1}{\cos\theta} - \dfrac{\cos\theta(\sin\theta + 1)}{\sin^2\theta - 1}$

$$= \dfrac{\sin\theta - 1}{\cos\theta} + \dfrac{\cos\theta(\sin\theta + 1)}{\cos^2\theta} = \dfrac{\sin\theta - 1}{\cos\theta} + \dfrac{\sin\theta + 1}{\cos\theta} = \dfrac{2\sin\theta}{\cos\theta} = 2\tan\theta$$

14. $\cot^2\alpha(\sec^2\alpha - 1) = \cot^2\alpha\tan^2\alpha = 1$

15. $\sin(x + y)\sin(x - y) = \dfrac{1}{2}[\cos(x + y - (x - y)) - \cos(x + y + x - y)]$

$$= \dfrac{1}{2}[\cos 2y - \cos 2x] = \dfrac{1}{2}[1 - 2\sin^2 y - (1 - 2\sin^2 x)] = \sin^2 x - \sin^2 y$$

16. $\sin^2 x \cos^2 x = \left(\dfrac{1 - \cos 2x}{2}\right)\left(\dfrac{1 + \cos 2x}{2}\right)$

$$= \dfrac{1}{4}(1 - \cos 2x)(1 + \cos 2x)$$

$$= \dfrac{1}{4}(1 - \cos^2 2x)$$

$$= \dfrac{1}{4}\left(1 - \dfrac{1 + \cos 4x}{2}\right)$$

$$= \dfrac{1}{8}(2 - (1 + \cos 4x))$$

$$= \dfrac{1}{8}(1 - \cos 4x)$$

17. $2\cos^2\beta - \cos\beta = 0$

$$\cos\beta(2\cos\beta - 1) = 0$$

$$\cos\beta = 0 \qquad 2\cos\beta - 1 = 0$$

$$\beta = \dfrac{\pi}{2}, \dfrac{3\pi}{2} \qquad \cos\beta = \dfrac{1}{2}$$

$$\beta = \dfrac{\pi}{3}, \dfrac{5\pi}{3}$$

Answer: $\dfrac{\pi}{3}, \dfrac{\pi}{2}, \dfrac{3\pi}{2}, \dfrac{5\pi}{3}$

18. $3 \tan \theta - \cot \theta = 0$

$$3 \tan \theta - \frac{1}{\tan \theta} = 0$$

$$\frac{3 \tan^2 \theta - 1}{\tan \theta} = 0$$

$$3 \tan^2 \theta - 1 = 0$$

$$\tan^2 \theta = \frac{1}{3}$$

$$\tan \theta = \pm \frac{\sqrt{3}}{3}$$

$$\theta = \frac{\pi}{6}, \frac{5\pi}{6}, \frac{7\pi}{6}, \frac{11\pi}{6}$$

19. $\sin^2 x + 2 \sin x + 1 = 0$

$$(\sin x + 1)(\sin x + 1) = 0$$

$$\sin x + 1 = 0$$

$$\sin x = -1$$

$$x = \frac{3\pi}{2}$$

20. $\sin u = \frac{12}{13} \implies \cos u = \frac{5}{13}$ and $\tan u = \frac{12}{5}$ since u is in Quadrant I.

$\cos v = \frac{3}{5} \implies \sin v = \frac{4}{5}$ and $\tan v = \frac{4}{3}$ since v is in Quadrant I.

$$\tan(u - v) = \frac{\tan u - \tan v}{1 + \tan u \tan v} = \frac{\frac{12}{5} - \frac{4}{3}}{1 + \left(\frac{12}{5}\right)\left(\frac{4}{3}\right)} = \frac{16}{63}$$

21. $\tan \theta = \frac{1}{2}$

$$\tan 2\theta = \frac{2 \tan \theta}{1 - \tan^2 \theta} = \frac{2\left(\frac{1}{2}\right)}{1 - \left(\frac{1}{2}\right)^2} = \frac{4}{3}$$

22. $\tan \theta = \frac{4}{3} \implies \cos \theta = \pm \frac{3}{5}$

$$\sin \frac{\theta}{2} = \sqrt{\frac{1 - \cos \theta}{2}} = \sqrt{\frac{1 - \frac{3}{5}}{2}} = \frac{\sqrt{5}}{5}$$

$$\text{or} = \sqrt{\frac{1 + \frac{3}{5}}{2}} = \frac{2\sqrt{5}}{5}$$

23. $5 \sin \frac{3\pi}{4} \cos \frac{7\pi}{4} = \frac{5}{2}\left[\sin\left(\frac{3\pi}{4} + \frac{7\pi}{4}\right) + \sin\left(\frac{3\pi}{4} - \frac{7\pi}{4}\right)\right]$

$$= \frac{5}{2}\left[\sin \frac{5\pi}{2} + \sin(-\pi)\right]$$

$$= \frac{5}{2}\left(\sin \frac{5\pi}{2} - \sin \pi\right)$$

24. $\cos 8x + \cos 4x = 2 \cos\left(\frac{8x + 4x}{2}\right) \cos\left(\frac{8x - 4x}{2}\right)$

$$= 2 \cos 6x \cos 2x$$

25. Given: $A = 30°, a = 9, b = 8$

$$\frac{\sin B}{8} = \frac{\sin 30°}{9}$$

$$\sin B = \frac{8}{9}\left(\frac{1}{2}\right)$$

$$B = \arcsin\left(\frac{4}{9}\right)$$

$$B \approx 26.39°$$

$$C = 180° - A - B \approx 123.61°$$

$$\frac{c}{\sin 123.61°} = \frac{9}{\sin 30°}$$

$$c \approx 14.99$$

26. Given: $A = 30°, b = 8, c = 10$

$$a^2 = 8^2 + 10^2 - 2(8)(10)\cos 30°$$

$$a^2 \approx 25.4359$$

$$a \approx 5.04$$

$$\cos B = \frac{5.04^2 + 10^2 - 8^2}{2(5.04)(10)}$$

$$\cos B \approx 0.6091$$

$$B \approx 52.48°$$

$$C = 180° - A - B \approx 97.52°$$

27. Given: $A = 30°$, $C = 90°$, $b = 10$

$$B = 180° - 30° - 90° = 60°$$

$$\tan 30° = \frac{a}{10} \Rightarrow a = 10 \tan 30° \approx 5.77$$

$$\cos 30° = \frac{10}{c} \Rightarrow c = \frac{10}{\cos 30°} \approx 11.55$$

28. $a = 4$, $b = 8$, $c = 9$

$$\cos C = \frac{4^2 + 8^2 - 9^2}{2(4)(8)} = \frac{-1}{64} \Rightarrow C \approx 90.90°$$

$$\sin A \approx \frac{4 \sin 90.90°}{9} \Rightarrow A \approx 26.38°$$

$$B \approx 180° - 26.38° - 90.90° = 62.72°$$

29. Area $= \frac{1}{2}(7)(12) \sin 60° \approx 36.4$ square inches

30. $s = \dfrac{11 + 16 + 17}{2} = 22$

Area $= \sqrt{22(11)(6)(5)} \approx 85.2$ square inches

31. $\mathbf{u} = \langle 3, 5 \rangle = 3\mathbf{i} + 5\mathbf{j}$

32. $\mathbf{v} = \mathbf{i} + \mathbf{j}$

$$\|\mathbf{v}\| = \sqrt{1^2 + 1^2} = \sqrt{2}$$

$$\mathbf{u} = \frac{\mathbf{v}}{\|\mathbf{v}\|} = \frac{1}{\sqrt{2}}(\mathbf{i} + \mathbf{j}) = \frac{\sqrt{2}}{2}(\mathbf{i} + \mathbf{j})$$

33. $\mathbf{u} = 3\mathbf{i} + 4\mathbf{j}$, $\mathbf{v} = \mathbf{i} - 2\mathbf{j}$

$$\mathbf{u} \cdot \mathbf{v} = 3(1) + 4(-2) = -5$$

34. $\mathbf{u} = \langle 8, -2 \rangle$, $\mathbf{v} = \langle 1, 5 \rangle$

$$\mathbf{w}_1 = \text{proj}_{\mathbf{v}}\, \mathbf{u} = \left(\frac{\mathbf{u} \cdot \mathbf{v}}{\|\mathbf{v}\|^2} \right) \mathbf{v} = \frac{-2}{26}\langle 1, 5 \rangle = -\frac{1}{13}\langle 1, 5 \rangle$$

$$\mathbf{w}_2 = \mathbf{u} - \mathbf{w}_1 = \langle 8, -2 \rangle - \left\langle -\frac{1}{13}, -\frac{5}{13} \right\rangle = \left\langle \frac{105}{13}, -\frac{21}{13} \right\rangle$$

$$= \frac{21}{13}\langle 5, -1 \rangle$$

$$\mathbf{u} = \mathbf{w}_1 + \mathbf{w}_2 = -\frac{1}{13}\langle 1, 5 \rangle + \frac{21}{13}\langle 5, -1 \rangle$$

35. $r = |-2 + 2i| = \sqrt{(-2)^2 + (2)^2} = 2\sqrt{2}$

$$\tan \theta = \frac{2}{-2} = -1$$

Since $\tan \theta = -1$ and $-2 + 2i$ lies in Quadrant II,

$\theta = \dfrac{3\pi}{4}$. Thus, $-2 + 2i = 2\sqrt{2}\left(\cos \dfrac{3\pi}{4} + i \sin \dfrac{3\pi}{4} \right)$.

36. $[4(\cos 30° + i \sin 30°)][6(\cos 120° + i \sin 120°)] = (4)(6)[\cos(30° + 120°) + i \sin(30° + 120°)]$

$$= 24(\cos 150° + i \sin 150°)$$

$$= 24\left(-\frac{\sqrt{3}}{2} + \frac{1}{2}i \right)$$

$$= -12\sqrt{3} + 12i$$

37. $1 = 1(\cos 0 + i \sin 0)$

$$\sqrt[3]{1} = \sqrt[3]{1}\left[\cos\left(\frac{0 + 2\pi k}{3}\right) + i \sin\left(\frac{0 + 2\pi k}{3}\right)\right], k = 0, 1, 2$$

$$k = 0: \sqrt[3]{1}\left[\left(\cos\left(\frac{0 + 2\pi(0)}{3}\right) + i \sin\left(\frac{0 + 2\pi(0)}{3}\right)\right)\right] = \cos 0 + i \sin 0 = 1$$

$$k = 1: \sqrt[3]{1}\left[\left(\cos\left(\frac{0 + 2\pi(1)}{3}\right) + i \sin\left(\frac{0 + 2\pi(1)}{3}\right)\right)\right] = \cos\frac{2\pi}{3} + i \sin\frac{2\pi}{3} = -\frac{1}{2} + \frac{\sqrt{3}}{2}i$$

$$k = 2: \sqrt[3]{1}\left[\left(\cos\left(\frac{0 + 2\pi(2)}{3}\right) + i \sin\left(\frac{0 + 2\pi(2)}{3}\right)\right)\right] = \cos\frac{4\pi}{3} - i \sin\frac{4\pi}{3} = -\frac{1}{2} - \frac{\sqrt{3}}{2}i$$

38. $x^5 + 243 = 0 \implies x^5 = -243$

The solutions to the equation are the fifth roots of

$-243 = 243(\cos \pi + i \sin \pi)$, which are:

$$\sqrt[5]{243}\left[\cos\left(\frac{\pi + 2\pi k}{5} + i \sin\frac{\pi + 2\pi k}{5}\right)\right], k = 0, 1, 2, 3, 4$$

$$k = 0: \ 3\left(\cos\frac{\pi}{5} + i \sin\frac{\pi}{5}\right)$$

$$k = 1: \ 3\left(\cos\frac{3\pi}{5} + i \sin\frac{3\pi}{5}\right)$$

$$k = 2: \ 3(\cos \pi + i \sin \pi)$$

$$k = 3: \ 3\left(\cos\frac{7\pi}{5} + i \sin\frac{7\pi}{5}\right)$$

$$k = 4: \ 3\left(\cos\frac{9\pi}{5} + i \sin\frac{9\pi}{5}\right)$$

39. Angular speed $= \dfrac{\theta}{t} = \dfrac{2\pi(63)}{1} \approx 395.8$ radians per minute

Linear speed $= \dfrac{s}{t} = \dfrac{42\pi(63)}{1} \approx 8312.7$ inches per minute

40. Area $= \dfrac{\theta r^2}{2} = \dfrac{(114°)\left(\dfrac{\pi}{180°}\right)(8)^2}{2}$

$\approx 20.267\pi \approx 63.67$ square yards

41. Height of smaller triangle:

$$\tan 16° \ 45' = \frac{h_1}{200}$$

$$h_1 = 200 \tan 16.75°$$

$$\approx 60.2 \text{ feet}$$

Height of larger triangle:

$$\tan 18° = \frac{h_2}{200}$$

$$h_2 = 200 \tan 18° \approx 65.0 \text{ feet}$$

Height of flag: $h_2 - h_1 = 65.0 - 60.2 \approx 5$ feet

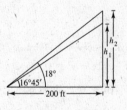

Not drawn to scale

42. $\tan \theta = \dfrac{5}{12} \implies \theta \approx 22.6°$

43. $d = a \cos bt$

$|a| = 4 \implies a = 4$

$\dfrac{2\pi}{b} = 8 \implies b = \dfrac{\pi}{4}$

$d = 4 \cos \dfrac{\pi}{4}t$

44. $\mathbf{v}_1 = 500\langle\cos 60°, \sin 60°\rangle = \langle 250, 250\sqrt{3}\rangle$

$\mathbf{v}_2 = 50\langle\cos 30°, \sin 30°\rangle = \langle 25\sqrt{3}, 25\rangle$

$\mathbf{v} = \mathbf{v}_1 + \mathbf{v}_2 = \langle 250 + 25\sqrt{3}, 250\sqrt{3} + 25\rangle$

$\approx \langle 293.3, 458.0\rangle$

$\|\mathbf{v}\| = \sqrt{(293.3)^2 + (458.0)^2} \approx 543.9$

$\tan\theta = \dfrac{458.0}{293.3} \approx 1.56 \implies \theta \approx 57.4°$

Bearing: $90° - 57.4° = 32.6°$

The plane is traveling
on a bearing of 32.6° at
543.9 kilometers per hour.

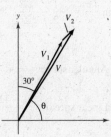

45. $\mathbf{w} = (85)(10)\cos 60° = 425$ foot-pounds

Chapter 7   Chapter Test Solutions

1. $\begin{cases} x - y = -7 \implies y = x + 7 \\ 4x + 5y = 8 \implies 4x + 5(x + 7) = 8 \end{cases}$

$9x + 35 = 8$

$9x = -27$

$x = -3 \implies y = 4$

Solution: $(-3, 4)$

2. $\begin{cases} y = x - 1 \\ y = (x - 1)^3 \end{cases}$

$x - 1 = (x - 1)^3$

$x - 1 = x^3 - 3x^2 + 3x - 1$

$0 = x^3 - 3x^2 + 2x$

$0 = x(x - 1)(x - 2)$

$x = 0, \quad x = 1, \quad x = 2$

$y = -1, \quad y = 0, \quad y = 1$

Solutions: $(0, -1), (1, 0), (2, 1)$

3. $\begin{cases} x - y = 4 \implies x = y + 4 \\ 2x - y^2 = 0 \implies 2(y + 4) - y^2 = 0 \end{cases}$

$0 = y^2 - 2y - 8$

$0 = (y + 2)(y - 4)$

$y = -2 \ \text{ or } \ y = 4$

$x = 2 \qquad\quad x = 8$

Solutions: $(2, -2), (8, 4)$

4. $\begin{cases} 2x - 3y = 0 \\ 2x + 3y = 12 \end{cases}$

Solution: $(3, 2)$

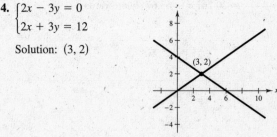

5. $\begin{cases} y = 9 - x^2 \\ y = x + 3 \end{cases}$

Solutions: $(-3, 0), (2, 5)$

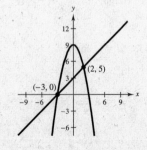

6. $\begin{cases} y - \ln x = 12 \implies y = 12 + \ln x \\ 7x - 2y + 11 = -6 \implies y = \frac{7}{2}x + \frac{17}{2} \end{cases}$

Solutions:
$(1, 12), (0.034, 8.619)$

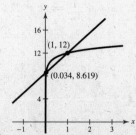
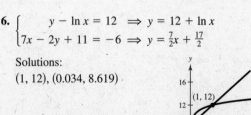

7. $\begin{cases} 2x + 3y = 17 \\ 5x - 4y = -15 \end{cases}$ $\begin{aligned} \Rightarrow \quad & 8x + 12y = 68 \\ \Rightarrow \quad & \underline{15x - 12y = -45} \\ & 23x = 23 \\ & x = 1 \Rightarrow y = 5 \end{aligned}$

Solution: $(1, 5)$

8. $\begin{cases} 2.5x - y = 6 \\ 3x + 4y = 2 \end{cases}$ $\begin{aligned} \Rightarrow \quad & 10x - 4y = 24 \\ \Rightarrow \quad & \underline{3x + 4y = 2} \\ & 13x = 26 \\ & x = 2 \Rightarrow y = -1 \end{aligned}$

Solution: $(2, -1)$

9. $\begin{cases} x - 2y + 3z = 11 \\ 2x - z = 3 \\ 3y + z = -8 \end{cases}$

$\begin{cases} x - 2y + 3z = 11 \\ 4y - 7z = -19 \quad -2\text{Eq.1} + \text{Eq.2} \\ 3y + z = -8 \end{cases}$

$\begin{cases} x - 2y + 3z = 11 \\ y - 8z = -11 \quad -\text{Eq.3} + \text{Eq.2} \\ 3y + z = -8 \end{cases}$

$\begin{cases} x - 2y + 3z = 11 \\ y - 8z = -11 \\ 25z = 25 \quad -3\text{Eq.2} + \text{Eq.3} \end{cases}$

$\begin{cases} x - 2y + 3z = 11 \\ y - 8z = -11 \\ z = 1 \quad \frac{1}{25}\text{Eq.3} \end{cases}$

$y - 8(1) = -11 \implies y = -3$

$x - 2(-3) + 3(1) = 11 \implies x = 2$

Solution: $(2, -3, 1)$

10. $\begin{cases} 3x + 2y + z = 17 \quad \text{Equation 1} \\ -x + y + z = 4 \quad \text{Equation 2} \\ x - y - z = 3 \quad \text{Equation 3} \end{cases}$

Interchange Equations 1 and 3

$\begin{cases} x - y - z = 3 \\ -x + y + z = 4 \\ 3x + 2y + z = 17 \end{cases}$

$\begin{cases} x - y - z = 3 \\ 0 \neq 7 \quad \text{Eq. 1} + \text{Eq. 2} \\ 3x + 2y + z = 17 \end{cases}$

Inconsistent
No solution

11. $\dfrac{2x + 5}{x^2 - x - 2} = \dfrac{2x + 5}{(x - 2)(x + 1)} = \dfrac{A}{x - 2} + \dfrac{B}{x + 1}$

$\quad 2x + 5 = A(x + 1) + B(x - 2)$

Let $x = 2$: $9 = 3A \implies A = 3$

Let $x = -1$: $3 = -3B \implies B = -1$

$\dfrac{2x + 5}{x^2 - x - 2} = \dfrac{3}{x - 2} - \dfrac{1}{x + 1}$

12. $\dfrac{3x^2 - 2x + 4}{x^2(2 - x)} = \dfrac{A}{x} + \dfrac{B}{x^2} + \dfrac{C}{2 - x}$

$\quad 3x^2 - 2x + 4 = Ax(2 - x) + B(2 - x) + Cx^2$

Let $x = 0$: $4 = 2B \implies B = 2$

Let $x = 2$: $12 = 4C \implies C = 3$

Let $x = 1$: $5 = A + B + C = A + 2 + 3 \implies A = 0$

$\dfrac{3x^2 - 2x + 4}{x^2(2 - x)} = \dfrac{2}{x^2} + \dfrac{3}{2 - x}$

13. $\dfrac{x^2 + 5}{x^3 - x} = \dfrac{x^2 + 5}{x(x + 1)(x - 1)} = \dfrac{A}{x} + \dfrac{B}{x + 1} + \dfrac{C}{x - 1}$

$\quad x^2 + 5 = A(x + 1)(x - 1) + Bx(x - 1) + Cx(x + 1)$

Let $x = 0$: $5 = -A \implies A = -5$

Let $x = -1$: $6 = 2B \implies B = 3$

Let $x = 1$: $6 = 2C \implies C = 3$

$\dfrac{x^2 + 5}{x^3 - x} = -\dfrac{5}{x} + \dfrac{3}{x + 1} + \dfrac{3}{x - 1}$

14. $\dfrac{x^2 - 4}{x^3 + 2x} = \dfrac{x^2 - 4}{x(x^2 + 2)} = \dfrac{A}{x} + \dfrac{Bx + C}{x^2 + 2}$

$\quad x^2 - 4 = A(x^2 + 2) + (Bx + C)x$

$ = Ax^2 + 2A + Bx^2 + Cx$

$ = (A + B)x^2 + Cx + 2A$

Equate the coefficients of like terms:

$1 = A + B, 0 = C, -4 = 2A$

Thus, $A = -2, B = 3, C = 0$.

$\dfrac{x^2 - 4}{x^3 + 2x} = -\dfrac{2}{x} + \dfrac{3x}{x^2 + 2}$

15. $2x + y \le 4$

$\quad 2x - y \ge 0$

$\quad\quad x \ge 0$

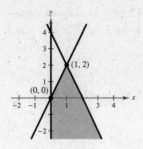

16. $y < -x^2 + x + 4$

$\quad y > 4x$

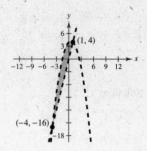

17. $x^2 + y^2 \le 16$

$\quad x \ge 1$

$\quad y \ge -3$

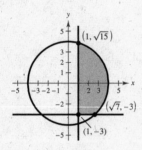

18. Maximize $z = 20x + 12y$ subject to:

$$\begin{cases} x \ge 0, y \ge 0 \\ x + 4y \le 32 \\ 3x + 2y \le 36 \end{cases}$$

At $(0, 0)$ we have $z = 0$.

At $(0, 8)$ we have $z = 96$.

At $(8, 6)$ we have $z = 232$.

At $(12, 0)$ we have $z = 240$.

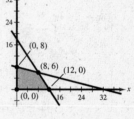

The maximum value, $z = 240$, occurs at $(12, 0)$.

The minimum value, $z = 0$ occurs at $(0, 0)$.

19. Let $x =$ amount in 8% fund.

Let $y =$ amount in 8.5% fund.

$x + y = 50{,}000 \implies y = 50{,}000 - x$

$0.08x + 0.085y = 4150$

Use substitution

$0.08x + 0.085(50{,}000 - x) = 4150$

$\quad 0.08x + 4250 - 0.085x = 4150$

$\quad\quad\quad\quad\quad\quad 100 = 0.005x$

$\quad\quad\quad 20{,}000 = x \implies y = 30{,}000$

$20{,}000 is invested at 8%.

$30{,}000 is invested at 8.5%.

20. $y = ax^2 + bx + c$

$(0, 6)$: $6 = c$

$(-2, 2)$: $2 = 4a - 2b + c$

$\left(3, \frac{9}{2}\right)$: $\frac{9}{2} = 9a + 3b + c$

Solving this system yields: $a = -\frac{1}{2}$, $b = 1$, and $c = 6$.

Thus, $y = -\frac{1}{2}x^2 + x + 6$.

21. Optimize $P = 30x + 40y$ subject to:

$$\begin{cases} x \ge 0, y \ge 0 \\ 0.5x + 0.75y \le 4000 \\ 2.0x + 1.5y \le 8950 \\ 0.5x + 0.5y \le 2650 \end{cases}$$

At $(0, 0)$: $P = 0$

At $(0, 5300)$: $P = 212{,}000$

At $(2000, 3300)$: $P = 192{,}000$

At $(4475, 0)$: $P = 134{,}250$

The manufacturer should produce 5300 units of Model II and not produce any of Model I to realize an optimal profit of $212,000.

Chapter 8   Chapter Test Solutions

1.
$$\begin{bmatrix} 1 & -1 & 5 \\ 6 & 2 & 3 \\ 5 & 3 & -3 \end{bmatrix}$$

$$\begin{matrix} -6R_1 + R_2 \rightarrow \\ -5R_1 + R_3 \rightarrow \end{matrix} \begin{bmatrix} 1 & -1 & 5 \\ 0 & 8 & -27 \\ 0 & 8 & -28 \end{bmatrix}$$

$$\begin{matrix} \\ -R_2 + R_3 \rightarrow \end{matrix} \begin{bmatrix} 1 & -1 & 5 \\ 0 & 8 & -27 \\ 0 & 0 & -1 \end{bmatrix}$$

$$\begin{matrix} \frac{1}{8}R_2 \rightarrow \\ -R_3 \rightarrow \end{matrix} \begin{bmatrix} 1 & -1 & 5 \\ 0 & 1 & -\frac{27}{8} \\ 0 & 0 & 1 \end{bmatrix}$$

$$R_2 + R_1 \rightarrow \begin{bmatrix} 1 & 0 & \frac{13}{8} \\ 0 & 1 & -\frac{27}{8} \\ 0 & 0 & 1 \end{bmatrix}$$

$$\begin{matrix} -\frac{13}{8}R_3 + R_1 \rightarrow \\ \frac{27}{8}R_3 + R_2 \rightarrow \end{matrix} \begin{bmatrix} 1 & 0 & 0 \\ 0 & 1 & 0 \\ 0 & 0 & 1 \end{bmatrix}$$

2.
$$\begin{bmatrix} 1 & 0 & -1 & 2 \\ -1 & 1 & 1 & -3 \\ 1 & 1 & -1 & 1 \\ 3 & 2 & -3 & 4 \end{bmatrix}$$

$$\begin{matrix} R_1 + R_2 \rightarrow \\ -R_1 + R_3 \rightarrow \\ -3R_1 + R_4 \rightarrow \end{matrix} \begin{bmatrix} 1 & 0 & -1 & 2 \\ 0 & 1 & 0 & -1 \\ 0 & 1 & 0 & -1 \\ 0 & 2 & 0 & -2 \end{bmatrix}$$

$$\begin{matrix} \\ -R_2 + R_3 \rightarrow \\ -2R_2 + R_4 \rightarrow \end{matrix} \begin{bmatrix} 1 & 0 & -1 & 2 \\ 0 & 1 & 0 & -1 \\ 0 & 0 & 0 & 0 \\ 0 & 0 & 0 & 0 \end{bmatrix}$$

3.
$$\begin{bmatrix} 4 & 3 & -2 & \vdots & 14 \\ -1 & -1 & 2 & \vdots & -5 \\ 3 & 1 & -4 & \vdots & 8 \end{bmatrix}$$

$$3R_2 + R_1 \rightarrow \begin{bmatrix} 1 & 0 & 4 & \vdots & -1 \\ -1 & -1 & 2 & \vdots & -5 \\ 3 & 1 & -4 & \vdots & 8 \end{bmatrix}$$

$$\begin{matrix} R_1 + R_2 \rightarrow \\ -3R_1 + R_3 \rightarrow \end{matrix} \begin{bmatrix} 1 & 0 & 4 & \vdots & -1 \\ 0 & -1 & 6 & \vdots & -6 \\ 0 & 1 & -16 & \vdots & 11 \end{bmatrix}$$

$$R_2 + R_3 \rightarrow \begin{bmatrix} 1 & 0 & 4 & \vdots & -1 \\ 0 & -1 & 6 & \vdots & -6 \\ 0 & 0 & -10 & \vdots & 5 \end{bmatrix}$$

$$\begin{matrix} -R_2 \rightarrow \\ -\frac{1}{10}R_3 \rightarrow \end{matrix} \begin{bmatrix} 1 & 0 & 4 & \vdots & -1 \\ 0 & 1 & -6 & \vdots & 6 \\ 0 & 0 & 1 & \vdots & -\frac{1}{2} \end{bmatrix}$$

$$\begin{matrix} -4R_3 + R_1 \rightarrow \\ 6R_3 + R_2 \rightarrow \end{matrix} \begin{bmatrix} 1 & 0 & 0 & \vdots & 1 \\ 0 & 1 & 0 & \vdots & 3 \\ 0 & 0 & 1 & \vdots & -\frac{1}{2} \end{bmatrix}$$

Solution: $\left(1, 3, -\frac{1}{2}\right)$

4. (a) $A - B = \begin{bmatrix} 5 & 4 \\ -4 & -4 \end{bmatrix} - \begin{bmatrix} 4 & -1 \\ -4 & 0 \end{bmatrix}$

$$= \begin{bmatrix} 1 & 5 \\ 0 & -4 \end{bmatrix}$$

(b) $3A = 3\begin{bmatrix} 5 & 4 \\ -4 & -4 \end{bmatrix} = \begin{bmatrix} 15 & 12 \\ -12 & -12 \end{bmatrix}$

(c) $3A - 2B = 3\begin{bmatrix} 5 & 4 \\ -4 & -4 \end{bmatrix} - 2\begin{bmatrix} 4 & -1 \\ -4 & 0 \end{bmatrix}$

$$= \begin{bmatrix} 15 & 12 \\ -12 & -12 \end{bmatrix} - \begin{bmatrix} 8 & -2 \\ -8 & 0 \end{bmatrix}$$

$$= \begin{bmatrix} 7 & 14 \\ -4 & -12 \end{bmatrix}$$

(d) $AB = \begin{bmatrix} 5 & 4 \\ -4 & -4 \end{bmatrix}\begin{bmatrix} 4 & -1 \\ -4 & 0 \end{bmatrix}$

$$= \begin{bmatrix} (5)(4) + (4)(-4) & (5)(-1) + (4)(0) \\ (-4)(4) + (-4)(-4) & (-4)(-1) + (-4)(0) \end{bmatrix}$$

$$= \begin{bmatrix} 4 & -5 \\ 0 & 4 \end{bmatrix}$$

5. $\begin{bmatrix} -6 & 4 \\ 10 & -5 \end{bmatrix}^{-1} = \dfrac{1}{(-6)(-5) - (4)(10)}\begin{bmatrix} -5 & -4 \\ -10 & -6 \end{bmatrix} = \begin{bmatrix} \frac{1}{2} & \frac{2}{5} \\ 1 & \frac{3}{5} \end{bmatrix}$

6.
$$\begin{bmatrix} -2 & 4 & -6 & \vdots & 1 & 0 & 0 \\ 2 & 1 & 0 & \vdots & 0 & 1 & 0 \\ 4 & -2 & 5 & \vdots & 0 & 0 & 1 \end{bmatrix}$$

$$\begin{array}{c} R_1 + R_2 \to \\ 2R_1 + R_3 \to \end{array} \begin{bmatrix} -2 & 4 & -6 & \vdots & 1 & 0 & 0 \\ 0 & 5 & -6 & \vdots & 1 & 1 & 0 \\ 0 & 6 & -7 & \vdots & 2 & 0 & 1 \end{bmatrix}$$

$$\begin{array}{c} -\frac{1}{2}R_1 \to \\ -R_3 + R_2 \to \end{array} \begin{bmatrix} 1 & -2 & 3 & \vdots & -\frac{1}{2} & 0 & 0 \\ 0 & -1 & 1 & \vdots & -1 & 1 & -1 \\ 0 & 6 & -7 & \vdots & 2 & 0 & 1 \end{bmatrix}$$

$$\begin{array}{c} -2R_2 + R_1 \to \\ \\ 6R_2 + R_3 \to \end{array} \begin{bmatrix} 1 & 0 & 1 & \vdots & \frac{3}{2} & -2 & 2 \\ 0 & -1 & 1 & \vdots & -1 & 1 & -1 \\ 0 & 0 & -1 & \vdots & -4 & 6 & -5 \end{bmatrix}$$

$$\begin{array}{c} -R_2 \to \\ -R_3 \to \end{array} \begin{bmatrix} 1 & 0 & 1 & \vdots & \frac{3}{2} & -2 & 2 \\ 0 & 1 & -1 & \vdots & 1 & -1 & 1 \\ 0 & 0 & 1 & \vdots & 4 & -6 & 5 \end{bmatrix}$$

$$\begin{array}{c} -R_3 + R_1 \to \\ R_3 + R_2 \to \end{array} \begin{bmatrix} 1 & 0 & 0 & \vdots & -\frac{5}{2} & 4 & -3 \\ 0 & 1 & 0 & \vdots & 5 & -7 & 6 \\ 0 & 0 & 1 & \vdots & 4 & -6 & 5 \end{bmatrix}$$

$$A^{-1} = \begin{bmatrix} -\frac{5}{2} & 4 & -3 \\ 5 & -7 & 6 \\ 4 & -6 & 5 \end{bmatrix}$$

7. $\begin{cases} -6x + 4y = 10 \\ 10x - 5y = 20 \end{cases}$

$$\begin{bmatrix} x \\ y \end{bmatrix} = \begin{bmatrix} \frac{1}{2} & \frac{2}{5} \\ 1 & \frac{3}{5} \end{bmatrix}\begin{bmatrix} 10 \\ 20 \end{bmatrix} = \begin{bmatrix} \frac{1}{2}(10) + \frac{2}{5}(20) \\ 1(10) + \frac{3}{5}(20) \end{bmatrix} = \begin{bmatrix} 13 \\ 22 \end{bmatrix}$$

Solution: $(13, 22)$

8. $\begin{vmatrix} -9 & 4 \\ 13 & 16 \end{vmatrix} = (-9)(16) - (4)(13) = -196$

9. $\begin{vmatrix} \frac{5}{2} & \frac{13}{4} \\ -8 & \frac{6}{5} \end{vmatrix} = \left(\frac{5}{2}\right)\left(\frac{6}{5}\right) - \left(\frac{13}{4}\right)(-8) = 29$

10. $\begin{vmatrix} 6 & -7 & 2 \\ 3 & -2 & 0 \\ 1 & 5 & 1 \end{vmatrix} = 2\begin{vmatrix} 3 & -2 \\ 1 & 5 \end{vmatrix} + \begin{vmatrix} 6 & -7 \\ 3 & -2 \end{vmatrix} = 2(17) + 9 = 43$

Expand along column 3.

11. $\begin{cases} 7x + 6y = 9 \\ -2x - 11y = -49 \end{cases}$ $D = \begin{vmatrix} 7 & 6 \\ -2 & -11 \end{vmatrix} = -65$

$$x = \frac{\begin{vmatrix} 9 & 6 \\ -49 & -11 \end{vmatrix}}{-65} = \frac{195}{-65} = -3$$

$$y = \frac{\begin{vmatrix} 7 & 9 \\ -2 & -49 \end{vmatrix}}{-65} = \frac{-325}{-65} = 5$$

Solution: $(-3, 5)$

12. $\begin{cases} 6x - y + 2z = -4 \\ -2x + 3y - z = 10 \\ 4x - 4y + z = -18 \end{cases}$ $D = \begin{vmatrix} 6 & -1 & 2 \\ -2 & 3 & -1 \\ 4 & -4 & 1 \end{vmatrix} = -12$

$$x = \frac{\begin{vmatrix} -4 & -1 & 2 \\ 10 & 3 & -1 \\ -18 & -4 & 1 \end{vmatrix}}{-12} = \frac{24}{-12} = -2$$

$$y = \frac{\begin{vmatrix} 6 & -4 & 2 \\ -2 & 10 & -1 \\ 4 & -18 & 1 \end{vmatrix}}{-12} = \frac{-48}{-12} = 4$$

$$z = \frac{\begin{vmatrix} 6 & -1 & -4 \\ -2 & 3 & 10 \\ 4 & -4 & -18 \end{vmatrix}}{-12} = \frac{-72}{-12} = 6$$

Solution: $(-2, 4, 6)$

13. $A = -\frac{1}{2}\begin{vmatrix} -5 & 0 & 1 \\ 4 & 4 & 1 \\ 3 & 2 & 1 \end{vmatrix} = -\frac{1}{2}(-14) = 7$

$$\begin{matrix} K & N & O \\ C & K & - \\ \textbf{14. } O & N & - \\ W & O & O \\ D & - & - \end{matrix} \begin{bmatrix} 11 & 14 & 15 \\ 3 & 11 & 0 \\ 15 & 14 & 0 \\ 23 & 15 & 15 \\ 4 & 0 & 0 \end{bmatrix} \begin{bmatrix} 1 & -1 & 0 \\ 1 & 0 & -1 \\ 6 & -2 & -3 \end{bmatrix} = \begin{bmatrix} 115 & -41 & -59 \\ 14 & -3 & -11 \\ 29 & -15 & -14 \\ 128 & -53 & -60 \\ 4 & -4 & 0 \end{bmatrix}$$

Message: $[11 \ 14 \ 15], [3 \ 11 \ 0], [15 \ 14 \ 0], [23 \ 15 \ 15], [4 \ 0 \ 0]$

Encoded Message: $115 \ -41 \ -59 \ 14 \ -3 \ -11 \ 29 \ -15 \ -14 \ 128 \ -53 \ -60 \ 4 \ -4 \ 0$

15. Let $x =$ amount of 60% solution and $y =$ amount of 20% solution.

$$\begin{cases} x + y = 100 \implies y = 100 - x \\ 0.60x + 0.20y = 0.50(100) \implies 6x + 2y = 500 \end{cases}$$

By substitution, we have

$$6x + 2(100 - x) = 500$$
$$6x + 200 - 2x = 500$$
$$4x = 300$$
$$x = 75$$
$$y = 100 - x = 25$$

Answer: 75 liters of 60% solution and 25 liters of 20% solution.

Chapter 9   Chapter Test Solutions

1. $a_n = \dfrac{(-1)^n}{3n + 2}$

$a_1 = -\dfrac{1}{5}$

$a_2 = \dfrac{1}{8}$

$a_3 = -\dfrac{1}{11}$

$a_4 = \dfrac{1}{14}$

$a_5 = -\dfrac{1}{17}$

2. $\dfrac{3}{1!}, \dfrac{4}{2!}, \dfrac{5}{3!}, \dfrac{6}{4!}, \dfrac{7}{5!}, \cdots$

$a_n = \dfrac{n + 2}{n!}$

3. $6 + 17 + 28 + 39 + \cdots$

$a_n = 11n - 5$

$a_5 = 50, a_6 = 61, a_7 = 72$

$S_5 = 6 + 17 + 28 + 39 + 50$

$\quad = 140$

4. $a_5 = 5.4, a_{12} = 11.0$

$a_{12} = a_5 + 7d$

$11.0 = 5.4 + 7d$

$5.6 = 7d$

$0.8 = d$

$a_1 = a_5 - 4d$

$a_1 = 5.4 - 4(0.8)$

$\quad = 2.2$

$a_n = a_1 + (n - 1)d$

$\quad = 2.2 + (n - 1)(0.8)$

$\quad = 0.8n + 1.4$

5. $a_n = 5(2)^{n-1}$

$a_1 = 5$

$a_2 = 10$

$a_3 = 20$

$a_4 = 40$

$a_5 = 80$

6. $\sum_{i=1}^{50} (2i^2 + 5) = 2\sum_{i=1}^{50} i^2 + \sum_{i=1}^{50} 5$

$\qquad = 2\left[\dfrac{50(51)(101)}{6}\right] + 50(5)$

$\qquad = 86,100$

7. $\sum_{n=1}^{7} (8n - 5) = 8\sum_{n=1}^{7} n - \sum_{n=1}^{7} 5$

$\qquad = 8\left[\dfrac{(7)(8)}{2}\right] - 7(5)$

$\qquad = 189$

8. $\sum_{i=1}^{\infty} 4\left(\dfrac{1}{2}\right)^i = \dfrac{2}{1 - \frac{1}{2}} = 4$

9. $5 + 10 + 15 + \cdots + 5n = \dfrac{5n(n + 1)}{2}$

When $n = 1$, $S_1 = 5 = \dfrac{5(1)(2)}{2}$, so the formula is valid.

Assume that $S_k = 5 + 10 + 15 + \cdots + 5k = \dfrac{5k(k + 1)}{2}$, then

$S_{k+1} = S_k + a_{k+1}$

$\qquad = \dfrac{5k(k + 1)}{2} + 5(k + 1)$

$\qquad = \dfrac{5k(k + 1)}{2} + \dfrac{10(k + 1)}{2}$

$\qquad = \dfrac{5k(k + 1) + 10(k + 1)}{2}$

$\qquad = \dfrac{5(k + 1)(k + 2)}{2}$

$\qquad = \dfrac{5(k + 1)[(k + 1) + 1]}{2}$

Thus, the formula is valid for all integers $n \geq 1$.

10. $(x + 2y)^4 = x^4 + 4x^3(2y) + 6x^2(2y)^2 + 4x(2y)^3 + (2y)^4$

$\qquad = x^4 + 8x^3y + 24x^2y^2 + 32xy^3 + 16y^4$

11. ${}_nC_r\, x^{n-r}y^r = {}_8C_5(2a)^3(-3b)^5$

$\qquad = 56(8a^3)(-243b^5)$

$\qquad = -108,864a^3b^5$

So, the coefficient of a^3b^5 is $-108,864$.

12. (a) ${}_9P_2 = \dfrac{9!}{7!} = 72$

(b) ${}_{70}P_3 = \dfrac{70!}{67!} = 328,440$

13. (a) ${}_{11}C_4 = \dfrac{11!}{7!4!} = 330$

(b) ${}_{66}C_4 = \dfrac{66!}{62!4!} = 720,720$

14. $(26)(10)(10)(10) = 26,000$ distinct license plates

15. $\underbrace{(1)}_{\text{owner}} \cdot \underbrace{(3)(2)}_{\substack{\text{bow} \\ \text{seats}}} \cdot \underbrace{(5)(4)(3)(2)(1)}_{\substack{\text{remaining} \\ \text{seats}}} = 720$ seating arrangements

16. $\dfrac{20}{300} = \dfrac{1}{15} \approx 0.0667$

17. $\dfrac{1}{{}_{60}C_8} \approx 3.908 \times 10^{-10}$

18. $P(E') = 1 - P(E)$

$\qquad = 1 - 0.75$

$\qquad = 0.25$ or 25%

Chapters 7–9   Cumulative Test Solutions

1. $\begin{cases} y = 3 - x^2 \\ 2(y - 2) = x - 1 \Rightarrow 2(3 - x^2 - 2) = x - 1 \end{cases}$

$$2(1 - x^2) = x - 1$$

$$2 - 2x^2 = x - 1$$

$$0 = 2x^2 + x - 3$$

$$0 = (2x + 3)(x - 1)$$

$$x = -\tfrac{3}{2} \ \text{ or } \ x = 1$$

$$y = \tfrac{3}{4} \qquad y = 2$$

Solutions: $\left(-\tfrac{3}{2}, \tfrac{3}{4}\right), (1, 2)$

2. $\begin{cases} x + 3y = -1 \Rightarrow 4x + 12y = -4 \\ 2x + 4y = 0 \Rightarrow \underline{-6x - 12y = 0} \end{cases}$

$$ -2x = -4$$

$$x = 2 \Rightarrow y = -1$$

Solution: $(2, -1)$

3. $\begin{cases} -2x + 4y - z = 3 \\ x - 2y + 2z = -6 \\ x - 3y - z = 1 \end{cases}$

Interchange equations.

$\begin{cases} x - 2y + 2z = -6 & \text{Eq.1} \\ -2x + 4y - z = 3 & \text{Eq.2} \\ x - 3y - z = 1 & \text{Eq.3} \end{cases}$

$\begin{cases} x - 2y + 2z = -6 & \\ 3z = -9 & 2\text{Eq.1} + \text{Eq.2} \\ -y - 3z = 7 & -\text{Eq.1} + \text{Eq.3} \end{cases}$

From Equation 2 we have $z = -3$. Substituting this into Equation 3 yields $y = 2$. Using these in Equation 1 yields $x = 4$.

Solution: $(4, 2, -3)$

4. $\begin{cases} x + 3y - 2z = -7 \\ -2x + y - z = -5 \\ 4x + y + z = 3 \end{cases}$

$\begin{cases} x + 3y - 2z = -7 & \\ 7y - 5z = -19 & 2\text{Eq.1} + \text{Eq.2} \\ -11y + 9z = 31 & -4\text{Eq.1} + \text{Eq.3} \end{cases}$

$\begin{cases} x + 3y - 2z = -7 & \\ y - \tfrac{5}{7}z = -\tfrac{19}{7} & \tfrac{1}{7}\text{Eq.2} \\ -11y + 9z = 31 & \end{cases}$

$\begin{cases} x + \tfrac{1}{7}z = \tfrac{8}{7} & -3\text{Eq.2} + \text{Eq.1} \\ y - \tfrac{5}{7}z = -\tfrac{19}{7} & \\ \tfrac{8}{7}z = \tfrac{8}{7} & 11\text{Eq.2} + \text{Eq.3} \end{cases}$

$\begin{cases} x + \tfrac{1}{7}z = \tfrac{8}{7} & \\ y - \tfrac{5}{7}z = -\tfrac{19}{7} & \\ z = 1 & \tfrac{7}{8}\text{Eq.3} \end{cases}$

$\begin{cases} x \phantom{+ -3y + \tfrac{1}{7}z} = 1 & -\tfrac{1}{7}\text{Eq.3} + \text{Eq.1} \\ y \phantom{- \tfrac{5}{7}z} = -2 & \tfrac{5}{7}\text{Eq.3} + \text{Eq.2} \\ z = 1 & \end{cases}$

Solution: $(1, -2, 1)$

5. $\begin{cases} 2x + y \geq -3 \\ x - 3y \leq 2 \end{cases}$

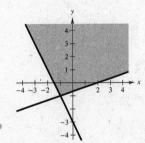

6. $\begin{cases} x - y > 6 \\ 5x + 2y < 10 \end{cases}$

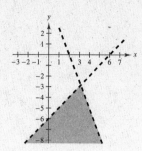

7. Objective function: $z = 3x + 2y$

Subject to: $x + 4y \leq 20$

$\qquad\qquad 2x + y \leq 12$

$\qquad\qquad x \geq 0, y \geq 0$

At $(0, 0)$: $z = 0$

At $(0, 5)$: $z = 10$

At $(4, 4)$: $z = 20$

At $(6, 0)$: $z = 18$

Minimum of $z = 0$ at $(0, 0)$

Maximum of $z = 20$ at $(4, 4)$

8. $\begin{cases} x + y = 200 \implies y = 200 - x \\ 0.75x + 1.25y = 0.95(200) \end{cases}$

$0.75x + 1.25(200 - x) = 190$

$0.75x + 250 - 1.25x = 190$

$\qquad\qquad -0.50x = -60$

$\qquad\qquad\quad x = 120$

$\qquad\quad y = 200 - x = 80$

Answer: 120 pounds of \$0.75 seed and 80 pounds of \$1.25 seed.

9. $y = ax^2 + bx + c$

$(0, 4)$: $4 = a(0)^2 + b(0) + c \implies c = 4$

$(3, 1)$: $1 = a(3)^2 + b(3) + 4 \implies 9a + 3b = -3$

$\qquad\qquad\qquad\qquad\qquad\qquad\quad 3a + b = -1$

$(6, 4)$: $4 = a(6)^2 + b(6) + 4 \implies 36a + 6b = 0$

$\qquad\qquad\qquad\qquad\qquad\qquad\quad 6a + b = 0$

Solving the system:

$\begin{cases} 3a + b = -1 \\ 6a + b = 0 \end{cases}$ yields $a = \frac{1}{3}$ and $b = -2$.

Thus, the equation of the parabola is $y = \frac{1}{3}x^2 - 2x + 4$.

10. $\begin{cases} -x + 2y - z = 9 \\ 2x - y + 2z = -9 \\ 3x + 3y - 4z = 7 \end{cases}$ $\qquad \begin{bmatrix} -1 & 2 & -1 & \vdots & 9 \\ 2 & -1 & 2 & \vdots & -9 \\ 3 & 3 & -4 & \vdots & 7 \end{bmatrix}$

11. $\qquad\qquad \begin{bmatrix} -1 & 2 & -1 & \vdots & 9 \\ 2 & -1 & 2 & \vdots & -9 \\ 3 & 3 & -4 & \vdots & 7 \end{bmatrix}$

$\begin{matrix} \\ 2R_1 + R_2 \rightarrow \\ 3R_1 + R_3 \rightarrow \end{matrix} \begin{bmatrix} -1 & 2 & -1 & \vdots & 9 \\ 0 & 3 & 0 & \vdots & 9 \\ 0 & 9 & -7 & \vdots & 34 \end{bmatrix}$

$\begin{matrix} -R_1 \rightarrow \\ \\ -3R_2 + R_3 \rightarrow \end{matrix} \begin{bmatrix} 1 & -2 & 1 & & -9 \\ 0 & 3 & 0 & & 3 \\ 0 & 0 & -7 & & 7 \end{bmatrix}$

$\begin{matrix} \\ \frac{1}{3}R_2 \rightarrow \\ -\frac{1}{7}R_3 \rightarrow \end{matrix} \begin{bmatrix} 1 & -2 & 1 & \vdots & -9 \\ 0 & 1 & 0 & \vdots & 3 \\ 0 & 0 & 1 & \vdots & -1 \end{bmatrix}$

$\begin{matrix} 2R_2 + R_1 \rightarrow \\ \\ \end{matrix} \begin{bmatrix} 1 & 0 & 1 & \vdots & -3 \\ 0 & 1 & 0 & \vdots & 3 \\ 0 & 0 & 1 & \vdots & -1 \end{bmatrix}$

$\begin{matrix} -R_3 + R_1 \rightarrow \\ \\ \end{matrix} \begin{bmatrix} 1 & 0 & 0 & \vdots & -2 \\ 0 & 1 & 0 & \vdots & 3 \\ 0 & 0 & 1 & \vdots & -1 \end{bmatrix}$

Solution: $(-2, 3, -1)$

12. $A + B = \begin{bmatrix} 4 & 0 \\ -1 & 2 \end{bmatrix} + \begin{bmatrix} -1 & 3 \\ 1 & 0 \end{bmatrix} = \begin{bmatrix} 3 & 3 \\ 0 & 2 \end{bmatrix}$

13. $-2B = -2\begin{bmatrix} -1 & 3 \\ 1 & 0 \end{bmatrix} = \begin{bmatrix} 2 & -6 \\ -2 & 0 \end{bmatrix}$

14. Use the result of Exercise 13.

$$A - 2B = A + (-2B) = \begin{bmatrix} 4 & 0 \\ -1 & 2 \end{bmatrix} + \begin{bmatrix} 2 & -6 \\ -2 & 0 \end{bmatrix} = \begin{bmatrix} 6 & -6 \\ -3 & 2 \end{bmatrix}$$

15. $AB = \begin{bmatrix} 4 & 0 \\ -1 & 2 \end{bmatrix}\begin{bmatrix} -1 & 3 \\ 1 & 0 \end{bmatrix} = \begin{bmatrix} (4)(-1) + (0)(1) & (4)(3) + (0)(0) \\ (-1)(-1) + 2(1) & (-1)(3) + (2)(0) \end{bmatrix} = \begin{bmatrix} -4 & 12 \\ 3 & -3 \end{bmatrix}$

16. $\begin{vmatrix} 8 & 0 & -5 \\ 1 & 3 & -1 \\ -2 & 6 & 4 \end{vmatrix} = 8\begin{vmatrix} 3 & -1 \\ 6 & 4 \end{vmatrix} - 5\begin{vmatrix} 1 & 3 \\ -2 & 6 \end{vmatrix}$

$$= 8(18) - 5(12)$$

$$= 84$$

Expand along Row 1.

17.

$$\begin{bmatrix} 1 & 2 & -1 & \vdots & 1 & 0 & 0 \\ 3 & 7 & -10 & \vdots & 0 & 1 & 0 \\ -5 & -7 & -15 & \vdots & 0 & 0 & 1 \end{bmatrix}$$

$$\begin{matrix} -3R_1 + R_2 \to \\ 5R_1 + R_3 \to \end{matrix} \begin{bmatrix} 1 & 2 & -1 & \vdots & 1 & 0 & 0 \\ 0 & 1 & -7 & \vdots & -3 & 1 & 0 \\ 0 & 3 & -20 & \vdots & 5 & 0 & 1 \end{bmatrix}$$

$$\begin{matrix} -2R_2 + R_1 \to \\ \\ -3R_2 + R_3 \to \end{matrix} \begin{bmatrix} 1 & 0 & 13 & \vdots & 7 & -2 & 0 \\ 0 & 1 & -7 & \vdots & -3 & 1 & 0 \\ 0 & 0 & 1 & \vdots & 14 & -3 & 1 \end{bmatrix}$$

$$\begin{matrix} -13R_3 + R_1 \to \\ 7R_3 + R_2 \to \end{matrix} \begin{bmatrix} 1 & 0 & 0 & \vdots & -175 & 37 & -13 \\ 0 & 1 & 0 & \vdots & 95 & -20 & 7 \\ 0 & 0 & 1 & \vdots & 14 & -3 & 1 \end{bmatrix}$$

$$\begin{bmatrix} 1 & 2 & -1 \\ 3 & 7 & -10 \\ -5 & -7 & -15 \end{bmatrix}^{-1} = \begin{bmatrix} -175 & 37 & -13 \\ 95 & -20 & 7 \\ 14 & -3 & 1 \end{bmatrix}$$

18. Let $x =$ total sales of gym shoes (in millions),

$y =$ total sales of jogging shoes (in millions),

$z =$ total sales of walking shoes (in millions).

$$\begin{bmatrix} 0.09 & 0.09 & 0.03 \\ 0.06 & 0.10 & 0.05 \\ 0.12 & 0.25 & 0.12 \end{bmatrix}\begin{bmatrix} x \\ y \\ z \end{bmatrix} = \begin{bmatrix} 442.20 \\ 466.57 \\ 1088.09 \end{bmatrix}$$

$$\begin{bmatrix} x \\ y \\ z \end{bmatrix} = \begin{bmatrix} 0.09 & 0.09 & 0.03 \\ 0.06 & 0.10 & 0.05 \\ 0.12 & 0.25 & 0.12 \end{bmatrix}^{-1}\begin{bmatrix} 442.20 \\ 466.57 \\ 1088.09 \end{bmatrix}$$

$$\approx \begin{bmatrix} 2042 \\ 1733 \\ 3415 \end{bmatrix}$$

Thus, sales for each type of shoe amounted to:

Gym shoes: \$2042 million

Jogging shoes: \$1733 million

Walking shoes: \$3415 million

19. $\begin{cases} 8x - 3y = -52 \\ 3x + 5y = 5 \end{cases}$, $D = \begin{vmatrix} 8 & -3 \\ 3 & 5 \end{vmatrix} = 49$

$$x = \frac{\begin{vmatrix} -52 & -3 \\ 5 & 5 \end{vmatrix}}{49} = \frac{-245}{49} = -5$$

$$y = \frac{\begin{vmatrix} 8 & -52 \\ 3 & 5 \end{vmatrix}}{49} = \frac{196}{49} = 4$$

Solution: $(-5, 4)$

20. $\begin{cases} 5x + 4y + 3z = 7 \\ -3x - 8y + 7z = -9, \\ 7x - 5y - 6z = -53 \end{cases}$ $D = \begin{vmatrix} 5 & 4 & 3 \\ -3 & -8 & 7 \\ 7 & -5 & -6 \end{vmatrix} = 752$

21. $A = \pm\dfrac{1}{2}\begin{vmatrix} -2 & 3 & 1 \\ 1 & 5 & 1 \\ 4 & 1 & 1 \end{vmatrix} = -\dfrac{1}{2}(-18) = 9$

$x = \dfrac{\begin{vmatrix} 7 & 4 & 3 \\ -9 & -8 & 7 \\ -53 & -5 & -6 \end{vmatrix}}{752} = \dfrac{-2256}{752} = -3$

$y = \dfrac{\begin{vmatrix} 5 & 7 & 3 \\ -3 & -9 & 7 \\ 7 & -53 & -6 \end{vmatrix}}{752} = \dfrac{3008}{752} = 4$

$z = \dfrac{\begin{vmatrix} 5 & 4 & 7 \\ -3 & -8 & -9 \\ 7 & -5 & -53 \end{vmatrix}}{752} = \dfrac{1504}{752} = 2$

Solution: $(-3, 4, 2)$

22. $a_n = \dfrac{(-1)^{n+1}}{2n + 3}$

$a_1 = \dfrac{1}{5}$

$a_2 = -\dfrac{1}{7}$

$a_3 = \dfrac{1}{9}$

$a_4 = -\dfrac{1}{11}$

$a_5 = \dfrac{1}{13}$

23. $\dfrac{2!}{4}, \dfrac{3!}{5}, \dfrac{4!}{6}, \dfrac{5!}{7}, \dfrac{6!}{8}, \cdots$

$a_n = \dfrac{(n + 1)!}{n + 3}$

24. $8, 12, 16, 20, \ldots$

$a_n = 4n + 4$

$a_1 = 8, \ a_{20} = 84$

$S_{20} = \dfrac{20}{2}(8 + 84) = 920$

25. (a) $a_6 = 20.6$

$a_9 = 30.2$

$a_9 = a_6 + 3d$

$30.2 = 20.6 + 3d$

$9.6 = 3d$

$3.2 = d$

$a_{20} = a_9 + 11d = 30.2 + 11(3.2) = 65.4$

(b) $a_1 = a_6 - 5d$

$a_1 = 20.6 - 5(3.2)$

$= 4.6$

$a_n = a_1 + (n - 1)d$

$= 4.6 + (n - 1)(3.2)$

$= 3.2n + 1.4$

26. $a_n = 3(2)^{n-1}$

$a_1 = 3$

$a_2 = 6$

$a_3 = 12$

$a_4 = 24$

$a_5 = 48$

27. $\displaystyle\sum_{i=6}^{\infty} 1.3\left(\dfrac{1}{10}\right)^{i-1} = \dfrac{1.3}{1 - \frac{1}{10}} = 1.3\left(\dfrac{10}{9}\right) = \dfrac{13}{9}$

28. $S_1 = 3 = 1[2(1) + 1]$

Assume that $S_k = 3 + 7 + 11 + 15 + \cdots + (4k - 1) = k(2k + 1)$.

Then, $S_{k+1} = 3 + 7 + 11 + 15 + \cdots + (4k - 1) + [4(k + 1) - 1]$

$$= S_k + (4k + 3)$$
$$= k(2k + 1) + (4k + 3)$$
$$= 2k^2 + 5k + 3$$
$$= (k + 1)(2k + 3)$$
$$= (k + 1)[2(k + 1) + 1].$$

Therefore, the formula is valid for all integers $n \geq 1$.

29. $(z - 3)^4 = z^4 - 4z^3(3) + 6z^2(3)^2 - 4z(3)^3 + (3)^4$

$$= z^4 - 12z^3 + 54z^2 - 108z + 81$$

30. $_7P_3 = \dfrac{7!}{(7 - 3)!} = \dfrac{7!}{4!} = 210$

31. $_{25}P_2 = \dfrac{25!}{(25 - 2)!} = \dfrac{25!}{23!} = 600$

32. $\dbinom{8}{4} = {}_8C_4 = \dfrac{8!}{(8 - 4)!4!} = \dfrac{8!}{4!4!} = 70$

33. $_{10}C_3 = \dfrac{10!}{(10 - 3)!3!} = \dfrac{10!}{7!3!} = 120$

34. B A S K E T B A L L

$$\dfrac{10!}{2!2!2!1!1!1!1!} = 453,600 \text{ distinguishable permutations}$$

35. A N T A R C T I C A

$$\dfrac{10!}{3!2!2!1!1!1!1!} = 151,200 \text{ distinguishable permutations}$$

36. $_{10}P_3 = \dfrac{10!}{(10 - 3)!} = \dfrac{10!}{7!} = 720$

37. The first digit is 4 or 5, so the probability of picking it correctly is $\frac{1}{2}$.
Then there are two numbers left for the second digit so its probability is also $\frac{1}{2}$.
If these two are correct, then the third digit must be the remaining number.
The probability of winning is:

$$\left(\tfrac{1}{2}\right)\left(\tfrac{1}{2}\right)(1) = \tfrac{1}{4}$$

Chapter 10   Chapter Test Solutions

1. $2x - 7y + 3 = 0$

$$y = \tfrac{2}{7}x + \tfrac{3}{7}$$

$$\tan \theta = \tfrac{2}{7}$$

$$\theta \approx 0.2783 \text{ radian} \approx 15.9°$$

2. $3x + 2y - 4 = 0 \Rightarrow y = -\dfrac{3}{2}x + 2 \Rightarrow m_1 = -\dfrac{3}{2}$

$$4x - y + 6 = 0 \Rightarrow y = 4x + 6 \Rightarrow m_2 = 4$$

$$\tan \theta = \left|\dfrac{4 - (-3/2)}{1 + 4(-3/2)}\right| = \dfrac{11}{10}$$

$$\theta \approx 0.8330 \text{ radian} \approx 47.7°$$

3. $y = 5 - x \Rightarrow x + y - 5 = 0 \Rightarrow A = 1, B = 1, C = -5$

$(x_1, y_1) = (7, 5)$

$$d = \dfrac{|(1)(7) + (1)(5) + (-5)|}{\sqrt{1^2 + 1^2}} = \dfrac{7}{\sqrt{2}} = \dfrac{7\sqrt{2}}{2}$$

4. $y^2 - 4x + 4 = 0$

$$y^2 = 4(x - 1)$$

Parabola

Vertex: $(1, 0)$

Focus: $(2, 0)$

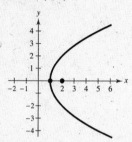

5. $x^2 - 4y^2 - 4x = 0$

$$(x - 2)^2 - 4y^2 = 4$$

$$\frac{(x - 2)^2}{4} - \frac{y^2}{1} = 1$$

Hyperbola

Center: $(2, 0)$

Horizontal transverse axis

$a = 2, b = 1,$

$c^2 = 1 + 4 = 5 \implies c = \sqrt{5}$

Vertices: $(0, 0), (4, 0)$

Foci: $\left(2 \pm \sqrt{5}, 0\right)$

Asymptotes: $y = \pm\frac{1}{2}(x - 2)$

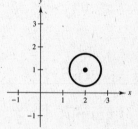

6. $9x^2 + 16y^2 + 54x - 32y - 47 = 0$

$$9(x^2 + 6x + 9) + 16(y^2 - 2y + 1) = 47 + 81 + 16$$

$$9(x + 3)^2 + 16(y - 1)^2 = 144$$

$$\frac{(x + 3)^2}{16} + \frac{(y - 1)^2}{9} = 1$$

Ellipse

Center: $(-3, 1)$

$a = 4, b = 3, c = \sqrt{7}$

Foci: $\left(-3 \pm \sqrt{7}, 1\right)$

Vertices: $(1, 1), (-7, 1)$

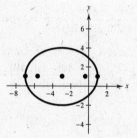

7. $2x^2 + 2y^2 - 8x - 4y + 9 = 0$

$$2(x^2 - 4x + 4) + 2(y^2 - 2y + 1) = -9 + 8 + 2$$

$$2(x - 2)^2 + 2(y - 1)^2 = 1$$

$$(x - 2)^2 + (y - 1)^2 = \frac{1}{2}$$

Circle

Center: $(2, 1)$

Radius:

$$\sqrt{\frac{1}{2}} = \frac{\sqrt{2}}{2} \approx 0.707$$

8. Parabola

Vertex: $(3, -2)$

Vertical axis

Point: $(0, 4)$

$$(x - h)^2 = 4p(y - k)$$

$$(x - 3)^2 = 4p(y + 2)$$

$$(0 - 3)^2 = 4p(4 + 2)$$

$$9 = 24p$$

$$p = \frac{9}{24} = \frac{3}{8}$$

Equation: $(x - 3)^2 = 4\left(\frac{3}{8}\right)(y + 2)$

$$(x - 3)^2 = \frac{3}{2}(y + 2)$$

9. Hyperbola

Foci: $(0, 0)$ and $(0, 4) \implies c = 2$

Asymptotes: $y = \pm\dfrac{1}{2}x + 2$

Vertical transverse axis

Center: $(0, 2) = (h, k)$

$\dfrac{a}{b} = \dfrac{1}{2} \implies 2a = b$

$c^2 = a^2 + b^2$

$4 = a^2 + (2a)^2$

$4 = 5a^2$

$\dfrac{4}{5} = a^2$

$b^2 = (2a)^2 = 4a^2 = \dfrac{16}{5}$

$\dfrac{(y - k)^2}{a^2} - \dfrac{(x - h)^2}{b^2} = 1$

$\dfrac{(y - 2)^2}{4/5} - \dfrac{x^2}{16/5} = 1$

$\dfrac{5(y - 2)^2}{4} - \dfrac{5x^2}{16} = 1$

10. (a) $x^2 + 6xy + y^2 - 6 = 0$

$A = 1, B = 6, C = 1$

$\cot 2\theta = \dfrac{1 - 1}{6} = 0$

$2\theta = 90°$

$\theta = 45°$

(b)

$x = x'\cos 45° - y'\sin 45° = \dfrac{x' - y'}{\sqrt{2}}$

$y = x'\sin 45° + y'\cos 45° = \dfrac{x' + y'}{\sqrt{2}}$

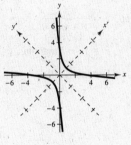

$\left(\dfrac{x' - y'}{\sqrt{2}}\right)^2 + 6\left(\dfrac{x' - y'}{\sqrt{2}}\right)\left(\dfrac{x' + y'}{\sqrt{2}}\right) + \left(\dfrac{x' + y'}{\sqrt{2}}\right)^2 - 6 = 0$

$\dfrac{1}{2}((x')^2 - 2(x')(y') + (y')^2) + 3((x')^2 - (y')^2) + \dfrac{1}{2}((x')^2 + 2(x')(y') + (y')^2) - 6 = 0$

$4(x')^2 - 2(y')^2 = 6$

$\dfrac{2(x')^2}{3} - \dfrac{(y')^2}{3} = 1$

For the graphing utility, we need to solve for y in terms of x.

$y^2 + 6xy + 9x^2 = 6 - x^2 + 9x^2$

$(y + 3x)^2 = 6 + 8x^2$

$y + 3x = \pm\sqrt{6 + 8x^2}$

$y = -3x \pm \sqrt{6 + 8x^2}$

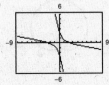

11. $x = 2 + 3 \cos \theta$

$y = 2 \sin \theta$

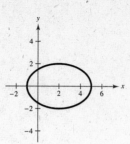

$x = 2 + 3 \cos \theta \implies \dfrac{x - 2}{3} = \cos \theta$

$y = 2 \sin \theta \implies \dfrac{y}{2} = \sin \theta$

$\cos^2 \theta + \sin^2 \theta = 1$

$\dfrac{(x - 2)^2}{9} + \dfrac{y^2}{4} = 1$

θ	0	$\pi/2$	π	$3\pi/2$
x	5	2	-1	2
y	0	2	0	-2

12. $(6, 4), (2, -3)$

$x = x_1 + t(x_2 - x_1) = 6 + t(2 - 6) = 6 - 4t$

$y = y_1 + t(y_2 - y_1) = 4 + t(-3 - 4) = 4 - 7t$

Answers are not unique. Another possible set:

$x = 6 + 4t$

$y = 4 + 7t$

13. Polar coordinates: $\left(-2, \dfrac{5\pi}{6}\right)$

$x = -2 \cos \dfrac{5\pi}{6} = -2\left(-\dfrac{\sqrt{3}}{2}\right) = \sqrt{3}$

$y = -2 \sin \dfrac{5\pi}{6} = -2\left(\dfrac{1}{2}\right) = -1$

Rectangular coordinates: $\left(\sqrt{3}, -1\right)$

14. Rectangular coordinates: $(2, -2)$

$r = \pm\sqrt{2^2 + (-2)^2} = \pm\sqrt{8} = \pm 2\sqrt{2}$

$\tan \theta = -1 \implies \theta = \dfrac{3\pi}{4}, \dfrac{7\pi}{4}$

Polar coordinates:

$\left(2\sqrt{2}, \dfrac{7\pi}{4}\right), \left(-2\sqrt{2}, \dfrac{3\pi}{4}\right), \left(2\sqrt{2}, -\dfrac{\pi}{4}\right)$

15. $x^2 + y^2 - 4y = 0$

$r^2 - 4r \sin \theta = 0$

$r^2 = 4r \sin \theta$

$r = 4 \sin \theta$

16. $r = \dfrac{4}{1 + \cos \theta}$

$e = 1 \implies$ Parabola

Vertex: $(2, 0)$

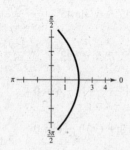

17. $r = \dfrac{4}{2 + \cos \theta} = \dfrac{2}{1 + \frac{1}{2} \cos \theta}$

$e = \dfrac{1}{2} \implies$ Ellipse

Vertex: $\left(\dfrac{4}{3}, 0\right), (4, \pi)$

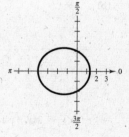

18. $r = 2 + 3 \sin \theta$

$\dfrac{a}{b} = \dfrac{2}{3} < 1$

Limaçon with inner loop

θ	0	$\dfrac{\pi}{2}$	π	$\dfrac{3\pi}{2}$
r	2	5	2	-1

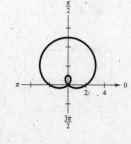

19. $r = 3 \sin 2\theta$

Rose curve ($n = 2$) with four petals

$|r| = 3$ when

$\theta = \dfrac{\pi}{4}, \dfrac{3\pi}{4}, \dfrac{5\pi}{4}, \dfrac{7\pi}{4}$

$r = 0$ when $\theta = 0, \dfrac{\pi}{2}, \pi, \dfrac{3\pi}{2}$

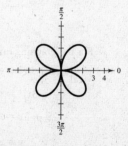

20. Ellipse, $e = \frac{1}{4}$, focus at the pole, directrix $y = 4$

For a horizontal directrix above the pole we have:

$$r = \frac{ep}{1 + e \sin \theta}$$

p = distance between the pole and the directrix $\Rightarrow p = 4$

Thus, $r = \dfrac{(1/4)(4)}{1 + (1/4) \sin \theta} = \dfrac{1}{1 + 0.25 \sin \theta}$.

21.

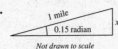
1 mile
0.15 radian
x
Not drawn to scale

Slope: $m = \tan 0.15 \approx 0.1511$

$\sin 0.15 = \dfrac{x}{5280 \text{ feet}}$

$x = 5280 \sin 0.15 \approx 789$ feet

22. $x = (115 \cos \theta)t$ and $y = 3 + (115 \sin \theta)t - 16t^2$

When $\theta = 30°$: $x = (115 \cos 30°)t$

$\quad\quad\quad\quad\quad y = 3 + (115 \sin 30°)t - 16t^2$

When $\theta = 35°$: $x = (115 \cos 35°)t$

$\quad\quad\quad\quad\quad y = 3 + (115 \sin 35°)t - 16t^2$

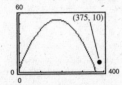

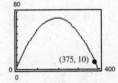

The ball hits the ground inside
the ballpark, so it is not a home run.

The ball clears the 10 foot fence
at 375 feet, so it is a home run.

Chapter 11   Chapter Test Solutions

1.

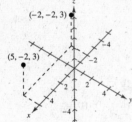

2. $AB = \sqrt{(8 - 6)^2 + (-2 - 4)^2 + (5 + 1)^2} = \sqrt{76}$

$AC = \sqrt{(8 + 4)^2 + (-2 - 3)^2 + (5 - 0)^2} = \sqrt{144 + 25 + 25} = \sqrt{194}$

$BC = \sqrt{(6 + 4)^2 + (4 - 3)^2 + (-1 - 0)^2} = \sqrt{100 + 1 + 1} = \sqrt{102}$

No. $\left(\sqrt{76}\right)^2 + \left(\sqrt{102}\right)^2 \neq \left(\sqrt{194}\right)^2$

3. Midpoint $= \left(\dfrac{8 + 6}{2}, \dfrac{-2 + 4}{2}, \dfrac{5 - 1}{2}\right) = (7, 1, 2)$

4. Diameter $= \sqrt{(8 - 6)^2 + (-2 - 4)^2 + (5 + 1)^2}$

$\quad\quad\quad\quad\quad = \sqrt{4 + 36 + 36} = \sqrt{76}$

Radius $= \sqrt{19}$

$(x - 7)^2 + (y - 1)^2 + (z - 2)^2 = 19$

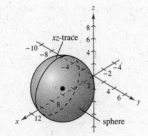

5. $\mathbf{u} = \langle 6 - 8, 4 - (-2), -1 - 5 \rangle = \langle -2, 6, -6 \rangle$

   $\mathbf{v} = \langle -4 - 8, 3 - (-2), 0 - 5 \rangle = \langle -12, 5, -5 \rangle$

6. (a) $\|\mathbf{v}\| = \sqrt{(-12)^2 + 5^2 + (-5)^2} = \sqrt{194}$

   (b) $\mathbf{u} \cdot \mathbf{v} = (-2)(-12) + 6(5) + (-6)(-5) = 84$

   (c) $\mathbf{u} \times \mathbf{v} = \begin{vmatrix} \mathbf{i} & \mathbf{j} & \mathbf{k} \\ -2 & 6 & -6 \\ -12 & 5 & -5 \end{vmatrix} = \langle 0, 62, 62 \rangle$

7. $\cos \theta = \dfrac{\mathbf{u} \cdot \mathbf{v}}{\|\mathbf{u}\| \|\mathbf{v}\|} = \dfrac{84}{\sqrt{76}\sqrt{194}} \approx 0.6918 \implies \theta \approx 46.23$ or 0.8068 radians

8. (a) $x = 8 - 2t, \ y = -2 + 6t, \ z = 5 - 6t$

   (b) $\dfrac{x - 8}{-2} = \dfrac{y + 2}{6} = \dfrac{z - 5}{-6}$

9. $\mathbf{u} \cdot \mathbf{v} = 0 - 2 - 6 \neq 0$ and $\mathbf{u} \neq c\mathbf{v} \implies$ neither

10. $\mathbf{u} \cdot \mathbf{v} = -2 + 3 - 1 = 0 \implies$ orthogonal

11. First two points: $\mathbf{v} = \langle 4, 8, -2 \rangle$

   Last two points: $\mathbf{w} = \langle 4, 8, -2 \rangle$

   Opposite sides are parallel and equal length.

   Adjacent sides: $\mathbf{v}$ and $\mathbf{u} = \langle 1, -3, 3 \rangle$

   Area $= \|\mathbf{u} \times \mathbf{v}\|$

   $\mathbf{u} \times \mathbf{v} = \begin{vmatrix} \mathbf{i} & \mathbf{j} & \mathbf{k} \\ 1 & -3 & 3 \\ 4 & 8 & -2 \end{vmatrix} = \langle -18, 14, 20 \rangle$

   $\|\mathbf{u} \times \mathbf{v}\| = \sqrt{18^2 + 14^2 + 20^2} = 2\sqrt{230} \approx 30.33$ square units

12. $\mathbf{u} = \langle 0, 8, -1 \rangle, \ \mathbf{v} = \langle 4, 5, -4 \rangle$

   $\mathbf{n} = \mathbf{u} \times \mathbf{v} = \begin{vmatrix} \mathbf{i} & \mathbf{j} & \mathbf{k} \\ 0 & 8 & -1 \\ 4 & 5 & -4 \end{vmatrix} = \langle -27, -4, -32 \rangle$

   Plane: $-27(x + 3) - 4(y + 4) - 32(z - 2) = 0$

   $-27x - 4y - 32z - 33 = 0$

   $27x + 4y + 32z + 33 = 0$

13. Let $A(0, 0, 5)$ be the vertex.

   $\mathbf{u} = \overrightarrow{AD} = \langle 4, 0, 0 \rangle, \ \mathbf{v} = \overrightarrow{AB} = \langle 0, 10, 0 \rangle,$

   $\mathbf{w} = \overrightarrow{AE} = \langle 0, 1, -5 \rangle$

   $\mathbf{u} \cdot (\mathbf{v} \times \mathbf{w}) = \begin{vmatrix} 4 & 0 & 0 \\ 0 & 10 & 0 \\ 0 & 1 & -5 \end{vmatrix} = 4(-50) = -200$

   Volume $= |-200| = 200$ cubic units

14. $2x + 3y + 4z = 12$

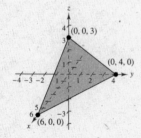

15. $5x - y - 2z = 10$

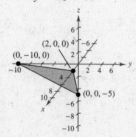

16. $\mathbf{n} = \langle 3, -2, 1 \rangle$, $Q = (2, -1, 6)$, $P = (0, 0, 6)$ in plane, $\overrightarrow{PQ} = \langle 2, -1, 0 \rangle$

$$D = \frac{|\overrightarrow{PQ} \cdot \mathbf{n}|}{\|\mathbf{n}\|} = \frac{|8|}{\sqrt{14}} = \frac{4\sqrt{14}}{7}$$

17. The normal vector to plane containing $(0, 0, 0)$, $(2, 2, 12)$ and $(10, 0, 0)$ is obtained as follows.

$\mathbf{v}_1 = \langle 2, 2, 12 \rangle$, $\mathbf{v}_2 = \langle 10, 0, 0 \rangle$

$$\mathbf{v}_1 \times \mathbf{v}_2 = \begin{vmatrix} \mathbf{i} & \mathbf{j} & \mathbf{k} \\ 2 & 2 & 12 \\ 10 & 0 & 0 \end{vmatrix} = \langle 0, 120, -20 \rangle$$

$\mathbf{n}_1 = \langle 0, 6, -1 \rangle$

The normal vector to the plane containing $(0, 0, 0)$, $(2, 2, 12)$ and $(0, 10, 0)$ is obtained as follows.

$\mathbf{u}_1 = \langle 2, 2, 12 \rangle$, $\mathbf{u}_2 = \langle 0, 10, 0 \rangle$

$$\mathbf{u}_1 \times \mathbf{u}_2 = \begin{vmatrix} \mathbf{i} & \mathbf{j} & \mathbf{k} \\ 2 & 2 & 12 \\ 0 & 10 & 0 \end{vmatrix} = \langle -120, 0, 20 \rangle$$

$\mathbf{n}_2 = \langle -6, 0, 1 \rangle$

The angle θ between two adjacent sides is given by

$$\cos \theta = \frac{|\mathbf{n}_1 \cdot \mathbf{n}_2|}{\|\mathbf{n}_1\| \|\mathbf{n}_2\|} = \frac{|-1|}{\sqrt{37}\sqrt{37}} = \frac{1}{37} \implies \theta \approx 88.45°.$$

Chapter 12   Chapter Test Solutions

1. $f(x) = \dfrac{x^2 - 1}{2x}$

$$\lim_{x \to -2} \frac{x^2 - 1}{2x} = \frac{(-2)^2 - 1}{2(-2)} = -\frac{3}{4}$$

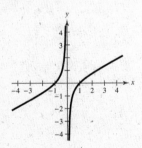

2. $f(x) = \dfrac{2x^2 - x - 3}{x + 1}$

$$= \frac{(2x - 3)(x + 1)}{x + 1} = 2x - 3, x \neq -1$$

$$\lim_{x \to -1} \frac{2x^2 - x - 3}{x + 1} = \lim_{x \to -1} \frac{(2x - 3)(x + 1)}{x + 1}$$

$$= \lim_{x \to -1} (2x - 3)$$

$$= -5$$

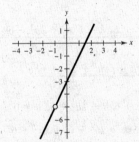

3. $f(x) = \dfrac{\sqrt{x} - 2}{x - 5}$

The graph has a vertical asymptote at $x = 5$.

$\lim\limits_{x \to 5} \dfrac{\sqrt{x} - 2}{x - 5}$ does not exist.

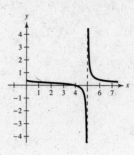

4.

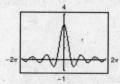

$\lim\limits_{x \to 0} \dfrac{\sin 3x}{x} = 3$

$f(x) = \dfrac{\sin 3x}{x}$

x	-0.02	-0.01	0	0.01	0.02
$f(x)$	2.9982	2.9996	?	2.9996	2.9982

5.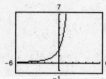

x	-0.004	-0.003	-0.002	-0.001	0	0.001	0.003	0.004
$f(x)$	1.9920	1.9940	1.9960	1.9980	?	2.0020	2.0060	2.0080

$\lim\limits_{x \to 0} \dfrac{e^{2x} - 1}{x} = 2$

$f(x) = \dfrac{e^{2x} - 1}{x}$

6. (a) $\dfrac{f(x + h) - f(x)}{h} = \dfrac{3(x + h)^2 - 5(x + h) - 2 - (3x^2 - 5x - 2)}{h}$

$= \dfrac{3x^2 + 6xh + 3h^2 - 5h - 3x^2}{h}$

$= 6x + 3h - 5$

$f'(x) = \lim\limits_{h \to 0} [6x + 3h - 5] = 6x - 5$

$f'(2) = 6(2) - 5 = 7$

(b) $\dfrac{f(x + h) - f(x)}{h} = \dfrac{[2(x + h)^3 + 6(x + h)] - [2x^3 + 6x]}{h}$

$= \dfrac{2x^3 + 6x^2h + 6xh^2 + 2h^3 + 6x + 6h - 2x^3 - 6x}{h}$

$= \dfrac{6x^2h + 6xh^2 + 2h^3 + 6h}{h}$

$= 6x^2 + 6xh + 2h^2 + 6, \ h \neq 0$

$f'(x) = \lim\limits_{h \to 0} [6x^2 + 6xh + 2h^2 + 6] = 6x^2 + 6$

$f'(-1) = 6(-1)^2 + 6 = 12$

7. $f'(x) = \lim\limits_{h \to 0} \dfrac{f(x+h) - f(x)}{h}$

$\qquad = \lim\limits_{h \to 0} \dfrac{4 - (3/4)(x + h) - [4 - (3/4)x]}{h}$

$\qquad = \lim\limits_{h \to 0} \dfrac{-(3/4)h}{h} = -\dfrac{3}{4}$

8. $f'(x) = \lim\limits_{h \to 0} \dfrac{f(x+h) - f(x)}{h}$

$\qquad = \lim\limits_{h \to 0} \dfrac{2(x+h)^2 + 4(x+h) - 1 - [2x^2 + 4x - 1]}{h}$

$\qquad = \lim\limits_{h \to 0} \dfrac{2x^2 + 4xh + 2h^2 + 4h - 2x^2}{h}$

$\qquad = \lim\limits_{h \to 0} (4x + 2h + 4) = 4x + 4$

9. $f'(x) = \lim\limits_{h \to 0} \dfrac{f(x+h) - f(x)}{h}$

$\qquad = \lim\limits_{h \to 0} \dfrac{\dfrac{1}{x + 3 + h} - \dfrac{1}{x + 3}}{h}$

$\qquad = \lim\limits_{h \to 0} \dfrac{(x + 3) - (x + 3 + h)}{h(x + 3 + h)(x + 3)}$

$\qquad = \lim\limits_{h \to 0} \dfrac{-1}{(x + 3 + h)(x + 3)}$

$\qquad = \dfrac{-1}{(x + 3)^2}$

10. $\lim\limits_{x \to \infty} \dfrac{6}{5x - 1} = 0$

11. $\lim\limits_{x \to \infty} \dfrac{1 - 3x^2}{x^2 - 5} = -3$

12. $\lim\limits_{x \to -\infty} \dfrac{3x^3}{x + 2}$ does not exist.

13. $0, \frac{3}{4}, \frac{14}{19}, \frac{12}{17}, \frac{36}{53}$

$\quad \lim\limits_{n \to \infty} a_n = \frac{1}{2}$

14. $0, 1, 0, \frac{1}{2}, 0$

$\quad \lim\limits_{n \to \infty} a_n = 0$

15. Width of each rectangle: $\frac{1}{2}$

$\quad$ Heights: $8, \frac{15}{2}, 6, \frac{7}{2}$

$\quad$ Area $\approx \frac{1}{2} \left[8 + \frac{15}{2} + 6 + \frac{7}{2} \right] = \frac{25}{2}$

16. Width: $\dfrac{4}{n}$, Height: $f\!\left(-2 + \dfrac{4i}{n}\right) = \left(-2 + \dfrac{4i}{n}\right) + 2 = \dfrac{4i}{n}$

$A \approx \sum\limits_{i=1}^{n} \left(\dfrac{4i}{n}\right)\!\left(\dfrac{4}{n}\right) = \dfrac{16}{n^2} \sum\limits_{i=1}^{n} i = \dfrac{16}{n^2} \dfrac{n(n+1)}{2}$

$A = \lim\limits_{n \to \infty} \dfrac{16}{n^2} \cdot \dfrac{n(n+1)}{2} = 8$

17. Width: $\dfrac{1}{n}$, Height: $f\!\left(\dfrac{i}{n}\right) = 1 - \dfrac{i^3}{n^3}$

$$A \approx \sum_{i=1}^{n}\left(1 - \frac{i^3}{n^3}\right)\!\left(\frac{1}{n}\right) = \sum_{i=1}^{n}\left(\frac{1}{n} - \frac{i^3}{n^4}\right)$$

$$= \frac{1}{n}\sum_{i=1}^{n}1 - \frac{1}{n^4}\sum_{i=1}^{n}i^3$$

$$= \frac{1}{n}(n) - \frac{1}{n^4}\!\left(\frac{n^2(n+1)^2}{4}\right)$$

$$= 1 - \frac{(n+1)^2}{4n^2}$$

$$A = \lim_{n\to\infty}\left(1 - \frac{(n+1)^2}{4n^2}\right) = 1 - \frac{1}{4} = \frac{3}{4}$$

18. (a) $y = 8.79x^2 - 6.2x - 0.4$

(b) Velocity = Derivative = $17.58x - 6.2$

At $x = 5$, velocity ≈ 81.7 ft/sec.

Chapters 10–12　　Cumulative Test Solutions

1. $\dfrac{(x-2)^2}{4} + \dfrac{(y+1)^2}{9} = 1$

Ellipse with center $(2, -1)$

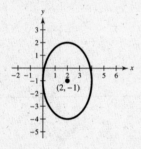

2. $\qquad x^2 + y^2 - 2x - 4y + 1 = 0$

$$(x^2 - 2x + 1) + (y^2 - 4y + 4) = -1 + 1 + 4$$

$$(x - 1)^2 + (y - 2)^2 = 4$$

Circle

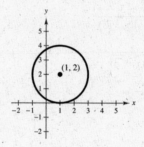

3. Ellipse

Vertices: $(0, 0)$ and $(0, 4) \implies a = 2$

Center: $(0, 2)$

Endpoint of minor axis: $(1, 2)$ and $(-1, 2) \implies b = 1$

Vertical major axis:

$$\frac{(x-0)^2}{1^2} + \frac{(y-2)^2}{2^2} = 1$$

$$\frac{x^2}{1} + \frac{(y-2)^2}{4} = 1$$

4. $x^2 - 4xy + 2y^2 = 6$

$B^2 - 4AC = 16 - 8 = 8 \implies$ Hyperbola

$$\cot 2\theta = \frac{1-2}{-4} = \frac{1}{4} \implies \theta \approx 37.98°$$

Graph as:

$$2y^2 - 4xy + (x^2 - 6) = 0$$

$$y = \frac{4x \pm \sqrt{16x^2 - 8(x^2 - 6)}}{4}$$

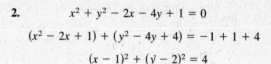

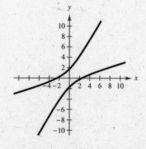

5. $x = 4 \ln t \implies t = e^{x/4}$

$y = \frac{1}{2}t^2$

$y = \frac{1}{2}(e^{x/4})^2 = \frac{1}{2}e^{x/2} = \frac{\sqrt{e^x}}{2}$

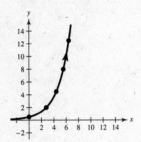

6.

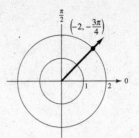

$\left(-2, \frac{5\pi}{4}\right), \left(2, \frac{\pi}{4}\right), \left(2, -\frac{7\pi}{4}\right)$

7.

$-8x - 3y + 5 = 0$

$-8\,r\cos\theta - 3r\sin\theta + 5 = 0$

$r(8\cos\theta + 3\sin\theta) = 5$

$r = \dfrac{5}{8\cos\theta + 3\sin\theta}$

8.

$r = \dfrac{2}{4 - 5\cos\theta}$

$4r - 5r\cos\theta = 2$

$4(x^2 + y^2)^{1/2} - 5x = 2$

$16(x^2 + y^2) = (5x + 2)^2 = 25x^2 + 20x + 4$

$9x^2 + 20x - 16y^2 + 4 = 0$

9. $r = -\dfrac{\pi}{6}$, circle

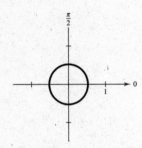

10. $r = 3 - 2\sin\theta$

Dimpled limaçon

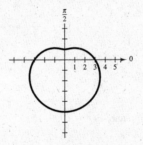

11. $r = 2 + 5\cos\theta$

Limaçon with an inner loop

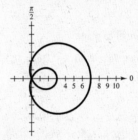

12. $(-6, 1, 3)$

13. $(0, -4, 0)$

14. $d = \sqrt{(4 - (-2))^2 + (-5 - 3)^2 + (1 - (-6))^2}$

$\quad = \sqrt{36 + 64 + 49}$

$\quad = \sqrt{149}$

15. $d_1 = 3, \ d_2 = 4, \ d_3 = \sqrt{4^2 + 3^2} = 5$

$d_1^2 + d_2^2 = d_3^2$

16. Midpoint: $\left(\dfrac{3 - 5}{2}, \dfrac{4 + 0}{2}, \dfrac{-1 + 2}{2}\right) = \left(-1, 2, \dfrac{1}{2}\right)$

17. Center $= (2, 2, 4)$

Radius $= \sqrt{2^2 + 2^2 + 4^2} = \sqrt{24}$

$(x - 2)^2 + (y - 2)^2 + (z - 4)^2 = 24$

18. *xy*-trace: $(z = 0)$

$(x - 2)^2 + (y + 1)^2 = 4$, Circle

yz-trace: $(x = 0)$

$4 + (y + 1)^2 + z^2 = 4$ or $(y + 1)^2 + z^2 = 0$, Point

$(0, -1, 0)$, Point

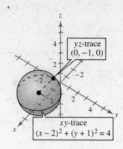

19. $\mathbf{u} \cdot \mathbf{v} = \langle 2, -6, 0 \rangle \cdot \langle -4, 5, 3 \rangle$

$= -8 - 30 = -38$

$$\mathbf{u} \times \mathbf{v} = \begin{vmatrix} \mathbf{i} & \mathbf{j} & \mathbf{k} \\ 2 & -6 & 0 \\ -4 & 5 & 3 \end{vmatrix} = \langle -18, -6, -14 \rangle$$

20. $\mathbf{u} \cdot \mathbf{v} \neq 0, \mathbf{u} \neq c\mathbf{v} \implies$ neither

21. $\mathbf{u} \cdot \mathbf{v} = -8 - 12 + 20 = 0 \implies$ orthogonal

22. $3\mathbf{u} = \langle -3, 18, -9 \rangle = -\mathbf{v} \implies$ parallel

23. $\overrightarrow{DA} = \langle 0, -2, 0 \rangle, \overrightarrow{DC} = \langle 2, 1, 0 \rangle, \overrightarrow{DH} = \langle 0, 0, 3 \rangle$

$$\begin{vmatrix} 0 & -2 & 0 \\ 2 & 1 & 0 \\ 0 & 0 & 3 \end{vmatrix} = 12 \text{ cubic units}$$

24. (a) Vector is $\langle 5 + 2, 8 - 3, 25 - 0 \rangle = \langle 7, 5, 25 \rangle$.

$x = -2 + 7t, y = 3 + 5t, z = 25t$

(b) $\dfrac{x + 2}{7} = \dfrac{y - 3}{5} = \dfrac{z}{25}$

25. $\mathbf{u} = \langle -2, 3, 0 \rangle, \mathbf{v} = \langle 5, 8, 25 \rangle$

$$\mathbf{u} \times \mathbf{v} = \begin{vmatrix} \mathbf{i} & \mathbf{j} & \mathbf{k} \\ -2 & 3 & 0 \\ 5 & 8 & 25 \end{vmatrix} = \langle 75, 50, -31 \rangle$$

Normal to plane

Plane: $75x + 50y - 31z = 0$

26.

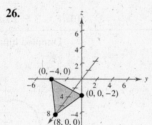

27. $\mathbf{n} = \langle 2, -5, 1 \rangle, Q = (0, 0, 25), P = (0, 0, 10)$ in plane, $\overrightarrow{PQ} = \langle 0, 0, 15 \rangle$

$$D = \frac{|\overrightarrow{PQ} \cdot \mathbf{n}|}{\|\mathbf{n}\|} = \frac{15}{\sqrt{30}} = \frac{\sqrt{30}}{2} \approx 2.74$$

28. Normal to plane containing: $(-1, -1, 3), (0, 0, 0)$ and $(2, 0, 0)$ is

$$\langle -1, -1, 3 \rangle \times \langle 2, 0, 0 \rangle = \begin{vmatrix} \mathbf{i} & \mathbf{j} & \mathbf{k} \\ -1 & -1 & 3 \\ 2 & 0 & 0 \end{vmatrix} = \langle 0, 6, 2 \rangle \text{ or } \mathbf{n}_1 = \langle 0, 3, 1 \rangle$$

Normal to front face is: $\langle 1, -1, 3 \rangle \times \langle 0, 2, 0 \rangle = \begin{vmatrix} \mathbf{i} & \mathbf{j} & \mathbf{k} \\ 1 & -1 & 3 \\ 0 & 2 & 0 \end{vmatrix} = \langle -6, 0, 2 \rangle \text{ or } \mathbf{n}_2 = \langle -3, 0, 1 \rangle$

Angle between sides: $\cos \theta = \dfrac{|\mathbf{n}_1 \cdot \mathbf{n}_2|}{\|\mathbf{n}_1\| \|\mathbf{n}_2\|} = \dfrac{1}{\sqrt{10}\sqrt{10}} = \dfrac{1}{10} \implies \theta \approx 84.26°$

29. $\lim\limits_{x \to 4} (5x - x^2) = 5(4) - 4^2 = 4$

30. $\lim\limits_{x \to -2^+} \dfrac{x + 2}{(x + 2)(x - 1)} = \lim\limits_{x \to -2^+} \dfrac{1}{x - 1} = -\dfrac{1}{3}$

31. $\displaystyle\lim_{x \to 7} \frac{x - 7}{(x - 7)(x + 7)} = \lim_{x \to 7} \frac{1}{x + 7} = \frac{1}{14}$

32. $\displaystyle\lim_{x \to 0} \frac{\sqrt{x + 4} - 2}{x} \cdot \frac{\sqrt{x + 4} + 2}{\sqrt{x + 4} + 2} = \lim_{x \to 0} \frac{(x + 4) - 4}{x\left(\sqrt{x + 4} + 2\right)} = \lim_{x \to 0} \frac{1}{\sqrt{x + 4} + 2} = \frac{1}{2 + 2} = \frac{1}{4}$

33. $\displaystyle\lim_{x \to 4^-} \frac{|x - 4|}{x - 4} = -1$

34. $\displaystyle\lim_{x \to 0} \sin\left(\frac{\pi}{x}\right)$ does not exist.

35. $f(x) = 3 - x^2$

$m = \displaystyle\lim_{h \to 0} \frac{f(x + h) - f(x)}{h}$

$= \displaystyle\lim_{h \to 0} \frac{3 - (x + h)^2 - (3 - x^2)}{h} = \lim_{h \to 0} \frac{-2xh - h^2}{h} = \lim_{h \to 0} (-2x - h) = -2x$

At $(1, 2)$, $m = -2$.

36. $f(x) = \sqrt{x + 3}$

$m = \displaystyle\lim_{h \to 0} \frac{f(x + h) - f(x)}{h}$

$= \displaystyle\lim_{h \to 0} \frac{\sqrt{x + h + 3} - \sqrt{x + 3}}{h} \cdot \frac{\sqrt{x + h + 3} + \sqrt{x + 3}}{\sqrt{x + h + 3} + \sqrt{x + 3}}$

$= \displaystyle\lim_{h \to 0} \frac{(x + h + 3) - (x + 3)}{h[\sqrt{x + h + 3} + \sqrt{x + 3}]}$

$= \displaystyle\lim_{h \to 0} \frac{1}{\sqrt{x + h + 3} + \sqrt{x + 3}} = \frac{1}{2\sqrt{x + 3}}$

At $(-2, 1)$, $m = \dfrac{1}{2}$.

37. $f(x) = \dfrac{1}{x + 3}$

$m = \displaystyle\lim_{h \to 0} \frac{f(x + h) - f(x)}{h}$

$= \displaystyle\lim_{h \to 0} \frac{\dfrac{1}{x + h + 3} - \dfrac{1}{x + 3}}{h}$

$= \displaystyle\lim_{h \to 0} \frac{(x + 3) - (x + h + 3)}{h(x + h + 3)(x + 3)}$

$= \displaystyle\lim_{h \to 0} \frac{-1}{(x + h + 3)(x + 3)}$

$= \dfrac{-1}{(x + 3)^2}$

At $\left(1, \dfrac{1}{4}\right)$, $m = \dfrac{-1}{16}$.

38. $f(x) = x^4$

$m = \displaystyle\lim_{h \to 0} \frac{f(x + h) - f(x)}{h}$

$= \displaystyle\lim_{h \to 0} \frac{(x + h)^4 - x^4}{h}$

$= \displaystyle\lim_{h \to 0} \frac{4x^3h + 6x^2h^2 + 4xh^3 + h^4}{h}$

$= \displaystyle\lim_{h \to 0} [4x^3 + 6x^2h + 4xh^2 + h^3]$

$= 4x^3$

At $(-1, 1)$, $m = -4$.

39. $\displaystyle\lim_{x \to \infty} \frac{2x^4 - x^3 + 4}{x^2 - 9}$

Does not exist

40. $\displaystyle\lim_{x \to \infty} \frac{3 - 7x}{x + 4} = -7$

41. $\displaystyle\lim_{x \to \infty} \frac{3x^2 + 1}{x^2 + 4} = 3$

42. $\displaystyle \lim_{x \to \infty} \frac{2x}{x^2 + 3x - 2} = 0$

43. $\displaystyle \sum_{i=1}^{50} (1 - i^2) = 50 - \frac{50(51)(101)}{6} = -42,875$

44. $\displaystyle \sum_{k=1}^{20} (3k^2 - 2k) = 3\frac{20(21)(41)}{6} - 2\frac{20(21)}{2}$

$$= 8610 - 420 = 8190$$

45. $\displaystyle \sum_{i=1}^{40} (12 + i^3) = 12(40) + \frac{40^2(41)^2}{4}$

$$= 480 + 672,400 = 672,880$$

46. Area $\approx \frac{1}{2}[1 + 2 + 3 + 4 + 5 + 6] = \frac{21}{2} = 10.5$ square units

47. Area $\approx \dfrac{1}{4}\left[\dfrac{1}{1 + \left(-\frac{3}{4}\right)^2} + \dfrac{1}{1 + \left(-\frac{1}{2}\right)^2} + \dfrac{1}{1 + \left(-\frac{1}{4}\right)^2} + \dfrac{1}{1 + 0} + \dfrac{1}{1 + \left(\frac{1}{4}\right)^2} + \dfrac{1}{1 + \left(\frac{1}{2}\right)^2} + \dfrac{1}{1 + \left(\frac{3}{4}\right)^2} + \dfrac{1}{1 + 1^2} \right]$

$$= \frac{1}{4}\left[2(0.64) + 2(0.8) + 2(0.941176) + 1 + \frac{1}{2} \right]$$

$$\approx 1.566 \text{ square units}$$

48. Width: $\dfrac{1}{n}$, Height: $f\left(\dfrac{i}{n}\right) = 1 - \left(\dfrac{1}{n}\right)^3$

$$A \approx \sum_{i=1}^{n} \left(1 - \left(\frac{i}{n}\right)^3\right)\left(\frac{1}{n}\right) = \frac{1}{n}\sum_{i=1}^{n} 1 - \frac{1}{n^4}\sum_{i=1}^{n} i^3 = \frac{1}{n}(n) - \frac{1}{n^4}\left[\frac{n^2(n + 1)^2}{4}\right]$$

$$A = \lim_{n \to \infty} \left[1 - \frac{1}{n^4}\left(\frac{n^2(n + 1)^2}{4}\right)\right] = 1 - \frac{1}{4} = \frac{3}{4} \text{ square unit}$$

49. Width: $\dfrac{6}{n}$, Height: $f\left(-3 + \dfrac{6i}{n}\right) = \left(-3 + \dfrac{6i}{n}\right) + 3 = \dfrac{6i}{n}$

$$A \approx \sum_{i=1}^{n} \left(\frac{6i}{n}\right)\left(\frac{6}{n}\right) = \frac{36}{n^2}\sum_{i=1}^{n} i = \frac{36}{n^2}\frac{n(n + 1)}{2} = \frac{18(n + 1)}{n}$$

$$A = \lim_{n \to \infty} \left[\frac{18(n + 1)}{n}\right] = 18 \text{ square units}$$

50. Width: $\dfrac{2}{n}$, Height: $f\left(-1 + \dfrac{2i}{n}\right) = \left(-1 + \dfrac{2i}{n}\right)^2 = 1 - \dfrac{4i}{n} + \dfrac{4i^2}{n^2}$

$$A \approx \sum_{i=1}^{n} \left[1 - \frac{4i}{n} + \frac{4i^2}{n^2}\right]\left(\frac{2}{n}\right)$$

$$= \frac{2}{n}\sum_{i=1}^{n} 1 - \frac{8}{n^2}\sum_{i=1}^{n} i + \frac{8}{n^3}\sum_{i=1}^{n} i^2$$

$$= \frac{2}{n}(n) - \frac{8}{n^2}\frac{n(n + 1)}{2} + \frac{8}{n^3}\frac{n(n + 1)(2n + 1)}{6}$$

$$= 2 - \frac{4(n + 1)}{n} + \frac{4(n + 1)(2n + 1)}{3n^2}$$

$$A = \lim_{n \to \infty} \left[2 - \frac{4(n + 1)}{n} + \frac{4(n + 1)(2n + 1)}{3n^2}\right] = 2 - 4 + \frac{8}{3} = \frac{2}{3} \text{ square unit}$$